HANDBOOK OF WESTERN PALEARCTIC BIRDS
Volume II

HANDBOOK OF WESTERN PALEARCTIC BIRDS

Volume II
Passerines: Flycatchers to Buntings

Hadoram Shirihai and Lars Svensson

H E L M
LONDON · OXFORD · NEW YORK · NEW DELHI · SYDNEY

Photographs by
Daniele Occhiato, Markus Varesvuo, Amir Ben Dov, Hanne & Jens Eriksen,
Vincent Legrand, Carlos Gonzalez Bocos, René Pop, Hugh Harrop,
David Monticelli, Mathias Schäf, Aurélien Audevard
and numerous others

Maps by Magnus Ullman

Editorial advice and assistance by
Nik Borrow, José Luis Copete, Guy Kirwan, René Pop and Nigel Redman

Identification consultants
David Bigas, Javier Blasco-Zumeta, Simon S. Christiansen, José Luis Copete,
Aron Edman, Marcel Gil, Alexander Hellquist, Magnus Hellström, Steve Howell and Peter Pyle

HELM
Bloomsbury Publishing Plc
50 Bedford Square, London, WC1B 3DP, UK

BLOOMSBURY, HELM and the Helm logo are trademarks of Bloomsbury Publishing Plc

First published in the United Kingdom in 2018

Copyright © Hadoram Shirihai and Lars Svensson, 2018
Photographs © named photographers (see credits list on p. 611), 2018

Hadoram Shirihai and Lars Svensson have asserted their right under the Copyright, Designs and Patents Act, 1988,
to be identified as Author of this work

Every effort has been made to trace copyright holders of materials featured in this book.
The publisher would be pleased to hear of any omissions or errors so they can be corrected.

All rights reserved. No part of this publication may be reproduced or transmitted in any form or by any means, electronic or mechanical,
including photocopying, recording, or any information storage or retrieval system, without prior permission in writing from the publishers

A catalogue record for this book is available from the British Library

Library of Congress Cataloguing-in-Publication data has been applied for

ISBN: HB: 978-1-4729-3737-7
ePub: 978-1-4729-6092-4   ePDF: 978-1-4729-3736-0
Printed in China by RR Donnelley

2 4 6 8 10 9 7 5 3 1

Publisher: Jim Martin
Design by Julie Dando, Fluke Art

To find out more about our authors and books visit www.bloomsbury.com and sign up for our newsletters

CONTENTS

Acknowledgements ... 7

Introduction
- Layout and scope of the book ... 8
- Species taxonomy ... 9
- Subspecies ... 10
- Sequence ... 11
- Nomenclature ... 11
- Photographs ... 12
- Maps ... 13
- Species accounts ... 14
- Glossary and abbreviations ... 17
- Selected gazetteer ... 19

An approach to moult and ageing birds in the field ... 20

General references ... 25

List of passerine families, traditional and new order ... 27

A brief presentation of passerine families ... 28

Species accounts
- Flycatchers ... 31
- Reedlings ... 61
- Babblers ... 64
- Tits ... 72
- Nuthatches ... 115
- Creepers ... 131
- Penduline tits ... 140
- Sunbirds ... 149
- White-eyes ... 160
- Orioles ... 163
- Shrikes ... 167
- Corvids ... 221
- Starlings and mynas ... 274
- Sparrows and allies ... 298
- Weavers and waxbills ... 336
- Vireos ... 351
- Finches ... 353
- New World warblers ... 462
- Tanagers ... 484
- New World sparrows ... 486
- Buntings ... 494
- Cardinals and icterids ... 572

Vagrants to the region ... 581

Checklist of the birds of the Western Palearctic – passerines ... 606

Photographic credits ... 611

Indexes ... 614

ACKNOWLEDGEMENTS

Anyone writing a detailed regional handbook is bound to be in debt to a very large number of helpful and knowledgeable people, and this book is no exception. It is impossible to produce a work like this without relying on the help and support of numerous persons, museums, institutions and libraries.

When one of us (HS) first came up with the idea of this book it was meant to be a somewhat more popular photographic guide, and early on photographers were asked to send in their best photographs. It started in the film era, but has subsequently moved into the digital world. This means that we have tested the patience of contributing photographers more than usual, and we thank all those who have stayed with the project despite its long gestation prior to publication. Although we have benefited from the cooperation of some 1200 photographers in Europe, Africa and Asia, of which about 750 are represented in the two passerines volumes, this is too many to be listed here (but all are credited in the list of photographers in each volume). Still, we must single out some who have either contributed significantly in numbers or quality, or who have made special efforts to fill gaps by travelling to remote parts of the treated region; hence our special thanks go to Abdulrahman Al-Siran, Rafael Armada, Aurélien Audevard, Amir Ben Dov, Carlos Gonzalez Bocos, Hanne & Jens Eriksen, Hugh Harrop, Lior Kislev, David Monticelli, Jyrki Normaja, Daniele Occhiato, René Pop, Mathias Schäf, Norman Deans van Swelm, Markus Varesvuo and Edwin Winkel. Many other photographers contributed commendably, and we would like to also mention Mike Barth, Arnoud B. van den Berg, Colin Bradshaw, Kris De Rouck, Axel Halley, Vincent Legrand, Alison McArthur, Werner Müller, Arie Ouwerkerk, Mike Pope, George Reszeter, Huw Roberts, Ran Schols, Ulf Ståhle, Gary Thoburn, Alejandro Torés and Matthieu Vaslin.

We have also benefited greatly from the use of a number of excellent internet collections of photographs, and their owners and staff have kindly cooperated with us to facilitate contacts with photographers and by publishing reports of progress and needs of the handbook project over the years. Those internet sites which we used the most are mentioned below under 'Photographs', but we would like to acknowledge here the significant help we received from Morten Bentzon Hansen (Netfugl), Dave Gosney and Dominic Mitchell (Birdguides), Emin Yoğurtcuoğlu (Trakus), Tommy P. Pedersen (UAE Birding), Tim Loseby and other members of the Oriental Bird Club (Oriental Bird Images), Askar Isabekov (Birds, Kazakhstan), Lior Kislev (Tatzpit) and David Bismuth (Ornithomedia). Dolly Bhardwaj, Prasad Ganpule and Abhishek Gulshan not only provided invaluable photographs of Indian birds, but also assisted with contacts with Indian photographers; we are very grateful to all these people.

There are a number of key persons whose contributions are of such importance for the successful completion of the book that they deserve warm thanks: Jim Martin (publisher), Guy Kirwan and Nigel Redman (text editing), Julie Dando (design), Magnus Ullman (maps), René Pop (photographic editor), José Luis Copete, Magnus Hellström and Alexander Hellquist (photographic consultants), and Nik Borrow (photographic assistance). For taxonomic input of various kinds we acknowledge valuable contributions by Per Alström, Nigel Collar, Joel Cracraft, Pierre-André Crochet, Edward Dickinson, Alan Knox, Mary LeCroy, Urban Olsson, Robert Prŷs-Jones and Frank Steinheimer. Alison Harding and Robert Prŷs-Jones helped on numerous occasions with tracing old references. Natalia Delvina and Alexander Hellquist helped with translation of some key Russian texts.

We are indebted to staff in the following museums, who have been cooperative, welcoming and helpful over the many years we have been preparing the manuscript, with often biannual visits to Tring, New York, Paris and Copenhagen. For covering travel expenses for one of many trips by LS to Tring, support received from the Synthesys project (financed by the European Community) is gratefully acknowledged. Three of several trips to New York by LS and one by HS were similarly kindly supported by the Chapman Collection Study Grant. Our particular thanks go to staff in visited museums: Robert Prŷs-Jones, Mark Adams, Hein van Grouw and Alison Harding (NHM, Tring), Joel Cracraft, Paul Sweet, Thomas J. Trombone and Peter Capainolo (AMNH, New York), Per Ericson, Göran Frisk and Ulf Johansson (NRM, Stockholm), Jon Fjeldså and Jan Bolding Kristensen (ZMUC, Copenhagen), Claire & Jean-François Voisin and Marie Portas (MNHN, Paris), Michael Brooke (UMCZ, Cambridge), Sylke Frahnert, Pascal Eckhoff and Jürgen Fiebig (ZMB, Berlin), Renate van den Elzen, Till Töpfer, Kathrin Schidelko and Darius Stiels (ZFMK, Bonn), Anita Gamauf and Hans-Martin Berg (NMW, Vienna), Steven van der Mije and Kees Roselaar (NBC, Leiden), Pavel Tomkovich, Eugeny Koblik and Sergey Grigoriy (ZMMU, Moscow), Andrey Gavrilov (ZIA, Almaty) and Daniel Berkowic and Roi Dor (TAUZM, Tel Aviv). We are also grateful to those museums which we had no opportunity to visit but who kindly arranged with loans of skins or sampling for DNA, and we thank Nicola Baccetti (ISPRA, Bologna), Georges Lenglet (IRSNB, Brussels), Jan T. Lifjeld (ZMUO, Oslo), Jochen Martens (JGUM, Mainz), Tony Parker (WML, Liverpool) and Kristof Zyskowski (YPM, New Haven).

We owe thanks to José Luis Copete, Nigel Cleere, Roy Hargreaves, Beth Holmes, Stephanie Peault, John Black and Claire Voisin for helping us to take some measurements, mainly in NHM, Tring. Guy Kirwan, Bo Petersson and Clive Walton provided other useful measurements.

For valuable help with double-checking identification, ageing and sexing of the selected photographs we are indebted to two independent groups of experts: (1) Magnus Hellström, Simon S. Christiansen and Aron Edman, and (2) José Luis Copete, David Bigas, Marcel Gil and Javier Blasco-Zumeta. Peter Pyle and Steve Howell kindly offered their expert advice on all American species as to captioned age and sex. Gabriel Gargallo helped with advice on moult and ageing and sexing of some species, while Yosef Kiat was consulted selectively regarding birds in Israel, and both are thanked for providing useful help. The input of all these helpful people has greatly improved the end result, but we should point out that we have in a very few instances opted to disregard their recommendations in favour of our convictions based on own research, and any lingering mistakes in how the photographs have been captioned are the full responsibility of the two authors.

Magnus Hellström contributed greatly to the section 'An approach to moult and ageing birds in the field'.

José Luis Copete and Marcel Haas kindly helped to compile a list of accidentals, generally with fewer records than ten, to be included in the passerine volumes.

We thank Norbert Teufelbauer, coordinator of the forthcoming Austrian breeding bird atlas, for helpfully checking several range maps concerning Austria.

Others who helped with various requests for information, or who provided other useful help, include: Mohammed Amezian, Raffael Ayé, Rosario Balestrieri, Mikhail Banik, Peter H. Barthel, Carl-Axel Bauer, Per-Göran Bentz, Anders Blomdahl, Leo Boon, Mattia Brambilla, Vegard Bunes, Ulla Chrisman, Alan Dean, Gerald Driessens, Gadzhibek Dzhamirzoev, Annika Forsten, Johan Fromholtz, Steve Gantlett, the late Martin Garner, Hein van Grouw, Morten Bentzon Hansen, Roy & Moira Hargreaves, Martin Helin, Niklas Holmström, Eugeny Koblik, Markus Lagerqvist, Antero Lindholm, Pekka Nikander, Barbara Oberholzer, David Parkin, the late David Pearson, Pamela Rasmussen, Yaroslav Redkin, Magnus Robb, the Scharsach family, Kathrin Schidelko, Maurizio Sighele, Brian Small, David Stanton, Darius Stiels and Menekse Suphi.

Finally, but most importantly, we thank our loving and supporting families for patience and encouragement throughout the years of preparation. Hadoram is grateful to María San Román, to his late daughter Eden and to all of his family, Lars to Lena Rahoult for endless support.

Hadoram Shirihai & Lars Svensson

INTRODUCTION

Layout and scope of the book

The original idea for this handbook was developed by HS in the late 1990s. Around 2000, LS was invited to be co-author. From the start, the aim for the project was to focus on identification and taxonomy, and to make it the most complete and profusely illustrated photographic guide to Western Palearctic birds. With the development of modern camera equipment, increasing numbers of birdwatchers travelling to distant parts of the region, and 'amateurs' making a significant photographic contribution through new digital camera equipment with image stabilisers and auto-focus, the time was ripe for producing the first comprehensive and complete photographic handbook to the birds of our part of the world.

The text aims to give a modern summary of advice on identification, ageing, sexing and moult, and to provide an up-to-date statement of the geographical variation within the treated range, describing in some detail all distinct subspecies. It also reflects the latest changes in taxonomy at lower levels and offers notes on these, and on unresolved problems. Whereas we have chosen to keep a traditional sequence, we still provide a summary of recent new insights in higher-level taxonomy. Each species account includes brief summaries of vocalisation and distribution, too. Maps produced especially for this book show the current summer and, often, winter ranges of nearly all treated species *and* subspecies.

The region covered is the Western Palearctic (Europe, Asia Minor, the Middle East and North Africa), but in this book we also treat the entire country of Iran (all, or the south-eastern part, are usually excluded) and all of Arabia (of which the southern parts are often omitted). Moreover, also covered are the Cape Verde Islands, the Azores, the Canary Islands, Madeira, Iceland (but not Greenland), Jan Mayen, Svalbard and Franz Josef Land. The limit in the south across the African continent follows the southern national or political borders of Western Sahara, Algeria, Libya and Egypt. We also include the disputed territory between Egypt and Sudan usually referred to as the Hala'ib Triangle (Gebel Elba). The eastern limit runs north from the Arabian Sea, excluding Socotra, along the eastern border of Iran through the Caspian Sea, then north along the Ural River and the 'ridge' of the Ural Mountains to the Kara Sea. A total of 408 species (plus seven extralimital ones afforded brief mention under species which they resemble) and 948 taxa are treated in the main section of the first two volumes covering the passerines (a further two volumes treating the non-passerines will follow). This first volume covers larks to warblers and crests, whereas Volume 2 will treat flycatchers to buntings and icterids.

Any species found within the covered range at least ten times is treated in the main section, whereas an additional 85 passerine species (34 of which are covered in volume 1) found on between one and nine occasions (in a few cases a little more) are given a brief presentation in an appendix. The grand total of passerine species treated therefore becomes 500, and the number of taxa 1033. The limit of ten records

Map showing the treated region.

must not be seen as firm. New records of rare species are made all the time, and it is impossible to be fully up-to-date. Our deadline for inclusion was the end of 2016. For a few of these vagrants the known number of records already exceeds nine, but either the species was a more frequent visitor in the past or it has become more regular only in the last few years.

We have included Iran in the Western Palearctic because we feel that its western parts have more in common with the Western than the Eastern Palearctic bird fauna, and since we only deal with whole nations or political territories rather than selected parts, eastern Iran is included as a bonus. The greater part of Iran belongs to the Palearctic fauna, and only for the south-eastern corner could a case be made for referring that to the Oriental region. The inclusion of southern Arabia seemed a logical change because, with the possible exception of the coastal corner of Yemen and Dhofar, the entire Arabian avifauna is more Palearctic than Afrotropical, and birdwatchers rarely visit only certain parts of these countries—they want to see every bird within the political borders once there.

It should be added that by expanding (or, to a much lesser extent—in southern Sahara—reducing) the conventional definition of the area known as 'the Western Palearctic' from a concept of natural and artificial borders to purely political ones, where they can be followed, we seek not to take issue with adherents of the former concept. Our decision is based purely on a pragmatic and birdwatcher-friendly approach.

During our work it became evident that with so many excellent photographs of nearly all possible plumages gathered in one handbook, photographs which often enable you to study feather wear and patterns in the finest detail, it would be a good time to expand and update the information offered on ageing and sexing. Similarly, we decided to expand the treatment of subspecies. Modern birdwatchers are knowledgeable and no longer interested only in species; they often seek to establish the subspecies, age and sex of birds, and undoubtedly the value of an individual record increases significantly if these further qualifications can be made.

Work was initially divided rather equally between the two authors, each writing first drafts of about half of the species. In the final stages HS did most of the work searching for and selecting photographs, drafting captions and checking the authenticity and data of the images, while LS undertook a final editing and review of the entire text including taking final decisions on geographical variation (which subspecies to be included), and on any contentious taxonomic issues, and also wrote the introductory and end matter. After the two authors were both content with the image selection and the detailed identification of the birds portrayed, all images with drafted captions were subject to a final check by two informal identification panels formed solely for this book (see Acknowledgements). Almost the entire text was also checked by Guy Kirwan.

Species taxonomy

Species concepts have been much discussed in recent years. The species is generally accepted as the basic unit in the natural hierarchy, yet its definition is still, to a certain extent, a matter of argument. As someone once put it, 'a species is what a majority of informed taxonomists agree about'. Just as family, genus and subspecies categorisations are somewhat diffuse and subjective in their definitions, so too is the species.

Today, several concepts exist side by side, the Biological Species Concept (BSC) which lays emphasis on reproductive isolation, the Phylogenetic Species Concept (PSC) which is defined as diagnosable populations that share a common ancestry, and the General Lineage Concept of Species (GLC) or the very similar Evolutionary Species Concept (ESC), which focus on the species as a lineage of directly related populations (and therefore focus more on the time aspect of evolution). There exist further concepts but generally speaking these are variations of the above-mentioned main concepts. In frustration over this unresolved situation leading to varying numbers of recognised bird species to protect, and in particular over the far-reaching and inconsistent influence of genetics in taxonomy (see also below), BirdLife International formed its own species concept based on five weighted criteria and noteworthy for excluding genetic evidence entirely (Tobias *et al.* 2010).

No concept can yet be said to be preferred by a majority of ornithologists, and in reality the main concepts, although theoretically very different, differ less than many believe in practical outcome. But there are differences. We basically favour the BSC, noting at the same time that its application has been influenced in recent years by the other concepts. For instance, if hybridisation with other species was previously seen

While attempting to apply a consistent species concept, the authors still at times ran into difficult borderline cases, and not all readers might agree with every decision made. A few examples will illustrate the dilemmas facing us. African Crimson-winged Finch *Rhodopechys alienus* (top left: Morocco, February) and Asian Crimson-winged Finch *Rhodopechys sanguineus* (top right: Turkey, June) have been split, whereas reasons for keeping Scottish *Crossbill scotica* (bottom left: Scotland, May) separate from Common Crossbill *Loxia curvirostra* were thought to be insufficient, in particular when considering that the Cypriot race *guillemardi* (bottom right: Cyprus, June), to take one example, is even more distinct than the Scottish population. (Top left, D. Occhiato; top right, H. Shirihai; bottom left, D. Whitaker, bottom right, H. Shirihai.)

as a barrier to species status under BSC, it can now be acceptable where limited, and if the resulting offspring are either less fit or do not alter the two involved species more than marginally and in a stable way. That allopatric populations, with a wide sea or a high mountain range between them, do not meet means that we cannot test their reproductive capacity. Still, this can generally be overcome by using good sense and observing and comparing them with similarly different populations that do live in contact. The problems of diagnosability in the PSC are often on the same level, so neither of these two concepts can claim to be 'superior'.

We have tried to find a sensible balance between, on the one hand, a conservative approach that favours stability and awaits solid proof from different independent sources before proposed splits or other changes are adopted, and on the other an ambition to mirror the latest developments in taxonomy. While it is good to be up-to-date, nobody wants to have to retreat after adopting changes too hastily. Reversals only cause confusion, and undermine the authority of the changer, so a cautious approach and good compromise are always sought.

When estimating whether a proposed new species split is warranted we have applied the following criteria: (1) it should be morphologically distinct in all individuals of at least one sex or age, and differences should not be minute; (2) it should live in biological segregation; (3) its vocalisations or behaviour ought to differ at least in one aspect; and (4) it is a further reassurance if it differs genetically. The first two requirements are compulsory, whereas the other two criteria are important but supplementary; they would strengthen a case but are not absolutely necessary. We are reluctant to accept new species that fulfil criteria 2–4 but not number 1. Similarly, for sympatric or parapatric populations we prefer to wait if points 1, 3 and 4 are fulfilled but not number 2. 'Minute differences', insufficient for separating two taxa at species level, are those which require advanced mathematical formulae to be proven, such as principal components analysis (PCA) or variance analysis (ANOVA); differences should be obvious without these methods. Subtle average differences are not enough. 'Biological segregation' is to be understood, as explained above, as a state where any hybridisation between two species is marginal (localised and involving relatively fewer pairs than would be expected if no barriers existed) and not known to increase in extent.

The Great Grey Shrike complex *Lanius excubitor* s.l. is one the most difficult taxonomic issues to deal with in the covered region. In our treatment we have been influenced by some of the findings in the phylogenetic study by Olsson *et al.* (2010), although this did not arrive at any firm conclusions or taxonomic recommendations. Our species taxonomy represents a cautious approach, defining only three species, Iberian Grey Shrike *Lanius meridionalis*, Northern Shrike *L. borealis* and Great Grey Shrike *L. excubitor*, the last mentioned including several races and race groups differing somewhat genetically, morphologically and geographically; one of these is shown here, *koenigi* of the Canary Islands (May). (H. Shirihai)

We considered at one stage to show for a few borderline cases, like the Siberian Chiffchaff *tristis* or the Eastern Black-eared Wheatear *melanoleuca*, the species epithet (the second word in a species' scientific name) within rounded brackets, thus e.g. '*Phylloscopus* (*collybita*) *tristis*', to signal that it is an incipient species on its (long!) way to becoming an undisputed separate species. Eventually, however, we decided not to do so. Any such practice, including the more formal one to recognise so-called allospecies (or even 'phylospecies') and show the species epithet of these within square brackets, involves subjective choices and a certain muddling of species taxonomy and nomenclature, at least to the layman. Consequently, we elected to make choices in each case and instead comment on the difficulties and the subjective nature of some of these decisions at the end of each relevant species account. The important thing is to describe all distinct taxa. The best taxonomic treatment is still, to a certain degree, a matter of taste and preference.

To conclude, we have tried to follow certain criteria and in general be conservative and cautious rather than bold in our choices. Someone may note that for the genus *Oenanthe* we have accepted or proposed more splits than previous authors, but this is just the result of consistently trying to apply the above-mentioned species criteria.

Subspecies

We have made an effort to independently assess all described subspecies that appear in modern ornithological literature within the range covered by the book. This has forced us to spend much time in museum collections, the only place where series of closely related but differently named subspecies can be compared side by side. Between the two of us we have spent many, many months in museums in order to write this handbook. One important object of this laborious work has been to decide whether a subspecies is sufficiently distinct to be recognised and treated as separate from other subspecies of the same species. We have, by and large, tried to apply the so-called 75% rule (e.g. Amadon 1949), meaning that at least three-quarters of a sample of individuals selected at random (in some cases of one sex or age) of a subspecies must differ diagnosably from other described subspecies within the examined species. The specimens selected for assessment should as far as possible be from similar age categories and seasons to allow for the most comparable plumages to be fairly assessed. The main limiting factor for this part of our work has been the scarcity of comparable material of some taxa in collections.

We have long felt the need to focus on *distinct* subspecies and to dismiss the many subtle or even questionable subspecies that were perhaps based on too short series or on skewed or inadequate material, or which in reality represented only minor tendencies within a population, far from the minimum 75%. Natural variation within any one population has often been underestimated when new subspecies have been described. In our opinion, it is better to concentrate on clear differences, and to account for slight clinal tendencies or deviations from the most typical under a single subspecies heading, rather than to create a confusing mosaic of subtle or dubious but formally named subspecies. Most of the final assessment of subspecies taxonomy was done by LS.

Compared to other handbooks and checklists, we accept about 15% fewer subspecies. Those deemed to be too similar to a neighbouring race to be upheld have been lumped with that race and treated as a synonym.

Much geographical variation within continentally distributed species is smooth and gradual, this variation usually being termed clinal. There are taxonomists who tend to see this as a problem and who prefer to treat the entire such range as one subspecies. It is true that clines create circumscription difficulties. However, we prefer to judge every case by its own merits; if the ends of a cline are very different, a minimum of two named subspecies seems reasonable, one at each end and with a diffuse centre of intergrading characters. If there are hints of stepwise change along the cline, these suggest subspecific borders and lead to recognising additional subspecies along the cline.

We want to stress that the definition of a subspecies is based on morphology. If size, structure, colours or patterns differ with sufficient consistency (in this work, as stated, by at least 75% of the examined specimens of at least one sex or age) within a geographically defined part of the range of a species, then it is a valid and distinct subspecies *irrespective* of whether an examination of its mitochondrial DNA or another genetic marker reveals no clear genetic difference from neighbouring subspecies. Geneticists are sometimes so impressed by their new tool that they forget these prerequisites and distinctions. If we are to re-define subspecies by their genetic properties, all existing subspecies need to be discarded and all work on geographical variation redone in laboratories. But in such a case, scientists need first to agree on exactly which genetic definition all variation is to be judged against. Since this is unlikely ever to happen, we had better stick to the morphology-based subspecies taxonomy we already have!

It is best to add that our approach to subspecies taxonomy is quantitative rather than qualitative. If a described taxon does not in our view reach the required level of distinctness, we place it in synonymy under a senior name. But that does not mean that we claim to know better than others regarding the recognition of a valid taxon, only that it is not in our view distinct enough according to our requirements. If others prefer to describe every tiny variation under a formal name it is up to them.

Subspecies deemed by us to be borderline cases as to whether they are warranted or not, and subspecies only differing very subtly from other valid subspecies, have been marked with an open circle in front of their names. For subspecies that are either for some reason questionable, or for which we failed to find sufficient or indeed

Museum collections are invaluable for the taxonomist, since examining a collection of bird skins (as bird study specimens are known by scientists) brought together from various parts of a species' range is the only way geographical variation can be properly studied. Here, as an example, four European subspecies of Coal Tit *Periparus ater* are compared, though many more have been described and are valid. These are just a few of the skins held in the Natural History Museum, Tring, England; the total is more than 700,000. (L. Svensson/Natural History Museum)

any material, a question mark has been placed in front of their names. Subspecies known or suspected to be now extinct carry a dagger mark in front of the name.

At the end of each subspecies entry a selection of synonyms has been listed alphabetically within brackets; these are subspecies names that appear in the literature or on some specimen labels in museums but which are either junior synonyms or are invalid for one reason or another. They are shown as a service to the reader and to make it clear that they have not simply been overlooked by us. We have as a rule only included the more recent and commonly seen synonyms, but a few older or more rarely used ones have also been included if considered helpful to the reader. Anyone seeking a more complete list is referred to Sharpe (1874–99), Hartert (1903–38), Peters *et al.* (1931–87) or Vaurie (1959, 1965).

Sequence

Within the passerine order, the sequence of families follows Voous (1977). The sequence within families and genera also generally follows Voous except where more recent insights have enabled revision. The Voous list was largely based on the so-called Peters Checklist (1931–87) and was the basis for influential handbooks like *The Birds of the Western Palearctic* ('*BWP*', Cramp *et al.* 1977–94) and is also largely used in *Handbook of the Birds of the World* ('*HBW*', del Hoyo *et al.* 1992–2012). Ornithology experienced a long period of relative taxonomic stability as long as it was based mainly on morphology and behaviour.

With the arrival of genetics as a new tool in systematics, starting with publications by Sibley & Ahlquist (1990) and Sibley & Monroe (1990), then followed by a stream of important contributions in the following two decades and more, it has become clear that the 'Voous order' does not correctly reflect the true relationship and evolutionary history of passerine families (or of non-passerines). A few regional faunas (e.g. Grimmett *et al.* 1998), world checklists (e.g. Dickinson 2003, Dickinson & Christidis 2014) and ornithological societies (e.g. British Ornithologists' Union 2009) have already applied dramatically different family sequences compared to the one worked out by Voous. Although differing in details, these new orders share several main features and are no doubt taxonomically more correct, but they will probably be further refined in years to come, clearly demonstrated by the fact that the passerine sequence suggested by Joel Cracraft for the fourth edition of *The Howard and Moore Complete Checklist of the Birds of the World* (Dickinson & Christidis 2014) differs in several aspects from what was thought by some to be definitive only about ten years ago. Considering this prevailing relative turbulence, and the advantage of using a familiar order which makes it easier for the reader to find a family or genus, we have decided to employ a sequence that is still mainly that of Voous, and which is similar to the one applied in the widely used *Collins Bird Guide* (Svensson *et al.* 2009), which treats nearly the same region. The family sequence used here can be compared to the new order in Dickinson & Christidis (2014) in a table on p. 27. With this background, it was also decided not to give family introductions in the main part of the book, since some of the long-established families have either been split or lumped, and some unexpected close relationships have now been revealed and vice versa. For an insight into modern passerine family taxonomy we refer the reader to Cracraft *in* Dickinson & Christidis (2014), and to the brief overview on pp. 28–30.

Nomenclature

Scientific names

The scientific names in this handbook mostly follow those adopted by Svensson *et al.* (2009) and Dickinson & Christidis (2014), with a few deviations due to different taxonomic choices made by the authors. In the end, the precise nomenclature chosen is that preferred by the authors.

A scientific species name consists of two words in Latin or in Latinised form, the first written with an initial capital letter signifying the genus (a group of closely related species), and the second being the species epithet ('species name' in everyday conversation, although a species name formally is both these words). To stand out better, scientific names below family level are recommended to be written in a different font, usually by using *italics*. A typical example would thus be *Parus major* for Great Tit. The word *Parus* ('tit') is the generic name and the species epithet *major* ('large') makes it clear that this is the Great Tit.

In the late 19th century the subspecies concept evolved and was subsequently commonly adopted, with eventually a third word being added to the scientific name to indicate the exact geographic origin of a bird based on its morphology. The Great Tits breeding in Britain have thus been named *Parus major newtoni*, based mainly on their subtly darker back and yellowish-white rather than pure white wing-bars and tertial-edges compared to the populations living in continental Europe, which are called *Parus major major*. For the first described subspecies within a species, *major* in this case, the subspecies name is created by simply doubling the species epithet, which happens only when more than one subspecies of the same species is described. Such first-described subspecies are often referred to as 'nominate' (or 'nominotypical'), and are sometimes mistaken as being 'more important'; however, the term only indicates that these taxa happen to be first described and have (by chance) the oldest names within the respective species. (To avoid such misunderstandings we have refrained from using the term.) If there is no conceivable geographical variation within a species it is said to be monotypic (being true for *c.* 20% of passerine species) and its scientific name is limited to two words.

We have included the name of the original author and the year of first publication of each approved scientific name, often useful to make a scientific name unambiguous, and which becomes important if any changes of scientific names need to be carried out. The family name of each author is written out with one traditional exception: the abbreviation 'L.' stands for the Swedish naturalist Carl von Linné, father of the binominal name system. (In the English-speaking world he is generally known by his birth name, Carl Linnaeus; however, he was ennobled in 1757 and changed his name thereafter[1].) Scientific nomenclature officially starts on 1 January 1758, the

---

[1] Since Linné's name still appeared as 'Linnæi' (genitive form of Linnaeus) on the title page of the important tenth edition of the *Systema Naturae* (1758), it is customary to refer only to this form of his name in nomenclature. However, several species names were first published in the twelfth edition (1766), and in this his name was written as 'a Linné' (Latinised form of von Linné). Formally, the Code issued by the International Commission on Zoological Nomenclature permits his later name to be used for such species. By abbreviating his name to 'L.' such potential confusion is avoided.

year in which the tenth edition of Linné's *Systema Naturae* was published. No earlier published names are permitted. If the author's name and year of publication appear within brackets it signifies that the species in question was originally described in a different genus. For instance it is *Alauda arvensis* L., 1758, for the Skylark, but *Lullula arborea* (L., 1758) for the Woodlark, since Linné originally described the latter as '*Alauda arborea*', thus in another genus than is the custom today.

Abbreviated first names, initials, are given only for authors like Brehm, Gmelin and Lichtenstein, where there existed more than one person with the same surname who each named several taxa. For people like Ernst Hartert or Otto Kleinschmidt, any namesake created only one or two names in all, so only the family name is given for the more famous person of the two.

Scientific names are governed by the rules laid out in the *International Code of Zoological Nomenclature* (ICZN, 4th edn. 1999, published by the International Commission on Zoological Nomenclature and available on the internet). Several amendments were published separately in 2012. The rules are logical and generally clear but require a wider knowledge of taxonomy, clear thinking and much practice to be properly understood. And even then you will often need advice from more experienced taxonomists or for particularly difficult questions a ruling by a commissioner linked to the ICZN.

Of the many rules in the Code only one needs to be commented on here, illustrating why the year of publication of a taxon name is sometimes important. The oldest available valid name has priority and should be used. This means that if an author decides to merge ('lump') two very similar subspecies (because they are not sufficiently distinct), as has happened in many places in this book, knowledge of which of the two merged subspecies was described and published first is essential, since the new 'larger' taxon must carry the older name.

English names

Choosing English bird names is a contentious subject, and although we are probably slowly approaching more international—or at least national or regional—agreement we are some way from achieving that, and it still seems impossible to please everybody. We have had the ambition to follow the attempts by the International Ornithological Congress (IOC) to create an international standard for English bird names (Gill & Wright 2006, with subsequent updates, also available online as Gill & Donsker 2016). In the end, we did not follow it 100%, but came very close. In a few instances we felt that selected modifiers in that work did not take into account recent taxonomic changes, and for some name choices we were sensitive to known criticism within Britain that several traditional and well-liked names (Rose-coloured Starling, Lapland Bunting) would be lost if the IOC standard was followed without exception. We also feel that for a small group of familiar birds, well known also to the layman, the modifier if needed should be 'Common' rather than 'Eurasian' or 'European' (e.g. Skylark, House Martin, Magpie, Raven, Starling and Bullfinch). Thus, in a few cases we followed our own conviction, always the privilege of independent authors.

A special problem in English names relates to whether to hyphenate compound words, write them as two or write them as one word. As one of us (LS) has been involved editorially in the fourth edition of *The Howard & Moore Complete Checklist of the Birds of the World*, it felt natural to adopt similar criteria for group names as those used in that checklist. In brief we avoid trying to create group names by the use of hyphens outside genera, and even within genera we avoid change to long-established names unless instructive gains are very obvious. Scientific names serve as indicators of relationship, therefore English ones do not need to attempt to do the same. If they do it is fine, but one should not introduce numerous new names in an effort to mirror the latest phylogenies published. It is better to have clear, unambiguous, brief, understandable and locally well-known names as far as possible. Since English indexes are arranged after the main (often last) word in a compound name, it is important to have this word unhyphenated. It is far more likely that you will find a sought-after lark under Lark than under Sparrow-Lark.

On the other hand we note that birdwatchers, patiently accepting and adopting the many changes of taxonomy and scientific names being imposed upon them at an ever-increasing rate, are surprisingly conservative and resistant to change when it comes to vernacular or English names. A little more flexibility and open-mindedness would not hurt if internationally accepted English names are eventually to be achieved.

Photographs

The search for images

One of the main features and strengths of this book is the comprehensive coverage of all important plumages with photographs taken in the field. It has been a delight to see the extremely high quality of many of the submitted photographs, and also to note the wealth of photographs now available for rarely photographed or even rarely seen species. The large 'team' of contributing photographers is to be warmly congratulated and thanked for its fine achievement.

The aim has been to include images of both sexes if they differ, young and older ages in as many stages as can clearly be distinguished, and often also the seasonal plumage stages, which usually means in spring and autumn. On top of this is geographical variation; we have strived to include at least one photograph, and often several, of each distinctive subspecies, or at least show examples of the variation if it is more subtle. There are on average more than ten photographs per species, with a maximum of 49 for Yellow Wagtail.

Despite being rather widespread and a well-known species, adult female Pied Flycatcher *Ficedula hypoleuca* has rarely been photographed in fresh autumn plumage. For closing such gaps, the project's photographers often targeted specific missing plumages. Sweden, October. (M. Hellström/Ottenby Bird Observatory)

To achieve all this, early on the publisher enrolled René Pop to act as photographic editor for the project. Later his task was taken over by HS, who has been responsible for the major part of the picture research. Further, the publisher set up a homepage especially for the project, devised and managed by then-project editor Jim Martin, where photographers could conveniently find out what was still missing, and to which site they could transfer their images for submission. In the later years of production the authors wanted to ensure that the remaining gaps were filled, and more photographs were brought into the project to raise the level of quality overall. They therefore enrolled three additional photographic researchers, José Luis Copete, biologist and freelance photographic editor for *Handbook of the Birds of the World*, Magnus Hellström, warden at Ottenby Bird Observatory, Sweden, and former chairman of the Swedish Rarities Committee, and Alexander Hellquist, current member of the Swedish Rarities Committee; all three also helped with important targeted searches of websites and journals to track down missing plumages. As mentioned above, all selected images and drafted captions were then carefully checked by two independent panels of experts (see Acknowledgements). In specific cases, additional experts were sometimes consulted to ensure the highest possible accuracy.

The following are some of the important internet collections of bird photographs which we have made much use of and of which the owners and staff kindly cooperated with us when searching for missing plumages, thus providing an important source for achieving complete photographic coverage: www.artportalen.se; www.birdguides.com; www.birdphoto.fi; www.birds.kz; www.birdsofsaudiarabia.com; www.birds-online.ru; www.kuwaitbirds.org; www.nature-shetland.co.uk; www.netfugl.dk; www.oiseaux.net; www.orientalbirdimages.org; www.ornithomedia.com; www.ouessant-digiscoping.fr; www.rbcu.ru; www.tarsiger.com; www.tatzpit.com; www.trakus.org; www.uaebirding.com.

As mentioned above, HS put in years of work checking out many tens of thousands of photographs, corresponding with photographers in many countries and searching websites to find crucial plumages and to check the identifications of many already submitted images, as well as drafting all of the captions. Without the combination of all these dedicated efforts this handbook would not have become as complete as it now is.

INTRODUCTION

Particular effort was made to illustrate the rarest species and subspecies, such as the range-restricted (and virtually inaccessible to European birdwatchers) Algerian Nuthatch *Sitta ledanti*, and the south-western Arabian population of Common Magpie *Pica pica*, the poorly known race *asirensis* (Saudi Arabia, July). (Left: R. Nehal; right: J. Babbington)

For the passerines, three photographers have between them contributed almost 25% of the total number of photographs, Daniele Occhiato, Markus Varesvuo and Hadoram Shirihai. Significant contributions have also been made by Hanne & Jens Eriksen and Amir Ben Dov. See also the List of photographers (p. 611).

Captions

Rather than keeping the captions as brief as possible and 'letting the images speak for themselves' we have opted for longer captions with selected identification advice and key points to note. From experience we know that many readers look at the images and read the captions first, long before they start penetrating the main text. By making the captions slightly longer and more interesting, we hope to bridge the difference between information conveyed by picture and word, and to activate and facilitate the way a photograph can be 'read'.

Much effort has been made to check not only identifications and ageing, but also to ensure that dates and localities for the photographs are correct. In a modern photographic handbook of birds it is no longer adequate to give only the species and the photographer's name for each picture. With a greater interest in taxonomy and subspecies, it is also necessary to give the location and time of year to help the reader independently evaluate whether a documented bird is correctly assigned. Therefore as a minimum we provide, whenever possible, country or smaller region and the month or season in the caption to every photograph.

It is still possible that the odd mistake has slipped through, and we will be happy to receive notification of any such claim via the publishers for possible future correction.

Maps

All of the maps have been commissioned specially for this book. We wanted clear and fairly large maps with up-to-date information about breeding range, winter range (if within the region covered) and main migration routes. We also wanted to show, for the first time for the majority of species, the ranges of the subspecies. The border between two fairly distinct subspecies has been indicated by an unbroken line, whereas a border between two very similar subspecies, or two that are clinally connected with intermediates over a wide hybrid zone, has been indicated by a dashed line. The borders are, for very natural reasons, approximate and do not claim to be precise, but they should give a fair overview of the geographical variation of a species.

The maps have been compiled by Magnus Ullman, well travelled in the entire region as a tour leader and experienced bird cartographer. He has had access to practically every known important published checklist or bird atlas when preparing the maps. Boundaries between subspecies have been largely worked out by LS based both on the literature and on museum collections.

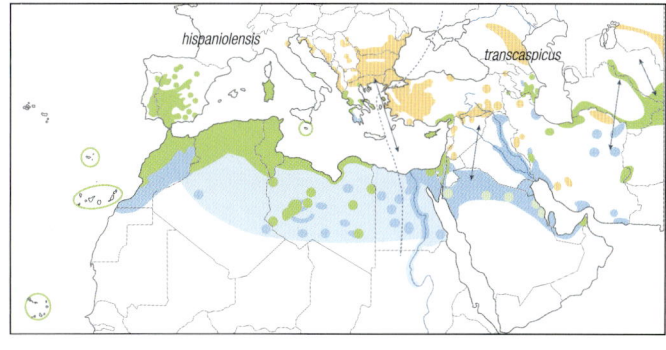

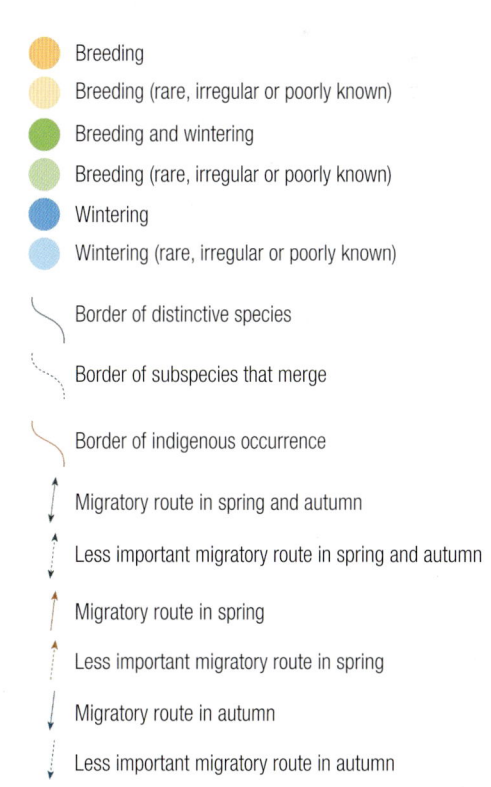

Example of range map, here showing that of the Spanish Sparrow *Passer hispaniolensis*. Colours and symbols are explained in the legend.

Species accounts

Names

Each species account is headed by a section giving the various names of the species. The principles for the selection of English and scientific names when more than one alternative exists has already been mentioned under 'Nomenclature'. A few alternative English or American names are given when these seemed called for. A short selection of foreign language names ends the heading.

Short introductions

Each species account opens with a brief outline of the species, a quick sketch to give a first presentation of the 'personality' of the bird, of what to expect from a first encounter and where to see it. These summaries are not meant to be complete or 'consistently' written. Therefore, do not expect to invariably find data on characteristic habits, favoured habitats or abundance here, to mention a few examples, although such information is frequently included. The larger type size (on tinted background) emphasises the preamble nature of these paragraphs.

Identification

Features helpful for identification of each species in the field are given under 'Identification'. Field characters thought to be more important than others *appear in italics*. Ageing and sexing are only briefly treated under this heading (but more thoroughly under a separate heading). If there is no significant geographical variation, the text is relevant for the entire area where the bird can be seen, but for species with more obvious geographical variation the text under 'Identification' generally relates to the most widely distributed subspecies in Europe. Whenever in doubt consult the information given under 'Geographical variation and range' further down in the species account.

The characters listed as important for identification 'in the field' are much about size, structure, colours and plumage details, and undeniably these often help in identifying a bird. But how a bird moves and behaves, and the 'personality' it conveys by this to the trained eye (often referred to as 'jizz') is almost equally important, only more difficult to describe in words and requiring more talent and training to use. Wherever such 'jizz' characters are obvious and useful, they are also included in the text, generally at the end.

Vocalisations

Of the various vocalisations of a bird, song is described first, followed by characteristic calls. By 'song' we mean not only what we conventionally think of as a pleasing and melodic repeated strophe, but any territorial or mate-seeking emittance of sound including the modest and irregular chuckles of a Fieldfare or feeble and 'pensive' notes by a White Wagtail. While the whole song, preceded and followed by a pause of varying length, can alternatively be called a *strophe*, a part of a song is referred to as a *phrase*, *syllable* or *note*.

Calls include contact calls, flight calls, alarm calls, anxiety calls and begging calls of young, but any calls are described only if they are characteristic enough to serve as aids for identification. Minor calls and minor call variations exist in many species, if not in most, but these variations have generally not been included unless they have some kind of significance. Voice descriptions are based largely on our own experience and on LS's personal recordings, but also on several published recordings (including those made available via free websites), and of unpublished recordings made available by courtesy of the recordists or by the National Sound Archive, British Library, London.

Anyone who has taken the trouble to memorise the songs and calls of birds knows how much easier field identification becomes. Waders, pipits and warblers, often so similar in appearance, suddenly become manageable due to their different calls and songs. True, just as when using 'jizz' as a tool in birding, sounds require a certain amount of gift to be successfully applied. But we believe that people often underestimate their own ability to learn and recognise bird vocalisations. All too often you hear resigned declarations that 'calls are just as variable as they are perceived and rendered by each observer'. This is simply not true! Bird sounds are rather more consistent than they are variable, and you too can probably learn to recognise bird sounds and describe in a meaningful way what you hear to others.

Rendering complex bird voices using letters and words can seem a blunt tool, but it is a start and often far better than what some sceptics tell you. Listen again and again to an unfamiliar call and write down your own rendering, and compare the call to other bird sounds (or other sounds than those given by birds) you are already familiar with. Little by little you will improve your skills and eventually be able to confidently sort out the entire chorus in a forest during a walk in May.

All sound transcriptions appear in *italics*, and **bold face** denotes emphasised (louder, stressed) syllables. An attempt has been made to render the pace at which syllables are uttered, *vivivi* being very quick, *vi-vi-vi* somewhat slower, *vi vi vi* still less rapid, *vi, vi, vi* yet again even slower, and *vi ... vi ... vi* more like individual calls given in a slow, hesitant series. The choice of vowels is meant to indicate roughly the pitch, from low to high *a, u, o, e, ü, ee, i*, thus *a* (as in 'after') is meant to be lower than *u* (as in 'use'), which is subtly lower than *o* (as in 'over'), the others in rising pitch being *e* (like in 'pet'), *ü* (like in German 'über'), *ee* (like in 'steel') and *i* (as in 'ring'). The inclusion of the German 'y' (ü, 'umlaut u') rather than the English letter is because the English y is pronounced in a variety of ways, e.g. like 'ju' (at the beginning of a word), often more like a diphthong ('ai') or, at the end of a word, in the same way as 'i' and therefore not well suited for call renderings. The way the consonants have been selected can help convey whether a call starts or ends with a hard, abrupt sound or if it is softer; *kek* or *tac* are sharper and harder sounds than *gep* or *bipp*.

Similar species

Reading the 'Identification' sections is a useful start when trying to put a name to an unfamiliar bird. The sections called 'Similar species' are then useful as a final check that you have arrived at the correct conclusion. Read through them with an open mind. Bird identification is not always simple, and only the knowledgeable realise that it is all too easy to fall victim to a delusion. Better to double-check the crucial characters once too often than not often enough, and compare them critically with what you have seen.

Ageing & sexing

This book is aimed mainly at ordinary birdwatchers who observe free-flying birds in nature and try to identify them with the help of eyes and ears, albeit usually aided by the use of binoculars, telescopes and perhaps cameras with powerful lenses. But it should prove useful also to ringers and museum workers with the bird in the hand. The sections on ageing and sexing are detailed enough to solve most problems facing a ringer or museum researcher. We have as a rule only left out certain specialised characters such as colour of gape, presence of tongue spots, brood patch, cloacal protuberance and growth bars on remiges or tail-feathers, all characters better reserved for specialist ringers' guides. Still, with optical equipment and cameras becoming ever better, and birders being increasingly knowledgeable (and demanding), we have strived to make the sections on ageing and sexing as complete as possible for both ordinary birdwatching and 'advanced birding' in the field.

Each section first provides a brief overview of what can be identified in the field as regards ageing and sexing. Then follows a summary of the moult strategy of the species, or, if moult differs geographically, the indicated subspecies. Often, knowledge of moult is essential for correct ageing, and sometimes also sexing. If you are not yet fully familiar with what moult is and how it is performed by birds we recommend that you consult specialist literature, since here is not the proper place to deal with it. A brief treatment is offered by Svensson (1992), a somewhat more detailed one in Ginn & Melville (1983), and quite comprehensive accounts are found for example in Stresemann & Stresemann (1966), Jenni & Winkler (1994) and Howell (2010). However, some comments on how knowledge of moult and feather wear can be useful for interested birdwatchers might be appropriate, and therefore the basics are covered in a separate chapter, 'An approach to moult and ageing birds in the field', which follows directly after the Introduction (p. 20).

After moult has been outlined in each species account, there follows a guide to ageing and sexing usually arranged in the two broad seasons of 'Spring' and 'Autumn' (though sometimes 'Summer' or other seasons or combinations of seasons are used). Spring here corresponds to plumages encountered in early spring to late summer (or mid autumn depending on timing of moult), whereas autumn refers to plumages worn in late summer or mid autumn to late winter or early spring. This may sound a bit loose, but since breeding, migration and moult all vary in details depending on species or population, the diffuse definitions are deliberate and unavoidable.

Juvenile plumage is treated separately, provided it differs from later plumages, and entries on this can appear either under Spring (rarely) or Autumn (often), if not even under a separate heading 'Summer', which is sometimes used. We have attempted to find practical and useful solutions that correspond to the plumage development of a certain species rather than adhere very strictly to a fixed formula of headings.

Biometrics

One of the reasons for the long gestation time of this book was the decision, wherever possible, to examine museum material (or in some cases live birds) rather than citing existing literature. We went back to the birds themselves as often as we could in our efforts to provide an independent view. Still, for some of the species or subspecies we rely on the thorough and careful work done mainly by C. S. Roselaar for *BWP*, and a few other fully referenced sources. Finally, for some species we commissioned the competent help of Nigel Cleere, José Luis Copete and Roy Hargreaves to take measurements, and a few others who helped take some measurements are credited in the Acknowledgements. The work is based on an extensive examination of museum specimens (mainly in Tring, New York, Stockholm, Copenhagen and Paris, but also to some extent in Berlin, Bonn, Leiden, Vienna, Tel Aviv, Almaty, St Petersburg and Moscow), and targeted fieldwork (in, for example, Armenia, the Caucasus, Israel, Kazakhstan, Siberia, Mallorca, the Pyrenees, Morocco, Tunisia and Turkey). Nearly every currently listed subspecies in major existing handbooks and checklists has been examined, measured (mainly by LS) and compared with related subspecies or populations in order to assess their distinctness and validity. In total, according to a rough estimate, *c.* 700,000 measurements of more than 40,000 specimens have been taken for the passerines alone.

Why so many measurements? Well, this book is meant to serve not only field ornithologists but also ringers and museum workers, and measurements and details of wing formulae are often indispensable tools when separating closely related and very similar taxa. Detailed measurements are required for a fair assessment of geographical variation and for establishing whether a described taxon really differs from those already known. And undoubtedly some basic size measurements also enable the successful identification of a bird in the field. Total length has been obtained from rather long series' of skins, calculating an average size span only after discarding all extremes and oddly prepared specimens. Only specimens prepared in such a way that the bill is pointing forward have been used.

The remaining measurements have been taken following the standard set out in Svensson (1992) with one exception: for measuring bill length to feathering we measure the exposed culmen, which means that we measure from the tip of the bill to where the culmen of the bill *disappears* under the forehead feathering (provided this is undamaged), thus not to the base of the forehead feathers. This is because pushing the calipers to these feather bases, frequently damaged on museum specimens and from time to time also on trapped live birds, gives on average more biased and variable values. In a few species with dense, hair-like nasal feathering covering the basal part of the culmen (e.g. redpolls, shrikes and some corvids, to name a few) these feathers have been discounted and the measurement taken to the border between the true forehead feathers and the nasal feathers.

Bill to feathering is the length from tip of bill to where culmen disappears under the true forehead feathers—thus visible culmen length. (L. Svensson)

The two most difficult measurements to obtain accurately, in our experience, are tail length and bill to skull. Added to this is tarsus length on skins if the feet are not well prepared. Tail length is easier to measure on live birds than on skins, because the latter can have glue or stained blood hidden under the tail-coverts, which will catch the end of the ruler before this reaches the true base of the tail-feathers, causing the measurement to be read misleadingly short. And the tail can be fixed in an unnatural way to the body by the taxidermist and cause either high or low values. Only by using a pair of compasses (rather than a thin ruler inserted between the tail and undertail-coverts) when such a skin is encountered can this problem usually be overcome. If tail length is important for identification and you arrive at a surprising value it is a good idea to re-do the measurement once or twice with various tools and, better still, to let someone else measure it independently.

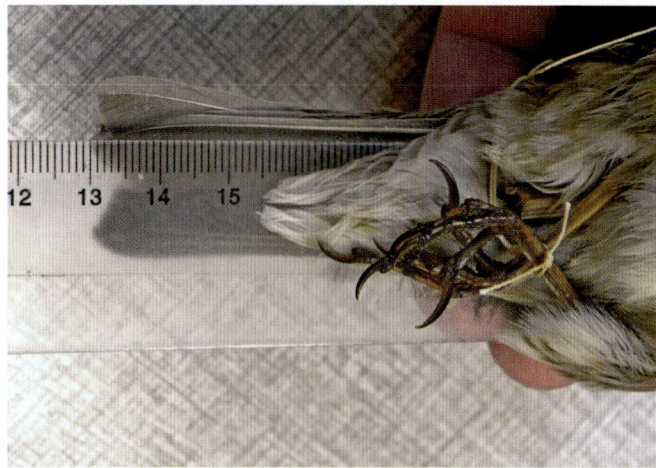

Total length is taken with a live bird in a natural relaxed position and only slightly stretched, bill pointing forward, the tip touching the zero-stop of a ruler. Measure to the tip of the longest tail-feather. When measuring skins it is important to only select specimens with a similar posture. (H. Shirihai)

Tail length is conveniently measured by inserting under the tail a thin ruler with the scale starting from the outer edge of one end and pushing it gently between the long tail-feathers (rectrices) and the longest undertail-coverts until it reaches the root of the central tail-feathers. Measure to the tip of the longest tail-feather. (L. Svensson)

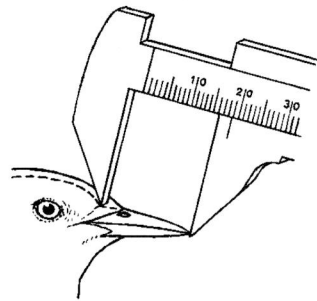

Bill to skull is best measured using sliding calipers of a sufficiently thick material (*c.* 3 mm). The outer caliper (the one pushed against the joint between skull bone and culmen of bill) should have a pointed profile. Push this caliper along the culmen until it reaches the start of the skull bone, then close the calipers until the inner one touches the tip of the bill. (L. Svensson)

The measurement of bill to skull shares the same basic limitation experienced with tail length: you cannot actually see the point to which you measure. There is sometimes an unevenness either on the foremost part of the skull bone or on the innermost part of the bill, and this creates a problem when locating (by feel only!) the groove or angle where the bill meets the skull. Always double-check any measurement troubled with such problems. In cases where we do not provide bill length to skull, only bill to feathering, this is because for these species in our experience the angle between bill and skull is too difficult to locate with accuracy for bill length to skull to be reliably applied.

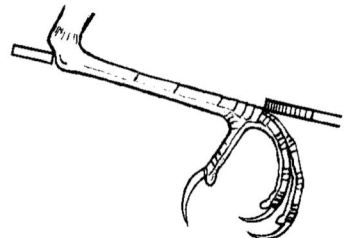

Technique for measuring the tarsus length on a skin. (L. Svensson)

Remember that the tarsus is measured from the groove or depression on the rear side of the ankle joint ('reversed knee') between the tarsus and the tibia, thus one should not include the lower knob of the tibia. However, it is more difficult to decide to where one should measure at the lower end of the tarsus. The rule-of-thumb to 'measure to the lower edge of the last complete scale on the front side before the toes diverge' is sometimes difficult to follow since you will encounter all kinds of intermediates between complete and divided scales at the base of the toes. The object should always be to obtain measurements that correspond to those given in the literature and ideally that are as similar as possible regardless of whether a skin or a live bird is measured. On a live bird it is natural to fold back the toes completely and measure to the lower end of the 'tarsus bone'. The toes of a dried skin are stiff so you need to figure out where the same point would be without being able to move the toes. We recommend that all complete scales are included and preferably one extra scale (towards the toes) rather than counting one too few if the scales are ambiguous.

The selection of measurements in this book varies somewhat between species reflecting their perceived usefulness for identification purposes. For all species wing and tail lengths are given with mean values (*m*) and sample sizes (*n*) in brackets. Measurements are divided between the two sexes if mean values differ in excess of 1–2%, or if sufficient specimens were available for examination. An average ratio between tail and wing lengths expressed as a percentage is also included for most taxa since this is often useful when assessing structure and for taxonomic considerations. Measurements of bill (sometimes several) and tarsus follow for nearly all species.

The section on biometrics is usually concluded with a description of the wing formula, i.e. the relative length of some key flight-feathers of the wing, usually expressed as the distance of the tip of each feather in relation to the wingtip (wt). One exception is the outermost short primary, which is measured in relation to the tip of the longest primary-covert (for a detailed explanation of the study and description of the wing formula, see for example Svensson 1992). Wing formula details have been limited to the basics thought to be particularly useful for identification.

Note that we number primaries from outside and inwards. Just as we do not walk backwards if we can avoid it we don't intend to describe the wing formula of a bird starting with p10, then p9 and so on backwards to p1. In moult studies the primaries are conveniently counted from inside and outwards as this is the same order in which they are replaced in 99.9% of the passerine species, but what works well for describing moult (just one specialised branch of ornithology) is not practical for dealing with the wing formula, where the interesting feathers are generally the outer ones. Just as it is possible for most people to cope with miles and kilometres at the same time, or euros and pounds, counting the primaries on a bird in two ways depending on circumstances presents no real obstacle.

We have included the distance between the two outermost primaries ('p1 < p2') for nearly all species except for those with a vestigial p1 (e.g. pipits, finches and buntings). The usefulness of this measurement for identification and for making taxonomic assessments was first noticed by HS in the 1980s. One advantage is that both these feathers, p1 and p2, are protected from excessive feather wear due to their sheltered position on the closed wing, allowing quite precise measurements to be taken even on heavily worn summer adults.

On the other hand, we do not include the so-called primary length (p3 length to base) since in our experience it is not unquestionably easier to take in a consistent way than the wing length according to the maximum method. But it is a fact that it ruffles the wing-feathers more, which is potentially harmful to the flight ability of handled live birds. In other words, we have independently arrived at partly similar conclusions as Gosler *et al.* (1995).

We nearly always provide both measurements and wing formula details for *each subspecies*, a novelty in this book. The current trend is to focus not just on species but on distinct subspecies as well—often the new species of tomorrow. Here we offer, possibly for the first time, fairly full coverage of all distinct subspecies of a large region.

Finally, it should be noted that total length is measured to the nearest half-centimetre, and feather-related measurements (wing and tail length, wing formula details) are measured to the nearest half-millimetre, and for these, whole-number values are not given with a decimal (thus we write '15' rather than '15.0'). For measurements of bare parts (bill, tarsus, claws, etc.), these are taken to the nearest decimal, i.e. tenth of a millimetre, and the decimal is invariably given also for even values (e.g. '6.0' rather than just '6'). These differences reflect in our experience a realistic level of accuracy that can be achieved and be repeated by others.

Geographical variation & range

This section opens with a summary of geographical variation within the species. If there is no geographical variation and no subspecies currently recognised, the species is said to be monotypic.

Under species with two or more recognised subspecies, each subspecies receives a separate entry. The entry starts with the scientific name plus author and year, followed by a brief summary of where it breeds and winters. Data on distribution is mainly based on available modern literature, but some information has also been taken from our extensive examination of museum specimens. A definition then follows of how a particular subspecies differs from others of the same species. We nearly invariably present separate biometrics and wing formula details for each subspecies. Where appropriate, comments are offered where our opinion differs from established handbooks and checklists. If our deviations are more complex to explain in the subspecies accounts, we have instead given them under 'Taxonomic notes' (see below). Each subspecies paragraph ends with a list of synonyms, mainly junior names which refer to the same population but which cannot be used under the present taxonomy and according to the rules of nomenclature. Since our research has led to a number of often cited subspecies being found insufficiently distinct and hence placed in synonymy, it is important to consult these lists to avoid the impression that they have been overlooked.

Our coverage of subspecies is complete only for the treated range, the Western Palearctic. For some species of particular interest we have also treated some (not necessarily all) extralimital subspecies to offer a better overview, or because occurrence of some of these within the treated range might be seen as reasonably likely in the future.

Taxonomic notes

Bird taxonomy has entered an era of great activity and exciting new results. This is largely due to the introduction of genetic analyses and new theoretical models. Some of the results and proposed changes appear to be solidly supported and are already followed by most authors, list editors and organisations. Others are more controversial and it seems a sound approach to let them rest a little, to be further discussed and tested independently by others before the proposals lead to changes in taxonomy. For studies relying only on mitochondrial DNA data, it is particularly desirable that the results are corroborated either by nuclear DNA or another 'independent' genetic method, or by morphology, vocalisations, behaviour, etc., before changes are implemented. Under 'Taxonomic notes' we give a few brief hints about ongoing work on the taxonomic scene and what currently is seen by us or others as a controversial question, or what remains to be discovered. We also hint at possible future splits of very distinct subspecies.

References

To save space we have restricted references under the species accounts to those in journals and a very few rarely seen books or sound recordings. We have omitted titles of papers, and only give the names of the journals (often abbreviated; see p. 18), volume or issue number, and page numbers. The inclusion of a reference does not mean that we necessarily agree with all of its conclusions (although this is frequently the case) but they are meant as a service to the reader to conveniently acquire a complete picture when studying a certain species. Handbooks, monographs and sound recordings have been used extensively throughout the preparation of this work and many would be cited under almost every species account if we were to include them there. Instead they are given in a comprehensive list of main references on pp. 25–26. Consulted atlases and checklists are also listed there.

Glossary and abbreviations

We have restricted the use of technical abbreviations in normal text to frequently utilised age categories, the four cardinal points in connection with geographical names (e.g. 'N Britain') and to numbering of certain wing-feathers or tail-feathers (e.g. 'p2' meaning 'second primary from the outside'). Abbreviations are listed below, and at the same time much of the terminology used is explained. In sections dealing with biometrics and wing formulae we have allowed for more abbreviations in order to gain space, since in these sections the same terms are repeated a multitude of times. They should still be easily understood given some practice. Some less common abbreviations are explained where they first appear. Note that we differ from several other books in focusing on the age of birds, not on plumages, since the latter vary considerably and often require specialist knowledge of moult.

Age and sex

ad	adult (mature bird in definitive plumage)
f.gr.	full grown(s) (fledged bird of unknown age)
imm	immature (young bird of unspecified age but still not adult)
juv	juvenile (young fledged bird in its first set of feathers)
1stS	first-summer bird (often = 2ndCY spring, a bird at age of about one year)
1stW	first-winter bird (strictly from start of replacement of juv plumage, often only partly, until when, in late winter or early spring, breeding/summer plumage starts to be attained; if exact plumage state cannot be ascertained, the term is used to refer to birds in their first autumn, winter and early spring)
1stCY	first calendar year (from hatching to 31 Dec of the same year)
1stY	first year of life, roughly speaking from hatching to first-summer
2ndCY	second calendar year (from 1 Jan the year after hatching to 31 Dec of the same year)
2ndS	second-summer bird (often = 3rd CY spring, at the age of about two years)
2ndW	second-winter bird (a bird which has attained winter plumage in its 2ndCY, either through moult or feather wear, often at the age of c. 14–16 months, and until the following spring)
♂, ♂♂	male, males
♀, ♀♀	female, females

Biometrics and wing formula

B	bill length (to skull unless otherwise stated; see opposite page)
B(f)	bill length to feathering (measured to where culmen is covered by real feathers, i.e. 'exposed culmen'; nasal hair or bristles not counted)
BD	bill depth (at feathering unless otherwise stated)
BD(n)	bill depth at inner (proximal) edge of nostril openings; when distal edge is meant this is stated as '**BD**(n dist)'
BW	bill width (at feathering unless otherwise stated)
BW(g)	bill width at gape ('corner of mouth', i.e. widest place where bill is hard)
emarg.	emargination of outer web of a primary
HC	hind claw length
L	length (total length from tip of bill to tip of tail, head/bill pointing forward but neck only gently outstretched)
m	mean value (often elsewhere abbreviated as '±')
MC	median (front) claw length
n	number (= sample size)
p, pp	primary, primaries (long outer wing-feathers forming the wingtip; abbreviation only appears combined with feather numbers, e.g. 'p3', 'pp3–5')
p1	first (outermost) primary
p10	tenth (innermost) primary
pc	primary-coverts (in wing formula descriptions referring to 'tip of longest')
r, rr	tail-feather(s) (= rectrix, -ices) (abbreviation only appears combined with feather numbers, e.g. 'r6', 'rr4–6')
s, ss	secondary(-ries) (somewhat shorter inner wing-feathers constituting the 'arm'; abbreviation only appears combined with feather numbers, e.g. 's1', 'ss3–6')
s1	first (outermost) secondary
T	tail length (from base of central tail-feathers to tip of longest feather)
TF	tail fork (distance between longest and shortest tail-feathers)
TG	tail graduation (the same definition as tail fork, only that the tail is rounded)
Ts	tarsus length (explained on opposite page)
W	wing length (distance from wing-bend, i.e. carpal joint, to tip of longest primary, wing-feathers flattened and straightened sideways against the ruler)
WS	wingspan (distance from wingtip to wingtip, wings gently stretched straight out)
wt	wingtip
>	more than, larger than
<	less than, smaller than

General

C	central (in connection with geographical name)
c.	circa (Lat.), approximately
cf.	confer (Lat.), compare with, see also
cm	centimetre
E	east, eastern (in connection with geographical name)
edn	edition
e.g.	exempli gratia (Lat.), for example
et al.	et alii (Lat.), and others
etc.	et cetera (Lat.), and so on
i.e.	id est (Lat.), that is to say
in litt.	in litteris (Lat.), in correspondence, written information
in prep.	in preparation, a reference underway but not yet published
Is	Islands
Jan, Feb	January, February, etc.
m	metre
mm	millimetre
Mts	mountains
N	north, northern (in connection with geographical name)

p., pp.	page, pages
pers. comm.	personal communication, verbally
S	south, southern (in connection with geographical name)
sec.	second(s)
s.l.	*sensu lato* (Lat.), in the broad sense
s.s.	*sensu stricto* (Lat.), in the narrow sense
ssp.	subspecies (geographically defined morphological variation within a species; the same thing as race, and these two words are used alternatively without signifying any difference of meaning)
syn.	synonym(s)
var.	*varietatis* (Lat.), variety, variation (usually = colour morph, in connection with scientific name)
vs	*versus* (Lat.), against; as opposed to
W	west, western (in connection with geographical name)
○	open circle in front of taxon name denotes a subtle subspecies that differs only in a very minor way from neighbouring subspecies (symbol appearing only under 'Geographical variation & range')
?	denotes a questionable subspecies, or one for which we have been unable to examine any or adequate material

Museums

AMNH	American Museum of Natural History, New York, USA
ISPRA	Istituto Superiore per la Protezione e la Ricerca Ambientale, Bologna, Italy
IRSNB	Royal Belgian Institute of Natural Sciences, Brussels, Belgium
JGUM	Johannes Gutenberg Universität, Mainz, Germany
MNHN	Muséum National d'Histoire Naturelle, Paris, France
NBC	Naturalis Biodiversity Centre, Leiden, the Netherlands
NHM	Natural History Museum, Tring, UK
NMW	Naturhistorisches Museum, Vienna, Austria
NRM	Naturhistoriska Riksmuseet, Stockholm, Sweden
TAUZM	Tel Aviv University Zoological Museum, Tel Aviv, Israel
UMCZ	University of Cambridge, Dept. of Zoology, Cambridge, UK
USNM	National Museum of Natural History, Smithsonian Institution, Washington, USA
WML	World Museum, Liverpool, UK
YPM	Yale Peabody University Museum, New Haven, USA
ZFMK	Zoologisches Forschungsmuseum Alexander Koenig, Bonn, Germany
ZIA	Zoological Institute, Almaty, Kazakhstan
ZISP	Zoological Institute, St Petersburg, Russia
ZMB	Museum für Naturkunde, Berlin, Germany
ZMMU	Zoological Museum, Moscow University, Russia
ZMUC	Zoologisk Museum, Copenhagen, Denmark
ZMUO	Zoologisk Museum, Oslo, Norway

Handbooks, journals and other references

The following are the most commonly used abbreviations of reference names.

Acta Orn.	Acta Ornithologica
Amer. Mus. Novit.	American Museum Novitates
BB	British Birds
BBOC	Bulletin of the British Ornithologists' Club
Biol. J. Linn. Soc.	Biological Journal of the Linnean Society
Bonn. zool. Beitr.	Bonner zoologische Beiträge
Bonn. zool. Monogr.	Bonner zoologische Monographs
Bull. AMNH	Bulletin of the American Museum of Natural History
Bull. OBC	Bulletin of the Oriental Bird Club (name changed to BirdingASIA in 2004)
Bull. OSME	Bulletin of the Ornithological Society of the Middle East
BW	Birding World
BWP	The Birds of the Western Palearctic
DB	Dutch Birding
HBW	Handbook of the Birds of the World
H&M 4	The Howard & Moore Complete Checklist of the Birds of the World, 4th edn
J. Avian Biol.	Journal of Avian Biology
J. Bombay N. H. S.	Journal of the Bombay Natural History Society
J. Evol. Biol.	Journal of Evolutionary Biology
J. f. Orn.	Journal für Ornithologie
J. of Orn.	Journal of Ornithology
Linn. Soc. Zool. J.	Zoological Journal of the Linnean Society
Mol. Phyl. & Evol.	Molecular Phylogenetics and Evolution
Notatki Orn.	Notatki Ornitologiczne
Novit. Zool.	Novitates Zoologicae
Orn. Beob.	Ornithologische Beobachter
Orn. Fenn.	Ornis Fennica
Orn. Jahrb.	Ornithologische Jahrbuch
Orn. Monatsber.	Ornithologische Monatsberichten
Orn. Vestnik	Ornitologicheskii Vestnik (*Messager ornithologique*)
PLoS ONE	Public Library of Science (open-access journal)
Ring. & Migr.	Ringing and Migration
Riv. ital. Orn.	Rivista italiana di Ornitologia
Ross. Orn. Zhurn.	Rosskii Ornitologicheskii Zhurnal (Russian Journal of Ornithology)
Sandgr.	Sandgrouse
Vår Fågelv.	Vår Fågelvärld
Zool. Med. Leiden	Zoologische Mededelingen, Leiden
Zool. Verh. Leiden	Zoologische Verhandelingen, Leiden
Zool. Zhurn.	Zooligicheskii Zhurnal
Zoosyst. Rossica	Zoosystematica Rossica

Selected Gazetteer

Arabia	Sometimes loosely used for the Arabian Peninsula, by and large south of the political borders of Jordan and Iraq.
Balearics	Group of islands in W Mediterranean Sea, including Mallorca, Menorca, Ibiza and adjacent smaller islands.
Balkans	Imprecise region in SE Europe usually comprising Croatia, Bosnia and Herzegovina, Albania, Greece, Macedonia, Montenegro, Bulgaria and E Romania. It takes its name from the mountain range with the same name.
Baltic States	The three countries of Estonia, Latvia and Lithuania.
Britain	See *Great Britain*.
Central Asia	(sometimes referred to as 'Russian Turkestan') includes the countries east of the Caspian Sea but does not include Iran, Pakistan, Tibet or other parts of China, nor Mongolia. The area thus includes Kazakhstan, Uzbekistan, Turkmenistan, Kyrgyzstan, Tajikistan and Afghanistan.
Cyclades	Group of Greek islands in the S Aegean Sea north of Crete.
Eurasia	The vast landmass formed by Europe and Asia.
Fenno-Scandia	Part of N Europe including Sweden, Norway, Finland and westernmost Russia (Murmansk Oblast, Karelia), thus excluding Denmark and Iceland.
Great Britain	England, Wales and Scotland, but not including Northern Ireland, Isle of Man or the Channel Islands. Also known simply as 'Britain'.
Gulf States	The Arab states around the Persian Gulf: Iraq, Kuwait, Bahrain, United Arab Emirates, Qatar, Saudi Arabia and Oman. Generally, only areas near the Gulf are included.
Hala'ib Triangle	Also known as Gebel Elba. Disputed territory on the border between Egypt and Sudan, on the Red Sea coast.
Iberia	The SW European peninsula comprising mainland Spain and Portugal.
Kalmykia	The plains NW of the Caspian Sea.
Kirghiz Steppe	Traditional name for the mainly vast grassy plains in N Kazakhstan, from east of the Volga River eastward to Zaisan Lake (thus not within Kyrgyzstan as the name might imply). Frequently now called the Kazakh Steppe, though not in HWPB.
Kurdistan	The region inhabited mainly by Kurdish people which encompasses parts of E and SE Turkey, N Syria, N Iraq and NW Iran.
Levant	A rather imprecise term here applied to the land on the eastern shores of the Mediterranean Sea. Commonly, Cyprus, a corner of S Turkey, Syria, Lebanon, Israel, the Palestine territory, Jordan and the Sinai Peninsula are included.
Macaronesia	Group of islands in the N Atlantic including Azores, Madeira, Canaries and Cape Verde.
Maghreb	A region of NW Africa traditionally encompassing the Atlas Mts and the coastal plains of Morocco, Algeria, Tunisia and Libya. It now comprises the whole of these four countries plus Western Sahara and Mauritania, but excluding Egypt (cf. North Africa).
Middle East	The same as Levant but with the addition of the whole of Turkey and Egypt, plus Iraq, Iran and the entire Arabian Peninsula. (Sometimes called the 'Near East'.)
Nordic Countries	Comprises NW Europe and includes Iceland, Norway, Denmark, Sweden and Finland.
North Africa	Here defined as the African countries mainly north of the Sahara, thus Morocco, Algeria, Tunisia, Libya and Egypt.
North Sea	Part of the N Atlantic Ocean located between Norway, Denmark, the British Isles, Germany, the Netherlands, Belgium and France.
Palearctic Region	('Arctic region of the Old World') One of eight ecological regions on Earth, covering Europe, N Africa, N Arabia and N Asia north of the Himalayas (but see elsewhere for specific definition of the Western Palearctic region for this handbook).
Sahel	Arid savanna region along the southern border of the Sahara Desert, Africa.
Scandinavia	The three countries of Denmark, Norway and Sweden (but not including Finland).
Siberia	Defined here in the broadest sense, thus reaching from the Ural Mts eastward to the Bering Strait and along the Pacific coast south to Ussuriland and also including Sakhalin, in the south to the borders of Kazakhstan, Mongolia and China. The eastern part of Siberia is sometimes known as the 'Russian Far East', but is here included in 'Siberia'. The Russian Far East is sometimes restricted to the extreme SE part of Siberia, i.e. Amurland and Ussuriland.
Transbaikalia	Area south and south-east of Lake Baikal in SC Siberia, including regions often referred to as Dauria and Chita.
Transcaspia	Area east of the Caspian Sea, mainly W Kazakhstan, W Uzbekistan and Turkmenistan.
Transcaucasia	Area south of the Caucasus, mainly extreme E Turkey, Georgia, Armenia and Azerbaijan.
Turkestan	See 'Central Asia'.
Tyrrhenian Islands	Corsica, Sardinia and smaller adjacent islands along the Italian coast.

AN APPROACH TO MOULT AND AGEING BIRDS IN THE FIELD

Many birds can be aged even in the field, using knowledge of plumage variation. One needs to take into account both the normal plumage colours and patterns, and seasonal variation due to moult, wear and bleaching. Knowing the age of a bird often increases the value of an observation. A brief outline of moult in passerines is therefore offered. Moult, the replacement of an old feather and growth of a new in its place, is essential for maintaining flight ability, heat and rain insulation, and sometimes serves to attain plumage patterns important for successful social interaction or camouflage. The examples of ageing Pine Grosbeak and Bullfinch in autumn based on moult, feather shape and feather wear explain well what to look for.

The shape and quality of tail-feathers differ in many passerine species between ad and 1stW, as this example of old (top; Sweden, Feb) and young (bottom; Sweden, Jan) Pine Grosbeak shows. Ad has on average broader feathers with more amply rounded tips, and the feathers are of a slightly denser and glossier texture enduring better abrasion and bleaching. 1stW tail-feathers are a trifle narrower with more pointed tips, and they are often subtly less glossy and wear quicker at the tips, sooner acquiring a ragged edge. Note that not all species show such obvious differences as in Pine Grosbeak, some requiring care and practice, and many even overlap in their extremes. (L. Svensson/NRM)

Ageing passerines according to complete or partial moult of greater coverts is practical in many species, in particular among thrushes and chats. However, the method is less often applicable for the families covered in this volume, either due to complete or non-existent moult of these coverts in the majority of young birds, or due to minute differences between ad and 1stW coverts pattern. Still, in a few species it can be applied at least for some birds, frequently among finches like here for Bullfinch, where a fair portion of young birds (bottom) leave a few outer juv greater coverts unmoulted differing markedly in being shorter and having tips tinged yellowish-buff or pale brown rather than greyish-white as in ad (top). Some 1stW Bullfinches leave only the carpal covert unmoulted, and this normally requires handling to be seen. Note that even ad Bullfinch has contrastingly greyer-looking primary-coverts and alula, but with experience one can notice that these feathers are slightly glossier than in 1stW. (L. Svensson/NRM)

Timing

The timing of the moult depends both on migrations and the breeding cycle. As is easily understood, moult involves production of numerous feathers and costs energy, and this is why it is performed either after breeding on the breeding grounds or in the winter quarters, thus not during any of the similarly energy-intensive migrations. Therefore, there are two main moulting periods—in *late summer* and during *winter*.

Terminology

The summer moult is often conducted close to the breeding grounds and generally completed before the start of the autumn migration (exceptions are few and insignificant). In adult birds that have just finished breeding, this moult is called the *post-nuptial moult* (literally 'after breeding moult'), while in young and newly fledged birds it is called the *post-juvenile moult*. In the post-juvenile moult the young birds replace all or most of their juvenile feathers that were grown in the nest. In a complete moult of remiges, primaries are nearly always replaced descendently (from inside out) and secondaries ascendently (from outside in), feathers being grown more or less symmetrically in both wings.

The winter moult is conducted sometime between the autumn and spring migrations, generally on the wintering grounds. It is called the *pre-nuptial moult* ('before breeding moult') regardless of the age of the bird. This is the moult in which many species acquire their breeding plumage. Most tropical migrants perform such a winter moult, while most short-distance migrants do not.

Keeping the above in mind, we now need to add the *extent* of the moult in order to describe it properly:

Complete moult includes the whole plumage and results in a uniform one-generation plumage.

Partial moult includes a part of the plumage, resulting in an often visible contrast between different feather generations. In some species the partial moult may be restricted to some head and body feathers, while other species moult more extensively and include all contour-feathers of the body, a variable number of wing-coverts, tertials, tail-feathers and sometimes even a few secondaries or primaries, but never the whole plumage.

With the above basic knowledge of passerine moult, it is convenient to combine the timing and extent of the moult in order to simplify the terminology. We follow Svensson (1992) using the following simple terms (upper case serves to signal complete, lower case partial):

Summer complete moult (**SC**) is a complete moult during late summer.

Summer partial moult (**sp**) is a partial moult during late summer.

Winter complete moult (**WC**) is a complete moult during winter.

Winter partial moult (**wp**) is a partial moult during winter.

Further, apart from the general rules outlined above and which hold true for the vast majority of species, it should be noted that a few species show a moult pattern that is more complicated, and in order to facilitate the description of these we use the following three additional terms or categories:

Arrested moult starts like a complete moult, involving the moult of some remiges, but is arrested (usually because of migration) before it is completed, and is not resumed later on. Hence, the rest of the unmoulted remiges will not be moulted until in the next complete moult.

Suspended moult starts like a complete moult, involving the moult of remiges, but is suspended (usually because of migration) before it is completed, but unlike in arrested moult it is subsequently resumed from the point of suspension, after arrival at winter or summer grounds. Thus, in the end all remiges will be renewed.

Partial moult of remiges is an energy-saving strategy to moult only some remiges, often the outer being most important for flight ability, while the rest are left unmoulted. This, for instance, is found in immatures of some species of warblers and shrikes wintering in the tropics.

The assessment of moult

Since adult and young birds often differ in the extent of the moult (complete vs partial), this provides a tool for separating these two categories. One could say that ageing a bird is much about assessing the moult history of the individual. Presence or absence of *moult limits* tells us whether the last performed moult was partial or complete, and combined with knowledge of both the normal plumage and the moult pattern of that very species, the age can frequently be derived with good certainty.

Finding a moult limit may be anything from straightforward to 'close to impossible'. In very many species, different feather generations normally differ from each other in colour, pattern, shape, structure and length. Our chances of finding such limits increase significantly if we know the (average) extent of moult used by the species we are looking at. Moult limits that are situated *within*, for example, the greater coverts, median coverts or tertials are generally much easier to find than those appearing *between* different feathers groups.

Note that in some groups of birds *both* adults and first-summers have a partial winter moult and therefore return in spring with a moult limit. It is thus essential to learn in which groups the method works in spring. Differences between the two age categories may still exist, but they are much more limited and often impossible to be certain of. Similarly, in some groups, juveniles perform a complete post-juvenile moult in the same season when their adults moult completely. These groups of birds accordingly cannot be aged using the moult limit method.

Juvenile feathers are grown rapidly and are inevitably of a slightly less endurable quality. This becomes especially important to know when examining spring birds with some retained juvenile feathers (often primary-coverts and remiges, sometimes also tail-feathers), since wear and bleaching tend to affect juvenile feathers more than feathers of later generations. This is helpful to know when trying to differentiate between worn adult feathers and worn juvenile feathers.

A warning is also called for: quite often '*false moult limits*' can appear, potentially leading to wrong conclusions about the age. Such false moult limits can occur when some inner (or outer) greater coverts naturally have a different colour or pattern from the rest, which therefore may look like a moult limit. Another common reason is when some feathers have been better preserved from wear and bleaching due to their more sheltered position in the wing, rendering them the appearance of being of a more recently moulted feather generation. Starvation or aberrations when feathers are grown might leave odd feathers shorter than their full length, again potentially appearing as a moult limit. Accidentally lost and replaced feathers may simulate a moult limit, so it is best to check both wings, since moult is nearly always symmetrical. The only way to avoid falling into these many traps is practice, and adopting a cautious approach in general.

The different main moult strategies

Below, comprehensive descriptions of the main moult strategies shown by Palearctic passerines are given. However, one should keep in mind that there are numerous exceptions from the rules, and individual variation may occur which is not described.

Strategy 1

Adult: SC – Young: sp (no winter moult in either age)

This moult strategy is found in, for example, some chats and most or all thrushes, tits, corvids and finches. The summer complete moult of adults results in a uniform one-generation plumage. The summer partial moult of young birds involves only a part of the plumage and results in two generations of feathers, with some contrast in wear or pattern between them. Since none of these groups conduct any moult during the winter, the same plumages are present also in spring, but by then are more worn.

When ageing by the pattern and colour of greater coverts, in some species a contrast among the innermost greater coverts can be misleading, as the example of this 1stS ♂ Chaffinch demonstrates (Germany, Mar). Whereas most central coverts are broadly white-tipped, the inner two or three are darker with more obvious yellow tinge on tip end edges. Still, all these are of same generation, the contrast constituting a so-called false moult limit. The true moult limit on this bird is to be found among the outermost greater coverts. The age is supported by the rather pointed and worn tail-feathers. (R. Martin)

Bramblings attain full summer plumage in post-nuptial (ad) or post-juv (1stY) moults in late summer, but at first pale buff or whitish tips to many feathers conceal much of the bright feather pattern of summer. Summer plumage is developed through successive feather wear during the winter season, not by moult. As can be ascertained by these two ♂♂, one ad (top; England, Feb) and one 1stW (bottom; Italy, Feb), the older bird must not by necessity appear neater than the young! Safest ageing in finches always after assessing colour and wear of primary-coverts, primaries, supported by shape of unmoulted tail-feathers. But here also on the 1stW bird by the retained outer three juv greater coverts being a little shorter and duller, and lacking rufous on tips. (Top: G. Thoburn; bottom: D. Occhiato)

Ageing of some finches can be challenging due to extensive moult of greater coverts in young birds, but in this example of ♂ Siskins, ad (top; Italy, Mar) has brighter yellow colour, broader greater coverts wing-bar and less whitish tips on black crown, whereas in the 1stW (bottom; France, Oct) at least one outer greater coverts is retained juv, duller and thinly tipped and edged off-white rather than yellow. Latter bird also has more subdued yellow colour and more extensive white tips on dark crown. That the ad has pointed and abraded tail-feathers is not unusual at this season. (Top: A. Audevard; bottom: D. Occhiato)

Strategy 2

Adult: SC, wp – Young: sp, wp

This moult strategy is found in, for example, pipits, wagtails, *Sylvia* warblers and buntings. The summer complete moult of adults results in a uniform one-generation autumn plumage. The summer partial moult of young birds involves only a part of the plumage and results in two generations of feathers, with some contrast in wear or pattern between them. In winter, both ages have a partial moult, which means that all birds will show a moult contrast during spring, regardless of age. When ageing such birds during spring, the following should be noted: (i) In some young birds the winter partial moult is *more extensive* than the summer partial. For those species both ages will show a single moult contrast in the wing, and ageing must be based on an assessment of the feathers that were not moulted during winter (often the outer greater coverts, all primary-coverts and remiges). One needs to decide whether they are of adult type (moderately worn) as in 3rdCY birds or older, or whether they are of juvenile type (more worn and bleached) as in 2ndCY birds. – (ii) In some young birds, the winter partial moult is slightly *less extensive* than the summer partial. In such cases a 2ndCY bird will sometimes show *two* moult contrasts (with three feather generations involved), while a 3rdCY bird or older always will show only one moult contrast (two feather generations present). – (iii) In some species the winter partial moult is restricted to body-feathers alone and does not affect the wing at all.

Flycatcher ♂♂ of genus *Ficedula* (Collared Flycatcher here) moult in late summer after breeding and attain a ♀-like brown-grey plumage. Still, ageing and sexing possible by details in plumage pattern. Age by white edges to tertials narrowing smoothly around tip in ad (top; Egypt, Sep), forming a step at shaft in 1stW, most obvious on central tertial (bottom; Netherlands, Oct), apart from size of basal white primary patch, larger in ad. Sex based on jet-black primaries, tail-feathers and bases to outer greater coverts (ad) or these feathers being at least quite dark brownish-black (1stW); ♀♀ have dark brown flight-feathers, not black. Note also that ad ♂ usually attains white tail-sides in autumn, while tail is usually all black in spring. (Top: E. Winkel; bottom: A. Ouwerkerk)

Although age of these two Collared Flycatcher ♂♂ in spring, ad (top; Denmark, May) and 1stS (bottom; Sweden, Aug), can immediately be ascertained by the jet-black primaries of ad with a very large white primary patch, differing from the brownish-tinged primaries with a variably small primary patch in 1stS, it is worth noting that both age categories replace tertials and greater coverts in a pre-nuptial late-winter moult and thus do not differ in the pattern of these. Ad has also on average a larger white forehead patch than 1stS. (Top: E. Foss Henriksen; bottom: L. Jonsson)

Strategy 3

Adult: SC – Young: SC

This moult strategy is applied by larks, starlings, some sparrows and a few other odd species. Both adult and young birds have a summer complete moult. Once this moult is completed, the ages are inseparable by plumage.

Just as with the example above of Brambling ♂ attaining its neat summer plumage solely through wear of feather tips, the same is true for the plumage development of Spanish Sparrow (above) and other species having the moult strategy 3. The freshly moulted autumn male (top; Israel, Nov) has some buff and white tips to many feathers making the pattern appear slightly dull, whereas in summer (bottom; Spain, Jun) all these tips are worn off and the plumage is very neat, bright and contrasting. Ageing impossible, though, due to complete summer moult of both age categories. (Top: A. Ben Dov; bottom: C. N. G. Bocos)

Strategy 4

Adult: sp, WC – Young: sp, WC

This moult strategy involves, apart from some swallows and many warblers wintering in the tropics, Golden Oriole and *Lanius* shrikes. Following the partial summer moult, all birds regardless of age category show moult contrasts. However, the ages are usually easily separated during the autumn since the remiges of adult birds are eight to ten months old and by now worn, while the same feathers in the young birds were grown in the nest only one to three months ago and are still fresh. During the winter, all birds have a complete moult after which the ages cannot be separated. It should also be said that quite a few species show prolonged or differentiated periods of moult during winter, which may result in moult contrasts that are of little or no use for ageing during spring.

The Red-backed Shrike is a typical example of a species following strategy 4 when moulting, which means that both age categories moult completely in their tropical winter quarters making ageing in the spring normally impossible. Extent of summer moult is slightly variable. The breeding summer ♀ has plain brown back and boldly barred breast and flanks (top; Iran, Jun), whereas some summer ♀♀ with remnants of upperparts barring and also of some immature pattern on tertials (centre; Kuwait, Aug) are presumed to be 1stS (though could also represent individual variation). 1stW in late summer and autumn (bottom; Israel, Oct) are readily recognised by having whole of upperparts boldly barred dark including dark subterminal bars on tail-feathers. (Top: E. Winkel; centre: M. Pope; bottom: A. Ben Dov)

Black-headed Bunting (Turkey, Jun) together with Red-headed Bunting are the only two buntings undergoing the complete moult (of both age categories) in winter quarters, hence cannot be aged in spring. (H. Shirihai)

Strategy 5

Arrested and suspended moults, partial moult of remiges

A general description of arrested and suspended moult can be found above under Terminology. It may be worth noting that both arrested and suspended moults are conducted in the same sequence as a regular complete moult, only being interrupted before completed (arrested moult) or introducing a lengthy pause halfway before completion (suspended moult), whereas the partial moult of remiges often starts about halfway out in the wing leaving the inner primaries unmoulted, and if any secondaries are moulted usually some inner ones are kept. Single species with either of these strategies are found in several diverse genera, good examples being Woodchat Shrike, northern populations of Isabelline Shrike and warblers such as several of the genera *Locustella* and *Sylvia*, and Great Reed Warbler. Barred Warbler and Ortolan Bunting are noteworthy for, as a rule, moulting their primaries in summer and secondaries in winter.

Woodchat Shrike typically moults incompletely in its first winter, returning in spring with moult contrasts in wing with some retained juv more brownish inner primaries, outer secondaries and (most) primary-coverts (♂ left; Oman, Mar). Young of the eastern subspecies *niloticus* often moult more extensively than ssp. *senator* or *badius* in Europe, many even replacing some outer primary-coverts to ad type as shown here (♂ right; Israel, Apr), a fact which can support racial identification. (Left: M. Varesvuo; right: Y. Kiat)

Masked Shrike has a complex and variable moult pattern and is clearly referable to Strategy 5, where suspended and arrested moult occur, and moult strategies include many odd variations. Both birds above are late summer non-ad birds (2ndW) based on having more than one generation of feathers in wing. At left, 2ndW ♀ (Israel, Aug) tertials are of two generations, and inner primaries have been more recently moulted than outer. The spread wing at right is of a similar ♀ (Israel, Aug), has three generations of feathers, outer three primaries and outer four secondaries (plus two outer primary-coverts) are retained juv, the rest of inner primaries and second longest tertial are post-nuptial (very recently grown), whereas rest of tertials and innermost secondaries are pre-nuptial feathers from late winter. (Left: E. Hadad; right: Y. Kiat)

GENERAL REFERENCES

The books, checklists, atlases and sound recordings below have been consulted throughout the preparation of this handbook and have been an indispensable source of knowledge and reference. Only references directly referable to the content of this volume have been included.

Ali, S. & Ripley, S. D. (1987) *Compact Handbook of the Birds of India and Pakistan*. 2nd edn. Oxford University Press, Oxford.

Alström, P., Colston, P. & Lewington, I. (1991) *A Field Guide to the Rare Birds of Britain and Europe*. HarperCollins, London.

Alström, P. & Mild, K. (2003) *Pipits & Wagtails of Europe, Asia and North America*. Christopher Helm, London.

Alström, P. & Mild, K. (in prep.) *Larks of Europe, Asia and North America*. Christopher Helm, London.

Amadon, D. (1949) The seventy-five per cent rule for subspecies. *Condor*, 51: 250–258.

Andrews, I. J. (1995) *The Birds of the Hashemite Kingdom of Jordan*. Privately published, Musselburgh.

Ash, J. & Atkins, J. (2009) *Birds of Ethiopia and Eritrea – An atlas of distribution*. Christopher Helm, London.

Ayé, R., Schweizer, M. & Roth, T. (2012) *Birds of Central Asia*. Christopher Helm, London.

Balmer, D. E., Gillings, S., Caffrey, B. J., Swann, R. L., Downie, I. S. & Fuller, R. J. (2013) *Bird Atlas 2007–2011; The breeding and wintering birds of Britain and Ireland*. BTO, Thetford.

Baumgart, W. (2003) *Birds of Syria*. 2nd edn. OSME, Sandy.

Beaman, M. (1994) *Palearctic Birds*. Harrier Publications, Stonyhurst.

Beaman, M. & Madge, S. (1998) *The Handbook of Bird Identification for Europe and the Western Palearctic*. Christopher Helm, London.

Beolens, B., Watkins, M. & Grayson, M. (2014) *The Eponym Dictionary of Birds*. Bloomsbury, London.

Bergmann, H.-H., Helb, H.-H. & Bauman, S. (2010) *Die Stimmen der Vögel Europas*. DVD. Aula, Wiebelsheim.

Bønløkke, J., Madsen, J. J., Thorup, K., Pedersen, K. T., Bjerrum, M. & Rahbek, C. (2006) *Dansk Traekfugleatlas*. Rhodos, Humlebæk.

Borrow, N. & Demey, R. (2001) *Birds of Western Africa*. Christopher Helm, London.

Brazil, M. (2009) *Birds of East Asia*. Christopher Helm, London.

Brewer, D. (2001) *Wrens, Dippers and Thrashers*. Christopher Helm, London.

Bundy, G. (1976) *The Birds of Libya*. BOU, London

Chappuis, C. (1987) *Migrateur et hivernants*. 2 cassettes. Grand Couronne.

Chappuis, C. (2000) *African Bird Sounds – 1. North-West Africa, Canaria and Cap-Verde Islands*. 4 CDs. SEOF/NSA, London.

Clarke, T. (2006) *Birds of the Atlantic Islands*. Christopher Helm, London.

Clement, P. & Hathway, R. (2000) *Thrushes*. Christopher Helm, London.

Clements, J. F. (2000) *Birds of the World. A Checklist*. 5th edn. Pica Press, Robertsbridge.

Cramp, S. & Perrins, C. M. (eds.) (1993–94) *The Birds of the Western Palearctic*. Vols. 7–9. Oxford University Press, Oxford.

Cramp, S., Simmons, K. E. L., Perrins, C. M. & Snow, D. W. (eds.) (2006) *BWPi*, version 2.0. [Includes film footage and sounds.] Oxford University Press, Oxford.

Dementiev, G. P. & Gladkov, N. A. (eds.) (1953–54) *Ptitsy Sovietskogo Soyuza*. Vols. 5–6. [In Russian.] (English translation, 1968, *Birds of the Soviet Union*. Jerusalem.)

Dickinson, E. C. (ed.) (2003) *The Howard & Moore Complete Checklist of the Birds of the World*. 3rd edn. Christopher Helm, London.

Dickinson, E. C. & Christidis, L. (eds.) (2014) *The Howard & Moore Complete Checklist of the Birds of the World*. 4th edn. Vol. 2. Aves Press, Eastbourne.

Dickinson, E. C., Overstreet, L. K., Dowsett, R. J. & Bruce, M. D. (2011) *Priority! The Dating of Scientific Names in Ornithology*. Aves Press, Northampton.

Dunn, J. L., Blom, E. A. T., Alderfer, J. K., Watson, G. E., Lehman, P. E. & O'Neill, J. P. (2006) *National Geographic Field Guide to the Birds of North America*. 5th edn. National Geographic, Washington.

Equipa Atlas (2008) *Atlas das aves nidificantes em Portugal (1999–2005)*. ICN & B. Assírio & Alvim, Lisbon.

Eriksen, H. & Eriksen, J. (2010) *Common Birds of Oman – An Identification Guide*. 2nd edn. Al Roya Publishing, Muscat.

Eriksen, J. & Reginald, V. (2013) *Oman Bird List*. 7th edn. Center for Environmental Studies and Research, Sultan Qaboos University, Muscat.

Estrada, J., Pedrocchi, V., Brotons, L. & Herrando, S. (eds.) (2004) *Atlas dels ocells nidificants de Catalunya 1999–2002*. Lynx Edicions, Barcelona.

Flint, P. R. & Stewart, P. F. (1992) *The Birds of Cyprus*. 2nd edn. BOU, Tring.

Fransson, T. & Hall-Karlsson, S. (2008) *Swedish Bird Ringing Atlas*. Vol. 3. Passerines. Naturhistoriska Riksmuseet & Sveriges Ornitologiska Förening, Stockholm.

Fry, C. H., Keith, S. & Urban, E. K. (eds.) (1992–2004) *The Birds of Africa*. Vols. 4–7. Academic Press, London.

Gallagher, M. D. & Woodcock, M. W. (1980) *The Birds of Oman*. Quartet Books, London.

Gavrilov, E. & Gavrilov, A. (2005) *The Birds of Kazakhstan*. Tethys, Almaty.

Gill, F. & Donsker, D. (eds.) (2016) *IOC World Bird List*. Version 6.2. Doi: 10.14344/IOC.ML.6.2.

Gill, F. & Wright, M. (2006) *Birds of the World: Recommended English Names*. Princeton University Press, Princeton.

Ginn, H. B. & Melville, D. S. (1983) *Moult in Birds*. BTO guide no. 19. Tring.

Glutz, U. N., Bauer, K. & Bezzel, E. (eds.) (1966–1998) *Handbuch der Vögel Mitteleuropas*. Vols. 1–14. Aula-Verlag, Wiesbaden.

Goodman, S. M. & Meininger, P. L. (eds.) (1989) *The Birds of Egypt*. Oxford University Press, Oxford.

Gorman, G. (1996) *The Birds of Hungary*. Christopher Helm, London.

Gosler, A. G., Greenwood, J. J. D., Baker, J. K. & King, J. R. (1995) A comparison of wing length and primary length as size measures for small passerines. *Ring. & Migr.*, 16: 65–78.

Grimmett, R., Inskipp, C. & Inskipp, T. (1998) *Birds of the Indian Subcontinent*. Christopher Helm, London.

Grimmett, R., Inskipp, C. & Inskipp, T. (2011) *Birds of the Indian Subcontinent*. Compact version. Christopher Helm, London.

Grimmett, R., Roberts, T. & Inskipp, T. (2008) *Birds of Pakistan*. Christopher Helm, London.

Gulledge, J. (ed.) (1983) *A Field Guide to Bird Songs of Eastern and Central North America*. 2nd edn. 2 cassettes. Cornell University Press/Houghton Mifflin, Boston.

Hagemeijer, W. J. M. & Blair, M. J. (1997) *The EBCC Atlas of European Breeding Birds*. T. & A.D. Poyser, London.

Handrinos, G. & Akriotis, T. (1997) *The Birds of Greece*. Christopher Helm, London.

Harrison, C. (1982) *An Atlas of the Birds of the Western Palaearctic*. Collins, London.

Hartert, E. (1903–21) *Die Vögel der Paläarktischen Fauna*. Vols. 1–2. Friedländer & Sohn, Berlin.

Hartert, E. (1921–22, 1932–38) *Die Vögel der Paläarktischen Fauna*. Vols. 3–4. ('Ergänzungsband', vol. 4, ed. by F. Steinbacher). Friedländer & Sohn, Berlin.

Hollom, P. A. D., Porter, R. F., Christensen, S. & Willis, I. (1988) *Birds of the Middle East and North Africa*. T. & A.D. Poyser, Calton.

Howell, S. N. G. (2010) *Molt in North American Birds*. Houghton Mifflin, Boston.

Howell, S. N. G., Lewington, I. & Russell, W. (2014) *Rare Birds of North America*. Princeton University Press, Princeton.

del Hoyo, J., Elliott, A. & Christie, D. A. (eds.) (2004–12) *Handbook of the Birds of the World*. Vols. 9–16. Lynx Edicions, Barcelona.

International Commission on Zoological Nomenclature (ICZN) (1999) *International Code of Zoological Nomenclature*. International Trust for Zoological Nomenclature, London. (www.iczn.org)

Isenmann, P. & Moali, A. (2000) *Birds of Algeria*. SEOF, Paris.

Isenmann, P., Gaultier, T., El Hili, A. Azafzaf, H., Dlensi, H. & Smart, M. (2005) *Birds of Tunisia*. SEOF, Paris.

Jännes, H. (2002) *Calls of Eastern Vagrants*. CD. Early Bird, Helsinki.

Jännes, H. (2002) *Bird Sounds of Goa & South India*. CD. Early Bird, Helsinki.

Jenni, L. & Winkler, R. (1994) *Moult and Ageing of European Passerines*. Academic Press, London.

Jennings, M. C. (2010) *Atlas of the Breeding Birds of Arabia*. Fauna of Arabia, Frankfurt & Riyadh.

Jobling, J. A. (2010) *The Helm Dictionary of Scientific Bird Names*. Christopher Helm, London.

Kennerley, P. & Pearson, D. (2010) *Reed and Bush Warblers*. Christopher Helm, London.

Kirwan, G. M., Boyla, K. A., Castell, P., Demirci, B., Özen, M., Welch, H. & Marlow, T. (2008) *The Birds of Turkey*. Christopher Helm, London.

Knox, A. G. & Parkin, D. T. (2010) *The Status of Birds in Britain & Ireland*. Christopher Helm, London.

Kren, J. (2000) *Birds of the Czech Republic*. Christopher Helm, London.

Larsson, L., Ekström, G., Larsson, E. & Gandemo, M. (2008) *Birds of the World*. 2nd edn. CD. Lynx Edicions, Barcelona.

Leibak, E., Lilleleht, V. & Veromann, H. (1994) *Birds of Estonia. Status, Distribution and Numbers*. Estonian Academy Publishers, Tallinn.

Lindell, L., Wirdheim, A. & Zetterström, D. (ed.) (2002) *Sveriges fåglar*. 3rd edn. [Official check-list of Swedish birds.] Vår Fågelv. Suppl. 32. SOF, Stockholm.

Marti, R. & del Moral, J. C. (eds.) (2003) *Atlas de las Aves Reproductoras de España*. SEO/BirdLife, Madrid.

Mild, K. (1987) *Soviet Bird Songs*. 2 cassettes. Privately published, Stockholm.

Mild, K. (1990) *Bird Songs of Israel and the Middle East*. 2 cassettes. Privately published, Stockholm.

Mitchell, D. & Young, S. (1997) *Photographic Handbook of the Rare Birds*. New Holland, London.

Palmer, S. & Boswall, J. (1981) *A Field Guide to the Birds Songs of Britain & Europe*. 16 cassettes. SR Phonogram, Stockholm.

Panov, E. N. (2005) *Wheatears of Palearctic: Ecology, Behavior and Evolution of the Genus Oenanthe*. Pensoft, Sofia.

Peters, J. L., Blake, E. R., Greenway, J. C., Howell, T. R., Lowery, G. H., Mayr, E., Monroe, B. L. Jr, Rand, A. L. & Traylor, M. A. Jr (1960–86) *Check-list of the Birds of the World*. Vols. 9–15. Harvard University Press, Cambridge, MA.

Phillips, A. R. (1991) *The Known Birds of North and Middle America*. Part 2. Denver Museum, Denver.

Porter, R. F. & Aspinall, S. (2010) *Birds of the Middle East*. 2nd edn. Christopher Helm, London.

Pyle, P. (1997) *Identification Guide to North American Birds*. Vol. 1. Slate Creek Press, Bolinas, CA.

Pyle, P., DeSante, D. F. Boekelheide, R. J. & Henderson, R. P. (1987) *Identification Guide to North American Passerines*. Slate Creek Press, Bolinas, CA.

Rasmussen, P. C. & Anderton, J. C. (2005, 2012) *Birds of South Asia. The Ripley Guide*. Vols. 1–2. 1st and 2nd edn. Smithsonian Institution, Washington & Lynx Edicions, Barcelona.

Redman, N., Stevenson, T. & Fanshawe, J. (2009) *Birds of the Horn of Africa*. Christopher Helm, London.

Ridgway, R. (1912) *Color Standards and Color Nomenclature*. American Museum of Natural History, New York.

Roberts. T. J. (1992) *The Birds of Pakistan*. Vol. 2: Passeriformes. Oxford University Press, Oxford.

Robson, C. (2011) *A Field Guide to the Birds of South-East Asia*. 2nd edn. New Holland, London.

Roché, J. C. (1990) *All the bird songs of Britain and Europe*. 4 CDs. Sittelle, Mens.

Roché, J. C. & Chevereau, J. (eds.) (1998) *A sound guide to the Birds of North-West Africa*. CD. Sittelle, Mens.

Roché, J. C. & Chevereau, J. (eds.) (2002) *Birds sounds of Europe and North-west Africa*. 10 CDs. Wildsounds, Salthouse.

Rogacheva, H. (1992) *The Birds of Central Siberia*. Privately published, Husum.

Roselaar, C. S. (1995) *Songbirds of Turkey: an atlas of biodiversity of Turkish passerine birds*. GMB, Haarlem & Pica Press, Robertsbridge.

Roselaar, C. S. & Shirihai, H. (in prep.) *Geographical Variation and Distribution of Palearctic Birds*. Vol. 1: Passeriformes. Christopher Helm, London.

Ryabtsev, V. K. (2001) *Ptitsy Urala*. UrGu, Yekaterinburg.

Schubert, M. (1979) *Stimmen der Vögel Zentralasiens + Mongolei*. 2 LPs. Eterna, Berlin.

Schubert, M. (1984) *Stimmen der Vögel. VII. Vogelstimmen Südosteuropas (2)*. LP. Eterna, Berlin.

Schulze, A. (2003) *Die Vogelstimmen Europas, Nordafrikas und Vorderasiens*. 17 CDs or 2 Mp3s. Edition Ample, Germering.

Sharpe, R. B. (1874–99) *Catalogue of the Birds in the British Museum*. Vols. 1–28. British Museum, London.

Shirihai, H. (1996) *Birds of Israel*. Academic Press, London.

Shirihai, H., Christie, D. A. & Harris, A. (1996) *Macmillan Birder's Guide to European and Middle Eastern Birds*. Macmillan, London.

Shirihai, H., Gargallo, G. & Helbig, A. J. (2001) *Sylvia Warblers*. Christopher Helm, London.

Sibley, C. G. & Ahlquist, J. (1990) *Phylogeny and Classification of Birds*. Yale University Press, New Haven, CT.

Sibley, C. G. & Monroe, B. L. Jr (1990) *Distribution and Taxonomy of Birds of the World*. Yale University Press, New Haven, CT.

Sibley, D. (2014) *The Sibley Guide to Birds*. 2nd edn. Knopf, New York.

Smithe, F. B. (1975, 1981) *Naturalist's Color Guide*. American Museum of Natural History, New York.

Snow, D. W. & Perrins, C. W. (eds.) (1998) *The Birds of the Western Palearctic*. Concise edition. 2 vols. Oxford University Press, Oxford.

Stresemann, E. & Portenko, L. A. (1960–98) *Atlas der Verbreitung palaearktischen Vögel*. Vols. 1–17. Akademie-Verlag, Berlin.

Stresemann, E. & Stresemann, V. (1966) Die Mauser der Vögel. *Journal für Ornithologie*, 107: Sonderheft. Berlin.

Strömberg, M. (1994) *Moroccan Bird Songs and Calls*. Cassette. Privately published, Sweden.

Svensson, L. (1984) *Soviet Birds*. Cassette. Privately published, Stockholm.

Svensson, L. (1992) *Identification Guide to European Passerines*. 4th edn. Privately published, Stockholm.

Svensson, L., Mullarney, K. & Zetterström, D. (2009) *Collins Bird Guide*. 2nd edn. HarperCollins, London.

Svensson, L., Zetterström, D. & Andersson, B. (1990) *Fågelsång i Sverige*. CD and booklet. Mono Music, Stockholm.

Thévenot, M., Vernon, R. & Bergier, P. (2003) *The Birds of Morocco*. BOU Checklist No. 20. BOU, Tring.

Thorup, K. (2004) *Bird Study*, 51: 228–238.

Ticehurst, C. B. (1938) *A Systematic Review of the Genus Phylloscopus*. British Museum, London.

Tomiałojć, L. & Stawarczyk, T. (2003) *Awifauna Polski*. 2 vols. PTPP, Wrocław.

Turner, A. & Rose, C. (1989) *Swallows and Martins of the World*. Christopher Helm, London.

Ueda, H. (1998) *283 Wild Bird Songs of Japan*. Yama-kei, Tokyo.

Urquhart, E. (2002) *Stonechats*. Christopher Helm, London.

Vaurie, C. (1959) *Birds of the Palearctic Fauna. Passeriformes*. H. F. & G. Witherby, London.

Veprintsev, B. N. & Leonovich, V. (1982–86) *Birds of the Soviet Union: A Sound Guide*. 7 LPs. Melodia, Moscow.

Veprintsev, B. N. & Veprintseva, O. (2007) *Voices of the Birds of Russia*. Mp3. Phonoteca, Moscow. (In Russian.)

Vinicombe, K., Harris, A. & Tucker, L. (2014) *The Helm Guide to Bird Identification. An In-depth Look at Confusion Species*. Christopher Helm, London.

Vinicombe, K. & Cottridge, D. M. (1996) *Rare Birds in Britain and Ireland*. HarperCollins, London.

Voous, K. H. (1977) *List of Recent Holartic Bird Species*. BOU, London.

Williamson, K. (1967) *Identification for Ringers 2. Phylloscopus*. 2nd edn. BTO, Oxford.

Williamson, K. (1968a) *Identification for Ringers 1. Cettia, Locustella, Acrocephalus and Hippolais*. 3rd edn. BTO, Oxford.

Williamson, K. (1968b) *Identification for Ringers 3. Sylvia*. 2nd edn. BTO, Oxford.

Witherby, H. F., Jourdain, F. C. R, Ticehurst, N. F. & Tucker, B. W. (1938–41) *The Handbook of British Birds*. H. F. & G. Witherby, London.

Wolters, H. E. (1979) *Die Vogelarten der Erde*. Paul Parey, Hamburg.

Zimmerman, D. A., Turner, D. A. & Pearson, D. J. (1996) *The Birds of Kenya and Northern Tanzania*. Christopher Helm, London.

Zink, G. (1973–85) *Der Zug europäischer Singvögel*. Vols. 1–4. Vogelzug-Verlag, Möggingen.

LIST OF PASSERINE FAMILIES: TRADITIONAL AND NEW ORDER

The order of families in this handbook is the traditional one found in Cramp et al. (1977–94), The Birds of the Western Palearctic (which used Voous 1977 as a basis), and with little variation in most field guides (e.g. Svensson et al. 2009). The choice to stick with this order is made solely to facilitate use; what is familiar to most readers is thought to be user-friendly. However, recent research using molecular biology clearly shows the traditional order to be in many instances wrong, largely based as it was on morphological similarity, which sometimes cannot separate homologous characters from those acquired through convergence or by chance. It is clear that soon we will all have to learn and use a new sequence of families. Interested parties are referred to vol. 2 of the recent fourth edition of The Howard and Moore Complete Checklist of the Birds of the World (Dickinson & Christidis 2014), where Joel Cracraft presents and explains the latest taxonomic developments. It should be stated that we are still a long way from full agreement over best taxonomy and sequence. The Howard and Moore checklist is just one of several options adopting a new order.

The table below offers a simple comparison between the traditional and new (sensu Howard and Moore, 4th edn) orders among the passerines. The number of families involved for the species in the main section of this handbook will increase under the new order by a quarter, from 34 to 44. As an example, the large family Sylviidae will be divided into several smaller families. Some of the changes are noteworthy, such as the fact that larks are closely related to cisticolas and prinias, the Bearded Reedling has proven not to be a timalid or parrotbill but closely related to the larks, further that swallows are surprisingly inserted among the various new warbler families, and that pipits and wagtails end up among sparrows and finches. Chats, redstarts and wheatears are not small thrushes, but part of the large flycatcher family. There is much new to digest and learn.

Traditional order
Sensu Voous / BWP

ALAUDIDAE – Larks
HIRUNDINIDAE – Swallows
MOTACILLIDAE – Pipits and wagtails
PYCNONOTIDAE – Bulbuls
BOMBYCILLIDAE – Waxwings and Hypocolius
CINCLIDAE – Dippers
TROGLODYTIDAE – Wrens
PRUNELLIDAE – Accentors
TURDIDAE – Nightingales, chats and thrushes
SYLVIIDAE – Warblers, crests and kinglets
MUSCICAPIDAE – Flycatchers
MONARCHIDAE – Paradise flycatchers
TIMALIIDAE – Reedling, babblers and parrotbills
AEGITHALIDAE – Long-tailed tits
PARIDAE – Tits
SITTIDAE – Nuthatches
TICHODROMIDAE – Wallcreeper
CERTHIIDAE – Treecreepers
REMIZIDAE – Penduline tits
NECTARINIIDAE – Sunbirds
ZOSTEROPIDAE – White-eyes
ORIOLIDAE – Orioles
LANIIDAE – Bush-shrikes and shrikes
CORVIDAE – Corvids
STURNIDAE – Starlings and mynas
PASSERIDAE – Sparrows and allies
PLOCEIDAE – Weavers
ESTRILDIDAE – Waxbills and allies
VIREONIDAE – Vireos
FRINGILLIDAE – Finches
PARULIDAE – New World warblers
THRAUPIDAE – Tanagers
EMBERIZIDAE – Buntings, New World sparrows and cardinals
ICTERIDAE – Bobolink and icterids

New order
Sensu Cracraft / H&M 4

VIREONIDAE – Vireos
ORIOLIDAE – Orioles
MALACONOTIDAE – Bush-shrikes
LANIIDAE – Shrikes
CORVIDAE – Corvids
MONARCHIDAE – Monarchs and paradise flycatchers
NECTARINIIDAE – Sunbirds
PRUNELLIDAE – Accentors
PLOCEIDAE – Weavers
ESTRILDIDAE – Waxbills, munias and allies
PASSERIDAE – Sparrows, snowfinches and allies
MOTACILLIDAE – Pipits and wagtails
FRINGILLIDAE – Finches
CALCARIIDAE – Longspurs and allies
EMBERIZIDAE – Buntings
PASSERELLIDAE – New World sparrows and allies
PARULIDAE – New World warblers
ICTERIDAE – Icterids
CARDINALIDAE – Cardinals, tanagers and allies
PARIDAE – Tits
REMIZIDAE – Penduline tits
ALAUDIDAE – Larks
PANURIDAE – Reedlings
CISTICOLIDAE – *Cisticola* warblers
LOCUSTELLIDAE – *Locustella* warblers
ACROCEPHALIDAE – Reed warblers
HIRUNDINIDAE – Swallows
PYCNONOTIDAE – Bulbuls
PHYLLOSCOPIDAE – Leaf warblers
SCOTOCERCIDAE – Bush warblers and allies
AEGITHALIDAE – Long-tailed tits
SYLVIIDAE – *Sylvia* warblers and allies
ZOSTEROPIDAE – White-eyes
LEIOTHRICHIDAE – Babblers and laughingthrushes
REGULIDAE – Crests and kinglets
BOMBYCILLIDAE – Waxwings
HYPOCOLIIDAE – Hypocolius
CERTHIIDAE – Wallcreeper and treecreepers
SITTIDAE – Nuthatches
TROGLODYTIDAE – Wrens
STURNIDAE – Starlings
CINCLIDAE – Dippers
MUSCICAPIDAE – Chats and flycatchers
TURDIDAE – Thrushes

A BRIEF PRESENTATION OF PASSERINE FAMILIES

As explained in the introduction to Volume I, the sequence adopted in this handbook is the traditional one among the passerines, starting with larks and ending with buntings and New World icterids. This is thought to be user-friendly, to facilitate finding a sought species quickly. Taxonomy is in a dynamic phase, with many new insights from molecular methods, and future lists and books will undoubtedly adopt a very different order. However, new findings are surfacing continuously, resulting in further recommendations for changes in taxonomy, and interpretations of results vary between taxonomists; we are far from seeing the end of such changes. This state of uncertainty is another reason to stick to a familiar sequence in this book.

Passerines comprise more than half of the global number of bird species. They are traditionally kept together in one large order, Passeriformes, but this might change in the future. They are often divided into suboscines and oscines, latter including the vast majority of species, and all the families listed below. Suboscines concern only a few extreme vagrants to the treated region (tyrant flycatchers, one pitta species; these are covered in Volume 1).

While the main section of the book contains accounts of the more regularly occurring 409 species (197 of which are treated in this second volume) arranged in one continuous sequence without division into family chapters, it is thought helpful to give brief family descriptions separately. The families below are those of the traditional order, but where appropriate comments on new thinking and alternative arrangements are offered.

Muscicapidae – Flycatchers

Small arboreal birds specialised in feeding on insects, often taken in flight. Flight is fast and direct, and the birds are capable of sudden twists and turns. Strong, broad bills, short legs, rather long and pointed wings. Plumage either streaked grey-brown and white (then sexes alike) or black-and-white (males) and brown-and-white (females). Song brief and rather simple, but characteristic. Nest in hole, either natural or a nestbox. Some species known for polygamous habits. Nocturnal migrants to tropics in winter. Nine regular species in the region, four more have occurred as vagrants. – Muscicapidae, the Old World flycatcher family as understood in the traditional taxonomy, was restricted to the 'true' flycatchers, belonging to such genera as *Muscicapa* and *Ficedula*. However, molecular methods have shown that the Muscicapidae also includes all the 'small thrushes', the chats, formerly referred to Turdidae. Thus robins, nightingales, redstarts, wheatears, etc. will in future all be relocated to Muscicapidae.

Monarchidae – Paradise flycatchers

Small or medium-sized birds with strong, broad-based and flattened bills and in some species elongated central tail-feathers, these then being extremely long in some adult males. Legs in comparison with bill rather weak. Often striking plumage pattern. Sexes similar or obviously different. Song of those species that have occurred within the covered region a rather simple whistled strophe. Nest a cup fixed in a fork of a branch high up in a tree. Usually sedentary. One species regular and two more accidental vagrants. – Although in traditional taxonomy usually placed after Old World flycatchers, modern research has shown this family to be more closely related to the corvids, and in the future it should preferably be listed immediately after them.

Timaliidae – Reedlings, babblers and parrotbills

Usually medium-sized (thrush-sized) but also including some small birds, widely distributed in tropical parts of Asia, but also many in Africa and some in Europe. Mainly insectivorous but will feed on large variety of food sources. Often sociable, both when breeding and at other seasons. Flight rather weak and fluttering due to short, rounded wings. Plumage often rather plain, feathering soft, sexes generally alike. Feet strong, bill often short and slightly decurved. Many species loud-voiced. Nest low in reeds or dense scrub, but also higher in trees. Sedentary as a rule. Five species regular in the region. – This family, where some rather different groups were included in traditional taxonomy, has not stood the test of time or molecular methods. Reedlings, of which one species, the Bearded Reedling, occurs in the Western Palearctic, is not related to the mainly east Asian parrotbills, but instead is closer to the larks. Timalids are closely related to *Sylvia* warblers (but are extralimital and unrepresented here). Babblers (*Turdoides*) are nowadays instead generally referred to a separate family, Leiothrichidae.

Aegithalidae – Long-tailed tits

Small, very long-tailed birds with tiny bill and short legs. Plumage soft and fluffy, colours subdued and mainly white, grey, rufous and black. Sexes alike. Weak flight with short, rounded wings. Often encountered in small parties. Does not seem to have a real song. Builds domed nest in fork of a tree, well camouflaged with lichens and pieces of bark, attached with cobweb. Mainly sedentary. One species, Long-tailed Tit, is found within the region. – In modern molecular-based taxonomy usually placed among the various new warbler families; thus, no longer close to the true tits.

Paridae – Tits

Small, rather robust, compact birds with strong bills and legs, capable of clinging up-side-down on thin branches or under food at bird feeders. Found in woods, parks and gardens, less often in more open country with tall bushes. Social when feeding, especially during non-breeding season, when several birds of different species often rove together in woods searching for food. Plumage variable, but many have a dark cap and bib. Sexes similar or entirely alike. Vocal, with large repertoire in some species, but song is rather monotonous and unremarkable. Nest in hole, natural or in a nestbox or hole in building. Mainly sedentary, but some movements in northern parts, at least in some years. 13 species. – In modern taxonomy, true tits (Paridae) together with penduline tits (Remizidae) are more or less basal to the large clade encompassing warblers, larks and swallows; they are not as closely related to nuthatches and treecreepers as their placement in traditional taxonomy implied.

Sittidae – Nuthatches

Small or medium-sized short-tailed birds with strong feet and straight, pointed, strong bill, capable of climbing downwards on tree trunks or under thicker branches. Movements of this kind performed without support of tail (like in treecreepers or woodpeckers). Some species prefer steep rock-faces for feeding and breeding. Rounded wings, but flight is often fast and direct. Feed on insects and spiders. Plumage rather similar in most species, lead-grey above, white or orange-buff below and with dark mask through eye. Sexes similar. Loud voice and characteristic song of several types. Nest in hole in tree or rock crevice, in both cases entrance hole often made smaller using mud. Sedentary. Six regular species and one vagrant. – As mentioned above, nuthatches are not as closely related to tits as their placement in traditional taxonomy suggests, but they are part of a clade containing both flycatchers and thrushes as well as dippers, waxwings and starlings.

Tichodromidae – Wallcreeper

Medium-large bird with striking appearance, in shape like a nuthatch but with longer, thinner and slightly down-curved bill, and plumage very different, with striking wing pattern in black, white and scarlet-red. Breeding male has extensive black bib. Moves with ease on vertical rock-faces, sometimes on walls like a nuthatch, but unlike a nuthatch often semi-opens wings for each hop, exposing red colour. Feeds on insects and spiders. Song, often in flight, a discreet, whistling strophe. Nests in crevice or under scree. Altitudinal movements and short-range migrant in winter. Only one species. – In traditional taxonomy usually afforded its own monotypic family inserted between nuthatches and treecreepers, but now seen as a subfamily within nuthatches, though its precise position is still uncertain.

Certhiidae – Treecreepers

Small arboreal birds feeding on insects or spiders taken by searching tree trunks and larger branches methodically from bottom upwards, thereby taking support from rather long, stiff tail in woodpecker fashion. Bill thin and down-curved, feet short but curved claws long. Plumage streaked brown and white, sexes alike. Song a short strophe with clear voice. Nest in cavity on tree trunk or under loose bark. Mainly sedentary. Two species within the treated region.

Remizidae – Penduline tits

Quite small birds found in open woodland, often near water. Insectivorous. Conical but fine, pointed bill, rather short tail, rounded wings. Acrobatic when feeding in canopy. Rather weak flight. Plumage variable, those of concern here rather similar with chestnut back, buffish underparts and a black pattern on head. Sexes similar. Song weak and high-pitched. Nest pouch-shaped suspended from thin branches, tunnel-entrance at top side. Presumed to be mainly nocturnal migrant, though southern breeders are resident. Three species occur within the treated region. – Together with the true tits (Paridae), according to modern taxonomic views the penduline tits form a basal part of a large and complex clade also including warblers, larks and swallows.

Nectariniidae – Sunbirds
Tiny or small birds specialised in nectar-feeding by having very thin, long bills, long tongue, short tail and small general size ('hummingbirds of the Old World'). Found in open forest, savanna, scrub, orchards, palm groves and gardens. Food also includes insects and spiders. Mainly arboreal. Plumage of males is often dark green or purple, iridescent and glossy, whereas females duller. Strong voice, clear, high-pitched or metallic song. Sedentary or seasonal shorter movements only. Four species occur within the covered region. – The traditional placement of the sunbirds, between penduline tits and white-eyes, does not hold when recent molecular research is considered. Thus, in modern taxonomy the sunbirds group with accentors (Prunellidae) in a rather basal clade where sparrows and finches and their allies also sit.

Zosteropidae – White-eyes
Small arboreal birds with a frayed tongue-tip, nearly all with greenish upperparts, yellow, buffish or white underparts, and prominent white eye-ring. Sexes alike. Numerous similar-looking species in Africa and southern Asia. Feed mainly on insects, berries, nectar and pollen. Social birds, often occurring in small parties. Vocal, although song and calls are rather simple and weak, song often a series of similar high-pitched notes. Nest is a cup within dense foliage of a tree. Sedentary or dispersive. Two species occur within the covered region. – Oddly, white-eyes have proven to be closely related to *Sylvia* warblers and babblers, not to sunbirds (also nectar-feeders) or *Phylloscopus* warblers (rather similar morphology).

Oriolidae – Orioles
Medium-sized birds with proportions recalling both starlings and thrushes but have narrower and longer wings than both. Body rather slender, tail medium long. Males of most species largely yellow and black, females and young duller greenish and streaked grey-white. Strong bill usually reddish. Highly arboreal requiring tall-growing trees, where they mainly keep to the canopy, being often surprisingly skulking and difficult to spot. Flight fast and strong, over longer stretches somewhat undulating. Song loud and characteristic, a flute-like whistle or yodelling. Nest a cup fixed in a branch fork high up in tree. Diurnal migrants that winter in tropical areas. One species a regular breeder, and one other recorded as a vagrant within the covered region. – Recent DNA-based research has confirmed the close relationship between orioles, shrikes and corvids.

Laniidae – Bush-shrikes and shrikes
Fairly small to medium-sized long-tailed birds of open habitats with scattered bushes or scrub at forest-edges, scanning for prey from highly visible lookout perch. Food insects, lizards, small rodents and birds. Will store surplus food by attaching it to thorny bushes (larder). Bill strong and sharply hooked with falcon-like tooth at side of upper mandible. Feet and claws also strong. Wings rather rounded, flight strong, over longer stretches undulating. Plumage chestnut or grey above (rarely black), whereas underparts are pinkish, buff or white. Usually a black (in females and young dark) mask through eye. Tail-sides white. Sexes similar or rather clearly different. Song of *Lanius* species either a simple territorial signal or a subdued prolonged warbling with some mimicry aimed at close-by female. Song in the only occurring bush-shrike, the Black-crowned Tchagra, is a remarkable loud series of whistles. Nest a cup in dense bush. Most species are migratory, some reaching tropical winter grounds. 13 species regular within the covered region. – While under traditional taxonomy bush-shrikes and true shrikes were often combined in one family, Laniidae, the former is nowadays generally afforded a separate family, Malaconotidae.

Corvidae – Corvids
Medium-sized to large birds with strong feet and very strong bills. Colours mainly black, in some with parts being grey or grey-white, though some smaller species have dark brown, rufous-grey or pinkish plumage with ornaments in bright blue, black and white, or are pied black-and-white. Omnivorous, some species specialised on large seeds, which are stored for winter season. Found in all kinds of habitats, woods, agricultural land, gardens, open heaths with scrub, etc. Sexes alike. Song primitive and for many species rarely heard. Nest built in tree of twigs, generally a flat cup but in magpies domed. Resident or short-range diurnal migrants in the north, sedentary in the south. 18 species occur regularly, another is a vagrant.

Sturnidae – Starlings and mynas
Medium-sized, usually short-tailed birds with strong and rather long feet. Often seen feeding on ground energetically, walking without stopping and head jerking. Social, will feed in small groups during much of year. Food insects, earthworms, berries, Flight strong and direct, often in dense flocks. Wings broad-based but rather long and pointed still. Plumage either blackish with metallic gloss, often finely spotted pale, or pink, brown or greyish with varying head and wing patterns. Bill rather long, pointed and generally slightly down-curved. Sexes similar. Accomplished singers, sometimes using expert mimicry. Nest in hole in tree, on house or in nestbox. Usually migratory, movements being diurnal. Ten species are regular and another two have been recorded as vagrants. – Starlings have traditionally been placed next to crows and their allies. However, molecular-based research shows them to be part of a large and diverse clade comprising also oxpeckers, mockingbirds, thrashers, dippers, flycatchers and thrushes.

Passeridae – Sparrows and allies
Small robust birds with strong conical bill adapted for feeding on seeds. Usually social, both feeding and breeding in small or large communities. Plumage rather plain, mainly brown or grey and white, sexes usually different (though in some species alike). Flight rather weak and direct, wingbeats fluttering or whirring, wings generally short and rounded. Mainly feeds on ground. Noisy, but song unremarkable, mostly consisting of repeated chirping notes. Nest usually a cup or domed construction of grass and fine twigs in bush or tree, sometimes in dense colonies. Alternatively in hole in tree, or on house. Most species resident, some perform short-range diurnal migration. 17 species. – This family remains much the same as it was seen in traditional taxonomy. The only surprising new insight is that it is, apart from with weavers (Ploceidae) and finches (Fringillidae), also rather closely related to pipits and wagtails (Motacillidae).

Ploceidae – Weavers
Small or moderately large, robustly built birds with strong, conical bill. Most species highly social, breeding in large, dense colonies, nests domed, in tree. Feeds on ground, usually in flocks. Wings short and rounded, still capable of strong flight. Often bright plumage, or with red, yellow, black or white markings. Females duller than males, also more streaked. Song simple, mostly chirping like in sparrows. Sedentary or make dispersive shorter movements outside breeding season. Two regularly occurring species. – See comment on taxonomy under Passeridae.

Estrildidae – Waxbills and allies
Tiny to small finch-like birds with conical bills and rounded wings. Found in open habitats, savanna, grassy fields, reedbeds, etc. Generally social habits, feeding in flocks and sometimes nesting in colonies, although solitary breeding is more common. Colours variable, some having red mask in otherwise rather plain grey-brown plumage, others being mainly all-red like male Red Avadavat. Bill red, blackish or grey. Sexes alike or at times clearly different. Song of variable length and quality, usually rather quiet and unremarkable. Mainly sedentary. Seven species occur naturally within treated region. – Closely related to sparrows and weavers.

Vireonidae – Vireos
Small to medium-sized flycatcher-like birds, which are mainly arboreal and feed on flying insects or other invertebrates found in tree foliage. Long pointed wings, strong fliers. In winter will also feed on berries and seeds. Bill rather strong, broad-based with tip finely hooked. Plumage usually grey or olive-grey above, off-white or slightly yellowish below. Some have contrasting marks on head. In some species iris reddish or white. Sexes alike or at least similar. Several species long-distance migrants, others are more sedentary. One species recorded so frequently as a vagrant to be treated in the main section, while another three are rarer vagrants. – Formerly often placed near the New World wood warblers (Parulidae), but molecular methods in recent taxonomies place the vireos more basal, close to orioles and corvids.

Fringillidae – Finches

Small to medium-sized birds with conical bills adapted for feeding largely on seeds. A peculiar adaptation is represented by the crossed mandible-tips of crossbills, serving to open conifer cones. Numerous species, inhabiting a rich variety of habitats from deserts to dense forests. Plumage and patterns also highly variable. Sexes similar or markedly different. Vocal ability quite evolved, and some are famous for their song and have been kept as cage-birds for this reason. Nest usually a cup placed in branch fork in dense bush or tree, rarely closer to the ground. Some northerly species or populations are prominent diurnal migrants, others in more southern and warm areas are sedentary or dispersive after breeding. 38 species are regular, a further four have been recorded as vagrants. – See note under Motacillidae for a comment on modern taxonomy.

Parulidae – New World warblers

Small, fine-billed, arboreal and insectivorous birds, the American counterpart to the Old World warblers (Sylviidae). Active, feeding in foliage with quick and agile movements. Some species adapted to walk on trunks and larger branches like a nuthatch. A rich variation of species, all with quite distinctive plumage patterns, often in lead-blue, yellow, orange, greenish, blue, black and white. Sexes similar or moderately different. Northern species migratory. Ten species are fairly regular vagrants and are hence treated in the main section, while another 15 are rarer and found in the vagrants section at the end of the book. – The Parulidae family is nowadays thought to be best placed after American sparrows (Passerellidae; formerly included in Emberizidae) and buntings (Emberizidae), and before icterids (Icteridae) and cardinals (Cardinalidae; formerly included in Emberizidae).

Thraupidae – Tanagers

Small to medium-sized, fairly long-tailed birds of the New World. Most are arboreal. Feed on berries, fruits, nectar, insects and seeds. Plumage of those which have been recorded as vagrants within the treated region invariably immatures (or females) with dull colours in olive-brown and yellow-grey, whereas adult males are largely deep red. Bill strong, culmen slightly decurved, colour mainly grey-horn. One fairly regular vagrant treated in main section, one further species rare. – Nowadays usually referred to Cardinalidae (cardinals, grosbeaks and allies), but rather confusingly keeping their vernacular name, tanagers.

Emberizidae – Buntings, American sparrows and cardinals

Mainly small birds with conical bills of somewhat varying size. Feed largely on seeds on ground and in grasses, but takes some insects and berries, too. Fairly slim shape with longish tail in some. Moderately pointed, but still rounded wings, strong fliers. Plumage variable, often streaked, males often having characteristic markings on head or throat. Sexes rather similar or alike in many species. Good singers, but strophe often short and simple. Nest on ground, in tussock or crevice or hole in ground. Northerly populations predominantly diurnal migrants. 26 regularly occurring species and another eight recorded as rare vagrants. – In traditional taxonomy, as here, Old World buntings lumped with New World sparrows (Passerellidae) and New World cardinals (Cardinalidae) in one large family. However, it is customary now to see these as representing three different families.

Icteridae – Bobolink and icterids

A New World family of birds of varying size, some being 'medium-small' and others quite large. Habitat choice and diet varies a lot within the family. Frequent size dimorphism between sexes. Long and pointed wings; strong, fast fliers. Majority of members black with metallic sheen; some have yellow, orange, red or other colour pattern added to the black. Males usually brighter than females. Two regular vagrants treated in the main section; to these are added three rarer species in the vagrants section. – In traditional taxonomy, often placed last in the sequence indicating a late evolution. In modern taxonomic thinking, Icteridae is part of the large so-called Emberizoidea clade, which comprises buntings, American sparrows, American wood warblers, cardinals and tanagers.

(ASIAN) BROWN FLYCATCHER
Muscicapa dauurica Pallas, 1811

Fr. – Gobe-mouches marron; Ger. – Braunschnäpper
Sp. – Papamoscas pardo; Swe. – Glasögonflugsnappare

A vagrant to the covered region, with most records from Scandinavia. Fortunate observers faced with such a bird should also consider the possibility of two other potential Asian grey-brown *Muscicapa* flycatchers. The Brown Flycatcher summers in wooded thickets over a wide area from the N Indian subcontinent to NW Thailand, and from S Siberia to NE China, Korea and Japan, and winters in the Indian subcontinent and further east to SE Asia, the Greater Sundas and the Philippines.

M. d. dauurica, presumed 1stS, Mongolia, May: note broad-based bill with typical straw-coloured base to lower mandible. Some have more dusky-grey wash on neck-sides, breast and flanks, leaving central throat more contrastingly white. Remiges, primary-coverts, alula and most or all greater coverts apparently juv and slightly bleached and brownish. (H. Shirihai)

IDENTIFICATION A *small, rather nondescript flycatcher* (almost as small as Red-breasted Flycatcher) with *ashy grey-brown upperparts, a large-headed appearance*, slightly darker face marked by a *dull white loral streak, large dark eye and pronounced white eye-ring* (especially prominent behind eye), and has long dark *grey-brown wings with narrow pale fringes, especially to the greater coverts, secondaries and tertials* (producing ill-defined whitish greater and, sometimes, median coverts bars in fresher plumage); *rather short*, mostly dark grey-brown *tail with narrow whitish tip and edges to outermost feathers*. *Underparts dull white*, variably sullied pale grey-brown on neck-sides, breast and flanks, appearing uniform at distance but sometimes faintly blotched or softly streaked at close range, when whitish edges to feathers visible. Some show *vague dark moustachial stripe and lateral throat-stripe*. Bill noticeably broad-based (triangular in ventral view with almost straight edges and rather 'full', not so pointed, tip) *and largely black, with rather extensive flesh-yellow base to lower mandible*. Legs dark horn to black. Postures and action resemble Spotted Flycatcher, though feeding sallies shorter, and its more compact form recalls those of *Ficedula* flycatchers, while, at first glance, overall appearance is intermediate between Spotted and ♀-like plumages of the black-and-white *Ficedula* species.

VOCALISATIONS Song consists of a thin, high-pitched series of varied squeaky notes, some being as high-pitched and thin as Goldcrest calls (thus partly difficult to hear for elderly people), interspersed by softer trills and whistles, and even some imitations. The pace and strophe length is variable, often stanzas are interrupted by brief pauses, and general tempo is rather slow. – Commonest calls are a high-pitched *tzi*, similar to Spotted Flycatcher, and a trilling *tzirrrr*.

SIMILAR SPECIES Potential confusion with Spotted Flycatcher covered under that species. – Perhaps equally likely to be confused with ♀-like plumages of *Ficedula* flycatchers, but easily separated by wing and tail patterns (both much bolder in these, but nearly uniform in Brown Flycatcher with at most a hint of paler tertial and greater coverts fringes). – Not dissimilar to *Dark-sided Flycatcher* (*M. sibirica* vagrant to the treated region), which is roughly equally large but differs in being somewhat slimmer (smaller-headed and smaller-eyed impression). It is longer-winged with longer primary projection (exceeding tertial length and reaching further down tail). Has rather darker grey-brown upperparts and usually much darker, blotchy or patchy streaks on breast-sides and flanks (unlike uniform wash with only vague streaking of Brown, though some Dark-sided have slightly more uniform underparts), creating a distinctive pale central belly line (not unlike some pewees *Contopus*). Bill, when seen from below, is narrower and edges slightly concave near tip, and is almost always all dark with a restricted orange-yellowish base to the lower mandible, normally not reaching beyond the nostrils. Pale throat less extensive but more contrasting, bordered by stronger lateral throat-stripe and striking pale moustachial stripe (and pale from throat continues as a half-collar behind the ear-coverts), but lores less white, and a whitish eye-ring is apparent only behind the eye. Undertail-coverts are blotched dark (unmarked in Brown). Unlike Brown, complete moult of Dark-sided occurs in winter quarters, thus adults are worn in autumn. Wing-length longer and wing shape more pointed than in Brown: W 77–85 mm; p1 < pc 3–6 mm; p2 =4/5 or 4. Emarg. pp3–4(5). – Extralimital *Grey-streaked Flycatcher* (*M. griseisticta*; NE & EC Asia, winters south to SE Asia & New Guinea; not treated, but see comparison on next page) is intermediate in size between Spotted and Brown, and is readily differentiated from latter by the more heavily, densely and boldly dark-streaked underparts to mid flanks, with only relatively restricted unstreaked belly, and the more striking pale eye-ring. Like Dark-sided has warmer and more rufous-tinged underwing-coverts and axillaries (washed brownish-white to creamy-white in Brown). Bolder dark lateral throat-stripe also useful when distinguishing Grey-streaked from Dark-sided, and former has duskier grey olive-brown head and rest of upperparts, contrasting with white wing fringes and vent. Has even longer wings and primary projection than Dark-sided. Bill generally rather long and stubby, less sharply pointed and largely black. Large dark eye enhanced by white eye-ring, but lores less clean with black streak between bill and eye. Undertail-coverts mostly white without dark blotchy centres (i.e. closer to Brown in this respect). W 78–86 mm; p1< pc 4–8 mm; p2 = 3/4 [=4]. Emarg. p3 (4) [5]. – These Asian *Muscicapa* flycatchers are further differentiated by their calls, behaviour and moult; see also comparison on next page.

M. d. dauurica, variation in spring/summer, presumed 1stS, Mongolia, May–Jun: mostly due to individual variation in feather wear (but also effects of light), some can appear more greyish (less brownish) like the left-hand bird, which also has more obvious pale wing fringes and rather well-developed dark lateral throat-stripe. Age of right bird suggested by apparently many retained and strongly beached juv wing-feathers. (H. Shirihai)

AGEING & SEXING Ageing requires a close check of moult and feather wear in wing, and is still difficult due to some overlap in characters. Sexes alike. Limited seasonal variation. – Moults. Complete post-nuptial moult in late summer (Jul–Sep), and partial post-juv moult at same time or a little later (Aug–Sep) including head, body and smaller secondary-coverts. Partial pre-nuptial moult in winter quarters (Feb–Apr), usually involving head and some of body, but apparently only occasionally a few tertials and/or some inner greater coverts, and perhaps also tail-feathers (but all of latter often retained, or only central pair replaced); extent of pre-nuptial moult apparently much restricted in ad. – **SPRING Ad** Less worn on spring migration but in summer bleaches browner, less grey above, and pale wing and tail-feather fringes reduced and whiter. May develop diffuse blotches on upper breast. Pre-nuptial moult can produce rather clear contrast, leading to incorrect ageing (cf. 1stS). **1stS** Retained juv primaries and tail-feathers more worn, and primary-coverts browner and narrower with very faded fringes, while greater coverts are much duller with paler fringes virtually lost. Moult contrast among tertials also often distinctive. – **AUTUMN Ad** As in spring but has more pronounced pale buff-brown fringes to secondary-coverts, secondaries and tertials, forming small wing-panel and slightly more obvious wing-bars. Tertial edges are usually pale buff-brown rather than whitish, narrower and slightly more diffusely demarcated than in 1stW, and usually

M. d. dauurica, 1stW, Scotland, Sep: when very fresh in early autumn, edges and tips to tertials and greater coverts broader and more obvious, and pale edges often end abruptly, with bolder and whiter tips to inner two tertials; wing-bars are more obvious than in later plumages. However, primaries and primary-coverts are (relatively) more worn than in ad. (M. Breaks)

M. d. dauurica, 1stW, Thailand, Jan: by winter, despite some wear, the edges to tertials and greater coverts are relatively broader, whiter and more distinct than ad. Readily identified by combination of dull white loral patch, large dark eye with broad white eye-ring, uniform ashy grey-brown above, wings only narrowly pale-fringed, underparts whitish, indistinctly sullied greyer at sides, and strong bill with clear yellowish base to lower mandible. (P. Ericsson)

fade off smoothly on inner webs. **1stW** Like ad, but edges to tertials (in particular inner two) and greater coverts have on average broader, whiter and more distinct edges, and pale edges often end abruptly on outer part of inner web of inner two tertials. Some have narrower and less distinct and less white edges and tips, being similar to ad with maximum amount of pale edges. Also, juv primary-coverts, primaries and tail-feathers, which are softer textured and already slightly worn, are often appreciably duller or browner-centred with more frayed fringes (from Sep onwards). Moult limits usually observable between moulted and unmoulted greater coverts (renewed inner greater coverts ad-like with narrow and diffuse fringes). Occasionally retains a few juv-feathers above, mainly uppertail-coverts. **Juv** Soft, fluffy body-feathers with extensive pale spots above, dusky-brown mottling below and clearer pale buff wing markings.

BIOMETRICS (*dauurica*) Sexes practically the same size. **L** 11.5–12.5 cm; **W** 67–78 mm (n 55, m 70.8); **T** 45–52.5 mm (n 46, m 48.6); **T/W** m 68.2; **B** 13.3–16.2 mm (n 49, m 14.3); **BD** 3.3–4.1 mm (n 34 m 3.7); **BW** 5.8–7.0 mm (n 38, m 6.2); **Ts** 12.0–14.3 mm (n 29, m 13.2). **Wing formula: p1** > pc 0–6.5 mm, < p2 31–37.5 mm; **p2** < wt 3.5–7 mm, =5/6 or =5; **pp3–4** about equal and longest; **p5** < wt 1–5 mm; **p6** < wt 6.5–10.5 mm; **p7** < wt 10–15 mm; **p10** < wt 16–22.5 mm; **s1** < wt 17.5–24.5 mm. Emarg. pp3–5.

GEOGRAPHICAL VARIATION & RANGE Very slight, involving colour, size and wing structure, but only *dauurica* is likely to reach the covered region. Three or four more extralimital races in SE Asia (not treated).

M. d. dauurica Pallas, 1811 (SC & SE Siberia, N Mongolia, NE China, E Asia between Sakhalin and Japan & Korea; winters India, S China, SE Asia including Philippines, Greater Sundas). Described above. Slightly larger and longer-winged than other races and has a strong bill. (Syn. *latirostris*.)

TAXONOMIC NOTES The scientific name *latirostris* Raffles, 1822, has sometimes been claimed to be the correct one (e.g. Mlíkovský 2012), but *dauurica* seems to fulfil all requirements in the Code and has priority.

Comparison of four streaked and brown Eurasian *Muscicapa* flycatchers, including Brown Flycatcher (top left: India, Dec), Dark-sided Flycatcher (top right: Hong Kong, Nov), extralimital Grey-streaked Flycatcher (bottom left: Hong Kong, Oct) and Spotted Flycatcher (bottom right: Italy, Aug), all 1stW: the three Asian species differ from Spotted Flycatcher by proportionately larger eye and more distinct white eye-ring; also, always pay attention to crown and breast streaking. – Brown Flycatcher is the least marked below, but beware that some may show a vaguely mottled breast in autumn, as here. – Dark-sided Flycatcher has much darker and blotchier streaking on breast and flanks, while its narrower dark bill has less extensive pale base to lower mandible. Dark lateral throat-stripes enhance pale throat patch and clearer pale moustachial stripe. Whitish eye-ring narrower and often apparent only behind eye. Undertail-coverts blotched dark, at least at bases and centres (unmarked in Brown). – Grey-streaked Flycatcher is characterised by dense and bold dark streaking on lower throat and breast, extending less boldly onto flanks. Largely black bill appears rather long and stubby. Undertail-coverts unmarked like in Brown but unlike Dark-sided. – Spotted Flycatcher readily told by finely streaked forecrown and breast. (Top left: N. Sant; top right: J. & J. Holmes; bottom left: M. & P. Wong; bottom right: D. Occhiato)

REFERENCES ALSTRÖM, P. & HIRSCHFELD, E. (1989) *Vår Fågelv.*, 48: 127–138. – BRADSHAW, C., JEPSON, P. J. & LINDSEY, N. J. (1991) *BB*, 84: 527–542. – DERNJATIN, P. *et al.* (2002) *Alula*, 4: 122–126. – LEADER, P. J. (2010) *BB*, 103: 658–671. – MLÍKOVSKÝ, J. (2012) *Zootaxa*, 3393: 53–54.

SPOTTED FLYCATCHER
Muscicapa striata (Pallas, 1764)

Fr. – Gobemouche gris; Ger. – Grauschnäpper
Sp. – Papamoscas gris; Swe. – Grå flugsnappare

Due to its habit of perching openly, often in parks or tall-grown gardens, and its sudden sallies to catch flying insects only to return to the same or similar perch, the Spotted Flycatcher has become a rather familiar bird to many. And this despite its quite plain grey-and-white plumage and unremarkable vocalisations. It is a widespread breeder in the W Palearctic arriving from its sub-Saharan winter quarters rather late, to N Europe only in mid May. Although almost nondescript, some obvious geographical variation can be noted, with Balearic breeders being oddly pale.

IDENTIFICATION A *dull-coloured, greyish-brown and white* flycatcher with a characteristic *slim build*, almost *upright stance* when perched on a branch, and *long-winged* appearance. Habitually perches in the open. Upperparts rather plain greyish-brown. *Crown rounded and narrowly dark-streaked* (forehead variably paler). *Brownish-grey streaks on throat and breast*, streaking becoming more ill-defined on breast-sides and flanks, otherwise largely dull white below. Rather short-necked, and head appears as if sunk slightly between shoulders, with dull brown face (sometimes with discreet dark moustachial and lateral throat marks), while *small-looking dark eye* and *indistinct whitish eye-ring* are useful field marks. Absence of extensive white in wings and tail important too: off-white fringes and tips to tertials and median and greater coverts produce *indistinct wing-bars*, and narrow pale edges to secondaries produce *slight inner wing panel*. Relatively long slim *tail appears slightly forked when folded*, and is only narrowly edged paler. Mostly *black, spiky bill-tip* but *quite broad base to bill*, particularly if viewed from below. Usually solitary. Feeding behaviour rather distinctive, sallying forth from low- to mid-height lookouts in a burst of wingbeats, returning via a gliding ascent to perch, often a new one but can select the same. Sometimes may hover briefly, but sallies are usually short and direct. Less frequently descends to take prey from ground. Never adopts cocked-tail stance of *Ficedula* flycatchers, but frequently flicks wings and tail, and will often accept rather close observer presence. Has rather direct flight with shallow undulating progress and elongated silhouette.

VOCALISATIONS Song rather weak and quiet, easily missed in woodland chorus, combining high-pitched, thin and sharp calls, and a series of faint short squeaks, the notes often delivered at rather slow and slightly uneven pace, e.g. *sip sip sree, sreeti sree-sip sree*, very rarely admixed with a low, sweet warble. – Characteristic call is a high-pitched, thin and slightly hoarse straight note, *zee*, which can be repeated a few times. When alarmed the call is combined with one or two short, dry clicking notes, *zee-tk* or *zee-tk-tk*.

SIMILAR SPECIES Given a reasonable view, should pose few identification problems, although the inexperienced could misidentify this species, especially in a brief view, as one of the ♀-like plumages of the black-and-white *Ficedula* flycatchers (mainly due to the slightly stronger wing markings

M. s. striata, NE Italy, Jun: dull-coloured flycatcher with almost upright stance and long-winged appearance. Habitually perches in the open. This race is dull greyish-brown above, with relatively broad and diffuse dark streaks on crown, and underparts dusky white, lower throat, breast and flanks variably dark-streaked (on the bird above rather diffuse and barely contrasting). Note dark lateral throat-stripe. Following complete pre-nuptial moult, ageing in spring impossible. (L. Sebastiani)

M. s. striata, Finland, May: streaking below varies both individually and even geographically—here a heavier-streaked bird, although streaks typically become more ill-defined on breast-sides and flanks. Ageing following complete pre-nuptial moult impossible. Not all birds perch upright; here is a more horizontal posture. (M. Varesvuo)

M. s. striata, 1stW, France, Sep: typical short-necked and long-tailed jizz. Note mostly fresh plumage, with no moult limits discernible, and with juv greater coverts being prominently white-tipped, forming well-marked wing-bar. (A. Audevard)

M. s. striata, 1stW, Extremadura, Spain, Sep: young in autumn can have much stronger, broader and more contrasting breast-streaking, and sometimes streaking even extends diffusely along grey-sullied body-sides. Note typical upright and open perch at top of branch. (H. Shirihai)

in some populations of Spotted Flycatcher), but larger size, different structure with longer wings and tail, streaky crown and breast, rather limited wing markings and an unpatterned tail should prevent such confusion. – A vagrant *Brown Flycatcher* is a possible pitfall, but this is smaller and slightly shorter-tailed with an unstreaked crown, almost unmarked pale brown-grey breast, a cleaner pale loral streak, prominent white eye-ring, and different bill shape, with a pinkish or straw-coloured base to the lower mandible (but does possess rather similar behaviour and habits). – Also consider two other potential vagrant, brownish-coloured and streaky flycatchers from Asia: Dark-sided Flycatcher *M. sibirica* and Grey-streaked Flycatcher *M. griseisticta*. Both are smaller than Spotted Flycatcher, differing in their (seemingly) larger eye and/or more prominent pale eye-ring, cleaner white central throat, often pronounced pale moustachial stripe (which may continue as a half-collar behind ear-coverts), stronger dark lateral throat-stripe and differently streaked breast and flanks (ill-defined and mostly on sides in *sibirica*, with distinctive pale central divide on underbody, but sharper and more pectoral-like in *griseisticta*); see also Brown Flycatcher.

AGEING & SEXING Ageing only possible in autumn, by moult and feather wear in wing. Sexes alike with limited seasonal variation. – Moults. Post-nuptial moult (partial) starting on breeding grounds (including part of head, body and secondary-coverts, often also one or more tertials, rarely odd secondaries and tail-feathers) in Jul–Sep, then either interrupted and resumed in northern Afrotropics, or, more normally, arrested. Post-juv moult less extensive than ad but similar in timing, mostly Jul–Aug, in Europe partial, including most or all of head, body, lesser and median coverts, but usually none or only 1–3 inner greater coverts (rarely a few more); no tertials or secondaries. More or less complete pre-nuptial moult performed by both age groups (but part or all of feathers renewed before autumn migration usually not renewed again in winter) Nov–Mar (to Apr/May in 1stY). Among European passerines the only species to habitually moult primaries ascendently, from outside of wing inwards. – SPRING **Ad** Less worn due to recent primary moult but some feathers (including sometimes odd secondaries, tail-feathers and tertials) moulted already in late summer/autumn, or immediately after migration, now already slightly worn. Plumage duller by late summer with reduced pale feather fringes and increased grey tones, wing-bars and pale secondary panel less conspicuous, and underparts streaking tends to coalesce, appearing blotchier on breast. Birds with

M. s. striata, juv, Sweden, Jul: typical for all juv flycatchers when still on breeding grounds (before post-juv moult commences) are heavily pale-spotted plumage and strongly buff-fringed wing. Plumage loose and more fluffy, especially noticeable here on throat, side of nape and lower flanks. Note also obvious yellow gape-flanges. (D. Jacobsen)

apparently uneven secondary moult may be ad with older worn post-nuptial feathers (not renewed again in pre-nuptial moult). Beware that both ad and 1stY can leave odd feathers unmoulted after pre-nuptial moult, often producing misleading tonal differences among different feathers in same tract. **1stS** Indistinguishable from ad. – AUTUMN **Ad** Worn. Check in particular for worn and bleached primaries and primary-coverts. Lacks any obvious bold buff-white spots on tips of greater coverts. Several show moult limits in secondaries. **1stW** Easily separated from ad by mostly fresh plumage, with no moult limits discernible in juv tertials, secondaries and tail. Juv greater coverts and uppertail-coverts have distinctive buffish-white spot at tips, forming well-marked wing-bar on greater coverts, and pale fringes to secondaries and tertials still rather fresh and much more conspicuous than those of ad. Some retain some juv rump-feathers and scapulars. **Juv** Distinctive, as, in addition to soft, fluffy body-feathers, has buffier upperparts with pale, buff-white spots, rufous-buff fringes to greater coverts and tertials, buff-white tips to uppertail-coverts and in general buffier underparts, with throat, breast and spotted or scaled dark brown.

M. s. neumanni, Turkey, May: a fairly normal appearance of this slightly paler subspecies compared to *striata*. Also in this race, streaking below varies quite strongly—here a heavier-streaked example. (D. Occhiato)

BIOMETRICS (*striata*) Sexes are practically the same size. **L** 14.5–15.5 cm; **W** 85–93 mm (*n* 60, *m* 88.7); **T** 57–65 mm (*n* 60, *m* 60.8); **T/W** 68.5; **B** 15.3–17.7 mm (*n* 27, *m* 16.5); **BD** 3.9–4.3 mm (*n* 17, *m* 4.2); **BW** 5.0–6.2 mm (*n* 27, *m* 5.6); **Ts** 13.2–15.5 mm (*n* 26, *m* 14.7). Wing formula: **p1** > pc 5 mm, to 2.5 mm <, < p2 44–52 mm; **p2** < wt 2–4 mm, =4/5 or 5 (90%) or =3/4 or 4 (10%); **p3** longest (rarely p3 = p4); **p4** < wt 0–1.5 mm; **p5** < wt 4–6.5 mm; **p6** < wt 11–13 mm; **p7** < wt 16–19.5 mm; **p10** < wt 23–29 mm; **s1** < wt 27–31 mm. Emarg. pp3–5.

GEOGRAPHICAL VARIATION & RANGE As many as 11 races have been described, and within the treated region six (e.g. in H&M 4). However, across most of the range variation is mostly slight and clinal, or too subtle to make separation practicable. There is a cline running from W Europe towards east, with dark and well-streaked birds in W Europe (*striata*), gradually becoming paler and less clearly streaked in the east (*neumanni*). The variation between the two is gradual and can only be appreciated with long series of breeding birds for comparison (birds on migration are mostly inseparable). Hence we found only reason to recognise the following four subspecies.

M. s. striata (Pallas, 1764) (Europe east to W Siberia and south to N Africa, but not on Balearics, and eastern limit partly unclear—see below; winters tropical Africa including in the south). Described above and compared to eastern paler and slightly plainer populations best characterised by being dull grey-brown to brownish-grey above, with forehead and forecrown sullied darker (fringed predominantly pale brownish-grey and indistinctly intermixed with white); dark streaks not so contrasting, being rather broad and diffuse. Underparts washed cream, lower throat, breast and flanks streaked on grey-buff suffusion. Birds with such characteristics generally occur in W & C Europe, but variation is inconsistent and mostly clinal. For example, birds from C & S England, but especially in the SE corner of the country, are particularly dark and brown, less grey above, and are dirty-looking below with broader streaking; they also tend to be slightly shorter-winged (**W** 81–86 mm, *n* 15, *m* 84.0), hence might warrant subspecific recognition, but here treated as a minor local variation within *striata* in want of a more comprehensive examination. Birds from N England, Scotland and especially in coastal continental Europe from Holland south to N Portugal are intermediate between classic *striata* (S Sweden) and these darker S English birds. – The situation is similarly unclear in the east. Vaurie (1959) and BWP extend the range of *striata* to just east of the Urals or to W Siberia. This may be practical, and is followed here, but it should be noted that, already from E Europe, birds gradually become paler, approaching *neumanni*, or appear intermediate; breeders in Levant and E Mediterranean are also to some degree intermediate and difficult to assign to one of the two races. – Breeders in Crimea ('*inexpectata*') were described from only four specimens and said to be distinctly darker than other forms, being browner above and more heavily streaked below than *striata*, but of similar size; such birds have never been examined by modern authors, nor have such birds ever been recorded on migration, and the evidence seems insufficient. Also, Hartert & Steinbacher (1933–38) cites Stegmann in stating that 'birds from Crimea do not differ from typical *striata*.' (Syn. *inexpectata*; *papamoscas*).

M. s. balearica von Jordans, 1913 (Balearics; winters W & SW Africa). Highly distinctive by combination of being paler and drabber, rather cream grey-brown above with (almost diagnostic) whitish crown, strongly enhancing dark streaking, and conversely has the poorest dark breast streaking (diffuse and pale, sometimes restricted to breast-sides, or appears almost washed out). Furthermore, it is distinctly smaller than surrounding populations (smallest of all races), has a proportionately broader but flatter bill, is rather long-tailed and has a more rounded wing. **W** 77–82.5 mm (*n* 22, *m* 80.1); **T** 55–62 mm (*n* 22, *m* 58.3); **T/W** *m* 72.7; **B** 14.7–18.2 mm (*n* 22, *m* 16.8); **BD** 3.5–3.9 mm (*n* 6,

M. s. neumanni, Mongolia, Jun: a particularly pale individual, paler overall, with on average paler forehead and crown (forehead off-white) and dark streaking sharper and more contrasting. Also, underparts generally cleaner, with thinner, less contrasting not so dark streaks which become very faint on lower chest and flanks. In spring ageing impossible. (H. Shirihai)

m 3.7); **BW** 5.3–6.5 mm (*n* 22, *m* 5.9); **Ts** 14.9–16.2 mm (*n* 6, *m* 15.4). Wing formula: **p1** > pc 1–5.5 mm, < p2 38–43.5 mm; **p2** < wt 2.5–6 mm, =5 or 5/6 (100%); **pp3–4** longest; **p5** < wt 1.5–3.5 mm; **p6** < wt 7.5–10 mm; **p7** < wt 12.5–14 mm; **p10** < wt 18–21 mm; **s1** < wt 20–23 mm.

○ *M. s. tyrrhenica* Schiebel, 1910 (Corsica, Sardinia, Elba, Italian west coast south to Campania; winters probably W Africa). Very close to *striata*, differing only by subtly smaller size (much overlap!) and being a little warmer brown above (but not as pale as *balearica*). Tentatively accepted, though certainly a weak race. The streaking of throat and breast is often rather diffuse and blotchy, and the chest can appear almost unpatterned, but many are inseparable from *striata* on this character alone. Wing shape intermediate between *striata* and *balearica*. Many migrant *striata* can be eliminated by long primary projection and more greyish-tinged upperparts. **W** 82–87 mm (*n* 13, *m* 83.1); **T** 58.5–65 mm; **T/W** 73.1; **B** 14.7–16.9 mm (*n* 12, *m* 16.1); **BD** 3.7–4.3 mm (*n* 12, *m* 4.0); **BW** 5.0–6.0 mm (*n* 12, *m* 5.4); **Ts** 14.5–15.8 mm. Wing formula: **p1** > pc 0–3.5 mm, < p2 41–46.5 mm; **p2** < wt 2–4.5 mm, =4/5 or 5 (rarely =5/6); **pp3–4** longest; **p5** < wt 2.5–4 mm; **p6** < wt 9–11.5 mm; **p7** < wt 14–16 mm; **p10** < wt 22.5–24 mm; **s1** < wt 22–26.5 mm.

M. s. balearica, presumed 1stS, Menorca, Jun: insular race distinctly smaller than other populations and also paler and drabber, cream-tinged grey-brown above with whitish linings on crown, strongly enhancing dark streaking, but much less marked dark breast-streaking (diffuse and pale, sometimes restricted to sides, or even almost washed out). Note heavily worn apparently retained juv primary-coverts. (P. Slade)

M. s. balearica, 1stW, Mallorca, Aug: again, note on this young bird overall pale plumage, barely developed breast-streaking (lower breast and flanks unmarked and much paler and cleaner impression), and especially whitish linings on crown. (S. Round)

Presumed *M. s. tyrrhenica*, 1stW, Italy, Aug: photographed in Firenze, a bit far inland for safe ssp. *tyrrhenica*, but during autumn migration and might not have set off due south; overall paler plumage with diluted breast-streaking suggests the paler WC Italian race. Fresh plumage with no discernible moult limits, and juv greater coverts have distinct buffish-white spots at tips, forming well-marked wing-bar. (D. Occhiato)

M. s. neumanni, ad, Ethiopia, Sep: partial post-nuptial moult commences on breeding grounds and usually includes secondary-coverts, often one or more tertials (here the longest on left wing, all on right), and sometimes a few secondaries and tail-feathers; primaries and primary-coverts are still unmoulted, being somewhat worn and bleached. (H. Shirihai)

M. s. neumanni Poche, 1904 (W & C Siberia east to N Mongolia and W Transbaikalia, Near East, and perhaps at lest part of SE Europe and Levant, or intergrades with *striata*, further Turkey, Caucasus, Transcaucasia south to SW Iran; winters E & S Africa). Compared to *striata* decidedly paler and greyer above, and has on average paler forehead and crown (forehead admixed with more off-white) with dark streaks being sharper and more contrasting. Pale fringes to wing-feathers tend to be broader and purer in fresh plumage, adding to the paler impression. Underparts also generally paler, with limited light greyish-brown wash and paler and more ill-defined streaks that barely extend to lower chest and flanks. Averages marginally larger. **W** 83–92 mm (*n* 31, *m* 87.0); **T** 55–70 mm (*n* 31, *m* 62.2); **T/W** *m* 71.5; **B** 15.8–18.8 mm (*n* 31, *m* 17.2); **BD** 4.0–4.3 mm (*n* 11, *m* 4.1); **BW** 5.0–6.3 mm (*n* 31, *m* 5.7); **Ts** 14.5–15.5 mm (*n* 11, *m* 15.0). **Wing formula: p10** < wt 21.5–25.5 mm; **s1** < wt 24.5–28 mm. – Breeders in E Iran, Transcaspia, Kazakhstan to Tien Shan, Pamir and N & W Pakistan ('*sarudnyi*') seem to fit into normal variation of *neumanni*, being only very subtly paler and sandier on upperparts, and perhaps a shade whiter on underparts, but far from consistent and hence best included here. – Breeders in SW Altai, N Mongolia and SE Transbaikalia ('*mongola*') have been separated on account of paler colours, perhaps similar to '*sarudnyi*' but greyer, less buff-tinged; the few birds examined from this region were inseparable from other paler *neumanni*. (Syn. *mongola*; *sarudnyi*.)

TAXONOMIC NOTE Considering the rather different morphology of the allopatric *balearica*, whereas most populations in the surrounding area differ rather clearly with little intraspecific variation, a case could be made for recognising this as a separate species. However, *tyrrhenica* of Corsica, Sardinia and apparently also the Tyrrhenian Sea coast over much of the Italian mainland appears intermediate (though closer to *striata*; see also Viganò & Corso 2015). Recently Pons *et al.* (2016) found a clear genetic difference between Balearic breeders (*balearica*) and breeders in 'mainland Europe' (*striata*; Spain, Italy except western coast, France, Denmark, Russia). Breeders of Corsica and Sardinia (*tyrrhenica*; two samples also from Tyrrenian coast) were intermediate. We disagree with their recommendation to separate *balearica* and *tyrrhenica* together as a separate species from *striata* for mainly two reasons: morphological distinction of *tyrrhenica* from Italian *striata* is poor (and as yet not comprehensively studied), and genetically it seems to bridge *striata* and *balearica* rather than group with latter; also, island populations very naturally develop higher genetic divergence faster than continental populations. We feel it remains to be demonstrated that *tyrrhenica* and *striata* on Italian mainland behave as two different species and that all individuals are morphologically distinct.

M. s. neumanni, 1stW, Israel, Oct: some individual variation in streaking below will occur, but most young *neumanni* show rather narrow and distinct streaks on throat and breast, becoming a little bit more ill-defined on breast-sides and flanks. Also typical is the light ground colour of the streaked crown. Note mostly fresh plumage, the juv greater coverts having whitish tips forming well-marked wing-bar. (A. Ben Dov)

REFERENCES Pons, J.-M. *et al.* (2016) *J. Avian Biol.*, 47: 386–398. – Viganò, M. & Corso, A. (2015) *Biodiversity J.*, 6: 271–284.

GAMBAGA FLYCATCHER
Muscicapa gambagae (Alexander, 1901)

Fr. – Gobemouche de Gambaga; Ger. – Savannenschnäpper
Sp. – Papamoscas Gambaga; Swe. – Akaciaflugsnappare

In Arabia, the Gambaga Flycatcher, named after the type locality in Ghana, is solely a summer visitor to wooded highlands, breeding from SW Saudi Arabia south to Yemen, whereas in Africa it is usually a partial migrant, with a very patchy distribution from the Horn of Africa west to Mali and south to Kenya. It is a plain-looking bird, quite similar to some of its relatives and needs to be identified with care in seasons when, in particular, the Spotted Flycatcher is present. It prefers dry highlands or foothills with acacia or other woods, not avoiding entering villages.

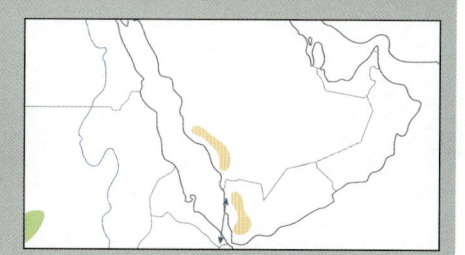

Ad, Yemen May: rather plain grey-brown flycatcher, with indistinct darker loral stripe, whitish throat, (variable) darker streaks on lower throat-sides, breast and upper flanks, wings fringed paler, and small but quite broad bill. Unlike in Brown Flycatcher, pale eye-ring is broken by hint of darker eye-stripe. Evenly feathered and less worn primary-coverts identify as ad. (W. Müller)

IDENTIFICATION A fairly small, compared to Spotted Flycatcher *more round-headed, rather plain grey-brown flycatcher*, with a uniformly dark grey-brown forehead and crown (only very faintly dark-mottled), *a dusky loral stripe and creamy upper lores and eye-ring*, a whitish throat, becoming pale greyish-drab on breast and flanks with diffuse darker streaks on lower throat-sides and breast (much less prominent streaking than in Spotted), and dusky-white lower belly and undertail-coverts. Wings predominantly grey-brown with narrow but rather distinct paler fringes, forming two slight wing-bars on larger coverts and rather obvious tertial linings (a pattern that resembles the paler and better-marked Mediterranean and eastern races of Spotted Flycatcher). Tail almost plain dusky brown. *Small but quite broad bill is black with extensive yellowish or pinkish-orange lower mandible.* Iris dark brown and legs black. Behaviour and habits very much like Spotted Flycatcher (at least in Arabia), but perhaps tends to make shorter and less frequent sallies, and perches less obtrusively than that species. Not particularly shy.

VOCALISATIONS Song is a high-pitched series of squeaking or creaking notes, interspersed by short trills. – Utters a Robin-like repeated *chik* or *zick* or a creaking *chee*, and *zickzick*, much sharper than any call of Spotted Flycatcher.

SIMILAR SPECIES Most serious confusion risk is with Spotted Flycatcher (which is a fairly common passage migrant through SW Arabia), but Gambaga is slightly smaller, shorter-winged (wingtip reaching only to base of tail; primary projection also shorter), rounder-crowned, often (but not invariably) perches less upright and has slightly browner and warmer upperparts. Additionally, most of the lower mandible is yellowish, there is virtually no forehead streaking, and the underparts streaking is also considerably less prominent than in Spotted. – Although hardly likely to ever come close to each other in nature, a museum ornithologist might confuse Gambaga Flycatcher with *Brown Flycatcher*, not least due to similar-looking broad-based bill, rather uniform brown-grey plumage and pale lores. Note that Gambaga has four emarginated primaries (Brown three), on average more obvious streaking or mottling on chest, and a shorter wing but proportionately longer tail. (The larger eye in Brown may be difficult to judge on study specimens!)

AGEING & SEXING Ageing difficult and requires close check of any moult contrast and amount of wear in wing. Sexes alike. Limited seasonal variation. – Moults. Extent and timing of moult is largely undescribed in the literature. Judging from specimens in NHM (32 specimens, of which 16 from Arabia), complete post-nuptial and partial post-juv moults in late summer or early autumn on breeding grounds (Aug–Nov?). Post-juv moult principally includes head, body, lesser and median coverts, and apparently all or most greater coverts and tertials. Pre-nuptial moult seems absent or very limited. – **SPRING Ad** In Feb–Mar some still have characteristic white tips to outer webs of primaries. **1stS** Difficult to separate but at least some have noticeably more worn and abraded primaries and primary-coverts; no white tips to outer webs of primaries (but in some ad these tips also already lost by Feb–Mar). – **AUTUMN Ad** Evenly fresh (no discernible moult limits). Note characteristic white tips to outer webs of primaries (tiny and bold, with abrupt step-like pattern at shafts). Primary-coverts and primaries relatively stronger textured, fresher and darker, and primary-coverts have even-width diffuse, narrow fringes. **1stW** From ad by broader pale tips to both webs of primaries, but these wear very rapidly and by Nov are often lost. Primary-coverts and primaries relatively weaker textured, more worn and paler or browner, and primary-coverts have sharply defined, narrow fringes and slightly broader and bolder tips, but differences from ad are rapidly reduced by increasing wear. **Juv** Distinct: soft, fluffy body-feathering, buff-spotted upperparts and heavier underparts streaking.

1stS, Yemen, May: narrow but rather distinct paler fringes to wing, forming wing-bars on larger coverts and rather obvious tertial linings, but tail almost plain brown. Some birds have plainer facial pattern. Here lack of breast-streaking is due to strong light and side view. Aged by thin white tips to primaries that extend to inner webs, and moult limit in greater coverts (outer ones juv). (W. Müller)

BIOMETRICS No sexual size difference. **L** 12–14 cm; **W** 73–79.5 mm (*n* 29, *m* 75.7); **T** 55–64 mm (*n* 29, *m* 59.0); **T/W** *m* 77.9; **B** 12.5–15.4 mm (*n* 29, *m* 13.9); **BD** 3.4–4.7 mm (*n* 29, *m* 4.0); **BW** 5.5–6.6 mm (*n* 29, *m* 6.0); **Ts** 14.0–16.0 mm (*n* 28, *m* 15.0). **Wing formula: p1** > pc 4–9 mm, < p2 28–36.5 mm; **p2** < wt 6–8 mm, =6/7 or 7 (90%) or =7/8 (10%); **pp3–5** about equal and longest; **p6** < wt 2.5–5 mm; **p7** < wt 7–9 mm; **p10** < wt 13–17 mm; **s1** < wt 14–19 mm. Emarg. pp3–6.

GEOGRAPHICAL VARIATION & RANGE Monotypic. – SW Arabia, W Yemen, N tropical Africa, S Sudan to Kenya; resident.

REFERENCES MÜLLER, W. (2010) *Sandgrouse*, 32:50–54.

1stS, Yemen, May: some have somewhat stronger face pattern, with even a slight, short, white supercilium, and more contrasting breast-streaking. Unlike Spotted Flycatcher, the smaller Gambaga has shorter primary projection and rounder crown. Much of lower mandible is yellowish, with only very tip black. Aged by diffuse white tips to primaries and moult limit in greater coverts. (W. Müller)

Ad, Djibouti, Sep: moult pattern automatically identifies this bird as ad (older than 1stW); note very thin but distinct white tips to outer webs of fresh inner *c*. 5 primaries, whereas the outermost long primaries are still unmoulted, by now very heavily bleached brown and abraded. (W. Müller)

Ad, Djibouti, Sep: some birds are hardly streaked below (though here exaggerated by strong light on chest), making separation from migratory Spotted Flycatcher easier. Note still unmoulted outer tail-feathers inferring a non-juv bird (i.e. ad/2ndW still in post-nuptial moult). (W. Müller)

RED-BREASTED FLYCATCHER
Ficedula parva (Bechstein, 1792)

Fr. – Gobemouche nain; Ger. – Zwergschnäpper
Sp. – Papamoscas papirrojo
Swe. – Mindre flugsnappare

This, the smallest flycatcher within the treated region, leads a largely anonymous life in dense, shady forests, behaving a little like a *Phylloscopus* warbler, always on the move in the canopy. Breeds primarily in E and C Europe, where it is a late-arriving summer visitor, wintering in Pakistan and India, probably also (uncommonly) in the Middle East, Arabia and Iran. The nest is placed in a hole or crevice, often in a broken or decaying tree trunk.

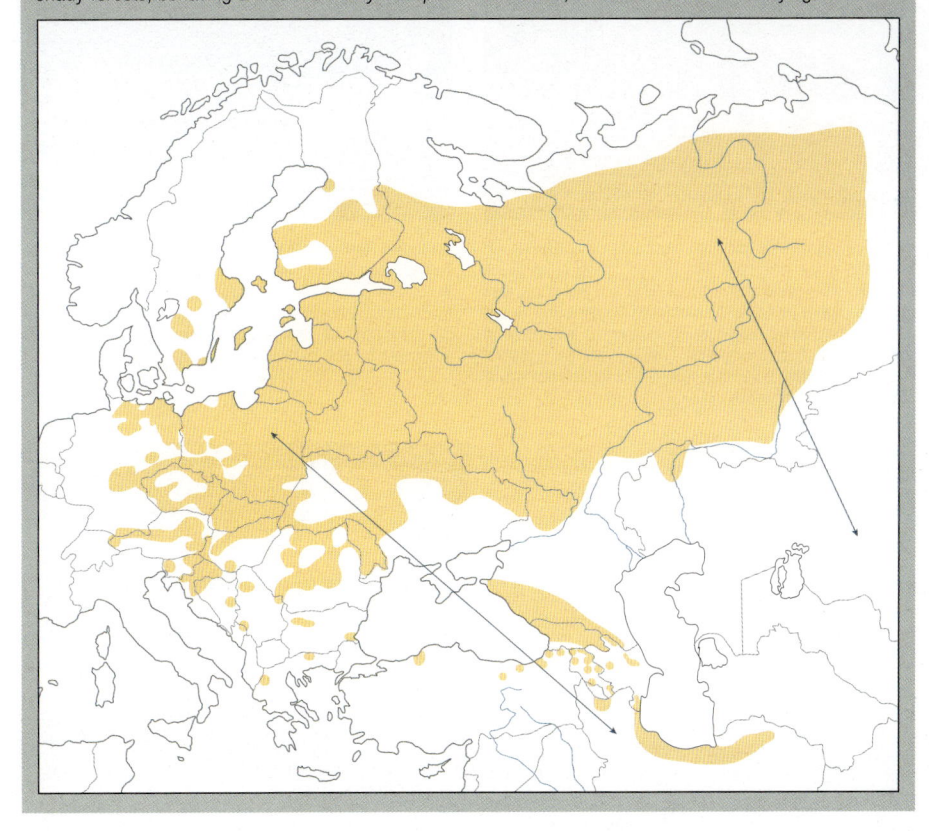

Ad ♂ summer, Germany, May: rusty-red throat sometimes extends to upper breast, thus (superficially) like much larger Robin, yet the bib is much smaller and less distinct in its lower part, and tail pattern is diagnostically black-and-white—especially obvious when tail is cocked. Evenly feathered wing and extensive rusty-red bib make this an ad. (T. Krumenacker)

♂♂ and older ♀♀ can attain a little red on chin and upper throat (young ♂♂ more so than old ♀♀), but never as extensive and full bib as adult ♂. Uppertail-coverts brown to dark brown-grey in ♀♀, dark brown-grey to blackish in ♂♂ (broadly tipped brown when fresh). Legs blackish. Bill dark but has largely pinkish-brown, rufous-brown or yellowish-brown lower mandible (only the tip is dark); note that in profile, the bill might still look nearly all-dark in normal lights and conditions. In winter, adult ♂ invariably retains the red bib (but grey head and neck become more brown-tinged), but at least some ♂♂ have a more subdued rust-buff bib due to extensive paler tips.

VOCALISATIONS Song loud for the bird's size, and consistent-sounding over the entire range. It opens with a few repeated high-pitched and sharp *zre* notes, followed by the 'real' song consisting of a few rhythmic motifs in Pied Flycatcher-fashion and is concluded by the most characteristic part, a series of descending, straight, clear whistling notes at moderate pace, e.g. (*zre… zre… zre…*), *see-chu see-chu see-chu weet weet*, *pee pee poo poo paa paah*. The descending series sounds rather mechanical and does not die off at the end, hence there is really no strong similarity with Willow Warbler (as sometimes claimed). – Has several calls. On migration and in winter most frequent is a short slurred rattle, *serrt*, a bit like Wren but less emphatic,

IDENTIFICATION Would be overlooked most of the time if it were not for its loud song and discreet but often-heard rattling call. *Small*, brown above, pale cream-buff below, with *black-and-white tail* forming a pattern like that in ♂ Red-backed Shrike. The tail is often cocked and slowly lowered again half-spread (to flash the black-and-white pattern) after the bird has landed. Young, first-summer ♂, and all ♀♀ are similarly cream-white on throat, and cream-buff (pale ochre, but with no or only limited grey) on breast and flanks, and brown on head and nape. Older ♂♂ attain a *rusty-red throat-bib*, somewhat variable in extent, and a lead-grey cast on head and sides of throat. Very rarely, both first-summer

Ad ♂ summer, Finland, May: the smallest flycatcher breeding in the treated region. Ad ♂♂ are unmistakable due to rusty-red throat and exposed white tail bases (visible even when tail is only partly open). Note typical posture with cocked tail and drooped wings. Aged by evenly ad wing feathers and unlike 1stS ♂♂ (which are ♀-like) has red throat and largely ashy-grey head. (M. Varesvuo)

Ad ♀ summer, Bulgaria, May: ad ♀ is duller and browner than ad ♂, lacking rusty-red bib, but still has diagnostic tail pattern. Note slight buff tinge on throat, which can be more obvious when fresh. Because most 1stS ♂♂ lack the rusty-red bib and are generally ♀-like, it is important to age the bird before sexing it: as this bird is ad (no juv wing-coverts, rather fresh and round-tipped tail-feathers), the lack of a red bib and subtly greyish-tinged head make it ♀. (C. Bradshaw)

and shorter. A dry clicking *tek* can be heard all year. When alarmed on breeding grounds both the *tek* call and the *zre* (the one introducing the song) can be heard, and a shy whistle *di* or, more often, disyllabic *dilü* (first note higher-pitched) or even notes transposed (last note higher-pitched), *düli*. Young beg with a sharp double *zri-zri*.

SIMILAR SPECIES In ordinary birding in Europe, the adult ♂ needs to be separated from *Robin*, which also has rusty-red throat and breast, but this is a larger bird with longer legs, spending much time on the ground, where it hops about like a little thrush. Note also the all olive-brown tail in Robin without any white (or black), and the considerably larger red breast-patch that reaches above bill to forehead, and almost encircles the eye. – More difficult (but less of a common problem!) is separation from *Taiga Flycatcher*: note that adult ♂ summer Taiga has a somewhat smaller red throat-patch delimited below by a pure grey broad band across upper breast (Red-breasted: the red on throat is often more extensive and fades off below, with blurred border to buff-white breast); brown-capped look (Red-breasted: variably lead-grey or brown-grey, not really 'brown-capped', but some are browner than others on crown); regardless of sex or age, Taiga has almost invariably jet-black uppertail-coverts, blacker even than central tail-feathers (Red-breasted: at most, some adult ♂♂ have blackish uppertail-coverts, uniform with the central tail-feathers, but usually the uppertail-coverts are slightly or clearly paler, more

1stS ♂, Finland, May: although most 1stS ♂♂ lack any red on throat, a few advanced, like this one, have rather a lot. Note also rather pure grey sides of head and seemingly dark uppertail-coverts revealing the sex. An ad ♀ with a little red on chin and upper throat (also rare) would have this less extensive and prominent. Though wing seems rather evenly worn, several central greater coverts still have remnants of the buff tips typical of young birds. (J. Normaja)

Presumed 1stS ♂, Germany, Jun: apparently 1stS ♂ by combination of ♀-like head and body, lacking any hint of red on throat, and quite blackish longest uppertail-coverts and central tail-feathers. Although angle of view makes it difficult to assess the entire wing, this seems rather brownish and worn, perhaps supporting the age as 1stS. (T. Krumenacker)

Breeding ♂ Red-breasted Flycatchers (left: ad, Germany, Jun; centre: 1stS, Finland, Jun) and Taiga Flycatcher (right: ad, Hong Kong, Feb): in Red-breasted, the red throat is often more extensive and can reach upper breast, merging gradually into buff-white chest, whereas Taiga tends to have smaller red throat with pure grey band across upper breast. However, rarely some 1stS ♂ Red-breasted (centre) have a small bib, too. Note tendency of ♂ Red-breasted to have lead-grey or brown-grey crown, versus browner in Taiga (but some overlap). On average Red-breasted has less black uppertail-coverts, brown-grey or brownish-black rather than jet-black as in Taiga. Also, Taiga has blacker bill, largely lacking pink-brown lower base. (Left: T. Krumenacker; centre: J. Vakkala; right: M. Hale)

Aberrant 1stS ♂, Finland, May: note remnants of pale tips on juv greater coverts, these being buff still in late May, a strong indication of Red-breasted (Taiga would have had white tips by this date); intense black of tail indicates ♂. The nearly all-black bill is a rare aberration and causes strong similarity with Taiga Flycatcher. Luckily, such odd birds are very rare (<0.5%?) but they serve to remind us that rare birds are best identified using several characters in combination. (J. Normaja)

Ad ♂, winter, England, Oct: another example of an almost aberrant plumage, this being an extreme as to amount of orange-buff tinge on belly and flanks and tawny-brown on upperparts. The extensive red bib is almost Robin-like. The latter, plus greyer-tinged forecrown and sides of head, and evenly feathered wing, make this an ad. (G. Thoburn)

brown-grey than uppertail); bill slightly deeper on average (by *c.* 6%), and all black, or nearly so, with only narrow paler brown rim ventrally outside feathering at base of lower mandible, though a very few exceptions occur (possibly including Taiga × Red-breasted hybrids?) with more extensive pale base below (Red-breasted: inner half or more of lower mandible pinkish- or yellowish-brown, at least when seen from below or handled). Note that bill colour of Red-breasted is often difficult to judge in normal lateral view, pale-based lower mandible frequently appears quite dark. Beware also of exceedingly rare melanistic Red-breasted with very dark bill. Immature Taiga has whitish outer edges and tips to tertials and greater coverts with a pattern more reminiscent of immature Pied Flycatcher than immature Red-breasted (though a few are intermediate or closer to *parva* in this respect). – There is a remote risk of confusion with a vagrant immature ♂ *Mugimaki Flycatcher* (*F. mugimaki*, extralimital, see vagrants section p. 583), of same size, red-throated, with some white at base of tail, but the white in Mugimaki is more restricted (outermost tail-feathers nearly all dark), greater coverts are pale-tipped, and often median coverts too, and tertials edged white.

AGEING & SEXING Ageing possible during 1stY. No seasonal differences. Sexes differ generally only after 1stY, rarely in 1stS. – Moults. Complete moult of ad, and partial of juv, in late summer (Jul–Sep). Partial pre-nuptial moult in winter quarters, apparently sometimes including tail-feathers, uppertail-coverts and some wing-coverts and tertials. – **SPRING Ad ♂** Chin, throat, and sometimes upper breast, rusty-red. Crown brown-grey, lores, ear-coverts, sides of neck/breast usually pure lead-grey. Not known whether variation in amount of red and grey reflects different ages. Uppertail-coverts usually dark brownish-black or blackish, about same colour as r1. **1stS ♂** Chin and throat usually pale cream or buff-white like ♀, though rarely has a little rusty-red on chin and upper throat. Usually some pale tips remaining on tertials and wing-coverts. Uppertail-coverts usually blackish as in ad ♂. **Ad ♀** As 1stS ♂, but even more rarely has red on chin/throat, and no pale tips on tertials or wing-coverts. Uppertail-coverts usually brown-grey, slightly paler than r1. **1stS ♀** As ad ♀, but generally some pale tips remaining on tertials and wing-coverts. – **AUTUMN Ad** As in spring. Tertials and wing-coverts uniformly grey-brown. Some ad ♂♂ attain a more subdued plumage with extensive buff tipping, the red bib becoming orange-ochre rather than rust-red. **1stW** As ad ♀, but tertials and wing-coverts have yellow-buff or rusty-buff tips. **Juv** Most of head and body-feathers diffusely spotted.

BIOMETRICS L 11–12 cm; W ♂ 66.5–72.5 mm (*n* 86, *m* 68.8), ♀ 63.5–70.5 mm (*n* 31, *m* 66.7); T 47–56 mm (*n* 129, *m* 50.5); T/W *m* 74.1; B 11.3–14.0 mm (*n* 126, *m* 12.6); BD 2.6–3.6 mm (*n* 111, *m* 3.1); Ts 15.5–17.8 mm (*n* 112, *m* 16.8). **Wing formula:** p1 > pc 1–8 mm, < p2 25.5–32.5 (once 35) mm; **p2** ~6 (=5/7); **p10** < wt 13–18.5 mm; **s1** < wt 14.0–20.5 mm. Emarg. pp3–5 (29%), pp3–5 and faint on p6 (52%), or equally strong on pp3–6 (19%).

GEOGRAPHICAL VARIATION & RANGE Monotypic. (Syn. *colchica*.) – E Europe from S Sweden, Germany, Austria, Bulgaria east to Russia and W Siberia, south-east to Transcaucasia and NW Iran; winters Pakistan, India.

Ad ♂, winter, England, Oct: all ad ♂♂ basically retain summer plumage during autumn/winter, with only slight reduction in size of orange bib, and sometimes also red colour subdued due to pale tips. Such birds, in particular as here with grey tinge on neck and sides of breast, plus very black uppertail, can be quite similar to some Taiga Flycatchers, but luckily ad ♂ Taiga in winter shows no or only very little red on throat, and has blacker lower mandible (base dull brownish on this bird). (G. Catley)

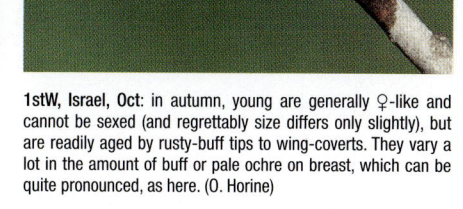

1stW, Israel, Oct: in autumn, young are generally ♀-like and cannot be sexed (and regrettably size differs only slightly), but are readily aged by rusty-buff tips to wing-coverts. They vary a lot in the amount of buff or pale ochre on breast, which can be quite pronounced, as here. (O. Horine)

TAXONOMIC NOTES Formerly treated as conspecific with Taiga Flycatcher, but plumage details, seasonal plumage development, bill colour, song, call and mtDNA differ clearly, and they are better separated as two species (Svensson *et al.* 2005). – Kashmir Flycatcher *F. subrubra* (extralimital, not treated), sometimes regarded as conspecific with *parva*, but quite distinct and clearly a separate species.

REFERENCES Cederroth, C., Johansson, C. & Svensson, L. (1999) *BW*, 12: 460–468. – Lassey, A. (2003) *BW*, 16: 153–155. – Svensson, L. *et al.* (2005) *BB*, 98: 538–541.

1stW, Israel, Oct: note typically drooped wings and partly cocked tail, as well as white basal tail-sides, which together with tiny size should clinch identification over much of the treated region. However, be aware of the possibility of a vagrant Taiga Flycatcher in autumn, and any with all-black bill, black uppertail and whitish underparts should be carefully checked. Aged by juv tertials and wing-coverts with ochre-buff tips. (A. Ben Dov)

Juv, Germany, Jul: in full juv plumage with typically loose feathering and profusely pale-spotted upperparts and diffusely mottled underparts. Note still quite pale largely pinkish-yellow bill with soft gape-flanges. Even young just out of nest will adopt typical posture with drooping wings and cocked tail. (W. Müller)

TAIGA FLYCATCHER
Ficedula albicilla (Pallas, 1811)

Fr. – Gobemouche de la taïga
Ger. – Taigazwergschnäpper
Sp. – Papamoscas de la taiga
Swe. – Taigaflugsnappare

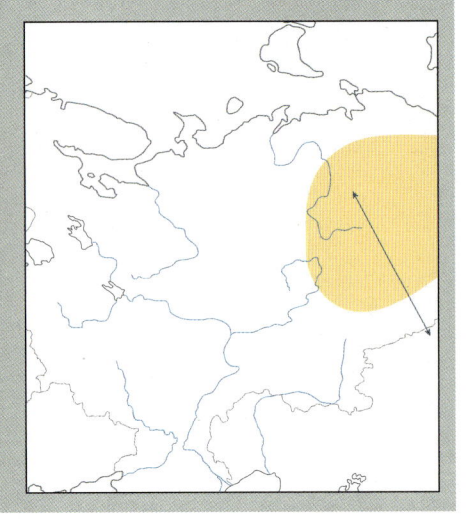

The sister species of Red-breasted Flycatcher, with similar appearance, but a different song and different plumage development. It replaces Red-breasted in the east from just west of the Ural Mountains all the way to the Pacific. It thrives in dense taiga with a mixture of coniferous trees and birch and other deciduous trees, but is also found close to settlements in more open forest edges. Winters in Nepal, E India and SE Asia, returning in late May to its breeding grounds. Habits much as for Red-breasted.

IDENTIFICATION As in Red-breasted, a *small bird, grey-brown above, pale below, with black-and-white tail*. ♂ in summer has a small *rusty-red throat-bib*, neatly *outlined below by a pure grey band across breast* (of variable prominence). ♂ often appears brown-capped due to brown tinge on crown, in contrast to pure lead-grey lores, supercilium and sides of neck (but beware that some are very similar as to head pattern). Ear-coverts often brownish. *Uppertail-coverts jet-black* in all plumages (including first-winter), *darker than uppertail*, which is probably safest clue to identity of birds in ♀-like plumage (but note that <1% have less black uppertail-coverts, these being equal in darkness to uppertail). Unlike in Red-breasted, 1stS ♂ habitually develops red bib on throat, thus breeding ♀-like ♂♂ do not occur, all are in principle ♂-coloured. The red bib of 1stY a is usually attained in early spring, but first traces can appear in Nov. Young, ♀ and ♂ in winter are off-white below with *limited amount of buff* or cream (some have none), and have a *dusky, grey-brown cast across breast*, and a tinge also on flanks. The red bib of ad ♂ is in winter to a variable degree lost, sometimes completely, more often partly, but it is always reduced in size and saturation. Often, the whitish throat stands out, surrounded by dusky colours. Head, nape and back are grey-brown, on average slightly greyer or darker above than Red-breasted. Legs blackish as in Red-breasted. *Bill appears all black*, blacker than in Red-breasted Flycatcher, but most have narrow rim of reddish-brown or other paler colour next to feathering on lower mandible (though this is generally visible only from below or in the hand). Exceptionally bill is more extensively reddish-brown basally on lower mandible, close to what is normal in Red-breasted. A few young Taiga may not develop dark bill until late autumn.

VOCALISATIONS Song differs markedly from that of Red-breasted. It recalls more the song of a bunting, or even a pipit, with its trills and repetition of fast notes on varying pitch, shuttling up and down the scale, e.g. *zri-zri-da zri-zri-da tü-tü-tü zrii-daa-zi*. Another common variation is a song that at some distance can recall Chaffinch song, but which immediately differs by the last phrase consisting of a fast and accelerating series of notes that rise slightly in pitch, ...*tu-to-te-ti*. The song never ends with a slow series of descending, clear notes like in Red-breasted Flycatcher. – Calls. The counterpart to the slurred and rather short rattle (*serrt*) by Red-breasted is usually a rather fast and longer dry trill, *trrrrrrr*, at a distance almost like the sound of a creaking branch in a windy forest, but when close the sound is more insect-like or machine-like. There is also a sharp, high-pitched *tzee*, used in anxiety.

Ad, ♂, China, May: similar to ♂ Red-breasted, being a very small flycatcher with a black-and-white tail pattern. Ad ♂ Taiga has a smaller rusty-red throat, and a grey area on breast (varies in width and boldness—see the two images below). All-black bill (including visible base of lower mandible) and very black uppertail are other important clues. Note also brown-capped impression. Bib coloration and evenly feathered wing make this an ad ♂. (R. Schols)

Variation in summer ♂♂, Mongolia, May: bird's posture, angle of view and light create variation in how birds appear, and to this is added natural plumage variation. Left-hand bird shows a classic grey breast (the grey being well defined, broad and pure), and a fairly extensive rusty-red throat (mainly due to more stretched neck). Right-hand bird seemingly has a more diffuse grey breast that does not form a band (probably due to angle of view), and a small rusty-red bib (neck more withdrawn). Note the brown-capped appearance of both birds (particularly the left one), while the right one shows the near-diagnostic jet-black uppertail-coverts (even blacker than uppertail). Both seem to show hint of whitish tips to outer coverts suggesting 1stS, and thus (unlike 1stS ♂ Red-breasted) the orange bib is already at this age well developed, which is a useful clue once age is confirmed. (J. Normaja)

SIMILAR SPECIES Separation of adult ♂ summer Taiga Flycatcher from similar *Red-breasted Flycatcher* treated in some detail under latter species. As to ♀♀ and young Taiga, note the practically invariably jet-black uppertail-coverts, darker than uppertail (Red-breasted: brown or brown-grey, rarely blackish-brown, never darker than uppertail); the dusky, brown-grey cast to breast and flanks (only rarely warmer buff); and the darker, and very slightly (by *c.* 6%) deeper, bill. On average there is a better emargination on the sixth primary in Taiga (see below).

AGEING & SEXING Ages and sexes differ. Ageing possible during 1stY. Ad ♂ changes summer plumage completely (or at least partly) to ♀-like in winter. Sexing possible from 1stS. – Moults. Complete moult of ad, and partial of juv, in late summer (Jul–Sep). Partial pre-nuptial moult in winter quarters. – **SPRING Ad ♂** Chin and throat rusty-red, completely encircled by lead- or ash-grey. Crown and ear-coverts brown, but lores, supercilium, sides of neck/breast usually pure lead-grey. Uppertail-coverts jet-black, darker than uppertail. Tertials and wing-coverts uniformly grey-brown without pale tips or edges. **1stS ♂** As ad ♂, but usually some pale tips and edges remain on tertials and wing-coverts. **♀** Chin and throat off-white, breast brown-grey. Ages often possible to separate as ad generally has uniformly grey-brown tertials and wing-coverts (any paler fringes very narrow), whereas 1stS usually has some rather obvious pale tips on these. – **AUTUMN Ad** Generally as ♀ in spring, thus sexes alike (red of ♂ disappears Sep–early Nov), but some ♂♂ retain a little red on chin (sometimes also on upper throat) through winter. Tertials and wing-coverts uniform grey-brown. **1stW** As ad but tertials and wing-coverts have buff-white or whitish tips and edges to outer webs. **Juv** Most of head and body-feathers diffusely spotted.

BIOMETRICS L 11–12 cm; **W** ♂ 66.0–74.0 mm (*n* 136, *m* 69.8), ♀ 65–73 mm (*n* 71, *m* 68.3); **T** 45–55 mm (*n* 221, *m* 50.3); **T/W** *m* 72.6; **B** 11.1–13.8 mm (*n* 209, *m* 12.7); **BD** 2.9–3.9 mm (*n* 184, *m* 3.3); **Ts** 15.4–17.9 mm (*n* 155, *m* 16.8). **Wing formula:** p1 > pc 0–9 mm, < p2 25.5–34 mm; **p2** < wt 3.5–9 mm, =6/7 or 6 (=5/6, rarely =7); **pp3–4** (5) about equal and longest; **p5** < wt 0–2 mm; **p6** < wt 2.5–6.5 mm; **p7** < wt 7.5–13 mm; **p10** < wt 10.5–19 mm; **s1** < wt 14.5–21 mm. Emarg. pp3–5 and faint on p6 (55%), or equally distinct on pp3–6 (43%), only very rarely emarg. restricted to pp3–5 (2%).

GEOGRAPHICAL VARIATION & RANGE Monotypic. – Extreme E European Russia east through Siberia to Sakhalin, Kamchatka, south to Altai, Mongolia, Amur; winters NE India, SE Asia.

♀, presumed ad, China, May: diagnostic jet-black uppertail-coverts, even blacker than uppertail. Also off-white underparts, with no hint of buff or cream, and darker and somewhat deeper bill than Red-breasted (blacker than in that species, except tiny area of paler brown at base of lower mandible). Presumed ad by fresh primary-coverts, but rest of wing difficult to assess on single image, hence the reservations as to ageing. (A. Kelly)

Ad ♂♂ autumn/winter (left: Thailand, winter; right: India, Jan): unlike Red-breasted Flycatcher, ad ♂ attains ♀-like winter plumage, but some ♂♂ retain a little red on chin and upper throat. Both show to some degree grey breast-band and at least the right one also brown-capped appearance typical of the species. Left bird shows typical black uppertail-coverts. Both are ad by mainly evenly feathered wings, the wing-coverts and tertials uniformly grey-brown with purer grey fringes. (Left: A. Soonas; right: H. Taavetti)

1stW, Hong Kong, China, Jan: showing well the diagnostic black uppertail-coverts (blacker even than uppertail), as well as hint of brownish cap in contrast to more greyish side of head. That the bill is blacker than in Red-breasted Flycatcher is difficult to see here. 1stW by white-tipped greater coverts and indented tips of outer webs of tertials (where white tips used to be). (M. & P. Wong)

1stW, United Arab Emirates, Dec: separated from Red-breasted Flycatcher in same plumage by same features as in fresh ad, but unlike ad note retained juv remiges (with worn primary tips) and especially wing-bar on juv greater-coverts formed by pale tips, which are white (or pale buffish-white at the most), not buff-orange or ochre like young Red-breasted. (H. & J. Eriksen)

Spring ♀ Taiga Flycatcher (top left: Mongolia, Jun) versus ♀ Red-breasted Flycatcher (bottom left: Sweden, May), and 1stW Taiga (top right: Finland, Nov) versus 1stW Red-breasted Flycatcher (bottom right: Netherlands, Oct): Taiga always has very black uppertail-coverts, darker than uppertail, whereas in Red-breasted these are brown or brown-grey, infrequently blackish-brown, never darker or blacker than uppertail. Underparts coloration is less consistent and thus less reliable, but Taiga often has a dusky grey suffusion to otherwise off-white breast (sometimes flanks, too) and a subtly deeper and blacker bill, rather than having warmer brown or even paler base to lower mandible, as in Red-breasted. Furthermore, retained juv tertials of Taiga tend to have whitish tips extending as pale fringe on outer web, whereas Red-breasted has clearly separated bolder and warmer pale tips (often ochre) and only indistinct paler fringes. The good quality of the wing-feathers indicate that both spring birds are ad, while both autumn birds are obvious 1stW birds by juv wing-coverts with prominent pale tips. (Top left: H. Shirihai; bottom left: M. Martinsson; top right: P. Hytönen; bottom right: B. van den Boogaard)

TAXONOMIC NOTE Formerly treated as an eastern subspecies of Red-breasted Flycatcher, but seasonal plumage development, morphology, bill colour, song, call and mtDNA (c. 7% or 6.3%; U. Olsson *in litt.*, Li & Zhang 2004) clearly differ, and they are better separated as two species. Ranges seem to overlap slightly in area west of Ural Mts, but unknown whether there is any sympatry. Hybrids have been claimed, but not established through observations, and very few intermediates (= possible hybrids) exist in collections.

REFERENCES Cederroth, C., Johansson, C. & Svensson, L. (1999) *BW*, 12: 460–468. – Lassey, A. (2003) *BW*, 16: 153–155. – Li, W. & Zhang, Y. (2004) *Zoological Research*, 25: 127–131. – Svensson, L. *et al.* (2005) *BB*, 98: 538–541.

1stW, Scotland, Sep: one of the best-documented records of Taiga Flycatcher in W Europe, showing off-white underparts and juv tertial pattern (white tips extending as white fringe on outer web). All-black bill possible to confirm in this angle of view. Aged apart from tertial pattern by juv greater coverts with whitish-cream tips forming obvious wing-bar. (H. Harrop)

SEMICOLLARED FLYCATCHER
Ficedula semitorquata (Homeyer, 1885)

Fr. – Gobemouche à demi-collier
Ger. – Halbringschnäpper
Sp. – Papamoscas semiacollarado
Swe. – Balkanflugsnappare

Perhaps best described as a mixture between Pied and Collared Flycatchers, the Semicollared replaces both in SE Europe, Turkey, the Caucasus and Iran. Has similar habitat requirements, being found breeding both in dense, shadowy forests and in glades or open patches in woods along rivers. Nowhere common, and frequently a species that is missed on trips to areas where it breeds. Summer visitor, wintering in E Africa and returning in late March and April.

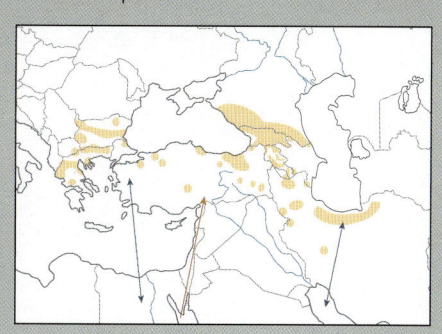

IDENTIFICATION The ♂ resembles a Pied Flycatcher, having *blackish nape and back of neck*, and only a *small white patch on forehead*, but differs in following: *much white in tail*, often with extensive white basally on both webs of outer feathers, easily visible when viewed from below; more *white on primary bases, forming a rounded patch* and extending somewhat outside tips of primary-coverts; and usually (but not invariably) broadly *white-tipped median coverts*. That the ♂ is 'semicollared' is variable: some indeed have a narrow white half-collar, but others look just like Pied in this respect. The ♀ is brown-grey above and whitish below, resembling most of all ♀ Collared due to predominantly greyish hue on upperparts, and a little more white on primary bases than in normal Pied, but differs from Collared when seen close by having *narrower white outer edges to tertials*, forming no large white wing patch (a small rounded one at the most), and on call. Still, often difficult to identify positively without the ♂ present! Note that some median coverts are whitish-tipped (unless heavily worn), but that rarely a hint of this also can be seen in ♀ Collared (and in autumn in all congeners!). Long-winged as Collared Flycatcher, but has proportionately very slightly longer tail than that species. Behaviour as other relatives. The adult ♂ in autumn has attained a more ♀-like plumage being brownish above and on head but can be told on blackish primaries and a fairly large white primary patch extending a little outside tips of primary-coverts.

VOCALISATIONS Song is like a mixture of Pied and Collared songs, a brief, squeaky, slightly monotonous strophe, with some rhythmical elements but also some 'strained', rather high-pitched notes, and frequently rolling sounds, *srre*. Slower pace than in Pied, more like Collared. It does not carry far (and is easily missed among other loud-singing species in the spring forest chorus), while the notes vary rather little in pitch. – Commonest call, often in alarm, is a repeated, clear, straight, piping note, *tüüp*, which can recall Siberian Chiffchaff (*tristis*) at a distance; a variation is the same call slightly upwards-inflected, *tüihp*. A completely different-sounding alarm is an almost straight, slightly 'cracked', intense *bersh* (despite very different rendering, still related to metallic *pik* alarm of Pied Flycatcher). A more subdued, dry clicking *tek* can also be heard.

SIMILAR SPECIES The ♂ in spring is separated from *Collared Flycatcher* by lack of broad white neck-collar, the white patch on forehead being small (can even be divided in two like standard Pied), the white mark at base of primaries being less extensive than in adult Collared, broadly white-tipped median coverts, and generally extensive white on outertail. Autumn birds very similar and best told on size of white primary patch and exact tertial pattern (although some

Ad ♂, Syria, Apr: resembles Pied Flycatcher in having blackish nape, but usually has substantial white 'semicollar' on neck-sides (in particular when stretching neck as here), more extensive white in tail and primary bases, and by diagnostic white median coverts wing patch extension (if present as here). Large white primary patch and evenly ad wing-feathers make this ♂ an ad. (A. Audevard)

Ad ♂, Greece, Mar: in this posture, with lowered neck, the 'semicollar' appears as a narrower white half-collar, but this bird also differs from the image above by its much smaller white forehead patches (just two small spots) and the seemingly blacker wing (partly due to angle of light). Again, note extensive white on outertail. Aged by evenly blackish ad wing-feathers. (P. Petrou)

Ad ♀, Turkey, Apr: unlike ♀ congeners, ad-type tertials have narrower and even-width white outer edges, with bases hardly merging to form white tertial patch. Note that there is often a narrow, at times uneven, white median coverts bar. Aged by rather extensive white primary patch and only ad wing-feathers, including primary-coverts, which are dark and fresh. (A. Öztürk)

HANDBOOK OF WESTERN PALEARCTIC BIRDS

Ad ♀, Israel, Mar: individual variation includes ♀♀ with a small white basal tertial patch (still smaller than those of congeners). This bird has rather obscure white median coverts bar with just two greyish-white tips. Note long, broad white primary patch (almost like Collared). Like all congeners, ad has partial winter moult, which can form misleading contrast. Note how primaries can appear worn and brownish by early spring, but primary-coverts still dark. (H. Shirihai)

1stS ♂, Israel, Mar: plumage like ad, but remiges, primary-coverts, outer greater coverts and alula are juv, obviously paler and browner than ad. At certain angles, 'semicollar' reaches close to central nape, giving impression of an almost complete collar. As here, the median coverts extension can be broken and less clear in some young ♂♂. Nevertheless, has typically small white forehead patches and much white in tail. (H. Shirihai)

1stS ♀, Turkey, Apr: unlike ad, basal white primary patch very small. This bird has rather atypical white basal tertial patch, though this still fairly small, and rather broad white tertial fringes. However, the white edges are still well separated by broad dark centres. White median coverts bar comprises just two white tips. Aged by apparently many juv wing-feathers. (H. Yıldırım)

probably inseparable without very close views or handling, unless call is heard). – From *Pied Flycatcher* ♂ by extensive white in tail, white-tipped median coverts, and often more white on primary bases. – Separated from adult ♂ *Atlas Flycatcher* by much white in tail, small white forehead patch and less extensive white mark at base of primaries. – The ♀ is trickier to separate from the other flycatcher ♀♀ (see Identification for some general guidance). To clinch the identification of a presumed vagrant ♀ Semicollared one needs to see a full set of characters and preferably hear the call; without the latter or handling, some must be left unidentified. The characters to note should include a slightly greyish cast to upperparts, a rather extensive (but not too extensive!) white primary patch reaching to or just outside tips of primary-coverts, a few white-tipped median coverts (white tips rather broad), quite narrow white outer edges to all tertials (ideally not widening much at base), tertials being rather pale brown, and, most importantly (but also most difficult to judge in the field without handling), rather pale brown outer tail-feathers lacking a white 'window' inside shaft, but with a pale (white in ad, buffish in 1stW) rim on outer web that runs all around the tip and continues onto inner web (nevertheless, a few Pied and Collared ♀♀ come close).

AGEING & SEXING Ageing possible during 1stY for ♂, and 1stW for ♀. (With experience and knowledge it is possible to age many ♀♀ in spring too; see below.) Sexes

Ad ♂, Ethiopia, Sep: like ♂ congeners, following complete post-nuptial moult, ♂ Semicollared attains ♀-like winter plumage, but can still be sexed by large white primary patch, quite well-developed white tertial patch, and blackish remiges (around whole tertial tips evenly narrow pale edges confirm ageing). Ad ♂ Collared similar but often eliminated by less jet-black primaries and on average smaller white primary patch (though some overlap). (H. Shirihai)

Ad, presumed ♀, Kuwait, Aug: showing diagnostic specific and ageing clues of even-width pale edges to tertials (no sharp widening of white fringe at shaft as in 1stY) which in this bird do not form white patch at base (cf. other images). The pale median coverts bar is rather obscured but still obvious (although beware that most congeners can show a similar bar in autumn). Sex rather controversial being either a dark-winged ♀ (perhaps most likely) or a ♂ with quite small primary patch. (P. Fågel)

1stW ♂, Ethiopia, Sep: fortunately many young Semicollared retain juv tertials with evenly narrow white fringes (well separated at base), thus white bases hardly merge to form white tertial patch. Note prominent median coverts bar, more prominent and pure white than seen in other species (in Collared or Pied thinner and often less pure white, though differences subtle). The two images seem to be of the same bird, a young ♂. The white primary patch is very small suggesting a ♀, though the blackish wing and uppertail and bold white wing pattern exclude a ♀. (T. Varto Nielsen)

FLYCATCHERS

differ clearly in summer plumage, but less obviously in winter. – **Moults.** Complete moult of ad, and partial of juv, in summer (Jun–early Aug). Partial pre-nuptial moult in winter quarters, including tertials (plus rarely s6), tail-feathers (affecting nearly all ♂♂, many ♀♀), and some inner greater and median coverts. – **SPRING Ad ♂** Upperparts black or blackish, except for usually small, single or double, white patch on forehead (1.5–6 mm high). Rump often grey. White wing patch on tertials and greater coverts large, and median coverts broadly tipped white. White at base of primaries obvious but variable in extent (starting on p3 or p4, or very rarely p5, reaching 3–8 mm > pc). Primaries and primary-coverts blackish, generally quite fresh. Tail with large white portions basally on both webs of rr5–6, with a smaller patch on outer web of r4. **1stS ♂** As ad ♂, but primaries, primary-coverts, alula and outer small wing-coverts clearly more brownish and abraded (primary-coverts also somewhat more pointed), creating obvious contrast to moulted and black rest of plumage. White at base of primaries fairly small (starting on pp5–6, or rarely p4, reaching 1 mm < or 0–5 mm > pc). White on forehead on average less extensive than in ad (1–4 mm high; exceptionally missing). **♀** Upperparts pale brownish-grey, grey as ♀ Collared Flycatcher, which it resembles, though uppertail-coverts and uppertail average paler grey-brown and are never blackish. White at base of primaries variable, usually visible as small mark, though rarely extensive (starting on pp3–4 in ad, and on pp4–6 in 1stS, and reaching 3–6 (9) mm > pc in ad, 0–3 mm > pc, or <1, in 1stS). (With practice a variable number of ♀♀ can be aged in spring based on same criteria as in ♂: 1stS has slightly narrower and more pointed primary-coverts, which are somewhat browner and more abraded; often slightly more obvious moult contrast in greater and median coverts; more pointed and abraded tips to tail-feathers. Note also average difference in extent of primary patch described above.) Tail-feathers slightly paler brown-grey than in congeners, and often lack pale 'window' on inner web of rr5–6; often has light margin around tip of r6, continuing along edge of inner web, and hint of darker subterminal bar around tip. White on outer edges of tertials narrower, not creating such a large white wing patch as in congeners, often none at all. No white hidden on nape held in Collared (which see). – **AUTUMN Ad** Both ♂♂ appear much as ♀ in spring, but upperparts browner, less grey, more like in Pied; thus, black upperparts, white on forehead, and some of the white on wing of ♂ disappear before migration, moulted to ♀-like appearance. Tertials grey-brown with narrow pale tips, evenly broad around tip of each feather. ♂♂ can be recognised by blackish wings with somewhat larger white patch at base of primaries (as in spring), and near-black outer greater and median coverts. **1stW** As ad, but tertials (especially central) have white tips that become much broader on outer webs, creating a 'step' in the white margin at the shaft. **Juv** Most of head and body-feathers diffusely spotted.

BIOMETRICS L 12–13.5 cm; **W** ♂ 79–87 mm (*n* 87, *m* 82.6), ♀ 76–84.5 mm (*n* 37, *m* 80.9); **T** ♂ 49–55 mm (*n* 79, *m* 52.2), ♀ 47–55 mm (*n* 35, *m* 50.9); **T/W** *m* 63.1; **B** 11.1–13.7 mm (*n* 94, *m* 12.4); **Ts** 15.2–17.7 mm (*n* 78, *m* 16.7). **Wing formula:** p1 > pc 0–3.5 mm (rarely 1 mm < or 5 mm >), < p2 37.5–46 mm; **p2** = wt 2–6.5 mm, =5 (45%), =4/5 (32%) or =5/6 (23%); **pp3–4** about equal and longest (though sometimes p4 0.5–2 <); **p5** < wt 2.5–6 mm; **p6** < wt 9–13 mm; **p10** < wt 20.5–26 mm; **s1** < wt 23–28 mm. Emarg. pp3–5.

GEOGRAPHICAL VARIATION & RANGE Monotypic. – SE Europe from Albania, Greece and Bulgaria east through N Turkey, Caucasus, Transcaucasia, N Iran, Zagros; winters central E Africa, apparently Kenya, Tanzania, Uganda and adjacent areas.

REFERENCES Curio, E. (1959) *J. f. Orn.*, 100: 176–209. – Haag, C. et al. (2009) *BW*, 22: 246–248. – Mild, K. (1994) *BW*, 7: 139–151; (with H. Shirihai) 231–240; 325–334. – Svensson, L. & Mild, K. (1992) *Alauda*, 60: 117–118.

Summer ♀♀ plumages (left-hand images ad, right-hand ones 1stS) of W Palearctic black-and-white flycatchers, with Collared Flycatcher (1st row, left: Switzerland, Jun; right: Israel, Apr), Semicollared Flycatcher (2nd row, left: Israel, Mar; right: Israel, Apr), Atlas Flycatcher (3rd row, left: Morocco, Apr; right: Tunisia, Mar), Pied Flycatcher ssp. *iberiae* (4th row, left: Spain, May; right: Spain, May), and Pied Flycatcher ssp. *hypoleuca* (5th row, left: Scotland, May; right: Finland, May): Collared: ad has most extensive white at base of primaries ('club-shaped'), and plumage above tinged grey with paler grey rump; 1stS can be tricky if not aged correctly first, white primary patch shorter and narrower than in ad, thus approaching ad ♀♀ of Pied or Atlas; note overall greyish plumage and paler rump, contrasting with blackish uppertail, and hint of paler semicollar, typical of Collared. – Semicollared: at all ages has relatively extensive white primary patch, with diagnostic narrow and almost even-width white tertial edges that do not merge to white patch at base (at the most a small patch), as well as spotty white median coverts line. – Atlas and Pied ssp. *iberiae*: resemble Collared, being often similarly tinged greyish above, with slightly paler rump. Also have quite substantial white primary patch, more or less midway between Collared and ad ♀ Pied (♀ *iberiae* differs only marginally from Atlas by on average smaller white areas in wing, and with 1stS ♀♀ only doubtfully separable in the field from Collared.) – Pied ssp. *hypoleuca*: dull brownish (less greyish) above, with white primary patch ending well short of edge of wing, nor does it extend outside tips of primary-coverts; 1stS often more readily identified due to very small or no visible white primary patch. (1st row, left: D. Saluz; right: A. Ben Dov. 2nd row, left: H. Shirihai; right: T. Krumenacker. 3rd row, left: P. Vantieghem; right: R. Pop. 4th row, left: F. López; right: J. Blasco Zumeta. 5th row, left: A. Hood; right: M. Varesvuo).

PIED FLYCATCHER
Ficedula hypoleuca (Pallas, 1764)

Fr. – Gobemouche noir; Ger. – Trauerschnäpper
Sp. – Papamoscas cerrojillo
Swe. – Svartvit flugsnappare

A common hole-nester in gardens, where it readily adopts nest boxes and—close to man—sings its merry tune, but is equally at home in remote boreal forests, then using natural tree cavities or old woodpecker holes. One of the best-studied passerines in Europe, and it is well known that varying proportions of ♂♂ lead a double life, with a secret second family away from the first nest. Summer visitor, returning in mid April and May from its winter grounds in W Africa south of the Sahara.

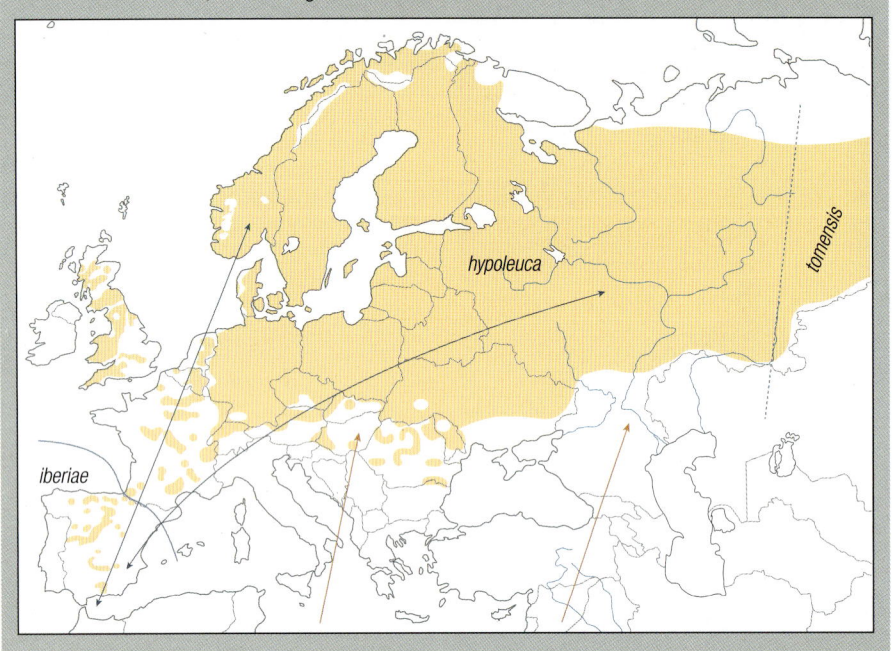

F. h. hypoleuca, 1stS ♂, Finland, May: young ♂ in spring almost acquires full ad breeding plumage, but remiges, primary-coverts and alula are juv, being rather obviously browner and a little paler than rest of wing. Tail and tertials have been renewed in pre-nuptial late winter moult. (M. Varesvuo)

vre-zi tsu tsu chu-we chu-we, zi zi zi. Now and then, a melodious figure is added, at least by certain accomplished singers. Paired ♂♂ mostly sing a shorter, more 'economical' song. – Commonest call, often in alarm, is a persistently repeated, loud, short metallic *pik, pik, pik,...* (also rendered *wit*). A more subdued, dry clicking *tek* can also be heard. From nocturnal migrants a high-pitched buzzing *bzz* (Robb 2015).

SIMILAR SPECIES In summer, the ♂ is separated from *Collared Flycatcher* by lack of broad white neck-collar, the white patch on forehead being small and generally divided in two, any white mark at base of primaries being tiny, and the primaries even in adult being a rather brownish-tinged dark grey. Normally, the black tail has white sides, but note that a very few have all-black tail like Collared and Atlas, and that this is the rule in the Iberian race (see Geographical variation). – Ssp. *hypoleuca* separated from ad ♂ *Atlas Flycatcher* by small white forehead patch, somewhat less black wings, and no or only tiny white mark at base of primaries. Much more difficult to separate from 1stS ♂ Atlas, and some are probably inseparable; any ♂ with small and clearly divided ('double') forehead patch is a Pied, not an Atlas. Ssp. *iberiae* ad ♂ of Pied Flycatcher is much more difficult to separate from ad ♂ Atlas Flycatcher, and sometimes inseparable unless DNA, detailed wing formula and/or call is noted. Take fullest notes if one is encountered outside its normal range. – From *Semicollared Flycatcher* by less white in tail, less white at base of primaries, absence of a white semicollar (though many Semicollared lack it too), and no pure white on median coverts in spring. (In autumn, *all* flycatchers can have pale-tipped median coverts.) – The ♀ is told by its upperparts being rather dark brown (not pale brown tinged greyish as Collared and Atlas, and see Iberian race below), and by the

F. h. hypoleuca, ad ♂, Wales, May: the most widespread black-and-white flycatcher in the treated region. Note the white forehead spots, typically divided into two, though rarely these merge into a larger one. White primary patch quite small. Ad remiges and primary-coverts make this ♂ an ad. Nevertheless, even ad ♂ has slightly more brown-tinged primaries and primary-coverts than rest of wing, albeit showing less contrast than 1stS ♂. (D. Pressland)

IDENTIFICATION A lively, small bird with rather large head, small bill, long wings and short black legs. The ♂ is black (or at least rather dark brown-grey) above, and white below, with a *white wing patch*, usually obvious white tail-sides, and a small *white spot on the forehead* (frequently divided into two small spots, one each side of the base of the bill; rarely the white on forehead is very small or missing). In the ♀, all the black (or dark brown-grey) is replaced by brown, and the forehead has no white spot, or is diffusely paler buff at most (but see Geographical variation & Range concerning Iberian race *iberiae*); the tail-sides are invariably white; and the wing-patch is somewhat smaller and at times less pure white. Behaviour important for identification too. Active and restless, moving around in the canopy. Sits still on a branch, dashes suddenly off to new perch, and *flicks tail and quickly lifts one wing or both upon alighting*.

VOCALISATIONS Song brief and rhythmical, tirelessly delivered, with minor alterations, a few motifs repeated in a jerky beat, rather in folk-dance fashion, e.g. *zi vre-zi vre-zi*

F. h. hypoleuca, ad ♂, more grey-brown bird, Italy, Apr: there is fair variation in darkness of ♂ plumage, many, in particular breeding away from the Atlantic, having the black of upperparts extensively admixed with grey-brown. The white on wing and forehead is as normal though. Ad by dark primaries and primary-coverts, and evenly feathered and still fresh wing. (D. Occhiato)

FLYCATCHERS

F. h. hypoleuca, 1stS ♂, Sweden, Apr: sometimes the two forehead spots merge to form a single larger (here almost heart- or peanut-shaped) patch. Many retained juv wing-feathers including some scapulars and wing-bend make this a 1stS. Note how small the white primary patch can be in a young ♂ Pied. (S. Johansson)

F. h. hypoleuca, 1stS ♂, grey-brown variation, England, Apr: in continental Europe, in particular in the east and in Russia, there is a fair number of brown-tinged ♂♂, and occasionally they occur (as here) in W Europe, too. Nearly all have at least some parts of upperparts blackish, and have white-spotted forehead. Note brownish and pointed juv alula and primary-coverts. (G. Sellors)

F. h. hypoleuca, ♀, Wales, May: ♀ usually easily sexed by lack of white forehead patch, by grey-brown upperparts, and lack of any black in wing. Easily separated from ♀ Collared by only tiny white mark at base of primaries (as here), or none at all. Ageing ♀♀ in spring, especially on photographs, is difficult and often impossible save for a few extremes as to freshness or heavy wear of primaries and primary-coverts. This bird is best left unaged. (R. Wilson)

and median coverts, plus rarely s6, and some tail-feathers (very rarely whole tail). – SPRING **Ad** ♂ Upperparts vary from black, or near-black, to grey or brown-grey, usually with small or moderate-sized white patch on forehead (single or divided, latter commonest), though this can rarely be very small or absent. White wing patch on tertials and greater coverts somewhat variable in size, generally large and involving almost entire outer web of inner two tertials. (Only exceptionally smaller patch as in ♀.) Non-white parts of tertials blackish or dark grey. Primaries and primary-coverts dark brown-grey, fairly fresh. A slight moult contrast present in greater and median coverts, but usually less pronounced than in 1stS. Amount of white in tail variable, usually partly white outer edges of rr4–6 (like in ♀); very rarely (< 3%) entire tail black. **1stS** ♂ As ad ♂, but primaries and primary-coverts on average more brownish and abraded, creating more contrast against moulted and blacker inner greater and median coverts (though difference from ad not always easy to see). ♀ Upperparts brown. No pure white patch on forehead (only a hint of a small pale buff-brown in some). Moderately large white wing patch, white on outer webs of inner tertials usually does not reach feather-shafts, and rest of tertials grey-brown, not blackish. (With practice a variable number of ♀♀ can be aged in spring based on same criteria as in ♂: 1stS has slightly narrower and more pointed primary-coverts, which are a little more brown and abraded; often slightly more obvious moult contrast in

F. h. hypoleuca, ad ♂, Sweden, Aug: in autumn, ad of both sexes attain ♀-like head and body plumage and sexing is less easy, but all ad safely aged by tertials having even-width white fringes that smoothly extend onto inner webs without a 'step' at the shafts, and wing evenly fresh. Sex obvious from near-black remiges, primary-coverts, alula and uppertail including uppertail-coverts, and quite large white tertial patch supports this. (L. Jonsson)

lack of white, or presence of only a tiny white mark, at base of primaries. If a white mark is present, it does not reach near the edge of the wing but stops halfway (four long outermost primaries lack any white at the base of outer webs). There is a risk of confusion in autumn between ad ♂ Pied and ad ♀ Collared, which at times have similar amount of white at primary bases, but the former should have darker flight-feathers and often more white basally on longest two tertials.

AGEING & SEXING (*hypoleuca*) Ageing nearly always possible during 1stY for ♂, and in 1stW for ♀. (With experience it is possible to age many ♀♀ in spring too; see below.) Sexes differ in summer plumage, but usually not in winter unless handled. Seasonal difference obvious in ♂♂. – Moults. Complete post-nuptial moult of ad in late summer (Jul–Aug, rarely Sep), though some leave a few innermost secondaries unmoulted (Jenni & Winkler 1994). Partial post-juv moult at the same time does not involve flight-feathers, tail-feathers, primary-coverts or many outer greater coverts. No moult usually of tertials (though rarely innermost replaced). Partial pre-nuptial moult in winter quarters, similar for both age categories, includes tertials, some inner greater

F. h. hypoleuca, ♀, Finland, May: a slightly paler brown bird than many, partly due to variation, partly to light. Wing-feathers renewed in pre-nuptial moult, here tertials and some inner greater coverts, are more greyish than the rest, but such moult contrasts occur in both 1stS and ad. Primary patch small and limited to innermost c. 4 primaries and does not reach tip of primary-coverts, eliminating ♀ Collared, which has more extensive patch. (D. Occhiato)

greater and median coverts; more pointed and abraded tips to tail-feathers.) – AUTUMN **Ad** As ♀ in spring, both sexes being brown above; thus, white on forehead, and large white wing patch of ♂, moulted to ♀-like appearance. Tertials grey-brown with narrow pale tips, evenly broad around tip of each feather. Some ♂♂ can be recognised by their very black tail-feathers, white either completely absent (very rare) or restricted to tiny patches on outer webs of rr5–6. Rarely, ad ♀ can have quite black uppertail-coverts and dark brown uppertail, so this is not sufficient for sexing in autumn. **1stW** As ad, but tertials (especially central) have white edges that become much broader on outer webs, creating a 'step' in the white margin at the shaft. Some (but not all) ♂♂ have blackish uppertail-coverts, ♀♀ never. **Juv** Most of head and body-feathers diffusely spotted.

BIOMETRICS (*hypoleuca*) **L** 12–13.5 cm; **W** ♂ 74–84.5 mm (*n* 181, *m* 79.5), ♀ 74–82 mm (*n* 67, *m* 78.0); **T** ♂ 48–56 mm (*n* 159, *m* 52.1), ♀ 46–54 mm (*n* 53, *m* 50.9); **T/W** *m* 65.3; **B** 10.9–13.6 mm (*n* 140, *m* 12.6); **Ts** 15.5–18.0 mm (*n* 139, *m* 16.8). **Wing formula: p1** > pc 0–5.5 mm, < p2 35–41 mm (once 42.5); **p2** < wt 3.5–8 mm,

F. h. hypoleuca, ad ♀, Sweden, Sep: combination of tertial pattern with evenly narrow pale tips, smoothly extending onto inner webs, evenly freshly moulted wing, dark brown-grey (rather than blackish) remiges, tail-feathers and uppertail-coverts identify this as ad ♀. Very small white primary patch and lack of white-tipped median coverts secure the species. (M. Hellström)

F. h. hypoleuca, 1stW ♂, France, Aug: aged by tertial pattern (especially inner two) with broader white edges becoming abruptly narrower at tips on inner webs, creating a 'step' at the shaft. Some ♂♂ have blackish uppertail-coverts and rectrices, as here. The very tiny white primary mark is the only species-specific feature. (T. Quelennec)

=5/6 (93%) or =5 (7%; once > p5 by 0.5 mm); **pp3–4** about equal and longest; **p5** < wt 2–5 mm; **p6** < wt 7.5–11.5 mm; **p7** < wt 12–16 mm; **p10** < wt 19–25 mm; **s1** < wt 21.5–27 mm (once 30). Emarg. pp3–5.

GEOGRAPHICAL VARIATION & RANGE There is a cline running from the Atlantic towards east in which ♂♂ become less black above, more grey, and increase slightly in size. Birds in south-west are invariably mainly black, with larger white forehead patch and more white in wing.

F. h. hypoleuca (Pallas, 1764) (Europe except Iberia; winters sub-Saharan Africa, mainly Senegal to Nigeria, but apparently some stay in SW & S Europe and NW Africa). Treated above. Usually blackish above in ♂ plumage, though some individual variation within its range, some being greyer, and far from a neat geographical pattern observed. White forehead patch of spring ♂♂ small (single or, more commonly, divided), usually 2–5 mm high. Single forehead patch usually slightly indented at centre. White on base of primaries (outer webs) from p6 or p7 (very rarely a small white mark on p5) and inwards, generally level with, or shorter than, tips of primary-coverts (but very occasionally 2.5 mm > pc; in this case still largely hidden on closed wing). Tail in ♂ usually with some white on outer 2–3 feathers, very rarely all black (< 1%). (Syn. *muscipeta*.)

○ **F. h. tomensis** (Herm. Johansen, 1916) (W Siberia; winters probably sub-Saharan Africa, but details poorly known). Similar to *hypoleuca* but is a trifle larger; spring ♂♂ are on average more grey-brown above, sometimes all grey-brown, and are thus invariably slightly paler. ♂♂ without any white on forehead occur. Grades into *hypoleuca* in the west over large area, and numerous birds are impossible to ascribe to subspecies. Although examined material of this race, and of birds breeding in C & E Europe (which can also be slightly browner and less black) is not extensive and therefore not fully conclusive, we have not found reason to distinguish a further and intermediate race ('*muscipeta*'). W ♂ 78–84 mm (*n* 33, *m* 81.0), ♀ 76.5–83 mm (*n* 10, *m* 80.0). (Syn. *sibirica*.)

F. h. iberiae (Witherby, 1928) (Iberia; winters presumably W Africa). Same size as *hypoleuca*, but plumage of ad ♂ in spring intermediate between *hypoleuca* and *speculigera*, or close to latter, usually being jet-black above, blacker on average than *hypoleuca* (small patches of grey-brown feathers in otherwise black upperparts occur only in c. 20%), has somewhat larger white patch on forehead (commonly 4–8 mm high, rarely 3–10 mm, shape rounded or oval, never divided), usually an all-black tail (exceptions rare, having a little white on edges of rr5–6), and has more white at base of

F. h. hypoleuca, 1stW, presumed ♂, Netherlands, Sep: aged as 1stW by the tertial 'steps' (white becoming abruptly broader on outer webs). Quite broad white edges to tertials and tips to greater coverts are typical for ♂♂, although wings and tail of this bird are not as dark brown-grey as they can be. Some 1stW have pale median covert tips that form a second wing-bar, like in Semicollared Flycatcher, but note brown upperparts and small white primary patch. (N. D. van Swelm)

F. h. hypoleuca, juv, Netherlands, Sep: note whitish spots above and some dark scaling below, typical of juvenile plumage in all flycatchers. Perhaps a late-hatched bird, since in Sep post-juv moult is usually completed and plumage more plain. (N. D. van Swelm)

F. h. iberiae, ad ♂, Spain, May: virtually identical to and usually inseparable from Atlas Flycatcher (here eliminated by locality), except by marginally less extensive white marks on forehead and wing. Black-and-white plumage, large white primary patch and jet-black ad wing-feathers age and sex this bird. (R. Fernández González)

primaries (in 85% from outer web of p3 or p4 and inwards, reaching 2–5.5 mm > pc, in 14% it starts on p5, and in 1% on p6; J. Potti *in litt.*). Some ad ♂♂ can attain a narrow white semicollar with age (Potti & Montalvo 1995; see also Taxonomic note), just like in Atlas Flycatcher. 1stS ♂ has more brownish and worn primaries and primary-coverts, and invariably a little white on outer retained juv rr (4)5–6; white on base of primaries from outer web of pp4–6 and inwards (very rarely a trace on p3), reaching to tip of primary-coverts. Size of white forehead patch does not seem to be closely correlated with age. Ad ♀ is rather greyish above with a dark-looking uppertail, resembling both Collared and Atlas Flycatcher ♀♀. It is a peculiar feature of this race that ad ♀ often attains a white forehead patch (2–7 mm high, rarely even more, in *c.* 40% of all 2ndY+, a higher percentage in older birds; Potti 1993.). **W** ♂ 75–82 mm (*n* 30, *m* 79.0), ♀ 74–82.5 mm (*n* 19, *m* 77.4); **T** ♂ 49–59 mm (*n* 30, *m* 52.0), ♀ 48–53 mm (*n* 19, *m* 50.3); **T/W** *m* 65.5; **Ts** 15.3–18.0 mm (*n* 49, *m* 16.9); **B** 11.4–14.9 mm (*n* 47, *m* 13.0). (The following biometrics of live birds in Segovia, C Spain, are from J. Potti *in litt.*: **W** ♂ 75–86 mm [*n* 1,393, *m* 79.9], ♀ 71–84 mm [*n* 1,695, *m* 77.8]; **T** ♂ 47–58 mm [*n* 597, *m* 51.9], ♀ 41–57 mm [*n* 757, *m* 50.6].) **Wing formula: p1** > pc 0–6 mm, < p2 34–41.5 mm; **p2** < wt 4.5–7 mm, =5/6 (85%) or =6 (15%); **p5** < wt 1–4 mm; **p6** < wt 7–11.5 mm; **p10** < wt 17.5–22.5 (once 26) mm; **s1** < wt 19–26 (once 28) mm.

F. h. iberiae, 1stS ♂, Spain, Jun: in full breeding plumage, but remiges and primary-coverts are juv and slightly browner than rest of black plumage, explaining near total lack of white primary patch. Such birds could invite confusion with ssp. *hypoleuca*, but note all-black tail and greyish rump (although both features can be rarely shown by *hypoleuca* too). White forehead patch smaller than in ad, but never divided in two small spots as in many *hypoleuca*. (R. Fernández González)

F. h. iberiae, variation in ad ♀, Spain, May–Jun: quite greyish above (though amount can vary individually and with light), with white primary patch rather large, intermediate in size between Pied and Collared. Many older ♀♀ (uniquely) attain white on forehead (left-hand image). (left: J. Blasco-Zumeta; right: J. L. Copete)

F. h. iberiae or *hypoleuca*, ad ♀, Spain, Apr: white primary patch rather prominent with small white spot on p5, which might indicate ssp. *iberiae*, while rather brownish-tinged upperparts including uppertail (lacking more greyish cast on back and blackish uppertail typical of *iberiae*) are better for *hypoleuca*. Neat wing seemingly only contains ad feathers. (C. N. G. Bocos)

F. h. iberiae or *hypoleuca*, 1stW, Spain, Sep: these taxa are inseparable in 1stW plumage, but the fact that this bird was photographed in Spain, and has a fairly large white primary patch suggest *iberiae*. (A. Torés Sanchez)

TAXONOMIC NOTE The status of Iberian race *F. h. iberiae* needs further study, considering its obvious external similarities to the recently split Atlas Flycatcher, and treating it as a subspecies of the latter is an alternative approach. However, on present knowledge best kept as a subspecies of Pied Flycatcher. Vocalisations are thought to be closer to *hypoleuca* than to *speculigera*, but studies still incomplete. – The name of the W Siberian race, often known as *sibirica* Khakhlov, 1915, was changed to *tomensis* in 1916 when the Pied Flycatcher was placed in *Muscicapa*. The Code stipulates (Art. 59.3) that such changes made before 1961 stand irrespective of whether preoccupation has later been removed due to change of genus.

REFERENCES Curio, E. (1960) *Vogelwelt*, 81: 113–121. – Mild, K. (1994) *BW*, 7: 139–151; (with H. Shirihai) 231–240; 325–334. – Potti, J. (1993) *Animal Behaviour*, 45: 1245–1247. – Potti, J. & Montalvo, S. (1995) *Ibis*, 137: 405–409. – Potti, J. et al. (2016) *Bird Study*, 63: 330–336.

ATLAS FLYCATCHER
Ficedula speculigera (Bonaparte, 1850)

Fr. – Gobemouche de l'Atlas; Ger. – Atlasschnäpper
Sp. – Papamoscas magrebí
Swe. – Atlasflugsnappare

Until recently regarded as a distinct NW African subspecies of the Pied Flycatcher, Atlas Flycatcher is now commonly treated as a separate species, as morphology is about as distinct as that of Semi-collared Flycatcher, and mtDNA and song apparently differ slightly but consistently. Breeds in dense oak forests and orchards, mainly on slopes and foothills of the Atlas ranges, commonly above 1000 m. Summer visitor, returning from W Africa surprisingly late—not until May, but vanguard of ♂♂ may arrive from second half of April or even a little earlier.

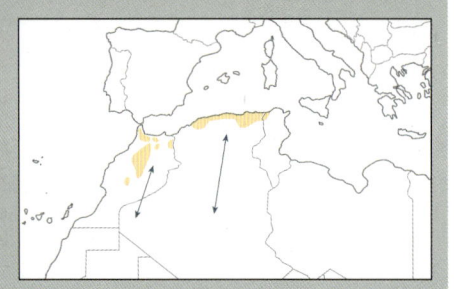

Ad ♂, Morocco, Jun: shares with Collared Flycatcher (and thus differs from ♂ Pied) extensive white primary patch and large white forehead as well as all-black tail, but lack complete, broad white neck-collar of Collared (though note narrow vague semicollar). White mid-wing patch particularly extensive. (R. Armada)

IDENTIFICATION Breeding adult ♂ is very similar to Collared Flycatcher, thus *quite black above* with much white, but lacks the complete, broad white collar around the neck (although it may have a semicollar, and—allegedly—rarely a near-complete collar, broken only in centre of hindneck). In common with Collared, has *much white on greater coverts and at base of primaries*, and a *large white single forehead patch* (rarely smaller patch). Also, the *tail is all black* in adults (with very few exceptions), sometimes also in first-years. Rump is variable, either *black as back*, faintly greyish *or prominently greyish-white*; all-black rump in combination with much white in wing is a good sign of this species. First-summer ♂ usually similarly black above as adult ♂ (albeit with brownish-tinged primaries), but has sometimes slightly less extensive white forehead patch and always much less white in wing (no more white than ♂ Pied Flycatcher, in fact), and sometimes shows white on tail-sides (when juv outer tail-feathers were not moulted in winter). Tricky first-summer ♂♂, some of which must be left unidentified, have rather small white forehead patch, which can almost (but not quite!) be divided in two (as it often is in Pied), and have some brown-grey admixed in the otherwise black upperparts, especially on lesser coverts and sides of back. In spring, upperparts of ♀♀ are often *tinged greyish* (like Collared), the rump slightly paler greyish, and the *uppertail near-black*, with a good white primary-patch in adults, a combination that may prove distinctive in relation to Pied (but not to some Collareds; it is desirable that these criteria are tested on a larger sample). Also, in spring, Iberian Pied population (*iberiae*) migrates through the normal range of Atlas Flycatcher, and some ♂♂ and many ♀♀ of these two are apparently inseparable. Spring migration of *iberiae* appears to be somewhat earlier than the arrival of Atlas Flycatchers (in April rather than May), which offers some help. About 40% of ad ♀ Iberian develop white forehead patch, while the proportion in ♀ Atlas is a mere 15%. – Autumn birds are very difficult (sometimes impossible) to separate from migrant Pied (especially of Iberian race *iberiae*), being similarly brown-grey above and whitish below with a whitish wing patch. Note in autumn on adult ♂, and some adult ♀♀, much *white at base of primaries reaching near edge of wing*, and beyond tips of primary-coverts. First-year ♂ can have as little white at bases of primaries as Pied Flycatcher, thus impossible to pick out on this character; but a few have a little more white, starting on the fourth primary and reaching just outside tips of primary-coverts. Outer tail-feathers have white portions, like Pied. Presumed to have on average slightly darker flight-feathers than first-year ♂ Pied, but difference subtle, and probably of little use in the field. First-year ♀ hardly separable from Pied. Habits as Pied Flycatcher.

VOCALISATIONS Song fairly similar to Pied, though differs consistently in being delivered at a rather slow tempo, almost in staccato, thereby at least in many strophes creating a slight similarity with the staccato-like song structure of Collared Flycatcher. Atlas Flycatcher has a slightly deeper voice than Pied; perhaps sounds 'less cheerful', more 'pensive' than Pied. Often contains on the one hand repeated phrases, on the other soft and melodic notes reminiscent of Blackcap song. (Based on unpubl. recordings by L. Wallin and J. L. Copete, and on own field studies.) – Alarm call is a repeated *veet* or *hiisk*, more similar to that of Pied (*pik*) than the rendering might imply, but still differing in being a trifle longer. The call is rather metallic and 'ends hard' (as in Pied), is not as drawn-out whistling à la Nightingale as in Collared. Has also a dry clicking *tek*, like that of Pied. Robb (2015) suggested fine differences in calls but these by and large require sonograms and specialist knowledge to be appreciated.

SIMILAR SPECIES See Identification for general differences from *Pied Flycatcher*. (Separation from the Iberian subspecies of Pied, *iberiae*, requires further study.) – Although ranges should differ, the rare occurrence of Atlas ♂♂ with

1stS ♂, Morocco, Jun: a quite advanced young ♂, in full breeding plumage, though remiges and primary-coverts are juv and tinged brownish, and primaries here lack visible white patch outside primary-coverts. Still readily told from ♂ Pied by large white forehead and mid-wing patches. All-black tail further supports this (though not all 1stS ♂♂ have it). (R. Armada)

FLYCATCHERS

Ad ♂♂ in summer of Collared (top left: Finland, May), Semicollared (bottom left: Kuwait, Mar), Atlas (top right: Morocco, Jun), 'Iberian' Pied ssp. *iberiae* (bottom centre: Spain, May) and Pied Flycatchers (bottom right: Switzerland, May): Collared is readily identified by its complete, broad white neck collar, very large white patch at base of primaries and large (and always single) white forehead patch. – Semicollared: by combination of white semicollar, small white patch on forehead usually divided into two spots, extensive white tail-sides (more white than congeners), extensive, rounded white primary patch and diagnostic white median coverts extension of white wing panel. – Atlas: distinctive by sharing many features with Collared, but lacks complete white collar (although it can show a semicollar), and unlike Pied it has large white primary patch and rather large, single white forehead patch, but usually no white in tail. – Race *iberiae* of Pied: differs only marginally from Atlas by on average smaller white areas in wings and on forehead. – Pied ssp. *hypoleuca*: never any white semicollar, smallest white primary patch of the five, usually two small white forehead spots (sometimes one a little larger) and nearly always some white on tail-sides. (Top left: J. Tenovuo; bottom left: A. Halley; top right: R. Armada; bottom centre: J. Blasco-Zumeta; bottom right: G. Schuler)

a semicollar, and more commonly with broadly white-tipped median coverts, will invite confusion with *Semicollared Flycatcher*. (A few claims of this from Algeria have not been substantiated following a careful check; cf. e.g. Svensson & Mild 1992.) Note that whereas adult ♂ Atlas in spring invariably has an all-black tail and a large white forehead patch, Semicollared has extensive white on outer tail-feathers but a rather small white forehead patch. ♀ Semicollared has paler uppertail being brown-grey rather than blackish. – Separation of ♀ from ♀ Collared Flycatcher in spring extremely difficult and often impossible in the field. Many but not all can be separated in the hand on fine details of biometry and wing formula, ♀ Atlas having a slightly shorter and blunter-tipped wing with p2 invariably falling between p5 and p6. ♀ Atlas also appears to have on average blacker uppertail in contrast to pale grey rump, but odd Collared can match this. – Finally, the rare occurrence of *hybrid* Pied × Collared must be considered, which might produce birds resembling Atlas Flycatcher.

AGEING & SEXING Apparently much as Collared Flycatcher, but little studied. – Moults. Similar to congeners. – SPRING Ad ♂ Upperparts black, except extensive white patch on forehead (6–11.5 mm long; once only 3 mm). Rarely limited brown-grey patches in the black of upperparts (either retained winter plumage or new feathers with this colour). Can apparently develop a narrow white semicollar, which in its extreme is only broken on hindneck (though the possibility that such—rare—birds may be hybrids with Collared cannot be ruled out). Rump can have some grey, even prominently greyish-white, but can also be black like

Ad ♀, Morocco, May: ♀ Atlas also resembles Collared, usually being similarly tinged greyish above, with slightly paler rump contrasting with blackish uppertail. Also has quite substantial white primary patch, quite like ad ♀ Collared. Therefore, safe separation of ♀ Atlas and Collared is very difficult and requires more research. Large primary patch and uniform ad wing-feathers make this ♀ an ad. (A. & G. Swash)

rest of upperparts. White patch on tertials and greater coverts large, with *c.* 5–7 inner greater coverts being all white, and white patch at bases of primaries rather large (starting on outer web of p3 or p4, reaching 4–10 mm > pc). Primaries and primary-coverts black, generally quite fresh. Tail practically invariably all black (only very rarely has some narrow white on r6). **1stS** ♂ As ad ♂, but primaries and primary-coverts clearly more brownish and abraded, creating obvious contrast with moulted and black rest of plumage. White at bases of primaries fairly small (starting on outer web of pp4–6, reaching 0–2 mm > pc, or 0–2 <). White patch on forehead usually fairly large (3–8 mm long) and rounded, rarely smaller and has hint of central divide. Usually black above as ad ♂, but commonly some brown-grey admixed on, e.g., wing-bend, scapulars and rump. Tail frequently has many outer retained juv feathers, showing much white, but rarely whole tail moulted to black ad type. ♀ Upperparts grey-brown, slightly greyer than in autumn, very similar to ♀ Collared. Rump slightly paler grey, uppertail blackish (unlike all ♀ Pieds and most ♀ Collareds). White mark at base of primaries small (starting on pp4–6 and reaching 0–4 mm > pc in ad; or on pp5–6, or exceptionally p7, 0–3 mm < in 1stY). (With practice a variable number of ♀♀ can be aged in spring based on same criteria as for ♂: 1stS has slightly narrower and more pointed primary-coverts, which are somewhat browner and more abraded; often slightly more obvious moult contrast in greater and median coverts; more pointed and abraded tips to tail-feathers. Note also average difference in extent of primary patch described above.) Frequently white-tipped median coverts. For reliable separation from ♀ Collared perhaps necessary to examine biometrics and wing formula in the hand, although ♀ Collared on average has less blackish uppertail. No white hidden on hindneck. – **AUTUMN Ad** Both sexes similar, much as ♀ in spring; thus, brown replaces black of upperparts, white on forehead, and much white on wing of ♂ (except white on primaries). Tertials grey-brown with narrow pale tips, evenly broad around tip of each feather. ♂ can be recognised by blackish wings with large white patch at base of primaries as in spring, and near-black inner greater and median coverts; also, central tail-feathers black. **1stW** As ad, but tertials (especially central) have white tips that become much broader on outer webs, creating a 'step' in the white margin at the shaft. Sexes generally differ in darkness of flight-feathers, ♂ very dark grey, ♀ slightly paler brown-grey. Note that ♂ has rather limited white at base of primaries like in spring (see above). **Juv** Most of head and body-feathers diffusely spotted.

BIOMETRICS L 12–13.5 cm; **W** ♂ 77–82.5 mm (*n* 38, *m* 79.8), ♀ 75–80 mm (*n* 10, *m* 78.0); **T** ♂ 48–56 mm (*n* 38, *m* 51.8), ♀ 46–52.5 mm (*n* 9, *m* 50.5); **T/W** *m* 64.8; **B** 11.6–14.2 mm (*n* 43, *m* 12.9); **Ts** 16.1–18.6 mm (*n* 45, *m* 17.3). **Wing formula: p1** > pc 0–5 mm, < **p2** 36–42 mm; **p2** < wt 4–7.5 mm, =5/6 (100%); **pp3–4** about equal and longest; **p5** < wt 1.5–4 mm; **p6** < wt 6.5–11 mm; **p6** < wt 11–14 mm; **p10** < wt 18–25 mm; **s1** < wt 20.5–28.5 mm. Emarg. pp3–5.

GEOGRAPHICAL VARIATION & RANGE Monotypic. – NW Africa from Morocco through N Algeria to NW Tunisia; winters W Africa.

TAXONOMIC NOTE The African population has always been known to differ from Pied Flycatcher on a level that might render it species status, but it was not until Sætre *et al.* (2001a, b) reported a fairly distinct level of genetic difference that the Atlas Flycatcher has been commonly separated. The intermediate character of Iberian Pied Flycatchers (*iberiae*) remains to be addressed in a consistent way. This race may well be better referred to Atlas Flycatcher than Pied based on several similarities. However, it is wise to await a more comprehensive study before a change is adopted.

REFERENCES VAN DEN BERG, A. B. (2006) *DB*, 28: 1–6. – COPETE, J. L. *et al.* (2010) *DB*, 32: 155–162. – CORSO, A. *et al.* (2015) *DB*, 37: 141–160. – ETHERINGTON, G. & SMALL, B. (2003) *BW*, 16: 252–256. – MILD, K. (1994) *BW*, 7: 139–151; (with H. SHIRIHAI) 231–240; 325–334. – ROBB, M. (2015) *DB*, 37: 161–163. – SÆTRE, G.-P., BORGE, T. & MOUM, T. (2001a) *Ibis*, 143: 494–497. – SÆTRE, G.-P. *et al.* (2001b) *Molecular Ecology*, 10: 737–749. – SVENSSON, L. & MILD, K. (1992) *Alauda*, 60: 117–118.

♀, Morocco, May: white primary patch quite substantial and suggesting ad. Whole plumage appears very neat, possibly indicating ad, but ageing of ♀♀ in spring notoriously difficult unless bird handled, thus best left unaged. (D. Barnes)

♀♀, Morocco, Jun: left-hand bird is strongly marked ad ♀, with white forehead, which occurs in *c.* 15 % of ad ♀♀ (like in ssp. *iberiae* of Pied); note darker grey centres to some coverts, pale rump and blackish tail-feathers, these having been replaced in winter and which contrast with unmoulted feathers. The right-hand bird, however, is a poorly marked 1stS ♀ with several unmoulted juv median and greater coverts. The white primary patch is quite substantial and bold in ad ♀, but much narrower in 1stS ♀. Both show larger white patches than most ♀ Pied *hypoleuca*, but it is usually impossible to separate ♀ Atlas from *iberiae* of the same age using this feature. (J. L. Copete)

COLLARED FLYCATCHER
Ficedula albicollis (Temminck, 1815)

Fr. – Gobemouche à collier; Ger. – Halsbandschnäpper
Sp. – Papamoscas acollarado
Swe. – Halsbandsflugsnappare

In mature deciduous woods in E Europe in summer you may find this elegant flycatcher with its strangely squeaky song. In the north it breeds on the large islands in the Baltic Sea of Gotland and Öland, as well as in Poland; in both areas some mix with the local Pied Flycatchers and produce hybrids. Has a similarly complicated sex life as the Pied Flycatcher, with males pairing with more than one ♀ simultaneously. Winters in SE Africa, returning to Europe from late April but mostly in May.

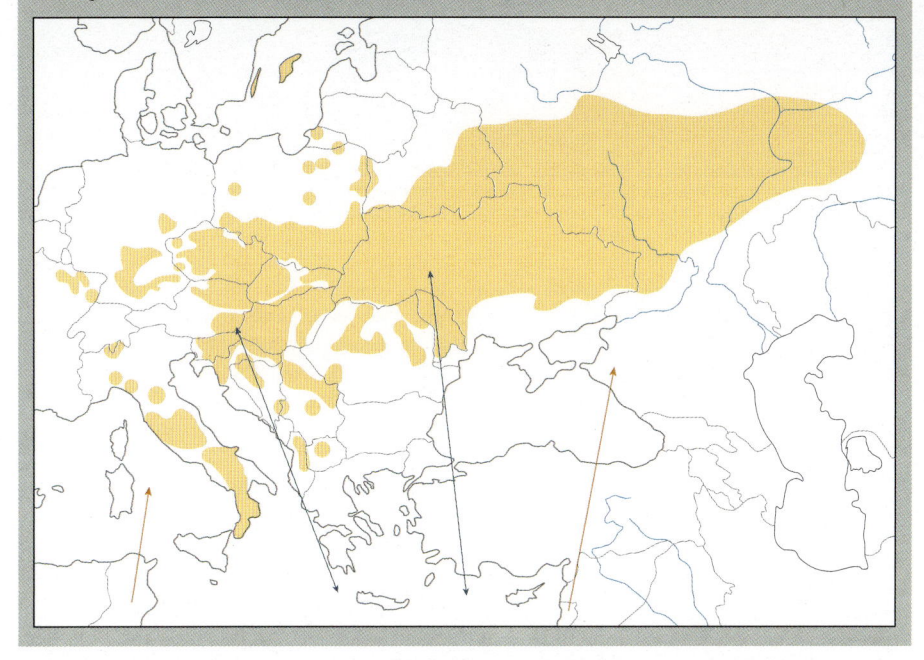

1stS ♂, Sweden, May: distinctive grey-white patch on lower back/rump. Unlike ad, white primary patch smaller and narrower; juv remiges, including primary-coverts, and outer greater coverts slightly brownish and confirm age as 1stS. The white forehead patch averages smaller than in ad, still is larger and rounder than in any Pied Flycatcher. (L. Jonsson)

Ad ♂, Germany, May: in full (immaculate) breeding plumage, and unlikely to be confused if complete white neck collar, extensive white primary patch and very large white forehead blaze are seen. Head pattern sometimes creates illusion of head being small. Aged by uniform ad wing-feathers. (R. Martin)

IDENTIFICATION In spring and summer, adult ♂ is perhaps easiest to describe as a Pied Flycatcher with a complete, *broad white neck collar*, much *more white at base of primaries* and a *larger white forehead patch*. Also, the rump averages paler, grey-white or even white. Due to the white collar, and large white forehead patch, can look small-headed. The tail is frequently nearly all black, or even entirely black, but can have some white on the sides. First-summer ♂ is similar, but has brownish-tinged primaries and outer wing-coverts, somewhat less white on forehead and wing, and more frequently shows white on tail-sides. Very rarely, the white collar can show some limited greyish-black on the central hindneck, but never a complete and broad dark break or 'bridge' in the white (which is a sign of the bird being either another species or a hybrid). ♀♀ and young, and autumn birds, of Collared and Pied are more similar: brownish-grey above and whitish below, with a whitish wing patch. But, notice on summer ♀ Collared: slightly paler and more *greyish upperparts*; rarely a hint of paler grey back of neck and rump (mirroring ♂ pattern); and, most importantly, more *white at the base of primaries reaching near edge of wing*, and almost invariably protruding beyond tips of primary-coverts, the white usually being *broadest over its distal part*, near the edge of the wing (sometimes described as 'club-shaped'). Very rarely, there can be a hint of pale or whitish tips to some median coverts, recalling ♀ Semicollared. Young, and autumn birds of both sexes, are browner above, much more similar to Pied. However, note amount and shape of white at primary bases (see above), which is nearly always a good clue. Adult ♂ in autumn, when seen well, differs by having quite *blackish outer wing-coverts, black wing-feathers* and central tail-feathers, and *a large pure white patch on primary bases*. Note that adult ♂♂ in autumn do not have all-black tail as in spring, but have some white on outer feathers, at times much (the tail is usually moulted twice a year). Habits as Pied. Restless and rather aggressive, generally dominating over Pied where the two occur together.

VOCALISATIONS Song a rather pathetic effort, the voice being strained and squeaky and not very pleasing. Shrill, harsh, whistling notes are delivered at moderate pace in a seemingly erratic pattern, and it can sound like 'laboured breathing'. Some notes in the song are like the call note or similar, *eehp*. The rhythm and pitch vary somewhat between strophes, one example can be rendered *zii zree, zet, zree, eehp zet-zet, zreeh zy*. – Call is a drawn-out, straight, thin

Ad ♀, Israel, Apr: large white primary patch (broadest over distal part, reaching edge of wing, appearing 'club-shaped'), and with characteristic pale, greyish-tinged upperparts, even with hint of paler collar like in ♂. Ad wing-feathers and extensive white primary patch age this bird. (A. Ben Dov)

— 55 —

Presumed 1stS ♀, Israel, Apr: here a slightly controversial bird demonstrating the difficulties often occurring in flycatcher identification. No white-tipped median coverts and large white tertial patch should eliminate Semicollared, and greyish-tinged upperparts seem to exclude Pied. Still, white primary patch for a ♀ Collared rather small, starting on p5 and ending just short of tips of primary-coverts, but rarely this can occur. Two or three outermost greater coverts seem retained juv, which would fit our assessment of species, sex and age. (G. Shon)

Ad ♂, Egypt, Sep: non-breeding ad ♂ in ♀-like plumage, but specific and ageing/sexing clues include the large white primary patch, extensive white tertial edges, blackish outer wing-coverts including some median and lesser ones, black remiges and central rectrices. Being ad, the white tertial tips are narrow and of even width around tip of feathers. The white forehead patch is either a remnant of summer plumage or a rarer variation. (E. Winkel)

whistle, *eehp*, as if 'inhaling', which recalls Thrush Nightingale alarm to the point of confusion. Has also a dry clicking *tek*, like that of Pied.

SIMILAR SPECIES Differences from *Pied Flycatcher* treated under Identification, but note for ♀♀ and autumn birds that some Pied can show a decent white bar on primary bases outside primary-coverts, but that the white does not reach so near to the edge of the wing as on ♀ Collared; still, the difference is not large, and caution is advocated. To separate a ♀ Collared in autumn with least white on primaries from an autumn adult ♂ Pied with maximum white, note that the latter usually has darker flight-feathers and more white basally on longest two tertials. – Collared summer ♂ can hardly be confused with other flycatchers, but the following pitfalls must still be mentioned: odd *Semicollared Flycatcher* ♂♂ can seem to have a near-complete white collar, reaching far back on sides of neck (though white never meets on nape); also, the collar of Semicollared is invariably narrow (nearly always broad in Collared); further, note in spring limited or no white in tail in Collared, but extensive white in Semicollared; forehead patch small on Semicollared (sometimes even divided at centre), large on Collared. More difficult still is to separate Collared from Semicollared in autumn. If correctly aged first, adults can usually be separated by size of white primary patch (larger in Collared; rather little overlap) and precise tertial pattern (more of a solid white patch on bases of tertials in Collared, narrower white edges on all three tertials and less of a white basal patch in Semicollared). First-winter birds are even more similar to first-winter Semicollared, require careful examination of size of primary and tertial patches, and some must probably be left undetermined without hearing the call or biometrics. – ♂ *Atlas Flycatcher* is very similar to ♂ Collared in all respects except collar, this apparently never becoming completely white on nape (though—allegedly—may come close), and many have no more hint of a white collar than has Pied. Also, ♀♀ of the two are very similar and often inseparable in the field, though many can be separated in the hand on biometrics and wing formula. On average, Atlas appears to have blacker uppertail in clear contrast to greyish rump, but at least some Collared come close or match this. – Even ♂ of the Iberian subspecies of *Pied Flycatcher* (*iberiae*) may approach Collared, with a fair amount of white on forehead and wing, an all-black tail, and very rarely even a rather broad and almost complete white collar, but both forehead patch and white in primaries are less extensive than in typical Collared, and there is invariably a dark 'bridge' across the white collar on nape. – Finally, one must be aware of the occurrence of Pied × Collared *hybrids*: look for contradicting set of characters, like full collar, and blackish primaries, but no or only very little white at primary bases, or much white at primary bases but incomplete collar.

AGEING & SEXING Ageing possible during 1stY for ♂, and in 1stW for ♀. Sexes differ clearly in summer plumage, but less obviously in winter, but still readily separable. – Moults. Complete post-nuptial moult of ad, and partial post-juv, in summer (mid Jun–Aug). Partial pre-nuptial moult in winter quarters, including head, body and tertials (plus rarely s6), tail-feathers (replaced by nearly all ♂♂, and by many ♀♀), and some inner greater and median coverts. Some 1stS ♂♂ replace tail to 'intermediate appearance', feathers being narrower, more pointed and less blackish, with more white on outer edge of r6. – **SPRING Ad ♂** Upperparts black, except broad and complete white collar, and usually extensive white patch on forehead (5.5–12 mm long). Rump often grey, sometimes white. White wing patch on tertials and greater

Ad, ♀, Turkey, Sep: narrow, even-width white tertial tips age this bird as ad. Given this, and dark brown-grey centred (rather than black) outer median and lesser coverts, brownish-grey rather than jet-black remiges and large white primary patch (albeit smaller than in ad ♂), it is possible to sex it as ♀, and separate it from Pied, but not from ad ♂ Atlas at this season (luckily, they seldom overlap in range). Told from Semicollared by tertial pattern and lack of median coverts bar. (F. Yorgancıoğlu)

1stW ♂, Netherlands, Oct: aged by pattern of white tips to tertials (especially central feather), with a broader outer web that forms a 'step' at the shaft. Given age and relatively large white primary patch, safely identified as Collared and sexed as ♂. Some 1stW Collared can show a faintly indicated white median coverts bar (retained juv feathers). (A. Ouwerkerk)

FLYCATCHERS

1stW ♂, Egypt, Oct: once aged as 1stW (by tertial pattern with much broader white edges on outer web than inner), the well-developed white primary patch and dark centres of tertials and greater coverts identify this bird as a 1stW ♂ Collared. (M. Heiß)

spring, but upperparts browner, less grey, more like Pied; thus, black upperparts, white on forehead, white collar and white in wing (except on primaries) of ♂ moulted to ♀-like appearance. Tertials grey-brown with narrow pale tips, evenly broad around tip of each feather. ♂ can be recognised by blackish wings with a large white patch at base of primaries as in spring, and near-black outer greater and median coverts; also, central tail-feathers black. Note that ad ♂ has large white portions in outer tail (unlike in spring). **1stW** As ad, but 1stW ♂ has not as jet-black primaries as autumn ad ♂, and tertials (especially central) have white tips that become much broader on outer webs, creating a 'step' in white margin at the shaft. Surprisingly, often considerably more white on outer tail-feathers than in 1stW ♂ Pied, many having very large white 'windows' on inner webs of rr5–6. Uppertail and uppertail-coverts often rather blackish in ♂♂, but some are brown like ♀♀. **Juv** Most of head and body-feathers diffusely spotted.

BIOMETRICS L 12.5–14 cm; **W** ♂ 77.5–87.5 mm (n 120, m 82.6), ♀ 76–84.5 mm (n 45, m 81.0); **T** ♂ 47–55 mm (n 97, m 50.6), ♀ 46–53 mm (n 36, m 50.1); **T/W** m 61.4; **B** 11.0–14.0 mm (n 127, m 12.6); **Ts** 15.7–18.0 mm (n 120, m 16.8). **Wing formula:** p1 > pc 0–4 mm (rarely 1 mm < or 5 mm >), < p2 (38) 39–46 mm; **p2** < wt 2.5–6.5 mm, =4 or 4/5 (63%), =5 (30%) or =5/6 (7%); **pp3–4** about equal and longest; **p5** < wt 3–7 mm; **p6** < wt 8–14 mm; **p7** < wt 12.5–19.5 mm; **p10** < wt 20–28.5 mm; **s1** < wt 23.5–30 mm. Emarg. pp3–5.

GEOGRAPHICAL VARIATION & RANGE Monotypic. – Local in France, Germany, Italy, on islands in Baltic Sea, more common C & E Europe, from Poland and S Russia and south-east; winters S Africa, mainly in Zambia, Zimbabwe and Malawi, possibly also adjacent areas (claimed records from W Africa usually refer to *F. speculigera*, from E Africa to *F. semitorquata*).

TAXONOMIC NOTE All black-and-white flycatchers are closely related, and have apparently diverged fairly recently. A certain amount of hybridisation still occurs between Collared and Pied in the Baltic, and rarely elsewhere. Furthermore, Semicollared Flycatcher was previously treated as a subspecies of either Pied or Collared, and Atlas Flycatcher was until recently treated as a subspecies of Pied.

REFERENCES MILD, K. (1994) *BW*, 7: 139–151; (with H. SHIRIHAI) 231–240; 325–334. – MILD, K. (1995) *BW*, 8: 271–277. – RIDDIFORD, N. (1991) *BB*, 84: 19–23.

coverts large, and white at base of primaries extensive (starting on p3, or very rarely p4, reaching 7–16 mm > pc). Primaries and primary-coverts black, generally quite fresh. Tail usually has small white portions on outer webs of some outer tail-feathers, or is all-black, or, very rarely has white tail-sides like ♀. **1stS** ♂ As ad ♂, but primaries, primary-coverts, alula and outer smaller coverts clearly more brownish and abraded, creating obvious contrast to moulted and black rest of plumage. White patch at base of primaries fairly small (starting on pp3–5, exceptionally p6, reaching 0–5 mm > pc). White on forehead similar to ad, or only subtly smaller (4–10 mm long). Tail usually moulted but is less black than in ad ♂ and commonly has a small white spot on outer web of r6 near tip; rarely more extensive white on r6. Rarely a hint of a greyish-black narrow 'bridge' across white neck at centre of nape (but a solid and broad black 'bridge' would probably invariably indicate hybrid origin). ♀ Upperparts pale brownish-grey, greyer than ♀ Pied Flycatcher, which it otherwise resembles. Uppertail, too, generally brown-grey, not blackish. White at base of primaries extends to near edge of wing (starting on pp3–5, exceptionally p6, and reaching 0–6 mm > pc in ad, 0–2 mm in 1stS, exceptionally to 2 mm <). On average, paler rump than in Pied. Rarely a hint of paler grey collar and thin whitish tips to a few median coverts. – AUTUMN **Ad** Both ♂♀ appear much as ♀ in

1stW ♂, Germany, Sep: another young 1stW ♂ with slightly smaller white primary patch, still darkness of remiges and prominence of white greater coverts bar, plus dark centres of tertials indicate it is a young ♂. (H. Schmaljohann)

1stW Collared Flycatcher (left: Turkey, Aug), Semicollared Flycatcher (centre: Armenia, Aug) and Pied Flycatcher ssp. *hypoleuca* (right: Netherlands, Sep): Collared readily distinguished by extensive white at base of primaries; Semicollared by combination of diagnostic tertial pattern (narrower and even-width white edges, with bases hardly merging to form white patch) and white primary patch rather small; and Pied has narrower and more limited white primary patch, and tertial pattern as Collared. All aged as 1stW by pattern of white tips to tertials (especially central one), broader on outer webs, becoming abruptly narrower at tips on inner webs, creating a 'step' at shaft. (Left: Ö. Necipoglu; centre: O. Z. Göller; right: N. D. van Swelm)

— 57 —

AFRICAN PARADISE FLYCATCHER
Terpsiphone viridis (Statius Müller, 1776)

Fr. – Tchitrec d'Afrique
Ger. – Graubrust-Paradiesschnäpper
Sp. – Monarca africano
Swe. – Afrikansk paradismonark

Though widespread in the Afrotropics, this flashy, long-tailed flycatcher is a very localised resident within the covered region, being restricted to foothills and montane woodlands of extreme SW Arabia and the Mahrah region of E Yemen and Dhofar in S Oman. To see this extremely beautiful bird you usually need to enter groves with rather open lower or middle parts, but still heavily shaded by the closed canopy. It is a quiet, gentle bird, not really shy though wary. Without warning it suddenly sits there on a branch, very still, only to soon dash out of sight again.

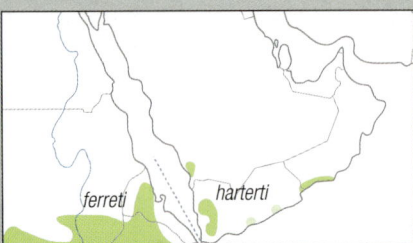

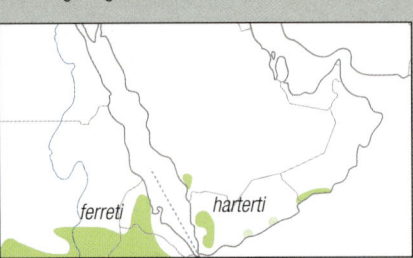

T. v. ferreti, ad ♂, Ethiopia, Aug: an extralimital race, widespread across Africa, and especially in Horn of Africa very similar to Arabian race *harterti*. Nevertheless, in both populations 'intermediate' morphs occur, among them quite frequent ♂♂ with white wing panel and central rectrices. Also note striking, bluish-coloured orbital-ring. (H. Shirihai)

IDENTIFICATION A large attractively coloured flycatcher, in adult ♂ plumage with *extremely long central tail-feathers*. *Glossy bluish-black head with short crest*, the black reaching down to chest and upper flanks, where it progressively becomes dark grey, and greyish-white to greyish-buff on undertail-coverts. *Rest of upperparts, tail, inner wing-coverts and tertials are russet-brown*, contrasting with mostly black wing-feathers, and variably extensive *white wing panel on greater coverts and fringes of secondaries* (on some, tertials also broadly fringed white). Several colour variations occur, mainly in ♂♂. Most common is so-called 'normal' or rufous-backed morph, described above, but there is also a rarer morph with more white on dorsal area and white portions on upperwings and tail, but head and underparts as 'normal' morph. (This is still quite different from 'white morph' of Indian Paradise Flycatcher; p. 584.) Various 'intermediate' morphs also occur, of which quite frequent are males with normal-

T. v. harterti, ad ♂, Oman, Jan: glossy blue-black head with short crest, chestnut upperparts, dark sooty-grey belly and large white panel on folded wing are typical features of ad ♂. Tail mainly chestnut, but streamers either chestnut or white, or mixture. Here tail-streamers are still growing, and are still short of full length. (H. & J. Eriksen)

T. v. harterti, ad ♂, Oman (both Jan): when not broken, tail-streamers of ad can be extremely long and elastic. This adult ♂ is of the white-tailed morph (while others are rufous, or mixed morphs). The short crest varies in size, individually and with the bird's behaviour and angle of view. (H. & J. Eriksen)

T. v. ferreti, ad ♀, Ethiopia, Aug: note combination of black outer greater coverts with narrow whitish fringes, contrasting with chestnut inner greater coverts, plainer russet tertials, small white wing panel (here just outer edges to secondaries), and lack of very long tail-streamers. (H. Shirihai)

T. v. harterti, ad ♀, Yemen, Nov: rather short, rounded crest, black of head and throat less intense and less glossed than in ♂, and underparts paler grey. Also has less blue bill, and narrower and duller bluish-grey orbital ring. (S. Kennernecht)

type plumage but restricted white linings to inner wings and white-centred central tail-feathers, the latter with black shafts. Also striking, in close views, is the *bluish-coloured orbital-ring* and rather *deep-based bill* (in all morphs). ♀ generally appears as 'normal' ♂ but *lacks tail-streamers*, has shorter crest, black of head to breast less intense and less glossy, underparts paler grey and most of russet plumage slightly duller (tail always russet, lacking white). ♀ also has a less blue bill and narrower and duller bluish-grey orbital ring. Usually *no white in wing*, or very restricted if any (white fringes to tertials and greater coverts). Note that some ad ♂♂ may lose or moult their tail streamers, and younger ♂♂ (at least first-summers) may have somewhat shorter tail streamers than older ones. Rather upright on perch, but despite rich plumage and extravagant tail, may be rather retiring and elusive in its often shady woodland habitats. Usually seen alone or in pairs. Hawks insects in quick sallies from a shaded perch, usually within a large tree.

VOCALISATIONS Song in Arabian subspecies *harterti* (there is considerable geographical variation in song types) is a loud, rapid, varied warble, containing pleasing whistles, often glissando notes, uttered while the bird fans and jerks its tail in sideways motion. – Calls consist of varied hoarse or raspy notes, e.g. a short, explosive, high-pitched *pcheeah* on falling pitch, a *zweet* or *zwayi*, and a quick rippling series of piping notes, *switty-switty-switty-*…

SIMILAR SPECIES Unlikely to be confused, especially the male, given a reasonable view and bearing in mind the species' highly restricted range in the covered region.

AGEING & SEXING Ad easily sexed. Limited seasonal variation. Moult, ageing and sexing poorly covered in the literature. Only 25 *harterti* examined (NHM), taken Nov–Mar (most), May and Sep. Freshest ad (apparently shortly after completing moult) are found in Nov–Dec, when 1stW has moderately worn primaries, primary-coverts and tail-feathers. Both age groups more worn from Feb–Mar, but no very worn specimens (just before moult) examined. Ad seems to undertake a complete post-nuptial moult probably between late summer and first half of autumn (the usual breeding season of Yemeni birds, correlated with the biannual seasonal rains, mostly Mar–Apr and Jul–Sep, but unknown if species breeds more than once in Arabia and how moult varies accordingly). Post-juv moult partial, and no evidence of any pre-nuptial moult. Post-juv moult involves all head, body,

T. v. harterti, 1stY ♂, Oman (left: Oct; right: Mar): unlike similar ad ♀, juv tertials have diffuse dark centres near shafts, juv outer greater coverts have brown centres (just visible on left bird) with russet fringes, and tertials and inner secondaries lack white fringes. (Left: D. Occhiato; right: G. Lobley)

T. v. harterti, 1stY ♀, Oman, Mar: young ♀♀ are dullest overall, with plainer wings, paler rufous upperparts, paler (more grey-black, less glossy blue-black) head, and paler grey underparts. Tail short, crest short and rounded. (M. Römhild)

lesser and median coverts but probably no or few innermost greater coverts, and no remiges, tertials, alula or primary-coverts, and apparently no tail-feathers. – SPRING **Ad ♂** Among the few *harterti* examined no white morphs were found, but two intermediate morphs had largely white tail streamers and one partly white. **Ad ♀** Unlike superficially similar 1stS, diagnostic black outer greater coverts present with narrow whitish fringes (contrasting with mostly russet inner greater coverts). White fringes to inner secondaries, plainer and deeper russet tertials, and central tail-feathers project slightly further (see 1stW ♂). Also, evenly feathered, unlike 1stS. **1stS** As 1stW, but feather wear makes moult limits more contrasting. Both sexes apparently retain 1stW plumage (including tail-feathers) but unknown (at least in Arabia) if they breed in such plumage. – AUTUMN **Ad** As in spring, but plumage generally more intensely coloured and glossy, and tail-streamers of ♂ more often complete (not broken as in some worn ♂♂). **1stW ♂** Unlike similar ad ♀, retained juv outer greater coverts dark brown with russet fringes, inner secondaries lack white fringes, and retained juv tertials have diffuse dark centres close to shafts. Juv primaries and primary-coverts are more worn and browner (blacker in ad). Tail graduation in 1stY (both sexes) mostly < 6 mm (in ad ♀ < 15 mm). **1stW ♀** Very like 1stW ♂ and separated from ad ♀ by same criteria, but has square tail, duller rufous-brown upperparts, a paler head and greyer underparts. Bill paler and browner in 1stY ♀ but blacker in 1stW ♂ and ad ♀, although this needs to be checked on live birds. Crest of most 1stW (both sexes) shorter than in ad ♀. **Juv** From ad ♀ by dark brown head, duller rufous-brown upperparts and brownish cast to underparts.

BIOMETRICS (*harterti*) **L** ad ♂ 26.5–45 cm, ♀♀/1stY 19–22.5 cm; **W** ♂ 83–94 mm (n 12, m 88.7), ♀ 81–91 mm (n 12, m 85.4); **T** (except streamers) ♂ 89.5–117 mm (n 10, m 100.2), ♀ 82–109 mm (n 12, m 95.9); **tail streamers** (r1>r2) ad ♂ 64–233 mm, ♀♀/1stY 1–35 mm; **T/W** ad ♂ m 292.6, ♀♀/1stY m 124.1; **B** 18.2–20.6 mm (n 24, m 19.5); **BD** 4.3–5.1 mm (n 19, m 4.6); **Ts** 14.3–17.2 mm (n 21, m 16.0). **Wing formula: p1** > pc 13–17 mm, < p2 19.5–25.5 mm; **p2** < wt 15–21 mm, = ss or < ss; **p3** < wt 4–6 mm; **pp4–5** about equal and longest; **p6** < wt 0.5–2 mm; **p7** < wt 5.5–8.5 mm; **p10** < wt 12–16.5 mm; **s1** < wt 14–17 mm. Emarg. pp3–6.

GEOGRAPHICAL VARIATION & RANGE Only one race occurs in the treated region, *harterti*, which is endemic to Arabia and is one of *c.* 10 generally recognised forms within this highly variable, otherwise African species, the closest extralimital being *ferreti* (Eritrea, Ethiopia, S Sudan).

T. v. harterti (Meinertzhagen, 1923) (SW Saudi Arabia, Yemen, S Oman; resident). Described above. Somewhat darker russet above than African races, tending towards ferruginous-brown, and black of head and chest tends to reach further down on underparts in ad.

Extralimital race: ○ *T. v. ferreti* (Guérin-Méneville, 1843) (Eritrea, Ethiopia, S Sudan, Somalia, Kenya west to Mali and Ivory Coast). Very similar to *harterti* but claimed to differ by their slightly smaller bill (at least in ♂♂) and on average in that breast is usually not as black, more slate-grey. Perhaps even more variable than *harterti* in that some ♂♂ are predominantly white on upperwing and tail (though not as white as white morph of Indian Paradise Flycatcher).

BEARDED REEDLING
Panurus biarmicus (L., 1758)

Alternative name: Bearded Tit

Fr. – Panure à moustaches; Ger. – Bartmeise
Sp. – Bigotudo; Swe. – Skäggmes

The Bearded Reedling forms a monotypic family (Panuridae), and is closely related to larks. It was previously thought to be related to parrotbills and babblers, but this has recently shown to be incorrect. It is patchily distributed from the British Isles and S Fenno-Scandia south to Turkey and Syria, and east to Transcaspia and NW China. Restricted to reedbeds, particularly extensive areas of *Phragmites*. In winter it may disperse quite widely (having even reached Kuwait) and use much smaller patches of suitable habitat. These irruptions can lead to the establishment of new colonies. Nevertheless, populations fluctuate greatly due to cold-weather losses.

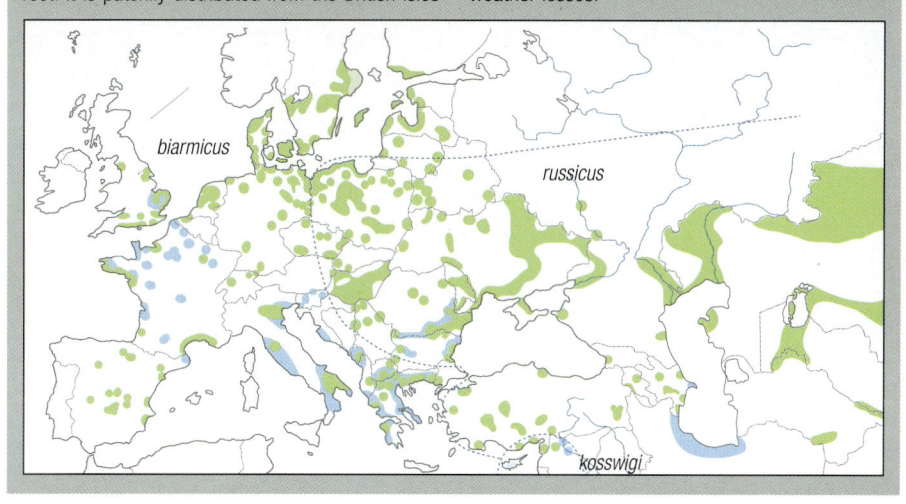

IDENTIFICATION A quite long-tailed, otherwise tit-sized passerine, *largely pale cinnamon-brown with white-edged well-graduated tail* and *striking white inner tertial edge* and *white on outer primaries*. There is also some black in wing. ♂ further characterised by striking *pale blue-grey head* and *long droopy black moustachial patch* (contrary to English name, the species lacks a 'beard'). *Eyes yellowish* and *bill bright orange-yellow*. Underparts largely ochre-buff with white throat and chest, and *black undertail-coverts*. ♀ paler with *head largely concolorous with upperparts* (some grey, mainly on ear-coverts) and *duller bill and eye*, while dark lores sometimes extend slightly upward. Whitish underparts have less contrasting buff flanks, and *lacks black undertail-coverts*. In many instances calls or just a brief glimpse are sufficient for identification. Typically shuffles nimbly through the reeds in short hops, constantly tail-flicking and occasionally moves up reed-stems to perch briefly in view. Whirring wingbeats and trailing tail also characteristic. Often in family parties or even larger groups comprising several broods, calling excitedly. Most easily seen on windless days, and gentle 'pishing' often helps bring birds into view.

VOCALISATIONS Song rather primitive, a series of 3–4 squeaky notes, e.g. *tschin dschik tschreh*, usually preceded by 3–5 introductory *ching* calls (see below) (song individually and perhaps geographically somewhat variable). – Unmistakable once learnt is a 'pinging' call, *pting* or *ching*, with a slight twanging quality, given both when perched and in flight, particularly as it moves in groups. The twanging call is sometimes more 'churring' and drawn out, *ptcherrr*. There is also a rolling *chirrr* and a soft clicking *plett* in 'conversation' or from feeding or agitated flocks.

SIMILAR SPECIES Practically unmistakable, even in least distinct juv plumage, given lack of other reedbed passerines with similar jizz, calls or plumage. Often, however, elusive and difficult to observe, or only heard. (Call can be mimicked by Reed and Marsh Warblers.)

AGEING & SEXING Ageing only possible of juv in early summer, before post-juv moult. Sexes easily separated. Little seasonal variation. – Moults. Both post-nuptial and post-juv moults are complete, mostly in Aug–Sep (hence ageing impossible thereafter). Pre-nuptial moult (of both ad and 1stY) apparently absent. Unconfirmed partial late-winter or early-spring body and tail moult has been reported, but unclear if merely a continuation of post-nuptial and post-juv moults, and whether only in 1stY or also ad. – **SPRING Ad** With wear head of ♂ becomes duller, less bluish-grey, but black face marks more pronounced. Both sexes pale cinnamon-brown above and on body-sides, with white below more extensive, and with pure white outer scapulars more prominent. By summer, pale tertial and tail-feather fringes almost worn off. **1stS** Indistinguishable from ad. Some ♀♀, presumably mostly 1stS, may have black blotches on mantle, and crown spotted or even heavily streaked black. – **AUTUMN Ad** See Identification for a general description of ♂ and ♀. **1stW** In general indistinguishable from ad. Birds with some black feathers on mantle, inner scapulars and crown probably young; such markings occur only in few ♂♂ and are then limited to lower mantle/back. **Juv** Mostly like ♀ but more straw-coloured with black saddle, extensively black primary-coverts and more extensive black centres to inner wing-coverts, including greater coverts, which lack broad cinnamon fringes and tips of ad. Tail-feathers also more extensively black, and some have dark lores (mostly juv ♂♂ which also have more orange bill, rather than greyish of juv ♀). **TG** in juv ♂ 33–41 mm (n 10, m 37.2), juv ♀ 28–33 mm (n 8, m 31.2) (Roselaar, in BWP).

P. b. biarmicus, ♂, Denmark, Jun: unmistakable given distinctive jizz, calls and plumage within reedbed habitat. Striking long black moustaches either side of pale blue-grey head (thus not 'bearded'), with complex white, chestnut and black wing pattern. The orange bill and straw-coloured eyes further make this bird among the most attractive in Europe. (H. Nussbaumer)

P. b. biarmicus, ♂, Belgium, Oct: in typical posture clinging to reed-stems in reedbed jungle, showing the long, droopy, black, moustachial patches either side of pale blue-grey head. (J. Fouarge)

P. b. biarmicus, ♂ (left) and ♀, Belgium, Oct: clear sexual dimorphism at all ages and seasons, but following post-juv moult ageing is usually impossible. All ages share the characteristic multicoloured, long, graduated tail, and the striking white panels on inner webs of tertials and outer primaries. ♀ lacks pale blue-grey head, long black moustaches and black undertail-coverts/vent of ♂, and has much duller bill. (J. Fouarge)

BIOMETRICS (*biarmicus*) **L** 15–16.5 cm; **W** ♂ 58–64.5 mm (*n* 15, *m* 61.2), ♀ 58–61 mm (*n* 13, *m* 59.9); **T** ♂ 73–85 mm (*n* 15, *m* 79.5), ♀ 72–78.5 mm (*n* 12, *m* 74.6); **TG** ad 33–46 mm (*n* 26, *m* 38.7); **T/W** *m* 127.7; **B** 9.8–11.4 mm (*n* 27, *m* 10.5); **B**(f) 6.8–8.2 mm (*n* 25, *m* 7.5); **BD** 3.4–4.4 mm (*n* 26, *m* 3.9); **Ts** 19.2–21.9 mm (*n* 26, *m* 20.2). **Wing formula: p1** in ad < pc 0–4 mm, minute and pointed, in juv > pc 4–9 mm, more rounded tip; **p2** < wt 2.5–7 mm, =6–7; **pp3–5** about equal and longest; **p6** < wt 1.5–4 mm; **p7** < wt 4–7 mm; **p8** < wt 5–9.5 mm; **p10** < wt 10–13.5 mm; **s1** < wt 11–15 mm. Emarg. pp3–6.

GEOGRAPHICAL VARIATION & RANGE Three races, all apparently rather poorly marked, afforded widespread recognition. All populations sedentary or partly short-range migrants, sometimes making irruptive movements.

P. b. biarmicus (L., 1758) (W & N Europe, in C Europe east to W Poland and W Czech Rep., in S Europe east to Greece and Thrace; northerly populations more apt to make autumn migrations, though many are resident). Described above. On folded wing of ♂, narrow white edges to primaries, narrow enough to usually also show blackish centres; secondaries and tertials edged rather dark cinnamon or rufous-ochre, and uppertail and flanks are darker cinnamon-tinged than in *russicus*, too.

P. b. russicus (C. L. Brehm, 1831) (EC & E Europe, east to E Asia, south to C Turkey, Caucasus region, NE Iran, Xinjiang and Qaidam Pendi). Similar to *biarmicus* but subtly larger and consistently paler, with ♂ paler and less bluish on head, and ♀ more rarely showing (very limited) black blotching and streaking above; more brownish elements in outer tail-feathers. On folded wing of ♂, primaries are edged prominently white and secondaries and tertials pale cinnamon-buff. Uppertail quite pale cinnamon-buff. Flanks

P. b. biarmicus, ♀, Finland, Feb: head of ♀ can be pale grey, mainly on ear-coverts, but crown is largely concolorous with buffish-brown upperparts, and compared to ♂ always has duller bill and pale undertail-coverts. (M. Varesvuo)

P. b. biarmicus, ♀, Belgium, Oct: the long tail is constantly flicked or raised when climbing a reed. Sometimes has dark marks on mantle. (J. Fouarge)

P. b. biarmicus, ♀, Netherlands, Mar: young are typically indistinguishable from respective ad after complete post-juv moult, though birds with some black feathers on mantle, inner scapulars and crown are probably young. (B. van den Boogaard)

P. b. biarmicus, juv ♂, Denmark, Jun: ♀-like, but has black saddle and more black in primary-coverts and tail-feathers than ♀♀. Unique to juv ♂ are the black lores and bright yellow-orange bill. (H. Nussbaumer)

P. b. biarmicus, juv ♀, Denmark, Jun: unlike ad ♀ has black saddle and all-black (rather than dark-and-white) primary-coverts, and differs from juv ♂ by plumage being overall more pale straw-coloured. Lores only pale grey, not black. Bill dusky yellowish-brown. (H. Nussbaumer)

P. b. biarmicus, juv ♀, Denmark, Jun: some juv ♀♀ have a darker dusky-grey loral patch, but never black like juv ♂, and always have greyish-brown (rather than orange) bill. (H. Nussbaumer)

rather pale cinnamon-buff, too. Tail proportionately somewhat longer, and tail graduation more pronounced. Middle toe + claw 15.0–15.5 mm (16.0–17.0 mm in *biarmicus*) (*BWP*). **W** ♂ 59–65 mm (*n* 12, *m* 62.6), ♀ 58.5–64 mm (*n* 12, *m* 60.9); **T** ♂ 79–92 mm (*n* 12, *m* 84.5), ♀ 72–87 mm (*n* 12, *m* 77.8); **TG** 35–48 mm (*n* 24, *m* 42.0); **T/W** *m* 130.9; **B** 9.5–11.1 mm (*n* 23, *m* 10.5); **Ts** 18.5–21.0 mm (*n* 24, *m* 19.7). (Syn. *raddei*.)

? ***P. b. kosswigi*** Kumerloeve, 1959 (Amik Gölü, S Turkey, if not now extinct, and presumably Syria; resident). Reportedly has deepest rufous upperparts and darkest lead-grey head in ♂. Breast and flanks said to be intense vinaceous-pink. ♀ reported to lack nearly all streaking on mantle and scapulars. **W** ♂ 61.5 and 63.5 mm, ♀ 61 and 61.5 mm (*BWP*). No material examined.

REFERENCES Buker, J. B., Buurma, L. S. & Osieck, E. R. (1975) *Beaufortia*, 23: 169–179. – Kumerloeve, H. (1958) *Bonn. zool. Beitr.*, 9: 194–199. – Pearson, D. J. (1975) *Bird Study*, 22: 205–227. – Saygili, F. et al. (2013) *Turkish Journal of Zoology*, 37: 149–156.

IRAQ BABBLER
Turdoides altirostris (Hartert, 1909)

Fr. – Cratérope d'Irak; Ger. – Rieddrossling
Sp. – Turdoide iraquí; Swe. – Irakskriktrast

This endemic of the Mesopotamia depressions, especially the lower Tigris and Euphrates valleys of SE Iraq and SW Iran, has recently been discovered along the entire Syrian Euphrates and as far north as Birecik in Turkey. Whether this range extension represents a recent colonisation (perhaps in response to habitat loss in Iraq) or is due to an observational lacuna is unclear, though its very recent discovery at a well-watched site in Turkey suggests the former. The Iraq Babbler frequents thickets, always in or around extensive waterside reedbeds, which is its principal habitat requirement.

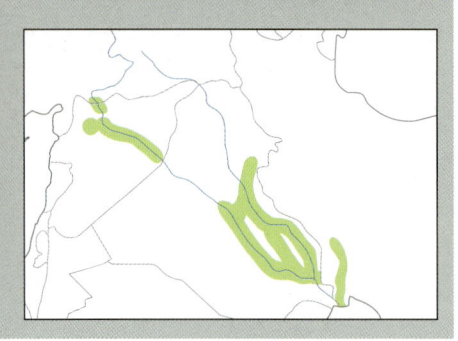

Turkey, May: fluttering flight appears laboured, on short rounded wings with long tail trailing, and typically ends in a straight glide, with one bird after another departing a bush or tree. (R. Armada)

Turkey, May: typical lookout posture, with delicate appearance, and small size diagnostic alone. Note bland face, bill shape, unstreaked buffish body-sides and long, deeply graduated tail. (H. Shirihai)

IDENTIFICATION *Smaller and slimmer* than all other *Turdoides* in the treated region, but plumage, structure and behaviour are typical of genus. *Predominantly drab buff-brown, with very long and well-graduated* (frequently cocked) *tail, and decurved, rather deep-based bill*. Plumage *indistinctly streaked above*, more obviously so on forehead, with *unstreaked buff-brown underparts, whiter throat, central breast and belly*, and warmest on flanks and undertail-coverts. Wings medium grey, whereas tail is slightly darker. Iris greyish-brown or more greyish. Bill horn-brown with flesh-coloured base to lower mandible. Legs horn-brown. Chiefly ground-dwelling, moves by hopping. Flight low and straight, appearing somewhat laboured on short rounded wings fluttering in bursts interspersed with straight glides, long tail trailing; often seen fluttering out of a bush, one after another, 'sailing' to bottom of the next bush, and so on. Other aspects of ecology and habits insufficiently described and studied, but probably very similar to other *Turdoides* species, including having a very complex 'anarchy group living' structure, with 10–20 or even more birds of different ages associating together.

VOCALISATIONS Like congeners, highly vocal year-round. Song is a series of cheerful and ringing short, high-pitched but musical trills, *birrrrrr*, remarkably similar to song of Little Grebe (at least if heard from a distance), but the ringing quality of the sound is actually more like that uttered by Pied Kingfisher, or even the ringing song notes often heard in Turkey from Lesser Whitethroat. – Broad variety of whistles and squeaks function as calls, but commonest are *tchit*, *whist* or *phist* given singly, sometimes repeated a few times, or a descending noisy chatter *pherrrrrreeee* or *pherrr pherrr pherrr*, etc., and in alarm a loud squeaking *phsioe* or *whsioe* (reminiscent of the typical alarm of Arabian Babbler in Israel, but lower and softer).

SIMILAR SPECIES Considering its overall shape and relatively small size, together with plumage details, should be fairly distinctive within covered range. Only known overlap with another *Turdoides* species is with Common Babbler in SE Iraq and neighbouring Iran; for differences see Common Babbler. – Given that Iraq Babbler breeds in reedbeds it may also require separation from the somewhat smaller Great Reed Warbler, but latter has shorter tail, straight bill and much warmer brown and completely unstreaked plumage, in addition to other structural, behavioural and vocal differences.

AGEING & SEXING Ageing possible by moult status and feather wear in wing. Iris colour appears useful, too, although full range of variation is still poorly known. Sexes similar apart from small average size difference. No seasonal variation. – Moults. Complete post-nuptial moult during late summer–autumn, at times protracted into winter. Post-juv moult is mostly complete like in adult but is even more protracted and variable, mainly occurring Jul–Nov/Dec (e.g. two birds toward the end of moult with still-growing remiges and tail-feathers in late Nov; NHM), but perhaps occasionally extended until Feb. Pre-nuptial moult (for both ad and 1stY) presumably lacking, but very few specimens or published data available. – **SPRING Ad** Slightly worn Feb–Mar, distinctly worn from May, when it also becomes slightly greyer above and whiter below. **1stS** Indistinguishable from ad if moult complete, but not known if some retain odd juv remiges

Turkey, May: in close views, the fine streaking above is visible. Note, despite the overall small size, the strong decurved and deep-based bill. Also unstreaked buff-brown underparts except whiter throat, supraloral and diffuse eye-surround. (D. Occhiato)

and primary-coverts. – AUTUMN **Ad** Many apparently retain some wing- and tail-feathers (old and distinctly worn), even in Dec. Those with pale iris are probably ad. Plumage warmer and slightly more buff-tinged than in 1stW. **1stW** Those in active wing moult show relatively less strong contrast between feather generations, unlike ad with some heavily abraded unmoulted feathers; often still has many fluffy juv body-feathers (characteristically replaced rather late during post-juv moult). Once moult completed inseparable from ad, but not known whether some retain any juv feathers. Those with dark brown iris are probably 1stY. **Juv** Soft, fluffy body-feathers, paler ground colour and less contrasting streaking; clear yellow gape on recent fledglings. (*BWP*; NHM; field observations in Turkey.)

BIOMETRICS (Birds with T down to 94 mm and TG to 38 mm have been excluded below since they were either deemed to be not fully grown or were very heavily worn.) **L** 19–23 cm; **W** ♂ 75–83.5 mm (*n* 12, *m* 78.4), ♀ 72–76 mm (*n* 7, *m* 74.4); **T** ♂ 98–120 mm (*n* 12, *m* 110.7), ♀ 99–105 mm (*n* 5, *m* 101.4); **T/W** *m* 138.6; **TG** 40–55 mm (*n* 19, *m* 47.0); **B** 20.0–26.1 mm (*n* 26, *m* 23.4); **B**(f) 14.9–20.0 mm (*n* 27, *m* 18.0); **BD** 6.0–7.3 mm (*n* 24, *m* 6.7); **Ts** 27.7–31.6 mm (*n* 27, *m* 29.5); **HC** 8.5–12.0 mm (*n* 24, *m* 10.0). **Wing formula: p1** > pc 16–24.5 mm, <

Syria, Jan: some birds are slightly warmer on head (more tawny or chestnut), or more strongly streaked above. (K. De Rouck)

p2 14.5–19 mm; **p2** < wt 9–15 mm; **p3** < wt 2.5–6 mm; **pp4–7** about equal and longest (though any, but usually p4 and/or p7, can be 0.5–2.5 mm <); **p8** < wt 1–3.5 mm; **p10** < wt 4–8.5 mm; **s1** < wt 6–11 mm. Emarg. pp2–7.

GEOGRAPHICAL VARIATION & RANGE Monotypic (but see Taxonomic notes). – SC Turkey, Syria, Iraq, extreme SW Iran; mainly resident or partly nomadic in off-season.

TAXONOMIC NOTES Although treated here as monotypic (following main sources like Vaurie 1959, Peters 1964, *BWP*, H&M 4), it should be noted that five specimens in NHM collected in or near Basra, SE Iraq, are slightly smaller, shorter-tailed, darker and greyer with more distinct dark streaking on upperparts than the majority, which are on average larger, longer-tailed, paler and more buffy or cinnamon-tinged (including the holotype from nearby Fao; AMNH). It is difficult to advance on this possible local and undescribed subspecies (species?) without access to longer series. A check of the mtDNA of the dark and smaller birds were performed on our request by U. Olsson and compared with a sample of the paler and larger birds (unpubl. 2011) but did not reveal any significant difference. Any comparison of colours and darkness needs to consider season and amount of wear, since abrasion of plumage tends to make it darker and greyer. The size difference is only moderate, and there is overlap for all measurements, but the main differences lie in shorter tail (97–102 mm versus 98–120 mm; both sexes), less graduated tail (38–49 mm versus 40–55 mm), and a shorter p1 in relation to primary-coverts (16–21.5 mm against 19–24.5 mm). Further investigations required.

REFERENCES Donaghy, N. (2006) *BW*, 19: 283–284. – Ticehurst, C. B., Buxton, P. A. & Cheesman, R. E. (1922) *J. Bombay N. H. S.*, 28: 381–427.

Iran, Jan: a warmer-toned bird with pinkish-cinnamon hue on head and flanks. (R. Felix)

Juv, Turkey, May: typical soft, fluffy body-feathers, paler ground colour and less contrasting streaking above, as well as yellow gape. Irrespective of age, some birds are duller. (H. Shirihai)

COMMON BABBLER
Turdoides caudata (Dumont, 1823)

Alternative name: Afghan Babbler

Fr. – Cratérope de l'Inde; Ger. – Langschwanzdrossling
Sp. – Turdoide indio; Swe. – Orientskriktrast

Slightly larger than the Iraq Babbler, with which it overlaps in Iraq, the Common Babbler is resident in semi-deserts and deserts with trees and bushes, and in cultivated lowlands and hills, from Iraq and S Iran east through S Afghanistan and the northern Indian subcontinent. The habitat accordingly differs from that of Iraq Babbler, but separation of the two should always be based on field marks as well.

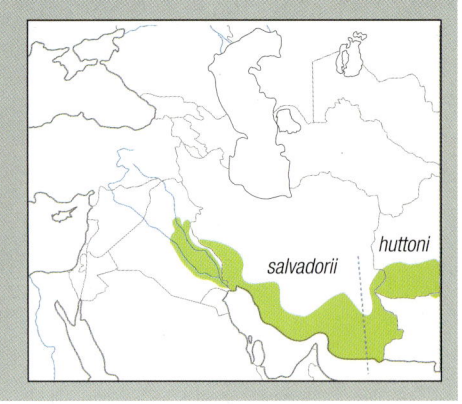

IDENTIFICATION A quite *large, bulky* babbler, with a long tail, *mainly dun-coloured* lacking any bold marks in yellow or black, but has *bold dark streaking on greyish sandy-brown crown, mantle and scapulars* (slightly flushed pinkish when fresh, much greyer plumage when worn), and quite *pronounced rusty-buff wash to cheeks*; rather distinctly streaked on pinkish sand-coloured lower breast and fore flanks (rest of underparts whitish-cream, brightest on throat and upper breast). In close views tail discreetly cross-barred. Has deep-based but *quite long brown bill with yellow base to lower mandible*. Iris yellowish-cream in adults, darker in young. *Legs typically dull yellow.* Lives year-round in small groups and is essentially terrestrial, moving by springing leaps and hops. Flies very low from bush to bush in characteristic straight flight, with fluttering rounded wings and glides typical of the genus.

VOCALISATIONS Not clear whether there is a song in conventional sense, but fast series of clear metallic notes on similar pitch, *pi-pi-pip-pi-pi-...*, rather reminiscent of Temminck's Stint display, seems to have territorial function. Most calls are uttered in social gatherings with more than two birds present. A common call of *huttoni* is a loud, fast and slightly descending series of spaced, clear, whistling notes, e.g. *ti-tee fyer fyeer fyeer fyeeer fyeeer...* (Rasmussen & Anderton 2005). As the notes progressively fall in pitch they also become more drawn-out. Another rendering of a common call is *tieh-too, tieh-too, tieh too-too-too-too-too-too* with a repeated piping note at end of about same pitch and length. Some calls accelerate into a rippling trill, *pi-pee-pee-pee-pee-pee-peerrrrrrr* (Beaman & Madge 1998). – In alarm utters *qwee qwe-e-e qwe-e-e*. There is also a distinctive multisyllabic whistle, *qwee-yur-qweeyur* (Roberts 1998).

SIMILAR SPECIES In the treated region only likely to be confused with similar *Iraq Babbler*, the two overlapping around N Persian Gulf. However, Common Babbler in this region has a longer and straighter (less decurved) bill, conspicuously stronger streaking on the crown and upperparts, streaked breast-sides and stronger, more dull yellow (rather than brownish) legs; and it tends to avoid extensive aquatic habitats (but is attracted to irrigated areas), unlike the reed-dwelling Iraq Babbler, although their behaviour is otherwise similar.

AGEING & SEXING Ageing requires a close check of moult and feather wear in wing, and use of iris colour, but variation and development of latter insufficiently studied to date. Sexes similar. No seasonal variation. – Moults. Complete post-nuptial moult, but is rather protracted and sometimes suspended, in summer or autumn. Post-juv moult also generally complete and takes place in summer–autumn, but extent, timing and possible degree of suspension vary with hatching date and region (mainly Jun–Nov, but sometimes lasts until Dec/Jan). Pre-nuptial moult (limited, involving some body and tail-feathers) generally in late winter to early spring. – **SPRING Ad** Fresh in Dec–Jan, slightly worn in March and distinctly worn in Apr–May. Worn birds become paler, less deeply saturated and have sharper streaks. Moult limits, if present, less contrasting. **1stS** Indistinguishable if no juv wing-feathers retained. Some may retain dark iris, but usually not clearly differentiated from ad. – **AUTUMN Ad** As moult is often protracted, many, even in late autumn, still have some worn wing- and tail-feathers. Those that have completed moult are evenly fresh, the pale iris being reliable for ageing. By and large, tinged warmer buff, especially on body-sides. **1stW** Some retain juv wing- and tail-feathers; those that have apparently moulted completely indistinguishable from ad, but if in active wing moult or suspended, less sharp differences between feather generations than moulting ad with many clearly worn feathers. Iris dark grey-brown (but variable, some being paler like ad). **Juv** Soft, fluffy body-feathers, streaks shorter and less intense (occasionally absent on underparts), and pink-buff fringes to wing- and tail-feathers more obvious.

BIOMETRICS (*salvadorii*) **L** 23.5–25 cm; **W** ♂ 89–93 mm (*n* 12, *m* 90.0), ♀ 85–91 mm (*n* 4, *m* 87.9); **T** ♂ 102–133 mm (*n* 12, *m* 121.3), ♀ 114–123 mm (*n* 4, *m* 119.5); **T/W** *m* 134.5; **TG** 35–66 mm (*n* 16, *m* 53.1); **B** 22.5–26.5 mm (*n* 18, *m* 24.2); **B**(f) 16.7–20.3 mm (*n* 18, *m* 18.6); **BD** 6.1–7.1 mm (*n* 18, *m* 6.6); **Ts** 29.1–31.7 mm (*n* 18, *m* 30.4). **Wing formula: p1** > pc 22–29.5 mm, < p2 17–21.5 mm; **p2** < wt 7.5–11.5 mm, =10 or <; **p3** 2–5 mm; **pp4–6**(7) about equal and longest; **p7** < wt 0–1.5 mm; **p8** < wt 1–3 mm; **p10** < wt 5–9 mm; **s1** < wt 7.5–11.5 mm. Emarg. pp2–7 (emarg. of p2 almost covers whole feather length).

GEOGRAPHICAL VARIATION & RANGE Variation is largely clinal across much of its Asian range, although at least within the covered region (Iraq and Iran) the variation is quite pronounced when combining size and coloration. Four races are recognised by most works, although only the following two occur within the treated range. All populations are sedentary.

T. c. salvadorii (De Filippi, 1865) (SE Iraq & SW Iran). Drab buff below with fine but distinct streaking on chest, diffusely continuing on flanks, and is generally a bit duller and browner above than *huttoni*. Underparts quite warmly tinged greyish buff-brown, deeper on the sides, vent and undertail-

T. c. salvadorii, Kuwait, Jan: a rather large, robust babbler that is mostly drab grey-brown but for extensive dark streaking above and on body-sides. Unlike sympatric (but distinctly smaller) Iraq Babbler, Common Babbler is diagnostically yellow-legged (rather than brownish or greyish). (D. Monticelli)

T. c. salvadorii, Kuwait, Dec: within the covered region, only likely to be confused with similar but distinctly smaller Iraq Babbler, although Common Babbler is also more heavily dark streaked above and on body-sides. Here an extreme example of a bird with very boldly streaked upperparts. (V. Legrand)

coverts. Upperparts grey-brown rather prominently streaked dark (a little fainter on lower back and rump). Rather bland face, and only obvious 'pattern' is a whitish throat. (Syn. *theresae*.)

T. c. huttoni (Blyth, 1847) (E Iran, S Afghanistan & S Pakistan). Very similar to *salvadorii* but is less distinctly and extensively streaked below (streaks on breast much finer and limited to fine shaft-streaks mostly on sides of breast and upper flanks). Also on average subtly paler overall (especially throat to chest are paler). Bill averages slightly shorter. **L** 24–25 cm; **W** ♂ 86–92 mm (*n* 13, *m* 88.3), ♀ 85–91 mm (*n* 10, *m* 87.4); **T** ♂ 104–128 mm (*n* 13, *m* 115.0), ♀ 107–127 mm (*n* 10, *m* 118.5); **T/W** *m* 132.7; **TG** 43–57 mm (*n* 20, *m* 50.8); **B** 21.5–25.6 mm (*n* 17, *m* 22.9); **B**(f) 16.3–20.0 mm (*n* 22, *m* 17.9); **BD** 5.5–7.1 mm (*n* 21, *m* 6.4); **Ts** 27.4–31.2 mm (*n* 23, *m* 29.4).

TAXONOMIC NOTE Rasmussen & Anderton (2005) gave species status to the two western subspecies mainly on account of differences in vocalisations and morphology; the vocal variation was pointed out already by Roberts (1998). Consequently, the above two subspecies were referred to as 'Afghan Babbler *T. huttoni*' as opposed to the Common Babbler *T. caudata*, including ssp. *caudata* of India and *eclipes* of N Pakistan and extreme NW India. IOC (2008) have since adopted this split; however, it has not been followed by sources like Collar (2006), *HBW* or Grimmett, Inskipp & Inskipp (2011), and here, too, inclusion of the two West Asian taxa in the Common Babbler is preferred until the case has been better investigated.

REFERENCES COLLAR, N. (2006) *Forktail*, 22: 85–112. – GREGORY, G. & NASRALLAH, K. (2007) *Sandgrouse*, 29: 218.

T. c. salvadorii, Kuwait, Jan: face often appears rather bland, with paler throat. Some are tinged more buffish-brown on cheeks and body-sides. Proportionately very long tail, which in close views is discreetly cross-barred. Note deep-based but quite long brown bill with a hint of yellow base to lower mandible. (D. Monticelli)

T. c. huttoni, Iran, Jul: this potential split from E Iran, S Afghanistan & S Pakistan is very similar to *salvadorii*, but averages more delicately built and smaller, paler (especially throat to chest are often whiter) and less distinctly and extensively streaked (streaks on breast much finer and limited to fine shaft-streaks, mostly at sides). Bill also slightly shorter. (H. van Diek)

ARABIAN BABBLER
Turdoides squamiceps (Cretzschmar, 1827)

Fr. – Cratérope écaillé; Ger. – Graudrossling
Sp. – Turdoide árabe; Swe. – Arabskriktrast

The Arabian Babbler is one of the most characteristic birds of *Acacia* deserts, and is also found at the edges of cultivation. Like its congeners it occurs year-round in small (familial or communal) groups. Easily located due to its far-carrying, high-pitched alarm calls, especially when a raptor, mammalian predator or snake is spotted. Endemic to the Middle East, where it occurs in S Israel south to Yemen and Oman.

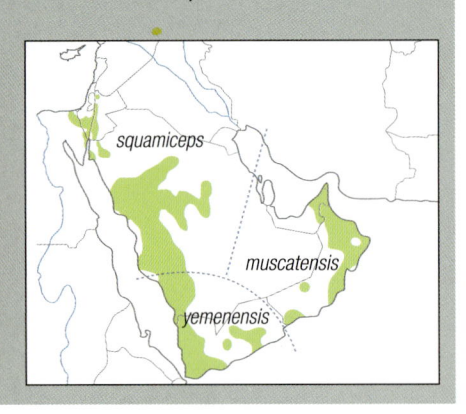

T. s. squamiceps, ♂ and ♀, Israel, Apr: well separated geographically from other *Turdoides*, and unmistakable within range due to its relatively large size, pale greyish-brown plumage, plain but for some mottling, and its shape and general behaviour. Ad can often be sexed: on left ♂ with greyish-black bill and whitish-yellow eyes, while ♀ on right has largely straw-coloured bill and ochre-brown eyes. (D. Forsman)

IDENTIFICATION   A ground-dwelling, *large thrush-sized babbler, stocky with a long, graduated tail.* Chiefly *dull grey-brown above,* though head and neck sometimes almost whitish-grey, finely streaked and mottled blackish-brown; streaking *blotchier on nape to mantle and scapulars.* Pale greyish underparts deeper on sides where very narrowly streaked, and *buffier on vent and undertail-coverts.* (Some are quite heavily spotted and mottled grey, especially on breast.) Often has pale surround to eye but duskier lores. *Bill rather long and clearly decurved* with obvious pale base. Strong legs dark yellowish-brown. Iris of adult whitish-yellow (darker in female and browner-grey in juvenile). Skulking, but groups often located by bursts of chattering notes and alarm calls. Moves by two-footed hops, with tail usually cocked, and usually flies low with fluttering wingbeats.

VOCALISATIONS   Song a series of high, piping *piu* notes, slowing at end. – Most characteristic call a short metallic *pii*, often repeated or combined with piercing, machine-like trill, *zuir'r'r'r'r'r'...* or *zeck-eck-eck-eck-eck-eck-...*; often utters chattering whistles and piping notes, e.g. *pyü pyuvü pii pyüvu pvie*, or a single abrupt high-pitched *peeeh*; also harsh Jay-like notes. Calls vary greatly, according to situation and social context.

SIMILAR SPECIES   Geographically well separated from other babblers, and unlikely to be confused given combination of larger size, much greyer overall plumage, strongly mottled appearance of paler head and neck, and lack of prominent underparts streaking. Plumage, shape and general character and behaviour also differ considerably from any passerine within range.

AGEING & SEXING   Ageing requires a close check of moult and feather wear in wing. Note, that due to the species' habitat and habits, ad soon acquires rather heavily worn primaries, and since primary-coverts and greater coverts are almost identically patterned at all ages, some practice is required to determine moult limits and to distinguish ad feathers from retained juv ones, especially in spring (and is not always possible). Iris colour appears useful, but detailed variation, also in relation to sex, is only partly known. Sexes very similar. No seasonal variation. – Moults. Complete post-nuptial moult in late summer and autumn (generally Jul–Dec). Partial (rarely complete) post-juv moult at about the same time, though exact timing and extent of post-juv moult highly variable, depending on hatching date; early-hatched birds may replace entire plumage, completed about same time as most ad, by Oct–Nov; most, however, replace all head, body, small wing-coverts, tail, usually some or all secondaries and tertials, sometimes outer primaries, and only some greater coverts; in contrast, late-hatched birds retain all primaries, primary-coverts, most or all secondaries, and often part of tail, and are still actively moulting in Dec. Few data available concerning pre-nuptial moult, in Nov–Feb, which involves part of body-feathers, perhaps some tertials, and some or all secondaries and tail-feathers; however, it is unclear to what extent this moult is truly pre-nuptial, i.e. not just complementary or continuation of previous interrupted moult. – **SPRING Ad** With wear becomes overall paler, and mottling more obvious. Older ♂♂ have greyish-black bill and whitish-yellow iris (in older ♀♀ straw-coloured bill and ochre-brown iris); however, many cannot be sexed. Some show pre-nuptial moult limits, but older ad wing-feathers are not as worn and brown as in 1stS. **1stS** Retained juv remiges and primary-coverts more worn with faded fringes. Minority apparently undertake complete post-juv moult and are indistinguishable from ad. Some, mostly ♀♀ and birds of any sex from late broods, retain dark iris. – **AUTUMN Ad** Moult often protracted, and many, even in late autumn, easily aged by old worn wing- and tail-feathers. Birds that completed moult are evenly feathered, fresh and have pale iris. At this season, overall warmer and slightly buffier, but streaks and mottling less obvious. **1stW** Those in active wing moult show less strong contrast among feather generations than ad with unmoulted feathers, and often still have some fluffy juv body-feathers, especially below. Very similar to fresh ad, but many retain remiges and primary-coverts, and often a few greater coverts, sometimes tertials and tail-feathers. Both retained juv primary-coverts and greater coverts are slightly weaker-textured, narrower and browner (less greyish), with slightly buffish and abraded, narrower margins (in ad: tips and fringes greyer, broader and more diffuse). A few 1stW, which apparently undertake complete post-juv moult, are indistinguishable from ad. Iris uniform dark grey-brown (but some develop partial paler areas and do not clearly differ from ad, especially not from ad ♀♀). **Juv** Soft, fluffy body-feathers, plainer above and buffier below. Iris dark grey-brown and bill black-horn.

T. s. squamiceps, Israel, Feb: typical posture with tail cocked. Some birds show more contrasting plumage with greyer head, whiter throat and warmer brown underparts, as well as more spotty head and blotches or streaks above and on chest. Ageing usually impossible in spring due to possibly complete moult also of young birds. At least some can be sexed by bill and eye colour, but birds which have not been possible to age, like here, are best left unsexed too. (H. Shirihai)

T. s. squamiceps, ad ♀, Israel, Oct: strong moult contrast in wing in Oct, dark iris and mainly yellowish bill identify this bird as an ad ♀. (A. Ben Dov)

T. s. squamiceps, 1stW, presumed ♂, Israel, Oct: less strong moult contrast in wing during autumn, and yellow gape, make this a 1stW, and the rather pale iris and mostly dark bill suggest a ♂. Juv/1stW tend to retain obvious yellow gape for some time, assisting ageing, especially as plumage differences from ad are not easy to appreciate. (A. Ben Dov)

BIOMETRICS (*squamiceps*) Sexes practically of the same size. **L** 27–29 cm; **W** 108–119 mm (n 26, m 113.7); **T** 122–147 mm (n 26, m 135.4); **T/W** m 119.2; **TG** 42–57 mm (n 25, m 52.0); **B** 23.9–28.7 mm (n 25, m 26.6); **B**(f) 19.0–24.0 mm (n 25, m 21.4); **BD** 7.0–8.3 mm (n 26, m 7.7); **Ts** 32.5–36.4 mm (n 26, m 34.6). **Wing formula: p1** > pc 27–38 mm, < p2 21–27 mm; **p2** < wt 8–15 mm, =9 or <; **p3** 1–3 mm; **pp4–6**(7) about equal and longest; **p7** < wt 0–3 mm; **p8** < wt 2–6 mm; **p10** < wt 6.5–14 mm; **s1** < wt 10–15 mm. Emarg. pp2–6 (often, p7 also slightly emarg.).

GEOGRAPHICAL VARIATION & RANGE Moderate, although three subspecies often recognised. All populations sedentary.

T. s. squamiceps (Cretzschmar, 1827) (Levant, Sinai, N & C Arabia, inland S Arabia). Treated above; rather pale and tinged pinkish sand-brown overall. Chest almost unmarked or shows very discreet blotches or streaks. Crown to upper mantle rather finely streaked dark brown. Long strong bill.

T. s. muscatensis Meyer de Schauensee & Ripley, 1953 (United Arab Emirates, Oman, SE Arabia). Almost as dark as *yemenensis*, but is slightly smaller and on average

T. s. muscatensis, ad ♂, Oman, Mar: this race is slightly smaller and on average somewhat paler and greyer above, but also less prominently streaked or blotched dark. Mainly dark bill and pure yellowish iris identify this bird as ad ♂. (M. Varesvuo)

somewhat paler and less prominently streaked or blotched on chest and upperparts. Grades into neighbouring races where they meet. Rather similar to *squamiceps* in coloration albeit somewhat purer grey above, and averages smaller. **W** 99–111 mm (n 12, m 107.3); **T** 119–137 mm (n 12, m 129.8); **T/W** m 121.0; **TG** 45–69 mm (n 11, m 53.5); **B** 22.5–26.7 mm (n 12, m 25.1); **B**(f) 17.3–20.9 mm (n 12, m 19.4); **BD** 6.7–8.3 mm (n 12, m 7.4); **Ts** 33.0–35.2 mm (n 13, m 34.3).

T. s. yemenensis (Neumann, 1904) (SW Saudi Arabia, W Yemen). Distinctly darker above and subtly shorter-billed than *squamiceps*; also, slightly smaller and has almost yellow-orange bill (both sexes) and whiter face. **W** 101–115 mm (n 25, m 105.8); **T** 112–138 (once 144) mm (n 26, m 130.9); **T/W** m 124.4; **TG** 42–65 mm (n 24, m 52.8); **B** 22.0–26.3 mm (n 26, m 24.5); **B**(f) 17.5–21.6 mm (n 26, m 19.5); **BD** 6.6–9.0 mm (n 24, m 7.4); **Ts** 32.0–36.0 mm (n 26, m 33.6). **Wing formula:** Emarg. pp2–7 (sometimes faintly also on p8).

TAXONOMIC NOTE See Fulvous Babbler for relationships.

T. s. yemenensis, Yemen, Jan: distinctly darker above, with contrastingly whiter face. Deep and bright (pure) yellow iris could suggest an ad (older) ♂, but note that the paler bill is a feature of both sexes of this race. (H. & J. Eriksen)

FULVOUS BABBLER
Turdoides fulva (Desfontaines, 1789)

Fr. – Cratérope fauve; Ger. – Akaziendrossling
Sp. – Turdoide rojizo; Swe. – Saharaskriktrast

A rather widespread, long-tailed and dull-coloured North African babbler of sandy lowland deserts with some bushes and *Acacia* trees, often appearing in noisy small parties. The Fulvous Babbler is resident from SC Morocco south to Niger, then breeds east across the Sahara to Eritrea. In Egypt now restricted to the Gebel Elba region, but formerly occurred north as far as Aswan, in the Nile Valley.

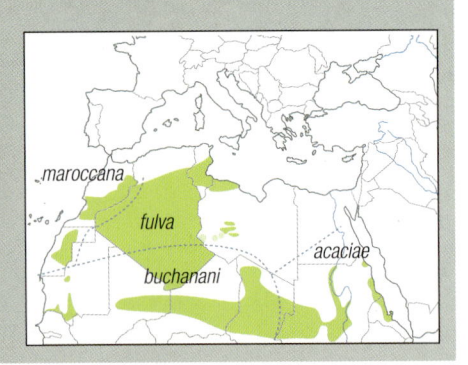

T. f. maroccanus, Western Sahara, Mar: ground-dwelling, long-tailed and plain, dull-coloured N African babbler, which moves by hopping and often holds tail slightly cocked. Decurved bill comparatively short but strong. Iris of ad pale yellowish to almost whitish. (V. & S. Ashby)

IDENTIFICATION A rather large babbler, typically in parties of up to ten. *Mainly a ground-dweller with very long (Magpie-like) tail often held somewhat cocked.* Plumage *warm sandy-brown, with faint dark scaling on crown and streaking on mantle and scapulars* (though often appears unstreaked), and *underparts mostly buff to cinnamon, with whitish throat.* Dark lores rather distinct. Iris of adult pale yellowish-brown to whitish-grey (but darker in younger birds). Slightly curved bill dark brown to almost black, occasionally with some yellow at base of lower mandible. Legs dark yellowish-brown. *Moves by leaping, hopping, and running.* Weak *flight, typical of genus, with straight path, spasmodic or mechanical fluttering wingbeats and brief glides on spread wings* (birds usually fly in a 'follow-my-leader' sequence).

VOCALISATIONS Song a series of piping diphthongs, *peeoo peeoo peeoo-peeoo-peeoo-peeoo*, accelerating slightly initially and each note descending perceptibly. – Rather noisy, commonest calls include a short, clear *chit*, various piping notes like *pee* and *peeuh*, and a sharp *pwit* in alarm. Also highly characteristic drawn-out and soft whirring trills *priür'r'r'r'r'r'r'*.

SIMILAR SPECIES Distinctive within its range.

AGEING & SEXING Ageing requires a close check of moult and feather wear in wing. Iris colour seems quite useful too, but variation requires further study. Sexes similar. Limited seasonal changes. – Moults. Complete post-nuptial moult, though sometimes protracted or even suspended (details poorly known), main moult period in summer and early autumn (mostly May–Oct). Partial post-juv moult (reportedly occasionally complete) at same time. Pre-nuptial moult limited but extended, generally in late autumn to early spring. Breeding period and timing of moult vary between populations, and moult seems highly variable individually. Moult in relation to age and sex also poorly understood; especially timing and extent of post-juv moult highly variable, apparently depending on hatching date. Some moult body-feathers and a very few wing-coverts, tertials and tail-feathers, while others moult these tracts more extensively, including some outer or all primaries and inner or all secondaries, and some may even replace entire plumage. Pre-nuptial moult involves part of head and body, sometimes a few secondaries, tertials, and some or all tail-feathers. – **SPRING Ad** With wear paler overall, less deeply saturated, with sharper streaks. Some show pre-nuptial moult limits in wing, but older ad feathers lack characteristic weak (juv) texture and are less bleached. **1stS** Retained juv primaries and primary-coverts more worn and abraded. Note that as wing-feathers are mostly identical to ad, some practice is required for correct ageing. Minority apparently undertake complete post-juv moult and are indistinguishable from ad. Some, apparently ♀♀ or young from late broods, may retain dark iris into spring. – **AUTUMN Ad** Moult often protracted and thus many, even in late autumn, easily aged by distinctly worn wing- and tail-feathers. Those that have performed a complete moult are evenly feathered, fresher, and have pale iris. Overall, warmer cinnamon-buff, and upperparts streaking less contrasting than in spring. **1stW** At least shortly after fledging, wings lack strong contrast between feather generations, unlike ad with some unmoulted feathers. Very similar to fresh ad, but any retained juv primaries slightly more worn, and primary-coverts and greater coverts slightly weaker textured and less fresh, with buffish and abraded fringes (broader and diffuser in ad). Some indistinguishable from ad. Iris dark grey-brown (but some have paler iris and are not obviously different from ad). **Juv** Soft, fluffy body-feathers, virtually unstreaked and yellower upperparts, paler underparts and blackish or horn-coloured bill.

BIOMETRICS (*fulva*) Sexes very nearly of the same size. **L** 24.5–27 cm; **W** 93–111 mm (*n* 26, *m* 99.4); **T** 112–138 mm (*n* 26, *m* 123.3); **T/W** *m* 124.1; **TG** 40–64 mm (*n* 23, *m* 52.7); **B** 21.8–26.3 mm (*n* 26, *m* 23.7); **B**(f) 17.3–19.8 mm (*n* 26, *m* 18.6); **BD** 6.6–7.8 mm (*n* 26, *m* 7.1); **Ts** 30.1–34.1 mm (*n* 24, *m* 32.5). **Wing formula: p1** > pc 26–31.5 mm, < p2 19–23.5 mm; **p2** < wt 8–13 mm, =9/10 or <; **pp3–6**(7) about equal and longest (though p3 and p6 sometimes 0.5–1.5 mm <); **p7** < wt 0–4 mm; **p8** < wt 3–7 mm; **p10** < wt 7–12.5 mm; **s1** < wt 9.5–13.5 mm. Emarg. pp2–7 (emarg. of p2 almost covering whole feather length).

GEOGRAPHICAL VARIATION & RANGE Four subspecies with variation expressed mostly in plumage differences (overall saturation of fulvous and buff, degree of brown and grey hues, and streaking above), but only slightly in size. Note

T. f. maroccanus, Western Sahara, Dec: this taxon is overall rather dark and warm brown (here slightly exaggerated by sunlight), with deep cinnamon-brown face and underparts (pure white restricted mainly to throat) and finely streaked dark crown. (C. N. G. Bocos)

T. f. maroccanus, Morocco, Mar: even older birds have quite pronounced yellow gape (which is especially contrasting due to this subspecies having mainly dark bill) that could suggest a juv. Note strong moult limit in wing which, given time of year, strongly suggests a 1stS. (D. Monticelli)

T. f. fulvus, ad (left) and juv, Tunisia, Apr: ssp. *fulvus* virtually lacks dark shaft-streaks on uniformly sandy upperparts, and is pale brownish cinnamon-buff below (body-sides a little deeper cinnamon). Juv at right has shorter bill and obvious yellow gape. (D. Occhiato)

that material of *maroccana* was limited—only six examined. Variation generally distinct (despite being claimed to be clinal or slight, e.g. in *BWP*). Ssp. *fulva* and *buchanani* are most similar, whereas *acaciae* and *maroccana* are rather distinctively differentiated both from each other and from the other two. All populations mainly resident, or make only shorter movements.

T. f. maroccana Lynes, 1925 (S & E Morocco, extreme W Algeria). Distinctive in being overall darker and warmer, with dark rufous-grey upperparts including uppertail, crown finely and diffusely streaked dark, sometimes extending to nape and upper mantle. Underparts deeper and more extensively cinnamon-brown than *fulva* (any white restricted mainly to chin, upper throat and mid belly), thus cheeks are saturated rufous-cinnamon, as are chest, flanks and undertail-coverts (flanks to undertail-coverts much whiter in *fulva*). Fairly small. The seven specimens examined (2 ♂, 4 ♀ and one unsexed, including holotype; NHM, ZFMK, AMNH, MNHN) had **W** 91–100 mm (*m* 95.9), **T** 103–130 mm (*m* 117.1), **T/W** *m* 122.3, **B** 22.5–25.7 mm (*m* 24.0), **B**(f) 16.8–20.0 mm (*m* 18.4), **BD** 6.5–7.3 mm (*m* 7.0) and **Ts** 31.1–34.0 mm (*m* 32.4). (Syn. *billypayni*.)

T. f. fulva (Desfontaines, 1789) (N Algeria, N Tunisia & NW Libya). Described above (see also Biometrics) and best characterised as being rather uniformly sandy or pale rufous-cinnamon above (tinged vinaceous in fresh plumage), with dark shaft-streaks above poorly developed, being thin and diffuse and evenly distributed throughout, or virtually lacking. Pale brownish cinnamon-buff below (not as pale as *buchanani*, but paler than other two), with deeper cinnamon tinge on breast-sides, flanks and undertail-coverts.

T. f. buchanani Hartert, 1921 (C Sahara from S Mauritania to C Chad; apparently also S & SE Algeria, but range in S Algeria unclear and needs clarification; breeders in Ahaggar and Tassili N'Ajjer in S & SE Algeria fit best this race). Pale rufous-buff like *fulva*, but is somewhat smaller. Very weakly streaked on crown, whereas breast is unstreaked. No sexual size difference. **L** 23–25 cm; **W** 90–100 mm (*n* 18, *m* 95.6); **T** 108–129 mm (*n* 18, *m* 119.8); **T/W** *m* 125.3; **TG** 43–60 mm (*n* 16, *m* 50.5); **B** 21.4–25.5 mm (*n* 18, *m* 23.1); **B**(f) 16.8–20.1 mm (*n* 18, *m* 18.1); **BD** 6.3–7.4 mm (*n* 18, *m* 7.0); **Ts** 30.5–33.4 mm (*n* 18, *m* 32.2). **Wing formula: p2** < wt 5.5–10 mm, =9 or < (rarely =8).

T. f. acaciae (M. H. C. Lichtenstein, 1823) (S Egypt, Sudan, Eritrea & NE Ethiopia). Smaller and proportionately longer-tailed than *fulva* and a little darker. Has diffusely but prominently dark-streaked crown and nape, greyish cheeks and loral area creating contrast with small but pale upper throat (nearly white, with only very faint pink-buff flush). On average, subtly more rufous uppertail, best visible when spread. Shorter (but subtly stouter) bill than *fulva*, which is diagnostically partly yellow, in some birds largely yellow tipped dark, but in others the yellow is less clean on the sides, being more restricted to the base (bill of other races: dark brown to almost black, occasionally with some limited yellow at base of lower mandible). **L** 24–26 cm; **W** ♂ 94–103 mm (*n* 17, *m* 98.3), ♀ 91–103 mm (*n* 13, *m* 96.5); **T** ♂ 119–139 mm (*n* 17, *m* 128.7), ♀ 120–136 mm (*n* 13, *m* 127.3); **T/W** *m* 131.4; **TG** 42–65 mm (*n* 29, *m* 55.3); **B** 20.8–24.8 mm (*n* 29, *m* 22.6); **B**(f) 16.0–19.2 mm (*n* 29, *m* 17.5); **BD** 6.4–7.7 mm (*n* 29, *m* 7.1); **Ts** 29.9–34.4 mm (*n* 30, *m* 31.8). **Wing formula:** Emarg. pp2–7 (emarg. of p2 almost covers whole feather length, of p7 sometimes less prominently; sometimes slight emarg. also of p8).

TAXONOMIC NOTES Forms species group (superspecies) with Arabian Babbler, Iraq Babbler, Rufous Babbler *T. rubiginosus* and Scaly Babbler *T. aylmeri* of Africa, and Striated Babbler *T. earlei* from the Indian subcontinent (three last-mentioned extralimital). – Two of the four subspecies of Common Babbler treated above, *buchanani* and *acaciae*, are fairly distinct and could, with further study, prove to be separate species (or at least distinctive subspecies on their way to becoming species). Further study required.

T. f. fulvus, Tunisia, Apr: some appear to have duskier lores, partly just an effect of light. Both ad and imm have quite pronounced yellow gape. (D. Occhiato)

T. f. acaciae, Sudan, May: note proportionately long and more rufous-brown tail, more extensively yellow bill and rather dark-toned plumage, with deeper rufous-buff underparts, diffusely but prominently dark-streaked crown and greyish head-sides contrasting with more whitish upper throat. (T. Jenner)

LONG-TAILED TIT
Aegithalos caudatus (L., 1758)

Alternative name: Long-tailed Bush-tit

Fr. – Mésange à longue queue; Ger. – Schwanzmeise
Sp. – Mito común; Swe. – Stjärtmes

The most common encounter with this widespread, charming little bird is brief and sudden, a little family party or group of birds noticed by their calls, emerging out of nowhere, nervously moving from perch to perch, gone before you really get a good glimpse of them. The Long-tailed Tit (or 'bush-tit' to signal that it is not closely related to the true tits at all) is a small, downy little bird with a long tail, looking cute and 'baby-like' due to its neckless rounded head and small bill. Resident, but northern populations will sometimes engage in invasion-like movements towards west and south. Has developed a number of different-looking subspecies, so different that it is hard to believe that the extremes are part of the same species.

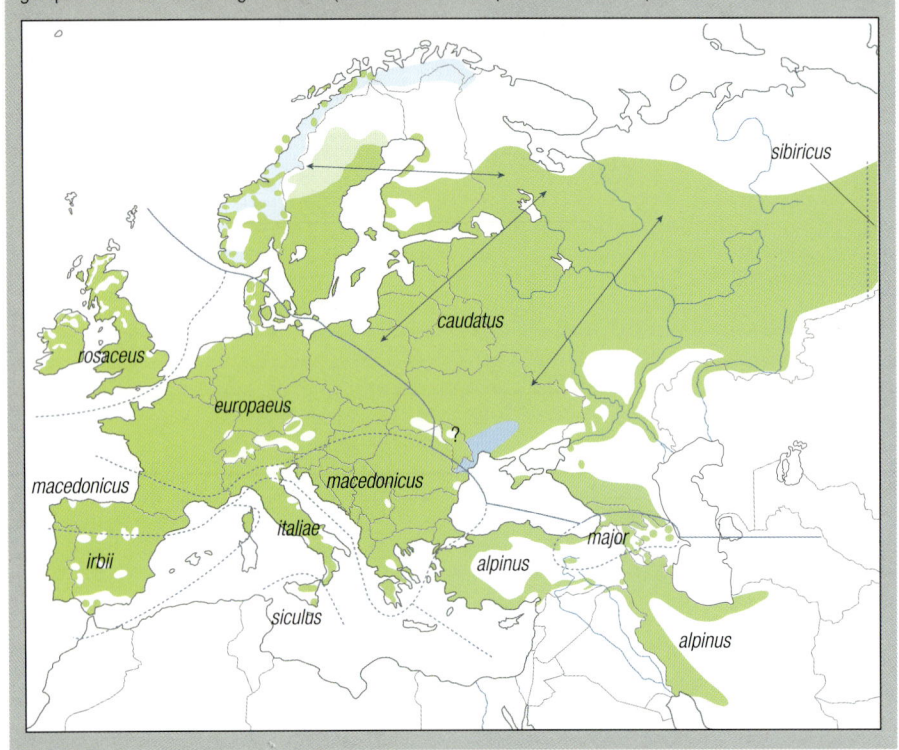

A. c. caudatus, Finland, Sep: long tail trails in typically short-distance but 'spur-of-the-moment' flight, which together with striking plumage and calls make the species always unmistakable. The wholly white head reveals it is a post-juv *caudatus*. (M. Varesvuo)

A. c. caudatus, Finland, Oct: a downy little bird with fluffy, loose plumage, a long tail and a 'cute' face. Unlike all other birds in treated region, ssp. *caudatus* is diagnostically white-headed. Flanks and lower belly tinged pinkish, scapulars pink-brown, and outer webs of tertials and secondaries form large white wing panel. Also has mostly black tail with white sides. Following complete moult after breeding, ageing is impossible. (M. Varesvuo)

IDENTIFICATION In much of its range the only small passerine with a strikingly long tail. Since it is one of the most variable species of all, and hence not easy to describe simply, it is at least always possible to liken it to a miniature toy kite or 'comet', or why not to a 'flying cotton ball' with a string attached! Often noticed in advance by its calls, then seen in a family group, arriving in hopping, hurried flight, one after another, then after a quick scan of the tree where they landed, changing perch a few times, off they go again, seemingly very restless. The *plumage is* always *fluffy*, woolly and loose, and both tertials and the *long tail are black with white sides*. Apart from black, grey and white, no colours other than dark pink and brown occur, the pink being found on scapulars and rump, and sometimes on belly and flanks. Head and underparts vary from *pure white* in N Europe to off-white with a brown or *black broad band on each side of crown*, from forehead and back, in rest of the range. In some areas dark-striped sides of head and a dark, rounded patch on centre of throat occur, or a necklace of diffuse dark blotches across upper breast. Mantle and back are black in northern populations but grey in southern, with some intermediates where these two types meet. When seen close, adults have an attractive brick-red or dark orange upper eyelid, which is easiest to detect on white-headed birds. *Black bill small and 'stubby'*, black feet small. Juveniles of all subspecies have much dark brown on sides of head, are swarthier than adults, and tail is somewhat shorter. Their eye-ring is dark purplish-red or dull dark red (rather than scarlet, brick-red or orange).

VOCALISATIONS The Long-tailed Tit provides the rare case of a European passerine without a proper territorial song (here understood as a reasonably loud and frequently repeated one). True, it can very rarely burst out into a Blue Tit-like high-pitched, clear trill, but this rare uttering appears to be more of an expression of excitement or alarm, hardly a normal song. No recurring song can be heard during nest building or from mating birds. Rarely, a subdued warble or twitter is heard at close range, interpreted as having a pair-bonding function. – Calls, on the other hand, are loud and uttered freely, and compensate for the 'missing song'. The most characteristic call is a vaguely Wren-like rolling, metallic, explosive *zerrr*, frequently repeated and used to keep the family party or flock together when feeding. When the flock is about to move, and from flying birds, another call is heard, a thin, short, sharp whistle, repeated three times, *see-see-see*, possible to confuse perhaps with some Blue Tit calls or Firecrest, although normally the tone is different. Also has a subdued dry clicking, conversational *pt*, which can be used in flight and sometimes is repeated in quick, rattling series, *pt-pt-pt-pt*. A fine, sharper *zit* or *chip* at anxiety.

SIMILAR SPECIES Can hardly be confused with any other species if seen reasonably well.

AGEING & SEXING (*europaeus*) Ageing not possible on

plumage after post-juv moult is completed, but eyelid colour generally differs into first autumn. Sexes alike. – Moults. Complete moult of both ad and juv in summer (Jun–early Oct, timing in juv obviously linked to fledging date). No pre-nuptial moult. – SUMMER **Ad** Lateral crown-stripes, if present, black. Back largely black, scapulars and rump with much greyish-pink. Central tail-feathers very nearly as long as tip of tail. Upper eyelid orange, scarlet or warm brick-red. **Juv** Broad lateral crown-stripes dark brown. Back dark brown. No pink in plumage. Central tail-feathers considerably shorter than tip of tail (Kipp 1968). Eye-ring (or at least upper eyelid) dull dark cold red or purplish-red.

BIOMETRICS (*europaeus*) **L** 13.5–15 cm; **W** ♂ 60–66 mm (n 28, m 62.8), ♀ 58–65 mm (n 28, m 61.4); **T** ♂ 78–94 mm (n 28, m 85.0), ♀ 75–92 mm (n 28, m 83.1); **T/W** m 135.5; **TG** 32–50 mm (n 53, m 43.7); **B** 6.7–8.2 mm (n 57, m 7.5); **BD** 3.3–4.4 mm (n 55, m 3.7); **Ts** 16.0–17.6 mm (n 56, m 16.8). **Wing formula: p1** > pc 5.5–12 mm, < p2 18–23 mm; **p2** < wt 10–13.5 mm, =9 or 9/10 (64%), =10 (24%), or =8 or 8/9 (12%); **p3** < wt 2–4 mm; **pp4–5**(6) about equal and longest; **p6** < wt 0–4 mm; **p7** < wt 2.5–8.5 mm; **p10** < wt 11–14 mm; **s1** < wt 11.5–15 mm. Emarg. pp3–7.

GEOGRAPHICAL VARIATION & RANGE Marked and complex variation with both moderate clinal changes forming some main patterns, and a few more abrupt changes with characteristic traits arranged more or less into subspecies groups. Several subspecies described, but not all have been possible to verify; only those that are reasonably distinct and meet the 75% rule requirement are listed below. Taxa within the coverage of this book often arranged in three groups, a policy followed here, although these groups are connected, and intergrading characters occur. Further distinct extra-limital subspecies exist apart from those listed below. Resident, but northern populations make shorter movements towards south and south-west, in some years even longer movements of invasion-like proportions.

'CAUDATUS GROUP'
A. c. caudatus (L., 1758) (Fenno-Scandia, Russia, Baltic States, NE Poland, W Siberia). Invariably white-headed (birds with hint of dark lateral crown-stripe show some degree of intergradation with *europaeus*), including white ear-coverts. Nape and upper mantle black with sharp demarcation to white head. Breast white without hint of dark cross-bar or dark blotches. Flanks and lower belly with rather restricted pink flush. Outer web of tertials broadly edged white and lower scapulars show much white, creating rather large

A. c. rosaceus, England, May: unlike northern races, has complete black lateral crown-stripe (becoming broad band when crown-feathers raised). This British and Irish race is differentiated from the main continental race, *europaeus*, by having narrower or more restricted white on centre of crown. (D. Tipling)

A. c. rosaceus, England, Feb: some are darker, others paler like this one; note dull white underparts with restricted vinaceous-pink tinge on rear flanks and strongly striped ear-coverts continuing into greyish-white, finely streaked throat and breast. (G. Thoburn)

A. c. europaeus, Switzerland, Jan: note broad black lateral crown-stripe starting from base of bill, off-white rest of head and underparts, and narrow white edges to tertials. Seen close to, the brick-red upper eyelid is visible even in dark-headed races. Very similar to British *rosaceus* but tail averages subtly longer and crown and breast a little whiter. (H. Nussbaumer)

A. c. europaeus, Germany, Apr: rather variable due to intergradation with neighbouring races, with in some birds broken dark lateral crown-stripe (here less than half developed), making separation from *caudatus* potentially less straightforward, but note streaky lateral crown-stripe behind eye and extensive pigmentation below, unlike *caudatus*. (M. Schäf)

white wing patch in fresh plumage. Dark pink on scapulars tinged brown. Intergrades in the east with ssp. *sibiricus* (see below). L 14.5–15.5 cm; W ♂ 62–69 mm (*n* 33, *m* 65.1), ♀ 63–66.5 mm (*n* 10, *m* 64.0); T ♂ 81–98 mm (*n* 33, *m* 89.9), ♀ 81–95 mm (*n* 10, *m* 88.7); T/W *m* 138.0. Wing formula: **p6** < wt 0–2 mm; **p7** < wt 2.5–6 mm. (Syn. *brachyurus*.)

Extralimital: ○ **A. c. sibiricus** (Seebohm, 1890) (C & E Siberia, Kamchatka, N Korea, Hokkaido; possibly occurring within treated range during invasion-type movements). Very similar to *caudatus*, and could alternatively be included in it with little difficulty. A fraction larger and longer-tailed, but much overlap. Pale parts on average very slightly cleaner white, but numerous inseparable on this character alone. No detectable difference in amount of white in wing. A subtle race. L 14.5–16 cm; W 62–70 mm (*n* 18, *m* 66.4); T 87–99 mm (*n* 18, *m* 93.6); T/W *m* 141.0. (Syn. *kamtschaticus*.)

'EUROPAEUS GROUP'

A. c. rosaceus Mathews, 1938 (Britain, Ireland). Very similar to *europaeus* (i.e. similar to those *europaeus* that have a complete black lateral crown-stripe), and of similar size or a fraction smaller, but differs subtly on proportionately slightly shorter tail, duskier greyish-white underparts, duller and more 'grey-striped' ear-coverts, usually narrower or more restricted white patch on crown, and more often has hint of dark bar across breast. Throat and breast sullied greyish and often diffusely blotched or streaked blackish. Belly tinged vinous-pink. L 13.5–15 cm; W ♂ 60–64 mm (*n* 15, *m* 61.9), ♀ 59–62 mm (*n* 14, *m* 60.2); T ♂ 80–90 mm (*n* 15, *m* 84.5), ♀ 76–84 mm (*n* 14, *m* 84.0); T/W *m* 135.0. Wing formula very similar to that of *caudatus*. (Syn. *chlamyrhodomelanos*; *rosea*.)

A. c. europaeus (Hermann, 1804) (Denmark, C Europe S of Baltic Sea, west to W France, south to S France, at least N & C Alps, Slovenia, N Croatia, Austria, N Bulgaria, possibly parts of Greece and NW Turkey). Treated above as to biometrics. Somewhat variable due to intergradation with neighbouring races, but in series typically differs from *caudatus* by variably broad black band across side of crown from base of bill, and white on head and underparts being on average slightly duller, less pure, and in white edges to tertials being narrower ('less white visible on closed wing'). A few (especially in Baltic States and E Poland) have broken-up lateral crown-stripe, or no stripe at all (thus more difficult to separate from *caudatus*, but are still more 'dirty' below and on head). Race *europaeus* is slightly smaller and shorter-

A. c. macedonicus, coastal NW Spain, Mar: this S European form is characterised by combination of broad black lateral crown-stripes (usually reaching base of bill), and by variable blotchy necklace-like breast-band, as well as pale parts extensively suffused pinkish-buff. This bird appears to show grey lower mantle/back, with blacker upper mantle, indicating some influence from neighbouring *irbii*, but by range it should be *macedonicus*. (A. Martínez Pernas)

A. c. macedonicus, Spain, Dec: blotchy necklace on upper breast obscured in some birds, but this race is also characterised by black mantle and back, much pinkish on scapulars and rump, and even over much of underparts (especially when fresh), with cheeks and neck-sides diffusely streaked grey-brown. Throat and upper breast off-white. (C. N. G. Bocos)

A. c. macedonicus, juv, Spain, Jul: note much browner (not black) head-sides and dark areas above, with dull dark red upper eyelids. No pink in plumage, while central tail-feathers are somewhat shorter than tip of tail. (F. Trabalon)

A. c. irbii, presumed 1stW, SW Spain, Sep: note extensive grey mantle/back, leaving only upper mantle black. Large pale cheek patch diffusely but extensively streaked grey-brown, with pale areas of throat and, especially, breast often mottled grey-brown. Also has extensive pinkish-brown wash on belly, but vent and undertail-coverts are even deeper brown. Well-defined black lateral crown-stripes rather narrow. In autumn, the bright dark red upper eye-ring suggests a 1stY. (A. M. Domínguez)

tailed (by c. 7%) than *caudatus*. – Birds in W France ('*aremoricus*') tend to be slightly smaller and darker, invariably with black lateral crown-stripes, and are thus intermediate between *europaeus* and *rosaceus*, but quite variable, and too subtle to warrant separation. Breeders in N Italy perhaps best referred to *italiae*, but some look very similar to *europaeus*. The situation in Greece is not clear either; some specimens from there are this race, others seem closer to *macedonicus*. – Birds in Japan—*trivirgatus* (syn. *japonicus*)—are extremely similar and apparently only differ on less pink flush to flanks and belly. (Syn. *aremoricus*; *tauricus*.)

A. c. macedonicus (Dresser, 1892) (N Iberia including N Portugal, Pyrenees, S France, Balkans, Hungary, Romania, possibly partly Alps, Greece and extreme W Turkey). Compared to *europaeus* slightly smaller, shorter-tailed and longer-billed. Black mantle and broad black lateral crown-stripes (stripes in vast majority reaching base of bill), and tends to have dark blotches across breast, forming a diffuse necklace. Much pink on scapulars and rump, the pink slightly brown-tinged. Cheeks/sides of neck diffusely streaked grey-brown. Throat variable, either off-white or sullied or mottled grey. – Birds in S France, Pyrenees and N Iberia traditionally separated as '*taiti*', but no detectable difference from *macedonicus* (which has priority) found when large series were compared. Not clear whether range of *macedonicus* divided in two parts or narrowly connected through

A. c. italiae, Italy, Jan: the race on mainland Italy is proportionately shorter-billed, usually has somewhat browner-tinged lateral crown-stripes and tends to have less pure or smaller grey area on mantle and back. (D. Occhiato)

A. c. italiae, Italy, Feb: some are heavily streaked on cheeks and thus resemble *rosaceus*, but *italiae* has grey back and rump, a largely brown lateral crown-stripe and is duller overall with pinkish areas somewhat less extensive. White inner wing panel can be reduced by wear. (D. Occhiato)

Alps. At least some Swiss breeders have dark blotches across breast and dusky streaks on sides of head tending towards *macedonicus*. **W** ♂ 57–64 mm (*n* 23, *m* 61.8), ♀ 56–62 mm (*n* 17, *m* 59.9); **T** ♂ 70–90 mm (*n* 23, *m* 82.1), ♀ 74–88 mm (*n* 16, *m* 80.6); **T/W** *m* 133.4; **B** 6.6–8.4 mm (*n* 39, *m* 7.8). Wing formula very similar to that of *europaeus*. (Syn. *taiti*.)

'ALPINUS GROUP'

A. c. irbii (Sharpe & Dresser, 1871) (C & S Portugal, C & S Spain, Corsica). Slightly smaller and shorter-tailed than *macedonicus*, with which it intergrades in N or C Iberia. Differs apart from smaller size and slightly blunter wing (with shorter primary projection) from *macedonicus* (including '*taiti*') by having lower mantle/back grey (only upper mantle is black). Cheeks tinged pinkish-brown and diffusely streaked grey-brown, and throat and breast diffusely but more boldly mottled grey-brown. Extensively pink on rump and over much of underparts (except on throat, sides of neck and upper breast). Close to *italiae* but is slightly smaller and has cleaner black lateral crown-stripes (with fewer brown tones). Bill a little longer but thinner than in northern populations. **L** 12–13.5 cm; sexes of very nearly same size; **W** 56–61.5 mm (*n* 34, *m* 58.6); **T** 71–84 mm (*n* 34, *m* 76.9); **T/W** *m* 131.3; **B** 7.2–8.5 mm (*n* 28, *m* 7.9); **BD** 3.1–3.9 mm (*n* 27, *m* 3.5); **Ts** 15.3–17.3 mm (*n* 30, *m* 16.5). **Wing formula: p1** < p2 16–19 mm; **p2** =10 (common), =9/10 (less common); **p6** < wt 0–1 mm; **p7** < wt 1.5–4 mm; **p10** < wt 9–12 mm; **s1** < wt 10–12.5 mm. (Syn. *tyrrhenicus*.)

○ **A. c. italiae** Jourdain, 1910 (mainland Italy except extreme north). Resembles *irbii* but is a little larger, proportionately shorter-billed, usually has a slight brown tinge to blackish lateral crown-stripes (on *irbii*, these are cleaner black), and has perhaps a shade paler grey back. Also, black on mantle often more extensive. Somewhat similar to *rosaceus* (with broad blackish lateral crown bands) but has grey back, is a little duskier overall, and has less pink, being more greyish and dull. Lower mantle/back mainly grey, rump grey with a little pink in most. Throat off-white, diffusely mottled grey on some. Often some irregular blackish marks across breast forming hint of necklace. Forecrown often tinged brown, centre of crown whitish. Sexes of similar size; **W** 56–64.5 mm (*n* 26, *m* 61.5); **T** 72–91 mm (*n* 26, *m* 79.9); **T/W** *m* 130.0; **B** 5.8–8.3 mm (*n* 23, *m* 7.4). **Wing formula: p1** < p2 18–21.5 mm; **p2** =9/10 or 10 (common), =9 or <10 (less common); **p10** < wt 10–13.5 mm; **s1** < wt 11.5–15.5 mm.

A. c. siculus (Whitaker, 1901) (Sicily). Quite characteristic. Small and short-tailed; rather pale rufous-brown lateral crown-stripes, starting diffusely from bill; often, whole forehead is tinged brown; centre of crown dusky-white with faint brown hue or ill-defined brown mottling. Head pattern stands out because of a black (or blackish) band across nape. Rest of upperparts dusky grey with slight brown tinge, but rump

A. c. italiae, Italy, Jan: forehead and forecrown often tinged brown, contrasting with whitish centre of crown. Note dark brown lateral crown-stripes and grey mantle and back. (D. Occhiato)

A. c. siculus, Sicily, Apr: this highly distinctive race is generally small-sized, with noticeably short tail. Also overall paler, including mostly dusky-white head with diffuse and mainly rufous-brown lateral crown-stripes (and forehead sometimes entirely washed rufous-brown). Mantle and back predominantly grey, but rump pinkish, and underparts greyish-white with limited brownish or pinkish hue on belly and vent, adding to the overall pale appearance. (R. Pop)

is pinkish. Underparts swarthy, dusky greyish-white with brown tinge, gradually becoming more pinkish on belly and vent. Throat diffusely blotched grey-brown on dusky-white. **L** 13–14 cm; sexes of very nearly same size; **W** 56–61 mm (*n* 15, *m* 58.4); **T** 69–77 mm (*n* 15, *m* 72.8); **T/W** *m* 124.7. – Strangely, this plumage is closely matched by Caucasian race *major*. Only differences seem to be that *major* is slightly larger, has on average a broader black band across nape (often invading upper mantle), paler and whiter underparts, cleaner pink colour on belly/flanks, and possibly also on average paler head.

A. c. major (Radde, 1884) (NE Turkey, Caucasus, W & C Transcaucasia). Very similar to *siculus* of Sicily (despite widely separated ranges, with populations of different appearance between them), also having brown lateral crown/nape stripes, largely grey back and dark blotches on throat.

Differs in being a little larger and longer-tailed, less blotchy and dusky, more clean and neat in colours. Black band across nape on average broader, often irregularly invading upper mantle. The grey of back often has more brown tones than in *siculus*. Centre of crown on average cleaner white, but overlap and variation make this character less useful. Ear-coverts striped brown on dusky-white ground. Breast mottled grey-brown on dusky ground. Belly and vent pinkish. Wings dark, white edges to tertials narrow, in worn plumage absent. Apparently no sexual size dimorphism; **W** 60–66 mm (n 20, m 63.0); **T** 75–85 mm (n 20, m 81.1); **T/W** m 128.7. (Syn. *caucasicus*.)

A. c. alpinus (Hablizl, 1783) (Turkey except in W & NE, also S & E Transcaucasica, N & W Iran). Rather small and markedly short-tailed. Has a characteristic blackish or dark grey rounded patch on throat (chin is pale, though), more solidly dark than in preceding two races. All pale areas in plumage tinged brownish (no pure whitish at all, though chin is nearly white), and breast has hint of thin brown crossbar.

A. c. major, Azerbaijan, Feb: distinctive, being proportionately long-tailed, with dark brown lateral crown-stripes (bordered behind by black nape) and mostly grey mantle and back. Ear-coverts, throat and chest are dusky white or mottled grey-brown (sometimes more distinctly so on breast), while belly and vent are tinged pinkish. (W. Müller)

A. c. alpinus, 1stW, Turkey, Dec: several unique characteristics of this race, especially a round blackish throat patch (when plumage worn in spring, patch becomes solid dark), and often has a vague dark breast-band below the patch (here a mere hint). Pale areas tinged brownish. The lateral crown-stripes are mainly black, well defined and usually start on forecrown. Ear-coverts typically heavily streaked brown. Mantle and back are largely grey. In Dec, the still deep red eye-ring indicates age. (F. Yorgancıoğlu)

A. c. alpinus, juv, Turkey, Jun: mid-June and already in post-juv moult, but body plumage mostly juv, showing that juv of this race is also distinctive with hint of dusky throat patch and largely brownish lateral crown-stripes. New central tail-feathers growing. (H. Shirihai)

Sides of head and neck diffusely streaked brown on dusky ground; crown similar. Mantle/back grey, but sometimes a narrow black band runs across upper mantle. Black lateral crown-stripes (with faint brown tinge), starting on forecrown and reaching mantle. Some pink often visible on rump. Juv has blackish throat patch replaced by rufous. **L** 10.5–13 cm; **W** ♂ 56–63 mm (n 22, m 60.1), ♀ 53–61 mm (n 12, m 57.4); **T** ♂ 60–77 mm (n 22, m 69.1), ♀ 60–75 mm (n 12, m 66.1); **T/W** m 114.6. – In N Iran, some birds are a little paler grey above and cleaner whitish below ('*passekii*'), but far from constant. Birds in Turkey ('*tephronotus*') on the whole a shade paler grey above than series of birds from S Caspian region (*alpinus* s.s.), but subtle and variable, and hardly sufficient ground for separation. (Syn. *passekii*; *tephronotus*.)

TAXONOMIC NOTE The fact that the population in the Caucasus is morphologically surprisingly similar to that on Sicily, both in plumage and structure, but widely separated by populations of different appearance, might merit a genetic study.

REFERENCES Downhill, S. (2012) *Ringing News*, 12: 7. – Jansen, J. J. F. J. & Nap, W. (2008) *DB*, 30: 293–308. – Kipp, F. (1968) *Vogelwarte*, 24: 284. – Päckert, M., Martens, J. & Sun, Y.-H. (2010) *Mol. Phyl. & Evol.*, 55: 952–967.

A. c. alpinus, juv, Turkey, Apr: a tight group of downy fledglings recently out of nest can appear especially 'cute'. This race has unique juv plumage, characterised especially by the rufous rather than blackish patch on lower throat. (I. Tunca)

MARSH TIT
Poecile palustris (L., 1758)

Fr. – Mésange nonnette; Ger. – Sumpfmeise
Sp. – Carbonero palustre; Swe. – Entita

The Marsh Tit is one member of a twin pair of rather colourless brown, black and white European tits, which are so similar in appearance that they would frequently be confused if it were not for their characteristic calls, the other being Willow Tit. The Marsh Tit prefers decaying deciduous or mixed woods on slightly damp ground, but is found in a variety of other habitats, including juniper thickets, and will come to garden feeders to take seeds. Nowhere numerous, and rarely joins the roaming packs of tits in winter, it is still widespread in Europe, missing only from N Fenno-Scandia, N Scotland, Ireland and much of Iberia. After a gap it also has a wide distribution in Asia.

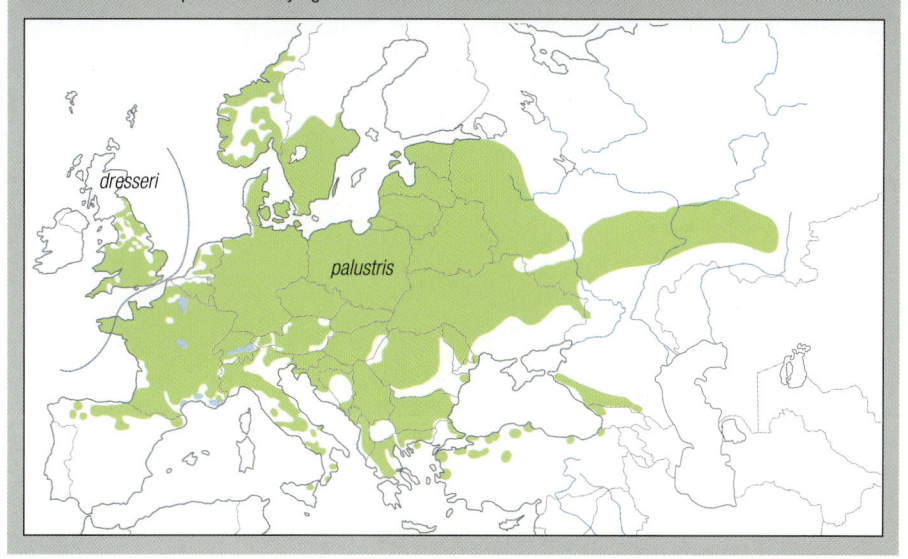

P. p. palustris, Denmark, Nov: two rather more consistent features eliminate Willow, albeit with variation and overlap. Firstly, note pale spot near base of upper mandible, and secondly, brownish-buff comma-shaped mark across lower and rear cheeks (between pure white of ear-coverts and brown mantle). In addition, the bib is generally small and more solid at its lower edge in most Marsh, and crown is more glossed. (O. Krogh)

IDENTIFICATION After noting *black cap* and *fairly small black throat bib*, *whitish cheeks*, *grey-brown upperparts* without any pale wing-bars, options have been narrowed down to either Marsh and Willow Tit. Then pay attention to any calls: vocalisation is the quickest and safest means of separating these two similar species, so should be noted first of all (see below). With silent birds you need to know what to look for: (i) Marsh Tit has a *more uniformly brown-grey wing*, lacking the whitish 'panel' along the folded secondaries usually seen in Willow (beware of light reflection on a very fresh wing in Marsh, which briefly can give a similar impression, and conversely Willow without a light 'panel' due to heavy feather wear or juvenile plumage, or certain Continental populations, having more brownish feather-edges); (ii) Marsh Tit has practically invariably *a pale spot on lower base of upper mandible*, below nostrils (Dewolf 1987, Broughton et al. 2008), whereas this is absent in nearly all Willow; (iii) nearly all Marsh Tits (at least in Britain and W Europe) have *a hint of a thin, dark* (brownish) *comma-shaped mark across the cheeks, cheeks being white in front of the 'comma' but slightly shaded behind it* (Willow: as a rule no such mark, at least not a brownish one, at the most a greyish hint, and whole cheek area more uniformly white; M. Pearson & P. Dunn *in litt*. 2008; Broughton 2009). To these frequently helpful characters come a few supplementary ones that show more overlap and which should be used with caution: (iv) the throat bib on Marsh is generally small and usually neatly outlined at its lower edge (on average larger and more broken up along its lower edge in Willow), but some overlap in this, especially some Willow Tits with smaller bib in Europe, but also Marsh Tits in Asia with large bib; (v) in good light the black cap of adult Marsh can be seen to be glossy, whereas it is more matt in adult Willow (although this character is often difficult to use in the field, is somewhat sexually biased with ♂♂ of both species having slightly more gloss on average, and does not hold well for juveniles or in Britain); (vi) the rear outline (behind the eye) of the black cap in Marsh is often more uneven or undulating, reducing pale area of 'rear cheeks' (Willow: cap behind eye is usually smoothly outlined in arch-shape, leaving more space for white 'rear cheeks'; Pearson & Dunn *in litt*.); and (vii) the tail is proportionately slightly shorter and ends more squarely in Marsh, is slightly longer and more graduated in Willow (for measurements, cf. Biometrics). – When comparing the two species in Britain or in rest of W Europe, remember that Willow Tits are more brown and warm-coloured below than in N Europe, and hence more similar to Marsh Tit.

VOCALISATIONS The song is a simple series of repeated notes, *chip chip chip chip chip* or *ziev ziev ziev ziev*. There is some variation in tone and pace, some sounding more resonant and then recalling a phrase from song of Tree Pipit, while faster variants can recall Greenfinch, *chüp-chüp-chüp-chüp-chüp*. Other variants with disyllabic notes recall Great or Coal Tit, *vita-vita-vita-vita-vita*. Although the song has here been likened to several other species, it is still usually easy—with some practice—to recognise the Marsh Tit by its particular voice. Rarely, a brief subdued warble is uttered, not that different from a similar alternative song of Willow Tit. – Calls are loud and characteristic and very different from those of Willow Tit. Commonest is an explosive, spirited and almost cheeky, disyllabic *pichay*. This call is sometimes trisyllabic, *zi-zi-chay*, or when agitated combined with a Blue Tit-like scolding series, *pichay-de-de-de-de-de-det*. Shorter variants *pichay-de-de* also heard. Conversational fine *zit* notes occur, too, but are not so different, if at all, from what you can hear from most feeding tits. Begging call of young a rhythmically repeated trisyllabic *dee-de-dah*, or, perhaps less often, a thin, sharp disyllabic *eehs-it* (Broughton 2009).

SIMILAR SPECIES The main problem, separation from *Willow Tit*, is dealt with under Identification, and see Biometrics and again under Willow Tit. – *Sombre Tit* overlaps in range with Marsh Tit in some parts of SE Europe and is superficially similar, but note on Sombre much larger size, more extensive dark bib and cap so that white cheeks are narrower and more wedge-shaped. – *Caspian Tit* is normally not a problem due to local and allopatric range, but is also immediately separated by brownish cap, much larger bib and shorter tail. Song and calls differ too. – *Siberian Tit*, which is not normally found in the same range as Marsh Tit, is also

P. p. palustris, Germany, Feb: very similar to Willow Tit, with the primary clue being that Marsh lacks Willow's whitish wing panel (but pale edges to the secondaries can coalesce and catch the light to form a panel as to some extent shown here). Ageing without handling difficult, but the overall fresh plumage (including primary-coverts) suggest an ad at this season. (A. Noeske)

Marsh (top two: left, Estonia, Apr; right, Netherlands, May) versus Willow Tits (bottom two, Finland: left, Feb; right, Jan): with silent birds there are several key identification clues, here in order of importance and degree of visibility. Even when quite distant, Marsh Tit lacks Willow's whitish secondary panel (though a few Marsh show a hint of this, and it is reduced or completely lacking in some Willow). Secondly (cf. right-hand images), there is a shiny pale spot near base of upper mandible in Marsh, which Willow lacks or at most shows only vaguely and rarely. Thirdly, Marsh has a subtle brownish-buff comma-shaped mark on lower and rear cheeks, between pure white ear-coverts and brown-tinged rear neck (thus pure white cheek area is smaller in Marsh). And fourthly note Marsh's often smaller bib with well-defined lower edge (usually larger with more ragged lower edge in Willow—but again, there is some overlap). Comparison here is between N European populations of the two species, which are generally paler/cleaner below. However, in other parts of Europe, Willow is browner and warmer below, and more similar to Marsh. (Top left: M. Varesvuo; top right: C. van Rijswijk; bottom left: H. Taavetti; bottom right: M. Varesvuo)

readily separated by its dark brown cap, tawny-tinged back, orange-brown flanks, broad white edges on folded secondaries (more even than on Willow Tit) and a much larger bib.

AGEING & SEXING (*palustris*) Ages differ slightly, young birds usually separable during 1stY, but ageing requires close views or handling. Sexes alike in plumage; ♂ is somewhat larger. On average, ♂♂ have W > 63 mm and ♀♀ < 66 mm, but this will not allow for atypically large ♀♀ or small ♂♂. – Moults. Complete post-nuptial moult of ad in summer after breeding (Jun–Aug). Partial post-juv moult in summer (Jul–early Oct) does not involve any primaries or secondaries but often several inner and rarely all greater coverts, odd tertials and part of or—rarely—whole tail. Rather large variation in extent of post-juv moult noted, even in same region, between studies. No pre-nuptial moult. – **AUTUMN Ad** Tail-feathers rather broad, tips more rounded than pointed, and fairly fresh through winter. Primary-coverts have rather broad and rounded tips with neat edges. All greater coverts uniform (but not always fresh after mid Sep). **1stCY** Tail-feathers (unless partly or wholly moulted in late summer) not so broad, tips more pointed than rounded, often abraded from late autumn or early winter. Primary-coverts have on average slightly more pointed tips and frayed edges. A few birds show contrast among greater coverts, a few outer ones being unmoulted, slightly longer, with paler buffish tips and being paler towards edge. Birds moulting all greater coverts rare. – **SPRING Ad** Same characters as in autumn apply, only now are more difficult to use due to more extensive wear.

Generally, though, tips of tail-feathers are less worn than in 2ndCY. **2ndCY** As autumn, but more abraded. In particular the tips of tail-feathers are more pointed and abraded than in ad. Still, many birds must be left un-aged.

BIOMETRICS (*palustris*) **L** 11.5–12.5 cm; **W** ♂ 60–70 mm (n 55, m 65.9), ♀ 60–67.5 mm (n 43, m 63.6); **T** ♂ 49–60 mm (n 55, m 55.2), ♀ 47–58 mm (n 43, m 53.3); **T/W** m 83.8; **TG** 1–5 mm (n 90, m 2.9); **B** 9.2–11.2 mm (n 95, m 10.2); **BD** 3.8–4.9 mm (n 74, m 4.4); **Ts** 15.0–17.5 mm (n 89, m 16.3). **Wing formula: p1** > pc 6.5–13 mm, < p2 17–25 mm; **p2** < wt 8.5–13 mm, =9 or 9/10 (81%), or =10 (19%); **p3** < wt 1.5–3 mm; **pp4–6** about equal and longest (p5 often 1 mm longer); **p7** < wt 1.5–4 mm; **p10** < wt 9.5–13 mm; **s1** < wt 10–13.5 mm. Emarg. pp3–7.

GEOGRAPHICAL VARIATION & RANGE Slight and clinal variation in W Palearctic, birds in west, and to lesser extent south, on average smaller and darker. After a careful comparison of material from most parts of the range in Europe, Turkey and Caucasus, only *dresseri* found to be sufficiently distinct from *palustris*, all other described taxa included in either of these two. No consistent plumage or biometrical differences noted between spring birds of, e.g., *palustris*, *stagnatilis*, *kabardensis* or *italicus* when series compared. Birds forming a different subspecies group in E Asia (so-called '*brevirostris* group'; extralimital) have generally proportionately longer tail and slightly different song. All populations sedentary.

P. p. palustris (L., 1758) (Europe except west, south to Sicily, Balkans, east to Turkey, Caucasus). Treated above. Usually rather light greyish-cinnamon below, cheeks somewhat paler and more whitish. Birds in W Europe and Italy on average slightly more saturated pinkish-brown below, in W France tending towards *dresseri*, but any differences subtle and variable, and none of several proposed subspecies is even close to 75% distinctness. (Syn. *brandtii*; *communis*; *italicus*; *kabardensis*; *longirostris*; *siculus*; *stagnatilis*; *tschusii*.)

P. p. dresseri (Stejneger, 1886) (Britain, W Brittany). Similar to *palustris*, and a few are inseparable from it, but when series are compared, at least 75% of *dresseri* are rather more evenly dusky-brown below with only somewhat lighter (off-white) cheeks. Flanks usually appreciably more cinnamon-tinged in fresh plumage. Upperparts cinnamon-tawny with slight grey tinge. Bill and tail tend to be subtly shorter. A vague sexual difference has been described (King 1990) with ♂ having on average more gloss on cap than ♀, but this appears to be very subtle and difficult to discern. As a rule, wing-length gives a fair guidance: ♂ > 62 mm, ♀ < 63 (Gosler & King 1989, King & Muddeman 1995), but leaves a few extremes wrongly sexed. Ageing relies primarily on shape and wear of primary-coverts and any retained juv tail-feathers, since most *dresseri* moult all greater coverts in post-juv moult (King & Muddeman 1995). **L** 11–12 cm; **W** ♂ 61–66 mm (n 15, m 63.4), ♀ 59–64 mm (n 13, m 61.3); **T** ♂ 47–54 mm (n 15, m 51.5), ♀ 45–53 mm (n 13, m 49.9); **T/W** m 81.3; **B** 9.0–10.3 mm (n 24, m 9.8). Rest of measurements and wing formula as *palustris*. (Syn. *darti*.)

Marsh Tit (left: Germany, Nov) versus **Willow Tit** (right: England, Nov): due to angle of view and light, the wing of the Marsh appears misleadingly to have a paler secondary panel (albeit brown-tinged), while on Willow the black bib is confusingly small with an atypically sharply defined lower edge. Complicating matters even further, the pale spot at base of upper mandible of the Marsh is much reduced, while the Willow has a faint hint of a brown comma on rear cheek, a character more associated with Marsh. In some areas, the two species can be equally warm-coloured below. Only the clearly whitish secondary panel in Willow helps to confirm species. This comparison demonstrates that all features can be affected by caveats, making it important to use as many as possible in combination, and underlining that silent birds may even prove impossible to identify. (Left: A. Noeske; right: DP Wildlife Vertebrates)

P. p. dresseri, England, Apr: this race has the underparts on average sullied evenly dusky buff-brown or cinnamon-tinged, highlighting the off-white cheeks. Note the pale bill spot (here closer to the cutting edges than in most) and faint brownish line across rear cheeks. The apparently heavily worn and pointed tail-feathers, and worn, brownish primaries suggest a 1stS. (J. Lewis)

P. p. dresseri, 1stS, England, Mar: primary-coverts and remiges are juv (making this a 1stS), and the worn and bleached browner primaries losing their fringes misleadingly suggest a Willow-like pale panel. However, this bird still has the brownish-buff comma-shaped mark on rear cheeks, pale bill spot and small bib of Marsh. (S. Round)

TAXONOMIC NOTES Gill *et al.* (2005) advocated a split of *Parus* into several genera, following a molecular analysis which revealed that the alternative to avoid non-monophyly of *Parus* would be to include some very different taxa. The split of *Parus* has since been adopted by, e.g., the BOU (cf. Sangster *et al.* 2005 for reasons). Johansson *et al.* (2013) reached the same conclusions when sampling all existing tit species including nearly every taxon.

REFERENCES Broughton, R. K. (2009) *BB*, 102: 604–616. – Broughton, R. (2010) *Ringers' Bull.*, 102: 106. – Broughton, R. K. *et al.* (2008) *Ring. & Migr.*, 24: 88–94. – Dewolf, P. (1987) *BUBO*, January: 10–11. – Eck, S. & Martens, J. (2006) *Zool. Med. Leiden*, 80: 10–12. – Gill, F. B., Slikas, B. & Sheldon, F. H. (2005) *Auk*, 122: 121–143. – Gosler, A. G. & King, J. R. (1989) *Ring. & Migr.*, 10: 53–57. – Johansson, U. *et al.* (2013) *Mol. Phyl. & Evol.*, 69: 852–860. – King, J. R. (1990) *BB*, 83: 510–511. – King, J. R. & Muddeman, J. L. (1995) *Ring. & Migr.*, 16: 172–177. – Sangster, G. *et al.* (2005) *Ibis*, 147: 821–826.

P. p. dresseri, Wales, Jan: variation within any population, even within this generally more brownish race, includes duller and somewhat greyer birds (or the light can create such impression). Note the lack of a whitish secondary panel and the species-specific whitish spot at base of upper mandible. Ageing without handling best avoided. (A. Williams)

WILLOW TIT
Poecile montanus (Conrad von Baldenstein, 1827)

Fr. – Mésange boréale; Ger. – Weidenmeise
Sp. – Carbonero montano; Swe. – Talltita

Closely recalls the Marsh Tit, thus brown-grey and white with a black cap and a black bib, and best told by its characteristic calls. Found in a variety of wooded habitats, in the north usually at rather higher altitudes in coniferous forest with a mixture of broadleaved trees, including decaying trunks in which it excavates its nest hole. In some parts of its range in W and C Europe it is often found at lower elevations and in more open habitats with a blend of trees and bushes. Not a frequent guest at feeders in gardens, but will come from time to time. Joins roaming packs of tits in winter more frequently. It is largely absent from southern Europe, but is common in N Fenno-Scandia and Russia, breeding in the taiga right up to the edge of the tundra or treeless mountains.

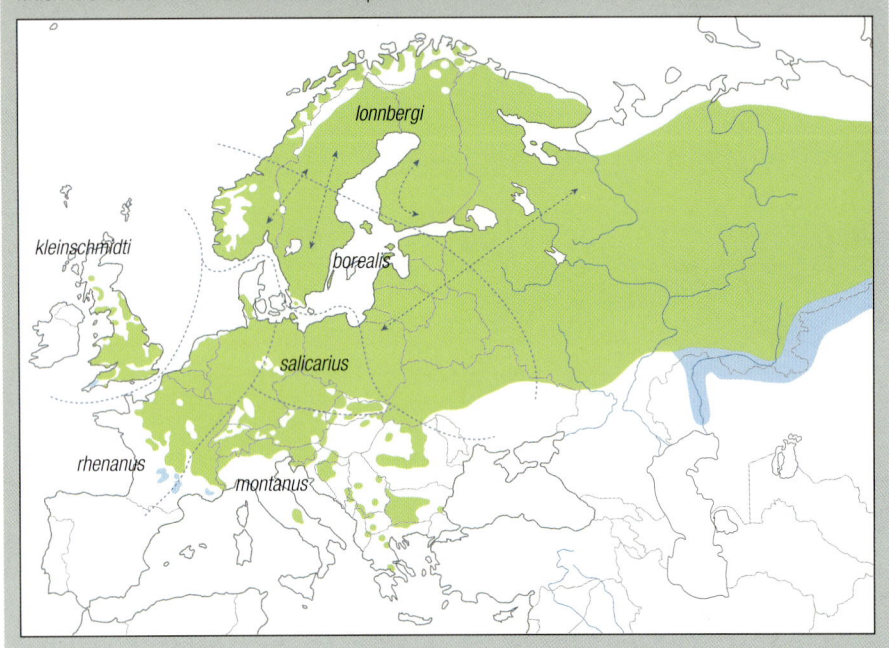

IDENTIFICATION See Marsh Tit for a detailed comparison between these two very similar species. Repeated here are the main points to note: vocalisations best means of telling the two apart (see below). Without the voice, remember: (i) Willow Tit has nearly always a *whitish or pale buff-brown 'panel' along the folded secondaries*, contrasting against darker brown-grey surround (beware of light reflection on very fresh wing of Marsh, which momentarily can give a similar impression, and conversely Willow without a light 'panel' due to heavy feather wear or juvenile plumage, or certain Continental populations of Willow, having more brownish than whitish feather-edges, affording less contrast); (ii) Willow Tit usually has an all-dark bill, rarely with a very thin pale line proximally on the cutting edges (Marsh: usually has a more obvious pale spot proximally on upper mandible between nostril and cutting edge); (iii) whole cheeks including rear part white (Marsh: only fore cheeks white, rear part sullied brownish or greyish, these two divided by hint of brownish comma-shaped mark across cheeks; Pearson & Dunn *in litt.*; Broughton 2009). Apart from these differences it is sometimes possible to see that the tail is a trifle longer and has more graduated end in Willow, versus slightly shorter and more square in Marsh. – Willow Tits in Britain and rest of W Europe are more brown and warm-coloured below than in N Europe, and hence more similar to Marsh Tits, whereas Willow Tits in Fenno-Scandia and Russia are much *paler and more greyish and white*.

VOCALISATIONS Willow Tit has several song types. The most common one, heard over much of the northern and western range, is a simple series of rather slowly repeated notes, each note descending slightly, rendering the song a melancholy ring, *piuh, piuh, piuh, piuh, piuh*. It has been likened to the alternative song of Wood Warbler (but in this the notes actually ascend, are repeated quicker, and the voice differs). Sometimes you hear one or more alternative songs with slightly different pitch and phrasing, perhaps then usually from different birds, but the same male can apparently also switch between a few variants. In the Alps (*montanus*), the normal *piuh* song is replaced by a dialect consisting of a slightly longer and faster series of more mechanical, short, straight and piercing notes, *düh düh düh düh düh düh düh düh*. There can be a slight deceleration, the notes becoming a trifle longer, towards the end. This song type, or at least a very similar one, is also heard over large parts of Siberia. Intermediate songs occur locally in northern and eastern parts of the Alps. Finally, in all populations a further song type is sometimes used, a brief, fast, high-pitched warbling outburst with a mixture of twittering and buzzing notes without a clear pattern, often ending in a trill. A similar alternative song can very rarely be heard also from Marsh Tit. – Calls are loud and luckily very different from those of Marsh Tit. Commonest call is a harsh and grating, repeated, drawn-out and stressed *taah*, 2–3 (4) in series and often preceded by 1–2 high-pitched, thin notes, *zi zi taah taah taah*. Conversational fine *see* or *zit* are heard, too, but are similar to what you can hear from most feeding tits. The *zi zi taah taah taah* call is used both for contact and mild alarm. When an owl is spotted, the *zi zi* notes are dropped and the *taah* call becomes even more stressed and drawn out, repeated in long series. Begging call of young a rather subdued and less frequently heard disyllabic trilling *chir-cherr*. – The harsh and 'angry' *taah* call is the easiest way to separate Willow and Marsh Tits in much of Europe. However, it should be noted that Siberian Tit has a very similar call, so in the northernmost taiga in N & NE Europe one has to be aware of this possibility as well. Similar problems occur with Songar Tit *P. songarus* (extralimital; not treated) in Central Asia.

SIMILAR SPECIES For separation from *Marsh Tit*, see that species, and under Identification and Biometrics. Note that Willow Tits in Continental Europe in worn plumage can have a less pale and contrasting secondary panel compared to N & E European breeders, hence they can be more similar to Marsh Tit in this respect, and all other characters must be considered as well. – Overlaps in range with *Sombre Tit* in SE Europe and Balkans, and could be confused if not seen well, but note Sombre's larger size, more extensive dark throat bib and cap so that white cheeks are narrower and more acutely wedge-shaped. Also, calls differ rather markedly. – *Siberian Tit* on the other hand has similar calls but is readily separated by its dark brown cap, warm tawny-tinged back, orange-brown flanks, and much larger throat bib. It also has an even whiter and more prominent secondary panel. Note, however, that Willow and Siberian exceptionally can hybridise where they meet, producing intermediates that can be very difficult to identify.

P. m. lonnbergi, Finland, Feb: the pale dusky grey-white underparts with hardly any brown suffusion identify this race of N Fenno-Scandia to N Russia, where it usually inhabits mixed coniferous and broadleaf forests. The most notable criteria separating this species from Marsh Tit are the whitish wing panel and extensive, clean white cheeks. The apparently evenly fresh wing suggests an ad. (H. Taavetti)

P. m. lonnbergi, Finland, Feb: rather purer white below than the otherwise very similar ssp. *borealis*. Note the well-developed white wing panel (most obvious in northern populations), rather large black bib with relatively indistinct borders, and completely white cheeks, making separation from Marsh Tit relatively straightforward. Without handling age is difficult to assess in this case. (M. Varesvuo)

P. m. lonnbergi, C Finland, Oct: photographed within the range of *lonnbergi*, but the degree of pinkish-brown suffusion on flanks is more typical of spp. *borealis*, suggesting that the two might intergrade in parts of Finland. Such warm flanks could invite confusion with Marsh Tit, but note the fairly large black bib, completely white cheeks and whitish wing panel. Evenly feathered wing suggests an ad. (M. Varesvuo)

P. m. rhenanus, France, Jan: in continental Europe the pale wing panel is often weakly developed, and flanks are usually warmer, thus more like Marsh Tit, while this bird even more confusingly has parts of cheeks tinged rufous-brown. However, other features, e.g. the larger black bib with rather ill-defined lower border, are typical of Willow. Also, Willow Tit has a very narrow pale line on the bill's cutting edges, whereas Marsh usually has a more obvious pale spot in front of the nostril and above the cutting edge. (A. Audevard)

P. m. rhenanus, Germany, Feb: rufous-tawny wash below, generally dirty whitish cheeks (faintly tinged rufous-buff, but not so obvious on this bird) and proportionately strong bill are features of this race. Note in this light and angle the very obvious whitish secondary panel. (F. Adam)

AGEING & SEXING (*borealis*) Ages differ subtly, young birds often separable during 1stY, but ageing requires close views or handling. Sexes alike in plumage; ♂ is slightly larger. – Moults. Complete post-nuptial moult of ad in summer after breeding (May–early Sep). Partial post-juv moult in summer (Jun–early Sep) does not involve any primaries or secondaries but sometimes odd tertials and central 1–2 pairs of tail-feathers. No pre-nuptial moult. – AUTUMN **Ad** All tail-feathers rather broad, tips more rounded than pointed, and fairly fresh through winter. Primary-coverts have rather broad and rounded tips with neat edges. All greater coverts uniform (but not always fresh after mid Sep). **1stCY** Tail-feathers not so broad, tips more pointed than rounded, often abraded from late autumn or early winter, but be aware of birds that replace a few central tail-feathers to ad shape. Some birds have slightly more rounded and ad-like shape to tail-feathers, and any intermediate bird should be left un-aged. Primary-coverts have on average slightly more pointed tips and frayed edges. A few birds show contrast among greater coverts, a few outer ones being unmoulted and slightly paler towards the edge. – SPRING **Ad** Same characters as in autumn apply, only now are more difficult to use due to more extensive wear. Generally, though, tips of tail-feathers are less worn than in 2ndCY. **2ndCY** As in autumn, but more abraded. In particular the tips of tail-feathers are more pointed and abraded than in ad. Many birds must be left un-aged.

BIOMETRICS (*borealis*) **L** 11.5–12.5 cm; **W** ♂ 62–68.5 mm (*n* 38, *m* 65.3), ♀ 60–66 mm (*n* 31, *m* 62.8); **T** ♂ 55–62 mm (*n* 38, *m* 58.5), ♀ 52–60 mm (*n* 31, *m* 56.5); **T/W** *m* 89.7; **TG** 4.5–9 mm (*n* 49, *m* 6.5); **B** 9.7–11.5 mm (*n* 48, *m* 10.6); **BD**(f) 3.8–4.7 mm (*n* 39, *m* 4.2); **Ts** 15.0–17.3 mm (*n* 49, *m* 16.2). **Wing formula: p1** > pc 9–13.5 mm, < p2 17–23 mm; **p2** < wt 8.5–12 mm, =9 or 9/10 (54%), =10 (32%), or =8/9 (14%); **p3** < wt 1.5–3.5 mm; **pp4–6** about equal and longest; **p7** < wt 1–5 mm; **p8** < wt 4–8 mm; **p10** < wt 9.5–12 mm; **s1** < wt 10–14 mm. Emarg. pp3–6, often faintly also on p7.

GEOGRAPHICAL VARIATION & RANGE Moderate and clinal variation over continuous range in lowland and taiga of W & N Europe, birds in the west becoming on average smaller and darker. Race in mountains of C & SE Europe, *montanus*, however, large and subtly greyer above, stronger-billed and has at least locally a different song, still kept as race with others on account of close resemblance and lack of firm evidence for status as separate species. Closely related to North American *P. atricapillus*, formerly regarded as conspecific but nowadays generally treated as separate species. Here separated also from closely related *P. songarus* in Central Asia and China (extralimital; not treated). Some 3–4 extralimital subspecies in Asia. All populations mainly sedentary, with some movements noted only in certain winters in northernmost part of range.

○ *P. m. lonnbergi* (Zedlitz, 1925) (N Fenno-Scandia north of *borealis*, including most of Finland, much of N Russia, lower Volga region, W Siberia). Similar to *borealis* but on average whiter below, lacking all or nearly all warm or pink-brown tones on flanks. Often, cheeks are purer white and upperparts a trace greyer, not as brown-tinged, but subtle. Some birds are inseparable, but series usually differ sufficiently to warrant separation of this race. Very pale birds collected near border between Russia and Estonia (Pskov) have same plumage, thus if these are genuine *lonnbergi* they are far outside normal range, but could be dislocated or migrant *lonnbergi* (invasion-type movements are known to occur), or are local colour aberrants. **W** ♂ 64–69 mm (*n* 25, *m* 65.5), ♀ 62–66 mm (*n* 16, *m* 63.5); **T** ♂ 56–62 mm (*n* 25, *m* 58.7), ♀ 57–61 mm (*n* 16, *m* 58.6); **T/W** *m* 90.3. **Wing formula: p2** =10 (71%), =9 or 9/10 (18%), or =8 or 8/9 (1%). Grades into slightly paler extralimital *baicalensis* in the east, starting from Yenisei Basin. (Syn. *rossicus*; *uralensis*. When series are compared of *uralensis* and *lonnbergi*, no consistent differences can be found.)

P. m. borealis (de Sélys-Longchamps, 1843) (S Fenno-Scandia, Baltic States, E Poland, SW Russia). Medium brown-grey above, dusky grey-white below with only slight brown element. Compared to *lonnbergi*, slightly duskier and more brown-tinged above and on flanks, at least evident when series are compared. Compared to *salicarius*, with which it intergrades on SE side of Baltic Sea, larger, less brown above and less ochre-tinged on flanks. Although representing only one step in a cline stretching from Britain to E Asia, *borealis* is rather constant over a large area. Grades into *lonnbergi* in the north and north-east, in Sweden between 63° and 65°N. – Birds with most pink-brown suffusion on flanks are found in S Norway ('*colletti*') and in SW Russia ('*tischleri*'), but normal variation within *borealis* makes it difficult to uphold such finer local differences as formal subspecies. (Syn. *colletti*; *tischleri*.)

P. m. kleinschmidti (Hellmayr, 1900) (Britain). The brownest race with entire underparts rather uniformly and dark dusky drab-brown (including centre of belly) with rufous-ochre tinge (flanks almost concolorous with upperparts), and cheeks also dusky, only subtly paler than underparts. Crown dark brown-black and has slight gloss, almost like in

P. m. salicarius, Czech Republic, Mar: this C European race is intermediate in colour between breeders in W and N Europe, but subtly closer to the pale birds in the latter area. Often its crown can be seen to have a sooty-brown rather than black colour, something which can be discerned here. Note whitish secondary panel, extensive pure white rear cheeks and all-dark bill typical of Willow Tit. (L. Mráz)

P. m. kleinschmidti, England, Dec: the brownest race with entire underparts rather drab-brown to buff-ochre. Note the confusingly small black bib, like Marsh Tit, especially when plumage still fresh and some black partly concealed by pale tips. Nevertheless, the pale wing panel, the fine all-dark bill, bulky head/nape and quite clean whitish rear cheeks in combination distinguish this Willow Tit. (J. Hawkins)

P. m. kleinschmidti, England, Mar: often has entire flanks and belly drab brown with ochraceous tinge and reduced contrast to duller brown upperparts. Note extensive clean white cheeks all the way to nape-side and edge of mantle and fairly extensive black bib. In this race, the pale wing panel is often buffish as long as plumage is fresh. Best left un-aged when seen at this angle. (L. Corbett)

P. palustris. Small like *rhenanus*, or even subtly smaller. **W** ♂ 57–61 mm (*n* 15, *m* 59.8), ♀ 56–60 mm (*n* 11, *m* 58.5); **T** ♂ 48–55 mm (*n* 15, *m* 50.9), ♀ 46–51 mm (*n* 11, *m* 49.2); **T/W** *m* 84.8. **Wing formula: p1** < p2 16–19 mm; **p2** =10 (85%), =9 or 9/10 (10%), or =8/9 (5%).

P. m. rhenanus (Kleinschmidt, 1900) (S Denmark, W Germany, Low Countries, France except east). Similar to *salicarius*, but underparts are warmer rufous-tawny, especially on flanks, while centre of breast and belly often slightly paler and more whitish. Cheeks very faintly tinged rufous-buff. Upperparts drab-tawny with rufous tinge, more rufous than *salicarius*. Secondary panel tawny-tinged, not very contrasting. Cap nearly black. Small, but bill proportionately strong. Wing somewhat more rounded, primary projection short. **L** 11 cm; **W** ♂ 59–64 mm (*n* 16, *m* 61.4), ♀ 57–61 mm (*n* 12, *m* 61.0); **T** ♂ 50–56 mm (*n* 16, *m* 52.3), ♀ 48.5–53.5 mm (*n* 12, *m* 50.9); **T/W** *m* 85.7; **B** 9.0–11.5 mm (*n* 27, *m* 10.4). **Wing formula: p1** < p2 16–20 mm; **p2** =10 (100%); **p7** < wt 1–3 mm; **p10** < wt 8–10 mm; **s1** < wt 9–11 mm. (Syn. *subrhenanus*.)

○ *P. m. salicarius* (C. L. Brehm, 1831) (E Germany, W Poland, Czech Republic, N Slovakia). Slightly smaller, proportionately shorter-tailed and somewhat browner above and below than *borealis*. Crown rather brownish-tinged. Compared to *montanus* a fraction more brownish above, less grey-tinged (but subtle!). **L** 11–12 cm; **W** ♂ 60–66 mm (*n* 11, *m* 63.0), ♀ 59–64 mm (*n* 13, *m* 61.6); **T** 48.5–57 mm

P. m. montanus, Switzerland, Nov: extensive dull pinkish-tawny tinge below, with restricted whiter central chest and belly among characteristics of this race. Note extensive black bib with mottled lower border (partially due to concealed black bases that, with wear, will form even larger bib in time). Grey-brown mantle and contrasting white panel in secondaries are also typical. Without handling ageing difficult. (L. Sebastiani)

(n 26, m 52.9); **T/W** m 85.1. **Wing formula: p10** < wt 9–11 mm; **s1** < wt 10–12 mm. Rest of measurements and wing formula as *borealis*. (Syn. *natorpi*.)

P. m. montanus (Conrad von Baldenstein, 1827) (mountains of E France, Alps, N & C Italy, C Europe east to W Black Sea, N Balkans). Very similar to *salicarius*, but slightly larger (as large as *borealis*, or even slightly larger) and stronger-billed, while underparts on average more evenly tinged dusky-tawny (less dominant ochre-tawny tinge to flanks contrasting with whiter centre of belly as in *salicarius* and *rhenanus*). Upperparts slightly more greyish-tinged brown than in *salicarius*, but this requires direct comparison of series. Larger than *rhenanus*, and less rufous-tawny on flanks. Crown perhaps very slightly more brownish-tinged than in either. In core area (Alps), song differs from rest of Willow Tit taxa in being a rapid and slightly longer series of straight *düh* notes (rather than the desolate-sounding downwards-inflected and usually fewer *piuh* notes at slower pace than the others). However, intermediate song is known, and *montanus* populations of E Europe and Balkans apparently invariably sing like taxa in N & W Europe. **W** ♂ 65–72 mm

P. m. montanus, Switzerland, Nov: evenly tinged pink-buff underparts, while upperparts have greyish-brown tone. Wing panel weakly developed in this race, and there can even be a vague buffish comma-shaped mark on the rear cheeks, making this race confusingly like Marsh Tit. Presumed ad by fresh remiges and tail, dark and well-rounded primary-coverts, and broad tail-feathers with rounded tips. (L. Sebastiani)

(n 21, m 66.6), ♀ 62–65 mm (n 14, m 63.5); **T** ♂ 54–61 mm (n 21, m 57.3), ♀ 51.5–57 mm (n 14, m 49.9); **T/W** m 85.9; **B** 10.3–11.7 mm (n 14, m 10.9); **Ts** 16.0–17.0 mm (n 13, m 16.3). **Wing formula: p1** < p2 18–21.5 mm; **p2** =10 (64%), or =9 or 9/10 (36%); **p7** < wt 1.5–3 mm; **p10** < wt 9.5–11 mm; **s1** < wt 9.5–11 mm. (Syn. *alpestris*; *alpinus*; *rhodopeus*; *transsylvanicus*.)

TAXONOMIC NOTES See under Marsh Tit regarding the generic arrangement of tits. – It has been proposed that the generic name *Poecile*, derived from Greek, is feminine, meaning that the scientific name should be '*P. montana*' (e.g. Sangster *et al.* 2007), but subsequently this was shown to be incorrect (N. David & M. Gosselin *in litt.*).

REFERENCES Broughton, R. K. (2009) *BB*, 102: 604–616. – Broughton, R. K. *et al.* (2008) *Ring. & Migr.*, 24: 88–94. – Dewolf, P. (1987) *BUBO*, January: 10–11. – Eck, S. & Martens, J. (2006) *Zool. Med. Leiden*, 80: 13–18. – Gill, F. B., Slikas, B. & Sheldon, F. H. (2005) *Auk*, 122: 121–143. – Sangster, G. *et al.* (2005) *Ibis*, 147: 821–826.

P. m. montanus, Switzerland, Jul: when very worn, as here, ad has black bib almost covering entire throat. Note stronger-billed and longer-tailed impression compared to other southern races. These characteristics, plus distinctive vocalisations, have raised questions concerning the taxonomy of European forms. (H. Shirihai)

CASPIAN TIT
Poecile hyrcanus Zarudny & Loudon, 1905

Fr. – Mésange hyrcanienne; Ger. – Kaspimeise
Sp. – Carbonero del Caspio; Swe. – Hyrkanmes

Previously commonly treated as a subspecies of the Sombre Tit, but based on clear differences in genetics, morphology and song, and some differences in habits and habitat choice, it is treated here as a species. It occurs in medium-high zones of broadleaved montane woods, and favours glades and edges with rich undergrowth. Like Willow Tit, but unlike Sombre Tit, excavates its nest hole in decaying trunks. Mainly resident, and restricted to the Talysh region in S Azerbaijan and Elburz Mts in NW Iran.

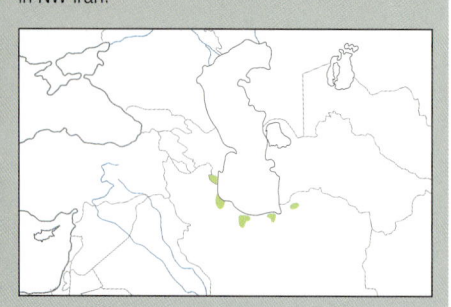

Iran, Apr: here at nest showing (compared to Willow Tit) brown cap and relatively uniform wing lacking a paler panel on secondaries. Also note (compared to Sombre Tit) the relatively broader and squarer whitish cheek patch. Strong feather wear to wings and tail, with pointed tips to latter, suggest 1stY. (B. Anderson)

Iran, May: very similar to Sombre Tit, but songs very different, and also differs in blackish bib that reaches to upper breast, but not to sides, with poorly marked lower border. The whitish cheeks appear wider. Also, cap essentially brown (not blackish), reaching well down hindneck, but rest of upperparts brownish-grey. Underparts off-white lacking strong buff wash. Seemingly evenly fresh wing suggests ad. (C. N. G. Bocos)

bushy habitats. Note in Caspian Tit a more uniform wing lacking obviously pale-tipped greater coverts and pale-edged tertials and secondaries; further its brown cap, broader off-white cheek patch (in Sombre usually quite narrow due to extensive dark bib invading lower cheeks) and ill-defined lower edge to bib (Sombre: nearly invariably sharp lower border). Song and alarm call differ markedly. – Hypothetical identification problems offered by *Marsh* and *Willow Tits*, which are widely allopatric and hardly could straggle to the range of Caspian Tit. Both have much less extensive throat bibs, and have black or blackish cap. – Although *Siberian Tit*

IDENTIFICATION Subtly smaller than Willow and Sombre Tits. Shape much as Willow Tit but is subtly *shorter-tailed* in comparison, hence a rather compact-looking species. *Cap dark brown* (not black), reaching down to upper mantle. Mantle greyish-brown. *Whitish edges to tertials narrow and less prominent* than in both ssp. *anatoliae* and *dubius* of Sombre Tit, soon wearing off, and tips to greater coverts not whitish, more pale brown; wing thus *more uniform dark brown* than in Sombre Tit. *Underparts off-white* lacking any buff hue in worn state, only a limited cream-buff tinge when fresh. *Throat patch blackish*, darker than cap and *reaching well onto upper breast but not far onto sides* (less extensive on sides than in Sombre). *Bib at times poorly marked at lower edge*, often with diffuse border. Song characteristic, similar to that of Willow Tit, thus a repeated mellow piping note rather than the quick scratchy, repetitive series of Sombre.

VOCALISATIONS Song described as a soft, simple short series of repeated notes, *tiu tiu tiu tiu*, quite similar to song of Willow Tit (Loskot 1982, 1987). – The alarm has been described as a single scratchy *chiev* or *chev* (like one note of Sombre Tit song) (Loskot 1982, 1987), quite different from the grating *zri-zri-zri* or rather Great Tit-like scolding *cheche-cheche...* of Sombre Tit.

SIMILAR SPECIES Needs mainly to be separated from *Sombre Tit*, although this has an allopatric range and generally occurs at lower altitudes and in more open maquis and

Iran, Apr: brown cap clearly paler than blackish bib. Note heavy bill and, when excited, peaked crown. Strong wear to primary-coverts may suggest 1stS, but safe ageing would require handling. (B. Anderson)

also has a brown cap (usually paler chocolate-brown, not as dark brown as Caspian) and an extensive blackish bib with ill-defined lower edge, confusion readily avoided due to Siberian having a longer tail, obvious whitish edges to greater coverts and tertials, plus much more orange-brown visible on flanks.

AGEING & SEXING Ageing and sexing not well studied but from few specimens examined appears similar to Sombre Tit. – Moults. Complete post-nuptial moult of ad in summer after breeding. Partial post-juv moult in summer. No pre-nuptial moult. – AUTUMN **Ad** All tail-feathers rather broad, tips more rounded than pointed, and fairly fresh through winter. Primary-coverts have rather broad and rounded tips with neat edges. All greater coverts uniform. **1stCY** Tail-feathers not so broad, tips more pointed than rounded, often abraded from late autumn or early winter. (As always, beware of birds that replace a few central tail-feathers to ad shape.) Primary-coverts have on average slightly more pointed tips and frayed edges. When still in juv plumage, feather structure on nape/upper mantle and on vent is looser. – SPRING **Ad** Same characters as in autumn apply, only are now more difficult to use due to more extensive wear. Generally, though, tips of tail-feathers are less worn than in 2ndCY. **2ndCY** As in autumn, but more abraded. In particular the tips of tail-feathers are more pointed and abraded than in ad. Many birds must be left un-aged.

Iran, Apr: extensive blackish bib reaching well down to upper breast, and bib also being clearly darker than browner cap. Here a bird with very obvious pale fringes to remiges, but wing still more uniform than Sombre Tit. Primaries dark, indicating ad. (W. Müller)

Iran, Aug: although separated geographically from the similar Sombre Tit, birds outside the normal range would require careful scrutiny for Caspian Tit's more uniform wing, browner cap and on average paler underparts. Beware that young Sombre can have a rather similar brownish cap. (A. Yekdaneh)

BIOMETRICS **L** 12.5–13.5 cm; **W** ♂ 67.5–71 mm (n 10, m 69.4), ♀ 64.5–69 mm (n 10, m 64.7); **T** ♂ 55–60 mm (n 10, m 56.7), ♀ 53–56.5 mm (n 10, m 54.4) (after Loskot 1978); **W** ♂♀ 69–72 mm (n 4, m 70.8); **T** ♂♀ 54–58 mm (n 4, m 56.0); **T/W** 77.2; **B** 10.5–12.2 mm (n 4, m 11.2); **BD** 4.7–6.0 mm (n 4, m 5.3); **Ts** 17.0–18.6 mm (n 4, m 17.8) (NHM, AMNH). **Wing formula: p1** > pc 10–13 mm, < p2 20–23 mm; **p2** < wt 11–15 mm, =9/ss; **p3** < wt 1–2 mm; **pp4–6** about equal and longest; **p7** < wt 1.5–3 mm; **p8** < wt 5–7 mm; **p10** < wt 9–12; **s1** < wt 10.5–12.5. Emarg. pp3–7. – Very limited material available for examination.

GEOGRAPHICAL VARIATION & RANGE Monotypic. – SE Transcaucasia (Talysh, S Azerbaijan), N Iran east to E Elburz; mainly resident but local movements noted. (Syn. *talischensis*.)

TAXONOMIC NOTES See under Marsh Tit regarding the generic arrangement of tits. – Due to combination of rather marked morphological differences between *hyrcanus* and closest populations of Sombre Tit subsp. *anatoliae* and *persicus*, different song in *hyrcanus* being more akin to Willow Tit (repetition of plaintive single note), habitat differences and habit of excavating nest hole rather than using existing one, *hyrcanus* has been suggested to constitute a separate species, more closely related to Willow and Songar Tits (Loskot 1977, 1982, 1987). Confirmation of this recommendation was recently offered by complete phylogeny (Johansson *et al.* 2013), placing Caspian Tit in same clade as Willow and Marsh Tits with strong support, but away from Sombre Tit.

REFERENCES Eck, S. (1980) *Zool. Abh. Staatl. Mus. Tierkunde, Dresden*, 36: 135–219. – Johansson, U. *et al.* (2013) *Mol. Phyl. & Evol.*, 69: 852–860. – Loskot, V. M. (1977) *Vestnik Ornithologii*, 4: 28–31. – Loskot, V. M. (1978) *Proc. Zool. Inst. Acad. Sci. USSR*, 76: 46–60. – Loskot, V. M. (1982) In: Gavrilov & Potapov, *Orn. Studies in USSR*, Moscow, pp 24–30. – Loskot, V. M. (1987) In: Stresemann *et al.*, *Atlas der Verbr. palaearkt. Vögel*, vol. 14.

Iran, Apr: here the clearly brown cap (rather than sooty brown-grey as in Sombre Tit) is very obvious, being warmer and paler than extensive blackish bib. Wing rather uniform, more so than in Sombre Tit. Plumage fresh and seemingly of one generation, indicating ad. (S. Rooke)

SIBERIAN TIT
Poecile cinctus (Boddaert, 1783)

Fr. – Mésange lapone; Ger. – Lapplandmeise
Sp. – Carbonero lapón; Swe. – Lappmes

The Siberian Tit is one of a few resident birds of the northern taiga, which for most of us requires a long trip to be seen; it rarely moves south to meet you. How such a small bird can cope with the severe cold during the long Arctic winter nights is difficult to understand, but a deliberate lowering of its body temperature is apparently a vital part of its strategy. Nowhere common, it is generally found in pristine boreal forests, preferably with much pine rich in lichens and some broadleaved trees mixed in, and—in summer—the nearness of running water.

P. c. lapponicus, presumed ad, Finland, Jun: has proportionately longish tail and diagnostic combination of orange-brown flanks, large black bib, dark brown-grey cap, rufous-tinged drab-brown mantle and back, and quite prominent whitish secondary panel. Tail-feathers and primary-coverts rather broad, tips rounded and fairly fresh, and greater coverts uniform, indicating an ad. (D. Occhiato)

IDENTIFICATION Best field marks are rusty or *orange-brown flanks* in combination with *very large black bib*, almost connecting with dark shoulders at sides of breast, supplemented by *dark brown-grey cap*. In a closer look, two more features add to the distinctive appearance of this species, the *warm drab-brown mantle and back* and the *prominent whitish panel on the folded secondaries*, the latter more obvious and contrasting than that in Willow Tit. Compared to Willow, Siberian is a trifle larger and has a proportionately *longer tail*. Subtler differences, still useful to know, are the slightly larger-looking head and the rich, 'fluffy' plumage (needed in the cold boreal winter). When agitated can raise crown-feathers to show hint of a crest. Brown-grey cap tends to be darker on sides, near-blackish on lores and behind eye. Behaviour very much like its congeners, seemingly restlessly active, feeding both in trees and, rather frequently, on the ground. Flight a bit jerky with bursts of wingbeats. Moves around in winter in small flocks, at times mixing with other species, mostly Willow Tits. Often quite tame.

VOCALISATIONS Like Willow Tit, has more than one song type. Although a more detailed study seems to be lacking, the most common song appears to be a grinding, buzzing, repetitive *chi-ürr chi-ürr chi-ürr…*, a little reminiscent of Sombre Tit or some odd utterings of Great Tit, or a simpler, *errr errr errr…* Another song type is a rather Marsh Tit-like repetition of one slightly harsh note, *che che che che…* Whether a melodic short outburst of notes, *zi-zi-dyetvuy* or similar, also has song function is more uncertain, but it could be a counterpart to the brief warble-type song of both Willow and Marsh Tits. – Calls basically resemble those of Willow Tit (but differ clearly from Marsh Tit). Commonest call is a grating, repeated, drawn-out and stressed, nasal *taah* or more often slightly bent at the end, *taa-eh*, a few in series and often like in Willow Tit preceded by 1–2 high-pitched, thin notes, *zi zi taa-eh taa-eh taa-eh*. Conversational fine *see* or *zit* are often uttered when feeding, and sometimes a harder *chik*, almost like knocking stones together, which can recall crossbill calls (Harrap 1996). Contact call is usually a fine, more drawn-out *tsih*.

SIMILAR SPECIES Due to its northerly range in taiga zone, needs only to be separated from *Willow Tit*, but this has black cap (very dark sooty-brown in juvenile), less extensive bib, pale grey mantle and back, and in N Europe greyish-white flanks, lacking the rusty tones of Siberian. Note also longer tail in Siberian, larger-looking head and fluffier plumage. – Further east, outside the range covered here, *Songar Tit* (*Poecile songarus*, extralimital, not treated) is a possible confusion species, but the two do not come into contact, and Songar Tit has paler flanks only faintly tinged ochre-buff, lacking the stronger rusty-brown tones of Siberian. Songar Tit also has smaller bib.

AGEING & SEXING (*lapponicus*) Ages differ subtly, young birds often separable during 1stY, but ageing requires close views or handling. Sexes alike in plumage; ♂ is very slightly larger. – Moults. Complete post-nuptial moult of ad in summer after breeding (late Jun–early Sep). Partial post-juv moult in summer (late Jul–mid Sep) does not involve any primary-coverts, primaries or secondaries, but sometimes odd tertials and central 1–2 pairs of tail-feathers. No pre-nuptial moult. – **AUTUMN Ad** All tail-feathers rather broad, tips more rounded than pointed, and fairly fresh through winter. Primary-coverts have rather broad and rounded tips with neat edges. All greater coverts uniform. **1stCY** Tail-feathers not so broad, tips more pointed than rounded, often abraded from late autumn or early winter, but beware of birds that replace a few central tail-feathers to ad shape. Some birds have slightly more rounded and ad-like shape to tail-feathers, and any intermediate bird should be left un-aged. Primary-coverts have on average slightly more pointed tips and frayed edges. A few birds show contrast among greater coverts, a few outer being unmoulted and slightly paler towards the edge. – **SPRING Ad** Same characters as in autumn apply, only

P. c. lapponicus, Finland, Mar: the large black bib covers entire throat and almost reaches the wing-bend. Close up, the brown-grey cap tends to appear darker at sides, near-blackish on lores and behind eye, probably largely the effects of light. Typical obvious contrast between cinnamon flanks and white breast and centre of belly. (H. Harrop)

are now more difficult to use due to more extensive wear. Generally, though, tips of tail-feathers are less worn than in 2ndCY. **2ndCY** As in autumn, but more abraded. In particular the tips of tail-feathers are more pointed and abraded than in ad. Many birds must be left un-aged.

BIOMETRICS (*lapponicus*) **L** 13–14 cm; **W** ♂ 65.5–72 mm (*n* 25, *m* 68.6), ♀ 64–71 mm (*n* 20, *m* 66.6); **T** ♂ 62–69 mm (*n* 25, *m* 65.2), ♀ 58–70 mm (*n* 20, *m* 63.2); **T/W** *m* 95.1; **TG** 6–13 mm (*n* 39, *m* 8.4); **B** 10.3–12.0 mm (*n* 43, *m* 11.2); **BD** 3.5–4.6 mm (*n* 40, *m* 4.1); **Ts** 16.0–17.6 mm (*n* 39, *m* 16.7). **Wing formula: p1** > pc 9–13 mm, < p2 19–22 mm; **p2** = wt 10–13.5 mm, =8/10 (75%), or =10 or 10/ss (25%); **p3** < wt 2–5 mm; **pp4–6** about equal and longest; **p7** < wt 1–5 mm; **p8** < wt 3–8 mm; **p10** < wt 11–14 mm; **s1** < wt 11–14.5 mm. Emarg. pp3–6, often less clear also on p7.

GEOGRAPHICAL VARIATION & RANGE Slight and clinal variation, plumage gradually becoming very slightly paler towards east and in Siberia towards the north. Resident. All populations sedentary. At least one more extralimital subspecies claimed from East Asia.

○ *P. c. lapponicus* (Lundahl, 1848) (N Fenno-Scandia, N Russia to near Urals). Treated above. On average slightly darker cap and warmer cinnamon mantle than in *cinctus*, but somewhat variable, and a certain overlap of colours with this race.

P. c. cinctus (Boddaert, 1783) (Ural region, Siberian taiga to Baikal area in SE). Very similar to *lapponicus* but is subtly paler and greyer above, with often slightly less clear contrast between crown and mantle (in particular mantle averages paler drab brown, less cinnamon), and less extensively rufous-ochre on flanks. On average proportionately slightly larger and longer-tailed. Many are very similar, and single birds not always separable even with series at hand for comparison. **W** ♂ 64–71 mm (*n* 14, *m* 68.7), ♀ 66–70 mm (*n* 9, *m* 68.2); **T** ♂ 63–70 mm (*n* 14, *m* 67.0), ♀ 62–70 mm (*n* 9, *m* 66.2); **T/W** *m* 97.2; **Ts** 16.0–18.0 mm (*n* 23, *m* 16.9). **Wing formula: p2** =9/10 (70%), =9 (15%), or =10 (15%). Emarg. pp3–7. (Syn. *kolymensis*; *obtectus*.) – Grades into *lapponicus* on western side of Urals.

TAXONOMIC NOTE See under Marsh Tit regarding the generic arrangement of tits.

REFERENCES Eck, S. & Martens, J. (2006) *Zool. Med. Leiden*, 80: 20. – Gill, F. B., Slikas, B. & Sheldon, F. H. (2005) *Auk*, 122: 121–143. – Taavetti, H. (2007) *Alula*, 13: 162–164.

P. c. lapponicus, Finland, Mar: unlike Willow Tit, the cap is sooty-brown, mantle sepia-brown or vinaceous-drab (not greyish) and flanks rusty-toned (greyish-white in N European Willow). Bib considerably larger than Willow Tit, but at this angle it can appear superficially smaller. Without handling age not certain. (H. Taavetti)

P. c. lapponicus, Finland, Mar: drab-brown mantle and back, whitish secondary panel, dark brown-grey cap, large blackish bib and orange-brown flanks. The primary-coverts (with pointed tips) and outer greater coverts are apparently juv, and worn tail-tip in Mar also suggests a young bird, but age not certain without handling. (H. Harrop)

P. c. lapponicus, Finland, Jun: often larger-looking head and seemingly richer and fluffier plumage are characteristics of this tit. Note that the orange-brown flanks can be less extensive in some birds, and the lower border to bib rather ragged. Without handling, ageing is not possible. (D. Occhiato)

P. c. lapponicus, juv, Finland, Jun: fluffy and dull plumage still lacking orange-brown on flanks and warmer vinaceous-drab colour on mantle, but has grey-brown cap and sooty bib. Tail and wings still not fully grown, while yellow gape is still prominent on this recently fledged young. (T. Muukkonen)

SOMBRE TIT
Poecile lugubris (Temminck, 1820)

Fr. – Mésange lugubre; Ger. – Trauermeise
Sp. – Carbonero lúgubre; Swe. – Balkanmes

'Sombre' is a translation of the scientific name 'lugubris', and although it may sound a little depressing it merely refers to the bird's dark and dull plumage. The Sombre Tit is a large and rather 'sluggish' tit, with its range restricted to SE Europe and W Asia. It lives mainly in open mountainous oak forests, is rather retiring and even shy, and nowhere common. It can also be found in olive or almond groves or taller maquis, juniper stands, etc. Resident.

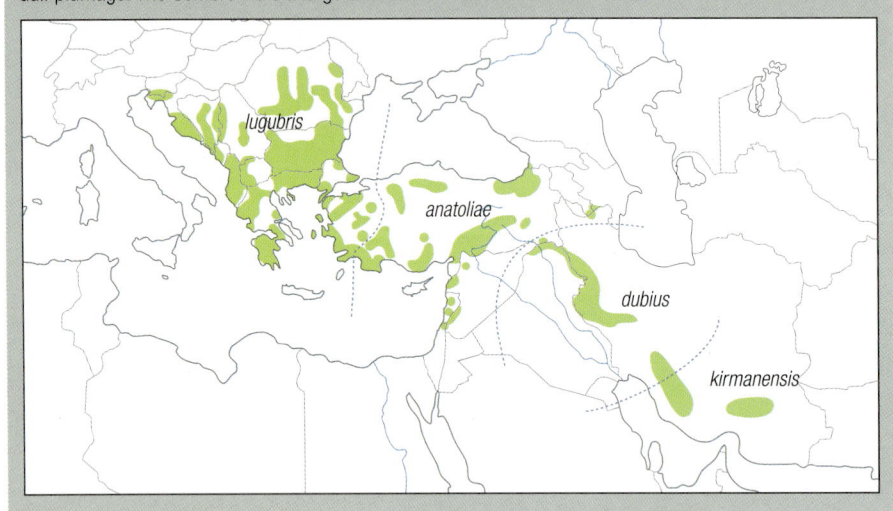

IDENTIFICATION *Large*, almost the size of a Great Tit, but with even *stronger bill* and proportionately slightly longer tail. Superficially like a large Willow or Marsh Tit, but differs in several respects, and in appearance is actually closer to Siberian Tit. The *cap is dark dull brown or sooty-black and rather extensive* (reaching just below eye), and the on average somewhat blacker *throat bib is very large* (much larger than in Marsh or Willow Tits), leaving quite *narrow white cheek patch* (narrowly wedge-shaped). Mantle and back are greyish-tinged tawny-brown in Europe but paler buff-grey further east. There is often a fairly obvious but *narrow pale panel* or line *on the wing* formed by light edges to the longest tertial and inner secondaries. In fresh plumage greater coverts are tipped paler forming a *hint of a wing-bar*. Underparts are dusky-white with a variable amount of pinkish-brown or drab brown hue on flanks. Dark grey feet are strong. Some of the calls and song variations are rather strident and buzzing in tone, different from other tits and helping to identify the species. Often sings from perch in tree-top.

VOCALISATIONS Like many other tits appears to have more than one song type, or at least displays much individual variation in the details. What appears to be the commonest song is a grating, simple series in Marsh Tit fashion, *chriv-chriv-chriv-chriv-...* (or rendered *bzz-bzz-bzz-bzz-...*). Another song type is similar but comprises multisyllabic notes, *chi-zre chi-zre chi-zre...* or *zrizri-zee zrizri-zee zrizri-zee...* Another song variation is two *che-ehv che-ehv*, then a brief pause, then two more such disyllabic notes, etc. – Calls variable but generally strong and a little coarse. Often heard is a series of fine notes, *si-si-si* (or more notes; can recall Long-tailed or Blue Tits), frequently combined with a sparrow-like, quarrelling or scolding *kerr'r'r'r'r* (a little like Blue or Marsh Tits), which is used in alarm. A variation is *zri-zri-zerrr*. Other calls are a sharp, grating *zri-zri-zri* and a rather Great Tit-like scolding *chechecheche...* Less often heard is a vaguely Willow Tit-like *zi-zi-zi-cheeh*, where the last harsh note is emphasized and a little drawn-out.

SIMILAR SPECIES Must first of all be separated from its close relative and similar-looking *Caspian Tit* (sometimes treated as a subspecies of Sombre Tit). However, separation should not present too much of a problem since they are allopatric (Sombre Tit missing in Azerbaijan and N Iran) and as Sombre Tit has a somewhat longer tail, more contrasting pale edges in wing (unless plumage heavily abraded), a more blackish cap, usually a better-defined throat bib and a warmer pink-buff flush on flanks. Song and some calls differ too. – *Marsh* and *Willow Tits* partly overlap with it in range, but both are smaller and have much less extensive throat bibs, leaving a fairly large pale cheek patch between black bib and a comparatively small cap. – As mentioned above, there is some resemblance to a dull *Siberian Tit*, but apart from being a resident species separated from Sombre by at least 2000 km, Siberian has usually much more orange-brown visible on flanks, and cap and bib are not so extensive as to create the same narrow pale cheek patch as in Sombre.

AGEING & SEXING (*lugubris*) Ages differ subtly, young birds often separable during 1stY, but ageing requires close views or handling. Sexes alike or very similar in plumage; the claimed sexual difference (BWP) in that ♂ has blacker cap and bib, ♀ browner, exist only as a very slight tendency with extensive overlap and is very difficult to use in practice except for a few extremes, or as an average difference when longer series are compared. ♂ is very slightly larger. – Moults. Complete post-nuptial moult of ad in late spring or summer after breeding (late May–early Sep). Partial post-juv moult in summer (Jul–mid Sep) does not involve any primary-coverts, primaries or secondaries but sometimes odd tertials and central 1–2 pairs of tail-feathers. No pre-nuptial moult. – **AUTUMN Ad** All tail-feathers rather broad, tips more rounded than pointed, and fairly fresh through winter. Primary-coverts have rather broad and rounded tips with neat edges. All greater coverts uniform. **1stCY** Tail-feathers not so broad, tips more pointed than rounded, often abraded from late autumn or early winter, but beware birds that replace a few

P. l. lugubris, Greece, Apr: a large, robust tit only superficially similar to distinctly smaller Willow or Marsh Tits. Note slightly browner cap, very large bib, due to this narrow whitish cheek patch, which is usually best clue to species apart from voice. Neck- and breast-sides and upper flanks washed pale brown. Upperparts dull grey-brown. Ageing without close inspection of wing difficult. (D. Tipling)

P. l. lugubris, 1stS, Bulgaria, Feb: in some birds, particularly of ssp. *lugubris*, cap and bib are clearly brown-tinged, not jet-black. In fresh plumage, the whitish fringes to greater coverts create a rather large wing panel. Moult limits among tertials and greater coverts, with retained juv feathers, worn and brownish juv primary-coverts, and worn tail-tip, identify this as 1stY. (M. Mendi)

P. l. lugubris, Greece, Aug: brownish-tinged cap, wide bib being dull brown-grey or blackish, underparts with more extensively buffish-brown wash. Fluffy feathers on flanks could suggest a juv moulting to 1stW. (N. Kontonicolas)

P. l. anatoliae, presumed 1stS, Turkey, Jun: a heavily worn bird in summer with very large black throat bib, reaching upper breast. Such strong wear of tail-feathers indicate that these are juv and the bird 1stS, but safe ageing requires full view also of wings. (H. Shirihai)

central tail-feathers to ad shape. Some birds have slightly more rounded and ad-like shape to tail-feathers, and any intermediate bird should be left un-aged unless other criteria can be employed. Primary-coverts have on average slightly more pointed tips and frayed edges. When still in juv plumage, cap and bib are browner and feather structure on nape/upper mantle and on vent is looser. – SPRING **Ad** Same characters as in autumn apply, only are now more difficult to use due to more extensive wear. Generally, though, tips of tail-feathers are less worn than in 2ndCY. **2ndCY** As in autumn, but more abraded. In particular the tips of tail-feathers are more pointed and abraded than in ad. Still, many birds must be left un-aged.

BIOMETRICS (*lugubris*) **L** 13.5–14.5 cm; **W** ♂ 70–78.5 mm (*n* 30, *m* 74.5), ♀ 70–77 mm (*n* 14, *m* 72.6); **T** 59–67 mm (*n* 43, *m* 63.2); **T/W** *m* 85.5; **B** 11.4–13.2 mm (*n* 43, *m* 12.3); **BD** 4.9–6.0 mm (*n* 37, *m* 5.4); **Ts** 18.0–22.0 mm (*n* 43, *m* 19.6). Wing formula: **p1** > pc 11–14 mm, < p2 20.5–26 mm; **p2** < wt 9–14 mm, =9/10 or 10 (74%), =9 (13%), or <10 (13%); **p3** < wt 1–4 mm; **pp4–6** about equal and longest; **p7** < wt 1–5 mm; **p8** < wt 4.5–8 mm; **p10** < wt 8–13.5 mm; **s1** < wt 10–14 mm. Emarg. pp3–7.

P. l. anatoliae, 1stS, Turkey, May: note typically quite narrow, wedge-shaped white cheeks between extensive sooty-black cap (reaching just below eye and onto hindneck) and very large bib (extending quite far onto upper breast). Characteristic is also the fairly obvious but narrow whitish secondary panel. Underparts dusky-white with usually limited pinkish-brown hue on flanks. Pointed and bleached primary-coverts and moult limits in greater coverts give the age. (G. Reszeter)

P. l. anatoliae, Turkey, Apr: a characteristic view of the species in its favoured habitat of dry hillsides with crops, such as pistachio, where it is often a highly characteristic avifaunal component. Note the distinctive, narrow, wedge-shaped white cheeks, large black cap, bib, stony grey-brown upperparts and dull white underparts. Both legs and bill are strong. (R. Martin)

GEOGRAPHICAL VARIATION & RANGE Mainly slight and clinal variation, although south-eastern populations are clearly paler. Resident.

P. l. lugubris (Temminck, 1820) (Slovenia, Croatia and Balkans south to Greece, Romania, Thrace and W Asian Turkey). Treated above. Cap dark blackish-brown, often appearing dull brown rather than blackish in strong light, but some are more blackish; cap does not reach upper mantle. Bib same or a little blacker, restricted to throat, often slightly irregularly outlined (though many are neat enough). Mantle greyish-tinged tawny brown. Underparts off-white with faint or obvious buff hue, on average more buff than in *anatoliae*. – Breeders in S & C Greece ('*lugens*') tend to be subtly more brown-capped, some having also a more washed-out and brown-grey bib, often a little warmer ochre-tinged mantle and more drab buff-grey, less clean whitish underparts. Still, breeders of S Balkans and Greece are variable, and plenty of darker birds inseparable from *lugubris* occur also, thus *lugens* is not recognised. (Syn. *lugens*; *splendens*.)

○ *P. l. anatoliae* (Hartert, 1905) (C & E Turkey including Taurus but not Izmir area in the west; W Caucasus, SW

Transcaucasia, NW Middle East). Very similar to *lugubris*. Cap and bib black or blackish, slightly blacker than in *lugubris*, blackish bib large and well defined, often just reaching upper breast (restricted to throat in *lugubris*). Upperparts on average slightly greyer, less brown than *lugubris*. Underparts off-white, usually with very little buff hue, but subtle difference from *lugubris*. Subtle race, but warranted when series are compared. **W** 69–75.5 mm (*n* 11, *m* 72.4); **T** 57.5–63 mm (*n* 11, *m* 60.9); **T/W** *m* 84.2; **Ts** 18.0–20.2 mm (*n* 11, *m* 19.4). Data for live birds from Mt Hermon, Israel (J. Langer & A. Rochman *in litt.*): **W** 68–76 mm (*n* 29, *m* 72.0); **T** 57–65 mm (*n* 12, *m* 61.2); **B** 10.8–14.4 mm (*n* 11, *m* 12.4); **Ts** 19.0–21.0 mm (*n* 6, *m* 19.6).

P. l. dubius Hellmayr, 1901 (SW Iran, NE Iraq; border in west towards *lugubris* not clear). Compared to *lugubris* smaller, proportionately shorter-tailed, with much paler and cleaner colours, extensive cap (reaching upper mantle) and neatly outlined bib jet-black, rest of upperparts pale sandy buff-grey, greater coverts tipped and tertials edged whitish, forming prominent pale line on closed wing. Underparts pale off-white with warm, cream-buff suffusion, bleaching whiter. Extension of bib on sides leaves only rather narrow off-white area on sides of head (narrower, more pointed wedge

P. l. anatoliae, juv, Israel, Jul: fluffy and duller plumage, with browner cap and bib, are typical of juv. (A. Ben Dov)

shape to whitish cheeks even than in European populations). **L** 12.5–13 cm; **W** ♂ 72–78 mm (*n* 14, *m* 74.5), ♀ 69–74 mm (*n* 12, *m* 72.3); **T** 54–64 mm (*n* 26, *m* 59.3); **T/W** *m* 80.7; **Ts** 18.4–20.3 mm (*n* 24, *m* 19.3). (Syn. *persicus*.)

P. l. kirmanensis (Koelz, 1950) (S Iran). Resembles *dubius* in extensive jet-black cap and bib contrasting with drab brown upperparts, but is consistently somewhat darker brown on mantle/back, albeit still a little paler than *lugubris* and *anatoliae*. **W** 74.5–78 mm (*n* 10, *m* 76.2); **T** 59–65.5 mm (*n* 10, *m* 62.5); **T/W** *m* 81.9; **B** 11.7–13.3 mm (*n* 10, *m* 12.7); **BD** 4.9–5.7 mm (*n* 10, *m* 5.3); **Ts** 18.5–20.5 mm (*n* 10, *m* 19.6).

TAXONOMIC NOTES See under Marsh Tit regarding the generic arrangement of tits. – As detailed under Caspian Tit, that taxon is split from Sombre Tit as separate species. (Loskot 1977, 1982, 1987, Harrap 1996, Johansson *et al.* 2013).

REFERENCES Eck, S. (1980) *Zool. Abh. Staatl. Mus. Tierkunde, Dresden*, 36: 135–219. – Johansson, U. *et al.* (2013) *Mol. Phyl. & Evol.*, 69: 852–860. – Loskot, V. M. (1977) *Vestnik Ornithologii*, 4: 28–31. – Loskot, V. M. (1978) *Proc. Zool. Inst. Acad. Sci. USSR*, 76: 46–60. – Loskot, V. M. (1982) In: Gavrilov & Potapov, *Orn. Studies in USSR*, Moscow, pp 24–30. – Loskot, V. M. (1987) In: Stresemann *et al.*, *Atlas der Verbr. palaearkt. Vögel*, vol. 14.

P. l. dubius, ad, SW Iran, Sep: this race, has much paler and cleaner plumage, large cap, sharply delimited blackish bib, rest of upperparts pale sandy buff-grey, innerwing edged whitish, forming prominent pale line (visible here only on recently replaced inner secondaries/tertials) and off-white underparts tinged cream-buff. Ad in post-nuptial moult. (M. Nemati)

CRESTED TIT
Lophophanes cristatus (L., 1758)

Fr. – Mésange huppée; Ger. – Haubenmeise
Sp. – Herrerillo capuchino; Swe. – Tofsmes

A tit nearly always closely linked to coniferous forests, and with a highly 'personal' look, being the only small passerine in the treated region with a pointed crest (Crested Lark, Thekla's Lark and Waxwing being medium-sized rather than small). Like Willow Tit, the ♀ often excavates the nest hole in a rotten tree-stump, but suitable nest boxes will suffice. Usually searches the lower branches of pines and spruces for larvae or spiders, but will also spend time feeding on the forest floor. Truly a resident bird that rarely leaves the safety of its breeding wood. Only locally is it found in deciduous forests in Europe.

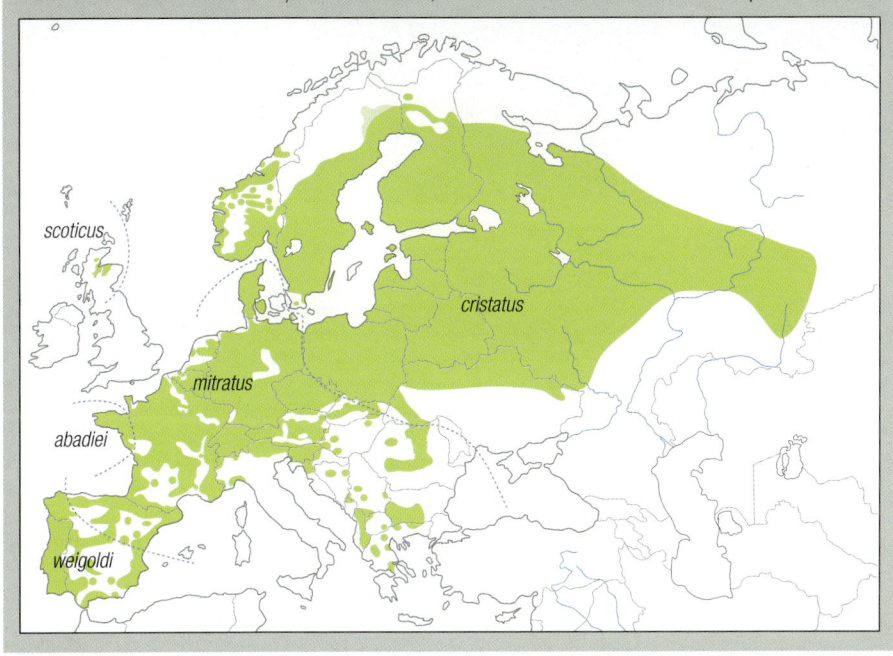

IDENTIFICATION Highly characteristic and not likely to be confused with any other bird within the region once seen well. Often noticed by its characteristic call. Small and rather compact, moving with small hops in canopy, often flicking wings and tail. Flight bouncing and quick, most recalling Marsh and Blue Tits. *Uniformly brown above* and *dirty-white below, often with ochre tinge*. Two features are striking, (i) the *pointed crest* and (ii) the *black-and-white-banded head pattern*. The crest is 'checkered' in black and white, and the *head is dirty-white with a small black bib* and a *black stripe running through eye and at an angle back below the ear-coverts*. The upper edge of the brown mantle is neatly but narrowly outlined in black, reaching the corners of throat bib. Bill black, legs grey.

VOCALISATIONS The song consists of repeated calls on two pitches given alternatively, or by calls interfoliated by high-pitched whistling notes: *burrurrit-seeseesit-burrurrit-seeseesit-burrurrit-seeseesit-*..., etc. – Normal call is the cheery trill *burrurrit*, which can recall one of the flight calls of Snow Bunting. When feeding, this call is mixed with fine, sharp *zit* calls, difficult to separate by themselves from those of other feeding tits.

SIMILAR SPECIES Should be distinct in all plumages once seen reasonably well!

AGEING & SEXING (*cristatus*) Ages very similar in autumn and generally inseparable in spring. Sexes alike in plumage, but size difference to some extent helpful once population known and studied. (Plumage differences mentioned in *BWP*, and repeated by Harrap 1996, have not been possible to confirm using museum material. The only and very slight difference that can be established is the on average darker, blackish, centres of crest-feathers with more contrasting dusky-white edges in ♂♂, whereas ♀♀ are more dull brown-grey, but much overlap and of limited value.) – Moults. Complete post-nuptial moult of ad in summer after breeding (mainly mid Jun–Sep). Partial post-juv moult in summer somewhat variable in extent depending on population, but generally does not involve any primary-coverts, primaries or secondaries. No pre-nuptial moult. – **AUTUMN Ad** Tail-feathers rounded and rather fresh through winter. All greater coverts moulted and uniformly fresh. Iris reddish-brown or hazel-brown. **1stW** Tail-feathers rather pointed, feathers in general slightly narrower, becoming worn towards end of autumn or in winter. Sometimes a visible moult contrast between inner new and a few unmoulted outer greater coverts. Iris greyish-brown or dull brown. **Juv** Differs in having slightly duller head pattern with shorter crest diffusely spotted in grey and buff-white, rather than neatly 'checkered' in black and white, bib is matt sooty-grey, eye-stripe narrower and less black, and black rim to mantle is lacking or narrow and broken.

BIOMETRICS (*cristatus*) **L** 11.5–12.5 cm; **W** ♂ 61.5–67.5 mm (*n* 42, *m* 64.7), ♀ 59.5–64 mm (*n* 18, *m* 61.8); **T** ♂ 48–55 mm (*n* 42, *m* 51.0), ♀ 46–51 mm (*n* 18, *m* 49.5); **T/W** *m* 79.5; **B** 10.0–11.5 mm (*n* 35, *m* 10.9); **BD** 3.2–4.0 mm (*n* 13, *m* 3.6); **Ts** 16.9–19.0 mm (*n* 35, *m* 17.8). **Wing formula: p1** > pc 7.5–12 mm, < p2 18–21.5 mm; **p2** < wt 8–14 mm, =8/10 (65%), =10 or <10 (25%), or =7/8 or 8 (10%); **p3** < wt 1–3 mm; **pp4–5** about equal and longest; **p6** < wt 0.5–2 mm; **p7** < wt 2.5–4.5 mm; **p8** <

L. c. cristatus, Finland, Jan: unmistakable given unique head pattern and pointed crest. N European race *cristatus* characterised by dull olive-brown upperparts and relatively limited buff-brown wash below. In head-on view, checkered crest and triangular black bib impart hourglass pattern. Ageing without closer inspection of wing difficult (this applies to most images). (M. Varesvuo)

L. c. cristatus, Finland, Oct: the black eye-stripe is connected with a black line below ear-coverts, while the upper mantle is narrowly outlined in black that connects with the bib. The crest can be raised almost at right angles. Some ssp. *cristatus*, especially when fresh, are warmer below. (M. Varesvuo)

L. c. scoticus, Scotland, Mar: this race is rather darker, more greyish-brown above (less tinged rufous-brown than other races), especially noticeable compared to paler *cristatus*. Here with typically steeply raised crest. (N. Blake)

L. c. scoticus, Scotland, Aug: a young bird in advanced post-juv moult, but head still essentially juv-like with weaker dark markings at sides, pale areas dirty white and crest short with only dotty white basal marks. (S. Round)

wt 6–9 mm; **p10** < wt 10.5–13 mm; **s1** < wt 11–13.5 mm. Emarg. pp3–7 (but emarg. partly rather poorly marked).

GEOGRAPHICAL VARIATION & RANGE Moderate and clinal variation, mainly affecting plumage colours. Sedentary.

L. c. cristatus (L., 1758) (N & E Europe, W Siberia, south to Czech Republic & N Ukraine). Treated above in the main account. Medium brown-grey above (with hint of olive-grey cast), dusky grey-white below with only very slight brown element. – Birds in E Russia and W Siberia ('*bashkirikus*') tend to be greyer and whiter, less brown, but when series compared difference very slight and insufficient for separation. (Syn. *bashkirikus*.)

L. c. scoticus Pražák, 1898 (NC Scotland). Small and rather dark in general. Slightly greyer brown above, not as rufous-brown as *mitratus* and *abadiei*. Wing rounded as in *mitratus* or on average even slightly more blunt. Sexes nearly the same size. **W** 59–65 mm (*n* 24, *m* 62.0); **T** 45–51 mm (*n* 24, *m* 48.4); **T/W** *m* 78.1. **Wing formula: p2** =9, 9/10 or 10 (87%), <10 (13%).

L. c. mitratus (C. L. Brehm, 1831) (W & C Europe from Denmark and Germany east to Romania, Austria and south to Balkans, west to N Spain and much of France). Warm tawny-brown above, dusky grey-white below with a variable amount of rufous-buff tinge on flanks and sometimes belly. Pale pattern of head rather dull, not as clean cream-white as in *cristatus*. Slightly larger and longer-legged than *cristatus* with fractionally rounder wing. A rather variable race, some being very similar to *abadiei* (see below), others to *cristatus*. **W** ♂ 62–70 mm (*n* 27, *m* 65.8), ♀ 60–66 mm (*n* 15, *m* 62.2); **T** ♂ 45–56 mm (*n* 27, *m* 50.9), ♀ 45–50 mm (*n* 15, *m* 47.5); **T/W** *m* 77.0; **Ts** 17.4–19.5 mm (*n* 40, *m* 18.4). **Wing formula: p2** =8/10 or 10 (88%), <10 (12%). – Birds of Romania, Austria, SE Germany, Switzerland and N Balkans often included in *cristatus*, but flanks in breeders from these areas usually saturated rufous-buff, and much closer to *mitratus*, in which they are included here. – Birds of Greece and W Bulgaria ('*bureschi*') tend to be less rufous-buff, more grey-brown like *cristatus*, but size and the more dusky head pattern conform to *mitratus*, thus included in this here. (Syn. *albifrons*; *alpinus*; *brunnescens*; *bureschi*; *heimi*; *poeninus*.)

○ *L. c. abadiei* (Jouard, 1929) (Brittany, perhaps also lower Loire). Very similar to *mitratus*, and they intergrade. When series are compared, underparts in *abadiei* are frequently both cleaner (less dusky) and more vividly rufous-tinged, and *abadiei* is slightly smaller. However, identical or near-identical birds occur in many parts of the range of *mitratus* (e.g. Calvados, Loire-et-Cher, Landes, Vosges, French Alps, Pyrenees), and not all birds from Brittany are that strongly rufous-tinged. Moreover, the holotype (examined; MNHN) is less typical, more like average *mitratus*. A subtle race, to say the least. **W** ♂ 61–65 mm (*n* 13, *m* 63.1), ♀ 57.5–62 mm (*n* 10, *m* 60.2); **T** ♂ 45–51 mm (*n* 13, *m* 47.8), ♀ 42–47.5 mm (*n* 10, *m* 45.7); **T/W** *m* 72.5. Wing formula as in *mitratus*.

○ *L. c. weigoldi* (Tratz, 1914) (Portugal, Galicia, C & S Spain south of León & Catalonia). Differs from *mitratus* in being slightly smaller and rather dark overall, more like *scoticus*. Still, a fraction paler brown above than *scoticus* (but hardly greyer as claimed), and whiter on belly. Very subtle race. Sexes of fairly similar size. **W** 56–66 mm (*n* 26, *m* 61.6); **T** 43–50 mm (*n* 26, *m* 47.0); **T/W** *m* 76.4. **Wing formula: p2** < wt 7–11.5 mm, =8/10 (56%), =10 or <10 (44%); **p10** < wt 9–11.5 mm; **s1** < wt 9–12.5 mm.

TAXONOMIC NOTE See under Marsh Tit regarding the generic arrangement of tits.

L. c. mitratus, Italy, Sep: the race widely distributed in W, C & S Europe is best characterised by its warmer and richer tawny-brown upperparts, with quite extensive rufous-buff on flanks and sometimes belly. The head pattern and black-and-white crest tend to be less neatly patterned than in *cristatus*, mainly because more black is exposed. (D. Occhiato)

L. c. weigoldi, Spain, May: the Iberian race differs only fractionally from *mitratus* in being darker than that, still it is slightly paler brown above than ssp. *scoticus*, the darkest race. The black-and-white crest is often raised to point even slightly forward. (C. N. G. Bocos)

COAL TIT
Periparus ater (L., 1758)

Fr. – Mésange noire; Ger. – Tannenmeise
Sp. – Carbonero garrapinos; Swe. – Svartmes

This, the smallest European tit, looks superficially like a smaller version of the Great Tit, being only larger-headed, more compact and generally lacks yellow below. It is a bird linked to conifers, both pine and spruce, and spends much time high up in these excavating seeds from cones or searching for insects and spiders. It is a less frequent guest at feeders near houses and therefore not as well known as some other tits. Resident, but like all species living partly on conifer seeds makes irruptive autumn movements in some years, often towards the south and south-west.

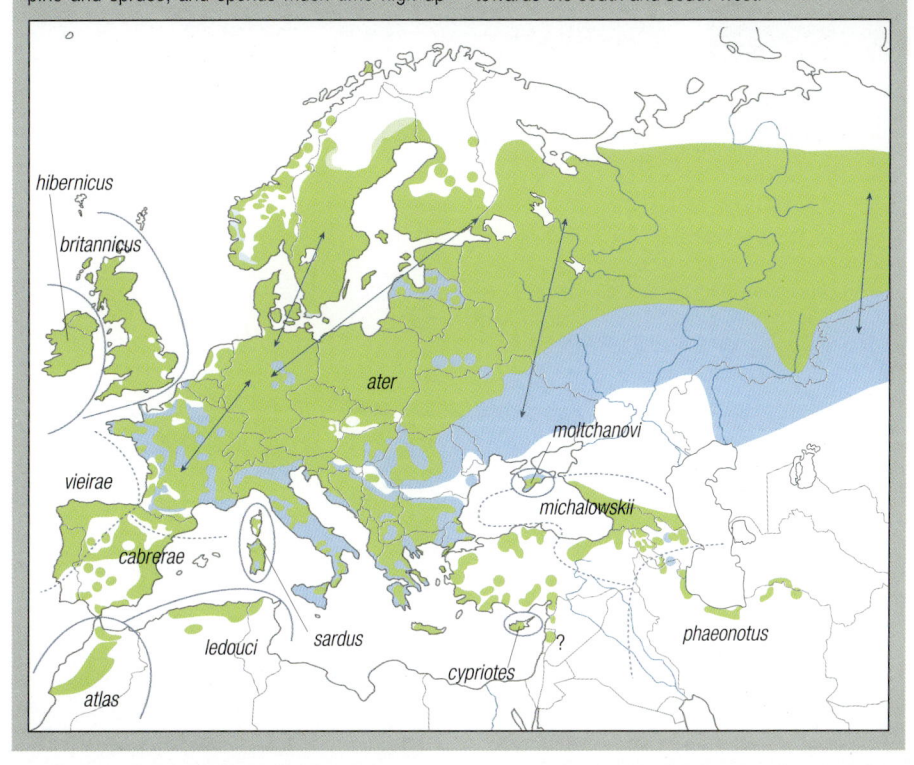

IDENTIFICATION Apart from *small size*, note *large-headed* but *short-tailed* appearance, and *black head* with three white patches: *white cheek patches* like in Great Tit, but also a characteristic oblong *white nape patch*. Mantle variably olive-brown or lead-grey, whereas *underparts* are often greyish-buff or *tinged ochre*, especially on flanks and vent. *Two prominent spotted white wing-bars* add to its characteristic looks. Can raise a hint of a crest when agitated, and elongated feathers can make the rear crown slightly angled in profile, even when flattened. Appears small and compact in its light, rather hopping flight. Always busy and energetic, frequently giving calls, at least some of which are fairly characteristic. A good deal of geographical variation affecting both colours and size. Birds in forests of N Algeria and N Tunisia are rather bright sulphur-yellow below instead of buff-white, and yellow instead of white on head patches.

VOCALISATIONS The song is basically similar to that of the Great Tit, only much faster and higher-pitched, with less amplitude between notes. It does not vary as much as the song of Great Tit, but there is still some individual variation in details. Nearly always the song is built up by repeated disyllabic or trisyllabic phrases, *seechu-seechu-seechu-seechu-seechu-...*, *seeweedü-seeweedü-seeweedü-seeweedü-...* or similar. – Most characteristic call has desolate ring, a fluty, slightly downwards-inflected *tüh(e)* or almost straight *tüh*. Many other fine, high-pitched but less characteristic contact calls heard from feeding birds, *swee-pee*, *tih*, *zi*, *zi-zi-zi*, *si-chu*, etc. Excited-sounding, fast series of squeaky and high-pitched notes also heard.

SIMILAR SPECIES If seen only briefly in the canopy of tall trees, possible to confuse with *Great Tit*, but much smaller, with proportionately larger head, and white (or yellowish) nape patch diagnostic, being narrow and vertically stretched. Note also lack of black central stripe on underparts, double white wing-bars with spotted appearance (solid single wing-bar in Great Tit), and lead-grey or olive-brown back, rather than moss green. Underparts never yellow except in some local populations (Ireland, NW Africa) or faintly in some juveniles.

AGEING & SEXING (*ater*) Ages differ subtly, young birds often separable during 1stY, but ageing requires close views

P. a. ater, ad, presumed ♂, Italy, Oct: small and compact, with rather large head, diagnostically black with large white cheek and nape patches. Also note lead-grey upperparts and two whitish wing-bars, and pale underparts with varying brownish-grey to ochre-brown tones on flanks. Aged by evenly fresh wing. Safe sexing often impossible and requires handling or comparison of both members of a pair, although in this case the deep black head with bluish sheen and large glossy black bib suggest a ♂. (D. Occhiato)

P. a. ater, 1stW, presumed ♂, Switzerland, Nov: lead-grey mantle, large black bib, white head markings, off-white underparts sullied brownish-grey and usually limited ochre-brown on flanks are important distinctions of race *cristatus*. Aged by moult limit with worn and browner outer (juv) at least four greater coverts; tentative sexing as in image to the left. (L. Sebastiani)

P. a. ater, 1stW, Poland, Feb: even in race *cristatus*, the underparts can be extensively sullied ochre-brown. Extent of moult extremely variable, and this individual has replaced very few greater coverts (the juv coverts show no bluish edges, thereby contrasting with the bluish-grey back). Sexing unsafe. (M. Szczepanek)

P. a. britannicus, ad, England, Mar: the British race is characterised by ochre-washed bluish-grey upperparts (ochre hue difficult to appreciate here, though) and by being rather strongly ochre-buff below. Glossy bluish-black crown and large bib suggest ♂, but much overlap and sexing best avoided. Evenly fresh wing typical of ad.

or handling. Sexes alike or very similar in plumage (despite Gosler & King 1989, King & Griffiths 1994); ♂ slightly larger. – Moults. Complete post-nuptial moult of ad in summer after breeding (mainly Jul–early Sep). Partial post-juv moult in summer (mainly Jul–mid Sep) somewhat variable in extent depending on population, but generally does not involve any primary-coverts, primaries or secondaries, but rarely odd or all alula-feathers, odd tertials and central 1–2 pairs of tail-feathers. No pre-nuptial moult. – ALL YEAR ♂ On average stronger bluish gloss on crown, but much overlap. Dark throat bib on average more extensive and blacker, and often has slight metallic gloss, but much variation and overlap (including in size and shape). Marginal and lesser coverts have on average darker grey (nearly black) and more contrasting bases, tipped bluish-grey. Note that reliable sexing is often impossible due to individual variation and extensive overlap. Plumage characters should be supported by long wing and tail. (The claimed sexual difference in that ♂ has more (dark) bluish-grey mantle and scapulars, ♀ more greenish-olive [*BWP*] could not be confirmed on a long series of Swedish specimens, NRM.) ♀ On average slightly less bluish tinge and gloss on crown (but invariably has some), and throat bib usually on average less extensive and more matt sooty-black, lacking gloss, but much variation and overlap. Marginal and lesser coverts centred dark grey, tipped dull lead-grey, with less contrast than in ♂. Sexing frequently difficult due to overlapping and faint differences. – AUTUMN **Ad** All greater coverts moulted and equally fresh, outer webs dark grey or blackish edged blue-grey and tipped white (tips often large and well-marked). Primary-coverts and alula fresh, rather dark grey edged blue-grey. Very slight colour difference between primary-coverts and rest of wing. **1stCY** Usually 1–6 outer greater coverts unmoulted (rarely 0–9), differing clearly or slightly from moulted ones of ad type by being slightly paler grey, edged olive-grey and tipped dusky-white, often with faint buff or olive tinge (pale tips often smaller and less distinctly demarcated than ad). Birds with one unmoulted can be overlooked as ad, and a very few moult all greater coverts to ad type. Primary-coverts are subtly paler and appear brown-tinged, and have on average slightly more pointed tips and duller edges. **Juv** Differs in being duller overall, matt sooty-black on crown and bib, latter being poorly developed, and in having pale head patches and underparts tinged yellowish. – SPRING **Ad** Same slight differences as in autumn apply, only now more difficult to use due to more extensive wear. **2ndCY** As in autumn, but greater coverts more abraded, and many birds intermediate and must be left un-aged.

BIOMETRICS (*ater*) **L** 10.5–12 cm; **W** ♂ 59–68 mm (n 73, m 62.4), ♀ 57–64 mm (n 47, m 60.3); **T** ♂ 43–51 mm (n 73, m 47.1), ♀ 42–49 mm (n 47, m 45.7); **T/W** m 75.7; **B** 9.5–11.9 mm (n 45, m 10.7); **BD** 3.2–4.0 mm (n 33, m 3.6); **Ts** 14.5–17.5 mm (n 46, m 16.0). **Wing formula: p1** > pc 5–11 mm, < p2 20–25 mm; **p2** < wt 6–10 mm, =7/9 (100%); **p3** < wt 0–2 mm; **pp4–5** about equal and longest: **p6** < wt 1–3 mm; **p7** < wt 4–7 mm; **p8** < wt 7.5–10.5 mm; **p10** < wt 11–14 mm; **s1** < wt 12–14.5 mm. Emarg. pp3–6, often a faint trace also on p7 (but all emarg. rather poorly defined).

GEOGRAPHICAL VARIATION & RANGE Partly slight or moderate and clinal variation (N & C Europe), partly rather abrupt and well marked (S Europe, N Africa, E Mediterranean), but traditional arrangement in racial groups not adopted here due to in our opinion less clear pattern and many shared traits across borders. The more marked variation in NW Africa may contain incipient species (see below). Many described races do not hold when longer series are compared. Those listed below are fairly distinct and therefore meaningful.

P. a. ater (L., 1758) (much of Europe, W Siberia, south to Pyrenees, Sicily, W & C Turkey; mainly sedentary, but in some years emigrates in large numbers towards southwest). Treated above. Bluish-grey mantle/back with slight olive or 'dirty' tinge in some. Large black throat bib. Rather pure white (or at least whitish) patches on sides of head and hindneck, lacking cream tinge. Underparts sullied brownish-grey on off-white ground, flanks with only slight ochre-brown tinge at most (lacking bright tawny or rufous colours of some

P. a. britannicus, presumed ad, England, Feb: note ochre-tinged lead-grey mantle and ochre-buff tinge to underparts typical of this race. The apparently evenly fresh wing with whitish-fringed primaries and bluish-grey primary-coverts suggest an ad. (G. Thoburn)

P. a. cabrerae, Spain, Apr: the east Iberian race is purer grey (less bluish) above and rather richer rufous-buff on flanks than nominate *ater*. In Spain, birds can be already heavily worn in Apr, so here the near lack of bluish gloss to crown and bib is not conclusive of a ♀; also, despite strong wear to outer primaries, there is no definite moult limit to indicate a young, meaning that age or sex should be left undetermined, at least without handling. (C. N. G. Bocos)

TITS

P. a. michalowskii, presumed ad, Russia, Nov: note strong feet, ochre-tinged blue-grey mantle, dusky buff-brown underparts, and rather extensive and sharply defined black bib extending to upper breast, all characteristics of this Caucasian and Transcaucasian race. Appearance of rather evenly fresh wing and bluish sheen to black of crown suggest age. (I. Ukolov)

P. a. phaeonotus, 1stW, Iran, Jan: strong feet, stout bill, shortish tail, darker and browner mantle without appreciable greyish or bluish tones, and flanks faintly tinged buff-brown are characters of this race. Apparently retained juv primary-coverts (rather pointed and only narrowly fringed paler), and moult limit in greater coverts indicate a young. (E. Winkel)

P. a. cypriotes, presumed ad ♂, Cyprus, Apr: the darkest race in the treated region, with black on head, especially that of bib, most extensive and glossy, leaving relatively narrow white cheek and nape patches. Mantle brownish tinged olive-grey, and underparts warm buff-brown. Bluish gloss to crown and evenly feathered wing suggest age and sex. (F. Trabalon)

other races). – A slight clinal trend in that birds in C & S Europe have less pure blue-grey mantle/back, more olive or buff tinge in the blue, and flanks and vent are slightly more saturated ochre-buff ('*abietum*'), but when series compared deemed too indistinct to be recognised. (Syn: *abietum*; *burgi*; *parisi*, *rapinei*, *rossosibiricus*.)

P. a. britannicus (Sharpe & Dresser, 1871) (Britain; resident). Small, short-tailed and somewhat darker than *ater*. Upperparts bluish-grey with ochre tinge, strongest in fresh plumage and on lower back/rump; mix of bluish and ochre tones create impression of olive hue. Pale patches on hind-neck and sides of head whitish with faint cream-white tinge in many. Flanks and vent rather strongly tinged ochre-buff. Rest of underparts rather 'dirty', sullied buffish-grey. Fewer than 5% have faint yellowish hue on belly, and very slightly on cheek patches, too. Sexing according to size and shape of throat bib claimed to be useful and reliable for many, with ♂♂ having larger and glossier black bibs (Gosler & King 1989, King & Griffith 1994), but much overlap and of very limited use. **L** 10–11 cm; **W** ♂ 58.5–64 mm (*n* 12, *m* 60.3), ♀ 58–62 mm (*n* 12, *m* 58.9); **T** ♂ 42–46 mm (*n* 12, *m* 43.7), ♀ 41–45 mm (*n* 12, *m* 42.6); **T/W** *m* 72.4. **Wing formula: p2** =7/9 (84%), =7 (11%), or =9 (5%). (Syn. *pinicolus*, which is not less tinged buff or ochre on upperparts as sometimes claimed, at least not in samples compared by us.)

P. a. hibernicus (W. Ingram, 1910) (Ireland; resident). Small as *britannicus* but has upperparts a mixture of olive-grey and umbra (warm ochre), giving both a slightly greenish and reddish-ochre impression (thus hardly any blue colour). Variable race, some being inseparable from *britannicus*, but *c.* 75% of post-juv retain some yellow tinge on cheek patches and underparts (thus considerably more than in *britannicus*). Bib colour of both sexes less black and glossy than in *britannicus* (King 1994), but difference subtle and difficult to use. Claims of on average somewhat smaller bib size than in *britannicus* could not be confirmed using the ample material at NHM, mainly due to substantial overlap and variation. **W** ♂ 58–62 mm (*n* 15, *m* 59.8), ♀ 56–61 mm (*n* 13, *m* 58.9); **T** ♂ 42–46 mm (*n* 14, *m* 43.8), ♀ 41–44 mm (*n* 13, *m* 42.3); **T/W** *m* 72.5. **Wing formula: p2** =7/9 (86%), =9 or <9 (14%).

○ **P. a. vieirae** (Nicholson, 1906) (NW Iberia; resident). Similar to *britannicus*, with same bluish olive-ochre upperparts, but even smaller, and stronger rusty-buff tinge on flanks and vent. Still, a very subtle race. **L** 9.5–10 cm; **W** ♂ 57.5–60 mm (*n* 12, *m* 58.6), ♀ 56–61 mm (*n* 13, *m* 58.9); **T** ♂ 40–44 mm (*n* 12, *m* 41.9), ♀ 41–44 mm (*n* 13, *m* 42.3); **T/W** *m* 71.5; **B** 9.0–10.3 mm (*n* 18, *m* 9.6); **Ts** 15.3–16.5 mm (*n* 17, *m* 15.9).

P. a. cabrerae (Witherby, 1928) (much of Iberia except north-west; resident). Less bluish-grey above than *ater*, more tinged ochre, though still has more bluish-grey tinge above than *vieirae* in direct comparison. Very slightly paler and cleaner than *vieirae*, which is a little duskier overall. Underparts are the same as in *vieirae*, slightly richer rufous-buff on flanks than in *ater*. Is clearly larger and longer-tailed than *vieirae*, and although some are very similar in plumage colours, cannot be included therein due to size and structure. Grades into *vieirae* in west and north-west, and into *ater* in the Pyrenees. Since this race is stable over a vast area it cannot be discarded as 'intermediates between *ater* and *vieirae*', both of which occur north of it. (Oddly, very similar to *michalowskii* of Caucasus and Transcaucasia, only slightly darker ochre-buff on flanks and vent.) **L** 10.5–11 cm; sexes of nearly same size. **W** 60–65.5 mm (*n* 24, *m* 62.6); **T** 44–48 mm (*n* 23, *m* 46.2); **T/W** *m* 73.8; **B** 10.0–11.5 mm (*n* 23, *m* 11.0); **Ts** 15.8–18.0 mm (*n* 23, *m* 17.1). **Wing formula: p2** =7/9 (96%), or =9 (4%).

○ **P. a. sardus** (Kleinschmidt, 1903) (Corsica, Sardinia; resident). Similar to both *ater* and *cabrerae*, but series differ from *ater* in having neatly outlined and rather small black throat bib (frequently leaving off-white gap between bib and side of nape, less commonly seen in *ater*), and by slightly stronger tawny or ochre colour on flanks. Mantle/back bluish-grey with limited ochre tinge (less ochre than in *cabrerae*), strongest on rump. Slightly smaller than *ater*. Generally slightly darker than *ater* overall, still variation, both individual and through wear, makes separation often difficult. **W** ♂ 59–67 mm (*n* 13, *m* 63.2), ♀ 59.5–64.5 mm (*n* 13, *m* 62.1); **T** 44–49 mm (*n* 26, *m* 46.4); **T/W** *m* 74.1. **Wing formula: p2** =7/9 (94%), =9 (6%).

P. a. moltchanovi (Menzbier, 1903) (S Crimea; resident). Similar to *ater* but is clearly larger with longer legs and stouter and longer bill. Blue-grey upperparts are a trifle paler with a very slight ochre-brown tinge, not as pure blue-grey as in *ater*, and underparts are subtly paler, flanks usually show reduced warm buff and grey tones. Also, bib is slightly less extensive, not invading upper breast as commonly as in *ater*. **W** 65–72 mm (*n* 26, *m* 68.3); **T** 47–54 mm (*n* 26, *m* 50.1); **T/W** *m* 73.4; **B** 11.0–13.0 mm (*n* 26, *m* 12.0); **BD** 4.1–5.3 mm (*n* 25, *m* 4.6); **Ts** 17.2–19.9 mm (*n* 26, *m* 18.8).

P. a. cypriotes, ♂, presumed ad, Cyprus, Oct: in fresh plumage warmer toned overall, and especially below; seemingly evenly fresh and ad wing-feathers and rounded tail-tips indicate possible age, while glossed blue-black crown and large bib firmly indicate ♂. (H. Shirihai)

P. a. cypriotes, presumed 1stW ♀, Cyprus, Oct: especially fresh ♀ of this distinctive race seems to be the strongest saturated with warmer buff pigments below. Duller black areas, small bib and less bluish (but more olive) tinged upperparts infer ♀, while more pointed and not so fresh primary-coverts suggest young age. (H. Shirihai)

P. a. cypriotes, juv, Cyprus, Apr: fluffy and dull plumage, matt sooty-black crown and poorly developed dusky bib, with pale head patches and underparts tinged yellowish-cream are features of juv of all races. Sexing impossible in this plumage. (R. Martin)

P. a. michalowskii (Bogdanov, 1879) (NE Turkey, Caucasus and Transcaucasia except in east; mainly resident). Large as *moltchanovi* with strong feet and stout bill, has same plumage colours as that, with mantle/back bluish brownish-olive, thus unlike *phaeonotus* further east has slight bluish-grey tinge, and flanks and vent tinged dusky grey-buff, usually with obvious buff-brown tinge, differing from *moltchanovi* only in the on average more extensive and more sharply defined black bib reaching upper breast. (Plumage surprisingly similar to that of Iberian *cabrerae*.) **L** 11–12 cm; **W** 64–72 mm (n 18, m 69.1); **T** 46–53 mm (n 18, m 49.1); **T/W** m 71.1; **B** 10.9–12.5 mm (n 18, m 11.4); **BD** 4.0–4.6 mm (n 15, m 4.3); **Ts** 17.0–19.7 mm (n 18, m 18.7). – Vaurie (1959) accepted also '*derjugini*' from the area, but apparently saw only two specimens, claimed to be long-billed and more greyish. However, comparison of long series in ZMMU from the entire Caucasus, Transcaucasus and NE Turkey shows that it is best to include *derjugini* of eastern Black Sea region in *michalowskii*. – Birds in C Caucasus eastwards grade into *phaeonotus* becoming successively less bluish or olive, more brown-tinged on mantle/back. (Syn. *derjugini*; *pageri*.)

P. a. phaeonotus (Blanford, 1873) (E Caucasus, Azerbaijan, N Iran south of Caspian Sea; mainly resident but may make shorter winter movements to lower altitudes). As large as preceding, with similar structure except that it has proportionately slightly shorter tail. Mantle and back darker and browner than in *michalowskii*, without any appreciable greyish or bluish tinge. Large black throat bib, still sometimes shows gap between crown and side of bib. Lower belly, flanks and vent tinged buff-brown. **W** ♂ 65.5–71 mm (n 16, m 68.9), ♀ 64–69 mm (n 9, m 66.2); **T** 44–51 mm (n 27, m 47.6); **T/W** m 70.2; **B** 10.2–12.0 mm (n 25, m 11.2); **BD** 4.1–4.9 mm (n 22, m 4.4); **Ts** 17.7–19.6 mm (n 26, m 18.7). – Tends to become subtly paler brown above, with a little more grey cast visible in the brown, towards east (Khorasan, '*chorassanicus*'), but slight, clinal difference that is hardly sufficient ground for separation. Claims of a 'narrower' bill in '*chorassanicus*' could not be confirmed on available material. (Syn. *chorassanicus*; *gaddi*—see Taxonomic notes.)

P. a. cypriotes (Guillemard, 1881) (Cyprus; resident). Darkest race of all, darker even than *sardus* and *vieirae*. Black on head more extensive and glossier than in other races, and hence white patches on cheeks and nape comparatively small. Mantle/back brownish with olive-grey tinge, rump slightly browner. Flanks rather dark dusky-brown. Medium large, with quite rounded wing. Sexes nearly same size. **W** 58.5–64 mm (n 24, m 60.7); **T** 41.5–47 mm (n 24, m 44.2); **T/W** m 72.7; **Ts** 16.5–18.2 mm (n 23, m 17.2). **Wing formula: p1** > pc 8.5–12 mm, < p2 16.5–21.5 mm; **p2** < wt 8–11 mm, =9/10 or 10 (53%), =8/9 or 9 (26%), or <10 (21%); **p7** < wt 1.5–4 mm; **p8** < wt 4–6 mm; **p10** < wt 8.5–10.5 mm; **s1** < wt 9–12 mm.

P. a. atlas (Meade-Waldo, 1901) (Atlas Mts in Morocco; resident). Large and strong-legged. Mantle greyish-green with very slight ochre-buff tinge, rump more greenish-ochre. Extensive sooty-black throat bib, wide near bill (black usually joins black cap either side of bill). Pale head patches and breast near-white, flanks olive-grey with slight brown tinge (but not rusty-buff). Wing rounded. **L** 11–12.5 cm; **W** ♂ 67–73 mm (n 12, m 69.5), ♀ 64–71 mm (n 12, m 66.4); **T** ♂ 47–52 mm (n 12, m 49.1), ♀ 45–49 mm (n 12, m 46.9); **T/W** m 70.7; **Ts** 16.5–18.2 mm (n 23, m 17.2). **Wing formula: p1** > pc 10.5–14 mm, < p2 20–25 mm; **p2** < wt 8–11 mm, =8/10 (71%), =7/8 or 8 (19%), or <10 (9%); **p3** < wt 1–2.5 mm; **p7** < wt 3–5 mm; **p8** < wt 6–9.5 mm; **p10** < wt 10.5–13.5 mm; **s1** < wt 11–15 mm.

P. a. ledouci (Malherbe, 1845) (N Algeria, NW Tunisia; resident). Almost as large as *atlas*. Mantle green, when fresh with slight blue-grey tinge, wearing to more dull greyish, rump always greener. Underparts and pale patches on sides of head lemon or pale sulphur-yellow (though patch on nape near-white on some). **W** ♂ 64–71 mm (n 12, m 66.8), ♀ 59–67 mm (n 13, m 62.9); **T** ♂ 45–51 mm (n 12, m 47.1), ♀ 42–49 mm (n 13, m 44.6); **T/W** m 70.6; **Ts** 16.5–18.2 mm (n 23, m 17.8). **Wing formula: p1** > pc 6.5–12 mm; **p2** =7/9 (82%), or =9 (18%), the rest much like in *atlas*.

TAXONOMIC NOTES See under Marsh Tit regarding the generic arrangement of tits. – The allopatric race *ledouci* is sufficiently distinct in morphology to raise the question of species status. A genetic study of this and *atlas*, and a comparison with as many as possible of the other Coal Tit races, is highly desirable. – Ssp. *phaeonotus* was described by Blanford based on three summer birds from the Zagros Mts, W Iran, and therefore thought to breed there, with another bird subsequently collected by Zarudny. However, it has apparently not been seen in the area since, and the hypothesised breeding can be questioned and could refer to migrants/stragglers from the north (Kirwan & Grieve 2010). A syntype in NHM (perhaps now the only remaining Zagros specimen) fits well with breeders of the S Caspian area, later described by Zarudny as *gaddi*, which name hence becomes a junior synonym. The *phaeonotus* syntype is in fresh post-juv plumage typical of autumn birds, casting doubt on the claim that it was collected in June.

REFERENCES Dickinson, E. & Milne, P. (2008) *BBOC*, 128: 267–268. – Gosler, A. G. & King, J. R. (1989) *Ring. & Migr.*, 10: 53–57. – King, J. R. (1994) *BBOC*, 114: 174–175. – King, J. R. & Griffith, R. (1994) *Bird Study*, 41: 7–14. – Kirwan, G. M. & Grieve, A. (2010) *BBOC*, 130: 83–87. – Markovets, M. Y. (1990) *Zool. Zhurn.*, 69: 127–129.

P. a. ledouci, presumed ad ♂, Algeria, Jun: a highly distinctive race due to its greenish grey-brown mantle (hardly visible here), yellow-tinged underparts and lemon-yellow head-sides, but nape patch white. Due to extensive black bib, the yellow cheeks are rather small. Bluish gloss to crown, large bib and apparently evenly feathered wing indicate ♂ and an ad, but without handling sexing and ageing uncertain. (V. Legrand)

P. a. ledouci, presumed ad, Algeria, Jun: a bird with stronger yellow tinge on underparts, these being equally yellow as head-sides. Wing-feathers including primary-coverts appear evenly fresh to suggest an ad. Since bib size and gloss varies individually, and sexual differences as to this are poorly studied, best to leave the bird unsexed. (V. Legrand)

(EUROPEAN) BLUE TIT
Cyanistes caeruleus (L., 1758)

Fr. – Mésange bleue; Ger. – Blaumeise
Sp. – Herrerillo común; Swe. – Blåmes

A much-loved bird even outside the ranks of bird-watchers, due both to its attractive colours and cute looks, but also perhaps to its bold manners among much larger birds on feeders. It is widely distributed in much of Europe and Turkey, but has recently been separated from its close relatives in N Africa and the Canaries, these nowadays regarded as constituting a separate species, the African Blue Tit. The Blue Tit is mainly resident, but in some years rather large numbers move south and west from northern parts of the range. Although usually a bird of woods, parks and gardens, in winter also often searches reedbeds for insects. Nests in holes.

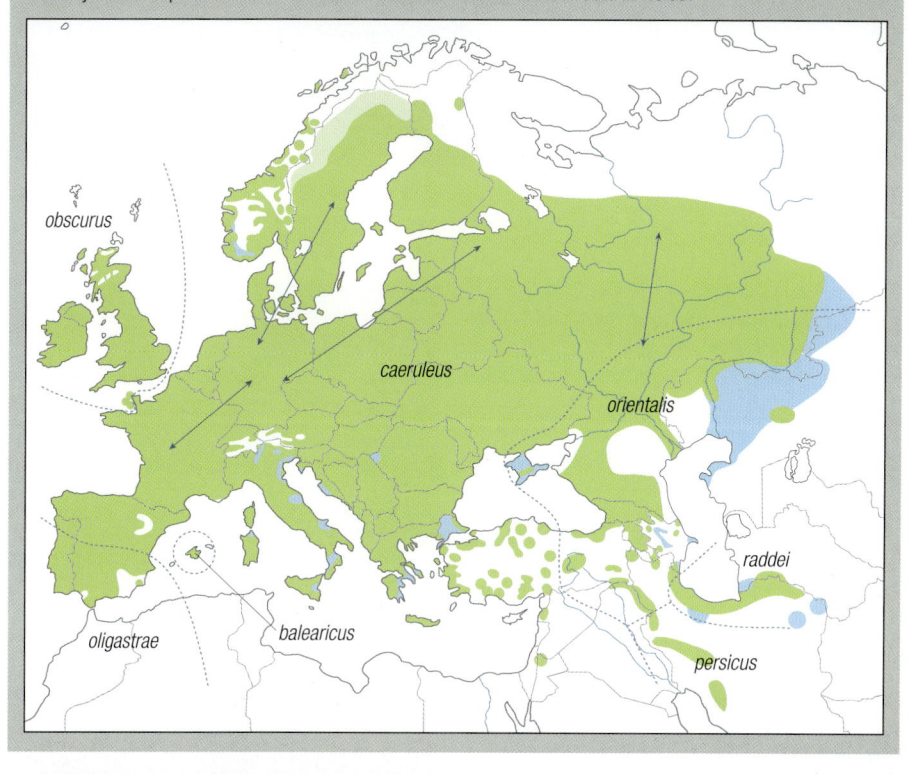

IDENTIFICATION A small tit with *rounded head* and *short bill*, giving it cute looks. Blue, yellow, white and green colours add to its attractive appearance. Best identified by head pattern, with *blue crown patch surrounded by white* ('blue beret'), white cheek patches outlined with a blackish throat bib, a narrow dark line below and a dark blue 'boa' across nape. *Wings and uppertail are bright blue* (duller and tinged greenish-grey in immatures), with *a single white wing-bar*. Mantle moss-green. *Underparts yellow* with a hint of a darkish central streak, flanks sometimes with a greyish or greenish tinge. Undertail-coverts whitish. Bill and legs dark grey. Juvenile has white of head tinged yellow, and dark markings lacking or indistinctly developed. Acrobatic and agile, often seen climbing upside-down in birch twigs or at bottom of seed feeders. Active and restless when feeding, movements quick, flight slightly 'hopping'. Not shy, and is a regular visitor to feeders.

VOCALISATIONS The song is high-pitched and clear, usually 1–3 fine notes followed by a clear trill ('crystal' clear voice) at a slightly lower pitch, *siih siih sürrrrrrrr*. A variation shortens the trill but quickly repeats the phrase twice, *si-si-sürr si-si-sürr*. Further song variants exist but with similar clear voice and fine notes, so usually readily recognisable. Each ♂ seems to stick to just one song type, consistently repeated. – Has a rich repertoire of calls, often identified by high-pitched, clear voice. Conversational fast *si-si-si* is perhaps most commonly heard (a little like Grey Wagtail; less intense and strident than similar call of Long-tailed Tit), but just as often the last note is on lower pitch, *si-si-si-su*. Single *si* also heard, then more difficult to pick out. When agitated (quarrelling, in anxiety) gives a scolding *kerrr'r'r'rek-ek-ek* with slowed and more stressed end.

SIMILAR SPECIES To the beginner somewhat resembles *Great Tit*, but is smaller with a more cutely rounded head, a blue cap surrounded by white rather than an all-black crown, and the dark central streak on lower breast and belly is indistinct and incomplete, not prominent and connected to black throat bib. – *Azure Tit* is white below, lacks dark bib and stripe across neck, and usually lacks any yellow; birds with mixture of white and yellow below could be hybrids. (Extra-limital race or separate species *flavipectus* of Azure Tit, in Central Asia, has yellow breast, but is otherwise like normal

C. c. caeruleus, ad ♂, Italy, Jan: unmistakable blue-white-and-black head pattern, including small blackish bib and dark blue nape. Bright blue wings and uppertail, wings with single white wing-bar, while back is greenish and underparts yellow with darkish central streak (usually visible only front-on). Aged by evenly fresh wing with bluish edges and no moult limits; ♂ by bright ultramarine-blue crown, deep blue edges to wing-coverts, and broad blue band on rear neck. (D. Occhiato)

C. c. caeruleus, presumed ad ♂, Finland, Apr: cute appearance emphasised by rounded head. The dark central streak to the yellow underparts is usually visible only in such views. The rather wide and more blackish-blue 'boa' on nape and broad and rounded tail-feathers (including outer ones) strongly suggest an ad ♂. (M. Varesvuo)

Azure Tit.) – Allopatric and closely related *African Blue Tit* separated by range, much darker bluish-black cap and more prominent dark head markings. Some races also lack the white wing-bar.

AGEING & SEXING (*caeruleus*) Ageing possible on colour and moult differences during 1stY. Sexes similar, but most are separable by combination of slight differences in plumage and size. – Moults. Complete post-nuptial moult of ad in summer after breeding (mainly mid Jun–mid Sep). Partial post-juv moult in summer generally does not involve any primary-coverts, primaries or secondaries. Usually replaces all greater coverts, tertials and alula, rarely also s6. Commonly, r1 is moulted, and some moult several pairs of tail-feathers, a few all rectrices. No pre-nuptial moult. – **ALL YEAR Ad** ♂ Crown bright ultramarine-blue. Edges to primaries, tail-feathers and wing-coverts intense cobalt-blue. Dark 'boa' across nape on average broader and darker indigo blue-black than in ad ♀, though some are intermediate. Wing > 63 mm (> 68 conclusive). **Ad** ♀ Blue of crown and edges to wings and tail less intense and dark, often slightly greyish-tinged. Dark 'boa' across nape on average narrower and less dark, bluish-grey. Wing < 69 mm (< 64 conclusive). **1stY** ♂ Often as ad ♂, but on average slightly paler and less brilliant blue colours. Primary-coverts not so clearly bluish-edged, slightly duller greenish or greyish-tinged. A few may be difficult to sex reliably. **1stY** ♀ Similar to ad ♀, but even duller, less clear blue colours. Primary-coverts edged dull greyish-green. A few may be difficult to sex reliably. – **AUTUMN Ad** Primary-coverts have rather rounded tips, edged bluish, uniform or nearly uniform with greater coverts and rest of wing. (Some ♀♀ have slightly duller greyish-blue edges, though no greenish cast.) Tail-feathers usually rather broad with rounded tips, keeping fresh through autumn and sometimes into early spring. **1stW** Primary-coverts have on average more pointed tips, edged dull bluish-green or greyish-green, after completing post-juv moult clearly duller than blue-edged greater coverts and rest of wing. Tail-feathers often slightly narrower and more pointed than in ad, and tips become abraded in autumn or early winter. **Juv** White of head in ad plumage replaced by pale yellow. Dark markings on head absent or only faintly suggested. – **SPRING Ad** As in autumn, although abrasion makes reliable ageing progressively more difficult (though generally still possible). **1stS** As in autumn. Beware that abrasion makes ageing more difficult later in the season.

BIOMETRICS (*caeruleus*, Scandinavia) **L** 11–12 cm; **W** ♂ (once 64) 65–71 mm (n 63, m 67.8), ♀ 60–68 mm (n 31, m 65.2); **T** ♂ 48–56 mm (n 61, m 52.9), ♀ 46–53.5 mm (n 31, m 51.7); **T/W** m 78.5; **B** 8.0–9.9 mm (n 92, m 9.0); **BD** 3.6–5.4 mm (n 66, m 4.8); **Ts** 15.0–17.6 mm (n 90, m 16.5). Birds of S Europe slightly smaller. **Wing formula: p1** > pc 4.5–10 mm, < p2 21–28 mm; **p2** < wt 7–11 mm, =8/9 or 8 (75%), =7/8 (24%), or =9 (1%); **p3** < wt 0.5–2 mm; **pp4–6** about equal and longest (p6 sometimes to 1.5 mm <); **p7** < wt 2–5 mm; **p8** < wt 5–10 mm; **p10** < wt 10–15 mm; **s1** < wt 11–15 mm. Emarg. pp3–7.

GEOGRAPHICAL VARIATION & RANGE Slight and clinal variation, mainly affecting plumage colours. A careful comparison of long series of specimens from relevant areas revealed slighter differences than often claimed, and a few subspecies are deemed here to be insignificant local or minor variations, and better viewed as junior synonyms of neighbouring taxa.

C. c. caeruleus (L., 1758) (much of Europe except west and south-west, but including Corsica, Sardinia, Sicily, Greece, Crete, W & C Turkey, Cyprus; mainly sedentary, but northern populations involved in large-scale exodus towards south-west in some years). Treated above. Underparts yellow with faint greyish-green tinge on flanks. Broad white wing-bar. Fairly large. – Birds of S Greece and Crete ('*calamensis*') said to be 'slightly smaller', but all birds in S Europe are slightly smaller than those of N Europe, and since the difference is both slight and clinal, with no plumage differences, insufficient for separation. (Syn. *calamensis*.)

○ *C. c. obscurus* (Pražák, 1894) (British Isles, Channel Is; resident). Very similar to *caeruleus* but subtly smaller, with less sexual size dimorphism, and has usually a trifle darker upperparts, but many are similar. Often has narrower, less white wing-bar, but somewhat variable, and is of restricted use. That underparts should be duskier, as often stated, is difficult to confirm on longer series. A very subtle race. See Sellars (1985) and Scott (1993) for reported sexing problems in this race, primarily in winter. **L** 10.5–11.5 cm; **W** 60–67 mm (n 31, m 63.3); **T** 45–51 mm (n 30, m 47.9); **T/W** m 75.7. Wing formula: **p1** < **p2** 21–25 mm; **p2** < wt 5–10 mm, =8/9 or 8 (83%), or =7/8 (17%); **p8** < wt 6–8.5 mm; **p10** < wt 10.5–13.5 mm; **s1** < wt 11–14 mm.

○ *C. c. oligastrae* (Hartert, 1905) (S Iberia; resident). Subtly smaller than *caeruleus*. Plumage appears very similar to that of *caeruleus*, but the blue of upperwing and uppertail is on average darker (and perhaps in some ad ♂♂ more intensely ultramarine, but rather variable). Said to be brighter yellow below and more grey-green, less bluish-tinged on mantle than *caeruleus*, but this is quite variable and subtle, and of little help. **W** ♂ 59–70 mm (n 14, m 62.4), ♀ 58–65 mm (n 9, m 61.7); **T** ♂ 45–54 mm (n 14, m 48.9), ♀ 45–51 mm (n 9, m 47.8); **T/W** m 78.4. Wing formula: **p1** > pc 7–10 mm, < **p2** 20–23 mm; **p2** < wt 8–10.5 mm, =8/9 or 9 (92%), or =8 (8%). (Syn. *harterti*.)

C. c. caeruleus, ad ♀, Germany, Feb: evenly-feathered wing and rounded tail-feather tips make this bird an ad, while slightly paler and duller blue crown patch and wing edges and paler blue and almost greenish-tinged uppertail suggest a ♀. The fairly narrow blue 'boa' across nape supports this. (S. Pfützké)

C. c. caeruleus, 1stY ♂, Germany, Oct: aged by greenish-blue juv primary-coverts, blue edges to remiges having a faint greenish hue (especially secondaries), and tips to outer rectrices pointed. Bright ultramarine-blue crown patch and broad blue band on neck typical of ♂. (M. Schäf)

C. c. caeruleus, 1stW ♀, Germany, Nov: dull blue juv primary-coverts and newer central rectrices (with rounded tips) identify this as a young bird; rather dull and pale blue crown patch, and narrow, paler blue band on neck indicate a ♀. However, such age and sex clues are rarely as striking as in this image, other birds can be more difficult. (A. Noeske)

C. c. caeruleus, juv, Italy, Jun: pale areas on head tinged yellow, dark bib is lacking and other dark markings on head are only vaguely developed. Body plumage still fully juv, being soft and fluffy. Note pointed tip of juv tail-feathers. (D. Occhiato)

C. c. obscurus, ad ♂, England, Feb: the British race is only subtly differentiated from continental European *caeruleus* by on average duller and darker mantle and back and on average a little narrower wing-bar. Evenly fresh wing with bluish edges and broad blue band on neck indicate age and sex, respectively. (G. Thoburn)

C. c. balearicus (von Jordans, 1913) (Balearics; resident). Differs from *oligastrae* in being slightly less intensely ultramarine-blue on head, wings and tail. Also a fraction paler on mantle, this being more yellowish-tinged greyish-green, not so bluish, than in *caeruleus*. There is apparently a tendency to be stronger yellow only on breast, whereas belly and flanks are paler and more whitish. Scant material examined. **W** 58–68.5 mm; **T** 47–52 mm; **B** 9.0–10.0 mm; **Ts** 15.1–16.8 mm (*n* 3).

○ **C. c. orientalis** Zarudny & Loudon, 1905 (SE Russia, lower Volga, SE Ukraine, Caucasus, Transcaucasia, E Turkey; mainly resident, but some south- and west-directed movements noted some years in north-eastern part of range). Resembles *caeruleus* but is a fraction larger on average and is usually slightly paler above, appearing to have a blue-grey 'sheen'. The yellow of underparts also tends to be very slightly paler. Rather subtle. **W** ♂ 64–71 mm (*n* 30, *m* 67.8), ♀ 64–68 mm (*n* 10, *m* 66.3); **T** 50–57 mm (*n* 40, *m* 53.4); **T/W** *m* 79.2; **B** 8.0–9.8 mm (*n* 39, *m* 9.0); **BD** 4.2–5.2 mm (*n* 36, *m* 4.7); **Ts** 15.8–17.6 mm (*n* 38, *m* 16.6). Wing formula very similar to *caeruleus*. – Populations of S Caspian ('*satunini*') inseparable (only very subtly smaller on average) and therefore included here in *orientalis*. (Syn. *georgicus*; *satunini*.)

○ **C. c. raddei** Zarudny, 1908 (NW Iran in coastal region south of Caspian Sea, SW Turkmenistan; mainly resident). Very similar to *caeruleus* but on average slightly darker above and below, and is somewhat smaller and subtly shorter-tailed than *caeruleus*. A weak subspecies. **L** 10.5–11 cm; **W** ♂ 61–68.5 mm (*n* 10, *m* 65.5), ♀ 59–64 mm (*n* 13, *m* 62.6); **T** ♂ 47–53.5 mm (*n* 10, *m* 49.1), ♀ 44–50 mm (*n* 13, *m* 47.2); **T/W** *m* 75.2; **B** 8.2–9.8 mm (*n* 23, *m* 9.0); **Ts** 15.7–17.7 mm (*n* 23, *m* 16.7). **Wing formula: p1** > pc 6–10 mm, < p2 20–25 mm; **p2** < wt 5.5–10 mm, =7/8 or 8 (75%), or =8/9 (25%); **p10** < wt 10–13 mm; **s1** < wt 10–13 mm.

C. c. persicus (Blanford, 1873) (Zagros Mts, Iran; resident). Very pale, upperparts a washed-out greenish-grey with very subdued blue element, on average slightly paler than *orientalis*. Blue colour of crown, wings and tail rather clear and pale, but blue of wing-bend and lesser coverts deeper ultramarine, more so than in *orientalis*. Dark line around white patches on sides of head blue-grey, often thin and can be partly broken or lacking. Underparts washed-out pale yellow, belly often more dusky-white than yellow. (Remarkably, this is extremely similar to ssp. *balearicus* of Balearic Is, only *persicus* is subtly more greyish on mantle!) **W** ♂ 65–68 mm (*n* 14, *m* 65.9), ♀ 59.5–64.5 mm (*n* 8, *m* 62.6); **T** ♂ 49–53 mm (*n* 14, *m* 51.6), ♀ 43–52 mm (*n* 8, *m* 48.1); **T/W** *m* 78.1; **B** 8.1–9.6 mm (*n* 25, *m* 8.8); **Ts** 15.0–17.3 mm (*n* 25, *m* 16.2).

TAXONOMIC NOTES See under Marsh Tit regarding the generic arrangement of tits. – Populations of N Africa and Canaries previously included in this species, but now commonly treated as a separate species, African Blue Tit.

REFERENCES Sellars, R. M. (1985) *Ringers' Bull.*, 6: 83–84. – Scott, G. W. (1993) *Ring. & Migr.*, 14: 124–128.

C. c. obscurus, 1stS ♀, England, Apr: aged by duller blue juv primary-coverts, moult limits in left wing tertials and moulted central rectrices, and sexed as ♀ by pale blue crown patch and narrower neck-band, as well as generally duller plumage of this race, combine to make this individual rather pale and washed out. (G. Thoburn)

C. c. raddei, presumed 1stS ♀, Iran, Jun: a bird in Mazandaran, on S coast of Caspian Sea, where the race *raddei* is known to occur, but by plumage seems nearer *persicus* of Zagros Mts, SW Iran. Note dull greenish-grey upperparts, with greyish tinge to blue of crown, wings and tail. Dark markings on head greyish too, while underparts are essentially dusky white with only little pale yellow. Surely a young ♀ with such dull plumage. (E. Winkel)

AFRICAN BLUE TIT
Cyanistes teneriffae (Lesson, 1831)

Fr. – Mésange nord-africaine; Ger. – Ultramarinmeise
Sp. – Herrerillo canario; Swe. – Koboltmes

Visitors to holiday resorts in N Africa and the Canaries may have noted that the blue tits there look a little unfamiliar, although basically the same as at home and readily recognisable as 'blue tits'. However, rather recently these slight differences in appearance, supported by genetic distance and some peculiarities in vocalisation, have led to the separation of the African Blue Tit. Common both in woods and nearer to humans in groves and gardens.

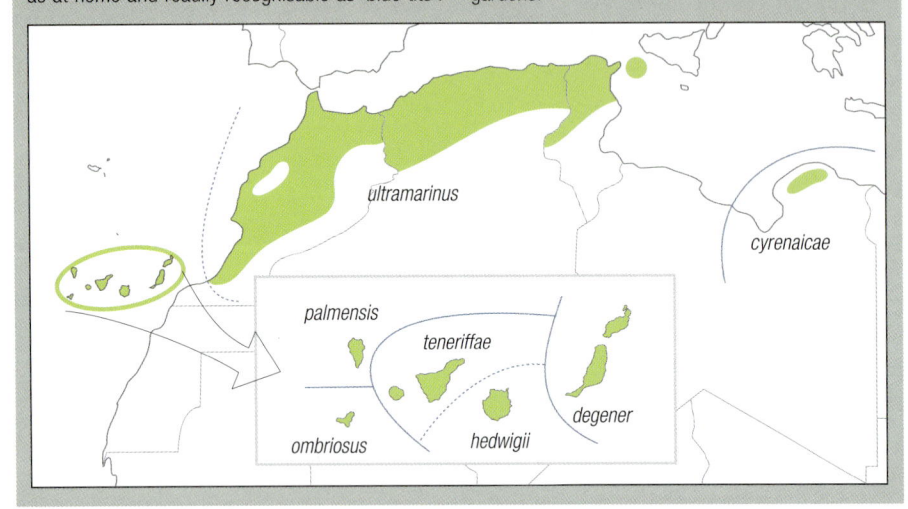

C. t. ultramarinus, presumed ♀, Morocco, Mar: less dark and greyer-blue crown, smaller, sooty-black (not jet-black, or glossed bluish) bib, and paler yellow underparts suggest a ♀. Ageing at this angle not possible. (H. van Diek)

IDENTIFICATION Very similar to Blue Tit but has *darker blue or blackish cap, narrower white border around it, broader and blacker stripe through lores and eye*, and, in some races, *lacks the white wing-bar*. Mantle colour somewhat variable, either bluish-grey or green. Yellow colour of underparts sometimes more intense, in one race (La Palma, Canaries) restricted to breast. Bill slightly longer and sometimes thinner. Song and several calls differ, too.

VOCALISATIONS Song differs clearly from that of Blue Tit in structure, recalling a mixture of Great Tit, Coal Tit and Marsh Tit in that a simple phrase is repeated in series. Fast variations resemble Coal Tit, slower Great Tit, although the voice is still high-pitched and clear like Blue Tit. Renderings of three common variations, among numerous known, could be *tseeya-tseeya-tseeya-tseeya-...*, *tsuweetsee-tsuweetsee-tsuweetsee-tsuweetsee-...* and *sítteritsee-sítteritsee-sítteritsee-sítteritsee-...* Songs usually consist of 3–8 phrases at a time, repeated with brief pauses between them. – A rich repertoire of calls as in Blue Tit, some similar to those of the European species, like fine *si-si-si-sü* and scolding *ker'r'r'err*, others markedly different. To the latter category must be referred the 'purring' calls, in the Canaries remarkably like those of Crested Tit, *perrerre-perrerre*, in N Africa not quite a copy but similar. Another peculiar call is a cheery, upwards-inflected *tsu-ee*, sometimes combined with the purring call.

SIMILAR SPECIES Hardly likely to be confused with any other species within its range, except in N Africa for *Great Tit*. Note smaller size of African Blue Tit, narrow white rim around dark cap and lack of complete black central stripe on breast (only a broken and indistinct one is present). – Differences from *Blue Tit* described under Identification and Geographical variation.

AGEING & SEXING (*ultramarinus*) Ageing usually possible on colour and moult differences during 1stY. Sexes very similar both in plumage and size, still often (but not always) separable. – Moults. Not closely studied, but appears to be similar to southern populations of Blue Tit. Complete postnuptial moult of ad in summer after breeding (May–Aug). Partial post-juv moult in summer generally does not involve any primary-coverts, primaries or secondaries. Usually replaces all greater coverts, tertials and alula. No pre-nuptial moult. – **ALL YEAR Ad ♂** Crown black with ultramarine-blue gloss. Throat bib jet-black with blue gloss. Black stripes on head broad and well-defined. Underparts rather intense yellow. Edges to primaries, tail-feathers and wing-coverts deep dark blue. Dark 'boa' across nape on average broader and darker blue-black than in ad ♀, though some are intermediate. Wing > 59 mm (> 64 conclusive). Note that there is some overlap in sexual characters and that many

C. t. ultramarinus, ad, presumed ♂, Tunisia, Apr: note blackish-blue cap only narrowly separated by thin white line from very broad blue-black eye-stripe and nape, and that yellow is most intense on breast. Aged by evenly fresh wing with bluish edges, and possibly a ♂ by deep ultramarine-blue crown patch, shiny blue edges to wing-coverts, and wide black-blue 'boa' on nape. (D. Occhiato)

C. t. ultramarinus, 1stW, presumed ♂, Tunisia, Nov: differs from European Blue Tit by blacker cap, with narrower white border, broader and blacker stripes around white cheeks, sometimes reduced white wing-bar, bluish-grey mantle, and tendency for yellow of underparts to be most intense on breast. Aged by dull blue juv primary-coverts; black crown patch faintly tinged blue, edges to wing-coverts shiny blue and black neck-band broad suggest a ♂, but there is much variation and overlap between sexes in this race. (A. Torés Sanchez)

— 100 —

cannot be reliably sexed. **Ad** ♀ Crown and edges to wings and tail less intense and dark blue, often slightly greyish-tinged blue. Bib often sooty-black, not jet-black with bluish gloss. Black stripes on head on average slightly narrower. Underparts often a trifle paler yellow. Dark 'boa' across nape on average narrower and less dark, bluish-grey. Wing < 65 mm. **1stY** ♂ Often as ad ♂, but on average slightly paler and less brilliant blue. Primary-coverts not so clearly bluish-edged, slightly duller greenish or greyish-tinged. A few aged as 1stY may be difficult to sex reliably. **1stY** ♀ Similar to ad ♀, but even duller, less clear blue colours. Primary-coverts edged dull greyish-green. – AUTUMN **Ad** Primary-coverts have rather rounded tips, edged bluish, uniform or nearly uniform with greater coverts and rest of wing. (Some ♀♀ have slightly duller greyish-blue edges, though no greenish cast.) Tail-feathers usually rather broad with rounded tips, keeping fresh through autumn and sometimes into early spring. **1stW** Primary-coverts have on average more pointed tips, edged dull bluish-green or greyish-green, after completing post-juv moult clearly duller than blue-edged greater coverts and rest of wing. Tail-feathers often slightly narrower and more pointed than in ad, and tips become abraded in autumn or early winter. **Juv** White of head replaced by pale yellow. Dark bib and stripe across neck-side only faintly suggested. – SPRING **Ad** As in autumn, although abrasion makes reliable ageing progressively more difficult (though generally still possible). **1stS** As in autumn; by late spring abrasion often makes ageing unreliable.

BIOMETRICS (*ultramarinus*) **L** 11–12 cm; **W** ♂ 60–67.5 mm (*n* 30, *m* 63.6), ♀ 59–64 mm (*n* 13, *m* 61.5); **T** ♂ 45–55 mm (*n* 30, *m* 48.9), ♀ 45–49 mm (*n* 13, *m* 47.3); **T/W** *m* 76.8; **B** 8.9–10.7 mm (*n* 45, *m* 9.7); **BD** 3.8–4.9 mm (*n* 45, *m* 4.3); **Ts** 16.0–18.0 mm (*n* 44, *m* 16.9). **Wing formula: p1** > pc 6–10.5 mm, < p2 19.5–25 mm; **p2** < wt 6.5–11.5 mm, =8/9 or 9 (79%), or =9/10 or 10 (21%); **p3** < wt 0–2 mm; **pp4–6** about equal and longest (p6 sometimes to 2 mm <); **p7** < wt 1.5–4 mm; **p8** < wt 4.5–7.5 mm; **p10** < wt 9.5–12.5 mm; **s1** < wt 11–14 mm. Emarg. pp3–7.

GEOGRAPHICAL VARIATION & RANGE Moderate variation in N African populations and E Canaries, more distinct on islands of C & W Canaries. All populations mainly sedentary.

C. t. ombriosus (Meade-Waldo, 1890) (W Canaries: El Hierro). Resembles *teneriffae*, but uppermost mantle is grey, lower mantle and back green. Upperwing dark, blue element subdued. Underparts darker bright yellow than other races,

C. t. degener, presumed 1stY ♂, Fuerteventura, Canary Is, Feb: slightly paler and greyer on mantle and back than *ultramarinus*. A bird with unusually strongly yellow underparts, while others often have slightly paler yellow belly. White wing-bar quite obvious. Apparently retained juv remiges and primary-coverts indicate age, while the deep ultramarine-blue crown patch and broad black-blue 'boa' across nape suggest a ♂. (M. Schäf)

C. t. degener, ad ♂ (right) and ♀, Fuerteventura, Canary Is, Feb: a perfect image showing sexual differences, with ♂ having much broader black-blue 'boa' on nape and neck-sides (but beware that size and shape also affected by posture), deeper bluish coloration above and, perhaps most conclusively, deeper yellow below. (T. Krumenacker)

C. t. degener, juv, Lanzarote, Canary Is, May: juv by pale areas on head being yellowish (not white), and lack of dark bib. Young age also evident from overall fluffy and soft-looking feathering. Head markings and crown patch somewhat darker and more prominent than in juv European Blue Tit. (H. Shirihai)

with only very slight dusky greenish-grey tinge on flanks. Very slight hint of a dusky-white, narrow wing-bar. Size and proportions similar to *teneriffae*, but tail proportionately slightly longer, and legs clearly longer. **W** ♂ 61.5–66.5 mm (*n* 15, *m* 63.4), ♀ 59–64 mm (*n* 7, *m* 60.7); **T** ♂ 48–55 mm (*n* 15, *m* 52.3), ♀ 47–54 mm (*n* 7, *m* 49.0); **T/W** *m* 81.9; **Ts** 17.1–19.4 mm (*n* 22, *m* 18.9).

C. t. palmensis (Meade-Waldo, 1889) (W Canaries: La Palma). Has head pattern much like *teneriffae*, but underparts differ markedly in that only breast, lower flanks and vent are yellow, whereas belly and upper flanks are dusky-white with no or only a faint trace of yellow; often a rather distinct border between vivid yellow and white. Sometimes a few black spots on centre of lower breast/belly. Blue colour of wings and tail rather subdued, closer to *degener* than to *teneriffae*. Mantle/back rather dark dirty bluish-grey. Wing rounded, bill rather thin, legs long and strong, tail long. **W** 57–65 mm (*n* 24, *m* 61.1); **T** 45–54 mm (*n* 24, *m* 51.0); **T/W** *m* 83.4. **Wing formula: p1** > pc 12–14.5 mm, < p2 15.5–18.5 mm; **p2** < wt 7–11 mm, =9/10 or 10 (70%), <10 (20%), or =9 (10%); **p10** < wt 6.5–10 mm; **s1** < wt 8–11.5 mm.

C. t. teneriffae (Lesson, 1831) (C Canaries: Tenerife, La Gomera). Similar to *ultramarinus* but black necklace usually narrower, and throat bib smaller and more sooty-black (lacking blue tinge). White band around dark cap

C. t. teneriffae, ad, presumed ♂, Tenerife, Canary Is, Mar: no white wing-bar and reduced dark central breast stripe in this race, which also usually has on average a rather small blackish bib. The rather broad black necklace possibly indicates a ♂. Note complete and neatly outlined white band around dark cap. Aged by evenly fresh wing with deep and shiny bluish edges. (V. Legrand)

C. t. teneriffae, presumed ad ♀, Tenerife, Canary Is, Oct: greyer blue plumage and less intense black head markings, suggest a ♀. Tentatively aged as ad by evenly fresh wing, but difficult to say for sure at this angle. (H. Shirihai)

TAXONOMIC NOTES See under Marsh Tit regarding the generic arrangement of tits. – Previously included as a group of distinct races in European Blue Tit, but now commonly treated as a separate species (Martin 1991). Sangster (2006) proposed species status for all Canary Islands races, but if consistently applied this would require a completely different taxonomic approach and in our opinion lead to undesired fragmentation, and thus it is not followed here. – Dietzen *et al.* (2008) lumped *degener* with *ultramarinus* on account of non-existent genetic difference in their study, but these two taxa still differ consistently morphologically outweighing the genetic argument (which only referred to a small part of the genome), and *degener* is maintained here. – Population on Gran Canaria originally named *hedwigii*, later emended to *hedwigae* by Manegold (2011) but such a change inadmissible according to the Code.

REFERENCES Dietzen, C. *et al.* (2008) *J. of Orn.*, 149: 1–12. – Kvist, L. *et al.* (2005) *Mol. Phyl. & Evol.*, 34: 501–511. – Manegold, A. (2011) *BBOC*, 132: 68. – Martin, J. L. (1991) *Ardea*, 79: 429–438. – Salzburger, W., Martens, J. & Sturmbauer, C. (2002) *Mol. Phyl. & Evol.*, 24: 19–25. – Sangster, G. (2006) *Mol. Phyl. & Evol.*, 38: 288–289.

neatly outlined and usually complete. No white wing-bar. Dark central breast-band missing or reduced to few spots. **W** ♂ 60.5–66 mm (*n* 14, *m* 62.7), ♀ 57–63 mm (*n* 16, *m* 59.9); **T** ♂ 45–55 mm (*n* 30, *m* 48.9), ♀ 46–52 mm (*n* 16, *m* 48.5); **T/W** *m* 80.3; **B** 9.4–10.8 mm (*n* 30, *m* 10.3); **BD** 3.7–4.5 mm (*n* 26, *m* 4.1); **Ts** 16.4–18.7 mm (*n* 28, *m* 17.6). **Wing formula: p1** > pc 9–11.5 mm, < p2 18–21 mm; **p2** < wt 6.5–11 mm, =8/9 or 9 (47%), = 9/10 or 10 (27%), or <10 (26%); **p7** < wt 0–2.5 mm; **p8** < wt 1.5–5.5 mm; **p10** < wt 7–11.5 mm; **s1** < wt 8–12.5 mm.

○ *C. t. hedwigii* (Dietzen *et al.*, 2008) (C Canaries: Gran Canaria). Inseparable from *teneriffae* on plumage (46 specimens directly compared with 21 *teneriffae*, NHM, refuting claims of paler grey-blue upperparts, a more extensive black bib with convex sides, and narrower white nape-band), but differs slightly genetically (Kvist *et al.* 2005, Dietzen *et al.* 2008) and has a subtly longer tail and stronger bill (latter by 4%). Said to have on average slightly longer song. Subtle race. **W** ♂ 61.5–66 mm (*n* 17, *m* 63.4), ♀ 59–62 mm (*n* 8, *m* 60.2); **T** ♂ 48–55 mm (*n* 16, *m* 51.6), ♀ 47–52 mm (*n* 8, *m* 50.0); **T/W** *m* 82.0; **B** 10.0–11.4 mm (*n* 27, *m* 10.7); **BD** 3.8–4.6 mm (*n* 25, *m* 4.3). (Syn. *hedwigiae*.)

C. t. degener (Hartert, 1901) (E Canaries: Lanzarote, Fuerteventura). Resembles *ultramarinus* but is slightly paler grey on mantle/back, with less or no blue tinge. Belly usually slightly paler yellow than breast. Blue of wings and tail slightly washed out, more greyish. Underparts purer pale yellow, with less greyish-olive cast. Bill slightly longer. Differs from *teneriffae* and *ombriosus* in presence of white wing-bar and paler colours. **W** ♂ 60–64 mm (*n* 12, *m* 62.4), ♀ 58–63 mm (*n* 10, *m* 59.6); **T** ♂ 45–54 mm (*n* 14, *m* 48.9), ♀ 42–48 mm (*n* 10, *m* 44.7); **T/W** *m* 75.7; **B** 9.4–10.7 mm (*n* 22, *m* 10.2).

C. t. ultramarinus (Bonaparte, 1841) (NW Africa east to Tunisia, Pantelleria). Treated above. Dark blackish cap with blue tinge, narrow (sometimes broken) white band between cap and broad blue-black necklace, connected to bluish-black triangular throat bib. Rest of underparts evenly yellow with dirty greyish-green tinge, cleanest yellow just below bib. Blackish central breast-band irregular or broken but present to some extent in all.

C. t. cyrenaicae (Hartert, 1922) (NE Libya). Smaller than *ultramarinus*, with darker greyish-blue mantle, and has less white on forehead. Underparts slightly darker and more saturated yellow, breast brighter. Usually somewhat longer-tailed and longer-legged. Wing more rounded, with shorter primary projection. **L** 11–11.5 cm; **W** 56–62 mm (*n* 12, *m* 57.8); **T** 46–52 mm (*n* 12, *m* 48.3); **T/W** *m* 83.5; **B** 10.2–11.1 mm (*n* 12, *m* 10.4). **Wing formula: p1** < p2 18–20 mm; **s1** < wt 7–11 mm.

C. t. ombriosus, ♂, presumed ad, El Hierro, Canary Is, Oct: resembles *teneriffae*, but has slight hint of whitish, narrow wing-bar, and is usually deeper bright yellow below, while upper mantle is grey, becoming greenish on lower mantle and back; blue upperwing generally duller. Aged by seemingly evenly fresh wing with purer bluish edges, and sexed as ♂ by broad jet-black head markings, with deep blue-black crown and very wide nape patch. (F. López)

C. t. palmensis, ad ♂, La Palma, Canary Is, Nov: rather uniquely, the underparts in this race are obviously yellow only on breast and body-sides, with contrastingly dusky-white belly patch. Central breast stripe usually absent or just a few black spots on lower breast/belly. (H. Nussbaumer)

C. t. palmensis, ad ♂, La Palma, Canary Is, Nov: the same bird as in image to left, here showing the rather subdued blue wings and dull bluish-grey upperparts, with rather narrow white wing-bar. Immaculate black-and-white head pattern also distinctive, with broad black necklace suggesting ♂. Aged by evenly fresh wing with rather pure bluish edges. (H. Nussbaumer)

AZURE TIT
Cyanistes cyanus (Pallas, 1770)

Fr. – Mésange azurée; Ger. – Lasurmeise
Sp. – Herrerillo azul; Swe. – Azurmes

The Azure Tit is an attractive, blue-and-white, close eastern relative of the widespread and well-known European Blue Tit. It is quite similar to the latter in appearance, only slightly bigger and with all yellow parts replaced by white and a different tail pattern. Resident in low-growing and open broad-leaved woods in Russia and Siberia, but there is also an isolated population in S Belarus. Records in W Europe are quite rare and may be linked to food shortage and irruptions originating further east. Favours willow stands along rivers and lakes, and is found in all sorts of thickets on damp ground including drier reedbeds mixed with trees, but also in oak forest on mountain slopes, etc.

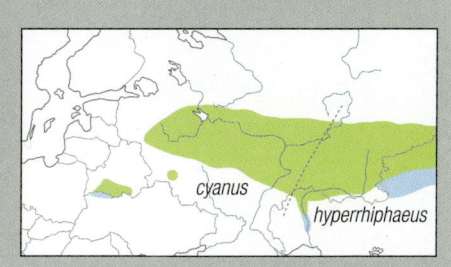

C. c. cyanus, presumed ad ♂, Belarus, Apr: unmistakable, smart-looking tit, with proportionately long tail and almost immaculate snowy-white and blue-and-grey plumage. Tentatively aged by evenly fresh wing, including primary-coverts with broad white tips, but ageing usually difficult in this species; apparently ♂ by rather broad, dark blue-grey nuchal band, evenly dark greyish-blue mantle, and shinier and bluer non-white areas of wing. (K. Blachowiak)

IDENTIFICATION Compared to Blue Tit a subtly bigger bird with proportionately *longer and a little more graduated tail*. The striking difference immediately identifying the species is the generous amount of white in the plumage: *crown and all of underparts including chin and foreneck are white*, and *tertials, greater coverts and outertail have extensive white parts*. Lower mantle and back are bluish-grey (not green). Like in Blue Tit there is a dark bluish-grey nuchal band with a narrow dark line running through the eye and joining the nuchal band. There is also a poorly-marked and indistinct thin dark central band on breast/upper belly visible in most birds, just like in Blue Tit. But apart from the white instead of yellow underparts, Azure Tit differs in *lacking the dark bib on the chin and the dark line across foreneck* below the cheeks. While it is true that a genuine Azure Tit has a white crown, a *very faint* bluish-grey shade, especially on the rear crown and in juveniles, may occur without this necessarily being a sign of hybrid origin. Since crosses with Blue Tits are well known and even have been given a separate name, 'Pleske's Tit', it is important to eliminate this option by going through a checklist of crucial characters (see Similar species). The white underparts can rarely be slightly stained or soiled buffish-cream and grey, especially in spring and summer on worn breeding birds, and in fresh autumn plumage there is often a very faint grey wash to the flanks.

VOCALISATIONS Has a rather rich and varied repertoire. Many utterings are reminiscent of European Blue Tit, others also of Great, Marsh, Crested or Long-tailed Tits. Song variable, can be quite Blue Tit-like with initial few high-pitched notes followed by a quicker series of lower-pitched ones, these concluding notes sometimes forming a clear trill, and then easily confusable with Blue Tit, e.g. *tse-tse tsih tsih-tserrrr*. Song variants that include repetition of a slightly longer phrase can somewhat recall Crested Tit in structure but with Blue Tit voice, *tsi-tsi-tserr-su-ih, tsi-tsi-tserr-su-ih, tsi-tsi-tserr-su-ih*. Simpler song structures also occur, like *tsirr-chi tsirr-chi tsirr-chi* or *tsowaytü tsowaytü tsowaytü* (more like a mix of Marsh and Great Tit). – Calls include Blue Tit-like fine *si-si-si*, or sharper variants, with voice more like Long-tailed Tit, *zi-zi-zi-zi*. When agitated or alarmed uses a Blue Tit-like scolding *ker'r'r'err*, often combined with high-pitched notes.

SIMILAR SPECIES The only risk of confusion with something other than the 'real thing' is with a hybrid between Azure Tit and Blue Tit, known as 'Pleske's Tit'. Such birds have been recorded many times in European Russia, a few times in Fenno-Scandia and in various places in E Europe. Therefore, although not formally 'a similar species', the following are useful points to remember when trying to eliminate the risk of mistaking a hybrid for a true Azure Tit: (i) tail should have extensive white, wedge-shaped corners involving much white distally on at least outer four feathers and broad white tips to rest; (ii) greater coverts should be all white or at least have white distally over more than half of their length, forming a very broad white wing-bar, broadest on the *inner* part, near the tertials; (iii) tips to primary-coverts should be broadly white making the white wing-bar on folded wing look angled or V-shaped; (iv) tertials should be largely white with only a bluish-black basal patch on outer webs (partially hidden), although juv-feathers may show less white; (v) chin and foreneck should be all white, without a dark bib or dark throat-line; and (vi) crown should be white or, if not pure white, with no more than a faint shade of blue-grey at rear.

AGEING & SEXING (*cyanus*) Extremely difficult to age, if at all possible. Sexes equally large and very similar in plumage. – Moults. Not closely studied. Complete post-nuptial moult of ad in summer after breeding (late Jun–Aug, at times early Sep). Partial post-juv moult in (late) summer generally does not involve any primary-coverts, primaries or secondaries. Usually replaces all greater coverts, tertials and alula, plus sometimes one or two pairs of central tail-feathers. Appar-

C. c. cyanus, Finland, Jan: the immaculate snowy-white and greyish-blue plumage is perhaps best admired from above, but depending on angle of view and light the attractive blue colours here look darker and duller. Ageing and sexing as for image above; very difficult, and best avoided when not handled. (J. Peltomäki)

ently no pre-nuptial moult. – Sexing generally not possible, but with series to compare, or with breeding pair together, some average difference show as slight tendencies and might be helpful. – **ALL YEAR Ad** ♂ Crown and centre of underparts (apart from short dark blue central line) pure white. Nuchal band rather broad and dark blue-grey. Lower mantle/back rather evenly dark blue-grey. Primaries rather clear blue (apart from white edges). **Ad** ♀ Crown and centre of underparts sometimes, but not always, faintly sullied cream or 'dirty-looking'. Nuchal band tends to be narrow and less well marked, and colour more grey than blue. Lower mantle/back often slightly less dark and blue, more dull greyish (though much overlap with ♂). Edges of primaries not so clear blue-tinged, have greyer element. **1stY** Sexing of imm even more difficult than of ad. On average, blue colours are more subdued, and although this refers to both sexes the difference in imm is even slighter than between ad. – **AUTUMN** In collections, plumage is very adult-like and neat in *all* birds, reasonably so also in 1stY. A very few seem to have more pointed tips to tail-feathers (= 1stY), others well rounded (= mainly ad?). Shape and colour of primary-coverts not very helpful.

BIOMETRICS (*cyanus*) **L** 12.5–14 cm; **W** 67–73 mm (*n* 20, *m* 69.8); **T** 59–67 mm (*n* 20, *m* 63.8); **T/W** *m* 91.3; **B** 9.4–10.5 mm (*n* 20, *m* 9.9); **BD** 4.5–5.5 mm (*n* 18, *m* 5.2); **Ts** 15.5–17.4 mm (*n* 19, *m* 16.3). **Wing formula: p1** > pc 8–13 mm, < p2 23–28 mm; **p2** < wt 8–11 mm, =8 (12%), =8/9 (76%) or =9 (12%); **p3** < wt 0.5–3 mm; **pp4–6** about equal and longest; **p7** < wt 1.5–3.5 mm; **p8** < wt 4.5–9 mm; **p10** < wt 9.5–14.5 mm; **s1** < wt 10–15.5 mm. Emarg. pp3–7.

GEOGRAPHICAL VARIATION & RANGE Slight variation over large areas of homogenous taiga habitats, somewhat more marked extralimitally in mountains of Central Asia (not treated here). As a rule resident, but in some years local winter movements towards south and south-west.

C. c. cyanus (Pallas, 1770) (Belarus, N Russia, European part). Treated above. Mantle rather dark greyish-blue, and dark portions of wings and tail deep blue and dark grey. Grades into W Siberian ssp. *hyperrhiphaeus* in Volga Valley and Ural Mts, mantle progressively becoming paler blue-grey, and dark portions of flight-feathers not quite so dark.

C. c. hyperrhiphaeus Dementiev & Heptner, 1932 (extreme SE Russia near Urals, SW Siberia, NW Altai). A trifle paler than *cyanus*, lower mantle/back being more grey than blue even in ad ♂. Rump almost white. White portions on wings tend to be subtly larger (but much overlap). The two races are connected by a wide zone of intergrading characters, and situation south-west of S Urals might need

C. c. cyanus, 1stW ♂, **Finland, Feb**: Azure Tits will visit bird feeders in winter just as other tit species, and this is how most are detected. Ageing and sexing difficult, but being one of three known young ♂♂ that reached Finland in winter 2006–07 helps. (T. Lindroos)

C. c. hyperrhiphaeus / tianschanicus, juv, **Kyrgyzstan, Jul**: greyer plumage, duller blue wing and tail, shorter tail and yellow gape, as well as fluffy feathering above and below. Juv of all races very similar. (H. & J. Eriksen)

C. c. hyperrhiphaeus, ad, **Altai, Russia, May**: further east, some populations are more grey than blue on mantle with white in wings usually more extensive (here an extreme example). Very broad white tips to primary-coverts and seemingly completely fresh wing indicate age, while sex is difficult to decide due to much overlap. (S. Pisarevsky)

C. c. tianschanicus, ♂, presumed 1stS, Mongolia, May: broad dark grey-blue bases to primary-coverts with narrow white tips apparently more common in extralimital race *yenisseensis*. This bird was singing and defending territory, hence it can safely be sexed. Age more difficult to decide, but slightly duller primary-coverts might infer 1stS. (H. Shirihai)

C. c. tianschanicus, ♀, Mongolia, Jun: sexing usually very difficult, but here behaviour in breeding activity helped. Note combination of weak eye-stripe and rather poorly defined and not so dark blue-grey nuchial band, plus not so bright blue wing, together often indicating ♀. (H. Shirihai)

further study. Birds collected in Orenburg fit this race, but so does an early March bird as far west as Odessa (a straggler?). Size and structure much as *cyanus*: **W** 67–74 mm (*n* 27, *m* 70.1); **T** 59–71 mm (*n* 26, *m* 63.3); **T/W** *m* 90.4. Wing formula: **p1** < p2 23–25.5 mm. – Grades into extralimital *tianschanicus* in S Siberia, NW Mongolia and S Altai.

C. c. flavipectus (Severtsov, 1873) (Central Asian mountains from Tien Shan to Pamir, extending through Afghanistan and apparently marginally into NE Iran, although exact range not known). Similar to *cyanus* and *hyperrhiphaeus* but in post-juv plumages readily recognised by large bright yellow patch on breast, brightest in ♂♂. Crown, cheeks and throat slightly sullied greyish. **W** 64–71 mm, **T** 55–62.5 mm, **Ts** 16–18.5 mm (Harrap & Quinn 1996).

Extralimital: **C. c. tianschanicus** Menzbier, 1884 (S Siberia, SE Altai, E Kazakhstan, Central Asia and Tien Shan region, N Mongolia eastward to Transbaikalia and Amur and Ussuri; N China). Blue of back deeper and darker than in *cyanus*, and crown often tinged pale grey-blue or grey. Size similar to that of *cyanus*. – Birds in the north and west, in S Siberia, Kazakhstan and NW Mongolia, sometimes separated as *yenisseensis*, tend to be slightly duller and paler on average, but much overlap and here tentatively included. **W** 62–73 mm, **T** 56–63 mm (Harrap & Quinn 1996). (Syn. *yenisseensis*.)

TAXONOMIC NOTES See under Marsh Tit regarding the generic arrangement of tits. Hybridisation with Blue Tit *C. caeruleus* where the two species come into contact occur, was summarised by Lawicki (2012), but frequency can merely be roughly estimated since many hybrids go unnoticed. Some hybrids (see above under Similar species) recall ssp. *flavipectus*, and securing detailed notes of tail, head and wing patterns is crucial. – Ssp. *flavipectus* by some regarded as separate species, 'Yellow-breasted Tit', but we follow Eck & Martens (2006) in regarding it as a distinct subspecies of Azure Tit.

REFERENCES Eck, S. & Martens, J. (2006) *Zool. Med. Leiden*, 80-5: 1–63. Harrap, S. (1995) *BW*, 8: 382–389. – Lawicki, L. (2012) *DB*, 34: 219–231.

C. c. hyperrhiphaeus / tianschanicus, 1stS, presumed ♀, Kazakhstan, May: narrow grey-white tips to primary-coverts clearly confirm age, while rather narrow and grey-tinged nuchial band and duller blue wing suggest a ♀. (C. Bradshaw)

C. c. flavipectus, presumed ad ♂, Kyrgyzstan, Apr: bright yellow breast patch and greyish tinge to white crown, cheeks and throat typical of this localised race in Central Asian mountains. Combination of seemingly quite fresh wing and tail and broad dark grey-blue nuchal band, and also dark bright blue in wing, infer possible age and sex. (M. Westerbjerg Andersen)

GREAT TIT
Parus major L., 1758

Fr. – Mésange charbonnière; Ger. – Kohlmeise
Sp. – Carbonero común; Swe. – Talgoxe

This is surely the best-known tit, widespread and common as it is, and well adapted to a life close to humans, in gardens and around houses. Its conspicuous plumage with strong colours and striking pattern makes it easily recognised. With this comes a snoopy and almost obtrusive nature paired with boldness; it is one of the few wild birds that dares to take food directly out of the hand of a human (although not all do so). Interestingly, and testimony to its success as a species, it is not confined to gardens and habitations, but is also found in the most remote taiga forests miles from the nearest road or house. It is a resident bird that rarely moves far from its breeding territory.

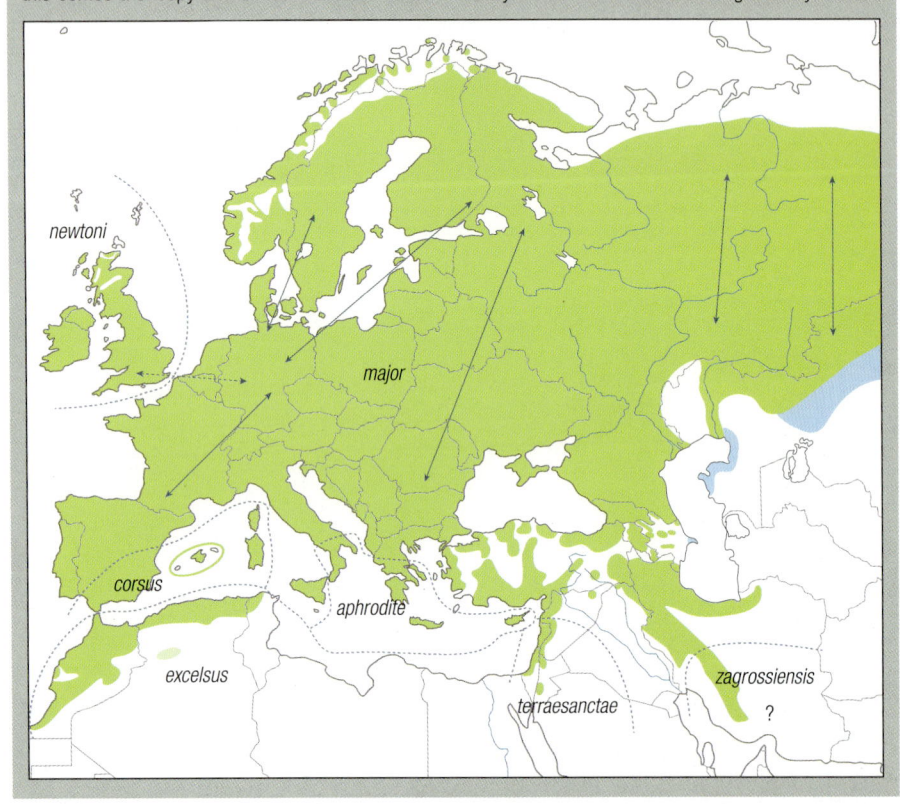

P. m. major, ad ♂, Finland, Oct: a large, robust tit, with black head, large white cheek patches, yellow underparts with black bib that narrows into ventral line, and bluish-grey wings with distinct white wing-bar make the species almost unmistakable. Aged by evenly-feathered wing (including primary-coverts) and sexed as ♂ by broad, unbroken, black ventral stripe. (M. Varesvuo).

IDENTIFICATION One of the *largest* tits, strongly built with a somewhat rounded head, a fairly long, square-ended bluish-grey tail, *strong grey legs* and a *short, strong and stubby, black bill*. Easily identified on its *yellow underparts* with a *black ventral line along the centre*, and its *glossy black head with large white patches on each side*, the white encircled by black margins. The *back* is an attractive *moss-green*, in some parts of Europe with bluish-grey rear part and rump. Wings are edged bluish-grey and have a *distinct white wing-bar* on greater coverts. There is generally a small yellowish-white patch on the nape, between the black rear crown and the green mantle. Sexes differ in that the ♂ has a complete ventral band widening to a broad black patch between the legs and on vent, whereas the ♀ has a narrower and often broken ventral band that does not broaden into a patch. – Juvenile separated by a yellow tinge to the pale cheek patches, and by lacking complete black borders to the pale cheeks. Also, all dark portions lack gloss and are dull and sooty rather than black.

VOCALISATIONS The song, although varying in presentation between ♂♂, is easily recognised by its loudness, moderate pace, simple structure and repetitive nature. Very commonly the song is just a series of the same two notes uttered alternatively, often rendered *tea-*cher *tea-*cher *tea-*cher *tea-*cher *tea-*cher. Another common variant repeats three notes, *see-see-chu*, in a series, and with a little imagination this can sound like a familiar tune by Mozart ('Non piú andrai' from 'La Nozze di Figaro', only with the final note missing). Less commonly, simple variants with just one repeated note also occur, *tsee tsee tsee tsee…*, and yet other variations can recall Marsh Tit song. – Has a very rich repertoire of calls, perhaps the richest of all European bird species. Here only a few examples are given. Among the commonest calls is a rather Chaffinch-like *ping ping*. Another frequent call is a cheerful *see-yuttee-yuttee* and *tee-tuuee*. In the non-breeding season you often hear a slightly melancholy *tee tee tüh* with last note lower. When agitated (quarrelling, in anxiety) gives a scolding, hoarse *che-che-che-che-che* (quite like a small Magpie!). Begging call of young, often heard in gardens in summer, is a persistent *ti-ti-te, ti-ti-te, ti-ti-te, …* etc.

SIMILAR SPECIES If in Europe or Turkey, once the yellow breast and the black-and-white head is seen there is usually no confusion possible. Blue Tit also has a yellow breast and a wing-bar (and occurs in gardens like Great Tit) but is smaller with a 'cuter' look created by smaller bill, shorter tail and more white on the head, with a blue cap or 'beret' on crown encircled by white, and with fewer or narrower dark markings on throat, around neck and on centre of underparts. – Birds more similar to Great Tit in size, shape and head pattern occur in E Iran, firstly the closely related *Cinereous Tit*, but this lacks yellow on the underparts and green on the back, being a blue-grey and off-white version of Great Tit. – In NE Iran the *Turkestan Tit* (which see) also occurs; it has the same plumage colours as Cinereous Tit, with blue-grey mantle and off-white underparts, and is thus readily separated from Great Tit by the lack of yellow. Turkestan Tit also has a proportionately longer tail, which is more graduated and has more white on its sides.

AGEING & SEXING (*major*) Ageing possible using colour and moult differences during 1stY. Sexes separable. No seasonal variation. – Moults. Complete post-nuptial moult of ad in summer after breeding (mainly late Jul–early Oct in N Europe, earlier in W & S Europe). Partial post-juv moult in summer (on average slightly earlier than ad moults) generally does not involve any primary-coverts, primaries or secondaries. All greater coverts usually replaced, but some retain a few outer; at least inner two tertials, often all, moulted, and often a few of the smaller alula-feathers (less often all). A variable number of tail-feathers (rarely none; almost invariably r1, but frequently several or all pairs) replaced. Very rarely a few inner primary-coverts are replaced as well (making ageing potentially more difficult), but no case of

P. m. major, ♀ (left) and ♂, **Finland, Apr** sexes differ in width, prominence and shape of black ventral band, which is usually glossy, broad and unbroken in ♂, becoming a broad patch between legs and on vent, narrower and often broken with no broad patch on vent in ♀. Crown and throat of ♂ are by comparison glossier black. (A. Juvonen)

P. m. major, ♀, **Finland, Apr**: matt sooty-black (less glossy black) dark parts on head and breast than in ♂, with ventral stripe narrow, partially broken or irregular, and which does not broaden into black patch on vent as in ♂. (A. Juvonen)

P. m. major, 1stY, presumed ♂, **Italy, Feb**: race *major* has saturated moss-green mantle and back and vivid yellow underparts with slight greyish-olive tinge on flanks and vent, and pure white wing-bar. Aged by more pointed, dull bluish-green primary-coverts, and sexed as most likely ♂ based on glossier black crown with metallic sheen, but central belly stripe is almost invisible. (D. Occhiato)

confirmed 1stW with moulted long outer primary-coverts. No pre-nuptial moult. – **ALL YEAR** ♂ Crown and throat glossy black, in some lights or angles with slight bluish, metallic sheen. Ventral stripe black and usually also glossy, wide and unbroken, broad first on upper breast and then also widening between legs and on vent to a large patch. Wing-length gives some support if studied population is well known (see Biometrics and Geographical variation). ♀ Dark colours on head and breast less jet-black and glossy, on average more matt and sooty-black. Difference most noticeable on throat and ventral band, whereas crown in ad ♀ can be quite black and glossy. Ventral band often narrow and can be broken or irregular lower down, not widening to broad black patch between legs and on vent as in ♂; often appreciably narrower on breast than in ♂. Wing-length can give some support when dealing with a borderline case. – **AUTUMN Ad** Longest primary-coverts have rather rounded tips, edged bluish-grey, uniform or nearly uniform with greater coverts and rest of wing. (Some ♀♀ have slightly duller greyish-blue edges, but no greenish cast.) Tail-feathers usually rather broad with rounded tips, keeping fresh through autumn and sometimes into early spring. **1stW** Primary-coverts have on average more pointed tips (but shape not infallible

P. m. major, 1stY ♀, **Netherlands, Jan**: primary-coverts just visible, revealing pointed and duller greyish-green edges to brown-grey juv feathers, while narrower, quite broken ventral stripe and lack of broad patch on vent safely sex it as ♀. (N. D. van Swelm)

P. m. major, 1stY ♀, **Netherlands, Mar**: duller young ♀♀ are often straightforward to age and sex. Note worn and pointed juv primary-coverts, pale yellow underparts and little blue tinge on wing edges, and duller black crown supports this. (N. D. van Swelm)

as age criterion), edged dull bluish-green or greyish-green, after completed post-juv moult often appreciably duller than blue-edged renewed greater coverts and rest of wing. Tail-feathers often slightly narrower and more pointed than in ad, and tips become abraded in autumn or early winter. **Juv** White of cheek patches on head replaced by pale yellow. Dark markings on head absent or only faintly suggested. — SPRING **Ad** As in autumn, although abrasion makes reliable ageing progressively more difficult (but generally still possible in many). **1stS** As in autumn, although abrasion makes reliable ageing progressively more difficult.

BIOMETRICS (*major*) **L** 13.5–15 cm; **W** ♂ 74–81 mm (*n* 32, *m* 76.7), ♀ 70–77 mm (*n* 38, *m* 73.9); **T** ♂ 60–69 mm (*n* 32, *m* 64.3), ♀ 57–66 mm (*n* 38, *m* 60.4); **T/W** 82.7; **B** 11.1–13.2 mm (*n* 64, *m* 12.2); **BD** 4.3–5.3 mm (*n* 61, *m* 4.8); **Ts** 18.7–21.3 mm (*n* 65, *m* 19.7). **White on r6:** wedge on inner web 4–33 mm (*n* 75, *m* 18.5) (79% had 7–28). **Wing formula: p1** > pc 8–12 mm, < p2 22–28 mm; **p2** < wt 9–15 mm, =8 or 8/9 (43%), =9 or 9/10 (47%) or <10 (10%); **p3** < wt 1–3 mm; **pp4–6** about equal and longest (though p6 often to 2 mm <); **p7** < wt 2–5.5 mm; **p8** < wt 6.5–10.5 mm; **p10** < wt 11–16 mm; **s1** < wt 12–16 mm. Emarg. pp3–7.

GEOGRAPHICAL VARIATION & RANGE Mainly slight and clinal variation. Forms superspecies with three other closely related and similar taxa in Asia, treated here as separate species. Within large part of range, variation is rather slight, affecting size, colour of mantle, underparts, amount of white on r6 and size of bill. Definition of races in the Mediterranean area poses many problems; if assessment is based on short series, or on not entirely comparable plumages, it is easy to conclude that several more races than listed here are sufficiently distinct. Here, focus is laid on post-juvenile ♂♂ in fresh plumage in an attempt to eliminate any variation not referable to populations. The following races are thought to be distinct. Plumage descriptions thus refer throughout to fresh post-juv ♂♂ unless otherwise stated. At least one more extralimital subspecies described, not treated (*kapustini*, Dzungaria, Mongolia and adjacent areas). Mainly sedentary, but some northerly breeders undertake winter movements towards south-west, at least in some years.

P. m. major L., 1758 (Europe except British Isles, south to Portugal, C Spain, France except Corsica, Italy except southernmost mainland but including Sicily, Greece except in the south, Turkey, Caucasus, Transcaucasia, W Iran south to about Esfahan in N Zagros). Treated above. Mantle/back saturated moss-green. Vivid yellow underparts with slight

P. m. major, juv, Finland, May: cheek patches tinged yellow, and dark markings on head and throat much duller sooty-black. Body plumage still mostly juv, being soft and fluffy, and gape is still yellow. (J. Peltomäki)

P. m. newtoni, ad ♂, England, Apr: the race breeding on the British Isles is only subtly differentiated from *major* by its darker greenish upperparts, and is rather strong-billed and long-legged. Also often has slightly yellow-tinged white wing-bar (difficult to see here). Aged by ad primary-coverts. A very typical ♂ by unbroken, broad ventral black stripe that widens between the legs. (G. Thoburn)

P. m. corsus, ad ♂, Spain, Apr: this race is on average subtly duller greenish above than *major*, with yellow below paler and suffused greyer on flanks. Sexing can be less straightforward as ad ♂ tends to have narrow and less solid black ventral stripe, but band usually still broadens between legs, and black of head is glossier (as here). (C. N. G. Bocos)

greyish-olive tinge on flanks and vent. Usually has yellowish patch on nape. Large. — We have not found compelling evidence to validate claims that birds of S Caspian coastlands ('*karelini*') or in NW Iran down to Zagros ('*blanfordi*') differ sufficiently from *major* to warrant distinction. The S Caspian population is only a trifle smaller than *major* (**W** ♂♀ 69–78 mm, *m* 73.9, *n* 18) and has slightly smaller white wedge on inner web of r6 (2–24 mm, *m* 10.3), but otherwise is identical in plumage and structure. Birds in W Iran do not differ from *major* in any appreciable way, except that they have slightly more white on r6 (10–38 mm, *m* 23.9, *n* 29). (Syn. *blanfordi*; *karelini*.)

○ *P. m. newtoni* Pražák, 1894 (Britain, Ireland). Plumage as *major*, or is subtly darker green on back. Nearly always has yellow-tinged edges to tertials and tips of greater coverts (usually whiter in *major*). Strong bill and long legs, but differences, if at all significant, are subtle and show much overlap; there is also seasonal variation in bill size (Gosler 1999). On average slightly less extensive white on r6. Same general size as *major*. **W** ♂ 74–81 mm (*n* 23, *m* 76.7), ♀ 70–75 mm (*n* 12, *m* 72.6); **T** ♂ 58–67 mm (*n* 22, *m* 63.1), ♀ 55–62 mm (*n* 12, *m* 59.4); **T/W** 82.1; **B** 10.7–13.6 mm (*n* 35, *m* 12.6); **BD** 4.4–5.9 mm (*n* 35, *m* 5.1); **Ts** 18.2–21.5 mm (*n* 35, *m* 20.3). **White on r6:** wedge on inner web 3–31 mm (*n* 32, *m* 16.0) (81% had

5–25). **Wing formula:** very similar to *major*, only subtly blunter wingtip: **p2** =8 or 8/9 (40%), =9 or 9/10 (30%) or ≤10 (30%).

P. m. corsus Kleinschmidt, 1903 (S & SE Spain, Balearics, Corsica, Sardinia). Has mantle/back on average subtly darker and duller green than in *major*, although many are inseparable. Underparts very similar to *major* but yellow perhaps subtly more suffused grey, especially on flanks. Has less white on inner web of r6 than *major*, although extremes overlap. Wing somewhat more bluntly tipped. Sexing sometimes not as straightforward as in *major*, black central band of ♂ frequently broken up like in ♀, but ♂ generally separated by more extensive and glossier black throat. **W** ♂ 71–77 mm (*n* 37, *m* 74.0), ♀ 68–75.5 mm (*n* 20, *m* 71.3); **T** ♂ 58–68 mm (*n* 37, *m* 63.2), ♀ 55–64 mm (*n* 20, *m* 59.5); **T/W** *m* 84.6; **B** 11.1–13.6 mm (*n* 57, *m* 12.6); **BD** 4.1–5.3 mm (*n* 56, *m* 4.7); **Ts** 18.5–22.3 mm (*n* 54, *m* 19.9). **White on r6:** wedge on inner web 0–25 mm (*n* 55, *m* 8.0) (84% had 2–19). **Wing formula: p1** < p2 20–27 mm; **p2** =8 or 8/9 (4%), =9 or 9/10 (39%) or ≤10 (57%); **p7** < wt 1.5–5 mm; **p8** < wt 4–10.5 mm; **p10** < wt 11–14.5 mm; **s1** < wt 11–16 mm. – We include birds of Balearics and Sardinia here since we could not validate claims of separate races for these ('*mallorcae*' and '*ecki*', respectively), but found them very similar to *corsus*. Among populations included here, those breeding on Corsica have least white on r6 (0–11 mm, *n* 33, *m* 4.5), whereas other populations have slightly more. (Syn. *alanorum*; *ecki*; *mallorcae*.)

P. m. excelsus Buvry, 1857 (NW Africa from Morocco to Tunisia). Plumage as *major*, but underparts on average a little deeper and purer yellow (and less tinged olive on flanks). On average much less white on r6 than in *major* and slightly less even than in *corsus sensu lato* (about the same small amount as in breeders on Corsica). Compared to *corsus*, slightly deeper, more saturated colours. Strong bill. **W** ♂ 71–77 mm (*n* 21, *m* 74.0), ♀ 70–75.5 mm (*n* 12, *m* 71.7); **T** ♂ 58–68 mm (*n* 21, *m* 63.9), ♀ 57–62 mm (*n* 12, *m* 60.1); **B** 11.5–14.1 mm (*n* 29, *m* 13.1); **BD** 4.4–6.1 mm (*n* 26, *m* 5.1); **Ts** 18.5–22.0 mm (*n* 30, *m* 20.5). **White on r6:** wedge on inner web 0–14 mm (*n* 33, *m* 4.5) (88% had 0–7). Wing formula very similar to *corsus*. (Syn. *lynesi*.)

○ **P. m. aphrodite** Madarász, 1901 (southernmost mainland Italy, S Greece, Crete, Cyprus). In the east (Greece, Cyprus), compared to *major* less bright moss-green on mantle, more tinged lead-grey, and grey of rump more

P. m. excelsus, ♂, presumed ad, Morocco, May: the NW African race is similar to *major*, but underparts are on average deeper and purer yellow. It is also a strong-billed race. Very broad and solid black ventral stripe visible on breast clearly indicates ♂. Ageing difficult to assess in this bird. (G. Olioso)

P. m. excelsus, 1stW ♀, Tunisia, Oct: retained juv primary-coverts lacking bluish-tinged edges, and narrower, quite broken ventral stripe that does not broaden between the legs indicate that this is a young ♀. (W. Suter)

P. m. aphrodite, ♂, Cyprus, Nov: this subtle and inconsistent race can be recognised only when comparing series of specimens next to series of neighbouring populations by its slightly duller, slightly olive-tinged yellow body-sides. ♂ by glossy, broad and unbroken black ventral stripe. (F. Trabalon)

P. m. aphrodite, presumed 1stW, Cyprus, Nov: this race is characterised by duller moss-green mantle. Note that flanks are less vivid yellow, being slightly tinged olive. Presumed 1stW due to apparently retained primary-coverts and some rectrices (but difficult to judge). Sex best left as undecided since ventral stripe is only partially visible. (A. McArthur)

P. m. terraesanctae, ad ♂, Israel, May: the Levantine race averages paler overall, with duller yellow underparts, more greyish-green upperparts and slightly duller blue wings. Age and condition of primary-coverts difficult to assess. Sex however easy as ♂ by glossy black parts and unbroken, broad ventral stripe that widens between legs. (H. Shirihai)

P. m. terraesanctae, ♀, presumed ad, Israel, May: even duller than ♂ with less glossed black head, and rather narrow and somewhat broken ventral stripe indicate a ♀. Note that bill in this race is proportionately quite strong. (H. Shirihai)

extensive, diffusely invading back. Flanks more tinged olive-grey. Birds in the west (S Italy) greener on mantle, closer to *major* and *corsus*. Slightly variable race, not always distinct. Biometrics close to *major*, only slightly smaller with shorter legs. **W** ♂ 70.5–76 mm (*n* 21, *m* 73.0), ♀ 67.5–72.5 mm (*n* 13, *m* 70.1); **T** ♂ 59–65 mm (*n* 21, *m* 61.9), ♀ 53–62 mm (*n* 13, *m* 57.3); **Ts** 18.0–20.6 mm (*n* 30, *m* 18.8). **White on r6**: wedge on inner web 7.5–30 mm (*n* 33, *m* 18.4) (82% had 10–30). **Wing formula: p2** =8 or 8/9 (16%), =9 or 9/10 (42%) or ≤10 (42%); **p7** < wt 1.5–3 mm; **p8** < wt 5.5–8 mm; **p10** < wt 10–13.5 mm; **s1** < wt 11–14 mm. – A long series from Crete ('*niethammeri*') was compared with birds from S Italy and mainland Greece, and virtually no difference could be detected. (Syn. *niethammeri*; *peloponnesius*.)

P. m. terraesanctae Hartert, 1910 (Levant). Very similar to *major* but slightly smaller and on average paler and cleaner yellow below, flanks not sullied olive-grey, and yellow-white nape patch larger and more prominent. Blue tinge on wings and tail slightly duller, more subdued. Compared to *aphrodite* slightly paler and cleaner in colours, but quite similar. **W** ♂ 69–77 mm (*n* 12, *m* 72.5), ♀ 66.5–71.5 mm (*n* 12, *m* 68.4); **T** ♂ 57–68 mm (*n* 12, *m* 61.5), ♀ 53–60 mm (*n* 12, *m* 57.0); **T/W** *m* 84.1; **B** 10.5–12.8 mm (*n* 24, *m* 12.0); **BD** 4.0–5.0 mm (*n* 24, *m* 4.5); **Ts** 18.0–20.7 mm (*n* 23, *m* 19.1). **White on r6**: wedge on inner web 15–30 mm (*n* 22, *m* 23.9) (85% had 20–30). **Wing formula**: close to *major*, **p2** =8 or 8/9 (22%), =9 or 9/10 (56%) or ≤10 (22%).

○ **P. m. zagrossiensis** Zarudny & Loudon, 1905 (Fars, S Zagros, SW Iran). Similar to *major*, which it grades into in N Zagros, but is slightly smaller and on average paler and cleaner yellow below, flanks not sullied olive-grey, and upperparts on average subtly paler, duller and less green, more tinged grey. **W** ♂ 74–78 mm (*n* 10, *m* 76.1), ♀ 69–78 mm (*n* 9, *m* 73.0); **T** ♂ 62–67 mm (*n* 9, *m* 64.4), ♀ 57–67 mm (*n* 9, *m* 59.9). **White on r6**: wedge on inner web 22–39 mm (*n* 18, *m* 30.6).

TAXONOMIC NOTES See under Marsh Tit regarding the generic arrangement of tits. – Best taxonomic arrangement of the Great Tit complex (comprising the superspecies *major–cinereus–minor–bokharensis*) has been much discussed and is a matter of some controversy. A common solution based on morphology is to single out Turkestan Tit *P. bokharensis*, but to keep the other three as distinct racial groups within a widely distributed and variable Great Tit. However, recent genetic evidence seems to indicate that Turkestan Tit is closely related to ssp. *major* of Great Tit, closer than to the more similar-looking Cinereous Tit. To avoid a paraphyletic arrangement, it seems better to afford all four groups in the Great Tit complex species status. Hybridisation in limited overlap areas is not a problem when adhering to the Biological Species Concept (*contra* Päckert & Martens 2008) provided there are no signs of weakening of the 'typicity' of the involved taxa. No such tendency has been noted according to Päckert & Martens (2008). Moreover, hybrids between Great Tit and Japanese Tit *P. minor* (extralimital; not treated) showed reduced fitness (Kvist & Rytkönen 2006). – As to the name *zagrossiensis*, this was published originally as *zayrossiensis* (sic), but since the type locality was given as Zagros Mts the correct spelling should be that first mentioned. Zarudny's intentions are clear as he later published a correction ('*zagrossiensis* Zarudny, 1911'). The senior name *blanfordi* Pražák, 1894, referred to a wider part of Iran, much of which (including the *blanfordi* type locality 'Tehran') we consider to be inhabited by birds inseparable from *major*, therefore name *blanfordi* cannot take precedence over *zagrossiensis*, and is instead regarded as a junior synonym of *major*.

REFERENCES Domènech, J., Senar, J. C. & Vilamajó, E. (2000) *Butlleti Anellament GCA*, 17: 17–23. – Eck, S. (1980) *Falke*, 27: 385–392. – Eck, S. (1988) *Zool. Abh. Mus. Tierk. Dresden*, 43: 101–134. – Eck, S. (2006) *Zootaxa*, 1325: 7–54. – Gosler, A. G. (1999) *BBOC*, 119: 47–55. – Kvist, L. *et al.* (2003) *Proc. Roy. Soc. London*, 270: 1447–1454. – Kvist, L. *et al.* (2005) *Mol. Phyl. & Evol.*, 34: 501–511. – Kvist, L. & Rytkönen, S. (2006) *Zootaxa*, 1325: 55–73. – Kvist, L. *et al.* (2007) *Biol. J. Linn. Soc.*, 90: 201–210. – Päckert, M. & Martens, J. (2008) *Ibis*, 150: 829–831.

CINEREOUS TIT
Parus cinereus Vieillot, 1818

Fr. – Mésange indienne; Ger. – Graumeise
Sp. – Carbonero cinéreo; Swe. – Gråmes

The well-known Great Tit has a wide distribution in Europe and Asia. It is divided into four rather well-defined subspecies groups sometimes maintained together in one species, the Great Tit, or split in two by recognising the Turkestan Tit as a separate species. Here we have split the complex into four species following recent genetic research and modern trends in taxonomy. The Cinereous Tit is a 'washed-out' copy of the European Great Tit, living in S Asia from E Iran and S Turkmenistan through India to SE Asia. It occurs in scrub, open forest and orchards at varying altitude, but in Central Asia mainly at 1000–2400 m.

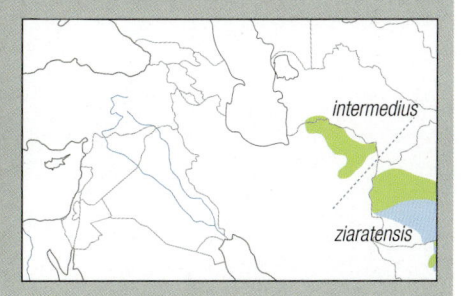

etc. Although to the human ear the song can sound quite similar to that of Great Tit, apparently the two differ acoustically when presented as sonograms in that individual notes in Cinereous Tit have steeper rises and falls in pitch, often forming 'Vs', whereas in Great Tit they form flatter traces (Harrap 1996). – Like Great Tit has a rich repertoire of calls, many of which recall those of the European species. A few are said to have a distinctive different quality.

SIMILAR SPECIES Realistically, within treated region, problems of identification will only occur in NE and E Iran and adjacent areas, where Cinereous Tit meets *Turkestan Tit*, which also lacks yellow on underparts and has similar blue-grey mantle. Turkestan Tit has a proportionately longer and more graduated tail, fringed with more white on its sides than in Cinereous Tit. This difference may not always be evident or easy to see, but supplementary characters exist: whereas Cinereous Tit has the same type of white cheek patch as in Great Tit, Turkestan Tit often has a slightly larger white cheek patch extending further towards the rear side of the neck, and the black border behind it is, as a rule, narrower;

P. c. intermedius, ♂, Iran, May: very similar in shape and plumage to Great Tit, with black crown and throat that continues as ventral stripe, large white cheek patches, but unlike Great Tit has bluish-grey upperparts and off-white underparts. Also very similar to Turkestan Tit, which also occurs in Iran, but note diagnostic dark centres to outer tail-feathers visible from below. Sexes differ as in Great Tit. (W. Müller)

P. c. intermedius, ad ♂, Iran, May: similarity to Great Tit obvious, but note lead-grey rather than moss-green mantle and off-white rather than yellow underparts. Wings edged greyish-blue with distinct white wing-bar on greater coverts. Unlike Turkestan, has much broader black band below and behind smaller white cheeks; wings also proportionately longer. Aged by evenly feathered wing (including primary-coverts) and sexed as ♂ by glossier black head, with ventral stripe broadening between legs. (W. Müller)

IDENTIFICATION Subtly *larger* than adjacent race *major* of Great Tit, but otherwise very similar in shape and general plumage, sharing the typical head pattern of *black crown and throat*, and *large white patches on sides of head* framed in black. Differs in post-juvenile plumages by *underparts being off-white* (with a slight, dull pink-buff tinge) instead of yellow, and *mantle/back bluish-grey* instead of moss-green. A few first-winter birds can have a very slight yellow tinge to sides of upper breast, but no yellow is seen in ad. Also, rarely a faint olive hue present on upper mantle. Outertail has rather little white, restricted to r6 and tip of r5 (at least in population within the covered region). *Wings are edged bluish-grey*, and as in Great Tit have a *distinct white wing-bar* on greater coverts. Sexes differ as in Great Tit in that the ♂ has a prominent ventral black band broadening to a patch between the legs and on vent, whereas the ♀ has a narrower and less solid black ventral band, and no broad black patch on vent. Juvenile is usually separated by a faint yellow tinge on the pale cheek patches, and by lacking solid black borders to the pale cheeks. Also, as in Great Tit, all dark portions lack gloss and are dull and sooty rather than black.

VOCALISATIONS Similar to those of Great Tit but perhaps consistently thinner and have been likened to Coal Tit (Harrap 1996). The song is a quick series of the same phrase of 2–4 notes repeated 5–8 times, e.g. *chew-a-ti chew-a-ti chew-a-ti chew-a-ti chew-a-ti* or *spi-tui spi-tui spi-tui spi-tui...*

P. c. intermedius, ♂, presumed 1stS, Iran, May: age inferred by slight contrast between apparently retained juv primary-coverts being slightly abraded and dull brown-grey with limited ash-grey edges, and newer greater coverts, darker and with neater lead-grey edges. Sexing straightforward by solid black ventral patch. Off-white underparts sometimes very slightly tinged pinkish-grey. (C. N. G. Bocos)

also, the large alula-feather is blacker and more contrasting in Cinereous Tit, with a neat whitish outer edge, whereas in Turkestan Tit this feather is paler and more bluish-grey, showing less contrast and lacking the same neat paler outer edge. It is worth remembering, too, that Turkestan Tit is a lowland species often found in close association with humans, while Cinereous Tit is more a bird of rough and mountainous country, often found far from settlements or cultivation.

AGEING & SEXING (*intermedius*) Ageing and sexing not closely studied, but appear to be very similar to what is known for Great Tit (which see). – Moults. For the treated subspecies, apparently much as in Great Tit. Several ad finish complete moult in mid or late Sep.

BIOMETRICS (*intermedius*) **L** 14–15 cm; **W** ♂ 71.5–81 mm (*n* 16, *m* 76.5), ♀ 69–75.5 mm (*n* 12, *m* 73.3); **T** ♂ 62–72 mm (*n* 16, *m* 66.9), ♀ 55.5–65.5 mm (*n* 12, *m* 61.5); **T/W** *m* 86.0; **TG** 2.5–6 mm (*n* 18, *m* 4.3); **B** 11.5–13.6 mm (*n* 28, *m* 12.8); **BD** 4.2–5.3 mm (*n* 23, *m* 4.8); **Ts** 19.0–21.7 mm (*n* 27, *m* 20.1). **White on r6:** wedge on inner web 13–29 (rarely 7–36) mm (*n* 26, *m* 21.7). **Wing formula:** close to that of Great Tit.

GEOGRAPHICAL VARIATION & RANGE Slight and clinal variation, mainly concerning size, paleness and amount of white in outertail. One race occurs within treated region, while a couple of more are extralimital. Apparently mainly sedentary.

P. c. intermedius Zarudny, 1890 (NE & E Iran, S Turkmenistan). Treated above. A few show a faint trace of green at uppermost mantle, but not known whether these represent hybrids with Great Tit (or carry genes from ancient hybridisation) or are part of normal variation. Apparently comes into contact with Great Tit ssp. *major* in SW Turkmenistan (north of Kopet Dag), and allegedly the two hybridise there (Formozov *et al.* 1993), but details poorly known.

Extralimital races: ○ *P. c. ziaratensis* Whistler, 1929 (Afghanistan between Herat in the west to Kabul in the east). Subtly paler above and a little smaller (**W** ♂ 71–76 mm [*n* 7, *m* 74.0]; Vaurie 1959) than *intermedius*. Also edges of tertials and inner secondaries purer white and therefore slightly more prominent.

P. c. decolorans Koelz, 1939 (easternmost Afghanistan east of Kabul and extreme north of Pakistan). A rather dark race with dusky-grey flanks.

TAXONOMIC NOTES See under Great Tit for reasons to treat Cinereous Tit as a separate species. – It has been suggested that *intermedius* has a mainly hybrid origin, but this was refuted by a genetic analysis (Kvist *et al.* 2007).

P. c. intermedius, presumed ♀, Iran, Sep: the appearance of a narrow and ragged black ventral band and undeveloped patch on vent suggest ♀. (M. Vejdani)

P. c. decolorans, ♀, Pakistan, Nov: an extralimital race but one that might straggle to Iran, differing by its darker colours above and below. Sex inferred from narrow ventral band which does not widen on vent. Ageing not safe. (T. Abbas)

REFERENCES Eck, S. (1980) *Falke*, 27: 385–392. – Eck, S. (1988) *Zool. Abh. Mus. Tierk. Dresden*, 43: 101–134. – Eck, S. (2006) *Zootaxa*, 1325: 7–54. – Formozov, N. A., Kerimov, A. B. & Lopatin, V. V (1993) *Arch. Zool. Mus. Moscow*, 30: 118–146. – Kvist, L. *et al.* (2003) *Proc. Roy. Soc. London*, 270: 1447–1454. – Kvist, L. *et al.* (2007) *Biol. J. Linn. Soc.*, 90: 201–210. – Päckert, M. & Martens, J. (2008) *Ibis*, 150: 829–831.

Aberrant Great Tit *P. major*, 1stW ♂, Spain, Feb: cream-white underparts with hardly any yellow, and greyer upperparts (almost no greenish) of this aberrant Spanish bird could cause confusion. Such diluted individuals can resemble the greyer and eastern forms of the Great Tit complex, or even Turkestan Tit *P. bokharensis*. Retained juv primary-coverts make this bird a 1stY, while glossy black crown and broad black patch between legs reveal a ♂. (R. Armada)

TURKESTAN TIT
Parus bokharensis M. H. C. Lichtenstein, 1823

Fr. – Mésange du Turkestan; Ger. – Turkestanmeise
Sp. – Carbonero Turquestano; Swe. – Turkestanmes

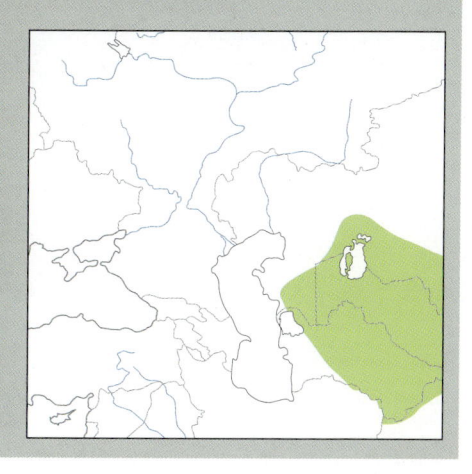

A Central Asian species, closely related to the Great Tit and the Cinereous Tit. Within the treated region, Turkestan Tit is only found marginally, in extreme NE Iran. From there its range stretches north-east, reaching as far as NW China and SW Mongolia. It is mostly found in towns and settlements, in gardens and valleys, but also in riparian vegetation and in various types of cultivation, requiring only holes to nest in and some tall bushes or trees. Habits resemble those of Great Tit. It is mainly resident, moving very little from its normal breeding areas.

IDENTIFICATION Of similar size to the Great Tit and Cinereous Tit but has proportionately slightly shorter, more rounded wings and somewhat *longer and more graduated tail* than both. The plumage has the same main pattern as Great Tit, only with all yellow and green replaced by off-white and bluish-grey. (The off-white underparts often have a faint pinkish-brown hue in fresh plumage.) This would seem a substantial difference, making separation easy, was it not for the existence of the similarly-coloured Cinereous Tit in S Turkmenistan and NE Iran, and further east into S Asia. Normally, the long and rather rounded tail gives Turkestan Tit a slightly different jizz compared to both Great and Cinereous Tits, with their shorter and more square-ended tails, but the difference is not always obvious and you often need to look for further but subtler characters. The *tail-sides have more white*, with much of the outermost tail-feather white and a deep white wedge on the inner web of the penultimate feather (Great Tit has white outer web and a narrow white wedge on the inner web of the outermost tail-feather, a wedge that reaches about halfway in, whereas the penultimate feather has only a narrow white tip and outer edge, or practically no white at all). This difference in amount of white

♂, Kazakhstan, May: essentially off-white and bluish-grey tit, lacking yellow and green of Great Tit, and unlike both Great and Cinereous Tits the tail is longer and tail-sides show more white (folded tail appears nearly all white from below with no visible dark centres to outer feathers). The white cheek patches are large and somewhat elongated, with much narrower black borders below and behind. ♂ has much broader black ventral stripe that broadens between legs. (V. & S. Ashby)

♂ (right) and ♀, Kazakhstan, May: a perfect image showing the sexual differences, with ♂ having all black marks broader and glossier. Still, note that the ♀ Turkestan has larger black patch on centre of belly and vent than in Great and Cinereous Tits. (A. Riley)

in the tail can sometimes be seen when a bird takes off or alights, but more often on a perched bird from below. The *folded tail* of a Turkestan Tit *appears nearly all white*, whereas on both Great and Cinereous Tits there is a dark centre with only the outer edges white. The *white cheek patch* is slightly more elongated in Turkestan Tit, *reaching slightly further back* towards the nape, and the *black band bordering it at the rear is usually narrower* than in Great and Cinereous Tit. Also, the *large alula-feather is not as dark and contrasting* as in the other two, less blackish and more bluish-grey, *while its outer edge is less neat* and less contrastingly white. This can be hard to see on a restless, moving tit, but if seen is a species diagnostic. Mantle/back is pale lead-grey or ash-grey, slightly paler on average than Cinereous Tit (which it otherwise resembles in general colours). Edges of wing-feathers and tail-feathers are pale greyish-blue (wing brightest, tail duller).

VOCALISATIONS The song consists of a short series of repeated simple notes, often two different ones in combination, *see-dee see-dee see-dee see-dee see-dee*, thus resembling that of Great Tit, only often slightly weaker or feebler. When the song is built up of the same note it resembles more Marsh Tit, *tiieh tiieh tiieh tiieh…* – Calls include a

Ad ♀, Kazakhstan, Feb: narrow black abdominal stripe and black band below and behind white cheeks characterise ♀, and evenly-feathered wing indicate age as ad. (A. Isabekov)

twittering *dididi*, sounding like a mixture of Great and Blue Tit, with a Blue Tit-like clear *see du-du* or *si-si-du-si*, and a Great Tit-like *see-chu see-chu*, often sparsely repeated when feeding. Mildly quarrelling *zre-zre-zre-zre* are more similar to corresponding irritated call of Great Tit, but less harsh in tone.

SIMILAR SPECIES The only other similar species within the treated region is the *Cinereous Tit*, which also lacks yellow and green. The discriminating characters are treated under Identification and under Cinereous Tit.

AGEING & SEXING Ageing has not been closely studied, but is assumed to be similar to Great Tit, with the possible difference that young Turkestan Tits probably moult more extensively in comparison, making ageing sometimes more difficult. Sexes differ in the same way as in Great Tit. – Moults. Probably follows same patterns as in Great Tit. One ad finished complete moult in mid Sep, outermost primaries still growing.

BIOMETRICS L 14.5–16 cm; **W** ♂ 67–80 mm (*n* 30, *m* 72.8), ♀ 67–74 mm (*n* 16, *m* 69.9); **T** ♂ 67–81 mm (*n* 30, *m* 73.9), ♀ 66–74.5 mm (*n* 16, *m* 70.2); **T/W** *m* 101.1; **TG** 8.0–15.5 mm (*n* 20, *m* 11.2); **B** 11.0–13.8 mm (*n* 44, *m* 12.3); **BD** 4.6–6.0 mm (*n* 44, *m* 5.2); **Ts** 18.0–20.8 mm (*n* 46, *m* 19.4). **White on tail-sides:** wedge on inner web of **r6** 35–52 mm (*n* 24, *m* 45.1); of **r5** 6–41 mm (*n* 24, *m* 27.5). **Wing formula: p1** > pc 9–15 mm, < p2 18–26 mm; **p2** < wt 7.5–13 mm, =8/9 (5%), =9 or 9/10 (35%) or ≤10 (60%); **p3** < wt 1–3.5 mm; **pp4–6** about equal and longest (though p6 often to 1.5 mm <); **p7** < wt 0.5–4 mm; **p8** < wt 3.5–7 mm; **p10** < wt 8–11.5 mm; **s1** < wt 9–13 mm. Emarg. pp3–7.

GEOGRAPHICAL VARIATION & RANGE Monotypic. – NE Iran through Central Asia to W Mongolia and Altai; resident. – At least three races described, but when comparing series we were unable to find any substantial difference that would warrant separation. (Syn. *ferghanensis*; *turkestanicus*.)

TAXONOMIC NOTE See under Great Tit for reasons to treat Turkestan and Cinereous Tits as separate species.

REFERENCES Eck, S. (1980) *Falke*, 27: 385–392. – Eck, S. (1988) *Zool. Abh. Mus. Tierk. Dresden*, 43: 101–134. – Eck, S. (2006) *Zootaxa*, 1325: 7–54. – Kvist, L. *et al.* (2003) *Proc. Roy. Soc. London*, 270: 1447–1454. – Kvist, L. *et al.* (2007) *Biol. J. Linn. Soc.*, 90: 201–210. – Päckert, M. & Martens, J. (2008) *Ibis*, 150: 829–831.

Presumed 1stY ♀, Kazakhstan, May: unlike Great and Cinereous Tits, has slightly shorter wing and proportionately longer and more graduated tail; extensive white tail-sides also visible. Note large white cheek patch, with much narrower black band below and behind (here sufficiently so to indicate a ♀). Faint pinkish-brown wash to off-white underparts still noticeable. Primary-tips and primary-coverts already quite abraded and brownish suggesting a young bird. (A. Audevard)

1stW, presumed ♀, Kazakhstan, Nov: typically very pale ash-grey and white tit with large white cheek patch and long tail, thus easily identified. Retained juv primaries and primary-coverts confirm young age, but as underparts only partly visible sexing uncertain. (A. Isabekov)

♀, Kazakhstan, May: black border below white cheek patch can be very narrow and even, as here, disconnected from black bib. Note also large alula (just visible) has grey inner web with narrower and less contrasting white edge to outer web, lacking contrasting dark alula with bolder white edge of Great and Cinereous Tits. Wing, tail and rest of plumage juv. (S. Harvančík)

KRÜPER'S NUTHATCH
Sitta krueperi Pelzeln, 1863

Fr. – Sittelle de Krüper; Ger. – Türkenkleiber
Sp. – Trepador de Krüper; Swe. – Krüpers nötväcka

This small, attractive nuthatch is confined to Turkey and Georgia, with one foot in Europe on the island of Lesbos in Greece. It belongs to a group of related smaller nuthatches that includes Corsican Nuthatch, Algerian Nuthatch, the North American Red-breasted Nuthatch and a couple of Asian species. These have all a common ancestor and have either dispersed or become isolated relict populations since. Krüper's Nuthatch lives mainly in mature spruce and pine forests in lower mountains, often between *c.* 800 and 2000 m, and is resident, apparently moving very little.

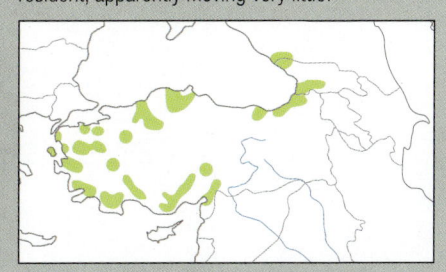

♂, presumed 1stS, Turkey, May: solid black cap and extensive rusty-red breast patch confirm sex. Dull grey-tipped primary-coverts and duller and worn large alula are apparently retained juv, suggesting 1stS, but nuthatches are very difficult to age without handling. (D. Monticelli)

♂, presumed ad, Turkey, May: combination of deep rusty-red breast, well-defined white supercilium, black forecrown and narrow black eye-stripe separate it from congeners, but solid breast and forecrown markings confirm that it is ♂. Seemingly evenly-feathered ad wing with pure grey primary-coverts indicate an ad. (D. Occhiato)

IDENTIFICATION About 85% the size of the common Eurasian Nuthatch, this lively nuthatch is easily identified by its *chestnut breast patch*, a rather square or oval patch on upper breast, somewhat diffuse and subdued in juvenile plumage, well-marked and deep rufous in later plumages. Upperparts are lead-grey except that both sexes in adult plumage have dark crown, the ♂ a *well-marked glossy jet-black crown patch* that clearly contrasts with the grey nape, the ♀ a smaller, less well-marked and less black patch restricted to forecrown and which often grades smoothly into grey nape. *Prominent white supercilium* is enhanced by the dark crown and a complete *black* (♂) *or grey* and less complete (♀) *eye-stripe*. Flanks are either pale bluish-grey without brown tinge (♂) or greyish with a buff-brown hue (♀). Grey-tipped tail corners practically lack any pure white. Identification is usually problem-free due to size and restricted breeding range in wooded habitat with few alternative species to eliminate. Sings from branch in tree, stretching up to expose chestnut breast patch, but can also sing while feeding on a trunk.

VOCALISATIONS Like most nuthatches, both vocal and loud-voiced. Song consists of a loud series of repeated nasal notes on two pitches, likened to the sound of old-fashioned car horns (those with a rubber ball), but perhaps more similar to Great or Coal Tit. Each bird can alternate between a few favoured variations in pace, the slower ones often clearly two-toned, *pju-di-pju-di-pju-di-pju-di-*..., while a slower and whining variety is an almost Hobby *Falco subbuteo*-like *pyew pyew pyew pyew pyew*..., whereas fastest sound like trills, *hididididi*..., still with a plaintive ring. – Call is an upwards-inflected *dyew-ee*, rather like Greenfinch, or in flight a clicking *chek*, a bit like Brambling. When alarmed gives a grating, hoarse, almost Jay-like *eehtch*, often repeated. This or a related sound can form a long quarrelling harsh series, *cheh-cheh-cheh-cheh-*...

SIMILAR SPECIES Only nuthatch within its range of its size, hence once smaller size than *Eurasian Nuthatch* is established the identification is done. If size is more dubious due to fleeting view, note dark crown and very distinct white supercilium, both often easier to ascertain than chestnut breast patch. Once red-brown breast patch seen all other species are immediately eliminated. – *Western Rock Nuthatch* is slightly larger (not always apparent though!), a shade paler grey above and lacks breast patch or dark crown patch. It is nearly always found in open montane habitats, not in closed woods like Krüper's Nuthatch usually is. – *Eastern Rock Nuthatch* is even larger than Western Rock Nuthatch

♂, presumed 1stS, Turkey, Apr: again, solid black forecrown and deep rusty-red breast of ♂. Rather well-marked rufous in ventral area and purer grey flanks are also features of ♂. Although primary-coverts seem to be juv (narrow and less pure grey, especially the tips), with already abraded slightly brownish primary-tips, and apparent moult limit in alula, handling is needed to confirm ageing. (R. Martin)

♀, presumed 1stS, Turkey, Apr: black on forecrown of ♀ tinged or mottled grey, patch smaller and less well defined at rear (unless heavily worn), eye-stripe in front of eye more broken up and grey, breast patch and sides of vent on average paler rufous, less deep rusty-red, sides of flanks tinged brown, less pure grey (still, some overlap in many of these characters). Primaries and primary-coverts are seemingly juv. (R. Martin)

♀, presumed ad, Turkey, Jun: already in post-nuptial moult with some inner primaries shed or growing. Rather mottled eye-stripe and seemingly diffuse rear border of black cap, as well as faint rufous tinge in greyish flanks and belly infer sex. Unmoulted outer primaries seem rather well kept and tinged lead-grey, hence most likely an ad. (H. Shirihai)

and has a stronger bill, which together with pale grey upperparts should immediately separate it.

AGEING & SEXING Ages practically alike after post-juv moult. Sexes differ at least after post-juv moult (from about Aug–Sep), possibly already after fledging in some. – Moults. Complete post-nuptial moult of ad in mid May–Aug (Sep). Partial post-juv moult in late summer or early autumn does not involve flight-feathers, primary-coverts or greater coverts. No pre-nuptial moult other than odd birds replacing scattered body-feathers (*BWP*). – **SPRING** ♂ Whole crown glossy black with sharp border to lead-grey nape. Eye-stripe black and prominent. Flanks tinged bluish-grey without brown tinge. ♀ Only forecrown black or dark sooty-grey, with little or no gloss, and grades into lead-grey hindcrown and nape (without sharp border). Eye-stripe grey, rather indistinct and sometimes broken. Flanks sullied grey, washed brown or buff. – **AUTUMN F.gr.** Sexes differ as in spring, both with uniform bluish-grey upperparts, black on crown, distinct supercilium and prominent rufous patch on upper breast. Ages may differ as in Eurasian Nuthatch in that some or all 1stW display a faint moult limit in median coverts, but not studied. **Juv** Similar to f.gr. but slightly duller grey above, with faint, thin whitish feather-tips. No black on crown, but many ♂♂ can show dark grey. Supercilium and eye-stripe often less well marked. Rufous patch on upper breast washed out and subdued due to paler rufous colour and pale feather-tips. ♂ sometimes recognised by purer grey underparts and rather obvious dark forecrown, ♀ being more buff below and usually lacking any darker cap.

BIOMETRICS L 11–12 cm. Sexes of nearly similar size, therefore combined: **W** 71–76 mm (n 19, m 74.1); **T** 35–42 mm (n 18, m 38.2); **T/W** m 51.5; **B** 17.2–19.3 mm (n 19, m 18.4); **BD** 3.6–4.5 mm (n 18, m 4.0); **Ts** 16.3–18.5 mm (n 18, m 17.5). **Crown patch** ♂ 13–16.5 mm, ♀ 8–13 mm. **Wing formula:** p1 > pc 5–10 mm, < p2 29.5–32.5 mm; **p2** < wt 6–8 mm, =6/7 (60%), =7 (30%) or =7/8 (10%); **p3** < wt 0–2 mm; **pp4–5** longest; **p6** < wt 1–2.5 mm; **p7** < wt 5.5–8 mm; **p8** < wt 10–12 mm; **p10** < wt 13–15 mm; **s1** < wt 13.5–16 mm. Emarg. pp3–6. – A different and slightly larger sample (Roselaar, in *BWP*) gave a better sexual differentiation (age classes combined): **W** ♂ 72–78 mm (n 20, m 74.6), ♀ 69–74 mm (n 14, m 71.9); **T** ♂ 36–40 mm (n 18, m 37.8), ♀ 34–38 mm (n 12, m 36.1).

GEOGRAPHICAL VARIATION & RANGE Monotypic. – Extreme E Greece (on Lesbos), Turkey, W Caucasus; resident.

REFERENCES HARRAP, S. (1993) *BW*, 6: 111–114.

♂, presumed 1stS, Turkey, Jun: whitish supercilium enhanced by black forecrown and black eye-stripe. Extensive rusty-red breast and near-solid black crown, but also purer grey flanks indicate ♂. Ageing difficult without handling, but apparent moult limit in alula and slightly brownish and mottled eye-stripe may suggest 1stS. (H. Shirihai)

Juv, Turkey, Jun: note fluffy feathering below, and much diluted red breast patch, while facial markings are obscured. Yellow mouth flanges at gape still visible. (H. Shirihai)

CORSICAN NUTHATCH
Sitta whiteheadi Sharpe, 1884

Fr. – Sittelle Corse; Ger. – Korsenkleiber
Sp. – Trepador corso; Swe. – Korsikansk nötväcka

Most birdwatchers who are not content with just their local patch, but want to get to know all of Europe's birds, must visit Corsica at least once. This is due to its two endemic birds, and especially the Corsican Nuthatch, a relict only found here. The total population is estimated to be *c*. 2000 pairs, dwelling in ancient Corsican pine forests with many lichens and dead trunks, at altitudes of 800 m and above. If you have not been there yet, plan a trip between late March and mid May when the ♂♂ still sing regularly. If you go in summer it will be harder work to locate the now often quiet birds, which are by then busy feeding their young.

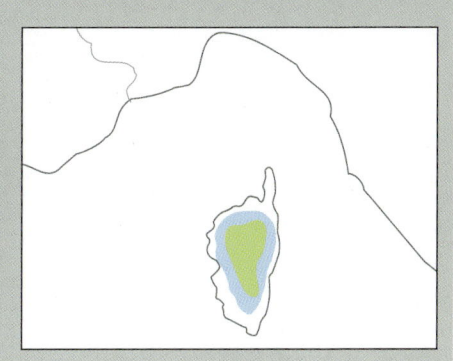

♂, Corsica, Mar: solid and glossy black crown and quite dark eye-stripe, with well-defined white supercilium are key species identification criteria, but also confirm that it is a ♂. Ageing difficult without handling. (L. Boon/Cursorius)

IDENTIFICATION Compared to the common Eurasian Nuthatch, Corsican Nuthatch is considerably smaller (roughly the same size as Krüper's Nuthatch) with a better-marked head pattern. Since it is the only nuthatch species found on Corsica, identification is uncomplicated. A small, *short-tailed* and compact *grey-backed* bird capable of climbing both up and down tree trunks (mainly mature or dead pines) on Corsica should be this species. Both sexes have a *well-marked white supercilium* enhanced by a well-defined *dark eye-stripe* and a *dark crown* (solid black in ♂, usually dark blue-grey in ♀). Seen close, grey of upperparts is pale lead-grey (thus with a faint bluish cast), while *underparts are sullied grey or buffish-grey*. Cheeks off-white, slightly paler than throat. Vent and undertail-coverts have diffusely darker grey feather-centres and paler tips, creating a slightly variegated pattern, visible in good light. There is *no chestnut* on lower flanks and vent like in Eurasian Nuthatch. Outer tail-feathers blackish with grey tips and minute white patches subterminally. Bill strong and pointed, grey with darker culmen and tip. Legs grey.

VOCALISATIONS Vocal in mating season and relatively loud-voiced (though not quite as much as Krüper's Nuthatch). Song is usually a fast repetition of one note on a constant pitch and pace, sounding like a shrill trill (at a distance can recall Alpine Swift *Tachymarptis melba* or even an insect or a frog), *didididididididi...* It somewhat recalls the fast song variety of Eurasian Nuthatch. Rarely, the song has a slower pace and consists of upwards-inflected notes, *dew dew dew dew dew dew...* – Call is either a monosyllabic *dee* or a brief song-like trill, *didid*. In alarm utters a Starling-like hoarse *pcheeh*, sometimes repeated in series, *cheh-cheh-cheh-...*, almost like a slow-motion Great Tit or a distant Jay. Fine whistling *weest* sometimes heard as well.

SIMILAR SPECIES For the not so experienced bird-watcher visiting Corsica, where Corsican Nuthatch is both endemic and the only nuthatch species, the main risk for confusion is *Treecreeper*. This is a similarly small bird that also climbs tree trunks, but note brownish upperparts, thin downcurved bill and habit of climbing mainly upwards using the stiff tail-feathers as support. Nuthatches do not use their tails and can climb in all directions, including downwards head first. – Museum ornithologists must consider several other closely related species occurring in scattered areas over the Northern Hemisphere (presumed to be relict species of a common widespread ancestor), including *Krüper's Nuthatch* (easily told by its chestnut breast patch), the very similar *Algerian Nuthatch* (with less extensive dark cap), the extralimital East Asian *Chinese Nuthatch* (*S. villosa*, not treated; slightly darker above and brighter below) and the North American *Red-breasted Nuthatch* (strongly orange-tinged underparts, more bold head pattern and supercilia joining on forehead).

AGEING & SEXING Ages practically alike after post-juv moult. Sexes differ slightly from fledging but more clearly after post-juv moult (from about Aug–Sep). – Moults. Complete post-nuptial moult of ad in summer (detailed timing not known). Partial post-juv moult in late summer or early autumn does not involve flight-feathers, primary-coverts or greater coverts. Apparently no pre-nuptial moult. – **SPRING** ♂ Crown glossy black. Eye-stripe black and prominent. ♀ Crown dark grey, when fresh with blue-grey feather-tips, slightly darker than back, especially when worn in summer, but never black. Eye-stripe grey and less well marked. – **AUTUMN F.gr.** As to ageing see Krüper's Nuthatch. Sexes differ as in spring. Mantle and back uniformly bluish-grey. 1y ♂ has on average less black crown than ad ♂ with more diffuse rear edge. **Juv** Like post-juv, but slightly duller grey above with faint whitish feather-tips. Dark cap of ad not fully developed, still ♂ usually has dark grey crown, while ♀ is mainly greyish.

Ad ♂, Corsica, Oct: the only nuthatch species on Corsica, but also identified by small size and well-marked plumage characters. Note conspicuous white supercilium contrasting with bold black cap and eye-stripe of ♂. Apparently evenly-feathered wing, including primary-coverts, suggests an ad. (G. Jenkins)

♂, presumed ad, Corsica, Oct: blue-grey upperparts and greyish pink-buff underparts (throat usually slightly paler). Bold white supercilium and black cap confirm sex as ♂, and seemingly evenly-feathered wing, including primary-coverts, suggest ad, but safe ageing difficult without handling. (G. Jenkins)

1stW ♂, Corsica, Oct: intermediate between ad ♂ and ♀ in having black cap restricted to forecrown, and cap being less solid. Note browner retained juv primary-coverts. (A. Easton)

♀, presumed ad, Corsica, Oct: unlike ♂, crown blue-grey, supercilium less well defined and eye-stripe grey and slightly broken-up. Age probably ad but difficult to decide without handling. (G. Jenkins)

BIOMETRICS **L** 11.5–12 cm; **W** ♂ 71–75 mm (n 13, m 73.2), ♀ 69–73 mm (n 12, m 70.8); **T** ♂ 38–43 mm (n 13, m 39.8), ♀ 37–42 mm (n 12, m 38.7); **T/W** m 54.5; **B** 16.5–18.7 mm (n 25, m 17.5); **BD** 3.2–4.0 mm (n 25, m 3.6); **Ts** 16.0–18.7 mm (n 23, m 17.4). **Wing formula:** **p1** > pc 7–9.5 mm, < p2 28–31 mm; **p2** < wt 5–7 mm, =6/7 (80%), or =7 or 7/8 (20%); **p3** < wt 0–1 mm; **pp4–5** longest; **p6** < wt 1–2.5 mm; **p7** < wt 5.5–8 mm; **p8** < wt 9.5–12 mm; **p10** < wt 14–16.5 mm; **s1** < wt 13–18 mm. Emarg. pp3–6.

GEOGRAPHICAL VARIATION & RANGE Monotypic. – Corsica; resident.

REFERENCES Harrap, S. (1993) *BW*, 6: 111–114. – Jenkins, G. & Jenkins, H. (2009) *BW*, 22: 22–27.

1stY ♀, Corsica, Mar: some young birds can be quite untidy in first summer due to some remnant of immaturity in the form of mottling on ear-coverts and retained brownish juv wing. Aged by moult limits in wing and tail. Lack of black on forecrown infers sex. (D. Occhiato)

ALGERIAN NUTHATCH
Sitta ledanti Vielliard, 1976

Fr. – Sittelle kabyle; Ger. – Kabylenkleiber
Sp. – Trepador de Kabilia; Swe. – Kabylnötväcka

The Algerian Nuthatch is a bird species that very few have seen despite its breeding on the southern side of the Mediterranean, very close to Europe. This is because it was unknown to science before 1975, when discovered and soon thereafter described for the first time, but also because it is very local and rare (estimated total population *c.* 600 pairs). To this can be added political and religious instability in Algeria ever since the bird was found, which has not encouraged birdwatchers to go there. However, there are signs of improvements in the country, and hopefully soon many more can visit the oak forests on the slopes of the Algerian Atlas Mountains between Algiers and Constantine to see this attractive nuthatch.

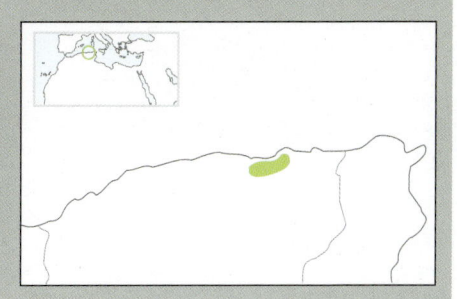

♂, Algeria, Jun: in some lights can appear considerably warmer below, being deeper salmon-pink, with hardly any contrast between pale throat and breast/belly. Prominent white subterminal band on some outer tail-feathers, just like Eurasian Nuthatch, but unlike Corsican and Krüper's. (D. Monticelli)

Ad ♂, Algeria, Jun: ♂ has characteristic black forehead and forecrown, forming a small cap not extending to rear crown, sharply delimited from broad cream-white supercilium. Note less solid black eye-stripe. Throat whitish, and remaining underparts cream-pink. Seemingly evenly-feathered and still rather fresh wing, including primary-coverts, indicate ad. (V. Legrand)

IDENTIFICATION Endemic to N Algeria, being restricted to a few montane oak forests, where it is the only nuthatch species, making identification straightforward. A *small* nuthatch with *bluish-grey upperparts* and *warm yellowish-buff* or ochre-buff *underparts*. Cheeks and upper throat are whitish, whiter than the cream-coloured lower throat. Both sexes have a *broad, prominent white supercilium* running from base of bill to well behind eye. Although the sexes are quite similar, the ♂ has a distinctly demarcated *jet-black forecrown patch* and a *blackish eye-stripe*, whereas the ♀ has an all-grey crown, or a mixture of grey and some blackish-grey patches, with a paler and less well-defined eye-stripe. The vent and undertail-coverts are uniformly buffish-white (or very slightly mottled grey), only slightly paler buff than belly. Several outer tail-feathers have broad white subterminal marks (the tips being grey), visible from below, rather like in Eurasian Nuthatch but unlike Corsican and Krüper's. Bill and legs as in Corsican Nuthatch.

VOCALISATIONS Song rather similar to Krüper's Nuthatch, a mechanically repeated disyllabic plaintive or whining note with nasal tone, *pieu-pi-pieu-pi- pieu-pi-pieu-pi-...*, which can recall a toy trumpet or old car horn. Pace sometimes slightly faster, but apparently never becomes trill-like. – Several calls, most commonly heard a short, harsh *chaeee*, which have been compared to the call of Jay. This call can also be repeated in short series, *cheeh-cheeh-cheeh*. A variation with more nasal voice recalls Trumpeter Finch. A merry-sounding upwards-inflected *dwu-ee* can sound like a Greenfinch and is obviously related to a similar call of Krüper's.

SIMILAR SPECIES Confusion with other nuthatches unlikely as this is the only *Sitta* species in Algeria. Compared to the similar *Corsican Nuthatch*, Algerian differs in having a more restricted dark cap confined to forecrown (easiest to see on ♂♂), a slightly narrower and less prominent dark eye-stripe, a slightly shorter white supercilium which does not reach quite as far back on the sides of the nape, often richer yellowish-buff or ochre-buff underparts (less dull pinkish-grey), and more extensive white patches subterminally on outer tail-feathers.

AGEING & SEXING No material seen. The following is based on the literature and available photographs. Ages

♀, Algeria, Jun: similar to ♂, but often completely lacks the black forecrown cap (though see image on next page, top left) and has even more indistinct or mottled dark eye-stripe. (D. Monticelli)

Presumed ad ♀, Algeria, Jun: some apparent ad ♀♀ have mottled black forehead with diffuse, uneven border to grey rest of crown, and also rather well-marked dark eye-stripe, hence can be difficult to separate from young ♂♂. Fresh primary tips in Jun fit best with ad ♀. (V. Legrand)

Juv, Algeria, Jul: variation in juv plumage is not fully established, but head usually all grey like rest of upperparts, with only indistinct mottled eye-stripe. Underparts paler and more buff than ad, bill base more pinkish-grey, and note soft, fluffy body-feathers. (A. van den Berg)

practically alike after post-juv moult. If plumage development is similar to closely-related Corsican Nuthatch, sexes could differ slightly after fledging, but probably more clearly after post-juv moult in summer. – Moults. Complete post-nuptial moult of ad in summer. Partial post-juv moult in summer does not involve flight-feathers, primary-coverts or greater coverts. Presumably no pre-nuptial moult. – **SPRING** ♂ Forecrown black, in clear contrast to grey hindcrown and nape. ♀ Forecrown often a mixture of black and grey, fading into grey hindcrown and nape without sharp border. Some appear blacker, approaching ♂, while others are nearly all grey on crown. – **AUTUMN F.gr.** Sexes differ as in spring, both having uniform bluish-grey upperparts, some black on crown in most, and distinct broad supercilium. **Juv** Not studied, but presumed to differ from post-juv in similar way as Krüper's and Corsican Nuthatches. Apparently some (all?) ♂♂ can show traces of black (or dark grey) on forecrown when in nest (Fosse 1992). But on the whole, juv probably invariably has more subdued head pattern than post-juv.

BIOMETRICS No material examined. The following is from *BWP*. **W** ♂ 80–83 mm (n 3, m 81.3), ♀ 77–81 mm (n 3, m 79.0); **B** 16.0–17.5 mm (n 6, m 16.6); **Ts** 19.0–22.0 mm (n 5, m 20.4).

GEOGRAPHICAL VARIATION & RANGE Monotypic. – NC Algeria; resident.

REFERENCES Van den Berg, A. B. (1982) *DB*, 4: 98–100. – Fosse, A. (1992) *DB*, 5: 234. – Harrap, S. (1992) *BW*, 5: 154–156. – Monticelli, D. & Legrand, V. (2009) *BW*, 22: 333–335. – Monticelli, D. & Legrand, V. (2009) *DB*, 31: 247–251. – Vielliard, J. (1976) *Alauda*, 44: 351–352. – Vielliard, J. (1978) *Alauda*, 46: 1–42. – Vielliard, J. (1980) *Alauda*, 48: 139–150.

Presumed ad ♀, Algeria, Jun: at nest entrance. Compared to the bird above (top left), has neater primary tips with well-defined pale edges. (R. Nehal)

(EURASIAN) NUTHATCH
Sitta europaea L., 1758

Alternative names: Wood Nuthatch, Western Nuthatch

Fr. – Sittelle torchepot; Ger. – Kleiber
Sp. – Trepador azul; Swe. – Nötväcka

A well-known bird even outside the ranks of birdwatchers. Being a common and widespread garden bird in Europe is one good reason for this, but it is above all its peculiar habit of climbing upside-down with head first on a vertical trunk or branch that has rendered it fame. Its loud and repeated calls also mean that it is often noticed, and its bold manners when visiting feeders can be striking, where it often chases away larger birds to be able to mess around with the sunflower seeds alone. Nests in a tree-hole, and often walls up the opening with mud for better protection.

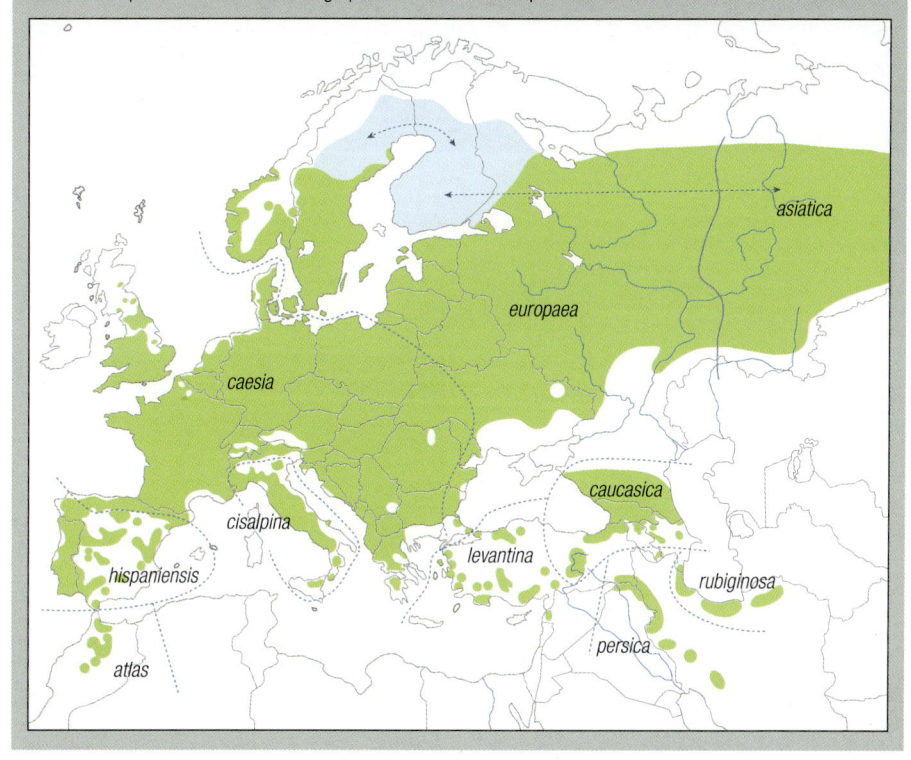

IDENTIFICATION Often noticed by loud calls and singular habit (shared with congeners) of being able to climb on vertical tree trunks without the support of the tail, even *moving downwards head first*. Its active and energetic movements and frequent sharp or loud whistled calls make it a real character. A small, compact bird with *strong legs and bill*, latter straight and pointed, and *short tail*. Tail is dark grey with a white subterminal band on outer feathers, which you can sometimes catch a glimpse of. *Upperparts uniform lead-grey* (as in most congeners), while underparts vary geographically, in Fenno-Scandia mainly off-white with variable chestnut (♂) or ochre-brown (♀) tinge on lower flanks and vent, in rest of Europe and SW Asia a warm orange-brown or rusty-buff colour on all or most of underparts. A *long black eye-stripe* runs from base of bill to sides of nape.

VOCALISATIONS Even though this not the most loud-voiced of nuthatches, the common Nuthatch of Europe is still often first noted by song or calls, and the song frequently appears remarkably loud and 'different' in the chorus of a park or garden. Song is given in a few variations, as so often in nuthatches, the commonest probably a rather slow series of whistling downwards-inflected notes with an almost wailing ring, *weeuu weeuu weeuu weeuu...*, but almost as frequently the notes are upwards-inflected, *vuuee vuuee vuuee vuuee...* The same bird can alternate now and then with very fast repetitions of one or two notes creating a trill, *vivivivivi...* These three song types can all be heard from a stationary ♂ using a high song post in crown of tall tree, and they can confidently be described as 'real' song. There is an alternative territorial call that is usually uttered by a feeding or moving bird (but still appears to have song function, at least partly), a repetition of three-note-combinations with rhythmic intonation, the last note often stressed, *dwet-dwet-**dwet**, dwet-dwet-**dwet**, dwet-dwet-**dwet**, ...* – Calls include conversational or agitated sharp *tsiit* and shorter *zit* notes, and a conversational softer *dwet* related to the above-mentioned song variation, only monosyllabic (but often much repeated). The *tsiit* call can also be slightly upwards-inflected, *tsoiit*.

SIMILAR SPECIES In much of Europe the only nuthatch, so once generic identity is established the species is known too. In SE Europe must be separated from *Western Rock Nuthatch*, most easily by different habitat (open lower mountains, bare rock faces, maquis but not closed forests) and vocalisations (louder and more striking trills with uninterrupted alternations between several variations, all at rather high speed). The slightly larger size of Western Rock Nuthatch is not always apparent; the paler grey upperparts are often easier to note, and Western Rock lacks chestnut on lower flanks of ♂ Eurasian Nuthatch, being merely diffusely pink-buff over much of lower underparts. Leg colour differs slightly and on average, in that Eurasian Nuthatch generally has some flesh-brown element in the grey, whereas the two rock nuthatches usually have cleaner slate-grey legs. – In Asia Minor the slightly smaller *Krüper's Nuthatch* overlaps locally with Eurasian Nuthatch, but habitats generally differ, with Eurasian Nuthatch at lower levels. The chestnut breast patch and dark crown will of course identify a Krüper's instantly once seen well. Krüper's Nuthatch also lacks the prominent white subterminal band on tail corners of Eurasian Nuthatch.

S. e. europaea, ♂, presumed ad, Norway, May: a medium-sized nuthatch with complex geographical variation, but in general characterised by long black eye-stripe, blue-grey upperparts and mainly white or orange-tinged underparts. ♂♂ of white-breasted group are characterised by chestnut flanks and lateral fringes to undertail-coverts. Seemingly evenly-feathered and fairly fresh wing, including primary-coverts, suggest an ad. (M. Winness)

S. e. europaea, ♀, presumed ad, Finland, Nov: ♀♀ often have cream-coloured wash on vent or belly, and chestnut on flanks always diluted and not contrasting. Seemingly evenly-feathered wing, including primary-coverts, suggests ad. However, ageing especially without handling of nuthatches is difficult and often unreliable. (T. Muukkonen)

S. e. asiatica, ♂, Mongolia, Jun: this race is predominantly white below, with bill fine and slender, and slightly upward-turned. Black eye-stripe well marked, accentuated by white cheeks and frequently also by narrow white supercilium narrowly extending across forehead. Sexual differences less pronounced, here a ♂ (seen to sing and defend territory) with no chestnut on flanks. (H. Shirihai)

S. e. asiatica, ♂, presumed 1stW, Finland, Oct: black eye-stripe prominent, often reaching further back than in *europaea*. When still fresh, greater coverts can show a faint wing-bar (probably most obvious in 1stW). Undertail-coverts with narrow dark chestnut fringes indicating ♂. Lower flanks variably tinged rufous-cinnamon, but area clearly smaller than in *europaea*, or all white (see image at left). Narrow juv primary-coverts better fit 1stW. (M. Varesvuo)

AGEING & SEXING (*europaea*) Ages alike or at least very similar after post-juv moult, and any difference subtle and difficult to use in normal field encounters. Sexes differ already among well-grown pulli. – Moults. Complete post-nuptial moult in summer, late May–Sep. Partial post-juv moult in summer or early autumn does not involve flight-feathers, tertials, primary-coverts or greater coverts (except very rarely a few innermost tertials and greater coverts). A varying number of median coverts are moulted, usually most and sometimes all. No pre-nuptial moult. – **SPRING** ♂ Lower flanks, vent and part of undertail-coverts saturated dark rufous (chestnut), sharply divided from whitish rest of underparts. ♀ Flanks, lower belly, vent and part of undertail-coverts warm rufous-buff fading into whitish breast and rest of underparts without sharp border. – **AUTUMN** Sexing as in spring. **Ad** Flight-feathers neat and fresh, rather dark grey with some gloss, edges tinged bluish-grey. All wing-coverts uniform bluish-grey. **Juv / 1stW** Similar to ad, but has slightly duller grey flight-feathers, which may attain slightly abraded tips earlier than in ad. Also, primary-coverts and all greater coverts are subtly duller and faintly paler grey on average (requires handling or close views, and many are still difficult to separate). Sometimes shows visible contrast between several moulted inner median coverts tipped lead-grey and some unmoulted outer ones that are subtly duller grey-brown, with hint of tips being tinged brownish-white and slightly more worn than inner (cf. Jenni & Winkler 1994). ♂♂ show hint of deep rufous ('chestnut') on lower flank and vent, ♀♀ are warm buff-brown lacking deep rufous.

BIOMETRICS (*europaea*) **L** 13.5–14.5 cm; **W** ♂ 86–94.5 mm (*n* 40, *m* 89.1), ♀ 83–88 mm (*n* 31, *m* 85.6); **T** ♂ 44–51 mm (*n* 40, *m* 47.1), ♀ 43–48 mm (*n* 31, *m* 44.8); **T/W** *m* 52.6; **B** 17.6–21.8 mm (*n* 69, *m* 20.0); **BD** 4.2–5.2 mm (*n* 69, *m* 4.6); **Ts** 18.3–22.0 mm (*n* 65, *m* 20.1). **Wing formula: p1** > pc 5–9 mm, < p2 37–40.5 mm; **p2** < wt 6–8 mm, =6/7 (95%) or =7 (5%); **p3** < wt 0–1 mm; **p4** longest (sometimes pp3–5); **p5** < wt 0–1 mm; **p6** < wt 1.5–4 mm; **p7** < wt 9–12 mm; **p8** < wt 13–17 mm; **p10** < wt 18.5–21.5 mm; **s1** < wt 20–23 mm. Emarg. pp3–6.

GEOGRAPHICAL VARIATION & RANGE Rather prominent variation, both in size and plumage colours. It is tempting to group taxa in two subspecies groups, 'white-breasted' in the north and east, 'red-breasted' in the west and south, but such an arrangement is superficial and probably does not reflect evolutionary history (Redkin & Konovalova 2006). Still it is useful when giving an overview of W Palearctic variation. Note that variation within the 'red-breasted group' is often slight and far from consistent within traditionally defined ranges, exhibiting all sorts of slight local variations in saturation of underparts and size of bill. Added to this is a substantial degree of individual variation, and variations linked to season, wear of plumage, age, etc. – Some of the problems experienced when attempting to validate generally accepted taxa are mentioned below under respective taxon, but some general problems and observations must also be stated. After a careful comparison of long series of most treated taxa, we have concluded that most claimed differences in darkness or exact hue of upperparts are either non-existent or of insignificant degree to serve for subspecific division. Staining of underparts, rendering the true colours a soiled or greyish cast, is a problem that affects especially birds in the breeding season that have still not moulted; it has not been possible to establish whether any of this variation is also innate and worthy of taxonomic treatment. For a consistent treatment, if accepting common practice to include British breeders ('*affinis*') in *caesia* on account of small and insignificant differences, a case could be made for lumping even

S. e. caesia, ad ♂, Germany, spring: a well-marked ♂, with deep orange-chestnut underparts (contrasting with white cheeks and chin). Not all ad ♂♂ across the range of *caesia* are as strikingly coloured as this bird, some being a little paler below (see below). Note ad primary-coverts with fresh lead-grey fringes. Darker chestnut edges on ventral feathering just discernible, inferring sex. (E. Kuchling)

S. e. caesia, ad ♂, Germany, Apr: irrespective of age, some ♂♂ have rather pale pinkish-brown underparts (apart from white cheeks and throat), but dark chestnut rear flanks and edges to vent confirm the sex. Ad primary-coverts have relatively broad and not heavily abraded lead-grey fringes in spring. (A. Noeske)

more races into *caesia* than we have done here. – Several additional extralimital distinct subspecies exist than those listed below. – All populations are mainly resident, but those in the north and north-east are prone to make movements in non-breeding season towards west and south, at least in some years.

White-breasted Group

S. e. europaea L., 1758 (S Fenno-Scandia, E Denmark, Baltic States, Russia east to region of Orenburg and Ustinov, Belarus, E Ukraine, possibly also west coast of Black Sea). Treated above. Large with underparts mainly off-white to white. Strong bill. Birds in S Latvia, E Poland, over much of Belarus and Ukraine, and possibly in E Bulgaria and Thrace, are either this race or intermediates with *caesia*. – There is a slight cline running from west to east, with birds becoming on average slightly whiter below ('*rossica*'), but differences slight and variable, and in our opinion do not warrant formal taxonomic treatment. (Syn. *homeyeri*; *norvegica*; *rossica*.)

S. e. asiatica Gould, 1835, 'SIBERIAN NUTHATCH' (easternmost Russia, Siberia north to c. 64°N, south to Altai, Tarbagatai, east to NW Mongolia, Sayan and W Baikal area). Considerably smaller than *europaea* and a trifle purer white below, but in contrast very slightly darker blue-grey above; furthermore, frequently has a rather well-marked white forehead and supercilium. Tends to have narrow pale wing-bar in fresh plumage. Bill often appears slightly up-turned due to near-straight culmen. **L** 12.5–13.5 cm; **W** ♂ 76–83 mm (n 28, m 79.8), ♀ 76–81 mm (n 13, m 78.7); **T** ♂ 39–43.5 mm (n 28, m 41.4), ♀ 39–42.5 mm (n 13, m 40.8); **B** 14.5–19.9 mm (n 41, m 18.0); **BD** 3.6–4.7 mm (n 40, m 4.2); **Ts** 16.8–19.4 mm (n 39, m 17.8). **Wing formula: p1** > pc 2–7.5 mm, < p2 34–37 mm; **p2** < wt 4.5–7.5 mm, =6/7; **p10** < wt 17–20 mm; **s1** < wt 18–21 mm. (Syn. *biedermanni*; *bifasciata*.) – Range overlaps partially with *europaea* in Orenburg region (with few mixed pairs), and in SW Yakutia with extralimital 'TAIGA NUTHATCH' *S. arctica*, which breeds in NE Siberian taiga east of c. 105°E (Redkin & Konovalovka 2006).

Red-breasted Group

S. e. caesia Wolf, 1810 (Denmark except in east, Britain, Continental Europe from Poland and Germany south to N Spain, France, Switzerland, Balkans including Greece, east at least to C Bulgaria, much of Romania, W Ukraine, extreme SW Belarus). Has warm cinnamon-buff underparts (of somewhat variable strength depending on age, plumage wear or staining, and individual variation) and is generally clearly smaller than *europaea*. Despite these differences, *europaea* and *caesia* intergrade in a zone of contact from S Latvia/E Poland through Belarus, Ukraine and apparently in region west of Black Sea, producing intermediate birds. **L** 13–14.5 cm; **W** ♂ 81–90 mm (n 33, m 86.5), ♀ 80–90 mm (n 38, m 83.4); **T** ♂ 41–48 mm (n 33, m 44.9), ♀ 40–47 mm (n 37, m 43.3); **T/W** m 51.9; **B** 18.7–22.1 mm (n 72, m 20.1); **BD** 4.3–5.6 mm (n 71, m 4.9); **Ts** 18.2–21.3 mm (n 71, m 19.9). **Wing formula: p1** > pc 2–7 mm, < p2 35–43 mm; **p2** < wt 5–10.5 mm, =6/7 (80%) or =7 (20%); **p10** < wt 17–22 mm; **s1** < wt 19–23 mm. – Specimens collected in Britain ('*affinis*') are often slightly duller and more 'dirty-looking' below, and the reason for this may well be staining. The difference from Continental *caesia* is anyway very slight, and we see no reason to uphold a separate race for these. – Birds of the E Adriatic coastland sometimes included in *cisalpina*, but although they appear as intermediates between this and *caesia*, they seem closer to the latter, being less saturated cinnamon below, and are included in *caesia* here. Birds of N Spain (e.g. León, Santander, Asturias) are clearly *caesia*, different from *hispaniensis*. Claimed local variations within *caesia* have led to the description of numerous separate races. We believe many or all of these were based on scant or inadequate material, and they are best treated as synonyms. (Syn. *affinis*; *domaniewski*; *extrema*; *harrisoni*; *hassica*; *loeppenthini*; *reichenowi*; *sordida*; *stolcmani*.)

S. e. caesia, 1stS ♂, England, Apr: typically alert and 'curious' look of a nuthatch. Diluted rufous below (still with chestnut rear flanks and edges to vent revealing sex) with slightly less pure black eye-stripe, and juv primary-coverts (with pointed tips and narrow grey fringes) inferring young age. (G. Thoburn)

S. e. caesia, 1stW ♀, Germany, Nov: a young bird in its first autumn (note juv primary-coverts with duller dark centres and relatively narrow grey fringes). Note rather pale orange-brown underparts lacking deep chestnut on undertail-coverts, rather ill-defined white cheeks and throat, plus less well-marked sooty-grey eye-stripe, which all confirm sex as ♀. (A. Noeske)

S. e. hispaniensis, ♂, Spain, Jan: very similar to (often inseparable from) ssp. *caesia*, but usually paler below, being more washed-out cinnamon-buff even in ♂ (more like ♀ of *caesia*), and is slightly smaller-billed. Deep chestnut rear flanks and edges to vent confirm sex. Ageing best left unresolved in this case. (M. Estébanez Ruiz)

S. e. hispaniensis, ♀, presumed 1stS, Spain, Jun: some young ♀♀ of this race are very pale below, being predominantly cream-buff, with sooty-grey eye-stripe. Such birds can recall Western Rock Nuthatch of SE Europe, but note yellowish-brown (instead of pure grey) legs; precise check of tail and undertail-coverts patterns will be required for safe separation. Brownish and abraded primary-tips suggest age. (R. Fernández González)

S. e. levantina, ad ♂, Turkey, Jan: this race tends to have a whitish forehead and sometimes a narrow white supercilium, otherwise many ♂♂ are deeply saturated cinnamon-red below, i.e. much like colourful *caesia*. Quite contrasting whitish cheeks and throat. Ad primary-coverts have broad and fresh lead-grey fringes. (M. Sözen)

S. e. levantina, ♀, presumed ad, Turkey, Oct: paler, more pinkish-brown below, with only slightly indicated whitish supercilium and forehead. Broad, round-tipped primary-coverts with fresh lead-grey fringes suggest ad, but ageing of nuthatches without handling inadvisable. Undertail-coverts faintly fringed rufous eliminates confusion with sometimes similarly orange-bellied juv Western Rock Nuthatch. (G. Ozgunlu)

S. e. hispaniensis Witherby, 1913 (Iberia except N Spain; northern border runs south of Asturias, León, Navarra). Similar to *caesia* but paler below, a more washed-out cinnamon-buff, and is a slightly smaller race with smaller bill. There seems to be a rather marked size variation, with birds tending to be smaller towards south and west, but apart from the fact that small birds are consistently found in C & S Portugal, there is much variation. **L** 12.5–13.5 cm; **W** ♂ 78–89 mm (*n* 27, *m* 84.6), ♀ 79–89 mm (*n* 17, *m* 83.1); **T** ♂♀ 39–47 mm (*n* 45, *m* 43.2); **B** 17.3–20.4 mm (*n* 45, *m* 18.9); **BD** 3.8–4.8 mm (*n* 44, *m* 4.4); **Ts** 17.1–20.4 mm (*n* 41, *m* 18.7). **Wing formula: p2** =6/7. (Syn. *minor*.)

○ **S. e. atlas** Lynes, 1919 (Atlas Mts in Morocco). Similar to *caesia* being similarly warm cinnamon-buff below and thus clearly more saturated on underparts than interposed *hispaniensis*. It could well be labelled *caesia*, but widely separated allopatric ranges of the same subspecies is never a desirable solution, and because it is somewhat larger on average than *caesia* with a proportionately longer tail, and a shorter tarsus and thinner bill (all these are only slight and average differences), it seems practical to recognise *atlas* as a separate taxon. Apart from the different coloration compared to *hispaniensis*, it also differs in having longer wing, tail and bill. **W** ♂ 85–90 mm (*n* 14, *m* 87.7), ♀ 83–87 mm (*n* 10, *m* 85.7); **T** ♂ 43–49 mm (*n* 14, *m* 46.5), ♀ 43–47 mm (*n* 10, *m* 45.3); **B** 18.2–21.4 mm (*n* 24, *m* 19.8); **BD** 4.1–4.7 mm (*n* 24, *m* 4.4); **Ts** 18.2–19.8 mm (*n* 23, *m* 18.9). **Wing formula: p1** > pc 5–9 mm, < p2 37–40 mm; **p2** < wt 6–9 mm, =6/7.

○ **S. e. cisalpina** Sachtleben, 1919 (Italy except southern slopes of Alps). Small and has underparts often a trifle deeper cinnamon-buff than in average *caesia*, but much overlap in this. A very few birds have a narrow greyish-white line on forehead above base of bill, or have forehead diffusely paler, but the vast majority not. Bill on average *c*. 1 mm shorter than in *caesia*. A subtle race. **L** 12.5–13 cm; **W** ♂ 82–88 mm (*n* 13, *m* 85.2), ♀ 80–87 mm (*n* 13, *m* 82.7); **T** ♂ 42.5–48 mm (*n* 13, *m* 44.5), ♀ 41–44 mm (*n* 13, *m* 42.7); **B** 17.2–20.4 mm (*n* 26, *m* 18.9); **BD** 4.2–5.0 mm (*n* 26, *m* 4.6); **Ts** 19.0–20.7 mm (*n* 24, *m* 19.8). **Wing formula: p2** =6/7 (80%) or =7 or 7/8 (20%). (Syn. *siciliae*.)

○ **S. e. levantina** Hartert, 1905 (Turkey except perhaps in N Thrace and in north-east; Levant). Small and saturated cinnamon-buff below, much like *cisalpina* (plumage differences from that require further study) but has a slightly thinner bill in dorsal view. Clear tendency to have whitish forehead and sometimes also narrow white supercilium. Claimed to have paler grey upperparts than, e.g., *caesia* and *caucasica*, but impossible to confirm when series are compared. **W** 80–88 mm (*n* 18, *m* 83.9); **T** 40–48 mm (*n* 18, *m* 44.4); **B** 17.5–20.0 mm (*n* 18, *m* 18.8); **BD** 3.9–4.4 mm (*n* 18, *m* 4.2); **Ts** 18.5–20.0 mm (*n* 17, *m* 19.4). **Wing formula: p2** =6/7 (70%) or =7 or 7/8 (30%).

○ **S. e. caucasica** Reichenow, 1901 (Caucasus, Transcaucasia except in east; apparently NE Turkey). Rather small and very deeply saturated cinnamon-buff below, like darkest variation of *caesia* or even a little deeper rusty colour. Chestnut on lower flanks in ♂♂ more restricted than in *caesia* or *europaea*. Often whitish forehead and sometimes thin white supercilium. Statements that it should have a short and blunt-tipped bill have not been possible to confirm on specimens. Has slightly broader bill than *levantina*. **W** 82–87.5 mm (*n* 12, *m* 85.1); **T** 41–48 mm (*n* 12, *m* 44.3); **B** 17.5–21.3 mm (*n* 12, *m* 19.0); **BD** 3.9–4.7 mm (*n* 12, *m* 4.5); **Ts** 18.8–21.3 mm (*n* 12, *m* 20.3).

○ **S. e. rubiginosa** Tschusi & Zarudny, 1905 (Azerbaijan, N Iran, SW Turkmenistan). Large as *caesia* (thus larger than *caucasica*), but has underparts slightly less saturated cinnamon-buff than that race. Markedly stronger legs and subtly shorter tail than *caesia*. Claims (Vaurie 1959; *BWP*) that the bill is deeper and wider at base, as well as longer, could not be verified. In fact, *rubiginosa* and *caesia* in material examined by us (NHM, MNHN, ZMC, NRM, AMNH) have very nearly identical bill sizes. A subtle race. **W** 81.5–90 mm (*n* 12, *m* 86.3); **T** 41–47 mm (*n* 12, *m* 43.8); **T/W** *m* 50.3; **B** 17.6–22.3 mm (*n* 16, *m* 20.3); **BD** 4.0–5.7 mm (*n* 15, *m* 5.0); **Ts** 18.5–22.7 mm (*n* 16, *m* 21.4).

S. e. persica Witherby, 1903 (NE Iraq, SW Iran, Zagros). Small and pale, especially above, palest in the group, but also below, being pale ochre-buff with only weak contrast to still paler buff throat (a similarly weak contrast is otherwise seen only in *hispaniensis*). Forehead invariably appears to have some cream-white. Together with *levantina* smallest race in this group. Sexual difference small in that ♂ has only very limited chestnut on lower flanks and vent. **L** 12–13 cm; **W** ♂ 82–87 mm (*n* 16, *m* 84.3), ♀ 78–85 mm (*n* 15, *m* 81.3); **T** ♂ 43–47 mm (*n* 16, *m* 44.6); ♀ 41–45 mm (*n* 15, *m* 43.1); **T/W** *m* 53.0; **B** 17.5–20.3 mm (*n* 30, *m* 19.0); **BD** 3.8–4.8 mm (*n* 30, *m* 4.2); **Ts** 17.5–20.1 mm (*n* 30, *m* 19.1). **Wing formula: p2** =6/7 (70%) or =7 or 7/8 (30%).

TAXONOMIC NOTE Various authors (e.g. Eck 1984, Redkin & Konovalova 2006, Zink *et al*. 2006) have suggested that extralimital *S. arctica* of NC and NE Siberia (TAIGA NUTHATCH) is quite distinct and perhaps better treated as a separate species, differing in mtDNA by *c*. 10% from *asiatica* and breeding sympatrically with it (admittedly over a very restricted area) with no or only limited hybridisation. It is not known whether *arctica* has ever occurred within the treated region, but it seems rather unlikely. For distinctions from *europaea* see Redkin & Konovalova (2006).

REFERENCES DICKINSON, E. C. (2006) *Zool. Med.. Leiden*, 80: 225–239. – ECK, S. (1984) *Zool. Abh. Mus. Tierk. Dresden*, 39: 71–98. – REDKIN, Y. & KONOVALOVA, M. (2006) *Zool. Med. Leiden*, 80: 241–263. – ZINK, R. M., DROVETSKI, S. V. & ROWHER, S. (2006) *Mol. Phyl. & Evol.*, 40: 679–686.

S. e. caesia, juv ♂, Netherlands, early summer: a recently fledged juv ♂ of the red-breasted race group. Note loose, fluffy body-feathers and uneven pink-brown and white feathering below, chestnut flanks signalling sex. Dark eye-stripe less well-marked and sooty-grey. Gape corners still swollen and yellow. (H. van Egdom)

EASTERN ROCK NUTHATCH
Sitta tephronota Sharpe, 1872

Fr. – Sittelle des rochers; Ger. – Klippenkleiber
Sp. – Trepador rupestre oriental
Swe. – Östlig klippnötväcka

This sister species to the Western Rock Nuthatch, although similar, is more spectacular in every respect. It has a louder voice, is larger in size, has a stronger bill and a better-marked black eye-stripe. It is resident in hills and lower mountains of E Turkey, Transcaucasia and Iran. A common place to get to know it has been in the neighbourhood of the Turkish city of Gaziantep or outside the town of Birecik. Although there is much overlap in habitat choice between the two closely related rock nuthatches, Eastern Rock tends to more often occur away from steep rock-faces in more varied and arboreal surroundings such as ravines or open valleys with rocks, boulders, groves and orchards.

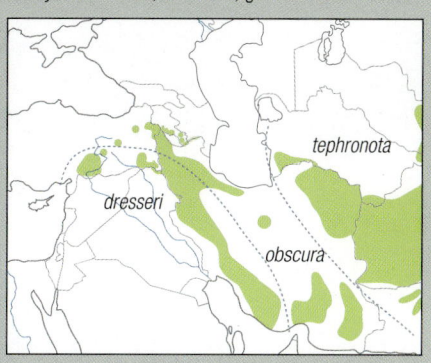

S. t. dresseri, presumed 1stS, Turkey, May: the largest nuthatch in the region, with powerful bill, more prominent and long, broad black eye-stripe ending on sides of mantle, and (especially in the western race) strong contrast between white throat to chest and cinnamon-rufous lower belly and flanks. Apparently strong wear in some outer wing-feathers indicates a young bird. (D. Occhiato)

IDENTIFICATION The largest nuthatch in the W Palearctic, although its slightly larger size than Western Rock Nuthatch is not always evident in the field. However, apart from being subtly *larger* overall, it has a proportionately *larger head* and a *heavier, longer bill*, which often gives it a more fierce and 'different' look. The plumage is very similar to that of Western Rock Nuthatch, in fact so similar that age-related, geographical and individual variations bridge nearly all claimed differences. The best means of separation are provided in most of the overlap area between the two (E Turkey, Transcaucasia, W & N Iran) by the eye-stripe, vocalisations and bill size. Eastern Rock Nuthatch in Turkey and Transcaucasia has a *very prominent black eye-stripe*, which broadens behind the eye, becoming *broader behind the eye than in front of it* (the reverse is true in Western Rock Nuthatch). From obliquely behind, both black stripes are visible at most angles (only one visible in Western Rock). The *song is considerably louder*, often *slower*, and *more varied* than in Western Rock (see below), and the *bill is somewhat stronger*, adding to the front-heavy look provided by the large head. Strangely, eastern populations, outside the overlap region with Western Rock, are slightly smaller and smaller-billed, with a more evenly broad black eye-stripe, therefore very similar to Western Rock Nuthatch and not always separable if it were not for provenance.

VOCALISATIONS All nuthatches are vocal and loud-voiced, but the Eastern Rock Nuthatch easily takes the prize within the W Palearctic. Song is varied but generally consists of accelerating or decelerating, loud, metallic, repeated notes running into mechanical trills, carrying very far as they echo against rock faces in valleys. Each ♂ alternates between 3–4 variations, some faster and clicking, others whistling, and still others more fluty and pleasing. The song resembles that of Western Rock Nuthatch but is generally lower-pitched, louder and more varied. – Calls are sharp and metallic, often single or short elements from the song, clicking sounds or shorter trills. As most nuthatches, it signals alarm or anxiety with Jay-like harsh notes, *cheeh-cheeh-cheeh*. When feeding or keeping contact with other birds a sharp *zeek*.

SIMILAR SPECIES Best separated from similar *Western Rock Nuthatch* within area of overlap by longer and broader rear part of black eye-stripe (both eye-stripes, one either side of head, often visible from behind, whereas in Western Rock at most one narrow and pointed eye-stripe can be seen from the same angle). The stronger bill and louder voice provide confirmation. – *Eurasian Nuthatch* is smaller, with either whiter or more orange-brown underparts and a much smaller bill.

AGEING & SEXING (*dresseri*) Ages usually alike after post-juv moult, but sometimes a subtle difference is discernible if handled. Sexes alike in plumage; size differs only slightly. – Moults. Complete post-nuptial moult in summer or early autumn, mid Jun–early Oct (Nov). Partial post-juv moult in late summer or early autumn generally does not involve flight-feathers, primary-coverts or greater coverts, but some birds replace outer primaries (*BWP*, Harrap 1996), occasionally also r1. No pre-nuptial moult. – **AUTUMN F.gr.** Flight-feathers neat and fresh, rather dark grey with some gloss. Upperparts including greater coverts uniformly grey. Eye-stripe black and prominent. **Juv** Similar to f.gr., but has slightly duller grey or grey-brown flight-feathers, which may show slightly abraded tips earlier than in ad. Some early-hatched young may moult some outer primaries into 1stW and then show moult contrast in mid-wing. When fresh, often has subtle pale feather-tips to grey of upperparts, and greater coverts finely tipped buff. Eye-stripe less well marked.

BIOMETRICS (*dresseri*) **L** 15–16.5 cm; **W** ♂ 87–97.5 mm (n 15, m 93.7), ♀ 89–94 mm (n 14, m 91.1); **T** ♂ 47–55.5 mm (n 14, m 52.4), ♀ 48–52 mm (n 13, m 49.7); **T/W** m 55.4; **B** 24.2–30.8 mm (n 30, m 27.1); **BD** 5.4–7.2 mm (n 30, m 6.2); **Ts** 26.0–30.5 mm (n 26, m 27.6). **Wing formula: p1** > pc 13–23 mm, < p2 25–32 mm; **p2** < wt 7–11 mm, =8 or 8/9 (70%), =7/8

S. t. dresseri, presumed ad ♂, Turkey, May: large and strong-billed. Pale lead-grey above, with black eye-stripe widening behind eye to form very prominent field mark. Note that bill often is obviously bicoloured due to paler grey base to lower mandible. Less worn wings in May suggest an ad, while very broad and intensely black eye-stripe indicates a ♂. (Y. Perlman)

Comparison of Eastern Rock *S. t. dresseri* (left two images) and Western Rock Nuthatches *S. n. syriaca* (right two) in Turkey (Jun): Eastern Rock has larger-headed and heavier-billed jizz, giving front-heavy appearance, and averages larger than Western Rock. At least in Turkey, the prominent black eye-stripe of Eastern Rock is useful, being very broad behind eye, ending rather squarely (bottom left). In Turkey the white throat to breast in Eastern tends to afford greater contrast with the usually quite deep cinnamon-rufous rear underparts. Nevertheless, some can have duller underparts (top left). In much of Eastern's range in Turkey, the only consistent feature versus Western apart from the broader black eye-stripe is the powerful bill with pale base to lower mandible. Western Rock has a notably slimmer and more delicate jizz, especially the relatively shorter, narrower and more pointed bill (which tends to show less pale basally, albeit with overlap). These images show that some Western Rock can have a rather wide black eye-stripe (top right), but the stripe usually narrows and fades at rear, and some have rustier bellies (bottom right), like Eastern Rock. Variation in appearance of rear eye-stripe also varies with gait and posture of head and neck, and partly with sex as well. Thus, observers should be aware that all such inconsistences can complicate identification. (Top left: K. Malling Olsen; bottom left: D. Occhiato; top & bottom right: H. Shirihai)

(25%) or =9 (5%); **p3** < wt 0–3 mm; **pp4–6** longest; **p7** < wt 1–6 mm; **p8** < wt 6–10 mm; **p10** < wt 11–15 mm; **s1** < wt 12–18 mm. Emarg. pp3–6, faintly also on p7 in some.

GEOGRAPHICAL VARIATION & RANGE Moderate variation. Three subspecies often recognised of which two, *dresseri* and *obscura*, are so similar that they could just as well be lumped. (If lumped, *obscura* is senior name with priority.) A cline of decreasing general size and bill size, and less broad black eye-stripe, runs from west to east.

S. t. dresseri Zarudny & Buturlin, 1906 (SC & SE Turkey, N Iraq, W Iran in Zagros to Fars). Described above. Said to be the largest with the strongest bill of the three races (although it seems only subtly larger and *not* heavier-billed than *obscura*). Pale bluish-grey above (marginally paler than *obscura*), white on throat and whitish or pale buff-white on breast and upper belly, offering good contrast with cinnamon-tinged vent. Black eye-stripe widens behind eye to prominent patch, significantly broader than in front of eye. (Syn. *kurdistanica*.)

○ **S. t. obscura** Zarudny & Loudon, 1905 (NE Turkey to SE Transcaucasica, N & C Iran east of Zagros including Elburz and south through Esfahan to Kerman region). Very nearly as large as *dresseri* with similar-sized bill, but has comparatively less prominent eye-stripe behind eye (still very broad and black!). Very similar to *dresseri* in most respects, except

***S. t. dresseri*, juv, Turkey, May:** juv separable only by shorter bill with more extensive pale base and fluffier rear underparts, as well as more pinkish-grey legs and still yellow gape when just fledged. The very broad black eye-stripe behind eye is already developed. (D. Monticelli)

***S. t. obscura*, Iran, Apr:** this race is like *dresseri* in most respects, but has less prominent eye-stripe behind eye, though still very broad, with fractionally duller upperparts. Further, underparts are more saturated (throat and breast duskier or more dirty white, lower belly and vent slightly deeper rufous) and legs proportionately marginally shorter. (B. Anderson)

Eastern Rock *S. t. dresseri* (right) **and Western Rock Nuthatches** *S. n. syriaca*, Turkey, May: where both species occur together, Eastern Rock usually has a larger body and head, with a proportionately longer and more powerful bill, and a more prominent black eye-stripe (here sufficiently broad and glossy to indicate a ♂). Nevertheless, differences between them are not always as obvious as in this image. (Y. Perlman)

S. t. tephronota, Iran, Jan: this race is slightly smaller, more delicately built and smaller-billed, with a more even-width and narrower eye-stripe, thus very similar to Western Rock Nuthatch and they are not always separable in their area of sympatry in NE Iran. (E. Winkel)

S. t. tephronota/obscura, Iran, Jun: photographed in NC Iran range (Mazandaran), where *obscura* should occur, but the intermediate-width black eye-stripe and underparts with little rufous suggest that the two races probably intergrade, or that racial differences are clouded by individual variation. (E. Winkel)

that upperparts tend to be fractionally darker grey, throat and breast a fraction duskier, not as clean white, lower belly and vent slightly darker and more rufous than cinnamon, and legs proportionately a little shorter. Statements that Armenian breeders are slightly paler on underparts could not be confirmed. **W** 85.5–94 mm (*n* 25, *m* 90.8); **T** 47–52 mm (*n* 25, *m* 50.1); **B** 25.7–29.3 mm (*n* 25, *m* 27.4); **BD** 5.4–6.8 mm (*n* 22, *m* 6.0); **Ts** 24.0–28.3 mm (*n* 25, *m* 26.8).

S. t. tephronota Sharpe, 1872 (E Iran, mountains of Central Asia). Very similar to *obscura* with which it intergrades, but is somewhat smaller with a finer bill and shorter legs, a slightly longer tail, and many have black eye-stripe narrower, especially noticeable behind eye (some overlap, though). Rather pale and pure grey above, though still (like most *obscura*) very slightly darker than *dresseri*. Rusty-buff lower belly/vent is rather pale ochre, contrasting little with paler upper belly/breast. **L** 14–15.5 cm; **W** ♂ 82–95 mm (*n* 19, *m* 87.8), ♀ 79–90 mm (*n* 19, *m* 85.4); **T** ♂ 48–55 mm (*n* 19, *m* 51.0), ♀ 46–52 mm (*n* 19, *m* 49.3); **T/W** *m* 57.9; **B** 22.2–27.3 mm (*n* 40, *m* 24.7); **BD** 4.8–6.3 mm (*n* 38, *m* 5.4); **Ts** 23.0–25.5 mm (*n* 35, *m* 24.5). (Syn. *iranica*.)

(WESTERN) ROCK NUTHATCH
Sitta neumayer Michahelles, 1830

Fr. – Sittelle de Neumayer; Ger. – Felsenkleiber
Sp. – Trepador rupestre occidental
Swe. – Klippnötväcka

There are three species of nuthatches in Europe of which this is the largest. The Rock Nuthatch is restricted to the Balkans in SE Europe, but its range continues through Turkey, the Levant, Transcaucasia and widely across Iran. It lives only where there are steep cliffs and exposed rock faces, but will now and then also feed among boulders and on smaller rocks in maquis or open woodland, especially in winter. Like all southern nuthatches it is sedentary. It builds a closed nest of mud and constructs an entrance tunnel. The nest is placed directly on a cliff-face, often in a crevice with overhanging rock as shelter.

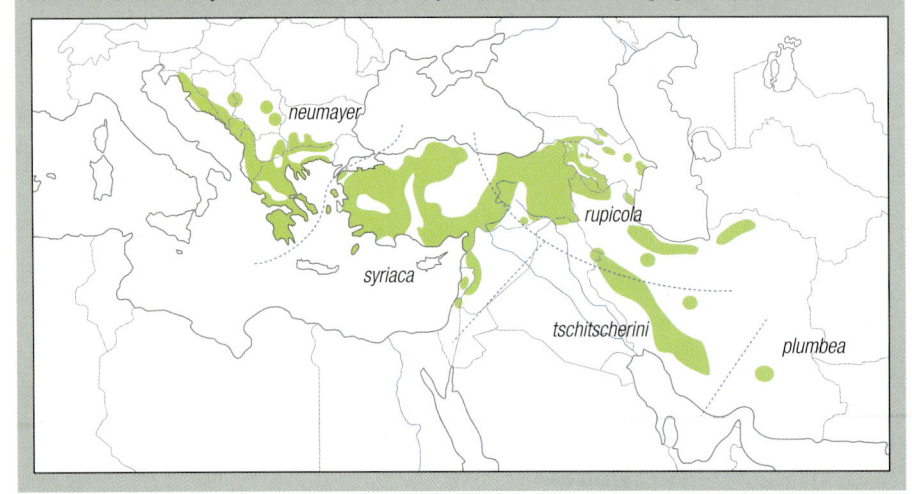

S. n. neumayer, presumed ad, Greece, Apr: in 'orange-bellied' Western Rock the warmer rear half of body contrasts with whiter throat and chest, unlike in Eurasian Nuthatch. Also unlike that species, and diagnostically, undertail-coverts lack white streaks or spots. Probably ad given evenly fresh wing-feathers (with pure lead-grey edges), including primary-coverts. (R. Pop)

IDENTIFICATION Slightly *larger than Eurasian Nuthatch*, but difference small enough to often pass unnoticed. There are usually few familiar birds nearby to compare with, and the species is often seen at some distance on rock faces making any detailed size evaluation difficult due to unfamiliarity to most observers with the habitat and distance. Compared to Eurasian Nuthatch it is *slightly paler and plainer* with subtly paler grey upperparts and a *white throat and breast, which grade into light rufous-buff belly and vent* (but without dark rufous on lower flanks and vent like ♂ Eurasian Nuthatch). Tail-feathers lack any white marks. *Head and bill proportionately larger* than in Eurasian Nuthatch, and Rock Nuthatch often differs in taking a more *upright stance when perched* on a rock or the ground. As in most nuthatches, there is a *long black mask* running from base of bill through eye to sides of nape. Legs strong, slightly longer than in Eurasian Nuthatch.

VOCALISATIONS Vocal and loud-voiced. Most sounds are shrill trills or fast series of repeated metallic or whistling notes, and can recall both Eurasian Nuthatch and Marsh Tit songs. Song is variable, often single or disyllabic notes repeated in long series, which may accelerate and increase in loudness, but also slow down and 'die' at the end. Several rather different series are alternated by the same ♂. If the singer is not close there can be a faint resemblance with a distantly-singing Woodlark. – Call comprises either short, sharp whistles when feeding and in contact, or short trills recalling the sound of the song. When alarmed or agitated gives harsh, short notes, often repeated in series, *cheeh-cheeh-cheeh*, doubtfully separable from similar call of Eastern Rock Nuthatch.

SIMILAR SPECIES For separation from *Eurasian Nuthatch* see Identification and under Eurasian Nuthatch, Similar species. Whether the bird perches on a rock or in a tree provides no fool-proof separation, so best to check size, proportions, stance, plumage colours and vocalisation. – *Eastern Rock Nuthatch* occurs in the same habitat and overlaps with Western Rock Nuthatch over a large area. Luckily, they differ most in the overlap area but are extremely similar elsewhere. In the overlap area, the very prominent black mask of Eastern Rock widens behind the eye (rather than narrowing slightly behind eye as in Western Rock). The head and the bill are also somewhat larger and heavier in Eastern Rock, and these differences can often be perceived. The song and calls are generally lower-pitched, louder and somewhat more varied in Eastern Rock than in Western Rock Nuthatch.

AGEING & SEXING Ages usually alike after post-juv moult, but sometimes a subtle difference discernible if handled. Sexes alike in plumage; size differ only slightly. – Moults. Complete post-nuptial moult in summer, mid Jun–Sep (Nov). Partial post-juv moult in late summer or early autumn generally does not involve flight-feathers or primary-coverts, but may involve r1, tertials and some inner greater coverts. No or only a limited pre-nuptial moult (restricted to body-feathers). – **AUTUMN F.gr.** Flight-feathers neat and fresh, rather dark grey with some gloss. Upperparts including greater coverts uniformly grey. Eye-stripe black and prominent. **Juv** Similar to f.gr., but has slightly duller grey or grey-brown flight-feathers, which may have slightly abraded tips earlier than in ad. Some early-hatched young can moult some outer primaries into 1stW and then show moult contrast in mid-wing. When fresh, often shows subtle pale feather-tips to grey of upperparts, and greater coverts finely tipped buff. Eye-stripe less well marked.

S. n. neumayer, presumed ad, Greece, Apr: usually appears somewhat larger and chunkier than Eurasian Nuthatch, and inhabits broken rocky terrain unlike that species. Underparts often whiter (as here) than 'cinnamon-bellied' races of Eurasian Nuthatch with which it co-exists. Ageing nuthatches without handling is usually unsafe, but in this case overall fresh feathers in Apr (and broadly grey-fringed primary-coverts) suggest an ad. (L. van Loo)

NUTHATCHES

S. n. neumayer, 1stS, presumed ♂, Greece, Mar: differs from Eurasian Nuthatch by rather long stout bill, long-legged appearance, and more uniform tail without black and white markings. Some have more extensive orange-tinged bellies, but unlike Eurasian Nuthatch breast whiter than belly, and undertail-coverts plainer. Much of visible wing is juv; a 1stY with a relatively broad and intense black eye-stripe suggests a ♂. (P. Petrou)

S. n. syriaca, Turkey, Jul: in classic pose in rocky habitat. It is usually impossible to sex a lone bird, and ageing without handling is often unwise, too. (A. Halley)

BIOMETRICS (*neumayer*) **L** 14.5–16 cm; **W** ♂ 77–86 mm (*n* 16, *m* 81.9), ♀ 73.5–84.5 mm (*n* 14, *m* 80.7); **T** ♂ 47–51 mm (*n* 16, *m* 48.8), ♀ 44–51 mm (*n* 14, *m* 47.6); **T/W** *m* 59.1; **B** 20.5–26.7 mm (*n* 36, *m* 23.8); **BD** 4.7–5.6 mm (*n* 35, *m* 5.1); **Ts** 21.3–24.5 mm (*n* 36, *m* 23.5). **Wing formula: p1** > pc 11.5–16 mm, < p2 21–27 mm; **p2** < wt 7.5–14 mm, =8/9 or 9 (53%), =9/10 or 10 (38%) or <10 or ≥8 (9%); **p3** < wt 0–2 mm; **pp4–6** longest; **p7** < wt 1.5–4 mm; **p8** < wt 3.5–8 mm; **p10** < wt 10–12.5 mm; **s1** < wt 9.5–14.5 mm. Emarg. pp3–6, faintly also on p7 in some.

GEOGRAPHICAL VARIATION & RANGE Limited variation, involving slight clinal changes in size and paleness. Smooth intergradation between races in areas of contact, hence many birds impossible to assign to a particular race. All populations resident.

S. n. neumayer Michahelles, 1830 (Balkans in SE Europe). Described above. Large, size almost as Eastern Rock Nuthatch, but bill proportionately somewhat smaller. Rather marked contrast between rusty-buff lower underparts and dusky-white upper half. Medium grey upperparts.

S. n. syriaca Temminck, 1835 (Turkey except east; Levant). Has upperparts paler grey than in *neumayer*. Tends to have broader black eye-stripe at rear in *tephronota* fashion. **L** 14–15 cm; **W** ♂ 75–82 mm (*n* 12, *m* 78.6), ♀ 74–80 mm (*n* 15, *m* 77.2); **T** ♂ 41–48 mm (*n* 12, *m* 45.5), ♀ 42–48 mm (*n* 15, *m* 44.8); **B** 20.5–24.3 mm (*n* 28, *m* 22.7); **BD** 4.5–5.6 mm (*n* 28, *m* 4.9); **Ts** 21.5–24.0 mm (*n* 27, *m* 22.9). Data from live birds at Mt Hermon, Israel (J. Langer & A. Rochman *in litt.* 1990, A. B. Dov & Y. Kiat *in litt.* 2013): **W** 76–85 mm (*n* 38, *m* 80.5); **T** 43–51 mm (*n* 25, *m* 47.1); **B** 19.0–25.0 mm (*n* 31, *m* 23.1); **Ts** 22.8–25.0 mm (*n* 13, *m* 23.7). (Syn. *zarudnyi*.)

○ *S. n. rupicola* Blanford, 1873 (E Turkey from *c.* 40°E, Transcaucasia, NE Iraq, N Iran south to *c.* 34°S). Somewhat smaller than—still very similar to—*neumayer*, differing mainly in paler rusty-buff lower underparts, creating less contrast with whitish breast. Upperparts slightly darker grey compared to *syriaca*, clearly darker than *tschitscherini*, and black eye-stripe behind eye better marked, not as broken and indistinct as *tschitscherini*. **L** 13.5–15 cm; **W** ♂ 76–85 mm (*n* 12, *m* 80.0), ♀ 74–82 mm (*n* 7, *m* 77.9); **T** ♂

S. n. syriaca, ♂ (left) & ♀, Israel, May: in Israel many pairs when seen together can show sexual dimorphism, with ♂ often having a considerably broader black eye-stripe, which is also intensely black; the ♀ has a much narrower eye-stripe that is less jet-black. Not all pairs show such clear differences. Both are apparently 1stS due to strong wear to remiges in May, but it seems that the ♀ is more strongly bleached and worn. (A. Ben Dov)

43–50 mm (*n* 12, *m* 46.6), ♀ 46–51 mm (*n* 6, *m* 48.2); **B** 18.9–25.7 mm (*n* 20, *m* 22.3); **BD** 4.3–5.4 mm (*n* 21, *m* 4.8); **Ts** 20.9–23.9 mm (*n* 19, *m* 22.6).

    S. n. tschitscherini Zarudny, 1904 (Zagros in SW Iran). Resembles *neumayer* in having rather distinctly bicoloured underparts, with fair contrast between whitish breast and rusty-buff lower underparts, but is considerably paler pure grey above. Black eye-stripe poorly developed behind eye. Even slightly smaller than *rupicola*. Thinner bill than other races. **L** 13–14.5 cm; **W** ♂ 73–82 mm (*n* 18, *m* 78.0), ♀ 74–79.5 mm (*n* 5, *m* 76.8); **T** ♂ 40–49 mm (*n* 18, *m* 45.0), ♀ 41–46 mm (*n* 5, *m* 43.6); **B** 19.0–24.2 mm (*n* 23, *m* 22.4); **BD** 4.0–5.3 mm (*n* 23, *m* 4.7); **Ts** 21.2–23.6 mm (*n* 22, *m* 22.4).

    ? ***S. n. plumbea*** Koelz, 1950 (Kerman in S Iran, east of *tschitscherini*). If three specimens in AMNH are representative (including holotype), is darker above, and throat and breast have a strong greyish wash (appearing more like a soil aberration). Size small (with data from a fourth specimen in Chicago provided by G. M. Kirwan): **W** 74.5, 76, 76.5, 78 mm; **T** 44.5, 45, 46, 46.5 mm; **B** 21.0, 22.1, 22.4, 22.6 mm. Scant material exists, and the race should be further confirmed in the field.

S. n. syriaca, juv, Israel, Jun: juv often show warm rufous-orange (albeit variable) over much of underparts except for white throat. Experience on Mt Hermon has shown that such rufous coloration fades rapidly in first three months post-fledgling. Best means of separation from superficially similar 'cinnamon-bellied' races of Eurasian Nuthatch are plainer undertail-coverts and grey (not brownish) legs. (A. Ben Dov)

S. n. rupicola, ad, Azerbaijan, Mar: differs from western races mainly by paler rear underparts, and black eye-stripe behind eye being narrower and less solid, but individual and sex-related variation render these differences of little use without comparative material. Aged by seemingly uniform (broadly fringed lead-grey) wing-feathers, including primary-coverts. (M. Heiß)

WALLCREEPER
Tichodroma muraria (L., 1766)

Fr. – Tichodrome échelette; Ger. – Mauerläufer
Sp. – Treparriscos; Swe. – Murkrypare

As evident from photographs in this book, the Wallcreeper is a spectacularly beautiful bird. It is also one of a kind, with no close relatives, being the only *Tichodroma* species in the world. A little surprisingly, it is more closely related to the nuthatches than to the treecreepers, although it resembles the latter more in shape and behaviour. Although widespread in high mountains of Eurasia, the Wallcreeper is rather elusive and difficult to see, and many birders have had to struggle hard to get it on their life lists. It breeds on steep rock faces at high altitudes, but often descends in winter to lower levels and can sometimes be seen at traditional sites, even in towns on cliffs or buildings.

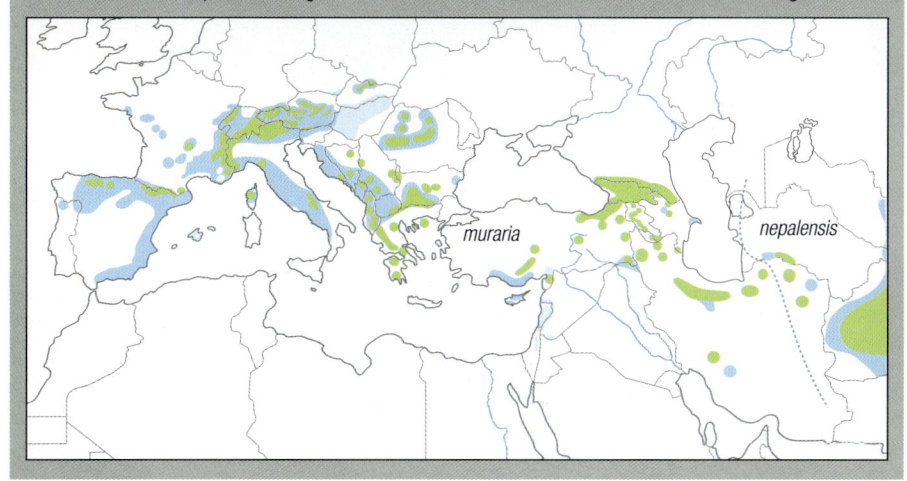

T. m. muraria, ♂, Turkey, Apr: the well-named Wallcreeper hitches upwards in small hops on a rock face usually using wing-flicks as support. Black bib extends to sides of head and breast in ♂, but note greyish belly, and rather bleached brownish wings, possibly juv feathers (but ageing still unsafe). Note characteristic scaly white tips to undertail-coverts. (R. Martin)

IDENTIFICATION It is difficult to misidentify this extremely attractive and striking bird, which lacks any relatives or potential confusion species. The body is only about the size of a Garden Warbler, but when it opens its *uniquely broad and rounded wings*, and flies along a rock face in its typically *unsteady and flapping, almost erratic, butterfly-like flight*, it looks more like a small thrush. The shape of the perched bird is nuthatch-like, i.e. *short-tailed and rather compact*, but with a *much thinner bill* (almost straight, very subtly downcurved only), perfect for picking insects and spiders from cracks in the rock faces it forages on. The Wallcreeper is basically a dark bird, rather uniformly *sooty-grey* above and below, with largely darker flight-feathers, but its unique feature are the colourful upperwings when spread: the *blackish upperwing has large crimson-red patches on forewing and bases of flight-feathers*, with large white spots on the primaries. White tail-corners add to the attractive flight pattern. In winter (late summer–early spring) the sexes are alike, with dark grey breast and belly, and a *rather well-marked off-white throat*. Upperparts are subtly paler grey than the belly. In early summer the ♂ becomes a little darker on the head and develops a *solid black throat bib* reaching the upper breast, whereas the ♀ acquires a more restricted and variable blackish or dark grey patch on centre of throat, surrounded by paler grey-white. A few ♀♀ attain very little dark on the throat even in summer (some none at all), while others have more, almost like ♂. The Wallcreeper has a peculiar habit of *flicking and semiopening its wings, often synchronised with each hop* when foraging on a vertical rock face. This exposes more of the red and the black-and-white wing pattern, which otherwise is largely concealed or inconspicuous on the perched bird. It may take a while to spot a Wallcreeper—rock faces attracting the species can be huge, and the bird is after all quite small in comparison. So look for something smaller than you might first anticipate! Remember that this is the only species that is grey-looking and moves in small treecreeper-like hops upwards on the rock face.

T. m. muraria, ♂, Bulgaria, Jun: usually creeps over vertical rock faces, displaying striking red flashes when flicking wings. At this time of year, ♂ easily recognised by extensive black bib. Ageing, however, mostly impossible without handling and sometimes difficult even then. (R. Wilson)

T. m. muraria, ♂, presumed 1stS, Switzerland, Apr: this unmistakable bird has unique appearance, including the long, thin, very slightly decurved bill, and short tail. Plumage generally grey, though ♂ in summer has black bib reaching breast, while rest of underparts are dark grey. Ageing difficult, but here most wing-feathers seemingly juv (especially note worn and bleached brown tertials). (R. Aeschlimann)

Sex variation in early spring, ♂ (left; Turkey, Apr) and ♀ (right; Slovakia, Mar): in addition to black bib, ♂ has more contrasting coloration, with brighter crimson primary patches and purer grey upperparts (especially crown and nape). By Mar, ♂ usually attains at least partial black bib of summer plumage, thus bird on right is clearly ♀, especially given the brown tinge to the grey upperparts. Due to great similarity between ad and juv wing-feathers, the species is as a rule difficult or impossible to age. (Left: R. Martin; right: S. Harvančík)

VOCALISATIONS Not often heard, except perhaps song in late spring. Song seems to occur in two variations, one a rather stereotyped brief strophe of three well-spaced whistling notes, the other a more varied warbling comprising whistling notes, and best labelled sub-song. The stereotyped song type might be described as a 'brief summary of Hoopoe Lark song'; it opens with one straight, low note followed by a louder rising glissando-whistle and finally a clear whistle that rises and falls in pitch, *tuu...ruuooeeh...ziii-üüh*. The varied, warbling song is difficult to describe, but could perhaps recall a singing Penduline Tit. The two song types are sometimes combined. – If enough time is spent at a breeding site, a few simple calls can often be heard. These include a buzzing *zerr*, a soft short whistle *tooee* and a call that has been likened to Tree Sparrow, *chup* or *cheu*.

SIMILAR SPECIES There are no risks of confusion—the Wallcreeper has a unique appearance!

AGEING & SEXING (*muraria*) Ages very similar or alike, and usually impossible to separate even when seen close. Sexes of ad differ in spring and summer. – Moults. Complete post-nuptial moult in summer, mainly Jul–Sep. Partial post-juv moult in late summer does not involve flight-feathers or, apparently, any greater coverts. Partial pre-nuptial moult of ad and 2ndCY in Feb–Apr involves head, neck, throat and breast, to enable breeding plumage to be attained. – **SPRING** ♂ Lores, chin, throat, sides of neck and centre of upper breast (sometimes entire breast) black. ♀ Lores grey, chin, throat and sides of neck white, upper breast pale grey. Lower throat/upper breast often has a variably large blackish or dark grey rounded (or irregular) patch, but some have more extensive black, approaching ♂ (rare), and others lack any patch (more common; variation not age-related; Löhrl 1976). – **AUTUMN** After post-nuptial moult of ad, sexes alike, both having white chin and throat, and grey breast without any black. Ageing rarely possible in the field after post-juv moult. – **Ad/1stW** Contrast between white (sometimes faintly buff-tinged) chin/throat and grey breast, border between these two rather well marked. In 1stW, on average slightly duller and greyer flight-feathers, greater coverts and primary-coverts (black parts) than ad, but rarely a distinct difference and therefore of limited use. **Juv** Very similar to ad, but often shows less contrast between pale grey (rather than white) chin/throat and only slightly darker grey breast.

BIOMETRICS (*muraria*) **L** 15–16.5 cm; **W** ♂ 95–105 mm (*n* 17, *m* 99.5), ♀ 94–103 mm (*n* 12, *m* 97.6); **T** ♂ 50–58 mm (*n* 16, *m* 52.9), ♀ 50–55 mm (*n* 9, *m* 52.3); **T/W** *m* 53.6; **B** 23.3–41.6 mm (*n* 32, *m* 32.2); **B/W**

T. m. muraria, ♀, Bulgaria, Jul: unlike summer ♂, lores greyish, chin to upper breast pale grey-white, with whiter throat and neck-sides, and varying number of blackish or dark grey feathers on lower throat and upper breast and especially belly (highly variable individually; practically none on this bird). (B. Nikolov)

T. m. muraria, ♀, Switzerland, Jun: in summer ♀ often has a black central throat patch, which varies in size and shape. (R. Kunz)

T. m. muraria, Slovakia, Feb: typically shows great agility in the air, here seen from below in characteristic downward-gliding flight. Underwing pattern is unique from below. Note quite well-marked white corners to tail. Since this is in winter sex cannot be decided. (S. Harvančík)

T. m. muraria, possible ♂, Spain, Nov: during autumn/winter, ♂ attains winter ♀-like plumage without black bib, making ageing and sexing largely impossible. Combination of dark sooty-grey, almost blackish-looking, underparts (except white throat/upper breast), pure grey upperparts and intense black ad wing-feathers infer a ♂. (F. López)

CREEPERS

T. m. muraria, possible ♂, Switzerland, Feb: in characteristic swooping flight, here with wings folded. Same assessment of sex as in previous image (previous page, bottom right) may apply here, too. (L. Howald)

T. m. muraria, ad, Spain, Sep: here an ad already in winter ♀-like plumage (at the end of postnuptial moult; unmoulted ad feathers detected in flight images). Sex difficult to decide safely. (H. Shirihai)

24.7–41.6 (*m* 32.6), usually >30 (87%); **Ts** 20.8–23.6 mm (*n* 30, *m* 22.5); **HC** 10.3–13.1 mm (*n* 27, *m* 11.6). **White on tail-tip:** length on **r6** 10–17 mm (*n* 20, *m* 14.0). **Wing formula: p1** > pc 12–16.5 mm, < p2 28–36 mm; **p2** < wt 11.5–16 mm, =9 or 9/10 (75%), ≤10 (20%) or >9 (5%); **p7** < wt 3–7 mm; **p8** < wt 7–11.5 mm; **p10** < wt 13.5–19 mm; **s1** < wt 15.5–21 mm. Emarg. pp3–5, faintly also on p6. – White on inner web of r6 measured from most basal spot to tip of feather irrespective of colour on tip (often grey).

GEOGRAPHICAL VARIATION & RANGE Morphological variation is surprisingly slight for a mainly sedentary species with a wide distribution in several unconnected mountain ranges. Just two subspecies, and even these are so similar that it could be argued that the Wallcreeper is best described as monotypic. However, Vaurie (1959) and *BWP* are provisionally followed in recognising two races.

 T. m. muraria (L., 1766) (Europe, Turkey, Caucasus, Transcaucasia, NW Iran; in winter may move to lower altitudes or towards west and south). Described above. (Syn. *longirostra*.)

○ **T. m. nepalensis** Bonaparte, 1850 (NE Iran, Turkmenistan, east to Tien Shan, Himalaya, N & C China, S Mongolia; resident or makes only short-range movements in winter, mainly to lower altitudes). Very similar to *muraria*, being only slightly larger and darker, with larger pale spots on flight-feathers, and ♂ on average is slightly more extensively black below in summer plumage, but much overlap. Although it is true that average bill-length is slightly shorter than in *muraria*, this varies considerably individually, making it a less useful character for separation of single birds. Sexes very similar in size: **W** ♂ 98–108.5 mm (*n* 24, *m* 102.7), ♀ 96–108.5 mm (*n* 32, *m* 101.1); **T** ♂ 53–60 mm (*n* 17, *m* 56.2), ♀ 50–60 mm (*n* 12, *m* 54.0); **B** 25.8–34.2 mm (rarely to 46.1; *n* 58, *m* 29.5); **B/W** 25.0–45.2 (*m* 28.9), usually <30 (83%); **Ts** 21.5–23.9 mm (*n* 34, *m* 22.8); **HC** 10.3–13.7 mm (*n* 36, *m* 12.0). **White on tail-tip:** length on **r6** 13–21.5 mm (*n* 36, *m* 17.7). **Wing formula: p1** > pc 12–19.5 mm; **p2** < wt 12.5–20 mm, =8 or 8/9 (50%) or =9 or 9/10 (50%); **p8** < wt 9–13 mm; **p10** < wt 18–23 mm; **s1** < wt 18–23 mm.

REFERENCES Bezzel, E. (1993) *Limicola*, 7: 35–48. – Löhrl, H. (1967) *J. f. Orn.*, 108: 221–223. – Löhrl, H. (1976) *Der Mauerläufer*. NBB no. 498. Wittenberg Lutherstadt. – Saniga, M. (1995) *DB*, 17: 141–145. – Saniga, M. (1999) *DB*, 21: 154–159.

T. m. muraria, ad, presumed ♂, Italy, Oct: relatively large, butterfly-like wings enable striking aerial manoeuvres. Intensely glossed black wing-feathers make this bird ad, while bluish-grey upperparts and pure grey nape and mantle suggest a ♂. Note blackish loral stripe. (L. Sebastiani)

T. m. muraria, Slovakia, Feb: age and sex often impossible to determine without handling, and difficult even then, but rather pale lores and pale grey underparts could suggest a ♀, rather than winter ♂. (S. Harvančík)

T. m. muraria, juv, Spain, Aug: evenly pale grey underparts (including chin/throat) with only slightly darker grey lower breast and belly; also overall fluffy juv plumage and yellow gape. (S. Bot)

(EURASIAN) TREECREEPER
Certhia familiaris L., 1758

Fr. – Grimpereau des bois; Ger. – Waldbaumläufer
Sp. – Agateador norteño; Swe. – Trädkrypare

A very discreet bird of tall, closed forests, parks and gardens over much of Europe, absent only from the south and on Iceland. It can be described as the counterpart of a mouse among birds, being small and brownish, creeping quietly along trunks and branches searching for food in a 'near-sighted' way. Unlike the nuthatches, it can only creep upwards, taking support from its long and stiff tail in woodpecker fashion. It often alights at the foot of a tree and then works its way upwards along the trunk into the crown, then suddenly flies down to the foot of the next tree, etc. It has a brief but clear, pleasant song, which often first signals its presence.

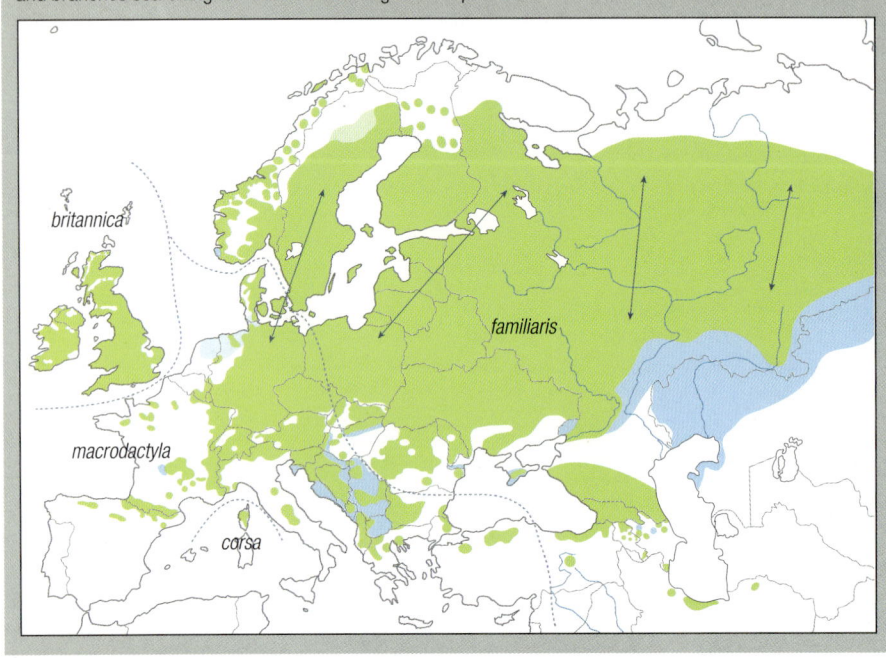

IDENTIFICATION There are two steps in the identification process. The first—to establish that it is a treecreeper of some kind—is easy. The second—to separate it from the Short-toed Treecreeper—is much more difficult. Treecreepers in general are recognised by their *small size*, their basically *white-speckled brown upperparts*, habit of *climbing trunks and branches upwards* in small hops or steps, and *long, fine and downcurved bills*. Then comes the more problematic part of the process, because what has been said so far fits equally well both of the European twin species, Treecreeper and Short-toed Treecreeper. To separate them, pay attention to tiny details in shape and plumage, but easiest is to use their different calls and songs (see below). Treecreeper is usually *slightly paler above* (an impression created by slightly larger and purer white spotting on the brown ground colour) and *cleaner white below*, with an on average *more prominent and purer white supercilium* that often reaches the forehead, a *slightly shorter bill* (but some overlap with Short-toed, and sexual differences in bill-length within each species must be kept in mind) and subtly longer claws. See Short-toed Treecreeper and under Similar species for more details of how to tell them apart.

VOCALISATIONS Has high-pitched, clear or fine trilling sounds. Song is a consistently repeated short and slightly irregular strophe of clear, soft whistling notes accelerating and falling in pitch, then ending with a little flourish or trill, *sree see-see-süü-soo sisisrreo-ee*. There is little local variation in the performance, all Treecreepers sound much the same. Due to the similar clear and high-pitched voice, could be confused only with some Blue Tit songs in the forest chorus and at some distance, but hardly possible when heard close. Also likened to Willow Warbler song, but this is more 'loveable' in tone, keeps a steady, moderate pace and lacks the terminal flourish. – Call is a very fine buzzing whistle, drawn-out a little and repeated a few times, *srree… srree… srree* (can recall similar call by Blackbird, but is usually softer, more drawn-out and more mechanically repeated, whereas Blackbird sounds a little stronger and rarely repeats). Furthermore, apparently mostly when alarmed or agitated,

C. f. familiaris, ad, Poland, Jan: small and mouse-like in its discreet behaviour when climbing up a trunk or along a branch, using the tail as a support. Strong and well-curved claws adapted to enable climbing even upside-down. Small, rounded white tips to primary-coverts, and uniformly dark wing-feathers, including alula and primary-coverts, make this an ad. M. Matysiak

C. f. familiaris, 1stW, Finland, Sep: at least in N Europe, whiter underparts, purer white supercilium (extending above lores, here even to forehead), more contrasting pale spots on crown and mantle, and somewhat shorter and less decurved bill, provide the first clues versus Short-toed Treecreeper. Broad white tips to primary-coverts confirm age. (H. Taavetti)

utters a less buzzing, high-pitched and shrill *zeee* (still differs from the clear call of Short-toed Treecreeper), which also can be repeated in series.

SIMILAR SPECIES Within the treated region only *Short-toed Treecreeper* is an alternative. See Identification for some general advice. Treecreepers in N & E Europe have more contrasting upperparts, with cleaner white supercilium (often reaching forehead) and white spotting or streaking in the brown ground colour, unlike Short-toed Treecreeper which is darker and duller brown above, with a more subdued supercilium. Flanks and vent are nearly pure white, whereas Short-toed Treecreeper invariably is slightly brown-tinged on these parts. However, Treecreepers breeding in C & S Europe are much more similar to Short-toed, and can even have a slight brownish wash on lower flanks. To separate them, one must learn the vocal differences and pay attention to bill length and the precise wing pattern. For details, see Short-toed Treecreeper.

AGEING & SEXING (*familiaris*) Ages often separable during 1stY if seen close or preferably handled (Cofta 1990, Soursa & Hakkarainen 2007), but some are still difficult. Sexes alike in plumage and differ only very slightly in size. – Moults. Complete post-nuptial moult of ad in mid Jun–Sep (Oct). Partial post-juv moult in late summer (onset often later than moult of ad) does not involve flight-feathers other than tertials. Usually replaces all median and greater coverts. No pre-nuptial moult. – **SPRING Ad** Alula and primary-coverts usually equally dark as rest of wing-feathers, with some surface gloss (though some birds more worn and slightly duller). White tips to outermost two long primary-coverts small and rather rounded, often only 1 mm long and can even be totally absent, very rarely slightly longer and somewhat more pointed. **1stS** Alula and primary-coverts usually subtly duller and paler than rest of wing-feathers, edges sometimes worn and slightly ragged. (Buffish-)white tips to outermost two long primary-coverts rather long and narrow, often wedge-shaped and even form a long 'channel' along shaft to base of feathers, only rarely small and more similar to typical spotting of ad, but then still more pointed, less rounded. – **AUTUMN** Ageing as in spring, only generally more obvious due

C. f. familiaris, 1stW, Finland, Mar: sometimes the white supercilium is less pure, while flanks and rear underparts are sullied drab or buff. Compared to Short-toed Treecreeper, has shorter less decurved bill, rather long hind claw, and wing-bar has diagnostic notch forming 'right-angle indentation' near edge of wing (arrow). Otherwise voice provides best means of identification in the field. Pale buffish tips to long primary-coverts prominent and oblong, confirming age. (M. Varesvuo)

C. f. macrodactyla, 1stW, Croatia, Jan: this continental European race averages slightly darker above than *familiaris*, with narrower and less well-defined supercilium. This individual shows less clear 'right-angle indentation' to wing-bar. Nevertheless, inner primaries have more washed-out buff-white tips covering entire edge, and gap between p6 and p7 is markedly larger, almost twice that between p7 and p8 (unlike Short-toed Treecreeper). Pale tips to outermost two primary-coverts are wedge-shaped and narrow, confirming age. (M. Matešić)

C. f. britannica, ad, England, Mar: a rather darker and richer-coloured race, often more rusty on lower flanks. Considering the relatively long bill, darkish fore-supercilium and reduced 'right-angle indentation' to wing-bar, this individual could be more prone to confusion with Short-toed Treecreeper. Nevertheless, has diagnostic off-white fringes to primary-tips, uneven spacing between p6, p7 and p8, and a long hind claw. Uniform wing-feathers and small, rounded white tips to primary-coverts confirm age. (G. Thoburn)

to fresher plumage. Tips of 1stW often more buff-tinged than in spring. **Ad** Pale tips to primary-coverts small or absent. Plumage neat, and flight-feathers glossy. **1stW** Pale tips to primary-coverts large and wedge-shaped, sometimes extensive reaching base of primary-coverts. Plumage neat but flight-feathers subtly less dark and glossy, and sometimes a faint difference in darkness of primary-coverts and rest of wing. **Juv** Similar to 1stW but usually less neat, more loose and woolly plumage, especially on vent and nape. Upperparts more profusely pale-spotted, and throat and breast finely vermiculated dark.

BIOMETRICS (*familiaris*) **L** 12–14 cm; **W** ♂ 62–68 mm (*n* 34, *m* 65.0), ♀ 60–64 mm (*n* 14, *m* 62.1); **T** ♂ 57–67 mm (*n* 27, *m* 61.4; heavily worn birds down to 53 mm), ♀ 55–65 mm (*n* 12, *m* 57.9; heavily worn birds shorter); **T/W** 94.1; **B** ♂ 14.0–20.0 mm (*n* 32, *m* 17.0), ♀ 13.7–15.7 mm (*n* 13, *m* 14.9); **Ts** 14.0–16.4 mm (*n* 44, *m* 15.2); **HC** 7.9–10.8 mm (*n* 44, *m* 8.8); **HC/B** 46.2–64.8 mm (*n* 41, *m* 54.5). **Wing formula: p1** > pc 7–11.5 mm, < p2 17–24 mm; **p2** < wt 8–11 mm, =8 or 8/9 (70%), =7/8 (16%) or =9 or 9/10 (14%); **p3** < wt 1–3 mm; **pp4–5**(6) about equal and longest; **p6** < wt 0–1.5 mm; **p7** < wt 4–7.5 mm; **p8** < wt 7–10 mm; **p10** < wt 10–13.5 mm; **s1** < wt 11–13.5 mm. Emarg. pp3–6. – A larger sample of live Finnish birds (Soursa *et al.* 2007) gave slightly larger measurements (age classes combined): **W** ♂ *m* 67.1 mm (*n* 375), ♀ *m* 63.6 mm (*n* 389); **B** ♂ *m* 18.0 mm (*n* 358), ♀ *m* 16.1 mm (*n* 360).

For difficult birds, resembling *C. brachydactyla*, check: (i) precise wing and alula pattern (see, e.g., Svensson 1992); (ii) hind claw length; and (iii) wing formula, where distance p6–p7 is about twice that of p7–p8, or even more, but less than twice in *brachydactyla* (a few birds are ambiguous, though). Ratio HC/B gives fairly good help in most populations: *familiaris* ≥ 46 (*brachydactyla* ≤ 49, thus overlap is 46–49). Large alula-feather usually has no white outer edge (92%), but sometimes a very thin or incomplete white edge (5%) or, rarely, a prominent one (3%). – A more elaborate formula than HC/B is described in Svensson (1992).

GEOGRAPHICAL VARIATION & RANGE Slight variation. There is a cline of decreasing colour saturation running from W Europe eastwards through Fenno-Scandia to Russia, with brown colour away from rump becoming slightly paler and more tawny-brown, less dark and rufous-tinged, and white spots being purer and better defined, but populations from Russia east through much of Siberia rather stable. Isolated population of Corsica differs in being subtly larger with obviously longer bill. Many more races described in Asia, but outside the scope of this book.

C. f. familiaris L., 1758 (Fenno-Scandia, Poland, Baltic States, Russia, Slovakia, E Hungary, Romania, Ukraine, Caucasus, Transcaucasia, N Iran; probably NE Turkey; short-range winter movements of northern breeders, in some years large-scale irruptions). Treated above. Whitish underparts without any brown tinge or with very restricted brown on lower flanks in some. Upperparts rather bright tawny-brown, with rusty or cinnamon tinge on rump, white streaks and spots fairly clean whitish and prominent. Whitish supercilium usually prominent. – Birds in northern and north-eastern part of range tend to be palest and have most obvious white supercilium. Scant material examined from Caucasus and adjacent areas but specimens seen seem to be very nearly identical to *familiaris*. (Syn. *caucasica*; *occidentalis*; *persica*; *rossica*.)

C. f. macrodactyla C. L. Brehm, 1831 (W & C continental Europe, including Denmark, Germany, Czech Republic, W Hungary, Bulgaria, Balkans, mainland Italy and west to France and Spain, possibly W Turkey; resident). Slightly darker on average than *familiaris*, pale streaks on upperparts more dusky brown-tinged and diffuse, also narrower on crown, and pale supercilium poorly developed. Underparts more dusky off-white, not so silky white. Usually more rufous tinge on lower flanks. **W** ♂ 61.5–68 mm (*n* 32, *m* 65.1), ♀ 60–65.5 mm (*n* 16, *m* 62.7); **T** ♂ 58–66.5 mm (*n* 13, *m* 63.0; heavily-worn birds shorter), ♀ 55–64 mm (*n* 12, *m* 59.5; heavily-worn birds shorter); **B** ♂ 15.1–19.3 mm (*n* 43, *m* 17.0), ♀ 13.5–17.8 mm (*n* 20, *m* 15.1); **Ts** 14.5–16.5 mm (*n* 23, *m* 15.5); **HC** 8.0–10.7 mm (*n* 64, *m* 9.1); **HC/B** 48.2–64.4 mm (*n* 64, *m* 55.9). (Syn. *rhenana*.)

○ **C. f. britannica** Ridgway, 1882 (Britain & Ireland; resident). Extremely similar to *macrodactyla*, only a trifle smaller and on average very slightly darker above and duskier below. Many are rustier on lower flanks, much like *C. brachydactyla*. A subtle race as many cannot be separated from *macrodactyla*. **L** 12–13.5 cm. **W** ♂ 59–67 mm (*n* 28, *m* 64.2), ♀ 59–64 mm (*n* 19, *m* 61.9); **T** ♂ 55–67 mm (*n* 24, *m* 59.7; heavily-worn birds shorter), ♀ 53–61 mm (*n* 17, *m* 57.3; heavily-worn birds shorter); **B** ♂ 14.8–18.9 mm (*n* 25, *m* 17.0), ♀ 14.4–18.0 mm (*n* 19, *m* 15.8); **Ts** 14.6–16.7 mm (*n* 27, *m* 15.5); **HC** 8.0–9.5 mm (*n* 46, *m* 8.7); **HC/B** 46.0–62.8 mm (*n* 46, *m* 52.9). (Syn. *meinertzhageni*.)

C. f. corsa Hartert, 1905 (Corsica; resident). Plumage very similar to *macrodactyla* (contrary to some statements), but underparts usually cleaner whitish, not so sullied dusky, and has rather obvious rufous tinge on flanks and vent, more rufous than is generally found in *macrodactyla* (but similar to many *britannica*). Size slightly larger than *macrodactyla*, and in particular bill is longer. **L** 13–14 cm. **W** ♂ 64–70 mm (*n* 25, *m* 67.8), ♀ 63.5–68 mm (*n* 14, *m* 65.3); **T** ♂ 61–69.5 mm (*n* 21, *m* 66.0; heavily-worn birds shorter), ♀ 58–66 mm (*n* 14, *m* 62.5; heavily-worn birds shorter); **B** ♂ 17.3–21.6 mm (*n* 25, *m* 19.9), ♀ 16.0–20.6 mm (*n* 14, *m* 18.0); **Ts** 14.0–17.0 mm (*n* 29, *m* 15.5); **HC** 8.0–9.6 mm (*n* 40, *m* 8.8); **HC/B** 38.8–53.8 mm (*n* 40, *m* 45.9). Note that ratio HC/B does not work as separation from *C. brachydactyla* (see Biometrics) in this race due to its long bill.

REFERENCES Cofta, T. (1990) *Notatki Orn.*, 31: 87–93. – Daunicht, W. D. (1991) *Limicola*, 5: 49–64. – Flegg, J. J. M. (1973) *Bird Study*, 20: 287–302. – Harrison, J. M. (1935) *Ibis*, 77: 437–438. – Mead, C. J. & Wallace, D. I. M. (1976) *BB*, 69: 117–131. – Soursa, P. & Hakkarainen, H. (2007) *Alula*, 13: 146–150.

C. f. britannica, 1stW, England, Oct: long hind claw and rather short bill. Despite that this is a rather warm-coloured race, this bird has relatively pale upperparts (due to slightly larger, purer white spotting), rather clean white underparts, and more prominent and purer white supercilium that just reaches the forehead. Pale tips to outermost long primary-coverts are wedge-shaped, long and narrow, typical of young age. (C. Bradshaw)

C. f. britannica, juv, Scotland, Jun: more woolly and loose plumage, especially on vent and nape, and plumage more profusely pale-spotted. Bill still very short, far from fully grown. (G. Thoburn)

SHORT-TOED TREECREEPER
Certhia brachydactyla C. L. Brehm, 1820

Fr. – Grimpereau des jardins; Ger. – Gartenbaumläufer
Sp. – Agateador común; Swe. – Trädgårdsträdkrypare

This species is extremely similar to the Treecreeper, and separation without the help of vocalisations or close views and expert knowledge is invariably difficult. Luckily, both song and most calls are distinct, and range sometimes helps a little. The short-toed species is absent (save a few stragglers) from the British Isles, most of Fenno-Scandia, the Baltic States and Corsica (where Treecreeper is present), but is common over much of the rest of S Europe, and occurs also in parts of N Africa. On average, it is found more often at lower altitudes than Treecreeper, but there is much overlap.

IDENTIFICATION Some general points are covered under Treecreeper. When it is obvious that one of the two European treecreeper species is involved, the best means of identification is to pay attention to vocalisations (see below). A silent bird can be very difficult to identify in the field, unless it allows prolonged and close observation. Short-toed Treecreeper has a *slightly longer bill* (an average difference, since there is some overlap with Treecreeper; bear in mind that bills of ♂♂ average slightly longer than of ♀♀), and this can be helpful with particularly long-billed ♂♂. That the *hind claw is subtly shorter* than in Treecreeper is usually difficult to ascertain in the field, but at least one should be aware of it when examining photographs. *Lower flanks and vent are brown-tinged* in Short-toed, and although British and Continental Treecreepers often show a hint of the same brown tinge, it is generally less obvious than in Short-toed. Finally there are some minute but important differences in wing pattern: (i) the *frontal side of the angled buff wing-band runs in rather even 'steps'*, one for each primary, whereas in Treecreeper there is a deep indentation forming a dark 'L-shaped' right angle that almost 'cuts through' the buff band; (ii) the *inner primaries have well-marked whitish tips*, in shape reminiscent of diamonds, contrasting against dark remainder of feathers, while Treecreeper has more washed-out buff-white tips that run along the entire edge of the tips; (iii) the *tips of the inner primaries are more evenly spread*, with fairly equal distances between pp6–8, whereas in Treecreeper there is a markedly larger step between pp6–7, often twice the distance found between pp7–8. Due to the pale tips on these feathers, these details are sometimes visible in the field.

VOCALISATIONS Nearly all calls and the song differ rather clearly from those of Treecreeper. The song is less fluent and sweet in tone, more staccato or jolting, and uneven in rhythm. It is also briefer, and the pitch does not fall so much or at all as in Treecreeper. In fact, the last one or two notes are often higher-pitched, faster and more stressed than the preceding ones. A fairly standard song can be rendered *tyyt e-too e-ty-teet*. Mimicry by Short-toed Treecreeper of Treecreeper song has been reported and much discussed, but it seems a quite rare occurrence, and it could be questioned whether not such birds are hybrids—which are known to occur—rather than pure Short-toeds. Song on Cyprus is described as slightly lower-pitched and longer, appearing slower (Förschler & Randler 2013). – The common call is very typical and easy to recognise by its very clear and straight tone, penetrating far in the woods due to its loudness and almost metallic ring, often given in short series at moderate pace ('between walking and jogging steps'), *tyyt… tyyt… tyyt…*, like 'drops of silver', much clearer than call of Dunnock (though otherwise similar in pitch and length). Rarely gives a more strident high-pitched *tsree*, often in series, and doubtfully separable from the common call of Treecreeper but apparently slightly lower-pitched.

SIMILAR SPECIES Can only be confused with *Treecreeper* within treated region. See Identification for a detailed discussion. Apart from differences in structure and voice, note the on average more blurred and dusky head pattern, the supercilium frequently being less white and prominent compared to Treecreeper, and the less clearly white-spotted upperparts, more dull brown in comparison. There is usually more rufous or dull brown on lower flanks and vent than in Treecreeper, but extremes are similar. Birds seen close and under ideal circumstances, or which are photographed, can often be separated using the wing pattern differences detailed under Identification.

AGEING & SEXING (*brachydactyla*) Ageing not studied in detail, but a preliminary check indicated that the same methods as for *C. familiaris* (which see) are applicable (Cofta 1990, Soursa & Hakkarainen 2007). Sexes alike in plumage, differing only very slightly in size. – Moults. Complete post-nuptial moult of ad in late Jun–Sep. Partial post-juv moult in late summer or early autumn does not involve flight-feathers other than tertials. Usually replaces all median and greater coverts. No pre-nuptial moult.

BIOMETRICS (*brachydactyla*) **L** 12–14 cm; **W** ♂ 60–67 mm (n 92, m 63.4), ♀ 56.5–66 mm (n 56, m 60.7); **T** ♂ 53–64 mm (n 79, m 58.3; heavily-worn birds down to c. 50 mm), ♀ 50–62.5 mm (n 48, m 55,7; heavily-worn birds down to c. 48 mm); **T/W** m 92.3; **B** ♂ 15.7–22.3 mm (n 89, m 18.8), ♀ 15.3–21.0 mm (n 58, m 17.4); **Ts** 14.1–16.7 mm (n 110, m 15.6); **HC** 6.7–8.4 mm (n 160, m 7.5); **HC/B** 33.5–49.0 mm (n 158, m 41.4). **Wing formula**: p1 > pc 7–11 mm, < p2 18–22 mm; p2 < wt 6–11 mm, =7/8 or 8 (50%), =8/9 or 9 (40%), or =9/10 or 10 (10%); p3 < wt 0.5–3 mm; pp4–5 about equal and longest; p6 < wt 0.5–3 mm; p7 < wt 3.5–6 mm; p8 < wt 6.5–9 mm; p10 < wt 9–13 mm; s1 < wt 9.5–14 mm. Emarg. pp3–6.

For difficult birds, resembling *C. familiaris*, check: (i) exact wing and alula pattern (see, e.g., Svensson 1992); (ii) hind claw length; and (iii) wing formula, where distance

C. b. brachydactyla, ad, Spain, Mar: easily confused with Eurasian Treecreeper, from which best separated by voice, but also by darker upperparts, warmer buff-brown flanks, dark-streaked lores (with shorter white supercilium), longer bill, and different pale marks in wing, all of which are judged only in close views and are not always valid throughout the range. Small, rounded white tips to darker primary-coverts suggest an ad, as do blackish areas around shafts of rectrices. (M. Saarinen)

C. b. brachydactyla, 1stW, Spain, Jan: unlike Eurasian Treecreeper has relatively longer bill and shorter hind claw. Broad white tips to greater coverts support ageing. (R. Rodríguez)

Short-toed (top three: Finland, Apr; Germany, Mar; Slovenia, Mar) and Eurasian Treecreepers (bottom three: Finland, Apr; Germany, Apr; Finland, Apr): silent birds can be very demanding to identify, unless a suite of features can be analysed in close views. Nine points should be considered. Often first noticed in Short-toed Treecreeper is its (1) slightly longer and more clearly decurved bill, then check (2) the anterior side of the buff wing-bar that forms rather even 'steps', one for each primary (lacking or having much-reduced 'right-angle indentation' of Eurasian). Further, (3) the large alula often has thin white edge along its entire length in Short-toed (linking broader white tip and base), whereas it is invariably broken in centre in Eurasian (differences best seen in two middle images), (4) the visible tips of inner primaries are more evenly spaced than in Eurasian, and (5) the pale primary-tips are whiter, bolder and smaller in Short-toed (broader, more diffuse and appear to 'wrap around' in most Eurasian). (6) The supercilium in Short-toed is frequently less white and prominent, and especially appears less prominent in front of eye than in most Eurasian. (7) The flanks of Short-toed are often washed brownish, but in Eurasian from UK and continental Europe there is often a hint of brown tinge. Note that (8) the upperparts of Short-toed are less obviously white-spotted, and duller brown than most Eurasian. As its name is meant to suggest, (9) the hind claw of Short-toed is shorter, but it is usually very difficult to see this in the field. Still, vocalisations are often the best means of identification. (Top left: M. Varesvuo; top centre: R. Martin; top right: S. Harvančík; bottom left: A. Below; bottom centre: T. Grüner; bottom right: M. Varesvuo)

p6–p7 is less than twice that of p7–p8 (sometimes nearly equal), but about twice or more in *familiaris*. Ratio HC/B offers fairly good help in most populations: *brachydactyla* ≤ 49 (*familiaris* ≥ 46, thus overlap is 46–49). Large alula-feather usually has an unbroken white outer edge, either narrow or broad (52%), but sometimes very thin or incomplete (23%) or even absent (25%). – A more elaborate formula than HC/B is described in Svensson (1992).

GEOGRAPHICAL VARIATION & RANGE Slight and clinal variation. Several races described, but all except three seem too slight and variable to be recognised, and even one of these three (*mauritanica*) is subtle and the other mainly based on slight vocal differences. All populations sedentary.

C. b. brachydactyla C. L. Brehm, 1820 (Europe, Turkey). Treated above. Dark brown above, sullied white below, upperparts with indistinct paler spots or streaks, rump rufous-tinged, underparts have variable rufous-buff tinge to lower flanks and vent. Somewhat variable individually and locally in size, bill-length and colour saturation, but all differences seem clinal and very slight. – A claimed colour difference in that birds in W Europe, especially those in Brittany ('*megarhyncha*'), are brighter and more rufous above could not be established despite examination of adequate material. Several other proposed subspecies proved inseparable or extremely vague not warranting separation. (Syn. *bureaui*; *harterti*; *lusitanica*; *megarhyncha*; *neumanni*; *nigricans*; *obscura*; *parisi*; *rossocaucasica*; *siciliae*; *spatzi*; *stresemanni*; *ultramontana*.)

○ *C. b. mauritanica* Witherby, 1905 (NW Africa). Very subtly larger and perhaps on average slightly darker brown and less rufous-tinged above than *brachydactyla*, but many are very similar and a few alike. Underparts in many have more extensive brown tinge on lower belly. Subtle race. **W** ♂ 63–70 mm (n 22, m 66.6), ♀ 60–66 mm (n 9, m 62.5); **T** ♂ 58–67 mm (n 19, m 61.7; heavily-worn birds down to 50), ♀ 54–64 mm (n 7, m 57.2; heavily-worn birds shorter); **B** ♂ 17.0–21.2 mm (n 22, m 19.6), ♀ 16.7–20.4 mm (n 9, m 17.8); **Ts** 15.0–16.8 mm (n 21, m 15.9); **HC** 6.9–8.3 mm (n 34, m 7.6); **HC/B** 36.4.2–44.4 mm (n 34, m 40.2). (Syn. *raisulii*.)

○ *C. b. dorotheae* Hartert, 1904 (Crete, Cyprus). Very similar to *brachydactyla*, only a fraction colder and less rufous-tinged brown above, and underparts are slightly whiter, flanks and vent lacking nearly any rufous. Small. Main difference appears to be a subtle difference in song, this being a little longer, containing more elements, and sounding subtly lower-pitched and slower (Förschler & Randler 2013). **W** ♂ 60.5–66.5 mm (n 18, m 63.6), ♀ 59–64 mm (n 12, m 61.0); **T** ♂ 57–64 mm (n 18, m 60.3), ♀ 50–59.5 mm (n 11, m 55.4); **B** ♂ 15.7–21.2 mm (n 18, m 18.9), ♀ 16.8–19.6 mm (n 12, m 18.0); **Ts** 14.5–16.0 mm (n 23, m 15.0); **HC** 6.6–8.0 mm (n 33, m 7.4); **HC/B** 34.9–46.4 mm (n 33, m 39.7).

REFERENCES COFTA, T. (1990) *Notatki Orn.*, 31: 87–93. – DAUNICHT, W. D. (1991) *Limicola*, 5: 49–64. – FLEGG, J. J. M. (1973) *Bird Study*, 20: 287–302. – FÖRSCHLER, M. I. & RANDLER, C. (2013) *Zoology in the Middle East*, 46: 37–40. – HIRSCHFELD, E. (1985) *BB*, 78: 300–302. – MEAD, C. J. & WALLACE, D. I. M. (1976) *BB*, 69: 117–131. – RODRIGUEZ DE LOS SANTOS, M. (1985) *BB*, 78: 298–300. – SOURSA, P. & HAKKARAINEN, H. (2007) *Alula*, 13: 146–150.

C. b. brachydactyla, ad, Germany, Jan: this bird has a wing-bar approaching Eurasian Treecreeper, with a slight 'right-angle indentation' on its anterior side. Still, note fairly long, well-curved bill, comparatively short hind claw and brown-tinged lower flanks. Dark and nearly unspotted primary-covert tips make this an ad. (R. Martin)

C. b. brachydactyla, 1stW, Spain, Oct: rather long bill and white supercilium mainly behind eye provide identification. However, this image also shows that some birds (mostly young) can have less well-marked whitish primary-tips. (F. Trabalon)

C. b. dorotheae, ad, Cyprus, Oct: very similar to race *brachydactyla*, but averages washed duller, being colder and less rufous-tinged brown above, and underparts are cleaner too, flanks and vent lacking nearly any rufous. This bird also shows the species' classic dusky lores, and anterior side of the angled buff remex-bar that runs in rather evenly spaced 'steps', one step for each primary. Small rounded pale tips to fresh and dark primary-coverts helps ageing. (H. Shirihai)

(EURASIAN) PENDULINE TIT
Remiz pendulinus (L., 1758)

Fr. – Rémiz penduline; Ger. – Beutelmeise
Sp. – Pájaro moscón europeo; Swe. – Pungmes

A free-hanging globed nest with a funnel entrance, suspended from a birch, willow or alder beside a wet area, often provides the first evidence of this species' presence. Breeds in S Scandinavia, C Europe and Iberia east to the Caspian Sea and N Kazakhstan. Northern populations move south in winter, but in the south the species is mostly sedentary. The Penduline Tit has a complex mating system, with most ♂♂ being polygamous, building several nests in a season to attract multiple ♀♀, whereas the latter are usually—but not invariably—monogamous.

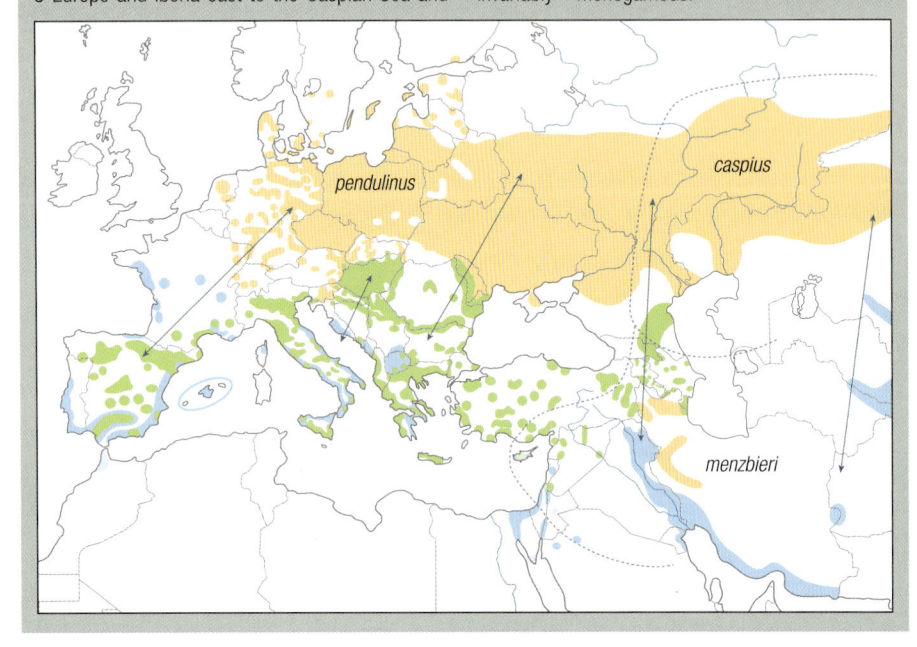

R. p. pendulinus, ♂, presumed ad, Hungary, Jun: with progressive feather wear, pinkish-chestnut below and on mantle becomes more exposed in ♂♂. Still not heavily worn plumage in Jun, including rufous-fringed primary-coverts, suggest an ad. (R. Martin)

IDENTIFICATION Very small, compact and short-winged tit-like bird of overgrown wet habitats, with a *conical and fine-pointed bill*, and distinctive far-carrying *high-pitched whistles*. Adults superficially recall a miniature ♂ Red-backed Shrike, with pale head, *black facial mask*, usually bordered above by a narrow chestnut band, *chestnut-brown upperparts* becoming paler posteriorly, *blackish tail- and flight-feathers*, whitish throat and pale buffish rest of underparts, *sometimes pinker on flanks* and often *brown-flecked on breast*. Iris dark reddish-brown. Bill and legs mostly grey. Sexing sometimes easy, especially if pair is seen together, but complicated by age-related variation; young are duller, including having no or only a poorly-marked mask. Gait and behaviour resemble true tits, often hanging at tip of fine branches of birch or willows, including upside-down, albeit rather less excitable and inquisitive compared to true tits, and flight action may recall a *Phylloscopus* warbler more than a tit. Often rather gregarious.

VOCALISATIONS Song is a simple ditty (slightly reminiscent of Goldfinch) comprising trills, the call, various twittering and strong buzzing notes (recalling flight-call of Tree Pipit), and a metallic rattling *r'r'r'…*, the elements well spaced, e.g. *tsiüü… sirrrr… twitwitwi… tsiüü…zver'r'r'… tsiüü…* Phrases used to build up the song according to one study being any of *pseeo-uit, psseo-srr, psee-shuit, pseeo-chip-chip, pseeo-tsee-tsee, pseee-u* (descending), *pseeo-wink-wink, pse-ee-u, tsu-i-u tsu-i-u* (latter in fast series; Persson & Öhrström 1980). – Commonest call, easily recognised, a soft, thin whistle, a wheezy, plaintive *tsiüü* or *seeoo*, recalling a Reed Bunting or distant Red-throated Pipit, but more drawn-out. Other variants include a trisyllabic *tsi-tsi-tsi*.

SIMILAR SPECIES Adult unmistakable due to very small size, pointed bill and diagnostic plumage. Juvenile, given a brief glimpse, might be mistaken for a *Bearded Tit*, but

R. p. pendulinus, ♂, Sweden, Apr: almost Red-backed Shrike-like plumage pattern, with broad mask. Unlike ♀, mask is solid, jet-black and very broad, extending to forehead, and wider behind the eye, too. Narrow chestnut band above black forehead, red-brown patches on mantle, scapulars and wing-coverts. Overall immaculate plumage probably indicates ad, but best not to attempt definite ageing. (D. Pettersson)

R. p. pendulinus, ♂, presumed 1stS, France, Mar: in early spring ochre-chestnut flecking on underparts sometimes mostly concealed, but this feature also varies individually. In this individual the chestnut band above the black forehead is much reduced, and chestnut on mantle is limited. Juv primary-coverts (dull brown-grey centres, rather pointed tips, diffusely paler edges) and tail-feathers (outer with sharply-pointed tips) infer age. (H. Harrop)

PENDULINE TITS

R. p. pendulinus, ad ♀, Slovakia, Apr: compared to ♂, mask visibly narrower and less intense black, and both black and chestnut on forehead reduced; also paler rufous above and on wings. At least in early spring, the crown is less white, more grey-brown, and underparts are also duller. Ad by uniform wing-feathers, including distinctly pale-edged primary-coverts. (S. Harvančík)

lacks long, graduated tail of latter, as well as extensive white fringes to primaries and primary-coverts. — Main identification challenge lies in separating *Black-headed Penduline Tit* and *White-crowned Penduline Tit* from the Eurasian species. (See these two for details.)

AGEING & SEXING Ageing requires close scrutiny of moult and feather wear in wing and tail. Sexes separable, especially in spring. — Moults. Complete post-nuptial moult of ad after breeding (mainly Jul–early Sep), and mostly partial post-juv moult (occasionally complete in some southern populations) at same time. Post-juv moult (by those undertaking partial moult) includes all head, body, lesser and median coverts, some or all greater coverts, tertials and often central to all tail-feathers. At least in Spain (G. Gargallo *in litt.*), Israel and Turkey, 1stW and ad undertake a pre-nuptial moult (mostly Nov–Dec), in 1stW largely complementary to post-juv moult in summer (i.e. most of the flight-feathers and greater coverts renewed during post-juv moult are not replaced again), but extent highly variable, and some populations (presumably mostly northern) have only limited or no winter moult (cf. *BWP*, Svensson 1992). — **SPRING Ad ♂** Sexed by broader mask, deeper chestnut lower mantle, purer grey crown (with only slight buff tinge), and breast often flecked extensively dark rufous. With wear, crown and neck much whiter, any chestnut on forehead reduced or lacking, chestnut of mantle somewhat deeper, with pale scapulars

R. p. pendulinus, ♀, Slovakia, Jul: with wear, head of ♀ often becomes contrastingly white and, unlike ♂, mask can be confined to behind eye, and underparts are plainer. Ageing unsafe unless handled or wing-feathers seen closer and at different angle. (S. Harvančík)

R. p. pendulinus, 1stS ♀, Italy, Feb: by late winter or early spring some young ♀♀ develop an ad ♀-like dark mask, but black and rufous on forehead is still rather subdued. Moult limit in greater coverts and tail, pointed juv primary-coverts and already worn and bleached browner primary-tips (compared to ad at same season) confirm age. (D. Occhiato)

R. p. pendulinus, ad ♂, Belgium, Jan: very fresh and crisply patterned ad ♂. Aside of the solid jet-black mask in winter, the already extensive amount of dark chestnut visible on mantle, the broadly white-tipped inner primaries, and rounded tips to all tail-feathers confirm age and sex. (J. Fouarge)

bar enhanced and rufous-chestnut flecking on breast-sides patchier and more extensive (albeit much individual variation). **Ad ♀** Mask usually narrower or less solid (black and chestnut on forehead reduced or virtually absent), while lower mantle and wing-coverts are pale rufous to yellowish red-brown, crown to nape slightly dusky grey-brown, and breast lacks or has only very limited ill-defined rufous flecking. **1stS** Diagnostically retains juv primary-coverts, usually some remiges, and sometimes a few outer greater coverts and tail-feathers, latter more worn and abraded browner with faded fringes, forming moult contrast easier to detect at this season. Some that moulted in winter may show three feather generations in wing. Some even more distinctive, having duller plumage, and conversely some 1stS ♂♂ resemble ad ♀; 1stS (especially ♀) with heavily-bleached juv greater coverts often has clear whitish wing-bar. – AUTUMN **Ad** Both sexes reliably aged by evenly very fresh wing- and tail-feathers (remiges and primary-coverts glossy greyish-black), and by sharper face markings. **Ad ♂** Compared to spring, lower mantle, scapulars and wing-coverts more buff-rufous, less deep chestnut, crown and nape grey (often tinged buff-brown), and chestnut and black forehead band more extensive. **Ad ♀** Generally more buff-tinged plumage, and mask less well marked than in spring. Sexed by similar criteria as in spring. **1stW** By and large as respective ad, but sexes differ less clearly, both having a smaller, less neat mask, crown to hindneck largely grey-brown (mostly ♂) or brownish-buff (♀); 1stW ♀ dullest, with drab grey-brown mantle and wing-coverts, and dark mask poorly delimited. Moult limits produced by retained juv outer greater coverts (contrasting with newly moulted inner ad-like greater coverts, which are deeper rufous); also, all or most primary-coverts and remiges are retained (weaker textured, paler and browner centred, and fringed more buffish), while juv primaries are more pointed, less square-ended and have reduced pale tips. A few 1stW in S Europe, however, undertake a complete moult, and are largely inseparable from ad (though usually most or all primary-coverts and central alula-feather are retained; Mariné *et al.* 1994). **Juv** Generally greyish-brown (with only limited cinnamon) above and pale buffish-white below, black of head replaced by ochre-grey. Supercilium and throat paler.

BIOMETRICS (*pendulinus*) **L** 11–11.5 cm; **W** ♂ 53.5–60.5 mm (*n* 17, *m* 56.6), ♀ 53–57 mm (*n* 12, *m* 55.1); **T** 44–48 mm (*n* 29, *m* 46.5); **T/W** *m* 83.2; **B** 9.7–12.5 mm (*n* 26, *m* 11.0); **BD** 4.1–5.0 mm (*n* 27, *m* 4.6); **Ts** 13.5–15.6 mm (*n* 28, *m* 14.4). **Wing formula: p1** > pc 0–2 mm, < p2 28–32 mm; **p2** = wt 2–5 mm, =6/7 (65%), =7 (25%) or =6 (10%); **p3** < wt 0–3 mm; **p4** (pp3–5) usually longest; **p5** < wt 0–1.5 mm; **p6** < wt 1–3.5 mm; **p7** < wt 4–6 mm;

R. p. pendulinus, 1stW ♂, Italy, Nov: some 1stY (especially in southern populations) attain very similar plumage to ad, like this young ♂. However, has narrower mask, slightly broken up above eye, brown, pointed and slightly worn primary-coverts and pointed central tail-feathers, moult limit in tertials and already slightly worn primary-tips, suggesting age. (D. Occhiato)

R. p. pendulinus, 1stW ♂, Italy, Dec: young ♂ inferred by combination of broad black mask, like ad, but overall duller plumage (though deeper chestnut on mantle than most ♀♀), thus appearing intermediate in upperparts coloration between ad ♂ and ♀. Ageing confirmed by juv primary-coverts and primary-tips that are already slightly worn. Tail and tertials replaced in post-juv moult. (D. Occhiato)

R. p. pendulinus, ♀, presumed ad, Italy, Nov: ♀ differs from ♂ by moderately sized dark mask being less solidly black and having some pale mottling on cheeks and forehead, and also by cinnamon-brown rather than deep chestnut mantle. Age suggested by evenly fresh wings and tail, but in S Europe one cannot firmly exclude possibility of a 1stW which had complete post-juv moult. (D. Occhiato)

R. p. pendulinus, ♀, Italy, Feb: a typical ♀ with narrow and broken-up mask and dull cinnamon-brown and buff upperparts. Tail-feathers fresh and of ad type (broadly rounded at tips), and all visible greater coverts seem moulted, but since a closer view of wings not possible ageing is best not attempted. What can be seen of primary-coverts seem juv. (D. Occhiato)

R. p. pendulinus, ♀, Italy, Dec: some young ♀♀ can possess a quite well-marked facial mask (here the rufous border above the dusky mask is especially well-marked), demonstrating that it is first important to age the bird by moult, but since this is not always possible without handling, birds like this are best left unaged. (D. Occhiato)

p10 < wt 8–11 mm; **s1** < wt 9–12 mm. Emarg. pp3–6 (on p6 sometimes a little fainter).

GEOGRAPHICAL VARIATION & RANGE Variation rather slight and mostly clinal, involving mainly head, upperparts and upperwing coloration (most markedly in ad ♂♂), and to some extent body size. Three races occur in the covered region (whereas a fourth, *jaxarticus*, is extralimital and not treated). Racial designation often clouded by marked individual variation and is further complicated by intergradation and hybridisation where some of the taxa meet.

R. p. pendulinus (L., 1758) (Europe from S Fenno-Scandia, Denmark, Netherlands and Iberia east to C European Russia, W Caucasus, N, C & W Turkey; short-range migrant in north and east, mainly resident in south). Treated above.

○ *R. p. menzbieri* (Zarudny, 1913) (SE Turkey, Cyprus, Transcaucasia, and Middle East to W Iran; altitudinal or shorter southward movements in winter). Similar to *pendulinus* but is somewhat smaller with subtly thinner bill, shorter tarsus, shorter tail, chestnut/black forehead averages narrower, and chestnut upperparts and wing-coverts contrast more markedly with whiter lower back and rump. **W** ♂ 52–58 mm (*n* 14, *m* 54.2), ♀ 51.5–56.5 mm (*n* 9, *m* 53.5); **T** ♂ 42–47 mm (*n* 14, *m* 44.3), ♀ 39–45 mm (*n* 9, *m* 42.4); **T/W** 80.8; **B** 9.3–12.2 mm (*n* 23, *m* 10.8); **BD** 3.9–5.2 mm (*n* 22, *m* 4.6); **Ts** 12.7–14.5 mm (*n* 20, *m* 13.8). (Syn. *persimilis*.)

R. p. caspius (Poelzam, 1870) (SE Russia at Volga and Ural rivers south to N Caucasus and W & N Caspian Sea; in winter to Iran, Iraq and, rarely, Levant). Rather variable (♂♂ especially), but classic ad ♂ has more chestnut on forecrown, often extending to central crown (even to nape), with a pale bar on hindneck. Chestnut mantle richer and—in summer—rufous flecking on underparts more intense and extensive; broader white fringes to secondaries, tertials and tail. ♀ approaches *pendulinus*, but often has paler and greyer crown, nape and upper mantle, with only traces of chestnut on forehead and crown-sides, while wings and tail are fringed whiter, and underparts are on average also whiter. 1stY very similar to *pendulinus* but chestnut on forehead and crown-sides rather pronounced, and crown often has some chestnut. Pale fringes to tail- and flight-feathers broader and paler. Size as *pendulinus*: **W** ♂ 56.5–60 mm (*n* 12, *m* 58.5), ♀ 54.5–58 mm (*n* 12, *m* 56.0); **T** ♂ 45–48.5 mm (*n* 12, *m* 47.3), ♀ 44.5–50 mm (*n* 11, *m* 46.4); **T/W** *m* 81.8; **B** 10.0–11.8 mm (*n* 23, *m* 11.0); **BD** 4.3–5.4 mm (*n* 22,

R. p. pendulinus, 1stW ♀, Germany, Oct: there is variation among young ♀♀, some being duller with very obscure dark face mask, as here. Aged as young bird by much of wing being juv, including seemingly very pointed primary-coverts. (M. Schäf)

R. p. pendulinus, juv, Italy, Aug: juv looks strikingly different due to lack of dark mask or any brown on mantle. Still has yellow gape and loose, fluffy plumage. (D. Occhiato)

Presumed *R. p. menzbieri*, ♂, Kuwait, Nov: birds wintering in Kuwait are thought to be *menzbieri*, a race that only differs subtly from *pendulinus*, by being on average a little paler. Sex indicated by fairly dark mask (though sometimes in winter not completely solid) and much chestnut flecking on breast. Ageing impossible without closer view of wing. (M. Pope)

Presumed *R. p. menzbieri*, ♀, Kuwait, Nov: again a bird in winter in Kuwait most likely of the race *menzbieri*. Sex obvious from poorly developed dark mask and medium brown mantle. Safe ageing would require a better view, hence not attempted. (M. Pope)

R. p. caspius, ♂, Russia, Apr: highly characteristic with the deep chestnut on forecrown extending across much of crown, leaving just a variable bar or patch on hindneck pale. Also has extensive (though somewhat variable) chestnut flecking on breast. Ageing at this angle not advisable. (I. Ukolov)

R. p. caspius, ♂, presumed 1stS, Kazakhstan, May: striking plumage with extensive black mask reaching to forecrown and then becoming deep chestnut on crown and sides of neck, leaving just a small white patch on hindneck. Sex further confirmed by extensive deep chestnut on mantle. Note in this race typical broad white fringes to secondaries, tertials and tail. Uniform ad type wing-feathers, still primary-coverts seem brownish and pointed suggesting 1stS. (A. Isabekov)

m 4.9); **Ts** 13.5–15.5 mm (*n* 24, *m* 14.3). (Syn. *castaneus*.)

TAXONOMIC NOTES Eurasian Penduline Tit is sometimes considered conspecific with Black-headed Penduline Tit *R. macronyx* and White-crowned Penduline Tit *R. coronatus*, but we follow Harrap & Quinn (1996) and Bot *et al.* (2011) in treating them as separate species. There is apparently no evidence of interbreeding between ssp. *jaxarticus* of Eurasian Penduline Tit (W Siberia, from Urals eastward) and ssp. *stoliczkae* of White-crowned Penduline Tit (Yenisei River eastward) in their contact zone of C Siberia and NW Altai, though *jaxarticus* tends towards *stoliczkae* in size and plumage, especially in the east (and the few *stoliczkae* examined from E Turkestan appear closer to *jaxarticus*). While giving both separate species status, we acknowledge the need for further analysis. – White-crowned Penduline and Black-headed Penduline Tits occur sympatrically over large areas apparently without interbreeding, but hybridisation between Penduline and Black-headed occurs within two apparently secondary intergradation zones, one in N Caspian and one in SW Caspian regions; at mouth of Ural River, Black-headed Penduline hybridises with *caspius* (Penduline), resulting in a variable population; between lower Kura River and Lenkoran in E Transcaucasia, *menzbieri* (Penduline) hybridises with '*neglectus*'-type (Black-headed), resulting in a polymorphic population. See Vaurie (1959), *BWP* and Harrap & Quinn (1996). Recently Barani-Beiranvand *et al.* (2017) confirmed substantial genetic divergence between Eurasian and White-crowned Penduline Tits, whereas Eurasian and Black-headed showed very modest genetic variation. Still, as these authors concluded, secondary contact and hybridisation resulting in extensive gene flow between Eurasian and Black-headed cannot be excluded, and therefore we prefer to keep them as separate species awaiting further research.

REFERENCES Barani-Beiranvand, H. *et al.* (2017) *J. Avian Biol.*, 48: 932–940. – Bot, S. *et al.* (2011) *DB*, 33: 177–187. – Mariné, R., Figuerola, J. & Gutiérrez, R. (1994) *Butlletí Grup Català d'Anellament*, 11: 11–13. – Persson, O. & Öhrström, P. (1980) *Anser*, 19: 219–226.

R. p. caspius, presumed ad ♀, Kazakhstan, May: surprisingly similar to a European *pendulinus* ♂, but locality and prominent whitish fringes to flight- and tail-feathers typical only for *caspius*. Both sexing and ageing difficult here, but chestnut on mantle rather limited, mask comparatively narrow lacking chestnut extension on crown, and breast unmarked whitish, hence probably a ♀. Wings and tail neat and of ad type, the most likely age of this bird. (A. Isabekov)

R. p. caspius, presumed 1stS ♀, Kazakhstan, May: very poorly developed mask and cinnamon-brown mantle clearly infer ♀. Less easy to judge whether this is a 1stS or a young of the year, already in post-juv moult starting to attain a more mature plumage. (A. Isabekov)

BLACK-HEADED PENDULINE TIT
Remiz macronyx (Severtzov, 1873)

Fr. – Rémiz à tête noire
Ger. – Schwarzkopf-Beutelmeise
Sp. – Pájaro moscón cabecinegro
Swe. – Svarthuvad pungmes

A close relative of the Penduline Tit of Europe, and sometimes treated as a racial group within that, but here treated as a separate species based on clearly different morphology and slightly different structure, while hybridisation between the two occurs but is mainly limited and local. Widespread in Central Asia but range rather fragmented, being restricted to reedbeds and bulrushes beside major river systems and inland seas and lakes, e.g. Caspian Sea and Lake Balkhash. Largely sedentary, but in winter some move further south.

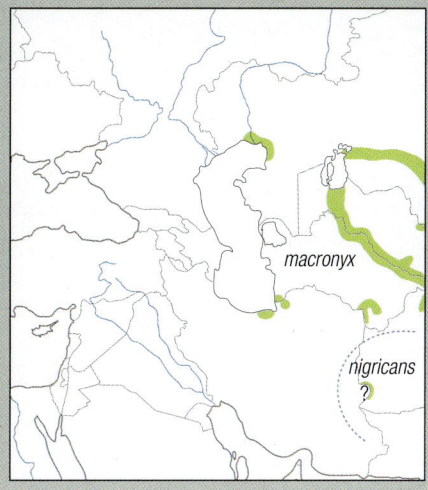

R. m. macronyx, ♂, presumed ad, Iran, Apr: interesting view of ♂ with spread wings, demonstrating bold pattern of chestnut patches on mantle and wings. Some buff-brown tips on nape are remnants of winter plumage not yet worn off. Whole plumage appears neat and ad-like, but image is slightly blurred, preventing ageing with certainty. (S. Papps)

R. m. macronyx, ad ♂, Iran, Apr: extensive black hood reaching breast, with characteristic and attractive pinkish-brown wash over rest of underparts, and deep chestnut upper mantle. Evenly-feathered wing, including primary-coverts, together with bold and colourful plumage are features of ad ♂. (C. N. G. Bocos)

IDENTIFICATION Much as described under Penduline Tit, though differs in being slightly longer-tailed. *Blackish hood of ♂*, reaching throat and even upper breast, is diagnostic, though extent rather variable; the black hood turns progressively into chestnut on mantle and throat and chest. The dark hood is often bordered by an ill-defined *whitish or buff neck-collar* emphasized by rich reddish-chestnut lower mantle and lesser coverts, and there is a *quite distinctly paler buffish shoulder bar*. Cinnamon-buff underparts (flecked rufous, especially in summer). ♀/young have duller head, only *forehead and mask dusky* (albeit extensive), *throat pale grey, variably mottled darker*, with characteristic *greyish-buff yellow-tinged crown to upper mantle*, rather *pronounced paler forecrown*, lower mantle pale rufous-chestnut, and breast and belly cream-yellow tinged brown. Wing and tail patterns, and bare-part colorations generally as in Penduline Tit, and behaviour also like that species.

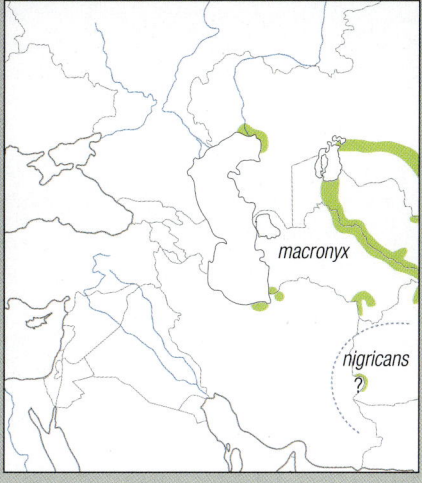

VOCALISATIONS Apparently similar to Penduline Tit, but reportedly louder and somewhat coarser (Harrap, in Harrap & Quinn 1996). Call reported to be longer and flatter (Bot & van Dijk 2009).

SIMILAR SPECIES Worn spring and early summer ♂ unmistakable. Beware that fresh ♂ (especially 1stW) has crown and throat tipped extensively cinnamon-buff and greyish, obscuring typical dark-hooded pattern, but dark elements visible on throat are still conclusive. Main identification challenge lies in separating ♀/young from those of *Penduline Tit* and *White-crowned Penduline Tit*, but black mask even in these duller plumages of Black-crowned is slightly more extensive on forehead and head-sides (extending well above eye and to lower cheeks), throat and breast often being mottled blackish-grey, and they often have quite pronounced pale forecrown above mask, a pale ochre cinnamon-buff 'shawl' from mid crown to upper mantle, and a warmer buff belly, a combination of characters never shown by White-crowned (although ssp. *caspius* of Penduline is rather well marked on crown, and hybrids between these two are extremely confusing). We recommend labelling only birds showing a convincing set of characters.

AGEING & SEXING Ageing possible in 1stY using moult pattern. Sexes separable, especially in spring and ad, with rather distinct seasonal variation in ♂♂ (due to feather wear). – Moults. Apparently largely as Penduline Tit, but only limited material examined in this respect. – SPRING **Ad ♂** By summer virtually entire head uniform blackish (sometimes with slight greyish or brownish tinge), throat and upper breast same or even blacker, and cream-buff collar narrower and better defined. Chestnut mantle deeper, with pale scapulars bar more obvious. Rufous-chestnut flecking on breast-sides more extensive. Fringes of tail and wings, including tertials, narrower. **Ad ♀** Clearly differs in paler throat and therefore masked appearance. Throat strongly mottled, and bib can be virtually absent. Mantle and wing-coverts usually paler rufous than in ♂, tinged tawny, and underparts have very limited and smaller rufous flecking. **1stS** As respective ad, but apparently most retain worn juv remiges, some wing-coverts and tail-feathers, while some are even more distinctive due to less-advanced plumage. – AUTUMN **Ad** Both sexes aged by evenly very fresh wing- and tail-feathers. They are also better marked with the species' characteristics than young. **Ad ♂** Crown and nape tinged dusky-

R. m. macronyx, ♂, presumed ad, Iran, Apr: characteristic, apart from all-black head (isolated tiny pale smudge on throat insignificant), is hint of slightly paler collar (narrow and indistinct in such worn plumage) separating black hood from chestnut mantle. Such immaculate plumage is usually a feature of ad ♂, but moult pattern cannot be ascertained to attempt ageing. (S. Papps)

brown and tipped buff (producing variable blackish mask), and dusky throat bib smaller and streakier than in spring. Chestnut mantle and wing-coverts tipped pale cinnamon-buff, while chestnut flecking on breast and flanks is mostly concealed. **Ad ♀** Compared to spring buffier, mask less intense and dark throat reduced. **1stW** Like Penduline Tit, those undertaking partial post-juv moult show moult limits among greater coverts, and have mostly retained juv primary-coverts and remiges. Some, however, apparently undertake a complete moult and are largely inseparable from ad. Both sexes resemble respective fresh ad, but have generally less intense blackish crown in ♂ and less intense mask in ♀, with less obvious dark throat bib. Some 1stW ♂♂ approach ad ♀ or appear intermediate between fresh ad ♂ and ♀ (thus, with less clear-cut examples, ageing should precede sexing). 1stW ♀ is dullest plumage, sometimes virtually lacking any dark on throat, with crown hardly darker and dark mask poorly defined, thus very similar to Penduline Tit; also more drab buffish grey-brown above, resembling juv. **Juv** Like juv Penduline Tit, but crown darker and greyer.

BIOMETRICS (*macronyx*) **L** 10.5–11.5 cm; sexes apparently same size. **W** 54–59 mm (n 10, m 57.0); **T** 43–52 mm (n 10, m 48.5); **T/W** m 84.9; **B** 10.9–12.2 mm (n 10, m 11.4); **BD** 4.6–5.1 mm (n 9, m 4.8); **Ts** 14.0–15.4 mm (n 9, m 14.7). **Wing formula: p1** > pc 0.5–4 mm, < p2 26–31 mm; **p2** < wt 2.5–4 mm, =7 or 7/8 (rarely =6/7); **pp3–5**(6) about equal and longest; **p6** < wt 0–2 mm; **p7** < wt 2–5 mm; **p10** < wt 7–10 mm; **s1** < wt 8–11 mm. Emarg. pp3–5(6).

GEOGRAPHICAL VARIATION & RANGE Only two subspecies recognised as being sufficiently distinct, and racial identification is probably only possible in ad ♂♂. Designation often complicated by individual variation and is particularly difficult in areas where hybridisation with *R. pendulinus* occurs.

R. m. macronyx (Severtzov, 1873) (lower R. Ural and N Caspian Sea, SE Caspian region, Aral Sea, Syrdar'ya, Amudar'ya, Lake Balkhash; mainly resident). Described above. Birds in SW Turkmenistan and the Iranian Caspian ('*neglectus*') sometimes separated on account of on average subtly smaller size and more extensive black hood of ad ♂ (sometimes reaching upper mantle and upper breast), thus approaching *nigricans*, but differences slight and inconsistent. – Birds in SW corner of Caspian Sea, in NW Iran and SE Azerbaijan, represent a mixed population from hybridisation between *macronyx* and Eurasian Penduline Tit (mainly ssp. *caspius*, but possibly also *menzbieri*), sometimes recognised as a separate subspecies of Black-headed Penduline Tit, '*altaicus*' (Radde, 1899), but not here. (Syn. *altaicus*; *aralensis*; *neglectus*; *ssaposhnikowi*.)

R. m. nigricans (Zarudny, 1908) (Iran in Seistan and along Helmand River on Afghan border, probably also adjacent Afghanistan; mainly resident or makes short-range movements). Differs in having a more extensive black hood reaching mantle. Both sexes apparently quite similar, but inadequate material examined to define sexual characters. Sometimes has lower mantle to rump blotched chestnut. Much of breast blackish-grey and thus often lacks even hint of pale collar. Underparts typically sullied warm buff-chestnut. **W** ♂ 55–59 mm (n 11, m 56.9); **T** ♂ 46–51 mm (n 11, m 47.8); **T/W** m 84.3; **B** ♂ 11.0–12.6 mm (n 11, m 11.9); **BD** ♂ 4.9–5.6 mm (n 7, m 5.2)); **Ts** ♂ 14.2–15.9 mm (n 11, m 14.8).

TAXONOMIC NOTES Black-headed Penduline Tit and White-crowned Penduline Tit are sometimes lumped with Eurasian Penduline Tit, but Black-headed often breeds sympatrically with White-crowned without interbreeding, and latter two are genetically well separated (Barani-Beiranvand *et al*. 2017). Considering not only rather clear differences in plumage patterns, but also in habitat choice, structure and size, treating all three as separate species appears justified. The mainly local hybridisation does not appear to challenge the overall pattern. – Subspecies taxonomy in Black-headed Penduline Tit is particularly complex due to hybridisation in some areas with Eurasian Penduline Tit, but also due to individual variation and the paucity of museum material of all named forms. Ssp. *neglectus* might be upheld, but we have examined too few specimens to be certain of its validity. The same is true for '*ssaposhnikowi*' (Lakes Balkhash, Alakol and Sasykkol, E Kazakhstan), reportedly of hybrid origin (*R. p. jaxarticus* × *macronyx*?) and at times (but not here) recognised as a separate subspecies. Black-headed Penduline Tit × Eurasian Penduline Tit hybrids, so-called '*altaicus*', from SW Caspian Sea, are variable but often have dark grey head with extensive white central crown to nape, and more obvious pale grey collar.

REFERENCES Barani-Beiranvand, H. *et al.* (2017) *J. Avian Biol.*, 48: 932–940. – Bot, S. & van Dijk, R. E.. (2009) *Sandgrouse*, 31: 171–176. – Bot, S. et al. (2011) *DB*, 33: 177–187.

R. m. macronyx, ♀, presumed ad, Iran, May: readily sexed by reduced black hood (pale rear crown and throat), thus superficially resembling ♂ Penduline Tit of race *caspius*. However, unlike this species, darker with less prominent white edges to wings and tail. Evenly-feathered wing, round-tipped tail-feathers, and no sign of any juv feathers in wing, together with not heavily worn plumage in May are features of ad ♀. (C. N. G. Bocos)

R. m. macronyx, var. '*ssaposhnikowi*', ♂, *presumed* ad, Kazakhstan, Jun: this variation, possibly the result of hybridisation between Eurasian (ssp. *jaxarticus*) and Black-headed Penduline Tits, is found in E Kazakhstan but could possibly straggle to within the treated region. Note lack of any white on nape or hint of paler collar, and also similarity with Eurasian Penduline Tit of ssp. *caspius*. (N. Bowman)

WHITE-CROWNED PENDULINE TIT
Remiz coronatus (Severtzov, 1873)

Fr. – Rémiz à couronné; Ger. – Kronenbeutelmeise
Sp. – Pájaro moscón coronado
Swe. – Vitkronad pungmes

This distinctive *Remiz* tit has traditionally been lumped with Eurasian and Black-headed Penduline Tits but is afforded separate species status here due to its breeding sympatrically with Black-headed Penduline Tit without mixing, its genetic distinctiveness and its distinctive plumage. It is restricted to Central and East Asia, where it frequents riverine forests and irrigated plantations, often in upland areas. White-crowned Penduline Tit is more migratory than its congeners, apparently habitually moving south and south-west in winter, with one claim from Austria at this season.

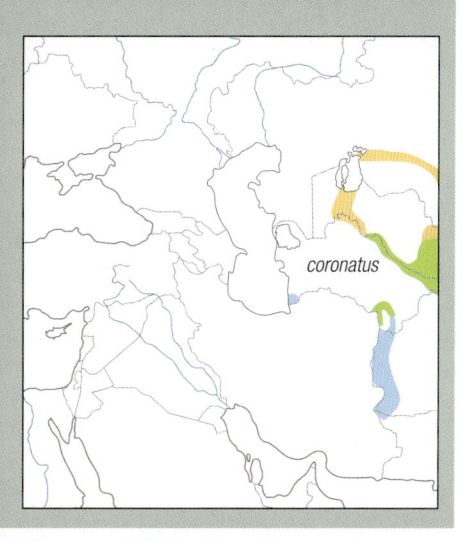

R. c. coronatus, ♂, presumed ad, Kazakhstan, May: dark nuchal band and deep chestnut mantle separated by white or pale grey collar. Lower mantle and scapulars pinkish-buff, gradually merging into pale grey-buff on back. Wings and tail have contrastingly white edges, and there is a deep chestnut-brown greater coverts patch. Evenly-feathered wing, including ad-like primary-coverts, suggest ad. (V. & S. Ashby)

IDENTIFICATION Much as Penduline Tit but is slightly *smaller* and proportionately *shorter-tailed*, with a similar *fine-pointed conical bill*. ♂ has very broad (albeit variable) *blackish mask broadly extending onto forehead and across rear crown and nape* (diagnostic), but separated from chestnut mantle by a *broad white collar*, leaving a *restricted white forecrown and ash-grey central crown*, sometimes forming oval-shaped patch. Underparts dull buffish-white tinged pink with, in worn plumage, chestnut on sides. *White fringes to tertials and secondaries broad and conspicuous*, and chestnut greater coverts form a dark band on wing, emphasized by pale buff scapulars. ♀/young usually distinctly duller, with less intense (even browner) mask that only narrowly or partially extends across nape, and pale grey crown and collar, but some adult ♀♀ are very similar to ♂♂, at times even being inseparable in the field. General behaviour as Penduline Tit.

VOCALISATIONS Presumed to be very similar to those of Penduline Tit. Common *pseeeu* call said to be 'perhaps slightly shorter and fuller, not trailing off so much at the end' (Harrap in Harrap & Quinn 1996). Also described is a fine twittering flight-call *ti-ti-ti-ti-ti*.

SIMILAR SPECIES Worn ♂♂ (spring–early summer) unmistakable due to extensive blackish mask extending to form substantial nuchal band. Fresh ♂ (especially 1stW) has black feathers of nuchal band tipped greyish-white, obscuring pattern, but usually has at least impression of 'bridle' on rear crown. Unlike Penduline Tit, also often has whiter upper border to mask, forming a supercilium and merging with ash-grey rest of crown. A white or pale grey collar is also distinctive. Main confusion risks lie in separating fresh ad ♀ and, especially, 1stY ♀, from very similar Penduline Tit (and even Black-headed Penduline Tit), but in at least some ad ♀♀, black on rear mask appears slightly more extensive with faint blackish-grey mottling on nape, and furthermore the whiter upper forehead and paler collar, if present, are conclusive. Most 1stW ♀♀ lack these diagnostic features and are inseparable from the other two species in the field. The slightly smaller size, shorter and thinner bill, and weaker legs are not useful for field identification.

AGEING & SEXING Ageing generally possible by close check of moult pattern. Sexes similar but usually separable, at least in spring. – Moults. Apparently largely as Penduline Tit, but few data available. – **SPRING** Sexing at times difficult; best character usually amount of chestnut blotching on chest (very few intermediates) and amount of deep chestnut on mantle (more overlap). **Ad ♂** With wear, forecrown and collar become virtually uniform white, and black nuchal band much broader and more solid. Also a more pronounced white throat develops, and prominent chestnut flecking on chest and sides becomes more exposed. Chestnut mantle deeper with pale scapular bar more obvious. Pale fringes to tail and wings, including tertials, wear narrower. **Ad ♀** Narrower mask and broken black nuchal band ill-defined and mottled. More greyish-buff crown and neck, with hardly contrasting collar, while mantle and wing-coverts are usually paler, and very limited and smaller rufous-chestnut flecks visible on chest and flanks. However, a few (perhaps older) ♀♀ approach ♂♂ (especially 1stS ♂♂). **1stS** Most retain distinctly worn juv primaries, primary-coverts and tail-feathers, and sometimes a few outer greater coverts. Some are even more distinctive, being considerably less advanced, with some 1stS ♂♂ intermediate between ad ♂ and ♀, while least advanced 1stS ♀♀ are very dull; 1stS with many retained greater coverts have stronger whitish wing-bar, and upperparts often appear variegated. – **AUTUMN** Sexing frequently difficult, sometimes impossible in the field. **Ad ♂** Compared to spring, nuchal band narrower and often mottled, and crown greyer, contrasting more with whiter forecrown. Mantle and wing-coverts paler, while chestnut flecking on underparts is mostly concealed. **Ad ♀** Mask less intense, brownish and hardly meets on nape. Both sexes have evenly fresh wing- and tail-feathers. **1stW** Moult pattern and contrast produced by post-juv moult mostly as Penduline Tit (and like latter, apparently many 1stW replace all greater coverts and can be harder to age, while it is unknown whether some undertake complete moult, which would make them inseparable from ad). Retained juv primary-coverts and remiges weaker textured, centred slightly browner and fringed buffish (less rufous), but these wear rapidly. 1stW ♂ generally intermediate between fresh ad ♂ and ♀, with less intense facial mask only extending slightly onto forehead, and nuchal band partial or spotty (often hardly reaches back). Also, has more obscure collar and duller chestnut upperparts. Many 1stW ♀♀ are very similar to Penduline Tit in same plumage: mask obscure and does not reach whitish forehead, while rest of head and nape is pale buff-brown. **Juv** Like juv Penduline Tit, but greyer above with characteristic white forehead and pale grey-brown crown.

R. c. coronatus, ♂, Kazakhstan, Jun: nuchal band is sometimes almost absent or very obscure, the white crown patch merging with whitish collar. Some ♂♂ are diffusely blotched pinkish-chestnut below. Age cannot be determined at this angle. (M. Vaslin)

BIOMETRICS (*coronatus*) **L** 9.5–10 cm; **W** ♂ 52.5–55 mm (*n* 14, *m* 53.5), ♀ 51–54 mm (*n* 13, *m* 52.3); **T** ♂ 40–44.5 mm (*n* 14, *m* 42.3), ♀ 39–44 mm (*n* 13, *m* 40.7); **T/W** *m* 78.5; **B** 9.2–10.8 mm (*n* 25, *m* 9.9); **BD** 3.7–4.4 mm (*n* 23, *m* 4.0); **Ts** 12.0–13.7 mm (*n* 23, *m* 12.9). **Wing formula: p1** > pc 0–3.5 mm, < p2 26–31 mm; **p2** < wt 1.5–4 mm, =6/7 (85%) or =7 (15%); **pp3–5** about equal and longest; **p6** < wt 0.5–1.5 mm; **p7** < wt 3–4.5 mm; **p10** < wt 8–10 mm; **s1** < wt 9–12 mm. Emarg. pp3–5 (6).

GEOGRAPHICAL VARIATION & RANGE Rather slight variation, with two races generally recognised, but within the covered region only *coronatus* occurs; ssp. *stoliczkae* (Yenisei River eastward) is commonly regarded as a race of the present species, but see Penduline Tit.

 R. c. coronatus (Severtzov, 1873) (Kazakhstan east of Aral Sea, Tajikistan, E Turkmenistan, N Afghanistan, Kyrgyzstan; winter movements may reach E Iran). Rather dark, and the nuchal band is practically invariably black and well marked in ♂♂.

 Extralimital: **R. c. stoliczkae** (Hume, 1874) (NE Zaisan Basin, where intergrades with *coronatus*, and NE Xinjiang, W, C, & N Mongolia, Tuva, S Krasnoyarsk region, S Yakutia, and Transbaikalia). Has more diluted colours than *coronatus* (particularly noticeable in ♂), and dark nuchal band is often missing or only partly developed at rear of crown. Rufous patch on upper mantle more buffish, and thus the deep chestnut panel on greater coverts is more strongly contrasting. Fringes of flight-feathers broader and purer white, those on tertials and secondaries forming a large obvious white wing panel when fresh. In worn plumage, the centre and rear crown may become heavily mottled black (though less so than in *coronatus* at a comparable stage of wear), the hindneck is still pure white (less greyish than in *coronatus*); rump fully white. In all plumages, bill shorter and more slender than in *coronatus*.

TAXONOMIC NOTES For some general comments on Remizidae taxonomy, see Eurasian Penduline Tit. – The breeding range of White-crowned Penduline Tit appears to overlap only to a limited degree or not at all (?) with Eurasian Penduline Tit, and these two seem ecologically separated: White-crowned breeds in riverine forests, or other forests and thickets, both in lowlands and hills, Eurasian Penduline chiefly in deciduous trees in direct contact with reed-beds, thus White-crowned is often found away from water, Penduline is more aquatic. Furthermore, White-crowned occurs over large areas in Eurasian sympatry with Black-headed Penduline Tit without any evidence of hybridisation. – As recently demonstrated (Barani-Beiranvand et al. 2017), White-crowned Penduline Tit is clearly separated genetically from Chinese Penduline Tit *R. consobrinus* (Manchuria, NE China, Amur; extralimital, not treated).

REFERENCES Barani-Beiranvand, H. *et al.* (2017) *J. Avian Biol.*, 48: 932–940. – Bot, S. *et al.* (2011) *DB*, 33: 177–187.

R. c. coronatus, ♂, presumed 1stS, Kazakhstan, Jun: most characteristic is ♂'s black mask that encircles whitish crown at rear through a nuchal band. The primary-coverts appear to be juv, but plumage rather crisply patterned like ad. (A. Audevard)

R. c. stoliczkae, ♀, presumed ad, Mongolia, Aug: dark mask dull and paler at loral area, and does not continue behind ear-coverts, typical of this race. Mantle has only hint of rufous, but no darker chestnut. Crown rather patchy grey-white with some buff element. Wings and tail freshly moulted and appear to be of ad type. (O. Baatargal)

R. c. stoliczkae, ♂, Mongolia, Aug: resembles *coronatus*, but black mask does not continue much behind ear-coverts. Differs from ♀ by the broader and more solid black mask and some chestnut flecks shining through on breast. Ageing not possible at this angle. (O. Baatargal)

R. c. stoliczkae, ♀, Mongolia, Jun: fluffy feathering on head and breast together with undeveloped general plumage pattern reveals young age. (H. Shirihai)

NILE VALLEY SUNBIRD
Hedydipna metallica (M. H. C. Lichtenstein, 1823)

Fr. – Souimanga du Nil; Ger. – Erznektarvogel
Sp. – Suimanga del Nilo; Swe. – Nilsolfågel

An African sunbird resident in *Acacia* scrub and semi-arid areas (as well as in gardens and cultivation near water) in NE Africa and the Middle East, from the Horn of Africa north to N Egypt, and SW Saudi Arabia east to SW Oman. There is some dispersal in winter. Egypt, where the species is locally rather frequent, is arguably the easiest country in the covered region to encounter it.

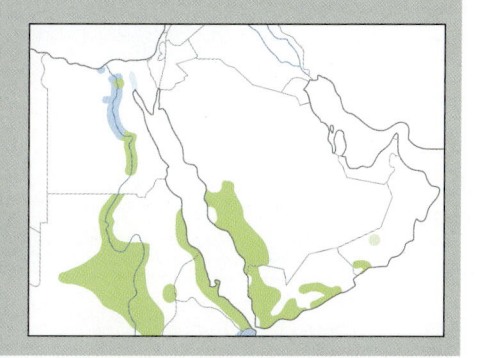

Ad-like ♂, Egypt, Apr: unmistakable sunbird of NE Africa and Arabia, at all ages showing a comparatively short, only slightly decurved bill, which separates it from all other sunbirds in region. Note the iridescent metallic purplish-blue crescent-shaped band on lower breast, but this is only visible when it catches light. Tail has markedly elongated central rectrices. Both age classes habitually moult completely in winter, making definitive ageing impossible. (D. Occhiato)

IDENTIFICATION Usually straightforward to identify in the covered region: *comparatively short, only slightly decurved bill* separates it from all other sunbirds, while highly distinctive *breeding-plumage ♂ has extremely long tail*, which may exceed body length. Breeding ♂ (c. mid Mar–end Jul) is *metallic glossy emerald-green above* and on throat/upper breast, but *rump and uppertail-coverts and a narrow band across upper breast is glossy purplish-blue*. Also, a small black bib on chin is present in most (but not all). Rest of *underparts intense yellow*, paler on vent and undertail-coverts. Elongated streamers narrow and blackish. ♀ lacks tail-streamers, and has dull brownish-grey upperparts with a slight greenish tinge, pale greyish-cream underparts, usually with variable yellow on breast and belly (sometimes quite saturated), and a subdued off-white or cream-coloured supercilium, paler throat, narrow brown loral stripe, and faintly pale fringes to wings and tail. Non-breeding ♂ acquires eclipse plumage in late summer (partially ♀-like, lacking streamers), but wings and tail are darker than in ♀, and greater coverts glossed metallic green. Chin and throat dull and pale yellow, with dark sooty-black central band or blackish blotches and some glossy green admixed. Upperparts mainly olive-drab, often with odd glossy green parts and some sooty-black patches or single feathers.

VOCALISATIONS Song a high-pitched series of trills and hisses, rendered *pruitt-prruitt-ptuitt-tiriririri-tiriri*. – Contact call is a piping *pee* or a series of repeated rising notes, e.g. *tschi*, *peee*, *pee-ee* and *pee-e-ee*, *cheeit cheeit*, *tee-weee*, or various thin squeaks and twittering, subsong-like calls.

SIMILAR SPECIES Breeding-plumage ♂ unmistakable. ♀/juvenile may require careful separation from *Palestine* or *Shining Sunbirds*, but all non-breeding ♂ plumages have yellow-washed underparts, which should separate Nile Valley Sunbird from either. Furthermore, in all plumages, comparatively short bill prevents confusion with Palestine Sunbird. – Extralimital Pygmy Sunbird (*H. platura*; not treated), which in size, structure, plumage and behaviour is exceedingly similar, apparently reaches close to the covered region (e.g. a small population in Tibesti, N Chad), but does not overlap in the W Palearctic proper. Furthermore, breeding adult ♂ Nile Valley Sunbird lacks Pygmy's bronzy-gold iridescence to the green upperparts, having instead a metallic bluish-purple band on the lower breast (absent in Pygmy), is more extensively violet dorsally (limited to uppertail-coverts in Pygmy), and has purer and more intense yellow underparts (often tinged orange in Pygmy). Separation of the two in other plumages is virtually impossible (though supercilium and eye-stripe somewhat more distinct in present species).

AGEING & SEXING Ageing requires understanding of seasonal variation according to age/sex and, often, close scrutiny of moult pattern in wing. Sexes strongly differentiated in breeding plumage due to marked seasonal variation in ♂. – Moults. Extent and timing poorly known; NHM mostly lacks specimens from moult periods and northerly latitudes, including Arabia and especially Egypt. Apparently the majority undertake a partial post-nuptial and post-juv moult, presumably mainly in May–Sep. Pre-nuptial moult in Dec–Mar seems largely complete for both age classes. Partial post-nuptial moult (when non-breeding ♂ plumage is acquired) usually involves renewal of head, body, a variable number of tertials, wing-coverts and r1. Post-juv moult more complex (see below). Breeding ♂ plumage attained during pre-nuptial moult, but many ♂♂ in such plumage have worn remiges that appear to be renewed well before acquiring breeding plumage, perhaps during or just after eclipse; ♀♀ prior to breeding appear evenly fresh throughout. Further studies of moult and seasonal variation required. – **BREEDING Ad ♂** (usually mid Mar–end Jul) With wear, bleaches dull green and more metallic blue above, and throat, wings and tail almost lack gloss, while tail-streamers become heavily

Ad-like ♂, Egypt, Apr: breeding-plumage ♂ is dark iridescent green on head, upperparts, smaller wing-coverts and upper breast, contrasting with dark purplish-blue rump and uppertail-coverts, paler blue gloss on dark uppertail, and bright lemon-yellow underparts below breast. However, the bluish-purple iridescence on lower breast appears black at many angles. (D. Occhiato)

Ad-like ♀, Egypt, Apr: unlike ♂, pale brown-grey above and on head-sides, with indistinct yellowish supercilium and sulphur- or lemon-yellow underparts. The yellow below should automatically separate ♀ and all non-breeding ♂ plumages of Palestine or Shining Sunbirds. (D. Monticelli)

Eclipse ad ♂, Egypt, Sep: ♂♂ show marked seasonal variation, as non-breeders are non-glossy, lack elongated tail-feathers and resemble ♀♀, although they often show a black patch on central throat and tend to have blacker tail- and flight-feathers. Eclipse plumage is attained during partial post-nuptial moult, when strong moult limits can be detected (as shown here). (V. & S. Ashby)

abraded; yellow underparts become paler. **Ad ♀** Wears to uniform medium brown-grey above (yellow-green tinge virtually lost), and face marks obscured; underparts lemon-yellow (can be rather faint), throat whiter, and tail and wings duller grey-brown. **1stS** Vast majority appear to undertake complete pre-breeding moult and cannot be aged thereafter. Some, however, apparently in first breeding plumage, retain juv remiges and primary-coverts, and perhaps a few outer greater coverts, which are brownish, worn and pale-fringed (differences most pronounced in ♂♂), while others are wholly distinctive, having only partial breeding plumage. – **NON-BREEDING Ad ♂** (eclipse; usually Aug–mid Mar, but sometimes acquired earlier) Partially ♀-like and lacks streamers, but has dull matt black chin and centre of throat forming a central band on otherwise yellowish underparts (sooty-black band sometimes with some bright green admixed). Also, has darker flight- and tail-feathers (still with a faint blue-green sheen), is less brown above and on wings than ♀, and in comparison has somewhat brighter yellow underparts. Upperparts mainly olive-drab, often with odd glossy green and some sooty-black patches or single feathers. Both when moulting in and out of eclipse, much

♂, probably ad, Oman, Dec: this ♂ is apparently in advanced moult, attaining eclipse (rather than moulting out of it). (M. Rouco)

Eclipse ad ♂ (early stage), Egypt, Jan: unlike sympatric Palestine and Shining Sunbirds, note yellow-washed underparts with clear-cut mesial-stripe and comparatively short bill. Also, combination of yellow below and rather broad, clear-cut mesial-stripe, and dark shoulder (against newly moulted and pale-fringed greater coverts) are features of ad ♂ in eclipse plumage. (L. Boon/Cursorius)

Eclipse ad ♂, Ethiopia, Sep: it is not always easy to detect if birds are acquiring or losing eclipse plumage, especially in regions where breeding is not associated with rains. (H. Shirihai)

plumage variation among ♂♂. **Ad ♀** Largely as breeding. Upperparts plain drab-brown, underparts very pale yellow lacking dark central band. **1stW** Extent of post-juv moult variable, perhaps linked to fledging dates, hence occurrence both of ♀-like ♂♂ lacking dark mesial streak on throat (or having only a hint of this) and any glossy greenish-blue tinge on upperparts, as well as much more advanced 1stW ♂♂ with partly attained breeding plumage in midwinter. Both sexes relatively fresh, i.e. lacking strong moult contrast and heavily worn wing-feathers of ad, but note after acquiring 1stW plumage they commence wing moult (producing moult limits), though primary-coverts and primaries not as contrastingly worn and heavily bleached as in ad). **Juv** Similar to ♀ but has soft fluffy appearance, paler throat and slightly greenish upperparts, with darker grey tail, and rest of underparts washed yellow.

BIOMETRICS L 14–18 cm (breeding ♂ including tail-streamers) or 9.5–10.5 cm (all other plumages); **W** ♂ 54–60 mm (*n* 16, *m* 57.5), ♀ 50.5–57 mm (*n* 13, *m* 53.7); **T** (excluding tail streamers) ♂ 37–43 mm (*n* 16, *m* 39.3), ♀ 33–39 mm (*n* 13, *m* 36.1); **T/W** (excluding tail-streamers) *m* 67.8; **tail-streamers** (in breeding ♂) 43–79 mm (*n* 12, *m* 58.9); **B** 12.2–14.5 mm (*n* 29, *m* 13.3); **BD** 2.8–3.6 mm (*n* 29, *m* 13.3); **Ts** 13.7–15.8 mm (*n* 22, *m* 14.9). **Wing formula: p1** > pc 1–4 mm, < p2 27–33 mm; **p2** < wt 0.5–3 mm; **pp3–5** about equal and longest; **p6** < wt 0.5–2 mm; **p7** < wt 2–5 mm; **p10** < wt 8–11 mm; **s1** < wt 9–12 mm. Emarg. pp3–6.

GEOGRAPHICAL VARIATION & RANGE Monotypic. – Nile Valley region in Egypt and Sudan, SW Arabia, E Chad, Eritrea, Somalia; resident. (Syn. *Anthodiaeta metallica*; *Anthreptes platura metallica*.)

TAXONOMIC NOTES Formerly often considered a race of Pygmy Sunbird *H. platura*, but no evidence of interbreeding where their ranges meet or overlap. – These two species were transferred to *Hedydipna* by Fry & Keith (2000), previously often referred to *Anthreptes*. Cheke & Mann (2006) claimed that *Hedydipna* is a junior synonym of *Anthodiaeta*, but this proposal was subsequently retracted (Mann & Cheke 2014).

REFERENCES Cheke, R. A. & Mann, C. F. (2006) *BBOC*, 126: 199–200. – Mann, C. F. & Cheke, R. A. (2014) *BBOC*, 134: 159–160.

1stW ♂, Oman, Nov: resembles ♀, but note some iridescent green lesser coverts just emerging, and plenty of green spots visible on head. Much of the wing apart from these green lesser coverts still hold retained juv feathers. The mesial-stripe is thin and poorly developed (hardly discernible on this bird). (H. & J. Eriksen)

1stW ♂, Saudi Arabia, May: a young ♂ in post juv-moult. (D. Alhashimi)

1stW ♀, Sudan, Nov: note lack of mesial stripe of young eclipse ♂. The yellow-washed underparts, clear pale supercilium and short bill prevent confusion with ♀ and non-breeding ♂ of sympatric Palestine or Shining Sunbirds. Nevertheless, in NE Africa, separation from extralimital Pygmy Sunbird is very difficult, but the more distinct pale supercilium, paler yellow rest of underparts, paler throat and greyer upperparts favour Nile Valley Sunbird. (T. Jenner)

Juv, Egypt, Apr: already in Apr in post juv-moult, but still differs mainly by less saturated colours and fluffier body plumage, especially on belly. (V. Legrand)

PURPLE SUNBIRD
Cinnyris asiaticus (Latham, 1790)

Fr. – Souimanga asiatique; Ger. – Purpurnektarvogel
Sp. – Suimanga asiático; Swe. – Purpursolfågel

Confined within the covered region to the extreme SE Arabian Peninsula, this small, elegant sunbird shares many plumage features with Palestine Sunbird but, fortunately, they are not known to overlap. It inhabits open woodland and thorn scrub, mainly in the lowlands, often in gardens, from N Oman, the United Arab Emirates and SE Iran east through S Asia. Some move south in winter, even within Arabia.

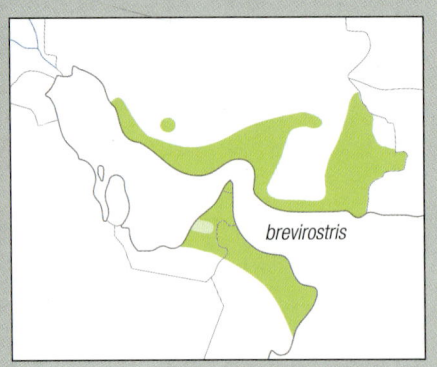

C. a. brevirostris, ♂, NW India, Aug: small, short-tailed and very dark sunbird of S Asia, much like Palestine Sunbird but of far more easterly distribution, although they almost meet in C Oman. Note relatively shorter and less decurved bill; also unlike Palestine, tufts on breast-sides (usually visible only in display) are predominately yellow with scarlet edges instead of mostly flame-orange. (S. Tiwari)

IDENTIFICATION A small, short-tailed sunbird with a *moderately long decurved bill* but *lacking elongated tail-streamers*. Identification usually straightforward: *breeding-plumage ♂ largely uniform deep greenish-blue* (appears all black in poor light) with characteristic *iridescent purple-blue sheen on chest*, although this is strongly reduced in some populations. Also, a *narrow deep red-brown breast-band* and *yellow-orange pectoral tufts* (located near wing-bend, in 'arm-pit'), though both usually hardly visible. Dark metallic blue or purple belly and vent often appear concolorous with rest of underparts. ♀ has largely *mouse-coloured or olive-brown upperparts*, and *underparts washed dull yellow*, often with a *narrow pale supercilium* and darker olive lores and ear-coverts, forming *poorly-defined mask*. Non-breeding ♂ acquires eclipse plumage, *underparts becoming yellow with a broad blackish stripe down centre*. Rarely, ♀ can show hint of dark mesial band on throat, though shorter and duller than in eclipse ♂. Like congeners, typically jerky and active (including hovering) when feeding. Rather noisy. Usually occurs singly, in pairs or small parties, being attracted to flowering trees, but more insectivorous than some relatives, and may perform short aerial sallies.

VOCALISATIONS Song is a busy *cheewit-cheewit-…* repeated 2–6 times, or *swee-swee-swee-swit zizi-ziz*, to some ears with Willow Warbler-like quality. Also, ♂ has sub-song (?) or alternate song consisting of a repeated strident note, *dzit dzit dzit* (recalling Zitting Cisticola), sometimes with added twittering elements. – In contact utters buzzing *zit* or *chip*, and a high-pitched, upward-inflected and somewhat Chiffchaff-like *sweep* or *tsweet*. Also at times utters a harsh monotonous whistle.

SIMILAR SPECIES No known overlap with only potential confusion species, *Palestine Sunbird*, of which ♀ has pale greyish (less yellowish-washed) underparts and a longer and more decurved bill. ♂ Palestine never shows narrow red-brown breast-band. Longer bill and extensive yellow underparts of ♀ separate it from Nile Valley Sunbird, which is also geographically separated from the present species.

AGEING & SEXING Ageing requires understanding of seasonal variation according to age and sex, and often close

C. a. brevirostris, ad-like ♂, United Arab Emirates, Feb: compared to Palestine Sunbird, despite superficially similar and almost uniform dark metallic-blue plumage of breeding ♂, when seen in direct sunlight, the gloss on head and upperparts is bluer, less green. However, the diagnostic narrow maroon band on lower breast is often obscured if the breast is not directly lit. (M. Barth)

C. a. brevirostris, ad-like ♂, United Arab Emirates, Jan: in the right light, purple-blue iridescence of throat and upper breast and narrow maroon breast-band may become visible. (T. Pedersen)

scrutiny of moult pattern in wing. Sexes strongly differentiated in breeding plumage, less so in non-breeding period. – Moults. Extent and timing of moult poorly known, and in Indian Subcontinent considered variable (few specimens available from Arabia and Iran). Post-nuptial and post-juv moults mostly partial, in Jul–Dec. Pre-nuptial moult appears mostly complete, in Nov–Mar, when ♂ breeding plumage attained. Partial post-nuptial moult (when non-breeding ♂ plumage acquired), apparently mostly involves renewal of head- and body-feathers, and it seems that remiges (both sexes) are mostly moulted during ♂'s eclipse period. – **BREEDING Ad ♂** (mostly Dec–Jul, e.g. in Oman 9 ♂♂ in full breeding plumage collected Feb–May, but in some areas breeding plumage apparently seen year-round) See Identification for a description. **Ad ♀** With wear, upperparts greyer and only a trace of yellow on underparts. **1stS** Apparently the vast majority moult completely, thus after moult inseparable from ad. However, some ♂♂ (apparently 1stS that did not moult completely prior to breeding) have remiges, primary-coverts and some tertials and outer greater coverts more worn, bleached browner and lacking bluish sheen (same pattern also in a few ♀♀, but moult limits much less contrasting). Some ♂♂ with mixture of breeding and eclipse plumage, when most are in breeding plumage, perhaps also 1stS. – **NON-BREEDING Ad ♂** (eclipse; at least Sep–Dec) Has partial ♀-like plumage, although compared to ♀ with brighter yellow underparts, conspicuous blue-black mesial line from throat to upper belly, darker head-sides, metallic blue-green wing-coverts and brownish-black remiges contrasting with greyish-green mantle, and metallic blue tail. Especially when in active moult, different stages of eclipse occur, with patchy iridescence on head, back and breast. **Ad ♀** Largely as breeding. **1stW** Both sexes aged by being relatively fresh (lacking strong moult contrast and heavily-worn wing-feathers of ad). Shortly after acquiring 1stW body plumage, they (especially ♂♂) moult their remiges, though resulting moult contrast differs from ad. Roberts (1992) states that young ♂♂ moult twice a year, i.e. once into ♀-like 1stW plumage and again into breeding plumage. Other literature also describes 1stW ♂♂ (after post-juv moult) as lacking black mark below, but we have not seen such ♀-like 1stW ♂♂. Instead, 1stW ♂♂ appear much as eclipse ad ♂♂ below, only with a thinner and shorter black mesial line, and upperparts as ♀ (also lack metallic blue-green wing-coverts and brownish-black remiges of eclipse ad ♂). **Juv** Similar to ♀ but has soft fluffy appearance to entire body plumage and paler yellow underparts, especially on flanks.

BIOMETRICS (*brevirostris*) **L** 10–11 cm; **W** ♂ 53.5–59 mm (*n* 12, *m* 56.4), ♀ 50–56 mm (*n* 12, *m* 52.8); **T** ♂ 34–38 mm (*n* 12, *m* 35.8), ♀ 30–34.5 mm (*n* 12, *m* 31.8); **T/W** *m* 61.9; **B** 15.1–19.4 mm (*n* 24, *m* 17.8); **BD** 3.0–3.4 mm (*n* 20, *m* 3.2); **Ts** 13.3–15.5 mm (*n* 19, *m* 14.3). **Wing formula: p1** > pc 0–3.5 mm, or to 2 mm <, < p2 27–33.5 mm; **p2** < wt 0.5–3.5 mm; **pp3–5** about equal and longest; **p6** < wt 1–3.5 mm; **p7** < wt 3.5–8 mm; **p10** < wt 7–13 mm; **s1** < wt 7.5–13 mm. Emarg. pp3–6.

GEOGRAPHICAL VARIATION & RANGE Moderate variation with 2–3 subspecies described but only one occurs within the covered range.

C. a. brevirostris (Blanford, 1873) (NE United Arab Emirates, N & C Oman, S Iran, E Afghanistan, Pakistan, W India; resident). Described above. There is a slight tendency for Indian birds to have stronger bill and a slightly more purple tinge on crown and mantle, but differences appear slight and clinal, and hardly ground for separation. Biometrics are the same.

C. a. brevirostris, ♀, Iran, Jan: yellowish wash on underparts clinches identification versus ♀ Palestine Sunbird; also often a vague pale supercilium (averages better developed than Palestine), and spread tail (or if viewed from below) shows wider and purer white corners. Also strongly resembles ♀ Nile Valley Sunbird, but bill of latter even shorter and they are geographically separated by a relatively tiny gap in Oman. (A. Ouwerkerk)

C. a. brevirostris, ♀, United Arab Emirates, Feb: with wear, pale greenish-grey upperparts and head-sides become purer grey. (M. Barth)

C. a. brevirostris, ♀, presumed 1stY, Oman, Jan: some ♀♀ show bicoloured underparts, with yellowish throat to chest and whitish belly to vent. Lack of dark mesial stripe eliminates ♂. (H. & J. Eriksen)

C. a. brevirostris, eclipse ad ♂, India, Jan: eclipse (non-breeding) plumage of ♂ develops gradually, producing variety of transitional patterns. Since eclipse normally occurs in Sep–Dec, such birds with new green and black feathers are likely to be ♂♂ moulting to breeding plumage. (S. K. Sen)

C. a. brevirostris, eclipse ad ♂, **United Arab Emirates, Jan**: here in full eclipse, and unlike ♀ or imm eclipse ♂ has notably brighter yellow underparts, conspicuous broad blue-black mesial line from throat to upper belly, darker head-sides, metallic blue-green wing-coverts and brownish-black remiges contrasting with greenish mantle. (M. Barth)

C. a. brevirostris, 1stY ♂, **Iran, Jan**: here in moult from imm eclipse (1stY) to breeding plumage, though still shows blackish stripe down centre of throat and breast. Chiefly when in active moult, different stages of eclipse occur, with patchy iridescence on head, back and breast. (A. Ouwerkerk)

C. a. intermedius, 1stW ♂, **NE India, Jan**: still in full imm eclipse with narrower and less solid blackish (admixed with blue) mesial stripe on centre of throat and chest. Unlike eclipse ad ♂, mesial stripe is narrower, and yellow below is more diluted, while upperparts are more greyish-olive and wings browner with metallic blue limited to small wing-coverts. In resting sunbirds, tongue often exposed outside bill, and face discoloured by pollen. (A. Samarpan)

C. a. brevirostris, 1stY ♂, **Kuwait, Jan**: during complete pre-nuptial moult, in Nov–Mar, ♂ (ad and 1stY) attains breeding plumage, and moulting remiges produce distinctive moult limits and gaps (as shown here). Brighter yellow breast, black mesial line concentrated on throat and upper breast, and rather few blue-green lesser coverts are features of 1stY ♂. (P. Fågel)

SHINING SUNBIRD
Cinnyris habessinicus (Hemprich & Ehrenberg, 1828)

Fr. – Souimanga luisant; Ger. – Glanznektarvogel
Sp. – Suimanga brillante; Swe. – Abessinsk solfågel

A relatively large sunbird that is reasonably common in better-vegetated wadis, usually at mid to higher altitudes, and is restricted to NE Africa and S Arabia, occurring in rocky thornbush in extreme SE Egypt, the Horn of Africa and south to C Kenya. Also found in coastal S Arabia east to S Oman, the latter country often that in which the species is first encountered by Western birdwatchers.

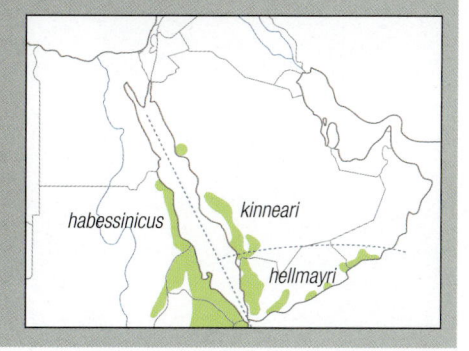

bluer tinge to rump and upper- and undertail-coverts. **Ad** ♀ With wear, Arabian birds become duller brown and Egyptian birds paler and greyer. **1stS** Apparently inseparable from ad if moult complete. – **NON-BREEDING** (autumn–winter) **Ad** Evenly fresh. **1stW** ♂ Readily differentiated by darker central throat, often has blacker breast patch and pale submoustachial stripe, and when commencing remiges moult to ad ♂ plumage may already show some green on head, back and breast, and red on breast-band. **1stW** ♀ Might be aged by presence of some juv wing-feathers. **Juv** Superficially resembles ♀, but has soft, fluffy appearance to entire body plumage, and chin to upper breast is darker (especially in ♂), also submoustachial stripe whitish-yellow, and in pale races (*habessinicus* and *hellmayri*) rest of underparts yellowish with ill-defined dark streaking (more pronounced in ♂); juv *kinneari* distinctive, as both sexes are mainly dark chocolate-brown.

BIOMETRICS (*habessinicus*) **L** 11.5–13 cm; **W** ♂

C. h. habessinicus, ad-like ♂, Egypt, Apr: the most northerly African race *habessinicus* just reaches the covered region in extreme SE Egypt. At this angle the violet-blue iridescence of the crown is just visible, and the blue feather fringes almost form a narrow lower border to the broad scarlet-red breast patch. (I. Moldován)

C. h. habessinicus, ad-like ♂, SC Ethiopia, Sep: mid-sized sunbird with a long strongly decurved bill, unmistakable due to combination of glossy green head, upperparts and chin to upper breast, becoming bluer just above the bright red breast patch, velvety black on belly, and uppertail-coverts being metallic blue. Blue feather fringes at base of scarlet-red breast patch very poorly developed, being limited or lacking compared to Arabian birds. (H. Shirihai)

IDENTIFICATION Readily separated from other sunbirds in the treated region by *larger size* and *comparatively longer, more decurved bill*. ♂ is bright metallic blackish-green, with broad carmine-red breast-band (as a rule rather inconspicuous, but width varies geographically, and prominence depends on light and angle of view) and darker throat to undertail-coverts and wings. In perfect light and close views, violet or bluish elements also visible. Yellow pectoral tufts visible in display. ♀♀ of Arabian subspecies difficult to confuse with other ♀ sunbirds, being *largely dark sooty-brown* with slightly paler underparts and *diagnostic whitish chevrons on vent and undertail-coverts*, clearly paler supercilium and dark eye-stripe, some fine pale tips on throat and small whitish tail-corners. African (Egyptian) ♀♀ much paler throughout, with *markedly pale vent and undertail-coverts* (pale fringes broader, dark centres narrower forming broad streaks). Immature ♂ very variable. Active, jerky movements (with gentle tail-flicking), and sometimes hovers. Often rather noisy and usually found alone or in pairs, occasionally in groups at flowering trees.

VOCALISATIONS Song notably fluty and fast, comprising repeated harsh *tuu-tuu-tuu-tuu*, *vita-vita-vita-vita*, *du-du-du-du* notes, closing with a Wren-like trill, and sometimes interspersed by jumbled whispering sounds and whirring notes (also used in subsong). – Varied calls, including a hard *dzit*, a *chewit-chewit* and rapid *che-che-che-che*.

SIMILAR SPECIES Not easily confused. It overlaps in SW Arabia with Palestine Sunbird, but this is distinctly smaller and lacks reddish breast-band of ♂ Shining, and ♀ is overall much paler on underparts with a shorter bill. Whitish fringes to vent of ♀ Shining provide another distinguishing feature. The manner in which it slowly flicks its quite long, broad tail, and its longer, more deeply undulating flight offer further identification features from Palestine Sunbird.

AGEING & SEXING Ageing requires knowledge of seasonal variation according to both age and sex, and often close inspection of moult pattern in wing. Sexes strongly differentiated, especially in breeding plumage. – Moults. Complete post-nuptial moult and partial post-juv moult in autumn (e.g. ad Arabian specimens in moult Aug–Oct). Limited data available on extent and timing of autumn moult: apparently, juv moults to intermediate 1stW plumage, but ad lacks true eclipse plumage, as it moults from one breeding plumage directly to next one. Ad pre-nuptial moult unknown, but 1stY pre-nuptial moult seems mostly complete, acquiring ad plumage, e.g. in Arabia worn young ♂♂ in moult Jul–Jan (mostly Sep–Dec); one from Egypt in Mar. – **BREEDING** (spring–summer) **Ad** ♂ Depth and extent of breast-band varies. Green iridescence to upperparts may have slight golden hue, while cap is sometimes tinged violet, wings violet, and narrow blue-green bands above and below carmine breast-band, as well as perhaps stronger coppery on mantle and back, or

C. h. habessinicus, ♀, SC Ethiopia, Sep: ♀ characteristic in being uniform grey-brown, lacking yellow tinge, but usually is at least a shade darker on chest and flanks, with white fringes to vent and undertail-coverts. (H. Shirihai)

C. h. hellmayri, ad-like ♂, Oman, Nov: two main features differentiate ♂♂ of Arabian races from African taxa: more violet-blue iridescence covering much of forehead and forecrown, and down head-sides and throat, and carmine-red breast patch is narrower and less solid, with more blue-turquoise fringes that form ragged but often quite substantial upper and lower borders. (M. Schäf)

C. h. hellmayri, ♂, presumed ad, Oman, Nov: violet-blue iridescence covers much of crown, and together with bluer throat gives impression of violet-blue face. Note the tip of a yellowish pectoral tuft just visible. In common with African races most of upperparts and small wing-coverts are iridescent green, contrasting with the violet gloss over rest of wing and metallic blue uppertail-coverts. Seemingly at end of post-nuptial moult, suggesting ad. (M. Schäf)

62–69 mm (*n* 12, *m* 67.1), ♀ 58.5–64 mm (*n* 12, *m* 52.8); **T** ♂ 44–50 mm (*n* 16, *m* 46.9); ♀ 40–44 mm (*n* 12, *m* 42.2); **T/W** *m* 69.6; **B** ♂ 20.7–24.1 mm (*n* 16, *m* 22.3), ♀ 19.0–24.3 mm (*n* 12, *m* 21.3); **BD** 3.0–3.9 mm (*n* 27, *m* 3.3); **Ts** 14.0–17.5 mm (*n* 26, *m* 15.9). **Wing formula:** **p1** > pc 4.5–7 mm, < p2 25–32.5 mm; **p2** < wt 3–5 mm, =6/7 (rarely =7); **pp3–5** about equal and longest; **p6** < wt 0.5–2 mm; **p7** < wt 3.5–7.5 mm; **p10** < wt 8.5–13 mm; **s1** < wt 9.5–14 mm. Emarg. pp3–6.

GEOGRAPHICAL VARIATION & RANGE Three subspecies occur in the covered region. These represent two distinct groups which may be specifically distinct (see Taxonomic notes). Two more races close in appearance to *habessinicus* in NE Africa are extralimital (not treated).

African subspecies group

C. h. habessinicus (Hemprich & Ehrenberg, 1828) (extreme S Egypt to N & C Ethiopia; resident). Has in ♂♂ much broader and more vividly carmine-red breast-band than in *kinneari* (with poorly and usually only partially developed metallic blue bands above and below). Metallic green of upperparts and breast tends to show golden-brown in certain light angles. Metallic gloss on crown purple rather than bluish as in the other races. ♀ distinctly paler than Arabian races, being greyish-brown above and paler cream-brown below. Pale fringes to vent and undertail-coverts broader, and size noticeably smaller than next two races.

Arabian subspecies group

C. h. kinneari Bates, 1935 (W Saudi Arabia; resident). Differs from *habessinicus* in generally narrower and less solid red breast-band of ad ♂, but the metallic blue bands around the red are usually better developed. ♀ dusky olive-brown (can appear almost blackish), with bold pale fringes to vent and undertail-coverts. Often shows more distinct facial pattern than African races, with a narrow but distinct pale supercilium, a dusky loral patch and a short submoustachial stripe. **W** ♂ 69–76.5 mm (*n* 15, *m* 71.8), ♀ 62.5–70 mm (*n* 8, *m* 65.2); **T** ♂ 47–52 mm (*n* 15, *m* 50.0), ♀ 44–48 mm (*n* 8, *m* 45.8); **B** ♂ 22.0–26.3 mm (*n* 15, *m* 23.3), ♀ 20.2–22.0 mm (*n* 8, *m* 21.6).

C. h. hellmayri Neumann, 1904 (extreme SW Saudi Arabia, Yemen, SW Oman; resident). Similar to *kinneari*, but ad ♂ sometimes has red breast-band admixed greenish-blue; ♀ differ more clearly than ♂ by being paler than *kinneari* but still darker than *habessinicus*; plumage mouse-grey with white primary-edges forming hint of paler wing panel. Juv, however, much like latter. **W** ♂ 70.5–74 mm (*n* 14, *m* 72.3), ♀ 61–66 mm (*n* 8, *m* 63.8); **T** ♂ 49–54 mm (*n* 13, *m* 51.6), ♀ 40.5–48 mm (*n* 9, *m* 44.4); **B** ♂ 23.0–27.0 mm (*n* 15, *m* 25.1), ♀ 21.9–24.9 mm (*n* 9, *m* 23.0).

TAXONOMIC NOTES The two Arabian races are rather close, but in turn differ markedly from *habessinicus* (and other extralimital subspecies in Africa) by the peculiarly dusky ♀ and juv plumages, and by their distinctly larger size. We suggest that *kinneari* and *hellmayri* may together constitute a species separate from the African races, but due to our limited familiarity with these forms in the field we await study of vocalisations and DNA analysis.

C. h. hellmayri, ♀, Oman, Nov: ♀♀ of Arabian races also differ rather markedly from those of African taxa, mainly in being much darker, almost dusky grey-brown. Note very short and thin pale supercilium, distinctive white scaly pattern to undertail-coverts, and some pale mottling on lower breast. Ageing unsafe without handling. (M. Schäf)

C. h. hellmayri, 1stW ♂, Oman, Oct: note dark central throat and upper-breast mark, bordered by broad pale submoustachial stripe; also a few isolated green feathers on breast. (D. Occhiato)

C. h. hellmayri, 1stY eclipse ♂, Oman, Mar: since ad lacks true eclipse plumage, ♀-like birds with some green on head, back and breast when commencing moult to ad plumage must be 1stY ♂♂. Unlikely to be confused with much smaller and shorter-billed Palestine Sunbird, which even in superficially similar plumage lacks present species' diagnostic whitish chevrons on vent and undertail-coverts. (J. Mayer)

PALESTINE SUNBIRD
Cinnyris osea Bonaparte, 1856

Alternative name: Orange-tufted Sunbird

Fr. – Souimanga de Palestine
Ger. – Jerichonektarvogel
Sp. – Suimanga palestino; Swe. – Palestinasolfågel

Also known as the Orange-tufted Sunbird (when merged as superspecies with similar-looking sub-Saharan *C. bouvieri*), although neither this name nor Palestine Sunbird is wholly suitable, the orange tufts being always extremely difficult to see, growing as they do near the axillaries and therefore often being covered by the wing-bend, and the range is hardly centred in or around Palestine. The species has a considerably broader range than suggested by its usual name, occurring in thorn savanna from W to C Africa, and is widespread in the Middle East and Arabia, where it reaches from extreme SW Syria across Arabia to S Oman. Sedentary or nomadic; favours sunny and warm lowlands and mountain slopes offering herbs and trees with flowers, on which it largely depends for its nectar and insect food. Reaches up to 3200 m in Yemen. In many areas it is also a common garden bird.

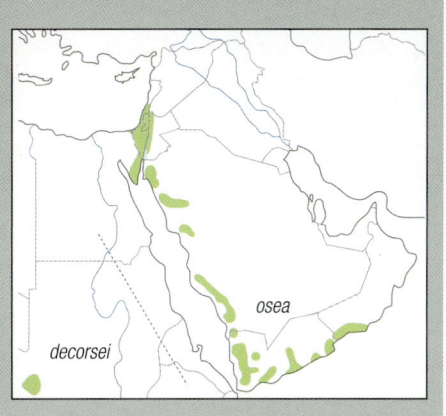

C. o. osea, ♂, Israel, Feb: ♂ can look almost all black at distance, but is glossy blue-green in full light at close range (the iridescence is not uniform, often appearing as scattered scales or patches). Ad and many 1stY undergo complete moult post-breeding, so birds with evenly-feathered ad wing cannot be aged. (H. Shirihai)

IDENTIFICATION A *small, dark sunbird* with a *short tail* and for its size *relatively long, decurved bill*. ♂ in full plumage is largely *dark metallic bluish* admixed with violet and green (at distance looks almost black). Its usually concealed *red-and-yellow breast-side tufts* may be visible at close range in display. The ♂ lacks ♀-like eclipse plumage of most other sunbirds. Adult ♀ is *rather dull and featureless*, being mainly *pale greyish olive-brown above*, including plain head (in close views may show very faint greyish supercilium); *below largely dull greyish-white with pale yellowish suffusion*, and faintly darker wash or mottling on throat, breast and sides (only visible in close views). Usually has darker wings and *contrastingly almost black tail*. Juvenile ♂ in various stages of its protracted post-juvenile moult distinctive, generally shows blotchy mixture of brown juvenile and glossy green and violet-blue black feathers, with brownish juv flight-feathers. Chin to chest and much of upperparts often moulted early. Restless, jerky and very active, with fast flights (on rapid flitting wingbeats) between flowers, and makes short tail-flicks when perched. Adapted to nectar-feeding and frequently hovers. Longer flights can reach quite high.

VOCALISATIONS Rapid (frantic and accelerating) song consists of Serin-like twitter, whistled notes and often includes a rattling trill, *tvui tvui tvui tirrrrrr* (can at distance recall part of Black Redstart song). – Has rich and loud repertoire of calls, including repeated descending, high-pitched whistled *viyu* (almost like Eurasian Nuthatch song) or upwards-inflected *tvueet*, sometimes with a short initial note, *che-wheeit*. Has several hard, dry contact notes, e.g. electric tongue-clicking *zet*, often repeated in an energetic series, *zut-zut-zut zut zut-…*

SIMILAR SPECIES For separation from *Purple Sunbird*, which is the only likely confusion species within the covered region (although they do not overlap), see that species. – Also unlikely to be confused with sympatric *Shining Sunbird*, even in ♀-like plumages, as Palestine is smaller with a proportionately slightly longer bill, and is distinctly paler in such plumages (especially compared to darker Arabian races of Shining), as well as lacking white-fringed and thereby 'patchy' ventral region of Shining. Absence of red breast-band prevents confusion with ♂ Shining. – Distinctly longer bill separates ♀-like plumages from *Nile Valley Sunbird*, which in such plumages also usually has much more extensive and brighter yellow underparts and a better-marked pale supercilium.

AGEING & SEXING Ageing of ♂♂ often straightforward when post-juv moult is underway, after that difficult or impossible. No seasonal or age-related differences in ♀. Sexes strongly differentiated in all post-juv plumages. – Moults. Complete post-nuptial and extensive post-juv moults (apparently leaving only remiges and tail-feathers), protracted due to long and variable breeding season, mostly May–Dec (less often Apr–Mar, thus moult can occur year-round). Seasonal moult peaks in S Arabian birds apparently earlier than in Levant. Note that this species (contrary to statements in most literature) lacks eclipse plumage in ad ♂. Juv ♂ moults directly to ad ♂ body plumage, and ad ♂ moults from full breeding plumage to next full breeding plumage. Whether any juv moults completely, thereby replacing also all flight-feathers, remains to be established. Many (or all) do not. – SPRING–SUMMER **Ad** ♂ Iridescence gives forecrown glossy violet sheen, becoming more blue on rear sides of crown, while rear cheeks, neck and much of upper body are glossy green; face to upper breast glossy violet-blue, rest of underparts dull black. Wings not glossy but dark brown-black, uppertail glossy bluish. Pectoral tufts orange and yellow, but usually concealed by wings and only occasionally visible in display. With strong wear, glossy areas become duller and bluer, even blacker due to more exposed bases. **1stS** ♂ Very difficult to separate from ad once body moult complete. Some ♂♂ still moulting into ad plumage show retained

C. o. osea, ♂, Israel, Jan: typical sunbird bill mid-length, blackish and rather fine, straight near base and strongly decurved over terminal part. Remiges, tertials, greater coverts, primary-coverts and alula reflect darker at certain angles, producing artificial effect of moult contrast. (H. Shirihai)

C. o. osea, ad ♂, Israel, Aug: in appropriate light and at close range, more purple-violet iridescence is reflected, especially on breast and forecrown. ♂ has yellow-and-orange pectoral tufts (right-hand image) that are usually hidden below wing-bend. Palestine Sunbird has no seasonal eclipse plumage (like some other species of sunbirds), thus ad ♂♂ moult straight to same plumage, like this ad ♂. (H. Shirihai)

brown juv feathers admixed. Primaries generally detectably more brownish than ad type, but due to individual variation some are still difficult to age. ♀ Brown-grey above, dusky grey-white below. Wings and tail slightly darker brown (pale fringes to tertials and wing-coverts almost worn off). **Juv** Soft fluffy body plumage, and underparts usually tinged yellowish, otherwise like ♀. Sometimes initially has subtly shorter bill. — AUTUMN–WINTER **Ad** ♂ Neat and freshly plumaged with strong metallic gloss. **Ad** ♀ As in spring–summer. Difficult or impossible to age. **1stW** ♂ Variable depending on hatching date and start of moult into ad ♂ but much as described under 1stS ♂ in spring–summer.

BIOMETRICS (*osea*) **L** 10–11 cm; **W** ♂ 53.5–56.5 mm (*n* 16, *m* 54.6), ♀ 49–53 mm (*n* 13, *m* 50.9); **T** ♂ 36–41 mm (*n* 16, *m* 38.8), ♀ 32–36 mm (*n* 13, *m* 33.9); **T/W** *m* 69.1; **B** ♂ 18.5–22.4 mm (*n* 16, *m* 20.1), ♀ 17.1–20.1 mm (*n* 13, *m* 18.9); **BD** 2.7–3.5 mm (*n* 27, *m* 3.1); **Ts** 14.2–16.8 mm (*n* 26, *m* 15.4). Wing formula: **p**1 > pc 6–8.5 mm, < p2 18.5–23 mm; **p2** < wt 3.5–5.5 mm, =6/7 (rarely =7); **pp3–5**(6) about equal and longest; **p6** < wt 0–2 mm; **p7** < wt 1.5–4.5 mm; **p10** < wt 5–8 mm; **s1** < wt 6–9 mm. Emarg. pp3–6.

GEOGRAPHICAL VARIATION & RANGE Rather clear variation, but generally only two subspecies recognised, one extralimital in sub-Saharan Africa (*decorsei*, not treated).
C. o. osea Bonaparte, 1856 (SW Syria, Lebanon, W Jordan, Israel, Sinai, W & S Saudi Arabia, Yemen, S Oman; resident). Described above. Distinctly larger than *decorsei* of mainland sub-Saharan Africa and lacks prominent blue-green patch on lower breast and upper belly present in *decorsei*. Some vocal differences compared to *decorsei* noted, latter having a metallic song consisting of a variety of *chwee* or *chwing* notes, while song of *osea* is more high-pitched and varied.

TAXONOMIC NOTE Given such marked morphological and acoustic differences (cf. above) between the two races, and their well-separated geographical ranges, it is possible that two species are involved.

C. o. osea, ♂, Israel, spring: in typical hovering flight when feeding at flowers. Note bluish iridescence on rump and uppertail-coverts (a colour which can also appear more purple depending on light and angle). (L. Kislev)

C. o. osea, ad ♀, Israel, Aug: unlike juv/1stW ♀ at this season, has no or very limited yellowish wash to rather plain, largely dull greyish-white underparts. From above, tail usually appears contrastingly dark. Distinctly paler than sympatric Shining Sunbird and lacks white-fringed ventral region of corresponding plumages of that species. Separable from ♀ Purple Sunbird by longer bill and lack of clear yellow on underparts, but the two are also allopatric. In post-nuptial moult. (H. Shirihai)

SUNBIRDS

C. o. osea, ♀, Israel, Feb: rather nondescript, mostly pale olivaceous grey-brown on head and upperparts, with very faint and short pale supercilium. Dull olive-grey underparts may show slight yellowish wash in fresh plumage with some mottling. Contrastingly black tail has clear pale fringes and tips to outer feathers visible from below. Birds without clear moult limits in wing impossible to age. (H. Shirihai)

C. o. osea, juv/1stW ♂, Israel, Aug: juv ♂ shows blackish bib and undergoes complete post-juv moult, while even in summer, as here, they can have some newly moulted green, blue and purple iridescent feathers, initially on dark bib, uppertail-coverts and small wing-coverts. As breeding can occur during *c*. 8 months of the year, such plumage can be encountered almost year-round. (H. Shirihai)

C. o. osea, juv/1stW ♂, Israel, Dec: young ♂♂ at various stages of post-juv moult were formerly erroneously thought to be non-breeders or in eclipse plumage. The right-hand bird is at more advanced stage of post-juv moult, already acquiring almost full ♂ plumage, with many dark metallic feathers above and on foreparts. Remiges and primary-coverts of both birds are still juv, while on left-hand bird the replaced tertials and most greater coverts form moult contrast. Also note the still-present yellow gapes. (A. Ben Dov)

C. o. osea, juv/1stW ♀, Israel, Dec: post-juv moult still incomplete (e.g. outer greater coverts retained). Note substantial yellow gape to relatively shorter bill (during first few weeks/months of life, bill usually does not reach full length). Unlike ad ♀ overall more brownish and olive (rather than greyish) above, with more dusky-yellow tones below; unlike same-age ♂, lacks dark bib. (A. Ben Dov)

ABYSSINIAN WHITE-EYE
Zosterops abyssinicus Guérin-Méneville, 1843

Alternative name: White-breasted White-eye

Fr. – Zostérops à flancs jaunes
Ger. – Somalia-Brillenvogel
Sp. – Anteojitos abisinio
Swe. – Abessinsk glasögonfågel

A common resident in SW Arabia, this typical white-eye occurs in all types of wooded habitats at altitudes up to 3000 m. For long thought to be the only *Zosterops* in Arabia, but recently a tiny population of Oriental White-eye was discovered in mangroves on an island in E Oman (apart from also occurring at one locality in S Iran). Fortunately, it is extremely unlikely that the two will ever come into contact, thereby ensuring almost no risk of confusion between the two. Elsewhere, Abyssinian White-eye extends from Eritrea south to Tanzania.

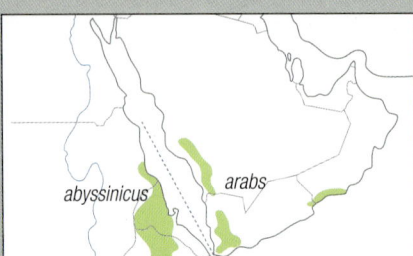

Z. a. arabs, ad, SW Saudi Arabia, Jul: warbler-like, mostly greenish-olive above, with the most obvious features being the broad white 'spectacles' and yellow throat. Mostly greyish and cream-tinged underparts, with buffish-brown wash on flanks. Also rather strong, slightly decurved bill with broad base. Ad by active post-nuptial moult at end of summer. (J. Babbington)

IDENTIFICATION A small warbler-like bird, readily identified by its *obvious white eye-crescents*, forming a near-complete ring. The combination of these white 'spectacles' with the *greyish-green upperparts*, marked only by slightly darker centres to the tertials, flight- and tail-feathers, *pale yellow throat and upper breast, slightly greyish-tinged lower breast, pale buffish flanks and belly*, and *yellowish cast to vent and undertail-coverts* is quite unmistakable within the covered region. The iris looks dark in the field, and the short bill is mostly black, but with a pale flesh-coloured base to the lower mandible, and similarly-coloured legs and feet. Usually found in small groups of up to 30 or so in the non-breeding season, which are very active when feeding and often easily approached.

VOCALISATIONS Song strangely machine-like (sounding like a 'geiger meter') or insect-like buzzing and finely rattling, the voice being rather strained, a series of oft-repeated thin phrases, each lasting c. 2 seconds, frequently with a very slight terminal flourish as the final note is doubled or trebled. Possible also to hear a reminiscence to Corn Bunting song in the voice. – Calls rather plaintive, though some notes sound slightly sharper, e.g. *teeyu*, *tew* or a sibilant *pseeyip*, slightly recalling Siskin. Groups continually utter contact calls. There are also reports of a purring call recalling Snow Bunting (*HBW*).

SIMILAR SPECIES Given a clear view of the eye-ring, Abyssinian White-eye should be unmistakable within the treated region. Oriental White-eye (no overlap, as in Arabia it only occurs on Mahawt Island, E Oman) is readily distinguished by brighter and darker yellow-green upperparts and head, yellow forehead, bright yellow throat and vent in contrast to rather pale pure greyish-white belly (Abyssinian slightly darker and faintly brown-tinged grey belly). Oriental also differs by often having a narrow lemon-yellow band along centre of belly.

AGEING & SEXING Ageing often difficult. Moult limits difficult to detect, and both ad and 1stW wear primaries and primary-coverts rather rapidly to abraded state. Sexes virtually identical in plumage (but see Biometrics). – Moults. Judging from examined specimens, ad apparently undergoes complete post-nuptial moult in late summer (mostly Aug–Sep), whereas post-juv moult is partial (involving all head, body, lesser and median coverts and most or all greater coverts, alula and tertials, but no remiges or primary-coverts, and no or few tail-feathers). Two ad in Dec had old unmoulted secondaries (apparently left in suspended moult). Extent of pre-nuptial moult unknown, but two (May, Aug) apparent 1stS had some new secondaries. – **SPRING Ad** Sexes very similar and usually not separable. In general, ♂ brighter/greener (less greyish) above and on wing fringes, with brighter yellow bib than ♀, but much overlap. **1stS** Very similar to ad, but primary-coverts and primaries more worn with narrower and more bleached fringes. Unknown if birds with new secondaries are always 1stS. – **AUTUMN Ad** As in spring but brighter greenish above and yellowish on bib. Compared to 1stW evenly fresh, and those with retained, very worn, secondaries distinctive. **1stW** As fresh ad, but juv primary-coverts and remiges (especially primaries) more weakly textured and less fresh, with thinner and sharper fringes to primary-coverts (less broad and diffuse fringes in ad). No data on iris colour with age, but both ad and 1stY described as brown (or fox-brown). **Juv** Closely resembles ad but perhaps distinguished by soft, fluffy body-feathering and paler yellow throat, greyer underparts, darker eyes and more extensively flesh-coloured bill.

BIOMETRICS (*arabs*) **L** 11–12 cm; **W** ♂ 55–61 mm (*n* 12, *m* 59.0), ♀ 55–59.5 mm (*n* 14, *m* 57.3); **T** ♂ 39–46 mm (*n* 12, *m* 42.9), ♀ 39–43 mm (*n* 14, *m* 40.9); **T/W** *m* 72.0; **B** 12.5–14.1 mm (*n* 25, *m* 13.4); **BD** 3.4–4.3 mm (*n* 25, *m* 3.7); **Ts** 16.0–18.0 mm (*n* 25, *m* 16.8). **Wing formula: p1** > pc 2–5 mm; **p2** < wt 3.5–6.5 mm, =7/9 (*c.* 90%)

Z. a. arabs, presumed ad, Yemen, Jan: when very fresh and in bright light, may appear saturated greener above, with warmer almost drab-coloured flanks. Seemingly evenly fresh wing suggests ad. (H. & J. Eriksen) or =9 or 9/10 (10%); **p3** < wt 0.5–2 mm; **pp4–5**(6) about equal and longest; **p6** < wt 0–2 mm; **p7** < wt 2–4 mm; **p8** < wt 3–8 mm, **p10** < wt 7–9 mm; **s1** < wt 8.5–10 mm. Emarg. pp3–6.

GEOGRAPHICAL VARIATION & RANGE Only one race occurs in the covered region, *arabs*, which is endemic to Arabia. This form is close to other races such as *abyssinicus* in E Sudan, Eritrea and further south, and others in the Horn of Africa or in E Africa (extralimital, not treated), all of which lack yellow on the belly and are further distinguished by the relatively paler yellow bib and generally duller plumage.

Z. a. arabs Lorenz & Hellmayr, 1901 (SW Saudi Arabia, Yemen, S Oman; resident). Described above. Rather dark greyish-green above.

TAXONOMIC NOTE Recently, breeders in Saudi Arabia (*arabs*) have been found to differ genetically from populations in the Horn of Africa and Socotra (mainly *abyssinica* and *socotrana*; Babbington, Boland, Kirwan, Shirihai & Schweizer, in prep.), on a level more typical for species than subspecies, although morphological differences seem to be slight or at least variable. More research should be carried out before the need for any taxonomic change can be evaluated.

REFERENCES Eriksen, H. *et al.* (2001) *Sandgrouse*, 23: 130–133. – Hamidi, N. (2006) *Podoces*, 1: 35–36.

Z. a. arabs, S Oman, Jan: habitually gives impression of curious look, probably because of the 'spectacled' eyes and otherwise plain head pattern. Without close check of moult pattern in wing (here invisible), ageing impossible. (H. & J. Eriksen)

Z. a. arabs, S Oman, Dec: some can have yellowish and greenish pigments rather diluted (here despite being rather fresh and in bright sunlight). Note also impression of slightly shorter and thinner bill on this bird. Best left unaged. (M. van den Schalk)

Z. a. arabs, presumed 1stS, S Oman, May: with wear, duller, paler and more greyish above, while white eye-ring wears subtly narrower. The rather dull coloration above, and already very worn and bleached primaries, may suggest a 1stS. (H. & J. Eriksen)

ORIENTAL WHITE-EYE
Zosterops palpebrosus (Temminck, 1824)

Alternative name: Indian White-eye

Fr. – Zostérops oriental; Ger. – Gangesbrillenvogel
Sp. – Anteojitos oriental; Swe. – Indisk glasögonfågel

This widespread and mainly Indian and SE Asian white-eye is also one of the rarest and most localised of all species occurring within the treated region. Recently, a tiny population was discovered in mangroves on the offshore island of Mahawt in E Oman, and there is also a small population in mangroves in Hormozgan, S Iran. Belongs to a genus with numerous similar-looking species (84 listed in *H&M* 4, and 74 in *HBW*), and identification would require great care if the majority of these were not endemic to islands or remote areas. In fact, it is quite rare that more than one species occurs in one particular place.

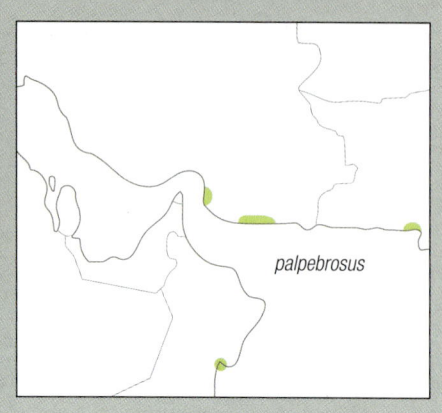

Z. p. palpebrosus, S Iran, Feb: small and short-tailed, yellowish olive-green above, with head and especially throat but also undertail-coverts yellow, otherwise dusky-white below, and this in combination with the prominent, snowy-white eye-ring readily identifies the species. Note clearly patterned black-and-grey bill (*cf. Abyssinian White-eye*). (A. Yekdaneh)

Z. p. palpebrosus, S Oman, Mar: prominent white eye-ring further enhanced by dusky-black feathering surrounding it, and bright lemon-yellow head that is even purer and brighter on throat. (H. & J. Eriksen)

IDENTIFICATION Similar to Abyssinian White-eye in having *olive-green upperparts* including unmarked yellowish-green wings, a *prominent white eye-ring* enhanced by dusky lower cheeks, *bright lemon-yellow throat and vent*, and *dusky greyish-white breast and belly*. It differs from Abyssinian mainly in having *stronger and purer colours*, being an altogether brighter-looking bird. On upperparts, *rump is brighter greenish-yellow*. On underparts, *flanks are grey* like the breast and belly, not brown-tinged as in Abyssinian, and *undertail-coverts are vivid lemon-yellow*, not dull dusky yellow-white as in Abyssinian. The *bright yellow forehead* continues narrowly in a short streak above lores, while *lores are blackish*. Iris dark brown. Bill dark and fine but strong-based with downcurved and acutely-pointed tip. *Legs medium to dark grey*, claws paler. Sexes alike including by size. Like other white-eyes, usually found in small groups or family parties at end of breeding or in the non-breeding season. Active, moving constantly. Often easily approached when feeding.

VOCALISATIONS Song a short series of call notes or call note-like sounds in irregular order. – Calls much as in Abyssinian White-eye, being rather plaintive, a persistently but rather slowly repeated piping *cheuw* or *tew*. A variant call is a purring downslurred, slightly guttural *djer'r'r'r*. Groups continually utter contact calls.

SIMILAR SPECIES Once the size, shape and general plumage colours are seen and the bird is established to be one of the white-eyes (small, short-tailed, olive-green, yellow and dusky-white with prominent white eye-ring) the only other species to be considered within the covered range is *Abyssinian White-eye*. This differs in being duller, less bright, and in having slightly paler and washed-out underparts, with brown-tinged flanks and without the bright yellow undertail-coverts of Oriental. Also, some Orientals have a hint of a yellow mesial band along centre of belly, not seen in Abyssinian. Bill and legs average darker in Oriental. Abyssinian White-eye is a slightly larger bird than Oriental, but without direct comparison this is difficult to appreciate.

AGEING & SEXING Ageing appears difficult due to subtle or lack of moult limits, and because both ad and 1stW wear primaries and primary-coverts rather rapidly. Sexes inseparable using plumage (but see Biometrics). – Moults. Not extensively studied, but appear similar to Abyssinian White-eye. – SPRING **Ad** Sexes similar or identical. **1stS** Very similar to ad, but primary-coverts and primaries more worn with narrower and more bleached fringes. – AUTUMN Not closely studied but presumed to be much as described under Abyssinian White-eye. **Juv** Resembles ad but distinguished by soft, fluffy body-feathering and overall duller colours.

BIOMETRICS (*palpebrosus*) **L** 10.5–11.5 cm; **W** ♂ 54–69 mm (*n* 12, *m* 57.4), ♀ 51–58 mm (*n* 15, *m* 55.1); **T** ♂ 38.5–43 mm (*n* 12, *m* 40.5), ♀ 35.5–42 mm (*n* 15, *m* 39.2); **T/W** *m* 70.9; **B** 11.0–13.4 mm (*n* 27, *m* 12.6); **BD** 3.0–4.0 mm (*n* 26, *m* 3.5); **Ts** 14.0–16.5 mm (*n* 24, *m* 15.5). Wing formula: **p1** minute; **p2** < wt 3–5.5 mm, =6/8; **pp3–5** about equal and longest; **p6** < wt 0.5–2.5 mm; **p7** < wt 3–6 mm; **p10** < wt 9–12 mm; **s1** < wt 10.5–13 mm. Emarg. pp3–6.

GEOGRAPHICAL VARIATION & RANGE Only one race occurs in the covered region, *palpebrosus*, but several extra-limital races breed in SE Asia. Note that range for treated race differs from that given in, e.g., Vaurie (1959), but follows that given in *HBW*.

Z. p. palpebrosus (Temminck, 1824) (Mahawt I, E Oman, local S Iran, NE Afghanistan, Pakistan, N & C India east to Burma; resident). Described above.

REFERENCES ERIKSEN, H. et al. (2001) *Sandgrouse*, 23: 130–133. – HAMIDI, N. (2006) *Podoces* 1: 35–36.

Z. p. palpebrosus, Mahawt Island, Oman, Dec: a rather dull individual. Separated both geographically and by plumage characteristics from the other *Zosterops* species found in the region, the Abyssinian White-eye, in being vividly coloured, with brighter greenish-yellow on face, bib, rump and undertail-coverts. Ageing unsure without handling and more experience with the local population. (H. & J. Eriksen)

Z. p. palpebrosus, Mahawt Island, Oman, Dec: in typical head-on view, and unlike the geographically separated Abyssinian White-eye, the yellow of face extends down to upper breast, and it has broader white 'spectacles'. (H. & J. Eriksen)

(EURASIAN) GOLDEN ORIOLE
Oriolus oriolus (L., 1758)

Fr. – Loriot d'Europe; Ger. – Pirol
Sp. – Oropéndola europea; Swe. – Sommargylling

In warmer parts of Europe this summer visitor dwells in mature deciduous woods, parks, riverine forests, etc. Tall oak trees, beeches, alders or poplars, preferably with a few glades and a pond or lake, are favoured habitats. There, in May and June, its lovely yodelling whistle can be heard, announcing the species' presence. Though the ♂ is about the brightest yellow there is, it is still difficult to see in the dense and often sunlit yellowish-green canopy. Golden Orioles are in fact most often seen when they fly over an opening from one perch to another. Winters in tropical Africa.

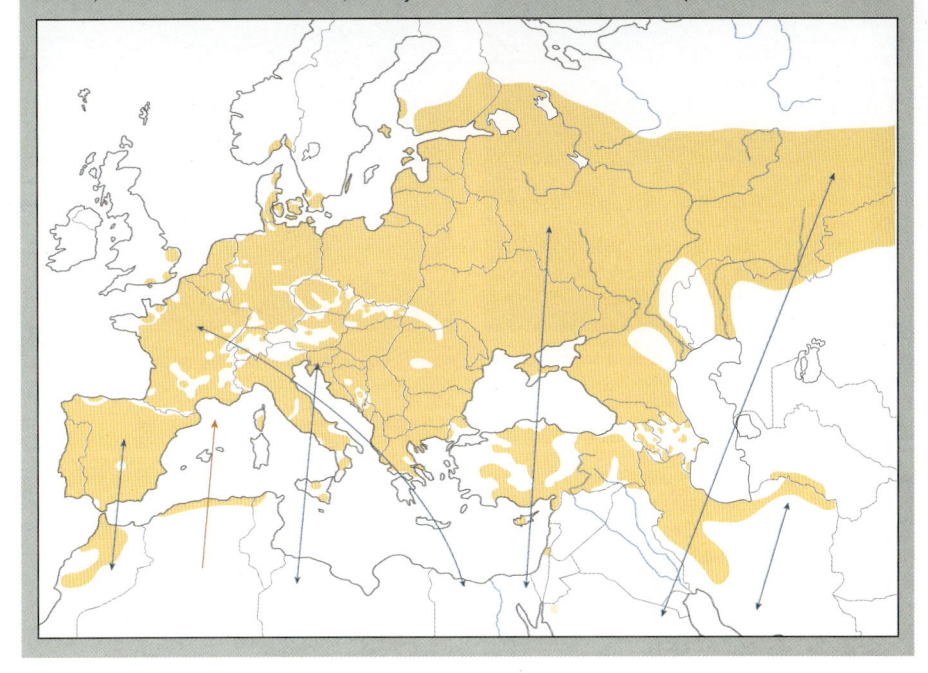

IDENTIFICATION A thrush-sized, slim, elegant bird with long, pointed bill, short legs and rather long wings. The adult ♂ is *bright yellow* with *black wings and tail*, all other plumages are greenish-grey with a variable yellowish tinge, and have dark green-grey wings and tail. *Bill dull reddish*, legs greyish. Adult ♂ has a dark loral stripe, a *yellow patch on the primary-coverts, yellow tail corners*, while other plumages have duller, less well-marked versions of these features. Young birds are more heavily streaked below than older. When perched in the canopy of a tree, behaves rather like a giant warbler, moving frequently, stretching for insects, etc. Often surprisingly difficult to see well when perched, despite its size, often striking colours and frequent calls. In flight, note fast progress, rather straight path or long shallow undulations (then somewhat thrush-like). During migration sometimes seen in small, loose flocks.

VOCALISATIONS Song a pleasing, yodelling, loud and rather deep-voiced whistle, *düd-lio-dih-dioo* and variants. Sometimes just a short *dioo*, at other times longer elaborations. Possible to confuse at some distance with Blackbird song (which is generally slower and more mellow), and comparatively easy to mimic by man (which will often attract the ♂). – Commonest call a rather raucous, emphatic *veeaahk* when anxious or agitated, sounding like a mixture of the calls of Collared Dove *Streptopelia decaocto* and Jay. Other calls include a falcon-like *gigigigig* in alarm (*BWP*) and subdued, meowing, slightly husky *veah*, related to the common raucous call.

SIMILAR SPECIES It is difficult to confuse the adult ♂ with any other species in the treated region. The ♀, and in particular younger birds, are more nondescript and, considering the greenish upperparts and the slightly paler and more yellowish rump, could, when flying away, possibly be confused with a *Picus* woodpecker, when the smaller size and slimmer body of the oriole is not apparent. However, *Picus* woodpeckers have larger heads, more rounded and slightly shorter wings, and fly in deeper undulations.

AGEING & SEXING (*oriolus*) Ages differ in 1stY, and for ♂♂ sometimes also slightly in 2ndY if seen close or handled (but beware numerous difficult intermediates between imm ♂♂ and ♀♀). Practically no seasonal differences. Sexes differ after second complete moult, performed in 2ndW. (A proposed one year longer immaturity for ♂♂ [*BWP*] is unconfirmed.) – Moults. Partial moult of ad, mainly of body, head and some inner wing-coverts, sometimes including also a few (1–3) inner secondaries, in late summer (mainly late Jun–Sep); complete moult in winter in Afrotropics. Rarely, 1–3 secondaries left unmoulted (possibly same as those moulted before autumn migration). Partial post-juv moult in summer involves no flight-feathers, and generally only some inner median and greater coverts. Complete moult in winter like ad (sometimes a few secondaries are left unmoulted, on average more than in ad). – **SPRING Ad** ♂ Head and body bright yellow, wings and tail jet-black (though sometimes a slight greenish tinge on basal 25% or less of r1). Rarely, yellow body-colour paler, which could imply ♀, but then note

Ad ♂, Spain, Jul: yellow plumage is pure, vivid and deep, eye-stripe jet-black, with wing-feathers glossed black, and primary-coverts have large yellow patch on tips; central tail-feathers black to base; deep red iris. Black eye-stripe limited to lores (any behind eye insignificant). Large yellow patch on primary-coverts, and jet-black remiges, indicate that this ♂ is an older bird. (C. N. G. Bocos)

Ad ♂, Hungary, May: unmistakable in flight given coloration and shape, as well as somewhat thrush-like flight with long shallow undulations. Note large yellow primary-coverts and tail-corner patches, and quite black flight-feathers, which together infer age and sex. (M. Varesvuo)

Ad ♀, **Bulgaria, Jun**: ad ♀, here with chick, can be rather bright yellow and quite similar to 1stS ♂ at same season, but note combination of grey loral patch, more greenish-tinged upperparts, and central tail-feathers greenish over basal half. Told from 1stS ♀ by diffuser and thinner streaking below, and fairly large pale yellow patch on tips of primary-coverts. (I. Hristova)

Ad ♀, **Poland, Jun**: compared to image to the left, this ad ♀ has bolder and more whitish patch on tips of primary-coverts. Note also that central tail-feathers have rounded tips and quite bold yellow tips to outermost rectrices. (M. Matysiak)

absence of any streaking below, and blackness of flight-feathers. Loral stripe jet-black. Yellow tip on inner web of r6 26–38 mm (or even slightly more in E Russian birds). Black primary-coverts tipped yellow on outer web by 7–16 mm. Iris reddish. **Ad ♀** Variable; either (rarely) resembling pale yellow ♂ (see above), but with greenish cast to outer webs of tertials and greater coverts, and on basal half (or more) of r1, on mantle and back, and dark greyish (not black) loral stripe; or (commonly) appearing more like imm birds, being greyish-white below mottled (on breast) or streaked (on belly and flanks) dark grey, with some yellow tinge on breast and flanks. Yellow tip on inner web of r6 23–30 mm. Dark olive-grey or almost blackish primary-coverts tipped yellowish on outer web by 3–10 mm (and often more extensively white on inner web). Iris reddish. **1stS** Like pale and streaked ad ♀, but on average with more distinct and darker streaking below, and with darker iris (rufous-brown rather than reddish). Sexes quite similar, usually impossible to separate in the field; ♂♂ tend to have more yellow across whole breast, and streaks on breast and flanks are usually narrow and weak (though many inseparable on this). Yellow tip on inner web of r6 24–35 mm in 1stS ♂♂ (once 20), 17–28.5 mm in 1stS ♀♀ (once 30). Dark primary-coverts tipped yellowish-white or white, on outer web 1–10 mm in ♂♂, 1–5 mm in ♀♀. Those (few) with very dark (blackish) large alula, some lesser coverts and primary-coverts combined with much yellow on breast and flanks and subdued streaking should be ♂♂, but majority are intermediates and inseparable. – **SUMMER Juv** As 1stS but readily separated when fresh by c. 0.5–1 mm-wide yellowish tips to median and greater coverts and to all or at lest shorter tertials. Note that these pale tips can wear off to a large extent as early as Sep. Recently fledged young have soft feathering on throat, neck and vent. Tail-feathers comparatively narrow and tips more pointed than in ad. Bill dark. – **AUTUMN Ad** As in spring. Tips of r1 rounded and worn. **2ndW** Sexes similar before complete

Ad ♀, **Israel, Apr**: some older ♀♀ can develop a plumage very close to ad ♂♂, but note grey-black loral patch, moderate-sized yellow tips to primary-coverts, deeper red bill and iris, and only very diffuse and thin dark streaks to quite rich yellow underparts. Also upperwing is greenish-grey, not black. (A. Ben Dov)

1stS ♂, **Hungary, Jul**: 1stS always has streaked underparts, but ♂ usually has primary-coverts tipped quite boldly white (not as narrow/spotty as retained juv coverts of 1stS ♀); dark loral patch is blackish-grey; iris reddish. Note worn unmoulted secondaries, with innermost feather growing. (M. Varesvuo)

moult. Greyish-white below, diffusely mottled grey on breast, streaked blackish on belly and flanks. Very similar to ad ♀ but sometimes separable by narrower and paler (off-white instead of yellow) tips to primary-coverts, being 1–7 mm (ad ♀ 4–10 mm). Tips of r1 rounded and worn (as in ad). No pale tips to median coverts. Bill dark pink-brown, iris reddish or darker rufous. **1stW** Sexes similar. Greyish-white below, streaked grey on throat, black on breast, belly and flanks. Tips of r1 slightly pointed and fresh. Habitually most (often all) juv median and greater coverts and all tertials remaining, tipped yellowish 0.5–1 mm wide. Bill dark, grey-black or dark pink-brown, iris dark brown or black. Pale yellowish tip of r6 20–30 mm in ♂♂, 19–25.5 mm in ♀♀. Dark primary-coverts tipped yellowish-white or white, 1–7 mm in ♂♂, 0.5–2 mm in ♀♀.

BIOMETRICS (*oriolus*) **L** 22–25 cm; **W** ♂ 143–161 mm (*n* 80, *m* 152.5), ♀ 141–158 mm (*n* 54, *m* 149.8); **T** 75–93 mm (*n* 135, *m* 82.8); **T/W** *m* 54.7; **B** 25.3–30.1 mm (*n* 59, *m* 27.7); **B**(f) 21.0–25.5 mm (*n* 133, *m* 23.8); **BD** (7.4) 7.8–9.7 mm (*n* 60, *m* 8.6); **Ts** 21.2–24.8 mm (*n* 59, *m* 22.7); **yellow tip longest pc** (outer web) **ad** ♂ 7–16 mm (*n* 48, *m* 11.1), **ad** ♀ 2–10 mm (*n* 21, *m* 5.0), **1stY** ♂ 1–10 mm (*n* 30, *m* 4.5), **1stY** ♀ 1–5 mm (*n* 29, *m* 2.2); **yellow tip r6** (outer web) **ad** ♂ 26–38 mm (*n* 48, *m* 33.5), **ad** ♀ 23–30 mm (*n* 19, *m* 26.8), **1stY** ♂ 20–35 mm (*n* 30, *m* 27.3), **1stY** ♀ 17–30 mm (*n* 29, *m* 23.7). Wing formula: **p1** > pc 17–30 mm, < **p2** 45–56.5 mm; **p2** < wt 7–14 mm, =5/6 (rarely =6); **p3** (pp3–4) longest; **p4** < wt 0–6 mm; **p5** < wt 14–20 mm; **p6** < wt 25–31 mm; **p7** < wt 33–39.5 mm; **p10** < wt 50–60 mm; **s1** < wt 51–64 mm. Emarg. pp3–4, often faintly also on p5.

GEOGRAPHICAL VARIATION & RANGE Monotypic. – Europe, NW Africa, Turkey, Caucasus, N Iran, W & C Siberia; winters tropical Africa south to South Africa, east to Kenya. – Compared to Indian Golden Oriole (see below) has long pointed wing and thin loral stripe, mask does not extend appreciably behind eye. Also, bill is subtly shorter with a slightly more attenuated and downcurved tip on average. – Breeders in Siberia ('*sibiricus*') are subtly larger and ad ♂♂ have on average slightly more yellow on tail-tip (in ad ♂ to 41 mm) than European breeders, but this is a comparably small difference, is apparently somewhat variable, is clinal from west to east without any clear boundary, and these breeders are therefore included here. **W** ♂ 152–162.5 mm (*n* 14, *m* 157.2), ♀ 151–161 mm (*n* 8, *m* 153.8); **T** 79.0–90.0 mm (*n* 22, *m* 85.5); **B**(f) 23.3–26.7 mm (*n* 22, *m* 24.7); **BD** 8.5–9.6 mm (*n* 22, *m* 9.2). – Breeders in E Caucasus and N Iran ('*caucasicus*') are as extensively yellow in ♂ plumage as are Siberian birds, but average somewhat smaller and shorter-winged. However, only a moderate difference presumably linked to shorter migrations with some overlap, hence again best included here. **W** ♂ (once 142) 146.5–158 mm (*n* 11, *m* 150.5), ♀ 145–154 mm (*n* 5, *m* 149.9); **T** 80.5–91.5 mm (*n* 16, *m* 84.7); **B**(f) 22.2–25.3 mm (*n* 16, *m* 24.0); **BD** 8.3–9.4 mm (*n* 15, *m* 8.7). (Syn. *caucasicus*; *galbula*; *sibiricus*.)

1stS ♀, Italy, Apr: sexing 1stY birds (note dark-streaked white underparts and fine white tips to tertials) is not always possible, but this bird with narrow, spotty white tips to primary-coverts must be a ♀. (G. Conca)

Presumed ad ♀, Israel, Apr: dark red iris, fully reddish bill, rather subdued mottling and thin grey streaking of breast and belly fit best with ad ♀. Rather extensive yellow below and seemingly broad and rounded tail-feathers with large yellow tips to outermost feathers further support this. (A. Ben Dov)

1stW, presumed ♀, Spain, Sep: note very fresh plumage, and generally almost blackish bill and iris (not reddish as 2ndW and ad ♀ at same season). Upperparts pale greenish-brown, and yellowish-white underparts distinctly streaked. Greater coverts brown-olive with thin yellow edges. Central tail-feathers have pointed tips (just visible). Primary-coverts tipped narrowly yellowish-white and very fresh compared to ad ♀. Most likely a ♀ with such thin whitish tips to primary-coverts. (C. N. G. Bocos)

Juv/1stW, Portugal, Aug: in partial post-juv moult, still showing bold dark streaking below typical of juv. Sex difficult to decide unless having very much or very little yellow on outer tail-feather tips and on primary-coverts (here invisible). This looks like a bird in the overlap range. (T. Tilford)

TAXONOMIC NOTE The extralimital taxon *kundoo* is sometimes treated as a race of Golden Oriole. However, considering the very marked morphological differences without appreciable intermediates where the ranges meet (Altai), the fact that the song differs clearly and that it winters exclusively in the Indian Subcontinent (*O. oriolus* as far as is known only in Africa) and performs complete moult in summer on breeding grounds rather than in winter, it is here (like in H&M 4, *HBW* and by IOC) regarded as a separate species. For convenience, and in the event that it straggles to the covered region, a brief description is provided below.

INDIAN GOLDEN ORIOLE *Oriolus kundoo* Sykes, 1832 (E Central Asia, Altai, SE Asia; short- or medium-range migrant wintering in Pakistan, India and Sri Lanka). Slightly larger and slimmer than Golden Oriole, with more rounded and therefore proportionately shorter wing and longer-looking tail, a longer and thinner-looking bill, the ad ♂ having broader black loral stripe clearly encircling eye, and mask extending visibly behind it. Outer tail-feathers entirely yellow in ad ♂ and the same or at least very extensive in ad ♀. In ad ♂ yellow wedge-shaped tips to tertials and secondaries, and tips to primary-coverts much more extensive than in Golden Oriole. Moult strategy differs from Eurasian Golden Oriole in

Indian Golden Oriole, ad ♂, India, Sep: extralimital Indian Golden Oriole (*O. kundoo*) is a potential vagrant to the treated region. Note slightly larger size but shorter wings, long-looking tail and, in ad ♂, extensive black mask reaching clearly behind eye. Also very large yellow patch on primary-coverts, yellow wedge-shaped tips to tertials and more yellow in outertail. Diagnostically undergoes complete post-nuptial moult in late summer/early autumn; this bird is halfway through primary moult. (R. Mallya)

Indian Golden Oriole, ad ♀, India, Oct: proportionately slightly shorter wing and primary projection (usually only 5–6 more evenly spaced primary-tips; in Eurasian 6–7, and more densely packed inwards). Has longer and stronger-looking bill than Eurasian, and dark eye-mask extends faintly behind eye. Reddish bill and iris eliminate 1stW, while rather strong yellow below and on tips of primary-coverts suggest ad ♀. (B. Monappa)

Indian Golden Oriole, 1stS ♂, India, Apr: unlike ad ♂, dark areas not jet-black, more sooty-black, and eye-mask only suggested. Yellow primary-coverts patch smaller. Also has thin underparts streaking, just like immature ♂ of European species. Note very strong and slightly more pointed and downcurved bill of this species. (A. Arya)

that ad has a complete post-nuptial moult in late summer and autumn, and only a limited partial winter moult. **L** 23–26 cm; **W** ♂ 135–148 mm (*n* 20, *m* 141.2), ♀ 134–144 mm (*n* 19, *m* 139.7); **T** 81–92 mm (*n* 39, *m* 87.0); **T/W** *m* 61.9; **B** 27.3–33.0 mm (*n* 37, *m* 30.7); **B**(f) 23.2–29.0 mm (*n* 37, *m* 27.1); **BD** 8.4–10.1 mm (*n* 36, *m* 9.2); **Ts** 21.2–24.0 (*n* 38, *m* 22.8); **yellow tip longest pc** (outer web) **ad ♂** 16–23 mm, **1stY ♂** 7–21 mm, **ad ♀** 5–10,5 mm, **1stY ♀** 1–8 mm; **yellow on r6** (either web) **ad ♂** entire inner web, **1stY ♂** whole inner web or 35–37 mm, **ad ♀** 27–36 mm (rarely entire inner web), **1stY ♀** 29–35 mm. **Wing formula: p1** > pc 21–30 mm, < p2 37.5–45 mm; **p2** < wt 8–12 mm; **pp3–4** longest; **p5** < wt 4–11.5 mm; **p6** < wt 12–24 mm; **p7** < wt 23–30 mm; **p10** < wt 36–46 mm; **s1** < wt 40–50 mm. Emarg. pp3–5, with a hint near tip also on p6 in many. (Monotypic; syn. *baltistanica*; *turkestanica*; *yarkandensis*.)

Indian Golden Oriole, 1stW ♀, India, Oct: very similar to same-age Eurasian Golden Oriole, but note diagnostic longer and thinner-looking bill, but relatively shorter primary projection; eye-mask faint but still extends well behind eye. Dark iris and bill and heavily-streaked underparts suggest age; thin yellow primary-covert tips and very strong breast streaking the sex. (D. Laishram)

ROSY-PATCHED BUSH-SHRIKE
Rhodophoneus cruentus (Hemprich & Ehrenberg, 1828)

Alternative name: Rosy-patched Shrike

Fr. – Tchagra à croupion rose; Ger. – Rosenwürger
Sp. – Bubú pechirrosado
Swe. – Rosabröstad busktörnskata

A quite unique E African bush-shrike that is a fairly common resident of dry thornbush and arid scrub in open acacia savanna, mostly below 1500 m in desert lowlands, but generally avoids bare-ground areas. Within the treated region, it is found only around Gebel Elba, within the Hala'ib Triangle, close to the SE Egyptian/NE Sudanese border. Performs spectacular loud-whistling duets and trios delivered from well-exposed bush- or treetops when mating or in territorial encounters. A calling bird usually seems to be very close but is actually far away! When duetting, birds are sensitive to close approach by observers, and can suddenly become quiet and disappear for long periods.

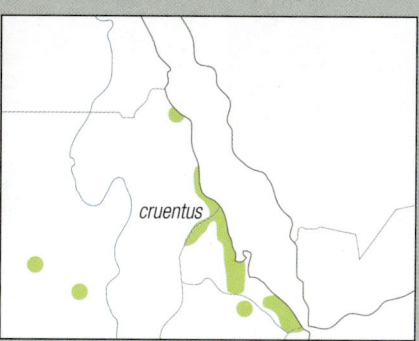

R. c. cruentus, ♂, Ethiopia, Nov: due to combination of size, shape and plumage, as well as voice and behaviour, unmistakable in the covered region. Note ♂'s conspicuous but typically ragged carmine-red central stripe from throat to upper belly (not reaching chin in this race). Ageing impossible without full knowledge of local moult pattern and timing. (I. Yúfera)

IDENTIFICATION Main characters include slim build, the *long and strongly graduated white-tipped tail* (forming broad white tail corners), dull drab brown upperparts and head-sides, a *contrasting rosy-red square patch on lower back*, and off-white underparts (with tiny creamy tinge). Has a short *white supercilium and broken white eye-ring*. Bill rather slim and black, legs brown-grey. ♂ has a conspicuous but irregular *carmine-red central stripe from throat to upper belly* (widest on throat and lower breast), ♀ has a white throat surrounded by a *black gorget* formed by black lower lores and lateral throat-stripe ending in a black bib on upper breast, the red being restricted to the middle of the lower breast and upper belly. Juv more 'normally shrike-like' in coloration, without red or black on underparts, instead having buff fringes to upperparts, upper wing-coverts, central tail-feather and tertials, bordered subterminally by dark brown, with pale central spots, but back rosy. Immatures are duller than adults and pale-fringed above. Frequently seen foraging on the ground below or near bushes, when it appears particularly slender and long-legged, like a truly terrestrial bird. May form small parties.

VOCALISATIONS Habitually sings antiphonal duets, the song having a totally unique quality within the treated region, although it bears some resemblance to the voices of some desert-dwelling larks. It consists of a varied series of high-pitched, mainly straight, whistles of piercingly metallic quality, *tuee-tueeuuee*, etc., sometimes dwindling at the end (details vary) with duetting birds facing each other on the same tree (see image p. 168), or occasionally a distant bird replies. ♀ seems to utter rather lower-pitched notes. – Various, harsher or highly complex musical calls also often heard in alarm or contact.

SIMILAR SPECIES Practically unmistakable if seen well, especially given carmine-red central stripe in ♂ and black pectoral patches in ♀. Also, if viewed from behind or above, note the diagnostic square-shaped reddish lower back patch in all plumages, also the larger size, and long, rather broad and strongly graduated tail with pattern reminiscent of a *Tchagra* species. – Beware that from a distance overall pale plumage and long tail may superficially resemble similar-sized and shaped babblers of desert habitats, especially before underparts patterns seen and calls heard. The species' often terrestrial behaviour offers another important clue for identification.

AGEING & SEXING (*cruentus*) Ageing only possible if post-juv moult incomplete. Sexes differ. – Moults. Complete (though prolonged and rather complex) post-nuptial moult of ad after breeding (Apr–Jun); post-juv moult usually partial and protracted (Jul–Nov), sometimes including some flight-feathers, often r1 or more tail-feathers, and some may even undergo complete moult. No evidence of pre-nuptial moult in late winter, but barely studied. – **ALL YEAR Ad** ♂ White chin and carmine-red central stripe from throat to upper belly. No black breast patch. **Ad** ♀ White throat with black surround, broadening to form broad black pectoral patch, with red central stripe below this. **Ad** Both sexes attain evenly-feathered ad plumage after post-nuptial moult. Shape of carmine-red central stripe of ♂, and black gorget of ♀, vary individually and depending on the bird's behaviour and viewing angle, influencing their apparent width and prominence. Black on the lower lores in ♀ may be concealed or altogether lacking. **1stY** Some young can be separated by at least some retained juv wing- and tail-feathers, but others indistinguishable if entirely moulted. At least during first months, many young are also visibly paler and greyer overall than ad or still have some buff fringes to many feathers, only gradually acquiring reddish and black feathering on underparts. **Juv** Much duller than ad and lacks striking underparts pattern (but has reddish patch on back), with buff-mottled crown, many buff fringes to upperparts and wing-feathers, a pale bill and very pale buff flanks.

BIOMETRICS (*cruentus*) L 22–25 cm; sexes nearly same size; **W** ♂ 92– 100 mm (*n* 12, *m* 95.3), ♀ 89–99.5 mm (*n* 7, *m* 95.1); **T** ♂ 107–123 mm (*n* 12, *m* 114.1), ♀ 110–115 mm (*n* 7, *m* 112.0); **T/W** *m* 119.0; **white on**

R. c. cruentus, ♀, SE Egypt, Apr: unlike ♂, has white throat surrounded by black gorget, which variably extends as lateral throat-stripe, with carmine-red restricted to centre of lower breast and upper belly. Ageing impossible. (B. Putsch)

R. c. hilgerti, ♂, C Ethiopia, Oct: this more southerly race is a potential vagrant to the covered region. Very similar to race *cruentus*, but slightly darker above and warmer below on flanks, and red stripe on underparts is broader and reaches the chin. (H. Shirihai)

R. c. hilgerti, ♀, C Ethiopia, Oct: ♀ of race *hilgerti* hardly differs from *cruentus*, except perhaps as in ♂♂, by slightly darker upperparts and warmer ochraceous flanks. The distinctive white-tipped tail is especially evident when tail is slightly fanned. (H. Shirihai)

R. c. hilgerti, ♂ (left) and ♀, C Ethiopia, Oct: singing characteristic duet, often fully exposed on treetop. (H. Shirihai)

inner web of r6 20–35 mm (n 19, m 27.0); **B** 22.7–26.3 mm (n 18, m 24.1); **B**(f) 17.5–20.8 mm (n 18, m 19.0); **BD** 7.7–8.7 mm (n 19, m 8.2); **Ts** 30.0–35.0 mm (n 19, m 32.9). **Wing formula:** **p**1 > pc 24–29.5 mm, < p2 17–23 mm; **p**2 < wt 4.5–10 mm, =8/10 (rarely =7/8, =8 or <10); **pp**3–6(7) about equal and longest (p3 and/or p6 rarely to 1.5 mm <); **p**7 < wt 0–4 mm; **p**8 < wt 2–6 mm; **p**10 < wt 7–13 mm; **s**1 < wt 10–16 mm. Emarg. pp3–6 (rarely also faintly on p7).

GEOGRAPHICAL VARIATION & RANGE Rather clear variation, with four subspecies recognised, of which only one occurs in the treated region, the others being extralimital and not treated in detail. Still a brief summary is useful. – Ssp. *hilgerti* (Djibouti, C & E Ethiopia, extreme SE Sudan, Somalia, N & C Kenya) is very similar to *cruentus*, but averages slightly darker drab grey above, has a more distinct vinaceous-chestnut tinge to the cap, which may also extend to mantle. Body-sides are deeper ochre, and the rosy-pink of the underparts is more extensive, reaching the chin in most ♂♂. – Ssp. *kordofanicus* (W Sudan) is also close to *cruentus* but is even paler and greyer above, with broader white markings on face. – Ssp. *cathemagmenus* (S Kenya, NE Tanzania) differs by darker upperparts and generally deeper carmine on underparts and back. ♂ differs markedly from ♂♂ of other races in having variable, almost ♀-like black gorget, but the throat within the gorget is carmine, not white, and often the black lateral throat-stripe is missing or broken, thus it appears to have red throat, black breast patch and then again a red lower breast patch.

R. c. cruentus (Hemprich & Ehrenberg, 1828) (Gebel Elba on border between SE Egypt and NE Sudan, south to Eritrea and extreme N Ethiopia). Described above. Differs from the other races in its rather pale drab grey upperparts and ear-coverts, with a slight vinaceous-pink wash on cap and the carmine-red of the underparts rather narrow and restricted, in ♂ usually not reaching chin.

TAXONOMIC NOTE Sometimes referred to the genus *Telophorus*, formerly also often to *Tchagra*. Recent molecular studies (Fuchs *et al*. 2012) showed that this species is more closely related to the genus *Chlorophoneus*, and these two in turn are sisters to *Telophorus zeylonus*, which Rosy-patched Bush-shrike somewhat resembles in plumage pattern (though the former is extensively yellow below). We follow Dickinson & Christidis (2014) in placing the Rosy-patched Bush-shrike in a monotypic genus, *Rhodophoneus*.

REFERENCES Fuchs, J. *et al*. (2012) *Mol. Phyl. & Evol.*, 64: 93–105.

BLACK-CROWNED TCHAGRA
Tchagra senegalus (L., 1766)

Fr. – Tchagra à tête noire; Ger. – Senegaltschagra
Sp. – Chagra del Senegal; Swe. – Svartkronad tchagra

The Black-crowned Tchagra is one of two representatives within the treated region of the so-called bush-shrikes, which are otherwise resident in the Afrotropics. Related to the true shrikes but differ clearly in plumage, vocalisations and behaviour. The generic name Tchagra was chosen by a French ornithologist, who described a South African member of the genus, the Southern Tchagra, referring to one of its calls when naming it. The Black-crowned Tchagra is resident in NW Africa, but is also distributed across Arabia, C and S Africa with a number of subspecies.

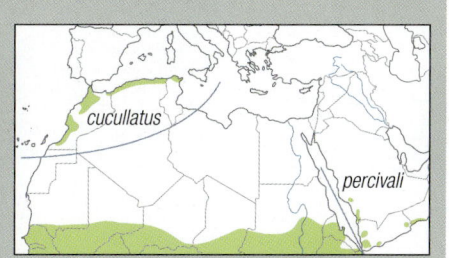

T. s. cucullatus, presumed ad, Morocco, Feb: unmistakable in the treated region once bold white and black stripes on head are seen; rufous wing panel conspicuous, too. Unlike the Arabian race, the N African taxon has black-centred tertials and scapulars. Also, the central tail-feathers are usually finely cross-barred. Ageing often requires closer inspection, but in this case the apparently rather fresh and evenly-feathered wing suggest an ad. (V. & S. Ashby)

IDENTIFICATION In shape and behaviour can superficially recall a Great Spotted Cuckoo, but is of course considerably smaller and differently plumaged. Similarities include a very *long and graduated, white-tipped tail*, at times sneaking or quiet habits low down in thickets and dense trees, or sudden appearance with glides and flapping wingbeats of the rounded wings, and *loud noises*. The tchagra is separated straight away by its rather large head with *prominent cream-white supercilium* between *black crown* and *black eye-stripe*. Most of the *underparts are grey*, with paler throat and centre of belly, whereas the *upperparts are a darker olive grey-brown*. The *upperwing is largely rufous* with some black on the scapulars and tertials (but see below). In flight, the *uppertail appears mainly black with white tip*, but when folded it becomes darker olive grey-brown due to the colour of the central tail-feathers, which then cover most of the tail. The central tail-feathers are usually finely cross-barred with darker lines, especially apparent at some angles. The *bill is black*, long and *quite heavy*. The *strong legs are yellowish-grey*. Sexes and ages alike. – The population on the Arabian Peninsula, *percivali*, differs considerably by being much smaller and shorter-tailed, with a thinner bill and lacks any black on scapulars and tertials, which are uniform with back.

VOCALISATIONS The song of the ♂ is a very loud and far-carrying series of drawn-out whistles (c. 7–12) of altering pitch, often connected via glissando passages, and usually each note has falling pitch, at least at the start. The voice sounds less like a bird, more like a human signal or a car's 'wolf-whistle', but if likened to a bird Tristram's Starling comes to mind. The song is quite variable, even from the same ♂. Sometimes during song, the ♀ joins in with a bubbling but dry tremolo sound, *tr'r'r'r'r'r'r'r'r'r*, which mixes with the louder whistles of the ♂. In duets the pair can also utter a variety of twittering and bubbling calls in excitement. – Calls are variable but includes above all a churring or grating *krrrr*, and an almost Blackbird-like *chok*.

SIMILAR SPECIES Can hardly be confused with any other species in the covered region, since all of its congeners occur south of Sahara. The similarity with *Great Spotted Cuckoo* is limited to general size and shape, and behaviour, but the two are readily separated as soon as the plumage is seen or the calls heard.

AGEING & SEXING (*cucullatus*) Ages very similar, usually possible to separate only in autumn and at closest range, or when handled. Sexes alike. – Moults. Complete post-nuptial moult of ad in summer after breeding (though no actively moulting birds seen). Post-juv moult usually partial and protracted (Jul–Nov), sometimes including some flight-feathers, rarely many (*BWP*, but proof of complete moult in 1stY lacking). Juv apparently often replace r1, or more tail-feathers. Apparently no pre-nuptial moult in late winter, but hardly studied. – **AUTUMN Ad** Mantle uniform brown. Flight-feathers blackish, tail-feathers distinctly tipped rather broadly white. Tail-feathers broad with neatly rounded tips. **1stW** Very similar to ad, but often recognised by retained parts of juv plumage. **Juv** Like ad, but mantle has diffuse buff edges, making upperparts less uniform brown. Flight-feathers browner, pale tips to tail-feathers narrower and buff-tinged, pale tips on rr5–6 separated from blackish rest of feathers by paler grey zone; tail-feathers also narrower and have more pointed, loosely-structured tips.

BIOMETRICS (*cucullatus*) **L** 22–25 cm; **W** 85–98 mm (*n* 29, *m* 92.3; ♀♀ < 93 mm); **T** 103–120 mm (*n* 28, *m* 111.3); **T/W** *m* 120.6; **TG** 24–34 mm (*n* 28, *m* 29.8); **B** 21.3–27.0 mm (*n* 28, *m* 24.7); **B**(f) 18.5–22.6 mm (*n* 28, *m* 21.3); **BD** 7.8–10.0 mm (*n* 26, *m* 9.1); **Ts** 29.6–34.5 mm (*n* 29, *m* 32.1). **Wing formula: p1** > pc 21–29 mm, < p2 16–21.5 mm; **p2** < wt 7–11.5 mm, =10 or 10/ss (45%), ≤ ss (31%), or =9 or 9/10 (24%); **p3** < wt 1–3 mm; **pp4–5**(6) about equal and longest; **p6** < wt 0–1.5 mm; **p7**

T. s. cucullatus, Morocco, Mar: striking head pattern comprising white supercilium bordered by black eye-stripe and median crown-stripe. The very long and graduated, white-tipped tail is also usually obvious. Grey tinge below varies in intensity. Whereas the bill is black, the strong legs are yellowish-grey. (D. Monticelli)

T. s. cucullatus, presumed 1stS, Morocco, Mar: this taxon characteristically shows bold black centres to scapulars and tertials. Also note striking supercilium and crown-stripe. Moult limits in wing and tail, with older and more worn feathers possibly juv, suggesting age. (R. van Rossum)

T. s. percivali, Oman, Mar: this Arabian endemic taxon is considerably smaller and slighter, as well as shorter-tailed, with a thinner bill; note the lack of black centres to scapulars and tertials, blacker uppertail and dusky undertail-coverts with white tips (uniform pale buff-grey in *cucullatus*). Wing-feathers seem to be mostly juv, but without handling difficult to be sure of ageing. (M. Römhild)

< wt 0.5–3.5 mm; **p10** < wt 6.5–11 mm; **s1** 5–12 mm. Emarg. pp3–7, sometimes even p8 faintly.

GEOGRAPHICAL VARIATION & RANGE Polytypic. At least ten races seem to be sufficiently distinct out of many more described, but extralimital taxa not examined in detail or treated here. The Arabian subspecies differs markedly, while other races are more similar.

T. s. cucullatus (Temminck, 1840) (Morocco, N Algeria, Tunisia, NW Libya; resident, but some local movements noted). Treated above. A large and dark race with invariably black crown and black-centred tertials and scapulars. Supercilium cream-yellow or pale ochre-buff unless very bleached. Legs yellow-grey. (Syn. *meinertzhageni*.)

T. s. percivali (Ogilvie-Grant, 1900) (Yemen, Oman, SW Arabia; resident). Quite different, potentially constituting a separate species. Much smaller, with blunter wing and proportionately shorter tail. Markedly thinner bill. No black on scapulars or tertials, which are uniform with back. Dusky undertail-coverts with white tips (*cucullatus*: uniform pale buff-grey). Blacker uppertail. Narrow supercilium, pure white in most (though juv has ochre-buff tinge). Crown not invariably black, in juv/1stW with brown admixed (not seen in *cucullatus*). Darker legs (can be blackish-looking).

T. s. percivali, Oman, Jan: no black centres to scapulars or tertials, and white supercilium is somewhat narrower than in *cucullatus*. Undertail-coverts greyish, typical of this race. Without handling difficult to age. (F. Trabalon)

L 18–21 cm; **W** 73–81 mm (*n* 19, *m* 77.2); **T** 79–103 mm (*n* 18, *m* 86.4); **T/W** *m* 111.5; **TG** 17–25 mm (*n* 18, *m* 20.8); **B** 19.7–23.6 mm (*n* 18, *m* 21.5); **B**(f) 17.1–19.7 mm (*n* 18, *m* 18.5); **BD** 6.2–7.3 mm (*n* 16, *m* 6.8); **Ts** 25.0–27.8 mm (*n* 19, *m* 26.4). **Wing formula: p1** > pc 15.5–24 mm, < p2 13–18 mm; **p2** < ss 2–7 mm; **p7** < wt 0–1 mm; **p10** < wt 2–6.5 mm; **s1** 3.5–8 mm.

TAXONOMIC NOTE Arabian race *percivali* differs rather dramatically from African ones, more than is generally found among races of polytypic species, and could well warrant separation as separate species. However, some differences between this and *cucullatus* are partially bridged by extralimital E African races *habessinica* (Eritrea, Ethiopia, SE Sudan, NW Somalia) and *warsangliensis* (N Somalia). A comprehensive study of genetics, vocalisation and behaviour would cast more light.

T. s. percivali, juv, Oman, Jan: similar to ad, but note overall soft texture typical of juv feathers, although tail is already in moult. Bill still greyish-tinged indicating a young juv. The presence of such fresh juv plumage in Jan indicates that the breeding season in Oman is related to rains. (M. Rouco)

BROWN SHRIKE
Lanius cristatus L., 1758

Fr. – Pie-grièche brune; Ger. – Braunwürger
Sp. – Alcaudón pardo; Swe. – Brun törnskata

An E Asian relative of the European Red-backed Shrike, and also resembles the Turkestan Shrike. Within its vast range, it inhabits similar mainly open habitats as Red-backed Shrike, but in northern parts it is more of a woodland bird, found in the taiga on bogs, at forest edges and in glades. Its main winter range includes India and SE Asia. Does not breed or winter within the range covered by this book, but it has straggled to W Europe many times (plus, e.g., Italy, Middle East, Oman) and needs to be considered when encountering any rufous-tailed or brown-backed shrike with a dark mask.

L. c. cristatus, ♂, Mongolia, May: prominent white supercilia often join on forehead; crown tinged vividly rufous, duller on rest of upperparts usually earth-brown or olive-brown, though rump and uppertail-coverts are again bright rufous. Washed ochraceous-buff on breast and flanks. Jet-black mask and no barring on breast-sides infer sex. Innermost primaries retained, bleached brown, could indicate 1stS, but variation poorly known. (H. Shirihai)

L. c. cristatus, ♂, Mongolia, May: could be confused with same sex Turkestan Shrike due to rather similar head pattern and upperparts. However, nearly always lacks white primary patch, and has stronger ochre-buff tinge on breast and flanks (in Turkestan underparts near-white with more pinkish, less yellow tinge). Also proportionately longer, narrower and more graduated tail, which is darker brown (less bright rufous than Turkestan). Head and bill somewhat heavier. Rather rounder wing with clear emargination on p5 in Brown Shrike. (H. Shirihai)

IDENTIFICATION Size similar to that of Red-backed and Turkestan Shrikes, or a fraction larger. Proportionately *larger head* and on average slightly heavier bill than Turkestan (but not a large difference when Siberian race *cristatus* is singled out, the race most likely to occur in Europe), and somewhat *longer, narrower and more graduated tail*. The legs, too, are subtly longer than in the other two species. Sexes differ slightly (in *cristatus*), ♀ being variably barred below, post-juv ♂ not. *Upperparts basically warm brown*, crown and rump/uppertail-coverts rufous, mantle and back more earth-brown or slightly tinged olive-brown. Forehead in adult ♂ usually paler and grey-tinged. *Uppertail dark brown-grey* with variable *rufous tinge* (quite subdued on some), undertail pale brown-grey, usually tinged rufous. When seen close, *tail-feathers are comparatively narrow*, and often appear *cross-barred above* (darker so-called growth-bars being prominent in many, but less so in 1st-years). *Mask complete and black* in adult ♂, usually less black in ♀ and immature ♂, and lores in these often show some pale elements. *Supercilium white, broad* and prominent. Underparts whitish with quite strong *ochraceous-buff tinge on breast and flanks* when fresh, bleaching more whitish in summer. *Tertials and greater coverts prominently fringed off-white* or rufous-buff. Primary projection rather short, but about same as on Turkestan Shrike, and many are not even that different from some Red-backed Shrikes. As a rule *no pale primary patch* in ♂ (and never in ♀), but exceptionally a small one may appear (indicating 'hybrid blood'?). First-winter is quite similar to both first-winter Red-backed and some young Turkestan Shrikes, being rufous-brown and largely barred dark above; see Similar species for details.

VOCALISATIONS Resembles those of the Red-backed Shrike. A common call has been described as a harsh, loud *shark*, also a more drawn-out *chr-r-r-ri* (BWP), but at least in Siberia call is very similar to Red-backed Shrike, a nasal, harsh-sounding, slowly-repeated *vehv*. Has similar warbling song full of mimicry to Red-backed, often delivered from high, visible song post. In display-flight utters hard, repeated *kriki* and harsher *tshef* calls (Lefranc & Worfolk 1997).

L. c. cristatus, ♂, United Arab Emirates, Apr: some ♂♂ are less bright overall, with especially less rufous on crown. Compared to Turkestan Shrike, no pale primary patch, and primary projection averages shorter (but differences not readily detected). Tail somewhat longer, narrower and more graduated compared to Red-backed and Turkestan Shrikes. Fresh tertials and greater coverts typically fringed off-white or rufous-buff. (M. Barth)

L. c. cristatus, ♀, Mongolia, May: unlike ♂, lores paler and mask less well defined, and ear-coverts brown-black rather than jet-black. Further, supercilium less distinct, sometimes narrower or shorter, and breast and flanks usually have variable amount of fine barring. Also slightly less bright rufous above than ♂. (H. Shirihai)

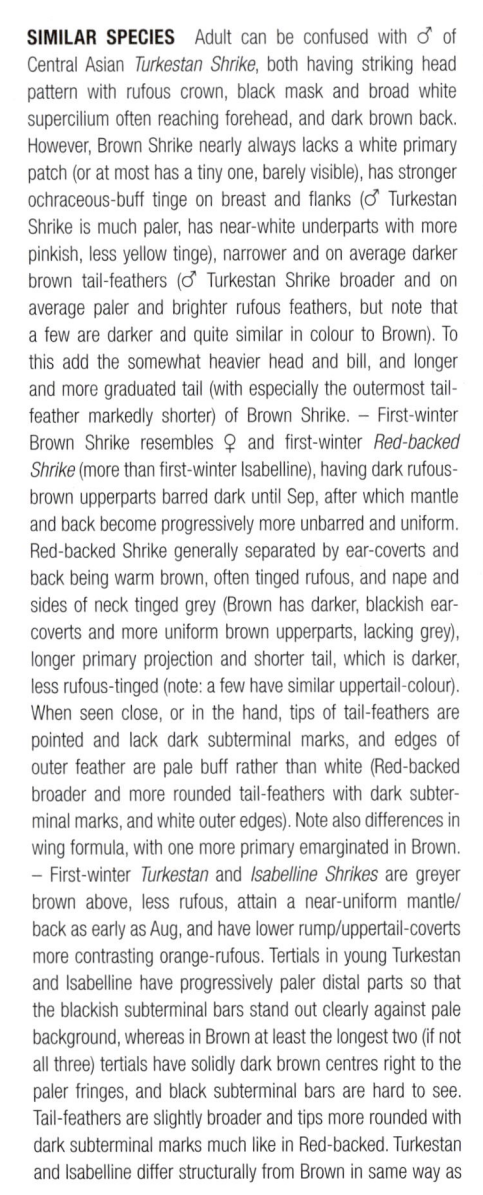

L. c. cristatus, ♀, Mongolia, May: some ♀♀ are less clearly differentiated from ♂, with only partly paler and faintly broken dark lores, and only faint barring on sides. The proportionately longer, narrower and more graduated tail is important versus Red-backed, Turkestan and Isabelline Shrikes. Note, when close, tail-feathers often appear characteristically faintly cross-barred above. (J. Normaja)

L. c. cristatus, ad ♀, Sri Lanka, Nov: young birds and ♀♀ can be surprisingly similar to corresponding plumages of Red-backed Shrike, but note short primary projection, proportionately larger head and stronger bill, and that mask on ear-coverts is sooty-brown or blackish (Red-backed: rufous-tinged). Note narrow tail-feathers and suspended moult of remiges (never occurring in Red-backed). (H. Shirihai)

SIMILAR SPECIES Adult can be confused with ♂ of Central Asian *Turkestan Shrike*, both having striking head pattern with rufous crown, black mask and broad white supercilium often reaching forehead, and dark brown back. However, Brown Shrike nearly always lacks a white primary patch (or at most has a tiny one, barely visible), has stronger ochraceous-buff tinge on breast and flanks (♂ Turkestan Shrike is much paler, has near-white underparts with more pinkish, less yellow tinge), narrower and on average darker brown tail-feathers (♂ Turkestan Shrike broader and on average paler and brighter rufous feathers, but note that a few are darker and quite similar in colour to Brown). To this add the somewhat heavier head and bill, and longer and more graduated tail (with especially the outermost tail-feather markedly shorter) of Brown Shrike. – First-winter Brown Shrike resembles ♀ and first-winter *Red-backed Shrike* (more than first-winter Isabelline), having dark rufous-brown upperparts barred dark until Sep, after which mantle and back become progressively more unbarred and uniform. Red-backed Shrike generally separated by ear-coverts and back being warm brown, often tinged rufous, and nape and sides of neck tinged grey (Brown has darker, blackish ear-coverts and more uniform brown upperparts, lacking grey), longer primary projection and shorter tail, which is darker, less rufous-tinged (note: a few have similar uppertail-colour). When seen close, or in the hand, tips of tail-feathers are pointed and lack dark subterminal marks, and edges of outer feather are pale buff rather than white (Red-backed broader and more rounded tail-feathers with dark subterminal marks, and white outer edges). Note also differences in wing formula, with one more primary emarginated in Brown. – First-winter *Turkestan* and *Isabelline Shrikes* are greyer brown above, less rufous, attain a near-uniform mantle/back as early as Aug, and have lower rump/uppertail-coverts more contrasting orange-rufous. Tertials in young Turkestan and Isabelline have progressively paler distal parts so that the blackish subterminal bars stand out clearly against pale background, whereas in Brown at least the longest two (if not all three) tertials have solidly dark brown centres right to the paler fringes, and black subterminal bars are hard to see. Tail-feathers are slightly broader and tips more rounded with dark subterminal marks much like in Red-backed. Turkestan and Isabelline differ structurally from Brown in same way as Red-backed does (slightly smaller head, finer bill, shorter and less graduated tail with broader tail-feathers than Brown), although primary projection is more similar, all except Red-backed having fairly short projection.

AGEING & SEXING (*cristatus*) Ages separable in 1stCY. Sexes similar but separable. – **Moults**. Moult pattern complex (differing partly also between races). Post-nuptial moult frequently starts on breeding grounds, often involving inner 2–3 primaries, several tertials and tail-feathers, then is suspended and finished on stop-overs or winter grounds in Oct–Nov (Dec). A few moult completely only after reaching the winter area (then often starting with any one of pp5–7, counted from outside), proceeding both strictly descendently and irregularly ascendently from this centre, or simultaneously from p10. Partial post-juv moult in Jul–Sep (Oct) involves head, body, most tertials and some smaller wing-coverts, rarely also r1, to a varying degree before autumn migration. 1stY birds then have a complete, or near-complete (arrested), moult in late winter (Feb–Apr). A second annual complete moult of ad has been proposed (Stresemann & Stresemann 1971) but also questioned and may be based on variation in ad moulting only once but in a wide time span (P. Pyle *in litt*.). Whether spring birds with arrested, incomplete moult of flight-feathers (often inner few pp older) solely represent 1stS remains to be clarified. – **SPRING Ad ♂** Mask invariably jet-black. No barring on sides of breast and flanks. White supercilium broad and prominent, often joining on greyish-white forehead. Crown to upper mantle vivid rufous. Lower mantle/back dark brown, less rufous, more olive-brown than crown. Uppertail-coverts bright rufous. Breast and flanks saturated ochraceous-buff, sometimes even faintly tinged rufous. Exceptionally a hint of white on primary bases. **Ad ♀** Resembles ad ♂. Generally (but not invariably!) lores less than all dark, and ear-coverts brown-

L. c. cristatus, presumed 1stW ♂, United Arab Emirates, Jan: typical young bird with large head, rufous crown, cream-white supercilium and long and apparently narrow tail. Greater coverts pattern and remnants of upperparts barring reveal age, while strong mask and well-marked supercilium in a young bird indicate sex. (I. Boustead)

L. c. cristatus, 1stY, India, Mar: depending on extent of post-juv moult, most 1stW show some vestiges of juv plumage, with crescent-shaped dark bars both above (here mostly visible on crown) and on body-sides, and rather incomplete supercilium. Sexing often difficult, but the poorly marked mask and bold scalloping below suggest ♀. Note heavy bill, relatively long narrow tail and short wings. (S. Heald)

L. c. cristatus, 1stW, Sri Lanka, Nov: depending on short primary projection, but also on posture and viewing angle, tail can appear longer on some birds. In late autumn, most young birds have replaced barred upperparts feathering with plainer brown feathers, but here faint traces of barring still remain on uppertail, while median and greater coverts have irregular buff pattern never seen in ad. (H. Shirihai)

$p7$ < wt 10–15.5 mm; $p10$ < wt 18–25 mm; $s1$ < wt 21–28 mm. Emarg. pp3–5.

GEOGRAPHICAL VARIATION & RANGE Polytypic. Four races described, two of which are quite similar and doubtfully separable unless compared directly (*cristatus* and *confusus*). The other two are distinct. All taxa are extralimital, but *cristatus* has straggled several times to Europe.

L. c. cristatus L., 1758 (Siberia from Ob to Pacific, south to Altai and Kamchatka; winters mainly in India and N Thailand). Treated above. Crown and rump/uppertail-coverts rufous, mantle and back more earth-brown or olive-brown. Has straggled to W Europe several times.

Extralimital races: ○ *L. c. confusus* Stegmann, 1929 (Extreme E Mongolia, Manchuria, SE Siberia in Amur and Ussuriland; winters SE Asia including Thailand, Borneo). Very close to *cristatus*, and of same size, but is subtly paler below and greyer above (some are very similar!). W 83.5–90.5 mm (n 13, m 87.3); T 81–90 mm (n 14, m 84.8); T/W m 96.8.

L. c. superciliosus Latham, 1801 (S Sakhalin, Hokkaido, N & C Honshu; winters Sundas). Small and slim with pointed wing (long pp projection), ♂ being bright rufous also on mantle/back, uniform with crown and rump, and fore-crown pure white, distinctly set off from deep rufous rest of crown. W 89–96.5 mm (n 10, m 91.6); T 83–93 mm (n 10, m 86.9); T/W m 94.8.

L. c. lucionensis L., 1766 (China, Korea; winters SE Asia). About same size as *cristatus*, subtly larger than other two races, ♂ has greyish crown and poorly or at least more diffusely marked supercilium. W 83–93 mm (n 11, m 88.7); T 79–89 mm (n 11, m 84.3); T/W m 95.1.

TAXONOMIC NOTE Judging from external morphology alone, both *lucionensis* and *superciliosus* appear quite distinct, sufficiently so to represent separate species. Future research will show whether this is correct.

REFERENCES Dean, A. R. (1982) *BB*, 75: 395–406. – Hellström, M., Kraft, D. & Svensson, L. (2007) *Vår Fågelv.*, 66 (8): 35–38. – Kryukov, A. P. (1995) *Proc. West. Found. Vert. Zool.*, 6: 22–25. – Lefranc, N. (2007) *Ornithos*, 14: 201–229. – Neufeld, I. A. (1978) *Trudy Zool. Inst. Akad. Nauk SSSR*, 68: 176–227. – Panov, E. N. (2009) *Sandgrouse*, 31: 163–170. – Stegmann, B. (1930) *Orn. Monatsber.*, 38: 115. – Stresemann, E. & Stresemann, V. (1971) *J. f. Orn.*, 112: 373–395. – Worfolk, T. (2000) *DB*, 22: 323–362.

black rather than jet-black. Some fine barring on sides of underparts, though extent variable, in some very little, and exceptionally none (then very difficult to sex). Upperparts on average slightly less brightly rufous than in ♂. Supercilium on average less distinct, sometimes narrower. – AUTUMN **Ad** As in spring. Tertials and greater coverts uniformly brown (fringed pale). **1stW** Late-moulting birds, still wearing much juv plumage, have upperparts mostly rufous with prominent crescent-shaped dark bars, and have poorly-developed pale supercilium and dark mask. More advanced birds moult in Jul–Aug (Sep) to more uniformly brown, unbarred upperparts and slightly more ad-like head pattern, but still smaller tertials and inner greater coverts have hint of dark subterminal bars or spots, and traces of dark barring remain on upperparts. **Juv** Like 1stW, but entire upperparts barred. Ear-coverts sometimes paler, concolorous with crown and rufous-tinged.

BIOMETRICS (*cristatus*) L 17.5–20 cm; W 83–92 mm (n 48, m 87.5); T 76–88 mm (n 41, m 82.9); T/W m 94.5; **TG** 15–25 mm (n 18, m 20.5); **B** 16.9–20.6 mm (n 35, m 18.8); **BD** 7.5–8.7 mm (n 35, m 8.1); **Ts** 24.0–26.5 mm (n 35, m 25.0). **Wing formula: p1** > pc 4–11 mm, < p2 26.5–33 mm, < wt 36–45.5 mm; **p2** < wt 7–13 mm, =6/7 (44%), =6 (29%), =5/6 (20%) or =7 (7%); **pp3–4** about equal and longest; **p5** < wt 1–4 mm; **p6** < wt 6–10 mm;

L. c. cristatus, 1stW, England, Sep: showing characteristic large-headed profile and heavy bill. Note also black rather than rufous-tinged mask, and typically heavy barring below. Upperparts also still barred dark, but as autumn advances upperparts will become plainer. (G. Reszeter)

L. c. cristatus, 1stW, England, Oct: some young moult later and retain extensive crescent-shaped dark bars on upperparts longer, by which they are more readily separated from Turkestan and Isabelline Shrikes. Such strongly-marked birds could be confused with Red-backed Shrike. Best separated by much shorter primary projection of usually up to four visible primary-tips (six in Red-backed), thus tail appears proportionately longer; also tail-feathers narrower. (S. Northwood)

L. c. cristatus, 1stW, Norway, Oct: as autumn progresses, dark barring of upperparts in young birds is gradually lost and replaced by rather plain rufous-brown mantle and back, and hence potentially difficult to separate from same-plumaged Turkestan Shrike, but has fairly short primary projection and almost solidly dark centres to tertials, and also note rather prominently ochre-buff body-sides and heavy bill. (F. Falkenberg)

TURKESTAN SHRIKE
Lanius phoenicuroides (Schalow, 1875)

Alternative name: Red-tailed Shrike

Fr. – Pie-grièche du Turkestan
Ger. – Rotschwantzwürger (Turkestanwürger)
Sp. – Alcaudón colirrojo
Swe. – Turkestantörnskata

The Turkestan Shrike has usually been treated as a subspecies of the Isabelline Shrike, but is regarded here as a separate species on account of distinct ♂ plumage, allopatric range with practically no mixing and substantial genetic distance.

It breeds in the major part of the so-called West (or Russian) Turkestan, which nowadays corresponds to Kazakhstan (except most of the steppe in the north), Uzbekistan, Turkmenistan, N and E Iran and other Central Asian countries north and west of the Tien Shan Range. It is mostly found in dry and warm habitats like deserts and semideserts with some elevated perches, as well as along rivers and on mountain slopes. It winters from C and E Africa to Arabia and S Middle East.

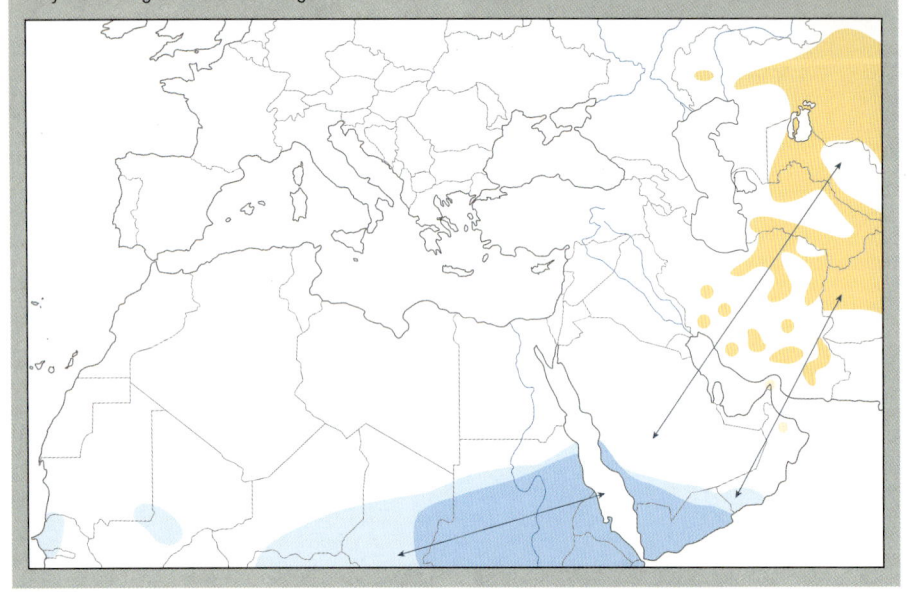

IDENTIFICATION A *brown* and buff-white bird with a *dark mask* and a *rufous-tinged tail*. Size same as or subtly larger than Red-backed Shrike, with proportionately somewhat *longer tail*. ♀ and young of Turkestan Shrike primarily resemble those of Isabelline Shrike, whereas adult ♂ is more similar to ♂ Brown Shrike. For a detailed comparison with Isabelline Shrike, see Similar species. – Adult ♂ Turkestan Shrike has *rufous crown*, complete *black mask* (usually c. 1 mm black over base of bill), and long and *prominent white supercilia* generally meeting on forehead. The supercilium typically is well demarcated from darker crown, not grading diffusely into paler crown. Further, it has *dark wings* (including quite dark wing-coverts) *with pale edges* when fresh, and a *small white primary patch*. Mantle and *back usually rather dark earth brown*, rarely somewhat paler sand-brown (never pinkish-grey). Uppertail is rufous, some being brighter, others darker and duller brown. The *underparts* are rather pale, *whitish* with variable light pinkish-buff tinge on sides and rarely on breast (throat never tinged pink). Sexes differ clearly: ♀ has whitish underparts variably 'scalloped' dark on breast and sides, and the mask is brown-grey rather than black, being faint or lacking in front of eye. Supercilium is rather prominent off-white, but is less distinctly outlined than in ♂, often has some fine barring and frequently blurs into pale lores. Crown variable, can be rather obviously rufous-tinged but also duller brown-grey. Primary patch invariably smaller than in ♂, and frequently absent. Plumage of ♀ variable, some having practically no dark barring below, others being extensively marked. First-winter resembles ♀ but has varying amount of subdued dark barring above, at least on tertials and inner greater coverts. Birds with reduced rufous on crown and brown on mantle (♂♂ easiest to detect), some so plain pale brownish-grey or even ash-grey above, have been named '*karelini*' and are thought to mainly be a colour morph (see Geographical variation and Taxonomic notes).

VOCALISATIONS Seem to resemble those of Red-backed, Isabelline and Brown Shrikes. Alternative song, or 'advertising call', a repeated harsh, loud *zech* or *vehv* from prominent perch. When alarmed, a drawn-out series of harsh notes, *tsche-tsche-tsche-tsche-tsche-…* (BWP). Has similar prolonged warbling sub-song full of mimicry as Red-backed Shrike.

SIMILAR SPECIES In many plumages strongly resembles *Isabelline Shrike*, especially ♀♀ and immatures. Adult ♂ usually differs from Isabelline on (i) more prominent, better-defined and evenly broad pure white supercilium, (ii) on average whiter throat and centre of breast/belly (though some birds are intermediate and tricky), (iii) on average more vivid rufous crown/nape (though Isabelline can have slight rufous tinge at least on crown), (iv) usually darker brown mantle/back which can be rufous-tinged, too, and only rarely or less typically is more greyish-tinged (invariably lacking pinkish flush found in many fresh Isabelline). Supporting details are (v) usually slightly broader black mask in its forepart, often extending narrowly to forehead (but similar pattern

♂, **Oman, Apr**: similar to ♂ Brown Shrike, mainly due to broad complete black mask, prominent white supercilium and rufous-tinged crown, but usually has white patch on primary bases, slightly shorter tail and smaller bill. Note that white supercilia generally meet on forehead. Rather dark earth-brown upperparts, but rump to uppertail contrastingly rufous. Birds that undergo complete winter moult (most!) cannot be aged in spring. (H. & J. Eriksen)

♂, **Kazakhstan, Jun**: a recent split from Isabelline Shrike, but has on average better-defined purer white supercilium, generally more vivid rufous crown and darker and duller brown rest of upperparts, while underparts are whiter. However, the size of a white primary patch is equally variable in the two species, this bird having only slight indication. (A. Audevard)

rarely found in Isabelline), and (vi) on average slightly darker flight-feathers. (Size of white primary patch seems very similar in the two.) Adult ♀ appears to subtly but consistently differ from ♀ Isabelline by slightly whiter underparts (at least throat and centre of body), better-marked dark barring or scalloping below in many, a tendency to have better-marked off-white supercilium behind eye and often darker brown-grey upperparts. Still, a few ♀♀ of the two seem very similar and are perhaps not separable in the field. Immatures are even more difficult to separate but tend to differ in the same way as ♀♀, only both species have pinkish-buff hue to much of underparts, eliminating or obscuring one of the differences between adult ♀♀. – Adult ♂ can be confused with *Brown Shrike*, both having striking head pattern with rufous crown, black mask, prominent broad white supercilium, and dark brown back. However, ♂ Turkestan Shrike has a white primary patch (Brown has none or at most a tiny one, barely visible); much paler, near-white underparts with pinkish-buff tinge on sides (Brown stronger ochraceous- or yellow-buff tinge on breast and flanks, though bleached summer birds are paler); broader tail-feathers that are on average lighter and brighter rufous (Brown has narrower and on average darker brown tail-feathers, in adults with fine dark cross-barring—so-called growth bars—visible at some angles, but a few Turkestan Shrikes have darker tail, quite similar in colour to Brown, although usually not obviously cross-barred). To this add the somewhat smaller head and bill, and shorter and less graduated tail, of Turkestan (outermost tail-feather of Brown generally clearly shorter than rest, with narrower, rather pointed tip). Also, lower mantle and back is more rufous-tinged in Turkestan, but slightly tinged olive-brown in Brown. – First-winter Turkestan Shrike resembles adult ♀ *Red-backed Shrike*, both being dull or warm brown above, with any fine barring of upperparts usually restricted to crown, scapulars and rump/uppertail-coverts. Adult ♀ Red-backed generally separated by ear-coverts and back being warm brown, often tinged rufous, and nape and sides of neck tinged grey (Turkestan has darker brown-grey ear-coverts and more uniform grey-brown upperparts, lacking contrasting grey nape), longer primary projection and shorter tail. The tail of Red-backed is darker, less rufous-tinged, above and paler greyish-white or cream-coloured (less russet-tinged) below. However, a very few have confusingly similar tail-colour, both species being rather similarly rufous-tinged dark brown. In such cases, it is often helpful to note precise pattern of outer tail-feathers, these having rather restricted dark subterminal bars but only very tips paler (Red-backed: more extensive dark subterminal bars and well-marked cream-white edges to distal parts of tail-feathers, especially on outer webs). Also, uppertail-coverts in Turkestan rufous-tinged, but duller brown or greyish in Red-

Presumed Turkestan Shrike, ♂, United Arab Emirates, Mar: a controversial bird showing mixed characters of both Turkestan and Isabelline Shrikes, still thought to be former. Note abnormal strong rufous-pink tinge on flanks and rather narrow white fore supercilium recalling Isabelline. Traits indicating Turkestan are rather well-defined pure white supercilium at rear and black mask extending to forehead (at least more common in Turkestan than in Isabelline). (M. Barth)

♀, Kazakhstan, Jul: unlike ♂, dark mask and off-white supercilium rather faint, upperparts dull brown-grey with crown similar or tinged slightly rufous, and whitish underparts variably 'scalloped' dark on breast and sides. White primary patch smaller than in ♂, frequently absent. These features hardly separate all ♀ Turkestan from ♀ Isabelline, but drabber mantle/back and strong barring on purer white underparts are average differences to note separating from Isabelline. (K. Haataja)

♀, Kazakhstan, May: differs from ♀ Isabelline by whiter underparts (chiefly throat and belly), better-marked dark bars on body-sides, usually has better-marked pale supercilium (difficult to see here) and duller brown-grey upperparts. Longer primary projection and small white primary patch eliminate ♀ Brown Shrike. Drab upperparts contrasting with rufous rump and uppertail separate from Red-backed. (C. Bradshaw)

backed. – Note that Turkestan and Red-backed are known to hybridise locally in Central Asia, further complicating matters. – First-winter Turkestan resembles first-winter *Brown Shrike*, but latter differs structurally from Turkestan in having somewhat larger head, heavier bill and narrower tail, although the difference in size of head and bill can be very slight in relation to ssp. *cristatus*, the race of Brown most likely to occur in W Europe. Brown is generally darker and more rufous above with more dark barring retained longer into first autumn (thereby resembling Red-backed more than Turkestan), and has *solidly dark brown longest two tertials* (if not all three), *dark right to well-marked pale fringes*, whereas young Turkestan has dark centres to all tertials and unmoulted greater coverts, which become progressively paler brown towards tips and edges, so that dark subterminal bars stand out clearly against pale background. – Juvenile is quite similar to juvenile *Bay-backed Shrike* but has less graduated tail with uniformly rufous feathers (Bay-backed: outer tail-feathers shorter and very pale, buff-white), and proportionally slightly stronger bill.

AGEING & SEXING Ages separable in 1stCY. Sexes separable from 1stS. – Moults. Post-nuptial moult on breeding

Ad ♂, Kuwait, Aug: in autumn, ad ♂ is like in spring and summer, showing striking head pattern with very broad black mask, broad white supercilium, dull brown upperparts, slightly rufous-tinged crown, and near-white underparts with hardly any pinkish-buff on sides in this case. (M. Pope)

Ad ♂, Oman, Sep: neat ♂♂ in Sep with no scalloped pattern below and intense black mask best fit ad that moulted completely, here rather early in autumn. Although very white body-sides readily exclude Brown Shrike, large white primary patch is a further confirmation. (H. & J. Eriksen)

Ad ♀, Kazakhstan, Sep: if aged first, then readily sexed. Moult limits in wing, e.g. still unmoulted ad primaries (with white bases) and primary-coverts, in combination with ♀-like plumage (e.g. weak mask and barred flanks) provide sex and age, preventing confusion with rather similar 1stW. Note new tertials and some greater coverts. White primary patch and rather small bill and head eliminate Brown Shrike. (R. Pop)

Presumed Turkestan Shrike, 1stW ♂, Kuwait, Jan: fairly dull above (with limited rufous on crown) and warm buff-brown underparts could invite confusion with Isabelline Shrike, and safe separation of young birds not always possible on current knowledge. However, pale supercilium fairly well-marked, with black mask reaching forehead, and dull greyish-brown mantle and back appear best for Turkestan. Rather intense black mask, only subdued barring of body-sides and small buff-white primary patch infer sex and age. (M. Barth)

grounds in Jul–Sep involves head, body, tertials and several tail-feathers. Complete pre-nuptial moult of ad in winter area (Arabia, Levant, E Africa) in (Nov) Dec–Mar (Apr). (Claims of suspended post-breeding moult involving some remiges, Demongin & Yosef 2009, based on one misidentified Isabelline Shrike.) Partial post-juv moult in Jul–Sep (Oct) variable, involves head, body, most tertials and some smaller wing-coverts, rarely also r1. 1stY moults completely, or nearly completely (can leave odd primaries or secondaries), in winter at same time as ad. – **SPRING** (From Apr?) Due to individual variation in moult, not safe to age birds based on differences in wear of flight-feathers. ♂ Mask (including lores) and bill to base entirely jet-black. No barring on breast or flanks. Usually a small to medium-large white patch on primary bases. ♀ Differs from ♂ by having partly pale lores, and rest of mask dark but not jet-black. Prominent but variable dark barring on sides of throat, breast and flanks, some being heavily barred, also on centre of breast, others being nearly unbarred. Primary patch absent or small. – **AUTUMN Ad** As ♂♀ in spring, though lores of ♂ partly pale like ♀. Tertials and greater coverts uniformly dark brown-grey (fringed pale). **1stW** Late-moulting birds, still mainly in juv plumage, have upperparts covered with prominent crescent-shaped dark bars, and poorly-developed dark mask. More advanced birds moult in Jul–Aug (Sep) to more uniformly brown, unbarred upperparts and slightly more ad-like head pattern, but unmoulted tertials and greater coverts are still pale, rufous-tinged with blackish subterminal bars or spots, and traces of dark barring remain on rest of upperparts. **Juv** Like 1stW, but crown to mantle less rufous, more buff-grey with dark barring. However, rufous on crown attained patchily from late Jun. Barring on underparts less regular and distinct than 1stW, more diffuse and irregular. Juv body- and head-feathers are also more 'woolly' in texture.

BIOMETRICS L 16.5–18 cm. Sexes of very nearly same size: **W** 88–98 mm (n 96, m 93.2); **T** 73–84 mm (n 99, m 78.5); **T/W** m 84.3; **TG** 7–15 mm (n 89, m 10.7); **B** 16.3–19.7 mm (n 96, m 18.1); **BD** 6.5–8.1 mm (n 81, m 7.3); **Ts** 21.7–25.3 mm (n 86, m 23.3). **White on pp** (outer webs) visible outside tips of pc (folded wing) 0–7 mm (more on ♂♂ and ad, but much overlap and any category can have 0 mm, ♀♀ max. 6 mm). **Wing formula: p1** > pc 0–9.5 mm, < p2 34–45 mm, < p3 42–52 mm; **p2** < wt 4.5–10 mm, =5/6 (58%), =6 (26%) or =6/7 (16%); **pp3–4** about equal and longest; **p5** < wt 1–5.5 mm; **p6** < wt 6–12 mm; **p7** < wt 10–16 mm; **p10** < wt 19–27 mm; **s1** < wt 22–29 mm. Emarg. pp3–5; on p5 often fainter (especially in juv/1stW), c. 12–15 mm deep.

GEOGRAPHICAL OR OTHER VARIATION & RANGE Since Turkestan Shrike is treated here as a separate species within the 'Isabelline Shrike superspecies complex', it becomes monotypic. The most obvious variation within the Turkestan Shrike's range refers to upperparts colours (both sexes), these being sometimes more grey than rufous-brown, in ♂♂ sometimes pure grey (var. '*karelini*'; see below and Taxonomic notes). It should be added that there is also a fair amount of less prominent individual variation in other characters.

L. phoenicuroides (Schalow, 1875) (N & E Iran, Afghanistan, Turkmenistan, Uzbekistan, S Kazakhstan; winters S Arabia, N & NE Africa, also less commonly C Africa west to Cameroon and Niger, and apparently also S Iran and east to NW India). Treated above. Entire upperparts brown, crown/nape and rump/uppertail-coverts of ♂ rufous. Mask completely black and prominent, often reaching narrowly to forehead. White supercilium broad and prominent, being well marked towards rufous crown and reaching forehead. ♀ more variable but often, too, has rufous-tinged crown/nape, and prominent but diffuse, off-white supercilium. In both sexes underparts as a rule quite pale, mainly whitish with a moderate pink-buff tinge on sides. Birds with more saturated underparts, but typical head pattern, occur; these are either atypical variants or possibly in a few cases represent intermediates with Isabelline Shrike, or show signs of gene flow from Red-backed. Juv/1stW has a distinct but rather short black subterminal bar or spot, and buff-white extreme tips to all tail-feathers. Uppertail-coverts invariably prominently barred through 1stW. (Syn. *canescens*; *montana*; *romanowi*; *varia*.)

L. phoenicuroides, var. '*karelini*', described from various localities in Central Asia, mainly from the border area between steppe and desert, differs as follows: ♂ has brownish-grey (topotypical birds even pure grey) crown,

1stW, presumed ♂, Oman, Oct: unlike 1stW/♀ Red-backed Shrike has darker brown-grey ear-coverts and more uniform duller grey-brown upperparts, lacking warm brown elements above and often contrasting grey nape of Red-backed, with any fine barring on upperparts usually restricted to crown, scapulars and rump to uppertail-coverts. Most of wing still juv, with juv pattern to primary-coverts and greater coverts including white tips and dark subterminal bars. (D. Occhiato)

1stW, Oman, Nov: immediately told from young Red-backed Shrike by rufous uppertail. Also, compared to Red-backed, outer tail-feathers have rather restricted dark subterminal bars and mostly just small paler tips. Further, black subterminal bars to tertials are narrower and less contrasting. Young shrikes have on average paler bills than ad with often, as here, just tip and culmen darker. (P. Alberti)

1stW, presumed ♀, W India, Oct: can be difficult to separate from Isabelline Shrike, although here duller, more brown-grey upperparts, whiter underparts with better-marked dark barring, and more noticeable off-white supercilium above and behind eye, point at Turkestan. Has smaller head and bill than Brown Shrike, and tertials and unmoulted greater coverts more clearly show dark subterminal bars against pale centres than in Brown. Rather extensive barring on head, breast and body-sides, and lack of white primary patch, indicate young ♀. (S. Singhal)

nape and mantle (but may have a very slight rufous tinge on forecrown). In fresh plumage also on average paler and more sandy-brown on mantle and back than typical *phoenicuroides*. Supercilium white, prominent and long, much as in *phoenicuroides*, but frequently more diffusely outlined in its forepart, this due the frequently pale forecrown. ♀♀ are much more difficult to identify reliably, as some ♀♀ *phoenicuroides* have very limited rufous on crown, too. Also, some that are less whitish below, more tinged cream or buff, strongly resemble *isabellinus*, and their safe separation seems problematic. Tarsus-length seems to be slightly longer than in *phoenicuroides*, approaching *isabellinus* (but as this is quite surprising, it needs confirmation on a larger sample): **Ts** 23.3–25.9 mm (*n* 22, *m* 24.2). Pp3–5 invariably emarginated as in *phoenicuroides*.

TAXONOMIC NOTES Not denying the occurrence of birds that seem intermediate between Turkestan and Isabelline Shrikes, or which at least create identification problems, such birds are few in relation to the large numbers that are typical, at least when ad ♂♂ are examined. The high mountain ridges which form a nearly unbroken border between the two taxa apparently represent effective barriers to unhindered gene flow, and samples of mtDNA show a difference of *c.* 4% (U. Olsson pers. comm.). This is sufficient ground to treat them as different species. In his monograph of Laniidae, Panov (2011) adopted the same taxonomy. – The true taxonomic status of the so-called 'var. *karelini*' remains unresolved. Geographical patterns and some morphological traits seem to support the theory postulated by Panov (e.g. 1983, 1996), followed by Kryukov (1995) and others, that '*karelini*' represents a more or less stable hybrid swarm between *phoenicuroides* and *collurio*. Undoubtedly *L. phoenicuroides* hybridises to a certain extent with *L. collurio* where they meet (e.g., N Iran, north of Aral Sea, Kirghiz Steppe, Zaisan Basin), producing a variety of intermediate plumages, some of which resemble the 'var. *karelini*' originally described by Bogdanov (1881). However, hybridisation does not seem to offer a complete explanation; '*karelini*' apparently has invariably the same wing formula as *phoenicuroides*, with pp3–5 fully emarginated, never intermediate between this and *collurio* (which has only pp3–4 emarginated and a slightly more pointed wing), and plumage colours are not consistent with hybridisation being the sole explanation. Like Stresemann & Stresemann (1972), we therefore regard '*karelini*' as a colour morph of

Var. '*karelini*', ♂, Oman, Apr: unlike typical *phoenicuroides* this colour morph has greyish crown, nape and mantle (although here slightly browner than extreme grey birds); supercilium white, prominent and long, much as normal *phoenicuroides*, but frequently more diffusely outlined in front of eye due to pale forecrown. (H. & J. Eriksen)

phoenicuroides, but accept that hybridisation with *collurio* may produce birds with quite similar appearance.

REFERENCES Bogdanov, M. N. (1881) *The shrikes of the Russian fauna and allies.* [In Russian.] St. Petersburg. – Dean, A. R. (1982) *BB*, 75: 395–406. – Demongin, L. & Yosef, R. (2009) *Sandgrouse*, 31: 160–162. – Kryukov, A. P. (1995) *Proc. West. Found. Vert. Zool.*, 6: 22–25. – Lefranc, N. (2007) *Ornithos*, 14: 201–229. – Lindholm, A. & Dernjatin, P. (2006) *Alula*, 00: 186–188. – Panov, E. N. (2009) *Sandgrouse*, 31: 163–170. – Pearson, D. J. (2000) *BBOC*, 120: 22–27. – Stegmann, B. (1930) *Orn. Monatsber.*, 38: 115. – Stresemann, E. & Stresemann, V. (1972) *J. f. Orn.*, 113: 60–75. – Worfolk, T. (2000) *DB*, 22: 323–362.

Presumed var. '*karelini*', ♀, Bahrain, May: ♀ of this variant is more difficult to identify due to slighter difference than in ♂, probably with some overlap, but on average visibly paler, more greyish-brown above than typical ♀ *phoenicuroides*; also purer whitish below with more limited barring (which could invite confusion with Isabelline Shrike). (A. Drummond Hill)

ISABELLINE SHRIKE
Lanius isabellinus Hemprich & Ehrenberg, 1833

Fr. – Pie-grièche isabelle; Ger. – Isabellwürger
Sp. – Alcaudón isabel; Swe. – Isabellatörnskata

The Isabelline Shrike breeds mainly in Mongolia and China, east of the closely related Turkestan Shrike, a species with which it has often been lumped in the past. However, here these two are regarded as separate species, and at least adult ♂♂ usually differ rather clearly. It occupies similarly dry and warm habitats as the Turkestan Shrike, being commonly found in deserts and semi-deserts, preferably along rivers and on mountain slopes. It is a summer visitor, wintering further south, from E Africa to NW India.

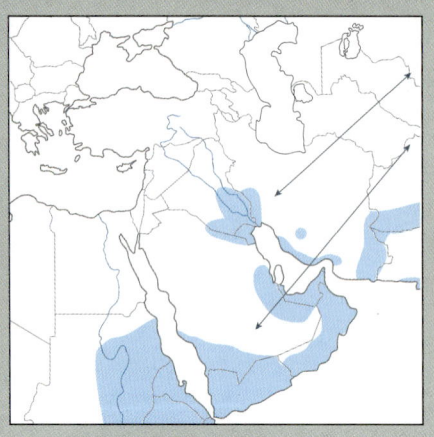

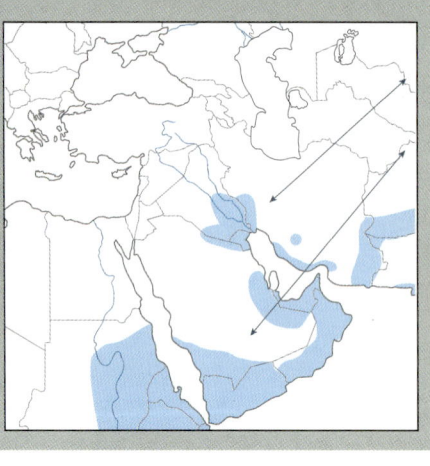

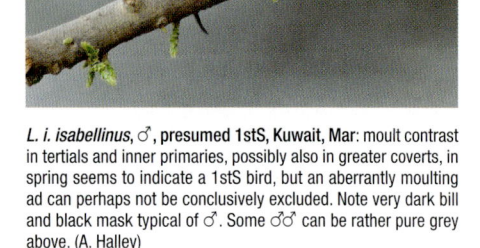

L. i. isabellinus, ♂, Mongolia, Jun: jet-black mask and no trace of bars on body-sides infer sex. Unlike Turkestan Shrike, has more indistinct and often buff- or rufous-tinged supercilium, less or no rufous on crown and nape, and sometimes shorter and narrower mask that either reaches bill or only marginally extends to lores and forehead. Warmer below with ochre-buff tinge, including faintly on throat and central breast (where typical Turkestan is white). Such warmer ♂♂ often have less pure grey upperparts. (H. Shirihai)

IDENTIFICATION Although geographically somewhat variable, this is generally a *greyish-brown and buff-white* bird with a *dark mask* and a *rufous-tinged uppertail*. Size same as, or subtly larger than Red-backed Shrike, with proportionately somewhat *longer tail*. Of the three subspecies, one is thought to occur more regularly within the treated region (*isabellinus*; breeds mainly in Mongolia and adjacent areas, wintering in S Uzbekistan, Turkmenistan, Middle East and Arabia, with a few reaching NE Africa), whereas the other two are believed to be winter visitors only to S Iraq and S Iran outside main wintering area in Pakistan and India (*arenarius* and *tsaidamensis*; breed in easternmost Central Asia and W China). Identification is hampered by a lack of knowledge of plumage variation of other than adult ♂♂ of both Isabelline and Turkestan Shrikes, and incomplete information concerning normal ranges outside breeding. – Adult ♂ *isabellinus* differs from Turkestan Shrike by combination of less rufous-tinged crown/nape (none or just a little, mainly on crown), an *indistinct and often buff- or pink-tinged supercilium* that does not reach forehead (except sometimes as a diffuse flush), and on average narrower black mask in its forepart, only rarely extending above nostrils and thinly onto forehead. Further, it has much *warmer pinkish-buff or ochre-buff tinge over entire underparts*, including faintly on throat and centre of breast (where typical Turkestan is white). Primaries are rather dark, often equally dark as Turkestan Shrike. White primary patch small to medium-large. Bill often all black, but at least some have a slightly paler base to lower mandible (not seen in breeding ♂ Turkestan Shrike). Sexes rather similar, more so than in Turkestan Shrike, but ♀ has at least some faint dark barring on breast, and an incomplete mask, lores being partly pale; blackish bill commonly has paler base to lower mandible. A very few more resemble Turkestan Shrike and are difficult to identify, having a narrow but rather more whitish supercilium and more rufous on crown and nape. – Adult *arenarius* and *tsaidamensis* resemble *isabellinus* but usually differ by lack

L. i. isabellinus, ♂, presumed 1stS, Kuwait, Mar: moult contrast in tertials and inner primaries, possibly also in greater coverts, in spring seems to indicate a 1stS bird, but an aberrantly moulting ad can perhaps not be conclusively excluded. Note very dark bill and black mask typical of ♂. Some ♂♂ can be rather pure grey above. (A. Halley)

L. i. isabellinus, ♂ (top) and ♀, Russia, May: some ♂♂ are quite pale below, with rusty-orange concentrated on flanks, and upperparts rather pure grey. Black mask on this ♂ complete (including lores), a normal and common pattern of this race. ♀ invariably has at least some faint dark barring on breast. Ear-coverts dark grey-brown but lores mostly pale and unmarked. There is a good deal of individual variation in plumage colours and patterns. (E. N. Panov)

of all-dark lores even in breeding ♂ (a few exceptions noted, though, displaying a complete mask), and have on average paler sand-brown upperparts and paler brown wings (tertials especially are paler brown compared to *isabellinus*). Uppertail is on average slightly paler and less vivid rufous. Sexes differ in the same way as in *isabellinus*, only less obviously, and some are difficult to sex. Ssp. *arenarius* has smaller white primary patch, which can be entirely hidden by coverts, and is the smallest of the three. – Note that numerous birds appear intermediate between *arenarius*, *tsaidamensis* and *isabellinus*, being difficult to label subspecifically away from breeding ranges.

VOCALISATIONS Resemble those of Red-backed and Brown Shrikes. Alternative song, or 'advertising call', a repeated harsh, loud *zech* or *vehv* from prominent perch. In alarm a drawn-out series of harsh notes, *tsche-tsche-tsche-tsche-tsche-…* (*BWP*). Has similar prolonged warbling sub-song full of mimicry as Red-backed.

SIMILAR SPECIES Strongly resembles *Turkestan Shrike*, especially in ♀ and immature plumages; see Identification and under Turkestan Shrike for details. – First-winter Isabelline Shrike resembles adult ♀ *Red-backed Shrike*, both being dull or warm brown above, with any fine barring on upperparts usually restricted to crown, scapulars and rump/uppertail-coverts. Adult ♀ Red-backed Shrike generally separated by ear-coverts and back being warm brown, often tinged rufous, and nape and sides of neck tinged grey (Isabelline has darker brown-grey ear-coverts and more uniform grey-brown upperparts, lacking contrasting grey nape), longer primary projection and shorter tail. The tail is darker, less rufous-tinged, above and paler greyish-white or cream-coloured, less russet-tinged, below. However, a very few have confusingly similar tail-colour, both species being rather similarly rufous-tinged. In such cases, often helpful to note precise pattern of outer tail-feathers, these having dark subterminal bars but only very tips paler (Red-backed: more extensive dark subterminal bars and well-marked cream-white edges to distal parts of tail-feathers, especially on outer webs). – Note that Isabelline and Red-backed are known to fairly frequently hybridise locally in Central Asia, further complicating matters. – First-winter Isabelline resembles first-winter Brown Shrike, but latter differs structurally from Isabelline in somewhat larger head, heavier bill and narrower and more graduated tail, although the difference in size of head and bill can be very slight in relation to ssp. *cristatus*, the race of Brown most likely to occur in W Europe. Brown is generally darker and more rufous above with more dark barring retained longer into first autumn (thereby resembling Red-backed more than Isabelline), and has solidly

L. i. isabellinus, ♀, presumed ad, Greece, Mar: a moderately marked ♀ in terms of barring but has quite warm orange-buff body-sides (unlike ♀ Turkestan Shrike, which is usually more heavily barred and whiter below). Being ♀, the rather large white primary patch indicates an ad; all visible wing-feathers, including primary-coverts, are ad-type (1stS often retains some juv feathers). (L. Stavrakis)

♀ Isabelline (top left: Mongolia, Jun), Turkestan (top right: Kuwait, Apr), Red-backed (bottom left: England, May) and Brown Shrikes (bottom right: Russia, May): first note that Isabelline and Turkestan usually show stronger contrast between rusty-red tail and rump and duller mantle, and nearly always have diagnostic white primary patch. Isabelline further separated from Turkestan by sandier (less greyish-brown) upperparts, poorly defined but warmer pale supercilium and warmer underparts with less barring at sides. ♀ Red-backed shows rich brown-tinged rufous upperparts, greyish rump and neck-sides, darker uppertail and heavy barring below, but no white primary patch; it also has the longest primary projection. Typical for ♀ Red-backed is also reddish mask (the others have darker mask). Brown Shrike is rather similar to Red-backed, but note larger head, narrow and deeply graduated tail (though not visible here), short primary projection, and darker and more complete mask. (Top left: H. Shirihai; top right: A. Al-Sirhan; bottom left: R. Ridley; bottom right: M. Hellström)

L. i. isabellinus, 1stS ♂, Israel, Apr: a rather unusual variation with strikingly pinkish-buff face and underparts in contrast to rather pure ash-grey upperparts, a combination never seen in Turkestan Shrike. Unbarred underparts and black mask give sex. Note retained juv (bleached brown) primary-coverts, and pale bill still in Apr, which infer age (J. Meyrav)

L. i. isabellinus, 1stS ♀, Kuwait, Mar: some ♀♀ have rather well-marked rear mask (ear-coverts), but unlike ♂♂ have less pure grey upperparts and invariably some faint barring on breast-sides and flanks. Strong wear to primary-coverts and to many unmoulted inner remiges and resultant moult contrast indicate age. (A. Halley)

dark brown longest two tertials (if not all three), dark right to well-marked pale fringes, whereas young Isabelline has dark centres that become progressively paler towards tips and edges of all tertials and unmoulted greater coverts, thus dark subterminal bars stand out clearly against pale background. – Juvenile quite similar to juvenile *Bay-backed Shrike* but has less graduated tail with uniformly rufous feathers (Bay-backed: outer tail-feathers shorter and pale, buff-white), and proportionately slightly stronger bill.

AGEING & SEXING (*isabellinus*) Ages separable in 1stCY. Sexes separable from 1stS. – Moults. Complex strategy, varying with race or even population. Timing of post-nuptial moult of ad variable, most replace some feathers on body and head plus smaller wing-coverts, some start their complete moult involving odd tail-feathers and remiges on breeding grounds in Jul–Sep, but apparently all finish after autumn migration, the majority moulting completely in midwinter. Partial post-juv moult variable, sometimes beginning in late summer, mainly performed in winter quarters. Many imm retain 2–5 inner juv primaries unmoulted into 1stS. – **SPRING Ad** Invariably has uniformly fresh wing-feathers. Note that birds with all wing-feathers freshly moulted might include a few 1stS that undergo a complete moult. **1stS** Birds with

L. i. isabellinus, ad ♂, United Arab Emirates, Oct: in fresh autumn plumage upperparts usually less pure grey due to fresh buffish tips. Also, unlike in breeding ♂♂, bill is somewhat paler basally. Not always easily separated from Turkestan (sharing white primary patch and contrasting reddish-orange uppertail and rump), but the warm ochre-buff underparts and lack of distinct white supercilium identify. (M. Barth)

L. i. isabellinus, ad ♂, Israel, Sep: an attractive ♂, compared to image above right having purer grey upperparts, but with typically diffuse supercilium and rich tinge of pink-orange on face and underparts. Some early post-nuptial moult has made for an unevenly-feathered wing, with moulted tertials notably contrasting against worn primaries. Bill now starting to become paler. (D. Forsman)

L. i. isabellinus, ad ♂, Sweden, Oct: again an ad ♂ in striking autumn plumage with combination of quite grey upperparts and rusty-pink flanks, breast-sides, supercilium and forecrown. Age based on total absence of barred feathers. (T. Bernhardsson)

L. i. isabellinus, ad ♂, Sweden, Nov: same bird as on previous page (bottom right). When wing fully stretched the white bases to the primaries form a large rounded or arch-shaped patch. Note heavily worn adult primaries, much more abraded at tips than would have been the case with a young bird about three months old. (S. Hage)

L. i. isabellinus, ad ♀, United Arab Emirates, Sep: similar to fresh ad in spring, but in autumn many ad have undergone partial post-nuptial moult; clear moult limits in wing include the still unmoulted primary-coverts, but recently renewed tertials and most greater coverts. Some ♀♀ show (unlike ♀ Turkestan) only very faint barring on flanks and breast-sides, while upperparts are more brownish-buff ('isabelline') than dull grey-brown. (I. Boustead)

pp6–10 more worn and paler brown than outer primaries (especially pp6–8, since pp9–10 are often hidden by tips of longest tertials) are this age. ♂ Mask (including lores) and bill to base entirely jet-black (though base of lower mandible might be slightly paler grey). No barring on breast or flanks. Usually a small white patch on primary bases. ♀ Although sexes differ slightly less on average than in Turkestan Shrike, ♀ differs from ♂ by having lores partly pale, and rest of mask dark but not jet-black. Barred dark prominently but variably on sides of throat, breast and on flanks, some being very lightly barred, in the field appearing unbarred. Primary patch absent or small, but odd birds have a larger one. – AUTUMN **Ad** As ♂♀ in spring, though lores of ♂ partly pale like ♀, and base of bill now pale. Tertials and greater coverts uniformly dark brown-grey (fringed pale). Uppertail usually rather vivid cinnamon-rufous (but some slight variation with some being a little duller brown). **1stW** Late-moulting birds, still mainly in juv plumage, have upperparts covered with prominent crescent-shaped dark bars, and a poorly-developed dark mask. More advanced birds start partial moult in late summer to more uniformly brown, unbarred upperparts and slightly more ad-like head pattern, but unmoulted tertials and greater coverts are still pale, rufous-tinged with blackish

L. i. isabellinus, ad ♀, Scotland, Oct: two images of the same bird. Only ad (worn and fresh) wing-feathers (and pure white primary patch just visible) infer age, while quite substantial barring below but weak face markings confirm sex. Note how angle of view and light conditions alter coloration, the same bird in right-hand image looking greyer above and less warm below, and thus more like Turkestan Shrike. (H. Harrop)

L. i. isabellinus, 1stW, presumed ♂, Kuwait, Dec: young bird following partial pre-nuptial moult, with primary-coverts (some still tipped whitish), inner primaries and outer secondaries juv, but rest of plumage, including tail, tertials, greater coverts, outer primaries and body pre-nuptial (some of it moulted earlier in autumn). The near lack of barring below and size of white primary patch fit 1stY ♂, but sexing of young birds in autumn complicated by some overlap. (R. Al-Hajji)

subterminal bars or spots, and traces of dark barring over rest of upperparts. Near-uniform tawny-brown tail-feathers with usually no or only vestigial dark subterminal spot, and faintly paler tips. Uppertail-coverts rufous, lightly barred, or even (from Sep) unbarred. **Juv** Like 1stW, but upperparts more evenly and prominently barred dark. Barring of underparts less regular and distinct than 1stW, more diffuse and irregular. Juv body- and head-feathers more 'woolly' in texture.

BIOMETRICS (*isabellinus*) **L** 17–18.5 cm; **W** ♂ 90–102 mm (n 56, m 95.7), ♀ 91–100 mm (n 20, m 94.6); **T** ♂ 76–88 mm (n 56, m 82.7), ♀ 77–84 mm (n 20, m 80.2); **T/W** m 85.8; **TG** 7–16 mm (n 68, m 11.1); **B** 15.5–21.5 mm (n 71, m 18.5); **BD** 6.8–8.3 mm (n 67, m 7.6); **Ts** 22.8–26.5 mm (n 68, m 24.5). **White on pp** (outer webs) visible outside tips of pc (folded wing) 0–12 mm (more on ♂♂ and ad, but much overlap and any category can have 0 mm, ♀♀ max. 7 mm). **Wing formula: p1** > pc 1–11 mm, < p2 32.5–43.5 mm, < wt 41–50 mm; **p2** < wt 5.5–12 mm, =5/6 or 6 (74%), or =6/7 or 7 (26%); **pp3–4** about equal and longest; **p5** < wt 1–5 mm; **p6** < wt 7–12 mm; **p7** < wt 11–16 mm; **p10** < wt 20–27 mm; **s1** < wt 22–30.5 mm. Emarg. pp3–5; on p5 often fainter (especially in juv/1stW), *c.* 12–15 mm deep.

GEOGRAPHICAL VARIATION & RANGE Three subspecies, all rather similar with a cline towards increasingly pale wing-feathers and slightly paler and more sandy colours overall running from north to south, and birds becoming larger in the east. Our knowledge of variation is still incomplete, and especially in the north (*isabellinus*) more variation is found than is usual in a single taxon. – Rarely forms mixed pairs with Turkestan Shrike (O. Belyalov *in litt*.; Panov 2009) where these two meet, e.g. at Charyn River, SE Kazakhstan, but probably also elsewhere north to the Altai–Dzungaria. Hybridisation with Red-backed Shrike have been rarely noted in a few overlap areas in Altai and NW Mongolia, such birds add to the difficulties of identifying less-than-typical Isabelline Shrikes.

L. i. isabellinus Hemprich & Ehrenberg, 1833. 'Isabelline Shrike' (sometimes referred to as 'Daurian Shrike') (S Altai, Mongolia, NW China, Transbaikalia; winters S Central Asia, E & S Middle East, Arabia, NE Africa). Has a generally poorly marked pale supercilium (often diffuse, narrow or short and tinged pink-buff, broader and whiter only above and just behind eye), crown variably grey, grey-brown, buffish-grey with a paler forepart or warmer ochre-brown (or even pale rufous), and usually quite saturated pink-buff underparts with only throat slightly paler or whiter. Bill

L. i. isabellinus, 1stW ♀, **Bahrain, Nov**: dark subterminal marks to juv primary- and greater coverts infer age; combination of relatively subdued face pattern, lack of white primary patch, and rather extensive barring on breast-sides and flanks indicate a ♀, apparently a rather extreme example (many are intermediate and thus impossible to sex in 1stW). Correct evaluation of age is often also vital in specific identification. (A. Drummond-Hill)

1stW ♀ Isabelline (top left: Oman, Nov), Turkestan (top right: Kuwait, Oct), Red-backed (bottom left: Germany, Sep) and Brown Shrikes (bottom right: India, Oct): illustrating the challenges facing you when separating young birds. 1stW ♀ Isabelline and Turkestan Shrikes often lack the white primary patch and are therefore less easily told from Red-backed and Brown Shrikes. First, note Isabelline's/Turkestan's paler (less rufous-brown) and plainer upperparts than Red-backed/Brown, with contrasting rusty tail and rump (though beware that some Turkestan, like here, can have surprisingly subdued reddish tinge). Isabelline and Turkestan can be extremely similar, with young ♀ Isabelline hardly differing in its subtly warmer, more buff underparts (at least throat and central body are generally whiter in Turkestan), with on average weaker dark barring below in many Isabelline (though hardly this one!), tendency to have washed-out supercilium (usually whiter behind eye in Turkestan) and often more sandy-brown above (darker brown-grey upperparts in Turkestan). Some young ♀♀ of these two might still have to be left unidentified in the field. Finally, Brown is similar to Red-backed, but has larger head and heavier bill, narrower deeply-graduated tail (here r6 falls distinctly short of tail-tip), and the shortest primary projection. The Red-backed is eliminated further, despite the young age, by its greyish-tinged neck-sides, reddish ear-coverts and pure white tail-sides. (Top left: A. Audevard; top right: R. Al-Hajji; bottom left: I. Boustead; bottom right: N. Devasar)

SHRIKES

♂ Turkestan (left: Kuwait, Sep) or Isabelline (right: Mongolian, ssp. *arenarius*, May) × Red-backed Shrikes: both hybrids have variably distinct black 'T' on tail but rufous instead of white tail-bases. The Turkestan × Red-backed, the commonest hybrid, has Red-backed's grey head and Turkestan's dull grey-brown upperparts. Isabelline × Red-backed hybrids have variably patterned head and body (but compared to Turkestan hybrids, the back is often paler brownish-grey and the face mask more restricted), often being surprisingly close to Isabelline, very rarely showing traces of chestnut mantle of ♂ Red-backed. Hybrids in young or ♀ plumages are much harder to detect, or to assign. The left bird is ad, appearing to show only ad-like wing-feathers. (Left: R. Al Hajji; right: M. Hellström)

size as in Turkestan Shrike, but tarsus very slightly longer. Breeding ad ♂ usually has complete black mask and bill (though a few have slightly paler base to lower mandible), and dark or moderately dark primaries, tertials and greater coverts (on average paler than most Turkestan Shrikes but darker than races *arenarius* and *tsaidamensis*; see below), primaries with small to medium-large white patch. Both lores and bill considerably paler outside breeding season. There is a fair and as yet not entirely understood plumage variation in ♂♂, some being richer rufous-tinged than others, and some develop quite pure grey mantle in worn summer plumage, at times both crown and mantle greyish (though forecrown often paler and tinged ochre-buff), whereas others remain more rufous-tinged. There is also a certain variation

L. i. arenarius, ♂, presumed ad, Iran, Jan: so-called 'Tarim Shrike' is rather distinctive in averaging smaller than ssp. *isabellinus* with a shorter, more rounded wing, and shorter primary projection. Typically peach-tinged buff-brown below and sandy-brown above, with inner wing, especially tertials, averaging paler, too. Black mask often broken at lores or at least narrower than in *isabellinus*. Age suggested by ad-like wing, while lack of white primary patch is diagnostic for ♂ *arenarius*. (E. Winkel)

Presumed *L. i. arenarius*, 1stW ♂, NW India, Jan: no white primary patch, short primary projection (just 3–4 primaries visible beyond tertials) and pale sandy-brown upperparts with fairly pale-centred tertials and greater coverts indicating this race (as does locality); also, uppertail on average slightly duller and less vividly rufous. But interestingly, note unmoulted inner primary just visible outside tertials thought to occur only in 1stY of race *isabellinus*. Sexed on black mask and apparently no barring below. (S. Singhal)

in the extent of the black mask, most having narrow forepart (lores) but a few somewhat broader reaching upper edge of bill-base, and rarely even onto forehead (< 1 mm). Normally, ♀ is lightly barred dark below on sides, but rarely virtually unbarred. – Extreme variations include birds that are less saturated pink-buff below, having slightly rufous-tinged crown and a fairly prominent whitish supercilium, and these resemble Turkestan Shrike. Birds with more greyish crown and mantle (especially when bleached in summer) can resemble Turkestan Shrike 'var. *karelini*', but generally separated by richer pink-buff tinge across whole breast, and less white and more indistinct supercilium. (However, the full range of variation within these two is not well known and needs further study.) More commonly, *isabellinus* difficult to tell from *arenarius* and *tsaidamensis*, and this is especially so for ♀♀ and imm. (Syn. *speculigerus*; see Taxonomic notes.)

L. i. arenarius Blyth, 1846 (Tarim Basin, NW China and adjacent areas to the north and east, apparently also Afghanistan and Kyrgyzstan, possibly Tajikistan; winters S Pakistan and W India, apparently west to S Iran, possibly also NE Arabia). Sometimes referred to as 'Tarim Shrike', it resembles *isabellinus* closely but is fractionally smaller and has slightly shorter, more rounded wing (due to shorter migrations). Underparts invariably pinkish-buff or pale buff-brown ('peach-coloured'), though throat can be slightly paler. Lores in ad spring ♂ generally not all black (though some come very close, and exceptionally might appear, or be, all black), ear-coverts not always black even in ♂♂, primaries on average paler brown than in *isabellinus* (but some overlap!), and a white primary patch is rarer and slightly smaller on average in ♂♂ (0–4 mm > pc; birds with 5–8 mm > pc noted, but might be *tsaidamensis* or intergrades) and often absent in ♀♀. Mantle and back slightly paler sand-brown in most. Tertials and greater coverts have somewhat paler centres and are more uniform with mantle and back (but some overlap with paler *isabellinus*). Uppertail is on average slightly paler or duller and less vivid rufous. These differences noted, still often frustratingly difficult to separate from migrant or wintering *isabellinus*. Sexes very similar, also in size, ♀ is mainly slightly duller than ♂; most show light

L. i. arenarius, ad ♀, Rajasthan, India, Dec: ♀ less clearly differentiated from ♂ in this race, though most show subdued facial pattern with poorly marked mask, and at least some light barring on breast and flanks. Note delicate impression with rounded head and smaller bill of this taxon, short primary projection and paler centres to tertials. Has moulted completely in late summer and shows no traces of immaturity, hence an ad. (J. Holmes)

Presumed *L. i. arenarius*, 1stW, presumed ♂, Rajasthan, India, December: similar to ssp. *isabellinus*, and many not separable without biometrics or away from breeding range. However, the location in winter (NW India), seemingly smaller size, including small bill, short primary projection and no visible prominent white primary patch, suggest ssp. *arenarius*. Remaining subtle traces of barring on upperparts and greater coverts reveal age. (J. Holmes)

barring on breast, which can wear off on breeders. Complete moult of flight-feathers by ad usually on breeding grounds before autumn migration. Young retain their flight-feathers one year. (An Isabelline Shrike in spring with two generations of flight-feathers should therefore in theory *never* be this race.) **L** 16–17.5 cm; **W** 86–97 mm (n 86, m 91.3); **T** 72–86 mm (n 85, m 79.4); **T/W** m 87.0; **B** 15.9–20.7 mm (n 63, m 18.3); **BD** 6.7–8.2 mm (n 63, m 7.3). **Wing formula: p1** > pc 3–12 mm, < p2 30–40 mm, < wt 38.5–46.5 mm; **p2** < wt 5–11 mm, =5/6 or 6 (54%), =6/7 or 7 (37%), or =7/8 or 8 (9%); **p6** < wt 4.5–11 mm; **p7** < wt 9–14.5 mm; **p10** < wt 19–24 mm; **s1** < wt 21–27 mm. Emarg. pp3–5; on p5 invariably distinct, $c.$ 14–18 mm deep.

Extralimital: **L. i. tsaidamensis** Stegmann, 1930 (Qaidam Pendi, Qinghai, NC China; apparently winters NE India west to SE Iran, but details poorly known). What is here called 'Tsaidam Shrike' resembles *arenarius* with usually incomplete black mask in ad ♂ and rather pale wing-feathers and tertials, but is slightly larger and has on average a larger white primary patch, at least as large as *isabellinus*. **L** 19–21 cm; **W** 92–99 mm (n 21, m 96.4; Stegmann 1930). Not studied in detail by us. One specimen in ZMMU had wing 100 mm. (Syn. *major*.)

TAXONOMIC NOTES Due to misidentification of the lectotype, two taxa of the Isabelline Shrike have had their scientific names changed (Pearson 2000). Accordingly, former '*speculigerus*' of Mongolia has become a junior synonym of *isabellinus*, and former '*isabellinus*' of Tarim Basin has changed to *arenarius*. The grounds for this change were questioned by Panov (2009), but repeated and expanded by Pearson et al. (2012). It should be noted that the *isabellinus* lectotype (MNB 1887, examined) untypically has narrow black over base of bill ($c.$ 1 mm), a feature more commonly found in Turkestan and Red-backed Shrikes (or hybrids between any of these two) but which can certainly also occur rarely in *isabellinus* of N & NE Mongolia; furthermore, in all other respects, the lectotype fits well with *isabellinus*, and two paralectotype ♂♂ in ZMB, collected on the same date and locality, are typical *isabellinus* in every respect. – Due to similarity between *isabellinus*, *arenarius* and *tsaidamensis* in imm and winter plumages, the exact range and winter grounds of these taxa is still inadequately known. Also, plumage variation within ssp. *isabellinus* appears to be surprisingly large and still inadequately understood and could indicate that a further taxon merits separation. – Ssp. *tsaidamensis* was provisionally treated as a synonym of *isabellinus* by Worfolk (2000) based on (limited?) field experience from Qaidam, but is stated by all others to be a larger version of *arenarius* with larger white primary patch (E. Panov *in litt.*), and this is corroborated by series of both in ZISP.

Presumed *L. i. arenarius*, 1stW, presumed ♀, NW India, Dec: location and impression of smaller and slighter build, including narrow bill, suggest *arenarius*. Because young *arenarius* retain their flight-feathers for one year, this bird's juv primary-coverts and seemingly only one generation of flight-feathers offer further support for the identification. Rather strong barring below suggests ♀. (S. Singhal)

REFERENCES Bogdanov, M. N. (1881) *The shrikes of the Russian fauna and allies*. [In Russian.] St. Petersburg. – Hellström, M. (2007) *Vår Fågelv.*, 66 (8): 30–34. – Dean, A. R. (1982) *BB*, 75: 395–406. – Kryukov, A. P. (1995) *Proc. West. Found. Vert. Zool.*, 6: 22–25. – van der Laan, J. (2008) *DB*, 30: 78–92. – Lefranc, N. (2007) *Ornithos*, 14: 201–229. – Panov, E. N. (2009) *Sandgrouse*, 31: 163–170. – Pearson, D. J. (2000) *BBOC*, 120: 22–27. – Pearson, D. J. et al. (2012) *BBOC*, 132: 270–276. – Stegmann, B. (1930) *Orn. Monatsber.*, 38: 115. – Stresemann, E. & Stresemann, V. (1972) *J. f. Orn.*, 113: 60–75. – Worfolk, T. (2000) *DB*, 22: 323–362.

RED-BACKED SHRIKE
Lanius collurio L., 1758

Fr. – Pie-grièche écorcheur; Ger. – Neuntöter
Sp. – Alcaudón dorsirrojo; Swe. – Törnskata

This small and neat shrike is widespread over much of Europe, although it has steadily declined in the north and north-west, and now no longer breeds in Britain. It is still common in SE Europe and can be seen there in considerable numbers from May to August in appropriate open habitats with thorny bushes and scattered trees. In early autumn it migrates over Greece and W Turkey, across the Mediterranean and through Egypt to its African winter quarters, which lie south of the equator, mainly in the south and east. On spring migration, returning birds take on a more easterly route and pass largely through Arabia and the Levant before returning to Europe.

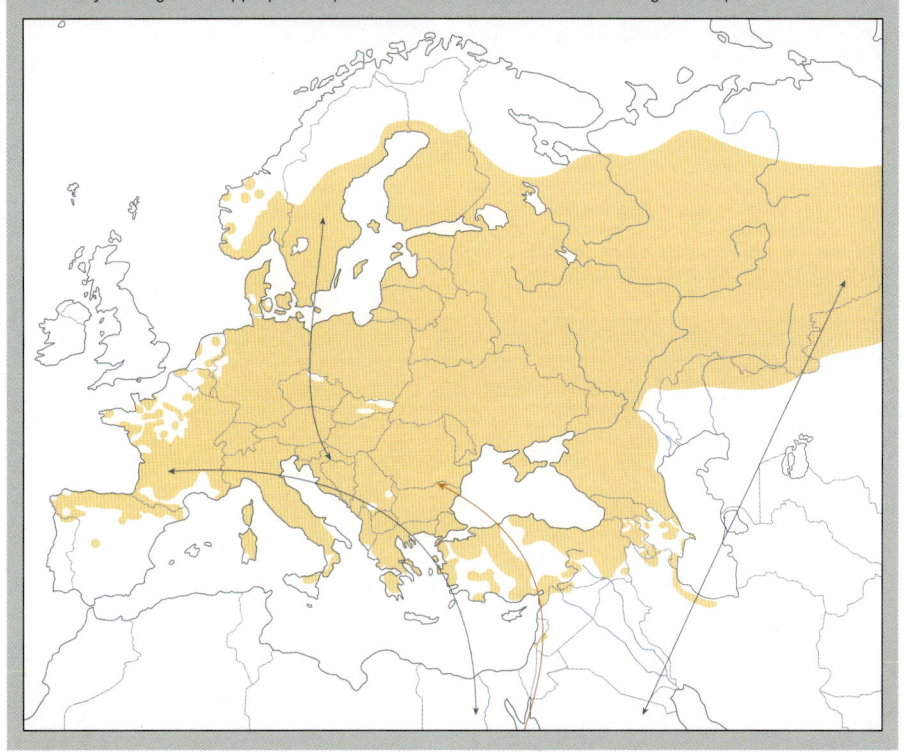

♂, presumed 1stS, Turkey, May: since all ages moult completely in winter, spring birds cannot be aged by moult pattern. However, the tertials of this bird have slight dark subterminal bars (almost intermediate between juv and ad feathers), and flanks are untypically barred that may infer immaturity. The black-and-white tail has an 'inverted T' pattern when spread. (D. Occhiato)

♂, Italy, Jun: this rather small, slim shrike, typically encountered on prominent perch, frequently twitching tail, is unmistakable in ♂ plumage. Note black mask, ash-grey crown and nape, contrasting with rufous back and salmon-pink breast, plus whitish throat and black-and-white tail. Undergoes complete winter moult, thus in spring the species cannot be aged. (L. Sebastiani)

♂, Greece, May: note black-and-white tail pattern from below. Black mask often reaches forehead above bill, and some birds show a vague white supercilium and forehead. Note on this bird presence of small white primary patch, although this is variable and in many birds and parts of the range it is lacking, while a few have a larger patch. (R. Pop)

IDENTIFICATION Behaviour as other shrikes, using prominent perch to scan for prey, often twitching tail. Drops to ground for food, or flies low, sweeping up to new perch. Longer flights slightly undulating, still low over ground. A small shrike with moderately long tail. In ♂ plumage easily told by *salmon-pink breast*, whitish throat, black mask, *ash-grey crown and nape*, and uppertail-coverts, *rufous back and black-and-white tail*. (The tail-pattern is similar to that of Red-breasted Flycatcher and many wheatears, with white sides basally on an otherwise black tail, forming an 'inverted T'.) The black mask sometimes reaches onto forehead above bill by 1–2 mm. Other variations in ♂ plumage affect the amount of rufous above, a few birds having less, with ash-grey invading much of mantle and scapulars. Some have a quite pale grey, or even white, forehead. Another variable character in adult ♂ is the presence of a small white primary patch or not. Generally, there is no visible patch, but some have a little white hidden. However, there is a percentage of birds with visible white outside primary-coverts, occurring more often in eastern parts of the range (although British, French and Finnish, too, are reported to rather commonly have a small white primary patch; J. Normaja *in litt.*). The ♀ requires more care as it resembles several other species, being brownish above and whitish below with crescent-shaped dark barring, especially on breast and flanks, but often also lightly on upperparts. Note that *ear-coverts and back* are basically *warm earth-brown or dull rufous-brown*, *uppertail dark brown-grey* and *undertail greyish* (generally lacking rufous tinge, but a few exceptions). Crown normally brown (rarely grey), whereas brown of nape, sides of neck and rump/uppertail-coverts is frequently obviously tinged grey. Supercilium diffusely paler, pale brown, cream-coloured or off-white. *Primary projection rather long*, and no hint of a light primary patch in ♀. There is quite some variation in ♀ plumage as to amount of barring and rufous above. Very rarely, ♀ attains a surprisingly ♂-like plumage (see Ageing & Sexing). Hybridises locally in Central Asia with Turkestan Shrike *L. phoenicuroides* and in Altai with Isabelline Shrike producing birds with intermediate characters.

VOCALISATIONS Most commonly heard is the alternative song ('advertising call') of the ♂, a fairly loud and far-carrying, repeated, nasal but harsh, monosyllabic *vehv… vehv…* The same call seems to be used also in contact or when mildly alarmed in the territory. Another peculiar sound with apparent song function is a repeated *ku-ehtch!*, surprisingly similar to take-off call by Common Snipe. When a ♀ has been attracted, the 'real' song is heard (albeit not frequently), a prolonged, rather subdued warbling, containing whistling notes, harsh clicking sounds and clever mimicry of other birds. Gifted singer, can even approach Marsh Warbler as to vigour and quality of imitations. – Normal call is the above-mentioned harsh, nasal *vehv*. A rapid, repeated *ke-vehv* is uttered in display-flights and territorial disputes. Full alarm is signalled by series of clicking but slightly harsh *tshek tshek tshek…*

SIMILAR SPECIES ♂ Red-backed separated from all other shrikes in region by combination of light grey head and deep rufous back and upperwing, still somewhat resembles *Bay-backed Shrike*, but clearly differs in having no or only very small white primary patch on brown-grey wing (Bay-backed has a large white patch on otherwise black wing). Also, black on forehead missing or not more than *c.* 2–3 mm (Bay-backed extensively black, reaching forecrown), and entire breast is salmon-pink (white with rufous-buff sides in Bay-backed). – ♀ and first-winter Red-backed very similar to same categories of both *Turkestan* and *Brown Shrikes*, unless close observation has established longer primary projection and lack of emargination on fifth primary in Red-backed. ♀ Red-backed separated by combination of (i) warm earth-brown or rufous-tinged upperparts, especially greater coverts, scapulars and back, but dark brown uppertail (Turkestan has more contrasting rufous-orange uppertail-coverts and usually, but not invariably, rufous-tinged uppertail); (ii) often not as dark rufous centres to longest two tertials so that black subterminal bars are clearly visible inside paler edges (same pattern found in young Turkestan, but differs from young Brown, which has darker brown centres to tertials making black subterminal bars difficult to discern and cream-white edges stand out well); (iii) ear-coverts brown, tinged rufous, being darker grey-black only in rare ♂-like plumage (ear-coverts of Turkestan and Brown typically darker, more blackish than warm brown); (iv) pale supercilium diffuse and tinged cream (whiter and more prominent in Turkestan); and (v) usually quite prominent dark barring on breast and flanks (fainter, more restricted barring in majority of Turkestan Shrikes). – First-winter Red-backed has upperparts generally rufous (in many hardly any grey) and is prominently barred from crown to uppertail-coverts throughout autumn (Turkestan, and some Brown, usually replace most of the barred juv feathers with more uniform upperparts feathers in Aug–Sep, thus attaining a more adult appearance earlier than Red-backed). Uppertail on average darker brown-grey with less rufous tinge than in Turkestan and Brown, and undertail (in few cases when seen in the field) less tinged orange-brown, being more cream-white, but a few are quite similar in tail

♀, **Greece, Apr**: very different from ♂, being brownish above with crescent-shaped dark barring on whitish underparts, especially on breast and flanks, plus weaker face pattern, with relatively indistinct rusty-brown ear-coverts. Cream-coloured lores may extend as indistinct pale supercilium. Variable grey tinge on crown, nape and neck-sides (here rather limited) and on rump and uppertail-coverts. Long primary projection. (R. Pop)

♀, **Iran, Jun**: some ♀♀ attain more ♂-like plumage, especially the amount of grey on crown and nape, and rufous on mantle and scapulars, but never acquire a clearly-defined black mask or unmarked underparts of ♂♂. (E. Winkel)

♀, **Turkey, May**: much individual variation in spring, with some ♀♀ having strongly-barred scapulars, and even dark subterminal lines on tertials apparently without this being linked to younger age. Note rufous mask and white tail-sides, typical of Red-backed. (D. Occhiato)

Ad ♂, **Ethiopia, Sep**: post-nuptial moult in late summer/early autumn has limited effect on ad ♂ plumage; most notable is the abraded remiges, which are not moulted until the birds reach winter quarters. (H. Shirihai)

colour, and caution is advocated when using it. For tricky birds important to note exact pattern of outer tail-feathers, these having dark subterminal bars and well-marked cream-white edges extensively on tips/distal parts, especially outer webs (Turkestan: more restricted dark subterminal bars and only tips paler). Further, primary projection slightly longer than in Turkestan and Brown, although difference small and not always easy to use. – Note that Red-backed hybridises locally in Central Asia with Turkestan Shrike, and in Altai with Isabelline Shrike, producing all kinds of mixtures. A common hybrid ♂ resembles Red-backed but has rufous tail with blackish T-pattern, others have reduced rufous on mantle and back. Hybrid ♀♀ frequently very difficult to separate from either parent. – From juvenile *Woodchat Shrike* by lack of obvious paler cream-white shoulder patches or of paler rump, and lack of broad pale tips to outer tail-feathers.

AGEING & SEXING Ages separable in 1stCY. Sexes differ in 2ndCY (at least from Feb) and thereafter. – Moults. Partial post-nuptial moult in some ad involves parts of head, body, some coverts and odd tertials in Jul–Aug. Any growing primaries or tail-feathers on breeding grounds are replacements for accidentally lost feathers rather than moult. Other ad moult no feathers before autumn migration. Complete moult (including flight-feathers) of ad in winter quarters (late Nov–Apr). Partial post-juv moult in Jul–Aug involves parts of head, body and some smaller coverts. Complete moult in winter quarters, as in ad. – **SPRING** ♂ Mask jet-black, loral part variably ending short of forehead or expanding onto it by 1–2 mm. Crown to upper mantle ash-grey, forecrown sometimes paler. Lower mantle/back, and much of upperwing, rufous. Uppertail-coverts ash-grey. Breast and belly unbarred salmon-pink. No white primary patch, or just a small one. Tail black with large white portions basally either side. (Ageing apparently impossible.) ♀ Ear-coverts warm brown; rarely darker, blackish-brown. Crown variable, generally dull brown but sometimes grey. Back and much of upperwing dull rufous-brown, not as vivid rufous as in ♂. Barring of upperparts variable, sometimes almost as extensive as in 1stW. Uppertail-coverts particularly variable, rufous with dark subterminal bars, unmarked dull grey-brown or rather

SHRIKES

Ad ♀, Israel, Sep: due to wear, ♀ can have purer grey on head and nape; note worn primary-tips. Wear has also by now removed nearly all barring on upperparts, if any existed in fresher spring plumage; this bird has only some faint marks on uppertail-coverts left. (L. Kislev)

Ad ♀, Kuwait, Aug: facial pattern, especially dark mask and grey crown and rump, can also be enhanced by feather wear, sufficient to slightly mirror ♂ plumage (but note heavily barred flanks). Beware that some Red-backed have rather rufous-tinged uppertail as here. (M. Pope)

pure grey, or a mixture of these. Tail dark brown-grey, slightly tinged rufous basally in some; outer tail-feathers narrowly edged and tipped buff-white. (Very rarely attains 'advanced', ♂-like plumage with crown unbarred ash-grey, ear-coverts near-black, back and upperwing saturated rufous, and tail brownish-black with rather extensive white basally on outer tail-feathers; however underparts invariably cream-white with some dark barring on sides.) Whether ♀♀ with many barred feathers on upperparts and retained, heavily-worn tertials with dark subterminal bars (juv pattern) are invariably 1stS is unknown. — AUTUMN **Ad** As in spring, but plumage abraded. Sometimes a few tertials moulted on breeding grounds, new ones being darker brown and fresher than unmoulted. **1stW** Upperparts mostly rufous with prominent crescent-shaped dark bars. Tertials have rather variable pattern but invariably white or buffish tips and blackish subterminal bars or spots. Sometimes also a dark line along shafts. **Juv** Like 1stW, but crown to mantle not as rufous, more buff-grey with dark barring. Barring of underparts not quite as regular and distinct as in 1stW, more diffuse and irregular. Juv body-feathers and head-feathers are also more 'woolly' in texture.

BIOMETRICS L 16–18 cm; **W** 89–100 mm (♂♂ nearly always >91 mm, ♀ < 99 mm; n 115, m 94.1); **T** 70–82 mm (n 115, m 76.1); **T/W** 76.4–85.4 (n 115, m 80.9); **TG** 5–14 mm (n 36, m 9.0); **B** 15.4–19.3 mm (n 58,

1stW, Israel, Sep: rather like ad ♀, but rufous upperparts have more prominent crescent-shaped dark bars from crown to uppertail-coverts, and tertials have pale tips and blackish subterminal bars. Further identification clues (versus Turkestan, Isabelline and Brown Shrikes) are long primary projection, dark brown (rather than vividly rufous) uppertail, white tail-sides, rufous-brown ear-coverts, diffuse pale supercilium, and moderately large head and bill. (A. Ben Dov)

1stW, Israel, Oct: some juv/1stW during autumn also superficially resemble same age Woodchat Shrike, but lack the diagnostic cream-white rump, the large pale primary patch (rufous-buff or whitish) and broad white tips to outer tail-feathers. Most 1stW Red-backed lack any hint of pale shoulder, or have as here pale bases to the lower scapulars forming vague patches, but never as contrasting as in young Woodchat. (A. Ben Dov)

m 17.4); **BD** 6.6–8.1 mm (n 35, m 7.3); **Ts** 21.7–24.5 mm (n 58, m 23.4). **White on pp** (outer webs) visible outside tips of pc (folded wing) 0–4 mm (white visible only in few %). **Black on forehead** ♂ 0–2 mm. **Wing formula: p1** > pc 0–6.5 mm (rarely 0.5 mm <), < p2 41–50 mm; **p2** < wt 2.5–7 mm, =4/5 (80%), =5 (11%), =5/6 (6%) or =4 (3%); **pp3–4** longest (p4 rarely 0.5–3 <); **p5** < wt 5–9 mm; **p6** < wt 11–14 mm; **p7** < wt 14.5–18 mm; **p10** < wt 24–30 mm; **s1** 27.5–35 mm. Emarg. pp3–4.

GEOGRAPHICAL VARIATION & RANGE Monotypic. — Europe from Iberia and Scandinavia east to SW Siberia, N Kazakhstan, south to Turkey, Caucasus, N Iran; winters mainly S Africa, in E Africa north to Kenya. — Six races have been described, but several authors (e.g. Lefranc & Worfolk 1997, and several references therein) have failed to confirm their required level of distinctness, and we, too, have found extensive variation within each proposed race, as well as much overlap in claimed separating characters, and prefer to lump all under *collurio*. (Syn. *fasciatus*; *jourdani*; *juxtus*; *kobylini*; *pallidifrons*.)

Juv, Spain, Aug: similar to 1stW, but rufous-brown upperparts often subtly paler, with dark barring above and below more diffuse and irregular than in 1stW; body- and head-feathers fluffy and softer textured. Note yellow gape still present. (M. Estébanez Ruiz)

BAY-BACKED SHRIKE
Lanius vittatus Valenciennes, 1826

Fr. – Pie-grièche à bandeau; Ger. – Rotschulterwürger
Sp. – Alcaudón dorsicastaño; Swe. – Indisk törnskata

This is a small and attractive-looking, long-tailed shrike with a primarily Indian distribution. Breeders in Central Asia (SE Iran, Afghanistan and SE Turkmenistan) are summer visitors, wintering in India and Pakistan, with a single recent breeding record from E Arabia, in N Oman. It is of about the same size and shape as the Masked Shrike, but its plumage is more akin to the Red-backed Shrike, though all but juveniles have a large white primary patch, a striking feature especially in flight.

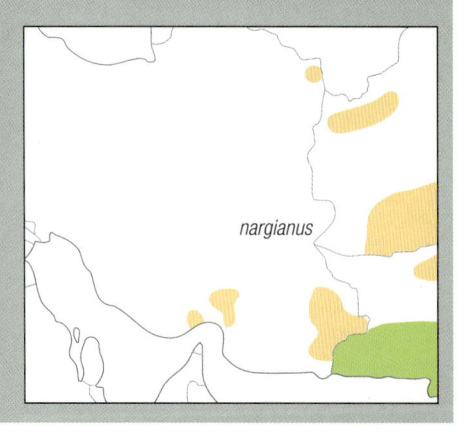

nargianus

IDENTIFICATION A 'medium-small' shrike that might be mistaken for a ♂ Red-backed Shrike at first glance, but has a *large white primary patch* on its otherwise *black wings*, and *much black on forehead*. Back and lower mantle *rufous*. Upper mantle and *nape ash-grey*, becoming pale *pearl-grey on crown*, next to the black forehead. *Rump grey-white*. Tail black with large white sides basally, rather like ♂ Red-backed Shrike. Underparts whitish with variable rufous-buff tinge on sides. Sexes similar, but ♀ is duller with less deep rufous on upperparts, and wings and tail are dark grey-brown rather than black. Dark mask of ♀ is blackish-grey, often with some paler feathers admixed. Juvenile is barred above and oddly has a rufous tail instead of the black-and-white of the adult, and its brown wing lacks any white wing patch (a few have a buffish hint). It needs to be separated from juveniles of other smaller shrikes (see below).

VOCALISATIONS Much like other shrikes, the vocalisation consists mainly of a variety of harsh simple calls and a more subdued, prolonged warbling song. Some calls, harsh but nasal in tone, serve as territorial signals and are repeated in series of twos or threes, e.g. *chew-chew-chew* or *keechew-keechew* (Harris & Franklin 2000). The warbling song consists of whistling notes, harsh sounds and mimicry, like in Red-backed Shrike. Alarm is a fast chattering rattle, *krrrrr…* and variants.

SIMILAR SPECIES Immediately separated from *Red-backed Shrike*, which it somewhat recalls, by very large white primary patch (Red-backed may have a tiny one at most), shorter wings and longer tail. – Main risk of confusion is with considerably larger *Long-tailed Shrike*, because the rather substantial size difference between these two is not always easy to appreciate on a single bird. Plumage-wise they are quite similar, especially ♂ Long-tailed Shrike ssp. *erythronotus* and ♀ Bay-backed. Notice less extensive black on forehead in Long-tailed; distinct buff-white or white edges to tertials (Bay-backed all dark); diffusely rufous-grey outertail (Bay-backed neatly black and white); and in flight or when seen from rear rufous rump/uppertail-coverts (Bay-backed greyish-white). Long-tailed has also, unsurprisingly, a proportionately longer tail. – Could theoretically appear like a *hybrid* between Red-backed and Lesser Grey Shrikes, but luckily such birds are exceedingly rare, and would perhaps show longer primary projection and deeper bill than Bay-backed. – Juvenile is quite similar to juv *Isabelline* and *Turkestan Shrikes*, being barred above with dark rufous uppertail, but has longer and more graduated tail, with much shorter and paler outer

L. v. nargianus, ad ♂, United Arab Emirates, Apr: superficially resembles Red-backed Shrike, but has a large white primary patch, blacker wings (tertials lack rufous fringes), broader black forehead and broad black eye-stripe extending to upper cheeks and ear-coverts, and pale ash-grey crown reaches further onto mantle. In this race, rufous lower mantle and scapulars less deep. Evenly fresh with only ad wing-feathers, and solid black areas sex it as ♂. (P. Arras)

L. v. vittatus, ad ♂, NW India, Jan: this extralimital race is similar to *nargianus* but has deeper, neater colours, especially the chestnut mantle and scapulars. There is also some resemblance to Long-tailed Shrike; however, Bay-backed is markedly smaller, with much more white in wing and tail, has a grey rump, and grey-white uppertail-coverts. Fresh ad plumage after complete post-nuptial moult. (N. Devasar)

L. v. nargianus, 1stS ♂, United Arab Emirates, Jun: worn 1stS of this race have paler grey cap, diluted orange-rufous mantle and buff-tinged flanks and are thus more similar to Long-tailed Shrike, especially ssp. *erythronotus*. Nevertheless, Bay-backed is noticeably smaller, with much more white in wing (in ad type remiges) and tail, and grey-white (not rufous) uppertail-coverts. Aged by juv primary-coverts and some other innerwing feathers. (M. Barth)

L. v. nargianus, 1stS ♂, Oman, Apr: unmoulted juv remiges, with a much-reduced diagnostic large white primary patch of ad. Unlike Long-tailed Shrike, note greyish-white (not rufous) rump and uppertail-coverts. (H. & J. Eriksen)

feathers, extensively buffish-white on tip and outer web, and has a weaker bill. (A bird that has lost its short, cream-white outer tail-feathers is even more similar to juvenile Isabelline, Turkestan or Red-backed Shrikes, but note long tail-feathers in combination with vivid rufous colour and extensive black subterminal marks on tips. Rump is usually barred pale greyish-white, paler than back and uppertail-coverts.) Practices long, elegant tail-dips not seen in the other species (Campbell et al. 2011). – It is easier to separate from juvenile Woodchat Shrike of local race niloticus by its rufous-tinged uppertail (Woodchat blackish with white sides and tip) and lack of primary patch (Woodchat has a large white patch on blackish wing).

AGEING & SEXING (nargianus) Ages and sexes differ subtly, ages in 1stY. – Moults. Complete post-nuptial moult (including replacement of flight-feathers) of ad generally after breeding but before autumn migration (late Jun–Sep). Apparently no pre-nuptial moult in ad. Partial post-juv moult surprisingly late and highly protracted (presumably due to long breeding season), judging from specimens of both nargianus and vittatus, Oct–Dec, rarely Sep–Feb, and very variable, usually involving tertials, secondary-coverts and much of body, also outer 6–9 primaries (rarely all), all or some inner secondaries, and all or most tail-feathers; head and body-feathers attain ad-type pattern. Some retain part of, or all, juv greater coverts, tertials and some outer tail-feathers. – SPRING **Ad ♂** Forehead and ear-coverts black. Black on forehead often reaches onto forecrown to level of eye. Crown very pale grey ('pearl-grey'). Nape to upper mantle slightly darker ash-grey. Lower mantle/back deep rufous. Rump/uppertail-coverts grey-white. **Ad ♀** As ad ♂, but black on forehead frequently reduced or mixed with grey, and back duller rufous-brown, not as vivid rufous as in ♂. Crown not as pale pearl-grey, more dirty grey. Wings and tail not jet-black, more dark grey-brown. **1stS** As ad, but usually has a few retained and heavily-abraded juv inner primaries, all or some outer secondaries, and generally all primary-coverts. – AUTUMN **Ad** As in spring, but plumage freshly moulted. **1stW** In early autumn, grows dark wing-feathers with large pale primary patch (though sometimes sullied buff, less distinct). Body plumage at first barred like juv, but progressively attains ad feathers, and largely ad appearance common from winter. **Juv** Completely different from later plumages, rather like juv Red-backed, Isabelline or Turkestan Shrikes although shorter-winged and longer-tailed, being grey-brown with darker barring above and diffuse scalloping on breast and flanks. Ground colour of crown and mantle pale grey-brown, rest somewhat darker and more sandy-brown, with some rufous on shoulders, greater coverts and tertials. Ear-coverts darker brown. Central tail-feathers dark brown tinged rufous, outer buff with broad near-white tips and outer edges. No or only lightly suggested white primary patch (on outer webs only).

BIOMETRICS (nargianus) **L** 16.5–18 cm; **W** 84–91 mm (n 27, m 87.5); **T** 82–93 mm (n 26, m 88.1); **T/W** 96.7–106.9 (n 22, m 101.2); **B** 16.5–18.3 mm (n 26, m 17.4); **BD** 6.2–7.0 mm (n 24, m 6.4); **Ts** 21.3–23.4 mm (n 22, m 22.6). **White on pp** (outer webs) visible outside tips of pc (folded wing) ♂ 12.5–21 mm, ♀ 9–14 mm. **Black on forehead** ♂ 7.5–12 mm (m 9.8), ♀ 5–7.5 mm (m 6.3). **Wing formula: p1** > pc 9–18 mm, < p2 28–33 mm; **p2** < wt 6–11.5 mm; **pp3–5** about equal and longest (or either or both of p3 and p5 0.5–1.5 mm <); **p6** < wt 3.5–7 mm; **p7** < wt 9–14.5 mm; **p10** < wt 17–22 mm; **s1** < wt 18.5–24 mm. Emarg. pp3–6, on p6 sometimes slightly less distinct.

GEOGRAPHICAL VARIATION & RANGE Two subspecies, differing in plumage and slightly in size. Only one appears within treated region.

L. v. nargianus Vaurie, 1955 (SE Iran, SE Turkmenistan, Afghanistan, Pakistan, with one recent breeding record from N Oman and sightings in United Arab Emirates; short-range movements in winter supposed to occur,

L. v. nargianus, 1stW ♀, NW India, Sep: ♀ in transition from juv to 1stW, facial mask only hinted; also, cap to mantle still predominantly grey, with border between grey upper mantle and rufous-cinnamon lower mantle diffuse. Large areas of wing and whole tail still retained juv. Sex inferred by dull colours lacking any deep chestnut or black. (S. Tiwari)

L. v. vittatus, 1stW ♂ (left: C India, Nov) and 1stW ♀ (right: C India, Sep): unmoulted juv remiges lack the diagnostic white primary patch of ad, and, when still fresh, black areas can have some pale tips. Young ♀ tends to lack the dark mask in first-year, or it is broken-up and indistinct. Also, rufous-brown back not as deep chestnut as in ♂. White primary patch lacking or only partly developed (see right bird) due to late or incomplete moult of juv primaries. (Left: S. Damle; right: N. Sant)

presumably into S Pakistan and NW India). Described above.

Extralimital: **L. v. vittatus** Valenciennes, 1826 (India, Nepal, Bengal; mainly resident). Similar but has deeper, neater colours, and sexes tend to be more similar, being frequently difficult to separate.

REFERENCES CAMPBELL, O., HARE, W. & MILIUS, N. (2011) Sandgrouse, 33: 7–11.

L. v. nargianus, juv, United Arab Emirates, Jun: like juv of other shrikes, with buff-brown to pale buff upperparts, pale buff to white underparts, closely barred black upperparts, body-sides and breast, rufous-brown tail, tertials and upperwing-coverts with buff and black marks on tips, but wings have white bases to primaries very limited and hardly visible on folded wing, or lacking entirely. Note long tail and short wings. (M. Barth)

LONG-TAILED SHRIKE
Lanius schach L., 1758

Fr. – Pie-grièche schach; Ger. – Schachwürger
Sp. – Alcaudón schach
Swe. – Rostgumpad törnskata

A rather large and long-tailed shrike with a primarily E Asian distribution. Breeders in Central Asia are summer visitors, wintering mainly in India, whereas birds in Pakistan, India and China generally are more or less resident. It does not currently breed within the treated region (formerly did so in NE Iran) but has straggled to Arabia and the Middle East, and rarely to Turkey and Europe. A bird of open habitats, warm-climate plains with cultivation or scattered bushes and trees, vegetated montane slopes, often associated with irrigation and plantations. It is also often found in parks and gardens.

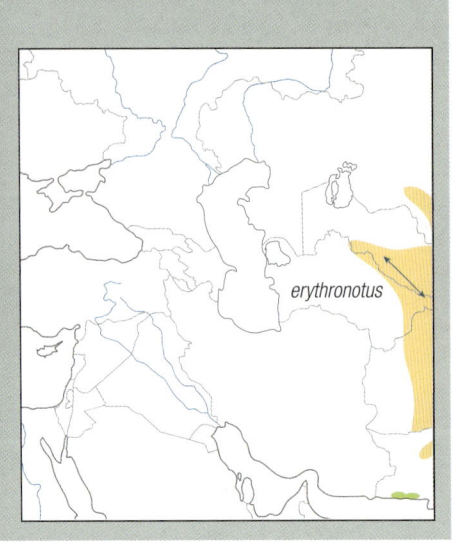

L. s. erythronotus, ad, presumed ♀, India, Jul: only ad (fresh or worn) wing-feathers. Presumed ♀ by rather dull plumage with rufous-buff less deep, dark areas of wings and tail tinged browner, mask narrower, smaller and more greyish-black, and rather dull grey crown and upper mantle, grading into ochre-rufous back. (A. Singh)

L. s. erythronotus, ad ♂, NW India, Feb: a large S Asian shrike with long and very slim but strongly graduated tail, with orange-buff to rufous scapulars, rump, flanks and vent. Small white primary patch often invisible on folded wings, while individual variation can include ad ♂♂ without patch, as here. Jet-black mask, reaching quite broadly above bill, and pure ash-grey areas but also quite deep rufous tones infer ♂; aged by only ad wing-feathers. (S. Singhal)

IDENTIFICATION Perhaps most significantly, lower back, most of *shoulders, rump, flanks and vent* are *rufous-buff* to deep orange-rufous. In flight, it resembles a (colourful) Great Grey Shrike due to its *long and strongly graduated tail*, and *rounded wings*. The tail is black with diffusely rufous-grey tip and sides, and the wings are dark with a *small white primary patch*, tertials broadly edged pale (rufous or white). Black mask runs slightly onto forehead (less in some ♀♀, and not in young birds). Crown and upper mantle grey, grading into ochre-rufous back. There is often a hint of a narrow whitish supercilium, and the forecrown can be slightly paler grey-white, between darker grey rest of crown and black forehead. Sexes similar, but ♀ is generally a little duller with less deep rufous on rump, and the wings and tail are dark grey-brown rather than blackish. Mask of ♀ is blackish-grey, not as pitch black as ♂. The juvenile develops rather adult-like upperparts early but has scalloped underparts, pale lores and forehead, and juvenile tail being rufous-tinged brown-grey with indistinct barring above.

VOCALISATIONS Much like other shrikes, consists mainly of a variety of harsh calls and a more subdued, prolonged warbling song. A repeated *kscha* or a disyllabic harsh *ger-lek* serves as territorial call ('alternative song'). Alarm is a harsh *kerrr kerrr kerrr...* Like Red-backed Shrike, the warbling song comprises whistling notes, harsh sounds and mimicry. It can be delivered in quite long sequences, lasting several minutes.

SIMILAR SPECIES Similarity with much smaller *Bay-backed Shrike* treated under that species. – Larger and longer-tailed than ♂ *Red-backed Shrike*, and has broadly pale-fringed tertials, rufous-tinged (not pink) flanks and lacks extensive white sides basally to tail. – Rufous-buff flanks, shoulders and rump separate it from any *Great Grey* or *Northern Shrike*. – The juvenile is rather extensively barred dark above and below, thereby resembling juveniles of *Brown, Red-backed, Isabelline* and *Bay-backed Shrikes* but is separated by larger size, longer tail and rufous tinge on scapulars, vent and lower back to uppertail-coverts. The mask is generally darker and better-developed than in the others. – Resembles also extralimital Grey-backed Shrike (*L. tephronotus*, E Himalaya, W China, wintering in India; not treated), which has shorter and largely rufous-tinged tail, generally lacks white primary patch, has less or no rufous tinge on scapulars and less on vent/undertail-coverts, as well as combination of dark grey upperparts (without any rufous away from rump) and moderate black on forehead.

AGEING & SEXING (*erythronotus*) Ages differ clearly in 1stY. Sexes differ subtly. – Moults. Complete post-nuptial moult (including replacement of flight-feathers) of ad after breeding but before autumn migration (mainly Jul–Aug). Partial post-juv moult coincides with this, but is more protracted and quite variable in extent, involving head and body, many wing-coverts, and sometimes a few tertials. Partial pre-nuptial moult occurs in late winter, during which

L. s. erythronotus, ad ♂, NW India, Feb: note pale-edged tertials; here white primary patch is mid-sized, but small in comparison to Bay-backed Shrike (of same age). Also diffusely pale (tinged rufous-grey) tip and edges to outer tail-feather, which fall well short of tail-tip. Rufous scapulars a feature of this westernmost race. Evenly fresh wing infers age, and deep black wing-bend and mask the sex. (S. Singhal)

SHRIKES

L. s. erythronotus, 1stS, presumed ♂, NW India, Feb: ageing usually not difficult and, with increased feather wear, moult limits with brown and abraded juv wing-feathers highly contrasting versus newly moulted black median and innermost greater coverts. However, sexing not always possible, especially with 1stY birds, although very black new wing-coverts and central tail-feathers indicate ♂, as does large white primary patch already in juv plumage. (S. Singhal)

L. s. erythronotus, ad ♂, India, Oct: following complete post-nuptial moult wing is evenly fresh and very black. The black wing and deep black mask also infer sex. Here an extreme example of a bird with rather large white primary patch and rufous edges to outer tail-feathers. (S. Sharma)

usually any barred juv feathers including tail-feathers are replaced, and 1stS will return appearing much like ad but apparently invariably has all primary-coverts, primaries and most secondaries unmoulted; of these unmoulted feathers, primary-coverts are brownest and easiest to use for ageing, strongly contrasting against new blackish greater coverts, whereas primaries are more variable in colour. – **SPRING Ad** ♂ Forehead and ear-coverts black, the black reaching 2.5–4 mm above bill. Crown and mantle usually quite pure ash-grey. Lower back/rump strongly tinged rufous. **Ad** ♀ As ad ♂, and some difficult to separate, but black mask duller (dark grey or blackish, not jet-black), black on forehead frequently reduced or even absent, and rest of plumage on average slightly duller. **1stS** As ad, but usually has retained juv primary-coverts, which are brown, contrasting with much darker greater coverts, and often slightly browner juv primaries, plus all or most secondaries. – **AUTUMN Ad** As in spring, but plumage freshly moulted. Tail-feathers have rounded tip. **1stW** A varying number of upperparts body-feathers barred dark, being juv, but from Aug these are usually replaced by uniform ad-type feathers, though some outer greater coverts and generally tertials still have dark subterminal bars. Moult of tail-feathers apparently variable in timing, some keeping barred juv tail-feathers until midwinter. New black central tail-feathers narrow and often quite pointed. **Juv** Entire plumage rather dusky and largely barred, also below. Vent, flanks and uppertail-coverts tinged ochraceous-rufous. Mask dark but duller, not blackish.

BIOMETRICS (*erythronotus*) L 21–23 cm; W ♂ 92–100 mm (*n* 22, *m* 95.4), ♀ 89–98 mm (*n* 16, *m* 93.0); T ♂ 105–116 mm (*n* 20, *m* 111.8), ♀ 100–112 mm (*n* 16, *m* 107.1); **T/W** *m* 116.5; **TG** 32–44 mm; **B** 18.3–22.5 mm (*n* 36, *m* 20.1); **BD** ♂ 8.3–9.2 mm (*n* 17, *m* 8.8), ♀ 7.6–8.8 mm (*n* 9, *m* 8.4); **Ts** 26.2–29.5 mm (*n* 37, *m* 27.8). **White on pp** (outer webs) visible outside tips of pc (folded wing) 0–7 mm. **Black on forehead** ad ♂ 3–6.5 mm, ad ♀ 0–4 mm. **Wing formula: p1** > pc 12–20 mm, < p2 24–29 mm; **p2** < wt 9.5–13 mm; **pp3–5** about equal and longest (or either or both of p3 and p5 0.5–2 mm <); **p6** < wt 2–5.5 mm; **p7** < wt 6.5–10 mm; **p10** < wt 17–21 mm; **s1** < wt 19–23 mm. Emarg. pp3–6, although usually much less distinct on p6.

GEOGRAPHICAL VARIATION & RANGE Several subspecies described. Due to its large range in Central and SE Asia, involving many islands and varying climates and habitats, morphological differences are not surprisingly fairly well marked, and even include a melanistic form. Two or three main racial groups can be discerned, but due to extensive hybridisation in areas of contact, geographical variation is complex and definition of such groups best avoided. Breeders in north-west and north-east have black mask and grey crown, those in-between and to the south have extensive black on crown, or even all-black head. Only one subspecies, *erythronotus*, has occurred within the treated region.

L. s. erythronotus (Vigors, 1831) (SE Turkmenistan, Afghanistan, Uzbekistan, S Kazakhstan, Pakistan, NW India; in winter short-range movements to south of range). Described above. Grey of crown/nape/mantle medium dark with faint brown hue. Scapulars and lower back strongly tinged rufous. Rather extensively rufous on flanks.

Extralimital races: *L. s. caniceps* Blyth, 1846 (W & S India, Sri Lanka; resident). Very similar to *erythronotus* but differs in slightly paler and purer grey upperparts, no or only very slight trace of rufous on scapulars, these being nearly all grey, and less extensive rufous on flanks.

L. s. schach L., 1758 (China; northern birds move south in winter, others resident). This too resembles *erythronotus* but differs by larger size, heavier bill, larger white primary patch, more extensive black on forehead often reaching forecrown, deeper chestnut on scapulars, lower back, rump, vent and flanks, and being slightly darker grey on crown/nape/mantle.

L. s. erythronotus, juv/1stW ♂, NW India, Oct: this plumage shows varying number of dark-barred juv feathers above and on body-sides, and much of wing is also still juv. Yellow gape still obvious. Rather pure ash-grey head and back and rather solid black mask infer sex. Strongly graduated tail, and rather deep orange-buff vent and uppertail-coverts important for identification. (S. Singhal)

L. s. erythronotus, 1stW ♂, NW India, Aug: some 1stW may lose many of the barred juv feathers, even by Aug. Usually, however, at least some 1stW greater coverts and tertials with dark subterminal lines still present. Blackish central tail-feathers reflect post-juv renewal. Large white primary patch on otherwise quite black wing indicates sex, and broad black mask and pure grey crown support this. (M. Mathur)

L. s. erythronotus, 1stW ♀, India, Dec: even in winter, some young ♀♀ still readily aged and sexed by more juv-like and extensively barred plumage, with strongly reduced white primary patch or, as here, having only a faint indication of one. Note still juv tail, appearing rusty-tinged and finely barred. (B. van den Boogaard)

LESSER GREY SHRIKE
Lanius minor J. F. Gmelin, 1788

Fr. – Pie-grièche à poitrine rose
Ger. – Schwarzstirnwürger
Sp. – Alcaudón chico; Swe. – Svartpannad törnskata

A medium-sized shrike with grey, black and white colours like one of the large species, e.g. Great Grey and Iberian Grey Shrikes. The main range of this warmth-dependent bird lies in Turkey and Central Asia, while it is absent or rare and local in W Europe, and only reasonably common in SE Europe. Often found in open low-altitude areas with scattered trees and bushes, orchards and lines of poplars, etc. Frequently perches on telephone wires along roads. Winters in S Africa, from where it returns to Europe mainly in April and early May.

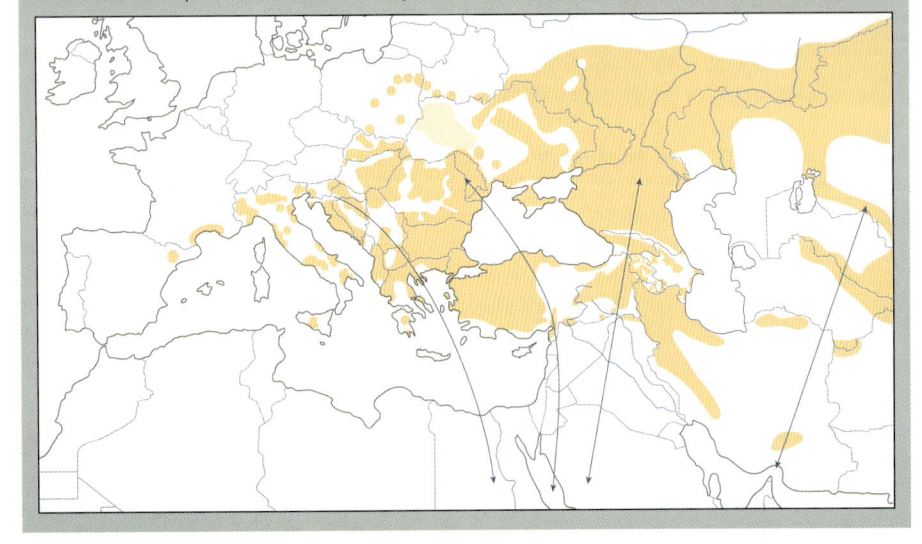

IDENTIFICATION Although its colours resemble those of the three larger *Lanius* species, being basically grey, white and black, one of the first things that strikes you in spring is the *salmon-pink breast* with a faint dirty-grey cast, contrasting with *pure white throat*. Adults in spring have a *black mask* also covering forehead and *reaching onto forecrown* (most prominently in ♂), but note that young birds have no black at all on forehead, and adults in autumn have a varying number of grey feathers mixed with the black (amount of grey poor indicator of sex). A *large white primary patch* is visible on the long and *pointed black wings*. Proportionately rather *large-headed* and *short-tailed*, the head often appearing rounded with peaked, or 'full', crown. The black tail has much white on sides, most basally. The bill is deep but short with rounded mandibles in profile, and is rather 'compressed' at the sides so that the culmen is sharply edged. Posture when perched is usually rather upright. Flight fast and direct, typically without the marked undulations of Great Grey Shrike (although this depends on distance and situation). Juvenile lacks black on forehead (but in late summer this provides limited help with ageing as adults by then can have started to moult and have more grey than black forehead), has entire crown grey finely barred darker (adults being more plain grey). Note also that mantle and scapulars in juveniles are barred, too, and greater coverts have white edges and tips forming a hint of a wing-bar.

VOCALISATIONS Rather quiet except when breeding. Alternative song ('advertising call') peculiar, a sharp, noisy *chillip!*, most recalling a call of a parakeet or budgerigar, given at intervals from high song post. This call is doubled in display-flight, *chillip-chillip!* (The chorus of calls from a small colony in a grove may sound like entering a pet shop!) The 'real' song is more subdued and continuous, a low, squeaky or rasping warble, quite like the sub-songs of Red-backed

♂, Greece, Apr: a fairly large shrike, superficially similar to the members of the Great Grey Shrike complex, being mainly grey, black, and white, and having conspicuous white primary patch, but has characteristic broad black mask that covers whole forecrown, and black rather than grey upperwing-coverts. In breeding plumage underparts washed pinkish. Sexes similar, but mask of ♂ jet-black, reaching well onto forecrown, and ♂ has on average deeper salmon-pink breast than ♀ (but some overlap). (R. Pop)

♂, Kuwait, May: black upperwing-coverts and conspicuous white primary patch (not extending to secondaries). Also square tail with broad feathers and an extensively white lateral base, while rump is not appreciably paler than back. Forehead, lores and ear-coverts jet-black, reaching well up on forecrown suggesting ♂. Ageing usually impossible in spring as both age classes undertake complete pre-nuptial moult. (R. Al-Hajji)

and Brown Shrikes (and questionable whether these three can be separated). – Anxiety and alarm calls are a series of hoarse, short notes, *che-che-che-che-...*, remarkably like Magpie.

SIMILAR SPECIES Superficially like *Great Grey* and *Iberian Grey Shrikes*, but generally smaller, with shorter tail and proportionally larger and more rounded head. Pink breast can invite confusion with *Iberian Grey Shrike*, but this has distinct narrow white supercilium reaching forehead, much darker grey upperparts and no black on forehead. Important to note long primary projection of Lesser Grey, equalling or usually exceeding tertial length (1/2 to 3/4 of tertial length in the larger species), and stout but short bill (more attenuated with more pronounced hook in the larger species). Juvenile Lesser Grey is largely unbarred below (has only fine barring on breast-sides and flanks) but is barred above (in general the opposite in northern populations of juv Great Grey, whereas southern have no or very little barring at all). Juvenile can also be similar to juvenile Great Grey of ssp. *pallidirostris* due to large white primary patch, but all-dark secondaries on folded wing. Note shorter bill with less developed hook, less strong feet, more black on lesser coverts near wing-bend, darker bill already by late summer and lack of any hint of paler scapular patch. – Juvenile *Masked Shrike* bears some resemblance to juvenile Lesser Grey, but is smaller with smaller head, smaller bill, has slightly longer and narrower tail, and a hint of the white shoulder patches typical of adults, although barred grey.

AGEING & SEXING Ages differ clearly in autumn, sexes usually in spring (sometimes all year). – Moults. Partial post-nuptial moult in late summer–autumn involves varying number of body-feathers, secondary-coverts and odd tertials. Complete pre-nuptial moult of ad (replacing flight-feathers) in winter quarters (Dec–Apr). Partial post-juv moult (Jul–Sep), variable in extent, mainly involving body and head. Complete pre-nuptial moult of 1stY in Africa as in ad (*BWP*). – **SPRING** Ageing impossible, unless odd brown and abraded juv secondaries retained (extremely rare). ♂ Forehead, lores and ear-coverts jet-black, black on forehead reaching well onto forecrown, 10.5–17 mm from base of bill (at feathering). On average deeper salmon-pink on breast than ♀ (but some overlap). ♀ As ♂, but black of forehead, lores and ear-coverts less extensive, reaching 6.5–11.5 mm (once 13) from bill, and being not so deep jet-black, more tinged dark brown-grey; sometimes black on forehead is admixed with paler grey spots or patches. On average less saturated pink hue on breast than ♂ (but some overlap). – **AUTUMN Ad** Much as in spring, but plumage abraded and

♀, **Iran, May**: unlike similar members of the Great Grey Shrike complex has rather long pointed wings and longer primary projection, a relatively shorter and less rounded tail, and a shorter, deeper, rather 'compressed' bill. Sexes similar, but black forehead of ♀ on average narrower, often partly mixed brown or grey, lores and ear-coverts tinged brown, and vinous-pink of underparts usually less deep, but in some ♀♀ (like that here) the pink wash can be substantial. (C. N. G. Bocos)

Ad ♂, **Egypt, Sep**: due to feather wear in summer, ad in autumn show virtually no pink below nor any pale fringes and tips to wing-feathers, thus wings look more solidly dark. Some ad keep black on forehead (albeit slightly broken by grey winter feathers) into early autumn, and are thus still possible to sex by width of black above bill (here sufficiently wide to suggest ♂). (E. Winkel)

Ad, presumed ♂ (left: Kuwait, Aug), and **presumed ♀** (right: England, Jul): the possible ♂ at left has already lost much of black forehead, but did moult its tertials (now very black tipped whitish). The right-hand bird has lost all of its black forehead already in Jul, and this together with brownish-grey rather than black ear-coverts and primaries suggest ♀. (Left: M. Pope; right: S. J. M. Gantlett)

black forehead admixed with varying number of grey feathers in both sexes (Jul–Mar). A very few retain all-black forehead. Nape and mantle unbarred, being pure pale grey. Greater coverts uniform blackish. Reliable sexing can be based only on how far onto forecrown black feathers reach (see Spring; still, some overlap, and many now impossible to sex due to extensive number of grey feathers). **1stY** Upperparts grey slightly tinged brownish, no black on forehead, crown, mantle and shoulders finely barred dark as long as juv feathers retained. From Sep often unbarred plain grey above. Greater coverts blackish, finely edged and tipped white. Sides of breast and flanks variably very finely barred grey. Sexes alike.

BIOMETRICS **L** 19–21.5 cm; **W** ♂ 113.5–126 mm (*n* 32, *m* 118.7), ♀ 113–120 mm (*n* 17, *m* 117.1); **T** ♂ 84–97 mm (*n* 33, *m* 89.1), ♀ 83–92 mm (*n* 17, *m* 87.0); **T/W** *m* 74.9; **B** 17.1–21.5 mm (*n* 51, *m* 19.7); **BD** 8.5–10.1 mm (*n* 51, *m* 9.5); **Ts** 23.0–25.6 mm (*n* 49, *m* 24.2). **White on pp** visible outside tips of pc (folded wing) 11–19 mm. **Wing formula: p1** < pc 0–4.5 or > pc 0–5 mm, < p2 57.5–70 mm; **p2** < wt 1–6 mm, about equally often =3, =3/4, =4 and =4/5; **p3** longest; **p4** < wt 3–6 mm; **p5** < wt 10–15 mm; **p6** < wt 14–22 mm; **p7** < wt 20–27 mm; **p10** < wt 36–44 mm; **s1** < wt 39–49 mm. Emarg. pp3–4, though p4 less distinctly in some.

GEOGRAPHICAL VARIATION & RANGE Monotypic. – Much of S, C & E Europe except Iberia, W France, Corsica, east to S Russia, SW Siberia, N Middle East, Turkey, Iran and much of Central Asia; winters S & SE Africa. – Populations in SE Russia, Siberia and Central Asia ('*turanicus*') described as larger, but difference slight, gradual and not distinct. 1stY '*turanicus*' said to be less grey, more buffish, but impossible to confirm this on a large sample. (Syn. *turanicus*; *yemenensis*.)

Juv/1stW, Egypt, Sep: 1stW in autumn and winter mostly still like juv, with some more or less fine dark barring on mantle and shoulders, lacking black on forehead, and with pale bill base. Grey upperparts can be slightly tinged brownish. Note very long primary projection and lack of any barring below, while upperparts are barred, which readily separates it from a young Great Grey Shrike. (E. Winkel)

Juv/1stW, Belgium, Aug: finely barred mantle, scapulars and rump and uppertail-coverts still quite pronounced here, and pinkish-buff tinge on breast-sides characteristic. Thus, juv is barred above but largely unbarred below (the reverse in northern populations of juv Great Grey, while southern populations have no or very little barring at all). Also characteristic of this age are the blackish greater coverts finely edged and tipped white, forming slight wing-bar. (V. Legrand)

1stW Lesser Grey (left: Kuwait, Aug) and Great Grey Shrikes ssp. *pallidirostris* (right: Kazakhstan, Oct): from Aug/Sep many young Lesser Grey Shrikes have moulted to more ad-like plumage, losing barring above. Juv/1stW (or dull winter ad) *pallidirostris* can appear rather similar to Lesser Grey due to pale bill base, pale lores giving open-faced impression, relatively longish wings and large white primary patch; young Lesser also lacks black forehead. Nevertheless, several key characters separate the two: (1) primary projection is longer in Lesser, hence showing more exposed primary-tips beyond tertials (at least six, in *pallidirostris* usually five or fewer); (2) Lesser has blacker upperwing-coverts with more black near wing-bend, whereas this area is pale grey in *pallidirostris*; (3) bill shorter with less developed hook in Lesser Grey; (4) at most, Lesser has hint of paler scapular patches, whereas in many *pallidirostris* these patches are well developed (but some overlap); (5) black on tail widest subterminally in Lesser Grey, instead of near base as in *pallidirostris*; and (6) lacks the cream-coloured rump patch of most young ♀ *pallidirostris*. These features should hold against dull or young plumages of most Great Grey Shrikes (not just *pallidirostris*). Both birds 1stW by juv wing-feathers. Sexing difficult due to much overlap. (Left: R. Al-Hajji; right: A. Wassink)

GREAT GREY SHRIKE
Lanius excubitor L., 1758

Fr. – Pie-grièche grise; Ger. – Raubwürger
Sp. – Alcaudón norteño; Swe. – Varfågel

The Great Grey Shrike, one of the largest shrikes, is a passerine adapted to a life much like a small bird of prey, patiently scanning its territory from an elevated perch, waiting for a careless mouse, small bird or large insect to show. It breeds in the north, primarily on open bogs in the taiga and on lower boreal mountains, and in the south in deserts and arid plains with some bushes or scattered trees to perch and build the nest in. Northern populations avoid the severe winter cold and move south in Europe to spend the winter months on pasture land, in fields and at lakesides wherever there are dense bushes and a few trees, whereas southern birds are resident. Rather shy and wary, flies off early, taking a new perch far away.

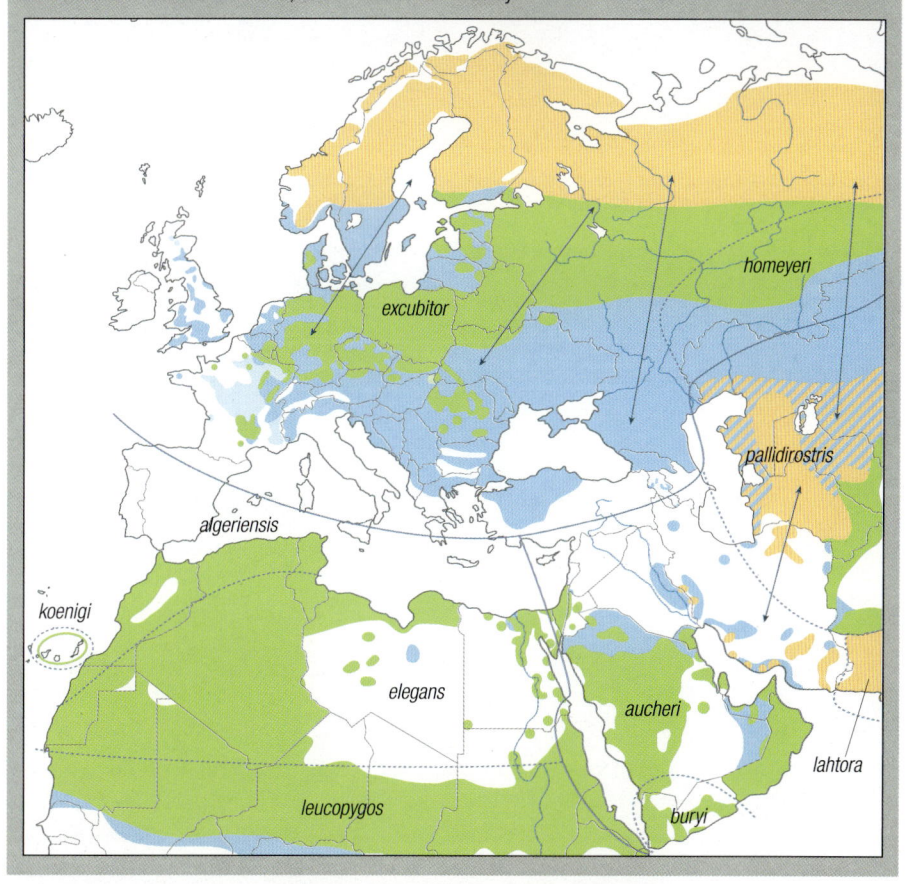

IDENTIFICATION The size of a thrush but with proportionately much longer tail, it is recognised by its *white and pale grey* colours relieved only by a *black mask through the eye*, and its largely black wings and tail. When spotted far away, perched atop a bush or tree, the *light plumage* and *long tail* are striking, but when a closer look is offered details like a *white wing patch*, *white-tipped tertials*, white tail-sides, often a diffuse whitish supercilium, and a slightly paler rump on the otherwise ash-grey upperparts can be seen. *Scapulars white* forming a prominent oblong patch between the grey and black. *Bill strong with raptor-like hooked tip* and even hint of 'toothed' cutting edge on upper mandible in falcon fashion. Bill varies from black (breeding ♂) to dark-tipped with paler pinkish-grey base (non-breeding immatures). Legs strong, dark grey. Young of northern populations have *underparts barred* or 'scalloped' grey, and ♀♀ and immature ♂♂ in Europe can show a hint of this pattern as well, while southern birds are invariably unbarred. In flight, note the long tail, the bursts of fluttering beats of rather *rounded wings* (in which the white primary patch flashes), and the resulting *undulating flight path*; undulations are often particularly deep just before landing on a new perch. Sometimes *hovers* like a raptor, at times prolonged and high up. Due to its vast and disjunct breeding range (as the species is treated here), geographical variation is marked with several subspecies showing different patterns, darkness, size and proportions as detailed below. There is also rather obvious individual variation within ssp. *excubitor* affecting the size and shape of white wing patch (being limited to primaries or extending to secondaries), amount of white in tail, prominence of mask, distinctness of whitish supercilium, shoulder patch and paleness of rump.

VOCALISATIONS Song is a 'meditative', slow and simple 'signalling', brief motifs of shrill, metallic or squeaky notes and often a doubled brief trill, *prre-prre* or more metallic *kleek-kleek*, these repeated similarly or identically for long periods with 2–4 sec pauses between them. Two or a few more variations can be heard from the same bird, but on the whole rather little variation. Like other shrikes, there is also a more subdued and prolonged warbling song type ('sub-song'), which includes some grating, hard notes and mimicry of other birds. In anxiety, including when spotting a Goshawk *Accipiter gentilis* or other large raptor, will give drawn-out, hoarse, nasal *vaaihk vaaihk* (or a few more notes), related to similar call of Red-backed Shrike but louder and coarser. Alarm (or just call?) can also be expressed with a trilling note, *prrrih*.

SIMILAR SPECIES Superficially similar to *Lesser Grey Shrike*, but separated by plumper, heavier build, larger and longer head with flatly rounded crown (Lesser Grey has steep

L. e. excubitor, ad ♂, Netherlands, Mar: ash-grey above and whitish below, with comparatively small but long black bill meeting black mask. Flight-feathers black with white bases to primaries, rather short primary projection, indistinct whitish patch on scapulars, and proportionately long black tail with white sides. Before subspecific identification is attempted, birds should be aged and sexed, because many racial characters also vary with age and sex. Aged by evenly fresh ad wing-feathers; ♂ by solid black lores and wing-feathers, and no trace of barring below. (W. Weenink)

L. e. excubitor, ad ♀, Denmark, Apr: very faint bluish tinge to ash-grey upperparts, whitish supercilium usually rather diffuse and short, moderately developed whitish shoulder and at least in N Europe 'single' white wing patch, rather restricted in comparison to some races in south and east, as well as pale grey wash to whitish underparts, are features of race *excubitor*. Sexed by some grey admixed on lores, and pale brown-grey barring on breast; only ad wing-feathers discernible. (E. Ødegaard)

L. e. excubitor, 1stS ♂, Sweden, Mar: aged by mostly retained juv greater coverts still having in spring remnants of pale tips forming thin wing-bar; only innermost greater covert moulted (blackish and ad-like). Sex inferred by solid jet-black mask, nearly all-black bill, very dark flight-feathers and any barring below very faint or already worn off. (P. Hvass)

L. e. excubitor, 1stS ♀, England, Mar: combination of less deeply black mask (in particular paler and narrower at loral area) and fine dark barring on much of underparts plus still rather pale bill infer that this is a 1stS ♀. (J. Richardson)

L. e. excubitor, ad ♂, England, Nov: when fresh, ad ♂ of ssp. *excubitor* has pale lead-grey upperparts, including crown and uppertail-coverts; underparts white without any vermiculations (usually shown to some degree by all other plumages in autumn). White patch at base of primaries often large, and sometimes also as here a second smaller patch on secondaries. Sex indicated by largely black bill and complete black mask, age by evenly fresh ad wing being glossy black. (D. Hutton)

L. e. excubitor, ad ♂, Finland, Nov: stunningly beautiful in flight. Being ad (all fresh and neat without any trace of barring below, plus black-looking wing) and having such broad and unbroken black mask and dark bill in late autumn infers sex. Quite large white primary patch, continuing narrowly on otherwise dark secondaries, supports subspecific identification. (J. Peltomäki)

forehead and frequently slightly peaked crown on rounded head), shorter and more rounded wings (primary projection 1/2 to 2/3 of exposed tertials, but equal to or exceeding tertials in Lesser Grey), broader and more graduated tail, and longer but often a fraction slimmer bill (Lesser Grey has shorter and deeper bill with more uniformly rounded mandibles in profile). Important to note in northern populations of young Great Grey Shrike barring *below*, but not above, as opposed to diffuse dark barring *above* in young Lesser Grey, but little below. — Much more difficult to separate from some *Northern Shrikes*, and some immatures and ♀♀ probably inseparable in normal field encounters. Due to extensive geographical variation in Great Grey Shrike especially, necessary to establish age, sex and then preferably subspecies before reliable identification can be attempted. The on average somewhat more barred underparts of Northern compared to *excubitor* Great Grey is not always evident or present. Upperparts of adult Northern on average slightly

L. e. excubitor, 1stW ♂, England, Nov: when fresh can show slight pinkish-buff wash below. Combination of only faint grey marks on breast, juv white-tipped greater coverts and rather solid black mask and pure bluish-grey upperparts make this a 1stW ♂. Slightly paler base to bill usual at this season. This bird has rather prominent white supercilium. Outer tail-feathers seem to have fairly extensive white portions. (P. Walkden)

paler grey than in *excubitor*, whereas immature Northern is slightly tinged yellowish-brown and perhaps a trifle paler above than Great Grey in corresponding plumage. Note that Northern Shrike invariably has a 'single wing patch', and in immatures often a slight rusty-brown tinge on breast-sides and flanks (brown hues absent or insignificant in immature Great Grey). White on outertail on average more restricted in Northern than in Great Grey (but much overlap). A first-winter ♀ *excubitor* can be so similar to a first-winter ♂ Northern that they are in practice inseparable without direct comparison or extensive experience, but latter usually has a more whitish and contrasting rump, and richer brown-tinged barring. – Separation from adult *Iberian Grey Shrike* is straightforward using combination of lack of strong pink flush below, much paler grey crown, nape and mantle, lack of well-marked, narrow white supercilium and smaller black mask (in Iberian extending far and broadly onto rear head-sides), but some juveniles may cause problems. However, juvenile *excubitor* generally is clearly more barred below, paler grey on crown and nape, often has a larger white primary patch, and has slightly shorter and less strong feet. Subspecies *algeriensis* of Great Grey is similarly dark grey on crown and mantle as Iberian Grey, but lacks the long and well-marked white supercilium and the strong pink flush on the underparts of adult Iberian Grey.

AGEING & SEXING (*excubitor*) Ages and sexes differ slightly in post-juv plumages. Seasonal variation insignificant. – Moults. Complete post-nuptial moult of ad in late summer (Jul–early Oct); partial post-juv moult of 1stY at same time does not involve primaries, secondaries, tail-feathers (except rarely r1) or primary-coverts, and none or few greater coverts and tertials. Partial pre-nuptial moult (Mar–Apr) of both age groups mainly involves some head- and body-feathers. – **SPRING Ad ♂** Underparts unbarred white (or off-white). Rarely, freshly moulted birds may have very faint traces of pale grey bars on breast (hardly visible in the field), but these bleach or wear off quickly. Wings and tail black or near-black. Bill and entire mask (including lores) jet-black. On average slightly more white in wing and tail than ad ♀, but much variation and of little use in practice (except perhaps when studying breeding pairs). **Ad ♀** As ad ♂, but underparts

L. e. excubitor, ♂, presumed 1stW, Finland, Nov: a bird with slightly larger primary patch which narrowly continues also on secondaries. Combination of being young (juv primary-coverts and greater coverts being subtly paler and more brownish than in ad) but still rather solid blackish mask and mostly dark bill, but also relatively much white in remiges, indicate 1stW ♂. (L.-O. Landgren)

L. e. excubitor, 1stW ♀, England, Oct: when fresh, young ♀ *excubitor* can be extensively barred below, mostly on chest and flanks, on cream-white ground with slightly pink or pale grey wash. Note pale-tipped greater coverts and pale grey lores. Beware that 1stY, especially ♀, with limited white in wing but extensive barring below, can approach Northern Shrike. (S. Ashton)

L. e. excubitor, juv, Sweden, Jun: variably barred brownish-grey below, greater coverts tipped whitish, plumage soft and fluffy, dark mask weakly developed (lores pale), and base of bill often pale pink-grey (though not so obvious here). As very recently fledged, wing-feathers and tail not yet grown to full length. (K. Johansson)

L. e. homeyeri, ad ♂, Kazakhstan, Nov: a variable race, but 'classic' examples in east of range have upperparts light grey, paler than ssp. *excubitor*, rump and uppertail-coverts light greyish-white and underparts pure white. Wing as a rule has larger white area than ssp. *excubitor*, invariably a 'double' patch with white at base of primaries usually extending 13–25 mm beyond primary-coverts, and white at base of secondaries 13–20 mm outside greater coverts; both r6 and r5 are often all white. (V. Fedorenko)

L. e. homeyeri var. *leucopterus*, ad ♂, Kazakhstan, Mar: as to amount of white in wing a highly variable race, and extreme examples are called var. *leucopterus* (common especially in east of range) and have very large white wing panel that continues broadly onto secondaries (as here), unlike in ssp. *excubitor*. Also characteristic is whiter rump and larger and purer white shoulder patches. Wing evenly feathered and black mask very broad inferring age and sex. (A. Timoshenko)

L. e. homeyeri, ad ♀, Kazakhstan Mar: race evident from locality, large white 'double' wing patch (extensive also on secondaries), as far as can be judged quite pale grey upperparts, and very large white shoulder patch. Age inferred by all visible wing-feathers seemingly intense black, whereas fine barring of breast, and hint of pale grey patch on lores, in an ad in spring fit only ♀. (A. Timoshenko)

L. e. aucheri, ad ♂, Israel, Jan: similar in size and proportions to N African taxa, but usually darker grey above and slightly washed grey on breast-sides and flanks (though difficult to see here) than *elegans*. Extent of white beyond coverts on primary bases quite large, but secondaries appear dark when wing folded, with only some whitish tips and edges forming a narrow line. Black mask extends well onto forehead, broadest in ♂. Outertail has moderately large white sides and corners. Evenly rather fresh wing-feathers and relatively solid black lores infer age and sex. (L. Kislev)

(breast to lower flanks) distinctly or faintly barred grey (exceptionally can wear off in worn summer birds). Wings and tail blackish-grey or dark brownish-black, less black than ♂ in direct comparison. Bill blackish with variable paler base (bluish-grey or horn), rarely all dark. Mask on average less black than ♂, lores often greyish or greyish-white, but some have virtually all-black mask. **1stS** As ad, but many show remnants of pale-tipped juv greater coverts before these entirely wear off. **1stS ♂** As ad ♂, but underparts usually faintly barred grey, and wings and tail not entirely black, more blackish-grey or dark brownish-grey. – AUTUMN **Ad** As in spring, but both sexes have bill and mask less black, base of bill and lores becoming variably paler, most pronounced in ♀. Greater coverts all black or have tiny wedge-shaped white spot at shaft (but not whitish tips across all greater coverts). **1stW** Resembles ad ♀ but has off-white tips to greater coverts, forming a narrow wing-bar. Sexes very similar, and usually not separable unless underparts are either only faintly barred (♂) or very heavily barred grey (♀), and wings and tail are either quite dark blackish-grey (♂) or relatively paler brown-grey (♀). **Juv** Like 1stW ♀, but more heavily barred (brown-)grey below, also on throat, and breast often tinged brown or pink-buff; dark mask poorly developed, base of bill extensively pale pink-grey. Pale tips to greater coverts buffish-white (later fading to whitish).

BIOMETRICS (*excubitor*) **L** 23.5–26 cm; **W** ♂ 108–121 mm (*n* 116, *m* 114.8), ♀ 108–116 mm (*n* 67, *m* 112.5); **T** ♂ 98–118 mm (*n* 116, *m* 110.0), ♀ 99–116 mm (*n* 67, *m* 107.5); **T/W** *m* 95.7; **TG** 12–30 mm (*n* 186, *m* 22.3); **width of r1** (measured 30 mm from tip) 13–18 mm (*m* 14.6); **white tip of r6** (inner web) 33–68 mm (*m* 48.6; rarely whole feather); **B** 19.2–25.3 mm (*n* 106, *m* 21.9); **B**(f) 15.1–19.2 mm (*n* 84, *m* 16.8); **BD** 8.0–9.7 mm (*n* 97, *m* 9.0); **Ts** 23.5–28.4 mm (*n* 148, *m* 26.8); **white on pp** visible outside tips of pc (folded wing) 4–21 mm (usually < 15 mm), **white on ss** visible outside tips of gc (though difficult to measure accurately) 0–19 mm (when present usually < 10 mm). **Wing formula: p1** > pc 11–19 mm, < p2 28.5–35 mm; **p2** < wt 10.5–17.5 mm, =6/7 (73%), =6 (16%), =7 (10%), or =5/6 (1%); **pp3–4**(5) about equal and longest (rarely p3 to 2 mm <); **p5** < wt 0–4.5 mm; **p6** < wt 8.5–14 mm; **p7** < wt 14–21.5 mm; **p10** < wt 21–29 mm; **s1** < wt 22–30.5 mm. Emarg. pp3–5.

GEOGRAPHICAL AND OTHER VARIATION & RANGE Extensive variation in plumage, size and proportions within the species' vast range. The variation is partly clinal, partly rather well-marked and disjunct without a clear pattern. Three racial groups or incipient species are combined and treated

L. e. aucheri, presumed ♀, Israel, Jan: here the typically grey-tinged breast-sides and flanks of this race can be clearly seen. At this angle, both age and sex are difficult to be sure of, but the black mask seems to narrow slightly in its forepart and only extend moderately onto forehead, while underparts both have a slight buff-brown cast and a trace of subtle barring indicating a ♀. (L. Kislev)

L. e. aucheri, 1stW ♂, Oman, Jan: typical for this race is rather dark grey crown and upperparts, broad black mask (in ♂♂ extending well up on forehead) and grey-washed flanks (not clearly visible here). Note large white primary patch, the same pattern as in *pallidirostris*, but young ♂♂ of that race do not develop black bill and complete black mask before late Mar or Apr, upperparts are paler grey, and underparts and scapulars are whiter. Age inferred by worn and brownish juv inner primaries. (H. & J. Eriksen)

as Great Grey Shrike; the three groups mainly discernible based on molecular data from mtDNA (Olsson *et al.* 2010), but these coincide poorly with morphological traits, and a split into three species is not followed here due to remaining uncertainties regarding degree of hybridisation and diagnosability of similar-looking taxa, which would end up in different species. (1) Northern group, GREAT GREY SHRIKE, consists of poorly differentiated and clinally connected *excubitor* and *homeyeri*, characterised by barred underparts in juv and imm plumages, and lightly barred underparts in ad ♀, broad tail-feathers, moderately graduated tail, moderately strong legs, in many birds—but far from all—visible white at base of secondaries on folded wing ('double wing patch'), rather narrow buffish or off-white tips to juv greater coverts soon bleaching to white, and tertials blackish and generally finely or moderately tipped white. Birds become slightly paler and have larger white areas from west to east. At eastern end of cline also fairly large individual variation, or the occurrence of a colour morph. Presence of 'single' or 'double wing patch', and some other traits, has led to separation of several subspecies, but listed characters very variable and not consistent within geographical areas, hence not recognised here. The two recognised subspecies *excubitor* and *homeyeri* are connected via a broad and diffuse zone of intergrading characters (by and large a cline), and lumping them is an alternative. However, they are provisionally maintained until variation in NW Asia is better known,

L. e. aucheri, ad ♂, Israel, Sep: combination of solid black mask covering forehead and largely brownish wing with new jet-black feathers emerging indicates ad ♂ in active wing moult (post-nuptial moult in this race can last well into autumn, during which time bill becomes paler). Note the grey-tinged lower flank and apparently limited white in outertail. (L. Kislev)

L. e. aucheri, 1stW, Israel, Aug: at end of post-juv moult. Large white primary patch, dark-based secondaries (with broad white tips, though) and rather dusky-grey upperparts. Most of wing still juv, but too young (and underparts invisible) to permit sexing. Apparently, partial tail-moult has started. (H. Shirihai)

L. e. pallidirostris, ad ♂, Mongolia, May: white on primaries extensive, forming very large white patch, while white on secondaries confined to tips and inner webs, forming a white 'grid' in flight, but which is usually not visible on folded wing or just shows as thin white linings on some (mostly inner) secondaries. Lesser upperwing-coverts pale grey. Age and sex as image above right. Note that breeding ♂ has black bill and complete black mask in this race. (J. Normaja)

and since the two ends of the cline are well differentiated. (2) Central Asian–Indo-Arabian group, or 'INDIAN GREY SHRIKE', consists of *lahtora*, *pallidirostris*, *aucheri*, *buryi* and *uncinatus* (last-mentioned extralimital: Socotra), characterised by invariably unbarred underparts, narrower tail-feathers and well-graduated tails, strong legs, no white on base of secondaries visible on folded wing (invariably 'single wing patch'), often-extensive black mask that in ♂ sometimes reaches onto forehead (sometimes extensively), and broad ochre-buff tips to juv greater coverts and tertials. A tendency to have peachy-buff tinge to breast in adults in parts of Central Asia. Ssp. *buryi* and *uncinatus* differ from the rest by being darker. (3) North African group, 'AFRICAN GREY SHRIKE', consists of *elegans*, *leucopygos*, *algeriensis* and *koenigi*, sharing with 'Indian Grey Shrike' unbarred underparts, narrow tail-feathers and well-graduated tail, strong legs, 'single wing patch', and tendency for some ♂♂ to have black on forehead. The main difference between groups 2 and 3 is genetic. The two first-mentioned races in this third group (see below) are pale and have large white portions on wings and tail, and white rump, the other two are darker with less extensive white. Said to intergrade with group 2 in E Egypt (Sinai) and along Red Sea coast, where *elegans*

L. e. pallidirostris, ad ♀, Oman, Mar: a migratory race, with an often quite notable primary projection (still, some overlap with other taxa). Tail-feathers typically narrow with much white on sides. Also characteristic are pale grey upperparts, with large white scapular patches, pale rump, prominent black mask, usually white underparts (but which can have strong pink-buff tinge). Very large white primary patch, but white on secondaries restricted to inner webs (visible in flight) and tips. Not entirely jet black wing-feathers, and black mask with grey loral spot, infer age and sex. (M. Varesvuo)

and *aucheri* meet, and at times these are indeed extremely difficult to separate morphologically; more study required to adequately resolve this issue.

(1) 'NORTHERN GROUP'
L. e. excubitor L., 1758 (Europe except Iberia and S France, NW Siberia; northern breeders make short- to medium-range winter movements, whereas southern birds are mainly sedentary). The GREAT GREY SHRIKE ssp. *excubitor* has ash-grey upperparts (in ad with very faint bluish tinge when fresh) from crown to lower back, forehead often faintly paler, and is white below, in all except ad ♂ with some degree of brown-grey or dull pink-brown barring, faint in 1stS ♂ and ad ♀♀, a little stronger in imm ♀ and usually fairly bold in juv. Both 'single' and 'double' white wing patches occur, latter more common in (but not restricted to) C & E Europe, rarer in N Europe. Variably large whitish shoulder-patch, variably distinct white supercilium (usually rather diffuse and short). Rump paler grey or greyish-white. White on tail-feathers varies from being restricted to tips of rr3–6, most extensively on r6, and white outer web of r6 dark, to much more extensive with all-white r6 (at most partly dark shaft), r5 with small dark portion on inner web alone, rest white (partly dark shaft); and extensive white tips on rr3–4, and smaller white tips on rest. See also Ageing & Sexing and Biometrics. – Grades steplessly into *homeyeri* in E Ukraine and SE Russia (lower Volga), but some birds with large double wing patch and pale forehead found in rest of Europe actually come very

close to genuine *homeyeri*, and correct subspecific labelling of these can be difficult or impossible. (Syn. *galliae*; *melanopterus*; *major*.) – Interaction with *sibiricus* Northern Shrike in C Siberia not well known and should be investigated. Great Grey Shrike and Northern Shrike allegedly intergrade in vast zone of Siberian taiga between Ob and Yenisei, but this may be mere assumption based on external similarity. Still, some specimens in collections appear to be intergrades between *excubitor* and *sibiricus*, others between *homeyeri* and *sibiricus*.

○ **L. e. homeyeri** Cabanis, 1873 (SE Russia to north slopes of Caucasus, S Siberia east to W Sayan, N Altai; short- or medium-range winter movements to S Central Asia). What is here called 'Homeyer's Grey Shrike' is a highly variable race as to amount of white in plumage, those with least white being very similar to *excubitor*: on average a trifle paler overall, forehead whitish, crown usually paler grey than *excubitor* (but some overlap), rest of upperparts slightly paler grey, white shoulder on average larger; invariably a 'double' and large white wing patch; tail on average has more extensive white, varying from r6 white with only

as a pale morph of *homeyeri* occurring in the eastern half of its range. Numerous very pale birds, showing full set of characters of 'var. *leucopterus*' according to available descriptions (notably plate in Bogdanov 1881), were obtained in breeding season within typical *homeyeri* range, e.g. Omsk, Novosibirsk, Tomsk, Krasnoyarsk, N Altai. (See also Stepanyan 1990.) An attractive morph, with black wing-feathers, even in juv ♂ (making ageing more difficult, especially in spring, when pale tips to greater coverts frequently wear off), very large white portions on wings and tail, and quite pale upperparts (pearl-grey with faint creamy hue). Forecrown whitish without distinct border to pale grey hindcrown. Lores often pale, practically never all black even in breeding ad ♂. Size and wing formula similar to *homeyeri* sensu stricto. White on pp > pc usually at least 17 mm, white on ss > gc usually at least 21 mm. On average somewhat more white on outertail than *homeyeri* sensu stricto. Just as separation between *homeyeri* and pale *excubitor* is complicated by the fact that variation is continuous, many birds cannot be labelled as one or the other.

(2) 'Central Asian–Indo-Arabian Group'

L. e. aucheri Bonaparte, 1853 (S Iran, Middle East, Arabia, shores of Red Sea, partly Nubia and Eritrea; possibly more extensively distributed in N Egypt and N Libya; largely resident, but northern breeders make short- to medium-range movements in winter). The 'Levant Grey Shrike' resembles *pallidirostris* and *elegans*, both of which it allegedly intergrades with in west and north-east of its range, and separation can be problematic. Differs from *elegans* in: (i) less white on outertail (often black portion on inner webs of all tail-feathers, or only r6 largely white); (ii) on average slightly shorter tail; (iii) in ♂ invariably 1–2.5 mm of black on forehead over base of bill (none or only very little in ♀; but note that odd ♂ *elegans* as far west as Tunisia, and several ♂ *pallidirostris*, also have some black on forehead); and (iv) usually a grey wash on sides of breast and flanks. White primary patch medium large. Differs from *pallidirostris* in: (i) more often black on forehead over base of bill in ad ♂ (though see above); (ii) usually smaller white primary patch (but some overlap); (iii) generally more obvious grey wash on breast-sides and flanks; (iv) underwing has some grey wash on primary-coverts and axillaries (*pallidirostris* usually cream-white); (v) white tertial-tips often smaller; (vi) on average less white on outertail (usually some dark basally on all tail-feathers, or only r6 nearly all white, whereas *pallidirostris* with few exceptions has rr5–6 nearly all white and reduced

L. e. pallidirostris, ad ♀, Kazakhstan, Mar: dark mask in ♀ often slightly brown-grey and usually incomplete, commonly broken in front of eye by grey-white patch on lores. The ♀ also has less jet-black remiges. Overall pale grey above, and rather white below with slight pink-buff tinge on breast-sides and flanks. White scapular and rump patches large. Bill can be all black in spring/summer ♀♀, too. Only ad wing-feathers, with no trace of pale-tipped greater coverts. (G. Dyakin)

dark portion on r4); (vii) much less white concealed on inner webs of secondaries (10–70% of inner webs of ss2–4 white in *aucheri*, 75–100% white in ad *pallidirostris*, although intermediate values occur in 1stY); (viii) proportionately longer tail; and (ix) more marked sexual size dimorphism. Juv cream-white below with greyish ochre-tinged breast. 1stS resembles ad but is usually distinguished by a few or many remaining juv greater coverts, and in most by contrast between faded and slightly brownish primaries and primary-coverts and new black tertials, median and some inner greater coverts. A few replace also some inner secondaries, and at times outer 3–5 primaries, and thus shows obvious moult contrast in 1stS. **W** ♂ 102.5–116 mm (*n* 35, *m* 110.0), ♀ 101–112 mm (*n* 29, *m* 107.6); **T** ♂ 101–117 mm (*n* 34, *m* 109.0), ♀ 96–115 mm (*n* 29, *m* 105.0); **T/W** 98.1; **TG** 21–35.5 mm (*n* 61, *m* 27.4); **width of r1** (measured 30 mm from tip) 8–13 mm (*m* 10.4); **white tip of r6** (inner

L. e. pallidirostris, ♂, presumed 1stS, Kazakhstan, May: pale plumage superficially resembles that of non-migratory ssp. *elegans* of N Africa, and some can be very difficult to separate even when handled. Still, ssp. *pallidirostris* often differs from *elegans* by on average longer primary projection and less white edges on secondaries and long tertials, these forming smaller white mark. Overall pale grey above, with large white scapular patches. Bill and mask often all black in spring/summer, even in 1stS ♂. Aged by seemingly juv primaries and primary-coverts. (A. Isabekov)

small dark portion on inner web reaching dark shaft, to all-white rr5–6, restricted amount of dark on inner web of r4, and extensive white tips and bases to rr(2)3–4. White at base of r1 21–42 mm (*n* 21, *m* 29.2) (in *excubitor* usually none). Grades into *excubitor*, and some birds impossible to distinguish. Biometrics: **W** ♂ 113–125 mm (*n* 71, *m* 118.7), ♀ 113–122 mm (*n* 26, *m* 116.9); **T** ♂ 108–126 mm (*n* 71, *m* 115.5), ♀ 100–121 mm (*n* 25, *m* 112.5); **T/W** 96.9; **TG** 14–35 mm (*n* 106, *m* 23.7); **width of r1** (measured 30 mm from tip) 13–17 mm (*m* 15.0); **white of r6** (inner web) invariably to base; **B** 19.9–25.7 mm (*n* 108, *m* 23.7); **B(f)** 16.3–19.7 mm (*n* 46, *m* 17.9); **BD** 8.7–10.2 mm (*n* 93, *m* 9.4); **Ts** 26.0–29.5 mm (*n* 108, *m* 28.0); **white on pp** visible outside tips of pc (folded wing) 11–29 mm (usually 13–25 mm), **white on ss** visible outside tips of gc (though admittedly difficult to measure accurately) 8–35 mm (usually 13–20 mm). – The palest birds have been described as '*leucopterus*' (*przewalskii* has priority if recognised as a subspecies), but until they have been established to have a separate breeding range, it is preferable to include them

L. e. pallidirostris, ad ♀, Kazakhstan, May: showing well the extensive white bases to primaries and very large white primary patch. Tips and inner webs to secondaries can be rather extensively white when wings fully spread, but are usually not visible on folded wing. Extensive white scapular patches. Breast and flanks strongly tinged pinkish-buff. No visible moult limits and blackish primary-coverts infer ad. Breeding ♀ typically has greyish lores and not quite black remiges. (A. Vilyayev)

L. e. pallidirostris, 1stS ♀, Kazakhstan, May: race indicated by date, locality, and large white primary patch but no white patch on secondaries. The wing pattern and pale lores in spring give a lead to the bird being young ♀ *pallidirostris* (or *elegans*). Indeed, no external character separates conclusively from ssp. *elegans*, only locality. Separation from *aucheri* helped by underparts lacking grey. Peachy tinge on flanks often seen in non-ad ♂ *pallidirostris*. Age obvious from strong moult contrast in wing. (S. Harvančík)

SHRIKES

L. e. pallidirostris, ad ♂, Kazakhstan, Sep: distinctive due to very pale ash-grey upperparts, single white very large primary patch, and also large white tertial tips, but secondaries black with only thin white edges and tips. In autumn, after breeding, black mask of ad ♂ becomes variably less complete, often with small pale mark on lores. Also, in Sep base of bill often paler. Evenly fresh after complete post-nuptial moult. (V. Federenko)

L. e. pallidirostris, ad ♂, Kuwait, Sept: when fresh, many ♂♂ have pinkish-buff tinge on face and underparts, at times rather strongly. Also note longish primary projection with well-spaced primary-tips, typical of this long-distance migrant, though much overlap with other taxa. When fresh, rump and uppertail-coverts are no whiter than rest of upperparts. This bird has a particularly strong and long bill. (A. Al-Sirhan)

L. e. pallidirostris, ad ♀, Oman, Nov: on average, ♀ has more extensive pale lores, paler bill-base and paler brown-grey remiges than ♂ of same age. Further, unlike similar 1stW ♂, ad ♀ tends to be more extensively buffish below, with less pure grey upperparts, but not as brownish- and buff-tinged or as weakly patterned on wings and face as 1stW ♀, and unlike 1stW of both sexes bill is more greyish-horn at base, less pinkish. (H. & J. Eriksen)

L. e. pallidirostris, 1stW ♂, United Arab Emirates, Nov: 1stW plumages are generally the most strikingly patterned, as they often show largely brownish wings with broad sandy-cinnamon to pale buff tips and fringes to median and greater coverts and tertials. Secondaries show contrasting mostly black outer webs and mainly white inner webs (the latter largely concealed). Pale area of bill greyish-pink; black mask incomplete, with just small dark area in front of eye and rest of lores pale, affording an open-faced expression. (Mattias Ullman)

L. e. pallidirostris, 1stW ♀, England, Nov: a surprisingly striking plumage considering its rather washed-out appearance, often tinged rufous-buff or sandy-pink. Wing- and tail-feathers and ear-coverts brown-grey. Large white primary patch and long primary projection characteristic, too. Most of the wing is juv. Note very pale and pink-tinged bill and pale lores. (G. Jenkins)

web) 30–75 mm (*m* 53.4) but frequently to base; **B** 19.0–24.8 mm (*n* 65, *m* 22.6); **B**(f) 14.9–19.0 mm (*n* 35, *m* 17.2); **BD** 8.7–10.6 mm (*n* 52, *m* 9.4); **Ts** 27.2–33.0 mm (*n* 62, *m* 30.1); **white on pp** visible outside tips of pc (folded wing) 10–24 mm. **Wing formula: p1** > pc 8–23 mm, < p2 24–36 mm; **p2** < wt 5–9.5 mm, =5/6 (90%), =6 (8%), or =5 (2%); **p6** < wt 7–16 mm; **p7** < wt 13–22 mm; **p10** < wt 24–34 mm; **s1** < wt 27–38 mm. – Meinertzhagen (1953) described '*theresae*' from N Israel as being darker above. The type (NHM) is severely stained and certainly darker due to this, while other spring birds from the same area are either paler grey or almost as dark grey as the type, so more likely only natural variation and to some extent staining involved. (Syn. *dubarensis*; *fallax*; *theresae*.)

L. e. pallidirostris Cassin, 1851 (Central Asia east of Caspian Sea, S Kazakhstan in Saxaul deserts, from C Iran to S Mongolia and N China; winters S Iran, Arabia, S Middle East, NE & E Africa). The 'DESERT GREY SHRIKE' (a name more appropriate than the often-used but misleading 'Steppe Grey Shrike') resembles both *aucheri* and *elegans*, and separation is at times problematic, surprisingly so especially in the case

L. e. pallidirostris, juv, Mongolia, Jun: typically has sandy and buff-brown pigments all over, with almost orange-buff rump. Note how broad and rufous ochre-coloured tips are to greater coverts when fresh, tips which then progressively will bleach to white and wear down to narrow fringes later in autumn and winter. (H. Shirihai)

of *elegans*. Birds in north and north-east are long-distance migrants (wintering in Arabia and NE Africa), but wing-shape still largely the same as *elegans* and *aucheri*, primary projection only a fraction longer on average. Differs from *elegans* in: (i) no, or less broad, white outer edges to secondaries (if any present mostly obvious near tips of secondaries); thus, whereas only few have prominent white outer edges to secondaries forming white 'panel', this pattern is the rule in *elegans*; (ii) often proportionately shorter and narrower tail; (iii) at times larger white primary patch; and (iv) a few have slightly longer primary projection. For differences from *aucheri*, see under that. Note mainly dark outer webs to all secondaries (producing dark innerwing when perched), but usually extensive white portions on inner webs often reaching shafts and leaving only a dark subterminal patch (giving white grid-shaped flash on innerwing in flight); this character is variable, though, and some ad have dark portions inside shafts of secondaries and thus somewhat less white on inner webs. Superficial resemblance to Great Grey Shrike *homeyeri* (especially pale morph var. '*leucopterus*'), but note (i) lack of white secondary patch on folded wing (only a large patch on primaries, plus large white tertial-tips; exceptional occurrence of 'double wing patch' in northern part of range presumably due to rare hybridisation with *homeyeri*); (ii) stronger feet; and (iii) narrower tail-feathers (see below). Upperparts pale grey with very faint brownish cast (breeding ad ♂ can appear pure grey). Underparts faintly tinged buff-pink in fresh plumage (many breeding ad ♂♂ become nearly pure white). No barring on underparts in any plumage or age. Sexual dimorphism in breeding season (c. Mar–Jul, sometimes Aug): ad ♂ has invariably (contrary to some statements in literature) jet-black lores, ear-coverts and bill; rarely, black mask reaches narrowly onto forehead; ad ♀ has somewhat paler lores, but near-black ear-coverts and bill. After breeding, many ad ♂♂ attain paler lores, and bill (especially base of lower mandible) gradually becomes less black. Sexes further differ in that ♂♂ have blacker flight-feathers (ad ♂ jet-black, ad ♀ dark blackish-grey). Imm have pale lores and markedly pale bill-base; ear-coverts dark, darkest on average in ♂♂. Easily aged through 1st autumn by broad rufous-buff tips and edges to greater coverts and tertials, often tail-feathers also tipped buff. From Oct, these buff edges and tips gradually bleach whiter and wear to become less evident. 1stS resembles ad (although ♂ may have less deep black lores, at least in Mar–Apr), but is usually distinguished by a few or many remaining juv greater coverts, and in ♂ by contrast between faded and slightly brownish primaries and primary-coverts and new black tertials, median and some inner greater coverts. Often some inner secondaries, and at times outer 4–6 primaries, renewed as well in (Sep) Oct–Nov of 1stCY, and thus such birds show moult contrast in 1stS. Sexes very nearly the same size. **W** 101–118 mm (n 156, m 110.8); **T** 88–115 mm (n 143, m 102.8); **T/W** m 92.9; **TG** 17–35 mm (n 100, m 24.9); **width of r1** (measured 30 mm from tip) 8–12.5 mm (m 10.1); **white tip of r6** (inner web) 52–76 mm (m 71.0) but more frequently to base; **B** 19.9–24.8 mm (n 126, m 22.4); **B**(f) 15.3–18.5 mm (n 33, m 16.9); **BD** 8.4–10.4 mm (n 91, m 9.5); **Ts** 27.2–32.5 mm (n 124, m 30.4); **white on pp** visible outside tips of pc (folded wing) 10–27 mm (very rarely < 14 mm). **Wing formula: p1** > pc 12–25 mm, < p2 27–38 mm; **p2** < wt 4.5–10 mm, =5/6 (80%), =5 (12%), or =4/5 (8%); **p6** < wt 6.5–18.5 mm; **p7** < wt 14–24.5 mm; **p10** < wt 26–34 mm; **s1** < wt 29–40 mm. (Syn. *assimilis*; *grimmi*.) – Birds of SE Iran and W Pakistan (Baluchistan) possibly undescribed race, differing in frequently peachy-brown or deep pink-buff tinge on underparts, variable but generally much stronger than in classic *pallidirostris* of further north in Central Asia. Secondaries either have much white on inner webs or with reduced white (latter more like *aucheri*). – As noted above, occasional interbreeding with ssp. *homeyeri* might hypothetically occur, since two *pallidirostris* had a hint of a double wing patch (Lop Nur in Oct, Ningxia late May, N China; ZISP).

L. e. buryi Lorenz & Hellmayr, 1901 (S Arabia, Yemen, apparently also coastal Djibouti and adjacent areas; resident). The 'YEMEN GREY SHRIKE' is almost as dark as *algeriensis*, being uniformly slate-grey above relieved only by a narrow whitish shoulder patch and a small white primary patch. Underparts including throat greyish, white only on centre of belly. White primary patch moderately large. Upperwing-coverts mainly black, lesser coverts grey-tipped near edge of wing. Sexes differ on average in that ♂ has 1–1.5 mm black on forehead over bill, ♀ not (or only diffuse hint of black). Extremely similar to extralimital *uncinatus* (see below), and safe separation other than by locality perhaps impossible; they might even be the same subspecies. Differs from similarly dark *algeriensis* in smaller size, usually not quite as dark slate-coloured crown, and greyish rather than whitish throat. Juv similar to ad, greyish below, strongest on breast and flanks (unlike juv *aucheri*, which is cream-white below with greyish-ochre breast). **L** 21–22 cm; **W** ♂ 101–111 mm (n 29, m 106.7), ♀ 100–109 mm (n 14, m 104.8); **T** ♂ 97–113 mm (n 29, m 105.7), ♀ 98–107.5 mm (n 13, m 102.7); **T/W** m 98.7; **TG** 22–34 mm (n 41, m 27.3); **width of r1** (measured 30 mm from tip) 8–12 mm (m 10.3); **white tip of r6** (inner web) 17–51 mm

L. e. buryi, ♂, Yemen, Jan: a dark and rather small race, with underparts including throat extensively sullied greyish, some being almost dusky. Palest birds can be quite similar to ssp. *aucheri*. Black mask very broad in ♂, extending slightly onto forehead over bill, and also to well below eye. No white supercilium. (H. & J. Eriksen)

L. e. buryi, ♀, Yemen, Jan: above slate-grey, with only indistinct whitish shoulder and small white primary patch; also extensively tinged greyish on breast and belly, and sometimes on throat, but others appear white-throated, as here. Black mask is broad but much narrower and paler on lores, and does not reach over bill, indicating a ♀. Brown alula seems juv, but safe ageing requires a better viewing angle. (H. & J. Eriksen)

L. e. lahtora, ad ♂, India, Dec: a large subspecies with a proportionately long tail and large head, but most characteristic is the very prominent and long black mask extending broadly over forehead and even extending to neck-sides (absent in other races), and considerable white in wing, including on secondaries even when wing folded, thus combined white panel on wing is largest among treated taxa. Only ad wing-feathers give the age. (J. & J. Holmes)

L. e. lahtora, ad ♂, NW India, Oct: very broad white wing panel in secondaries, with inner five feathers appearing all white. Also, unlike most other races, lesser and marginal coverts blackish (sometimes a grey spot on 'wrist'), a character also shown by ssp. *buryi* from SW Arabia, which is otherwise much greyer and smaller. Age as bird at left. (S. Singhal)

L. e. lahtora, ad ♀, India, Aug: here in advanced post-nuptial moult. Unlike ♂, black mask narrower, extending only slightly over bill, still quite broad and long compared with other races. (G. Venkataraman)

L. e. uncinatus, presumed ad ♂, Socotra, Dec: an insular extralimital race very similar to ssp. *buryi* in its rather dark grey plumage, but is smaller and has proportionately longer bill; relatively limited white in wing. (A. Al-Sirhan)

(m 33.1); **B** 18.6–23.3 mm (n 45, m 21.3); **B**(f) 14.5–18.1 mm (n 41, m 16.5); **BD** 8.8–10.3 mm (n 41, m 9.4); **Ts** 26.5–30.6 mm (n 48, m 28.3); **white on pp** visible outside tips of pc (folded wing) 6–16 mm. **Wing formula: p1** > pc 12–23 mm, < p2 26–35 mm; **p2** < wt 6–10 mm, =5/6 (33%), =6 (33%), or =6/7 (33%); **p6** < wt 6.5–13 mm; **p7** < wt 12–18 mm; **p10** < wt 21–29 mm; **s1** < wt 25–33 mm. (Syn. *arabicus*.)

Extralimital: **L. e. lahtora** (Sykes, 1832) (India; presumed resident, but claimed to sometimes winter in E Arabia). The 'INDIAN GREY SHRIKE' is rather large with well-graduated tail, is rather dark grey above and has a broad black mask (very similar in both sexes) usually reaching well above bill on forehead (up to 5 mm in ad ♂, to 3.5 mm in ♀ and 1Y ♂) and far down on sides of neck, where it often broadens and becomes diffuse, has no or only faint pale supercilium, and very broad white distal outer edges and tips to secondaries and tertials (tips 9–16 mm), forming very wide white wing patch or 'panel' on folded secondaries. Fairly large white primary patch on which black primary shafts penetrate the entire white patch (at least on many outer primaries), and black on each primary forms a wedge into the white. In ad, inner webs of inner five secondaries all white, or nearly so. Lesser and marginal coverts black or blackish (usually grey in *aucheri* and *pallidirostris*, but perhaps some overlap). Moults rather late, often between (Jul) Aug–Nov. Sexes nearly the same size. **L** 23–26 cm; **W** ♂ 106–118 mm (n 28, m 111.8), ♀ 104.5–117.5 mm (n 17, m 110.5); **T** ♂ 104–125 mm (n 29, m 112.5), ♀ 104–121 mm (n 16, m 111.0); **T/W** m 100.7; **TG** 24–41 mm (n 46, m 32.5); **B** 19.2–23.8 mm (n 49, m 22.0); **B**(f) 15.9–19.0 mm (n 19, m 17.1); **BD** 9.1–11.0 mm (n 42, m 10.0); **Ts** 28.9–33.3 mm (n 50, m 31.2); **white on pp** visible outside tips of pc (folded wing) 8–23 mm. **Wing formula: p1** > pc 14–24 mm, < p2 25.5–35 mm; **p2** < wt 5.5–9.5 mm, =5/6 (90%) or =6 (10%); **p6** < wt 8–14 mm; **p7** < wt 14.5–19 mm; **p10** < wt 27–29 mm; **s1** < wt 29–36 mm.

○ **L. e. uncinatus** Sclater & Hartlaub, 1881 (Socotra; resident). The 'SOCOTRA GREY SHRIKE' is very similar to *buryi* (rather than *aucheri* as has been stated; Kirwan 2007) having similarly dark grey plumage (only paler throat by comparison), but is c. 5% smaller and has proportionately longer bill and legs (as so often found in island populations), on average less extensive white tips to tertials and smaller white primary patch. Since it differs by c. 2.5% in mtDNA from *aucheri*, being sister taxon to all remaining Great Grey Shrike taxa, and is isolated on an island, it has been suggested as a separate species, but we prefer to keep it as a race of the already polytypic and highly variable *L. excubitor*. **L** 21–23 cm; **W** 96.5–103 mm (n 12, m 100.3); **T** 94–103 mm (n 12, m 97.6); **T/W** m 97.4; **TG** 25–31 mm; **B** 21.3–25.8 mm (n 10, m 23.7); **B**(f) 18.0–21.0 mm (n 10, m 19.4); **BD** 8.6–10.3 mm (n 11, m 9.5); **Ts** 27.8–30.6 mm (n 12, m 29.4); **white on pp** visible outside tips of pc (folded wing) 1–9 mm. **Wing formula: p7** < wt 13–16 mm; **p10** < wt 22.5–27 mm; **s1** < wt 23–28 mm.

(3) 'NORTH AFRICAN GROUP'

L. e. elegans Swainson, 1832 (interior of N Africa from Morocco to Egypt, N of Sahara, in SE to N Sudan; resident). What is here called the 'ELEGANT GREY SHRIKE' is paler than *algeriensis*, with somewhat paler grey upperparts and whiter underparts (breast often faintly tinged cream, flanks sometimes with some pale grey), and the white primary patch generally is much larger, creating superficial resemblance to Great Grey Shrike. Note absence of 'double wing patch', but in fresh plumage white outer edges to secondaries—and edges and broad tips to long tertials—form a white 'panel'

L. e. elegans, ad ♂, Tunisia, May: one of the palest races, but very variable, and extreme individuals can show white supercilium and large white wing panel, encompassing most remiges. Those with most extensive white outer edges to secondaries (like here) can appear to have a second wing patch or even recall Indian *lahtora*. The scapular patches are also extensively white, and sometimes the rump, too. Extensive white areas, plus solid jet-black mask that extends well above bill confirm ♂, while lack of moult limits in wing suggests ad. (R. Pop)

L. e. elegans, ad ♂, Morocco, Mar: age and sex as bird to the left, but white areas smaller, including white line above eye-stripe, which is almost absent. Nevertheless, extensive white primary patch characteristic, and white edges to outer webs of secondaries also present, only are narrower on this bird. Also typical are rather pale grey upperparts, underparts white with a faint cream-tinged hue, and outer tail-feathers extensively white. (D. Monticelli)

L. e. elegans, 1stS, presumed ♀, Western Sahara, Apr: as central bird, but browner and less solid dark mask that does not reach above bill equals a ♀. Quite characteristically, rump patch can be whiter, especially with wear. Note apart from white primary patch also long white panel or line on innerwing formed by white edges and tips to secondaries, typical of this taxon. (R. Armada)

on folded wing (which can appear as a second wing patch, only sited further back). Tends to have outer webs of outer secondaries basally white (often forming concentrated patch on folded wing just outside tips of greater coverts), not seen in other taxa in this group or in 'Indian Grey Shrike'. In flight or in the hand extensive white portions visible on inner webs of secondaries. No barring below in any plumage. White supercilium variable, usually narrow and short, but some have long and distinct one, even reaching above bill. Sexes alike or very similar. Black mask sometimes reaches to 2 mm above base of bill in ♂ (never in ♀). Much white on outertail, rr5–6 usually all white (except dark shafts), but r5 sometimes has limited black basally or on inner web. Complete moult sometimes well into Nov. 1stY frequently moults some outer primaries and inner secondaries during 1stW, making at least some 1stS easy to age using contrast between darker outer and browner, more worn, inner primaries. Juv has tertials, r1, greater coverts and uppertail-coverts paler and tinged brown. Often extremely similar to *pallidirostris*, which see for slight average differences. A rather variable race, which increases the identification difficulties. **L** 22–24 cm; **W** 100–116.5 mm (*n* 81, *m* 106.9); **T** 95–119 mm (*n* 81, *m* 106.1); **T/W** *m* 99.3; **TG** 21–37 mm (*n* 76, *m* 30.4); **width of r1** (measured 30 mm from tip) 8.5–12.5 mm (*m* 10.3); **white on inner web of r6** invariably to base of feather; **B** 19.3–24.3 mm (*n* 79, *m* 21.9); **B(f)** 15.5–19.6 mm (*n* 37, *m* 17.0); **BD** 8.5–10.3 mm (*n* 77, *m* 9.4); **Ts** 27.5–32.6 mm (*n* 81, *m* 30.1); **white on pp** visible outside tips of pc (folded wing) 14–27 mm, **white on ss** usually not visible on folded wing except. **Wing formula:** p1 > pc 14–24 mm, < p2 27–36.5 mm; **p2** < wt 5–10 mm, =5/6 (92%) =5 (4%), =4/5 (2%), or =6 (2%); **p6** < wt 7.5–16 mm; **p7** < wt 12–21.5 mm; **p10** < wt 24–33 mm; **s1** < wt 25–36 mm. – Apparently grades into *aucheri* in Sinai and (?) E & N Egypt, but true status and genetics of birds in Libya, Egypt and N Sudan should be further investigated. – Grades into *algeriensis* in the interior of NW Africa, and intermediates from zone of intergradation sometimes separated, e.g. '*oasis*' (Biskra, Algeria) or '*batesi*' for birds approaching *elegans* but often a trifle darker, as in S Tunisia; however, too variable to be recognised as a separate subspecies. (Syn. *dealbatus*; *oasis*.)

L. e. algeriensis Lesson, 1839 (coastal NW Africa; resident). The 'Algerian Grey Shrike' is darker than *elegans* but the same size. It is similar to *koenigi*, being greyish on breast and flanks, but is slightly larger and often even darker grey on crown, mantle and upper shoulders (the darkest *algeriensis* are darker even than ssp. *buryi*); it usually lacks a white supercilium (but may have traces of grey-white above black mask near eye). Sometimes faint ochre tinge below and faint grey barring on breast. Rather restricted white on outertail, and tends to have black spear-shaped marks along shafts of outer tail-feathers. Sexes differ on average in that ♂ has 1–2 mm black on forehead over bill, ♀ not (or only diffuse hint of black). **L** 23–25 cm; **W** ♂ 103–112 mm (*n* 16, *m* 108.8), ♀ 101–111 mm (*n* 17, *m* 106.8); **T** ♂ 101–115 mm (*n* 16, *m* 109.9), ♀ 99–114 mm (*n* 17, *m* 106.4); **T/W** *m* 100.1; **TG** 24–40 mm; **B** 20.0–23.6 mm (*n* 44, *m* 21.9); **BD** 9.4–10.8 mm (*n* 33, *m* 10.0); **Ts** 28.7–32.5 mm (*n* 38, *m* 30.6); **white on pp** visible outside tips of pc (folded wing) 4–17 mm. **Wing formula:** p1 > pc 13–22.5 mm, < p2 27–37 mm; **p2** < wt 5.5–12 mm, =5/6 (89%), =6 (8%), or =6/7 (3%); **p6** < wt 8–13 mm; **p7** < wt 13–18.5 mm; **p10** < wt 25–29 mm; **s1** < wt 26–34 mm. – Grades into *elegans* in the interiors of NW Africa. Intermediates from zone of intergradation given various names, e.g. '*dodsoni*', '*batesi*'; however, too variable to be recognised as separate subspecies.

Intergrade *L. e. elegans* × *algeriensis*, 1stS ♂, NC Morocco, Feb: the pale *elegans* is normally found in inland deserts and mountainous habitats, whereas the darker *algeriensis* is confined to coastal lowlands. Somewhere between these the two races intergrade. Here a bird with rather dark grey crown and nape, lacking white supercilium, still overall paler than typical *algeriensis*. This 1stY bird has recently moulted tertials, innermost secondaries and longest primaries, these being black and tertials having extensive white edges and tips. Sex inferred by solid black mask that extends just above bill. (C. Gouraud)

L. e. elegans, young ♂, SE Morocco, May: southern populations are early breeders, and young of the year can appear already in spring in such advanced plumage. Race inferred by locality but also by quite pale colours, while young age obvious from unmoulted juv greater coverts broadly fringed and tipped buffish-white. Sex indicated by such dark and complete mask at young age, but also by near-black flight-feathers. (J. Thalund)

L. e. elegans, juv (left: Western Sahara, Apr) and 1stW (right: Italy, Nov): juv mostly characteristic with still small bill and pale plumage, with pronounced broad rusty-brown fringes to wing-coverts and tertials (left), often even more striking in autumn (right) if juv feathers not moulted. On 1stW note white fringes to inner secondaries forming broad white panel, and bill being fully grown and already more uniform darkish, unlike bill in ssp. *pallidirostris*. The short primary projection is also noteworthy. (Left: R. Armada; right: I Maiorano)

L. e. algeriensis, ad ♂, Morocco, Mar: darker than neighbouring ssp. *elegans* but almost as dark as *koenigi*, and usually lacks any trace of white supercilium (or has only vague pale line above eye). Note rather restricted white in wing and outer tail-feathers. Jet-black mask extending above bill as thin line, confirming sex, while lack of moult limits in wing infers age. (D. Monticelli)

L. e. algeriensis, 1stS, presumed ♂, Morocco, Feb: almost dusky-grey underparts enhance white-throated impression, and sometimes shows faint ochre tinge on breast-sides in certain lights. ♂ by some black on forehead, rather large white primary patch, and dark-looking flight-feathers. 1stS by moult limits in wing. (D. Occhiato)

L. e. algeriensis, 1stS ♀, Western Sahara, Apr: as image to the left, but a paler individual with marked ochre tinge in the dusky grey below (which might be due to staining). Typically lacks any white supercilium and the black mask being narrow in its fore part. (P. Adriaens)

L. e. koenigi Hartert, 1901 (Canary Islands; resident). The 'CANARY ISLANDS GREY SHRIKE' is slightly smaller than *elegans* and appears large-headed and small-bodied in comparison. It is dark grey above, especially on crown, and largely pure grey on breast and flanks. Black face mask extensive, white supercilium narrow and often indistinct. Sexes of very nearly the same size in examined sample. **L** 21–23 cm; **W** 95–106 mm (*n* 32, *m* 101.1); **T** 93–112 mm (*n* 32, *m* 103.0); **T/W** 101.9; **TG** 22–34.5 mm (*m* 27.5); **width of r1** (measured 30 mm from tip) 7.5–11.0 mm (*m* 9.5); **B** 20.1–24.5 mm (*n* 32, *m* 22.7); **B(f)** 15.9–19.6 mm (*n* 27, *m* 17.2); **Ts** 27.7–31.2 mm (*n* 31, *m* 29.8); **white on pp** visible outside tips of pc (folded wing) 5–14 mm. **Wing formula: p1** > pc 11.5–19 mm, < p2 26–31.5 mm; **p2** < wt 7–9.5 mm; **p10** < wt 18–27 mm; **s1** < wt 21–31 mm. – Sexes virtually identical in appearance, but slight sexual size dimorphism found in sample of live birds (Gutiérrez-Corchero et al. 2007): **W** ♂ 96–103 mm (*n* 44, *m* 99.3), ♀ 92–102 mm (*n* 17, *m* 97.0), **T** ♂ 100–113 mm (*m* 106.3), ♀ 97–109 mm (*n* 17, *m* 102.7), **B** ♂ 12.2–14.6 mm (*n* 44, *m* 13.4), ♀ 12.0–13.4 mm (*n* 17, *m* 13.3), **BD** ♂ 8.5–9.8 mm (*n* 44, *m* 9.1), ♀ 8.4–9.3 mm (*n* 17, *m* 8.8).

Extralimital races: **L. e. leucopygos** Hemprich & Ehrenberg, 1833 (Sudan, much of sub-Saharan Africa, possibly S Algeria but information lacking; resident). Here named the 'SAHEL GREY SHRIKE', similar to *elegans* in coloration, although

L. e. koenigi, ad ♂, Canary Is, Feb: this insular race is rather small, with relatively short wings, long tail, and proportionately strong bill and legs, rather dark grey plumage with contrasting white throat, and attractive black-and-white wing pattern. Evenly fresh wing suggests ad, while jet-black lores and wing-feathers confirm ♂. (M. Schäf)

L. e. koenigi, ad ♀, Canary Is, Feb: almost as dark as Iberian Grey Shrike, with rather similar white pattern in flight-feathers and tail, being rather limited, but unlike Iberian supercilium less pronounced or almost absent; underparts also pure grey, not pinkish. Colour and pattern further resemble ssp. *algeriensis*, but *koenigi* is smaller, with shorter wings and tail, and bill proportionately longer. Age due to wholly ad wing, ♀ by narrower and browner mask that does not reach above bill. (M. Schäf)

L. e. koenigi, 1stS ♂, Canary Is, May: large-headed impression, with small white primary patch and white on inner secondaries and tertials confined to outer edges and tips forming small and isolated white mark. The visible primaries are juv, while the extensive solidly jet-black mask that reaches to forehead infers sex. (H. Shirihai)

on average even paler above, and often has more obviously white rump, and underparts tinged stronger cream. Somewhat smaller than *elegans*, with smaller bill and more rounded wings. Often has narrow black on forehead above bill in ♂. **L** 20–22 cm; **W** 93–105 mm (once 108; *n* 27, *m* 100.0); **T** 92–108 mm (*n* 27, *m* 101.6); **T/W** *m* 101.6; **TG** 22–35 mm (*m* 30.8); **width of r1** (measured 30 mm from tip) 8.5–10.2 mm (*m* 9.1); **B** 16.8–21.7 mm (*n* 27, *m* 19.8); **B**(f) 14.0–16.2 mm (*n* 17, *m* 15.4); **BD** 7.9–9.2 mm (*n* 27, *m* 8.4); **Ts** 25.3–29.7 mm (*n* 25, *m* 27.8); **white on pp** visible outside tips of pc (folded wing) 9–20 mm. **Wing formula: p1** > pc 12–23 mm, < p2 23–34 mm; **p2** < wt 4–11 mm, =5/6 (common), or =5 or 6 (rare); **p6** < wt 7.5–12 mm; **p7** < wt 13–18 mm; **p10** < wt 23–30 mm; **s1** < wt 24.5–33 mm. (Syn. *orbitalis*; *pallens*.)

? *L. e. jebelmarrae* Lynes, 1923 (breeding range unknown; described from winter in Jebel Marra, W Sudan). A controversial subspecies known only from four Nov specimens (NHM). These differ from only fractionally larger *leucopygos* in having grey, not white rump, and proportionately slightly longer tail, but otherwise are very similar. Smaller than both *elegans* and *aucheri*, which they also resemble in several respects. Could represent intermediate population between all three subspecies mentioned, but difficult to assess definitively based on so few specimens. **L** 21.5–22 cm; **W** 97–100 mm (*n* 4, *m* 99.3); **T** 103–108 mm (*n* 4, *m* 105.1); **T/W** *m* 105.9; **TG** 29–34.5 mm; **B** 18.3–20.2 mm (*n* 3, *m* 19.3); **BD** 7.0–8.4 mm (*n* 4, *m* 8.0); **Ts** 27.0–28.3 mm (*n* 4, *m* 27.7); **white on pp** visible outside tips of pc (folded wing) 8–14 mm.

TAXONOMIC NOTES The taxonomy of the Great Grey Shrike complex, including the closely related Northern Shrike and Iberian Grey Shrike, was examined based on mtDNA and its phylogeny was estimated (several independent studies, Mundy & Helbig 2004, Gonzalez et al. 2008 and Klassert et al. 2008, but most extensive coverage of taxa in Olsson et al. 2010). This showed that the usual modern division into a northern and a southern species, with Iberian referred to the Southern, does not comply with genetic evidence. Iberian Grey proved to be closely related to Northern Shrike, and the latter not be closely related to Great Grey Shrike despite the apparent clinal connection in the taiga belt between these two. Other surprising results were that the very similar and allegedly intergrading taxa *elegans* and *aucheri* are in different clades, and that *pallidirostris*, *aucheri* and *lahtora*, which formed a tight group, proved to be genetically relatively close to Great Grey Shrike of N Europe, and comparatively more distant from the N African taxa (which they phenotypically resemble more closely). It is not easy to interpret these findings into a clear and functional taxonomy, but to avoid non-monophyletic species we have provisionally split the complex into three rather than two species. However, it should be noted that the Great Grey Shrike as presented here is a mixed and heterogeneous assemblage, and might require further splits into two or more species as further data become available. It is also fair to mention that the genetic findings cited above have been questioned (e.g. by Panov & Bannikova 2010) since they are based solely on mtDNA (with possibly resulting artefact relationships) and are in such obvious conflict with morphologically-based phylogenies. Only more research will show which taxonomic model is preferable. – One of the paralectotypes of *leucopygos* (ZMB 1802) has wing 108 mm, slightly grey-tinged flanks, a rather large bill and long primary projection, and is probably rather an intermediate between *aucheri* and more typical *leucopygos* from centre of its range than an ideal type specimen. However, another paralectotype (ZMB 1800) from same locality and series is typical in every respect. – Tajkova & Redkin (2014) suggested that *homeyeri* is a colour morph or variation within *excubitor* and synonymised these two, and instead recognised *leucopterus* as the race breeding in SW Siberia and N Kazakhstan. However, apart from overlooking that *przewalskii* has priority as subspecies name (whereas *leucopterus* is a valid morph name), this view does not explain how breeders with *homeyeri* traits occur in SW Siberia and N Kazakhstan, sharing this range with birds with *leucopterus* traits. It seems best to view *leucopterus* as a colour morph of *homeyeri*, latter the older name of the two.

REFERENCES Bogdanov, M. N. (1881) *The Shrikes of the Russian Fauna and Allies*. [In Russian.] St. Petersburg. – Gutiérrez-Corchero, F., Campos, F. & Hernández, M. A. (2007) *Ardeola*, 54: 327–330. – Gonzalez, J. et al. (2008) *J. of Orn.*, 149: 495–506. – Kirwan, G. M. (2007) *Sandgrouse*, 29: 135–148. – Klassert, T. E. et al. (2008) *Mol. Phyl. & Evol.*, 47: 1227–1231. – Meinertzhagen, R. (1953) *BBOC*, 73: 72. – Mundy, N. I. & Helbig, A. J. (2004) *Molecular Ecology*, 59: 250–257. – Olsson, U. et al. (2010) *Mol. Phyl. & Evol.*, 55: 347–357. – Panov, E. N. & Bannikova, A. A. (2010) *Sandgrouse*, 32: 141–146. – Perttula, P. & Tenovuo, J. (2002) *Alula*, 8: 54–60. – Stepanyan, L. S. (1990) *Conspectus of the ornithological fauna of the USSR*. Moscow. – Tajkova, S. U. & Redkin, Y. A. (2014) *Journal of the Natural Museum, Prague*, 183: 89–107. – Tenovuo, J. & Vamela, J. (1998) *Alula*, 4: 2–11.

L. e. koenigi, juv, Canaries, May: juv of this taxon is similar in coloration to juv ssp. *excubitor*, but it is darker grey above and usually shows no or only very insignificant and limited dusky barring below as here. Also, the dark mask is rarely if ever as prominent and developed as here in ssp. *excubitor*. (H. Shirihai)

L. e. leucopygos, ad ♂, NE Chad, Oct: the least-photographed taxon, overall very similar to ssp. *elegans* although averages smaller, with smaller bill, proportionately slimmer body and larger head, relatively shorter wings, on average even paler above, and often has more obviously white rump. Evenly fresh ad wing, while deeper mask reaching above bill and jet-black lores and wing-feathers confirm ♂. (P. Vangiersbergen)

NORTHERN SHRIKE
Lanius borealis Vieillot, 1808

Alternative name: Siberian Grey Shrike

Fr. – Pie-grièche boréale; Ger. – Sibirienraubwürger
Sp. – Alcaudón boreal; Swe. – Sibirisk varfågel

Recent genetic studies of the Great Grey Shrike complex showed that the barred forms of Siberia, the Central Asian mountains, and of Arctic and boreal North America, are all closely related, but on the other hand rather distant from those living in much of Europe, the Middle East and N Africa. These barred forms are therefore combined here into a separate species, the Northern Shrike (which is the established name for the North American birds, with the range in America constituting a large part of this species). Like the Great Grey Shrike in Fenno-Scandia and Russia, it lives mainly on bogs in taiga, but one race has adapted to more alpine habitats and mountain slopes. It is a medium- to short-range migrant, known to have straggled to Fenno-Scandia, and may easily be overlooked in autumn when young Great Grey Shrikes look frustratingly similar.

L. b. sibiricus, 1stW ♂, SC Siberia, Dec: very similar to Great Grey Shrike but separated by locality, date, comparatively small white primary patch, all-black secondaries, contrasting whitish rump (though not visible here), faint buff-brown tinge both above and below, and having fine dark barring below even though it is a ♂. Pale-tipped greater coverts reveal that these are juv and define the age, while moderate, thin barring of underparts and darkness of remiges give the sex. (E. Shnayder)

IDENTIFICATION Of similar size to the *excubitor* race of the Great Grey Shrike, with similar proportions, the size of a small thrush, and has rounded wings and long, well-graduated tail. Like that species, it has mainly rather *pale grey* colours relieved only by a *black mask through the eye*, and largely black wings and tail. Unlike Great Grey Shrike in N Europe it is *barred dark below in all plumages* (including in ad ♂) except in worn summer plumage, and has the entire plumage faintly tinged buff-brown when fresh. However, these differences are subtle, and in practice safe separation of Northern Shrike from a young or ♀ of *excubitor* is extremely challenging. Upperparts in Northern are light grey, a trifle paler than in *excubitor*. It has a '*single wing patch*' of variable size, frequently *quite small* (but slightly larger in a few). *Secondaries are invariably very dark*, also on inner web. Pale supercilium narrow and sometimes rather well marked (recalling Iberian Grey Shrike) but in others more diffuse (much like *excubitor*). The dark mask tends to be larger and extend slightly further back in many *sibiricus* versus *excubitor*, but there may be quite extensive overlap if sufficient material is compared. Shoulder patch variable, sometimes rather large (and in some birds faintly tinged buff-brown). *Tail* has usually no white basally, and *rather limited white on sides and outer corners*, often with black shafts reaching far into the white (but note that extralimital *bianchii* (Sakhalin) indeed has white bases to outer tail-feathers). A few birds noted to have much of r6 white rather than only outer half or less, as is the usual pattern. *Lower rump and uppertail-coverts nearly always contrastingly light grey or greyish-white*. First-winter is usually not possible to separate from first-winter *excubitor* Great Grey, but upperparts average slightly paler grey-tinged yellowish, underparts barring is comparatively subtly darker on slightly yellowish-brown off-white basis, and primary patch is mainly confined to outer webs of outer 4–6 primaries, lacking white on inner webs as in normal *excubitor*. Adults in autumn (no pale tips to greater coverts, on average slightly darker bill) with much barring below from throat to undertail-coverts, and with faint rusty-brown hue on breast-sides and flanks, are good candidates for Northern, but do take full documentation!

VOCALISATIONS Judging from the American race, the song and calls are very similar to those of Great Grey Shrike. Not sufficiently studied to enable a closer comparison.

SIMILAR SPECIES As mentioned under Identification, frequently confusingly similar or even inseparable from immatures and some ♀♀ of *excubitor* Great Grey Shrike. Combination of small white primary patch (with in young birds white mainly confined to outer webs), contrasting whitish rump, all-black secondaries, limited white on outertail, prominent barring over entire underparts and faint yellowish-brown cast on upperparts and sides might help to identify a vagrant Northern. Any adult in autumn with completely barred underparts should be easier, since in Great Grey *excubitor*, adult ♂ is entirely unbarred below, and adult ♀ is lightly barred on breast and sides alone, and can look unbarred at a distance. However, heavily-worn breeders of Northern are likely to lose all barring below, hence even summer birds could potentially resemble Great Grey Shrike! – Superficially similar to immature *Lesser Grey Shrike*, but apart from being somewhat larger and heavier, note shorter and more rounded wings on Northern (primary projection 2/3 to 3/4 of exposed tertials, but equal to or exceeding tertials in Lesser Grey) with smaller primary patch, and barring *below*, not above (young Lesser Grey: diffuse dark barring *above*, but not much below).

AGEING & SEXING (*sibiricus*) Ages and sexes differ slightly. – Moults. Complete post-nuptial moult of ad in late summer (Jul–early Oct), partial post-juv of 1stY at same time not involving primaries, secondaries, tail-feathers (except rarely r1) or primary-coverts, and none or few greater coverts and tertials. Partial pre-nuptial moult (Mar–Apr) of both age groups mainly involves some head- and body-feathers. – **SPRING Ad ♂** Underparts lightly barred grey on off-white ground. Wings and tail black or near-black. Bill and entire mask (including lores) jet-black. May wear to unbarred underparts in summer (requires confirmation). **Ad ♀** As ad ♂, but underparts distinctly barred grey or grey-brown. Wings and tail blackish-grey or dark brownish-black, less black than ♂ in direct comparison. Bill blackish with variable paler base (bluish-grey or horn), rarely all dark. Mask on average less black than in ♂, lores often greyish or greyish-white, but some have virtually all-black mask. As with ♂, may attain nearly unbarred underparts in summer. **1stS** As ad, but many show remnants of pale-tipped juv greater coverts before these entirely wear off. – **AUTUMN Ad** As in spring, but in both sexes bill and mask are less black, base

L. b. sibiricus, ad ♂, SC Siberia, Nov: in spite of being an ad ♂ in fresh plumage is barred dark below, a combination excluding Great Grey Shrike. Degree of barring varies individually and with viewing conditions, and is not always easy to detect, and in worn summer plumage can be entirely missing. White wing patch restricted to primaries. Also unlike pale races of Great Grey, note limited white in tail (right). Aged by lack of moult limits in wing and absence of white-tipped greater coverts. (E. Shnayder)

L. b. sibiricus, 1stY ♂, SC Siberia, Jan: two images of the same bird. Note combination of diffusely brown-barred underparts, dirty brownish hue to grey upperparts, and limited white in wing and tail, which should exclude any Great Grey Shrike (though in theory odd 1stW ♀ Great Grey could be very similar). The comparatively modest barring below on a *sibiricus* indicates a ♂. Also quite typical of 1stY ♂ is the weaker and browner mask and pinkish-based bill. Much of wing still juv. (V. Ivushkin)

of bill and lores become variably paler, most pronounced in ♀. Greater coverts all black or have tiny white spot at shaft (but not buffish or whitish tips across whole greater coverts). **1stW** Resembles ad ♀ but has buffish or off-white tips to greater coverts, forming a narrow wing-bar. Sexes very similar, and usually not separable unless underparts are either rather faintly barred (♂) or very heavily barred grey-brown (♀), and wings and tail are either quite dark blackish-grey (♂) or relatively paler brown-grey (♀). **Juv** Like 1stW ♀, but grey upperparts rather more obviously tinged yellowish-brown, while underparts are even more heavily barred grey-brown, also on throat, and breast and flanks often tinged brown or pink-buff; dark mask poorly developed. Pale tips on greater coverts buffish (later fading to whitish).

BIOMETRICS (*sibiricus*) **L** 23–25 cm; **W** ♂ 113–125 mm (*n* 32, *m* 117.0), ♀ 111.5–122 mm (*n* 21, *m* 115.3); **T** ♂ 105–119 mm (*n* 32, *m* 111.7), ♀ 103–118 mm (*n* 21, *m* 110.0); **T/W** *m* 95.6; **TG** 12–28 mm (*n* 58, *m* 21.7); **width of r1** (measured 30 mm from tip) 13.5–17 mm (*m* 15.0); **white tip of r6** (inner web) 33–53 mm (*m* 42.9); **B** 19.3–25.0 mm (*n* 49, *m* 22.5); **B(f)** 16.2–18.6 mm (*n* 29, *m* 17.5); **BD** 8.5–9.6 mm (*n* 39, *m* 9.0); **Ts** 22.7–27.6 mm (*n* 52, *m* 25.7); **white on pp** visible outside tips of pc (folded wing) 0–17.5 mm; usually no white on ss, but rarely 1–15 mm (possibly intergrades with *L. excubitor homeyeri*). **Wing formula: p1** > pc 10–22 mm, < p2 29–36 mm; **p2** < wt 12–16.5 mm, =6/7 or 7 (96%), =5/6 or 6 (2%), or 7/8 (2%); **pp3–4**(5) equal and longest; **p5** < wt 0–3 mm; **p6** < wt 8–14.5 mm; **p7** < wt 13.5–17.5 mm; **p10** < wt 22–27.5 mm; **s1** < wt 23–30 mm. Emarg. pp3–5.

GEOGRAPHICAL VARIATION & RANGE Mainly moderate clinal variation among birds in taiga belt, but some more abrupt or marked changes in plumage pattern in Far East and in Central Asian mountains. All races are extralimital, but at least one has straggled, and may straggle regularly, to the treated region. Claimed to be a not uncommon vagrant to Ukraine in winter (Tajkova & Redkin 2014).

○ ***L. b. sibiricus*** Bogdanov, 1881 (C & E Siberia roughly from Yenisei eastwards, Transbaikalia, Russian Far East, possibly N Mongolia; winters S Siberia, Central Asia in Kyrgyzstan, E Kazakhstan, Mongolia, Manchuria, NE China; vagrant in Europe, with few records in Ukraine and E Russia, a few unsustained reports from Fenno-Scandia but at least one in Norway confirmed by DNA). Treated above. Some have a little white on bases of secondaries, concealed by coverts on closed wing, rarely more and visible on at least outer secondaries (perhaps the result of cross between *sibiricus* and Great Grey Shrike *homeyeri*). – Apparently intergrades with *mollis* in NW & NC Mongolia. Often quite difficult to separate from some *mollis*, but usually lacks the rufous lower flanks generally found in that race, or at least has this colour much reduced, and especially ♀♀ have darker

(blackish) primaries and secondaries, even in juv plumage (*mollis*: primaries brown) with purer white and more distinctly defined wing patch, also in juv plumage (*mollis*: edges often diffuse, patch sometimes tinged buff or rufous). (Syn. *major*.) – Birds deemed to be hybrids with *homeyeri* have large white primary patch, some white visible also on secondaries, and 3 mm white on forehead. One such bird, at least, recorded in Sweden.

○ ***L. b. mollis*** Eversmann, 1853 (Russian Altai, Sayan, NW & NC Mongolia, Tarbagatay, E Tien Shan, Chinese Turkestan; winters S Central Asia). The darkest form, actually quite similar to *sibiricus* but is slightly larger, has significantly longer legs and is subtly shorter-tailed, proportionately. In fresh plumage it is on average even more heavily barred below, and has a rufous tinge, especially on shoulder patches, uppertail-coverts, breast-sides and lower flanks. Ad ♂ is greyish on mantle (still with brown tinge), but may bleach to purer grey in summer, while heavily worn breeders may lose all barring and become pure grey above and uniform pink-buff below, a both surprising and rarely documented transformation (specimens in ZMMU). Ad ♀ is usually somewhat browner above, but will also become purer grey on back in summer and lose most or all underparts barring. Wing-coverts and secondaries in both sexes rather blackish, but primaries more brown-grey. 'Single wing patch' often less clearly delimited, and faintly ochre-tinged. Sexes very nearly the same size. **W** 110–123 mm (*n* 52, *m* 118.7); **T** 98–120 mm (*n* 50, *m* 110.8); **T/W** *m* 93.3; **TG** 14–27 mm; **B** 19.4–24.8 mm

L. b. sibiricus, 1stW ♀, SC Siberia, Oct: strongly barred below with grey of plumage subdued and rather dominated by brown, buff and gingery hues. Limited white in wing and tail, pale rump and incomplete and brownish mask are other features of young ♀ Northern Shrike. Much of wing still juv. (V. Ivushkin)

(*n* 46, *m* 22.5); **Ts** 25.8–29.5 mm (*n* 50, *m* 27.9). Amount of white in wing and wing formula very nearly the same as *sibiricus*. – Birds in Tarbagatay, E Tien Shan and Chinese Turkestan ('*funereus*') claimed to be darker and duller than *mollis*, but a comparison of fairly extensive material from

L. b. sibiricus/borealis, 1stW, Finland, Dec: one of the few Finnish Northern Shrike records. On current knowledge, it is impossible to distinguish NW Siberian *sibiricus* from N American *borealis* unless handled and biometrics examined. Strongly barred underparts washed pinkish-brown, limited white in wing and tail, and face pattern (more juv-like, or recalling *pallidirostris* Great Grey) in Dec identify this bird as Northern Shrike. Already bill colour infers age. Sex best left open as blackish flight-feathers seem to indicate ♂, while head pattern is better for ♀. (A. Uppstu)

L. b. mollis, 1stS, presumed ♂, Russian Altai, Apr: by range could be either ssp. *sibiricus* or *mollis*, but this bird should be *mollis* being very likely a young ♂ in Apr (black bill, almost complete mask) and considering it has such extensive rufous barring below, brown primaries and white wing patch with slightly diffuse or buff distal edge. Also, note obviously brown-tinged grey upperparts. Aged by prominent pale wing-bar along tips of retained juv greater coverts. (E. Shnayder)

L. b. borealis, 1stW ♂, Azores, Oct: the first confirmed record of this taxon within the covered region. Compared to ad ♂ at same season, note warmer (pinkish-brown) tinge and more extensive barring below, also bleached brown juv outer wing, including primary-coverts, and larger pinkish base to bill. Unlike same age ♀, new wing-feathers black. Aside of the barred underparts, the limited white wing patch and tail-corners, and whiter rump patch, readily distinguish it from Great Grey Shrike. (D. Monticelli)

both areas revealed no such distinction, hence best included here. (Syn. '*funereus*'; see Taxonomic notes.)

Two further races, also extralimital, which will only rarely occur within the treated range but treated for completeness:

L. b. borealis Vieillot, 1808 (E Canada; makes short- to medium-range movements south in winter). Very similar to *sibiricus*, but differs in slightly darker and more brown-tinged upperparts, a fraction darker barring below, on average smaller white primary patch and less white on outertail. Imm frequently has practically no visible primary patch on folded wing, and the distal edge of the patch is broken or irregular. Rump/uppertail-coverts as contrasting whitish as *sibiricus*. Sexes nearly same size. **W** 108–122 mm (*n* 34, *m* 116.6); **T** 92–117 mm (*n* 34, *m* 107.3); **T/W** *m* 92.0; **width of r1** (measured 30 mm from tip) 13–17 mm (*m* 15.5); **white tip of r6** (inner web) 26–50.5 mm (*m* 35.7); **B** 18.7–24.3 mm (*n* 34, *m* 22.1); **B(f)** 16.6–19.5 mm (*n* 34, *m* 17.9); **BD** 8.6–9.7 mm (*n* 33, *m* 9.2); **Ts** 26.0–28.5 mm (*n* 34, *m* 27.1); **white on pp** visible outside tips of pc (folded wing) 1–10 mm. (Syn. *invictus*.)

L. b. bianchii Hartert, 1907 (Sakhalin, S Kurils; winters Hokkaido, Japan). Quite different from rest in being largely unbarred below in ad plumage, pure grey above lacking any brown tones, and therefore is very similar to Great Grey Shrike *excubitor*, differing only in on average whiter rump and having white bases to outer tail-feathers. Some have a slightly more pronounced hook on tip of bill (Hartert 1910); inadequate material examined. **W** 111–116 mm (*n* 10, *m* 113.6); **T** 102–109 mm (*n* 9, *m* 106.3); **T/W** *m* 93.5; **B** 20.4–24.7 mm (*n* 10, *m* 22.4); **B(f)** 17.9–18.6 mm (*n* 10, *m* 18.4); **BD** 8.7–9.5 mm (*n* 10, *m* 9.2); **Ts** 24.5–26.3 mm (*n* 10, *m* 25.5); **white on pp** visible outside tips of pc (folded wing) 6–13.5 mm; very rarely 1–3 mm white on some outer ss. **Wing formula: p1** > pc 14–18 mm, < p2 30–33 mm.

TAXONOMIC NOTES A recent genetic study reconstructed the phylogeny of the entire Great Grey Shrike complex (Olsson *et al.* 2010) and showed that one clade consists of the Northern Shrike, Iberian Grey Shrike, Loggerhead Shrike (*L. ludovicianus*; extralimital), Chinese Grey Shrike (*L. sphenocercus*; extralimital) and Somali Fiscal (*L. somalicus*; extralimital), whereas the forms in Europe, N Africa, Middle East and Central Asia, thus including *excubitor* Great Grey Shrike, are more distant and form a separate clade. Other studies had previously arrived to similar conclusions (Mundy & Helbig 2004, Gonzales *et al.* 2008, Klassert *et al.* 2008). To avoid non-monophyletic species, and preferring not to lump all above-mentioned taxa into one large and rather heterogeneous species, we split the Northern Shrike from the rest of the Great Grey Shrike complex. The apparent smooth morphological transition in W Siberia between Great Grey *excubitor* and Northern *sibiricus*, where many birds are quite similar and

L. b. borealis, ♂, presumed 1stW, USA, Dec: this taxon is very similar to ssp. *sibiricus*, but differs in being slightly duller above, with fractionally darker barring below, on average smaller white primary patch and less white on outertail. Rump and uppertail-coverts are contrasting whitish like in ssp. *sibiricus*. The alula appears juv, suggesting 1stW. (D. Speiser)

difficult to separate, implying some or even much gene flow, requires closer study. Available material from supposed area of contact is scant. See Taxonomic notes under Great Grey Shrike. – Northern Shrike (*mollis*) breeds sympatrically locally with Great Grey Shrike (*pallidirostris*) in Mongolia without mixing (Neufeld 1986, Panov 1995). – Ssp. *mollis* includes the synonym *funereus*, supposed to differ in darker colours and to breed south of *mollis* in E Tien Shan and Chinese Turkestan, and possibly in Tarbagatay and Dzungarian Alatau. According to Hartert, the original description of *mollis* by Eversmann was based on juv plumage (although the type specimen, examined in ZISP, is an ad ♀). As detailed above, there is a fair amount of sexual dimorphism in *mollis*, and whether this played a part or not when Menzbier (1894) described *funereus* as a new taxon, based on a winter ad ♂ from Yining, Tien Shan, is not known. Available specimens of *mollis* and '*funereus*' including types of both (ZISP) show no consistent differences when sex, age and season are separated. – Ssp. *bianchii* groups genetically with *borealis* and *sibiricus* despite appearing morphologically like a relict isolated population of Great Grey Shrike *excubitor*. Worthy of more study. – Several winter records of *sibiricus* in Ukraine reported (Tajkova & Redkin 2014) based on morphology alone.

REFERENCES Gonzalez, J. *et al.* (2008) *J. of Orn.*, 149: 495–506. – Klassert, T. E. *et al.* (2008) *Mol. Phyl. & Evol.*, 47: 1227–1231. – Mundy, N. I. & Helbig, A. J. (2004) *Molecular Ecology*, 59: 250–257. – Neufeldt, I. A. (1986) *Coll. Pap. Zool. Inst. USSR Acad. Sci.*, 150: 3–178. – Olsson, U. *et al.* (2010) *Mol. Phyl. & Evol.*, 55: 347–357. – Panov, E. N. (1995) *Proc. West. Found. Vert. Zool.*, 6: 26–33. – Perttula, P. & Tenovuo, J. (2002) *Alula*, 8: 54–60. – Stepanyan, L. S. (1990) *Conspectus of the ornithological fauna of the USSR*. Moscow. – Tajkova, S. U. & Redkin, Y. A. (2014) *Journal of the Natural Museum, Prague*, 183: 89–107. – Tenovuo, J. & Vamela, J. (1998) *Alula*, 4: 2–11.

L. b. borealis, 1stW, Canada, Sep: if such bird was to reach W Europe it could be very difficult to separate from some young ♀♀ Great Grey Shrike. Note combination of very extensive and rufous-tinged barring on breast, belly and flanks (in Great Grey barring would be more grey-brown), and obvious brown tinge to the grey of upperparts (less brown tinge in Great Grey), which often is on average slightly paler overall. It will also be vital to document the precise wing and tail patterns. Separation from *sibiricus* difficult without biometrics. Juv by pinkish base to lower mandible. (J. Jantunen)

IBERIAN GREY SHRIKE
Lanius meridionalis Temminck, 1820

Fr. – Pie-grièche méridionale; Ger. – Iberienraubwürger
Sp. – Alcaudón real; Swe. – Iberisk varfågel

Long lumped with all the other large grey shrikes, subsequently usually with the so-called Southern Grey Shrike, but recent research has shown that the Iberian Grey Shrike is distinct enough to deserve treatment as a separate species. Its range is widely separated from its closest related forms, which—surprisingly—are those of Siberia and Arctic North America, not the N African populations. It is found over much of Iberia on arid plains with scattered trees, and locally also in S France, where it can be found year-round on calcaric heaths near, e.g., La Camargue and at many other places near the Mediterranean coast. In S France formerly both Great Grey and Iberian Grey Shrike occurred (although in the breeding season not together), but recently Great Grey has become rare and has withdrawn northward.

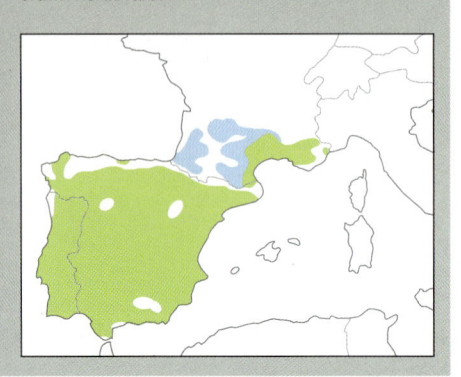

IDENTIFICATION At first glance resembles Great Grey Shrike, especially the N African race *algeriensis*, which is similarly sized and long-tailed with rounded wings, these and tail being black with white marks. However, in a closer look the Iberian Grey Shrike is quite characteristic: underparts dusky, *breast and belly pinkish* with varying greyish or buffish tinge, chin/upper throat and vent diffusely whitish or buffish-white; some birds are more grey than pink on flanks; very *dark slate-grey on crown, nape and upper mantle*, quite dark grey also on rest of mantle and back; *extensive black mask* (narrower on lores) widening to broad black patch behind eye and reaching a good 2 mm below eye; *very narrow but distinct white supercilia*, usually running from rear eye above lores and *joining over bill*; supercilium seen head-on 'undulating', running down over narrow black lores and upwards again over bill and eye; usually quite *small white primary patch* (never any white on secondaries), and medium-large to large (4–8 mm) white tertial-tips patch; rump/uppertail-coverts only slightly paler grey than rest of upperparts (not whitish); fairly large (but frequently narrow) pure white shoulder patches. Somewhat more rounded wing than in Great Grey Shrike. *Strong, long legs*. Bill blackish in most plumages (darkening quickly even in juv). Juvenile matures to adult-like plumage surprisingly quickly, only being usually diffusely barred grey on pink breast and belly-sides; difficult to separate in the field from adult from c. October.

VOCALISATIONS Song appears very similar to that of Great Grey Shrike (which see), and no consistent differences detected. Just like Great Grey Shrike, it has also a subdued, protracted warbling song variation. – Anxiety and alarm calls, too, apparently resemble those of Great Grey Shrike.

SIMILAR SPECIES For separation from superficially similar *Lesser Grey Shrike*, both having pinkish breast in adult summer, note: lack of black forehead expanding onto forecrown (as in ad summer Lesser Grey); slightly duskier throat (Lesser Grey has contrasting white throat); larger size; larger, more elongated head; longer bill with often longer hook on upper mandible; shorter, more rounded wing; longer tail; and usually much smaller white primary patch. Furthermore, adult Iberian Grey invariably has a distinct white supercilium, lacking in Lesser Grey. – In juvenile plumage similar to juvenile *Great Grey Shrike* but differs in stronger buffish-pink tinge, and on grey barring below being paler and less distinct.

AGEING & SEXING Ages differ slightly, but generally less than in Great Grey Shrike. Sexes very similar, often inseparable in the field. – Moults. Feather replacement as in Great Grey Shrike with the difference that the moult period stretches longer into autumn (often until Nov) and that 1stY often moult flight-feathers partly, after which they show contrast between old and newer feathers. – SPRING **Ad ♂** Chin/upper throat, and vent/undertail-coverts diffusely whitish (or buff-white), lower throat, breast and sides of belly salmon-pink tinged grey or buff-grey. Wings and tail black or near-black. Bill and mask (including lores) jet-black. **Ad ♀** As ad ♂, but breast to lower flanks on average slightly less pure pink, more tinged buff-grey, and dark mask sometimes not entirely black but has some grey admixed. Wings and tail on average slightly less blackish than ♂ in direct comparison; in practice, many are inseparable. **1stS** As ad, but some have slightly more abraded and greyish primaries, and most show slight contrast between black inner greater coverts and a few retained outer juv greater coverts that are paler and browner. – AUTUMN **Ad** As in spring, but in both sexes bill and mask are a little less black. Sometimes faint barring on chin and sides of throat (both sexes). Greater coverts can be all black, or black with distinct small wedge-shaped white tips. **1stW** Separated by some outer juv greater coverts being less black with diffuse buff-white edges, or edges and tips, in contrast to blacker inner greater coverts. In early autumn, dark mask often less black and distinct, and often has some barring on chin and sides of throat. Underparts often greyish without pink flush. In late autumn increasingly difficult to separate from ad unless retained juv outer greater coverts visible. **Juv** Grey upperparts tinged brown, face mask poorly developed. Usually more extensively barred (brown-)grey below (except on upper throat and vent). Pale edges, or edges and tips, on all greater coverts are buffish.

BIOMETRICS L 23–25 cm; **W** ♂ 103–113 mm (n 30, m 107.3), ♀ 101–110 mm (n 18, m 105.6); **T** ♂ 105–118 mm (n 29, m 111.0), ♀ 101–113 mm (n 16, m 109.2); **T/W** m 103.4; **TG** 20–39 mm (n 52, m 29.6); **width of r1** measured 30 mm from tip (10) 11–13 mm; **B** 20.2–25.3 mm (n 50, m 23.3); **BD** 8.5–10.0 mm (n 47, m 9.3); **Ts** 27.5–31.8 mm (n 52, m 29.8); **white on pp** visible outside tips of pc (folded wing) (0) 2–10 mm, **white on ss** absent. **Wing formula: p1** > pc 11–19.5 mm, < p2 25–31.5 mm; **p2** < wt 10–15.5 mm, =6/7 (62%), =7 (30%) or =7/8 (8%); **pp3–4** about equal and longest; **p5** < wt 0.5–3 mm; **p6** < wt 5.5–9 mm; **p7** < wt 11–16 mm; **p10** < wt 21–25 mm; **s1** < wt 22–29 mm. – In a study of live birds in Spain (Infante & Peris 2004) practically no sexual size difference was found: **W** ♂ m 107.6 mm (n 21), ♀ m 107.2 mm (n 14), **T** ♂ m 116.8 mm, ♀ m 114.6 mm. In another study of live birds in Navarra, N Spain (Gutiérrez-

Ad ♂, Spain, Jul: very characteristic with underparts extensively sullied salmon-pink, broad black mask and short, narrow but distinct white supercilium, running from above bill to slightly beyond eye. Upperparts rather dark grey (particularly crown), and white primary patch usually small. Inner web of r6 basally black, with distinctive black shaft; rest of tail-feathers show generally less white than most forms of Great Grey Shrike. All visible wing-feathers ad, and intense black of mask and wing-feathers infer age and sex. (A. M. Domínguez)

Ad, presumed ♂, Spain, Jan: due to light and angle, the two wings look very different but surely the right wing appears paler mainly due to light reflection; as far as can be judged all wing-feathers are very fresh, and small whitish tips to greater coverts are often seen in ad. Sexing often difficult in winter, as ♂ to some degree approaches ♀ in mask and underparts coloration as in this example. Very pure grey upperparts and jet-black tail and wings (only the left one possible to judge) perhaps indicate ♂. (A. M. Domínguez)

Ad ♀, Portugal, Apr: sexes very similar, but ♀'s mask sometimes tinged slightly browner or greyer, or narrows obviously near bill-base. Also tends to have underparts duller and slightly less pure pink, tinged buff-grey, especially on sides, while wing-feathers are less intense black. Evenly-feathered wing (as far as can be judged!) suggests ad. (M. Lefevere)

Ad, presumed ♀, Spain, Oct: in autumn, bill often becomes slightly paler at base. An extreme example in having much-reduced salmon-pink below, with much thinner black loral-stripe near bill base, indicating ♀; wing evenly fresh. Note broad black mask bordered by narrow but well-defined white supercilium, which often appears 'undulating' in shape. A much darker grey bird than most populations of Great Grey Shrike. (M. Schäf)

Corchero et al. 2007), the similar size of the sexes was again confirmed: **W** ad ♂ m 106.3 mm (n 51), ad ♀ m 105.6 mm (n 44), **T** ad ♂ m 116.8 mm, ad ♀ m 114.0 mm.

GEOGRAPHICAL VARIATION & RANGE Monotypic. – Iberia, S France; resident.

TAXONOMIC NOTE The Iberian Grey Shrike is similarly distinct as several other allopatric forms recently afforded species status (Madeira Firecrest, Atlas Flycatcher, Corsican Finch, Striated Bunting, to name a few). The combination of pinkish-grey underparts, narrow and distinct white supercilium, the dark upperparts, and the proportionately long tail makes it quite characteristic. In a recent comprehensive genetic study (Olsson et al. 2010), the phylogeny of a large part of the Great Grey Shrike complex was estimated, confirming earlier findings (Helbig in Lefranc & Worfolk 1997, Mundy & Helbig 2004, Gonzalez et al. 2008, Klassert et al. 2008) that the Iberian Shrike is not part of the so-called Southern Grey Shrike, but is instead genetically grouped with the Siberian and Arctic North American clade. As meridionalis is geographically widely separated from these, is resident and has developed a characteristic and invariably diagnosable morphology, and although genetic distance is still small, we see no strong reason not to treat it as a separate species.

REFERENCES Gonzalez, J. et al. (2008) J. of Orn., 149: 495–506. – Gutiérrez-Corchero, F. et al. (2007) Ring. & Migr., 23: 141–146. – Infante, O. & Peris. S. J. (2004) Ardeola, 51: 455–460. – Klassert, T. E. et al. (2008) Mol. Phyl. & Evol., 47: 1227–1231. – Mundy, N. I. & Helbig, A. J. (2004) Journal of Molecular Evolution, 59: 250–257. – Olsson, U. et al. (2010) Mol. Phyl. & Evol., 55: 347–357.

Ad ♀, Spain, Nov: in early autumn, bill often becomes slightly paler at base. An extreme example in having much-reduced salmon-pink below, with much thinner black loral-stripe near bill base, indicating ♀; wing evenly fresh. Note broad black mask bordered by narrow but well-defined white supercilium, which often appears 'undulating' in shape (see also images above as to this). White on lower scapulars narrow and limited, but can still be rather contrasting. (S. Fletcher)

Ad, presumed ♂, Spain, Sep: in moult with outer tail-feathers growing and rather tatty body plumage. Still, salmon-pink flush below (save whiter throat) obvious, indicating ♂, although mask is less solid due to moult. Long and rather narrow tail typical for the species. (H. Shirihai)

1stW, presumed ♂, Spain, Sep: aged by bill colour, and juv greater coverts, these having still prominent but diffuse buff-white tips forming complete wing-bar. In early autumn, dark mask often less black and distinct in both sexes; underparts less smoothly flushed pink, but the extensive amount of pink perhaps favours ♂. (C. N. G. Bocos)

Juv, Spain, Jul: grey upperparts less pure, often tinged brownish, and mask poorly developed; pale edges and tips to greater coverts rufous-buff. Bill-base and lower mandible pinkish. Note that salmon-pink below in juv can be very limited. (A. M. Domínguez)

Juv, Spain, Jul: juv often highly charaarcteristic due to strong buffish-pink tinge below, and does not develop grey bars below. Broad dark mask and narrow white supercilium developed early. Bill consistently two-coloured with lower mandible pinkish-brown and upper more blackish. (A. M. Domínguez)

WOODCHAT SHRIKE
Lanius senator L., 1758

Fr. – Pie-grièche à tête rousse; Ger. – Rotkopfwürger
Sp. – Alcaudón común; Swe. – Rödhuvad törnskata

A colourful and attractive, mid-sized shrike of partly wooded but mainly open, warm habitats in S Europe. Like all shrikes, the Woodchat Shrike is often seen perching on telephone wires or on exposed tall bush or a tree, waiting for a beetle or other large insect to show on the ground below. The loud and varied song often contains mimicry. Has become scarcer, and the range reduced in northern parts. Winters in Africa south of Sahara, and to a lesser extent in S Arabia.

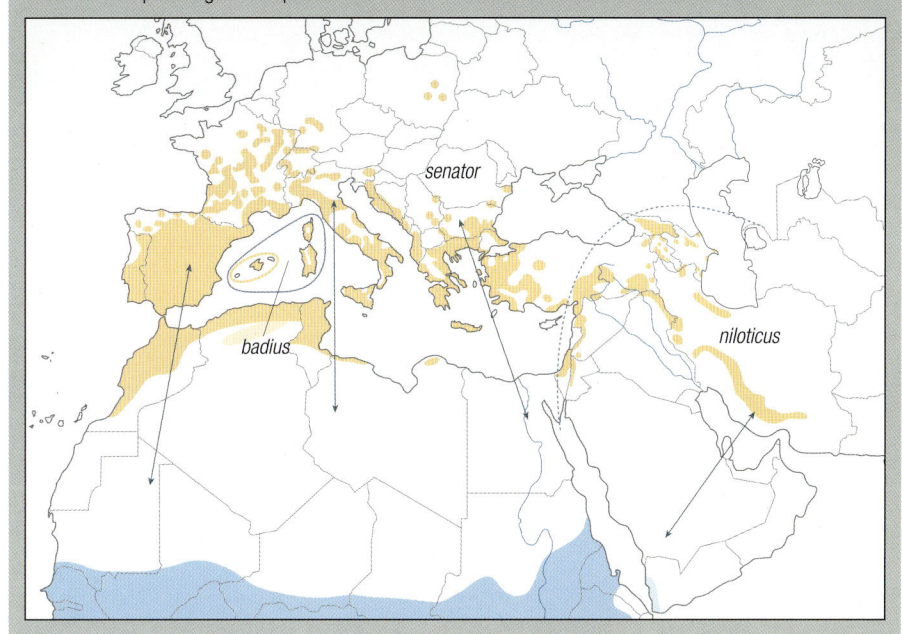

IDENTIFICATION The adult is a striking bird and will hardly cause any identification problems. The combination of black forecrown joining a black mask through eye and continuing down neck-sides, *rufous hindcrown, nape and upper mantle*, *black lower mantle and back* (♀: brown-grey), *whitish uppertail-coverts* and *prominent white shoulder patches*, *black tail with white sides*, and often a peachy-buff hue on breast and flanks makes the adult unmistakable. The ♂ also has ash-grey lower back and rump, and pure white uppertail-coverts and sharply outlined large primary patch. The ♀, apart from the brown-grey lower mantle and back (can rarely be darker blackish-brown but not have jet-black feather centres), has usually a slightly—or markedly—less complete dark forecrown and mask broken up with much pale feathers. ♀ also often has the pale primary patch slightly tinged rufous-buff and less well-marked distally. The juvenile can be more difficult; resembles juvenile Red-backed Shrike but apart from being a trifle larger and proportionally larger-headed, note *hint of pale shoulder patches* and *pale rump* already in juvenile plumage (cream-white or off-white, heavily 'scalloped' or barred dark), and in most a hint of a pale rufous primary patch, or rufous-buff outer webs basally on primaries (exception: race *badius* of large W Mediterranean islands lacks large pale primary patch, has none or only vestigial in ♂). Also, usually paler ochraceous-buff or off-white ground colour on heavily barred crown and nape than in juvenile Red-backed, which is darker and more grey-brown or dark rufous.

VOCALISATIONS Song rather loud, given at somewhat varying but (at least in SW Europe) generally moderate pace, often with phrases well articulated and repeated a few times, usually 2–3, thereby sometimes creating a certain superficial resemblance to Blyth's Reed Warbler. At other times, perhaps as a rule in SE Europe and W Asia, the song is more varied, with few repetitions, but of a similar character. Voice rather sharp, metallic and squeaky. Mimicry of other birds can be incorporated, whole imitations or fragments. Alternative song is more subdued and continuous, a low, squeaky warble, quite like the sub-songs of Red-backed and Lesser Grey, and questionable whether these can be separated; sometimes given from hidden song post unlike the full song, which is delivered from exposed perch. – Anxiety and alarm calls are a series of hoarse, short notes, *veh-veh-veh-veh-veh-...*, or a faster and drawn-out, harsh trill, *chehrrrrrrr*.

SIMILAR SPECIES Juvenile resembles juvenile Red-backed Shrike—cf. Identification. – Shares pattern of white shoulder patches, white primary patches (in most) and black tail with white sides only with adult Masked Shrike, but this is smaller, proportionately slimmer and longer-tailed, has black-and-white head without any rufous, and a dark rump. Juvenile can be confused with juvenile Masked, but apart from structural differences already mentioned, juvenile Masked is basically a grey-and-white bird (brown element subordinate), whereas Woodchat has obvious rufous and ochre-brown elements, notably on tertials and greater coverts, often also on nape.

AGEING & SEXING (*senator*) Ages and sexes differ though not always clearly. – Moults. Strategy complicated, rather like in, e.g., Great Reed Warbler. Complete post-nuptial moult (including replacement of flight-feathers) of ad generally in winter quarters (late Aug–Dec), but frequently partial moult of body-feathers in breeding area, and occasionally 1–3 inner primaries moulted already in Mediterranean region or N Africa (Jul–Aug). 2ndW moults as ad although

L. s. senator, ad ♂, Spain, May: easily identified to species by rufous crown and nape, and broad black mask and forecrown. Mantle and flight-feathers mainly black, indicating sex, with large white primary patch, and another oblong (and even larger) white patch on scapulars. Ad in spring has mostly fresh plumage following recent largely complete moult in winter, but secondaries and wing-coverts still fringed whitish or buff. (M. Römhild)

L. s. senator, ad ♂, Spain, May: conspicuous in flight is the large white uppertail-coverts patch. Black tail has white sides, but unlike in the eastern race *niloticus* the tail-feathers have no white base. Also unlike *niloticus* the white primary patch averages smaller (still large!) and tapers outwards to become narrower at wing-edge. Age and sex as for image to the left. (J. Lidster)

any retained, by then much abraded, juv primary-coverts, inner primaries, and often a few tertials and tail-feathers are moulted in Mediterranean region or N Africa (Jul–Sep). Apparently then suspends, migrates to C African winter quarters to conclude moult. Partial (arrested) post-juv moult in winter quarters (Oct–Feb), involves varying number of outer primaries (commonly outer 7–8 primaries, rarely 6–9), 0–5 outer secondaries, and some to all tail-feathers. – **SPRING Ad ♂** Forehead (except sides over bill, which are generally buff-white) and ear-coverts black, rarely with some limited buff or brown mottling; typically, black reaches down to culmen of bill. Crown to upper mantle deep rufous. Lower mantle/upper back black (or very dark brown-grey with obvious blackish tinge), in fresh plumage tipped rufous-buff. Lower back/rump ash-grey. Uppertail-coverts white or cream-white (except longest, which is somewhat darker). White primary patch when fresh often tinged rusty but this soon bleach to pure white; patch well defined also distally. Underparts nearly invariably unbarred (exceptions very rare). **Ad ♀** As ad ♂, but frequently shows reduced or no black on forehead, black if any as a rule not reaching down to culmen of bill, and lower mantle/upper back dull grey-brown or dark brown-grey (only very rarely has faint trace of black tinge; never jet-black feather centres). Lower back/rump as lower mantle/upper

L. s. senator, ♂ (left) and ♀, Spain, Apr: sexing not always straightforward, but when pair seen together, as here, ♀ at right obviously different by dark mask being incomplete and broken up by white, and crown/nape being a paler rufous (♂ has purer and more solid black mask with white concentrated on lores, and crown/nape darker rufous). (A. M. Domínguez)

L. s. senator, 1stS ♂, Spain, Apr: moult in young is partial, then suspended before autumn migration and resumed on wintering grounds, but pre-nuptial also partial, and all young birds return in spring with moult contrast in wing. Here some central remiges and all primary-coverts were left unmoulted, the normal pattern. (M. Lefevere)

L. s. senator, 1stS ♀, France, Apr: a classic-looking ♀ with poorly developed dark mask and brown-grey mantle and back. Aged by at least juv primary-coverts being brown and abraded in contrast to blacker renewed greater coverts; difficult to say anything firm about age of remiges, but most likely a few inner are also retained juv, only not visible here. (A. Audevard)

L. s. senator, ♀, presumed 1stS, Spain, May: a dull and presumably young ♀ in spring, which has still poorly developed head pattern. Age controversial, and some ad ♀♀ in spring can appear very similar, but apparently central tail-feathers retained juv, brown and worn, and moult contrast in wing of ♀♀ not always easy to see without handling. Note also grey-brown mantle and rufous-buff tinge on primary patch, both typical of young ♀♀. (J. L. Muñoz)

back, or somewhat greyer (though not pure ash-grey as in typical ♂). Pale primary patch white or, more often, partly slightly tinged rufous-buff even in spring, distal delimitation frequently slightly blurred. Rarely some faint barring on breast-sides and flanks (but unbarred birds do occur). **1stS ♂** As ad (more definite plumage develops from late autumn), but usually has a few (commonly 1–3) retained and heavily abraded juv inner primaries, some or many outer secondaries and generally all primary-coverts (though rarely replaces some or all primary-coverts). Often some (rather faint) barring on breast-sides and flanks. **1stS ♀** As ad ♀, but generally possible to age in same way as 1stS ♂. However, difference in colour between old and new primaries, secondaries and wing-coverts less obvious, and some are difficult to age; on such birds note difference in wear, and generally paler shafts to retained juv flight-feathers. Rather commonly has some dark barring on sides below (but barring as such is not a good clue to age, and some ad ♀♀ have bold barring still in spring). – **AUTUMN Ad** As in spring, but plumage abraded. **2ndW** Many recognised by a few retained heavily worn and bleached juv inner primaries, outer secondaries, tertials and tail-feathers. **1stW** Depending on date of hatching and season, either still largely in juv plumage, or much of body moulted to more ad appearance. **Juv** Upperparts off-white or cream-buff to buff-brown, heavily barred from crown to uppertail-coverts. Rear crown and nape often show discernible rufous tinge (in both sexes). Tertials and greater coverts show much rufous and are irregularly barred. Hint of pale primary patch tinged cream or rufous-buff and diffusely outlined, often absent (only ochre outer webs on outer primaries). Rump/uppertail-coverts cream-white, paler than back. No black on upperparts or head.

BIOMETRICS (*senator*) **L** 16.5–18 cm; **W** 93–104 mm (n 114, m 97.8); **T** (70) 72–85 mm (n 118, m 77.0); **T/W** m 78.8; **TG** 8–15 mm (rarely 6.5–18 mm; n 56, m 10.8); **B** 16.7–20.0 mm (n 58, m 18.0); **B**(f) 12.5–14.7 mm (n 40, m 13.6); **BD** 6.3–8.6 mm (n 57, m 7.7); **Ts** 22.0–24.9 mm (n 54, m 23.6). **White on pp** (outer webs) visible outside tips of pc (folded wing) 4–14 mm (once 2 mm on 'retarded' 2y bird). **Black on forehead** ♂ (4) 6.5–15 mm (m 9.5), ♀ 0–12.5 mm (m 4.3). **Wing formula: p1** > pc 4–13 mm, < p2 35–46 mm; **p2** < wt 5.5–9.5 mm, =5/6 (79%), =5 (14%) or =6 (7%); **pp3–4** about equal and longest; **p5** < wt 2.5–7.5 mm; **p6** < wt 8–15 mm; **p7** < wt 14–20 mm; **p10** < wt 24–31 mm; **s1** < wt 26–34 mm. Emarg. pp3–4, and p5 slightly in many.

GEOGRAPHICAL VARIATION & RANGE Three rather distinct forms, differing in plumage and slightly in size. The three subspecies have separate ranges with little or no tendency to clinal connection.

L. s. senator L., 1758 (Europe except Balearics, Corsica, Sardinia; also W Turkey; winters sub-Saharan Africa from Senegal to Sudan). Described above. Moderately large white primary patch. No or only little white basally on tail-feathers, invariably hidden below tail-coverts. Birds in Macedonia and SE Europe commonly have some white hidden at base of tail, on average whiter underparts and larger white primary patch, tending towards *niloticus*. (Syn. *erlangeri*; *flückigerii*; *hensii*; *italiae*; *pomeranus*; *rufus*; *rutilans*; *weigoldi*, see Taxonomic note.)

L. s. badius Hartlaub, 1854 (Balearics, Corsica, Sardinia; winters W Africa east to Cameroon). Resembles *senator* but differs in lacking white primary patch (although rarely may have a tiny one, barely visible). Black on mantle/upper back averages more extensive, ash-grey on lower back/rump is darker, black on forehead averages narrower than in other races (although much overlap). General size a little larger, and bill stronger. Sexes on average more similar in plumage, thus ♀ more ♂-like and invariably has black on forecrown. **L** 17–18.5 cm; **W** 97.5–105 mm (n 30, m 100.7); **T** 74–83 mm (n 30, m 77.5); **T/W** m 77.1; **B** 17.3–20.5 mm (n 21, m 18.9); **BD** 8.3–9.0 mm (n 20, m 8.6). **White on pp**

L. s. senator, ♂, presumed 2ndW, Spain, Aug: post-nuptial moult of 2ndW often starts in summer in breeding areas (note already renewed several inner primaries and some wing-coverts), then is suspended before migration. By and large, 2ndW in early autumn appears much like ad in spring and summer, though unmoulted remiges and coverts are very worn and abraded. (M. Mendi)

L. s. senator, ad ♀, Spain, Jul: just prior to autumn migration, just like ad ♂, ad ♀ is generally very similar to the appearance in spring/summer. The grey-brown mantle and ragged mask and broken-up forecrown band indicate sex, while some dark subterminal marks on breast-sides support it. All wing-feathers seem of ad type and same age, the dark primary-coverts excluding a young ♀. (C. N. G. Bocos)

L. s. senator, juv, possible ♂, Spain, Aug: clearly a juv with profusely barred plumage, best separated from juv Red-backed by large pale primary patch and pale buff-white scapulars being paler than rest of innerwing and mantle. The large white primary patch suggests a ♂, but sexing of juv not easy. (C. N. G. Bocos)

L. s. senator, juv, presumed ♀, Spain, Sep: clearly a juv, and very small pale primary patch suggests a ♀, while mask is indistinct and usually confined to the ear-coverts. Underparts mainly white with obvious dark scaling on breast and flanks. Such greyish young Woodchat Shrikes resemble juv Masked Shrike but are obviously bulkier and shorter-tailed, and has diagnostic pale uppertail-coverts (not really visible here). (M. Goodey)

L. s. badius, 1stS ♂, Mallorca, Spain, May: differs from ssp. *senator* by having white primary patch either small and concealed by primary-coverts, or entirely absent. Also, black forecrown of breeding ad is narrower, rufous of crown is darker, and bill averages slightly deeper-based. The few unmoulted browner and more worn tail-feathers (one central) and wing-feathers (primary-coverts, inner primaries and several secondaries) are juv. (A. Audevard)

L. s. badius, ♀, presumed 1stS, Corsica, Jul: no white bases to primaries visible on folded wing, inferring race. Clearly a ♀, and age presumably 1stS considering quite brown and worn wing and also some juv-like dark subterminal markings to breast-side feathers. Moult has started with some new wing-coverts growing, dark-centred with rusty tips. (U. Paal)

L. s. badius, ad ♂, Mallorca, Spain, Aug: worn ad ♂ during post-nuptial moult (patchy orange-stained feathers are freshly moulted and typically deeper cinnamon-buff than in *senator*). Note black mantle just visible. Also, at this time of year bill tends to become paler than during breeding. (D. Monticelli)

L. s. badius, juv/1stW, Mallorca, Spain, Aug: similar to juv *senator*, but upperparts, head-sides and flanks sometimes slightly darker, with heavier dark bars, while cream-white at base of primaries is absent (but can be replaced by rufous, as here, at least on outer webs), or at least if present not extending beyond primary-coverts on closed wing. Note also the for identification important white shoulder patches. (J. Bazán Hiraldo)

L. s. niloticus, ad ♂, Azerbaijan, May: compared to previous race, generally has more extensive and purer white areas, including larger white primary patch and tail-base (see bird with spread tail at far right). Bill often slightly less heavy at base. ♂ by solid black mask (and note that pale forehead patches are small), crown to upper mantle deep rufous, with mantle/upper back black. Immaculate plumage and evenly-feathered wing infer age. (K. Gauger)

L. s. niloticus, ad ♀, Israel, Mar: ♀♀ told by grey-brown mantle and less developed head pattern with extensive cream-buff patches on face, dark mask often subdued or limited, and chestnut crown slightly duller, too. That some ad ♀♀ have almost as large white primary patch as ♂ is evident here. Wing evenly fresh ad. (L. Kislev)

L. s. niloticus, 1stS, ♂, Oman, Mar: at all ages *niloticus* characterised by extensive white at base of all tail-feathers, here clearly visible from above. Young age obvious from very clear moult contrasts in wing and brownish primary-coverts. Certainly a ♂ considering extensive black mask and very black tail and black new wing-feathers with pure white and sharply delimited primary patch. The seemingly mainly brown-grey mantle and extensive white flecking of forehead can confuse. (M. Varesvuo)

L. s. niloticus, 1stS, ♂, Kuwait, Mar: at all ages *niloticus* characterised by extensive white at base of all tail-feathers, here just visible at the side of tail-base. Young age obvious from very clear moult contrasts in wing and brownish primary-coverts. Certainly a ♂ considering extensive black mask and very black tail and black new wing-coverts. Note that in early spring, the seemingly mainly brown-grey mantle and extensive white flecking of forehead can confuse. (A. Halley)

L. s. niloticus, 1stS ♀, Israel, Mar: often returns very early in spring, then sometimes with quite fresh wing that has wing-coverts and tertials broadly fringed rufous-buff, though as can be seen here at least primary-coverts are juv. White primary patch largely covered by drooping secondaries. ♀ by grey-brown mantle and dull head pattern. (P. Alberti)

L. s. niloticus, ad ♂, Israel, Sep: typical for race *niloticus* is its early complete (or quite extensive) post-nuptial moult that starts on breeding grounds, with some birds as early as in Aug/Sep already having completely replaced their body- and wing-feathers, as here. Black mantle partially concealed by rufous tips to fresh feathers, but black mantle and flight-feathers, and extensive mask, distinguish it from ad ♀. (L. Kislev)

L. s. niloticus, ad ♀, Ethiopia, Sep: easily told as ♀ by the grey-brown mantle and the fairly subdued head pattern, and as ad by very dark primary-coverts and by being completely fresh and with no traces of immaturity. Note that it has already finished complete post-nuptial moult. When very fresh, rusty-buff fringed wing-coverts and tertials can recall those seen in fresh Northern Wheatear. (H. Shirihai)

0 mm (rarely 1–3) [exceptionally 5] > pc. **Black on forehead** ♂♀ 4–10 mm (*m* 6.6).

L. s. niloticus (Bonaparte, 1853) (SE Turkey, Middle East east to SE Iran; winters extreme S Arabia, NE Africa from S Sudan to Kenya in south and to NE Rep. of Congo in south-west). Differs in having white base to all tail-feathers (15–32 mm on central pair, mostly hidden below tail-coverts but often some visible on outermost), usually larger white primary patch (7.5–19 mm > pc; primary patch large and generally mainly pure white already in juv, unlike in juv *senator*, which has it tinged cream or rufous-buff, diffusely edged, and often smaller or even lacking), whiter underparts and different moult strategy (due to earlier breeding, moults earlier, beginning complete post-nuptial moult in Jul and generally finish primaries on breeding grounds, but leaves a few or all secondaries until after autumn migration). 1stY moult extensively but probably never complete, leaving a few inner primaries, outer secondaries and variable number of primary-coverts unmoulted, but in rare cases renew whole wing except primary-coverts. 1stY ♂ often has ♀-like head and mantle until Mar–Apr, its sex revealed only by new wing-coverts being more blackish-based. Black on forehead much like in *senator*, ♀ often having little or none. (Several ♂-like specimens in collections labelled '♀' surely erroneous.) –

L. s. niloticus, 1stW ♂, Israel, Aug: although 1stW *niloticus* can undertake rather extensive post-juv body moult before or during migration, this bird has not come that far. It can still tentatively be sexed as ♂ based on extensive rufous nape/neck and unusually large white primary patches. Especially in this plumage, *niloticus* can appear paler than any *senator*, with white rump and much white on tail-base and primary bases. (H. Shirihai)

Birds in SE Europe and W Turkey are intermediates between *senator* and this race. **W** 93–104 mm (*n* 96, *m* 98.9); **T** 71–87 mm (*n* 94, *m* 79.7); **T/W** *m* 80.5; **B** 17.5–19.1 mm (*n* 25, *m* 18.3); **BD** 7.1–8.6 mm (*n* 20, *m* 7.7). **Wing formula: p1** > pc 0.5–10.5 mm, < p2 39–47 mm. (Syn. *paradoxus*; *pectoralis*.)

TAXONOMIC NOTE Birds in Iberia have a slightly shorter wing than *senator*, W *m* 95.1, and these and birds of NW Africa have been described as ssp. *rutilans*. However, late spring and summer birds of NW Africa, in all likelihood mainly breeders, have wing *m* 97.5 (*n* 30), very nearly the same as breeders in Germany and Poland (*m* 97.7). Between Iberia and Germany/Poland, breeding birds of S France (*m* 98.6), Italy and Croatia (*m* 99.5) have slightly longer wings. Since plumage is the same in all of these areas, the slight and far from geographically clear differences in wing-length seem insufficient grounds for accepting *rutilans*.

REFERENCES Rowlands, A. (2010) *BB*, 103: 385–395. – Small, B. J. & Walbridge, G. (2005) *BB*, 98: 32–42.

L. s. niloticus, juv/1stW, Israel, Sep: clearly a young bird with considerable remnants of dark scalloping above and on upperwing. Large pure white primary patch (even in young) typical for eastern race *niloticus*. Limited rufous on nape or flanks can invite confusion with young Masked Shrike, but that species firmly eliminated by emerging rufous on crown and nape/mantle, cream-white rump and strong bill. (L. Kislev)

MASKED SHRIKE
Lanius nubicus M. H. C. Lichtenstein, 1823

Fr. – Pie-grièche masquée; Ger. – Maskenwürger
Sp. – Alcaudón núbico; Swe. – Masktörnskata

One of the smallest shrikes, long-tailed and slim, almost the shape of a strongly-built wagtail—maybe seemingly a somewhat far-fetched similarity, but enhanced by its undulating flight and long dark tail with white sides. Thrives in more closed habitats than most other shrikes, in woods with glades and undergrowth, at forest edges, etc., and will often perch in the canopy of large deciduous trees, where difficult to spot. Winters in E Africa south of Sahara, and, rarely, in SW Arabia, from where it returns to Europe, Turkey and Central Asia mainly in late March to early May.

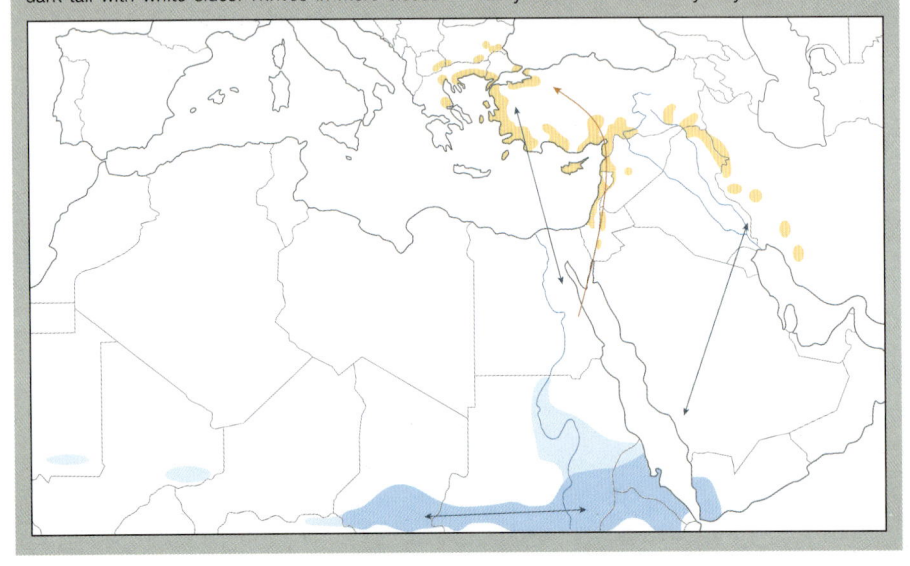

IDENTIFICATION The black (or dark grey), white and 'peachy' colours, the general shape with *narrow long tail*, and large *white shoulder and primary patches* generally identify the species straight away in Europe and Asia. On a closer look, note *white or buff-white forehead extending back over eye*, above black mask. Once status as a shrike is established based on stationary 'surveying' behaviour and bill-shape, adults are generally unmistakable, with ♂ neater and largely *glossy black upperparts*, whereas ♀ has greyish or at most dull blackish upperparts (crown and nape darkest). Both have rusty or *rufous-pink* (peachy-coloured) *belly and flanks*, and some have the same colour on breast but usually less saturated. The juvenile, however, might be a little confusing to the inexperienced, being *greyish above* with dark barring, barred also on the *pale shoulder patches and forehead*; underparts are off-white with faint buff tinge, and some irregular grey barring, especially on sides. As with adults, size and shape are good clues to species, not least including the rather slim shape and delicate bill for a shrike. Note that some early-hatched young can attain surprisingly adult-like plumage by Sep. In general, only first calendar-year birds have any dark barring below, but very rarely a few older ♀♀ can have faint traces on breast-sides.

VOCALISATIONS Song peculiarly resembles a mixture of song of Olive-tree and Upcher's Warblers, being a prolonged raucous and scratchy staccato-like phrase. However, it differs in being more monotonous staying on one pitch, also quite rugged and jolting in structure, and compared to Upcher's is slower and drier without any high-pitched, squeaky notes. It is not quite as deep-voiced and grating as Olive-tree, and with practice it is not difficult to separate. – In anxiety a hoarse, vibrating *chaihr* or *ch'chaihr*, superficially like Common Snipe *Gallinago gallinago* (!), and when more alarmed a series of scratchy notes that become a dry rattle, *chre-chre-chre-chre-chre...* (or *zer'r'r'r'r'r't*), a little like Mistle Thrush but often more extended.

SIMILAR SPECIES Adults unmistakable within treated range. To separate juvenile Masked from reasonably similar juvenile *Woodchat Shrike*, note Masked's slimmer general shape, longer, narrower and darker tail with white sides, smaller head and bill, grey ground colour above (lacking any rufous or ochraceous on tertials, greater coverts or nape), and all-dark rump. – At extreme south of range theoretically possible to encounter vagrant of either of extralimital sub-Saharan and reasonably similar *Somali Fiscal* (*L. somalicus*) or *Common Fiscal* (*L. collaris*), but the former, or the northern races of the latter, lack white forehead and rufous on flanks (though present on southern races of Common Fiscal). Also, note grey or white rump of these African species.

AGEING & SEXING Ages differ subtly, sexes usually clearly (although a few ♀♀ approach ♂ plumage and can be tricky). Size between sexes differs only very slightly (c. 1%). – Moults. Moult strategy rather variable. Complete post-nuptial moult of ad either before (late Jun–Sep) or after migration to winter quarters (Aug–Dec); a third category, apparently including most 2ndY, starts replacing some inner primaries, tertials and tail-feathers before migration, then suspends and resume moult in winter quarters. A variation is to moult all primaries and tail-feathers before migration, but all or most secondaries after, in winter quarters. A few ad (2ndY+) leave odd secondaries unmoulted. Partial post-juv moult protracted (Jul–Dec) and variable, sometimes commences in late summer with moult of body- and head-feathers (new feathers of upperparts sometimes still barred, but pattern then coarser and feathers of better quality), plus tertials and all or most of rectrices (some outer tail-feathers may be left unmoulted), in others (perhaps from late broods) no moult in Oct or Nov; thus, more advanced can start to resemble ad already in mid Sep, others (commoner) retain

Ad ♂♂, Kuwait, May (left) and Greece, Apr (right): unmistakable given jet-black from crown to tail with bold black mask, white 'forehead blaze' and scapular patch, white outer tail-feathers, and conspicuous white patch at base of primaries. Underparts white, except distinctive rufous-orange flanks. Evenly fresh and intense black areas including mantle infer age and sex. (Left: R. Al-Hajji; right: M. Schäf)

many barred feathers (especially on upperparts and vent) until reaching winter grounds, whereupon head and body are renewed to ad appearance (often from Dec). Most 1stY replace tertials and one or a few outer secondaries, and odd tail-feathers in autumn–winter (but no primaries), returning in spring with two generations of feathers, whereas a minority retain all juv flight-feathers. (Juv flight-feathers, especially in ♂♂, last well due to their dark colour.) Generally no pre-nuptial moult in late winter, but a few replace a few tertials and tail-feathers, perhaps just accidentally-lost feathers. — **SPRING Ad ♂** Upperparts (crown to rump) glossy jet-black (but often some grey basally on lower back, which may be partly visible). Wing-feathers dark brownish-black. Some lesser coverts often broadly grey-tipped. **Ad ♀** As ad ♂, but upperparts either brownish-grey or greyish-black (rarely dull blackish, then similar to ♂ in the field). Usually darkest (near-black) on crown, nape and uppertail-coverts, whereas mantle and back are greyer. Wing-feathers grey-brown, not as blackish as in ♂, but a few are darker, almost as in ♂. A very few ad ♀♀ have faint traces of barring on breast-sides. **1stS** Differs from ad in both sexes by showing contrast between newer, darker flight-feathers (notably tertials) and retained juv, which are paler, browner and more abraded. As a rule, all or most primaries and primary-coverts, plus all or a few outer secondaries (and odd other feathers)

Ad ♀, Israel, Mar: similar to ♂ but black of mantle and back, wings and tail replaced by duller grey-brown with variable black suffusion, often showing some contrast between dull black cap and browner mantle, while white of forehead, supercilium, scapulars, primary bases and underparts tend to have cream tinge and often being less sharply delimited from dark adjoining tracts. Age inferred from evenly-feathered ad wing. (M. Varesvuo)

1stS ♂, Jordan, Apr: some 1stS ♂♂ more closely resemble ad ♀♀, with black areas less intense (but still too blackish on mantle and back to be ♀) and pale facial area less pure white. Also characteristic of this plumage is contrast between brown-grey eye-stripe and jet-black crown and mantle. Being 1stS (largely brownish-tinged, worn wing), it is unlikely to be ♀ with such ♂-like plumage. Note that juv primaries especially in ♂♂ remain fresh well into spring due to their dark colour. (H. Shirihai)

1stS ♀, Turkey, Apr: young ♀♀ are dullest with dark areas brownish-tinged. Strong moult contrast between newer, darker tertials and older, browner and more abraded juv primaries and primary-coverts. Diagnostic white 'forehead blaze', long slim tail with white sides, and rufous-orange flanks make this species easy to identify. (P. Leigh)

are old, whereas at least tertials, often also a few inner secondaries are new and black(er). — **AUTUMN Ad** Much as in spring, thus sexes differ. No barring on body or secondary-coverts (except very rarely restricted on breast-sides in ♀). Plumage either partly abraded and partly freshly moulted, or (from late Oct?) moulted altogether without contrast. Uppertail-coverts variable, all black or, more commonly, with extensive white tips or edges. **2ndW** Those 'non-ad' with heavily worn tips to brownish, still unmoulted primaries and tail-feathers presumed to be 2ndW, still wearing many juv feathers; borderline cases do occur, and only obvious birds should be aged. **1stW** Late-hatched still have many barred juv feathers on upperparts and vent in Nov, but when post-juv moult advanced (at times already from mid Sep) becomes quite similar to ad in the field (but invariably separable in the hand). On average browner flight-feathers with more prominent buff-white fringes in fresh plumage. **Juv** Upperparts mainly greyish (actually with buff-brown tinge, but brown colour often unobtrusive in the field) barred dark. Underparts off-white with faint yellowish or buff hue, coarsely barred dark on sides of throat and breast, slightly on flanks, and on undertail-coverts. Large white primary patch as in ad.

Ad ♂, Israel, Aug: sex inferred by black mantle and back, even though face and underparts are quite rufous-tinged. During autumn, plumage by and large resembles spring but is often more patchy due to highly variable post-nuptial moult. Note that bill can become paler from late summer. The characteristic long, slim and rather square-ended tail is noticeable in this image. (D. Laredo)

Ad ♀, Israel, Aug: sex obvious from dark brown-grey rather than black mantle and wing-bend. Still, from a brief glance such dark ♀♀ can appear like imm ♂, though a closer look will prevent a mistake. Note typical jizz and posture, with pronounced tail-waving in this long-tailed species. (H. Shirihai)

Juv, presumed ♂, Turkey, Sep: late-hatched birds are conspicuously barred and highly characteristic due to combination of slim, long-tailed shape, and especially greyish and white plumage. Unlike any other plumage the off-white underparts are coarsely barred dark at sides. The very large white primary patch indicates a ♂ and blacker flight-feathers support this. (R. Debruyne)

Most can be sexed by darkness of flight-feathers, ♂ having blackish, ♀ much or somewhat paler grey (only few being ambiguous).

BIOMETRICS L 17.5–19 cm; **W** 83–96.5 mm (n 138, m 90.2); **T** 80–92 mm (n 137, m 86.9); **T/W** m 96.3; **B** 16.1–19.0 mm (n 100, m 17.8); **BD** 6.0–7.7 mm (n 89, m 6.6); **Ts** 20.5–23.4 mm (n 94, m 22.0). **White on pp** (outer web) visible outside tips of pc (folded wing) ♂ 11–24 mm (n 39, m 17.2), ♀ 7–20 mm (n 52, m 12.8). **Wing formula: p1** > pc 4.5–13 mm (once 15), < p2 30–41.5 mm; **p2** < wt 6.5–12 mm (once 14), =6/7 or 7 (82%), <7 (10%), or =6 (8%); **pp3–4**(5) about equal and longest; **p5** < wt 0–4 mm; **p6** < wt 4–9 mm; **p7** < wt 8–14 mm; **p10** < wt 16–23 mm; **s1** < wt 17–26 mm. Emarg. pp3–5.

GEOGRAPHICAL VARIATION & RANGE Monotypic. – Balkans, Bulgaria, Turkey, Levant, W & SW Iran; winters mainly NE Africa, also west to Chad, probably also S Arabia. (Syn. *atticus*.)

Juv/1stW Masked Shrike (top: Israel, Aug), Woodchat Shrike ssp. *senator* (bottom left: England, Oct) and Red-backed Shrike (bottom right: Israel, Sep): in Masked, ill-defined mask and whitish patch on scapulars produce a pattern rather similar to that of young Woodchat, but Masked should still prove readily distinguishable by smaller size (with slimmer, more delicate build, appearing small-headed and characteristically longer-tailed), greyer upperparts (without rufous-brown on wings and nape), diagnostic dark rump/uppertail-coverts, which are concolorous with rest (normally lacking cream-white patch on rump and uppertail-coverts of Woodchat), and at least two outer tail-feathers largely white. Most 1stW Red-backed lack any hint of pale shoulder patches that are often well developed in Masked and Woodchat, or at most sometimes, as here, shows pale bases to lower row of scapulars in form of vague patches. Red-backed is usually further eliminated by warmer rufous ear-coverts (not visible here though), mantle, uppertail and uppertail-coverts that contrast with greyer nape and rump. All three are also separated by different jizz and tail pattern. (Top: L. Kislev; bottom left: M. Schmitz; bottom right: A. Ben Dov)

(EURASIAN) JAY
Garrulus glandarius (L., 1758)

Fr. – Geai des chênes; Ger. – Eichelhäher
Sp. – Arrendajo común; Swe. – Nötskrika

A widespread and familiar corvid of the Palearctic that exhibits much geographical variation and breeds mostly at temperate and Mediterranean latitudes, as well as east through Siberia and the Himalayas, to Japan and Indochina. Inhabits many woodland types, coniferous, evergreen and open broadleaf deciduous forests, even large wooded parks and suburban areas, including orchards and large gardens. Prefers lowlands, but occurs to the tree-line in some countries. Resident, though subject to some local and altitudinal movements; in some years northern populations undertake large-scale irruptions. One peculiar habit is that of burying food in autumn for winter use, which has probably enabled it to utilise such a wide range of habitats.

G. g. glandarius, ad, Finland, Nov: characteristic domed head shape exaggerated when crown-feathers erected. Striking wing pattern also includes white patch at base of otherwise jet-black secondaries and greyish outer webs to primaries. Innermost tertial partly chestnut but not easy to see. Aged as in image below, and also by quite blackish, broad tail-feathers with broadly rounded tips. (M. Varesvuo)

IDENTIFICATION Unmistakable and rather noisy, its *screeching call* is often heard and recognised before you see the bird. About same size as Jackdaw but has longer tail. Rotund body, domed head, and *short, stout bill*. When perched strikes you at first as 'a large *pinkish bird*', which is unusual (Hoopoe *Upupa epops* has very different structure and frequents more open habitats). Often seen *flying longer distances in slow, straight flight with jerky flaps* of its broad wings, when pattern often striking, especially *bold black-and-white secondaries* and *white rump against blackish tail*. Only in closer views are *glossy azure-blue* (barred black) *primary-coverts and inner greater coverts* visible, but diagnostic when seen. Head pattern variable over wide range, but invariably includes *bold black moustachial stripe or rounded patch*. Crown either black-streaked on white or pale rufous ground, or solidly black. *Iris bluish-white*, bill black and tarsi dull flesh-coloured. No sexual difference as to plumage. Generally shy and wary, and when exited crown is ruffled or raised. Hops and 'bounces' along branches and on ground, often twitching and jerking tail or flicking wings. Several birds can assemble when owl or cat is spotted and noisily mocked.

VOCALISATIONS Song comprises elements of calls, a rather odd mixture of clucking, knocking, mewing and raucous sounds, not far-carrying. – Most familiar call an intense, hoarse *kschaach*, often repeated twice, and used in warning or advertisement (can appear 'hysterical' when several call together, usually on sighting a raptor or other perceived danger). Gives varied low mewing, clicking or chirruping sounds, e.g. Common Buzzard *Buteo buteo*-like *piyeh*, in different social contexts. Mimicry well developed, including many avian and non-avian sounds, which may also be used in the song. Can even mimic its arch-enemy the Goshawk's *Accipiter gentilis* cackle (kya-kya-kya!) or almost any other bird species, e.g. Crane's *Grus grus* flight-call, Blackcap song, and human whistles or machinery.

SIMILAR SPECIES If seen well unmistakable. If seen less well—a glimpse in a forest glade or in distant flight—separation from *Nutcracker* can be a theoretical problem, since both have broad, rounded wings and straight flapping flight

G. g. glandarius, ad, Finland, Nov: a widespread corvid with striking wing pattern, especially glossy azure-blue cross-barred primary-coverts, inner greater coverts and alula. Aged by large number of black crossbars on the bright blue patch, with black bars usually thinner, denser and more evenly spaced, resulting in more regular pattern. (J. Peltomäki)

G. g. glandarius, ad, Spain, Nov: much variation within *glandarius*, some being less grey and more pinkish-buff than typical birds in Scandinavia and N Europe, as shown by this bird in NC Spain. Note whitish black-streaked forecrown, strong black bill and bold black moustachial patch. Finely and densely barred alula-coverts typical of ad. (C. N. G. Bocos)

G. g. glandarius, 1stW, France, Jan: after post-juv moult, very similar to ad, but black-barred blue wing-bend comprises retained juv primary-coverts and some duller greater coverts and alula of more blue-grey appearance; fewer black crossbars variable in spacing and prominence, creating less regular pattern. (A. Audevard)

which can appear somewhat laborious. Note longer tail of Jay, and usually the pink-buff body colour is visible even in brief views. The white rump patch also certainly identifies Jay, as does the large white patch or bar on the secondaries. – Hoopoe shares the pink-buff general body coloration and has similarly broad, rounded wings and 'flappy', unsteady flight. However, note smaller head and body, narrower neck, long, narrow, downcurved bill and more dainty movements of Hoopoe.

AGEING & SEXING (*glandarius* group) Ageing can be challenging, careful inspection of moult, width and shape of tail-feathers and pattern in wing often needed (some can be difficult to determine reliably). Sexes differ only slightly in size (wing-length sometimes useful—see Biometrics—though substantial overlap). – Moults. Complete post-nuptial and partial post-juv moult in summer (mostly Jun–Sep, but sometimes into Oct). Post-juv renewal includes head- and body-feathers, all lesser and median coverts, but only limited and variable number of greater coverts and alula (according to locality). Reports of pre-nuptial (winter) moult need verification, and may represent a continuation of post-nuptial and post-juv moults, as essentially limited to some southern populations that might moult extensively and over prolonged period, including part of head- and some body-feathers, wing-coverts and tertials, perhaps even a few remiges (apparently mostly 1stY). – **AUTUMN** Post-moult ad lacks moult limits, whereas 1stY at least when handled may show some contrast, most obvious in those 1stY that replaced only some greater coverts and tertials, but more difficult to judge in those (few?) that moulted extensively, especially if all greater coverts moulted to ad-like pattern and length. On 1stY that moulted all greater coverts, differences still detectable (with experience) in feather quality and pattern of greater coverts (ad-like) and primary-coverts (retained juv feathers), plus width and shape of outer tail-feathers. **Ad** Black crossbars on blue primary-coverts, alula and greater coverts denser and more numerous, (7) 8–12 bars on outermost greater covert; black tips not counted; typically 9–11. Black crossbars usually slightly narrower and more evenly spaced, imparting more regular overall pattern. Large alula habitually subtly duller blue (at times almost tinged greyish-blue) even in ad. Two alula-coverts bright blue and typically finely and densely barred (11–14 bars). If barring still ambiguous, check if r5 is noticeably rounded at tip (tip usually slightly obtuse) and broad, being 25–29 mm wide (measured *c.* 40 mm from tip), while dark parts of secondaries and tertials are blacker, and glossier. **1stY** Retained juv primary-coverts and alula, and often some outer greater coverts, have fewer black crossbars, (5) 6–8 (9) bars on outermost greater covert if this still juv, barring more irregular and pale areas broader. Since many retain some outer greater coverts, they can be reliably separated in the hand using the contrast these create; juv greater coverts usually also slightly shorter and weaker textured. Black bars often less even in width, and some bars are fainter or variably spaced, creating a more irregular pattern. On average, 1stY have broader black crossbars than ad (though some overlap). Two alula-coverts rather sparsely barred (8–11 bars). It is often advisable to check r5 (21–26 mm wide), while dark of secondaries and tertials is more brownish-grey (less blackish) and outer webs and tips of primaries generally less fresh, looser textured, pointed and sometimes abraded. Furthermore, birds that renew only a few uniformly dark inner greater coverts show contrast between new velvet black and retained juv brownish-black coverts. Feather quality and wear or bleaching differences generally more pronounced towards spring. Winter-moulted birds not studied by us, but 1stY that undergo late moult should show very distinctive moult limits. **Juv** Generally resembles adult (see Geographical variation) but body-feathers characteristically fluffy. Juv *glandarius* and *minor* groups have ground colour of head and body distinctly more reddish, white facial area reduced, with reduced and fainter dark streaks on crown, and no vinous and grey bars on hind-crown, thus somewhat recalling *brandtii* group.

BIOMETRICS (*glandarius*) **L** 32–35 cm; **W** ♂ 169–197 mm (*n* 95, *m* 182.8), ♀ 168–193 mm (*n* 61, *m* 178.9); **T** ♂ 133–162 mm (*n* 95, *m* 150.5), ♀ 133–164 mm (*n* 61, *m* 148.1); **T/W** *m* 82.5; **B** ♂ 29.5–37.8 mm (*n* 57, *m* 33.3), ♀ 28.7–35.2 mm (*n* 43, *m* 32.3); **BD** 11.0–14.5 mm (*n* 100, *m* 12.6); **Ts** 37.5–46.0 mm (*n* 104, *m* 41.7). **Wing formula: p1** > pc 23–33 mm, < p2 35–49 mm; **p2** < wt 30–41 mm, =10/ss or =ss(rarely =10); **p3** < wt 3–16 mm; **pp4--6** about equal and longest (although any of these can be 0.5–3 mm < wt); **p7** < wt 1–8 mm; **p8** < wt 9.5–17 mm; **p10** < wt 26–40 mm; **s1** < wt 29–42 mm. Emarg. pp3–7 (also usually near tip on p8).

GEOGRAPHICAL VARIATION & RANGE Highly complex, with as many as *c.* 45 subspecies proposed just in the W Palearctic, belonging to several more or less distinctive groups, though following our current revision just 11 diagnosable races recognised in four main groups. The arrangement in groups is meant to facilitate the overview, but it

G. g. rufitergum, ad, England, Dec: the British race is generally richer coloured than *glandarius*, more brownish vinaceous-pink, with deeper vinaceous ear-coverts and hindneck that contrast little with rest of upperparts, with grey tone of the nominate virtually absent. Regular barring to light blue wing-bend indicates age. (G. Ferrari)

should be noted that not all groups are clearly defined, and previous authors like Vaurie (1959) or Roselaar (in *BWP*) came to partly different conclusions. Some groups may be products of long-term isolation, which in some cases future molecular (and field) work may prove to be specifically distinct (see Taxonomic notes). – Defining subspecies within the Jay (and indeed the Siberian Jay) is possibly the most subjective and 'blurred' field any taxonomist can venture into. If the intention is to see a tiny difference and name it, ample opportunities exist. Practically all morphological traits vary clinally *and* individually, creating a quagmire of variation, and even when comparing long series (not always the practice in the past) it is possible to arrive at different conclusions depending on the underlying taxonomic philosophy. As Voous (1953) put it in his admirable survey of the variation of the Jay in Europe, 'the intergradations are so gradual and the individual variation is so unexpectedly large that the application of subspecific names as a method of expressing geographical variation has proved to meet with serious difficulties'. The following is based on extensive examination of several large collections, in particular those in NHM, AMNH, MNHN, ZFMK and NRM. Focus has been laid on reasonably distinct variations that affect a clear majority of individuals in any population, meaning that a number of very subtle, variable or questionable subspecies have been treated as junior synonyms. – All populations are mainly sedentary, but some movements towards south and south-west occur in northernmost populations at least in some years.

EUROPEAN JAYS, THE *GLANDARIUS* GROUP
(Europe east to E European Russia; S Caspian Sea, Cyprus) Forehead and forecrown whitish or pinkish-buff, narrowly streaked dark, head and body pale pinkish vinaceous-brown, often with strong grey elements (with rufous undertone) on mantle and back, in contrast to purer rufous tones on hindneck. Underparts extensively saturated vinaceous-buff, but face and bib relatively pale. Variation in Europe mainly clinal, differences often slight, frequently slighter than individual variation, and recognition of several races apart from *glandarius* doubtful and not followed here. Generally becomes gradually pinker (less grey) from C Europe towards west, while southern birds are variable and difficult to fit into a neat pattern. No consistent biometric differences within the group. Size generally large, but two isolated forms south of Caspian Sea and on Cyprus smaller.

G. g. glandarius (L., 1758) (Europe except Ireland, Britain, Brittany, Crimea and easternmost European Russia). Described in main account above. There is extensive variation in plumage colours, some having quite white black-streaked forecrown, others a much more saturated pink-rufous ground colour on much of head, and saturation in pink and rufous colours of neck and body also vary to some extent. It is impossible to say that one type is the 'real' *glandarius*, the other not, since both types—and all shades of intermediates—appear in same areas (albeit with slightly variable proportions in various parts of the range). Some exhibit a rather strong pink-rufous flush on rear head-sides, neck and nape, whereas others are slightly duller greyish-white with reduced rufous or pink. Underparts are generally a rather full greyish-pink, providing contrast to whitish bib, but some are paler below with ensuing less obvious bib contrast. All have a greyish cast on back, although even this is variable, with especially birds from Netherlands (and also E European breeders) frequently having reduced grey on back, being more pink- than grey-tinged. Breeders in Netherlands also tend to be subtly paler on underparts, but a very fine difference and not consistent. – Other variations, sometimes the basis for subspecific separation (but not here), include birds in C & S Iberia said to be greyer below and heavier-billed ('*fasciatus*') or in N Portugal to Galicia and N Pyrenees slightly paler and creamier below ('*lusitanicus*'), but extensive individual variation and many inseparable from *glandarius*. Populations on Corsica ('*corsicanus*') and Sardinia ('*ichnusae*') have been claimed to be distinct, but they seem very close or identical to *glandarius*, Corsican birds sometimes subtly darker but much

Presumed *G. g. glandarius*, ad, England, Oct (left) and *rufitergum*, England, Nov: flies short distances with jerky flaps of broad rounded wings. Striking features in flight are cobalt-blue wing patch, bold white patches on secondaries and rump, and black tail. Left bird has quite grey-tinged back and shoulders, and is therefore most probably not a local English breeder. Right bird is darker rufous overall, typical of the English race. Left bird has broad and blunt-tipped tail-feathers and 11 crossbars on outermost greater coverts inferring age, while the right bird is best left unaged. (Left: R. Steel; right: N. West)

***G. g. rufitergum*, juv, England, Jul**: duller and paler than ad, both pink and grey colours undeveloped, with fluffy body-feathers, and reduced white face and dark streaking on crown still sparse and thin. (A. Warner)

***G. g. glaszneri*, ad, Cyprus, Oct**: a small race with narrowly black-streaked forehead and crown, broadly fringed greyish rufous-brown, lacking white on face. Throat dusky buff-white, often sullied pink-grey. Back and shoulders obviously greyer. Aged by regularly and rather densely black-barred azure-blue wing-bend. (H. Shirihai)

overlap, Sardinian very subtly smaller on average and a trifle paler above, but majority inseparable from *glandarius*. Birds from mainland Italy, Sicily and coast of W Balkans, sometimes all of Balkans and W Turkey ('*albipectus*'), but sometimes restricted to Sicily ('*jordansi*'), have been described as having a quite white head and a paler back, approaching populations in N Africa, but we find them variable and impossible to separate from the quite variable *glandarius*. Populations in SE Europe and Thrace (including '*graecus*' of Greece, S Balkans and S Bulgaria, '*cretorum*' from Crete, and '*ferdinandi*' of E Bulgaria and NW Turkey) are all rather variable and poorly differentiated from each other and other populations of *glandarius* further west or north. Birds of Serbia and Croatia often subtly paler and more greyish-tinged pink below than *glandarius*, but subtle and variable, with overlap, and best included in latter. Attempts to separate populations on some other Aegean islands ('*chiou*', '*samios*', '*zervasi*') seem poorly founded. – Intergrades with *brandtii* in Russia west of Ural Mts over a wide zone, variably-looking hybrids sometimes recognised as separate race ('*severtzowii*') but too variable and better included in either race or defined as intermediate. (Syn. *albipectus*; *athesiensis*; *corsicanus*; *cretorum*; *fasciatus*; *ferdinandi*; *graecus*; *hilgerti*; *ichnusae*; *jordansi*; *lusitanicus*; *septentrionalis*; *yugoslavicus*.)

○ **G. g. hibernicus** Witherby & Hartert, 1911 (Ireland). A very subtle subspecies but recognisable when series of same season and age are compared. Upperparts saturated more darkly rufous, less brown, with somewhat deeper-toned underparts, but otherwise as *rufitergum*. Usually sides of head and neck, and crest are darker rufous than *rufitergum*. Structure closer to *glandarius* with proportionately longer tail than *rufitergum*. **W** ♂ 177–187 mm (*n* 12, *m* 181.8), ♀ 170–183 mm (*n* 13, *m* 175.2); **T** ♂ 145–161 mm (*n* 12, *m* 152.0), ♀ 143–160 mm (*n* 13, *m* 149.8); **T/W** *m* 84.5; **B** ♂ 30.4–36.0 mm (*n* 12, *m* 33.1), ♀ 30.0–33.8 mm (*n* 13, *m* 31.6); **BD** 11.0–14.0 mm (*n* 24, *m* 12.4); **Ts** 40.7–45.5 mm (*n* 26, *m* 42.3).

G. g. rufitergum Hartert, 1903 (Britain, Brittany). Marginally smaller on average and proportionately shorter-tailed than *glandarius* and *hibernicus*, and differs from *glandarius* by more vinaceous-pink (usually much less grey) back, thus no contrast between pink-rufous nape and similar-coloured back (latter lacks clear grey element obvious in other two races). On average slightly more saturated vinaceous pinkish-brown below (warmer cinnamon tone, less greyish and pink) than *glandarius*, but slightly less so than *hibernicus*. Following Ridgway (1912), upperparts are 'Army Brown' to 'Etruscan Red', whereas underparts are 'Light Corinthian Red'. Forehead and crown whitish to buffish-white with blackish streaking appearing on average broader and shorter, and somewhat more rounded, but much individual variation and overlap. A few are difficult to separate from *glandarius*, which in winter may occur in the British Isles. – Populations of Brittany ('*armoricanus*'), and to some extent also of rest of coastal W France, are intermediate between *rufitergum* and *glandarius*, but closer to former, hence included here. **W** ♂ 172–191 mm (*n* 15, *m* 179.3), ♀ 169–184 mm (*n* 14, *m* 174.2); **T** ♂ 135–156 mm (*n* 15, *m* 143.3), ♀ 132–146 mm (*n* 14, *m* 140.5); **T/W** *m* 80.3; **B** ♂ 29.5–36.0 mm (*n* 15, *m* 33.3), ♀ 29.9–37.1 mm (*n* 14, *m* 32.6); **BD** 11.5–14.2 mm (*n* 29, *m* 12.9); **Ts** 39.4–45.4 mm (*n* 30, *m* 41.8). (Syn. *armoricanus*; *caledoniensis*.)

G. g. hyrcanus Blanford, 1873 (Elburz Mts and S Caspian shore in N Iran). The darkest form in the *glandarius* group (from which it is geographically well separated), darker even than *hibernicus*, especially below, though difference is not large. (Note that ground colour of head-sides and fore-crown in *hyrcanus* usually is whitish with subdued pink-buff tinge, much darker rufous-vinaceous in *hibernicus*.) Practically no grey elements above, being saturated dull pinkish-drab (less grey than *hibernicus*). Throat bib sullied greyish pink-buff, less contrasting than in other members of group. Relatively small and proportionately short-tailed, though much overlap with *krynicki*. Sexes appear to be same size. **W** 160–180 mm (*n* 21, *m* 169.9); **T** 124–144 mm (*n* 21,

G. g. brandtii, ad, Mongolia, Jun: relatively small and typically foxy on crown to neck. Forehead and crown densely streaked black, almost coalescing on crown. Upper mantle buff-brown, grading into darker pinkish-grey lower mantle, scapulars and back. Blue tone to pale iris. Azure-blue wing-bend regularly barred, barring fairly dense with 12 bars on alula indicating ad. (H. Shirihai)

m 132.0); **T/W** *m* 77.7; **B** 29.3–36.6 mm (*n* 20, *m* 33.1); **BD** 12.0–13.6 mm (*n* 18, *m* 12.7); **Ts** 37.4–42.5 mm (*n* 21, *m* 40.2). – Intergrades or hybrids with *krynicki* (of '*atricapillus* group') in W Caspian region (SE Transcaucasia) are apparently basis for description of '*caspius*' (type examined; NHM), these having variably black crown and are larger. (Syn. *caspius*.).

G. g. glaszneri Madarász, 1902 (Cyprus). A small and isolated race (nearest population of this group is on Crete). Crown and sides of head and neck darkly rufous-tinged, especially forecrown. Crown streaked black, streaks not merging into dark patch like Turkish *krynicki*. Throat bib dusky buff-white, often sullied pink-grey. Only moderate grey wash on mantle and scapulars. A controversial form that could have evolved in a number of hypothetical ways, but most likely is a relict isolate or that colonised Cyprus from Aegean Is. **W** ♂ 163–175 mm (*n* 15, *m* 170.3), ♀ 158–173 mm (*n* 12, *m* 168.4); **T** ♂ 137–154 mm (*n* 15,

m 144.7), ♀ 135–152 mm (*n* 12, *m* 143.8); **T/W** *m* 85.1; **B** 28.5–34.3 mm (*n* 27, *m* 30.7); **BD** 11.2–13.0 mm (*n* 20, *m* 12.0); **Ts** 37.0–43.0 mm (*n* 27, *m* 39.5).

ASIAN JAYS, THE *BRANDTII* GROUP
(NE European Russia, Urals through Siberia to Mongolia, Japan and China)
Several races described but only one occurs in W Palearctic. Highly distinctive genetically (see Taxonomic notes) and morphologically, being relatively small and characteristically foxy on crown to neck. Forehead and crown narrowly streaked black, often particularly heavily spotted at front, sometimes black almost coalescing. Upper mantle buff-brown, gradually merging into mid grey lower mantle, scapulars and back. Breast rufous as head, contrasting strongly with blackish moustachial stripe and variably pale bib. Belly more greyish-tinged. Reported to have blue tone to iris.

G. g. brandtii Eversmann, 1842 (NE Russia, Ural region, Siberia south to Sayan and Altai, east to Amur and Ussuriland). Markedly rufous-tinged on crown, nape, sides of head and neck. Forehead and crown narrowly streaked black, often particularly heavily spotted at front, sometimes black almost coalescing. Upper mantle buff-brown, gradually merging into mid grey of lower mantle, scapulars and back. Breast rufous as head, contrasting strongly with blackish moustachial stripe. Reported to have blue tone to iris. Throat bib variable, usually dusky greyish-pink and not very prominent, though rarely more contrasting whitish like *glandarius*. Upper breast usually saturated rufous, grading to more pinkish-grey on belly as *glandarius* (thus often quite dusky greyish-tinged below). Comparatively short-legged. – An area in N European Russia (with odd birds also in W Siberia) is inhabited by birds of variable intermediate appearance between *brandtii* and *glandarius*, sometimes referred to as '*severtzowii*'; range difficult to define due to variability of characters, some being close to *brandtii*, others more like *glandarius*. Alternatively could be described as a hybrid swarm. Biometrics of *brandtii*: **W** ♂ 173–186 mm (*n* 17, *m* 179.0), ♀ 171–185 mm (*n* 12, *m* 175.7); **T** ♂ 145–163 mm (*n* 17, *m* 153.2), ♀ 143–162 mm (*n* 12, *m* 152.1); **T/W** *m* 85.9; **B** 30.7–35.2 mm (*n* 29, *m* 32.4); **BD** 10.1–13.6 mm (*n* 27, *m* 12.1); **Ts** 37.5–42.4 mm (*n* 29, *m* 40.3). (Syn. *bambergi*; *kurilensis*; *okai*; *severtzowii*; *taczanowskii*; *ussuriensis*.)

MIDDLE EASTERN JAYS, THE *ATRICAPILLUS* GROUP
(Turkey, Levant, W Asia)
Generally large and long-legged but comparably short-tailed. Contrastingly black-capped, often with rather elongated feathers forming loose crest on hindcrown. Distinctly pale-

G. g. brandtii, presumed 1stW, E Kazakhstan, Oct: this race, which in autumn may move south into Kazakhstan, is typically tinged rufous-orange on head and neck, which coloration extends variably onto upper breast, sometimes faintly also on flanks. Blue patch on wing-bend has rather few dark bands suggesting a young bird, but safe ageing would require handling. (V. Vorobyov)

G. g. krynicki, ad, NE Turkey, May: one of two races in this group, characteristic in being distinctly black-capped and pale-faced. It is also overall pale, with pale pinkish-grey upperparts. Quite evenly and densely barred on bright azure-blue wing-bend, in particular on alula, indicating ad. (C. N. G. Bocos)

G. g. krynicki, ad, Georgia, Apr: note rather elongated crown-feathers, characteristic of the Middle Eastern taxa. Is somewhat darker than *atricapillus*, with less clean white face, and head-sides tinged darker pink-buff. Also slightly richer drab pinkish-brown nape and upper mantle, becoming greyer over rest of mantle and scapulars. Shiny blue wing-bend, with all primary-coverts, greater-coverts and alula evenly and densely barred black implying ad. (V. & S. Ashby)

faced (even whitish, but slight rufous tinge in some populations). Nape and mantle almost uniform. Like in *glandarius* group several indistinct populations have been named, but we recognise just two.

G. g. krynicki Kaleniczenko, 1839 (Rhodes, Lesbos & E Aegean Is, Turkey, Caucasus, Transcaucasia, N Middle East including N Syria, N Iraq, east to Azerbaijan and extreme NW Iran and Zagros Mts; apparently Crimea). Has black-streaked white forehead to fore or mid crown, rearwards becoming solid black. Head-sides variously tinged darker, or paler vinaceous or pink-buff or almost whitish (though not clean white as *atricapillus*). Drab pinkish-brown above, with slightly richer vinaceous hindneck and upper mantle, becoming indistinctly greyer over rest of mantle and scapulars. Underparts usually heavily saturated dirty drab grey-brown with slight pinkish hue. In comparison somewhat darker or more saturated colours than *atricapillus*. Proportionately shorter-tailed but longer-legged than *glandarius*. – Breeders of W, C & S Turkey ('*anatoliae*') often separated on account of intermediate characters between *krynicki* and *atricapillus*, and on average more pinkish forecrown and sides of head than typical *krynicki*, but birds in most of Turkey quite variable and better included in former. Biometrics of *krynicki*: **W** ♂ 177–200 mm (*n* 16, *m* 186.9), ♀ 175–196 mm (*n* 18, *m* 185.7); **T** ♂ 141–167 mm (*n* 16, *m* 150.6), ♀ 135–164 mm (*n* 18, *m* 151.1); **T/W** *m* 80.9; **B** 31.4–37.2 mm (*n* 36, *m* 33.6); **BD** 12.5–15.0 mm (*n* 24, *m* 13.6); **Ts** 40.5–47.0 mm (*n* 36, *m* 43.6). **Wing formula: p1** > pc 31–37 mm, < p2 41–46 mm. (Syn. *anatoliae*; *caspius*; *chiou*; *hansguentheri*; *iphigenia*; *lendli*; *nigrifrons*; *rhodius*; *samios*; *zervasi*.)

G. g. atricapillus Geoffroy Saint-Hillaire, 1832 (Lebanon, S Syria, N & C Israel, NW Jordan, S Iraq, SW Iran). Shares with *krynicki* a complete black hindcrown but has much whiter rest of head with extensive almost pure white throat, white unstreaked forecrown and white (rather than pinkish-tinged) head-sides. It is also considerably paler pinkish-grey below and above. Some slight variation with odd birds showing slight trace of darker *krynicki* colours. **W** ♂ 173–195 mm (*n* 14, *m* 182.7), ♀ 170–188 mm (*n* 12, *m* 179.4); **T** ♂ 133–161 mm (*n* 14, *m* 146.1), ♀ 134–153 mm (*n* 12, *m* 143.5); **T/W** *m* 79.9; **B** 31.0–36.8 mm (*n* 27, *m* 34.0); **BD** 12.3–14.9 mm (*n* 25, *m* 13.6); **Ts** 40.0–45.5 mm (*n* 27, *m* 43.1). **Wing formula: p1** > pc 30–40 mm, < p2 38–43 mm. (Syn. *melanocephalus*; *susianae*.)

NORTH-WEST AFRICAN JAYS, THE *MINOR* GROUP
(Atlas Mts in Morocco and Algeria)
The following two subspecies share streaked crown lacking full black cap, quite marked rufous tinge on nape and neck-sides, at times creating a faint suggestion of a rufous-pink half-collar, subtly darker than rest of upperparts with some contrast to near-white throat and head-sides (though half-collar much less prominent than in *cervicalis*; see below). Complete moult usually finished at end of Oct.

○ **G. g. whitakeri** Hartert, 1903 (Atlas Mts in N Morocco and NW Algeria). Differs from *cervicalis* further east in less extreme pattern, more approaching *glandarius* group. Rufous on nape and neck-sides much less intense and often not forming hint of dark necklace (though some have a trace), white on head and throat less pure and less

G. g. atricapillus, ad, Israel, Mar: broad white face and forehead extending almost to mid crown (dark streaks, if any, narrow and much reduced) and onto head-sides, leaving rather narrow black cap, with contrasting dark bill and black moustachial patch. Both upperparts and underparts almost uniform vinaceous-pink. Aged by ad type primary-coverts. (D. Occhiato)

Presumed *G. g. whitakeri*, Morocco, May: subspecies by range, otherwise virtually identical to *minor*, the other NW African race. Streaky black crown (broad black feather centres and narrow pink-white fringes), vinaceous-brown half-collar reaching rear head-sides, and underparts dirty pinkish-grey, often with vinaceous-brown chest and sides. (V. Legrand)

G. g. whitakeri, juv/1stW, Morocco, Jul: blue wing-bend with irregular and partly well-spaced black barring indicative of juv, as is washed-out body plumage. (C. Gouraud)

extensive. Black broad streaks on crown do not merge into uniform black cap, rufous ground colour on hindcrown visible between streaks, and black ends on rear crown, does not cover nape as *cervicalis*. Tail lacks nearly all visible grey barring on base. Underparts pinkish-grey, slightly paler on belly, more like *glandarius* group. A little larger than *minor* to south, which has otherwise near-identical plumage. Long-tailed. **W** ♂ 169–188 mm (*n* 14, *m* 177.2), ♀ 162–175 mm (*n* 10, *m* 169.1); **T** ♂ 147–166 mm (*n* 14, *m* 155.0), ♀ 141–156 mm (*n* 10, *m* 148.5); **T/W** *m* 87.5; **B** 29.2–40.1 mm (*n* 26, *m* 31.7); **BD** 11.7–13.6 mm (*n* 27, *m* 12.6); **Ts** 39.0–43.6 mm (*n* 27, *m* 41.0). **Wing formula: p1** > pc 26.5–38 mm, < p2 36–42 mm; **p8** < wt 9–13 mm; **p10** < wt 22–33 mm; **s1** < wt 25–34 mm.

G. g. minor Verreaux, 1857 (local in Atlas Mts of Algeria and Morocco south of *whitakeri*). Close to *whitakeri*, and probably intergrades with it where they meet, but is smaller and usually has more saturated rufous-tinged nape and sides of head, not white (but variable, and some are close to *whitakeri* with a hint of whiter face). Upperparts also variable, as a rule slightly darker than in *whitakeri*, often with rather strong grey element and pinkish flush, but some have more rufous-tinged upperparts. Crown streaking also variable, often bold but some have narrower streaks. – Birds from Middle Atlas in Morocco ('*theresae*') are best included as the type (NHM) is inseparable from *minor*, and any average difference must be minute. Birds from Greater Atlas in Morocco ('*oenops*') are likewise inseparable from typical *minor*. No major differences between Algerian and Moroccan birds evident, except that Algerian breeders sometimes are tinged a little more pink below, while Moroccan birds are more uniform dull grey-buff. Moroccan birds also appear a trifle purer grey above, but much overlap between the two areas. **L** 29–32 cm; **W** ♂ 156–172 mm (*n* 14, *m* 164.4), ♀ 151–167 mm (*n* 13, *m* 158.8); **T** ♂ 136–149 mm (*n* 14, *m* 142.6), ♀ 132–147 mm (*n* 13, *m* 138.1); **T/W** *m* 86.7; **B** 27.2–31.9 mm (*n* 27, *m* 29.8); **BD** 10.3–12.4 mm (*n* 26, *m* 11.4); **Ts** 36.0–41.0 mm (*n* 26, *m* 38.5). **Wing formula: p1** > pc 21–33 mm, < p2 32.5–44 mm; **p8** < wt 6.5–11 mm; **p10** < wt 22–28 mm; **s1** < wt 25–30 mm. (Syn. *oenops*; *theresae*.)

UNRESOLVED FORM

G. g. cervicalis Bonaparte, 1853 (Tunisia, NE Algeria). Most distinctive of all forms in the complex. Pure white sides of head and throat contrast against black moustachial and black rear crown (feathers well elongated into loose crest). Head pattern further enhanced by chestnut or ferruginous 'collar' encircling entire neck (though weakest or broken at front). Forehead white, prominently spotted and streaked black. Mantle and back 'Cinnamon Drab' to 'Light Cinnamon Drab', underparts dull 'Vinaceous Brown' (Ridgway 1912). Base of tail has bold grey barring more prominent than in other taxa, reaching far outside tip of wing. Breast tends to be finely mottled or striated grey-brown. Centre of belly near white, contrasting with saturated vinaceous-rufous lower flanks. Long-tailed. Oddly different from geographically close *whitakeri* and hardly a natural part of the '*minor* group', hence placed here in a separate group of uncertain relationships. (Only molecular analysis can resolve this.) Originally described from Algeria; Tunisian birds sometimes named '*koenigi*' but clearly synonymous. **L** 33–34 cm; **W** ♂ 173–192 mm (*n* 20, *m* 182.8), ♀ 170–182 mm (*n* 11, *m* 177.5); **T** ♂ 146–168 mm (*n* 20, *m* 158.4), ♀ 147–161 mm (*n* 11, *m* 154.9); **T/W** *m* 86.9; **B** 29.3–36.7 mm (*n* 35, *m* 32.3); **BD** 10.8–14.3 mm (*n* 34, *m* 13.0); **Ts** 39.0–43.9 mm (*n* 34, *m* 42.1). **Wing formula: p1** > pc 25–34 mm, < p2 38.5–48 mm. (Syn. *koenigi*; *lambessae*.)

TAXONOMIC NOTES Some of the above groups or distinctive isolated forms may prove to be better treated as separate species, but it is too early to speculate further. A preliminary molecular study by Akimova *et al.* (2007) found *brandtii* to be genetically significantly different from *glandarius* and *krynicki* (but note that *brandtii* was only sampled from Primorskie, Russian Far East, hence distance may play part). Differences between *glandarius* and *krynicki* were lower, perhaps suggesting free current or historic gene flow between the *glandarius* and *atricapillus* groups (interbreeding reported from several localities, and Akimova *et al.* 2007 mentioned hybrids in their analysis). However, the extent of contact in Near East and SE Europe between *glandarius* and black-headed *atricapillus* groups not fully understood, and whether they commonly hybridise, while similar uncertainties exist regarding possible hybridisation between other groups. Vaurie (1959) suggested that *severzowii* is a very variable race of hybrid origin (*glandarius* × *brandtii*), and Mayr & Greenway (in Peters 1962) mentioned secondary intergradation between *krynicki* and *hyrcanus* around Lenkoran (mixed population named '*caspius*'). Akimova *et al.* (2007) also pointed out the tiny genetic distance between *glandarius* and '*iphigenia*' of Crimea, despite that the latter is closer morphologically to the *atricapillus* group (and included here in *krynicki*). Evidently, there are several unresolved relationships between groups or taxa in the Jay complex. Only a comprehensive taxonomic revision using a full range of tools including more than one genetic marker will resolve some of these questions.

REFERENCES Akimova, A. *et al.* (2007) *Ross. Orn. Zhurn.*, 16(356): 567–575. – Voous, K. H. (1953) *Beaufortia*, 2(30): 1–41.

G. g. cervicalis, ad, Tunisia, May: highly distinctive. Despite being geographically close to *whitakeri*, has *atricapillus*-like solid black cap and largely whitish rest of head, but also a unique broad chestnut collar reaching across hindcrown to breast-sides. Upperparts predominantly pale vinaceous-grey. Quite evenly and densely barred wing-bend suggests ad. (G. Olioso)

SIBERIAN JAY
Perisoreus infaustus (L., 1758)

Fr. – Mésangeai imitateur; Ger. – Unglückshäher
Sp. – Arrendajo siberiano; Swe. – Lavskrika

In the W Palearctic this bird of remote conifer forests is restricted to northernmost latitudes, where it favours Norway Spruce and Scots Pine festooned with beard lichen, as well as larch and Downy Birch. Readily tolerates human approach, presumably because the bulk of its range is sparsely inhabited. In places, e.g. parts of Finland, its range has decreased in recent decades due to expansion of managed forests, cultivation and settlements. Usually occurs in small territorial flocks with complex hierarchy year-round. Pairs may breed alone or in small groups. Some winter dispersal occurs, and food is stored for winter consumption. Main food is berries and insects.

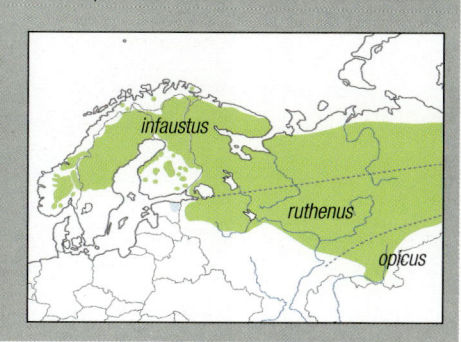

P. i. infaustus, Finland, Apr: this small (roughly thrush-sized), fairly long-tailed jay has a restricted range in W Palearctic, in coniferous forests at northernmost latitudes. Can be rather tame. Note rufous-orange patches on forewing, rear body and tail-edges, and dark brown cap extending below eye. Ageing without handling very difficult. (J. Normaja)

IDENTIFICATION *Small jay* (only somewhat larger than a thrush), *fairly long-tailed* (slightly rounded at tip) and typically soft drab brown-grey, relieved by *dull rufous-orange patches at base of primaries, primary-coverts, on rump and uppertail-coverts and some outer tail-feathers*. Paler bushy nostril feathers form contrastingly *small tufts* either side of bill. *Crown variably darker brown, especially around eyes*, contrasting slightly with paler buff-grey throat. Breast buffish-brown and rest of body dull buffish-rufous. May slightly raise crown-feathers in excitement. Rather large head even when crown-feathers not raised, and has thick neck. All sexes and ages alike. Iris brown. Tarsi and proportionately short but strong, pointed, bill black. Gait comprises short hops and some sidling movements. Manages to cling upside-down, tit-like. *Flight in forests often low*, a burst of wingbeats followed by *long, straight glides. Moves silently through the woods*, in follow-the-leader file. In summer can be rather shy and elusive, unlike in winter when it is much tamer and can approach foresters' or hunters' camp sites looking for food. Also, the incubating ♀ is renowned for its total tameness (similar to Dotterel).

VOCALISATIONS Song, audible only at close range and therefore rarely heard, comprises twittering and chattering notes mixed with melodious whistles and mewing sounds, varying in pitch and speed. Can include mimicry of various other birds. – Subdued *chet* when feeding and climbing, or in flight. Occasionally utters a harsh Jay-like *tchair* (singly or repeated), or a mewing *geeaah*, suggesting Jay or a *Buteo*; the mewing sound can be combined with a brief clicking note, *kla-yeeah*. Also a brisk, cheerful *kyu ke-ke* repeated a few times, a woodpecker-like *kek*, a hawk-like, slightly deeper *ka* and multisyllabic *kvi-kvi-kvi* and *ti-ti-ti-teah*. Bill-snapping also noted.

SIMILAR SPECIES Noticeably smaller and slighter than *Jay*, with much shorter, more pointed bill and longer tail, as well as rather drab plumage, while dusky head and rufous in wings, rump and tail-sides should make the species unmistakable.

AGEING & SEXING Ageing sometimes possible using moult pattern and shape of tail-feathers and alula (but beware some intermediate cases). Sexes differ only in average size (see Biometrics). – Moults. Complete post-nuptial and partial post-juv moult in summer (mostly late May–Aug, rarely mid May–Sep). Post-juv renewal limited to head- and body-feathers, lesser and median coverts (apparently never tertials, greater coverts or alula). No pre-nuptial moult. – **AUTUMN Ad** Evenly fresh, and dark parts of remiges and primary-coverts greyer/blacker and glossier, less brownish. Fringes and tips of primaries fresher and stronger textured. Tip of r6 and alula-feathers smoothly rounded and fresher. **1stY** As late as Sep some retain soft, fluffy juv feathers on throat and neck. Retained juv remiges, primary-coverts, tertials and greater coverts duller/browner, less blackish and glossy, and outer webs/tips of primaries generally less fresh, weaker textured and more pointed, and sometimes abraded. Especially check for contrast between fresh lesser and median coverts, and retained juv rest of wing. Tips of r6 and alula-feathers rather tapering, more pointed, and edge of former slightly frayed. Towards spring, with heavy wear, these differences can become difficult to evaluate. **Juv** Like ad but differs mainly in darker ground colours above, more buff-tinged on forehead and cheeks, paler and duller belly, and much shorter and looser body-feathers, especially on underparts, while rufous in wing and tail is less bright. – **SPRING** Ageing as in autumn but due to bleaching and abrasion more difficult, requiring practice and care, and will leave more birds undecided than in autumn.

BIOMETRICS (*infaustus*) **L** 27–30 cm; **W** ♂ 138–148 mm (*n* 38, *m* 142.9), ♀ 133.5–144.5 mm (*n* 45, *m* 138.9); **T** ♂ 123–145 mm (*n* 38, *m* 132.5), ♀ 118–137 mm (*n* 45, *m* 129.5); **T/W** *m* 93.0; **B** 24.7–31.0 mm (*n* 42, *m* 26.9); **BD** 8.7–11.2 mm (*n* 40, *m* 9.7); **Ts** 34.3–39.2 mm (*n* 43, *m* 36.9). **Wing formula: p1** > pc 26–34 mm; **p2** < wt 23–31 mm, < p10 (usually = ss, rarely =10); **p3** < wt 8–11 mm; **p4** < wt 2–5 mm; **pp5–6** about equal and longest; **p7** < wt 2–4 mm; **p8** < wt 7–11 mm; **p10** < wt 21–25 mm; **s1** < wt 23–31 mm. Emarg. pp3–8.

GEOGRAPHICAL VARIATION & RANGE Mainly only slight geographical variation, description of which is complicated by a certain amount of individual variation and the clinal nature of any slight colour difference, which run in partly different directions. A general trend is seen for colours to become slightly paler and greyer and size larger from west to east, but both in Fenno-Scandia and Far East populations also become increasingly darker and more rufous-tinged from north to south. There are considerable problems to delimit any named subspecies along these clines. We have opted to recognise only reasonably different races occurring within the treated region. (Several more extralimital subspecies have been described, the validity of which has not been examined here, mainly due to scarcity of relevant material in museums visited.) All populations resident.

P. i. infaustus (L., 1758) (Fenno-Scandia and extreme NW & N European Russia). Described above. Relatively pale-bodied with extensive buffish or pale rufous-ochre hue. Rufous in wing averages less bright than in some other races, but still quite extensive. Very slight average tendency for birds to become a little more saturated towards south, but when long series of breeders from N & C Fenno-Scandia are compared no grounds for recognising a different subspecies in the south evident. Grades into *rogosowi* in east (extralimital). – Birds in NE Russia and NW Siberia (N Urals and

P. i. infaustus, Finland, Mar: may suddenly dive straight down from a perch, as shown here. Note diagnostic rufous-orange patches at base of primaries, primary-coverts, on rump and uppertail-coverts and some outer rectrices. (M. Varesvuo)

P. i. infaustus, presumed ad, Sweden, Dec: individual variation includes birds with rather extensive rufous-tinged body, contrasting with greyish breast. Primary-coverts rather broad and rounded at tips suggesting ad. (T. Lundquist)

P. i. infaustus, ad, Norway, Jul: sometimes solid dusky head can appear almost blackish. Note pale bushy nostril feathers forming small whitish tufts either side of bill. Buffish-tinged looser rear body-feathers a feature of ad, as are broad, obtusely tipped outer tail-feathers. (K. Dichmann)

lower Ob) sometimes separated as 'ostjakorum', but apparently only very subtle and far from consistent differences from *infaustus*, and best included here. (Syn. *manteufeli*, *ostjakorum*.)

○ ***P. i. ruthenus*** Buturlin, 1916 (E Estonia, WC to SC European Russia, from St. Petersburg region to Kazan and Perm, and WC Siberia to middle Ob). A little darker and richer coloured on average than *infaustus*, especially has darker brown head, with extensive rufous-buff suffusion to body and brighter rufous rump and larger rufous primary-coverts patch. Still, intergrades smoothly into *infaustus* at northern border, and subspecific determination frequently difficult over a rather large area. **W** (sexes combined) 136–146 mm (n 52); **T** 133–143 mm (n 52); **Ts** 36–39 mm (n 52) (BWP, NRM).

○ ***P. i. opicus*** Bangs, 1913 (SE Russia at middle Volga and Orenburg, SW Siberia through Barnaul and Krasnoyarsk east to Altai). Resembles *ruthenus* but has rather pale grey breast but warm orange belly. Dull, dark cinnamon-brown cap. Quite large reddish wing patch at base of pp6–10 and

P. i. infaustus, Finland, Feb: in flight, note rufous underwing-coverts concolorous with flanks, belly and undertail. Plumage colours blend surprisingly well with bark on trunks of pine trees. (H. Taavetti)

ss1–2. Outer tail-feathers have very little or no grey. Generally more rufous-tinged throughout than *ruthenus*. **W** (sexes combined) 135–145 mm (n 38); **T** 127–142 mm (n 38); **Ts** 35.7–38.4 mm (n 38) (BWP, NHM, NRM).

Extralimital: ○ ***P. i. rogosowi*** Sushkin & Stegmann, 1929 (C Siberia from Tomsk to middle Yenisei). Very similar to *infaustus* but averages a shade paler and greyer. Much smaller orange wing-patch than *opicus*. Scant material examined, but size appeared similar to *infaustus*.

TAXONOMIC NOTE Often considered to form a superspecies with Sichuan Jay *P. internigrans* of W China and Grey Jay *P. canadensis* of North America, which latter replaces it in equivalent habitats there.

P. i. infaustus, presumed 1stW, Sweden, Sep: already seemingly worn tail, not very fresh wing-coverts and overall dull plumage indicate young bird. Sometimes crown in young birds is medium grey-brown with only area around eye blackish. (V. Dell'Orto)

P. i. infaustus, presumed juv, Norway, Jul: ageing Siberian Jay is difficult, not least due to rather loose underbody feathering also in ad. Unkempt feathering of neck indicates juv, and brown of back is rather dull. Still, broad and bluntly rounded tail-feathers more typical of ad complicate ageing, but tips show hint of poor quality indicating juv. (K. Dichmann)

AZURE-WINGED MAGPIE
Cyanopica cyanus (Pallas, 1776)

Fr. – Pie bleue; Ger. – Blauelster
Sp. – Rabilargo; Swe. – Blåskata

A small, elegant corvid found in Spain and Portugal, where it occurs up to *c.* 700 m altitude. Arboreal, it frequents overgrown but never truly dense natural woodlands (both deciduous and coniferous), as well as open plantations in grassland and scrubby areas, and olive groves and cork oak. A European relict of an otherwise E Asian species, occurring from Lake Baikal to Japan, and due to wide gap the Iberian population sometimes split (although here they are kept as one species considering morphological and vocal similarities). Usually nests in loose colonies, with a single nest in each tree. Often parasitised by Great Spotted Cuckoo. Diet consists mainly of seeds of oak and pine, supplemented by fruits, berries, invertebrates and scraps. Stores food in ground. Rather gregarious and, outside breeding season, roams in family parties or larger groups (sometimes numbering a few tens), when occasionally wanders north-east, even to the Pyrénées Orientales (S France). The Portuguese range has increased slightly in recent times.

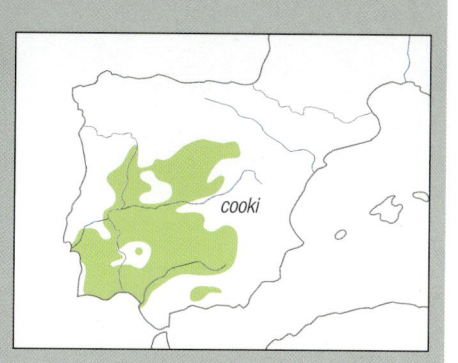

C. c. cooki, ad, Spain, Jul: relatively small body and very long, graduated tail afford Magpie or Great Spotted Cuckoo-like silhouette, but unmistakable due to azure-blue wings and tail. Contrasting black cap against clean white throat, grading into dull pinkish olive-brown underparts. Above, vinaceous drab-brown. Ad by evenly-feathered wings and tail. Tail-feathers lack white tips, and black cap intense and glossy. (C. N. G. Bocos)

IDENTIFICATION A palish, medium to large, and slim magpie, with a relatively small body and very *long, graduated, pale blue tail*. Typically *pale vinaceous drab-brown above*, with contrasting *black cap* and *mainly azure-blue wings*. Soft bluish tones to wings and tail often only visible in close views and good light. Clean white throat gradually grades to cream suffused dull pinkish-brown over rest of underparts. The short wings when folded have a dark lower border formed by dark outermost primaries, and inside this *a white panel formed by outer edges to primaries*. Iris brown, bill and tarsi black. Flight recalls Magpie but is lighter and more elastic, often looser and progress more sustained, birds typically sweeping up into canopy following low flight. Gait more like Jay, with bouncy hops. Forages freely on ground under canopy. Often flicks wings and tail. Can be rather shy and secretive, but often occurs in noisy groups, the birds following one another from tree to tree.

VOCALISATIONS Song seldom heard and poorly known, soft high-pitched chattering and squeaky notes by ♂ when displaying to ♀. – Calls complex, but commonest is a high, grinding, slightly nasal, gently upslurred *vrruih*, an ascending *schrie*, and trilled *screeep*, dry and trembling, all constantly given by flocks on the move. A husky rising whistle *zhreee* or *wee-wee-wee-u*, high *kui* and whining *vih-e* also heard. Alarm a drawn-out, high-frequency, rolling or rattling *krrrrrr*.

SIMILAR SPECIES Given reasonable views unmistakable, but at first glance or in poor views its silhouette could resemble Magpie or even Great Spotted Cuckoo.

AGEING & SEXING Aged reliably by moult pattern. Sexes alike except marginally in size (see Biometrics). – Moults. Complete post-nuptial moult, and post-juv (partial to near-complete) moult, mostly May–Sep. Post-juv renewal includes entire head- and body-feathers, lesser and median coverts, also most to all greater coverts; tertials and alula also often moulted fully (rarely innermost or outermost tertial is excluded, and 1–2 alula feathers); invariably r1 is moulted, while s6 is also quite frequently replaced. Some moult more extensively: occasionally including r2 and, rarely, r3, or even most or all tail-feathers, and sometimes 1–3 additional secondaries. Very rarely a few outer primaries and apparently an irregular number of primary-coverts replaced too. Post-juv moult in ♂♂ more extensive (e.g. ♀♀ practically never moult any primaries and fewer greater coverts). – AUTUMN **Ad** Evenly fresh/worn, and all wing-coverts and tail-feathers uniform azure-blue; tail-feathers also broad, almost square-ended and lack white tips (except, rarely, vestigial ones on r1); cap glossed black. **1stY** Blue of retained juv tail-feathers and wing-feathers duller, and black cap duller and less glossed. Check for retained juv primaries and primary-coverts, as well as some unmoulted tail-feathers, secondaries, tertials, alula, and maybe outer greater coverts, which are relatively less fresh, weaker textured and duller, sometimes with abraded outer webs/tips. These contrast in colour and abrasion, especially against newly moulted and fresher ad-like adjacent feathers. Ageing rather straightforward if at least some secondaries and primaries are replaced (moult limits most obvious). Furthermore, retained tail-feathers are narrower, more pointed and less fresh, and, if not heavily worn, tipped off-white, clearly contrasting with new central feathers (but some moult all tail-feathers). Generally, differences in feather texture and moult contrast become more pronounced with wear. **Juv** Resembles ad but has shorter and looser body-feathers, especially below, and tail is often shorter, plumage generally duller and somewhat browner, with pale fringe to black of cap, buffy wash on mantle, narrow whitish median coverts bar, and whitish fringes to tail-feathers.

BIOMETRICS (*cooki*) L 32–35 cm; **W** ♂ 129–140 mm (*n* 14, *m* 134.3), ♀ 124–138 mm (*n* 15, *m* 130.8); **T** ♂ 165–207 mm (*n* 14, *m* 187.4), ♀ 157–198 mm (*n* 15, *m* 176.3); **TG** 73–123 mm (*n* 13, *m* 102.3); **T/W** 136.7;

C. c. cooki, ad, Spain, Apr: general shape somewhat recalls Magpie, but flight lighter, and combination of rounded wings and long graduated tail—both bluish-grey—and black cap readily identify flying birds, even if seen briefly. (G. Reszeter)

Ts 32.0–36.5 mm (*n* 32, *m* 34.7); **B** ♂ 23.1–29.0 mm (*n* 14, *m* 26.5); ♀ 24.0–29.5 mm (*n* 15, *m* 26.2); **B**(f) 18.5–22.4 mm (*n* 18, *m* 20.8); **BD** 8.3–9.7 mm (*n* 29, *m* 8.9). **Wing formula:** p1 > pc 24–36 mm; **p2** < wt 20–29 mm, < ss (=10 to = ss); **p3** < wt 5–9 mm; **p4** < wt 0–3 mm; **pp(4)5–6** about equal and longest; **p7** < wt 1–4 mm; **p8** < wt 7–10.5 mm; **p10** < wt 20–24 mm; **s1** < wt 22–26 mm. Emarg. pp3–8.

GEOGRAPHICAL VARIATION & RANGE Given that the Iberian population is a relict, being separated for a presumably long time from its E Asian relatives, morphological differences are surprisingly slight, mainly consisting of subtly longer tail and more extensive white tips to tail-feathers in Asia. Only SW European *cooki* treated here, while seven more extralimital subspecies occur in Asia. The Asian races appear very similar and largely clinally connected, with only *cyanus*, *japonicus* and *interpositus* being reasonably distinct (Kryukov *et al.* 2004; own study), but a closer examination of their validity lies outside the scope of this book.

C. c. cooki Bonaparte, 1850 (Iberia; resident). Treated above. Differs from *cyanus* (at least Lake Baikal area and N Mongolia, but probably much of Russian Far East as well if some subtle races are synonymised) in proportionately slightly shorter tail lacking the extensive white tips on central tail-feathers of all Asian populations, and in darker vinaceous-pink and less greyish hue to body plumage, creating stronger contrast with white throat. – As in the Asian population, Iberian birds show similar subtle variation, with a tendency towards paler and greyer birds in the north and inland, darker and browner birds in the south and near coasts. Furthermore, Iberian birds from northern and inland localities average slightly larger than those in coastal and southern localities, but overlap in size (and plumage) too great to warrant recognition. (Syn. *gili*.)

TAXONOMIC NOTE It has been proposed (e.g. Kryukov *et al.* 2004, Haring *et al.* 2007) that genetic distance between *cooki* and its Asian relatives is more typical of two species. Still, this is subjective and one could argue that with the retention of such strong similarity between the two populations despite their long separation (*c.* 2–3 million years) it is more natural to keep them together.

C. c. cooki, Spain, May: rather gregarious, often appearing in family parties. The Iberian endemic *cooki* differs from Asian *cyanus* in lacking extensive white tips to central tail-feathers and has more obvious white throat. (H. H. Larsen)

C. c. cooki, 1stS, Spain, Apr: 1stY very similar to ad but black cap duller and less glossy, and blue of juv tail- and wing-feathers duller. In this bird, the unmoulted primary-coverts are relatively less bright blue-grey, with some brownish tinge on inner webs showing, forming moderate moult limits. (G. Reszeter)

REFERENCES Alarcos, S. *et al.* (2007) *Ring. & Migr.*, 23: 211–216. – Cooper, J. H. & Voous, K. H. (1999) *BB*, 92: 659–665. – Cruz Solis, C., Lope, F. & Sanchez, J. M. (1992) *Ring. & Migr.*, 13: 27–35. – Haring E., Gamauf, A. & Kryukov A. (2007) *Mol. Phyl. & Evol.*, 45: 840–862. – Kryukov, A. *et al.* (2004) *J. Zool. Syst. Evol. Research.*, 42: 342–351. – Zhang, R. *et al.* (2012) *Mol. Phyl. & Evol.*, 65: 562–572.

C. c. cooki, 1stW, Spain, Sep: juv can undertake rather extensive post-juv moult, resulting in rather fresh ad-like appearance in autumn. However, note moult limits among tertials, and some crown-feathers not yet moulted completely, revealing white bases, providing clues to age. (H. Shirihai)

C. c. cooki, juv, Spain, Jul: note looser and rather patchily patterned juv body-feathers, especially below, and extensive pale fringes to black crown-feathers. Also has quite dull-coloured inner greater coverts and narrow buff-white median and greater coverts bars. (C. N. G. Bocos)

(COMMON) MAGPIE
Pica pica (L., 1758)

Fr. – Pie bavarde; Ger. – Elster
Sp. – Urraca común; Swe. – Skata

Together with Blackbird, Robin and several of the tits, the Magpie is among the most familiar European birds. This widespread Palearctic crow breeds from the boreal taiga to the Mediterranean region, and in steppe and semi-desert zones in the south. Found in most lightly wooded habitats, including farmland. It also thrives in suburban and urban habitats.

Sedentary, with limited dispersal. Essentially territorial year-round, pairs being usually monogamous for life. Its broad diet of small mammals and insects masks a specialisation to take the chicks and eggs of small passerines, though it also takes fruit, a wide spectrum of carrion, refuse and scraps. It stores food for short periods and is well known for 'stealing' shiny objects (although this is apparently grossly exaggerated).

P. p. pica, 1stW, England, Oct: compared to ad, juv primaries have shorter white fields, leaving broader black tips. Juv head- and body-feathers duller and slightly brown-tinged, less deep black. Cream-brown rump patch. (R. Steel)

IDENTIFICATION Medium-large and markedly *long-tailed* with striking *black-and-white plumage*. Head and upperparts, throat, breast, wings and strongly graduated tail black with metallic gloss (bluish, purple and green sheen visible at the correct angle). *Belly and scapular patch white*. Short rounded *wings have large white primary flash* with only tips and outer webs of remiges black. Variable amount of white on rump (most obvious in northernmost populations). Bare parts mostly black, but 'African Magpie' has bare blue skin behind the eye (postocular patch). Solitary but can gather in noisy flocks of up to *c.* 25 birds, known as 'parliaments'; *several often fly in follow-the-leader succession, chattering as they go*. In winter may roost in larger flocks. Wary but not timid, though invariably keeps healthy distance from man. Usually has confident walk with swaying body movement, but may hop quickly forward or sideways with wings barely open, usually with tail held up (flicked and fanned in excitement). Often patrols roadsides, lawns and flowerbeds in centre of cities. Usually flies short distances, dashing and sweeping, with *bursts of wingbeats* interspersed by *short or long glides*, glides sometimes almost vertically to ground from high perch.

VOCALISATIONS Song, linked to display, is weak and rarely heard (unless at very close range) and formed by mixture of various calls, harsh subsong-like twittering and soft warbles, high-pitched babbling notes, and some mimicry. – Most familiar call is far-carrying alarm, a long-drawn-out, fast, very hoarse staccato rattle *tsche-tsche-tsche-tsche-tsche* or *chak-chak-chak-chak*, machine-gun-like (or like 'amplified Great Tit alarm'). Other calls rather varied, chiefly

P. p. pica, ad, England, May: unmistakable and among the best-known European birds. Long, strongly graduated tail, and white scapulars and belly identify the species, but sheen of various metallic colorations visible only at right angles. Ad by evenly-feathered wing. (M. Fowler)

P. p. pica, ad, England, Mar: the attractive iridescence of the flight-feathers is revealed at close range and in right light. In Europe, race *pica* has relatively less white on primaries than populations further east, still much—as here—in ad. This race is also characterised by usually narrow pale grey rump patch. Metallic gloss predominantly deep blue, but on uppertail also green and purple. (R. Steel)

hoarse and whining sounds admixed, depending on context, e.g. disyllabic *cha-ka* or *chiah-cha* or *shrak-ak* or *ch'chak* and ascending *che-uk* in conversation; also yelping, clicking, etc. Sounds alarm against domestic cat.

SIMILAR SPECIES Unmistakable, no other species has similar combination of size, plumage pattern, characteristic shape and behaviour, making the Magpie easily identified at first glance, or by call.

AGEING & SEXING Easiest to age by length of black tips to otherwise white outermost two primaries, but also check moult pattern for any ambiguous bird. Sexes alike but differ slightly in size; still, racial variation must be borne in mind (see Biometrics and Geographical variation). Ad ♂♂ have on average proportionately slightly longer tail than ♀♀. – Moults. Complete post-nuptial moult in ad, and partial post-juv moult, in summer into autumn (May–Nov, mostly Jun–Sep). Post-juv renewal includes entire head- and body-feathers, lesser and median coverts; also variable number of greater coverts, usually at least inner ones replaced (if not all); sometimes also some tertials and r1. – **ALL YEAR Ad** Plumage (depending on season) evenly fresh or worn, and dark parts of remiges and primary-coverts intensely black and glossy; fringes and tips of primaries, primary-coverts, greater coverts and alula relatively fresh and firm-textured. Relatively small black tips (rarely absent) to outer otherwise white primaries, and p1 markedly narrow for 3/4 of its distal length, black tips of p1 <13 mm, p2 5–14 mm (*pica*, Sweden). Border between black tips and white bases sharper than in 1stY (very little overlap). Tail-feathers rather broad with rounded tips (unless heavily worn). **1stY** Dark parts of juv remiges, primary-coverts and alula (plus usually some unmoulted outer greater coverts, and often all or just outer one or two pairs of tertials) are duller and slightly brown-tinged blackish, less deep black and glossy. On close inspection, outer webs and tips are relatively less fresh, weaker textured and more pointed, and sometimes more abraded (i.e. compared to ad at same season). Check for contrast created by moult limits between fresh median and lesser coverts, and retained juv primaries and primary-coverts; if not all greater coverts moulted, there is slight contrast between new inner and retained juv outer coverts (number of replaced greater coverts varies between populations). Relatively large blackish tips to otherwise white outer primaries, and p1 broader and more wedge-shaped, not as attenuated over distal 2/3 as ad; black tips of p1 12–24 mm, p2 18–34 mm (once 47; *pica*, Sweden). Border between dark tips and white bases diffuser than in ad (very little overlap). Tail-feathers average narrower, more pointed and less fresh, and if central feathers moulted clear contrast formed (unlike ad). Differences in feather texture and moult contrast more pronounced with wear. **Juv** Resembles ad, but has shorter, looser body-feathers, especially below, tail often shorter, and dark plumage duller, with black more sooty-tinged and dull, and white of scapulars, rump and underparts suffused pale buff or dirty cream, thus pied pattern less sharp. Feather gloss confined to wings and tail and is less brilliant. At least on fledging, iris pale grey or greyish-blue (instead of dark brown), bill greyish-black, cutting edges and flanges pale yellow or whitish.

BIOMETRICS (*pica*) **L** 40–51 cm, of which tail *c.* 25 cm; **W** ♂ 187–211 mm (*n* 50, *m* 196.2), ♀ 180–195 mm (*n* 33, *m* 188.5); **T** ♂ 235–296 mm (15% had shorter tail, to 200 mm, partly due to wear, feather loss or moult; *n* 49, *m* 250.8), ♀ 193–260 mm (*n* 33, *m* 232.3); **T/W** ♂ *m* 127.8, ♀ *m* 123.2; **TG** 80–142 mm (*n* 73, *m* 109.1); **B** ♂ 35.7–42.5 mm (*n* 37, *m* 38.5), ♀ 33.5–38.7 mm (*n* 31, *m* 35.6); **B(f)** ♂ 30.0–36.1 mm (*n* 35, *m* 33.3), ♀ 28.3–33.1 mm (*n* 31, *m* 30.6); **BD** 11.8–15.0 mm (*n* 70, *m* 13.6); **Ts** 43.5–54.2 mm (*n* 72, *m* 48.6). **Wing formula: p1** > pc 9–32 mm, < p2 50–65.5 mm; **p2** < wt 27–49 mm, =9/10 or 10 (53%), =8/9 or 9 (34%) or =10/ss or =ss (13%); **p3** < wt 4–13 mm; **pp4–5** about equal and longest; **p6** < wt 1–4 mm; **p7** < wt 6–13 mm; **p8** < wt 18–25 mm; **p10** < wt 34–44 mm; **s1** < wt 41–55 mm. Emarg. pp3–7 (sometimes very faintly also on p8).

GEOGRAPHICAL VARIATION & RANGE Variation largely clinal and moderate over temperate zone of continental Europe and N Asia, but more complex and marked further south, with Iberian *melanotos* and NW African *mauritanica* differing more clearly, and the isolated Arabian mountain-living *asirensis* perhaps meriting specific treatment. In want of a comprehensive study of *asirensis* we prefer to keep it as a distinct race of Common Magpie; see Taxonomic notes. Three racial groups proposed for better overview, but differences between the first two are partly clinal and not large. All populations sedentary and longer movements exceptional.

THE *PICA* GROUP
(Much of Europe, Turkey, Siberia, Central Asia) Clinal and only moderate variation. Size increases towards north and east, rump turns from rather dark to more whitish and amount of white in wing increases in the same direction, from southwest to north and east. Due to individual variation and partly subtle differences (and even lack of knowledge or agreement concerning most important traits), and due to only scant material available from some critical areas, borders between all four subspecies difficult to define. See Taxonomic notes.

P. p. pica, 1stS, England, May: overall duller, with brownish-tinged juv primaries on which smaller white area is revealed when wing not completely folded. Also vestigial bluish bare spot behind eye, a feature of some juv *pica*, which sometimes remain a year later. (G. Thoburn)

P. p. pica (L., 1758) (British Isles, continental Europe south to Pyrenees and east to C Ukraine, in S Scandinavia north to *c.* 65°N, Turkey except in extreme east, Cyprus, N Levant). Medium-sized (see Biometrics) with relatively reduced white on primaries, in ad being fully concealed on folded wing (black tips on inner web of p1 in ad ♂ [0] 1–6.5 mm, in ad ♀ 2–12 mm, in 1stY 12–24 mm, on p2 in ad ♂ 5–14 mm, in ad ♀ 8.5–14 mm and in 1stY 18–34 mm [once 47]). Rump usually pale grey, infrequently purer white, rarely dark grey and seldom blackish; metallic gloss predominantly deep and bright blue, but on uppertail also green and purple. (Syn. *galliae*; *germanica*.)

P. p. pica, 1stW, England, Oct: in flight from below, note species-specific white wing patches and white belly, and long, strongly graduated tail; broad dark primary tips infer age. (R. Steel)

P. p. pica, juv, Italy, Jul: looser body-feathers, especially below, with black areas more sooty-tinged, and white areas suffused pale buff or dirty cream. Note bluish bare skin below and behind eye. (D. Occhiato)

○ ***P. p. fennorum*** Lönnberg, 1927 (Scandinavia north of *c.* 65°, Finland, Baltic States, Russia north of *c.* 55° east at least to region of Perm and Ufa). Close to *pica* but averages somewhat larger, and connected with neighbouring races via intergrading populations of intermediate size (still consistently large over a vast area). Rump patch on average purer white and larger, but overlap. Amount of white in wing, and size of black tips on inner webs of outer primaries, much the same as ssp. *pica*. **W** ♂ 190–218 mm (*n* 44, *m* 204.7), ♀ 182–202 mm (*n* 30, *m* 193.6); **T** ♂ 215–292 mm (*n* 44, *m* 259.9), ♀ 200–270 mm (*n* 30, *m* 240.1); **T/W** ♂ *m* 126.9, ♀ *m* 124.0; **TG** 75–146 mm (*n* 62, *m* 108.0); **B** ♂ 37.7–47.7 mm (*n* 37, *m* 41.1), ♀ 35.0–40.8 mm (*n* 30, *m* 37.8); **B**(f) ♂ 32.3–42.0 mm (*n* 33, *m* 35.1), ♀ 28.7–35.9 mm (*n* 28, *m* 32.3); **BD** 12.5–15.4 mm (*n* 63, *m* 13.9); **Ts** 43.3–55.4 mm (*n* 63, *m* 49.0). Wing formula very similar to *pica*.

P. p. bactriana Bonaparte, 1850 (SE Russia, E Ukraine, Crimea, easternmost Turkey, Caspian region, Iraq except south, lower Volga from about Samara and lower Ural River east across much of Central Asia through Kirghiz Steppe to Tarbagatai, Fergana and Tarim Basin, south to Khorasan and Baluchistan in Iran, also in Afghanistan and to Gilgit; see Taxonomic notes). Close to both *pica* and *fennorum*, but when series compared averages larger than *pica* in most

P. p. fennorum, ad, Finland, Nov: this race is somewhat larger, with white areas on average larger than in *pica* (but much overlap). Evenly-feathered wing infers age. (M. Varesvuo)

P. p. fennorum, Finland, ad (left: Nov) and 1stS (Apr): ad of this population often has very narrow black wingtips, but much overlap with *pica*. Like *pica*, the retained juv primaries of 1stS (on right) have smaller white areas, hence broader dark wingtips. (M. Varesvuo)

measurements and has proportionately longer tail and larger white area in primaries than both; white visible on at least pp7–8 on folded wing, thus more restricted black tips (on inner web of p1 in ad 0–5 mm, in 1stY 9–22 mm, on p2 in ad ♂ 0–8 mm, in ad ♀ 0–10 mm [once 24.5, once 30] and in 1stY 15–30 mm). Rump patch nearly always purer white and larger, and dark areas above generally more glossed metallic green, especially on secondaries (though some variation). **W** ♂ 198–216 mm (*n* 17, *m* 205.9), ♀ 181–203 mm (*n* 12, *m* 194.2); **T** ♂ 237–300 mm (*n* 16, *m* 268.8), ♀ 228–278 mm (*n* 12, *m* 252.8); **T/W** *m* 130.4; **TG** 93–153 mm (*n* 28, *m* 119.4). Tarsus, bill and wing formula very similar to *pica* and *fennorum*. (Syn. *kot*.)

○ ***P. p. leucoptera*** Gould, 1862 (Urals east of Perm and Ufa, W & C Siberia, Sayans, Altai, Zaisan, S Transbaikalia, N & W Mongolia; see Taxonomic notes). Averages the largest race in the group and differs from almost as large *fennorum* mainly by slightly more extensive white in wings (showing more white on primary tips in folded wing). Rump usually mainly pure white (though sometimes narrow and with slight suffusion of brown-grey). Glossy sheen on primaries and tertials mainly green, on secondaries blue. – Ssp. *leucoptera* (originally described from 'E Siberia', later defined as 'SE Transbaikalia') is only very subtly larger with on average a little more white on remiges than birds described from SC Siberia ('*hemileucoptera*'), differences of subtle clinal

P. p. bactriana, ad, Kazakhstan, Sep: this eastern race differs from *pica* and *fennorum* by having proportionately longer tail and larger white area in primaries. Evenly-feathered wing infers age. (R. Pop)

P. p. leucoptera, ad, Mongolia, May: the largest race. Differs only subtly from similar *fennorum* by on average slightly larger white area in primaries, which can be exposed even on folded wing. Evenly-feathered wing infers age. (M. Putze)

P. p. leucoptera, ad, Mongolia, May: rump sometimes as here narrow and suffused brown-grey, more often larger and white. Aged as previous bird. (M. Putze)

nature and hence latter best included here. **W** 199–235 mm (n 18, m 211.6); **T** 240–310 mm (n 18, m 271.8); **T/W** m 128.6; **TG** 101–158 mm (n 18, m 128.6); **B** 33.9–45.6 mm (n 17, m 39.6); **B**(f) 30.5–37.0 mm (n 15, m 33.8); **BD** 12.2–15.5 mm (n 17, m 13.9); **Ts** 44.1–51.0 mm (n 17, m 47.9). **Wing formula: p1** > pc 16–34 mm, < p2 57–70 mm; **p2** =9/10 or 10 (61%) or =8/9 or 9 (39%); **p7** < wt 8.5–18 mm; **p8** < wt 22–29 mm; **p10** < wt 40–52 mm. (Syn. *hemileucoptera*.)

THE MAURITANICA GROUP
(Iberia and NW Africa) Includes two distinctive races, grouped together due to both having (variable amount of) bare bluish skin around and in particular behind eye, reduced white in wing, and lacking a pale rump patch, but they differ quite considerably in overall size and tail-length. That they should have a proportionately slightly thinner bill (*BWP*) could not be confirmed. Metallic gloss purplish-green to dull bronzy-purple, especially on tail.

P. p. melanotos A. E. Brehm, 1857 (Iberia). Averages smaller and darker than *pica*, with more extensive black tips to primaries (on inner web of p1 in ad 6.5–14 mm, in 1stY 13–28 mm, on p2 in ad 10–23.5 mm, in 1stY [once 19]

P. p. leucoptera, 1stS, Mongolia, Jun: sometimes due to feather wear, moult limits in wing and abraded tail very obvious, making ageing straightforward. Even in young birds of this race, the white primary areas are extensive. (H. Shirihai)

29–37 mm). Rump dark grey-brown, occasionally blackish, any white feather bases concealed. Sometimes has vestigial bluish skin behind eye, chiefly S of Madrid (but note that such a bare spot exceptionally occurs in C European *pica* and is typical of all juv magpies). **W** ♂ 179–203 mm (n 13, m 189.5), ♀ 175–191 mm (n 12, m 182.5); **T** ♂ 226–283 mm (n 13, m 250.0), ♀ 203–259 mm (n 12, m 231.0); **T/W** ♂ m 131.9, ♀ m 126.4; **TG** 81–140 mm (n 27, m 111.7); **B** ♂ 32.3–42.1 mm (n 13, m 36.9), ♀ 33.3–38.5 mm (n 12, m 35.9); **B**(f) ♂ 27.9–35.1 mm (n 12, m 31.6), ♀ 29.0–33.0 mm (n 8, m 30.8); **BD** 11.3–13.7 mm (n 27,

P. p. melanotos, ad, Spain, Apr: compared to *pica* rather small, with less white in wing and no pale rump. Also has variable bare bluish skin around or behind eye, varying from none (as here) to quite large, almost as in NW African *mauritanica*. Evenly-feathered wing infers age. (B. Marnell)

P. p. melanotos, ad, Spain, Apr: pale rump patch lacking, or is very small. Also, note rather broad dark wingtips, despite being apparently ad. (J. Laborda)

P. p. melanotos, presumed ad, Portugal, Mar: a few birds have rather extensive bluish skin around and behind eye, like race *mauritanica*. Glossy black plumage and advanced bare-parts colours suggest ad. (M. Lefevere)

P. p. mauritanica, ad, Morocco, Feb: the relatively smaller white areas make this race very distinctive, as well as the purple-blue bare skin behind eye and on lower orbital ring. Evenly-feathered ad wing infers age. (D. Occhiato)

m 12.7); **Ts** 45.2–51.9 mm (*n* 27, *m* 48.5). **Wing formula: p1** > pc 13–30 mm, < p2 47.5–60 mm; **p2** < wt 26–46 mm; **p7** < wt 7–14 mm; **p8** < wt 17–25 mm; **p10** < wt 31–43 mm; **s1** < wt 38–49 mm.

P. p. mauritanica Malherbe, 1845 (NW Africa). The 'African Magpie' is small and markedly shorter-winged and longer-tailed than *melanotos*. Typically, ad has rather large cobalt-blue bare spot behind eye, and rump is always black (or at least very dark grey-black). Slightly less extensive white in primaries with correspondingly larger black tips to primaries (on inner web of p1 in ad 21.5–37 mm, in 1stY 21.5–28 mm, on p2 in ad 16–32 mm, in 1stY [once 14] 19–38 mm). With such extensive overlap in amount of black on primary tips, ageing becomes considerably more difficult than in *pica* or *melanotos*. Concentrate on amount of metallic gloss, on width of tail-feathers, whether tips of tail-feathers are well rounded or more pointed, on narrowness of p1, and whether border between black and white on outer primaries is sharp or diffuse. **W** ♂ 157–173 mm (*n* 12, *m* 163.7), ♀ 149–164 mm (*n* 12, *m* 156.9); **T** ♂ 233–290 mm (*n* 12, *m* 256.3), ♀ 210–274 mm (*n* 12, *m* 240.3); **T/W** ♂ *m* 156.4, ♀ *m* 153.2; **TG** 91–146 mm (*n* 24, *m* 117.2); **B** ♂ 35.5–40.0 mm (*n* 12, *m* 38.3), ♀ 34.5–39.1 mm (*n* 12, *m* 37.6); **B(f)** ♂ 30.6–34.4 mm (*n* 11, *m* 33.1), ♀ 30.0–33.0 mm (*n* 12, *m* 31.5); **BD** 11.0–13.6 mm (*n* 24, *m* 12.7); **Ts** 44.4–50.8 mm (*n* 25, *m* 47.1). **Wing formula: p1** > pc 11–20 mm, < p2 41–52 mm; **p2** < wt 25–35 mm, =10/ss or =ss (68%), <ss (18%) or =9/10 or 10 (14%); **p3** < wt 3–11 mm; **pp4–5**(6) about equal and longest; **p6** < wt 0–5 mm; **p7** < wt 3.5–8 mm; **p8** < wt 11–17 mm; **p10** < wt 23–36 mm; **s1** < wt 27–39 mm.

The *ASIRENSIS* GROUP
(Asir Mts, SW Arabia) Just one very distinctive race, widely allopatric and characterised mainly by its proportionately larger and thicker bill, and stronger legs (see Taxonomic notes).

P. p. asirensis Bates, 1936 (Asir massif of SW Saudi Arabia). The 'Arabian Magpie', is a large form with markedly thick bill, rather short tail and strong feet. Shows reduced white on wings and scapulars, with moderate amount of white on primaries and correspondingly broad black tips, especially in 1stY. Dark plumage dull black, with strongly reduced metallic gloss, but tail glossed purple. Rump all dark. Observers report that the main call is very different, a mournful prolonged screech. Limited material available for examination, only 3 ♀♀ (NHM): **W** 205–226 mm (*m* 214.3); **T** 228–250 mm (*m* 239.7); **T/W** *m* 111.8; **TG** 77–97 mm

(*m* 87.0); **B** 45.8–47.8 mm (*m* 47.0); **B(f)** 41.0–42.1 mm (*m* 41.6); **BD** 15.2–16.2 mm (*m* 15.8); **Ts** 49.9–52.9 mm (*m* 51.6). **Wing formula: p1** > pc 11.5–33 mm, < p2 58–63.5 mm; **p2** < wt 28–50 mm; **p3** < wt 7–12 mm; **pp4–5** about equal and longest; **p6** < wt 1–3 mm; **p7** < wt 4–7 mm; **p8** < wt 11–17 mm; **p10** < wt 32–35 mm; **s1** < wt 36–41 mm.

TAXONOMIC NOTES By some the Magpie is thought to constitute at least two if not three separate species, the three-species scenario represented by the above division into three subspecies groups. However, a comprehensive study including of genetic distance is lacking, and there are apparently no clear differences in vocalisations between *melanotos* and *mauritanica* on the one hand and the other European populations on the other (J. L. Copete pers. comm.). The blue bare skin behind the eye, thought to be a defining character for *mauritanica*, shows a certain clinal variation in that a hint of it also appears in a minority of *melanotos* and even in some south-western populations of *pica*. In the case of the 'Arabian Magpie' some clear vocal differences are known, but considering the small and widely isolated range of this population, morphological differences are still surprisingly moderate. All taxa therefore maintained in one species. – Due to paucity of relevant specimens in visited collections, but also because of the subtle differences and clinal nature of most of the morphological characters used to define them, it has been difficult to firmly establish the borders between the three subspecies *fennorum*, *bactriana* and *leucoptera*. The borders shown on the map represent an 'educated guess', but it could be that *bactriana* also breeds over much of W Siberia including S Ural region, and that *leucoptera* is found only in the Yenisei Basin and eastward, thus is purely extralimital. – Note that the taxon '*hemileucoptera*' is placed here as a junior synonym of *leucoptera*, whereas other sources (HBW) treat it under *bactriana*. Such discrepancies are an effect of the difficulties referred to above.

REFERENCES BÄHRMANN, U. (1958) *Vogelwelt*, 79: 129–135. – LEE, S. et al. (2003) *Mol. Phyl. & Evol.*, 29: 250–257. – LÖNNBERG, E. (1927) *Fauna och Flora*, 22: 97–110. – SEEL, D. C. (1976) *Ibis*, 118: 491–527.

P. p. asirensis, 1stS, Saudi Arabia, Mar: the 'Arabian Magpie' is distinctive in having proportionately larger and thicker bill, but subtly shorter tail, with reduced white in wings and scapulars (the latter often only partly exposed or invisible). Dark plumage dull black, with much-reduced metallic gloss and all-dark rump. Aged by moult limits in wing. (M. Al Fahad)

P. p. asirensis, juv–1stY, Saudi Arabia, Jul: juv also very distinctive, with no visible white in scapulars and stronger crow-like bill. Note moult limits in wing and pale gape line. (J. Babbington)

PLESKE'S GROUND JAY
Podoces pleskei Zarudny, 1896

Fr. – Podoce de Pleske; Ger. – Pleskehäher
Sp. – Arrendajo terrestre iraní
Swe. – Pleskes ökenskrika

This species, the sole representative within the covered region of its remarkable genus, is considered endemic to Iran, although it seems probable that the species also occurs in S Afghanistan and perhaps W Pakistan. A ground-dweller, the Iranian Ground Jay as it is also known, appears to be confined to semidesert sandy plains with *Zygophyllum atriplicoides* bushes, which it uses as lookout perches. Fortunately, the Pleske's Ground Jay is reportedly quite frequent within its range and appropriate habitat, and may even lose its natural fear of humans if the observer remains quiet

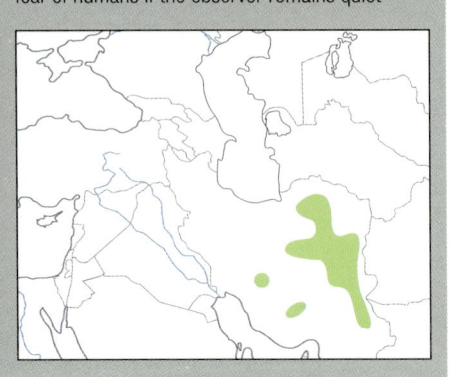

IDENTIFICATION Smallish, *warm orange-buff ground jay with a moderately long, slightly down-curved bill*. Head and much of both upper- and underparts orange-buff with sandy or pinkish tones, but becoming *whitish on throat and around eyes*, thereby emphasizing *well-defined black eye-stripe* that extends slightly behind eye. Variable oval *black patch on lower throat* (just touching upper breast) is a good character, but is absent or reduced in juvenile. Striking *black-and-white wing pattern* very obvious in flight and in rear view of perched bird. Remiges black with broad white central patch on primaries, narrowing to a band on trailing edge of secondaries. Median and greater coverts black with broad white tips forming bars (less distinct on median coverts); lesser coverts as rest of upperparts. *Glossy blue-black tertials* (broadly edged white) and *tail highly contrasting* in fresh plumage. Bill black, legs greyish-white and iris dark brown. Freely runs on ground among low vegetation, usually flying only short distances.

VOCALISATIONS Call described as a clear, rapid (ten notes per second) and high-pitched *pee-pee-pee-pee-pee* given from bush top, recalling distant song of Rock Nuthatch (Madge & Burn 1993).

SIMILAR SPECIES The only ground jay in the covered region and therefore unmistakable. – Given brief or poor views, might be mistaken for *Hoopoe Lark*, but that differs considerably in overall coloration and shape, and wing and tail pattern.

AGEING & SEXING Just eight specimens examined (NHM, AMNH). Ageing apparently possible at least in autumn by slight moult limits, but requires close views or handling. Sexes alike, but extremes might prove distinct on a larger sample, ♂♂ identifiable by combination of longer wings and bill, as well as larger black throat and loral patches. – Moults. Ad has complete post-nuptial moult in summer, young a partial post-juv moult. Ad with growing remiges in Jul–Aug. Not known whether there is a partial pre-nuptial moult. – **Ad** Uniformly fresh or worn plumage lacking obvious moult limits in wing. Bill black or dark. Black breast patch and loral mark. **Juv** Rather like ad, but differs mainly in looser body-feathers and duller overall coloration, as well

Possible ♂, Iran, Jul: in typical upright stance. The immaculate plumage (plus rather freshly and evenly-feathered wing), with large black throat and loral patches suggest an ad ♂, although sexes are mainly inseparable. The mostly black wings and all-black tail present a nice contrast with the otherwise orange-buff plumage. (M. Westerlind)

♂, Iran, Jul: song often given from top of bush. Large black breast and loral patches suggest ♂. Partly open wing revealing large white patches that are otherwise largely concealed. (J. Thalund)

Possible ♀, Iran, Mar (same individual): comparatively small size and ground-dwelling habits surprisingly may recall a lark at first fleeting sight. Moderately long, slightly down-curved bill. Head and much of body orange-buff. Whitish around eyes, on fore-cheeks and throat, emphasising well-defined black eye-stripe that broadens in front of eye. Glossy black, broadly white-edged tertials highly contrasting when fresh. Note that black throat patch can appear smaller when body-feathers puffed up and neck hunched down. Judged by rather short bill, pale colours and small throat patch this could possibly be a ♀, but sexing is merely tentative. (Left: H. van Diek; right: M. Hornman)

Possible ♂♂, Iran, Apr: sexes have very nearly same plumage, although black throat patch averages slightly larger in ♂♂, and general orange-buff colour may prove somewhat darker by comparison. ♂ is somewhat larger and has longer bill, and judging by all these fine differences both these birds could well be ♂♂. (Left: A. Sangchouli; right: A. Yekdaneh)

Juv, Iran, Jul: duller and more yellow-tinged sandy-buff than ad, lacking black throat-patch. (J. Thalund)

as in lack of black breast and loral patches, and by having a pinkish bill.

BIOMETRICS L 24–25.5 cm; W 118–126 mm (n 8, m 120.9); T 82–88 mm (n 8, m 84.0); T/W m 69.5; B 38.0–43.8 mm (n8, m40.5); B(f)32.6–36.3mm(n8,m34.3); BD 8.7–9.9 mm (n 7, m 9.3); Ts 43.1–47.0 mm (n 8, m 44.2). **Wing formula:** p1 > pc 31–37 mm, < p2 22–26.5 mm; p2 < wt 7–9 mm, =7 or 7/8; pp3–5(6) about equal and longest; p6 < wt 0–4 mm; p7 < wt 4–7 mm; p8 < wt 11–13 mm; p10 < wt 17–22 mm; s1 < wt 22–26.5 mm. Emarg. pp3–6.

GEOGRAPHICAL VARIATION & RANGE Monotypic. – Breeds in C Iran from Khorasan east to Baluchistan; suspected to breed also in adjacent W Pakistan and SW Afghanistan; resident.

TAXONOMIC NOTE Often suggested to form a superspecies with the other three *Podoces*, being especially similar to Pander's Ground Jay *P. panderi* of Turkmenistan north to Kazakhstan.

REFERENCES HAMEDANIAN, A. (1997) *Sandgrouse*, 19: 88–91.

(SPOTTED) NUTCRACKER
Nucifraga caryocatactes (L., 1758)

Fr. – Cassenoix moucheté; Ger. – Tannenhäher
Sp. – Cascanueces norteño; Swe. – Nötkråka

A rather unique corvid of northern and central latitudes (from N and C Europe east to Japan, also Himalayas to SW and C China), in which food-storing is highly developed, and an ability to find cached food even below deep snow enables it to be sedentary even in far north. Two distinctive populations, European *caryocatactes* ('Common' or 'Spotted Nutcracker'), which depends on seeds of Norway Spruce or Arolla Pine (according to region), occasionally larch, silver fir and nuts of Hazel, while Siberian *macrorhynchos* ('Slender-billed Nutcracker') primarily feeds on seeds of Siberian Stone Pine and spruce. The latter is known for its irruptive movements to the south-west in years of poor food supply. Following influxes, some may remain to breed in wintering areas. Such irruptions have reached Britain and Italy, but it is uncertain whether they are responsible for recent colonisations in Belgium and France, and range expansions in Fenno-Scandia and Czech Republic. Sociable, sometimes appearing in large flocks during eruptions, and can be very tame. Accidentals have reached Turkey and Spain.

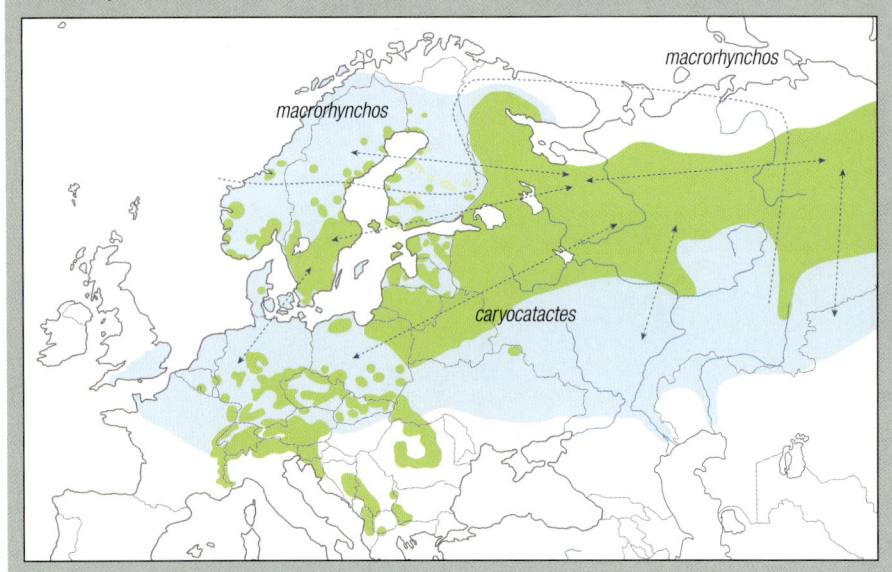

IDENTIFICATION A Jay-sized, characteristically stocky corvid with *quite short tail, long sharp-pointed bill* (quite powerful) and *dark-capped*, large head. *Dark chocolate-brown plumage, heavily and boldly sprinkled with white drop-shaped spots* (profuse and small on head-sides to upper throat, becoming larger on mantle and belly, though rump all dark, with narrow white tips to upperwing-coverts). *Undertail-coverts, vent and tail-corners are contrastingly white.* Crown sometimes narrowly streaked or spotted buffish. Wings and much of tail glossy brownish-black and unspotted, but *underwing-coverts heavily white-tipped* forming several irregular bands. European and Siberian birds differ as follows: former has bill thicker based and heavier looking, and tail tipped narrowly white, while latter's bill is somewhat longer and more uniformly narrow, with broader white tip to tail. Sexes alike and ages by and large similar. Iris dark brown, and bill and tarsi black. Feeds in trees and on ground, where moves by hops and sidling jumps. Flight is direct but slow and fluttering rather like Jay or Magpie, with no undulations when travelling longer distances, but still readily identified by silhouette with long bill and short tail.

VOCALISATIONS Territorial signal uttered by both sexes is the commonest call, a machine-like even-pitched buzzing *krrrrreh*, often repeated 2–5 times. The 'real' song, seldom heard and apparently uttered only by ♂, is a variable mix of different calls, but phrases, executed in calm pace, can be rather long and subdued (subsong-like); at times ventriloquial. Included in the song are harsh chattering notes, some brief warbling, squeaky, whistling and miaowing sounds, as well as melodious babbling and piping notes; some mimicry can occur, too. – Generally rather silent outside breeding season, but presence often revealed by the characteristic loud and hard, drawn-out, even-pitched buzzing, almost machine-like call, usually repeated several times, *krrrrreh krrrrreh krrrrreh* etc. Has several other less characteristic calls, such as a Jackdaw-like *kya* and a harder variant *kiak*. Alarm either a whistling *pyuh* or a hard, knocking doubled *kyek-ek*.

SIMILAR SPECIES If seen well, unlikely to be confused with any other similar-sized bird; shape and behaviour also highly characteristic, making the species easy to identify, even at a glance. More challenging is of course to differentiate between the two races of Nutcrackers (see above and below).

AGEING & SEXING Aged by moult pattern, and by amount and shape of white tips to inner remiges and some wing-coverts (largely after Svensson 1992). Sexes differ only in average size, and difference minute (see Biometrics). – Moults. Complete post-nuptial moult of ad and partial post-juv moult in summer, mostly Jun–Aug. Post-juv renewal includes entire head- and body-feathering, lesser coverts and none to all median coverts (usually at least inner ones replaced); rarely a few inner greater coverts also moulted, but apparently never replaces tertials and alula-feathers. No regular pre-nuptial moult. – **AUTUMN Ad** Evenly fresh, remiges and primary-coverts black and glossy, not brownish. Edges and tips of primaries, primary-coverts, greater and median coverts relatively fresh and firm-textured. Remiges and alula generally lack white tips, but tiny white dots on some (especially inner primaries and outer secondaries). A few outer greater coverts and primary-coverts have small white tips forming a narrow wedge at the shaft. All or at least some

Presumed *N. c. caryocatactes*, ad, Austria, Jan: dark-capped stocky corvid with long pointed bill. Unmistakable by dark chocolate-brown plumage, boldly sprinkled with white spots. However, racial separation in the field is difficult; in this bird, the fairly thick-based bill suggests *caryocatactes*. Wing without white tips to wing-coverts makes this bird ad. (T. Grüner)

Presumed *N. c. caryocatactes*, 1stS, Poland, Apr: note large-eyed impression and very strong bill adapted for opening conifer cones or hazel nuts. White tips to secondary-coverts (small), as well as to primary-coverts and alula, indicate 1stS. Thicker-based bill suggests race *caryocatactes*. (A. Faustino)

inner median coverts blackish without white tips: if white tips present, they are wedge-shaped. Tail-feathers wider with on average more rounded tips, width of r1 (measured 2 cm from tip) 25–28 mm. **1stY** Dark retained juv remiges, primary-coverts, tertials and alula, and usually most or even all greater coverts (sometimes a few outer median coverts) duller and browner, more grey-brown with only slight gloss. Outer webs and tips of juv primaries generally less fresh, weaker textured and rather pointed, and sometimes abraded. Especially look for contrast between fresh median and lesser coverts on the one hand, and retained rest of juv wing on the other. Remiges also clearly and extensively tipped white (check especially inner primaries and outer secondaries). Retained juv median and greater coverts have rather extensive white tips (not just wedge-shaped spots near shaft as in ad), but some moult most or all to ad-type feathers. Primary-coverts and alula similarly tipped white on most. (Note: a few young have more limited white on tips of coverts and alula, sometimes almost as little as those ad with much white on tips; thus always check moult limits in wing, and presence and extent of white tips on primary-coverts and remiges). Tail-feathers average narrower: width of r1 (measured 2 cm from tip) 20–27 mm. Also white tips to tail-feathers less square-cut, forming shallow notch at shaft on r1 (in ad, more sharply-defined tip

N. c. caryocatactes, 1stW, Finland, Aug: much of wing, including primaries, primary-coverts and greater coverts, and at least some median coverts, are juv feathers with duller bases, and large white tips. Note short but thick-based bill and seemingly little white on tail-tip. (M. Varesvuo)

N. c. caryocatactes, Finland, Sep: rather square-shaped broad wings, as well as diagnostic boldly white-spotted underparts and underwing pattern make the species wholly distinctive. Thick-based bill infers race. (M. Varesvuo)

N. c. caryocatactes, juv, Switzerland, Jun: rather like ad, but duller and browner, with more diffusely spotted underparts and much looser body-feathers. (R. Kunz)

of even width). Some retain soft fluffy juv feathers on lower neck and vent until Sep. **Juv** Rather like ad, but duller and browner overall with less gloss on wing-feathers, and is diffusely spotted, while white-spotted greater coverts form noticeable wing-bar. Much shorter and looser body-feathers, especially below. – **SPRING** The same ageing criteria as in autumn applicable, but by spring, with wear white tips considerably reduced (narrower, more rounded and more frayed) or entirely lost, and differences in shape of white tips therefore often less clear, making caution essential when attempting ageing. Nevertheless, differences in feather texture and moult contrast usually become more pronounced with wear.

BIOMETRICS (*caryocatactes*) **L** 30–34 cm; **W** ♂ 177–195 mm (n 18, m 186.5), ♀ 179–189 mm (n 15, m 184.5); **T** ♂ 114–130 mm (n 18, m 121.3), ♀ 114–125 mm (n 15, m 120.0); **T/W** m 65.0; **white tip of r6** (measured on inner web parallel with and as near as possible to shaft) 13–26 mm (n 33, m 20.2); **B** ♂ 45.5–55.1 mm (n 18, m 51.0), ♀ 44.7–52.7 mm (n 15, m 48.9); **B(f)** ♂ 40.5–48.3 mm (n 18, m 45.0), ♀ 40.0–46.6 mm (n 15, m 43.1); **BD** (measured just inside gonys bulge) 14.0–16.2 mm (n 33, m 15.1); **BD/B** 26.5–33.1 (n 33, m 30.2); **Ts** 40.0–43.9 mm (n 32, m 41.8). **Wing formula: p1** > pc 25–40 mm, < p2 40.5–48.5 mm; **p2** < wt 22–33 mm, =7 or 7/8 (*c.* 60%) or =8 or 8/9 (*c.* 40%); **p3** < wt 1–8 mm; **pp4–5**(6) about equal and longest; **p6** < wt 0–4 mm; **p7** < wt 16–21 mm; **p10** < wt 37–47 mm; **s1** < wt 40–51 mm. Emarg. pp3–6 (also sometimes near tip on p7).

GEOGRAPHICAL VARIATION & RANGE At least nine subspecies recognised, forming complex pattern of three main groups (some perhaps species), but only the northern *caryocatactes* group, with two races, occurs in the covered region and is hence the only one treated here. All populations are mainly sedentary, but especially northern ones can make large-scale eruptions in years with meagre food supplies, reaching far west and south of their normal breeding range.

N. c. caryocatactes (L., 1758) (Europe to E European Russia, where grades into *macrorhynchos* somewhere west of Urals). The common European form. Its main distinctions from 'Slender-billed Nutcracker' are slightly larger overall size, slightly thicker-based bill, less extensive white on tail-tip and somewhat longer legs (see Biometrics). Beware of intermediates, thus only clear examples should be assigned to subspecies. It is always best to combine several measurements for a reliable identification. (Syn. *wolfi*.)

N. c. macrorhynchos C. L. Brehm, 1823 (extreme E Russia W of Urals east across much of Siberian range, south to Altai, Tarbagatai, N Mongolia and Manchuria; frequently

Presumed *N. c. macrorhynchos*, 1stW, Finland, Sep: typical jizz in flight. Ageing possible based on 1stY having juv primary-coverts, tertials and alula, and most or even all greater coverts (sometimes a few outer median coverts, too), extensively tipped white (not restricted to small wedge-shaped spots near shaft of ad). Race probably *macrorhynchos* based on large amount of white on tail-tip, but bill seems slightly thick-based, still within normal variation. (M. Varesvuo)

N. c. macrorhynchos, ad, N Mongolia, Jun: unlike *N. c. caryocatactes* has fairly slender and invariably rather long bill, broader white band on tail-tip and shorter legs, and race confirmed by location. Evenly quite fresh wing, with no white tips on secondary-coverts, indicating ad. (H. Shirihai)

Presumed *N. c. macrorhynchos*, ad, Netherlands, Nov: subspecies based on fairly slender, long bill compared to *caryocatactes*. Amount of white at tail-tip impossible to evaluate at this angle. Fresh glossy wing with hardly any white tips indicates ad. (R. Pop)

involved in large invasions to west, following which some breed in N & C Europe, now for many consecutive years in NE Sweden and N Finland). Has fairly slender bill compared to *caryocatactes* (though bill-length is only fractionally longer), and white band on tail-tip is broader (some overlap though). It is slightly smaller on average with a little shorter legs. **W** ♂ 172–191 mm (*n* 14, *m* 181.5), ♀ 172–183 mm (*n* 13, *m* 177.8); **T** ♂ 110–127 mm (*n* 14, *m* 118.6), ♀ 111–121 mm (*n* 13, *m* 115.9); **T/W** *m* 65.0; **white tip of r6** (measured on inner web parallel with and as near as possible to shaft) 21–33 mm (*n* 37, *m* 27.0); **B** ♂ 47.6–55.0 mm (*n* 14, *m* 51.8), ♀ 44.9–54.0 mm (*n* 13, *m* 49.8); **B(f)** ♂ 42.1–48.5 mm (*n* 14, *m* 45.7), ♀ 40.5–46.3 mm (*n* 13, *m* 44.5); **BD** (measured just inside gonys bulge) 12.0–13.7 mm (*n* 36, *m* 12.9); **BD/B** 22.9–28.3 (*n* 36, *m* 25.4); **Ts** 36.6–41.8 mm (*n* 33, *m* 39.4). **Wing formula: p1** > pc 23.5–37 mm; **p2** =7 or 7/8 (*c.* 95%) or =8 (*c.* 5%); **pp4–5** about equal and longest; **p6** < wt 2–9 mm; **p7** < wt 16.5–24 mm. (Syn. *altaicus*; *kamchatkensis*; *sassii*.)

REFERENCES Jenni, L. (1983) *Orn. Beob.*, 80: 136–137. – Lehikoinen, E. (1979) *Orn. Fenn.*, 56: 24–29.

N. c. macrorhynchos, Azerbaijan, Sep: subspecies based on locality, long, slender and very pointed bill, and very broad white band on tail-tip. (M. Heiß)

ALPINE CHOUGH
Pyrrhocorax graculus (L., 1766)

Fr. – Chocard à bec jaune; Ger. – Alpendohle
Sp. – Chova piquigualda; Swe. – Alpkaja

In summer this graceful crow frequents the highest mountains (especially boulder-strewn slopes and cliffs, chiefly above tree-line), regularly to 4000 m, but descends in winter, to valleys and around villages and ski stations. Within the covered region it breeds in the Atlas, Sierra Nevada, Pyrenees, Alps, Apennines, Balkans, Caucasus, Turkey south to about Lebanon, and in Iran, while elsewhere it reaches Central Asia and the Himalayas. The Alpine Chough is most regularly confused with the partly overlapping Red-billed Chough, which, however, tends to frequent lower altitudes, as well as rocky coasts in W Europe. Nests on inaccessible rock faces. Forages over a wide variety of open habitats. Pairs may form loose colonies of up to a few tens. In spring to autumn feeds mainly on insects, but at other times berries, refuse and scraps are busily sought by picking or digging. Social year-round, with much larger flocks forming after breeding and in winter.

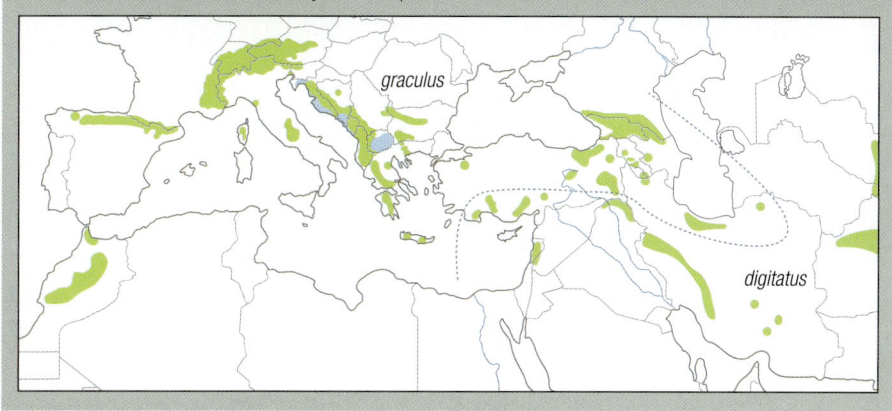

IDENTIFICATION Small or medium-sized *jet-black* corvid, appearing quite slim when perched but somewhat more compact in flight (similar size to Jackdaw, or fractionally larger). In close views, *wings and tail slightly glossed bluish-green*. Sufficiently close in size and structure to Red-billed Chough to sometimes cause identification problems, but *vocalisations, longer tail with narrower base* and *slightly less broad and 'fingered' wingtips* usually help. The *shorter pure yellow bill* is diagnostic, but colour and length not always easy to see, and note that bill of recently fledged Red-billed Chough is shorter than adult and tinged orange, which can appear yellow at long range (see below). Also typical is the *rather small head and rather straighter bill* (decurved mainly at tip). The long tail becomes quite rounded at the end in soaring flight. Iris dark brown and legs variably reddish (brighter orange-red in adult). Gait shuffling with short hops. *Wingbeats rather loose but fast and deep*, and tail often fanned widely. Habitually in flocks, which are vocal, wheeling and gliding effortlessly close to cliffs, and often indulges in spectacular aerobatics and playful chases.

VOCALISATIONS Song and subsong seldom heard; described as a succession of warbling, squeaky, chittering and churring sounds (*BWP*). – Calls loud and often heard, generally higher-pitched than those of Red-billed Chough. Typically gives a very sharp, piercing, slightly downslurred *ziieh* (or *ziiah*) and a high-pitched rolling trill *shrrrrree* (with an almost electric quality, likened to the sound of a stone thrown across thin ice); the two calls sometimes given together, *zrrrrii-eh*. (Typical call of Red-billed Chough is a loud, throaty *kiaach* or *chough*, more reminiscent of Jackdaw, although there is some slight overlap between the two choughs as to this particular call.)

SIMILAR SPECIES Even at long range can be reliably separated from *Red-billed Chough*, especially if pitfalls are considered. At rest adult Alpine Chough has a much shorter, pale yellow bill (may look whitish at distance), with rather straight lower but clearly decurved upper mandible, and the feathering usually extends further onto upper mandible than lower (about equal on Red-billed). In Red-billed, the bill is long, scarlet-red and strongly decurved (especially upper mandible). Alpine often has a less obvious blue and purple gloss to the plumage than Red-billed, and can even appear rather brownish, especially on head. Useful at longer range is tail-length: on Alpine, it is longer and projects well beyond wingtips; on Red-billed, shorter than or equal to wingtip. Juvenile Red-billed has a shorter bill than adults, which is rather pale orange-brown to pinkish-yellow, but always slightly longer than Alpine and never pure yellow, while young birds are usually accompanied by adults. – Despite of structural differences already mentioned, differences in flight silhouette are surprisingly often difficult to see, thus voice is often key. Nevertheless, the following combined differences should produce correct identifications. Alpine has proportionately slightly narrower wings (which are narrower than the tail is long) than Red-billed (in which they are as broad as the tail is long), affording impression of long 'arm' but short 'hand' (opposite in Red-billed). Alpine normally shows five, shorter and less obviously 'fingered' primaries (frequently six well-spaced deeply 'fingered' in Red-billed), but beware effects of moult. Alpine rarely has the typically vulture-like well-spread and longer outer primaries of Red-billed, which appear to have a distinct step between them and the inner primaries and secondaries. Alpine also tends to have a bulging trailing edge and slightly narrower wing base, making the wings appear less rectangular or square than Red-billed, with a more rounded and shorter 'hand'. It also has a slightly narrower (chiefly at base) and longer tail which, when closed, looks less square-ended than that of Red-billed (differences impossible to judge when fanned), and appears somewhat smaller-headed. All of these features, however, require experience of both species, and many individuals at longer range cannot be separated using structural features alone. – Both choughs are very dark-plumaged and, in flight, differ from the somewhat smaller *Jackdaw* by their prominently 'fingered' wingtips, and Alpine Chough has narrower-based and slightly longer tail (in Jackdaw tail barely equals wing width). Also, Alpine Chough has a smaller and narrower head. In certain lights, the jet-black underwing-coverts appear obviously darker than the dark grey flight-feathers (entire underwing uniform dark grey in Jackdaw). Although Jackdaw is a skilful flier and can be acrobatic, it lacks the incredible diving and soaring abilities of both choughs, or at least does not demonstrate them often.

AGEING & SEXING Ageing not easy using moult differences due to limited wing moult in juv, and because ad and juv feathers are very similar in colour and pattern; shape and condition of tail-feathers most reliable character. Sexes alike but differ marginally in size (see Biometrics). – Moults. Complete post-nuptial moult of ad in (May) Jun–Oct. (Interestingly, a 1stW ♀ in Dec in Geneva had dropped inner two primaries in both wings [AMNH 675.744], but could be aberrant or the label date is wrong.) Simultaneous partial post-juv moult includes entire head- and body-feathers and lesser coverts, but only few median coverts; no greater coverts, tertials, alula, primary-coverts, remiges or tail-feathers

P. g. graculus, presumed ad, France, Aug: a pair seen in typical alpine habitat. Plumage looks uniformly jet-black suggesting ad. Lemon-yellow short bill diagnostic. (J. Wynn)

are replaced. – ALL YEAR **Ad** In late summer/early autumn recognised by active moult of flight-feathers. Tail-feathers comparatively broad, tips rounded (cf. Svensson 1992). Evenly-feathered wing, tail and body plumage, underparts black with gloss. **1stY** Any moult contrast between replaced lesser and a few median coverts, and rest of retained juv wing-feathers, often difficult to detect, but close inspection of primary-coverts, greater coverts and outer primaries may reveal that these are duller and less intense black (more brownish-tinged), relatively weaker textured and less fresh, sometimes with abraded outer webs and tips (i.e. more worn than ad at same season). Retained tail-feathers more brownish-black (less glossy) and narrower, with more pointed and less fresh tips and fringes, becoming quite worn by late winter (cf. Svensson 1992). Underparts tend to be duller/browner and less glossed. In worn plumage (first spring/summer), body has more grey feather-bases exposed than ad, often appearing tinged grey-brown, less uniform black, especially underparts. Bill and legs slightly duller than in ad, at least in some. **Juv** Resembles ad but has looser body-feathers, especially below, and is duller, browner and less glossy, lacking metallic gloss of ad, with duller greyish or brownish-red legs (red only by first spring), and duller yellow bill (with horn-coloured base), though soon turns yellow.

BIOMETRICS (*graculus*) **L** 36–39 cm; **WS** 66–75 cm; **W** ♂ 247–274 (once 281) mm (*n* 15, *m* 262.4), ♀ 245–265 mm (*n* 12, *m* 254.3); **T** ♂ 146–168 mm (*n* 15, *m* 157.1), ♀ 141–162 mm (*n* 12, *m* 152.1); **T/W** *m* 59.9; **B** 30.9–38.8 mm (*n* 29, *m* 34.6); **B**(f) 22.3–29.5 mm (*n* 29, *m* 25.5); **BD** (measured just inside gonys bulge) 9.0–11.2 mm (*n* 29, *m* 9.9); **Ts** 41.6–47.7 mm (*n* 29, *m* 44.8). **Wing formula: p1** > pc 38–55 mm, < p2 61–72 mm; **p2** < wt 26–48 mm, =6/7 (*c*. 90%) or =5/6 or 6 (*c*. 10%); **p3** < wt 1–11 mm; **p4** longest; **p5** < wt 2–9 mm; **p6** < wt 25–35 mm; **p7** < wt 48–58 mm; **p10** < wt 83–93 mm; **s1** < wt 93–109 mm. Emarg. pp3–5.

Alpine Chough (top left, centre left and centre right: Morocco, Feb, and bottom left: Switzerland, Jul) versus Red-billed Chough in flight (top right, centre left and centre right: Morocco, Feb, and bottom right: Switzerland, Jul): the two choughs partially overlap in range and even mingle in same flocks (as illustrated by central images, where both species are seen together), but are unlikely to be confused if bill is seen well. Alpine has rather short bill (longer, more decurved bill in Red-billed). In Alpine, nostril feathers usually extend further onto upper mandible than lower. Note flight silhouette with Alpine's proportionately narrower wings with five shorter, less obviously 'fingered' primaries than Red-billed (frequently six well-spaced, longer primaries). Alpine tends to have slightly longer tail than wing is broad (Red-billed about equal) and slightly narrower base, and is smaller-headed. However, moult can considerably alter wing shape and number of 'fingered' primaries. (Top left, top right, centre left and centre right: M. Varesvuo; bottom left and bottom right: H. Shirihai)

GEOGRAPHICAL VARIATION & RANGE Rather slight, with racial designation based solely on size. Two races recognised here. All populations are resident and make only local or altitudinal movements in winter.

P. g. graculus (L., 1766) (S Europe, NW & NE Turkey, Caucasus, Transcaucasia, N Iran, Morocco). Smaller than *digitatus* but otherwise very similar (see Identification and Biometrics).

P. g. digitatus Hemprich & Ehrenberg, 1833 (S Turkey, Levant, Zagros in W Iran, much of Central Asia, north to SC Siberia). Slightly larger than *graculus* with heavier bill and stronger feet. Wing formula differs somewhat. **W** ♂ 250–291 mm (n 12, m 281.8), ♀ 252–285 mm (n 12, m 270.5); **T** ♂ 155–187 mm (n 13, m 170.8), ♀ 152–179 mm (n 12, m 163.4); **T/W** m 60.5; **B** 33.2–42.2 mm (n 25, m 37.0); **B(f)** 22.7–31.0 mm (n 25, m 27.3); **BD** (measured just inside gonys bulge) 8.9–12.0 mm (n 25, m 10.3); **Ts** 38.5–46.8 mm (n 23, m 42.2). **Wing formula: p1** > pc 43–58 mm, < p2 57–76 mm; **p2** < wt 40–57 mm, =6/7 or 7 (*c.* 50%) or =7/8 or 8 (*c.* 50%); **p3** < wt 6–13 mm; **pp4–5** about equal and longest; **p6** < wt 15–30 mm; **p7** < wt 37–52 mm; **p10** < wt 88–112 mm; **s1** < wt 97–137 mm. (Syn. *alpinus*; *forsythi*.)

P. g. graculus, 1stW (left) and ad, Morocco, Feb: about same size as Jackdaw, but jet-black, and thus most like Red-billed Chough, with which it often overlaps in range. Pure yellow and shorter bill diagnostic, while proportionately longer tail and narrower wings with slightly less 'fingered' tips clinch identification. Note 1stW has juv tail-feathers, these being less glossy and narrower, while underparts of body show more grey feather-bases than ad, appearing less uniform black. (M. Varesvuo)

P. g. graculus, ad, Italy, Feb: quite slim black corvid with an elongated rear, the tail projecting well beyond wingtips, unlike Red-billed Chough. Also typical is the rather small head and shorter, only slightly curved yellow bill. Legs often appear brighter orange-red. Evenly feathered jet-black wing ages the bird. (L. Gabrielsen)

P. g. graculus, 1stS, Switzerland, Jul: 1stY has browner tone to black plumage and often shows moult limits in wing (here in median coverts), which can be seen even over reasonable distances. (H. Shirihai)

TAXONOMIC NOTE Variation is clinal and limits between the two recognised races are difficult to define and to some extent arbitrary. Breeders in the Alps are rather large and overlap in size with eastern birds. Even in the eastern range size is not consistent. Further study is needed.

REFERENCES Königstedt, D., Langbehn, H. & Frede, M. (1990) *Limicola*, 4: 22–27. – Madge, S. (1994) *BB*, 87: 99–105.

P. g. graculus, 1stW, Morocco, Jan: young can retain a dusky tinge to bill, at least at tip, and darker legs well into first winter. Also has moult limits in wing, evident at close range and in good light. Shorter yellow bill and clearly longer tail separate from Red-billed Chough (C. N. G. Bocos)

P. g. graculus, juv, Switzerland, Aug: duller, browner and less glossy plumage, lacking full metallic gloss of ad, with duller brownish-red legs, and duller pinkish-yellow to horn-coloured bill with darker tip. (H. Shirihai)

(RED-BILLED) CHOUGH
Pyrrhocorax pyrrhocorax (L., 1758)

Fr. – Crave à bec rouge; Ger. – Alpenkrähe
Sp. – Chova piquirroja; Swe. – Alpkråka

Despite its rather fragmented range in W and S Europe as well as in the Near East and the Caspian region, unlike the Alpine Chough the Red-billed Chough is not restricted to rocky mountains and is thus often more widespread than its congener. In W Europe (Britain, Ireland, Iberia and France) and the Canaries (on Palma) it favours rugged, rocky coasts with nearby grassy areas. Although the two species partly overlap in high mountains, Red-billed is mostly found below Alpine, in the south mainly between 1200 and *c.* 3000 m in the breeding season, but in the Himalayas is found breeding up to *c.* 5000 m and foraging as high as at 7950 m. There is also a relict population in the Ethiopian highlands. Breeds alone or in loose colonies; resident but performs altitudinal movements. Some W Palearctic populations have recently declined, but elsewhere it seems to adapt quite well to land-use changes.

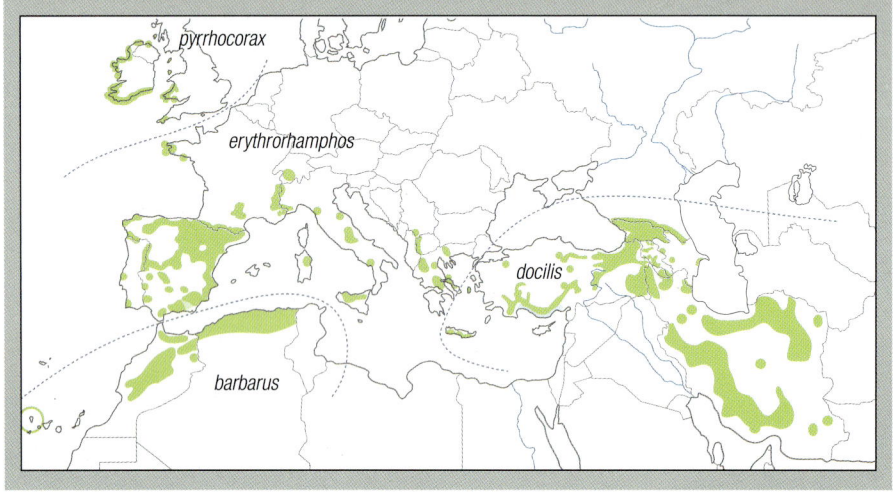

IDENTIFICATION Dashing, *slim, medium-sized black crow* with *diagnostic, pointed, decurved bright red bill* and *reddish legs*. Entirely velvet-black, slightly glossed bluish-purple or greenish on body, but more so on wings and tail. Best separated from Alpine Chough by bill colour, though the longer, more slender and curved outline of the bill are often sufficient clues. Note also the proportionately *broader wings with longer and squarer wingtip, and wider spread and deeply 'fingered' primaries*. Tail rather short (equals wing width) and more square-ended, and, *when perched, wingtip reaches tail-tip* (tail protrudes clearly beyond wings in Alpine Chough). Note extent of feathering on bill base: dense bristles at base of both mandibles are equal in extent. Further differences between them, and potential pitfalls, are discussed at length under Alpine Chough. Like Alpine Chough, in flight (and in good light) blackish underwing-coverts contrast with silver-grey flight-feathers (unlike more uniform grey Jackdaw). Gregarious and very vocal, and most calls are diagnostic. Walks confidently, runs or hops, and feeds almost entirely on ground (on insects, in winter chiefly on seeds and berries), the bill being used to dig and overturn stones (such behaviour often useful in long-range identification). Very agile and buoyant in flight, 'playing' and soaring on thermals, and performing spectacular aerobatics around cliffs, including sweeping dives, often ending in hurtling rolls and tumbling manoeuvres. Fearless and rather approachable.

VOCALISATIONS Song and subsong seldom heard but described as a succession of low warbling, chittering and churring sounds, resembling 'songs' of some other corvids (*BWP*). – Most typical, and often uttered in advertisement in flight, is a cutting, almost whizzing, slightly descending *chiach* or *kiaach* (in the opinion of some the origin of its name Chough; sometimes accompanied by wing- and tail-flicking if given on ground), which superficially resembles some calls of Jackdaw, but is more piercing and explosive, with a 'thicker' finale (but beware *kyow* of young Jackdaw). A wide spectrum of related calls are known, e.g. *chiaa, chrai, chi-ah, tschraah, chwee-ow*, etc.

SIMILAR SPECIES Sometimes loosely associates with Jackdaw and Alpine Chough, but even at long range both of these can be reliably eliminated. The uniform blackish plumage (lacking grey on rear of head) and different structure, with strongly 'fingered' wingtip, longer red bill and red legs, among other features, prevent confusion with *Jackdaw*. – Separation from *Alpine Chough* is briefly outlined above and discussed at length under that species. Beware of hybrids between the two species.

AGEING & SEXING Ageing based on subtle moult limits difficult and requires very close observation or handling. Sexes alike butt differ marginally in size (see Biometrics). – Moults. Complete post-nuptial moult of ad in Jun–Sep (Oct). Simultaneous partial post-juv moult includes entire head- and body-feathers and lesser coverts, but only few median coverts; no greater coverts, tertials, alula, primary-coverts, remiges or tail-feathers are replaced. – **ALL YEAR Ad** In late summer/early autumn recognised by active moult of flight-feathers. Tail-feathers comparatively broad, tips rounded (cf. Svensson 1992). Evenly-feathered wing, tail and body plumage, underparts black with gloss. Downcurved bill long and bright red, legs bright red too. **1stY** Most wing- and tail-feathers are retained juv feathers; especially check primary-coverts, greater coverts, long primaries and tail-feathers, which are more often detectably duller, and relatively weaker textured, less fresh and less glossy than freshly moulted ad feathers (at least some are more brownish-tinged than pure blackish). These fade more rapidly than ad feathers, and sooner acquire abraded outer webs and tips. Retained tail-feathers narrower and more pointed (as in Alpine Chough). Tip of p1 pointed or narrowly rounded, less broadly rounded than ad (cf. Svensson 1992). Until Oct–Nov a few late-moulting birds may retain some scattered juv body-feathers and lesser and median coverts, these being markedly browner and hardly glossed. Similarly, underparts tend to be duller, browner and less glossed. In worn plumage (1st-spring/summer), body often appears less uniform black, especially on underparts. Bill often slightly shorter and dull orange-red in early autumn. Legs, too, often duller red. **Juv** Resembles ad but has looser body-feathers, especially below, and is duller, lacking obvious metallic gloss, has pinkish-brown or blackish (rather than red) legs, and has shorter and less curved pinkish-brown to orange-brown bill.

BIOMETRICS (*pyrrhocorax*) **L** 37–40 cm; **WS** 71–80 cm; **W** ♂ 254–287 mm (*n* 13, *m* 270.9), ♀ 253–277 mm (*n* 12, *m* 263.8); **T** ♂ 119–139 mm (*n* 13, *m* 128.6), ♀ 120–131 mm (*n* 12, *m* 125.8); **T/W** *m* 47.6; **B** ♂ 48.0–61.0 mm (*n* 11, *m* 54.0), ♀ 45.7–57.3 mm (*n* 12, *m* 50.9); **B**(f) 38.4–51.0 mm (*n* 23, *m* 45.0); **BD** (measured just inside gonys bulge) ♂ 11.0–13.2 mm (*n* 13, *m* 12.2), ♀ 10.5–12.2 mm (*n* 12, *m* 11.6); **Ts** 47.3–57.0 mm (*n* 24, *m* 51.9).

P. p. pyrrhocorax, ad, Wales, Sep: diagnostic bright red decurved bill, dark red legs and all-black plumage. Tail is relatively short, at most reaching wingtip, whereas tail clearly protrudes beyond wings in Alpine Chough. Recently moulted with seemingly only new feathers and bright red bare parts in Sep suggest ad. (M. Lane)

P. p. erythroramphos, ad, France, Mar: note dense bristles at base of both mandibles are about equal in extent (cf. Alpine Chough). This continental European race differs only by slightly larger size from *pyrrhocorax*. Aged by bright red bill and feet and fresh, glossed plumage. (A. Audevard)

P. p. erythroramphos, 1stS, France, May: heavily worn and brownish primary tips and less bright red bill suggest age (but remember dark on tip of bill can be dirt). (A. Audevard)

Wing formula: p1 > pc 56–69 mm, < p2 57–68 mm; **p2** < wt 19–33 mm, =6/7; **p3** < wt 0–7 mm; **pp**(3)**4–5** about equal and longest; **p6** < wt 9–20 mm; **p7** < wt 41–52 mm; **p10** < wt 87–98 mm; **s1** < wt 94–107 mm. Emarg. pp3–5.

GEOGRAPHICAL VARIATION & RANGE Four races within the treated region differing in size and to a lesser extent in colour and intensity of gloss. Only ad can generally be reliably assigned to race, and exact boundaries between the taxa might require further elucidation. There is a general trend towards smaller size in the north and on coastal cliffs or inland hills, with larger birds being found in high mountains and in the south. Gloss is generally strongly purplish in populations of humid climates, but more greenish in arid regions. All populations essentially sedentary.

P. p. pyrrhocorax (L., 1758) (Britain, Ireland). Relatively rather small (see Biometrics), and ad has bluish-green gloss (faintly purplish at some angles) on upperwing and tail, whereas body is duller and more sooty-black. Breeders in Brittany, France, tend towards this race, but still best included in next race.

○ **P. p. erythroramphos** (Vieillot, 1817) (W & S France, Iberia, Alps to Austria, Italy, Sicily and Sardinia, Balkans, Greece except extreme south-east). Very similar to *pyrrhocorax*, being only slightly larger on average. Said to have greener gloss than *pyrrhocorax*, but impossible to confirm when series directly compared. **W** ♂ 270–310 mm (*n* 15, *m* 288.3), ♀ 270–286 mm (*n* 11, *m* 280.4); **T** ♂ 127–146 mm (*n* 15, *m* 136.1), ♀ 125–139 mm (*n* 11, *m* 132.4); **T/W** *m* 47.3; **B** ♂ 52.8–59.9 mm (*n* 14, *m* 56.0), ♀ 49.2–58.6 mm (*n* 11, *m* 53.7); **B**(f) 40.8–52.8 mm (*n* 26 *m* 47.1); **BD** ♂ 11.0–14.3 mm (*n* 15, *m* 12.5), ♀ 11.4–12.9 mm (*n* 11, *m* 12.1); **Ts** 49.4–59.6 mm (*n* 27, *m* 53.4). **Wing formula: p1** < p2 64–77 mm; **p6** < wt 13–24 mm; **p7** < wt 45–59 mm; **p10** < wt 94–107 mm; **s1** < wt 104–116 mm.

○ **P. p. barbarus** Vaurie, 1954 (NW Africa and La Palma, Canary Is). A little larger than *erythroramphos*, thus clearly larger than *pyrrhocorax*, but has slightly longer bill than both, which is also subtly deeper based. Glossed green on body and upperwing-coverts, and more bluish-green, less purplish, on mantle and scapulars than both. **W** ♂ 274–312 mm (*n* 20, *m* 296.1), ♀ 258–298 mm (*n* 12, *m* 278.4); **T** ♂ 125–148 mm (*n* 20, *m* 135.9), ♀ 122–143 mm (*n* 12, *m* 132.2); **T/W** *m* 46.5; **B** ♂ 53.8–64.1 mm (*n* 20, *m* 60.9), ♀ 50.6–62.8 mm (*n* 12, *m* 55.2); **B**(f) 43.6–58.2 mm (*n* 32, *m* 51.0); **BD** (measured just inside gonys bulge) ♂ 12.1–14.1 mm (*n* 20, *m* 13.3), ♀ 11.6–14.0 mm (*n* 12, *m* 12.5);

P. p. barbarus, ad, La Palma, Canaries, Jul: compared to previous subspecies, slightly more robust and longer billed, with subtly greener gloss on body and wing-coverts, and more bluish-green tinge on mantle and scapulars. Strongly glossed and less worn at this season, indicating ad. (B. Rodríguez)

P. p. barbarus, 1stW, Morocco, Feb: compared to Alpine Chough, note proportionately broader wings with longer and squarer tips and wider spread, deeply 'fingered' primaries. Relatively short and more square-ended tail than Alpine Chough (which see). Greyish feathers below and abraded wingtips indicate 1stW. (M. Varesvuo)

Ts 48.3–58.2 mm (*n* 32, *m* 53.3). **Wing formula: p1** < p2 55–79 mm; **p10** < wt 80–106 mm; **s1** < wt 94–115 mm. (Syn. *pontifex*.)

○ ***P. p. docilis*** (S. G. Gmelin, 1774) (extreme SE Greece, Crete, Asia Minor, Levant to N Arabia, Caucasus, Turkmenistan and Iran to Afghanistan). A trifle larger on average than *barbarus*, thus is largest race but has relatively smaller bill. Claimed differences in gloss ('greener than other races') difficult to confirm. A very subtle race. **W** ♂ 285–326 mm (*n* 12, *m* 311.1), ♀ 282–302 mm (*n* 12, *m* 292.1); **T** ♂ 135–152 mm (*n* 12, *m* 144.6), ♀ 133–151 mm (*n* 12, *m* 139.3); **B** ♂ 53.4–59.6 mm (*n* 12, *m* 56.1), ♀ 49.7–59.8 mm (*n* 12, *m* 52.4); **B**(f) 42.7–51.3 mm (*n* 27, *m* 46.2); **BD** (measured just inside gonys bulge) ♂ 12.2–14.2 mm (*n* 12, *m* 13.3), ♀ 11.6–13.5 mm (*n* 12, *m* 12.4); **Ts** 47.5–56.9 mm (*n* 27, *m* 52.3). **Wing formula: p1** < p2 63–90 mm; **p10** < wt 101–123 mm; **s1** < wt 117–137 mm.

REFERENCES KÖNIGSTEDT, D., LANGBEHN, H. & FREDE, M. (1990) *Limicola*, 4: 22–27. – MADGE, S. (1994) *BB*, 87: 99–105.

P. p. docilis, Turkey, May: largest race but has relatively small bill. Here a presumed breeding pair. The bird on the right possibly 1stS, since it has rather worn primaries and apparent moult limits in wing. (R. Pop)

P. p. docilis, Georgia, Jun: typical flight silhouette, with broad, deeply 'fingered' wings. Note shorter bill of this race. (K. Malling Olsen)

P. p. pyrrhocorax, juv, Wales, May: resembles ad but bill duller red and a little shorter and less down-curved, and legs not as bright, more brown-tinged. Note also prominent pale gape. The fact that plumage is slightly duller with looser body-feathers requires good light and close observation. (M. Potts)

P. p. erythroramphos, Spain, Sep: often patrolling barren mountainsides in small parties, noticed first by their characteristic calls. In flight readily told from Alpine Chough by the longer, more pointed and red bills, but also by broader wings and shorter tails. The first bird of the three here is in the final stage of primary moult. (H. Shirihai)

(WESTERN) JACKDAW
Corvus monedula L., 1758

Fr. – Choucas des tours; Ger. – Dohle
Sp. – Grajilla común; Swe. – Kaja

One of the most familiar of European crows, Jackdaws are often found around homes, breeding in holes in buildings or in their chimneys. However, its natural breeding habitat, long before man provided convenient nest sites, is rugged coastal and inland cliffs. Favours open country with scattered trees, parkland and woods, and agricultural land. Within the covered region it is widespread, albeit locally in decline. A short-distance migrant in the north, otherwise resident. In winter it prefers open cultivated valleys, damp pastures and sea shores. Gregarious and colonial by nature, Jackdaws often form small flocks that chiefly seek invertebrates, fruit, seeds, carrion and scraps on the ground. The pair bonds for life, and the two are seen together most of the time. Large roosts form in autumn and winter at traditional sites in favoured woods or avenue trees in towns.

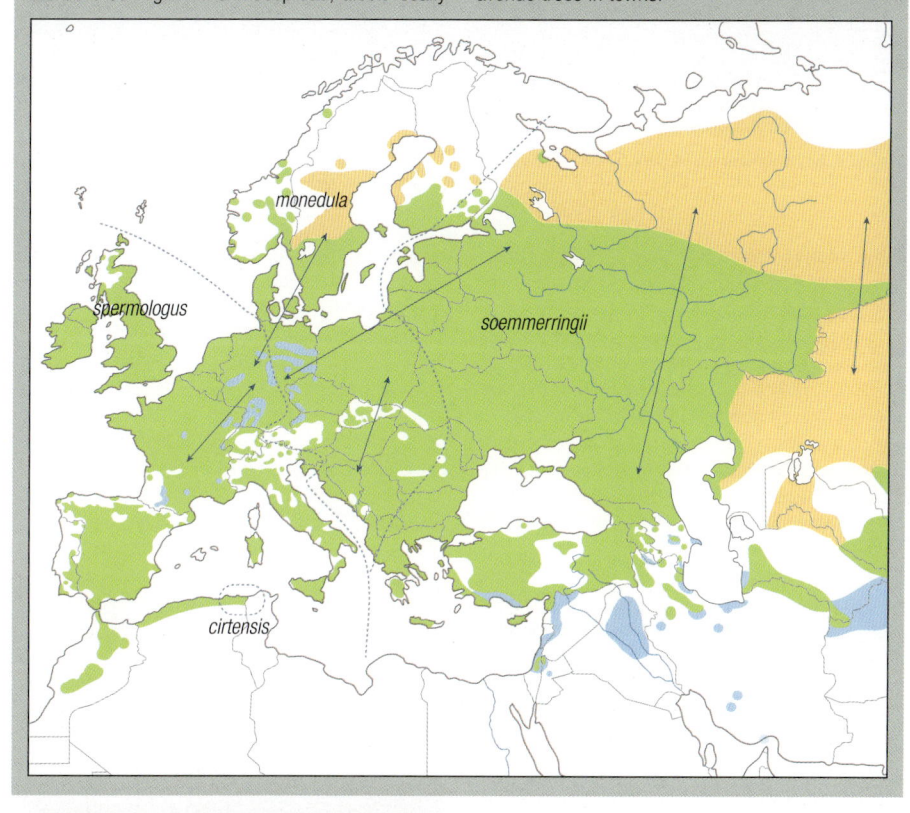

IDENTIFICATION *Small, stocky, black-and-grey corvid.* Usually unmistakable given rather compact appearance, characteristic *silky grey neck-sides and nape*, contrasting *blackish face*, *stubby, short bill* and *pale eye* (greyish blue-white in adult) visible in close views. Mantle, scapulars and smaller wing-coverts dark to mid grey, and underparts even paler, being light dull sooty-grey. A few (chiefly in the east) possess ill-defined whitish neck-side patches that almost form half-collar. At close range fresh autumn plumage shows faint purple to bluish sheen, or faint green and blue gloss (mainly on wings). *Uniformly dark grey underwing* generally lacks contrast, at most has slightly darker coverts. On ground adopts rather upright stance and appears short-legged with jaunty or wagging, rapid walk, often accompanied by head-bobbing, and occasional short sidling jumps. Spread wings appear oval-shaped, and tail is comparatively long. Flight is buoyant with *quick jerky wingbeats* (much quicker than in Carrion Crow or Rook, two other all-dark corvids of open habitats). At long range, flight silhouette and action occasionally somewhat pigeon- or roller-like. Frequently performs acrobatics, with tight circling and tumbling close to cliffs, though usually less agile than choughs. *Sociable and vocal*, and often forms mixed feeding flocks with Rooks and other corvids.

VOCALISATIONS Song a continual low chatter of call-type units with much variation in loudness, but seldom heard. – Conversational and advertisement calls rather short but quite pleasing, some also hoarse and harsh. Most frequent is a short, rather high-pitched, abrupt or even explosive and resounding k*eak!* or *kyack!*, and a shorter jolting *kya* (usually repeated in energetic series and mostly given in flight; frequently extended into characteristic cackle). Other calls include yelps, a short, indistinct but high *kyow*. A low, drawn-out *kyaahr* or *carr-r-r-r*, or *chaiihr* used in alarm (sometimes quite loud and harsh), and harsher *tschreh*. A 'cackling' in chorus is given by large roosting flocks as they settle for the night. Also varied monosyllabic clicking, clucking and hissing sounds, which vary in volume and with mood and context.

SIMILAR SPECIES Unlikely to be confused with other corvids, even if observed at long range, but bear in mind the possibility of a vagrant *Daurian Jackdaw* or a *House Crow* (latter normally only found in extreme south-eastern part of region), both of which possibilities are discussed under their respective accounts.

AGEING & SEXING Ageing generally possible if seen close or handled. Separation of ad and 1stY from late autumn relies principally on moult pattern, shape of unmoulted tail-feathers and sometimes iris colour. Sexes alike but differ marginally in size (see Biometrics). – Moults. Complete post-nuptial moult of ad in Jun–Sep (Oct). Simultaneous partial post-juv moult includes entire head- and body-feathers and lesser coverts, but only few median coverts; no greater coverts, tertials, alula, primary-coverts, remiges or tail-feathers are replaced. – **ALL YEAR Ad** Initially, in late summer and early autumn, many ad easily separated by heavy moult and whitish iris (pale silvery-grey). Later in winter and spring more difficult to age but often has glossier flight-feathers and broader tail-feathers with broadly rounded less abraded tips. No detectable moult limits. **1stY** Most juv wing- and tail-feathers retained, thus no moult of remiges in late summer or early autumn. Juv feathers (in particular, check primary-coverts, greater coverts, long primaries and tail-feathers) often detectably duller, and relatively weaker textured, less fresh and less glossy than fresh-moulted ad feathers (at least some juv feathers are more brownish-tinged). Juv feathers also fade more rapidly, and sooner acquire abraded outer webs and tips. Occasionally replaces some central tail-feathers. Retained juv tail-feathers narrower and more pointed than new of ad type. Iris initially bluish or brown, then successively becomes paler in autumn, light brown or pale greyish with brown smudge. Some attain ad appearance in mid autumn, others much later. **Juv** Has looser body-

C. m. spermologus, ad, Netherlands, Mar: race characterised by being rather dark with mid-grey shawl and head-sides, albeit with much individual variation. Species unmistakable in being a small stocky corvid, with black-and-grey plumage, including contrasting blackish face, and a rather stubby bill and pale eye. Evenly fresh wing infers age. (R. Pop)

C. m. spermologus, ad, Netherlands, Feb: when fresh and at right angle can show more contrasting paler grey neck-sides, even almost forming a half-collar at lower edge, with a faint purple, green or bluish sheen, mainly on wings. Note characteristic facial pattern. Aged by the fresh, glossed wing. (N. D. van Swelm)

C. m. spermologus, 1stS, England, Apr: many juv wing- and tail-feathers, which are visibly worn and brownish, reveal age, but in first spring iris often already white like ad. (I. Fisher)

C. m. monedula, Finland, Oct: in flight, rather small and compact all-dark corvid, and usually readily identified, even before the plumage diagnostics are visible. (M. Varesvuo)

feathers, especially below, and is duller overall, lacks obvious metallic gloss, with dark iris (brown, dull bluish or dark grey).

BIOMETRICS (*monedula*) **L** 31–34 cm; **WS** 65–73 cm; **W** ♂ 228–245 mm (*n* 17, *m* 235.1), ♀ 218–233 mm (*n* 17, *m* 225.5); **T** ♂ 118–135 mm (*n* 17, *m* 123.8), ♀ 113–127 mm (*n* 17, *m* 118.6); **T/W** *m* 52.6; **B** 30.0–37.4 mm (*n* 33, *m* 33.7); **BD** 11.0–15.0 mm (*n* 33, *m* 13.5); **Ts** 38.4–46.0 mm (*n* 33, *m* 43.1). **Wing formula: p1** > pc 36–48 mm, < p2 56–69 mm; **p2** < wt 12–23 mm, =5/6 (50%), =5 (38%) or =4/5 (12%); **pp3–4** about equal and longest (p4 rarely < wt 0.5–4 mm); **p5** < wt 10–18 mm; **p6** < wt 30–38 mm; **p7 <** wt 44–54 mm; **p10** < wt 71–87 mm; **s1** < wt 78–94 mm. Emarg. pp2–5.

GEOGRAPHICAL VARIATION & RANGE Rather slight and clinal variation without sharp borders between described subspecies, frequently making subspecific identification difficult or speculative. Differences mainly involve depth of grey on rear of head and neck-sides (with variably-developed whitish patches or half-collar on latter) and colour of underparts. However, these also vary with age, sex, and season (including strongly with feather wear, especially by spring), and young should never be included in subspecific assessments. Also body colour (including grey of neck) is influenced by altitude and latitude, being darker in western and northern, more humid and higher areas, but paler further south, in drier climates and in lowlands. Most populations are mainly sedentary, but northern birds may winter somewhat further south and west as indeed some ringing recoveries prove.

C. m. spermologus Vieillot, 1817 (British Isles, W & SW Europe east to Italy and extreme NW Balkans, also Morocco, NW Algeria). Rather dark with mid grey nape and head-sides. Underparts blacker than in *monedula*, which it otherwise closely resembles. On average subtly darker nape shawl than *monedula*, but many are just as pale as *monedula*, and this claimed difference is very subtle and of little use to separate them. Variation in underparts sometimes referred to as representing morphs, but not adopted here due to much individual variation. – Birds in S Denmark, Germany, Netherlands and Belgium, and in much of C & S Europe, intermediate between this race and *monedula*, or very close or identical to latter, and included here within it. **W** ♂ 223–249 mm (*n* 15, *m* 239.0), ♀ 223–242 mm (*n* 12, *m* 231.3); **T** ♂ 115–133 mm (*n* 15, *m* 126.5), ♀ 119–134 mm (*n* 12, *m* 123.8); **T/W** *m* 53.2; **B** 32.1–38.5 mm (*n* 27, *m* 35.2); **BD** 12.9–15.7 mm (*n* 26, *m* 14.1); **Ts** 42.1–47.0 mm (*n* 27, *m* 44.4). Wing formula very nearly identical to *monedula*.

C. m. monedula L., 1758 (Fenno-Scandia, Denmark, W Germany, Poland, W Ukraine, Czech Republic, Slovakia, W Romania, Hungary, Austria, Switzerland, W Balkans). Averages paler than *spermologus*, with grey on head/neck paler than underparts (both equally dark in *spermologus*). Underparts medium ash- or plumbeous-grey, paler than *spermologus*. Nape and neck-side patches rather pale grey, but often restricted to small whitish crescent on lower neck-sides, though (even in worn plumage) hardly ever as broad and contrasting as birds further east or true *soemmerringii*.

C. m. monedula, Finland, Jan: highly sociable, forming large, noisy flocks with buoyant flight and characteristic quick, jerky wingbeats. (M. Varesvuo)

C. m. monedula, Finland, Jan: characteristic 'bluntly pointed' wings but proportionately rather long tail. Head pattern with blackish face, pale neck-sides and whitish eye clinch identification. (M. Varesvuo)

CORVIDS

– Birds in S Denmark, Germany, Netherlands and Belgium often a little darker below and tend towards *spermologus*, still majority identical or very close to *monedula* and best included herein. (Syn. *brehmi*; *hilgerti*; *turrium*.)

○ **C. m. soemmerringii** J. G. Fisher, 1811 (Russia, Belarus, much of Ukraine, E Romania, Bulgaria, Macedonia and adjacent parts of Balkans, Turkey, Near East, N Iraq, much of Iran, Caucasus, Transcaspia, Central Asia, much of Siberia; possibly also parts of Finland, Baltic States, E Poland, Czech Republic, Hungary). Similar to *monedula* but subtly larger on average with slightly stronger bill. Has dark grey rather than near-black underparts (thus close to *monedula*). Upperparts in fresh plumage often blackish with pale grey scalloping on mantle and back. Paler grey shawl variable in prominence, some like darkest *monedula*, others very prominently pale grey. Majority have grey-white mark or half-collar, but a prominent and long whitish mark only occurs in *c*. 50%. Often better contrast between black crown and rather pale grey surround. Birds without grey-white collar patch can be inseparable from *monedula*. – Intergrades with *monedula* in N Sweden, Finland and in E Europe & Balkans. **W** ♂ 222–253 mm (*n* 13, *m* 237.2), ♀ 222–242 mm (*n* 13, *m* 230.2); **T** ♂ 110–134 mm (*n* 13, *m* 126.0), ♀ 117–128 mm (*n* 13, *m* 121.1); **B** *m* 34.7 mm (*n* 27); **BD** *m* 13.9 mm (*n* 24); **Ts** *m* 43.5 mm (*n* 27). Wing formula very nearly the same as *monedula*. (Syn. *collaris*; *pontocaspicus*; *schlueteri*; *sophiae*; *tischleri*; *ultracollaris*.)

○ **C. m. cirtensis** (Rothschild & Hartert, 1912) (Constantine, NE Algeria; possibly also adjacent parts of N Africa). Resembles *spermologus* but is slightly larger and clearly paler smoke-grey below, near *monedula* but more uniform with less contrasting blackish bib. Less glossy metallic colours, more matt and evenly smoke-grey above. Only slightly paler grey shawl poorly developed (thus apparently invariably like darker *spermologus*). **L** 33–36 cm; **W** ♂ 228–243 mm (*n* 9, *m* 237.9), ♀ 212–229 mm (*n* 11, *m* 220.4); **T** ♂ 119–136 mm (*n* 9, *m* 126.2), ♀ 110–121 mm (*n* 11, *m* 117.8); **T/W** *m* 53.1; **B** 31.9–38.0 mm (*n* 22, *m* 35.2); **BD** 12.9–15.1 mm (*n* 16, *m* 14.0); **Ts** 39.8–48.3 mm (*n* 24, *m* 44.3).

TAXONOMIC NOTE A recent molecular analysis (Haring *et al*. 2007) found jackdaws to form a clade basal to other *Corvus*, but referring them to a separate genus *Coloeus* is arbitrary, and a larger genus *Corvus* is preferred here.

REFERENCES Haring, E., Gamauf, A. & Kryukov, A. (2007) *Mol. Phyl. & Evol.*, 45: 840–862. – Harrop, A. (2000) *BW*, 13: 290–295. – Henderson, I. G. (1991) *Ring. & Migr.*, 12: 23–27. – Offereins, R. (2003) *DB*, 25: 209–220.

C. m. monedula, ad, Finland, Nov: some birds of this race are as dark as *spermologus*, and lack any trace of a whitish crescent on lower neck-sides. Note purple sheen to wing. Evenly fresh and glossy wing infers ad. (M. Varesvuo)

Presumed *C. m. soemmerringii*, ad, Finland, Nov: this race winters sparsely in Fenno-Scandia, averaging slightly paler grey on head with distinctly whiter half-collar in most birds (but much individual variation, and extremes of *monedula* can be very similar). Evenly fresh and glossy wing infers ad. (M. Varesvuo)

G. m. soemmerringii, ad (right) and 1stW, Israel, Jan: often overall slightly paler grey than *monedula*, especially the shawl and grey-white half-collar (left bird), although this is highly variable, some being like the darkest *monedula* (right). Presence of grey-white neck-sides unrelated to age—it being here better developed in the young bird (aged by juv, brown, wings). (H. Shirihai)

G. m. soemmerringii, ad, Cyprus, Apr: on palest birds, in which grey-white shawl is very obvious, it is difficult to see whitish half-collar. Note pale grey scalloping on mantle and back. White iris immediately silence any thoughts of Daurian Jackdaw. Evenly-feathered ad. (F. Trabalon)

DAURIAN JACKDAW
Corvus dauuricus Pallas, 1776

Fr. – Choucas de Daourie; Ger. – Elsterdohle
Sp. – Grajilla dáurica; Swe. – Klippkaja

The Daurian Jackdaw replaces Western Jackdaw in S Siberia, Transbaikalia, on the E Tibetan Plateau, and in parts of China and Mongolia. It occurs in quite open habitats, e.g. trees in river valleys, fields, open woodlands and clearings, cultivated areas, and in mountains, including rock faces, and the species can visit suburban areas with large trees and fields for foraging. Migratory or resident, thus northern breeding areas are mostly vacated in winter, while the species is resident or dispersive in the south. Winters south to Russian Turkestan, Korea, Japan (rare) and S China, but within the treated region known solely as a vagrant and is likely to join flocks of Western Jackdaws. Records come from Netherlands, France, Sweden and Finland.

Ad, Mongolia, May: a smallish crow of Asian origin, in which invariably pied ad is readily separated from Western Jackdaw of similar size and shape. Note also dark eye. (H. Shirihai)

IDENTIFICATION A *smallish pied crow* that resembles Western Jackdaw in size and structure. Two distinctive age-related plumages, the adult is pied whereas juvenile/immature are dark. *Adult bicoloured* almost like a small Hooded Crow. *Upperparts and head glossy black* (with slight bluish-purple sheen) except *greyish-white neck collar*. *Underparts mainly grey-white*, too, except *black vent and bib*. *Blackish head-sides* behind eye variably *striped silvery-white*. Dark juvenile usually shows hint of pied pattern, but strongly obscured and restricted. *Immatures mainly all-dark* (darker even than Western Jackdaw), with most pronounced *pale area being rear ear-coverts to hindneck*, where variably striped whitish. (1st-year plumage thus mainly black, as juvenile moults to dark plumage in first autumn and remains blackish until first complete moult when *c.* 1 year old.) Therefore, young birds might easily be confused with Western Jackdaw. Bill and legs black, iris dark brown in adult and immature. Gait, behaviour and flight action similar to Western Jackdaw.

VOCALISATIONS Mostly very similar to Western Jackdaw, but sometimes slightly more nasal and even ventriloquial in tone. Also said to possess some more raw and harsher notes recalling Carrion Crow, e.g. a short *kaah*.

SIMILAR SPECIES Pied adults are striking and usually unmistakable, even compared to the palest eastern races of *Western Jackdaw*. However, take into account potential for confusion with leucistic Western, which should, however, still lack the silvery-white striped rear ear-coverts of Daurian. Juvenile and immature Daurian invariably darker and more uniform than adult, thereby inviting confusion with Western Jackdaw, but still diagnostically diff in being blacker (less blackish-grey) and having, at least in juvenile plumage, to some degree (albeit strongly obscured) pied pattern of adult, i.e. some contrast between black of broad bib to upper chest, and at least a hint of faint grey to pale silver ear-coverts streaking in close views (in *monedula* smaller bib grades into dark grey underparts, without contrasting darker patch, and no pale ear-coverts striping). Daurian Jackdaw is dark-eyed (at least many immatures and all adult Western have whitish iris or partially so). Beware, however, that the pale 'silver stripes' on ear-coverts often wear off rapidly, especially in darker individuals, with heavily worn 1st-summers sometimes fully black-headed. – Adult at a distance somewhat recalls *Hooded Crow* (especially due to Hooded Crow's contrastingly blackish head, breast and wings, and pale grey collar effect). However, Hooded readily eliminated by being distinctly larger and differently shaped, especially the longer and stronger bill, and by its grey back (black in Daurian). – *House Crow* (a species with a track record of ship-assisted vagrancy, but no overlap in natural ranges) could superficially recall Daurian Jackdaw due to its paler grey collar and underparts. Again, the larger size, slimmer shape and longer and heavier bill of the crow, among other characters, should identify it.

AGEING & SEXING Ageing straightforward during 1stY. Apart from obvious plumage differences between 1stY and ad, ageing also supported by moult pattern, shape of unmoulted tail-feathers. Sexes alike but differ marginally in size (see Biometrics). – Moults. Complete post-nuptial moult of ad in Jun–Oct. Simultaneous partial post-juv moult involves entire head- and body-feathers and lesser coverts, but only few median coverts; no greater coverts, tertials, alula, primary-coverts, remiges or tail-feathers are replaced. – **ALL YEAR Ad** Invariably pied (boldly patterned in black and near-white), but note that 2ndY (i.e. 2ndW to 2ndS) has variably less pale body parts, these often being more greyish-tinged white. Note, often considered dimorphic with respect to presence and amount of silvery-white stripes on head-sides, with 'dark-headed birds' having entire head blackish except slightly greyer ear-coverts, and 'silver-eared birds' having densely striped silvery-white head-sides, especially behind eye, but much individual variation and wide range of intermediates. Whitish underparts and collar often tinged pinkish-grey in fresh plumage, but usually when breeding washed ochre or isabelline (by soil contamination). Apart from wing-, tail- and bill-length (and especially if pair seen together), ad ♀ tends to have black areas less glossed than ad ♂. It has not been possible to confirm the existence of a 'black morph' ad (sometimes mentioned in the literature). **1stY** Following post-juv moult (from about Aug) markedly different from ad, being virtually all dark, but often still has greyish-striped ear-coverts and nape, and broad black bib contrasts slightly with paler and greyer surround or mottled underparts (but these vary individually). Like ad, some are markedly striped silvery-white on rear head-sides and some virtually lack this, or stripes wear off. Individual variation means that some 1stY are less blackish, but tinged slightly more grey, and pied pattern of ad is 'ghosted', but this seems more strongly developed in 1stS, mainly due to wear and beaching. Black remiges, tail and body duller (more greyish brown-black) with reduced gloss (or latter more bluish, less purplish), and also relatively less fresh, sometimes with abraded outer webs and tips. **Juv** Has looser body-feathers, especially below, and though can appear faintly pied, is distinctly duller than ad, with black parts duller and greyer (and variably mottled) and pale areas ill-defined and dirty light grey. Moults into blacker plumage soon after fledging. Iris grey-brown (dark to mid brown in adult).

BIOMETRICS L 31–34 cm; WS 63–74 cm; W ♂ 226–248 mm (n 15, m 236.1), ♀ 213–233 mm (n 15, m 222.7);

Ad, Mongolia, Jun: pied pattern also distinctive in flight. At distance can recall palest races of Hooded Crow but is distinctly smaller, with more compact shape, and shorter bill; also all-dark underwing (Hooded Crow has paler grey foremost underwing-coverts). (J. Lidster)

1stS, Japan, Mar: unlike Western Jackdaw diagnostically overall blacker, only sides of head being striped in silver-grey. Daurian Jackdaw (at all ages) is dark-eyed, unlike Western. Much of wing is juv. (T. Shimba)

1stS, Mongolia, May: all-dark imm could easily be confused with Western Jackdaw, but is often more uniformly darker. (M. Putze)

T ♂ 123–137 mm (*n* 15, *m* 128.1), ♀ 113–128 mm (*n* 15, *m* 119.5); **T/W** *m* 53.9; **B** 30.1–35.0 mm (*n* 30, *m* 32.9); **BD** 12.0–14.1 mm (*n* 29, *m* 13.1); **Ts** 39.0–46.7 mm (*n* 30, *m* 43.0). **Wing formula: p1** > pc 36–52 mm, < p2 58–71 mm; **p2** < wt 13–24 mm, =5/6 (90%) or =5 (10%); **pp3–4** about equal and longest; **p5** < wt 10–18 mm; **p6** < wt 30–38 mm; **p7** < wt 44–54 mm; **p10** < wt 71–87 mm; **s1** < wt 78–94 mm. Emarg. pp2–5.

GEOGRAPHICAL VARIATION & RANGE Monotypic. – C & E Siberia, Mongolia, Manchuria, SW, NC & E China; mainly resident, but may make short-range movements in winter, at least in northern parts. (Syn. *khamensis*; *neglectus*.)

TAXONOMIC NOTES See Western Jackdaw. – Daurian Jackdaw is sometimes considered conspecific with Western Jackdaw, but molecular differences (Haring *et al.* 2007), as well as distinctive plumage and eye colours, and unique plumage sequence (pied juv, black 1stY and pied ad) support their separation. – In older literature (e.g. *BWP*) the species is sometimes stated to have two colour morphs even as adults, one pied and one dark, but like Svensson (1992) we have only found a tendency to two morphs among young birds. – Note that some, especially in extreme west, have obscure, greyer stripes on hindneck, and even more extensive grey suffusion over the flanks and lower belly, suggesting gene flow with *C. monedula*. Interbreeding with *monedula* reported from several parts of Asia.

REFERENCES Haring E., Gamauf, A. & Kryukov A. (2007) *Mol. Phyl. & Evol.*, 45: 840–862. – Leader, P. J. (2003) *BB*, 96: 520–523. – Madge, S. (1994) *BB*, 87: 99–105.

Left: 1stS, Mongolia, Jun: especially during 1stS can develop paler areas and thus appear superficially like Western Jackdaw. Note diagnostic combination of blackish bib (though difficult to see due to brownish-grey background) and bib extending to breast. Also, note marbled pattern over rest of underparts. (H. Shirihai)

1stS Daurian Jackdaw (left: Mongolia, Jun) vs. Western Jackdaw (right, *C. m. soemmerringii*, Israel, Jan): at this age, three characters should always separate Daurian: broad blackish bib to upper chest (in Western smaller bib grades into dark grey underparts); pale marbled underparts below dark breast (lacking in Western, or only show some streaking); dark eyes (most young Western have whitish-grey iris). However, there is some degree of variation within (partially due to moult and wear) and overlap between the two species in all these characters. (H. Shirihai)

HOUSE CROW
Corvus splendens Vieillot, 1817

Fr. – Corbeau familier; Ger. – Glanzkrähe
Sp. – Cuervo indio; Swe. – Huskråka

This small crow, native to the Indian subcontinent, SE Asia and SW China, is a successful ship-borne colonist that has spread widely across East Africa north to the Levant and Kuwait and east to Australia, as well as to various islands. Often closely associated with humans, foraging around settlements, and outside natural range usually breeding close to ports and towns with trees for nesting and food in the form of refuse; locally inland (below c. 1500 m). Often considered a pest, but in Eilat, Israel, helps eliminate other pests and specialises in predating feral pigeon's nests. Ship-assisted accidentals have reached Ireland, the Netherlands and Spain, and a small breeding population has been established in the Netherlands.

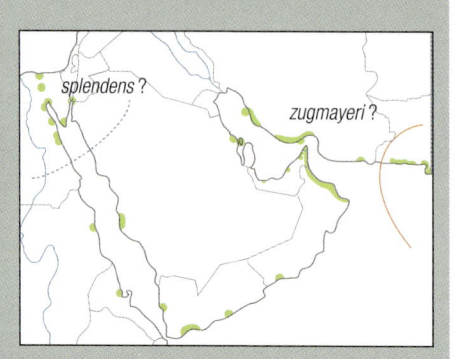

C. s. zugmayeri, ad, Oman, Nov: this race has paler grey collar, contrasting with black parts. When excited species often raises crown-feathers to form peculiarly domed head profile. Evenly-feathered and strongly glossed wing infers age. (D. Occhiato)

IDENTIFICATION A *medium-sized, slim*, two-toned crow with a domed crown, *relatively long and proportionately strong bill*, and longish tail. It sometimes also appears long-necked and long-legged. Simple plumage pattern, with *blackish crown and forehead to throat*, abruptly becoming *grey on head-sides and neck, imparting diagnostic broad greyish collar extending to chest, bordered below by dusky underparts*, with black on back, wings and tail. Superficially similar to Hooded Crow or a pale Jackdaw, but at close range adult shows gloss to black plumage, with stronger blue gloss particularly on forewing. At some distance or in certain lights can appear all dark, the grey colour becoming blackish. Iris brown, bill and legs black. Sexes similar. This conspicuous crow is often encountered around habitation. On ground gait as Hooded Crow, but appears more agile due to smaller size and slimmer build with longer legs (especially tibia). In flight, characteristic silhouette of longish neck, narrow head and long bill; wings broad and rounded, and tail long with round corners, which can render it a certain jizz like a small Raven. Highly social and can be rather noisy; usually alert and wary.

VOCALISATIONS Song and subsong seldom heard (presumably due to limited study). – Rather vocal (frequently calls in flight), often giving a harsh or rasping *kraar kraar kraar*, which can recall Nutcracker a little. There is also a flatter, toneless *kwar kwar kwar* or dry *quah quah*. Other softer and more subdued calls noted, e.g. guttural *caw*, a low *kurr* or protracted *krrrraaar*, a low-pitched *kowk*, etc.

SIMILAR SPECIES Superficially recalls Hooded Crow (only limited overlap of ranges, e.g. N Egypt), but is c. 10% smaller, with proportionately heavier bill (more parallel upper and lower edges and, especially, deeper at the tip, with more curved upper mandible). House Crow also has diagnostic dark mantle and scapulars to uppertail-coverts (grey in Hooded), but blackish hood is less complete (grey collar extends to ear-coverts), and black of throat does not extend as 'splash' marks on breast as often in Hooded. The latter, plus usually dusky (with lead-grey hue) underparts, makes grey collar more complete than in Hooded. Beware occasional hybrid Carrion × Hooded Crow, which can even more closely resemble House Crow in plumage, but latter has clearly better-defined and broader grey collar than any hybrid, with slimmer build, different head profile and bill shape. Furthermore, such hybrids are unlikely to occur in the same range as House Crow. Plumage may also somewhat recall eastern populations of Western Jackdaw and especially Daurian Jackdaw, but both are at least c. 20% smaller, with especially much shorter bill, different pattern of grey, paler iris (adult Western Jackdaw) and different call.

AGEING & SEXING Ageing generally possible if seen close or handled. Separation of ad and 1stY from late autumn relies mainly on moult pattern, shape of unmoulted tail-feathers and sometimes iris colour. Sexes alike but differ marginally in size (see Biometrics). – Moults. Complete post-nuptial moult of ad in Jun–Sep (depending on population/ time of breeding; some indigenous Asian populations moult later). Simultaneous partial post-juv moult includes entire head- and body-feathers and lesser and median coverts, but usually only few inner greater coverts (varies between none and all); tertials and alula also moulted partially (but few data); no primary-coverts or remiges moulted, but central tail-feathers (r1) often replaced. – **ALL YEAR** Initially, at end of breeding season, many ad still in heavy moult while juv fresh. Once moult completed, check following pattern. **Ad** Evenly fresh or worn wings, tail and body with no obvious moult limits. Tail-feathers broad with rather square tips. Blue gloss more uniform and intense. **1stY** Moult contrast best seen (when bird seen close or handled) between moulted median coverts, some tertials and (inner) greater coverts on the one hand being darker and glossier, and rest of juv wing-feathers (usually including at least some retained outer greater coverts); retained greater coverts, remiges and primary-coverts relatively weaker textured and less fresh, sometimes with abraded outer webs and tips (generally more worn than ad at same season). Retained tail-feathers more brownish-tinged, less glossy and narrower, with more tapering, rounded and less fresh tips and fringes, already quite heavily worn when a few months old. In worn plumage (1st-spring or summer) moult limits often more pronounced, but in both age classes grey collar better defined and paler with wear. **Juv** Resembles ad but has looser body-feathers, especially below, and is duller; grey areas darker than in ad (collar sometimes duskier and less well defined) and black of plumage dull, not glossy.

BIOMETRICS (*zugmayeri*) L 39–42 cm; WS 72–86 cm; W ♂ 241–286 mm (n 12, m 267.3), ♀ 241–270 mm

C. s. zugmayeri, 1stS, Oman, Mar: in sunlight and favourable angle, moult contrast in 1stY easily detected (juv wing feathers are worn and bleached brownish). (M. Varesvuo)

(n 12, m 256.7); **T** ♂ 151–178 mm (n 12, m 165.3), ♀ 140–174 mm (n 12, m 155.6); **T/W** m 61.2; **B** 44.0–54.4 mm (n 24, m 48.5); **BD** (at lower mandible feathering) 15.0–17.7 mm (n 23, m 16.5); **Ts** 43.2–52.5 mm (n 24, m 47.1). **Wing formula: p1** > pc 43–64 mm, < p2 52–73 mm; **p2** < wt 18–36 mm, =6 or 6/7 (85%), =5/6 (10%) or =7/8 (5%); **pp3–4** about equal and longest (though p3 often to 3 mm <); **p5** < wt 2.5–13 mm; **p6** < wt 14–28 mm; **p7** < wt 40–62 mm; **p10** < wt 73–101 mm; **s1** < wt 79–104 mm. Emarg. pp3–5.

GEOGRAPHICAL VARIATION & RANGE Rather slight and largely clinal; racial designation mostly involves tone of grey on collar and chest. Several races recognised, but two relevant within treated region (either introduced or self-established at many coastal localities in E Africa and along Red Sea and Persian Gulf). See Taxonomic note. All populations including introduced apparently resident.

C. s. zugmayeri Laubmann, 1913 (SE Iran to Pakistan, NW India and S Kashmir; presumably this race introduced or ship-assisted to Oman and Europe). Described above; when series compared with *splendens*, quite distinctly paler overall, grey collar pale ash-grey or pale smoky-grey when fresh, to almost dirty grey-white when worn (collar becomes almost dull sandy-white when very worn), and strongly contrasts with black parts. Subtly smaller than *splendens*.

C. s. splendens Vieillot, 1817 (India except in west, also Nepal and Bangladesh; possibly predominantly this race introduced to Israel and adjacent parts of Middle East). Has darker (medium) ash-grey collar, contrasting less with duller blackish head, underparts and mantle. Proportionately slightly shorter tail but thicker bill. Sexual size dimorphism larger than in *zugmayeri*. **W** ♂ 253–292 mm (n 12, m 270.3), ♀ 238–260 mm (n 12, m 249.0); **T** ♂ 149–175 mm (n 12, m 161.8), ♀ 137–168 mm (n 12, m 151.5); **T/W** m 60.4; **B** 44.5–56.0 mm (n 23, m 49.2); **BD** (at lower mandible feathering) 15.5–19.3 mm (n 23, m 17.5); **Ts** 43.0–52.8 mm (n 24, m 47.8).

TAXONOMIC NOTE It is usually impossible to know origin of introduced birds and difficult to assign them to race, either because they appear to be derived from intermediate populations or are a mixture of various introduced races. Most in Middle East and E Africa appear to be *C. s. splendens*, but those in, e.g., Iran, Kuwait and Oman could be *zugmayeri* or intermediates. Further, *protegatus*-like bird recorded in Aden and Yemen. For further information see Ryall (2010) and Roselaar (*in BWP*). For birds introduced to Israel, see Shirihai (1996), for those in Egypt, see Bijlsma & Meininger (1984), and for Arabia, see Jennings (2010).

REFERENCES Bijlsma, R. G. & Meininger, P. L. (1984) *Gerfaut*, 74: 3–13. – Ottens, G. & Ryall, C. (2003) *DB*, 25: 312–319. – Ryall, C. (2010) *BBOC*, 130: 246–254.

C. s. zugmayeri, ad (left) and 1stY, Oman, Mar: with wear paler areas become dirty grey-white and sharply contrast with black parts, making this race even more distinctive. However, possibility of confusion with Hooded Crow is greater, although present species usually appears smaller, with diagnostic dark mantle to uppertail-coverts (grey in Hooded), pale collar extends to ear-coverts and black of throat does not extend to breast (as often in Hooded). 1stS shows mainly juv wing and tail, contrasting with glossy black median and lesser coverts. (M. Varesvuo)

C. s. splendens, ad, Israel, Mar: underwing-coverts usually darker than remiges (in Hooded Crow underwing-coverts are pale grey). Note typical flight silhouette, which once learnt can identify the species even at long range. Compared to Arabian birds this race usually has clearly darker grey plumage, notably on collar, contrasting less with head, underparts and mantle. (H. Shirihai)

C. s. splendens, Israel, Mar: note typical jizz and posture, long-legged with thin 'trousers', as well as proportionately heavy, long bill with quite curved upper mandible. Metallic blue sheen to plumage visible in close views. (H. Shirihai)

C. s. splendens, 1stS, Israel, Mar: an extremely dark example of this race, with pinkish dusky-grey collar, thus blackish head and belly are less contrasting. Note proportionately slightly thicker bill than in Arabian race. Aged by many juv wing-feathers, which are worn and bleached brownish. (H. Shirihai)

ROOK
Corvus frugilegus L., 1758

Fr. – Corbeau freux; Ger. – Saatkrähe
Sp. – Graja; Swe. – Råka

An elegant crow of open country, the Rook is locally rather frequent and widespread across Europe to the northern Middle East and through Asia. Highly gregarious, it nests in noisy colonies and keeps in flocks, either when feeding or at large communal roosts. Frequently forms mixed feeding flocks with Jackdaws and Starlings, and sometimes forages with Lapwings *Vanellus vanellus* or even Calandra Larks. Usually it forages on fields and plains (taking invertebrates, plant material, small vertebrates, scraps and carrion), and inhabits broad valleys with clumps of woodland and extensive pasture or agricultural landscapes with ploughed fields, often near farms and habitations. Its preference for agricultural areas with some tall trees has permitted the species to expand in modern times, but a strong decline (due variously to persecution, changes in agricultural methods, use of pesticides, etc.) has been reported in some countries. The Rook is partially migratory.

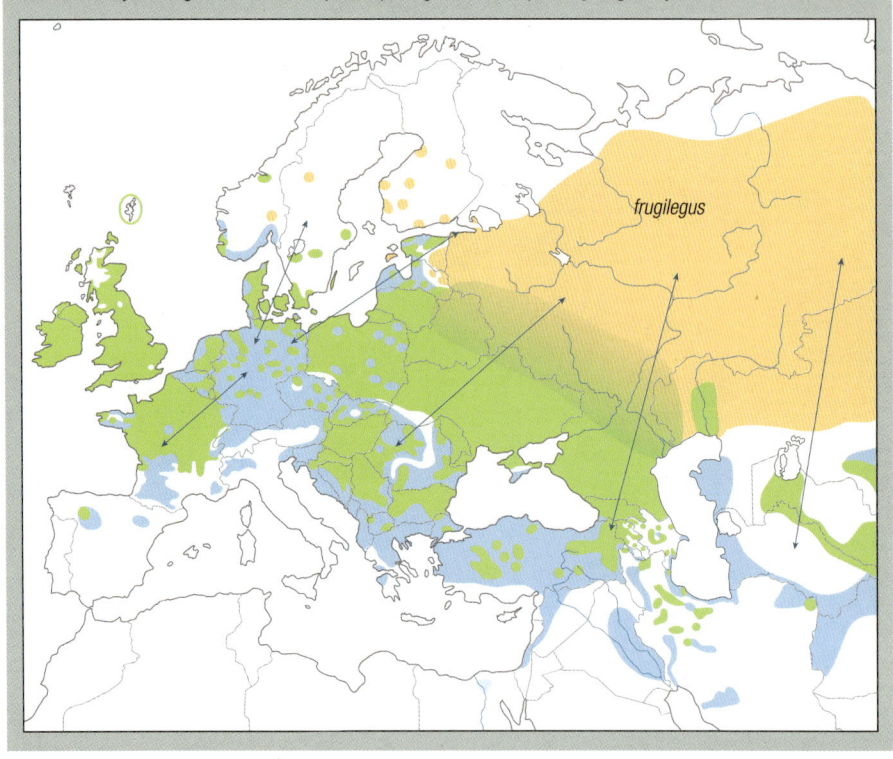

IDENTIFICATION Rather large, size as Carrion or Hooded Crow with *all-black glossy plumage*, which at certain angles gives metallic reflection in reddish-lilac, blue and green (varying with the light). Most adults within a flock show diagnostic *pale whitish-grey bare skin at base of bill, over lores and on chin*, but no nostril feathering at bill-base. Characteristic profile compared to Carrion Crow includes proportionately *smaller head* with *steeply rising forehead, peaked or domed crown* (heightened by bare face), and more *conical and sharply-pointed bill*. Bill also has straighter ridge to culmen (i.e. less curved tip of upper mandible) than Carrion. Often appears to have short nape, flat breast, deeper body and ample belly due to *drooping, loose belly-feathers and 'baggy' thighs*. Juveniles have strongly reduced gloss and brownish cast to black plumage, are feathered fully black on face, nasal tuft covers basal half of upper mandible (diagnostic bare bill-base develops Feb–May of 2nd calendar year), and loose thigh-feathers usually less evident. Can be prone to confusion with Carrion Crow (see below). In flight, compared to Carrion Crow, Rook appears slightly longer winged with longer 'hand', deeper 'fingered' wingtip with narrower primaries, while secondaries tend to become narrower inwards (wings more evenly broad in Carrion); head and bill project further, and tail is proportionately slightly longer and more wedge-shaped. Flight also appears more 'elegant', more direct and regular, with wingbeats often appearing quicker and deeper, more regular and less laboured ('hands' tend to be more backswept). Iris dark brown, bill black but whitish at base, becoming dusky over terminal half, and tarsus black. Upright gait typically more waddling than in Carrion Crow.

VOCALISATIONS Song and sub-song (by ♂) seldom heard but comprise mixed medley of various calls, including various soft cawing sounds, buzzing rasps, and rattling, gurgling and crackling calls, some of which resemble a loud Starling. – Commonest call, in advertisement or contact but usually in alarm, a cawing *kaah*, or high-pitched almost plaintive *kraa-a*, both more deliberate, prolonged and relatively harsher and flatter (less raucous and rolling) than Carrion Crow; loudest calls accompanied by bowing and tail-fanning. A deep, harsh, sonorous *korr* or *krah*, as well as 'caws' in duet, during bowing-display, and higher-pitched *kar-kor kar-kor kar-kor kar-kor* also recorded, where each *kar* is given by one mate and the *kor* is immediate response from the other. Other calls, variable in timbre and pitch but less loud, though of similar quality, are used in various social contexts. Mimicry apparently rare. Alarm is a repeated raucous *kaah*.

C. f. frugilegus, ad, Finland, Apr: black plumage glossed metallic blue, green and purple, with diagnostic pale bare skin at base of bill and on chin. Nostril-feathers at bill base wear off when maturing. Also characteristic are proportionately small head with steep forehead, peaked or domed crown, and conical and sharply pointed bill. Evenly-fresh plumage and extensive bare facial skin age this bird. (M. Varesvuo)

C. f. frugilegus, ad, Finland, Apr: long wings in flight, with proportionately longer 'hand', which is deeply 'fingered' with long narrow primaries. Tail also proportionately long and slightly wedge-shaped. Typical head and bill shape with diagnostic pale bare fore-face. (M. Varesvuo)

SIMILAR SPECIES Adults (see above) readily separated from other black-plumaged crows by bare greyish-white face near bill-base, and relatively long, pointed, pale-based bill, as well as peaked appearance to crown and sloping forehead. Darker juveniles and immatures with fully-feathered face (including bristly feathers at base of upper mandible) can be much more problematic to separate from *Carrion Crow* than often imagined. At close range, experienced observers should be able to detect the structural differences mentioned above. Further, in close inspection, bill-base in young Rooks is sometimes already slightly paler by 1st-winter (uniform in Carrion). The bristly feathers at base of upper mandible tend to curve up and taper forward in Rook (feathering and cutting edges more parallel and fore edge rounder in Carrion); lower bill-base and chin also often more exposed or slightly scruffy (commonly from Jan, at times from Nov). Plumage generally glossier than Carrion Crow. In Rook, tail-tip projects slightly further beyond wingtips (when wings completely folded) Also, Rooks are gregarious, tending to gather in large flocks at all times, whereas Carrion Crow is typically a solitary nester but sometimes congregates in small post-breeding flocks. – A Rook flying overhead at some altitude and giving single rolling *kor* can momentarily be mistaken for *Raven* before true size becomes evident. Note also less protruding neck,

C. f. frugilegus, Germany, Jan: often forms large winter roosts, prior to which species is often seen in tight flocks, when jizz and silhouette are important for identification. Flight rather 'elegant', direct and not so 'lazy' or laboured as of Carrion Crow, the wingbeats appearing slightly more rapid and elastic with more backswept wings. (M. Putze)

C. f. frugilegus, ad, Turkey, May: when collecting food, nearly always on the ground, the dry grey-white skin around the base of bill can be extended like a sack below the lower mandible. Evenly-fresh wing confirms age. (D. Occhiato)

C. f. frugilegus, ad, Israel, Dec: note typical small, slightly domed head, long wings and heavy, long-looking bill. Combination of whitish bare skin at bill base, with no nostril feathers in Dec, and evenly feathered and not heavily worn wing are indicative of an ad. (A. Ben Dov)

smaller bill, slightly shorter wings and quicker wingbeats but slower flight, etc. – Even the darkest young *Jackdaw* should never cause real confusion, being overall at least 30% smaller than Rook, while bill and overall shape, and some plumage elements usually differ.

AGEING & SEXING Aged in late spring and summer by presence of bare skin or not and coloration at bill-base, lores and chin. Moult pattern in wings and tail useful too (though contrast in gloss and ground colour of new and old feathers not always easily seen in fresh plumage). 1stY shows moult limits but ad not; shape and condition of tail-feathers also helpful. Sexes alike but differ marginally in size (see Biometrics). – Moults. Complete post-nuptial moult of ad and partial post-juv moult mostly Jun–Oct. Post-juv renewal includes entire head and body, all lesser coverts, variable number of median coverts (none to almost all), and usually only a few inner greater coverts; tertials and alula also occasionally partially replaced, but only rarely some primaries and (apparently) corresponding primary-coverts (few data); central tail-feathers (r1) or several more tail-feathers often replaced as well. – **ALL YEAR Ad** Evenly fresh (or worn depending on season) wings, tail and body plumage, lacking obvious moult limits; when fresh, black plumage has strong bluish-purple and green gloss. Nostrils, cheeks and chin bare. Tail-feathers broad with rounded tips. **1stY** Moult contrast best seen between moulted median coverts, some tertials and (inner) greater coverts on the one hand, and rest of juv wing-feathers on the other (usually including at least some retained outer greater coverts). Retained greater coverts, remiges, primary-coverts and tail-feathers relatively weaker textured and less fresh, duller and less intense black (more brownish) and less brightly glossed, already from winter or at least early spring having abraded outer webs and tips. Retained tail-feathers narrower, with more tapering and less fresh tips and edges. Bare face may not be fully acquired until 1stS, with bristles and feathers on forehead and covering nostrils lost last, feathers being partially present until Apr–Jun, but cheeks and chin often bare (or partially so) by Jan, sometimes even from mid or late Dec. Correspondingly, bare facial skin changes from soft and pink to hard and whitish. Especially from mid or late spring check moult pattern for safe ageing; in worn plumage (1st-spring/summer) moult limits often more pronounced. At fledging, iris pale grey-blue, with brown to grey-brown or hazel-brown as ad attained in 1st autumn. **Juv** Resembles ad but has looser body-feathers, especially below, and is duller. Also lacks bare whitish face and has dull, not glossy, plumage.

BIOMETRICS (*frugilegus*) **L** 43–48 cm; **WS** 83–97 cm; **W** ♂ 314–340 mm (*n* 12, *m* 325.0), ♀ 292–315 mm (*n* 16, *m* 304.3); **T** ♂ 161–180 mm (*n* 12, *m* 170.0), ♀ 151–172 mm (*n* 16, *m* 159.8); **T/W** *m* 52.4; **B** 48.8–65.0 mm (*n* 28, *m* 57.5); **B**(f) 48.3–61.5 mm (*n* 28, *m* 53.7); **BD** 16.6–20.0 mm (*n* 28, *m* 18.5); **Ts** 50.0–58.9 mm (*n* 28, *m* 53.5). **Wing formula: p1** > pc 55–77 mm, < p2 62–88 mm; **p2** < wt 18–33 mm, =5/6 (100%); **pp3–4** about equal and longest (though p3 often to 8 mm <); **p5** < wt 3–13 mm; **p6** < wt 28–45 mm; **p7** < wt 58–79 mm; **p10** < wt 100–132 mm; **s1** < wt 110–140 mm. Emarg. pp2–5 (sometimes a hint also on p6). – See also Fletcher & Foster (2010) for less well-marked sexual size difference based on a large sample of freshly dead birds in N England.

GEOGRAPHICAL VARIATION & RANGE Two well-differentiated subspecies, but only *frugilegus* confirmed to have occurred in the covered region, *pastinator* being extralimital and is not thought to appear in winter either, hence is not treated. The difference between the two races consists of *frugilegus* acquiring bare skin around base of bill and on chin in adults, whereas in *pastinator* these parts remain largely feathered. Intermediates said to occur in Sayan and Altai.

C. f. frugilegus L., 1758 (Europe, Crimea, Caucasus, N Iraq, Iran, much of Central Asia; northern populations make short- or medium-range movements to W, C & SE Europe, Turkey, S Middle East). Described above. Bare skin develops in ad around base of bill and on chin. Minor variations in average bill size occur in Central Asia but are local and show much overlap, hence poor grounds for subspecific recognition. (Syn. *tschusii*; *ultimus*.)

REFERENCES FLETCHER, K. & FOSTER, R. (2010) *Ring. & Migr.*, 25: 47–51.

C. f. frugilegus, 1stS, Finland, Apr: bare face may not be fully attained until 1stS, with feathers covering nostrils lost last, but fore cheeks and chin by this time often partially bare, as shown here. Note moult contrast between moulted median coverts and juv greater coverts and remiges, the latter bleached and more brownish. (M. Varesvuo)

Rook (left: Finland, Oct) versus Carrion Crow (right: Germany, Oct): in young (especially 1stW) bare greyish-white bill-base is limited or lacking, and nostril-feathers still present, meaning that other characters must be used. Note Rook's slightly peaked crown with relatively longer, pointed bill, and more triangular profile. Further, the nostril-feathers tend to appear more bulky in Rook (slightly flatter and less bulky in Carrion Crow). Lower bill-base and chin often slightly scruffy, and may reveal slightly paler base. In Rook plumage generally appears glossier than Carrion Crow. Both show moult pattern of 1stW. (Left: J. Normaja; right: M. Schäf)

C. f. frugilegus, 1stS, Finland, Apr: compared to ad in flight, 1stS has limited whitish bare skin at sides of bill, but feathers on nostrils still cover large area. Note typical head and bill structure and long outerwing and tail, adding to overall quite distinct jizz. (M. Varesvuo)

C. f. frugilegus, 1stY, Israel, Feb: from mid or at least late winter young can develop quite substantial bare skin on fore cheeks/bill-base, but nostril-feathers still present. Triangular head shape with long conical bill diagnostic, too. Note much of wing subtly abraded and brownish, slightly contrasting with renewed lesser and median coverts (A. Ben Dov)

CARRION CROW
Corvus corone L., 1758

Fr. – Corneille noire; Ger. – Rabenkrähe
Sp. – Corneja negra; Swe. – Svartkråka

The Carrion Crow, together with the Hooded Crow *C. cornix*, forms a fascinating but highly controversial topic in Palearctic taxonomy. Despite extensive study in C Europe of the hybrid zone between the two, no consensus has been reached whether to treat them as separate species or as a single species. The reason for them being treated here as two species is partly practical: in more than 95% of their respective ranges they appear and act as different species, and separate accounts facilitate description. Carrion Crow is a truly widespread, numerous and well-known bird in W and SW Europe. It prefers open country with scattered trees and woodlands, is a ground-feeding scavenger in agricultural landscapes, typically in pastures or arable fields. Often found in proximity to towns with large gardens, parks and farms with woods, etc. It mostly takes invertebrates but is fond of carrion, frequently including roadkills. Scavenging at rubbish tips is another trait, while in some coastal areas birds often forage at tidal estuaries.

IDENTIFICATION *Large all-black strong-billed crow*, with entire plumage faintly glossed metallic green, bluish-lilac or purplish, especially when fresh; dullest on mantle and underparts. Sexes similar. Juvenile duller, almost lacking gloss and tinged blackish-brown. Iris dark brown, bill and tarsus black. Frequently interbreeds with Hooded Crow, mostly in C Europe and the C Mediterranean region, with hybrids showing variable features (see below). In Europe, only likely to be confused with young Rook (both having black-feathered nostrils and surround of bill-base), but distinguished by *sleeker and less intensely glossy black plumage*, and *heavier and deeper bill, which is blunter-tipped, with a more curved culmen in its outer part*. Also has slightly larger, flatter head, a little squarer tail, and lacks the loose thigh-feathers ('baggy trousers') of Rook. Rather slow flight with regular but loose and almost indolent, floppy wingbeats. Intermittent short glides rare or absent. Gait usually less erect than Rook, a steady walk, with clumsy hops and sidling jumps. Singles, pairs and flocks occur year-round. Gathers in communal roosts and flocks at rich food sources, especially post-breeding, but nests singly, not colonially like Rooks. Wary, even in urban areas, though can be bolder when not persecuted.

VOCALISATIONS Song is a triple *kraah kraah kraah*, slowly repeated, a hard croaking, difficult to separate from Hooded Crow but perhaps on average a little harder, less rolling and more down-bent, *kraaeh*, thereby imparting a certain malevolent tone. Sub-song seldom heard, but apparently are low-intensity versions of other calls, as well as croaking and bubbling sounds, or croaking, in succession; soft crackling also heard, and some mimicry too. – Vocal year-round, commonly uttering abrupt, oft-repeated, rasping *kraah* (harsh but vibrant) or *kraaer*—often dropping in pitch at end (generally more rolling, harsher, more vibrant and resonant than common call of Rook). Also fast-repeated knocking *klok-klok-klok*, a nasal strained *keerk*…, or *konk*. Gives repeated, angry, hard rattles *krrrr-krrrr-krrrr* when mobbing a raptor. Sometimes gives deep *konk-konk*, somewhat resembling Common Raven, but these are usually followed by typical thinner calls.

SIMILAR SPECIES Must be carefully distinguished from young Rook. For detailed separation from Rook in juvenile and immature plumages, see that species. See also *Jackdaw*. – At a distance separation from *Raven* needs some attention. Raven is not only considerably larger but also has a different shape with longer neck, a larger head and very heavy bill, which combined with slightly longer (wedge-shaped) tail gives it a more slender and less compact appearance. In flight, wingbeats are powerful, deeper and slower, and wings usually come through as longer and more angled at the carpals. Often soars high up like a large raptor, whereas Carrion Crow is not much of a soarer and generally stays at lower level.

AGEING & SEXING Aged by moult pattern in wing and tail; separation of ad and 1stY from late autumn relies mainly on moult pattern, amount of gloss and shape of unmoulted tail-feathers. Sexes alike but differs marginally in size (see Biometrics). – Moults. Complete post-nuptial moult of ad and partial post-juv moult mostly May–Sep. Post-juv renewal includes entire head- and body-feathers and usually all lesser and median coverts; variable number of inner greater coverts also replaced (none to all recorded). Tertials perhaps also moulted partially (few data); no primary-coverts, alula or any remiges moulted, but central tail-feathers (r1) often replaced. – **ALL YEAR** Close to end of breeding season, many ad still in heavy moult, but juv/1stW relatively fresh. **Ad** Wings, tail and body evenly fresh (or slightly worn depending on season); deeper and purer black with more intense gloss (remiges bluish, purple and green); tail-feathers broad with bluntly truncated tips, and still fresh at least through Dec. **1stY** Moult contrast best detected (especially in the hand) between slightly darker and above all more glossy

C. c. corone, ad, Switzerland, Mar: birds with shorter, stubbier bill easily separated from young Rook. When head-feathers relaxed the crown is rather flat and almost level with bill. Some feather tracts show purple and green gloss, strongest on scapulars and wing-coverts. Evenly-feathered, rather fresh wing infers age. (H. Shirihai)

C. c. corone (left: Germany, Jan; right: France, Mar): flight jizz enables identification. The left-hand bird, especially, shows the typical compact, round-winged and short-tailed jizz (compared to Rook), augmented by the slow, almost sluggish, floppy wing action. (Left: M. Putze; right: H. Michel)

moulted median coverts, some (inner) greater coverts and sometimes odd renewed tertials, and rest of juv wing-feathers (including retained alula and usually at least some outer greater coverts). Retained greater coverts, remiges and primary-coverts relatively weaker textured, less fresh, duller, and sometimes have abraded outer webs and tips (generally more worn than ad at same season). Plumage in general, especially primaries and tail-feathers, less intensely black (more brownish) and much less brightly glossed (primaries have limited gloss when fresh, but much less from Oct). Retained tail-feathers also narrower, with more tapered, less fresh tips and fringes, even quite worn in birds few months old. In worn plumage (1st-spring/summer) moult limits often more pronounced, and hence clearer contrast apparent between juv abraded and browner tertials and neighbouring longer scapulars. In both age classes, especially 1stS, with wear plumage becomes noticeably duller and much drabber. In recently fledged juv iris brown-grey, becoming cinnamon-brown in 1stW and darker brown-black with less cinnamon tinge in ad. **Juv** Resembles ad but has looser body-feathers, especially below, and is overall duller with limited gloss.

Head brown-black with slight oily gloss; nape, upperparts and upperwing-coverts duller (sooty-black or brown-black), with rather restricted glossy purplish-black tones. Underparts sooty-black, virtually unglossed. Yellow gape often still evident in many recently fledged juv.

BIOMETRICS (*corone*) **L** 44–48 cm; **WS** 85–100 cm; **W** ♂ 316–348 mm (*n* 12, *m* 330.9), ♀ 297–327 mm (*n* 13, *m* 314.0); **T** ♂ 170–190 mm (*n* 12, *m* 179.3), ♀ 158–183 mm (*n* 13, *m* 172.7); **T/W** *m* 54.6; **B** 48.1–60.5 mm (*n* 25, *m* 53.3); **BD** 17.0–21.5 mm (*n* 24, *m* 18.6); **BD** (at gonys) 16.3–18.5 mm (*n* 13, *m* 17.4); **Ts** 51.4–62.0 mm (*n* 25, *m* 57.0). **Wing formula: p1** > pc 47–75 mm, < p2 73–98 mm; **p2** < wt 29–40 mm, =6/7 (100%); **pp3–5** about equal and longest (though either of p3 and p5, or both, often to 9 mm <); **p6** < wt 13–27 mm (in *C. frugilegus* 28–45 mm); **p7** < wt 56–70 mm; **p8** < wt 78–98 mm; **p10** < wt 100–120 mm; **s1** < wt 115–137 mm. Emarg. pp3–5 (sometimes faintly also on p6).

GEOGRAPHICAL VARIATION & RANGE Rather slight variation mainly affecting size and bill proportions. See Taxonomic notes for explanation of adopted taxonomy within the crow complex. The relationships are complicated and far from unanimously agreed.

C. c. corone L., 1758 (W Europe including S Scotland, England, Germany west of Elbe, Czech Republic, W Austria, France, Iberia; largely resident but some shorter winter movements to south of range and to N Africa). Described above. Slight and clinal variation, with subtly smaller and thinner-billed breeders often found in Iberia, but much overlap and variation, and best included in *corone*. (Syn. *pulchroniger*.)

Extralimital: **C. c. orientalis** Eversmann, 1841 (Transcaspia north to Aral Sea, Altai and east through much of Asia including N Afghanistan, Kashmir; winter movements reach S Afghanistan and NW India; stragglers claimed from Turkey and Lebanon). Very similar to *corone*, differing only in larger size and proportionately slightly longer tail and thinner bill. Claimed to overlap and to some limited extent hybridise with Hooded Crow *C. cornix sharpii* in Siberia (Knox *et al.* 2002). **W** ♂ 324–370 mm (*n* 13, *m* 349.8), ♀ 313–348 mm (*n* 13, *m* 329.7); **T** ♂ 185–225 mm (*n* 13, *m* 199.2), ♀ 177–207 mm (*n* 13, *m* 191.6); **T/W** *m* 57.4;

C. c. corone, ad, Germany, Mar: like other crows, when excited the crown-feathers can be raised and forehead misleadingly appears steep and crown domed, but such posture is rarely held for long. Note overall more compact shape than Rook. Evenly-feathered rather fresh wing in spring infers age. (T. Krüger)

C. c. corone, 1stW, Germany, Nov: unlike ad, black plumage less clearly glossed (bluish-lilac or purplish and green), instead tinged brownish, visible in certain light. Individual variation includes birds with rather long and more pointed bill, which are less readily separated from young Rooks. Note moult limit between moulted lesser coverts and retained median and greater coverts. (M. Putze)

B 51.0–67.6 mm (*n* 28, *m* 60.2); **BD** 17.4–21.0 mm (*n* 28, *m* 19.3); **BD** (at gonys) 16.3–19.7 mm (*n* 7, *m* 17.6); **Ts** 53.2–65.3 mm (*n* 28, *m* 60.2). Wing formula very similar to *corone*. (Syn. *interpositus*; *saghalense*; *yunnanensis*.)

TAXONOMIC NOTES Two disjunct and subtly different black crow populations exist, *corone* in C & SW Europe and *orientalis* in much of Asia. Together they form the Carrion Crow as defined here. The interposed grey-and-black forms here tentatively and for practical reasons treated separately (Hooded Crow *C. cornix*, which see). Although *corone* and *orientalis* may not be sister taxa, despite close phenotypic similarity, they are still kept as one species pending further study. There is already a rich literature dealing with the Carrion and Hooded Crow complex (see References), but the picture is not yet quite elucidated, and we have favoured an interim solution based more on phenotypic than genetic characters, the latter anyway being partly contradicting and far from clear.

REFERENCES Brodin, A. & Haas, F. (2009) *Evolutionary Ecology*, 23: 17–29. – Brodin, A., Haas, F. & Hansson, B. (2013) *J. Avian Biol.*, 44: 281–287. – Cook, A. (1975) *Bird Study*, 22: 165–168. – Haas, F. & Brodin, A. (2005) *Ibis*, 147: 649–656. – Haas, F. *et al.* (2009) *Molecular Ecology*, 18: 294–305. – Haas, F., Knape, J. & Brodin, A. (2010) *J. Avian Biol.*, 41: 237–247. – Haring, E., Gamauf, A. & Kryukov, A. (2007) *Mol. Phyl. & Evol.*, 45: 840–862. – Knox, A. G. *et al.* (2002) *Ibis*, 144: 710. – Kryukov, A. P. & Blinov, V. N. (1994) *J. f. Orn.*, 135: 47. – Kryukov, A. *et al.* (2012) *Zoological Science*, 29: 484–492. – Parkin, D. T. *et al.* (2003) *BB*, 96: 274–290. – Risch, M. & Andersen, L. (1998) *J. f. Orn.*, 139: 173–177. – Saino, N. (1992) *Ornis Scandinavica*, 23: 111–120. – Saino, N. & Scatizzi, L. (1991) *Bollettino di Zoologia*, 58: 255–260. – Saino, N. & Villa, S. (1992) *Auk*, 109: 543–555. – Saino, N. *et al.* (1992) *Biochemical Systemativs & Ecology*, 20: 605–613. – Wolf, J. B. W. *et al.* (2010) *Molecular Ecology*, 19 (Suppl. 1): 162–175.

C. c. corone, 1stW, England, Nov: rather compact with short, rounded wings and short tail, especially compared to Rook. Aged by slight contrast between moulted lesser coverts and subtly duller rest of wing, but also note pointed primary-coverts. (S. Round)

C. c. corone, France, Feb: ageing in the field often difficult and depends on light and reflections. Here a bird with mixture of green and purple gloss all over suggesting ad, but unless handled ageing unsafe. (A. Audevard)

C. c. corone, Netherlands, Nov: typical group foraging on open ground. Note smaller 'trousers' than Rook indicating species even before details of bill and head can be seen. (N. D. van Swelm)

C. c. corone, juv, Netherlands, Jul: yellow gape often still evident in recently fledged juv. Note duller plumage with limited gloss and looser body-feathers, especially below. (R. Pop)

HOODED CROW
Corvus cornix L., 1758

Fr. – Corneille mantelée; Ger. – Nebelkrähe
Sp. – Corneja cenicienta; Swe. – Gråkråka

A well-known large passerine in N and E Europe whose range continues south-east to the Middle East, occurring both in agricultural and pasture lands, open forests and in and around habitations. It is not always well liked due to its catholic feeding habits, including frequently robbing other birds' nests or killing chicks and young, but it also takes insects, rodents, cereal grain and scraps. Mainly sedentary in the south, whereas northern populations are partially or wholly migratory. Its relationship to the Carrion Crow is much debated, and the two are certainly very closely related, but the Hooded Crow is treated here as a separate species, not least for practical reasons. Intergradation occurs with Carrion Crow in a peculiarly narrow and rather stable hybrid zone, which runs basically from S Denmark and N Germany south through C Europe to N Italy.

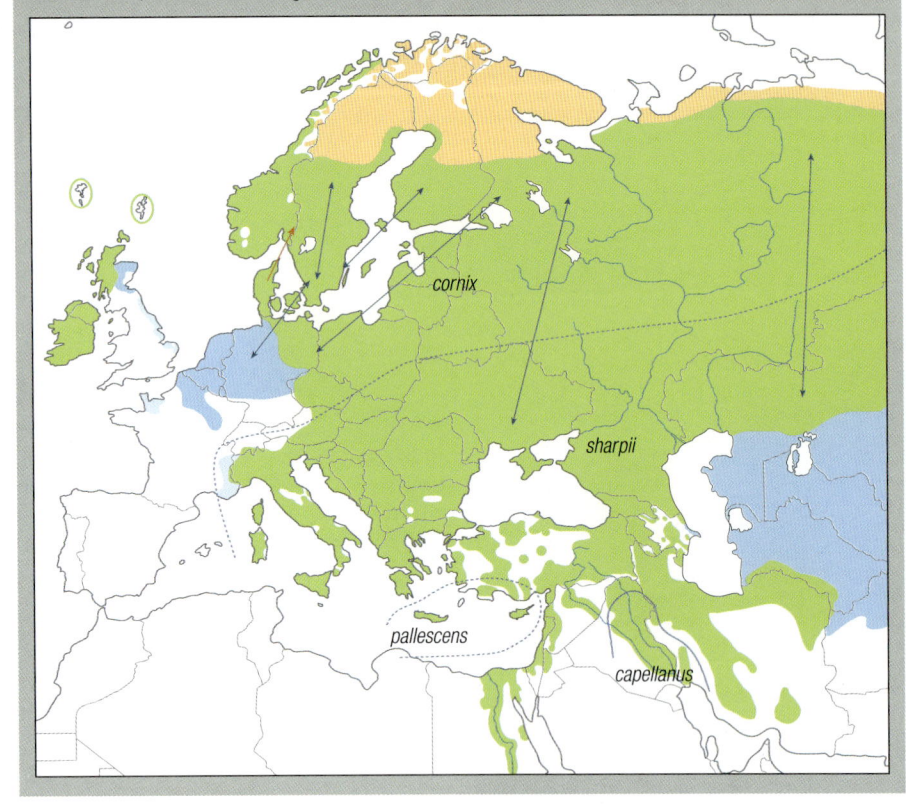

IDENTIFICATION Large grey-and-black crow, with *shape and size exactly as Carrion Crow*, including *strong black bill*, but unmistakable due to its bicoloured plumage, with *dirty grey body* (varying in paleness geographically) *contrasting with black head and smudgy black bib*, at its lower edge usually ending in bold streaks and uneven border (like a splash of black colour), with *glossy black wings and tail*. Juvenile has black parts duller, almost unglossed. In flight from below, note *grey underwing-coverts contrasting with black flight-feathers*. Iris dark brown, and bill and legs black. *Broad 'fingered' wings obvious in flight*, which means that a flying crow can recall a small buzzard at first glance. Rather slow but powerful flight, with rather regular but quite loose, almost lazy wingbeats. *Absence of frequently interspersed gliding moments* without wingbeats safely separates from a distant raptor. Shares many behavioural characteristics with Carrion Crow, including occurrence of small flocks feeding on fields, often in the company of more numerous Jackdaws and Rooks, but larger gatherings of Hooded Crows may occur locally, outside breeding season and at roosts. Gait steady, with direct and rather quick walk interspersed by sudden clumsy hops or sidling jumps. Generally shy and wary, but in some areas rather bold when taking food, even in middle of streets in towns. Often mobs birds of prey, and while it will boldly pursue a Sparrowhawk *Accipiter nisus* closely it keeps a healthy distance from Goshawk *A. gentilis* (which helps to separate these two at times similar raptors). In return often mobbed intensely by small passerines like White Wagtail and Fieldfare.

VOCALISATIONS Noisy, calls mostly hoarse, hard and croaking. Song is a repeated well-spaced series of usually 3–5 identical croaks, *krrrah krrrah krrrah*, delivered from high perch, usually in a tall tree, with body crouching, neck extended with ruffled feathers and head lowered for each call. Each croak is straight (or with faint drop in pitch at end), raw and resounding but with some individual variation between birds or situations. No major differences in vocalisations between Carrion and Hooded Crows, but latter on average has straighter pitch and more 'rolling r', is less malevolent-sounding than former (though much overlap and of subordinate importance in separation). – Song-type croaks also used as call, but then less consistently repeated, less emphatically delivered. Various *krrahr*, *krra-krra-kraa* and more Rook-like *gaah* (without rolling r) noted. Birds bicker among themselves, or mob Sparrowhawk or an owl, with stifled but stubborn grating *krrrr-krrrr-krrrr…*; against feared Goshawk, give full vent to loud, hard and indignant *krrah* notes. Various other more subdued calls noted in interaction between mates or with young.

SIMILAR SPECIES Generally unmistakable, but in poor light and distant views the diagnostic bicoloured pattern is not always obvious. Sometimes rather darker N European form, *cornix*, can in certain lights appear duskier and more uniform, at least briefly. Hybrids with Carrion Crow variably show features of both parents, but sometimes have restricted grey foreparts and thus appear superficially similar to several other bicoloured crows, e.g. *House Crow* or *Daurian Jackdaw* (which see). – Iraqi race *capellanus* of Hooded Crow can appear confusingly much paler, almost black and white, especially in bright light, making it superficially rather similar to extralimital *Pied Crow* (sub-Saharan Africa, vagrant to Western Sahara, Algeria, once to Libya), and similarity enhanced since *capellanus* also has rather heavy bill and strong feet. Still, Pied Crow readily told by black rump, upper-tail-coverts, lower belly and vent, and all-black underwing.

AGEING & SEXING Aged by moult pattern in wings and tail (after post-juv and post-nuptial moults young shows moult limits but ad does not), by shape and condition of tail-feathers, and by colour of bare parts, including inside of upper mandible—all much as in Carrion Crow (which see for details). Sexes alike but differ marginally in

C. c. cornix, Finland, Mar: grey-and-black crow, including grey underwing-coverts contrasting with black flight-feathers. Also contrasting black hood and smudgy black bib, ending in bold streaks. (M. Varesvuo)

size (see Biometrics); no obvious seasonal plumage variation other than effects of wear and bleaching, which strengthen plumage contrast. – Moults. Rather similar to Carrion Crow, but moult on average more prolonged, with greater number of wing-coverts retained (number of moulted feathers highly variable). – **ALL YEAR** At end of breeding season, many ad still in heavy moult, but juv/1stW relatively fresh. **Ad** Wings, tail and body evenly fresh (or somewhat worn depending on season); deeper and purer black with more intense gloss (remiges bluish, purple and green); tail-feathers broad with bluntly truncated tips, and still fresh at least in Dec and reasonably fresh into spring. **1stY** Moult contrast best detected between slightly darker and above all glossier moulted median coverts, some (inner) greater coverts and sometimes odd renewed tertials, and rest of juv wing-feathers (including retained alula and usually at least some outer greater coverts). Retained juv greater coverts, remiges and primary-coverts relatively weaker textured, less fresh, duller, and sometimes have abraded outer webs and tips (generally more worn than ad at same season). Flight-feathers less intense black (subtly more brownish) and much less brightly glossed (primaries have some limited gloss when fresh, but much less from Oct). Retained tail-feathers also narrower, with more tapered, less fresh tips and fringes, and can be already quite worn in birds only a few months old. In worn plumage (1st-spring/summer) moult limits often more pronounced, with clearer contrast apparent between abraded and browner juv tertials and neighbouring longer scapulars. In both age classes wear makes plumage noticeably duller, and much drabber. Iris colour development much as in Carrion Crow. **Juv** Resembles ad but has looser body-feathers, especially below, and is duller. Grey areas darker and less uniform than in ad (especially above, which at close range shows brown or buff hue to feathers, above all fringes, with darker tips), while black of plumage is dull too, almost lacking gloss. Black bib on central breast often smaller and more greyish-black.

BIOMETRICS (*cornix*) **L** 44–48 cm; **WS** 85–100 cm; **W** ♂ 316–348 mm (*n* 12, *m* 330.9), ♀ 297–327 mm (*n* 13, *m* 314.0); **T** ♂ 170–190 mm (*n* 12, *m* 179.3), ♀ 158–183 mm (*n* 13, *m* 172.7); **T/W** *m* 55.7; **B** ♂ 50.0–59.3 mm (*n* 21, *m* 55.3); ♀ 49.0–57.3 mm (*n* 15, *m* 52.4); **BD** 17.4–20.5 mm (*n* 37, *m* 19.0); **Ts** ♂ 53.0–61.0 mm (*n* 20, *m* 58.0); ♀ 51.5–60.2 mm (*n* 15, *m* 57.1). **Wing formula: p1** > pc 50–67 mm, < p2 73–92 mm); **p2** < wt 24–39 mm, =6/7 (common) or =6 (rare); **pp3–5** about equal and longest (though either of p3 and p5, or both, often to 7 mm <); **p6** < wt 16–28 mm; **p7** < wt 53–67 mm; **p8** < wt 73–88 mm; **p10** < wt 97–116 mm; **s1** < wt 108–128 mm. Emarg. pp3–5 (sometimes faintly also on p6). – In a study in NW Italy (Giammarino *et al.* 2012), length

C. c. cornix, Finland, Mar: in typical group. Grey-and-black plumage pattern (especially grey back and underparts) unique among crows in region. Ageing requires closer inspection of moult pattern. (M. Varesvuo)

C. c. cornix, ad, Scotland, Mar: in neutral light, it is possible to appreciate the fairly dark and dull grey coloration of the pale parts of *cornix*. Evenly-fresh wing infers age. (G. Jenkins)

C. c. cornix, 1stS, Finland, Mar: note typical moult pattern in wing of 1stY, with moult contrast intensified by wear, and rather obvious in appropriate light. (M. Varesvuo)

C. c. cornix, presumed 1stS, Scotland, Jun: combination of grey back and black hood diagnostic. Shape, size and flight mode as Carrion Crow. At start of post-nuptial moult (innermost primary growing), much of wing seems still mostly juv (including apparently pointed primary-coverts), but the replaced median and lesser coverts form a subtly darker patch. (R. Ridley)

C. c. sharpii, ad, Israel, Mar: compared to race *cornix* grey parts paler, often with some brown tones and dark shaft-streaks. All wing-feathers evenly fresh. (H. Shirihai)

C. c. sharpii, 1stS, Italy, Apr: not all *sharpii* appear paler than *cornix* due to variation, light or plumage wear. Here a slightly darker-looking bird. Note clear moult limits in wing, indicating age. (D. Occhiato)

C. c. sharpii, juv, Egypt, May: like ad, but plumage duller with looser body-feathers; shortly after fledging the pinkish-yellow gape is still evident. (E. Winkel)

C. c. pallescens, presumed ad, Cyprus, Aug: differs subtly from similar *sharpii* by slightly paler and warmer drab grey colour (tinged cream), less pure grey, and slightly stronger bill. Somewhat worn and in moult, but no visible juv feathers in wing suggests ad at end of complete post-nuptial moult. (A. Fadeev)

of skull + bill provided good support for sexing, albeit still with some overlap: **H+B** ♂ 90.2–114.0 mm (*n* 70, *m* 96.5), ♀ 84.2–96.1 mm (*n* 75, *m* 90.9).

GEOGRAPHICAL VARIATION & RANGE Rather slight and largely clinal variation over a large area, but one subspecies stands out rather more. The text below differs from most traditional checklists in that it affords *sharpii* a wider range than is usually the case. After careful comparison in several museum collections it is evident that breeders very similar or identical to populations in the Caucasus, Central Asia and SW Siberia, areas normally associated with *sharpii* (the type specimen is from Punjab in winter), breed also in Italy, Corsica and over large part of SE Europe. There are no sharp geographical borders between subspecies, which instead are connected by wide zones of intergrading characters. Racial designation mostly involves darkness and tone of grey areas, as well as size of wing, bill (length and depth) and tarsus. Separation safer with birds in fresh plumage. Also beware that birds in arid southern and eastern areas fade much more rapidly to pale greyish-white, and as size is to some extent related to sex (and age), preferably only specimens of same sex and age should be compared.

C. c. cornix L., 1758 (Scandinavia, Finland, Russia north

C. c. capellanus, presumed ad, SW Iran, Jan: highly distinctive race, with grey parts much paler, almost whitish, and bill markedly heavy, legs long, and tail long and rather strongly graduated. Note black-feathered tibia, more extensively so than in other races. Plumage and wing appears fresh and immaculate, suggesting ad. (R. Felix)

Hybrids *cornix* × *corone*, Germany, Sep–Jan: a spectrum of hybrids, from almost Hooded-like to Carrion-like. These vary in amount of grey versus black and in amount of black streaking both above and below. The challenge is detecting those hybrids that are very similar to one or other species. Those most like Hooded (top two) usually show some dark on flanks and/or vent, and streaks at border between black hood and grey areas, while the most Carrion-like (bottom right) usually show vestiges of blacker breast or have uneven black plumage. Ageing should follow same criteria as for non-hybrids. (Top right: M. Schäf; rest: M. Putze)

of c. 54°N, NW Siberia, Ireland, Isle of Man, NW Scotland, Germany east of Elbe, Poland, Belarus, N Romania; populations in north-east migrate in winter to S Scandinavia and W Europe, breeders in south and west are mainly resident. Treated above (in fresh plumage has grey parts rather dark ash-grey (limited brown tinge), rather like 'Pale Payne's Grey' (Ridgway 1912, Pl. 49). Size rather large.

C. c. sharpii Oates, 1889 (Provence, Corsica, Italy, Balkans, S Romania, Bulgaria, S Ukraine, Slovakia, S Poland, Russia south of c. 54°N, SW Siberia, much of Turkey, Levant south to Egypt, Caucasus, N Iran to Kazakhstan; southern populations mainly resident, northern partly short-range migrants in winter). Has grey parts distinctly paler and less ashy, more brown-tinged, than *cornix*, but size similar or slightly smaller and bill a little slimmer. Intergrades smoothly with *cornix* in wide zone over S Germany through Czech Republic, Hungary, S Poland, Belarus and S Russia to middle Ob and lower Yenisey. (Part of this intergradation zone sometimes separated as '*subcornix*', but not followed here.) – Breeders on Corsica and Sardinia ('*sardonius*') not separable when individual variation is accounted for by using sufficiently long series. Samples from Black Sea region ('*christophi*', '*valachus*') and S Russia ('*khozaricus*') are paler than typical *cornix*, and fit much better with *sharpii*, many being identical with birds from Caucasus and Iran, thus included here. Birds in Levant ('*judaeus*', '*syriacus*') and Egypt ('*egipticus*') tend to be slightly smaller, and those in Egypt also average subtly darker, but slight difference and much overlap. Similarly, breeders in much of Turkey and Caucasus ('*kaukasicus*') are too similar to average *sharpii* to warrant separation. – Race *sharpii* also reported to overlap and partly hybridise with all-black *orientalis* in Turkmenistan and S Kazakhstan. – **L** 44–47 cm; **W** ♂ 290–344 mm (*n* 42, *m* 313.7), ♀ 285–323 mm (*n* 35, *m* 302.0); **T** ♂ 162–192 mm (*n* 42, *m* 175.4), ♀ 157–183 mm (*n* 35, *m* 171.2); **T/W** *m* 56.1; **B** 45.6–58.7 mm (*n* 87, *m* 53.3); **BD** 15.0–19.9 mm (*n* 85, *m* 17.7); **Ts** 51.5–61.0 mm (*n* 89, *m* 56.3). Wing formula similar to *cornix*. – (Syn. *christophi*; *egipticus*; *judaeus*; *kaukasicus*; *khozaricus*; *sardonius*; *syriacus*; *subcornix*.)

○ **C. c. pallescens** Madarász, 1904 (Crete, S Aegean Is, Cyprus, S Turkey; possibly Syria). Very similar to *sharpii* but differs subtly when series are compared by slightly paler and warmer drab-grey colour (tinged cream), not appearing as pure grey as that race (although even *sharpii* has a slight tinge of brown in the grey). Some specimens appear paler on lower nape and upper mantle, where *sharpii* is a little darker grey. Averages the smallest race, although much overlap with *sharpii* of Levant and Egypt. Bill slightly stronger than *sharpii*. **W** ♂ 294–329 mm (*n* 12, *m* 310.3), ♀ 271–297 mm (*n* 5, *m* 286.8); **T** ♂ 160–185 mm (*n* 12, *m* 172), ♀ 150–170 mm (*n* 5, *m* 162.8); **T/W** *m* 56.0; **B** 50.2–62.3 mm (*n* 17, *m* 55.2); **BD** 16.8–19.3 mm (*n* 16, *m* 18.3); **Ts** 52.0–61.0 mm (*n* 17, *m* 56.0). (Syn. *minos*.)

C. c. capellanus Sclater, 1877, 'Mesopotamian Crow' (Iraq and SW Iran; vagrant to United Arab Emirates). A very distinct race, having grey parts much paler, almost whitish (often with slight vinous-pink or buff hue when fresh), wearing and bleaching to near-white. In particular, upperparts are paler; a few *sharpii* can approach *capellanus* in paleness of underparts, but upperparts consistently clearly paler in latter. Also, tibia-feathering more extensively black than in other races. Large, but much variation, and some small birds are as large *sharpii*. Bill markedly heavy and legs long, tail long and rather strongly graduated, approaching *orientalis*. Throat-feathers rather long. – Said to intergrade with *sharpii* at foot of Zagros Mts in Iran. **W** ♂ 298–353 mm (*n* 14, *m* 329.8), ♀ 300–335 mm (*n* 11, *m* 319.4); **T** ♂ 177–222 mm (*n* 14, *m* 197.9), ♀ 168–205 mm (*n* 11, *m* 187.8); **T/W** *m* 59.6; **B** 56.3–67.3 mm (*n* 31, *m* 61.7); **BD** 18.9–22.8 mm (*n* 30, *m* 20.5); **Ts** 61.2–71.5 mm (*n* 30, *m* 67.1).

TAXONOMIC NOTES The question whether Hooded and Carrion Crows should be treated as races of one species or as two different species has been much studied and discussed without reaching a consensus; both views have their pros and cons. The two are connected in Europe by a narrow and rather stable hybrid zone. The narrowness of this zone seems to indicate that selective forces favour only one form on each side, but despite much research, it has not been possible to define such a selective force, and the two are genetically very close. We refrain from taking a side in the ongoing debate, and rather let a practical aspect decide: it is easier to describe Hooded and Carrion Crows if they are afforded separate accounts. – Oddly, the type specimen of the very pale race *capellanus* appears darker than the type of the medium-pale *sharpii*, both in NHM. However, when a feather was cleaned (by H. van Grouw) this proved to be a typically pale whitish-grey. It turned out that the type had been brought alive from present-day Iraq to London and was held in the London Zoo for some time, where it became heavily stained.

REFERENCES Giammarino, M., Quatto, P. & Soglia, D. (2012) *Ring. & Migr.*, 27: 38–42. – Haring E., Gamauf, A. & Kryukov A. (2007) *Mol. Phyl. & Evol.*, 45: 840–862. – See also references listed under Carrion Crow.

PIED CROW
Corvus albus Statius Müller, 1776

Fr. – Corbeau pie; Ger. – Schildrabe
Sp. – Corneja pío; Swe. – Svartvit kråka

A medium-sized bicoloured large Afrotropical crow, which only marginally enters the extreme SW Palearctic in Western Sahara. It is apparently also recorded quite regularly along the Nile to *c.* 20°N. Observed in Libya and S Algeria, and may breed there, too. Otherwise it occurs more regularly in coastal and central Mauritania, the sub-Saharan Sahel and savanna regions right across the continent. In the Afrotropics an often quite widespread and abundant resident, and always attracted to cultivation and habitation.

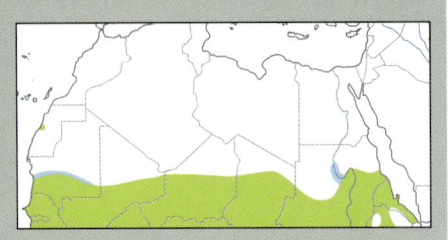

1stW, Comoro Is, Nov: following partial post-juv moult (some lesser and median coverts renewed, as here), while much of rest of wing is juv, slightly duller and more brown-grey. Note typical bill shape. (H. Shirihai)

Morocco, Mar: raven-like crow, with diagnostic pied body, notably the white breast and upper belly and broad white hind-collar (see images below). Also has raven-like lanceolate throat-hackles. Note very broad, deeply 'fingered' wings. (V. Legrand)

IDENTIFICATION Distinctively plumaged raven-like crow, with *black-and-white pattern* recalling Hooded Crow of the race *capellanus*, although unlike that species has *black nape, shoulders, vent and lower belly*, in other words a larger part is black. About the *same size as Carrion and Hooded Crows*, or subtly larger, with moderately long but rounded wings and proportionately *slightly shorter tail*, with *quite long and deep bill*, and *long legs*, thus in a way rather strongly built like a true raven in shape and general appearance. At rest, primaries reach tail-tip, and tail is only slightly graduated at the tip. Look for the diagnostic pied body, created by the *white breast and upper belly, and broad white hind-collar across mantle and neck-side*, sharply set-off from black head, throat and rear body (latter with blue or purple gloss). Also has raven-like well-developed lanceolate throat-hackles. Bare parts essentially black, but legs and toes often appear paler and greyer. Frequently associates with other scavengers. Flight and behaviour mostly as Brown-necked or Fan-tailed Ravens.

VOCALISATIONS Gives varied long and short cawing notes in flight and at rest, which might recall both Raven and Rook. Can also utter a much quieter and somewhat musical *clork*.

SIMILAR SPECIES Unmistakable if pied pattern with all-black head and lower belly/vent confirmed, and if size and shape are noted, but beware leucistic Brown-necked Raven or Common Raven. – If viewing distance is longer and observation shorter, Hooded Crow needs to be eliminated. Apart from more extensive black hood reaching to mantle and black lower belly and vent, note much heavier Raven-like bill in Pied Crow, and that black bib ends neatly at lower edge, not streaky or irregularly as in Hooded. Also, in flight Pied has much more restricted grey-white on underwing, mainly comprising 'armpits' (in Hooded also lesser and median coverts grey).

AGEING & SEXING Ageing not closely studied, but from a check of museum specimens does not seem to differ from other corvids using moult limits in upperwing and shape and wear of tail-feathers. Sexes inseparable, though ♂ averages very slightly larger, albeit overlap extensive (see Biometrics). No seasonal variation, though wear affects degree of gloss to black plumage. – Moults. Not well studied, but it seems that the species undertakes complete post-nuptial moult (can be confusingly protracted) and partial post-juv moult (latter including head, body, and some or all lesser and median coverts). However, breeding season and the related moult periods within the covered region remain to be established, but based on freshness of feathers and active moult in a few specimens examined from nearby areas, e.g. W Africa and Sudan, it seems that moult occurs in late summer and autumn, at least into Nov. Pre-nuptial moult apparently absent. – **SPRING–SUMMER Ad** With wear gloss of black feathers is reduced and white pattern becomes more striking. **1stS** Effect of feather wear stronger than in ad, and black areas may become browner. Also shows more contrasting moult limits, with browner juv upperwing-coverts and tertials. – **AUTUMN–WINTER Ad** When very fresh, black areas glossed violet to purple with some blue sheen. **1stW** All or at least most flight- and tail-feathers, primary-coverts, greater and some lesser and median coverts are juv, less strongly glossed and slightly weaker textured or browner, while black areas below may appear more scalloped. Retained flight- and tail-feathers, but also primary-coverts, are narrower and somewhat more sharply pointed. **Juv** Basically similar to ad but has slightly duller and less contrasting plumage pattern, with black parts unglossed and white parts mottled darker.

BIOMETRICS L 45–49 cm; **WS** 88–103 cm; **W** ♂ 328–382 mm (*n* 15, *m* 366.7), ♀ 332–378 mm (*n* 15, *m* 353.5); **T** ♂ 160–186 mm (*n* 15, *m* 178.8), ♀ 161–185 mm (*n* 15, *m* 173.9); **T/W** *m* 49.0; **B** 52.7–62.4 mm (*n* 29, *m* 56.7); **BD** 19.0–23.0 mm (*n* 28, *m* 20.9); **Ts** 55.3–66.8 mm (*n* 27, *m* 60.3). **Wing formula: p1** > pc 76–92 mm, < p2 70–98 mm; **p2** < wt 13–32 mm, =5/6 (100%); **pp3–4** about equal and longest (though either rarely to 8 mm <); **p5** < wt 7–18 mm; **p6** < wt 44–62 mm; **p7** < wt 77–98 mm; **p10** < wt 124–149 mm; **s1** < wt 135–162 mm. Emarg. pp3–5 (sometimes a hint near tip of p6 also).

GEOGRAPHICAL VARIATION & RANGE Monotypic. – Sub-Saharan Africa from Western Sahara, Mauritania, and east to Chad and Sudan, south of this over much of subtropical Africa; occasional in S Algeria, vagrant to Libya; mainly sedentary. No geographical variation. – Reported to hybridise locally with Dwarf Raven *C. edithae* in E Ethiopia and Somalia. (Syn. *C. scapulatus*.)

Ad, Western Sahara, Dec: when neck stretched, white of belly extends as broad white hind-collar across upper mantle. Note huge, heavy Raven-like bill. Evenly-fresh wing with metallic gloss infers age. (C. Batty)

Ad, Gambia, Nov: aged as in image to the left. Often the wing-bend conceals the fact that white collar joins pale underparts. Sometimes shaggy throat hackles can be quite obvious when the bird is excited. (V. & S. Ashby)

BROWN-NECKED RAVEN
Corvus ruficollis Lesson, 1831

Fr. – Corbeau brun; Ger. – Wüstenrabe
Sp. – Cuervo desertícola; Swe. – Ökenkorp

The desert counterpart of the Common Raven, which it largely replaces across N Africa and in the Middle East, the Brown-necked Raven is largely resident in semi-deserts but can survive even in the harshest of environments. It is often found in desert oases and cultivation, including on the outskirts of towns and villages. A scavenger, it scans the ground for food during slow but steady flight at moderate heights. Usually encountered singly or in pairs, but small or even large flocks not rare, and very large flocks can gather at favoured roosts or feeding sites. It has wandered north to S Syria and extreme SE Turkey from the main Middle Eastern range.

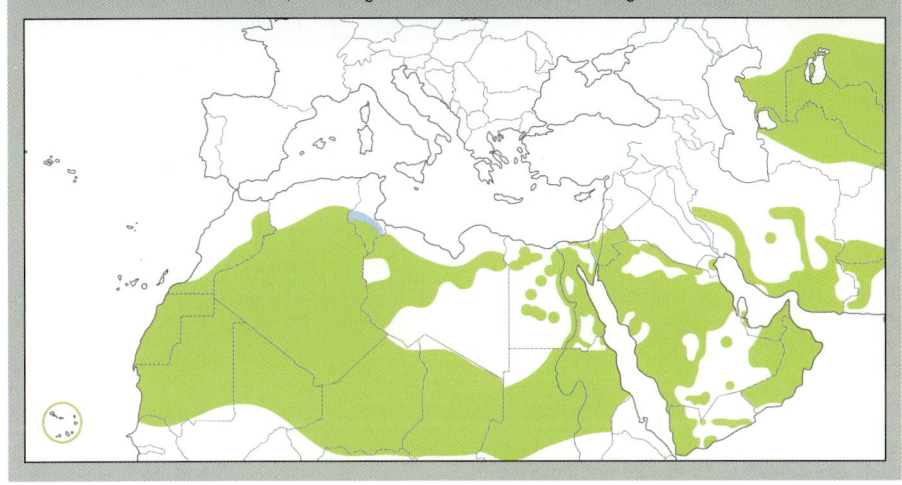

IDENTIFICATION A large corvid with *general size and structure most like Common Raven*, but *slighter and sleeker*, especially the head, which has *less shaggy throat hackles*, and the *slenderer bill*, longer and more pointed wings, and slight differences in tail proportions and structure. Looks all black at a distance (closer and in good light strongly glossed purplish-violet or purplish-bronze, especially above), but has a *brownish gloss*, especially *on the rear of the head*, which is not always easily detected. In general crow-like when perched due to smaller head and somewhat more crow-like than raven-like bill, whereas the opposite is true in flight: long, slightly angled, narrow wings, long tail and slow but powerful wingbeats give it a rather raven-like look. Bare parts essentially black, but legs and toes often appear greyer. Frequently associates with other scavengers, and is attracted to habitation. Behaviour mostly as Common Raven.

VOCALISATIONS Has either a high-pitched trilling crowing with falling end, *krrri-eh*, or a slightly rising harsh *kraao*, often repeated and varying in pitch and strength, but with experience typically sounds much less deep and croaking than most calls of Common Raven, being somewhat more akin to Carrion or Hooded Crows, and, in some cases, Rook. As a rule, lacks the deep resonance and rolling quality of Common Raven's wooden croak.

SIMILAR SPECIES Brown-necked Raven is always difficult to distinguish from *Common Raven* in areas where they overlap. Direct size and structural comparison is rarely possible because each tends to avoid areas inhabited by the other. Even seen close, the eponymous brown neck may be virtually impossible to see (and some Common Ravens in the Middle East can also show a hint of this feature). Although structural differences permit certain identification, careful observation is required, especially of an out-of-range vagrant. N African populations of Common Raven are smaller and shorter-billed than European forms and, as in all populations in arid habitats and strong light, may appear browner in worn plumage. With experience, the following should enable their separation, but voice is the easiest distinction. On the ground, the folded wingtips in Brown-necked Raven reach near or beyond the tail-tip, whereas the wingtips of Common Raven usually fall well short. Brown-necked is less robust, with a slimmer neck and head, and slightly less stout bill, often with a downward droop, and generally lacks the obvious shaggy-throated appearance and prominent nostril feathering of Common Raven (but calling or agitated Brown-necked can certainly show an 'untidy throat'). In addition, Brown-necked tends to have longer-looking legs and a more direct, quite free-stepping walk, varied with occasional jumps and sidling movements. In flight Brown-necked appears slighter than Common, with a less heavy head; wings generally narrower, less broad-based with more tapered 'hand' ('fingers' shorter, less splayed), and tail less wedge-shaped and less full, with a narrower base (sometimes with slightly projecting central feathers). Flight action appears less powerful and majestic at long range, and the bird often seems to hang more freely in the wind, when structural silhouette differences are best appreciated. A surprisingly useful clue is that Brown-necked has a frequent habit of holding its bill slightly down in flight. Furthermore, Brown-necked in good light often shows stronger contrast between the dark grey-black flight-feathers and jet-black coverts on the underwing.

AGEING & SEXING Ageing possible using moult limits in upperwing and shape and wear of tail-feathers. Sexes inseparable, though ♂ averages slightly larger (see Biometrics). No seasonal variation, although wear affects degree of gloss to black plumage. – Moults. Complete post-nuptial moult and partial post-juv moult in summer–autumn,

Israel, Feb: range sometimes overlaps with similar Common Raven, but in close views and appropriate light the extensive brown hue of head, neck and breast is diagnostic. At least in Israel, the clearly shorter tail is diagnostic, too. Ageing impossible from this image. (H. Shirihai)

generally Jun–Oct. Post-juv moult includes head, body, and some or all lesser and median coverts. Pre-nuptial moult absent. – SPRING **Ad** With wear becomes drab with patchy coppery gloss and even browner shawl, which may incline to rusty-brown, contrasting with blacker wings and tail. **1stS** Plumage progressively browner, especially neck and underparts. More contrasting moult limits, with browner juv upperwing-coverts and tertials, but difficult to assess in the field. – AUTUMN **Ad** When very fresh, glossed violet to purple on head and underparts, and violet elsewhere, with purplish-bronze sheen to head-sides, nape, hindneck and upper mantle. **1stW** All flight- and tail-feathers, primary-coverts, greater and some lesser and median coverts juv, less strongly glossed and textured, and browner, while underparts may appear more scalloped (beware of possible confusion with ad that still has not moulted completely in autumn, showing moult limits with distinct brown feathers). The retained flight- and tail-feathers, and especially primary-coverts, are also narrower and more sharply pointed. **Juv** Generally dull black with faint oil-blue gloss, less intense purplish-violet and no brown shawl until 1st-autumn moult. Underparts more fluffy with slight mottled appearance.

BIOMETRICS **L** 49–54 cm; **WS** 103–120 cm; **W** ♂ 352–417 mm (n 25, m 387.7), ♀ 335–405 mm (n 23, m 379.5); **T** ♂ 185–270 mm (n 25, m 217.6), ♀ 194–246 mm (n 23, m 212.5); **T/W** 56.1; **B** 60.3–77.9 mm (n 49, m 66.8); **BD** 18.7–24.0 mm (n 44, m 21.6); **Ts** 59.3–72.1 mm (n 49, m 64.5). Wing formula: p1 > pc 76–115 mm, < p2 69–110 mm; **p2** < wt 17–45 mm, =5/6 or 5 (75%) or =4/5 (25%); **pp3–4** about equal and longest (though either rarely to 12 mm <); **p5** < wt 5–32 mm; **p6** < wt 42–91 mm; **p7** < wt 82–130 mm; **p10** < wt 145–192 mm; **s1** < wt 147–207 mm. Emarg. pp2–5.

Ad, Israel, Sep: in closer view some plumage details may be detected. Note blacker eye-surround (mainly due to differences in reflection), throat markings (shadow between elongated feathers), and unevenness over rest of underparts (fringes darker than centres). All wing evenly fresh and glossy, indicating ad. (A. Ben Dov)

GEOGRAPHICAL VARIATION & RANGE Monotypic (see Taxonomic note). – Saharan N Africa from Cape Verdes, N Senegal, Mauritania, Western Sahara and Morocco east to Egypt, Sudan and Chad, then through Arabia, on Socotra, S Levant, Kuwait, S & E Iraq, S Iran, S Pakistan and S & extreme N Afghanistan; also Transcaspia, Aral Sea region east through S & C Kazakhstan and adjacent areas; all populations mainly resident.

TAXONOMIC NOTE The population breeding in the Horn of Africa, *edithae* Phillips, 1895 (N Kenya, Somalia, Djibouti, E Ethiopia), is very similar in plumage to Brown-necked Raven *ruficollis* but distinctly smaller. Usually it has been treated as a small subspecies of Brown-necked Raven, but Ash (1983) suggested it might merit separate species status, and Jønsson et al. (2012) demonstrated that it is sister to Pied Crow and that these two in turn are sisters to Brown-necked Raven. To avoid polyphyly, *edithae* is best treated as

Oman, Oct: slenderer, more crow-like shape and jizz and shorter tail are important for separation from Common Raven. However, such clearly lighter build or brown plumage only rarely apparent in field. (D. Occhiato)

United Arab Emirates, Sep (left) and Mar: throat hackles can produce a shaggy-throated appearance, but never as prominent as in Common Raven. Left bird also shows that Brown-necked in certain light can appear glossy black. Partly due to the wind, the species' long legs are obvious in this image. The right image shows that wing is narrower and innerwing a little shorter than in Common Raven. (Left: D. Clark; right: T. Pedersen)

a separate species, 'DWARF RAVEN'; the alternative to merge Pied Crow with Brown-necked Raven is less appealing. Dwarf Raven, being entirely extralimital, is not further treated here. The close relationship between Dwarf Raven and Pied Crow is further demonstrated by the frequent hybridisation between the two in some areas where they meet. Similar hybridisation has not been noted where Pied Crow meets Brown-necked Raven further west in Africa.

REFERENCES ASH, J. S. (1983) *Scopus*, 7: 54–79. – JØNSSON, K. A., FABRE, P.-H. & IRESTEDT, M. (2012) *BMC Evol. Biol.*, 12: 72.

1stS, Egypt, May: throat hackles and nasal bristles weakly developed compared to Common Raven. This image perfectly shows how tail falls short of wingtips. Slightly abraded primary-coverts and worn flight feathers, while greater coverts are also juv. (V. Legrand)

Brown-necked Raven (top two images: Apr, Sep) versus Common Raven *C. c. laurencei* (bottom two: Apr, Feb), Israel: in parts of Middle East, Brown-necked occurs sympatrically with Common Raven. Although the local race of Common Raven (*laurencei*) is among the largest and its tail usually clearly projects beyond the wingtips (the opposite in Brown-necked), in certain lights Brown-necked can look almost blackish (top left), while *laurencei* is often tinged slightly brownish, sometimes obviously (bottom left). Common Raven's throat hackles do not always produce a shaggy throat (bottom left), and the converse can be true in Brown-necked (top right). Common Raven's N African race *tingitanus* is smaller and shorter-billed, but jizz of Brown-necked can still be useful, especially the less heavy head and neck, slender bill, less broad-based wings with more tapered hand (five 'fingers' rather than six), and less full tail with a narrower base. Furthermore, in good light Brown-necked often shows more contrasting and blacker underwing-coverts and impression of eye-surround. (Top and bottom right: A. Ben Dov; Top and bottom left: H. Shirihai)

COMMON RAVEN
Corvus corax L., 1758

Fr. – Grand corbeau; Ger. – Kolkrabe
Sp. – Cuervo grande; Swe. – Korp

The largest crow (and passerine) of the Northern Hemisphere, widespread in Eurasia, reaching east to the NW Indian subcontinent, China and Japan, and also occurring in North America. Mostly resident, but subject to local movements, it frequents hilly or mountain terrain with cliffs, vast coniferous forests and woody riverine landscapes or rugged coastlines and inshore islands, as well as steppe and semi-desert, and occurs up to *c.* 5000 m. Pairs for life, and territories are large, nest built in tall tree or on inaccessible cliff. Food varied, from carrion and scrap to smaller birds and mammals, or their eggs or young. Although mainly solitary, the pair may join others to form small aggregations in winter, if food is abundant. Often soars high up like a buzzard.

C. c. corax, ad, Finland, Mar: size comparable to Common Buzzard but has slightly longer wings, very powerful and agile flight, during which birds often soar with long primaries well splayed and wings held flat or slightly lowered unlike Buzzard. Contrasting blacker body and underwing-coverts; protruding neck, and heavy bill. (M. Varesvuo)

IDENTIFICATION *Large, all-black*, crow with beautifully *glossed plumage* when seen close. Size comparable to Common Buzzard *Buteo buteo*, but it has even slightly longer wings and thereby wider wingspan, noticeable when they are seen together. The *deep, croaking calls* are also characteristic, and often betray its presence even at some distance. The *shaggy throat-feathering* with prominent throat-hackles, nasal bristles and *long and heavy bill* impart a notably bulky appearance to the head and neck, not only in flight but also when perched. *Wingtips usually fall just short of tail* when at rest. The metallic green and bluish-lilac sheens are visible in close views but vary with light conditions. Flight is powerful and agile, birds often seen soaring and, around nesting areas, engaging in impressive acrobatics, tumbling, diving on closed wings and rolling onto the back in mid-air. Even at long distance readily separated in flight from buzzards by *protruding neck, head and heavy bill*, long wedge-shaped tail, and *wings when gliding and soaring held flat or slightly down-bent*. Active direct flight fast on powerful, deep wingbeats, *outerwing slightly angled back*. Gait is a bold walk with some forward leaps or sidling hops.

VOCALISATIONS Song is a variety of ventriloquial, subdued knocking and resounding notes without any clear structure (to the human ear), including a peculiar hollow *klong*, a disyllabic *kvorrl-ka!*, a fast trisyllabic note with dry wooden sound *ke-ko-ka*, and a dashing *krrü-a*. – Calls loud, jarring and rolled (with hard, rolling 'r'), and a strangely percussive quality lacking in smaller *Corvus*. Frequently uttered in flight, sometimes in alarm or excitement, is a deep *korrp*, often rather quickly repeated 3–4 times, *korrp-korrp-korrp*, or a higher-pitched and drier knocking *toc-toc-toc*. A variety of related calls, often more conversational, including a deep, rather hollow *pruk ... pruk*, and more drawn-out, harsher *krrah*. Even a miaowing call noted from flying bird, but function unknown.

SIMILAR SPECIES Unmistakable if seen well. Differs from all other W Palearctic relatives in its deep, massive bill, flat head, shaggy throat, long powerful body, broad and long wings, strong legs and feet, and rather long wedge-shaped tail. However, in N Africa and the Levant, care must be taken to separate this species from *Brown-necked Raven*, as in certain lights both may show brownish napes. Their separation is discussed under Brown-necked Raven. In Common Raven, flight is rather majestic, often at great height and includes more gliding and soaring than other corvids, which may provoke confusion with some raptors, if only seen briefly or at great distance. However silhouette and flat wing posture helps separate a distant Raven.

AGEING & SEXING Ageing requires close scrutiny of moult and feather wear in wing and tail, and like its relatives is based mainly on any moult contrast on upperwing and on

C. c. corax, ad, Finland, Mar: at certain angles and at close range can show green, purple and metallic blue gloss. Note shaggy throat hackles, as well as nasal bristles on very powerful bill. Evenly-feathered glossed wing infers ad. (M. Varesvuo)

wear and shape of tail-feathers. Sexes indistinguishable, though ♂ subtly larger (albeit much overlap, sexes nearly same size). – Moults. Complete post-nuptial moult of ad in spring, summer and autumn, May–Nov (thus protracted or variable) and partial post-juv moult in summer and autumn, but timing varies with latitude. Post-juv moult (starts at age of *c*. 2 months), includes all or most of head and body, all or part of lesser and median coverts, but no greater coverts and only occasionally some tertials. Pre-nuptial moult absent. – **SPRING Ad** No moult contrast obvious in upperwing. Central tail-feathers broad and tips truncate, only moderately worn. With wear becomes duller and less glossed, especially on wings and tail, and these parts can appear browner in early summer. **1stS** Best aged by rather contrasting moult limits with browner, retained juv feathers in wings and tail. Central tail-feathers somewhat narrower on average, and tips more tapering, not as obtuse as in ad, comparatively more worn at tips. Also progressively more bleached and brown, less black and glossed. – **AUTUMN Ad** Fresh plumage glossed stronger purple around head and nape, mainly purple-blue above and on shaggy throat-feathers, and reddish-purple on tail and underparts. **1stW** Like fresh ad but all flight- and tail-feathers, primary-coverts and greater coverts, and some lesser and median coverts juv, slightly weaker textured, less strongly glossed and a little browner, affording moult contrast, while overall less brightly glossed than ad. **Juv** Plumage generally dull sooty-black, with restricted purplish-blue gloss and sometimes a dull grey-mottled appearance and fluffy-feathered rear underparts.

BIOMETRICS (*corax*) **L** 56–67 cm; **WS** 115–130 cm; **W** ♂ 404–459 mm (*n* 14, *m* 428.8), ♀ 400–446 mm (*n* 13, *m* 425.4); **T** ♂ 228–275 mm (*n* 14, *m* 242.3), ♀ 225–254 mm (*n* 13, *m* 240.4); **T/W** *m* 56.4; **B** 69.4–81.7 mm (*n* 28, *m* 74.5); **BD** 25.0–28.8 mm (*n* 26, *m* 26.9); **Ts** 60.0–68.9 mm (*n* 27, *m* 65.0). Wing formula: p1 > pc 87–115 mm, < p2 70–113 mm; p2 < wt 19–41 mm, =5/6 (common) or =5 (rare); pp3–4 about equal and longest; p5 < wt 5–15 mm (rarely 1–22); p6 < wt 41–73 mm; p7 < wt 79–115 mm; p10 < wt 146–185 mm; s1 < wt 163–205 mm. Emarg. pp3–6 (on p6 sometimes less prominent).

GEOGRAPHICAL VARIATION & RANGE Four races generally recognised in W Palearctic, but variation very slight, mainly concerning plumage gloss, colour of feather bases and length of throat-feathers, but more pronounced in size. Individual variation and intergradation obscure most geographical differences in the treated region. All populations mainly sedentary, movements rather irregular and hardly seasonal. Several extralimital subspecies in Asia and North America, not treated.

○ ***C. c. varius*** Brünnich, 1764 (Iceland, Faeroes). Subtly larger than *corax*, in particular has stronger bill and feet (as so often with island populations). On average a little duller, not as glossy. Subtle differences that overlap with *corax*. A pied morph (black and largely grey in irregular pattern) existed on Faroes but extinct since early 1900s. **W** ♂ 410–464 mm (*n* 13, *m* 430.9), ♀ 414–450 mm (*n* 12, *m* 425.5); **T** ♂ 225–257 mm (*n* 13, *m* 241.5), ♀ 225–253 mm (*n* 12, *m* 240.4); **T/W** *m* 56.3; **B** 70.0–85.6 mm (*n* 27, *m* 77.9); **BD** 26.0–31.0 mm (*n* 27, *m* 28.1); **Ts** 62.9–72.5 mm (*n* 27, *m* 67.1).

C. c. corax L., 1758 (Europe and Mediterranean islands to Siberia). Described under Identification and Biometrics. Note strong purple-blue gloss and well-developed throat hackles. Quite large, only subtly smaller on average (much overlap) than *varius*, but bill and feet a little smaller. – Birds in SW Europe, notably in Iberia and on Balearics ('*hispanus*'), tend to average slightly smaller than breeders in, e.g. Sweden, but difference is clinal and slight, and no steps or breaks in the cline have been found that would motivate recognition of a separate subspecies. (Syn. *cretae*; *cyprius*; *dardaniensis*; *hispanus*.)

C. c. corax, ad, Norway, Apr: long wedge-shaped tail varies in shape, depending how tightly it is folded. In spite of holding large territories when breeding, Raven is a social bird which can be seen in small groups when scavenging or feeding. (M. Varesvuo)

C. c. corax, ad, Norway, Apr: often seen soaring and gliding, or engaging in impressive mid-air acrobatics. Note typically flattish wing posture when gliding, with outerwing slightly lowered. (M. Varesvuo)

C. c. corax, ad, Spain, Sep: throat hackles relaxed, but still has thick-necked impression. Nasal bristles cover large area along the heavy bill, heightening bulky appearance to head. The wingtips sometimes appear to reach (rather than fall short of) tail-tip, and even in sunlight the metallic green and bluish-lilac sheen is often only slightly reflected. All wing ad. (H. Shirihai)

C. c. corax, juv, France, May: pinkish-orange gape and dull sooty-black head and body plumage. Fluffy throat hackles not yet developed. (A. Audevard)

C. c. corax, juv, Spain, Jul: in post-juv moult. Some juv can be very brown-toned on head and body and could be confused with Brown-necked Raven in relevant parts of range. (F. Trabalon)

C. c. laurencei, 1stY, Israel, Feb: a rather large race with proportionately shorter tail (though usually still projects somewhat beyond wingtips). Unlike *corax*, throat hackles shorter and less shaggy, and these feathers often glossy and can appear slivery-grey (though not visible here). Aged by many contrasting juv brown and worn wing-feathers. (A. Ben Dov)

CORVIDS

C. c. laurencei, Israel, Apr: typical flight silhouettes, especially the left bird in the right-hand image, with its anchor-like shape. Race *laurencei* is one of largest subspecies, and the largest individuals can appear only a little smaller than the male Bonelli's Eagle *Aquila fasciata* it is mobbing. (H. Shirihai)

C. c. tingitanus, presumed 1stY, Morocco, Feb: the smallest race in the covered region, with relatively stout, deep bill and well-arched culmen. Much of wing seemingly juv. (D. Occhiato)

C. c. tingitanus, ad, Canary Is, Feb: especially in N Africa and Canaries, care is needed to separate this species from Brown-necked Raven, as it can show brownish head and body, a tail slightly shorter than the wings, and this race also has less obvious throat hackles. However, nasal bristles on the very powerful bill still help to identify. (M. Schäf)

C. c. laurencei Hume, 1873 (Crete, Turkey, Cyprus, Middle East, N & W Iraq, S Caucasus, Iran, W Indian subcontinent). Has slightly duller gloss and is more oily blue than fresh *corax*. Head often tinged coppery-brown (though never as pronounced as in Brown-necked Raven), with shorter, less shaggy throat-hackles. Basal neck-feathering variable, often almost whitish. Rather large with proportionately slightly shorter tail than previous two races. – Birds in E Greece often included in this race, but fit better with *corax*. – **W** ♂ 403–477 mm (*n* 11, *m* 436.5), ♀ 390–451 mm (*n* 13, *m* 429.4); **T** ♂ 210–257 mm (*n* 11, *m* 238.5), ♀ (178) 215–262 mm (*n* 13, *m* 230.5); **T/W** *m* 54.0; **B** 60.7– 82.5 mm (*n* 25, *m* 74.6); **BD** (21.7) 23.3–30.0 mm (*n* 24, *m* 26.2); **Ts** 63.0–73.5 mm (*n* 25, *m* 67.2). (Previously often named *subcorax*, but this referred to Brown-necked Raven; see Hartert 1910, 1922.)

C. c. tingitanus Irby, 1874 (Canaries, N Africa). Smallest, with bill short but markedly stout and culmen strongly arched. Quite glossy in ad plumage, but little or subdued metallic colour, more 'oily' black-brown, when very fresh with slight purple tinge. Becomes browner in worn plumage; throat hackles distinctly shorter than in other races. Call often different from in Europe, a more subdued and hard rolling *krrahb*, but other calls are more like those in Europe. **W** ♂ 350–430 mm (*n* 23, *m* 386.7), ♀ 340–409 mm (*n* 23, *m* 377.6); **T** ♂ 176–243 mm (*n* 23, *m* 208.5), ♀ 175–230 mm (*n* 23, *m* 208.5); **T/W** *m* 54.5; **B** 547.9– 73.8 mm (*n* 46, *m* 65.8); **BD** 23.3–29.5 mm (*n* 46, *m* 25.4); **Ts** 59.5–73.0 mm (*n* 46, *m* 65.1). **Wing formula: p1** > pc 78–100 mm, < p2 78–98 mm; **p10** < wt 145–165 mm; **s1** < wt 149–183 mm. – Birds on Fuerteventura ('*jordansi*') average subtly smaller than breeders on Tenerife, but no difference obvious when compared with birds from NW Africa and best included in a slightly wider subspecies concept. (Syn. *canariensis*; *jordansi*.)

FAN-TAILED RAVEN
Corvus rhipidurus Hartert, 1918

Fr. – Corbeau à queue courte; Ger. – Borstenrabe
Sp. – Cuervo colicorto; Swe. – Kortstjärtad korp

This remarkable-looking raven of rocky or montane deserts is mainly resident and often gregarious, feeding in mountains or deserts, but also visits refuse tips, cultivations and oases in small parties. It shares several affinities with some African ravens, and occurs across sub-Saharan Africa, thereafter also through Arabia north to Jordan and Israel, with a recent claim from southernmost Turkey. The Dead Sea Depression is one of the best areas in our region to encounter the species.

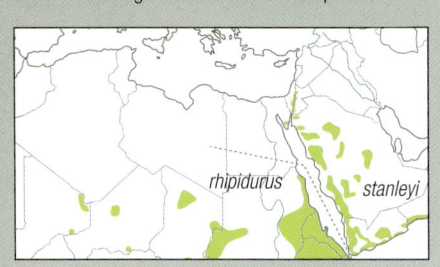

IDENTIFICATION Rather large, mainly all-black crow, with *folded wingtips extending well beyond tail-tip and a stumpy, wedge-shaped tail* affording a unique flight profile among the region's corvids. Instantly identified by its *peculiar flight silhouette, comprising tail-less appearance and very broad bat-like wings*. Underwing-coverts blacker than underside of the remiges, much like in a chough. *Bill rather short and thick*, but head appears fairly small in flight. At close range, oily bluish-purple sheen with *some pale mottling due to some visible grey feather bases on neck and body*, and has slightly elongated throat-hackles and forked nasal bristles. Forehead-feathering peculiarly 'tufted' and uneven when seen close. On the ground makes sidling jumps and occasional leaps, when the *very short tail makes the legs appear taller*. Often seen soaring, and is acrobatic in display.

VOCALISATIONS Song rarely heard and little studied, described as a medley of croaking calls, soft clucks and high-pitched squeals, plus a frog-like tremolo (Goodwin 1986). – Commonest call a short, resounding, rolling *korr*, more similar to Common Raven than to Brown-necked Raven (latter can recall Hooded Crow), often given in short series. Also a rolling *trrrü* with dry wooden ring, somewhat recalling Rook. Soft clucks, high-pitched squeals, and loud tremolos suggestive of some frogs. Various knocking sounds also noted.

SIMILAR SPECIES Size and shape quite different to larger and long-tailed *Common Raven*. – Head and body close in size to *Brown-necked Raven*, but Fan-tailed has shorter, broader wings and a much shorter, nearly 'non-existent' tail. With experience, unmistakable given its unusual flight silhouette.

AGEING & SEXING (*stanleyi*) Ageing sometimes possible using amount of gloss and wear in plumage, and width and shape of tail-feathers. Sexes inseparable, except rarely by size, in breeding pair ♂ averages larger (but much overlap). No seasonal variation (except in gloss to plumage). – Moults. Complete post-nuptial moult of ad protracted, noted during Jul–Feb (Mar). Partial post-juv moult chiefly Jul–Nov, involves head, body, most or all lesser and some median coverts. Pre-nuptial moult absent (but note late finish to complete moult in many ad). – SPRING **Ad** With wear becomes slightly duller; blue and purple tones become oily blue-bronze or copper, and appear browner, especially on head and neck. Tail-feathers average broader with more obtuse tips, and due to late moult remain fresh well into spring. **1stS** As ad but moult limits sometimes visible, with browner juv wing- and tail-feathers; also generally more bleached and less glossed, or browner. Tail-feathers average slightly narrower, and wear

C. r. stanleyi, ad ♂ (right) and ♀, Israel, Mar: quite a character, mainly due to its small compact size, with very short tail (often wholly cloaked by the wings). When a pair is seen side by side, the duller ♀ (on left) can be quite obvious, as well as smaller size, lighter build (especially head) and less glossy plumage. All wing-feathers ad. (H. Shirihai)

C. r. stanleyi, Israel, Mar and Apr: unmistakable due to unique shape, with short but very broad wings and short tail, which is fanned when soaring and just projects beyond the wings' trailing edges—in such views, the species can be identified instantly, even at very long range! Blacker underwing-coverts highly contrasting. (H. Shirihai)

is obvious on tips in spring. – **AUTUMN Ad** Evenly fresh (no moult limits), and plumage glossed strongly blue-black, with purple and bronze iridescences in some lights. Note that some birds moult remiges all through winter. **1stW** Like fresh ad, but all flight- and tail-feathers, primary-coverts, apparently all greater coverts and a few lesser and median coverts are retained juv feathers, slightly weaker textured, less strongly glossed and a little browner; occasionally some body-feathers also kept (less strongly glossed than ad). The retained flight- and tail-feathers are also to some degree narrower and more sharply pointed. **Juv** Plumage has much less gloss, being more brown-tinged.

BIOMETRICS (*stanleyi*) **L** 42–46 cm; **WS** 103–114 cm; **W** ♂ 353–378 mm (*n* 12, *m* 365.1), ♀ 342–368 mm (*n* 8, *m* 356.9); **T** ♂ 145–157 mm (*n* 12, *m* 150.5), ♀ 134–158 mm (*n* 8, *m* 143.8); **T/W** *m* 39.6; **B** ♂ 47.7–54.6 mm (*n* 12, *m* 51.3), ♀ 46.6–56.6 mm (*n* 8, *m* 49.8); **BD** 20.5–23.0 mm (*n* 22, *m* 21.3); **Ts** 54.0–61.5 mm (*n* 23, *m* 57.3). Wing formula: **p1** > pc 74–106 mm, < p2 74–93 mm; **p2** < wt 20–37 mm, =5/6 (100%); **pp3–5**(6) about equal and longest; **p6** < wt 0–15 mm; **p7** < wt 39–59 cm; **p8** < wt 70–92 mm; **p10** < wt 115–138 mm; **s1** < wt 121–148 mm. Emarg. outer web pp2–5.

C. r. stanleyi, ad, Israel, Feb: tail falls well short of wingtips. In close views, some paler streaking and marbling may be evident, especially around head. Evenly-fresh wing infers age. Note spine-tipped tail-feathers. (A. Ben Dov)

C. r. stanleyi, Oman, May: in flight can almost resemble a black plastic bag in a desert sandstorm (!). Pale scaling on mantle and scapulars can be visible in close views. (H. & J. Eriksen)

GEOGRAPHICAL VARIATION & RANGE Moderate variation concerning only size, with northern and eastern breeders being slightly smaller.

C. r. stanleyi Roselaar, 1993 (Sinai, Israel, Jordan, Arabia south to Yemen; resident). Described above. The smallest race.

C. r. rhipidurus Hartert, 1918 (sub-Saharan Africa from Niger to Sudan, S & C Egypt, Eritrea, Ethiopia, Somalia; resident). Somewhat larger, longer-legged and heavier-billed than *stanleyi* but otherwise identical. **L** 44–50 cm; **WS** 105–126 cm; **W** ♂ 349–421 mm (*n* 15, *m* 385.9), ♀ 358–399 mm (*n* 13, *m* 379.3); **T** ♂ 136–169 mm (*n* 15, *m* 152.5), ♀ 138–161 mm (*n* 12, *m* 148.8); **B** 52.2–61.7 mm (*n* 28, *m* 56.9); **BD** 21.0–24.5 mm (*n* 26, *m* 22.4); **Ts** 58.0–70.5 mm (*n* 28, *m* 64.4).

REFERENCES Roselaar, C. S. (1993) *DB*, 15: 258–262.

C. r. stanleyi, 1stY (with Common Raven behind) Israel, Feb: 1stY with many worn and brown-bleached juv wing-feathers. In such a perfect comparison between the two species, Fan-tailed (on left) can be seen to be much smaller. (A. Ben Dov)

TRISTRAM'S STARLING
Onychognathus tristramii (P. L. Sclater, 1858)

Alternative name: Tristram's Grackle

Fr. – Rufipenne de Tristram; Ger. – Tristramstar
Sp. – Estornino irisado; Swe. – Sinaiglansstare

Tristram's Starling, an attractive if at times somewhat noisy species, is a Middle Eastern endemic and the sole Palearctic representative of the Afrotropical *Onychognathus* starlings. The species has increasingly adapted to urban areas, even nesting on buildings and taking advantage of scraps at picnic sites. Nevertheless, it also symbolises the unspoilt rocky deserts that it shares with species such as Sand Partridge *Ammoperdix heyi*, Fan-tailed Raven, White-crowned Wheatear and Sinai Rosefinch, and is often found in association with Nubian Ibex.

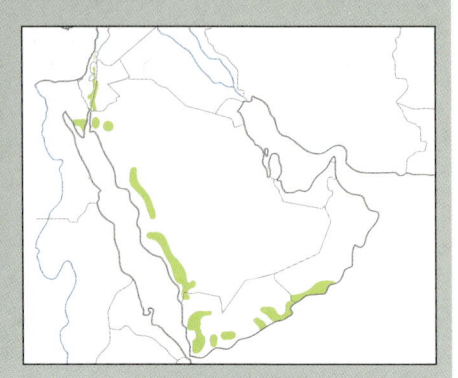

IDENTIFICATION A rather long-tailed and long-winged starling, *mainly blackish* with a square-ended tail and the *basal two-thirds of the primaries bright rufous, forming a conspicuous panel on the folded wing and a bold colourful patch on both wing surfaces in flight*. Typically occurs in rather *noisy flocks, uttering diagnostic whistles* audible even at long range. ♂ has iridescent plumage particularly when fresh, and ♀ readily recognised by the less uniform, browner plumage with greyer (slightly streaked) head and duller wing patches. Bare parts generally black, but iris reddish-brown, imparting rather fierce look at close quarters, and bill and legs strong and more grey-black. Flight rather fast, with action recalling a large thrush rather than a starling, but flocks often dash past with sweeping, diving and climbing manoeuvres, and sometimes engage in aerial play accompanied by far-carrying calls. On the ground, proceeds with bouncing hops and a striding, purposeful walk.

VOCALISATIONS Song a series of whistles and fluting notes, with a strangely melancholic quality, and interspersed squeaky and rasping notes. – Calls loud and often heard, very characteristic. Gives loud glissando, fluting whistle *chee-oo-wee* or *dee-oo-ee-o*, unlike any other bird call in the region, especially typical when uttered in chorus from approaching flock. Alarm is a hoarse scream or irritated sound, *veeech* (somewhat resembling alarm of Golden Oriole).

SIMILAR SPECIES Given the lack of congenerics in the treated region, should prove unmistakable. No other red-winged starling has been recorded on the Arabian side of the Red Sea or so far north in Africa. The reddish-brown primaries, particularly striking in flight, offer a ready distinction from other starlings in the covered region.

AGEING & SEXING Ages and sexes readily separable. Very little seasonal variation. – Moults. Complete post-nuptial and partial post-juv moult, generally Jul–Sep (can finish Oct). Post-juv moult includes all head, body, lesser and most or all median coverts and a variable number (none to all) of greater coverts, tertials, alula, and often central tail-feathers, but apparently only rarely odd remiges or primary-coverts. Pre-nuptial moult absent. – **SPRING Ad** ♂ Black head and body, black tail, wings with large rufous primary patches (which become less bright with wear, but no big change). Primary-coverts all black. **Ad** ♀ Differs from ♂ in having dark brown body (rather than black) and paler brown-grey head. Upper breast, neck and nape diffusely but boldly streaked dark brown-grey. All-dark brown primary-coverts. **1stS** Differs from similar ad primarily in having bicoloured

♂ (left) and ♀, Oman, Jan: this Middle Eastern endemic may habitually follow livestock, a behaviour even more commonly seen in Oman. Aside of the diagnostic orange in wing, ♂ is mostly glossy black and ♀ is browner and greyer and streaked in front and on neck. (H. & J. Eriksen)

Ad ♂, Israel, Dec: a stunningly beautiful starling often first noticed by its loud whistling voice. The black plumage of ♂ has faint greenish and purple gloss when seen close and in good light. Evenly-feathered ad wing with all-black primary-coverts. (H. Shirihai)

Ad ♀, Israel, Apr: conspicuous large rufous-orange primary patches framed by dark wingtips. Unlike ♂, ♀ has greyish head and neck and duller sooty-black body. Ad of both sexes have all-dark primary-coverts, lacking rufous-buff bases. Note longer tail compared to Common Starling. (I. Butler)

primary-coverts, these having a subterminal rufous patch and contrasting dark tips. Also (but less easy to see) has contrasting moult limits, with more bleached and browner juv wing- and tail-feathers, and body plumage duller and less smoothly pigmented. – **AUTUMN Ad** Largely as in spring but more pronounced purplish- and greenish-blue iridescence. ♀ has head and neck brownish-grey, and typically looks mottled (with faint narrow black shaft-streaks in close views). **1stW** Retained juv primary-coverts have rufous or cinnamon central portion (forming easily seen pale subterminal patch bordered by the dark basal area and dark tips of the coverts). Each sex like respective ad (but generally less strongly glossed and varyingly browner), but all flight- and tail-feathers, primary-coverts and some outer greater coverts still juv, a little weaker textured, less glossed and somewhat browner, especially at edges and tips. **Juv** Soft fluffy appearance to entire body plumage, which is even less glossy than ad ♀. Head, neck and breast are darker, being dull black or sooty-brown like rest of body; iris duller brown.

BIOMETRICS L 26–28 cm; **W** ♂ 145–156 mm (n 18, m 150.8), ♀ 137–148.5 mm (n 12, m 144.5); **T** ♂ 102–112 mm (n 18, m 107.1), ♀ 93–110 mm (n 13, m 102.2); **T/W** m 70.9; **B** 25.6–32.8 mm (n 30, m 29.8); **BD** 8.7–10.0 mm (n 28, m 9.2); **Ts** 28.0–34.1 mm (n 31, m 31.6). **Wing formula: p1** small, > pc 0–5 mm or to 2 mm <; **p2** < wt 0–7 mm; **pp**(2)**3–4** about equal and longest; **p5** < wt 1–8.5 mm; **p6** < wt 11–20 mm; **p7** < wt 20–29 mm; **p10** < wt 34–46 mm; **s1** < wt 39–51 mm. Emarg. pp3–6 (less strongly near tip of p6).

GEOGRAPHICAL VARIATION & RANGE Monotypic. – Israel, Jordan, Sinai, W & SW Arabia, Yemen, S Oman; resident. – Southern birds have been separated as '*O. t. hadramauticus*', but differences not well described, and all previous handbooks and checklists have synonymised it with *tristramii*, a sound policy followed here. (Syn. *hadramauticus*.)

Ad ♀, Israel, Feb: usually readily sexed by duller and browner (never deep black) plumage with even paler brown-grey head. Upper breast, neck and nape diffusely but boldly streaked dark brown-grey. Aged by evenly-feathered wing, including primary-coverts. (H. Shirihai)

Israel, Jan: very noisy flocks often dash past high over head or in climbing or sweeping flight. Conspicuous rufous wing panels are also well visible from below. Characteristically quite long-tailed. (H. Shirihai)

Ad ♂ (left) and ad ♀, Sinai, Egypt, Nov: metallic blue iridescence mostly evident in shade, especially if plumage fresh (both sexes), when ♀ also often appears to have greyer head. Depending how tightly wings are folded, size of rufous primary patches varies. Evenly-fresh wings, including primary-coverts, in both birds infer age. (C. Bradshaw)

1stW ♀, Oman, Nov: desert dwellers often best seen or photographed where there is fresh water. Young birds resemble ad, but much of wing, including primary-coverts, juv, being browner and here already slightly worn. Note moult limit in median coverts, but also between juv tail and moulted uppertail-coverts. (D. Occhiato)

1stS ♂, Israel, Mar: compared to ad, note black-tipped pale-based primary-coverts forming small isolated pale patch apart from large primary patch, readily visible in flight but often also on perched birds. (H. Shirihai)

(COMMON) STARLING
Sturnus vulgaris L., 1758

Fr. – Étourneau sansonnet; Ger. – Star
Sp. – Estornino pinto; Swe. – Stare

Although it has decreased substantially in the last three decades, the Starling has recovered recently and is still one of the most widespread and numerous birds in Europe with an estimated population in the region of 40 million pairs. It is also a well-known and sufficiently popular bird to have been brought by emigrant Europeans to other continents, and is therefore now established in North America, Australia and South Africa. It can form enormously large, dense and noisy flocks, roosting in reedbeds and frequenting orchards or vineyards by day. Locally, it can also become something of a pest roosting in thousands in city centres or parks.

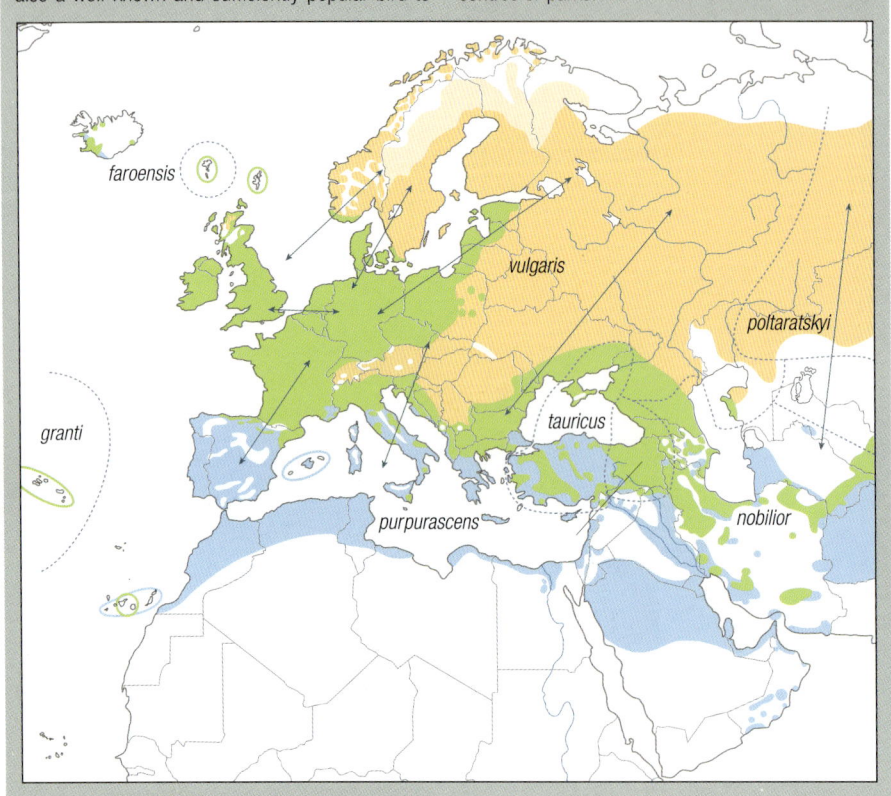

IDENTIFICATION A *dark*, medium-sized, chunky bird with flat forehead, *pointed bill*, *short tail*, but rather pointed, *triangular wings*. In adult plumage glossy black with metallic (variably purple, green or bluish) sheen and, in fresh plumage, *sprinkled with tiny buff-white feather-tips*. Bill in spring and summer largely *yellow*, legs dirty-pink, in autumn and winter both bill and legs are dark. Breeding birds best sexed by bill colour, ♂♂ having bluish-grey base to lower mandible, ♀♀ pinkish-white. There is also an average difference in length of throat-feathers ('hackles') and size and shape of pale tips to these, ♂♂ having longer and narrower feathers with smaller and thinner or rectangular pale tips, ♀♀ slightly broader and shorter feathers with broader and more rounded pale tips. Young are dull brown lacking paler feather-tips, and have dark brown bill and legs. In all plumages told by characteristic energetic, restless gait (e.g. when feeding on a lawn) with rapid legwork and *constant jerking head movements*; capable also of sudden quick rushes. In flight, note fairly fast progress, straight path, occasional short glides and clumsy landing, thudding down on fluttering wings. Often seen in large flocks, then notice dense formation and simultaneous turns. For a bird frequently living near man and nesting on houses or in gardens, it is surprisingly wary, taking off early with a short purring call.

VOCALISATIONS Song special: often opened with a joyful drawn-out, bent whistle *tseeoouh*, then a frantic series of expertly performed mimicry of other birds, interfoliated with clicking or creaking sounds and squeaky notes, all seemingly at the same time and at just as hectic a pace! It is difficult to understand how the Starling manages it, but the song sounds as if coming from at least two different birds at the same time. The song is not loud (save the whistles, or when the bird is particularly agitated), and has a regular rhythm with recurring clicking or squeaky sounds. The song's delivery, with the bird seemingly in a trance, is accompanied by wing-flaps and erect throat-feathers. – Calls include a short, purring *chürr* when taking off or in flight; a hoarse, often repeated, grating *tcheehr*, or a hard, clicking *spet*, when alarmed. Young are very noisy, giving repeated, dry rolling or rippling calls, at times from flocks of grown immatures sounding like a very loud, metallic rattle.

SIMILAR SPECIES For the layman must be separated from Blackbird, for the expert from Spotless Starling. From *Blackbird* by: shorter tail; at least some pale spotting on rump and vent (even in summer, when pale tips become largely worn off in ♂♂); dull pinkish or reddish-brown legs; and energetic and continuous walk with jerking head (not hopping, or walking, stopping and 'freezing' all the time like Blackbird). – From *Spotless Starling* on the following: (i) in spring/summer, adults still spotted, though ♂♂ unspotted on head, throat and breast (Spotless has no spots at all), in autumn/winter all plumages profusely spotted (Spotless: only immatures and adult ♀ finely spotted on rear parts); (ii) has more obvious metallic gloss, this being more clearly divided in green (or blue) and purple (Spotless: more subdued and uniformly-shaded greenish-purple, almost 'oily' gloss, always lacking any blue); (iii) shorter and less narrow 'hackles' (elongated feathers on throat and upper breast), particularly obvious in ♂♂; (iv) darker and duller pink legs (Spotless brighter pink, but some overlap); (v) on average very subtly longer and more thin-based bill (Spotless: tends to have thicker-based, more downcurved bill, but considerable overlap). – There is the remote risk of confusion between Starling and ♂ *Black Lark*, especially as eastern Starlings are less pale-spotted, the two being fairly similar in flight. Note more pointed wings and bill, and less rounded head, of Starling.

AGEING & SEXING (*vulgaris*) Ageing possible if seen close or handled. Some seasonal plumage differences due to wear, and juv readily separated in summer. Sexes (after post-juv moult) differ, more so in spring than autumn. –

S. v. vulgaris, ♂, Scotland, Jun: racial characteristics include green gloss over much of plumage, with variable purple or blue on head, throat and flanks; ochraceous-buff tips on upperparts and vent in breeding plumage. ♂ has base of lower mandible tinged bluish-grey, glossier plumage and fewer and smaller pale spots. Very narrow and long throat-feathers. In full breeding plumage ageing impossible. (I. H. Leach)

S. v. vulgaris, ♀, England, Apr: some *vulgaris* have a little more purple gloss on breast-sides. ♀ by whitish base to lower mandible. Pale spots on average slightly larger and remain longer into spring. Species' chunky jizz with flat forehead, pointed bill and short tail but mostly blackish plumage well shown here; any prominently spotted feathers at this season immediately eliminate Spotless Starling. (G. Thoburn)

STARLINGS AND MYNAS

S. v. vulgaris, ♂ (right) and ♀, Netherlands, Mar: ♂ courting a ♀ on roof top. In early spring, because ♀♀ retain larger pale spots later, sexual differences are more pronounced. Very narrow throat-feathers of ♂ and sex-related coloration to base of lower mandible visible, greyish in ♂ and whitish in ♀. (R. Schols)

S. v. vulgaris, ♂, Estonia, May: rather pointed, triangular wings in flight. Note sex-related coloration to base of lower mandible, greyish-blue in ♂, and dark plumage by late spring almost unspotted. From Spotless Starling, note a few large spots on rear underparts, and longer and thinner-based bill. (M. Varesvuo)

Moults. Complete moult of both ad and juv in late summer (mainly Jun–Sep), though moult of juv somewhat later and more protracted (late Jun–early Nov). – SPRING **Ad ♂** Plumage has more gloss but fewer and smaller pale spots. Head, throat, breast and upper belly become unspotted with wear. Elongated tips of throat-feathers very narrow and long. (Longest throat-feathers from base to tip 20–26 mm.) Base of lower mandible tinged bluish-grey. Iris uniform dark brown. **Ad ♀** Pale spots on average slightly larger and remain longer in summer. Elongated tips of throat-feathers slightly broader than and not quite as long as in ♂. (Longest throat-feathers 15.5–21.5 mm.) Base of lower mandible whitish with pink hue. Iris dark or medium brown with a whitish or yellowish inner or outer circle. **1stS ♂** As ad ♂ but averages slightly more spotted, and throat-feathers slightly shorter (16–20 mm). Note that these are tentative age characters separating only a few. **1stS ♀** As ad ♀ but averages somewhat more spotted, and throat-feathers slightly shorter (13–17 mm). – SUMMER **Ad** Plumage dark with metallic gloss, pale spots remaining at least on back, rump, wings and belly (Europe). Bill pale yellow. **Juv** No metallic gloss; largely uniform dull grey-brown with paler throat and pale, greyish- or buffish-white edges to wing-feathers and tail-feathers. Unspotted. Bill dark. Sexes differ already in that

S. v. vulgaris, ad ♂ (top left: Netherlands, Oct), ad ♀ (top right: England, Feb), 1stW ♂ (bottom left: Finland, Oct) and 1stW ♀ (bottom right: Finland, Oct): both ages undergo complete moult at the end of breeding season, still ageing and sexing is possible on tail shape and tail pattern (requires close views or handling), general gloss and detailed shape of pale spots (best judged on throat and breast). Ad ♂ is most glossed in winter with relatively small and narrow pale spots, but has longest black shaft-streaks penetrating *c.* 3/4 into pale spots. Ad ♀ and 1stW ♂ rather similar in having reduced gloss, larger spots, and shorter black shaft-streaks, thus these two difficult to separate, but ad ♀ has even browner plumage, narrower and more arrowhead-shaped pale spots with long shaft-streaks, almost like ad ♂. 1stW ♂ blackish and glossy but pale spots more broad and rounded than in ad ♂. 1stW ♀ dullest, with little gloss, pale spots large and rounded with insignificant shaft-streaks, and has generally browner ground colour. (Top left: R. Pop; top right: G. Jenkins; bottom left and bottom right: M. Varesvuo)

♂♂ typically have all-dark or uniform brown iris, ♀♀ a paler yellowish-grey or pale brown iris (see references in Svensson 1992; Smith *et al.* 2005). – AUTUMN **Ad ♂** Much gloss in plumage. Pale spots small and narrow, with long black shaft-streaks penetrating *c.* 3/4 of pale spots. Tips of tail-feathers rather pointed, central feathers with clear-cut buff margins, and dark subterminal bar distinct and glossy. **Ad ♀ / 1stW ♂** Slightly less strong gloss than in ad ♂, and spots a little larger and less narrow, black shaft-streaks shorter, penetrating about half of the pale spots. (Difficult to separate in the field.) Ad ♀ has very nearly same shape and pattern of tail-feathers as ad ♂ (only subtly less typical), whereas 1stW ♂ has more rounded tips and indistinct pattern by and large as described for 1stW ♀. **1stW ♀** Very little gloss. Pale spots large and rounded, almost dominating dark ground colour. Black shaft-streaks insignificant. Tips of tail-feathers rounded, central feathers with ill-defined edges and diffuse medium dark, matt subterminal bars.

BIOMETRICS (*vulgaris*) **L** 19.5–22 cm; **W** ♂ 126–137 mm (*n* 30, *m* 133.0), ♀ 124–135 mm (*n* 29, *m* 128.7); **T** ♂ 61–67 mm (*n* 30, *m* 63.9), ♀ 59–65 mm (*n* 29, *m* 61.9); **T/W** *m* 48.0; **B** 26.7–30.0 mm (*n* 57, *m* 28.4); **BD** 6.7–8.4 mm (*n* 56, *m* 7.6); **BW** 7.2–8.4 mm (*n* 56, *m* 7.7); **BW**(n dist) 6.1–7.3 mm (*n* 56, *m* 6.6); **Ts** 26.5–30.6 mm (*n* 56, *m* 28.8). **Wing formula: p1** minute, < pc 10–17 mm (ad) or 4–11 mm (juv); **p2** longest (sometimes pp2–3 longest); **p3** < wt 0.5–2 mm; **p4** < wt 4–10 mm; **p5** < wt 12–19 mm; **p6** < wt 20–27 mm; **p7** < wt 26–35 mm; **p10** < wt 43–50 mm; **s1** < wt 46–54 mm. Emarg. pp3–4.

GEOGRAPHICAL VARIATION & RANGE Rather slight variation, mainly involving colour of metallic gloss, and amount of spotting, but some structural differences also in north-west and south-west of range. Most subspecies grade continuously into each other over large areas, hence not all birds can be subspecifically labelled. Three more extralimital races in Asia unlikely to reach W Palearctic (and therefore not treated).

S. v. vulgaris L., 1758 (Europe east to Urals, except as detailed below; winters in W & S Europe, N Africa, Levant). Treated above. Green tinge on ear-coverts, strong green gloss on upperbody (hindneck to uppertail-coverts), wing-coverts (although these often bluish), breast and centre of belly, but also some purple on lower throat and breast-sides. Lower flanks often tinged purple. Rest of head and throat usually purple, crown with some green admixed. Retains ochraceous-buff tips on upperparts while breeding. – Birds of Shetlands and Outer Hebrides have same plumage and only a fraction thicker-based bill, with much overlap, too small a difference to recognise as separate subspecies. (Syn. *balcanicus*; *britannicus*; *ferdinandi*; *graecus*; *jitkowi*; *ruthenus*; *zetlandicus*.)

S. v. vulgaris, juv, Netherlands, Jun: prior to or during post-juv moult, birds with complete or mostly juv plumage are dull brown without paler feather-tips, and have dark bill and reddish-brown legs. Note characteristic dark loral patch, pale fore-supercilium and buff-brown wing edgings. (R. Pop)

S. v. vulgaris, 1stW, Italy, Jul: note start of complete post-juv moult on underparts and in wing, thus 1stW plumage emerging. (D. Occhiato)

S. v. vulgaris, Finland, Dec: in non-breeding season, forms enormous, dense and noisy flocks, known as murmurations, especially prior to roost, often in reedbeds. These large flocks often perform remarkable aerobatics. About 9000 birds visible on this photograph. (M. Varesvuo)

S. v. purpurascens, Israel, Jan: large flocks can also form during day at rich food sources, e.g. at garbage dumps, as here in Jordan River Valley, where most birds in winter are known to be or approach *purpurascens*. (H. Shirihai)

S. v. faroensis Feilden, 1872 (Faeroes; resident). Larger, with stronger feet and bill, and in juv plumage is on average somewhat darker than *vulgaris*, being sooty-black with only slightly paler throat. Ad has somewhat less gloss than *vulgaris*, and pale spots are smaller and wear off sooner. **L** 21.5–23 cm; **W** ♂ 135–142 mm (*n* 16, *m* 138.3), ♀ 131–142 mm (*n* 10, *m* 134.9); **B** ♂ 29.5–33.5 mm (*n* 16, *m* 31.8), ♀ 29.0–32.4 mm (*n* 10, *m* 31.3); **BD** 7.6–9.3 mm (*n* 26, *m* 8.4); **BW** 7.7–8.7 mm (*n* 26, *m* 8.2); **BW**(n dist) 6.4–7.6 mm (*n* 26, *m* 7.1); **Ts** 27.2–31.4 mm (*n* 21, *m* 30.0).

○ **S. v. granti** Hartert, 1903 (Azores; resident). Very similar to *vulgaris* but has on average slightly shorter and stouter bill. (General size the same, including wing, tail and tarsus.) Mantle sometimes tinged purplish, but more often green as in *vulgaris*. A very subtle race. **W** ♂ 129–136 mm (*n* 9, *m* 131.9), ♀ 125–134 mm (*n* 12, *m* 129.0); **T** ♂ 58–64 mm (*n* 9, *m* 61.5), ♀ 57–64 mm (*n* 12, *m* 60.1);

S. v. granti, ad ♂, Azores, Oct: very slightly shorter and stouter bill, and purplish-tinged mantle, otherwise very similar to *vulgaris*. Aged by extensive gloss in winter with relatively small and narrow, arrowhead-shaped spots with long black shaft-streaks. (D. Occhiato)

S. v. poltaratskyi, ♂, Kazakhstan, May: compared to *vulgaris*, note more purple sheen on head. Also tends to shed nearly all pale spots in late spring and summer. (P. Dubois)

Presumed *S. v. tauricus*, ♂, probably ad, Israel, Dec: birds approaching *tauricus* frequent wintering areas in parts of Israel, and are characteristic in being almost entirely glossed green, with largely dark grey underwing-coverts, only finely fringed paler. Note typical jizz in flight. ♂ by extensive plumage gloss in winter and presumed ad by relatively small, more arrowhead-shaped spots with long black shaft-streaks. (A. Ben Dov)

S. v. purpurascens, ♂, Turkey, Jun: characteristic is lack of pale spotting, very like Spotless Starling, in summer, and also extensive purple sheen on sides of head and body, especially on lower breast and scapulars. Mix of green and purple-blue metallic gloss on head and breast (rather than uniform purplish-blue and slightly subdued gloss) distinguishes species from Spotless. (G. Conca)

T/W *m* 46.6; **B** 25.0–29.0 mm (*n* 21, *m* 27.0); **BD** 7.3–8.7 mm (*n* 21, *m* 7.9); **BW** 7.0–8.3 mm (*n* 19, *m* 7.7); **BW**(n dist) 6.5–7.5 mm (*n* 19, *m* 7.0); **Ts** 28.1–30.5 mm (*n* 21, *m* 29.2).

S. v. poltaratskyi Finsch, 1878 (SE Russia, C Urals, W & C Siberia, Altai, W Mongolia, N & E Kazakhstan; winters E Iran, Afghanistan, N India). Like *vulgaris*, including most measurements, but head is more purple than green, and flanks usually more purplish-blue. Underwing-coverts broadly fringed buff, much more so than in *vulgaris* or *tauricus*. **W** ♂ 128–136 mm (*n* 16, *m* 131.1), ♀ 124.5–132 mm (*n* 13, *m* 128.5); **T** ♂ 59.5–68 mm (*n* 16, *m* 63.2), ♀ 58.5–68 mm (*n* 13, *m* 63.2); **T/W** *m* 48.6; **B** 27.5–31.0 mm (*n* 29, *m* 29.1); **BD** 6.3–8.5 mm (*n* 26, *m* 7.2); **BW** 6.9–8.5 mm (*n* 29, *m* 7.5); **BW**(n dist) 5.7–7.2 mm (*n* 29, *m* 6.3); **Ts** 28.0–30.7 mm (*n* 27, *m* 29.6). (Syn. *dzungaricus*; *menzbieri*; *zaidamensis*.)

S. v. tauricus Buturlin, 1904 (SE Ukraine, Crimea, Sea of Azov, W & C Turkey; partly resident, but some move south and south-west in winter). Glossy green on much of head, neck and throat/upper breast, with dull green tinge on flanks, whereas lower breast, mantle, scapulars, rump and uppertail-coverts are purple; sometimes a bluish gloss on back. From about early May, ad is generally unspotted above, but a few are finely spotted until Jun. Underwing-coverts largely dark grey, only finely fringed white. Size as *vulgaris*. Limited material examined. **W** ♂ 134–139 mm (*n* 5, *m* 136.1), ♀ 121–132.5 mm (*n* 3, *m* 128.2); **T** ♂ 60–68.5 mm (*n* 5, *m* 65.3), ♀ 58.5–62.5 mm (*n* 3, *m* 60.5); **T/W** *m* 46.8; **B** 27.5–31.5 mm (*n* 8, *m* 27.4); **BD** 7.5–9.2 mm (*n* 8, *m* 8.0); **BW** 7.9–8.7 mm (*n* 8, *m* 8.1); **BW**(n dist) 6.7–7.0 mm (*n* 8, *m* 6.9); **Ts** 29.2–31.3 mm (*n* 8, *m* 30.2).

○ **S. v. purpurascens** Gould, 1868 (E Turkey, E Black Sea, N Iraq; winters in south of range and through S Middle East south to Egypt). Has green head like *tauricus* but purple sheen on ear-coverts (variable, often much), belly and flanks; strong purple gloss on lower breast and scapulars; green on throat/upper breast and wing-coverts; upper body a mixture of purple and blue-green. From about Apr or early May ad often unspotted above. – Intergrades with both *tauricus* and *caucasicus*. **L** 21–22.5 cm; **W** ♂ 127–138 mm (*n* 12, *m* 133.1), ♀ 125–134.5 mm (*n* 14, *m* 129.6); **T** ♂ 59.5–70 mm (*n* 12, *m* 64.3), ♀ 59.5–66.5 mm (*n* 14, *m* 62.9); **T/W** *m* 48.4; **B** 28.3–32.9 mm (*n* 26, *m* 30.1); **BD** 6.7–8.7 mm (*n* 24, *m* 8.0); **BW** 6.9–8.5 mm (*n* 24, *m* 7.8); **BW**(n dist) 5.4–7.3 mm (*n* 24, *m* 6.6); **Ts** 26.8–31.5 mm (*n* 26, *m* 29.7). (Syn. *oppenheimi*.)

○ **S. v. nobilior** Hume, 1879 (N & E Caucasus, areas N, W & SW of Caspian Sea, NW Iran, S Turkmenistan, W & S Afghanistan; mainly resident, vacating only higher altitudes). Similar to *purpurascens* but whole head and throat are more a mixture of green and purple gloss (thus on average less green). Mantle to uppertail-coverts and scapulars also a mix of green and purple. Upperwing largely purplish-blue and purple. Ad becomes entirely unspotted above when breeding. **W** ♂ 122–135 mm (*n* 14, *m* 130.2), ♀ 125–133 mm (*n* 12, *m* 127.5); **T** ♂ 61–68 mm (*n* 14, *m* 64.3), ♀ 62–66 mm (*n* 12, *m* 63.6); **T/W** *m* 49.6; **B** 28.1–35.9 mm (*n* 26, *m* 31.0); **BD** 7.5–8.3 mm (*n* 26, *m* 7.6); **BW** 7.0–8.1 mm (*n* 26, *m* 7.5); **BW**(n dist) 6.0–6.8 mm (*n* 26, *m* 6.5); **Ts** 28.4–33.0 mm (*n* 24, *m* 30.1). – Birds in Caucasus ('*caucasicus*', described from Kislovodsk, N Caucasus) average more green and blue than purplish on head, but subtle with much overlap and individual variation, thus included here. (Syn. *caucasicus*; *dresseri*; *heinrichi*; *persepolis*.)

REFERENCES SMITH, E. L. *et al.* (2005) *Ring. & Migr.*, 22: 193–197.

Presumed *S. v. purpurascens*, ♂ (left) and ♀, Israel, Jan: many winterers in E Israel match *purpurascens*, having extensive purple sheen to ear-coverts and body-sides, especially lower breast; green dominates throat/upper breast and wing-coverts, while upper body, including scapulars are mix of purple and blue-green. This race intergrades with other races, including *tauricus*. Both ad by small, narrow, more arrowhead-shaped spots with long black shaft-streaks, with ♀ browner and less glossy than ♂. (H. Shirihai)

S. v. nobilior, ♂, Azerbaijan, May: like *purpurascens* but head to upper breast has purer purple gloss, while mantle and scapulars show mix of green and purple, and upperwing largely purplish-blue and purple. Ad becomes entirely unspotted above when breeding, as shown here. (M. Heiß)

SPOTLESS STARLING
Sturnus unicolor Temminck, 1820

Fr. – Étourneau unicolore; Ger. – Einfarbstar
Sp. – Estornino negro; Swe. – Svartstare

The Spotless Starling replaces the Starling in Iberia, NW Africa and on larger Mediterranean islands (except the Balearics), and has fairly recently expanded into S France, where it overlaps with Starling. Similar to Starling in most respects, but usually does not form such huge flocks in non-breeding season, although it will mix with large flocks of Starling. Breeds in holes in a variety of habitats including on houses and near man. Resident or only short-distance migrant.

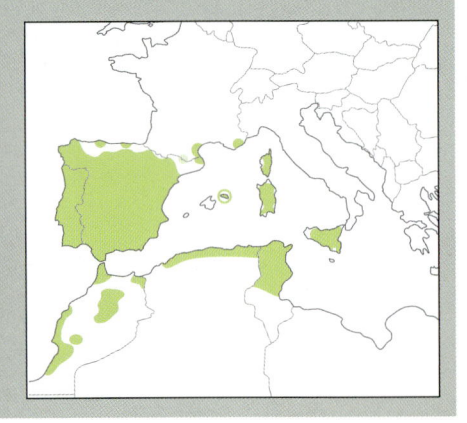

IDENTIFICATION Very similar to Starling, only differing in considerably fewer light spots in plumage, and in a few other more subtle ways, detailed under Similar species. In spring and summer virtually *unspotted*, *uniformly black* with a slight metallic gloss. It often has *less gloss* than in Starling (but fresh-plumaged ♂♂ can be glossy, too), and is *more evenly purplish-green all over*, not green in one part and purple in another like Starling. Similarly, it lacks any of the blue gloss visible at some angles on Starling. In winter may have very fine pale spotting on upper tail-coverts and vent, and faintly elsewhere on rear body, especially in first-winter ♀. In breeding season on average brighter and paler pink legs, but some are duller brownish-pink like Starling. The bill is yellow in summer, dark brown in winter. Gait, flight and behaviour much as Starling (which see). Juvenile usually inseparable from juvenile Common Starling.

VOCALISATIONS Song much the same as in Starling, but often the commonly-heard loud whistle differs in having a fizzing quality, and being more drawn-out and more penetrating with an upward-bent, *tsooeesh*; some whistles sound more like Starling, though. Another sound that seems to be species-specific is a variety of soft upwards-inflected (or downwards) glissando whistles, *tuuooee* (*teeooh*), pleading or miaowing in quality, and which can vaguely recall Tristram's Starling in tone, or a sound of human origin. – Calls like Starling.

SIMILAR SPECIES For some general advice, see same section under Starling. From *Starling* in following: (i) in late spring/summer, adults entirely unspotted (Starling: ♂ usually unspotted only on head, throat and breast), in autumn/winter immatures and adult ♀ finely spotted on rear parts, whereas adult ♂ has very tiny and narrow pale spots over whole plumage (Starling: all plumages profusely pale-spotted, spots rather broad except on chin and throat); (ii) on average has less metallic gloss, which is more uniformly-shaded greenish-purple all over (Starling: gloss strong and clearly divided into green or blue and purple); (iii) much longer and narrower hackles (elongated feathers on throat and upper breast); (iv) when breeding often brighter and paler pink or pinkish-red legs (Starling: dirty pink or pinkish-brown), but some are darker-legged and do not differ from Starling; (v) on average very subtly shorter, thicker-based, more down-curved, 'mean-looking' bill (Starling: often slightly thinner-based, straight bill, but considerable overlap).

♂, Spain, May: ad ♂ is entirely unspotted year-round, unlike Common Starling (though some Common become largely unspotted with wear in summer, especially in east). Also unlike Common Starling, uniformly black plumage with impression of oily purplish-green gloss that is more evenly distributed, not clearly separated in mixed green and purple parts, and lacks any blue gloss found in Common Starling. (R. Tidman)

♂, Spain, May: elongated tips to much longer and narrower throat-feathers (hackles) are typical, and these even reach lower breast, while bill is shorter, thicker-based and sometimes more down-curved than Common Starling. Legs average brighter and paler pink. Both this and left-hand bird sexed by bluish-grey base of bill and very narrow and extremely long hackles. (C. N. G. Bocos)

♀ (in front) and ♂, France, Apr: unlike ♂, ♀ is often spotted when still fresh (autumn to early spring) with fine pale tips, most visible on mantle, scapulars and rear underparts; hence can be confused with ♂ Common Starling, but note more even purplish or greenish gloss (depending on light and angle) on whole head and body. (L. Petersson)

AGEING & SEXING Ageing straightforward in summer, but after post-juv moult requires close views and attention to subtle feather details. Very slight seasonal differences caused by wear. Sexes (after post-juv moult) differ slightly. Juv differs markedly. — Moults. Complete moult of both ad and juv in late summer (mid Jun–Oct), though moult of juv somewhat later and more protracted (Jun–early Nov). — SPRING **Ad ♂** Entirely unspotted (all year). Plumage glossy. Elongated tips to throat-feathers very narrow and extremely long, 28–41 mm. Base of lower mandible bluish-black. Iris uniformly dark. **Ad ♀** On the whole unspotted, though fine pale tips visible on mantle and underparts when bird seen close. Plumage has some gloss. Elongated tips to throat-feathers slightly broader than and not as long as in ♂, 21–27 mm. Base of lower mandible pale pinkish-brown or whitish. Iris bicoloured. **1stS ♂♀** As ad, and often inseparable, but averages slightly duller, and long throat-feathers slightly shorter, 1stS ♂ 24–28 mm, 1stS ♀ 17–23 mm. — SUMMER **Ad** Plumage dark with metallic gloss. Bill yellow. **Juv** No metallic gloss; largely uniform dull grey-brown with paler throat and pale, greyish- or buffish-brown edges to wing- and tail-feathers. Unspotted. Bill dark. — AUTUMN **Ad ♂** Much as in spring, though gloss less obvious. Long throat-feathers as in spring. Tips of tail-feathers tend to be pointed (see Common Starling for details). **Ad ♀** On the whole unspotted, though fine

♀, Spain, Jun: with strong feather wear, ♀ (especially 1stS) often becomes almost uniformly dusky-brown, with hardly any gloss, and some remnant pale feather-tips on undertail-coverts and vent. ♀ also has broader and shorter hackles than ♂. Base of lower mandible pale pinkish-brown to whitish, and iris bicoloured. (C. N. G. Bocos)

Spotless Starling (top left: Spain, May; bottom left: Spain, Dec) versus Common Starling (top right: ssp. *nobilior*, Azerbaijan, May; bottom right: ssp. *purpurascens*, Israel, Jan): due to variation in both species, Spotless (top left; 1stS ♀) and Common Starlings (top right; ♂) during late spring and summer can be very similar, especially as spots in Common can be completely lost (especially in eastern races). Consequently, the less metallic gloss of Spotless, which is more uniform greenish-purple, becomes important, whereas Common is more strongly glossed green or blue and purple on different body parts. Spotless has on average a very subtly shorter, thicker-based, more down-curved bill (but much overlap as can be seen here). The most heavily spotted Spotless (bottom left; 1stW ♀) requires careful separation from dullest and least glossed and spotted Common Starling in winter (bottom right; ad ♀). In combination, note duller and duskier blackish coloration of Spotless, homogenously and moderately glossed purplish-green (vs. brightly but patchily glossed green and purple in Common), narrower pale feather-tips (especially to undertail-coverts), and any bill differences. (Top left: C. N. G. Bocos; bottom left: M. Varesvuo; top right: M. Heiß; bottom right: H. Shirihai)

Ad ♂, Spain, Sep: after complete post-nuptial moult. ♂ is uniform black, so in fresh plumage (autumn to early spring) is automatically separable from Common Starling. Extremely long hackles on throat and typically 'oily' plumage. Bill becomes solid black or shows a minute yellowish tip, as here. All-dark iris confirms sex. (H. Shirihai)

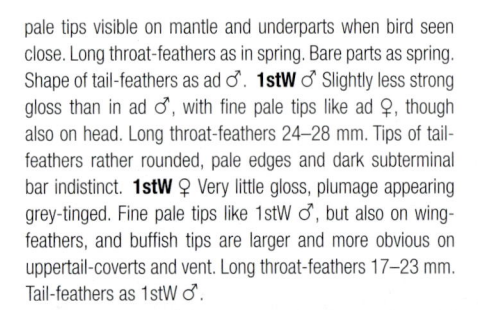

Presumed 1stW ♂, Spain, Dec: following complete post juv-moult, 1stW ♂ is obviously glossed, but typically shows very fine pale tips and greyish shaft-streaks like winter ad ♀. By midwinter hardly shows any pale fringes to wing-feathers, unlike Common Starling, which is furthermore profusely pale-spotted. However, definite ageing/sexing impossible without clear view of tail-tip. (M. Varesvuo)

pale tips visible on mantle and underparts when bird seen close. Long throat-feathers as in spring. Bare parts as spring. Shape of tail-feathers as ad ♂. **1stW ♂** Slightly less strong gloss than in ad ♂, with fine pale tips like ad ♀, though also on head. Long throat-feathers 24–28 mm. Tips of tail-feathers rather rounded, pale edges and dark subterminal bar indistinct. **1stW ♀** Very little gloss, plumage appearing grey-tinged. Fine pale tips like 1stW ♂, but also on wing-feathers, and buffish tips are larger and more obvious on uppertail-coverts and vent. Long throat-feathers 17–23 mm. Tail-feathers as 1stW ♂.

BIOMETRICS L 20–22.5 cm; **W** ♂ 129–140 mm (n 36, m 133.8); ♀ 123–136 mm (n 29, m 129.9); **T** ♂ 60–69 mm (n 36, m 64.0); ♀ 58–67.5 mm (n 29, m 62.7); **T/W** m 48.0; **B** 26.0–31.5 mm (n 65, m 28.7); **BD** 6.7–9.3 mm (n 64, m 7.9); **BW** 7.1–8.7 mm (n 65, m 7.8); **BW**(n dist) 5.8–7.8 mm (n 65, m 6.7); **Ts** 27.1–32.1 mm (n 65, m 29.6). **Wing formula: p1** minute; **pp2–3** longest (sometimes p2 0.5–1.5 mm <); **p4** < wt 2.5–7.5 mm; **p5** < wt 9–17 mm; **p6** < wt 16–31 mm; **p7** < wt 24–32.5 mm; **p10** < wt 40–49 mm; **s1** < wt 43–52 mm. Emarg. outer webs of pp3–4.

GEOGRAPHICAL VARIATION & RANGE Monotypic. (Birds of N Africa have on average subtly longer legs, but too tiny a difference to warrant separation.) – Iberia, extreme S mainland France, S & E Corsica, Sardinia, Sicily, NW Africa; resident.

REFERENCES HIRALDO, F. & HERRERA, C. M. (1974) *Doñana Acta Vertebrata*, 1: 149–170.

1stW ♀, Spain, Dec: the dullest plumage (juv excluded), with very little gloss. Note how dense spotting forms fine pale grey-buff streaking on crown, head-sides and upper mantle. Also, clearer pale edges to wing-feathers, and more prominent buffish tips to uppertail-coverts and vent. Such birds invite confusion with the least spotted Common Starlings. (M. Varesvuo)

Juv, Spain, Jun: often inseparable from juv Common Starling, except perhaps by stubbier bill. (M. Schäf)

STARLINGS AND MYNAS

PIED MYNA
Gracupica contra (L., 1758)

Alternative name: Asian Pied Starling

Fr. – Martin pie; Ger. – Elsterstar
Sp. – Estornino pío; Swe. – Svartvit stare

An open-country bird of damp grasslands, lightly wooded and cultivated lowlands, native to S and SE Asia but since the late 1980s has also become established in parts of the United Arab Emirates and perhaps also in at least one city in Saudi Arabia, presumably in all cases due to introduction by man. Like all starlings, omnivorous and feeds on ground, often seen in pairs or small groups in spring, and frequently forms larger flocks outside breeding season.

IDENTIFICATION Clearly more *a typical starling in structure and behaviour* (compared to other mynas), and within the covered region unmistakable. Medium-sized. Adult has crown, upperparts, throat, upper breast and tail black, contrasting with *clean white ear-coverts patch and rump*. Rest of *underparts slightly dirty greyish-white*. White edges to scapulars and inner coverts form *diagonal wing-bars*, which appear as two tramlines at edge of the mantle in flight. Adding to its characteristic head pattern, there is a *grey-white loral patch* and *bright orange skin around the eye*. Base of bill bright orange-red. Iris orange-buff and legs pinkish. Sexes alike; ♂♂ average blacker and glossier, but complete overlap with some ♀♀. Juvenile is a duller version of adult. Slow and rather butterfly-like flight, but mainly terrestrial in habits like other starlings, feeding on ground, and usually appearing rather sociable, especially outside the breeding season.

VOCALISATIONS Song is prolonged, being sometimes

G. c. contra, India, spring: with bill open to cool down in hot weather. Note contrasting white ear-coverts patch narrowly extending backwards at upper edge, conspicuous white rump, orange orbital ring and orange-red bill base, while tip of bill and loral patch are whitish. Vast overlap in plumage pattern makes sexing generally impossible. (D. Robinson)

G. c. contra, India, Feb: bold white diagonal wing-bar and white cheeks, and bright orange-red and yellow on eye and bill make identification easy, even in flight. (D. Tipling)

G. c. contra, presumed ad ♀ (left) and ♂, India, Sep: apparently near end of complete post-nuptial moult, which occurs earlier than moult in young; bright orange bare parts at this season usually indicate ad. ♀ has slightly paler blackish upperparts, including crown. Still, ages and sexes usually very similar and inseparable. (C. N. G. Bocos)

likened to that of a Skylark and by others to a myna, reasonably varied, comprising high-pitched warbles, wheezing and chuckling notes, and long whistles, as well as some mimicry. – Calls generally myna-like, a descending *treek-treek-treek*, and disyllabic *cheek-cheurk*. Flocks also give high-pitched musical notes. Alarm a harsh series, *shree-shree-…*

SIMILAR SPECIES This starling-sized bird with black-and-white plumage should prove unmistakable, especially given its limited range in W Palearctic.

AGEING & SEXING Ageing impossible after post-juv moult. Sexes alike, though ♂ averages very slightly larger and really black and glossy plumage seems commoner among ♂♂; still, total overlap in characters between sexes. – Moults. Both ad and juv have complete moult in late summer/early autumn. Apparently no pre-nuptial moult in late winter. — **ALL YEAR Ad** Bare parts vary in colour and slightly in extent, especially orbital skin (orange to red). Especially when very fresh, feathers of forehead pale forming narrow band to nostrils (sullied white), crown (glossy dark green and purple) hackled, and some birds appear slightly streaked on neck and nape (white shafts and greyish webs), while feathers on hindcrown and nape show some iridescence, which may extend over sides of neck to throat. **1stY** Probably indistinguishable unless some juv wing- and tail-feathers retained. **Juv** Considerably duller but same general pattern, with dark parts drab brown and no iridescence. Lacks white on forehead, has dirty white ear-coverts browner at lower and rear edges, and buffish-white underparts smudged or streaked brown. Pale scapular lines narrow and ill-defined. Bare parts duller.

BIOMETRICS (*contra*) **L** 20–23 cm; **W** ♂ 112–125 mm (*n* 18, *m* 118.6), ♀ 111–122 mm (*n* 14, *m* 116.9); **T** ♂ 65–72 mm (*n* 18, *m* 68.1), ♀ 62–73 mm (*n* 14, *m* 68.2); **T/W** *m* 57.1; **B** 29.0–35.5 mm (*n* 48, *m* 31.7); **BD** 7.1–9.0 mm (*n* 43, *m* 8.1); **Ts** 29.7–35.3 mm (*n* 50, *m* 33.2). **Wing formula: p1** minute; **p2** < wt 0.5–5.5 mm, =5 or 5/6 (77%), =4 or 4/5 (14%) or =6 (9%); **pp3–4**(5) about equal and longest; **p5** < wt 0–4.5 mm; **p6** < wt 4–8 mm; **p7** < wt 9.5–15 mm; **p8** < wt 13.5–20.5 mm; **p10** < wt 20–30 mm; **s1** < wt 24–32 mm. Emarg. pp3–6.

GEOGRAPHICAL VARIATION & RANGE Moderate variation, mainly involving shades of metallic gloss, but also presence or not of pale supercilium above base of bill on forehead. Several extralimital subspecies in SE Asia, not treated since they are apparently not involved in known introductions in Middle East.

G. c. contra (L., 1758) (N & C India, Bangladesh, W Assam, introduced Middle East; mainly resident). Treated above. Uniform dark crown. Narrow pale forehead band reaches nostrils.

G. c. contra, India, Mar: ageing and sexing in spring as a rule impossible, due to much overlap and variation. Still, the matt brownish-black back and wings make it fair to speculate that this might be a ♀. (S. Sharma)

G. c. contra, juv, India, Jul: general pattern reminiscent of ad, but dark areas browner, throat and upper breast streaked white and rest of underparts marbled brown, and bare parts darker and duller. (S. K. Sen)

ROSE-COLOURED STARLING
Pastor roseus (L., 1758)

Alternative name: Rosy Starling

Fr. – Étourneau roselin; Ger. – Rosenstar
Sp. – Estornino rosado; Swe. – Rosenstare

Nomadic and of characteristically erratic occurrence, in response to occurrence of swarms of locusts and other insects, a huge flock of these beautiful starlings is surely one of the most charismatic sights of the steppes and other open areas of West and Central Asia. Rose-coloured Starling breeds from west of the Black Sea (at least in some years) to the Ukraine, on the Kirghiz Steppe and in SW Siberia, as well as through Turkey and N Iran to W China. It winters primarily in India and SE Arabia, but wanders widely and has overwintered in W Europe, where it is a regular vagrant, often joining flocks of Common Starlings.

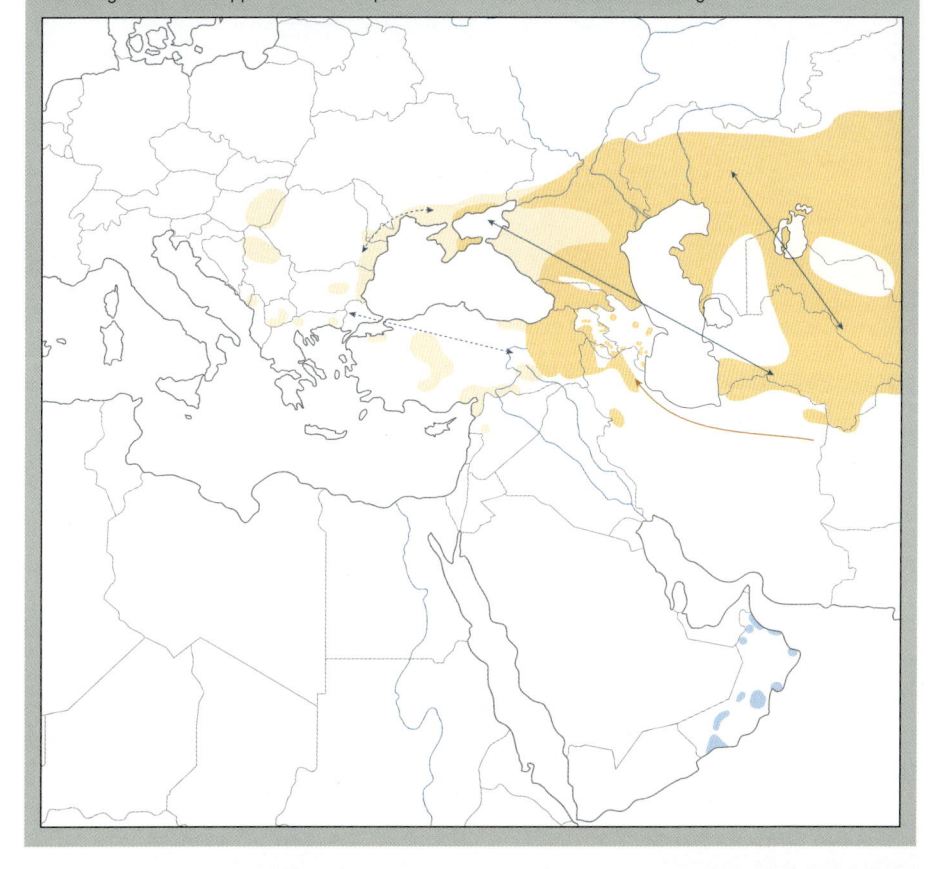

Ad ♂, Kazakhstan, May: pink mantle, scapulars and much of underparts, contrasting with shiny black head, wings and tail, and shaggy crest often erected when breeding. Neatness of plumage and bicoloured bill, plus length of crest-feathers, suggest this ♂ is ad. (J. Normaja)

IDENTIFICATION Adult unmistakable given *pink mantle, scapulars and underparts* (in worn plumage forming striking pink-jacketed appearance) *contrasting with black head, wings and tail*, and has long and drooping shaggy crest. Plumage highly contrasting in flight, when *pink rump* also striking. Bill rather short with a curved culmen, mostly yellowish-pink with a black base. Iris black and tarsus pinkish to pale orange-pink. Adult ♂ usually very neat with pure pink parts contrasting with green-glossed black areas and long crest. The shaggy crest is very evident in display, but in other situations the bird can appear crestless. In general, ♀/non-breeding adult duller, with a shorter crest and browner-toned pink upperparts with in particular scapulars tinged grey-brown, though adult ♂ usually recognised by all-black (or nearly so) undertail-coverts and longer crest-feathers (adult ♀ broadly buff-fringed dark undertail-coverts and shorter crest). Juvenile in early autumn *generally dull*

Ad ♂, Mongolia, Jun: two images of same bird showing typical features, including all-dark underwing (coverts can appear almost blackish-brown). Both pink and black tones can change depending on angle of view and light. (H. Shirihai)

sandy-brown, with darker flight-feathers and tail, pale bars and fringes to wing-feathers, pale rump patch, often some mottling on breast, whitish throat, yellowish lower mandible and pale eye-ring, making it easily separated from juv Common Starling. Compared to latter in flight, wings less pointed and triangular, slightly more rounded and appear to beat more loosely. Tail seems to be roughly the same length but have more rounded corners, at least in some. Flocks also appear to sweep, circle and land more slowly than Common Starling.

VOCALISATIONS Song a jingling, noisy chorus of knocking, squeaky and 'silvery' sounds, but unlike Common Starling does not mimic; song often from large colony a loud chorus of squeaky, knocking and whistling notes, which can recall noisy young Common Starlings. ♀ sometimes utters a loud *zilij-zilij-zilij* in later stages of courtship. — Calls include a frequent, harsh *tschirr*. A raucous *kritsch* noted in alarm (rather like Common Starling). Feeding flocks produce a rapid, high-pitched chatter.

SIMILAR SPECIES Given reasonable views, even the dull juvenile plumage should prove unproblematic, as distinctions from similar-age *Common Starling* are large and obvious, notably the pale rump, the in general paler brown ground colour, the hinted wing-bars or pale edges to many upperwing-feathers, and the shortish bill being straw-coloured rather than dark. — For separation from *Wattled Starling* see that species. — Unexpected pitfall, at least at quick glance, is with immature ♂ *Black-throated Thrush* versus moulting 1stY Rose-coloured Starling with darker and mottled head and bib. Once shorter tail, different more curved shape of bill, pale-fringed upperwing and pale rump are seen this mistake should be readily avoided.

AGEING & SEXING Ageing unproblematic in early summer but difficult following complete post-juv moult, after which both age and sex need to be considered based on differences in plumage and bare parts. Sexes slightly dimorphic and generally separable after post-juv moult. — Moults. Both ad and juv undertake complete moult in late summer and autumn (much variation depending on breeding time), often mid Aug–Nov (Dec). No pre-nuptial moult. — **WINTER–SPRING Ad ♂** When still very fresh duller pink due to brownish-buff fringes, with greyish fringes to black of head, nape and breast; from Mar wear reduces nearly all fringes, and bird acquires immaculate glossy black and rosy-pink plumage. Undertail-coverts and tibia feathering ('trousers') at first thinly pale-fringed, when worn in late spring becoming nearly or virtually all black. Shaggy crest on hindcrown long and extensive, longest feathers 27–45 mm. General size averages larger than ♀, hence useful to compare with other birds or take wing- and tail-lengths of any handled bird to support plumage characters. **Ad ♀** Like ad ♂ but usually duller black, and pink parts often have more brownish-buff

Ad ♀, Turkey, May: note combination of no indication of a crest (or at the most a very short one), pink parts duller and less pure, and undertail-coverts and 'trousers' broadly pale-fringed. (D. Occhiato)

Mongolia, Jun: like most starlings, commonly seen in large, dense flocks. Immediately identified by its bicoloured plumage, and the only remotely similar species (if views distant and brief) is Golden Oriole, but even if pink rather than yellow colour cannot be appreciated, dense flock formation and direct flight of the starling should still readily differentiate them. (H. Shirihai)

Ad ♂ (right) and 1stS ♀, Azerbaijan: ad ♂ has pink mantle, scapulars and underparts, and some blue sheen on wing-coverts. In contrast, 1stS ♀ usually has all of the pale areas sullied brownish-grey, but still will appear clearly bicoloured with pinkish rump in most lights and views. (M. Heiß)

fringes when very fresh (in some, fringes remain until at least May). Undertail-coverts and tibia feathering ('trousers') when fresh broadly pale-fringed, remaining so through spring. Still, a very few difficult to separate on plumage colours alone. Elongated crown-feathers 17–33 mm (mostly 18–22 mm). See ♂ for size difference. Rear neck or nape feathers usually basally purplish-black with much metallic gloss (broadly tipped brown). Still, a few are very brown-looking and difficult to separate from 1y ♀ in the field. **1stS ♂** Impossible to age reliably unless in wing moult with browner and pale-edged juv wing-feathers still present, contrasting strongly against much blacker new remiges. Yet, many have plumage generally somewhat duller and duskier than ad ♂, usually more like a dull ad ♀, and some show other vestiges of immaturity, e.g. some white basal feathering to throat, and partial juv bill coloration (not all pink but with variably large dark smudge at tip). Elongated crown-feathers 24–34 mm. Rear neck or nape feathers rather dull dark brown with limited metallic gloss. Rarely odd outer juv secondaries retained being brown and slightly worn. **1stS ♀** Plumage variable but usually recognised on subdued pink, pale parts instead being sullied greyish buff-brown, and on dark parts being dark brown rather than dark grey or blackish. Bill has dark

Presumed 1stS ♂, Kazakhstan, May: both ad and juv undertake complete moult, thus moult is of no use for ageing, and sexing is also difficult, especially the separation of ad ♀ from 1stS ♂. Here, combination of vestiges of immaturity (some brownish on pink parts, especially on scapulars), the shorter crest and some dusky elements near bill-tip could suggest 1stS ♂, still not very different from some ad ♀♀. (P. Dubois)

1stS ♀, Turkey, May: the dullest plumage, with pink areas sullied brown, and dark areas of head, neck, throat and breast blackish-brown, while pink of mantle is dusky-looking due to much brown-grey admixed. Note duller and less pure pink area to bill. (R. Armada)

Ad, presumed ♀, Kazakhstan, Sep: during early autumn in early stages of complete post-nuptial moult (here inner primaries are growing), ad is bleached and worn, and both sexes appear more like breeding ad ♀ (with shorter or no crest, dirty pinkish upperparts and browner-toned underparts, plus less neat pattern of bill). Here, some pale tips to dark ventral area suggest ad ♀. (R. Pop)

1stW, Oman, Dec: during complete post-juv moult, with browner and pale-edged juv wing-feathers, contrasting strongly against much blacker new remiges. Head, neck and bib also have newer black feathers, but mostly tipped paler and mixed with brown juv feathers. Prolonged transition produces wide variety of patterns. Note still more juv-like yellowish bill. (H. & J. Eriksen)

smudge at tip. Elongated crown-feathers 11–19 mm. Rarely odd outer juv secondaries retained being brown and slightly worn. – **SUMMER–AUTUMN Ad** Largely as spring but when heavily worn (prior to autumn migration), in Jul–Sep, rarely until Oct, gloss to black parts duller, pink parts diluted, some brown basal feathering exposed and overall plumage rather untidy; bill brownish-pink. Sexing usually possible by all-black undertail-coverts and 'trousers' (any remaining pale fringes now insignificant), and long crest in ♂, still broadly pale-fringed undertail-coverts and 'trousers', and shorter crest in ♀. **1stW** This plumage generally lacking, as juv moults directly to ad plumage (usually in winter quarters). However, in autumn to early winter there is a quite prolonged transitional period (often due to moult suspension), when birds wear mixed ad and juv feathers, with fewer pure pink feathers, which are tipped extensively brownish (especially in ♀♀). Such birds resemble pale and worn autumn ad ♀♀, but

Juv, Germany, Oct: usually dull greyish sandy-brown with slightly darker brown wings, yellowish-tinged bill, slightly paler rump and hint of pale wing-bars. Here seen with two 1stW Common Starlings and a flock of Dunlins *Calidris alpina*. (T. Langenberg)

Juv, Netherlands, Sep: juv can in certain lights appear overall colder and greyer. Note individually variable mottling on breast, whitish throat, pale loral area, bill with yellowish lower mandible and cutting edges, and with darker tip and curved culmen, which all separate this plumage from juv Common Starling. (A. Ouwerkerk)

are often duller still, fringed broadly fawn-brown on head and breast, pink hardly apparent on mantle, with buffish fringes to coverts and tertials, and underparts pinkish-buff. **Juv** Washed-out, mainly sandy-brown to buff grey-brown above with darker grey-brown tail and wings (the latter distinctly but narrowly pale-fringed forming hint of wing-bars). Underparts off-white with brown-streaked chest. Bill pale horn or yellowish-brown. Juv plumage usually moulted from Aug or Sep (sometimes starts even later) in winter quarters, but frequently all or most feathers retained until Nov by vagrants to W Europe.

BIOMETRICS L 20–22 cm; **W** ♂ 124–140 mm (*n* 37, *m* 131.8), ♀ 116–133 mm (*n* 44, *m* 126.8); **T** ♂ 63–74 mm (*n* 31, *m* 68.8), ♀ 61–72 mm (*n* 29, *m* 66.4); **T/W** *m* 52.2; **B** 21.5–25.9 mm (*n* 58, *m* 24.0); **B**(f) 17.1–21.0 mm (*n* 56, *m* 18.9); **BD** 6.7–8.6 mm (*n* 55, *m* 7.5); **Ts** 28.5–33.1 mm (*n* 60, *m* 30.7). **Wing formula: p1** minute; **p2** (pp2–3) longest; **p3** < wt 0–4.5 mm; **p4** < wt 4–11 mm; **p5** < wt 11–19 mm; **p6** < wt 18–26 mm; **p7** < wt 24–34 mm; **p10** < wt 40.5–49 mm; **s1** < wt 42–52 mm. Emarg. pp3–4 (on p4 somewhat less prominently).

GEOGRAPHICAL VARIATION & RANGE Monotypic. – SE Europe, Ukraine, Crimea, Caucasus, Turkey, Transcaucasia, Near East, Iran, much of Central Asia; partly migrates towards south and south-west, partly makes invasive mass movements in various directions.

TAXONOMIC NOTE For placement in monotypic genus *Pastor* to avoid non-monophyletic Sturnidae, see Lovette *et al.* (2008) and Zuccon *et al.* (2008).

REFERENCES VAN DEN BERG, A. B. (1982) *DB*, 4: 136–139. – LOVETTE, I. J. *et al.* (2008) *Mol. Phyl. & Evol.*, 47: 251–260. – ZENATELLO, M. & KISS, J. B. (2005) *Ring. & Migr.*, 22: 163–166. – ZUCCON, D., PASQUET, E. & ERICSON, P. G. P. (2008) *Zoologica Scripta*, 37: 469–481. – ZENATELLO, M. & KISS, J. B. (2005) *Ring. & Migr.*, 22: 163–166.

Juv, England, Aug: in shape and flight mode similar to Common Starling (left) but immediately told by off-white and brown colours. Note largely yellow-based rather thick bill, brown-centred undertail-coverts and underwing-coverts, and brown-streaked breast. (P. Morris)

WATTLED STARLING
Creatophora cinerea (Meuschen, 1787)

Fr. – Étorneau caronculé; Ger. – Lappenstar
Sp. – Estornino caranculado; Swe. – Flikstare

A resident of open dry savannas in sub-Saharan Africa, where it nests in huge colonies, but in the treated region it is generally only a scarce and irregular visitor that has straggled to Oman and Yemen. Prefers areas with short grass, where it forages on ground, characteristically probing the ground with open bill. Follows cattle or wild grazing animals to take advantage of disturbed insects. Also lands on them to take ectoparasites. Highly nomadic and opportunistic following fluctuations in food supply.

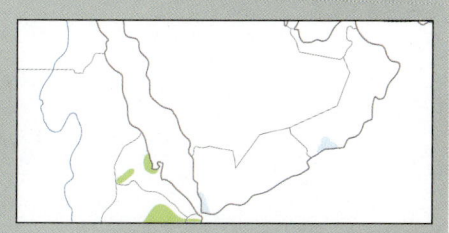

Ad ♂ breeding, South Africa, Sep: bald crown and large wattles develop seasonally in older ♂♂, with bright yellow occiput skin, black forecrown and long wattles. The long wattles dangle from the lower mandible, but their formation, and the various wattles on forecrown, above the eyes and upper mandible vary greatly. The bald patch and long wattles, together with the large and pure white primary-coverts patch, indicate an ad ♂. (V. & S. Ashby)

♂ breeding, Namibia, Nov: white primary-coverts patch suggests ad ♂. Only partly bald-headed with still rather short wattles further suggest a younger ad. Unique head pattern, and cream-grey body contrasting with blackish wings and tail make the species unmistakable. (H. Shirihai)

SIMILAR SPECIES Non-breeders are most likely to be encountered in the treated region, of which the dullest may require careful separation from immature *Rose-coloured Starling* as, especially in flight, they can appear superficially similar due to their contrasting dark wings and tail and whitish rump patch. However, Wattled Starling has a dark lateral throat-stripe, among other diagnostic facial marks, and bear in mind that Rose-coloured Starling is a considerable rarity in most areas of potential overlap.

AGEING & SEXING Juv readily separated from ad in summer. Sexes differ in breeding plumage, but (at least in Africa) neither breeding nor moulting behaviour is obviously seasonal. – Moults. In general complete post-nuptial and post-juv moults, but some juv have irregular primary moult with odd feathers retained in wing (NHM). – **BREEDING Ad ♂** White patch on leading edge of primary-coverts, visible in flight, only acquired in ad ♂. Some lose most feathers on head, chin and throat, developing variable pendulous wattles at bill-base, occasionally forming a large dewlap, with other wattles on crown. Completely bald crown and

IDENTIFICATION Distinctive starling, with older ♂♂ in full breeding plumage unmistakable having *cream-grey body plumage*, becoming whiter on inner scapulars, wing-coverts, underparts and rump, contrasting with *blackish rest of wings and tail*. A large part of the rear head becomes bare, with *occiput skin bright yellow*, while *rest of head is black with long, black wattles dangling from the throat, and two upstanding black wattles present above the eyes*. However, non-breeding and young ♂♂ and all other plumages are *grey to pale buffish-brown with blackish wings and tail*, with *barely contrasting white rump in flight*, some yellowish behind eye, black lores and lateral throat-stripe (the latter highly variable). Bill pale pink in breeding birds, otherwise tinged slightly brownish, eye dark brown and legs dark brown. A highly gregarious and somewhat flighty starling, whose rather slow flight recalls that of Rose-coloured Starling.

VOCALISATIONS Song unstructured, rather low-pitched metallic or liquid notes given in long strophes, usually rather quiet, being accompanied by tail movements in certain contexts. Sometimes sings in groups, and ♀ may solicit ♂ using high-pitched twittering notes. – In flight often noisy and can continuously utter mixed creaking, gurgling, rasping and clicking notes.

♀ breeding, South Africa, Oct: note lack of white primary-coverts patch and small bright yellow occiput skin behind eye, but blackish lores and lateral throat-stripes. Precise ageing impossible without locating moult limits in wing, as both age classes essentially moult completely. (H. Shirihai)

enlarged wattles occur only in older ♂♂, and are retained just 3–4 months, mostly in Apr–Dec. Younger ad ♂ (second breeding season) usually at most only partially bald and wattled. **Ad ♀** Like non-breeding ♂ but lacks white primary-coverts patch. Some older ♀♀ may develop slight wattles and even bald yellow areas on head, thus resembling imm/subad ♂♂. **1stY** Probably indistinguishable from ad (young ♂ especially close to ad ♀) unless has some retained juv wing- and tail-feathers. Some imm ♂♂ may already have slight or even substantial wattles, but they apparently shed rather few head-feathers. – **NON-BREEDING Ad** Both sexes resemble breeding ad ♀ (with no enlarged wattles or bald area on head), but ad ♂ aged and sexed by white primary-coverts patch, tends to have more yellowish-white around eye and better-developed blackish lores and lateral throat-stripes, and sometimes has vestigial wattles. In general, black areas of ad (especially when fresh) are glossed bronzy-green. **1stY** Often separable from non-breeding ad if moult incomplete, by presence of some retained juv feathers. **Juv** Differs from ♀ by overall browner body-feathers, with cream-coloured underparts and rump patch, highly restricted face marks with mostly dark lores and yellowish orbital ring. Remiges brownish.

Ad ♂ non-breeding, Kenya, Aug: apparently in full post-nuptial moult (outer primaries still unmoulted), when bald crown becomes feathered and wattles already lost. White primary-coverts patch confirms age and sex. (V. & S. Ashby)

1stY ♂, non-breeding, South Africa, May: at end of post juv-moult (outer primaries juv). Differs from non-breeding ad ♂ by reduced yellow occiput skin and white primary-coverts patch involves only central and innermost feathers (number increases with age, but most 1stY after first post-juv moult have two/three outermost coverts blackish). (P. Ryan)

1stY ♂, Ethiopia, Sep: during long transition from juv to 1stW, ♂ is like ♀, especially if primary-coverts not yet replaced, but still characteristic in having irregular but rather broad pale (grey and yellow) skin around and behind eye, blackish loral mark and lateral throat-stripes, and dirty base to lower mandible. Note species' diagnostic white rump on right-hand bird. (H. Shirihai)

1stY ♀, South Africa, spring: during prolonged transition from juv to 1stW, young ♀ is characteristic in having only tiny area or no exposed yellow skin around and behind eye, with indistinct or virtually no blackish loral mark and lateral throat-stripes, resulting in rather bland head pattern. (J. Hager & R. Harding)

Kenya, Jul: mostly young birds in post-juv moult. Note distinctive white rump patch. In the covered region, this starling is most likely to be encountered in such dull plumage, which may require careful separation from young Rose-coloured Starling. (J. Jourdan)

BIOMETRICS L 20–22 cm; **W** ♂ 118–127 mm (n 15, m 122.4), ♀ 114–123.5 mm (n 15, m 119.3); **T** ♂ 62–70 mm (n 15, m 66.3), ♀ 60–69 mm (n 15, m 65.2); **T/W** m 54.4; **B** 22.0–27.1 mm (n 30, m 24.9); **B**(n) 12.3–15.8 mm (n 30, m 14.5); **BD** 6.2–8.4 mm (n 28, m 7.3); **Ts** 28.1–32.0 mm (n 29, m 29.6). **Wing formula: p1** minute; **pp2–3**(4) about equal and longest; **p4** < wt 0–3 mm; **p5** < wt 2–8 mm; **p6** < wt 8–14 mm; **p7** < wt 12–21 mm; **p10** < wt 28–37 mm; **s1** < wt 30–39 mm. Emarg. pp3–5.

GEOGRAPHICAL VARIATION & RANGE Monotypic. – SW Arabia, S Sudan, Somalia, Eritrea, Ethiopia and further south in E & S Africa; sedentary. (Syn. *Gracula carunculata*.)

STARLINGS AND MYNAS

BRAHMINY STARLING
Sturnia pagodarum (J. F. Gmelin, 1789)

Fr. – Étourneau des pagodes; Ger. – Pagodenstar
Sp. – Estornino de las pagodas; Swe. – Pagodstare

Breeds in the Indian subcontinent from the Himalayan foothills to Sri Lanka; also found in Pakistan and E Afghanistan, straggling to Iran. Mostly resident and partial short-range migrant, usually encountered in dry and open lowland country, not shunning towns, villages and cultivation with groves and scattered trees; in some areas rather local and patchily distributed. Any passage or dispersal takes place late September to early March. Has established a (probably feral) breeding population in United Arab Emirates; a fairly regular visitor to Oman.

♂, United Arab Emirates, Jan: thick-necked impression due to dense plumes, while black crown, contrasting pure ash-grey rest of upperparts and cinnamon or rufous-orange underparts (with paler shaft-streaks) make this starling very distinctive. Sexed by long, spiky, black crest with very long blue-black hackles (reaching mid-back); also bluer bill-base, brighter in breeding season. As both age classes moult completely, ageing is impossible. (M. Barth)

Presumed ♂, non-breeding, India, Aug: in excitement, hairy crest can be puffed out. Bold white corners to tail, but flight-feathers blackish. Also characteristic are bright white iris and small pale orbital skin behind eye. Latter, yellow bill with slate-blue base and yellow tarsus are all duller when not breeding. Apparently ad at end of moult cycle. (P. Maheria)

IDENTIFICATION A medium-sized starling (size as Common Starling or averages a bit smaller), appearing thick-necked, stout-billed and with rounded wings, giving *myna-like jizz*. Crown black extending down to nape with the feathers elongated into a long, lax, hairy crest (often raised in excitement), with rest of head, neck and underparts cinnamon rufous-orange (with paler shaft-streaking). Upperparts pale drab with ashy-grey tinge contrasting with brownish-black primaries. Secondaries brown, broadly edged pale drab-grey above, uniform with upperparts. In flight, spread *tail shows bold white corners*, but primaries lack white patch as in mynas, and underwing-coverts are contrastingly pale. *Iris bright white-blue* and has *slate-blue orbital skin patch behind eye*. Yellow bill has slate-blue base, and tarsi are yellow. Feeds in flowering and fruiting trees; also feeds on ground, especially among grazing animals.

VOCALISATIONS Song described to be rather short, comprising a gurgling drawn-out cry followed by a louder, more vehement bubbling yodel, rendered as *gu-u-weerh-kwurti-kwee-ah*. – Among calls a rapid series noted of harsh parakeet-like shrieks that may begin with a nasal note. In alarm gives series of short grating *churrs*.

SIMILAR SPECIES Easily distinguished from any native or similarly introduced starlings, including from the superficially similar Rose-coloured Starling, by the absence of a black bib.

AGEING & SEXING As in all starlings, ageing impossible after post-juv moult is completed in late summer (thus juv readily separated in summer). Some seasonal plumage differences occur due to wear. After post-juv moult, sexes differ. – Moults. Complete post-nuptial moult in Oct–Nov. Post-juv moult at same time (or slightly earlier) is complete, and 1stW birds are thus indistinguishable from ad. – **Ad** Sexes rather similar but ♂ usually shows longer, spikier, blacker crest, longer neck-hackles, and paler/purer grey mantle than ♀ (in extreme cases the ♂'s neck-hackles extend halfway down mantle). – **Juv** In general duller above, crown-feathers duller, browner and shorter (lacking crest), underparts duller orange-buff and plainer (lacking white shaft-streaking of ad), grey upperparts tinged buffish and wing-coverts and tertials and secondaries fringed buff. Further, bill is duller, and it has an olive-coloured patch behind eye. – **1stY** Can only be separated as long as some characteristic juv feathers are retained.

BIOMETRICS L 18–20 cm; **W** ♂ 102.5–112 mm (*n* 12, *m* 106.9), ♀ 98–108.5 mm (*n* 12, *m* 101.9); **T** ♂ 60–70 mm (*n* 12, *m* 66.0), ♀ 59–68 mm (*n* 12, *m* 62.7); **T/W** *m* 61.7; **B** 18.0–23.5 mm (*n* 26, *m* 20.6); **B**(f) 13.5–18.5 mm (*n* 26, *m* 15.8); **BD** 5.2–6.8 mm (*n* 25, *m* 6.0); **Ts** 26.0–30.3 mm (*n* 26, *m* 28.4). Wing formula: p1 minute; **p2** < wt 0–3 mm; **pp**(2)**3–4** longest; **p5** < wt 1–5 mm; **p6** < wt 6–10 mm; **p7** < wt 10.5–15.5 mm; **p10** < wt 20–27 mm; **s1** < wt 21.5–31.5 mm. Emarg. pp3–5 (rarely faintly also on p6).

GEOGRAPHICAL VARIATION & RANGE Monotypic. – NE Afghanistan, N Pakistan, S Nepal, India, Bangladesh, Sri Lanka; resident or short-range movements to India, occasional in SE Iran; vagrant N Oman, feral population in United Arab Emirates. (Syn. *Temenuchus pagodarum*.)

♀, C India, early spring: note that hackles on neck of ♀ are rather short, while black feathers of hindcrown are not elongated to expand over mantle. Also, grey of upperparts subtly tinged sandy-brown. (V. Nair)

Juv, India, Sep: rather dull brown above, including shorter crown-feathers without crest. Grey-tinged upperparts pinkish-buff, wing-coverts and tertials fringed buff, and underparts dull orange-buff and plainer (shaft-streaks obscure or lacking). Bill dull, and olive-coloured patch behind eye. Such plumage, or transition towards ad, can persist for some months in autumn and winter. (P. J. Saikia)

COMMON MYNA
Acridotheres tristis (L., 1766)

Fr. – Martin triste; Ger. – Hirtenmaina
Sp. – Miná común; Swe. – Brun majna

The Common Myna frequents cultivation and gardens in villages, towns, wooded areas and even deserts, being generally a rather common and widespread resident. It is mostly found in lowlands and occurs naturally from SE Iran and Transcaspia east to SE Asia, but recent scattered records from many Middle Eastern countries are all assumed to emanate from escapes or are in some way man-assisted. Very characteristic when seen in flight between trees, the rounded wings displaying large white patches.

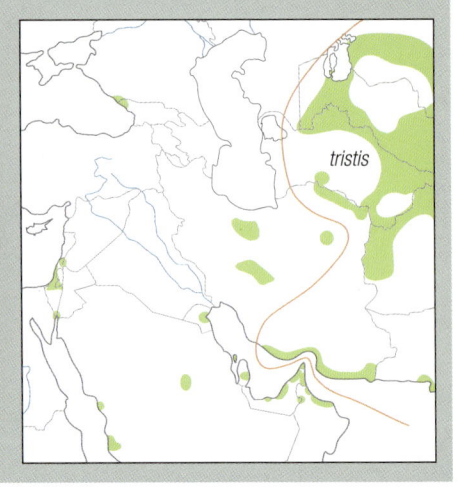

VOCALISATIONS Very vocal year-round. Pairs may duet and perform quite elaborate vocal displays. Song repetitive with much mimicry, one variation can be described as *chilp chilp chilp chirk-a chirk-a chirk-a, chwee chewee*, but many different notes and combinations can be given, and ecstatic singing involves a continuous stream of varied sounds. Capable of mimicking a wide range of birds, and human voices. – *Chip* calls given during aggressive encounters and a scolding *chake chake*. A rough *traaah* in alarm (often serving to alert other birds to potential danger).

SIMILAR SPECIES Only likely to be confused with Bank Myna. Their separation is discussed under that species.

AGEING & SEXING Ageing impossible after post-juv moult. Sexes normally inseparable in the field (but see Biometrics); tempting to believe that neatest-looking birds are only ♂♂, but some are ♀♀, too; overlap in characters seems complete. – Moults. Both post-nuptial and post-juv moults complete, generally late Aug–early Dec, but timing

yellow. Flight consists of noticeably flapping wingbeats, and it is a very noisy, jaunty bird as it walks restlessly on open ground, taking advantage of every possible food source. Typically roves in parties through palm groves.

A. t. tristis, Seychelles, Nov: bare skin behind and below eye, deep-based and slightly decurved bill and rather long bright yellow legs readily eliminate confusion with other starlings. Darkest tracts are sooty-black head, neck and bib, and palest are cream-buff undertail-coverts. Note white in tail and wings. Because both age classes undergo complete moult post-breeding, ageing is impossible. (H. Shirihai)

A. t. tristis, Oman, Nov: characteristically very noisy when encountered in small groups. Note white-tipped part of tail broader on corners. (D. Occhiato)

IDENTIFICATION Rather plump and robust-bodied starling with deeper-based, short and slightly decurved bill, and relatively longer legs than other starlings in W Palearctic. Generally glossy *black on crown, lores and nape*, whereas *throat, neck and upper breast are more matt sooty-grey* (can look blackish in certain lights), while *rest of body is deep vinaceous-brown*, interrupted only by *bare yellow skin behind and below eye* adjoining upper ear-coverts, the yellow patch creating a 'large-eyed' impression. The bristled forehead-feathers form a flexible frontal crest. Also has *striking white wing patch* formed by bases to primaries and entire primary-coverts, and *white tail corners* that become progressively broader towards outermost pair of feathers. In flight these conspicuous white patterns, and *the pure white underwing-coverts*, are obvious and provide useful identification aids. Undertail-coverts greyish-white, contrasting with buffier belly and vinaceous-brown breast and flanks. *Bill mainly yellow* (often connected to yellow wattle above ear-coverts by narrow yellow skin from gape). Iris reddish-brown and *legs*

A. t. tristis, Oman, Nov: during display, when pair seen side-by-side, ♂ often appears visibly larger-headed with stronger-glossed black areas, but there is much overlap between the sexes. Note the short, bristled feathers on forehead forming a slight frontal crest, but never as bulging as in Bank Myna. (M. Schäf)

STARLINGS AND MYNAS

A. t. tristis, Oman, Mar: striking white wing patch encompassing basal primaries and whole primary-coverts; from below the patch is larger, extended inwards onto underwing-coverts. The white-tipped tail becomes progressively broader towards the outermost pair of feathers. (M. Varesvuo)

A. t. tristis, Oman, Mar: flight is rather slow and flapping on rounded wings, giving it a typical jizz. Note white patterns in tail and wings. (M. Varesvuo)

Presumed *A. t. melanosternus*, Israel, Dec: although only *tristis* is definitely known to occur in the treated region, some birds are clearly darker, perhaps suggesting that the darker race *melanosternus* also occurs, a race which originates from Sri Lanka being overall darker and having narrower white tail-tip. (A. Ben Dov)

and strategy apparently adaptable, especially in introduced populations, and data from covered region very scant. – **ALL YEAR Ad** When fresh, black parts glossier, iris perhaps deeper reddish-brown. Some appear warmer chestnut on mantle and back. **1stY** Probably inseparable from ad unless (rarely) retains some juv feathers, perhaps even some remiges and tail-feathers, which are browner and less glossed, with less pure white tail-tip (sometimes with dusky tips and shafts). Also, primary-coverts tipped black, especially on inner webs (reduced or lacking in ad), and has less uniform secondary-coverts and tertials with some black bases visible. **Juv** Duller than ad with soft fluffy appearance to lower body, greyish brown-black head, paler brown throat and breast, and lacks crest or fully-coloured wattle. Also, unlike ad, lower back to uppertail-coverts buff-brown, faintly mottled darker, and longest tertials may have off-white fringe close to tip.

BIOMETRICS (*tristis*) **L** 23–25 cm; **W** ♂ 142–152 mm (*n* 12, *m* 145.2), ♀ 132.5–143 mm (*n* 12, *m* 137.7); **T** ♂ 83–91 mm (*n* 12, *m* 87.4), ♀ 79–87 mm (*n* 12, *m* 82.0); **T/W** *m* 59.9; **white tip of r6** (inner web) 13–22 mm (*m* 17.6); **B** 25.1–28.7 mm (*n* 35, *m* 27.0); **B(f)** 17.0–21.2 mm (*n* 24, *m* 19.2); **BD** 7.5–9.5 mm (*n* 22, *m* 8.2); **Ts** 34.3–40.0 mm (*n* 24, *m* 37.4). **Wing formula: p1** minute; **p2** < wt 0–4 mm; **pp(2)3–4**(5) about equal and longest; **p5** < wt 0–3 mm; **p6** < wt 5.5–11 mm; **p7** < wt 13–19 mm; **p10** < wt 29–36 mm; **s1** < wt 32–38 mm. Emarg. pp3–5 (rarely a hint also near tip of p6).

GEOGRAPHICAL VARIATION & RANGE Limited variation.

All birds occurring within the treated region, of introduced origin, are probably ssp. *tristis*. Only one other race, ssp. *melanosternus* (endemic to Sri Lanka), generally recognised, and those in W Kerala (S India) appear to represent intergrades between this and *tristis*. All populations resident.

A. t. tristis (L., 1766) (Transcaspia, S Kazakhstan, much of Central Asia south to SE Iran, India, east to S China and SE Asia; introduced e.g. in Middle East, Arabia, Africa, Australia). Treated above. Differs from the local Sri Lanka race in being overall somewhat paler and having broader white tail-tip.

293

BANK MYNA
Acridotheres ginginianus (Latham, 1790)

Fr. – Martin riverain; Ger. – Ufermaina
Sp. – Miná ribereño; Swe. – Brinkmajna

An inhabitant of open country in the N Indian subcontinent, from Pakistan east to Bangladesh. It is now also an established breeder in the United Arab Emirates, and has more recently bred in Kuwait and Oman, and has been recorded in Saudi Arabia, all as a result of escapes or other human influence, although Omani records may represent a natural expansion from the Emirates. Bank Mynas breed colonially in riverbanks (hence their name) and feed on all kinds of animals, fruit and scraps around habitations or near water or in wet meadows.

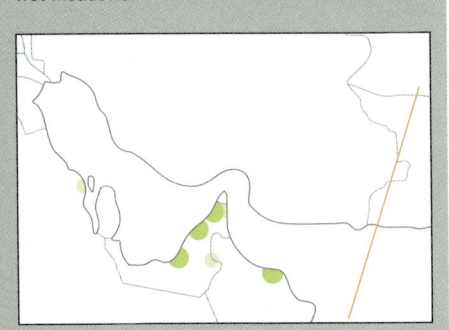

Bank Myna (right: ad) and Common Myna (left: in post-juv moult), Kuwait, Dec: unlike Common Myna, Bank has smoky-grey body, contrasting with black crown and head-sides, brick-red or orange-red orbital skin and better-developed bristled forehead, often creating a small tuft against the seemingly concave crown. Ageing and sexing impossible on current knowledge. (M. Pope)

IDENTIFICATION Medium-sized, plump-bodied starling, in many respects much like slightly larger Common Myna, with an *overall deep smoke-grey body*, which becomes slightly washed pinkish on centre of belly and vent, the rather uniform grey contrasting with *black crown and head-sides*, *brick-red orbital skin* around eye and small tuft of slightly erect, bristled forehead-feathers making the *crown appear peculiarly concave* in profile. *Primary patches and tail corners pale rufous-pink*. In flight, wing and tail pattern a useful identification pointer, with the base of the primaries white, becoming pale rufous-pink on the inner feathers and primary-coverts, and rufous-tipped tail-feathers (broadest on the outermost feathers), while the *underwing appears basically two-coloured* (mostly black with buffish-rufous coverts and white primary bases). Bill mostly orange-yellow, iris orange-red (adult) and legs orange-yellow. Young duller and less distinctive, very similar to juvenile Common Myna but note greyer upperparts, darker belly-centre and grey rather than blackish chin/upper throat. Gregarious year-round and has the typically jaunty gait and flight pattern of other mynas, and like Common Myna is noisy and habituated to man.

VOCALISATIONS Rather like Common Myna, albeit slightly softer, shriller and higher-pitched: song a typical myna suite of high-pitched whistles, low creaks and various warbling noises. – Utters a slightly harsh *wheek* in contact or flight, and a raucous alarm call.

SIMILAR SPECIES Main confusion risk throughout range is with *Common Myna*, as they share many structural and morphological similarities. However, Bank Myna is slightly smaller, with a dark-capped appearance (throat greyish like rest of underparts), rather than dark-hooded pattern of Common Myna (in which dark on head extends to throat and breast). Also essentially greyer overall (even bluish-grey in adults, not brown). Considerably smaller wing-patches and tail corners, which are rufous-tinged, not large and white like Common Myna, and underwing pattern is wholly distinctive. The two are further separated by the deeper and redder bare skin around the eye of Bank Myna. Juvenile perhaps less distinctly differentiated being overall duller, but is still identifiable using the same characteristics.

AGEING & SEXING Ageing impossible after post-juv moult. Sexes indistinguishable on plumage, but differ somewhat in size, ♂ being slightly larger. – Moults. Extent and timing of moult seem to differ little from Common Myna. Both ad and juv undertake protracted complete post-nuptial and post-juv moults in summer or early autumn (earlier on average in ad). No pre-nuptial moult known. – **ALL YEAR Ad** When very fresh, breast and flanks usually strongly washed cinnamon, and mantle suffused brown, otherwise as described under Identification. **1stY** Probably indistinguishable from ad unless retains some juv wing- and tail-feathers. **Juv** Considerably browner upperparts and head than ad, and underparts mostly paler buff, including underwing-coverts. Wings duller (with narrow pale fringes to rufous-buff greater coverts and tertials), and wing panel and tail edges tinged whitish-buff (instead of deep orange buff). Bare parts duller.

BIOMETRICS **L** 20–22.5 cm; **W** ♂ 121–132 mm (*n* 7, *m* 125.7), ♀ 116–125 mm (*n* 12, *m* 120.4); **T** ♂ 68–74 mm (*n* 7, *m* 71.1), ♀ 62–71 mm (*n* 12, *m* 66.4); **T/W** *m* 57.5; **rufous tip of r1** 0–7 mm (*m* 2.8); **B** 22.5–27.8 mm (*n* 25, *m* 24.8); **B**(f) 14.5–20.1 mm (*n* 25, *m* 18.1); **BD** 6.5–8.0 mm (*n* 25, *m* 7.3); **Ts** 32.5–37.9 mm (*n* 25, *m* 35.8). **Wing formula: p1** minute; **pp 2–4** about equal and longest; **p5** < wt 2.5–9 mm; **p6** < wt 9–16 mm; **p7** < wt 15–21 mm; **p10** < wt 26–35 mm; **s1** < wt 28–37 mm. Emarg. pp3–5 (rarely a hint also near tip of p6).

GEOGRAPHICAL VARIATION & RANGE Monotypic. – Pakistan, N India, Bangladesh; introduced or man-assisted range expansion to United Arab Emirates, Kuwait, Oman, Saudi Arabia; resident.

India, Apr: pale orange-buff (rather than pure white) tail corners and undertail-coverts differentiates Bank from Common Myna. Note how appearance of orange-red orbital skin and black forehead crest differ depending on angle of view. A markedly long-legged starling. (A. Deomurari)

STARLINGS AND MYNAS

India, Jun (left) and United Arab Emirates, Mar: compared to Common Myna, white primary patch considerably smaller and variably tinged pale rufous-pink. The underwing shows smaller white primary bases, with the lesser and median coverts tinged buffish-rufous, but greater coverts are mostly black, creating rather narrow pale forewing panel. (Left: S. N. H. Mhatre; right: M. Barth)

Kuwait, Jan: in non-breeding period, at least some ad have slightly paler and duller bare parts. Diagnostic pale orange-buff tail corners are well shown on this bird. (V. Legrand)

Juv, India, Aug: in summer, post-juv moult starts with head and body, but here wing- and tail-feathers seem still juv. Note deep red iris like ad, but other bare parts more yellowish than red, grey of plumage duller and tinged brownish on mantle, and head sullied greyer. Tail-tip still whitish instead of orange-buff as in ad. (A. Deomurari)

Juv Bank Myna (left: India, Aug) versus Common Myna (right: United Arab Emirates, Sep): identification of juv in areas where both species occur requires close views. Useful to note iris colour, which is reddish even in nestlings of Bank and bluish-tinged in Common. Also, bill is subtly thinner and longer and more orange-tinged in Bank, shorter, stubbier and yellower in Common. Unlike Common Myna, Bank has almost uniformly grey upper- and underparts (browner in most Common, but varies geographically) and lacks the extensive pale belly. In juv Bank dark of head encompasses crown and head-sides, but not throat and neck as in Common. Juv Bank has tail corners tinged whitish-buff (whiter than in ad Bank), but still not as pure white as in Common. (Left: N. Devasar; right: T. Pedersen).

AMETHYST STARLING
Cinnyricinclus leucogaster (Boddaert, 1783)

Alternative name: Violet-backed Starling

Fr. – Étourneau améthyste; Ger. – Amethystglanzstar
Sp. – Estornino amatista; Swe. – Ametiststare

Breeds in savannas and sparsely-wooded mountains over large parts of C and S Africa, with a limited occurrence also in SW Saudi Arabia (Hejaz and Asir) and W Yemen. Nests in tree hole, in Arabia in early spring. Often nomadic behaviour, small flocks moving around in search of fruiting trees. Leaves Arabia for Africa in non-breeding season. Vagrant to Israel, United Arab Emirates and Oman.

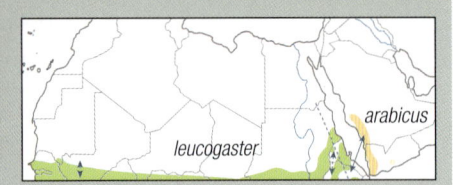

IDENTIFICATION A *small, rather compact starling* with short but strong dark bill. The adult ♂ is a strikingly beautiful bird with *glossy violet-blue upperparts and hood including upper breast*. The iridescent violet colour can become more bronze-pink at certain angles, or even look black. *Belly pure white*. The ♀, first-year ♂ and juvenile are plain and colourless, having grey-brown upperparts, and juveniles and immatures have pale rufous or buff feather-edges and darker centres, producing a streaky effect when still fresh, while *underparts are off-white and boldly streaked dark*. Bill and legs blackish, iris mainly dark brown. An indistinct very thin pale eye-ring visible only at close range. When not breeding forages in flocks, often in foliage or practise fly-catching in the air. Flight fast and direct. In flight sometimes shows some rufous basally on inner remiges, and this can, when extensive, create a rufous panel. Has a habit of flicking wings when alighting.

VOCALISATIONS The song is variable and rather unmusical with components of repeated chattering notes, *chet-chet-chet* or *clep-clep-clep*, at times with inserted dry trills, *cherrr*. – A variety of calls including rather sharp and penetrating whistles on different pitch, contemplative slow chatter, buzzing sounds and harsh quarrelling notes.

SIMILAR SPECIES Post-juv ♂ should be unmistakable, although extralimital Abbott's Starling (*Pholia femoralis*; Kenya and Tanzania; not treated) has same size and similar basic colours, although differs in being darker and bluer, less pink-tinged. The bluish colour reaches further on lower breast than in Amethyst Starling. Abbott's Starling also has a white eye-ring (in Amethyst only a very thin rim at most). ♀ Amethyst is paler than ♀ Abbott's Starling, especially on throat, and, again, lacks a whitish eye-ring. – In a poor and fleeting view the ♀ could possibly be confused with juvenile *Golden Oriole*, although this species in this plumage would hardly ever be encountered on the Arabian Peninsula. Golden Oriole is also considerably larger, more long-winged and elongated in shape, has at least some greenish tinges above and yellowish-tinged rump and tips to primary-coverts and outer tail-feathers, all lacking in the grey-and-white ♀ Amethyst.

AGEING & SEXING (*arabicus*) Ages similar and ageing not always possible, although juv generally readily separable from older generations. Sexes differ clearly from late 1stW on. – Moults. Complete post-nuptial moult of ad in late summer. Sequence of feather replacement unusual and

C. l. arabicus, ad ♂, **Saudi Arabia, Mar**: a compact and beautiful starling, especially ad ♂ with its glossy violet-blue upperparts and hood, sharply contrasting with silk-white underparts. Fore-face and throat can appear blackish at some angles. Evenly-feathered wing infers age. (J. Babbington)

C. l. arabicus, ♀, presumed 1stY, **Saudi Arabia, May**: despite lacking colourful iridescent plumage, ♀ still distinctive in being boldly streaked dark below, with blotchy dark markings. Apparent moult limits in wing suggest 1stY. (J. Babbington)

C. l. arabicus, ad ♂, **Saudi Arabia, Jun**: in certain lights the glossy violet-blue of ad ♂ looks paler and more reddish-purple. Note also how shape changes when neck is withdrawn to give a more compact look. (J. Babbington)

slightly variable, flight-feathers being moulted first and body (plus often tertials and some inner secondaries) later; some birds manage only flight-feathers before migration, whereas others moult completely while still in the breeding area. Partial post-juv moult (unusual for a starling!) at same time only involves head, body and many wing-coverts, and usually also whole tail. Seasonal variation very slight. – SPRING **Ad ♂** Glossy violet above and on throat and breast, belly pure white, violet colour becoming more purplish-pink with wear. **Ad ♀** Upperparts, greater coverts and tertials uniform grey-brown (only very slightly mottled darker on nape). Underparts off-white boldly streaked dark grey-brown on throat and breast, streaks gradually becoming broader and rounder on lower breast and flanks. **1stS** Upperparts as ad ♀ or slightly more blotched or diffusely streaked dark, but best separated by any retained juv coverts having rufous-buff or whitish edges before too worn. Streaks on underparts average slightly thinner, thus throat appears a little paler. Sexes hardly separable unless 1stS ♂ has attained the first violet glossy feathers (in Dec or Jan), or the stronger abrasion of primaries will reveal the younger age of a 1stS ♂. – AUTUMN **Ad** Very similar to spring plumage for both sexes, but on average ♂ darker purplish-blue, and ♀ has broader pale feather-edges above. **1stY** Doubtfully separable from non-breeding ad unless separated by stronger wear to retained juv feathers. **Juv** Differs from ♀ in prominently streaked crown and nape, and broadly rufous-edged upper wing-coverts and tertials. Remiges have broad rufous edges to inner webs.

C. l. arabicus, ♀, Saudi Arabia, Jul: in ♀ plumage differs from African *leucogaster* and *verreauxi* by its plainer grey-brown upperparts with very subdued pale feather-edgings only, and rounder and blotchier streaking below. Overall jizz, especially the short, strong dark bill, important in separation from superficially vaguely similar Song Thrush, or even young Golden Oriole. Age difficult to ascertain from this image. (J. Babbington)

C. l. verreauxi, 1stY ♂, South Africa, Nov: ad ♂ of this S African race differs from *arabicus* in having white base to outertail (not visible here). Seemingly duller purple upperparts, greyer-tinged underparts and unmoulted juv primaries indicate age. (H. Shirihai)

C. l. verreauxi, ♀, South Africa, Nov: in ♀ plumage this taxon (like *leucogaster* of Sahel belt) characteristically shows obvious pale feather edges above and narrow dark streaking below. Furthermore, the continental African races often show a rusty-tinged head in both ad ♀ and young. (H. Shirihai)

BIOMETRICS L 16.5–18 cm; **W** ♂ 103–111 mm (n 9, m 105.8), ♀ 98–105 mm (n 8, m 101.8); **T** ♂ 59–65 mm (n 10, m 62.1), ♀ 59–67 mm (n 8, m 63.9); **T/W** m 60.5; **B** 16.8–21.4 mm (n 18, m 18.4); **B**(n) 11.4–14.3 mm (n 18, m 12.6); **BD** 4.8–6.0 mm (n 17, m 5.3); **Ts** 19.8–23.0 mm (n 18, m 21.2). **Wing formula**: **p1** minute; **pp2–3**(4) about equal and longest; **p4** < wt 0–4 mm; **p5** < wt 4–8.5 mm; **p6** < wt 11–15.5 mm; **p7** < wt 15–21.5 mm; **p10** < wt 27–33 mm; **s1** < wt 29–34 mm. Emarg. pp3–5.

GEOGRAPHICAL VARIATION & RANGE Widespread in sub-Saharan Africa and S Arabia. Moderate variation with only three subspecies described, of which only one occurs within the covered range, the other two being extralimital.

C. l. arabicus Grant & Mackworth-Praed, 1942 (SW Arabia, SE Sudan, N & E Horn of Africa; migratory, apparently departs Arabian Peninsula post-breeding, wintering in adjacent Africa and further south). Treated above. Differs from ssp. *leucogaster* (Sahel from Senegal to SW Sudan and S & W Horn of Africa) mainly in that ♀ is plainer grey-brown above, not boldly streaked, and has fewer paler feather-edges in fresh plumage. ♀ typically has rounded black blotches or broad streaks below. – Birds in S Ethiopia sometimes separated as '*friedmanni*', but differences seem negligible. (Syn. *friedmanni*.)

Extralimital: *C. l. verreauxi* (Bocage, 1870) (S Congo, Angola, S Kenya south to South Africa). Very similar to *arabicus*, but has white bases to outer rectrices, and in ♀ plumage rather bold pale edges to upperparts feathering (at least when fresh).

SAXAUL SPARROW
Passer ammodendri Gould, 1872

Fr. – Moineau des Saxaouls; Ger. – Saxaulsperling
Sp. – Gorrión del saxaul; Swe. – Saxaulsparv

Within the covered region restricted to extreme NE Iran, at least as rare winter guest, but presumably also as a breeder. It is by and large restricted to Central Asia (Turkmenistan, Kazakhstan, Uzbekistan, Kyrgyzstan, Tajikistan, Mongolia and W China). Lives in sandy desert and semi-desert with thickets and Saxaul trees (Haloxylon), nests placed in tree holes or in the base of stick nests belonging to raptors. Breeds alone or in small colonies, and can form small flocks in non-breeding season. Predominantly resident, some also being dispersive or partial migrants.

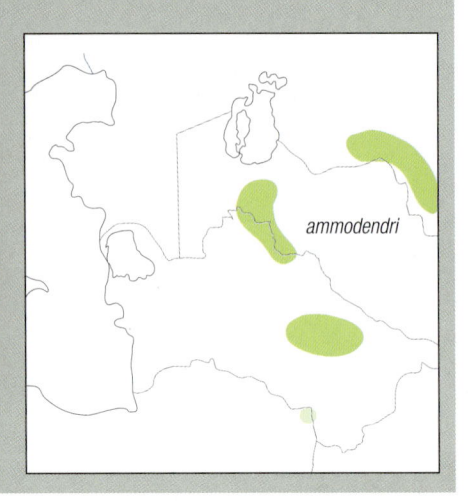

P. a. ammodendri, ad ♂, Kazakhstan, Jun: broad black crown-stripe, black mask with thin stripe behind eye and long black bib, plus conical black bill. Whitish fore supercilium, rufous-cinnamon rear head-sides, and conspicuous black-and-white 'shoulder' are further characteristics. Western populations have generally colder plumage, otherwise subspecific variation limited. Age and sex as for next image. (N. Bowman)

IDENTIFICATION A medium-sized sparrow with conspicuously coloured ♂, being basically *pale drab and white* with *strikingly beautiful head pattern, crown and upper mantle black, sides of rear head* (from eye back, and above thin black eye-stripe) *rufous-cinnamon*, rest of *face and head including supercilium whitish with black lores, eye-surround* and *long black bib* (reaching upper breast). Lower mantle brown-grey streaked black, back and rump same but more diffusely. Wings overall pale with *black centres to tertials broadly fringed white. Double white wing-bar.* Conical bill *black* in breeding ♂, pale horn-brown in winter ♂ and ♀. ♀ much plainer, lacking head pattern of ♂, except diffuse dark bib generally developed (can be very faint); apart from hint of dark bib, rather House Sparrow-like, though paler. White of underparts sullied buff-grey, while upperparts more sullied brown-grey. However, in close views some of the specific diagnostic features should be visible, which mirror the head and wing patterns of ♂. The wings are relatively short and the tail long and narrow; head typically square in profile, otherwise jizz very similar to House Sparrow.

VOCALISATIONS Squeaky song typical of sparrows, a rhythmic repetition of chirps, e.g. *tchip* or more drawn-out *chirrip*; sometimes higher-pitched *siep* or soft *swee* are also inserted. Song can at times be more complex and slow, sounding somewhat like that of Dead Sea Sparrow, at least from a distance. The song generally does not carry far. – Calls include e.g. a soft *cheg-cheg-cheg* (again reminiscent of some notes uttered by congeners), which can become louder and more tremulous in alarm.

SIMILAR SPECIES Given reasonable view of head and wing, ♂ easily separated from any other sparrow within the covered region. – ♀ can be confused with ♀ of either *Spanish* or *House Sparrows*, especially if seen alone, and if head and wing patterns are obscured, or if dealing with poorly-marked individuals. If any doubt exists, ♀ Saxaul Sparrow should show the following diagnostic combination: (i) bicoloured supercilium, white in front and above the eye, becoming wider, more diffuse, buffier (and often slightly rusty) behind the eye; and (ii) tricoloured wing pattern with cinnamon lesser coverts panel and black-and-white median coverts bar. – ♀ can also be confused with ♀ *Dead Sea Sparrow*, here compared with local Central Asian race *yatii* ('Afghan Sparrow'), but latter is distinctly smaller. – Separation from the more closely similar ♀ *Sind Sparrow* always requires a close check of the following two key features: (i) the bicoloured supercilium (see above), whereas in Sind Sparrow this is very obscure/greyish or absent above the lores, but above and behind the eye is distinctly paler, buff-cream to purer white, better defined and often further emphasized by darker lateral crown-stripes, lacking in Saxaul Sparrow; and (ii) the basal whitish primary patch is lacking or tends to be rather ill-defined in Saxaul, whereas in Sind Sparrow it is always present, and tends to be bolder marked.

AGEING & SEXING Ageing generally impossible after post-juv moult concluded. Sexes differ sharply after post-juv moult, sometimes also subtly before. Some limited seasonal changes due to feather wear, especially in ♂. – Moults. Complete post-nuptial and post-juv moults, apparently starting at least from Jul. Pre-nuptial moult in winter absent. – **SPRING Ad ♂** See above for a general description. With wear in spring, pattern of head and wings becomes better-marked. Mantle to uppertail-coverts more sandy-grey, bleaching much greyer, and black streaking becomes more prominent when pale feather-fringes almost wear off; in extremely worn birds, underparts, too, become dirty white, and the white wing-bars more or less wear off. Often worn birds appear overall darker. However, black spotting of rump is less affected by wear. **Ad ♀** Pattern also becomes bolder and paler with wear. Bill may become dusky-horn in breeding season. Some older and 'more advanced' ♀♀ can develop stronger head and wing markings, and even crown

P. a. stoliczkae, ad ♂, Mongolia, Jun: with rufous-cinnamon rear head-sides and otherwise neat head pattern a distinctive and attractive sparrow of Central Asian desert and semi-desert environments. Due to always complete moult, ageing not possible. Age and sex inferred by black bill and typical mature plumage. This extralimital race is very similar to *ammodendri*, only warmer brown on mantle and back. (H. Shirihai)

SPARROWS AND ALLIES

P. a. stoliczkae, ♀, Mongolia, Jun: mostly drab grey-brown and House Sparrow-like, but close views should reveal the diagnostic bicoloured supercilium, white in front and above eye, becoming wider and more cinnamon at rear, and emphasised by (variably dark) grey eye-stripe. (H. Shirihai)

P. a. stoliczkae, ♀, Mongolia, Jun: note broad cream-white wing-bar on otherwise black median coverts, as well as cinnamon-tinged lesser coverts. In such poor views separation from House Sparrow is challenging. (H. Shirihai)

and lesser coverts may darken somewhat, appearing almost intermediate between ♂ and ♀, but they are always much duller than any ♂ at same season. Legs in both sexes bright reddish-pink when breeding. **1stS** Both sexes inseparable from respective ad, but see 1stW. – **AUTUMN Ad ♂** Appears much as in spring/summer, but due to fresh feather tips has warmer ginger wash to pale areas above and below; the distinctive wing pattern is less striking, and streaking above less apparent, while black bib is slightly paler due to white tips. Black of crown partly concealed by pale buff feather-fringes. **Ad ♀** Warmer sandy-buff than in spring. **1stW** Much as respective ad, but at least some ♂♂ have comparatively broader sandy-buff fringes than ad ♂, with poorly-developed dark marks on crown, face and bib, and the diagnostic wing pattern is to some extent more washed-out. At least in early autumn, bill is less pure black. **Juv** Resembles ♀ but face and wing markings even duller, while black streaks on cap and mantle are browner and less sharply defined, underparts paler and have fluffier feathering. Bill and legs pale flesh, and yellow gape may still be apparent. Many juv ♂♂ can be sexed by black lores and eye-stripes, and pale rufous supercilium.

BIOMETRICS L 15–16 cm; **W** ♂ 74–85 mm (*n* 16, *m* 79.4), ♀ 74–80 mm (*n* 7, *m* 76.6); **T** ♂ 57–68.5 mm (*n* 16, *m* 64.1), ♀ 61–66 mm (*n* 7, *m* 63.9); **T/W** *m* 81.6; **B** 12.2–14.5 mm (once 11.0; *n* 23, *m* 13.4); **B**(f) 8.8–12.1 mm (*n* 23, *m* 10.4); **BD** 7.3–8.7 mm (*n* 23, *m* 7.7); **BW**(g) 6.9–8.0 mm (*n* 23, *m* 7.5); **Ts** 18.8–21.0 mm (*n* 23, *m* 19.8). **Wing formula: p1** minute; **p2** < wt 0–2 mm; **pp**(2)**3–4** about equal and longest; **p5** < wt 0.5–2.5 mm; **p6** < wt 4–8 mm; **p7** < wt 8.5–12 mm; **p10** < wt 17–21.5 mm; **s1** < wt 18–23 mm. Emarg. pp3–5, sometimes slightly also near tip of p6.

GEOGRAPHICAL VARIATION & RANGE Polytypic, with three subspecies generally recognised, all in Central Asia, Mongolia and W China. Within the treated region only ssp. *ammodendri* has been recorded (NE Iran). The three subspecies differ in colour saturation, by width of black streaking above, and in extent of black streaking on rump; also differ in wing- and tail-length, and by depth and width of bill. As these differences are clinal and some populations are intermediate in size and amount of streaking, assessment of subspecies generally requires sufficient specimens at hand of known origin and in comparable stage of wear. Many are indeterminable.

P. a. ammodendri Gould, 1872 (NE Iran, NW Afghanistan, E Turkmenistan, N Uzbekistan north to S & C

P. a. ammodendri, ♀, presumed ad, Kazakhstan, May: some older ♀♀ may appear almost intermediate between the two sexes (here note well-developed orange-tinged supercilium and some black mottling on crown and lesser coverts), still always much duller than ♂ at same season. (V. & S. Ashby)

Kazakhstan; mainly resident, but some dispersal in winter noted). Described above. Typically, ground colour of ♂ above light drab-grey or sandy-grey, slightly brighter yellowish-pink on mantle and scapulars when plumage fresh, and rufous-cinnamon stripe behind eye also less bright than in races further east and north. Some ill-defined dull grey marks on central back and uppertail-coverts can be present, but rump unmarked sandy-grey. Upperparts of ♀ generally tinged light drab-grey, less brown than in races further east and north. Relatively small. – Breeders in Turkmenistan and N Afghanistan ('*korejewi*') tend to have more unstreaked rump and uppertail-coverts, but some variation and clinal transition, hence provisionally included. (Syn. *korejewi*.)

Extralimital: *P. a. stoliczkae* Hume, 1874 (W Mongolia, NW & W China to SW China). Similar to *ammodendri* but mantle and back more cinnamon-buff, not as greyish. (Syn. *plecticus*.)

REFERENCES ROSELAAR, C. S. & ALIABADIAN, M. (2009) *Podoces*, 4: 1–27. – BOURNE, W. R. P. (1957) *Ibis*, 99: 182–190. – VAURIE, C. (1958) *Ibis*, 100: 275–276.

P. a. ammodendri, juv ♂, Kazakhstan, Jul: already at this age (before post-juv moult seems to have started) some can be sexed—note black eye-stripe and pale rufous rear head-sides. Bill still pale flesh, and yellow gape also apparent. (K. Haataja)

HOUSE SPARROW
Passer domesticus (L., 1758)

Fr. – Moineau domestique; Ger. – Haussperling
Sp. – Gorrión común; Swe. – Gråsparv

One of the commonest and most widespread passerines in the world, the all-too-familiar House Sparrow is a native of Eurasia south to N Africa and east to China, but has also been widely introduced to many areas throughout the globe. Indeed, given the current decline in some NW European populations, it is perhaps doing better in other parts of its range than it is 'back home'. Mostly resident, breeding close to man, often near farms and villages but also in parks and gardens of large towns. Often quite gregarious even in the breeding season, the chirping chorus from a larger flock in a hedgerow can be almost deafening.

IDENTIFICATION A heavy-billed rather large-headed sparrow, which can appear a little clumsy, especially in flight and when landing. Slightly larger than Tree Sparrow, but marginally smaller than Spanish Sparrow. ♂ easily recognised by *grey crown broadly bordered chestnut-brown on both sides, the chestnut colour also covering entire nape*. A *black mask continues as a bib on centre of throat and widens over upper breast*. In fresh plumage, *cheeks are sullied pale greyish*, becoming almost off-white when worn. *Upperparts rufous and ochre streaked black*. A bold *white median-coverts wing-bar* is noticeable, otherwise wings mostly rufous and dark grey with obvious dark-centred tertials. Lower back to uppertail-coverts and underparts mostly pale grey. *Bill black in breeding plumage*, yellowish-horn in other seasons. In contrast, ♀ *largely nondescript, being dull brown above and dusky yellow-grey below with an indistinct but rather long, pale supercilium behind eye* enhanced by a *dark eye-stripe below*. Mantle and scapulars streaked dark like ♂, but less rufous-tinged. There is a *buff-white median coverts wing-bar*, and *ill-defined buffish-grey tertial fringes*. In none of these characteristics is the ♀ very dissimilar from congeners. Bill generally greyish straw-coloured with a darker culmen and tip. In both sexes iris is blackish-brown and tarsus pale grey-brown. Hops boldly on ground. Flight slow but direct on whirring wings, can appear laborious. Lands rather clumsily with sudden collapse.

VOCALISATIONS Vocal, almost constantly uttering various chirping calls. The song is an emphasized series of chirping calls strung together, which is arguably one of the least-developed songs among Palearctic birds. No clear structure, recognised mainly by voice and energetic repetition of notes of slightly varying type, *chirp cheev chirp cherp cheve chirp...* – Basic *chirp*, *chee-ip* or *chirrup* vary according to situation and behaviour; also a drawn-out *siep* or soft *swee*, and in excitement a hard rattling series, e.g. *churr-r-r-r-it-it*. Sometimes a low, husky *chreek* when flushed. A louder and more strident *chweep* can recall Richard's Pipit to a surprising degree.

SIMILAR SPECIES ♂ unmistakable given a reasonable view, but still must be safely separated from ♂ *Spanish Sparrow*. Note grey centre to crown, no bold black streaking on flanks and less black on mantle and shoulders. – Is even more similar to ♂ *Italian Sparrow*, but differs again by its grey central crown, by no or practically no white supercilium and

P. d. domesticus, ♂, Germany, Apr: distinguished from congeners by grey cap bordered by chestnut-brown bands, and dull greyish-white cheeks lacking black central patch. Has rather large black bib becoming mottled on upper breast. Also rather conspicuous white wing-bars, and bold blackish streaks to largely rufous-chestnut mantle and scapulars. Late summer moult complete in both ages, preventing ageing thereafter. (M. Schäf)

P. d. domesticus, ♀, Netherlands, Apr: compared to ♂, lacks contrasting head and throat patterns, and is predominantly greyish and dirty brown, but has pale creamy supercilium, behind eye enhanced by dark eye-stripe. Mantle and scapulars streaked dark, patterned like ♂ but much less rufous-tinged. (R. Pop)

SPARROWS AND ALLIES

P. d. domesticus, ♂, Netherlands, Oct: when fresh has dusky-greyish cheeks and underparts, while both chestnut head-sides and black bib are mottled with pale tips, and bib is often restricted to chin and upper throat. Upperparts also warmer brown when very fresh. (N. D. van Swelm)

P. d. domesticus, ♂♂, presumed ad (left) and 1stW, Netherlands, Oct: both ages moult completely in late summer/early autumn, but many young ♂♂ show vestiges of immaturity, with more extensive pale feather-tips creating less clear-cut head pattern and more obscure bib, and bill mostly yellowish. (N. D. van Swelm)

much duskier greyish cheeks. – With less experience, or in a poor and fleeting view, could be confused with *Tree Sparrow*, but latter has smaller, rounder head, all-rufous crown and a characteristic black spot or 'comma sign' on whitish cheeks. It also has a near-complete narrow white necklace. – ♀/juvenile might easily be confused with *Spanish Sparrow*, and these two are possibly indistinguishable in poor views. This pitfall is fully discussed under Spanish Sparrow. – The situation is even more complicated and challenging when encountering hybrids between House and Spanish Sparrows, and with ♀ *Italian Sparrow*, especially in overlap zones. There is some advice offered under Italian Sparrow, but ♀-coloured of the two are basically inseparable. – Other sparrows covered should not pose a real problem, unless observed poorly. For inexperienced observers, ♀/juvenile House Sparrow might be confused with *Rock Sparrow*, but latter has bolder head pattern with dark lateral crown-stripes leaving an obvious pale central stripe. It also is slightly larger and has proportionately larger head and bill. – ♀-like *Common Rosefinch* should present no real problems being larger and slimmer, with streaked underparts, greyer plumage and different bill shape, in addition to many other minor differences.

P. d. domesticus, ♀, Netherlands, Apr: when very fresh, pale areas overall tend to have deeper and warmer colours. Note moderately sized conical bill, smaller than in Saxaul Sparrow. (N. D. van Swelm)

P. d. domesticus, juv, Netherlands, May: soft, fluffy body-feathers, with rather obscure head pattern compared to older ♀. Pale tip to bill and yellow gape-flanges still rather obvious. Lack of greyish chin and throat suggests ♀. (N. D. van Swelm)

P. d. domesticus, juv, Germany, Aug: at least until late summer, even into autumn, many juv are ♀-like with obvious yellow gape. Many have rather blotchy and uneven underparts, which could suggest ♀ Spanish Sparrow, thus beware of this pitfall. Juv with more clearly dusky throat (like left-hand bird) could be ♂. (M. Schäf)

AGEING & SEXING Ageing only possible in summer before post-juv moult finished. The black mask in ♂♂ is known to, on average, slightly increase in size in older birds, but much overlap (Nakagawa & Burke 2008). Sexes generally easily differentiated once post-juv moult under way, and when seen close or handled many juv (even pulli) can be sexed. Moderate seasonal variation in ♂ (due to feather abrasion). – Moults. Both ad and juv moults completely, including wings and tail, in late summer, but timing varies with latitude, number of broods and hatching date (span Jun–Oct). – SPRING **Ad** ♂ See under Identification. **Ad** ♀ With wear may develop tiny dark-mottled bib and, rarely (presumably older ♀♀), quite a substantial blackish bib and some rufous in lesser coverts, but rest of plumage like normal ♀. **1stS** Inseparable from ad. – AUTUMN **Ad** ♂ Overall duller and less solidly patterned than in breeding plumage: bib largely concealed, often almost limited to throat and only vaguely visible on upper breast. Dull olive-brown tips to grey parts of crown, back, rump and cheeks, and some pale drab-buff to grey parts of breast and flanks. **Ad** ♀ Much as in spring, but deeper grey below with warmer tawny-buff fringes above and on wings. **1stW** Both sexes assume respective ad plumage by Sep–Oct (retained juv feathers until then diagnostic), but more extensive pale feather-tips in 1stW ♂ produce less clear-cut pattern with more obscure bib, but much overlap and this cannot be used for ageing; iris perhaps more olive-brown than richer brown of ad, but difficult to detect. **Juv** Soft, fluffy body-feathers. Bill has pale tip and yellow gape-flanges. Crown and rump brown-mottled, throat and belly are whiter. Wing-bars very poorly defined or lacking, and supercilium strongly reduced. About 95% of juv ♂♂ develop vague greyish chin and throat early (even as pulli), whereas juv ♀ lacks this in c. 92% (Harrison 1961, Cheke 1967, Cordera 1990).

BIOMETRICS (*domesticus*) **L** 14.5–15.5 cm; **W** ♂ 76–83.5 mm (*n* 18, *m* 80.1), ♀ 74–80 mm (*n* 13, *m* 77.2); **T** ♂ 55–62 mm (*n* 18, *m* 58.7), ♀ 54–61 mm (*n* 13, *m* 57.0); **T/W** *m* 73.6; **B** 13.2–16.0 mm (*n* 30, *m* 14.6); **B**(f) 11.0–13.2 mm (*n* 30, *m* 12.3); **BD** 8.1–9.5 mm (*n* 30, *m* 8.7); **BW**(g) 7.4–9.3 mm (*n* 30, *m* 8.4); **Ts** 17.9–20.3 mm (*n* 30, *m* 18.9). Wing formula: p1 minute; pp2–4 about equal and longest; p5 < wt 1–3 mm; p6 < wt 4.5–9 mm; p7 < wt 8–14 mm; p10 < wt 18–22.5 mm; s1 < wt 19–23 mm. Emarg. pp3–5.

GEOGRAPHICAL VARIATION & RANGE Clinal and rather slight variation overall, but eastern populations clearly smaller with finer bill and paler colours, and are migratory, unlike western populations. Western and eastern races

P. d. tingitanus, ♂, Morocco, Feb: compared to most European races, subtly smaller-billed and paler overall. Here still in fresh plumage with black on head partly admixed with grey-white and cheeks sullied grey, but note characteristic whiter frontal and lower edges making cheeks look almost bicoloured. Later when worn may develop blackish streaking on grey crown. (D. Occhiato)

P. d. niloticus, ♂ (two in front) and ♀, Egypt, May: similar to *biblicus* or slightly paler, differing mainly in much smaller size, but unlike similarly small-sized 'Indian Sparrow' group, cheeks always sullied greyer, grey cap usually extends to nape (rather than being restricted to crown), and uppertail-coverts uniform grey-brown. (J. Hering)

P. d. biblicus, ♂, Cyprus, Apr: this race is overall pale but not as pale or small as *niloticus*, and pale areas extensively washed sandy-buff when not heavily worn. Rufous fringes above pale, buffish cinnamon; cheeks pale ash-grey and wear almost to whitish-grey. (F. Trabalon)

P. d. biblicus, ♀, Cyprus, Apr: ♀ of this race also overall pale, still dusky brown but with more greyish tinge. Note also the contrasting pale areas in wing. (F. Trabalon)

SPARROWS AND ALLIES

therefore best seen as two racial groups. Virtually only ♂♂ are separable, using a combination of size (though much overlap) and plumage (against comparative material, and bear in mind effects of strong seasonal feather wear). The following races are deemed sufficiently distinct to be recognised within the covered area. At least two more extralimital subspecies (not treated).

DOMESTICUS GROUP
Generally large, ♂♂ having dusky greyish cheeks and underparts (but some overlap with 'Indicus Group'), rump usually pure grey. All populations sedentary.

P. d. domesticus (L., 1758) (Europe [see map], most of Turkey, Ukraine except Crimea, east across N Central Asia to Kamchatka, N Sakhalin, NW Manchuria; introduced Americas, Azores, Cape Verdes, Africa, Australia; resident). Described above. A rather large and dark subspecies. Brown of head in ♂ deep chestnut, cheeks rather dusky-grey (tinged olive-yellow when fresh), only rarely a little purer whitish. – Birds in parts of the Mediterranean region sometimes separated ('*balearoibericus*') on account of being subtly paler, but very slight difference indeed and much overlap, thus not followed here. (Syn. *baicalicus*; *balearoibericus*; *hostilis*; *semiretschiensis*; *sibiricus*.)

P. d. tingitanus Loche, 1867 (NW Africa; resident). Differs only slightly from *domesticus*, being almost same size, but proportionately subtly smaller-billed and shorter-tailed, and paler overall. Worn ♂♂ often become rather noticeably streaked black on crown, and when heavily abraded some acquire almost entirely blackish crown. Cheek patch whitish, but slightly grey usually on rear centre, while frontal and lower edges are pure white making cheek patch look bicoloured. **W** ♂ 73–83 mm (n 17, m 77.5), ♀ 72–80 mm (n 17, m 76.3); **T** ♂ 52–60 mm (n 17, m 55.5), ♀ 49–59 mm (n 17, m 54.6); **T/W** m 71.6; **B** 14.0–15.9 mm (n 26, m 14.7); **B**(f) 10.4–13.3 mm (n 26, m 11.9); **BD** 7.8–9.0 mm (n 26, m 8.4); **BW**(g) 7.7–8.9 mm (n 26, m 8.1); **Ts** 17.5–20.4 mm (n 26, m 19.0). (Syn. *maroccanus*.)

P. d. niloticus Nicoll & Bonhote, 1909 (N Egypt, south along Nile to Wadi Halfa, Sudan; resident). Similar to *biblicus* or slightly paler, differing mainly in smaller size (even matching *indicus*). **W** ♂ 69–79 mm (n 12, m 74.0), ♀ 69–74 mm (n 11, m 72.4); **T** ♂ 50–60 mm (n 12, m 54.7), ♀ 50–56 mm (n 11, m 53.6); **T/W** m 74.0; **B** 12.2–15.5 mm (n 23, m 13.9); **B**(f) 10.3–12.2 mm (n 23, m 11.3); **BD** 7.1–8.3 mm (n 23, m 7.8); **BW**(g) 7.0–8.3 mm (n 23, m 7.6); **Ts** 16.1–19.5 mm (n 23, m 17.9). (Syn. *halfae*.)

P. d. biblicus, ♂, presumed 1stW, Israel, Sep: despite being overall pale, when fresh typically extensively washed sandy-buff with faint grey hue. After complete moult of both ages, ageing usually impossible, but at least some young show more extensive yellowish tone to bill and gape, and broader pale tips to head concealing more of the summer pattern. (H. Shirihai)

P. d. bactrianus, ♂, NE Iran, May: one of the races in the 'Indian Sparrow' group and, unlike *domesticus*, decidedly smaller, cheeks and underparts average whiter, grey of crown is more restricted, and bill relatively more slender (still can look big as here). (C. N. G. Bocos)

P. d. bactrianus, ♀, NE Iran, May: rather small bill and overall pale plumage in this race, like it has been newly washed. (C. N. G. Bocos)

The following two subspecies show traits of both groups, thus represent an intermediate link between them, but are kept in the 'Domesticus Group', which they most resemble, for convenience.

P. d. biblicus Hartert, 1904 (Crimea, E Turkey, Cyprus, Caucasus, Transcaucasia, Levant including Sinai and east through Iraq and Iran except north and south-east; resident). Resembles *domesticus*, and has same large size, but is somewhat paler, with crown, cheeks and rump ash-grey in worn ♂ plumage, extensively washed sandy-buff when fresh; rufous-chestnut less deep. **W** ♂ 75.5–84.5 mm (n 16, m 79.9), ♀ 72–83 mm (n 16, m 77.0); **T** ♂ 54–62.5 mm (n 16, m 57.7), ♀ 51–61.5 mm (n 16, m 55.8); **T/W** m 72.4; **B** 13.3–16.1 mm (n 32, m 14.7); **B**(f) 10.7–13.6 mm (n 32, m 12.3); **BD** 7.6–9.6 mm (n 32, m 8.7); **BW**(g) 7.5–9.0 mm (n 32, m 8.2); **Ts** 17.2–21.2 mm (n 31, m 19.0). (Syn. *persicus*.)

○ *P. d. hyrcanus* Zarudny & Kudashev, 1916 (SE Azerbaijan, NW Iran; mainly resident). Close to *biblicus* but averages slightly smaller. Mantle subtly darker chestnut than *indicus*. Crown of ♂♂ often slightly dark-streaked in worn plumage. Breast and flanks rather darkly washed grey. Sexes seem similarly sized. Sometimes placed in the 'Indicus Group', but morphology slightly better fits the 'Domesticus Group'. **W** 70–80 mm (n 16, m 75.3); **T** 52–59 mm (n 16,

P. d. indicus, ♂, India, Dec: in fresh plumage cheeks still slightly sullied greyish (more pure white when worn). Note small grey crown, and sometimes rusty-brown wash to rump/uppertail-coverts. Generally recalls *bactrianus* but breast and flanks more grey-tinged and upperparts subtly paler, less dark rufous. (A. Banerjee)

P. d. indicus, ♂, India, Jun: here in worn summer plumage, when black bib of ♂ is better marked against whiter cheeks and underparts. (B. Castelein)

— 303 —

m 55.2); **T/W** *m* 73.3; **B** 13.2–15.2 mm (*n* 17, *m* 14.0); **B**(f) 10.5–12.3 mm (*n* 17, *m* 11.3); **BD** 7.3–9.1 mm (*n* 17, *m* 8.0); **BW**(g) 7.1–8.2 mm (*n* 17, *m* 7.7); **Ts** 17.1–19.5 mm (*n* 17, *m* 18.5).

INDICUS GROUP
Generally small, ♂♂ having on average whiter cheeks and underparts, and rump more brown-tinged, less pure grey. Still, some overlap of characters between groups. Northern populations migratory. Wing formula by and large the same as in the 'Domesticus Group'.

P. d. bactrianus Zarudny & Kudashev, 1916 (NE Iran, W Turkmenistan, from Aral Sea region east to S Kazakhstan south through N & E Afghanistan to NW Pakistan; migrant, winters Sind and adjacent parts of India, but apparently some populations short-range migrants only or even resident). A small race with upperparts bright and dark rufous streaked black, much like *domesticus*, but underparts and cheeks much whiter (often wear to pure white). **W** ♂ 75–87 mm (*n* 14, *m* 77.9), ♀ 72–77.5 mm (*n* 12, *m* 74.3); **T** ♂ 52–58 mm (*n* 12, *m* 55.2), ♀ 49–57 mm (*n* 12, *m* 52.7); **T/W** *m* 71.3; **B** 12.7–14.7 mm (*n* 25, *m* 13.7); **B**(f) 9.8–12.7 mm (*n* 26, *m* 11.2); **BD** 7.5–8.5 mm (*n* 25,

P. d. indicus, ♀, India, Feb: like other members of the 'Indian Sparrow' group, note general compact appearance, with short, stubby bill. (R. Mallya)

P. d. hufufae, ♂, Oman, Nov: duller and greyer than *indicus*, with cap, rump, and uppertail-coverts purer grey, and chestnut of upperparts and wing of ♂ distinctly reduced, being paler and less cinnamon. (D. Occhiato)

P. d. hufufae, ♀, Oman, Nov: ♀ of this race represents pale version of 'Indian Sparrow', some being characteristically whitish below and on face. (P. Alberti)

m 8.0); **BW**(g) 6.9–7.8 mm (*n* 24, *m* 7.4); **Ts** 17.4–20.1 mm (*n* 24, *m* 18.6). (Syn. *griseogularis*.)

P. d. indicus Jardine & Selby, 1831 (much of Arabia except east, S Afghanistan, Pakistan, N India south of Himalayas and SE Asia to Laos; introduced SE Africa, Indian Ocean islands; mainly resident). Resembles *bactrianus* but breast and flanks more grey-tinged, and upperparts subtly paler, less dark rufous. Black bib of ♂ slightly smaller and better-marked against pale underparts. **W** ♂ 70–82 mm (*n* 13, *m* 75.4), ♀ 69–77 mm (*n* 12, *m* 72.3); **T** ♂ 49–60 mm (*n* 13, *m* 55.2), ♀ 50–56 mm (*n* 12, *m* 53.3); **T/W** *m* 73.6; **B** 13.2–16.0 mm (*n* 30, *m* 14.6); **B**(f) 11.0–13.2 mm (*n* 30, *m* 12.3); **BD** 8.1–9.5 mm (*n* 30, *m* 8.7); **BW**(g) 7.4–9.3 mm (*n* 30, *m* 8.4); **Ts** 17.9–20.3 mm (*n* 30, *m* 18.9). (Syn. *buryi*; *confucius*; *enigmaticus*; *nigricollis*; *soror*.)

P. d. hufufae Ticehurst & Cheesman, 1924 (NE & E Arabia, N Oman; resident). A small, pale and greyish version of *indicus* with reduced rufous in ♂. **W** ♂ 66–74 mm (*n* 12, *m* 71.7), ♀ 64–73 mm (*n* 12, *m* 69.2); **T** ♂ 50.5–57 mm (*n* 11, *m* 54.0), ♀ 44–55 mm (*n* 12, *m* 51.3); **T/W** *m* 74.6; **B** 11.4–15.0 mm (*n* 24, *m* 12.7); **B**(f) 9.1–11.8 mm (*n* 25, *m* 10.2); **BD** 6.8–7.9 mm (*n* 23, *m* 7.3); **BW**(g) 6.0–7.3 mm (*n* 23, *m* 6.8); **Ts** 16.4–18.5 mm (*n* 25, *m* 17.6).

TAXONOMIC NOTES Hybridisation between House Sparrow and Spanish Sparrow is extensive in several well-studied areas (e.g. S Italy, Tunisia, Algeria) producing hybrids of mixed appearance, NW African of which having been afforded 'scientific' names such as '*flückigeri*', '*ahasver*' and '*bergeri*'. Just as closely-related pairs with extensive hybridisation like Carrion and Hooded Crows, and Yellowhammer and Pine Bunting, are treated here as two different species, respectively, House and Spanish Sparrows are similarly regarded as different species. Over most of their ranges they appear and act as species, but they have apparently not yet developed functional reproductive barriers where they meet. – Formerly, Italian Sparrow was similarly seen as 'a stabilised hybrid swarm' between House and Spanish Sparrows, but recently it has been proposed that Italian Sparrow has reached a status not obviously different from a separate species, and is treated as one here (cf. Italian Sparrow). – The situation over large parts of Central Asia is especially complex and interesting in that representatives of the 'Domesticus Group' (*domesticus*) and 'Indicus Group' (*bactrianus*) appear to overlap when breeding, with resident *domesticus* already breeding when migrant *bactrianus* arrives, effectively limiting interbreeding, while they also differ somewhat in ecology, timing of moult and in behaviour, suggesting that they are candidates for separate species. However, *indicus* reportedly intergrades with *biblicus* (of the 'Domesticus Group') in parts of Iran. Thus, in some areas, populations of both racial groups occur in sympatry without evidence of hybridisation, while in others there is extensive interbreeding.

REFERENCES CHEKE, A. S. (1967) *Ringers' Bull.*, 3 (2): 7–8. – CORDERO, P. J. (1990) *Butlletí Anellament GCA*, 7: 3–6. – ELGVIN, T. O. *et al.* (2011) *Molecular. Ecology*, 20: 3823–3837. – HARRISON, J. M. (1961) *BBOC*, 81: 96–103. – HERMANSEN, J. S. *et al.* (2011) *Molecular Ecology*, 20: 3812–3822. – JENSEN, H. *et al.* (2013) *Molecular Ecology*, 22: 1792–1805. – NAKAGAWA, S. & BURKE, T. (2008) *J. Avian Biol.*, 39: 222–225. – SÆTRE, G.-P. *et al.* (2012) *J. Evol. Biol.*, 25: 788–796. – TÖPFER, T. (2006) *Zootaxa*, 1325: 117–145.

ITALIAN SPARROW
Passer italiae (Vieillot, 1817)

Fr. – Moineau cisalpin; Ger. – Italiensperling
Sp. – Gorrión Italiano; Swe. – Italiensk sparv

The sparrows of Italy, including related islands such as Sicily and Malta, and also Crete in S Greece, represent something of a taxonomic enigma. They have been variously considered as a hybrid phenotype without taxonomic status, or as a subspecies of either House or Spanish Sparrows, but also (including here) as a different species. These sparrows represent a stabilised hybrid population between House Sparrow and Spanish Sparrow, and although still controversial it has increasingly become accepted as a viable method of speciation. In the extreme north of its range, *italiae* continues to hybridise with *domesticus*.

♂, Italy, Nov: in fresh plumage when chestnut and black on head less solid, including the bib, but head coloration still resembles Spanish, and bib that of House. Both age classes undergo complete moult post-breeding. Bill-base yellowish at this time. (D. Occhiato)

♂, Italy, Jun: resembles ♂ Spanish on head, with chestnut-red crown to nape (no grey centre) and bright white cheeks, but underparts and rump are as House. Also has less black on mantle and shoulders, with narrower and clearly broken white supercilium unlike typical Spanish. Due to complete autumn moult of both age classes, ageing thereafter is impossible. (D. Occhiato)

♂ (left) feeding juv, Italy, Aug: has less black on mantle than Spanish, but pattern still closer to that species than to House. Extent of black on lores and bib varies, here it is rather extensive. Underparts more like House, lacking or having much less extensive black markings, even in late summer. Juv inseparable from either Spanish or House. (E. Winkel)

IDENTIFICATION Closely resembles both House and Spanish Sparrows, perhaps closest to latter, given head pattern of ♂, with *chestnut-red crown to nape (no grey centre)* and *bright white cheeks*, but usually has *narrower and clearly broken white supercilium* (nearer in this respect to House Sparrow). *Upperparts have less black on mantle than Spanish*, but *stronger pale tramlines than House*, and is much *richer chestnut on small wing-coverts* with brighter double wing-bars, but *rump purer grey* like House Sparrow (not mottled brownish as Spanish). Underparts have House Sparrow-like pattern of *black lores and bib* (albeit more extensive), but has *only limited or no black streaking on lower breast and flanks* characteristic of Spanish (at most has fine streaks on breast, but almost invariably lacks flank markings), while *rest of underparts greyer* like House Sparrow. 'Typical' ♀ Italian tends to share characters of both species (which are already very similar!), and can be inseparable from either. Tends to *lack diffuse grey streaking on breast and flanks* present in at least some ♀ Spanish, but shares with that species presence of *pale tramlines on mantle*.

♀, Italy, late spring: ♀ not safely identifiable outside native range. It closely resembles ♀ Spanish Sparrow in having on average stronger face pattern than average House Sparrow (especially better-defined supercilium) and more prominent pale tramlines on mantle, but like House usually lacks obvious dark streaking on underparts. (A. Rossella)

VOCALISATIONS Mostly as House Sparrow and probably as a rule inseparable.

SIMILAR SPECIES See above for differences from *House* and *Spanish Sparrows*. It should be emphasized that outside known range even classic birds of both sexes will usually prove inseparable, since hybrids between Spanish and House Sparrows can possess Italian Sparrow-like characteristics.

AGEING & SEXING As in Spanish Sparrow. Sexes seem to differ rather obviously in size.

BIOMETRICS **L** 14.5–16 cm; **W** ♂ 79–82.5 mm (*n* 12, *m* 80.0), ♀ 75–78 mm (*n* 13, *m* 76.5); **T** ♂ 54–59 mm (*n* 12, *m* 57.0), ♀ 52–59 mm (*n* 13, *m* 55.4); **T/W** *m* 71.8; **B** 13.4–16.4 mm (*n* 25, *m* 14.7); **B**(f) 11.1–13.8 mm (*n* 25, *m* 12.3); **BD** 8.1–9.3 mm (*n* 24, *m* 8.8); **BW**(g) 7.8–9.2 mm (*n* 25, *m* 8.4); **Ts** 18.5–20.0 mm (*n* 24, *m* 19.3). **Wing formula: p1** minute; **pp2–4** about equal and longest; **p5** < wt 0.5–3 mm; **p6** < wt 4–8.5 mm; **p7** < wt 9–13 mm; **p10** < wt 18–23 mm; **s1** < wt 19–24 mm. Emarg. pp3–5.

GEOGRAPHICAL VARIATION & RANGE Monotypic. – Italy south of foothills of Alps and related islands, but apparently not on Sardinia, also adjacent parts of Provence, Corsica, Crete; sedentary. (Syn. *schiebeli*.)

♂, presumed young, Italy, Nov: autumn ♂ with broad pale tips and heavily obscured head pattern suggests 1stW. Also extensive yellow on bill. Specific identification is less straightforward, but note Spanish-like head, and underparts and rump more like House. (D. Occhiato)

♂, presumed ad, Italy, Nov: some ♂ Italian Sparrows have red crescent-shaped breast-band, sometimes called 'red morph'. (D. Occhiato)

♀♀ in fresh plumage, Italy, Dec–Jan: most Italian ♀♀ in fresh plumage are like left-hand bird, being more like House Sparrow in lacking streaky underparts of Spanish, but a few, like that on right, are sullied and streaked grey below, and could be confused with latter. Outside usual range any variation will prove inseparable from House or Spanish. (Left: D. Occhiato; right: G. Conca)

TAXONOMIC NOTES After much debate regarding the true status of Italian Sparrow (subspecies of either House or Spanish Sparrows, hybrid swarm created by extensive mixing between these two, or a separate species with hybrid background, so-called 'homoploid hybrid speciation'), many authorities now opt for the last-mentioned model (Hermansen *et al.* 2011). Some proof is offered by the fact that Italian Sparrow occurs sympatrically with Spanish Sparrow on the Gargano Peninsula, SE Italy. Further, if *italiae* is a hybrid it would be illogical to include it as a race of either species (and *italiae* would have priority over *hispaniolensis* if subsumed within Spanish Sparrow; Töpfer 2006). Given that Italian Sparrow is morphologically distinct over a large area with little recent genetic inflow from either parent species, it seems defensible to treat it as a separate species.

REFERENCES Elgvin, T. O. *et al.* (2011) *Molecular Ecology*, 20: 3823–3837. – Hermansen, J. S. *et al.* (2011) *Molecular Ecology*, 20: 3812–3822. – Hermansen, J. S. *et al.* (2014) *Molecular Ecology*, doi: 10.1111/mec.12910. – Sætre, G.-P. *et al.* (2012) *J. Evol. Biol.*, 25: 788–796. – Töpfer, T. (2006) *Zootaxa*, 1325: 117–145. – Trier, C. N. (2014) *PLoS Genet.*, 10: e1004075.

♂ House × Spanish Sparrow, Spain, Jun: in Spain but especially in North Africa, House Sparrow hybridises with Spanish, producing Italian-like variants. (C. N. G. Bocos)

SPANISH SPARROW
Passer hispaniolensis (Temminck, 1820)

Fr. – Moineau espagnol; Ger. – Weidensperling
Sp. – Gorrión moruno; Swe. – Spansk sparv

A sociable species that breeds colonially, usually in trees (unlike House Sparrow that often breeds in man-made structures), but can also nest in lower parts of twig nests of larger birds such as raptors and White Storks *Ciconia ciconia*, in telegraph poles, etc. Spanish Sparrow is widespread from the NE Atlantic islands to Central Asia and extreme SW China. The European range is strangely disjunct, with strongholds in Iberia, Sardinia, Balkans and north to Romania. A resident or partial migrant in the west, some populations, mainly in the east, are migratory, and it is an erratic winter visitor further south, reaching the N Sahara, Nile Valley, N Arabia and Indian Subcontinent.

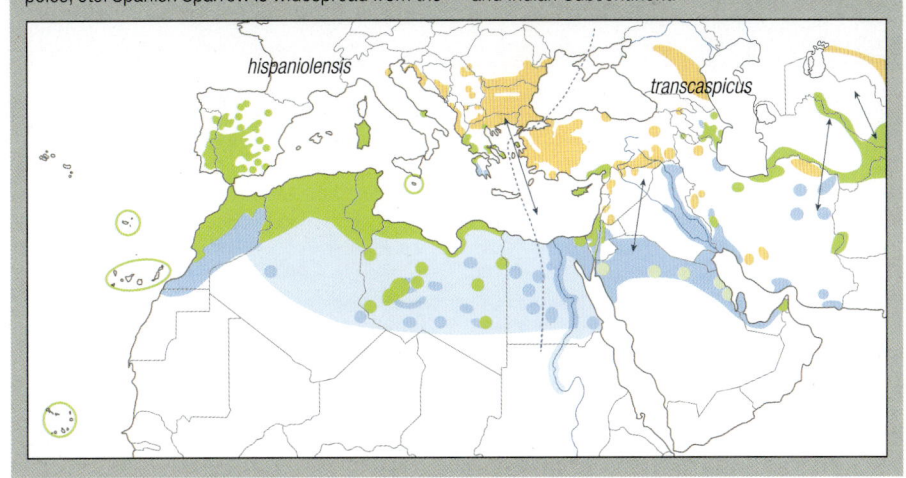

IDENTIFICATION Similar size to House Sparrow with on average a fraction heavier bill, but structural differences insignificant compared to plumage differences in ♂. This has a *dark chestnut crown and nape*, and *whiter cheeks*. Especially in spring, extensively *black throat and breast continue blotchily onto scapulars and mantle*, and as *coarse black arrowhead streaks on flanks* to undertail-coverts. ♂ also has a *short white supercilium*, black lores and eyestripe, and *white belly*. ♀ usually indistinguishable at distance from House Sparrow. Iris dark brown, bill mostly black in breeding ♂, dark-tipped yellowish-horn in non-breeding ♂ and ♀. Legs brown. Flight action perhaps less whirring than House Sparrow, a little more powerful, but hardly separable on this. General behaviour as House Sparrow. Spanish is distinctly more adapted to natural habitats (e.g. dry open country with tall bushes and trees or open woodland, riverside trees), and is much less frequently found in towns. It is equally as gregarious as House Sparrow and, in addition to forming single-species flocks, also mixes commonly with that species. In some areas gathers into dense, large foraging or migratory flocks, which sometimes number thousands. Chorus from large colony impressive and has characteristic metallic and high-pitched ring.

VOCALISATIONS Similar to House Sparrow, but chirps more metallic, e.g. *chirrip*; also higher-pitched *chweeng*; both have monosyllabic *siep* or soft *swee*, but typically more drawn-out in Spanish; also a characteristic squeaky, far-carrying *chee-chee-chee* or *chiree*, a distinctly abrupt *chup* and, in excitement, a shorter, deeper rattle, *churr-r-r-r-it-it*, with softer *ch* sounds. – Calls richer but mostly given in faster monotonous series, e.g. *chilli-chilli- chilli-…*, which is far-carrying, especially from large colonies.

SIMILAR SPECIES ♂ easily separated from *House Sparrow* given reasonable view of head and underparts patterns, although hybrids are varied, and can be very difficult to distinguish from pure Spanish Sparrow. – ♂ *Italian Sparrow* can resemble Spanish closely, and variation in both means that some odd birds are best left unidentified. – ♀ Spanish is extremely difficult to separate from ♀ *House Sparrow*, especially if seen alone, and previous experience is often needed. Even in-the-hand identification is not always possible. Hybridisation further complicates the issue. Five slight differences are important to know, and if used in combination identification is often possible: Spanish has (i) faint darkish streaks and spots on breast and (more diffusely) on flanks (though House Sparrow, especially in bleached plumage, can show some darkish streaks on breast and, slightly, on flanks, but lacks 'dappled' appearance of Spanish); (ii) more pronounced pale mantle and scapular tramlines; (iii) on average somewhat better-defined, longer and paler supercilium; most Spanish tend to have (iv) faint darker shaft-streaks on crown (paler, more uniform in House); and (v) whiter median-coverts tips (but some overlap). Other, very subtle differences perhaps also useful, although much overlap: (vi) tertial fringes in Spanish more sandy-grey (less rufous-brown), less contrasting with centres; (vii) uppertail-coverts and rump normally show faint darkish shaft-streaks (indistinct or lacking in House); (viii) undertail-coverts have more diffuse greyish centres (in House, centres usually narrower, often almost restricted to shaft, and darkish); and most Spanish have (ix) a darker and more contrasting eye-stripe, and (x) paler ear-coverts. Slight structural differences are sometimes useful: (xi) Spanish's bill averages slightly longer and broader-based, contributing to the less square-headed profile. Vocal differences may prove important means of separation, at least sometimes and with familiarity.

♂ (second from the top) and three ♀♀, Israel, Apr: showing sexual differences and variation in degree of streaking in ♀♀. Degree of streaking or mottling below varies individually and is enhanced with wear. Complete autumn moult in both age classes makes ageing impossible thereafter. (A. Ouwerkerk)

♂, Spain, Jun: easily recognised by vivid head pattern, with chestnut-brown crown and nape, large black-splashed tracts above and below (especially in spring) and full black gorget. Also has short, prominent white supercilium above black loral patch. (C. N. G. Bocos)

♂, possibly 1stS, Israel Mar: ♂♂ that retain almost fresh winter-like plumage into spring, with remaining extensive fringes to darker centres (especially on crown and below), and still yellowish base to bill, are perhaps 1stS. (H. Shirihai)

♀, Bulgaria, Jun: a bird with unusually dark-streaked breast and flanks. Compared to ♀ House Sparrow, apart from underparts streaking, on average has slightly larger bill, somewhat longer and clearer supercilium and better-defined pale 'braces' on mantle. (A. Wilson)

AGEING & SEXING Ageing only possible in summer before post-juv moult finished. After this, ageing of a few typical 1stW ♂♂ only sometimes possible in the hand. Sexes generally easily differentiated once post-juv moult under way. Moderate seasonal variation in ♂ (due to feather abrasion). – Moults. Both ad and juv moults completely, including wings and tail, in late summer (often Jul–Sep, sometimes Oct); on average, juv moults slightly earlier than ad. – SPRING **Ad ♂** Crown to mantle wears to uniform dark chestnut. All-black bib and breast with pronounced black arrowheads on flanks narrowly fringed whitish. In early summer, before moult starts, black marks even bolder and coalesce over large areas. **Ad ♀** Very similar to ♀ House Sparrow, but a trifle larger with on average a little longer bill (at times with blackish upper mandible). Often has slightly more pronounced head pattern. With wear fine dark streaking on crown and underparts more pronounced. **1stS** Inseparable from respective ad, but ♂ often less solidly coloured (retains pale feather-tips longer) and at least some differ from ad by median-coverts pattern, as in autumn. – AUTUMN **Ad ♂** Duller and less contrastingly patterned than in breeding plumage: chestnut crown tipped pale cinnamon-brown, and black areas below largely concealed, with much white admixed, thus black bib and flank markings smaller and sparser. Unlike 1stW ♂, black

♀♀ Spanish Sparrow (top left: Israel, Apr; bottom left: Cape Verdes, Mar) versus House Sparrow (top right: Germany, spring; bottom right: and Israel, Feb): obviously streaked (top left) and weakly streaked (bottom left) Spanish compared with unstreaked (top right) and vaguely streaked (bottom right) House. Given reasonable views the two are unlikely to be confused as in the two top birds. However, some birds require prolonged views and others are probably inseparable. Spanish often has clearer stippled effect, especially on breast, clearer dark shaft-streaks on crown, better-developed pale tramlines above, more sandy-grey (less rufous-brown) tertial fringes, faint darkish shaft-streaks on uppertail-coverts and rump (indistinct or lacking in House), diffuse but broader dark centres to undertail-coverts (confined to shafts in House), and bill slightly longer and deeper-based. (Top left: V. Legrand; top right: I. Schultz; bottom left: D. Occhiato; bottom right: A. Ben Dov)

SPARROWS AND ALLIES

♂, presumed ad, Israel, Nov: in fresh autumn plumage, when blackish centres above and below are broadly pale fringed, with overall pattern still like spring/summer birds with very extensive black centres both below and above, indicating ad. (A. Ben Dov)

♂, presumed 1stW, Israel, Dec: at this season ageing usually impossible by moult patterns, but at least some presumed young have black or rufous centres more concealed, obscuring chestnut crown, face pattern and bold streaking above and below. (L. Kislev)

bases to median coverts more sharply defined against white tips. **Ad ♀** Much as breeding ad ♂, but warmer tawny-buff above and on wings. Below warmer greyish, and streaking less pronounced. **1stW ♂** Assumes ad plumage by Sep/Oct, but more extensive pale feather-tips tend to lend a duller pattern (including more obscured black marks). Presumed reliable criterion from most ad (but allow for some variation): pattern of blackish bases to median coverts uneven, with longer dark area on inner webs, and dark portions merge diffusely with white tips. Also, iris less rich brown compared to ad, especially in early autumn, but frequently difficult to judge. **1stW ♀** Perhaps usually indistinguishable from ad ♀ except sometimes by iris colour. **Juv** Resembles least-marked ♀. Has pale gape-flanges and soft, fluffy body-feathers. Largely inseparable from same-age House and Italian Sparrows, and various hybrids.

BIOMETRICS (hispaniolensis) **L** 14.5–16 cm; **W** ♂ 74–82 mm (n 60, m 78.2), ♀ 75–81 mm (n 16, m 77.8); **T** ♂ 52–59 mm (n 60, m 55.8), ♀ 53–58 mm (n 16, m 55.4); **T/W** m 71.3; **B** 13.2–17.0 mm (n 74, m 14.9); **B**(f) 11.5–14.8 mm (n 75, m 12.8); **BD** 8.1–10.0 mm (n 76, m 9.0); **BW**(g) 7.7–10.5 mm (n 76, m 8.5); **Ts** 18.3–22.3 mm (n 76, m 19.7). **Wing formula: p1** minute; **pp2–4** about equal and longest; **p5** < wt 1–2.5 mm; **p6** < wt 5–8 mm; **p7** < wt 9–12 mm; **p10** < wt 17–23 mm; **s1** < wt 19–25 mm. Emarg. pp3–5.

GEOGRAPHICAL VARIATION & RANGE Very slight variation, mainly concerning size and saturation of plumage colours, eastern birds tending to be paler.

P. h. hispaniolensis (Temminck, 1820) (Iberia, NW Africa east to Libya, Sardinia, Balkans, Thrace; introduced Cape Verdes, Canaries; mainly sedentary). Described above. On average subtly smaller and darker than *transcaspicus*, but difference slight and many are indistinguishable. – Southern populations ('*arrigonii*') stated to be very slightly smaller-billed and shorter-legged, but practically no difference, extensive overlap and clinal character of variation, thus included here. (Syn. *arrigonii, canariensis; salicaria*.)

○ *P. h. transcaspicus* Tschusi, 1902 (Asian part of Turkey, Near East, N Caucasus, Transcaucasia, Iraq, Transcaspia, Afghanistan and to foothills of Tien Shan, Tarim Basin; mainly migrant, wintering in Punjab and Sind, but also west through S Iran, S Iraq, Arabia to Sudan). Subtly paler above than *hispaniolensis*, but a very slight difference requiring direct comparison of series, and still not always discernible. Same size as *hispaniolensis*. **W** ♂ 75–82 mm (n 15, m 79.3), ♀ 74–79 mm (n 12, m 76.8); **T** ♂ 53–60 mm (n 15, m 56.6), ♀ 52–58.5 mm (n 12, m 54.0); **T/W** m 70.9; **B** 13.7–16.3 mm (n 26, m 14.6); **B**(f) 11.5–13.6 mm (n 27, m 12.4); **BD** 8.7–9.7 mm (n 25, m 9.1); **BW**(g) 7.8–9.3 mm (n 27, m 8.5); **Ts** 18.5–21.2 mm (n 27, m 19.8). (Syn. *terekius*.)

TAXONOMIC NOTE Se notes under House Sparrow concerning hybridisation between Spanish and House Sparrows.

REFERENCES ELGVIN, T. O. et al. (2011) *Molecular Ecology*, 20: 3823–3837. – HERMANSEN, J. S. et al. (2011) *Molecular Ecology*, 20: 3812–3822. – JENSEN, H. et al. (2013) *Molecular Ecology*, 22: 1792–1805. – SÆTRE, G.-P. et al. (2012) *J. Evol. Biol.*, 25: 788–796. – TÖPFER, T. (2006) *Zootaxa*, 1325: 117–145.

♀, Kuwait, Nov: in fresh plumage, with pale areas sullied buffish, including supercilium. Note characteristic streaking of much of underparts. (M. Pope)

Juv, Spain, Aug: yellow gape-flanges and soft, fluffy body-feathers reveal young age, otherwise like ♀. (C. N. G. Bocos)

SIND SPARROW
Passer pyrrhonotus Blyth, 1845

Alternative names: Jungle Sparrow, Sind Jungle Sparrow

Fr. – Moineau à calotte grise; Ger. – Dschungelsperling
Sp. – Gorrión del Sind; Swe. – Sindsparv

This localised species is resident in scrub and tall grass of extreme SE Iran and in lowland riparian *Tamarix* and *Acacia* scrub, usually away from habitation, along major rivers of Pakistan and extreme NW India. Range is apparently slowly expanding. Loosely colonial, feeding on ground, mainly on grass seeds.

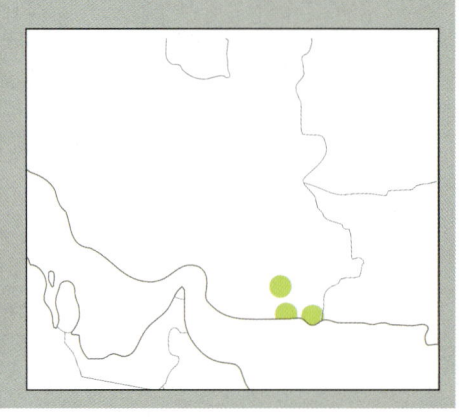

♀, India, Apr: mantle 'braces' pale buff, with greyer cast to crown and ear-coverts, contrasting with whiter chin and throat. Strong wing pattern, with broadly white-tipped median coverts, obvious dark centres to tertials, pale secondary panel and basal primary patch, and chestnut-brown lesser coverts. Bill black when breeding. (N. Devasar)

♂, India, Mar: much like a small House Sparrow with finer bill. Somewhat similar head pattern in ♂ plumage, with grey crown and nape. Note more reddish chestnut-brown sides to nape, in form of broad band surrounding medium grey cheeks, smaller black bib and almost uniform greyish underparts. (A. Arya)

♂, India, Dec: some pale tips on chestnut rear crown-sides and blackish bib when very fresh. Note well-developed pattern above and double wing-bars. Complete autumn moult of both age classes makes ageing impossible thereafter. (A. Arya)

IDENTIFICATION Rather closely recalls House Sparrow but is *smaller and more round-winged*, which makes tail look a little longer. ♂ immediately recognised by *grey, not chestnut, nape, grey cheeks* and *chestnut rump*. ♂ furthermore has *small black bib that does not extend to breast*. Other features are the same as those found in ♂ House Sparrow: lores and a small mask black, and *crown grey with a broad chestnut-brown band behind eye* surrounding the medium grey lower ear-coverts and cheeks. Mantle and scapulars much as House Sparrow, *heavily streaked buff* and some *chestnut on lower back and rump*, while median coverts broadly tipped white, forming *obvious wing-bar*, and tertials have bold dark centres and buff fringes. There is an obvious *pale buffish basal primary patch*, and *chestnut lesser coverts*. Underparts pale off-white to slightly greyish. Bill black in breeding season. ♀ very closely resembles House Sparrow. Not particularly attached to habitations or cultivations, and forms only comparatively small flocks in non-breeding season, sometimes mixing with House Sparrows. At times rather confiding.

VOCALISATIONS Song comprises long arrays of typical House Sparrow-like chirps, interspersed with a Pied Wagtail-like call note and various twittering sounds. The song is said to be slightly softer than House Sparrow. – Calls like those of House Sparrow, varied chirping notes in long series. Also gives a soft *chup* (lower-pitched than House Sparrow).

SIMILAR SPECIES Distinguished from quite similar *House Sparrow*, close to which it sometimes nests, by virtue of ♂ having grey cheeks (local House Sparrows: white), smaller black bib (not reaching upper breast) and chestnut lower back and rump (rather than greyish). Note also from behind that grey of centre on crown runs down nape onto upper mantle (House Sparrow: chestnut nape). ♀ Sind Sparrow usually has a more prominent buffish-white supercilium (reaching nape-sides), warmer buffish-brown lower back and rump, greyer ear-coverts, contrasting with white throat, usually more rufous on wing-coverts (chiefly lesser), and better-marked wing-bars and bolder tertial pattern, somewhat like pattern in ♂♂. However, both sexes are just as likely to be distinguished from House Sparrow by their smaller size, overall slimmer appearance and smaller bill. – Cf. also similar-sized ♀ Dead Sea Sparrow.

AGEING & SEXING Ageing only possible in summer before post-juv moult finished. Sexes generally easily differentiated once post-juv moult underway. Moderate seasonal variation in ♂ (due to feather abrasion). – Moults. Both ad and juv moults completely, including wings and tail, in late summer. – SPRING **Ad ♂** With wear, head and wing patterns better-marked, chestnut parts bright, underparts paler but whitish median coverts wing-bar much reduced. **Ad ♀** Resembles a small House Sparrow, but has more prominent pale supercilium that reaches nape-sides. **1stS** Inseparable from respective ad. – AUTUMN **Ad ♂** Grey of crown, cheeks and nape tinged pale sandy-brown, chestnut band behind eye duller, bib reduced and has whitish tips, duller culmen and bill tip, and lower mandible yellow-brown. **Ad ♀** Generally warmer than in spring, with broader fringes, and more buffish supercilium and fringes to coverts. **1stW** Indistinguishable from respective ad, but presumably least-marked ♂♂ in autumn (with extensive pale fringes/tips) are 1stW. **Juv** Like ♀ but has soft, fluffy body-feathers.

BIOMETRICS L 12–13 cm; **W** ♂ 66–69.5 mm (n 15, m 67.7), ♀ 61–65 mm (n 12, m 63.3); **T** ♂ 47–53 mm (n 15, m 50.8), ♀ 45–51 mm (n 12, m 48.2); **T/W** m 75.6; **B** 11.1–12.2 mm (n 12, m 11.7); **B**(f) 8.3–10.1 mm (n 11, m 9.1); **BD** 5.9–7.0 mm (n 16, m 6.4); **BW**(g) 4.6–6.0 mm (n 12, m 5.3); **Ts** 15.4–18.1 mm (n 13, m 16.4). **Wing formula: p1** minute; **pp2–5** about equal and longest (p5 rarely to 3 mm <); **p6** < wt 3–8 mm; **p7** < wt 6–9.5 mm; **p10** < wt 13–18 mm; **s1** < wt 14.5–18 mm. Emarg. pp3–6.

GEOGRAPHICAL VARIATION & RANGE Monotypic. – SE Iran, Pakistan, NW India; sedentary, or makes only short-distance movements outside breeding season.

♀♀, India, Dec: in fresh plumage, supercilium sullied rufous-buff, long, broad and well-defined. Comparatively pale bill in non-breeding season. (Left: A. Arya; right: N. Devasar)

DEAD SEA SPARROW
Passer moabiticus Tristram, 1864

Fr. – Moineau de la Mer Morte; Ger. – Moabsperling
Sp. – Gorrión del Mar Muerto; Swe. – Tamarisksparv

A local resident or partial migrant in lowland SC Turkey and Syria east to SW Iran, also found locally around the Dead Sea and in the Arava Valley of Israel/Jordan, and at least formerly bred on Cyprus. It also has an isolated population in the borderland between SE Iran and SW Afghanistan. It prefers tall trees or large bushes near water. The winter range is rather poorly known, but it has been noted south to Bahrain and Saudi Arabia, as well as in S Israel, but it appears to largely vacate the extreme north. At this season it is sometimes recorded far from water and occasionally associates with Spanish Sparrows.

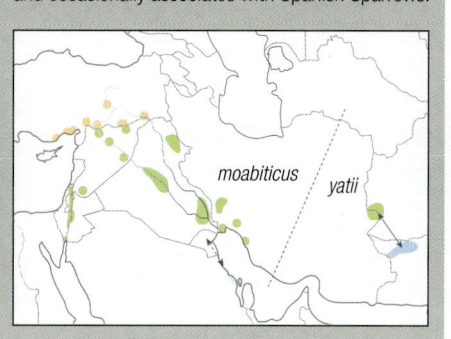

P. m. moabiticus, ♂, Israel, May: typical display posture with spread wings displaying chestnut lesser coverts. Complete autumn moult of both age classes makes ageing impossible thereafter. (H. Shirihai)

P. m. moabiticus, ♂, Iran, Apr: compact with distinctive head pattern, especially when worn, with grey crown and ear-coverts, white supercilium becoming cinnamon at rear, short black eye-stripe, black bill and bib, and white submoustachial stripe that broadens into bright yellow neck-side patch. Wing and upperparts patterns no less distinctive. (Magnus Ullman)

IDENTIFICATION *Diminutive, small-billed sparrow.* Breeding ♂ colourful with *bluish-grey crown, cheeks and nape, white supercilium* (becoming yellow-buff or even rufous-cinnamon behind eye), *white submoustachial stripe that broadens into a bright yellow half-collar.* The short *eye-stripe is black*, as are bill and bib, and there is a small white mark below eye. Upperparts mostly chestnut and cinnamon, boldly streaked black, with *chestnut-orange scapular fringes and wing-coverts forming a chestnut panel* between mantle and rest of wing. There is a *whitish primary patch* outside tips of primary-coverts, and narrow *pale wing-bars*. Underparts largely greyish-white (sometimes tinged pinkish-cinnamon), but undertail-coverts faintly rusty. Legs flesh-brown and iris dark sepia-brown. ♀ usually clearly differentiated (though some have partial ♂ pattern), resembling ♀ House Sparrow with streaky upperparts, but has *whitish supercilium, buff-white submoustachial stripe*, chin and throat, and sometimes *yellowish lower throat-feathering*. Also note small size and *rounded wings* making *tail look proportionately longer*. Action and behaviour more like Tree Sparrow, presumably due to small size, but even more restless and jerky, and flight has fast whirring wing action. Rather elusive, and even large flocks can remain unseen in dense bushes. However, builds a unique, huge round nest (often overlooking water and constructed of bare hard twigs) that is visible at some distance.

VOCALISATIONS Song a distinctive fast, rolling or sawing che-*rirp* che-*rirp* che-*rirp* che-*rirp* or a monotonous chihp-chihp-chihp-chihp-chihp-... or a slower and more varied series, particularly in colonial chorus, tri-rirp chrirp, tlir-tlir-tlir chlip chrilp chrirp chilp. Voice a little feebler than House or Spanish Sparrows, but is perhaps most similar to latter (though can be perceived as slightly sharper and harder). – Calls resemble Spanish Sparrow with similar metallic ring but perhaps sharper and more liquid, tchrelp and a repeated disyllabic chep-chew or tcheep-tcheep. Rattle call also rather Spanish-like, but feebler and higher-pitched, trrirrrp or more rattling tret-et-et-et-et.

SIMILAR SPECIES Unmistakable provided small size and ♂'s unique head and wing patterns are visible. ♀, however, is by and large a replica of ♀ *House* and *Spanish Sparrows* and requires careful separation (especially in mixed flocks with latter species), but like ♂ is much smaller with a tiny bill and shorter primary projection, less than half tertial length (with three occasionally four, more closely-spaced primary tips visible; four or five widely-spaced tips in the other species). Also, note pronounced buffish-white supercilium above and behind eye, broadening posteriorly. Upperparts paler sandy-grey and more regularly and sharply dark-streaked. Pale-tipped wing-bars (especially on median coverts) more contrasting, and variably tinged chestnut. Underparts largely whitish. May show yellowish tinge to supercilium and throat-sides. Note also undertail-coverts strongly patterned with dark centres and pale fringes.

AGEING & SEXING Ageing only possible in summer before post-juv moult finished. After moult, 1stY similar to respective ad but some young ♂♂ separable in the hand, as are some older ♀♀. Sexes generally easily differentiated once post-juv moult under way. Moderate seasonal variation in ♂ (due to feather abrasion). – **Moults.** Both ad and juv moults completely, including wings and tail, in late summer. Pre-nuptial moult absent. – **SPRING Ad ♂** With wear (Apr–Jul) plumage pattern becomes more clear and better-marked, including bluish ash-grey crown and cheeks, black mask and (larger) black bib, and chestnut mid-wing panel. Also, underparts become whiter, rufous centres to undertail-coverts more exposed, and pale wing-bars almost lost. **Ad ♀** Note that some, especially older ♀♀, may develop slightly greyer heads and even rufous-orange tips and fringes to median coverts, even greater coverts and scapulars, but rest of plumage still essentially ♀-like. **1stS** Inseparable from respective ad, but it seems 1stS ♂♂ are overall duller, while ♀♀ apparently never acquire chestnut coverts like some older birds. – **AUTUMN Ad ♂** Grey of crown, cheeks and nape tinged pale sandy-brown, bib reduced and slightly flecked whitish, while underparts warmer and bill brownish. **Ad ♀** Less greyish with broader fringes and buffier supercilium and fringes to coverts. **1stW** At least some ♂♂ have much less

P. m. moabiticus, ♂, Turkey, May: rufous-chestnut median and greater coverts with narrow whitish tips, lesser coverts black with broad cinnamon to white tips, forming wing-bar. Primary-coverts largely black, tertials black with broad pale cinnamon fringes. Note well-patterned buff-brown upperparts with brighter cinnamon and black-streaked mantle and scapulars. (M. Schäf)

P. m. moabiticus, ♀, Turkey, May: recalls ♀ House or Spanish Sparrows, even in wing and face patterns, but much smaller, with shorter primary projection. Also, generally paler and usually has hint of buffish in post-ocular supercilium. Diagnostic yellow patch on neck-sides and rusty-brown undertail-coverts are not always visible. (R. Armada)

P. m. moabiticus, ♀, Israel, May: faint yellow patch on neck-sides (visible only on some birds, and visibility varies with light and angle), and rusty-brown undertail-coverts. Often looks long-tailed due to short primary projection. (H. Shirihai)

P. m. moabiticus, ♀, presumed ad, Turkey, May: chestnut tips to greater coverts indicate older ♀. Here chestnut-tinged ventral area and dark chestnut centres to undertail-coverts show well. (M. Schäf)

grey on head, and chestnut mid-wing panel less solid (due to partially-exposed blackish centres to wing-coverts, unlike ad); young ♀♀ apparently never acquire chestnut fringes to coverts. **Juv** Like ♀ but has soft, fluffy body-feathers. Iris more olive-grey (dark brown or sepia-brown in imm and ad, respectively).

BIOMETRICS (*moabiticus*) **L** 11–12.5 cm; **W** ♂ 58–63 mm (n 13, m 60.6), ♀ 55–60.5 mm (n 12, m 58.2); **T** ♂ 46–51 mm (n 13, m 48.1), ♀ 43–48 mm (n 12, m 45.8); **T/W** m 79.0; **B** 10.2–12.0 mm (n 24, m 11.1); **B**(f) 7.6–9.6 mm (n 25, m 8.8); **BD** 5.7–6.4 mm (n 22, m 6.0); **BW**(g) 4.5–5.5 mm (n 25, m 4.9); **Ts** 15.0–17.0 mm (n 25, m 15.9). **Wing formula: p1** minute; **pp2–5** about equal and longest (p5 rarely to 3 mm <); **p6** < wt 2–5.5 mm; **p7** < wt 4.5–8 mm; **p10** < wt 11–16 mm; **s1** < wt 13–16.5 mm. Emarg. pp3–6.

GEOGRAPHICAL VARIATION & RANGE Two allopatric subspecies with clear differences in plumage colours and structure, but little in size. Separate species status for these have been proposed (see Taxonomic note) but is not followed here.

P. m. moabiticus Tristram, 1864 (SE Turkey, Syria, Israel, Jordan, Iraq, SW Iran; winters in south of range). Described above. Underparts dusky-white with at most only very slight yellow wash in some. Birds in Turkey, Iraq and SW Iran ('*mesopotamicus*') sometimes separated on account of subtly larger size, but difference very slight with much overlap, thus best included here. (Syn. *mesopotamicus*.)

P. m. yatii Sharpe, 1888, 'Afghan Sparrow' (E Iran, SW Afghanistan; short-range migrant in winter to W Pakistan). Differs rather clearly in being longer-winged and longer-legged, and in that ♂ has underparts washed clear lemon-yellow, only undertail-coverts being whitish. The yellow is strongest on lower flanks and upper breast. Upperparts streaking averages narrower and are fringed slightly paler, more sandy-brown, while the lower back and rump are paler, and the grey head is paler, too. Central tail-feathers have somewhat narrower dark centres and broader pale fringes.

P. m. moabiticus, Israel, Feb: diagnostic head and wing patterns show up well on this flock in typical dense reedbed habitat. (L. Kislev)

Undertail-coverts less well saturated, and has almost no chestnut flecking on rear flanks. ♀ differs less clearly, but is rather extensively washed pale yellow, mainly on lower throat and upper breast and, in some, yellowish-buff on body-sides. Slightly paler and sandier above (with paler buff-grey fringes). Tends to have better-marked supercilium. **W** ♂ 62–68 mm (n 12, m 65.0), ♀ 61–65 mm (n 12, m 62.9); **T** ♂ 48–53 mm (n 12, m 50.4), ♀ 46–50 mm (n 12, m 48.6); **T/W** m 77.4; **B** 10.8–13.1 mm (n 23, m 11.6); **B**(f) 8.3–9.7 mm (n 22, m 9.1); **BD** 6.0–7.2 mm (n 22, m 6.5); **BW**(g) 4.7–6.5 mm (n 22, m 5.6); **Ts** 16.0–17.6 mm (n 23, m 17.1).

SPARROWS AND ALLIES

TAXONOMIC NOTES Race *yatii* considered by Boros & Horváth (1954) as a separate species due to strong differences in colour and, to some extent, size. Kirwan (2004) provided detailed analysis to support this assumption. At least morphologically, the two forms are highly distinctive, perhaps suggesting long separate evolutionary trajectories, a premise that demands testing using molecular methods. Given that *yatii* and *moabiticus* are allopatric, it is impossible to test the degree to which they are reproductively isolated. They are not known to differ in vocalisations. A cautious policy is therefore adopted, treating them as conspecific.

REFERENCES Boros, I. & Horváth, L. (1954) *Acta Zool. Acad. Sci. Hungary*, 1: 43–51. – Kirwan, G. M. (2004) *Sandgrouse*, 26: 105–111.

P. m. moabiticus, ♂, Israel, Nov: when fresh in autumn and winter, grey of crown, cheeks and nape tipped pale sandy-brown, bib reduced and slightly flecked whitish, and bill paler. (L. Kislev)

P. m. moabiticus, presumed 1stW ♂, Kuwait, Nov: following complete moult (late summer) ageing usually impossible, but at least some young ♂♂ show less solid chestnut mid-wing panel than ad, and have much less grey on head, restricted mostly to ear-coverts, while blackish bib is smaller and yellow tinge to rear supercilium fainter. By early spring differences from ad are lost. (M. Pope)

P. m. moabiticus, ♂ at its nest, Israel, Jun: anyone that travels in the Jordan River Valley or the Dead Sea Depression looking for the enigmatic Dead Sea Sparrow should be aware that often well before the bird is spotted or heard, its presence will be revealed by its distinctive large oval-shaped nest, often a few clustered together in one area. The nest can be as much as 100 times larger than the bird itself. (H. Shirihai)

P. m. yatii, ♂, Iran, May: unlike *moabiticus*, underparts extensively washed lemon-yellow, and upperparts streaking averages narrower. Also broader and warmer sandy-brown fringes to wing-feathers and mantle, and undertail-coverts much less saturated with almost no chestnut flecking on rear flanks. (C. N. G. Bocos)

P. m. yatii, ♀, Iran, May: differs less clearly from *moabiticus* than ♂, but has uneven pale yellow-olive to yellowish-buff wash below, tends to have better-marked supercilium, and upperparts sandier. Specifically distinct (like *moabiticus*) in being much smaller than ♀ House and Spanish Sparrows with smaller bill, shorter primary projection, and pronounced supercilium above and behind eye. (C. N. G. Bocos)

IAGO SPARROW
Passer iagoensis (Gould, 1838)

Alternative names: Cape Verde Sparrow, Rufous-backed Sparrow

Fr. – Moineau à dos roux; Ger. – Rostsperling
Sp. – Gorrión de Cabo Verde; Swe. – Kap Verdesparv

The Iago Sparrow is endemic to the Cape Verde Islands, where it is generally common and widespread (absent only from Fogo). It inhabits desolate lava fields, gorges with cliffs, arid plains, and areas near cultivation. Closer to habitation, it will mix with Spanish Sparrow and even enters villages and towns in some areas. Usually observed in small parties, it can form large flocks where abundant.

IDENTIFICATION Compact sparrow with a strong conical bill, short wings and longish tail. ♂ unmistakable due to *blackish crown, rufous-cinnamon wedges above and behind eye* (broadening on nape-sides), narrow white line above lores, black eye-stripe continuing behind *whitish cheeks*, and *narrow black bib* restricted to centre of throat. Dusky shawl contrasts with *bright chestnut mantle* (streaked black), *scapulars, back and lesser coverts*. The rump is more brownish-grey. Wings have a *bold white median coverts bar*, and less distinct buff fringes and tips to greater coverts. Underparts white, tinged cream in centre and grey on breast and flanks. ♀ much duller, resembling ♀ House Sparrow but has slightly *more contrasting whitish supercilium emphasized by dark eye-stripe*, and (on some) a tiny dusky bib and dusky cheeks, and *black-streaked mantle-sides and scapulars*. Compared to ♀ House and Spanish Sparrows appears *neater*, with broader white median coverts bar (when fresh), *better-defined face pattern* and *obvious pale fringes to tertials* (more ill-defined in other species). Bill black in breeding ♂, horn-coloured in non-breeding plumage and ♀. Legs lead-grey. Highly gregarious, as typical of genus.

VOCALISATIONS Song by ♂ at nest a repeated slightly trilling high-pitched chirping, *chirrp*, more slurred than in House and Spanish Sparrows (Bourne 1955), with some variation in details. – Calls much like in other *Passer* species. Basic *chirrp* call clearly slightly higher-pitched than equivalent call of House Sparrow (and subtly higher-pitched than Spanish Sparrow), and is generally doubled or trebled. Also noted are a twangy *cheep* and a doubled *chew-weep*.

SIMILAR SPECIES On Cape Verdes could only be confused with the much larger and chunkier *Spanish Sparrow*, but the main plumage criteria for both sexes should readily separate the two.

AGEING & SEXING Ageing only possible before and during moult (provided some old flight-feathers retained). Sexes differ markedly already as juv. Limited seasonal variation related to feather wear. – Moults. Both ad and juv undergo complete moult, mainly in winter to early summer, but moult is protracted and probably can last even longer in juv. (Breeding normally in Aug–Jan.) Pre-nuptial moult presumably absent. – **BREEDING Ad ♂** With wear, head and wing patterns become more well-defined and neat; especially crown gets blacker, and tips to greater coverts whiter and narrower. **Ad ♀** See under Identification. **1stS** Inseparable from respective ad. – **NON-BREEDING Ad ♂** When fresh head pattern slightly less clear-cut, being tipped pale sandy-brown, and bib reduced. Bill horn-coloured (or even paler on lower mandible). **Ad ♀** Much as in worn

Ad ♂, Raso, Cape Verdes, Oct: blackish crown merges into grey nape, bordered by broad chestnut supercilium. Upperwing has obvious chestnut lesser coverts patch and bold white median coverts wing-bar. Plain sullied off-white underparts. Moult varies with breeding, but when only ad feathers visible and bill is black on a nest-building bird, age should be ad. (E. Winkel)

Ad ♂, Santiago, Cape Verdes, Feb: dark bib not always easy to see. Much individual and seasonal variation (here fresh and moulting to ad plumage). (V. Legrand)

♀, Santo Antão, Cape Verdes, Mar: similar to House and Spanish Sparrows, but has much shorter primary projection and is neater, with well-defined pale supercilium enhanced by dusky crown and dark eye-stripe. White median coverts bar can wear off almost completely as here. (H. Shirihai)

♀, Santo Antão, Cape Verdes, Feb: sometimes dusky crown appears as dark patch on forecrown. Note cinnamon-brown lesser coverts and white median coverts bar somewhat mirroring ♂ pattern. After moult completed ageing is impossible. (H. Shirihai)

1stY ♂, Santo Antão, Cape Verdes, Feb: subdued head pattern and pale bill indicate young age. Otherwise, ageing rarely possible after moult completed. (H. Shirihai)

spring plumage but buffier on pale parts, including supercilium and fringes to coverts. **1stW** Indistinguishable from ad, but owing to extended moult period, even months after breeding season has ended, many still retain juv wing- and body-feathers. **Juv** Resembles ♀, with soft, fluffy body-feathering, but ♂ already has black-streaked crown and partly rufous rear supercilium. Bill yellowish horn-coloured to dull pinkish-grey.

BIOMETRICS L 12–13 cm; W ♂ 61.5–68 mm (*n* 12, *m* 64.0), ♀ 55–61 mm (*n* 12, *m* 58.4); T ♂ 48–57 mm (*n* 12, *m* 51.3), ♀ 41–53 mm (*n* 12, *m* 45.9); T/W *m* 79.4; B 13.3–15.7 mm (*n* 22, *m* 14.5); B(f) 9.8–13.3 mm (*n* 22, *m* 11.2); BD 6.0–8.7 mm (*n* 23, *m* 7.4); BW(g) 5.3–7.1 mm (*n* 24, *m* 6.4); Ts 15.5–19.6 mm (*n* 23, *m* 17.6). **Wing formula: p1** minute; **p2** < wt 1–2 mm; **pp3–5** about equal and longest; **p6** < wt 2–5 mm; **p7** < wt 5–8 mm; **p10** < wt 10–15 mm; **s1** < wt 11–16 mm. Emarg. pp3–6.

GEOGRAPHICAL VARIATION & RANGE Monotypic. – Cape Verdes (except on Fogo); sedentary. – There are some slight average and clinally varying differences in plumage colour, bill size and length of wing and tarsus within the Cape Verde Is (Bourne 1955, 1957), but not distinct enough to warrant separation (Vaurie 1958). (Syn. *brancoensis*.)

TAXONOMIC NOTE Sometimes united with Rufous Sparrow *P. motitensis* (extralimital; not treated) in a single polytypic species, but the Iago Sparrow, the Rufous Sparrow and their closest relatives are nowadays usually considered to represent six allospecies.

REFERENCES Bourne, W. R. P. (1955) *Ibis*, 97: 548–549. – Bourne, W. R. P. (1957) *Ibis*, 99: 182–190. – Vaurie, C. (1958) *Ibis*, 100: 275–276.

♀, São Nicolau, Cape Verdes, Oct: very fresh plumage. Typical greyish hue to dark parts of head. Note proportionately heavy bill, cinnamon-brown scapular fringes and lesser coverts, and relatively thin white median coverts bar. (E. Winkel)

Juv (with ♂ on right), Santo Antão, Cape Verdes, Feb: juv has same plumage as ♀, only with slightly warmer brown upperparts and more buffish supercilium, fluffy body-feathers, and yellow gape. (H. Shirihai)

DESERT SPARROW
Passer simplex (M. H. C. Lichtenstein, 1823)

Fr. – Moineau blanc; Ger. – Wüstensperling
Sp. – Gorrión sahariano; Swe. – Ökensparv

A resident of sandy deserts (often found in scrub and *Acacia* around oases) in SE Morocco to C Libya, Mauritania and from C Mali to C Sudan. The species appears to wander in response to rainfall and the availability of grassy areas. The easiest place to see it remains the EC Moroccan Merzouga close to the Algerian border. Formerly, the similar-looking Asian isolate *zarudnyi* was considered a subspecies of this species, but is treated here as a separate species, Zarudny's Sparrow, for reasons explained therein.

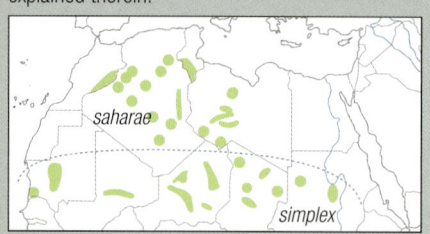

P. s. saharae, ♂, Morocco, Feb: most ♂♂ of this race are variably washed ash-grey above, with cream-white rump and uppertail-coverts. Generally unmistakable by short, broad blackish eye-stripe. Well-developed black bib to centre of breast is striking, especially head-on. Rest of head- and neck-sides whitish, and underparts cream-buff to white. Also note diagnostic wing pattern. (D. Occhiato)

IDENTIFICATION A very pale, dainty sparrow with a stubby bill. Attractive ♂ plumage *almost uniform pale greyish creamy-drab above*, though usually crown to mantle and lesser coverts silvery grey, with *whiter rump and uppertail-coverts*. Head has *black loral stripe* (in shape of mask), *whitish cheeks* and a *black narrow bib* widening to a patch on upper breast. Variegated wing pattern comprises *bold whitish median coverts bar, grey-black bases to greater coverts* which, together with *mostly black alula and primary-coverts*, create *narrow V-shaped dark mark* bordering *buffish-white greater coverts band*. Extensively pale cream-pink remiges form conspicuous pale wing area, which contrasts against *grey-black centres to tertials and dark primary tips*. Tail dark-centred and further darkens towards tip. Underparts dull white, often faintly tinged pink or buff. ♀ *has head and upperparts isabelline to pink-buff*, with faint (or no appreciable) pale supercilium, *warmer ochre-toned crown, mantle, breast and flanks*. Nearly invariably *lacks black bib* (has vague hint at most) and has only indistinct dark wing markings, with usually no or only diffuse pale wing-bars. *Bill when breeding black* in ♂, variably dark brown-grey in ♀, but dull brown or straw-coloured at other seasons. Tarsus flesh-brown. Lives in small groups, often attracted to nearness of villages.

VOCALISATIONS Song much more metallic and musical than House Sparrow, comprising a melodious series of trills, often a mixture of a quick double note and more drawn-out chirps, *chee-wee, cheerp, chee-wee, cheerp, chee-wee,...* etc. – Calls more liquid and metallic sonorous than chirps of House Sparrow, but also some more like House Sparrow but usually higher-pitched, *chip-chip*, or subdued lower-pitched, repeated *chu*, or *chüp-chüp-chüp* recalling Greenfinch in flight, any of these often mixed with more normal sparrow-like chirps. When alarmed or in agony utters typical chattering series for sparrows, *cher'r'r'r'r'r'r'r*.

SIMILAR SPECIES ♂ wholly distinctive at all seasons, but a lone ♀ or juv seen briefly and poorly might cause confusion with other desert passerines, e.g. Bar-tailed Lark, or ♀/young Trumpeter Finch. Nonetheless, the wing and tail patterns of Desert Sparrow once seen are diagnostic. – Very similar to Zarudny's Sparrow, but the two are mainly resident, live in widely separated areas and are unlikely to ever come into contact. Still, in case the unexpected happens, or as guidance for museum workers, Desert Sparrow is shorter-tailed (see Biometrics). Judging from specimens it is about the same size as Zarudny's Sparrow or fractionally smaller (not larger, *contra* Kirwan et al. 2009). It has a more subdued facial pattern, and ♀♀ are much paler and washed out, being more buffish (stronger sexual dimorphism). ♂ plumage averages slightly more buff-tinged grey (Zarudny's: purer grey and white), and dark bib tends to be broader and more extensive (Zarudny's: narrower, but probably some overlap). ♀ Desert Sparrow is clearly smaller than the ♂, whereas in Zarudny's Sparrow sexes are nearly the same size. Bill averages slightly longer and more pointed, less bulbous in Desert Sparrow.

AGEING & SEXING (*saharae*) Ageing only possible before and during moult (provided some old flight-feathers retained), but see below regarding some 1stW ♂♂. Sexes differ sharply in plumage, slightly in size. There is some seasonal change due to feather wear, especially in ♂. – Moults. Both ad and juv undergo complete moult, mainly in late summer (apparently often starting Jul and ending in Sep). Pre-nuptial moult usually absent, but odd ♂♂ replace one or more tertials. – **SPRING Ad** ♂ See under Identification. With wear patterns of head and wing become neater. Especially bib is blacker and broader, and black bases to greater coverts somewhat more exposed. **Ad** ♀ See general description under Identification. **1stS** Inseparable from respective ad, but see 1stW. – **AUTUMN Ad** ♂ Due to fresh feather-tips has warmer, ginger wash to forehead, crown-sides, mantle and lesser coverts, but extent varies (some already quite pure greyish). Also, often has buff tinge to upper breast, flanks, greater coverts and wing panel. Black of bib and bases to greater coverts largely concealed, and bill brown with prominent yellowish-horn base. **Ad** ♀ Warmer sandy-buff than in spring. **1stW** Much as respective ad, but at least some ♂♂ are comparatively more sandy-buff (much less grey) than ad ♂, with poorly-developed dark marks on face and bib. **Juv** Resembles ♀ but face and wing markings even duller, upperparts less buff, underparts uniform sandy-white. Bill and legs pale flesh.

BIOMETRICS (*saharae*) **L** 13.0–14.5 cm; **W** ♂ 74–83 mm (*n* 17, *m* 78.9), ♀ 72–81 mm (*n* 15, *m* 75.7); **T** ♂ 52–59 mm (*n* 17, *m* 55.1), ♀ 50–58 mm (*n* 15, *m* 53.4); **T/W** *m* 70.2; **B** 11.0–14.5 mm (*n* 25, *m* 12.8);

P. s. saharae, ♂, Morocco, Dec: ♂♂ when very fresh are tinged buffish sandy-brown above, and deeper buff below. Diagnostic wing pattern with 'angle-shaped' dark mark comprising black-centred median coverts, alula and primary-coverts. Also has black wingtip. Lesser and median coverts variably tipped paler. (R. Messemaker)

B(f) 8.1–11.8 mm (*n* 25, *m* 9.9); **BD** 6.8–8.2 mm (*n* 25, *m* 7.5); **BW**(g) 6.1–8.0 mm (*n* 25, *m* 7.1); **Ts** 17.7–21.3 mm (*n* 32, *m* 19.6). **Wing formula: p1** minute; **p2–4** about equal and longest (either or both of p2 and p4 often to 1.5 mm <); **p5** < wt 0.5–4.5 mm; **p6** < wt 5–10 mm; **p7** < wt 9–14 mm; **p10** < wt 18–23.5 mm; **s1** < wt 19–24.5mm. Emarg. pp3–5.

GEOGRAPHICAL VARIATION & RANGE Clinal and moderate variation in N Africa.

○ ***P. s. saharae*** Erlanger, 1899 (NW Sahara from SE Morocco to SW Libya; resident or makes shorter local movements). Described above. The largest and palest form. Upperparts in ♂ pale ash-grey, in ♀ pale pink-buff. Underparts more pale cream-white (less pink-buff). Probably grades into *simplex* in the south where they meet.

Mainly extralimital: ***P. s. simplex*** (M. H. C. Lichtenstein, 1823) (C & NE Mali, Niger, N Chad, NW & C Sudan; perhaps marginally enters the covered region in S Algeria; sedentary). Very similar to *saharae* but a trifle smaller, and ♂ slightly darker above and has underparts subtly more tinged pink-buff, less whitish (but needs direct comparison to discern). ♀ is similarly slightly darker above and more pinkish-buff, less whitish below. Scant material available. **W** ♂ 76–78 mm (*n* 5, *m* 76.8), ♀ 73–76 mm (*n* 4, *m* 74.3); **T** ♂ 50–55 mm

P. s. saharae, ♀, Morocco, Mar: lacks dark face and bib of ♂, with only alula and tips to primaries and primary-coverts blackish. This bird has clearly paler-tipped lesser and median coverts, forming vague wing-bars. (R. van Rossum)

P. s. saharae, ♀, Tunisia, Feb: especially in ♀, the black bill of a breeder is very contrasting. Wing pattern with dark alula, tips to primaries and primary-coverts, and dark inner web to tertials, as well as dark-centred tail, eliminates confusion with other desert passerines. (G. Olioso)

(*n* 5, *m* 53.0), ♀ 46–52 mm (*n* 4, *m* 49.0); **T/W** *m* 67.7; **B** 11.4–13.4 mm (*n* 9, *m* 12.4); **B**(f) 8.6–10.2 mm (*n* 9, *m* 9.7); **BD** 6.3–8.0 mm (*n* 9, *m* 7.1); **BW**(g) 6.5–7.6 mm (*n* 9, *m* 7.0); **Ts** 18.4–19.5 mm (*n* 9, *m* 18.8). **Wing formula:** Sometimes a hint of emarg. also on p6.

TAXONOMIC NOTE Desert Sparrow is treated here as a separate species versus Zarudny's Sparrow, for reasons given under latter species.

REFERENCES Densley, M. (1990) *BB*, 83: 195–201. – Kirwan, G. M. *et al.* (2009) *DB*, 31: 139–158.

P. s. saharae, ♀, Western Sahara, Jan: in fresh winter plumage, when bill is pale. Specific wing pattern is again well visible. (C. N. G. Bocos)

P. s. saharae, juv, Western Sahara, Mar: like ♀, but dark areas of wing browner. Note fluffy juv body-feathers and coverts, and yellow gape. (V. Legrand)

ZARUDNY'S SPARROW
Passer zarudnyi Pleske, 1896

Fr. – Moineau de Zaroudny; Ger. – Sarudnysperling
Sp. – Gorrión del karakum; Swe. – Turkestansparv

This rare Central Asian desert-dwelling sparrow is a close relative of the Desert Sparrow of Saharan Africa, a species with which it has often previously been lumped. Zarudny's Sparrow occurs in the Karakum Desert of Turkmenistan and in the Kyzylkum Desert of Uzbekistan, and apparently at least formerly also bred in NE Iran but now seems extinct there. According to recent reports, numbers elsewhere appear to be declining. Just like its African relative, it seems to live sedentarily in small parties, but is partially nomadic depending on rains and seed supply.

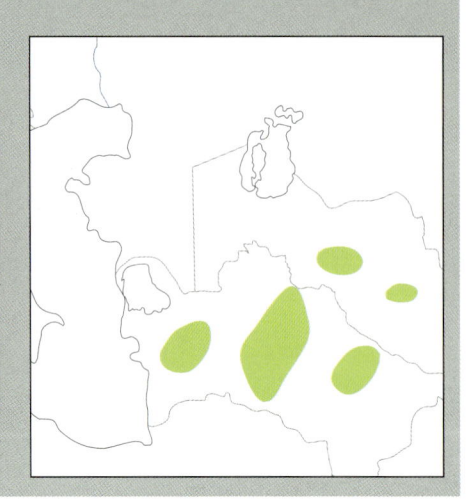

♂ (top left) and ♀, Turkmenistan, May: same pair as in two images below, nest building in tamarisk. The ♂ shows the broad black tertial centres, and the ♀ its more sandy-brown upperparts. (N. Redman)

IDENTIFICATION Like its African relative, a *small, pale sparrow* with a *chunky bill*. In general very similar to Desert Sparrow, and the two have probably by and large the same body size (if anything Zarudny's seems the larger of the two), a *large pale panel on wing*, white median-coverts wing-bar and black (or dark) in wing is restricted to primary tips, tertial centres and a *black* V-*shaped wing mark* formed by primary-coverts, alula and bases to greater coverts. Zarudny's differs from Desert Sparrow in its proportionately *longer tail*, strongly reduced sexual dimorphism (♀ more similar to ♂), a *better-marked facial mask* and a more globular-shaped and therefore slightly shorter-looking bill. ♂ in bleached plumage is *purer grey above*, and generally *whiter below*, whereas in fresh plumage the difference is less significant (though it probably averages more grey-tinged even then). ♀, unlike Desert Sparrow, is closer in appearance to ♂, with at least blackish-grey bib and mask (sometimes near-black), and is greyer overall. In both sexes the *dark bib seems to be on average broader and more extensive*, especially in its lower part (but certainly some overlap).

VOCALISATIONS No differences from Desert Sparrow known, but largely unstudied. Whether it is true, as claimed by some Russian authors, that the call resembles those of Goldfinch *Carduelis carduelis* remains to be better established.

SIMILAR SPECIES See Desert Sparrow for differences from latter, and from other similar species.

AGEING & SEXING Not closely studied, mainly due to scant material available in collections, but appear to be similar to what is known for Desert Sparrow. Birds finishing complete moult have been recording in Aug and early Sep.

BIOMETRICS L 13.0–14.5 cm; **W** ♂ 69–74 mm (*n* 17, *m* 72.0), ♀ 68–72 mm (*n* 9, *m* 69.9) (Dementiev & Gladkov 1954). – **W** ♂ 69–76.5 mm (*n* 17, *m* 72.5), ♀ 69.5–72.5 mm (*n* 10, *m* 71.2); **T** ♂ 59–65.5 mm (*n* 17, *m* 62.3), ♀ 52.5–67 mm (*n* 10, *m* 62.6); **T/W** *m* 86.6; **B** 9.8–12.6 (*n* 27, *m* 11.2). (G. M. Kirwan *in litt.* 2013; note that when comparing measurements of same specimens of other taxa, GMK measured tail longer by *c.* 5% than LS.) **Wing formula:** Not closely studied but presumed to be similar to that of Desert Sparrow.

GEOGRAPHICAL VARIATION & RANGE Monotypic. – C Turkmenistan, local C Uzbekistan, at least formerly also marginally in NE Iran; sedentary, with nomadic tendencies noted.

TAXONOMIC NOTE Considering that Zarudny's Sparrow is a sedentary-living widely allopatric isolate with differences evident both in size, structure and plumage morphology, including near-absent sexual dimorphism, and knowledge that the mtDNA difference between Desert and Zarudny's Sparrows is *c.* 5% (U. Olsson pers. comm. 2014), the two are better treated as separate species.

REFERENCES Densley, M. (1990) *BB*, 83: 195–201. – Kirwan, G. M. *et al.* (2009) *DB*, 31: 139–158. – Redman, N. (1993) *BB*, 86: 131–133.

♂, Turkmenistan, May: unlike Desert Sparrow, black centres to wing-feathers are more exposed, including on median coverts wing-bar, producing bolder black and grey-white pattern, while upperparts are overall purer grey. (M. Raes)

♀, Turkmenistan, May: unlike Desert Sparrow, ♀ has rather ♂-like plumage, including blackish-grey bib and mask, but upperparts still drabber, being sandy-brown. The images included here are among the very few available of the species, which is among the least documented of the treated region. (M. Raes)

TREE SPARROW
Passer montanus (L., 1758)

Fr. – Moineau friquet; Ger. – Feldsperling
Sp. – Gorrión molinero; Swe. – Pilfink

Although distinct enough when seen close, still commonly mistaken by the layman for 'yet another House Sparrow', a chirping brown-streaked sparrow that often enters gardens and frequents feeders in winter. A widespread resident of farmland and open, wooded country across much of Eurasia, south to Morocco, and in the south-east across the N Middle East to SE Asia. Tree Sparrow has been introduced to the Canaries, many Pacific islands and even locally in the USA. In W Europe, it is often less 'urban' than House Sparrow, just as its name implies, but in Asia and parts of Europe it replaces the latter species as *the* town sparrow. A partial migrant in some northern areas.

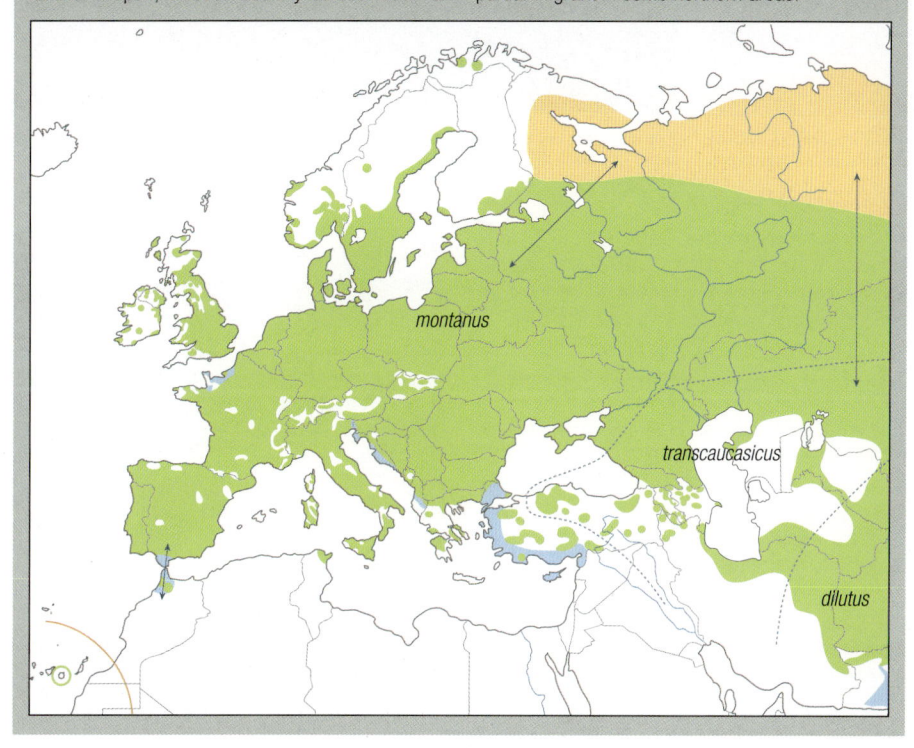

IDENTIFICATION Medium-sized, mainly brown-and-white sparrow with slim, relatively dainty appearance and *rather small, rounded head*. Sexes alike. Striking combination of *dark pinkish-chestnut crown and nape*, *white cheeks with diagnostic isolated black cheek-spot* just below and behind eye. *Black lores* and *small black bib* and *black bill* are obvious when seen head-on. There is also a narrow *white half-collar*, only just broken at rear in most birds. Mantle and back markedly rufous and ochre, boldly streaked black, rump and uppertail-coverts yellowish-brown unstreaked or only diffusely mottled. Brownish-chestnut wings with *two white wing-bars*. Breast dusky grey-white and flanks washed brown, but belly white and undertail-coverts pale buff-white. In flight, wingbeats and direct progress characteristically rather rapid, adding to the overall compact appearance. Gregarious, but usually less so than House and Spanish Sparrows, at least over much of Europe. Frequently forms mixed flocks with House Sparrow. Like all sparrows feeds mostly on ground with constant short hops and movements.

VOCALISATIONS Song (unlike House Sparrow) more musical, higher-pitched and consists of a rhythmic repetition at relaxed pace of fuller *tsvit* notes, only slightly varied, *tsvit, tsweet, tsvit, tsvit, tveit, tsvit,...* etc. – Commonest call a pleasant-sounding *tsuwitt*, slightly nasal and disyllabic, with a cheery tone; also a rather nasal clicking *teck* or *tett*, given several times on taking flight, and in flight. When alarmed at nest a nasal chattering or 'bubbling' *pet-pet-pet-pet-...*

SIMILAR SPECIES Once pinkish-chestnut crown, isolated conspicuous black ear-covert spot within white cheek, white neck collar, and smallish black bib are seen this species is unmistakable. Tree Sparrow is easily separated from *House Sparrow*, given the obvious differences in head pattern, structure and voice. (Beware, however, that these two very rarely hybridise, and birds of mixed parentage lack the well-marked cheek patch, possess a larger bib and even have some grey on the forecrown.)

AGEING & SEXING Ageing generally only possible before and during moult (provided some old flight-feathers retained). Sexes alike. Limited seasonal variation. – Moults. Both ad and juv moults completely, including wings and tail, in late summer and early autumn, mainly Jul–Oct, but timing varies with latitude, number of broods and hatching date. Like in other *Passer* species, pre-nuptial moult absent. – **SPRING Ad** See under Identification for general description. With wear, crown becomes uniformly deep vinaceous-chestnut, upperparts slightly duller and white of cheeks and collar purer. Breast-sides and flanks less extensively sullied buffish-grey, and black chin and upper throat broader. **1stS** Inseparable from ad. – **AUTUMN Ad** When very fresh, narrow buff tips to crown-feathers, bill browner with yellowish-horn lower mandible (rather than mostly black when breeding). Also more brightly fringed rufous-buff above and warmer below. Cheeks less pure white and black bib more restricted. **1stW** Identical to ad following complete post-juv moult. **Juv** Greyish centre to forecrown merging with paler rusty-brown sides. Lacks or has only diffuse dusky cheek spot and bib. Yellow base to bill. Eye-stripe greyish. Wing-bars less white, more buff.

P. m. montanus, Scotland, Apr: two ad copulating. Easily distinguished from several congeners by white cheeks with conspicuous black spot, pinkish-chestnut crown and black lores and bill. The only member of the genus *Passer* lacking sexual dimorphism. As with all sparrows ageing impossible once late-summer moults completed. (P. McLean)

P. m. montanus, England, Jan: delicate appearance and rounded head often obvious. Upperparts brownish-buff and brown-grey, streaked brown-black on mantle and scapulars, though head pattern is always the most distinctive feature. (R. Tidman)

P. m. montanus, Finland, Feb: crisply patterned when fresh. Note relatively small bib not reaching lower throat. Small rounded head obvious, as is nearly complete white collar (usually broken only at back of neck). (V. Legrand)

P. m. montanus, Finland, Oct: in non-breeding season, bill-base variably yellow. Still in post-nuptial or post-juv moult of remiges, but since no unmoulted remiges visible age cannot be decided. (M. Varesvuo)

BIOMETRICS (*montanus*) **L** 13.5–14.5 cm; **W** ♂ 70–75 mm (*n* 15, *m* 72.7), ♀ 67–72 mm (*n* 18, *m* 69.5); **T** ♂ 52–58 mm (*n* 15, *m* 55.2), ♀ 49–56 mm (*n* 18, *m* 52.6); **T/W** *m* 75.8; **B** 11.8–13.5 mm (*n* 28, *m* 12.7); **B**(f) 9.6–11.3 mm (*n* 32, *m* 10.3); **BD** 6.9–8.0 mm (*n* 31, *m* 7.4); **BW**(g) 6.4–7.8 mm (*n* 28, *m* 7.0); **Ts** 16.5–18.8 mm (*n* 29, *m* 17.5). **Wing formula: p1** minute; **pp2–4**(5) about equal and longest; **p5** < wt 0–2 mm; **p6** < wt 3.5–6.5 mm; **p7** < wt 8.5–11 mm; **p10** < wt 16–20 mm; **s1** < wt 18–20 mm. Emarg. pp3–6.

GEOGRAPHICAL VARIATION & RANGE Slight variation only. Three races generally recognised within the covered region, but separation often difficult. Several more subspecies described from Asia.

P. m. montanus (L., 1758) (Europe to W & SC Turkey, NW Asia; mainly resident, but some northern populations move shorter distances towards south-west in winter). Described above. Subtle and inconsistent differences are the basis for several more proposed subspecies, but none seems warranted. (Syn. *boetticheri*; *catellatus*; *ciscaucasicus*; *hispaniae*; *margeretae*; *stegmanni*; *volgensis*.)

○ *P. m. transcaucasicus* Buturlin; 1906 (N & E Turkey, Astrakan and lower Volga region, Caucasus, Transcaucasia, NW Iran; mainly resident). Subtly smaller, shorter-tailed and slightly paler above than *montanus*, but not quite as pale as *dilutus*. Underparts (especially belly) pale, more whitish than in *montanus*. Sexes nearly the same size. **W** ♂ 67.5–74 mm (*n* 12, *m* 70.1), ♀ 66–71 mm (*n* 9, *m* 68.3); **T** ♂ 50–58 mm (*n* 11, *m* 53.0), ♀ 46–56 mm (*n* 9, *m* 52.3); **T/W** *m* 72.3; **B** 11.4–13.3 mm (*n* 21, *m* 12.3); **B**(f) 9.2–10.6 mm (*n* 21, *m* 10.0); **BD** 6.2–7.7 mm (*n* 21, *m* 68); **BW**(g) 6.0–8.1 mm (*n* 21, *m* 7.0); **Ts** 16.0–17.9 mm (*n* 21, *m* 16.8).

P. m. dilutus Richmond, 1896 (E Iran in Khorasan and Kerman, Afghanistan and east through Kazakhstan to Mongolia and N China; mainly resident). Paler above than other races, chestnut and rufous colours more 'washed-out'. Underparts tend to be warmer sandy-buff, less white. Very slight sexual size difference. **W** 69–76.5 mm (*n* 25, *m* 72.2); **T** 51–59.5 mm (*n* 24, *m* 54.6); **T/W** *m* 75.4; **B** 11.5–15.3 mm (*n* 24, *m* 12.3); **B**(f) 9.3–11.7 mm (*n* 24, *m* 10.2); **BD** 6.5–7.9 mm (*n* 25, *m* 7.0); **BW**(g) 6.5–7.4 mm (*n* 25, *m* 7.0); **Ts** 15.1–19.5 mm (*n* 25, *m* 17.5).

P. m. montanus, juv, Spain, Jul: duller than ad, with subtly greyish-tinged centre to crown grading into paler rusty-brown sides and has only diffuse dusky cheek spot and 'bib'. Also yellow base to bill and gape are still obvious shortly after fledging. (C. N. G. Bocos)

P. m. transcaucasicus, NW Turkey, Jan: in N & E Turkey birds are subtly smaller and paler above, otherwise look very much like *montanus*. Photographed in Bolu, NW Turkey, near the border between the two races. No ageing criteria known following complete post-nuptial moult of both age classes. (T. Çağlar)

P. m. dilutus, Mongolia, May: still slightly paler above than *transcaucasicus*, chestnut and rufous more washed-out. Underparts tend to be warmer sandy-buff, less white. (H. Shirihai)

ARABIAN GOLDEN SPARROW
Passer euchlorus (Bonaparte, 1850)

Fr. – Moineau d'or du Yémen; Ger. – Jemen-Goldsperling
Sp. – Gorrión árabe; Swe. – Arabguldsparv

The Arabian Golden Sparrow is resident in savannas and agricultural coastal lowlands of SW Saudi Arabia to E Yemen. It also occurs in Djibouti and N Somalia. At times considered conspecific with the very similar Sudan Golden Sparrow, that species having a much larger distribution across the entire Sahel belt in Africa. A small, highly gregarious sparrow that nests in large colonies.

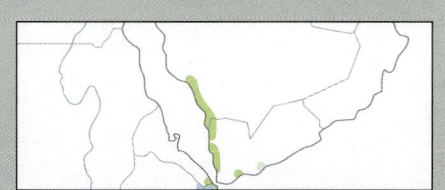

♂, Saudi Arabia, Jan: unlike Sudan Golden Sparrow, Arabian is deeper yellow with no chestnut-brown on upperparts. Overall golden, except partly concealed dark flight-feathers, primary-coverts and alula, fringed narrowly whitish (dark tertial centres and tips to primaries often distinctive). Black bill and dark eye striking in breeding plumage. (A. Al-Aumari)

IDENTIFICATION Breeding ♂ *canary yellow* (back and scapulars subtly washed olive-green), *except contrasting dark grey flight-feathers including tertials, primary-coverts and alula, all fringed whitish. Black bill and dark eye are striking.* Non-breeding ♂ superficially recalls ♀. The ♀ is paler and duller yellow and rather nondescript, *pale yellow on head and underparts, greyish-brown on back* (often with olive tinge), *mantle and back faintly dark-streaked, although crown, nape and cheeks often rufous-tinged.* Dark flight-feathers and wing-coverts have creamy-grey fringes (generally creating *diffuse pale wing-bars*). Non-breeding ♀ on average duller yellow, more brown-grey. Restless. Flight fast and agile, with action recalling small African finches. Associates with man and habitations in many areas of Arabia, where often forms mixed flocks with House Sparrow and Rüppell's Weaver.

VOCALISATIONS Song apparently undescribed. – The commonest call is a typical *Passer* chirp, a strident short note, higher-pitched than House Sparrow. Flocks utter a constant resounding twittering.

SIMILAR SPECIES Can only be confused with *Sudan Golden Sparrow*, with which there is no known overlap in range in W Palearctic (however, they may be in contact in Djibouti). Adult ♂♂ in breeding plumage readily separated, as Arabian has mantle and wing-coverts mainly yellow (rather than chestnut), and tertials and rectrices broadly white-fringed (the tail appearing largely white when closed). ♀-like plumages more difficult to separate, but Sudan has obvious dark-streaked mantle/scapulars (only faint shaft-streaks, or appearance of being unstreaked in Arabian), a prominent pale median coverts bar and post-ocular supercilium, warmer sandy-brown (not olive-tinged grey-brown) upperparts, and lacks yellow underparts and face. Presence of ♂♂ should help eliminate any confusion. – Young might also be mixed up with ♀/immature *House Sparrow* or *Rüppell's Weaver*, but both are larger and always show much bolder mantle streaking among other characteristics.

AGEING & SEXING Ageing generally only possible before and during moult (provided some old flight-feathers retained). Sexes differ, especially in breeding plumage. ♂♂ show seasonal variation due to bleaching, but unknown whether ♂ has additional pre-nuptial moult; still, transformation to full breeding plumage appears surprisingly sudden. Thus, plumage development and seasonal variation, and moult, are still only partly known. Judging from specimens (NHM), some ♂♂ appear to develop an intermediate, less mature plumage in 1stS; requires confirmation using larger sample. – Moults. Both ad and juv presumably undertake a complete moult after breeding (both breeding season and moult seem protracted being rain- and food-dependent). Some appear to have arrested or suspended moult. – **BREEDING Ad ♂** Yellow on head to back, and on wing-coverts and underparts most intense at this season. Bill black. **Ad ♀** Yellow paler and confined to head, neck and underparts. Bill dark brown. **1stS** Inseparable from respective ad. Some ♂♂ apparently paler, with at least partly ♀-like plumage coloration, and apparently breed in such plumage (to be confirmed). – **NON-BREEDING Ad ♂** Much like ♀ due to greyish-brown feather tips (partly concealing yellow feather-bases) to crown, mantle/back (faintly streaked), scapulars and wing-coverts. Usually, however, head has rufous-cinnamon wash and some exposed yellowish on forehead. Underparts also washed yellowish, deeper on throat. Bill as ♀. **Ad ♀** At most, pure yellow restricted to face and throat, while upperparts more tawny-brown than in spring. **1stW** Indistinguishable from respective ad once moult is advanced or finished, but in the event some juv feathers retained in wing and on body (can be seen even long after breeding season) ageing possible. Sexing largely impossible, but ♂ often averages slightly more intense yellow on head-sides and underparts,

♂ (above) and ♀, Saudi Arabia, Jan: two birds mating. In breeding plumage, ♀ acquires russet tinge above, and underparts have more yellow wash, though never as vivid as ♂, and bill is never blackish. (A. Al-Aumari)

♂ (right) and ♀, Saudi Arabia, May: in spring with wear, ♂ wing-feathers lose most of the whitish fringes, and bill becomes paler. ♀ is in moult to winter plumage, which is paler and drabber, albeit with darker centres to primary-coverts and alula, and tips to primaries, still distinctive (but lacks pale wing-bars of fresh plumage). (J. Babbington)

♀, Yemen, Jan: ♀♀ become slightly more yellow on head and underparts in breeding season (which can start early), and crown and nape can attain a slight rufous tinge. (J. Oláh)

♂♂, Saudi Arabia, Nov: non-breeding birds when fresh have overall golden plumage, with dark grey-brown centres to wing-feathers largely concealed, and bill is paler. (M. Al Fahad)

and is tinged more rufous-cinnamon above with darker remiges. Legs of young reportedly browner. **Juv** Much as non-breeding plumages, but has soft, fluffy body-feathers and is greyer and more mottled on mantle than ad ♀, virtually without any yellow pigment.

BIOMETRICS **L** 11–12 cm; Size differs very little between sexes. **W** 54–64 mm (n 28, m 59.5); **T** 40–48 mm (n 28 m 45.2); **T/W** m 76.0; **B** 10.6–12.0 mm (n 27, m 11.3); **B**(f) 8.1–9.9 mm (n 27, m 8.9); **BD** 5.8–7.0 mm (n 25, m 6.3); **BW**(g) 4.9–6.8 mm (n 28, m 5.7); **Ts** 15.4–17.9 mm (n 28, m 16.3). **Wing formula: p1** minute; **pp2–5** about equal and longest; **p6** < wt 2.5–5 mm; **p7** < wt 5–8 mm; **p10** < wt 10.5–15 mm; **s1** < wt 12–17 mm. Emarg. pp3–5. – For measurements of 27 live birds (sexes combined) from N Yemen, see Bowden (1987).

GEOGRAPHICAL VARIATION & RANGE Monotypic. – SW Saudi Arabia, Yemen, N Somalia, Djibouti, E Ethiopia; resident or nomadic.

REFERENCES BOWDEN, C. G. R. (1987) *Sandgrouse*, 9: 94–97.

♂, Saudi Arabia, Oct: another non-breeding ♂ with largely yellow head and body but silvery yellowish-grey wings and back. Bill, though dark, is not black. Impossible to age due to usually complete moult of both age categories. (D. al Hasimi)

SUDAN GOLDEN SPARROW
Passer luteus (M. H. C. Lichtenstein, 1823)

Fr. – Moineau doré; Ger. – Sudan-Goldsperling
Sp. – Gorrión dorado; Swe. – Guldsparv

Sudan Golden Sparrow is a resident of arid thorn savannas across a narrow zone of the Sahel, from Eritrea west to Mauritania and Senegal. It breeds mainly just south of the covered region, occurring in the Tibesti of extreme N Chad and in W Mauritania, but is apparently an established resident in the extreme southern corner of Algeria, hence its inclusion in this book. Rather abundant in places and sometimes attracted to cultivated areas, it is highly gregarious and breeds colonially.

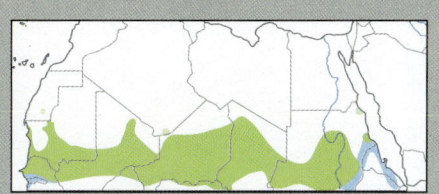

♂, Western Sahara, Feb: unmistakable given bright lemon-yellow head, rump and underparts, rich chestnut-brown mantle, scapulars, edges to secondaries and tertials, and blackish wing-coverts edged rufous with double whitish wing-bars. Also unlike Arabian Golden Sparrow, bill does not become black in breeding season. (T. Jones)

IDENTIFICATION ♂ unmistakable during the breeding season having *underparts, head and rump entirely canary-yellow*, the *mantle, back, scapulars and inner wing-coverts rich chestnut-brown*, while outer wing-coverts, primaries, tertial centres and tail are dark brown to black, tertials broadly edged and tipped rufous. Note also *conspicuous whitish greater coverts wing-bar*, and *black bill*. Much duller ♀ is *pale sandy-brown, finely streaked darker on upperparts*, but head is almost plain yellow-buff with *quite pronounced, yellowish-buff supercilium behind eye*, narrow dark eye-stripe and dingy yellowish-buff face. *Throat pale*, and submoustachial stripe tinged pale yellowish, sides of neck and breast to flanks yellowish-buff. Belly and undertail-coverts whitish. Dark grey *wings have less striking wing-bar than ♂* (though still an important field character). ♀ and non-breeding ♂ have brownish-grey culmen and pinkish-horn lower mandible; tarsus pinkish-brown. Very similar to Arabian Golden Sparrow (which see) in habits and behaviour.

VOCALISATIONS Song contains some elements recalling House Sparrow, with a sibilant twitter emanating from colonies. – Flight-call is a fast *che-che-che*, repeated 7–8 times and can suggest that of Redpoll.

SIMILAR SPECIES Closely recalls *Arabian Golden Sparrow* but adult ♂ of latter has extensive and deeper yellow plumage, and lacks any chestnut on upperparts, while ♀ is greyer and unstreaked above. – A remote risk of confusion is provided by ♀ *Yellow-breasted Bunting*, which is superficially similar and could stray as far as the range of Sudan Golden Sparrow; also beware latter's frequent escapes from captivity, e.g. at Eilat, Israel. Yellow-breasted Bunting is readily separated by its partly white outer tail-feathers, its plainer grey-brown head-sides, all pale pinkish bill, etc.

AGEING & SEXING Ageing generally impossible after post-juv moult. Sexes differ, especially when breeding. – Moults. Complete post-nuptial and post-juv moult, generally in May–Aug in N Chad (but post-juv probably suspended and may continue until much later), apparently further south mainly much later. ♂ shows seasonal variation due to bleaching. Unknown whether ♂, especially 1stY, has additional pre-nuptial moult, but transformation to ad-like breeding plumage quite marked. – **BREEDING Ad** ♂ Yellow and chestnut parts most intense at this season. **Ad** ♀ Yellow paler and confined to underparts and face. **1stS** ♂ Inseparable from ad unless some juv feathers retained. Some distinctive, being less advanced in colour, and can breed in such plumage. Differs from ad ♂ in less pure yellow crown, which looks dirty or mottled, and chestnut area on mantle and scapulars slightly less deep or patchily tipped brownish or sandy, while rest of wing duller, as is yellow of face and underparts. **1stS** ♀ Duller, but differences from ad ♀ hard to detect. In both sexes and at all ages, wing-bars may be reduced or absent in extremely worn plumage. – **NON-BREEDING Ad** ♂ Duller than breeding plumage and can show paler lores and greyish wash to some yellow parts, e.g. ear-coverts and uppertail-coverts. **Ad** ♀ Yellow further reduced, while upperparts more tawny-brown. **1stW** ♂ Rather similar to ♀, but has greyer head and rich brown tips to mantle and scapulars, is more intensely washed pale yellow on head-sides and underparts, and wing- and tail-feathers are all blacker, with whiter median coverts wing-bar, perhaps more rufous fringes to inner wing-feathers, and duskier bill. May retain juv feathers in wing and body for several months. **1stW** ♀ Like ad ♀ but even duller; ageing sometimes possible by moult pattern (as 1stW ♂). **Juv** Much as ♀ but has soft, fluffy body-feathers, lacks any yellow below and has grey flecks on head and nape, with a pale whitish to yellow-buff supercilium and greyer legs.

BIOMETRICS L 11.5–12.5 cm; W ♂ 62–67 mm (n 12, m 64.4), ♀ 58–65.5 mm (n 12, m 64.4); T ♂ 43–51 mm

♀, Sudan, Apr: pale sandy-brown indistinctly streaked darker above, with dingy yellowish-buff face, except brighter supercilium bordered by narrow dark eye-stripe. Indistinctly patterned wings have pale buff wing-bars and some dark centres, including to tertials. (Z. Celebic)

(*n* 12, *m* 47.5), ♀ 42–51.5 mm (*n* 12, *m* 45.7); **T/W** *m* 73.6; **B** 10.6–12.7 mm (*n* 24, *m* 11.6); **B**(f) 8.1–10.0 mm (*n* 24, *m* 9.1); **BD** 5.7–7.0 mm (*n* 23, *m* 6.2); **BW**(g) 5.1–7.0 mm (*n* 24, *m* 5.9); **Ts** 15.7–17.5 mm (*n* 24, *m* 16.4). **Wing formula: p1** minute; **pp2–4**(5) about equal and longest; **p5** < wt 0–2 mm; **p6** < wt 3–5.5 mm; **p7** < wt 5–9 mm; **p10** < wt 13–17 mm; **s1** < wt 14.5–18.5 mm. Emarg. pp3–5.

GEOGRAPHICAL VARIATION & RANGE Monotypic. – Sahel, from Eritrea west to Mauritania and Senegal, including extreme S Algeria; resident.

REFERENCES Summers-Smith, D. (1988) *The Sparrows*. Calton.

Senegal, Apr: a perfect image showing ad ♂ (top right) and below it a 1stY ♂ with diluted plumage, plus two ♀♀ (on far left) from front and behind, respectively. The second bird from the left on the bottom row is probably an ad ♂ due to much yellow below, while the far right lower bird could be a young ♀. Unlike ♀ Arabian Golden Sparrow, supercilium better defined and mantle streaking more obvious. (T. Haslam)

♂, Burkina Faso, Jan: at end of post-breeding moult, with all feathers seemingly new. Ageing not possible. (V. van der Spek)

1stW ♂, Western Sahara, Feb: 1stW ♂ has dirty greyish-yellow head, creamy-yellow underparts and brown-marbled mantle and scapulars, plus duskier bill. Many primaries and primary-coverts are juv. (T. Jones)

Presumed ♀, Mali, May: plumage variation and timing and extent of moult are still insufficiently well known to definitely age and sex all individuals. (T. Helsens)

PALE ROCK SPARROW
Carpospiza brachydactyla (Bonaparte, 1850)

Alternative name: Pale Rockfinch

Fr. – Moineau pâle; Ger. – Fahlsperling
Sp. – Gorrión pálido; Swe. – Blek stensparv

This charismatic but unassuming species summers on grassy hillsides with rocky outcrops and stony valleys in E Turkey, Armenia and Iran, with scattered occurrences elsewhere in the Middle East including Israel and N Saudi Arabia. It winters in NE Africa, the Near East and Iraq south to SW Arabia. Numbers fluctuate in response to annual grass development and thus rainfall. In some years it may breed either much further south or north, in arid areas that have experienced unusually wet winters, but in others it appears very scarce and is confined to certain core areas.

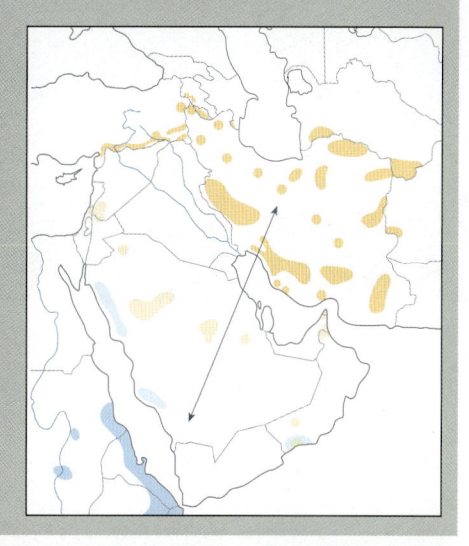

Israel, May: this anonymous-looking, unstreaked, sandy grey-brown sparrow is long-winged (with long primary projection) and has paler wing-bars (most obvious on median coverts), with a dingy-white panel on secondaries. Here in typical posture emitting the insect-like buzzing song. (A. Ben Dov)

IDENTIFICATION An unobtrusive sparrow-sized bird with a slim body and relatively small head lacking obvious field marks. Usually encountered either in brief fly past, when its shallowly undulating lark-like flight and diagnostic calls like a distant bee-eater permit identification, or by first hearing its insect-like high-pitched buzzing song. Overall plain grey-brown (but despite its English name hardly very pale), with an indistinct dusky cream-whitish supercilium and throat, *paler eye-ring and small but prominent dark eye*. Heavy finch-like bill with convex culmen and *flat crown* form nearly unbroken line, the *bill rather dark grey* dorsally when breeding with slightly paler base to lower mandible (in non-breeding bill entirely paler grey-brown). Poorly-defined dark moustachial stripe and lateral throat-stripe, dark grey-brown wings with hint of *narrow double pale wing-bars*, dull white secondary panel and *ill-defined paler tertial fringes*. Generally paler dusky-white below, only tinged slightly warmer on sides. Undertail-coverts have some dark centres. Rather short-tailed, an appearance accentuated by the *long primary projection*. In flight, *small white spots on tail-tip* can be obvious, especially from below or against dark background. Legs rather pale pinkish-brown. Usually gregarious, especially in non-breeding season, but flocks only rarely large.

VOCALISATIONS Song, usually given from ground, is a one-second-long throaty, piercing buzz, quite insect-like, sometimes prefaced by a few short Corn Bunting-like introductory notes, *bz-bz-bz-bzzrrüüüiz*. – Flight-call a soft *pluip* or *twee-ou* or a low trill, *trrrü*, which can recall a distant Bee-eater *Merops apiaster*.

SIMILAR SPECIES Anonymous-looking, but this is what makes its overall appearance so distinctive. Differs from ♀ and juv *Yellow-throated Sparrow* in smaller size and somewhat slimmer shape, darker grey, deeper-based and proportionately larger bill, lack of any hint of cinnamon or chestnut on lesser coverts and the indistinct double wing-bars, rather than boldly white-tipped median coverts creating very obvious upper wing-bar. ♂ of latter should not prove a significant confusion risk. – In Arabia and Africa confusion with *Bush Sparrow* is a possibility, but this is considerably smaller, colder grey, less tinged buff-brown, has a somewhat slimmer bill, lacks white tail-tip and has different flight call. – Larger and bulkier *Rock Sparrow* has bold dark crown bands either side of a narrow pale median crown-stripe, faint yellow throat spot (in adults), obvious underparts and mantle streaking, more massive bill and generally darker upperparts. – Inexperienced observers should bear in mind ♀-like plumages of *Sinai Rosefinch* or juvenile or non-breeding *Trumpeter Finch*, and even an unusually greyish and unstreaked *Short-toed Lark* might require elimination.

AGEING & SEXING Ageing generally only possible before and during moult (provided some old flight-feathers retained), sometimes also during 1stY. Sexes alike. Limited seasonal variation. – Moults. Ad and 1stY undergo protracted moult commenced on breeding grounds, suspended during autumn migration and resumed in winter quarters, usually being completed before end of year (but see BWP, and below about some 1stS having retained primary-coverts and a few greater coverts and remiges), thus post-nuptial moult complete and post-juv moult probably best regarded as usually complete but with unknown percentage retaining some juv wing-feathers. Apparently no pre-nuptial moult in early spring. – **SPRING Ad** For a general description see Identification. Fresh in early spring; with wear becomes browner with much-reduced face and wing markings. Unlike 1stS probably lacks moult limits. **1stS** Many inseparable from ad, but some retain all or some juv primary-coverts and some greater coverts, others retain some remiges (preferably outer secondaries and innermost primaries), which are browner, paler and sometimes slightly more worn. – **AUTUMN Ad** Prior to completion of moult those with unmoulted remiges and wing-coverts distinctive (such

United Arab Emirates, Mar: largely nondescript, but has white spots on inner webs of tail-tip (especially obvious when tail is spread, but when folded only visible from below). In spring and summer, birds without moult limits cannot be aged. (M. Barth)

♂, presumed 1stS, Turkey, May: spring birds often appear rather featureless with just a hint of pale wing-bars, and note rather dark-looking conical bill of this ♂ (singing). Considering strong moult limits with some heavily worn and bleached inner wing-coverts retained presumably a 1stS. (L. Svensson)

Oman, Mar: this bird has unusually well-developed whitish wing-bars and secondary panel. Entire plumage has been moulted. (M. Varesvuo)

1stS, Turkey, May: slim body and small head, with narrow pale supercilium, ill-defined pale submoustachial and dark lateral throat-stripes. Throat centre whitish and rest of underparts mainly white with warmer/greyer sides and breast. Note stout pale bill with dark culmen, swollen at base with more rounded culmen than Rock Sparrow. Retained juv remiges and primary-coverts (sharply pointed) reliably age as 1stY. (D. Occhiato)

United Arab Emirates, Mar: often gregarious in non-breeding season. Long pointed wings but proportionately shorter tail with triangular shape in flight afford somewhat lark-like appearance. The diagnostic white spots on tail-tip and boldly white-tipped coverts are often visible in flight. (K. Al Dhaheri)

feathers much older, worn and bleached, with brownish centres and fringes almost worn off). **1stW** Similar to ad but has juv wing-feathers, and probably some juv tail- and body-feathers; latter less worn than unmoulted ad feathers. **Juv** Soft, fluffy body-feathers. Plumage overall paler, with especially well-marked pink-buff tips to wing-coverts, but less obvious supercilium. Dusky-brown tip to rather pale bill and greyish-pink legs.

BIOMETRICS L 13–14.5 cm; W ♂ 94–100 mm (n 13, m 96.4), ♀ 88–94 mm (n 17, m 91.8); T ♂ 47–53 mm (n 13, m 50.5), ♀ 45–53 mm (n 17, m 49.8); T/W m 53.4; B 12.0–15.8 mm (n 30, m 13.9); B(f) 10.1–12.1 mm (n 30, m 11.0); BD 6.6–8.5 mm (n 29, m 7.4); BW(g) 7.5–8.5 mm (n 7, m 8.2); Ts 16.8–19.5 mm (n 30, m 18.3). **Wing formula: p1** minute; **pp2–3** about equal and longest; **p4** < wt 3–8 mm; **p5** < wt 10–15 mm; **p6** < wt 14–20 mm; **p7** < wt 19–26 mm; **p10** < wt 32–39 mm; **s1** < wt 34–42 mm. No emarg. of pp.

GEOGRAPHICAL VARIATION & RANGE Monotypic. – SE Turkey, Levant north to Lebanon, N & C Arabia, Armenia, Iran; winters NE Africa, SW Arabia, probably also S Middle East. (Syn. *psammochroa*.)

TAXONOMIC NOTES Relationships within Ploceidae not clear and require molecular research, which apparently still is lacking. The species' breeding habits do not suggest a particularly close relationship with other *Petronia* (with which it has sometimes been subsumed) or *Gymnoris*.

REFERENCES KIRWAN, G. (1998) *Sandgrouse*, 20: 8–12.

Oman, Jan: when very fresh (in winter) often appears colder/greyer (less brownish) above, and bill less bicoloured and more uniform grey-horn. (H. & J. Eriksen)

Ad, Israel, Sep: prior to completion of prolonged post-nuptial moult, when typically abraded feathers produce whitish patches, most noticeably on scapulars and median coverts, giving an overall scruffy appearance; note also very worn primaries. (A. Ben Dov)

YELLOW-THROATED SPARROW
Gymnoris xanthocollis (Burton, 1838)

Alternative name: Chestnut-shouldered Bush Sparrow

Fr. – Moineau à gorge jaune; Ger. – Gelbkehlsperling
Sp. – Gorrión cuelligualdo
Swe. – Gulstrupig stensparv

This Asian sparrow mainly occurs in lowlands with scattered trees, orchards, dry open forests with bushes and oases, over a wide area from SE Turkey to NW India. A migrant, wintering in S Pakistan and India, but also in S Middle East and Arabia. The Yellow-throated Sparrow has reached S Sri Lanka and in the west to Israel, Jordan, Egypt and even Malta. Often first detected when the ♂ is singing, a loud, rapid flow of chirping notes delivered from perch in a tree. It nests in holes, usually in trees.

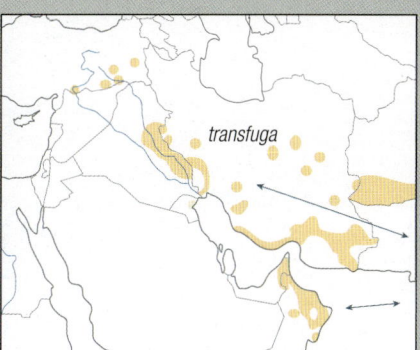

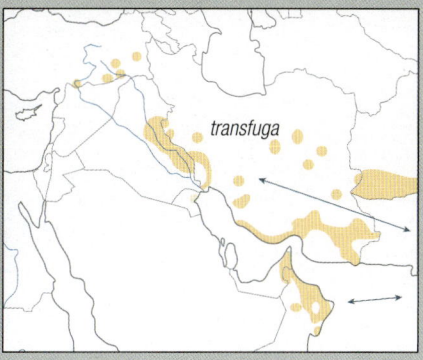

G. x. transfuga, ♂, **May**: rather unique long, fine and pointed, conical bill, which is black in breeding ♂. Note rather bland head pattern except dark lores and yellow lower-throat spot, bordered by pale olive-grey chest, while chin and abdomen are purer white. (H. Shirihai)

IDENTIFICATION Rather nondescript, being largely unstreaked olive grey-brown, with a rather characteristic *long, pointed conical bill* (with almost straight culmen), which is black in breeding ♂. The ♂ has dark lores, very faint supercilium, *yellow lower-throat-spot* (often difficult to see), olive-grey breast, purer white chin, upper throat and rest of underparts, *diagnostic deep rufous lesser coverts* (also difficult to see) and *obvious pale double wing-bars* (much broader, bolder and whiter on median coverts). ♀ similar but generally duller, lacking yellow on lower throat (or has faint wash), with browner lesser coverts, pale buff wing-covert tips, making distinctive bill shape and whitish wing-bars even more important for identification. Bill of ♀ and non-breeding ♂ dusky or horn-coloured. Iris usually black-brown, legs grey. Typically inhabits orchards and other lightly wooded habitats, where it can be quite common but is rarely truly gregarious, except in non-breeding season and at roosts. Flight consists of dipping undulations accompanied by diagnostic calls.

VOCALISATIONS Song recalls House Sparrow, an 'endless' series of chirping notes, but with a softer and more liquid tone (still can be perceived as loud) and much faster tempo, *chip-cheeve-chip-chock-chirp-chet-chiv-...*, etc., usually delivered from a relatively high perch in a tree. – Calls less suggestive of House Sparrow, a piercing *chiah*, soft and tuneful, somewhat like a chough.

G. x. transfuga, ♀, **Turkey, May**: no chestnut lesser coverts patch, but has pale double wing-bars (bolder on median coverts). Yellow spot on lower throat very faint and not always visible. A pale supercilium over lores to eye, or reaching slightly behind it. (H. Shirihai)

SIMILAR SPECIES ♂ unlikely to be confused given range and a good view. ♀ and juvenile may be confused with *Pale Rock Sparrow*, but that should not present a significant identification problem, especially given combination in Yellow-throated of unique bill with finer shape and straight culmen, distinctive pale median coverts wing-bar, uniform tail (lacking white tips) and dull-coloured legs. The two species also have rather different habits and habitat preferences.

AGEING & SEXING Ageing generally only possible before and during moult (provided some old flight-feathers retained). Sexes moderately differentiated and mainly separable in spring and summer. Seasonal variation generally limited. – Moults. Both ad and juv moult completely in late summer, mainly Jul–Sep. Pre-nuptial moult absent. – **SPRING Ad** ♂ Sexed by black bill, deeper yellow patch on throat and brighter wing markings. **Ad** ♀ Bill dusky

G. x. transfuga, ♂, **Turkey, May**: overall pale plumage with deep bright rufous (rather than chestnut) lesser coverts patch are features of the south-western race. Rufous lesser coverts patch (can be concealed by scapulars) contrasts with bold white median coverts bar, whereas greater coverts bar is much less obvious. (D. Occhiato)

or horn, has ill-defined or mere hint of yellow on throat (sometimes none), with duller wing pattern. With extreme wear, both sexes may lack wing-bars. **1stS** Probably indistinguishable from ad. – AUTUMN **Ad** Sexes less clearly differentiated, as both have horn-coloured bill and paler lores, reduced or wholly-concealed yellow throat spot, and less deep rufous lesser coverts. However, most ♂♂ have yellow on throat and more rufous on shoulders. Both sexes evenly very fresh with some light cinnamon on paler parts, including supercilium, ear-coverts, wing-bars, and fringes of tertials and flight-feathers. **1stW** Identical to ad following complete post-juv moult. Unconfirmed if some retain juv wing-feathers. **Juv** Similar to ♀ but has soft, fluffy feathering and sandy-brown upperparts, a rather vague, pale buffish supercilium, indistinct wing-bars and lacks any yellow on throat.

BIOMETRICS (*transfuga*) **L** 13–14 cm; **W** ♂ 80.5–88 mm (*n* 14, *m* 82.8), ♀ 75–80 mm (*n* 12, *m* 77.3); **T** ♂ 47–55 mm (*n* 14, *m* 51.9), ♀ 42–52 mm (*n* 12, *m* 47.8); **T/W** *m* 62.3; **B** 11.8–16.5 mm (*n* 25, *m* 14.5); **B**(f) 8.7–12.6 mm (*n* 25, *m* 11.3); **BD** 5.1–6.8 mm (*n* 26, *m* 6.1); **BW**(g) 5.2–6.5 mm (*n* 26, *m* 5.9); **Ts** 16.2–17.7 mm (*n* 24, *m* 16.9). **Wing formula: p1** minute; **pp2–4** about equal and longest (p4 sometimes to 3 mm <); **p5** < wt 3–7 mm; **p6** < wt 7–12.5 mm; **p7** < wt 12–17 mm; **p10** < wt 22–29 mm; **s1** < wt 24–31 mm. No emarg. of pp.

GEOGRAPHICAL VARIATION & RANGE Two races generally recognised, but variation is slight and clinal. Only *transfuga* known to occur within the covered region.

G. x. transfuga Hartert, 1904 (SE Turkey, Iraq, Kuwait, NE Arabia, S Iran east to C & S Pakistan; winters in south of range and in India and S Arabia). Described above. (Syn. *occidentalis*.)

G. x. transfuga, ♀, Turkey, May: unlike ♂, bill never black in breeding season, but lores variably darker. Yellow on throat impossible to see at this angle, but note distinctive bill shape and wing-bars. Given complete seasonal moult of both age classes, ageing is impossible thereafter. (D. Occhiato)

G. x. xanthocollis/transfuga, ♀, India, Dec: in fresh plumage, when lack of rufous lesser coverts patch confirms sex. During winter in India, both local *xanthocollis* and migrant *transfuga* occur, and these are only doubtfully separable in the field, especially in ♀-like plumages. (P. Ryan)

G. x. transfuga, juv/1stW, ♂/♀, United Arab Emirates, Sep: much of plumage and entire wing apparently juv and already worn. The dullest, least-marked and most variable plumage, but pale double wing-bars, together with faint pale supercilium and diagnostic bill shape, identify. (K. Al Dhaheri)

G. x. transfuga, juv (fed by ♀), Turkey, Jun: juv similar to ♀, but sandy-brown above, lacks yellow throat spot, and has indistinct median coverts bar (narrower and not pure white). Note fluffy flanks feathering. Unlike Pale Rock Sparrow lacks white tail-tip. (Y. Perlman)

BUSH SPARROW
Gymnoris dentata (Sundevall, 1850)

Alternative name: Bush Petronia

Fr. – Moineau soulcie à ventre blanc
Ger. – Buschsperling
Sp. – Gorrión chico; Swe. – Bukstenssparv

This unassuming and easily-overlooked sparrow occurs marginally within the covered region in dry, semi-arid areas of SW Arabia, mostly in *Acacia* woodland. Large flocks may form to exploit abundant food sources. Mainly Afrotropical, its range extends across sub-Saharan Africa, from S Mauritania to Ethiopia and Eritrea.

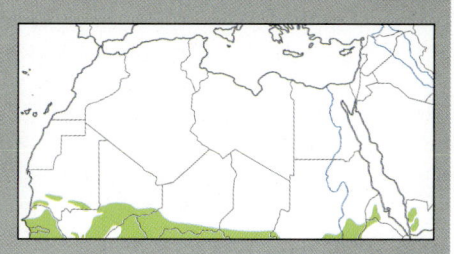

♂, Ethiopia, Sep: note cinnamon-brown supercilium broadening at rear, almost encircling ash-grey cap, with ill-defined pale wing-bars and whitish eye-ring. These features combined with short, black, clearly conical and sharply pointed bill, are diagnostic. (H. Shirihai)

IDENTIFICATION A relatively featureless species with arguably even fewer field characters than other related species in the treated region. *Upperparts are largely plain grey-brown, only diffusely mottled darker. Chin and central throat white* (forming a narrow pale band along centre) with an obscure yellow spot on the lower border in breeding plumage. Cheeks and rest of underparts dusky-grey, though belly more whitish. On folded wing note hint of paler patch on primary bases. Legs greyish-black. ♂ has a *broad, warm cinnamon-brown supercilium starting behind eye*, which variably extends to neck-sides and nape. Wing-bars ill-defined and somewhat variable buff-white (can be so dull brownish as to be very inconspicuous). Tertial fringes and secondary panel pale brown and can contrast with darker primary-coverts and outer greater coverts. Whitish edges and tips to tail-feathers indistinct. ♀ strongly recalls ♀ House Sparrow, sometimes with *some dark streaks on mantle* and well-defined wing markings. Compared to House Sparrow, *supercilium is somewhat bolder and paler*. Whitish eye-ring usually broader on upper half, and eye blackish. *Short bill is conical and sharply pointed* with only subtly curved culmen. Usually occurs in pairs or small groups, often initially noticed perched quietly near top of a low tree, flicking tail in a sparrow-like manner. Flight also direct and sparrow-like.

VOCALISATIONS Song bunting-like with a rapid twitter and, mainly, sparrow-like chirps, *triup-triup-triup-triup*, or undulating *chup-chee-chup-chee-chup-...*, but generally appearing rather formless, and notes given, singly, doubly or trebled. Occasionally speeded-up, and individual notes sometimes lengthened and given at even but slightly more rapid pace than normal song. – A sparrow-like *chewee* in flight.

SIMILAR SPECIES Main confusion risk is *Yellow-spotted Sparrow* (*G. pyrgita*, extralimital and not treated), but unlikely to be encountered in the Palearctic. Yellow-spotted is whiter on belly, less dusky, and white throat is broader, not a narrow central band as in Bush. It also has plain brown primaries, lacking hint of pale primary bases patch as in Bush. – *Pale Rock Sparrow* is about 25% larger, overall warmer-tinged yellowish buff-brown (less cold greyish) and drabber, with plainer upperparts, a longer primary projection, brighter reddish-brown legs, better-defined pale submoustachial stripe and dark lateral throat-stripe. Also, head pattern is much plainer and unlike either sex of Bush Sparrow, especially ♀, which has bold pale supercilium. The two species also have different tail patterns and flight-calls. – ♀-like *House* and *Spanish Sparrows* may also offer potential confusion with ♀ Bush Sparrow, but aside from very different shape, behaviour and voices, the Bush Sparrow's supercilium is much bolder, the whitish throat better defined by the darkish lateral throat-stripe and breast, and mantle and scapulars lack obvious buff-brown 'braces' of the others. Additional differences can be detected by close inspection of the wing patterns.

AGEING & SEXING Ageing only possible before post-juv moult and post-nuptial moult of ad become well advanced. Sexes usually strongly differentiated. Rather limited seasonal variation. – Moults. Both ad and juv undergo complete moult after breeding, but few data available from Arabia on timing or extent. Birds from Sep and Oct (NHM) very fresh and recently moulted, whereas Apr birds heavily worn. – **BREEDING Ad ♂** See Identification for a general description. Bill usually all black. Some have underparts slightly tinged buffish. Both sexes wear to more uniform wing pattern, as paler fringes, especially on tertials, almost wear off in spring. **Ad ♀** See Identification for a general description. Additionally, yellow throat-spot less noticeable than in ♂, and bill has flesh-coloured base to lower mandible, thus not all black. Note

♂, Yemen, Jan: in worn plumage, overall duller, and wing-bars almost worn off. (H. & J. Eriksen)

that cream-white supercilium extends above and just in front of eye, making whitish eye-ring inconspicuous. **1stS** Inseparable from respective ad. – **NON-BREEDING Ad ♂** Bill dark brown with pale horn-coloured base, and plumage fresher than breeding. **Ad ♀** Supercilium more buff-tinged, and yellow throat-spot largely concealed. **1stW** Inseparable from respective ad, though some 1stW ♂♂ can be a little duller on head, while least-advanced ♀ appears almost like juv. **Juv** Paler and more nondescript than ♀ but also consistently warmer and browner, with warm buff fringes to wing-feathers. Supercilium buff-tinged, no yellow on throat. Underparts slightly more whitish. Bill as ad ♀ and sometimes has a row of dark spots on lower throat.

BIOMETRICS L 12–13.5 cm; **W** ♂ 75–87 mm (n 19, m 80.1), ♀ 73–80 mm (n 15, m 75.8); **T** ♂ 43–52 mm (n 19, m 46.8), ♀ 40–46 mm (n 15, m 43.2); **T/W** m 57.8; **B** 11.5–14.3 mm (n 35, m 13.2); **B**(f) 9.5–12.3 mm (n 32, m 11.0); **BD** 6.0–7.5 mm (n 32, m 6.8); **BW**(g) 6.9–8.3 mm (n 11, m 7.5); **Ts** 15.0–17.5 mm (n 33, m 16.4). **Wing formula:** p1 minute; pp2–4 about equal and longest; p5 < wt 0.5–4 mm; p6 < wt 5–11 mm; p7 < wt 10–15 mm; p10 < wt 20–25 mm; s1 < wt 20.5–27 mm. Emarg. pp3–5.

GEOGRAPHICAL VARIATION & RANGE Monotypic. – SW Arabia, and from Senegal south to C Nigeria and across the Sahel to S Sudan, Eritrea, Ethiopia; sedentary. – A separate subspecies has been described from S Niger ('*buchanani*'), but is poorly differentiated and recognition is hardly warranted.

♀, Ethiopia, Jan: ♀ has rather inconspicuous coloration, but still shows pale wing-bars, dark-streaked mantle and ill-defined wing markings, including tertial fringes and pale secondary panel, but most prominent is broad pale supercilium. Note bicoloured, short, pointed and conical bill. (C. N. G. Bocos)

♀, Ethiopia, Sep: in distant views, clear whitish supercilium from above eye and to rear, throat patch, and ill-defined pale wing-bars, as well as fairly slim conical bill serve to identify. Once moult is completed ageing is no longer possible. (H. Shirihai)

ROCK SPARROW
Petronia petronia (L., 1766)

Fr. – Moineau soulcie; Ger. – Steinsperling
Sp. – Gorrión chillón; Swe. – Stensparv

Rock Sparrow inhabits montane country and is also often found in open terrain with small eroded precipices, such as ravines and rocky wadis, quarries, walls, ruins and even occupied buildings. It occurs on some NE Atlantic islands east through S Europe and N Africa to Central Asia and C China, wintering in the breeding range and also further south in Asia to N Pakistan.

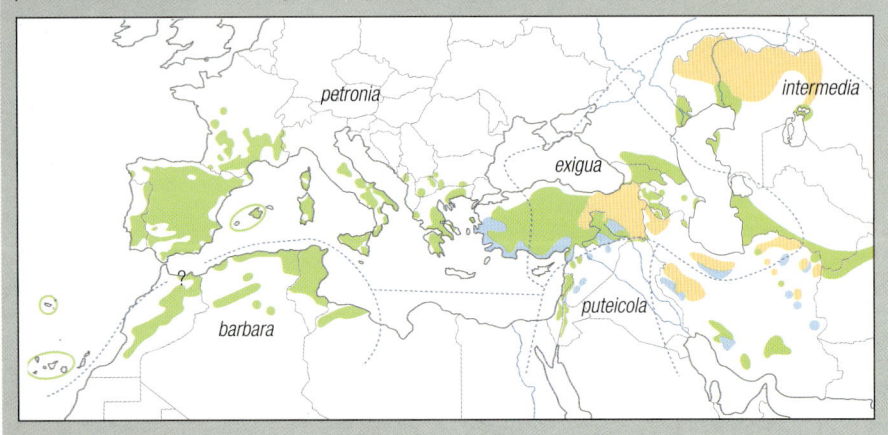

P. p. petronia, Spain, Feb: a sparrow-like, mid-sized, front-heavy and square-tailed bird with boldly-striped head, broad pale supercilium, dark lateral crown- and eye-stripes, pale below streaked dark on breast and flanks, and conspicuous pale spots on tail-tip. Pale buff 'braces' on mantle, and dark-centred undertail-coverts. Conical bill is rather heavy at base and always bicoloured. (C. N. G. Bocos)

IDENTIFICATION A bulky, long-winged, square-tailed and mostly dull-looking sparrow, albeit has some characteristic features. *Bill heavy*, with *conical shape* typical of finches and sparrows, horn-brown with a yellowish base to lower mandible. *Head appears large with broad dark lateral crown-stripes* either side of a pale median crown-stripe, *buffish-white supercilium*, *dusky ear-coverts* (pale drab, slightly darker than supercilium) and *a faint cream-coloured submoustachial stripe*. Sometimes vague dark moustachial and lateral throat-stripes visible too. *Lower throat has a pale yellow spot* (not always visible). Rest of *underparts dusky or buffish-white, faintly streaked or spotted brown* on breast and belly, stronger on flanks and undertail-coverts. *Upperparts brown-grey heavily streaked* and mottled *dark*. A *spotty wing-bar* on median coverts (a second wing-bar on greater coverts is less well marked), *boldly-tipped tertials*, and a small dark primary-coverts patch add to the species' characteristics. *Tail dark brown with obvious white tips*; in flight, this is often an important aid to identification. Very active, especially on the ground, where it possesses a somewhat lark-like manner, but also hops boldly, and is often encountered in large noisy flocks that sometimes make long escape-flights.

VOCALISATIONS Noisy, giving a variety of loud and rather melodious calls. Song apparently comprises a loose collection of drawn-out calls 'with many vowels' (glissando-type calls), e.g. *sle-veeit* or *tve-yuitt*, and a shorter *tvüitt*. – Calls include a short nasal *vüi* or *chwee*, rather penetrating and far-carrying; in anxiety a hard trilling *tii-tür'r'r'r'r'*, other notes more squeaky, but a sweet *peeuh-ee* recalls Greenfinch.

SIMILAR SPECIES Unlikely to be confused with ♀ *House* or *Spanish Sparrows*, even if seen poorly, as Rock Sparrow has stronger head and bill, a striped crown pattern, longer primary projection, and proportionately shorter and broader, white-tipped tail. However, these features may require prolonged scrutiny before they are detected, but experienced observers should easily identify the species by its powerful appearance in flight and unmistakable calls. – For separation from *Pale Rock Sparrow* see that species.

AGEING & SEXING Ageing generally only possible before and during moult (provided some old flight-feathers retained). Sexes alike. Limited seasonal variation. – Moults. Ad and 1stY undergo complete moult in late summer, mostly Jul–Sep. Pre-nuptial moult apparently absent. – SPRING **Ad** A general description is found under Identification. With wear, more contrastingly patterned, and yellow throat-spot noticeably bright, especially in ♂, but if strongly bleached (late summer) may lose most pale tips and fringes above, and striped head pattern and streaks above and below become more obscure, thus darkens immediately before onset of moult. **1stS** Inseparable from ad. – AUTUMN **Ad** Prior to completion of moult those with unmoulted (heavily worn and bleached) remiges and wing-coverts distinctive. When still very fresh, ground colour of body rather noticeably tinged tawny or buff. **1stW** Similar to fresh ad but before moult completed may still possess juv wing-feathers, which are less worn than unmoulted old ad feathers. Yellowish feathers on throat just developing or still lacking (usually distinctive in ad). **Juv** Soft fluffy body-feathers and lack of yellow on lower throat make this plumage readily distinguishable. Additionally, note bolder dark crown- and eye-stripes, dark feather-centres above and to wing-coverts, overall buffier tinge to plumage, and more pinkish-yellow lower mandible.

BIOMETRICS (*petronia*) **L** 14.5–16 cm; **W** ♂ 93–100 mm (*n* 16, *m* 96.6), ♀ 90–97 mm (*n* 13, *m* 93.3); **T** ♂ 50–57 mm (*n* 16, *m* 53.5), ♀ 46.5–54 mm (*n* 13, *m* 51.7); **T/W** *m* 52.7; **B**(f) 12.6–14.7 mm (*n* 31, *m* 13.7); **BD** 9.3–12.0 mm (*n* 31, *m* 10.2); **BW**(g) 7.4–9.5 mm (*n* 31, *m* 8.5); **Ts** 17.1–19.5 mm (*n* 29, *m* 18.0). **Wing formula: p1** minute; **pp2–3**(4) about equal and longest; **p4** < wt 0–2.5 mm; **p5** < wt 5–9.5 mm; **p6** < wt 8–17.5 mm; **p7** < wt 19–23 mm; **p10** < wt 30–37 mm; **s1** < wt 31–39 mm. Emarg. pp 3–4, sometimes faintly also on p5.

P. p. petronia, Greece, Apr: yellow patch on throat not always visible unless neck is stretched or feathers are fluffed out. This race has greyish-brown upperparts, stripes on cap and upperparts being darker, yellow spot deep sulphur and bill rather thick-based. (R. Pop)

GEOGRAPHICAL VARIATION & RANGE Rather moderate and clinal variation. Six races recognised, five of which breed within the treated region.

P. p. petronia (L., 1766) (Canaries, Madeira, S Europe to W Turkey; possibly N Morocco; sedentary). Comparatively dark and small. A general description is found under Identification. Minor subtle average differences within the range sometimes the basis for separation of further races, but none considered sufficiently distinct. (Syn. *balearica*; *hellmayeri*; *macrorhynchos*; *madeirensis*.)

P. p. barbara Erlanger, 1899 (N Africa, except possibly N Morocco; sedentary). A fraction larger and longer-tailed than *petronia*. Also slightly paler brown, and less heavily streaked than *petronia*, streaks dark brown rather than blackish; dark stripes on cap browner. Has on average larger bill than other races (except perhaps *puteicola*). **W** ♂ 96–105 mm (*n* 18, *m* 99.3), ♀ 90–98 mm (*n* 12, *m* 95.0); **T** ♂ 53–60 mm (*n* 18, *m* 56.4), ♀ 50.5–56 mm (*n* 11, *m* 53.5); **T/W** *m* 56.5; **B**(f) 13.5–15.9 mm (*n* 33, *m* 14.7); **BD** 9.9–12.0 mm (*n* 33, *m* 10.9); **BW**(g) 8.1–10.3 mm (*n* 29, *m* 9.3); **Ts** 17.2–20.2 mm (*n* 33, *m* 18.8).

○ ***P. p. exigua*** (Hellmayr, 1902) (SE Ukraine, S Russia, C & E Turkey, Caucasus, S Transcaspia, N Iran; sedentary). Very similar to *petronia* in plumage colours and pattern, and could perhaps just as well be included therein, but is very slightly larger and heavier-billed, thus provisionally accepted. Streaks on underparts more distinct than in *barbara*. **W** ♂ 97–103 mm (*n* 12, *m* 99.6), ♀ 92–100.5 mm (*n* 9, *m* 95.9); **T** ♂ 54–58 mm (*n* 12, *m* 56.3), ♀ 51–57 mm (*n* 9, *m* 54.3); **T/W** *m* 56.5; **B**(f) 12.9–15.4 mm (*n* 22, *m* 14.1); **BD** 9.9–11.5 mm (*n* 22, *m* 10.7); **BW**(g) 8.3–10.5 mm (*n* 22, *m* 9.1); **Ts** 17.0–19.5 mm (*n* 22, *m* 18.5).

P. p. puteicola Festa, 1894 (SE Turkey, Syria, Israel, Jordan, Iraq; sedentary). Large overall with a strong bill, the largest race (together with *intermedia*). A distinctly pale race, slightly paler than others and rather sandy-toned at all seasons, with rather fine dark streaking; streaks on mantle and crown-stripes greyish olive-brown; underparts extensively whitish or sandy-white, while throat-spot paler yellow than in previous races (in which spot is deep sulphur-yellow). **W** ♂ 97.5–105 mm (*n* 13, *m* 101.0), ♀ 94.5–103 mm (*n* 14, *m* 98.8); **T** ♂ 55–61 mm (*n* 13, *m* 58.0), ♀ 53–61 mm (*n* 14, *m* 56.6); **T/W** *m* 57.4; **B**(f) 13.5–16.2 mm (*n* 28, *m* 15.0); **BD** 10.2–12.0 mm (*n* 28, *m* 11.0); **BW**(g) 8.1–10.2 mm (*n* 28, *m* 9.4); **Ts** 17.8–20.6 mm (*n* 27, *m* 19.3).

○ ***P. p. intermedia*** Hartert, 1901 (Lower Volga and Ural Rivers, Transcaspia, E & S Iran, including Zagros, much of S Central Asia, reaching into China; northern populations winter in south of range). The largest race (together with *puteicola*), strangely similar to widely allopatric *barbara* (NW Africa),

P. p. petronia, presumed 1stW, Spain, Sep: in fresh plumage at end of apparent post-juv moult (some unmoulted coverts seem to be juv); once moult is completed will be inseparable from fresh ad. (C. N. G. Bocos)

P. p. petronia, juv, Spain, Jul: as ad but plumage on average subtly tinged more pale brown, streaking less well defined and lacks yellow spot on throat. Gape still yellow. (C. N. G. Bocos)

P. p. exigua, Azerbaijan, Sep: in crisp, freshly moulted autumn plumage (and therefore impossible to age). Note rather thick-based bill of this large race. (M. Heiss)

similarly pale above with pale brown lateral crown-stripes. Upper mantle often unstreaked drab brown, lower mantle and back rather boldly streaked, but somewhat variable. **W** ♂ 95–107 mm (*n* 22, *m* 101.3), ♀ 93–102 mm (*n* 13, *m* 97.4); **T** ♂ 51–61 mm (*n* 22, *m* 56.5), ♀ 52–58 mm (*n* 13, *m* 55.0); **T/W** *m* 56.0; **B**(f) 12.6–15.9 mm (*n* 36, *m* 14.2); **BD** 9.5–11.5 mm (*n* 35, *m* 10.4); **BW**(g) 8.2–10.5 mm (*n* 35, *m* 9.2); **Ts** 17.5–20.2 mm (*n* 33, *m* 18.6). (Syn. *haermsi*, *kirhizica*.)

P. p. exigua, Azerbaijan, May: showing tail pattern well when preening. Subspecies based on locality. (E. Didner)

P. p. puteicola, Israel, May: this population gets worn and bleached rather early, with much-reduced pale fringes above. Despite being worn, racial features still visible, including sandy-brown upperparts, duller stripes on head, streaks on mantle and scapulars narrower and more brown-grey, and throat spot pale yellow and often difficult to see. (A. Ben Dov)

P. p. intermedia, Mongolia, Jun: showing advanced wear and bleaching. Racial features include rather pale upperparts with drab brown streaking and lateral crown-stripes, and pale, ill-defined streaking on underparts. (H. Shirihai)

P. p. intermedia, Mongolia, May: showing diagnostic white tip and dark subterminal band to tail. Note chunky body, triangular-shaped wings and large-headed jizz in flight. (J. Hornbuckle)

(WHITE-WINGED) SNOW FINCH
Montifringilla nivalis (L., 1766)

Fr. – Niverolle alpine; Ger. – Schneefink
Sp. – Gorrión alpino; Swe. – Snöfink

A hardy passerine of high altitudes, restricted in S Europe to the highest mountains, where it is mainly sedentary, only moving to somewhat lower levels in winter. Commonly first encountered around restaurants at ski resorts, sometimes in the company of Alpine Accentors. Can form rather large flocks in winter. Apart from Europe, it occurs on suitable mountains in Turkey and the Caucasus, and eastwards locally to W China and W Himalayas.

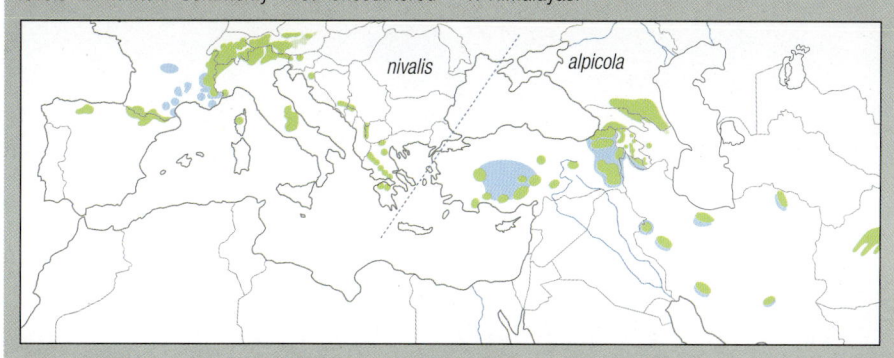

VOCALISATIONS Song a rhythmic repetition of both squeaky, hard notes (like Dipper song) and dry rolling notes (a bit like Redpoll song), e.g. *zrit zrit zi-zerrrrrr-zi zi zi-zerrrrrr-zi zi…*, etc., often given in fluttering display-flight. Variation can include a rather liquid repeated double note, *zre-litt zre-litt*, etc. – A variety of calls noted, including a grating or miaowing *weah* or strident high-pitched *pschieu*, frequently given by flocks in flight. A rattling *chet-et-et-et-et* also noted. When alarmed a related rattling or drawn-out purring *zer'r'r'r'r'r'r*, etc., somewhat recalling Crested Tit. Also a series of short piping notes, *pe pe pe pe pe…* (begging young?), and a rolling *zr'r'r'ruit*. A clear whistling note *pyu* can recall Bullfinch.

SIMILAR SPECIES Differs from *Snow Bunting* in plainer brown upperparts lacking variegated pattern caused by pale-tipped dark-centred scapulars and tail-coverts. Also, primaries are black to base, leaving slightly less extensive white wing patch visible in flight than in adult ♂ Snow Bunting. Differs in several calls and in habitat and range, and the two should not readily be confused.

AGEING & SEXING Ageing generally only possible before and during moult (provided some old flight-feathers retained), but when handled several can be told based on detailed pattern of primary-coverts and innermost primaries (see below, and Strinella et al. 2013). Sexes very similar but sometimes separable in breeding plumage, at least in the hand. Limited seasonal variation. – Moults. Ad and 1stY undergo protracted complete moult in late summer, mostly Jul–Oct (rarely Jun–Dec). Pre-nuptial moult apparently absent or very limited. – **SPRING** Sexing criteria described here probably sexes correctly at least 75%, perhaps 90%. One should be aware of the occurrence of odd ♀♀ with nearly same plumage as ♂, and vice versa. (Strinella et al. 2013; own data.) **Ad ♂** Head cold ash-grey with irregular but prominent black bib (most have some white tips admixed, especially on chin) and blackish loral mask. Central tail-feathers and long uppertail-coverts jet-black (latter sometimes with odd white feathers), outer primaries dark brownish-black. Innermost primary (p10) all white (except some dark at base, usually hidden by coverts), p9 having inner web all white in its outer part, and often also some white inside shaft on outer web. Primary-coverts more extensively white (longest pc3 tipped 2–5.5 mm black, pc4 3–6 mm while pc6 is either all white or has only isolated dark streak or patch at shaft). Smallest lesser coverts grey-tipped (longer not). Bill sometimes all black but usually has small yellowish base to lower mandible. **Ad ♀** Head brown-grey (not pure grey, but extremes of both sexes come close). Bib and loral mask

M. n. nivalis, ad ♂, Switzerland, Jul: large, plump and strikingly patterned, especially the conspicuous black-and-white wings and tail. Sexes not always easy to separate, but purer grey head, relatively extensive black bib and lores, and uniform black bill indicate a ♂, most certainly an ad, as post-juv moult would just have started in July. (H. Shirihai)

IDENTIFICATION Large, long-winged and rather attenuated snow finch with striking plumage. Conspicuous *white panels on wings and tail-sides* (on average less extensive in ♀/juvenile) contrast with *black primaries, alula and central tail*, which latter also has narrow black terminal band. Pied wing and tail pattern recalls Snow Bunting, but these two probably never meet, neither geographically nor altitudinally. Sexes nearly similar, only rarely separable. Adult has pale *brownish-grey or ash-grey head and olive sepia-brown mantle* with ill-defined darker feather-centres and broad blackish tertial centres. *Blackish bib* and small loral mask especially developed in breeding ♂ (but can be similarly strong in odd ♀♀), otherwise mainly white below. Rather long pointed bill largely black with limited pale base to lower mandible (breeding) or has extensive yellowish base to lower mandible (at other seasons). Legs dark grey-brown. Juvenile like adult, but overall duller and lacks darker bib, instead has rather bright yellow-orange bill with just extreme tip darker. Tame, yet often appearing nervous, can suddenly take off for no apparent reason. Does not perch on trees, hops and walks or runs across open ground or between low alpine cover, frequently flicking wings and tail upwards. Flight action fast and powerful on flickering but stiff wingbeats, more clearly undulating over longer distances.

M. n. nivalis, ad ♀, Switzerland, Jul: compared to ♂ has less pure grey head with black bib and lores indistinct or virtually lacking. Mantle slightly duller brown, with less immaculate wing pattern, partially due to browner primaries, and primary-coverts have more extensive black tips. Dark bill has brownish-yellow lower mandible. Dark bill and neat plumage infer sex and age at this date. (H. Shirihai)

M. n. nivalis, ♂, possibly ad, Austria (Jan) and France (Aug): strikingly patterned, especially the conspicuous white portions in wings and tail. That primaries and upper central tail are black (rather than brownish-black) indicate ♂♂. Upper primary-coverts of right bird largely white with small black tips only, indicative of ad. Wing pattern very similar to Snow Bunting but primaries black to base, thus has less extensive white in wing. (Left: K. Lane; right: T. Krumenacker)

M. n. nivalis, ♂, presumed ad, Italy, Feb: in winter, bill is orange. Purer grey head, deep orange-brown iris, black bases to bib and lores, and apparently extensive white bases to longest primary-coverts infer sex. Ageing difficult, but vivid reddish-brown colour of iris and quite grey head might indicate ad. (L. Sebastiani)

M. n. nivalis, presumed ad ♀, Italy, Mar: can remain in winter plumage well into spring. Combination of dirty grey head and only very faint black bases to bib in Mar, still fair amount of white at base of primary-coverts and deep orange iris indicate ad ♀; 1stS ♀ would have had all-dark primary-coverts; while primaries are too brownish for 1stS ♂. (D. Occhiato)

generally less well defined and extensive than in typical ♂ (often more narrow mottled patch on chin and upper throat, rather than full bib). Central tail-feathers dark brown-grey or blackish (but not deep jet-black, still subtle difference, if any, when extremes compared), and longest uppertail-coverts a mixture of blackish, brown and off-white. P10 white as in ad ♂, p9 variable, often with all-white inner web but no white on outer web, rarely with some dark also on outer part of inner web or with white extending slightly onto outer web distally. Primary-coverts have more extensive black tips (longest pc3 tipped 5–10 mm black, pc4 3–7.5 mm while pc6, as in ad ♂, is either all white or has only isolated dark streak or patch at shaft). Most of the longer lesser coverts have small grey tips. Dark bill has extensive brownish-yellow lower mandible. See Biometrics. **1stS** ♂ As ad ♂ or intermediate between this and ad ♀, grey head often slightly brown-tinged. Many are inseparable, but at least typical ones possible to pick out based on pattern of primary-coverts and innermost primaries. 6–7 outer primary-coverts tipped black (2.5–12 mm) against 4–5 in ad ♂. Often black shafts on outer four primary-coverts. P10 white as in ad; p9 has no or only very thin white streak inside shaft on outer web and also a variably large dark smudge or dark patch near tip on inner web. **1stS** ♀ As ad ♀ or even slightly duller, central tail-feathers dark brown-grey, and tertials and outer primaries also on average less dark compared with ad ♀ or 1stS ♂. Primary-coverts typically have all dark outer webs, or dark webs with white outer edge, thus have dark 'bands' running from dark tips to bases. Black on inner webs of outer primary-coverts often 7–12 mm. P10 often sullied darker along shaft and subterminally, and p9 has much dark at tip of inner web and never any white on outer web (pale fringe not counted). – **AUTUMN Ad** Both sexes more ♀-like, with dark lores and bib obscured or lacking, and bill dull yellowish-brown with black tip. Chin of ♂, however, tends to have darker feather-tips.

Best sexed by combination of amount of white/black in wings (as spring, see above) and at times by size; ♂ also tends to have greyer head, and blacker uppertail and uppertail-coverts generally visible. **1stW** Very similar to ad once post-juv moult completed, but see differences in amount of white

Snow Finch ssp. *nivalis*, ♂♀ (e.g. bottom left), and Snow Bunting (e.g. centre in full profile, and far right-hand bird), Italy, Mar: despite both having orange bills, rather similar wing pattern and often occupying similar habitats, they are still readily separated, even in winter when they are most similar. Snow Finch lacks Snow Bunting's strongly patterned head-sides and well-fringed upperparts, among other details. (D. Occhiato)

M. n. nivalis, presumed 1stW ♂, Italy, Mar: still mainly in winter plumage. Combination of 'medium small' dark tips to primary-coverts, and a hint of bib and still mostly olive-brown iris suggest 1stW ♂. Quite richly rufous-tinged back and very little black on tips of outer tail-feathers seem to exclude ♀. Note also slightly dusky yellow bill, often seen in young in winter. (D. Occhiato)

M. n. nivalis, France, Jan: typical winter flock with mixed ages and sexes. In flight, unlike Snow Bunting, the primaries are black to base, leaving less extensive white wing patch. Birds with the most striking pattern are ad ♂♂ with jet-back wingtips and very limited black tips to primary-coverts. (S. Remijn)

in wing as described above under spring, which will separate majority though not all. For much of autumn, iris dull olive-grey (rather than reddish-brown of ad). **Juv** Soft fluffy body plumage duller, head usually tinged buffy-brown, pale grey chin (♂) and white areas tinged sandy. Bill yellow-orange with only extreme tip darker.

BIOMETRICS (*nivalis*) **L** 17.5–19 cm; **W** ♂ 118–127 mm (*n* 17, *m* 121.6), ♀ 108.5–123.5 mm (*n* 17, *m* 117.6); **T** ♂ 67–76 mm (*n* 17, *m* 71.3), ♀ 61–71 mm (*n* 17, *m* 67.3); **T/W** *m* 58.0; **B** 13.9–18.0 mm (*n* 35, *m* 15.6); **B**(f) 11.1–14.0 mm (*n* 35, *m* 12.7); **BD** 7.8–9.0 mm (*n* 34, *m* 8.4); **BW**(gape) 7.0–8.6 mm (*n* 35, *m* 7.9); **Ts** 21.4–23.5 mm (*n* 35, *m* 22.4). **Wing formula: p1** minute; **pp2–3** about equal and longest; **p4** < wt 2.5–5 mm; **p5** < wt 13–18 mm; **p6** < wt 22–27 mm; **p7** < wt 28–34 mm; **p10** < wt 45–51 mm; **s1** < wt 46–52 mm. Emarg. pp 3–4 (sometimes a hint also on p5).

GEOGRAPHICAL VARIATION & RANGE Moderate variation involving size, bill-length, upperparts coloration, especially head, and amount of black on upperwing-coverts and secondaries. Two races recognised.

○ **M. n. nivalis** (L., 1766) (S Europe; sedentary). Described above. Breeding ♂ has crown and neck ash-grey or bluish-grey, contrasting with deep sepia-brown mantle and scapulars. Breast and flanks washed grey. Upperwing-coverts and secondaries white, except black at extreme bases. Large, but bill relatively short.

○ **M. n. alpicola** (Pallas, 1811) (S & E Turkey, Caucasus and Iran east to W Pamirs; sedentary). Differs from *nivalis* in its brown-tinged rather than ash-grey crown, more or less concolorous with back, and by being overall somewhat paler and subtly smaller than *nivalis*, but with a proportionately longer tail and bill. Also, underparts are whiter, it has more black on lesser coverts and outermost secondaries, and bib is more solidly black. **W** ♂ 114–124 mm (*n* 20, *m* 119.3), ♀ 108.5–118 mm (*n* 15, *m* 113.9); **T** ♂ 67–79 mm (*n* 20, *m* 71.4), ♀ 62–73 mm (*n* 15, *m* 67.4); **T/W** *m* 59.5; **B** 14.5–18.5 mm (*n* 33, *m* 17.0); **B**(f) 12.2–16.0 mm (*n* 34, *m* 14.2); **BD** 8.2–9.8 mm (*n* 34, *m* 8.9); **BW**(g) 7.5–9.0 mm (*n* 34, *m* 8.3); **Ts** 19.3–22.8 mm (*n* 34, *m* 21.6). – Turkish breeders ('*leucura*') tend to be a little shorter-winged, but difference slight and insufficient ground for separation. (Syn. *gaddi*; *groumgrzimaili*; *leucura*; *prosvirowi*, *tianshanica*.)

REFERENCES STRINELLA, E. *et al.* (2011) *Ring. & Migr.*, 26: 1–8. – STRINELLA, E. *et al.* (2013) *Avocetta*, 37: 9–14. – WINKLER, R. & WINKLER, A. (1985) *Orn. Beob.*, 82: 55–66.

M. n. nivalis, juv, Switzerland, Jul: head tinged brownish, bill yellow-orange, still obvious yellow gape visible, and has duller, soft fluffy body-plumage. (H. Shirihai)

M. n. alpicola, ♂, Turkey, May: note brown-tinged grey crown, paler brown mantle and scapulars and whiter underparts than *nivalis*. Also apparently shows less white in wing due to having more black on bases of remiges, while bib averages more solid black. (D. Occhiato)

M. n. alpicola, ♀, Turkey, May: especially in ♀ of this race, crown often extensively sullied brownish, while young ♀ more frequently lacks any visible white on bases of primary-coverts. (D. Occhiato)

STREAKED WEAVER
Ploceus manyar (Horsfield, 1821)

Fr. – Tisserin manyar; Ger. – Manyarweber
Sp. – Tejedor estriado; Swe. – Streckad vävare

A native of Asia which within the covered region is confined to the Nile Delta, where a non-native population became established as recently as in the 1970s based on escapes from Alexandria Zoo. Recently it appears also to have become established in Riyadh (Saudi Arabia), the United Arab Emirates and Qatar, also via escapes.

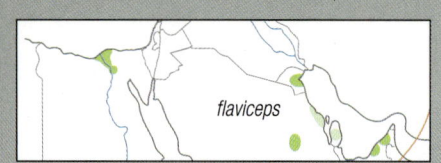

♂ breeding, Egypt, Apr: identification and sexing easy due to yellow crown, dark-streaked breast and flanks, and dark or blackish-brown upperparts, feathers fringed buff and white (creating tramlines). Ageing unsure as many 1stY achieve full breeding plumage after complete wing moult. (V. Legrand)

IDENTIFICATION Being a weaver, it has a heavy conical bill and some strong colours and patterns, and lives socially in dense groups. A rather *small* species. Breeding ♂ has *golden-yellow crown contrasting with blackish-brown head-sides* and dusky throat, and diagnostic combination of *bold blackish streaks over much of upperparts and breast*; otherwise fringed buffy-grey above and mainly dirty white below. The *thick conical bill* is brownish-black. ♀ lacks yellow crown, but has heavy, arrowhead streaks on crown, upperparts and underparts, with pronounced yellowish-white supercilium and a yellowish ill-defined crescent-shaped patch behind dark ear-coverts, on each side of nape. Furthermore, has a pale submoustachial stripe and a pronounced dark moustachial and (sometimes) lateral throat-stripe. Head-on appears pale-faced (often yellowish). Bill dark brown or horn-coloured, often with pinkish or yellowish tinge to lower mandible. Non-breeding ♂ like ♀ but has a hint of yellow cap and dark cheeks, and a brighter yellow face. Wing pattern in all plumages comprises dark-centred coverts and flight-feathers, the former with white fringes and tips, and latter with greenish or yellowish fringes. Gregarious, nesting colonially in dense reedbeds. Flight, general behaviour and gait recall *Passer* sparrows.

VOCALISATIONS The short song consists of *c.* 6 high-pitched whistled *see* notes, culminating in a prolonged wheezing. The overall sound produced by a colony may recall that of a group of House Sparrows. – Call is a wheatear-like *chack*.

SIMILAR SPECIES No congeners occur within tiny W Palearctic range, hence risk of confusion is limited. Still, ♀-like plumages could perhaps be confused with other *Ploceus* species that have escaped, especially Lesser Masked Weaver (*P. intermedius*; feral colony in Sharjah, United Arab Emirates) and Baya Weaver (*P. philippinus*; has also recently bred in Riyadh and the United Arab Emirates).

AGEING & SEXING Ageing apparently impossible following post-juv moult, but little studied. Sexes strongly differentiated in breeding plumage. – Moults. Few data from Egypt, but most likely ad has complete post-nuptial moult, and juv usually moults completely too, or at least almost so, some retaining a few or many juv secondaries and perhaps additional feathers; moults take place mostly Jul–Oct (Jun–Dec). Pre-nuptial moult in Feb–Apr is partial, and is apparently more extensive in 1stY. – **BREEDING Ad** For a general description and sexual differences, see Identification. Pale fringes to upperparts and wings bleach to pale buff-white, or largely wear off, with plumage becoming more solidly dark above and more heavily streaked below. **1stS** Those which moult completely are inseparable from ad, but many retain at least some juv remiges, primary-coverts, and perhaps outer greater coverts and tail-feathers. Some ♂♂ distinctive, being less advanced, and appear to retain partial winter plumage. – **NON-BREEDING** Fresh (Nov–Feb/Mar). In both sexes, cinnamon fringes to upperparts and wings distinctive, and bill paler, light horn-brown. **Ad ♂** Generally ♀-like but tends to have brighter yellow supercilium, cheeks and submoustachial stripe, usually some dark grey triangular spots on throat, sometimes with some yellow on crown, contrasting with darker head-sides. Streaking on underparts highly variable, but usually broader than in ♀, covering much of chest (rather than being mostly confined to sides). **Ad ♀** Like breeding, but pale facial parts and neck whiter, less yellow (bright yellow tinge almost restricted to mark on neck-side, just behind ear-coverts), while dark streaks below are partly concealed and narrower. **1stW** Perhaps indistinguishable from ad if moult completed. However, many retain juv wing-feathers, especially some primary-coverts, outer primaries and secondaries. At this season birds in many different stages (intermediate between juv and non-breeding) present near colonies. **Juv** Like ♀, but often has reduced (or even lacks) streaking on buffy underparts; face markings

♂ breeding, United Arab Emirates, Sep: ♂♂ in breeding plumage can occur year-round, even in autumn, and there is evidence that only some undertake partial moult to winter plumage. This ♂ moulted its body plumage completely, but not the head, while bill remains black. (H. Roberts)

♂ breeding, United Arab Emirates, Sep: unlike previous ♂, this bird's moult was very limited, but some dark feathers on neck-sides suggest that it is about to moult directly to breeding-like head plumage; bill pale orange, as non-breeding. (M. Barth)

inconspicuous and streaks on upperparts duller. Bill yellow-brown, with flesh-coloured tone to lower mandible, and tail-feather tips more sharply pointed.

BIOMETRICS (*flaviceps*) **L** 13–14 cm; **W** ♂ 68–71 mm (*n* 12, *m* 69.6), ♀ 66–72 mm (*n* 12, *m* 68.4); **T** ♂ 38–45 mm (*n* 12, *m* 41.9), ♀ 41–47 mm (*n* 12, *m* 43.8); **T/W** *m* 62.2; **B** 16.8–19.9 mm (*n* 24, *m* 18.1); **B**(f) 13.6–17.2 mm (*n* 24, *m* 15.6); **BD** 9.1–11.0 mm (*n* 24, *m* 10.1); **BW**(g) 7.4–9.4 mm (*n* 23, *m* 8.6); **Ts** 18.5–22.8 mm (*n* 24, *m* 20.4). **Wing formula:** p1 minute; **pp2–4**(5) about equal and longest; **p5** < wt 0–3 mm; **p6** < wt 1.5–6 mm; **p7** < wt 4–9 mm; **p10** < wt 11–15 mm; **s1** < wt 10–15.5 mm. Emarg. pp 3–6.

GEOGRAPHICAL VARIATION & RANGE Rather well-marked variation in SE Asia, but apparently only one subspecies, *flaviceps* of the Indian Subcontinent, involved among the introduced breeders in the Nile Delta.

P. m. flaviceps (Lesson, 1831) (Pakistan, India, Sri Lanka; resident). Slightly less saturated yellow and chestnut in ♂ breeding plumage than *manyar* of Java.

♀ breeding, Egypt (left) and United Arab Emirates, Sep: heavily streaked above and below, with pronounced yellowish-white supercilium and crescent-shaped yellow patch behind ear-coverts. (Left: V. & S. Ashby; right: H. Roberts)

♂ non-breeding, Oman, Jan: generally ♀-like plumage, but has more yellow in supercilium and nuchal patch. Entire plumage replaced, so the bird could be either an ad or 1stY, although there is some evidence that at least some ad ♂♂ (older such?) skip this plumage. (I. Lowson)

Non-breeding, India, Dec: typical non-breeding, ♀-like plumage. Ageing and sexing usually not feasible. (C. Francis)

Juv/1stY, India, Aug (left) and Egypt, Oct: shortly after fledging (left) heavily streaked above, but can remain partly unmoulted even into late autumn (right), when fairly distinctive in appearing uniform below. However, note diagnostic buffish-yellow nuchal patch in both birds. (Left: V. V. Mishra; right: M. Heiß)

RÜPPELL'S WEAVER
Ploceus galbula Rüppell, 1840

Fr. – Tisserin de Rüppell; Ger. – Gilbweber
Sp. – Tejedor de Rüppell; Swe. – Rödahavsvävare

One of the more familiar and numerous E African weavers. In the covered region its range is restricted to SW Saudi Arabia, from about Jiddah and south (where introduced), but also further south in Yemen and S Oman, with recent records (again of escapes) in the Persian Gulf. It frequents savanna-like habitats, oases or foothill copses, breeding in colonies near water and is often attracted to cultivation.

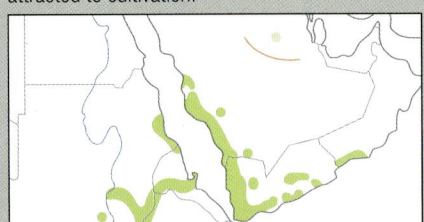

IDENTIFICATION Like a cross between a Siskin and a sparrow, with a *stout, noticeably pointed bill with slightly decurved culmen*, short tail and wings, and typically colourful weaver plumage. Breeding ♂ is mainly rich *golden-yellow* with a *dark chestnut mask*, has *blackish-brown lores and upper chin*, metallic blackish-grey bill, virtually unstreaked bright yellow crown to back, and *orange-red eyes*. ♀ (and non-adult) has almost *unmarked head*, dark-streaked olive-brown upperparts, predominantly whitish underparts, washed buffish-yellow on (unstreaked) throat to upper flanks, with paler, dark red-brown iris and horn-grey bill. In all plumages upperwings narrowly fringed paler, including the tertials. Gregarious, often forming large flocks and, like most weavers, very noisy around colonies.

VOCALISATIONS Song a prolonged series of high-pitched strained and metallic notes and rattles, now and then interrupted by a drawn-out wheezing in Barn Swallow or Siskin fashion, e.g. *chek-zik-zik-zezeze-trirrr-chek-zweeeeeehz-chek-zeze-kakaka…*, etc. One scratchy sound of the song has been likened to paper tearing. Usually starts as a low grating chatter and accelerates into a double swizzle, with brief insect-like hissing chatter at end. – Other harsh notes in alarm and contact include a quite explosive *chucka*, an upslurred whistle *chewy* and a squeaky *cheee-cheee*.

SIMILAR SPECIES In the treated region requires separation only from ♀ *Arabian Golden Sparrow* (no range overlap with Streaked Weaver), from which it is readily distinguished by its bulkier appearance, longer, slightly drooping bill, shorter tail and streaked upperparts. – ♀/juvenile could be confused with a *Passer* sparrow, but overall shape, bill structure and finer upperparts streaking on more olive-yellow ground colour are usually sufficient to overcome such problems.

AGEING & SEXING Ageing not well studied, and probably often difficult after post-juv moult concluded. Sexes strongly differentiated in breeding plumage, much less so when younger; full breeding-plumage ad ♂♂ occur year-round in Arabia. – Moults. Few birds in moult from Arabia examined. Based on limited experience of the species' annual breeding cycle in the covered region, appears to breed virtually year-round, mainly correlated to biannual rainy seasons, and fresh-plumaged birds encountered from late autumn, fresh or slightly worn in Dec, moderately worn in Jan–Feb, rather worn in Mar–Apr and very worn in May–Sep, with complete post-nuptial moult apparently mostly in Aug–Sep, when ad

♂ **breeding, Ethiopia, Oct**: unmistakable, being mainly rich golden-yellow with dark chestnut mask, blackish-brown lores and chin, dusky mottling on olive-yellow mantle, and orange-red eyes. Upperwings narrowly fringed yellowish-green, producing double wing-bars. (H. Shirihai)

♀, **Oman, Jan**: stout, noticeably pointed bill with typically slightly decurved culmen. Also normal ♀-like weaver plumage with dull olive-brown and streaked upperparts. Underparts paler and unstreaked, washed buffish-yellow, and eyes very dark red-brown. Short wings with pale wing-bars and tertial fringes. (F. Trabalon)

WEAVERS AND WAXBILLS

1stY ♂, Oman, Nov: note combination of ♀-like lack of mask, dull orange iris and pale buffish-brown underparts. This bird is in post-juv moult. (D. Occhiato)

♀, Oman, Nov: at end of post-juv moult. Yellowish throat slightly contrasting. Bill and legs appear pale pinkish-brown, and upperparts fringed duller and brownish. (D. Occhiato)

♂ breeding plumage replaced by next one (previous literature suggests a winter/non-breeding plumage attained by ♂♂, but this seems incorrect, at least for ad). Timing and extent of post-juv moult unclear, but probably complete or almost so, with both sexes achieving ♀-like plumage, but at a few months age some ♂♂ acquire variable number of yellow feathers below, but not chestnut mask. Unknown whether they breed in this plumage. – **BREEDING Ad** See under Identification. **1stY** Those moulting to ad-type breeding plumage inseparable from ad. Iris apparently less pure red-brown than ad (field observations). – **NON-BREEDING Ad** When very fresh, ♂ has paler chestnut face with reduced or virtually no black on lores or chin. **1stY** Both sexes more ♀-like. Shortly after moult, ♂ still lacks chestnut face and rich yellow underparts, but compared to ♀ is greener above with yellower wing and tail fringes, and has more yellowish on throat and breast. Some have darker bill. 1stY ♂ in ♀-like plumage occurs mostly Nov–Dec (–Mar), when some have finished primary moult; note that some retain some worn secondaries. **Juv** Similar to ♀ but has soft fluffy appearance to entire body plumage, and olive-brown iris. In Arabia one fresh juv collected 9 Mar and another on 11 Jul.

BIOMETRICS L 13–14 cm; **W** ♂ 70–76 mm (n 13, m 74.2), ♀ 66–73 mm (n 12, m 69.1); **T** ♂ 44–53 mm (n 13, m 48.6), ♀ 41–49 mm (n 12, m 45.3); **T/W** m 65.6; **B** 15.5–18.4 mm (n 24, m 16.9); **B(f)** 12.0–15.8 mm (n 24, m 14.0); **BD** 8.1–10.1 mm (n 24, m 9.0); **BW(g)** 7.2–9.3 mm (n 24, m 8.0); **Ts** 18.6–22.1 mm (n 24, m 20.5). **Wing formula: p1** minute; **pp2–5** about equal and longest; **p6** < wt 1–4 mm; **p7** < wt 2–7 mm; **p10** < wt 10–15 mm; **s1** < wt 10.5–14.5 mm. Emarg. pp 3–6.

GEOGRAPHICAL VARIATION & RANGE Monotypic. – E Sudan, Eritrea, N Ethiopia, N Somalia, introduced SW Saudi Arabia, Yemen, S Oman, Kuwait; sedentary. – Although probably best kept as monotypic, Arabian ♂♂ appear generally slightly paler than African populations, with plainer upperparts (only indistinctly streaked or virtually unstreaked) and have a somewhat more pointed (less compressed) bill.

REFERENCES AL-SAFADI, M. M. (1996) *Sandgrouse*, 18: 14–18.

Juv, Yemen, Jan: a juv in still neat and quite yellow-tinged plumage. Iris colour difficult to judge in this angle, but eye looks overall dark. Median coverts appear rather softly textured. (H. & J. Eriksen)

Presumed juv, Oman, Apr: dull plumage with very obscure greenish- or buffish-yellow on face, and similarly subdued wing and tail fringes (perhaps unmoulted and bleached juv feathers), are best indication of young ♀. (R. Kunz)

COMMON WAXBILL
Estrilda astrild (L., 1758)

Fr. – Astrild ondulé; Ger. – Wellenastrild
Sp. – Estrilda común; Swe. – Helenaastrild

Common Waxbill, a frequently kept cagebird, has been widely introduced in many parts of the world, and is now established in Portugal (since 1967, with perhaps 10,000–100,000 pairs), Spain (since 1977), the Azores (mid 1980s but not since) and the Cape Verdes (confined to Santiago); local introductions on Gran Canaria and Madeira, and in other countries apparently in progress. In its native range over much of sub-Saharan Africa, it is generally found in tall-grass savannas, marshes, cultivations, gardens, villages and towns, often near water.

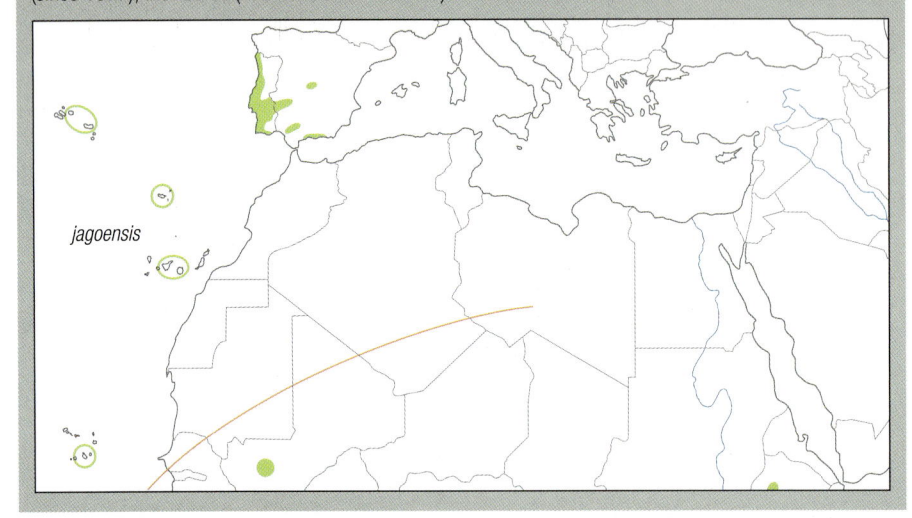

E. a. jagoensis, ♀ (left) and ♂, Portugal, Sep: note subtly more orange (less deep red) bill and mask, and unbarred crown, as well duller and less pink-tinged underparts of ♀. (J. Viana)

tieh-prt-tieh-prte-tieh-... – Calls include a short but slightly buzzing monosyllablic *tzep*, and flocks utter a muted twittering, becoming more urgent and excited when disturbed.

SIMILAR SPECIES The only significant confusion risks are with escapes of two common cagebirds: Black-rumped Waxbill *E. troglodytes* (which has distinctive black uppertail-coverts and tail, and has occurred in Iberia, where it has even been claimed to hybridise with the present species), and Crimson-rumped Waxbill *E. rhodopyga* (which has a black bill and red fringes to the wing-coverts, tertials and tail). – No overlap with *Arabian Waxbill* or *Red-billed Firefinch*, and very different in all plumages from *Red Avadavat*. From all of the above and several other closely-related small finches, usually safely differentiated by combination of bright red bill and mask in both sexes (but not juvenile), lack of black or bright red rump (but ♂♂ of some races have rather obvious rufous-red patch). It also lacks red in wings, and its barred tail is never uniform black. Instead, adult ♂ has almost blackish vent and undertail-coverts.

AGEING & SEXING Ageing possible in the hand by moult pattern. Sexes rather similar, though usually separable. Insignificant seasonal variation (caused by feather wear). – Moults. Complete post-nuptial moult and partial post-juv moult mainly in late summer or early autumn (Jul–Oct). Post-juv moult usually involves head, body and wing-coverts,

E. a. jagoensis, ♂, presumed ad, Spain, Sep: tiny finch with wax-red bill and bright red mask tapering at rear, plumage almost entirely vermiculated dark brown on buffish-grey upperparts and warmer pinkish-tinged underparts. Note solidly black vent and red patch in centre of belly (darker and more extensive in ad ♂). Graduated tail barred dark brown. (M. Rojas Ruiz)

IDENTIFICATION A quite *small* and *small-headed* finch-like bird, with short, rounded wings, slim body and *long graduated tail*, which typically hops close to the ground, climbs stems, and flies rapidly on whirring wingbeats. Gregarious, usually in small, rather tame but restless, tight flocks. Readily identified by combination of *wax-red bill* and *bright red stripe through eye* tapering to a point above rear ear-coverts, otherwise *predominantly pale brown-grey finely vermiculated dark above*, and usually *warmer pinkish-buff below, finely but extensively barred on sides*, but *cheeks, throat and upper breast usually whiter* and some have rose-red central belly (then a sign of ♂ sex) and *almost black undertail-coverts* (darker and more extensive in ♂). Graduated tail boldly barred dark brown. Tail appears comparatively long in flight.

VOCALISATIONS Song a repeated three- or four-syllable phrase that rises in pitch, *te-tete-prizz* (*prizz* note sounding extended), sometimes culminating in an excited series *prt-*

E. a. jagoensis, presumed ad ♀, Spain, Oct: lacks any pink-cinnamon tones on flanks or reddish-rose patch on belly, while rear underparts are tinged browner than ♂ (of any age). Colourful mask and bill at this time of year fit well, and indication of moult suggests a bird completing post-nuptial moult, thus probably ad ♀. (M. Rojas Ruiz)

some or all tail-feathers and tertials. Partial pre-nuptial moult (extent poorly known), mostly Jan–Feb. Data available from Portugal, but other populations differ due to environmental factors; e.g. in Cape Verdes, breeding commences at onset of rains, in Jul–Aug, and moult occurs from Dec. — SPRING **Ad ♂** For general description, see Identification. Note extensive black ventral area and often prominent red central patch on belly. **Ad ♀** Duller than ♂, especially on underparts, with reduced pink-cinnamon tinge (and reduced if any rosy patch on belly), almost lacks blackish vent and undertail-coverts, or has these dark brownish (usually blacker and more extensive in ♂). Face patch extends less behind eye, and both face patch and bill orange (less pure red), though some approach dullest ♂. Races with rufous-red rump and uppertail-coverts (especially *rubriventris*) have this colour much reduced in ♀. **1stS** As respective ad, from which best separated by retained juv wing-feathers (more worn and abraded browner). Some moult substantially in winter, but unknown if all juv feathers ever replaced. — AUTUMN **Ad** Sex differences as spring. Plumage fresh. **1stW** As respective fresh ad, but sexual differences less obvious (plumage duller and more buffish, approaching ♀). Extent of post-juv moult quite variable, but primary-coverts and apparently at least most remiges are retained (being slightly weaker textured, duller and fringed buffier), forming moderate moult limits. **Juv** Even duller than ♀, with bill black, and lacks obvious cross-barring and sometimes even red facial mask. Underparts entirely buff-brown (no black on vent), though often has pale red loral line or some red on underparts.

BIOMETRICS (*jagoensis*) Sexes of same or very similar size. **L** 10–11 cm; **W** 44–48 mm (*n* 19, *m* 46.2); **T** 38–47 mm (*n* 16, *m* 42.5); **T/W** *m* 92.1; **B** 8.9–10.9 mm (*n* 19, *m* 9.5); **B**(f) 7.0–8.9 mm (*n* 19, *m* 8.0); **BD** 5.3–6.7 mm (*n* 19, *m* 6.0); **BW**(g) 4.5–5.8 mm (*n* 19, *m* 5.2); **Ts** 12.4–14.5 mm (*n* 19, *m* 13.5). **Wing formula: p1** minute; **pp2–5** about equal and longest; **p6** < wt 0.5–3 mm; **p7** < wt 2–5.5 mm; **p10** < wt 7–10 mm; **s1** < wt 6–10.5 mm. Emarg. pp 3–6.

GEOGRAPHICAL VARIATION & RANGE Up to 14 races described, but intergradation and subtle, variable differences probably would invalidate several if comprehensively revised. Portuguese populations rather variable, too, apparently because various races from Angola were introduced. However, majority perhaps nearest to *jagoensis* (described from Cape Verdes, where also introduced, but this population originates from coastal Angola, possibly south to extreme NW Namibia), albeit with more extensively red underparts, though also approach *rubriventris* (Equatorial Guinea, Gabon to extreme N Angola) and *angolensis* (W Angola), which races are often considered to have been introduced elsewhere. Other introductions relate to *astrild* (S Namibia, S Botswana, W Orange Free State and Cape Province). Of these, *jagoensis* is paler and greyer, ♂♂ being washed with limited brown above, reddish breast and belly patch poorly developed, relatively whiter on chin and throat, and weakly barred throughout. ♂ *rubriventris* is reddest or richest pink of all, especially above, including on rump and uppertail-coverts, with lower breast and belly warmer cinnamon. ♀♀ generally harder to assign.

E. a. jagoensis, 1stS ♂, Cape Verde, Mar: a few very worn and bleached juv tail-feathers and secondaries suggest age. Underparts rather extensively sullied pink but not as rich as in ad. Sex inferred by dark red bill and mask, and by pinkish underparts. (D. Occhiato)

E. a. jagoensis, presumed 1stW ♂, Azores, Oct: as ad ♂ but reddish bill slightly duller. Underparts have a pinkish flush inferring ♂, but still no darker red patch on central belly indicating young ♂. (D. Occhiato)

E. a. jagoensis Alexander, 1898 (W Angola; resident; introduced Cape Verdes, Gran Canaria, Madeira, Portugal, Spain). Described above and discussed under Geographical variation, a comparatively rather pale and greyish race.

REFERENCES Vowles, G. A. & Vowles, R. S. (1987) *Ring. & Migr.*, 8: 119–120.

E. a. jagoensis, 1stW ♀, Cape Verdes, Oct: hardly any pink below and dull yellowish-orange bill confirm sex and age. (E. Winkel)

E. a. jagoensis, juv, Portugal, Oct: note black bill and weak cross-barring, as well as still partly fluffy plumage, still growing tail and yellow gape, which combined identify this bird as juv. (N. Loverock)

ARABIAN WAXBILL
Estrilda rufibarba (Cabanis, 1851)

Fr. – Astrilde d'Arabie; Ger. – Jemen-Astrild
Sp. – Estrilda árabe; Swe. – Jemenastrild

A resident of better-vegetated grasslands and scrubby *Acacia* areas in the foothills and mountains of SW Saudi Arabia and Yemen, the Arabian Waxbill is still a rather poorly-known bird, despite recent fieldwork in the remote parts of Arabia it inhabits. Occurs both in arid country and in cultivated areas.

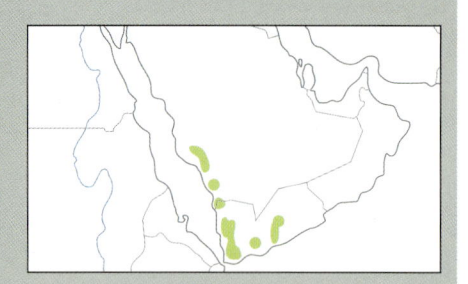

Presumed ♂, Yemen, May: finely vermiculated greyish upperparts and only indistinct warmer wash, except faint buff vermiculations on sides of whitish underparts. Bill is mostly dark, not all red. Dark crimson mask, and tail and rump are black with narrow white edges to outer rectrices. Sexes largely alike, but some reddish on bill, purer grey upperparts and faint pinkish tinge below indicate ♂. (W. Müller)

Yemen, May: note variation in amount of red on black bill and black in red mask. The species should be readily recognised in most views, having very fine vermiculations (visible only when close) on distinctly grey upperparts, by mask and bill colours, largely black and deeply graduated tail with whitish edges to outermost feathers, and rather evenly pale and indistinctly barred underparts. (W. Müller)

Presumed ♀, Yemen, May: partly black mask, buff instead of pinkish below and very limited red on bill indicate ♀. Apparently in post-nuptial moult with at least a central tail-feather growing. (W. Müller)

Presumed ♀, Yemen, Jan: buff instead of pinkish underparts indicate ♀. Especially in non-breeding plumage, bill is all black. (H. & J. Eriksen)

IDENTIFICATION Typical waxbill, a small delicate grey-brown finch with a *deep red mask*. Upperparts including wings predominantly *dull brownish-grey and finely vermiculated*. Rump black. Tail *comparatively long and well graduated, dark brown* with fine off-white sides. Whitish cheeks, throat and neck-sides contrast with slightly darker pale greyish-buff or pale brownish rest of underparts. That underparts, too, are finely vermiculated is not usually obvious in normal encounters, except on lower flanks, where it is most prominent. *Undertail-coverts off-white*. Short, stubby *bill usually all dark*, although at least some breeding adult ♂♂ have red cutting-edges and base to lower mandible. Like most waxbills typically forms small, fast-moving flocks, often mixing with Zebra Waxbill and frequently found near water. Flicks and waves tail when perched. Flight rather jerky.

VOCALISATIONS Song apparently unrecorded (Christensen & Porter 1987; Hollom *et al.* 1988). – Flocks utter a *tse-tsee* or *chee* virtually constantly, and in flight give harsh series, *chee-chee-chee-chee-chee*, and *che-che* or *chee* (1–2 syllables). Other (contact?) notes include a hard buzzing *dzit*, *dzeet* and *chzit*. Several other minor calls noted from feeding or roosting flocks.

SIMILAR SPECIES Overlaps (and frequently associates) with *Zebra Waxbill* (which see). – No other waxbills occur in the SW Arabian highlands, and it should be readily separated from any potentially-escaped Estrildidae in Arabia by having much finer vermiculations on distinctly greyer upperparts, the near lack of any pink, rather distinct whitish cheeks, and the largely black bill instead of red. The dark tail (with whitish edges to outermost feathers visible from below), black rump, and white vent and undertail-coverts, are also important features.

AGEING & SEXING Sexes largely indistinguishable. Seasonal variation rather limited. – Moults. Few data on moult, and annual cycle poorly known. Breeding recorded once in Jun, juv noted in Oct–Nov when one ad was in advanced (presumably complete post-nuptial) moult (Christensen & Porter 1987); seven very worn birds, some already starting to moult, in Dec–Feb, and in Jul (three), while fresh ad (or birds at end of moult) and 1stY recorded in Dec–Mar and Aug (NHM). Breeding and moult strongly linked to twice-yearly seasonal rains, mostly in Mar–Apr and Jul–Sep, suggesting that moult and ageing are at least as complex as in Common Waxbill. Post-nuptial moult apparently complete (possibly suspended, at least in some), and post-juv moult mostly partial. Pre-nuptial moult unrecorded but still possible. – ALL YEAR **Ad** For general description, see Identification; some, apparently ♀♀, have mask almost black, with warmer upperparts and buff wash to belly and vent. Bill bluish-black in non-breeding season, but others show pale grey or whitish bills. **1stY** Most apparently retain at least some remiges and primary-coverts, and in early stages are relatively fresher than heavily-worn breeding ad (but later slightly more worn and bleached compared to freshly-moulted ad). Primary-coverts of several, apparently 1stY, unbarred grey or brown (unlike ad), but this needs to be confirmed. Some, also perhaps 1stY, appeared to be in extensive moult, and it seems probable that some undertake complete post-juv moult. **Juv** Soft fluffy body-feathering and blackish mask, brown upperparts (often has dark-capped appearance), darker wings, buff wash to underparts and dark grey bill.

BIOMETRICS Sexes of same or very similar size. **L** 10–11 cm; **W** 46–50.5 mm (n 24, m 48.1); **T** 40–49 mm (n 24, m 44.9); **T/W** 93.5; **B** 9.6–11.9 mm (n 23, m 10.3); **B**(f) 7.5–10.0 mm (n 23, m 8.4); **BD** 5.5–6.7 mm (n 22, m 6.0); **BW**(g) 5.3–6.3 mm (n 23, m 5.7); **Ts** 12.2–13.7 mm (n 24, m 13.2). **Wing formula: p1** minute; **pp2–5** about equal and longest; **p6** < wt 0.5–3 mm; **p7** < wt 2.5–5 mm; **p10** < wt 7–10 mm; **s1** < wt 8–10.5 mm. Emarg. pp 3–6.

GEOGRAPHICAL VARIATION & RANGE Monotypic. – SW Saudi Arabia, Yemen; sedentary.

REFERENCES CHRISTENSEN, S. & PORTER, R. F. (1987) *Sandgrouse*, 9: 98–101.

RED AVADAVAT
Amandava amandava (L., 1758)

Alternative names: Red Munia, Strawberry Finch

Fr. – Bengali rouge; Ger. – Tigerfink
Sp. – Bengalí rojo; Swe. – Röd tigerfink

A small, dark red finch-like bird of S Asia which has established local thriving populations in SW Europe in recent times. All populations in the W Palearctic stem from escapes, and this avadavat is now locally common in SW Spain and Portugal, as well as in NE Italy, N Egypt and E Arabia. Found primarily in tall, dense grass, reeds and sugarcane, but also in bushes and low scrub. Often seen in small flocks when breeding, forming larger flocks at other seasons.

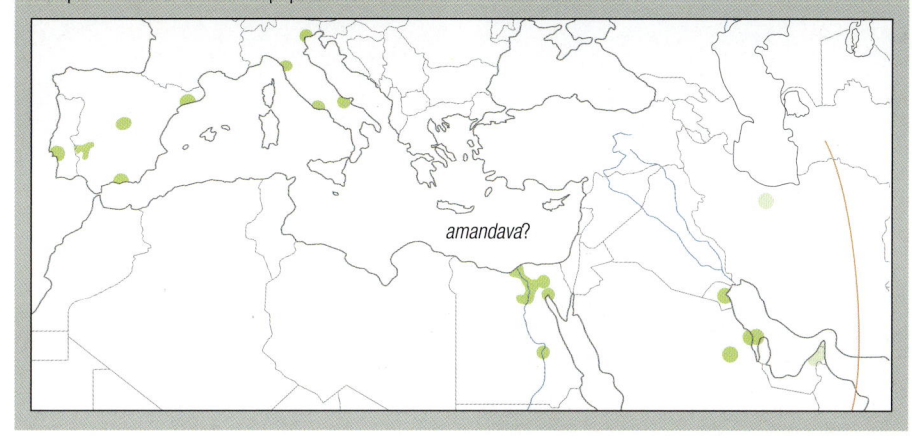

A. a. amandava, ad ♂, India, Dec: resembles ♀ but throat whitish, and red uppertail-coverts speckled white. Tail blackish, and iris red. Combination of small size, restless behaviour and conspicuous white-spotted rufous-red rump patch make identification easy, even in such duller plumage. (S. Singhal)

IDENTIFICATION A nervous and flighty, short-tailed finch, with a distinctly conical bill, and often merely heard as small parties fly past. ♂ in breeding plumage very attractive, being *predominantly carmine-red with numerous irregular pearly-white spots*, primarily on chest, flanks, rump, scapulars and blackish wings (spotted on all coverts and tertial tips), while *undertail-coverts and vent are blackish*, and bill coral-red. ♀ almost lacks red in plumage, predominantly *dusky grey-brown above and paler grey-buff below*, with fulvous-yellow flanks, and has *blackish lores, red rump and uppertail-coverts, and small spots restricted to wings*. Additional features in all plumages are the blackish tail and pinkish legs, whereas eye colour varies with age and sex.

VOCALISATIONS Song a weak, high-pitched warble, given by ♂. – Both sexes utter varied squeaks and chirps as calls.

SIMILAR SPECIES Readily identified even in a brief view. Combination of small size, restless behaviour and bright red plumage (fast-flying birds are conspicuous by the crimson-red rump patch) with numerous white spots ensure easy identification.

AGEING & SEXING Ageing not well studied and seems difficult in most seasons. Sexes usually differ markedly. Rather marked seasonal variation in ♂. – Moults. Complex and poorly understood, with very limited data from birds established in the covered region. Breeding season varies (e.g., in SW Spain nests mainly Jul–Oct, in N Italy mainly Nov–Dec, in N Egypt Apr–Jun but also Aug and Nov, while in native Indian Subcontinent main breeding season is during monsoon, continuing into dry season), and hence moult timing and extent of moult also vary. Moult of ♀ generally occurs slightly later than ♂. – **BREEDING** Following complete pre-nuptial moult of ad (remiges can be renewed completely but often partly, perhaps complementarily, as continuation of partial post-nuptial moult), in Spain mainly Apr–Aug. In W Palearctic birds in full breeding plumage mostly in Jul onwards. **Ad ♂** For general description, see Identification. Extent of white spotting varies individually, as does line of small white spots below eye. Iris deep red-brown. **Ad ♀** As ad ♂ but wings often tinged reddish, and tail and uppertail-coverts occasionally have small white spots. Also, unlike ♂, has duller pinkish-brown bill, and iris is often duller. **1stS** Many inseparable from respective ad, but at least some retain some juv primary-coverts and remiges, and some ♂♂ breed in intermediate plumage (between non-breeding and breeding). – **NON-BREEDING** Ad attains eclipse ♀-like plumage following partial post-nuptial moult of head, body, lesser and median coverts, and sometimes inner greater coverts, often a few remiges (generally Jul–Dec, in Spain chiefly Nov–Dec, then suspend Jan–Apr when in non-breeding plumage). **Ad ♂** resembles ♀, but face and throat duskier, and red uppertail-coverts speckled white, with larger white tips to greater coverts, deeper buff-brown breast, sometimes with scattered red feathers on breast and belly, and perhaps on head, lores black, usually with red stripe above; tertials and wing-coverts blacker (less brown). Ad non-breeding ♂ plumage rare in some populations, perhaps because non-breeding is partly or largely suppressed. Moult of ♀ appears to be more restricted or suppressed when moulting to non-breeding plumage. **Ad ♀** As breeding but tends to have reduced or no red pigmentation and white body spotting; also lacks any pink-red on

A. a. amandava, ad ♂ breeding, Italy, Aug: very small, brightly coloured finch, which in breeding plumage is easily sexed: ♂ is carmine-red with numerous small white spots, including on dark wings, while crown to mantle are mottled olive-brown, and undertail-coverts and vent are blackish. Both sexes have coral-red bill (except black culmen ridge). Ad at end of post-nuptial moult, outermost primaries still growing. (D. Occhiato)

A. a. amandava, ♀, presumed ad, Italy, Sept: almost lacks red in plumage save for reddish tinge to rump and uppertail-coverts. Note characteristic blackish lores, and that white spots are small and few and restricted to wing. Red iris and apparently ad wing-feathers suggest age. (D. Occhiato)

belly and has no or very indistinct white spotting on red uppertail-coverts, eliminating confusion with non-breeding ♂. Iris in both sexes reddish. **1stY** Most juv moult partly (but can suspend at any stage) into ad non-breeding plumage and before acquiring full breeding plumage (e.g. in Spain, body moult recorded Oct–Jan, sometimes suspended Feb–Apr, then followed by moult of body, wings and tail, Apr–Jul). Most 1stY aged by combination of browner and more worn juv primary-coverts, remiges, tertials, perhaps some alula, outer greater coverts, and even some tail-feathers, and dull olive-brown iris. ♂ (but not ♀) often has some red feathers on head, especially in supercilium. A few moult directly from juv to breeding plumage, or acquire plumage intermediate between breeding and non-breeding. **Juv** Soft fluffy body-feathers, with fawn-brown upperparts, blackish-brown bill and no red rump. Cheeks pale grey-buff, chin and mid-belly off-white, brown-buff throat, chest and flanks (thus juv is considerably more dusky than post-juv plumages). Buff tips to greater and median coverts create wing-bars. ♂ can have some red around eye.

BIOMETRICS (*amandava*) Sexes same or very similar size. **L** 10–10.5 cm; **W** 44–50 mm (*n* 24, *m* 48.1); **T** 34–41 mm

A. a. amandava, 1stY ♂, **India, May**: after partial post-juv moult with obvious moult contrasts in wing. Rufous uppertail-coverts with small white spots fit well with 1stW eclipse ♂. Bill almost pure red, but iris red-brown. (A. Arya)

A. a. amandava, 1stY ♀, **India, Jan**: very similar to 1stY ♂, but plumage averages still duller with fewer or smaller white spots, and bill not as red, has more darkish portions. Note also darker red-brown iris and retained juv primary-coverts. (R. Mallya)

(*n* 23, *m* 37.3); **T/W** *m* 77.8; **B** 9.0–10.9 mm (*n* 24, *m* 10.0); **B**(f) 7.4–9.3 mm (*n* 24, *m* 8.3); **BD** 5.4–6.4 mm (*n* 24, *m* 5.8); **BW**(g) 5.0–6.3 mm (*n* 24, *m* 5.6); **Ts** 13.0–14.0 mm (*n* 23, *m* 13.4). **Wing formula: p1** minute; **pp2–5** about equal and longest; **p6** < wt 1–3 mm; **p7** < wt 2–4.5 mm; **p10** < wt 6–8.5 mm; **s1** < wt 6–9.5 mm. Emarg. pp 3–6.

GEOGRAPHICAL VARIATION & RANGE Moderate variation, generally three subspecies recognised. Ssp. *amandava* is probably the form introduced in the W Palearctic.

A. a. amandava (L., 1758) Pakistan, India, S Nepal; introduced in SW & S Europe, Egypt, E Saudi Arabia, Bahrain; sedentary). Described above.

REFERENCES Costa, H., Lobo Elias, G. & Farinha, J. C. (1997) *BB*, 90: 562–568. – De Lope, F., Guerrero, J., De La Cruz, C. & Da Silva, E. (1985) *Alauda*, 53: 167–180.

A. a. amandava, 1stW ♂, **Italy, Oct**: typically fawn-brown with brownish-orange pale areas on bill and no red rump. Also buffish wing-bars on juv coverts, but spotty white on moulted inner greater coverts. Rest of wing and tail juv. In ♂ lores are blackish. (D. Occhiato)

A. a. amandava, 1stW, presumed ♀, **Italy, Nov**: wing still juv, but iris already dull red. Lack of black mask indicates ♀. (D. Occhiato)

A. a. amandava, juv, **Italy, Nov**: note fluffy body-feathers, and rather dusky olive to buffish-brown plumage, blackish-brown bill and no red rump. Buff-tinged wing-bars are the most obvious character. (D. Occhiato)

ZEBRA WAXBILL
Amandava subflava (Vieillot, 1819)

Alternative name: Orange-breasted Waxbill

Fr. – Bengali zébré; Ger. – Goldbrüstchen
Sp. – Bengalí cebra; Swe. – Guldbröstad tigerfink

This tiny, brightly coloured seed-eater is widely distributed across much of sub-Saharan Africa, but within the treated region is virtually confined to S Yemen and C Saudi Arabia, where it overlaps locally with the endemic Arabian Waxbill. Usually found feeding in flocks in grassy savannas but also along river-edges in reeds, rice fields and other cultivation. Energetic and nervous, constantly on the move. Partly nomadic in response to rains and food supply.

A. s. subflava, ♂, Ethiopia, Dec: unmistakable by yellow throat and cheeks, deep orange-red rest of underparts, with boldly barred flanks, crimson-red supercilium, and bill waxy-red with blackish culmen. (C. N. G. Bocos)

IDENTIFICATION *Minute size*, small even for a waxbill, and very brightly coloured. ♂ readily identified by strong *yellow and orange underparts*, with *duskier and pale-barred flanks*, *crimson-red supercilium* (almost forming mask), *rump and undertail-coverts*, and *waxy red bill* (some blackish on culmen and gonys). ♀ much paler and lacks red mask and breast coloration, has rather dark lores and *restricted red-orange rump and vent*, while rest of *underparts are yellowish-buff* and *flanks less boldly barred*. In all plumages graduated tail is dark brown to blackish, and *iris of adult reddish*. Usually occurs in small flocks, larger in non-breeding season. Often difficult to observe in vegetation as it quickly disappears, sliding down cereal stems. Fast flight with short, jerky undulations.

VOCALISATIONS Song in Africa a jumbled series of high-pitched notes, often sustained for several minutes and interspersed by a low *cheup*, e.g. *chit-cheet-chink-cheup-chink-chink-chip-cheet…* – Calls varied, including a rather soft metallic twittering, *zink zink*, or *trip-trp-trp-trp*, piping *ptik* given in flight, and a short *cheep* in other situations; also a hard, short metallic *zzrrep*, repeated 2–3 times, and *tip, tif* or *tith-tith*, etc.

SIMILAR SPECIES In Yemen frequently associates with *Arabian Waxbill*, but easily separated by reddish bill and rump, bright red supercilium and richly coloured underparts of ♂, as well as less buzzing calls.

AGEING & SEXING No Arabian specimens examined and no published data on moult or seasonal ageing or sexing available from elsewhere. Specimens from Africa indicate that sexes differ year-round (once post-juv moult complete) with limited seasonal variation. – Moults. Apparently complete post-nuptial moult in ad. Post-juv moult more controversial in extent, perhaps variably partial or complete. Many apparently suspend renewal of outer primaries and sometimes secondaries. Extent of pre-nuptial moult unknown, but some in non-breeding period renewing inner remiges and some coverts. Further study required. – **BREEDING Ad ♂** See Identification for general description. However, strength of underparts and bill coloration vary, with some largely yellow, suffused orange on breast and chin, in others almost entire underparts washed red-orange; flank barring varies from olive-green through yellow to white, and rump and uppertail-coverts from bright scarlet to orange. Upperparts wear slightly duller. **Ad ♀** Some have distinct white throat. **1stS** Inseparable from ad if moult complete. Iris colour unknown. – **NON-BREEDING Ad ♂** Upperparts warmer, otherwise as breeding. **Ad ♀** As breeding. **1stW ♂** Apparently many moult completely, thus inseparable from ad ♂, but others moult only partly, retaining at least some remiges and primary-coverts (these being rather worn and bleached), contrasting especially with new greater coverts and tertials. There is a fair amount of variation in plumage, some being much duller, especially on underparts. Bill dull red with dark brown culmen already in rather early stages. Iris dull olive-brown. **1stW ♀** Note dull yellowish-olive iris. At least some can be aged by same moult pattern as 1stW ♂. **Juv** Soft fluffy body-feathering. Generally browner, less olive and yellow, lacks ♀'s red on rump and flanks barring, is browner above, and bill is duller.

BIOMETRICS (*subflava*) Sexes same or very similar size. **L** 8.5–9.5 cm; **W** 39–46 mm (*n* 24, *m* 42.5); **T** 27–34 mm (*n* 24, *m* 30.6); **T/W** *m* 72.0; **B** 7.8–9.8 mm (*n* 24, *m* 8.7); **B**(f) 6.0–8.2 mm (*n* 24, *m* 7.3); **BD** 4.7–5.9 mm (*n* 24, *m* 5.3); **BW**(g) 4.7–5.6 mm (*n* 24, *m* 5.0); **Ts** 11.0–11.9 mm (*n* 23, *m* 13.4). **Wing formula: p1** minute; **pp2–5** about equal and longest; **p6** < wt 0.5–2 mm; **p7** < wt 2.5–4 mm; **p10** < wt 5.5–7.5 mm; **s1** < wt 6–8 mm. Emarg. pp 3–6.

GEOGRAPHICAL VARIATION & RANGE Fairly well-marked variation, but only two subspecies separated, one of which occurs within the covered region.

A. s. subflava (Vieillot, 1819) (C Saudi Arabia, Yemen, W Africa east across Sahel to S Sudan & Ethiopia; mainly sedentary but can make nomadic movements in search of food). Described above. Differs from more southerly African race *clarkei* in being somewhat less colourful with less extensive orange-red below.

A. s. subflava, ♂, Yemen, Mar: one of very few images of a species rarely photographed in the region. Some ♂♂ are much duller overall, and especially below, but still have the red mask. (R. Porter)

A. s. subflava, ♂, Ethiopia, Feb: bright crimson-red rump and uppertail-coverts contrast with almost uniform dull greyish-brown upperparts. Barred flanks, head pattern and bill colour add to characteristic appearance. (T. V. Nielsen)

A. s. subflava, ad ♀ (third from top) and three 1stY ♂♂, South Sudan, Mar: ♀ lacks bright crimson-red supercilium and deep orange underparts of ♂, but has darkish lores and rich yellowish-buff underparts with olive slightly barred flanks. Three 1stY ♂♂ show variation in development of red feathers on supercilium and rump. (T. Jenner)

INDIAN SILVERBILL
Euodice malabarica (L., 1758)

Fr. – Capucin bec-de-plomb
Ger. – Indischer Silberschnabel
Sp. – Capuchino picoplata indio
Swe. – Indisk silvernäbb

This small seed-eater has its core range in India, but also breeds as far west as E Saudi Arabia and N Oman. The Indian Silverbill has been introduced in Israel and recently been recorded in Sinai. Escapes have also established small population in SE France. It is usually found in arid thorn savanna and scrub with grasses, where it feeds in small flocks, energetically searching seedheads for food, accompanied by tail-flicking and sharp calls. Introduced birds can occur in town parks and gardens.

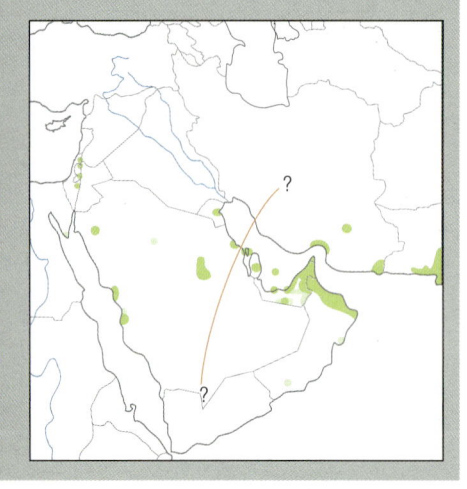

IDENTIFICATION Small and slender, with a *large conical bill* and *long, pointed tail*. Predominately pale fawn grey-brown above, with *whitish rump and uppertail-coverts*. The long and sharply pointed *tail is blackish*. Cream-white underparts variably washed or faintly barred buff on flanks. The deep-based, finch-sized *bill* (with clearly arched culmen) *is glossy pale silver or grey-blue*; the outline of the head and bill seem to run as one *unbroken curved line from crown to bill-tip*. Iris black, orbital ring grey and tarsus pale purplish-grey to rich vinous-red. Sociable, especially at roosts, but frequently only heard in dense scrub or tall grass, and flocks are often quite nervous and fast-moving; occasionally hitches up stems to perch briefly in view, otherwise moves by hops and jumps, occasionally flicking tail. Flight whirring with short undulations, but less jerky and agile than in *Estrilda*.

VOCALISATIONS Similar to African Silverbill, but song much shorter (less tuneful), an abrupt trill with drawn-out whistles and tripled chirps. – In flight, flocks very vocal, ceaselessly giving a rapid *cheet-cheet-cheet*, interspersed by short, weak sparrow-like chirrups, and *tr-r-r-rt* or *trit-trit* sounds; also recorded a high-pitched *zip-zip*, a Linnet-like but more tinkling, conversational *seesip-seesip*, and a harsh *tchwit*.

SIMILAR SPECIES Differs from *African Silverbill* in all plumages by wholly or mostly white (not all-black) rump, and by lack of vermiculations on wing-coverts and tertials, though some have delicate barring on flanks. Also, usually appears to have a slightly shorter bill and paler face, chin and throat, with narrow, short and faint whitish supercilium.

AGEING & SEXING Plumage variation, ageing and moult as for African Silverbill, but few data available from populations within the covered region. Ageing after post-juv moult completed often impossible, but some juv moult partly and can probably be aged if handled, using moult contrasts. Sexes largely inseparable, but ad ♂ tends to be finely barred on lower back, with whiter lores, and is suffused (and perhaps barred) stronger buff on sides. Also, bill has stronger bluish tone than ♀. Seasonal variation rather limited. – Moults. Complete post-nuptial moult of ad, whereas post-juv moult apparently can be either complete or partial. Partial pre-nuptial moult unconfirmed. – **BREEDING Ad** Bill varies from pale slate-blue to silvery pink (with darker culmen and tip), but can be darker, and with wear becomes duller than

Breeding ♂, Bahrain, Jun: differs from African Silverbill in having white (not black) rump and by lack of vermiculations on wing-coverts and tertials. Has diffusely barred flanks, and usually appears to have slightly shorter bill and paler face, with whitish lores and supercilium from bill to over eye. Ageing difficult, but such immaculate plumage best fits ad. (A. Drummond-Hill)

♂ (right) and ♀, breeding pair, Israel, Jul: sexual dimorphism not always obvious, but ♀ generally has slightly less clear light rufous barring on flanks, and tends to have face somewhat mottled. No obvious age-related moult pattern visible. (L. Kislev)

Non-breeding ♂, Israel, Dec: in winter, many (of both age classes) have completed moult making ageing and sexing very difficult, but clean white face and barred flanks with rufous-sandy markings suggest ♂. (L. Kislev)

in non-breeding plumage. Also, darker brown feather centres on forehead and crown may create speckled appearance. ♀ tends to have face washed browner and somewhat darker, with dark-mottled chin. **1stS** At least some retain some heavily-worn juv wing-feathers, but many moult completely, being inseparable from ad. – **NON-BREEDING Ad** Warmer cream-buff tone to fresh plumage. Some pale fringes to coverts, and especially to tertials. **1stY** Like fresh ad (those that moulted completely are inseparable), but may retain some juv remiges, primary-coverts, possibly outer greater coverts and tail-feathers, which are weaker textured and browner (contrasting with new ad-like darker or even blacker feathers). Such less-advanced birds average duller, with strongly reduced barring on upperparts including upperwing. **Juv** Soft, fluffy feathering, with slightly warmer brown upperparts, pale buff fringes and tips to upperwing (which lack vermiculations), and less spotted head; underparts plainer with slight barring on chin, and has less graduated, slightly shorter tail.

BIOMETRICS Sexes same or very similar size. **L** 11–12 cm; **W** 50–57 mm (n 24, m 53.7); **T** 37–55 mm (n 24, m 46.4); **T/W** m 86.3; **B** 9.7–11.7 mm (n 24, m 10.6); **B**(f) 7.7–10.5 mm (n 24, m 9.3); **BD** 6.5–8.3 mm (n 24, m 7.6); **BW**(g) 5.9–7.5 mm (n 24, m 6.6); **Ts** 12.0–13.5 mm (n 23, m 12.8). **Wing formula: p1** minute; **pp2–5** about equal and longest; **p6** < wt 2–5 mm; **p7** < wt 4–8 mm; **p10** < wt 10–14 mm; **s1** < wt 11–15 mm. Emarg. pp 3–5 (sometimes faintly also on tip of p6).

GEOGRAPHICAL VARIATION & RANGE Monotypic. – E Saudi Arabia, United Arab Emirates, N Oman, S Iran, Pakistan, India and north to Himalayas, Sri Lanka; introduced Levant, Sinai, W Saudi Arabia, SE France; sedentary.

TAXONOMIC NOTES Previously considered conspecific with African Silverbill, but genetic work has confirmed their distinctiveness, and they now occur allopatrically in Oman and have quite different song structures. They apparently hybridise in Arabia, although other authors have not reported evidence of this, except in captivity. The ranges of both are currently spreading, and it will be interesting to observe their behaviour in new contact areas. Both species frequently placed in *Lonchura*, but see Sorenson *et al.* (2004).

REFERENCES Harrison, C. J. O. (1964) *Ibis*, 106: 462–468. – Sorenson, M. D., Balakrishnan, C. N. & Payne, R. B. (2004) *Systematic Biology*, 53: 140–153.

Non-breeding ♀, Israel, Dec: sullied buff-brown and unbarred underparts, with throat and face—including supercilium—dirty white are features of ♀. (L. Kislev)

1stY ♂, Israel, Jan: note moult limits in wing and, despite age, quite clean whitish face and buff-barred flanks, which indicate ♂. (A. Ben Dov)

♀, presumed 1stW, Oman, Nov: overall dull and featureless, with hardly any suggestion of white supercilium or whitish throat, plain flanks, and rather darkish bill, which are all features of young ♀. (M. Varesvuo)

Juv, India, Feb: hint of soft, fluffy feathering on scapulars and lower belly, warmer brown upperparts, pale buff fringes and tips to upperwing, plain underparts, and bill darkish. (R. Mallya)

AFRICAN SILVERBILL
Euodice cantans (J. F. Gmelin, 1789)

Fr. – Mannikin bec d'argent
Ger. – Afrikanischer Silberschnabel
Sp. – Capuchino picoplata africano
Swe. – Afrikansk silvernäbb

Widespread in arid thorn savanna and *Acacia* scrub of sub-Saharan Africa, north of the Congo forest belt, but within the treated region the African Silverbill occurs only in SW Arabia, where it comes close in range to Indian Silverbill. There is also a recent record in Egypt. A confiding bird, easy to approach, which is often seen in flocks feeding on the ground among low grasses.

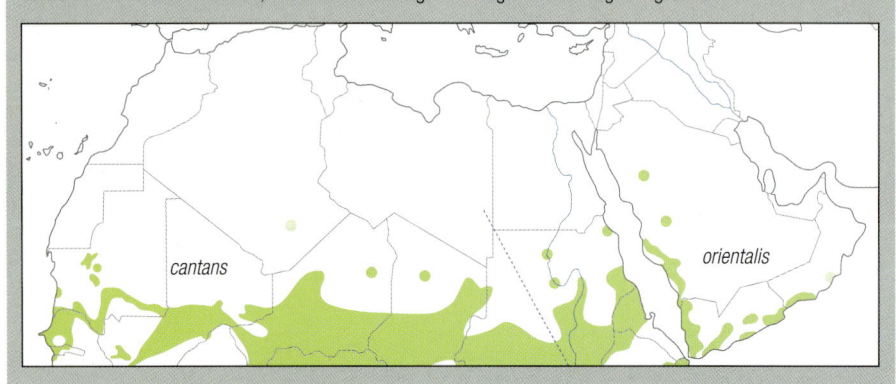

E. c. orientalis, non-breeding ♀, presumed 1stW, Oman, Nov: fresh, but suggested moult limits in tail and possibly wing could indicate a young bird. Rather pale and narrower barring above, and browner rump and uppertail-coverts infer ♀. (M. Schäf)

E. c. orientalis, ♂ (right) and ♀, breeding pair, Ethiopia, Sep: sexes differ in that ♂ has jet-black, ♀ blackish grey-brown rump and uppertail-coverts. Unlike Indian Silverbill has more buffish sandy-brown upperparts and faint to moderate barring or vermiculations, boldest on wing-coverts, tertials and across lower back. Lacks Indian's barred flanks, has darker mottling on chin and throat, lacks pale supercilium and is marginally larger-billed. (H. Shirihai)

IDENTIFICATION General appearance and behaviour much like Indian Silverbill (which see), but has contrastingly *all-black rump and uppertail-coverts* (never any white). Compared to Indian Silverbill more buffish sandy-brown upperparts, *faintly barred or vermiculated wing-coverts, tertials, lower mantle and scapulars* (usually visible at close range), and *darker, browner chin, throat and face*. No pale supercilium, lacks barring on flanks and is marginally larger-billed and shorter-tailed than its Indian relative.

VOCALISATIONS Like Indian Silverbill, but short song consists of high-pitched single notes, followed by double notes, each descending and then rising. Calls appear similar to those of Indian Silverbill.

SIMILAR SPECIES See Identification and *Indian Silverbill* for distinguishing features from that species.

AGEING & SEXING Ageing after completed post-juv moult often impossible, but some juv moult partially and can probably be aged if handled using moult contrasts. Sexes largely inseparable. Seasonal variation rather limited. – Moults. Ad has complete post-nuptial moult, whereas post-juv moult apparently can be either complete or partial, but few data on timing or extent available from the treated region. Partial pre-nuptial moult unconfirmed. Thus, post-moult ageing usually impossible except for some 1stY with retained juv feathers (usually requires examination in the hand). – **BREEDING Ad** Bill varies from pale slate-blue to silvery pink (with darker culmen and tip), but can be darker. Plumage becomes somewhat duller with wear than non-breeding plumage, and darker brown centres to feathers of forehead and crown may give speckled appearance. ♀ tends to have face washed browner and darker, with dark-mottled chin. **1stS** At least some retain some juv, heavily-worn, wing-feathers, but many moult completely, being thereafter inseparable from ad. – **NON-BREEDING Ad** Warmer cream-buff tone to fresh plumage; some pale fringes to coverts, especially tertials. **1stW** Like fresh ad (those that moulted completely are inseparable), but may retain some juv remiges, primary-coverts, possibly outer greater coverts and tail-feathers, which are weaker textured and browner (contrasting with new ad-like darker or even blacker feathers). Such less-advanced birds average duller, with strongly reduced barring on upperparts including wings. **Juv** Soft, fluffy feathering, with slightly warmer brown upperparts, pale buff fringes and tips to upperwings (which lack vermiculations), and less spotted head; underparts plainer with slight barring on chin, and has less graduated, slightly shorter tail.

BIOMETRICS (*orientalis*) Sexes same or very similar size. **L** 10–11 cm; **W** 49–55 mm (n 24, m 51.5); **T** 35–47 mm (n 24, m 41.4); **T/W** m 80.2; **B** 9.4–10.9 mm (n 24, m 10.2); **B**(f) 8.4–9.9 mm (n 24, m 9.1); **BD** 6.6–8.3 mm (n 23, m 7.4); **BW**(g) 6.3–7.5 mm (n 24, m 6.8); **Ts** 12.0–13.3 mm (n 24, m 12.4). **Wing formula:** p1 minute; **pp2–5** about equal and longest; **p6** < wt 1–5 mm; **p7** < wt 4–8 mm; **p10** < wt 10–13 mm; **s1** < wt 10.5–13 mm. Emarg. pp 3–5 (sometimes faintly also on tip of p6).

WEAVERS AND WAXBILLS

GEOGRAPHICAL VARIATION & RANGE The variation appears very slight with only two subspecies generally recognised, and only *orientalis* (NE Africa, Arabia) breeds within the treated region.

E. c. orientalis (Lorenz & Hellmayr, 1901) (SW Arabia, Yemen, S Oman, E Sudan, Eritrea, Ethiopia, Somalia south to Kenya; sedentary). Differs from *cantans* in slightly paler underparts but darker sandy-brown upperparts, more distinct close barring on upperwing, breast-sides and flanks, and less buff-tinged face. Many are, however, intermediate and indistinguishable. Furthermore, breeders in N Sudan to Red Sea ('*inornata*') close to *orientalis*, but claimed to have darker brown to fuscous uppertail-coverts with variable dark reddish fringes to these feathers and those of tail, but differences slight, variable and not sufficient grounds for recognition. (Syn. *inornata*.)

Extralimital: **E. c. cantans** (J. F. Gmelin, 1789) (Senegal, Mauritania east through Sahel region to S & W Sudan). Very similar to *orientalis* but averages a little paler above and more saturated below. Fine barring of upperwing less distinct.

TAXONOMIC NOTE See notes under Indian Silverbill.

E. c. orientalis, presumed non-breeding ♂, Oman, Oct: fresh, but despite appearance of moult limits in tail (so perhaps a young bird) age uncertain. Strength of barring best fits ♂. (D. Occhiato)

E. c. orientalis, non-breeding flock, Oman, Jan: mix of ages and sexes, but in winter differences are less obvious. The forward-facing central bird has fewer and very faint buff barring on central flanks, but obvious barring above, and mottled chin and upper throat. (H. Michel)

E. c. orientalis, non-breeding ♀, presumed 1stW, Oman, Nov: duller, almost featureless plumage with darkish bill are features of young autumn ♀. (M. Schäf)

E. c. orientalis, juv, Oman, Apr: upperparts and head brown and rather plain. Underparts pale with slight mottling. Note rather short tail and yellowish-white gape. (R. Kunz)

RED-BILLED FIREFINCH
Lagonosticta senegala (L., 1766)

Fr. – Amarante du Sénégal; Ger. – Senegalamarant
Sp. – Amaranta senegalesa
Swe. – Rödnäbbad amarant

This small African estrildid finch is both common and often found near settlements and in gardens, hence is well known within its range. It is, of course, also a frequently kept cagebird. In the wild, it is widespread across all of sub-Saharan Africa and was introduced in S Algeria since the 1940s. It is otherwise only an erratic straggler to the W Palearctic, known from a single Moroccan record in the mid 1960s, and from an introduction in Egypt that failed.

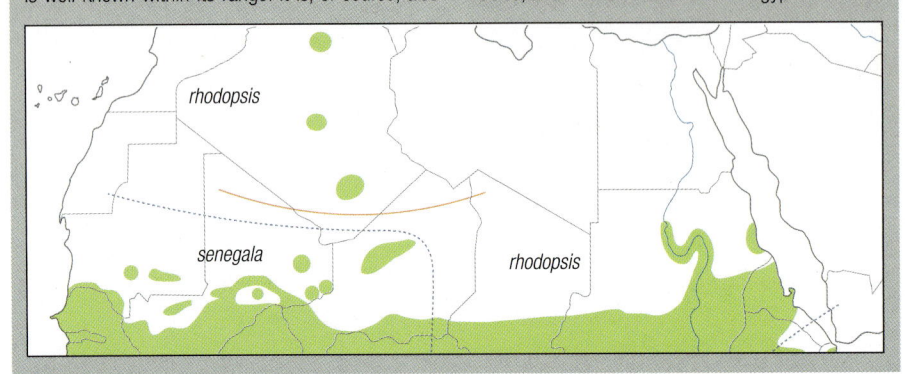

L. s. senegala, breeding pair, ♂ (front) and ♀, Mauritania, Jan: note differences in colour and width of eye-ring between sexes. ♂ has only single white spot on side of breast (number varies individually). If seen well, unlikely to be confused, except perhaps with escaped congeners or Red Avadavat. Ageing impossible in this view. (K. de Rouck)

IDENTIFICATION A tiny waxbill-like finch that usually frequents *Acacia* grassland, but also often enters gardens and cultivations. ♂ has head and body to posterior underparts, rump and uppertail-coverts *unmarked red*, except for *tiny white spots on breast-sides and upper flanks*. Throat, rump and uppertail-coverts are more pinkish-red, while red on rest of body has a slight orange-brown tinge. Mantle, back and upperwings are variably duller russet-brown. Flight-feathers are dark brown, *tail mostly blackish* and undertail-coverts rather dull grey-brown. Bill in ♂ *coral-red*, with ill-defined black culmen emphasizing sharp-pointed shape; colours duller in ♀. ♀ is by and large a rather unassuming uniform buff-brown, though still has characteristic *pale spotting on chest near wing-bend, red lores and rump, and red uppertail-coverts patch* contrasting with darker tail. Eyes dark red-brown, with narrow but *conspicuous yellowish-white orbital ring*. Diminutive, hops on ground restlessly and forms small, roving flocks that are fast-flying, but is sometimes rather approachable when feeding or at drinking spots.

VOCALISATIONS Song a short, lively musical signal, repeated a few times, opening with two or three short, sharp notes followed by a loud, clear glissando-whistle, e.g. *chick-see-tsuwooey*. A variation may be rendered *tu-tweet-tweet*. – Call involves a short whistle *zweea* (single or double), high-pitched *dwee* or *pfweet*. In alarm a weak piping *tweet-tweet* or *chick*, often repeated.

SIMILAR SPECIES Within the covered region, the colourful ♂ is only likely to be confused with escaped congeners or *Red Avadavat*. Both sexes of latter have very dark wings with some white spotting contrasting with the rest of upperparts, and have dark lores. ♂ Red Avadavat has much more extensive and bolder white underparts spotting continuing onto scapulars, and red coloration in both sexes is much brighter, more intense and crimson-red rather than scarlet. Much plainer young already shows wing-pattern differences, albeit these are less well marked.

AGEING & SEXING Ageing after post-juv moult completed often impossible, but some juv moult partly and can probably be aged if handled, using moult contrasts. Sexes differ following post-juv moult. Seasonal variation rather limited. – Moults. Ad has complete post-nuptial moult, whereas post-juv moult apparently can be either complete or partial. Moult in northernmost areas (and hence perhaps in W Palearctic) presumably in May onwards, but generally protracted, and timing and extent seem dependent on latitude, environmental conditions and breeding season. Pre-nuptial moult unknown. – **ALL YEAR Ad ♂** See Identification, but with wear, pink-red of head and upperparts become duller, even browner, underparts also paler and have reduced white spotting on breast-sides. **Ad ♀** Even less change occurs than in ♂. **1stY** Only separable from ad if some juv wing-feathers retained. Plumage—especially of 1stY ♂—variably developed, some attain ad ♂ colours, but usually has less extensive red, and some are mostly ♀-like with vinaceous-red restricted to forecrown or throat, or red wash on cheeks and breast, but others show no red at all. **Juv** Resembles ♀ but has soft, fluffy body-feathers and black bill, dull grey lores and is drabber, lacking any white spots on breast-sides. Reddish colour restricted to rump and uppertail-coverts.

BIOMETRICS (*rhodopsis*) **L** 9–10 cm; **W** ♂ 48–52 mm (*n* 12, *m* 49.6), ♀ 44–50 mm (*n* 12, *m* 47.9); **T** ♂ 33–38 mm (*n* 12, *m* 36.1), ♀ 32–36 mm (*n* 12, *m* 33.7); **T/W** *m* 72.5; **B** 8.8–10.7 mm (*n* 24, *m* 9.7); **B**(f) 7.1–9.3 mm (*n* 24, *m* 8.3); **BD** 5.0–6.4 mm (*n* 24, *m* 5.6); **BW**(g) 4.7–6.2 mm (*n* 24, *m* 5.4); **Ts** 10.8–12.9 mm (*n* 23, *m* 12.2). **Wing formula: p1** minute; **pp2–5**(6) about equal and longest; **p6** < wt 0–2.5 mm; **p7** < wt 1–4 mm; **p10** < wt 6–9 mm; **s1** < wt 6.5–9.5 mm. Emarg. pp 3–6 (on p6 slightly less prominently).

GEOGRAPHICAL VARIATION & RANGE Moderate variation. Often about seven races recognised, just one of which following introduction occurs within the covered region.

L. s. rhodopsis (Heuglin, 1863) (local S & C Algeria, SE Niger, S Chad, east to S Sudan, Eritrea, Ethiopia; sedentary). Slightly paler than *senegala*, warm brown parts being more cinnamon-tinged or tawny. (Syn. *erythreae*; *flavodorsalis*.)

Extralimital: **L. s. senegala** (L., 1766) (S Mauritania, Senegal, Gambia, Mali; sedentary). Slightly darker red than *rhodopsis*, brown parts a little darker, too.

L. s. rhodopsis, breeding ♂, Ethiopia, Sep: this race is among the palest, still is strikingly red with earth-brown back. Throat, rump and uppertail-coverts pinkish-red, grading into orange-red chest, and lower belly and undertail-coverts greyish olive-buff. Note small white spots on breast-sides, while bill is pinkish-red with grey culmen. Ageing impossible in this view. (H. Shirihai)

L. s. rhodopsis, breeding ♀, Ethiopia, Sep: ♀ largely brown with short reddish loral mask, reddish rump and duller bill. Note small whitish spots at breast-sides. Ageing not possible without handling. (H. Shirihai)

RED-EYED VIREO
Vireo olivaceus (L., 1766)

Fr. – Viréo à oeil rouge; Ger. – Rotaugenvireo
Sp. – Vireo chiví; Swe. – Rödögd vireo

One of the most frequent transatlantic vagrant passerines, the Red-eyed Vireo can probably appear at first glance to a European birdwatcher like a cross between a Spotted Flycatcher and an Icterine Warbler. It has been recorded in Iceland, Britain & Ireland (where it is now almost expected in October), France, the Netherlands, Germany, Spain, Malta and Morocco. It breeds over much of North America, and in South America south to N Argentina.

V. o. olivaceus, 1stW, England, Oct: large-headed and warbler-like, with rather distinctive quite long, deep-based and slightly blunt-tipped bill. Note long pointed wings but fairly short, square-ended tail. Remiges and primary-coverts juv, latter subtly contrasting to more bright green-edged moulted greater coverts. This bird already has reddish-brown iris, almost ad-like but not quite. (G. Thoburn)

IDENTIFICATION An Icterine Warbler-sized bird, with *rather long, deep-based* (and blunt-ended) *bill*, somewhat flat-crowned head (which looks proportionately elongated or large), and—to continue the comparison with Icterine Warbler—has similarly *long and pointed wings, dark legs* and a fairly short, *square-ended tail*. Most obvious is the *bold white, dark-bordered supercilium* and *bluish-grey crown with an olive cast*; *amber to red eye* emphasized by white lower crescent. Upperparts mainly olive-green, brighter on wings and tail, and underparts off-white. Some first-year birds possess a slight yellow tone below. Bill black with a grey or horn-coloured base to lower mandible, and the strong legs are bluish or lead-grey. Mostly arboreal. Takes insects and berries (especially in autumn), and is usually solitary and rather tame, though can make dashing sallies and be rather secretive, sometimes perching quietly for long periods.

VOCALISATIONS Song, unlikely to be heard in the W Palearctic, is a steadily repeated, short and slightly varied note, a little recalling House Sparrow in tone. – Commonest call a nasal, soft but strained *quee* or *chway* with a distinctive complaining 'compound' timbre. Also a *zherr* and descending *rreeea*.

SIMILAR SPECIES In poor views might be confused with smaller, plain Old World warblers. However, unlike any European birds, Red-eyed Vireo has a stronger bill and diagnostic combination of long whitish supercilium, dark lateral crown- and eye-stripes, grey-tinged crown and (in adult) red eye. – Two other vireos have been recorded in W Europe and should be borne in mind: Yellow-throated Vireo (UK, Sep 1990, and Germany, Sep 1998) and Philadelphia Vireo (Eire, Oct 1985 and UK, Oct 1987); see vagrants section, p. 589.

AGEING & SEXING Can be aged by moult and feather wear prior to completion of moult, and iris colour is especially reliable in autumn. Sexes alike, and seasonal plumage variation limited to wear and bleaching. – Moults. Complete post-nuptial moult and partial post-juv moult mostly from Jul onwards, suspended Sep–Dec and resumed Jan–Apr. Pre-nuptial moult poorly understood, probably largely complementary to post-breeding and juv moults, but perhaps has genuine pre-nuptial moult as well. Prior to migration, ad often replaces some wing-coverts and even some outer secondaries and inner primaries, as well as some tail-feather (usually including r1). 1stW may extensively renew head, body, secondary-coverts, (rarely outer greater coverts retained). However, some migrate in mostly juv plumage. – **SPRING–SUMMER Ad** Prior to completion of moult, generally duller, with a more dusky greyish-olive (less greenish) tinge to upperparts, and purer grey crown and whiter underparts, albeit some variation with the occurrence of brighter birds (also depending on state of moult). Once moult completed entire plumage fresh and bright (recalling fresh 1stW). **1stS** Differs in moult pattern, degree of feather wear and iris colour (see autumn–winter). Apparently never recorded in W Palearctic in spring plumage. – **AUTUMN–WINTER Ad** Relatively duller than fresh 1stW, and any unmoulted remiges and primary-coverts rather worn, wing showing moult limits during suspension even if not in actual moult. **1stW** (most frequently recorded plumage in Europe) Differs from ad by retained juv remiges and primary-coverts, which are relatively little worn (but more so than any recently renewed ad feathers). Typically brighter and greener, with more yellowish tinge to underparts. Fresh covert-tips may produce faint wing-bars and a pale wing panel on fresh secondaries. Retains dark brown to reddish-brown iris (reddish to bright red in ad), but colour closer to ad by spring. **Juv** Soft, fluffy body-feathering noticeably drabber than in ad, with dull brown crown (only indistinct lateral stripes developed), duller

V. o. olivaceus, 1stW, Azores, Oct: note long primary projection. Diagnostically, prominent whitish supercilium is emphasised by dark lateral crown- and eye-stripes. Bluish-grey crown and silk-white throat are striking, and note green upperparts and off-white rest of underparts. Uniformly fresh primary tips and primary-coverts are best age criteria. (V. Legrand)

supercilium suffused olive at rear, while face, neck-sides and breast appear slightly mottled. Iris dark brown. This plumage may be retained rather late in autumn.

BIOMETRICS (*olivaceus*) **L** 13–14.5 cm; **W** ♂ 81–88 mm (*n* 12, *m* 83.7), ♀ 76–87 mm (*n* 12, *m* 80.6); **T** ♂ 53–59 mm (*n* 12, *m* 55.7), ♀ 47–58 mm (*n* 12, *m* 51.9); **T/W** *m* 65.5; **B** 15.3–18.3 mm (*n* 24, *m* 17.8); **BD** 4.3–5.2 mm (*n* 24, *m* 4.7); **BW** 4.0–5.6 mm (*n* 24, *m* 5.1); **Ts** 17.0–18.5 mm (*n* 24, *m* 17.8). **Wing formula: p1** minute, < pc usually 5–9 mm, < p2 *c.* 52–55 mm; **p2** < wt 3–6 mm, =4/5, =5 or =5/6 (rarely =4); **p3** longest (sometimes pp3–4); **p4** < wt 0–3 mm; **p5** < wt 2–5 mm; **p6** < wt 7.5–12 mm; **p7** < wt 12–16 mm; **p10** < wt 19.5–24.5 mm; **s1** < wt 22–27 mm. Emarg. pp3–5.

GEOGRAPHICAL VARIATION & RANGE Apparently only *olivaceus* recorded in W Palearctic, the only subspecies breeding in North America, other races being described from Central and South America.

V. o. olivaceus (L., 1766) (Canada, North America; winters South America). Described above.

REFERENCES Bradshaw, C. (1992) *BW*, 5: 308–311.

V. o. olivaceus, 1stW, Azores, Oct: mostly arboreal, sometimes perching quietly for long periods. Dark bill has grey-coloured base to lower mandible, and note how chunky bill can appear, especially in birds with quite short bills. Primary-coverts juv. (V. Legrand)

V. o. olivaceus, 1stW, Azores, Oct: often, fresh 1stW (the age most frequently recorded in Europe) shows bright greenish-olive rump and uppertail-coverts, and hint of pale wing panel. Head-sides often slightly tinged buff. At this age, and especially in woodland shade, any red in iris is rarely visible in the field. (V. Legrand)

V. o. olivaceus, 1stW, Azores, Nov: a rather dull bird still largely in juv plumage. Especially Philadelphia Vireo can be rather similar, but that species always has less obvious dark lateral crown-stripes, more yellow-tinged underparts and different bill shape. (V. Legrand)

(COMMON) CHAFFINCH
Fringilla coelebs L., 1758

Fr. – Pinson des arbres; Ger. – Buchfink
Sp. – Pinzón vulgar; Swe. – Bofink

Over much of Europe the familiar Chaffinch is one of the commonest birds wherever there are trees, its presence announced from early spring to summer by its energetically repeated, characteristic song. Widespread in all kinds of woodland, farmland and 'green' areas in towns from Iceland to W Siberia, south to the NE Atlantic islands and east to the Caspian Sea. Northern populations move south and south-west in large numbers in winter, mainly to W and S Europe. The rather different-looking African and Atlantic Islands populations are described at length below.

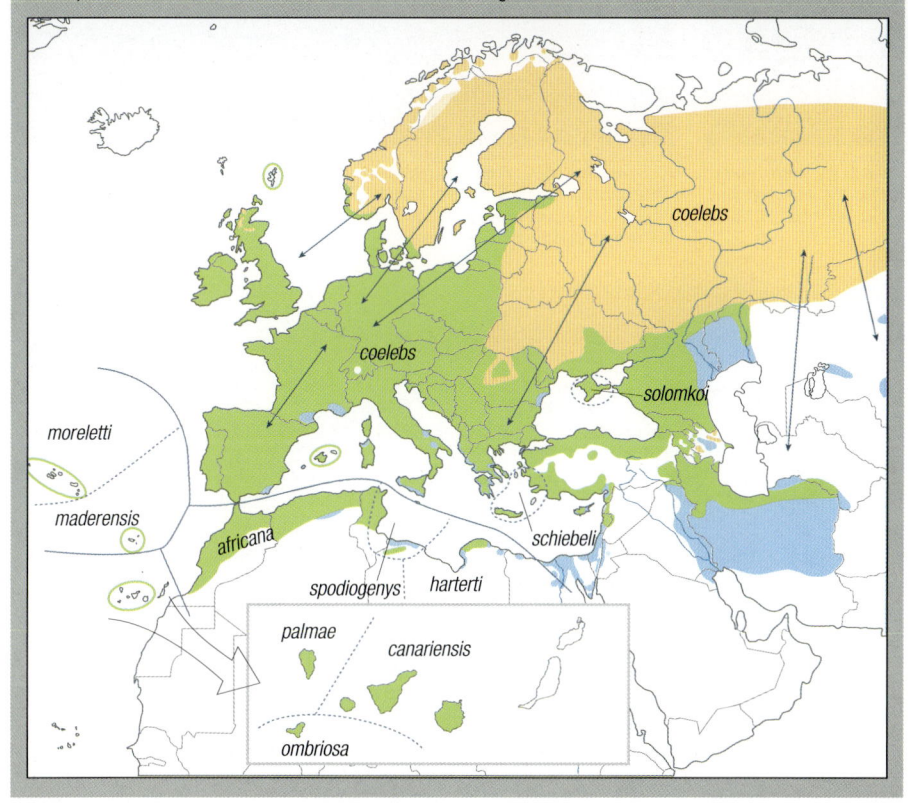

IDENTIFICATION Medium-sized finch with a *conical, pointed bill*, flattish crown (sometimes with hint of peak at rear) and slightly forked tail. Both sexes have a *bold whitish greater-coverts wing-bar*, and *white median and lesser coverts* forming a striking 'shoulder' patch. In flight easy to see *white outer tail-feathers*. ♂ has *black forehead, pale grey-blue crown to upper mantle and scapulars*, and moss-green lower back to uppertail-coverts. Cheeks to upper breast vinaceous-pink or pinkish-brown, *becoming pinkish-white on belly* and cream-white on undertail-coverts. Bill blue-grey with dusky tip and paler base to lower mandible. Most conspicuous features of ♀ are the pale wing and tail markings, almost identical to those of ♂. The head of ♀ has a *broad pale grey band behind eye continuing on neck-sides, contrasting with darker dusky-grey crown-sides and nape*. Rest of upperparts brownish grey-olive with *slight greenish tinge, especially on rump*, while underparts are greyish-white, darker on flanks and breast. Bill pale grey-brown with dusky tip. Small eye looks blackish. Usually forages on ground with characteristic jerky hops. Long-undulating, purposeful flight, with sharp turns and descent, but short escape-flight (usually into nearest cover) is with half-spread tail. Easily flushed, but rather tame around bird tables. Highly gregarious on migration and in winter, when often occurs with other finches, buntings and sparrows.

VOCALISATIONS Song easily remembered: a brief, loud, almost rattling, descending series of sharp notes usually ending in a multisyllabic flourish, *zitt-zitt-zitt-zitt-sett-sett-sett-sett-sett-chatt-chiteriidia* (constantly repeated with brief pauses). Degree of individual variation in song, but each ♂ sings by and large only one type, again and again. Some songsters have a habit of adding a Great Spotted Woodpecker *Dendrocopos major*-like *kick* at the end, rather oddly. Apparently to save energy, also has brief territorial signal, sometimes called 'rain song'. This varies more notably geographically to form dialects. In C Europe and S Scandinavia often a rolling or discordant, straight whistle *rrhü*, slowly, but incessantly repeated, in N Scandinavia and Finland an upslurred forceful *huitt*, in S Europe a Nightingale-like straight whistle *hiiht*. Other variations noted. – Of several calls probably commonest are a sharp *fink*, rather like Great Tit but more metallic, and a low, timid *yupp* in flight and on migration. Gives alarm with a very high-pitched *ziiit*. – Song and calls of Atlantic Is races rather distinct, briefly described under Geographical variation below.

SIMILAR SPECIES Generally unmistakable given bold white wing markings and unique plumage, distinctive shape and calls. If not seen well, might only be confused with Brambling, which has a similar wing pattern, but is easily separated by its conspicuous white rump, all-black tail, much whiter belly and, in ♂♂, deeper orange-red shoulder and breast. Migrating flocks tend to be looser, whereas flocks of Brambling are on average denser, but much overlap, hence flock density only a hint at species.

AGEING & SEXING (*coelebs*) Ageing possible when seen close or handled based on presence of moult contrast or not in wing, and on shape and wear of tail-feathers, sometimes supported by degree of wear of tertials. Sexes differ clearly after post-juv moult. Some seasonal variation due to feather wear. – Moults. Complete post-nuptial moult in ad and partial post-juv moult in Jul–Sep (Jun–Oct). Post-juv moult involves head, body, all lesser and median coverts, most or all greater

F. c. coelebs, ad ♂, Spain, Jun: one of the most widespread woodland birds in Europe, readily distinguished by bold white greater coverts bar and triangular white 'shoulder', the ♂ having blue-grey crown and nape, chestnut mantle, and pinkish- or vinaceous-brown face and breast. Aged by still rather fresh wing without visible retained juv coverts. (C. N. G. Bocos)

F. c. coelebs, ad ♀, Finland, Apr: clearly more nondescript than ♂, being rather uniform olive-brown above with darker brown lateral crown-stripes, greener rump, and greyer face, breast and flanks grading to white mid-belly and undertail-coverts. Only rather fresh ad wing-feathers visible. (M. Varesvuo)

F. c. coelebs, 1stS ♂, Germany, Mar: moss-green rump and uppertail-coverts just visible between wings. Aged by moult limits in wing, juv primary-coverts and alula being paler and greyer than in ad. Note that crown often appears gently peaked, and that white-sided tail is slightly forked. (R. Martin)

F. c. coelebs, 1stS ♀, France, March: olive-yellow secondary panel obvious on some ♀♀, but white 'shoulder' patch can be partly concealed by scapulars, as here. Ageing of ♀♀ often requires handling, but here seemingly a moult limit in greater coverts, worn and pointed tertials and pointed tail-feathers indicate age. (A. Audevard)

coverts, sometimes some alula-feathers and tertials, central tail-feathers, but no primary-coverts or primaries, and only very rarely odd inner secondaries. Pre-nuptial moult (regardless of age) absent. – **SPRING Ad** (for very obvious sex differences see Identification) At least ♂♂ aged by lack of moult limits in wing (primaries, primary-coverts and tail-feathers of same age and relatively fresh), whereas in ♀♀ moult limits are subtle and at times difficult to see making ageing impossible. Tertial fringes broader (less abraded) and tipped rufous (♂) or rufous-yellow (♀). **1stS** As ad, but has relatively more worn and duller juv wing-feathers, including all remiges and primary-coverts, usually some outer greater coverts, and all or most tertials, alula and tail-feathers (moult limits generally detectable, especially in ♂♂). Tertial fringes narrower (more abraded) and paler or whiter. 1stS ♂ sometimes still has slight brownish wash, e.g. on crown, and is overall relatively less immaculate and less intensely coloured than ad. – **AUTUMN Ad ♂** Fresh buff and olive tips obscure spring plumage pattern: forehead band greyish-black (instead of black), grey crown tinged brownish, mantle darker brown, lower back to uppertail-coverts tinged brownish-grey, and pink-brown of face and underparts duller. Wing-bars more yellowish-white, and fringes of tertials and remiges brighter greenish-buff. Once moult completed, evenly fresh and lacks moult contrast in wing of 1stY. **Ad ♀** Head and upperparts tinged browner, underparts more brownish-grey and wing-bars buffier. Both sexes, prior to completing moult, may possess some heavily-worn unmoulted remiges. Once moult completed, evenly fresh and lacks moult contrast in wing of 1stY. **1stW** General plumage pattern in each sex similar to that of fresh ad. 1stW ♂ has pale or buff tips somewhat broader, with grey of cap sullied browner, mantle and scapulars greener buffish-brown, and vinaceous of underparts even less intense, while white areas on remex bases and edges/tips of coverts average narrower (though much variation and overlap with ad). Both sexes aged by retained juv remiges, primary-coverts and often some or all tertials, some outer greater coverts, odd alula and tail-feathers (weaker textured and duller, contrasting with fresh, blacker or at least darker-centred feathers in these tracts). Ageing best performed noting slightly paler and more greyish large alula and primary-coverts in 1stY (not by trying to establish moult contrast in greater coverts, which rarely exists). Tail-feathers slightly narrower and more pointed, especially central pair, which tends to lack dark subterminal spot of ad. Moult limits in 1stW ♀ more difficult to ascertain, and southern populations may replace more in post-juv moult and are thus less clearly differentiated from ad (as always, making ageing more difficult, both in spring and autumn, especially if no juv outer greater coverts are retained). **Juv** Like non-breeding ♀ but has greyish-white nuchal spot, browner mantle, dull brownish-green rump, paler underparts and soft, fluffy body-feathers. Juv ♂ generally richer coloured compared to ♀.

BIOMETRICS (*coelebs*) **L** 14.5–16 cm; **W** ♂ 86–95 mm (*n* 60, *m* 89.5), ♀ 81–87 mm (*n* 30, *m* 84.0); **T** ♂ 62–71 mm (*n* 60, *m* 67.4), ♀ 59–68 mm (*n* 30, *m* 62.7); **T/W** *m* 75.1; **B** 13.0–16.1 mm (*n* 86, *m* 14.3); **B**(f) 10.3–12.9 mm (*n* 85, *m* 11.8); **BD** 7.0–8.5 mm (*n* 81, *m* 7.6); **Ts** 17.0–19.3 mm (*n* 79, *m* 18.1). **Wing formula: p1** minute; **p2** 1–4 mm < wt, =5 or 5/6; **pp3–4**(5) about equal and longest; **p5** < wt 0–1.5 mm; **p6** < wt 5–8 mm; **p7** < wt 12–16 mm; **p10** < wt 20–26 mm; **s1** < wt 21–28.5 mm. Emarg. pp3–6.

GEOGRAPHICAL VARIATION & RANGE Only slight variation in the north, but marked and complex in the south and south-west. Usually three main subspecies groups recognised based on geographical region but also on rather clear

F. c. coelebs, ♂ (left) and ♀, Sweden, Sep: in flight, note white forewing ('shoulder') patch and prominent wing-bar across tips of greater coverts that continues onto primary bases, while white on outer tail-feathers often visible too. (M. Varesvuo)

F. c. coelebs, ad ♂, Israel, Dec: in winter, fresh buff and olive tips blur the otherwise distinctive plumage pattern, grey crown being less pure, and black forehead largely concealed. Wing-bars and fringes to tertials and remiges have yellowish tinge. Bill mostly pale pinkish-grey. No moult limits in wing, and primary-coverts quite dark, all inferring age. (A. Ben Dov)

F. c. coelebs, ad ♀, Spain, Dec: slightly warmer colours when fresh, otherwise as in spring and summer. Head, breast and flanks dull brownish, lacking ash-grey and rufous-pink of ♂. Note greenish tinge on rump. Once moult completed in early autumn plumage is evenly fresh and lacks moult contrasts in wing of 1stW. (M. Varesvuo)

F. c. coelebs, 1stW ♂, Israel, Oct: in autumn, broad pale tips generally conceal or obscure plumage coloration in young ♂. Aged by retained juv primary-coverts and outermost greater covert, these being a shade paler grey than moulted inner greater coverts. (H. Shirihai)

differences in ♂ plumage, the Eurasian Group ('the *coelebs* group'), the North African Group ('the *spodiogenys* group') and the slightly more variable Atlantic Islands Group ('the *canariensis* group').

EURASIAN GROUP ('Common Chaffinch' – Europe, W Siberia, Turkey, Middle East) comprises only three subspecies differing only marginally in plumage or bill size. There is a subtle cline also in wing-length becoming shorter in the south, but small differences, much overlap and poor grounds for recognition of separate subspecies. Any fair assessment of geographical variation needs also to take into account a certain amount of individual plumage variation among ♂♂, and this is particularly true for the widely-distributed ssp. *coelebs*. The degree of vinaceous tinge on cheeks, chin, throat and breast, normally present to some degree in *coelebs*, varies even within Sweden (the type locality), and in some birds is more or less replaced by more rufous or dull cinnamon-brown hues. Likewise, the darkness of the underparts varies (though very pale pink-rufous birds are rare and untypical), as are the upperparts as to darkness of chestnut mantle and brightness of green rump. Without an awareness of this variation, and perhaps working with shorter samples of specimens (not all of which are breeders), it is all too easy to end up recognising more subspecies than is warranted.

F. c. coelebs L., 1758 (Europe except Crimea and Crete; also Turkey, Near East, Caucasus region, W Siberia; winters in south and west of range, or, to lesser extent, in N Africa). Described above. – Note individual variation within any subpopulation as described above, this being the main reason for not recognising many claimed subspecies as listed below. Breeders in Britain ('*gengleri*', '*scotica*') and Ireland ('*hibernicus*') claimed to be slightly darker brown (rufous), less vinaceous-tinged on cheeks and throat/breast compared to *coelebs*, but no difference evident when series of ♂♂ of both from mid Mar to late Aug compared. Corsican ♂♂ ('*tyrrhenica*') average a shade paler and more pinkish below, but much overlap and subtle. Breeders in Levant ('*syriaca*') similarly seem inseparable from *coelebs*, and claimed paler plumage could not be confirmed using available material. At least some birds from C Caucasus ('*caucasica*') tend to have less clean lead-grey crown and nape in ♂♂, but others do not differ from *coelebs*, and perhaps best included here on present evidence. A series from N Iran ('*alexandrovi*', '*transcaspia*') was compared with *coelebs* and found inseparable in plumage and only c. 2–3% smaller with vast overlap. (Syn. *alexandrovi*; *balearica*; *caucasica*; *cypriotis*; *gengleri*; *hibernicus*; *hortensis*; *iberiae*; *sarda*; *scotica*; *spiga*; *syriaca*; *transcaspia*; *tyrrhenica*; *wolfgangi*.)

○ **F. c. solomkoi** Menzbier & Sushkin, 1913 (Crimea; mainly sedentary, but winter movements poorly known). Similar to *coelebs* and could perhaps be included. Differs from latter mainly by subtly longer wing and tail and slightly stronger bill. Also, a vague tendency in ♂♂ to have purer lead-grey scapulars causing the chestnut mantle to be better defined laterally, but a slight difference only, and similar birds can sometimes be found among European *coelebs*. Claimed

F. c. coelebs, 1stW ♀, Italy, Dec: even in young autumn ♀ pale wing markings are sufficiently conspicuous to ensure identification, along with the typically shaped head with peaked hindcrown and conical bill. Moult limits in young ♀ rather difficult to ascertain, but stronger wear to primaries, and more pointed and browner juv primary-coverts and alula reveal age, while central tail-feathers are moulted. (D. Occhiato)

to be slightly paler above, back being paler chestnut brown, but this is not correct. The underparts are also claimed to be paler than in *coelebs* (Vaurie 1959), but this must be extremely slight if true. Still, scant material examined. **W** ♂ 89–95.5 mm (n 8, m 91.8), ♀ 81.5–91 mm (n 3, m 86.8); **T** ♂ 67–74 mm (n 8, m 69.6), ♀ 63–71 mm (n 3, m 66.5); **T/W** m 76.0; **B** 14.3–17.4 mm (n 10, m 15.8);

F. c. coelebs, two juv, ♂ (left: Spain, Jul) and ♀ (right: Spain, Jun): sexual differences already clear in very young plumage. (C. N. G. Bocos)

F. c. africana, ♂, presumed ad, Morocco, Apr: N African population distinctive by lead-grey head-sides in ♂, similar to cap. Greenish above and pinkish below; black forehead and loral area large. Broken white eye-ring and/or short white supercilium behind eye, absent in *coelebs*, while wing and tail show a little more white. Seemingly ad by evenly feathered wing. (R. Armada)

F. c. africana, ♀, Morocco, Mar: very similar to ♀ *coelebs*, but on average slightly more vividly green above. Sometimes has clear white spot above eye. Also on average broader white areas in wing and tail, and has heavier bill. Ageing difficult in this case. (H. van Diek)

B(f) 12.0–14.6 mm (*n* 10, *m* 13.2); **BD** 8.2–9.6 mm (*n* 8, *m* 9.1); **Ts** 17.8–19.5 mm (*n* 11, *m* 18.7). – No material available from W Caucasus, claimed to be this subspecies. C Caucasus populated by *coelebs*.

○ **F. c. schiebeli** Stresemann, 1925 (Crete; possibly extreme SW Turkey; mainly sedentary, but winter movements poorly known). Very similar to *coelebs*, but in ♂♂ lower back and rump greener, chestnut of mantle less dark, a little washed out and more heavily invaded with green, cheeks and underparts average paler. Scapulars dark lead-grey invaded with brown. **W** ♂ 84–91 mm (*n* 16, *m* 86.7), ♀ 80–85 mm (*n* 13, *m* 82.1); **T** ♂ 60–69 mm (*n* 16, *m* 65.8), ♀ 59–67 mm (*n* 13, *m* 63.1); **T/W** *m* 76.3; **B** 14.1–16.2 mm (*n* 20, *m* 15.0); **B**(f) 11.5–13.2 mm (*n* 21, *m* 12.3); **BD** 7.3–8.5 mm (*n* 20, *m* 7.8); **Ts** 17.2–19.0 mm (*n* 20, *m* 18.1).

NORTH AFRICAN GROUP ('African Chaffinch' – NW Africa) comprises three similar-looking races that differ rather clearly from the Eurasian Group. Breeding ♂ blue-grey on crown, ear-coverts and nape, appearing greyish-hooded, with paler throat. Forehead and lores black, contrasting with broken white eye-ring. Some have bold white nape spot, common in the east, rare in the west. Rest of upperparts pale green to blue-grey, sometimes tinged warm brown, forming marked saddle. More extensive white on wing-bars, tertials and secondaries fringes, and tail-edges. Compared to Eurasian Group, ♀ generally greyer brown above, with a grey-green rump, bluish-grey uppertail-coverts, and has some pinkish-buff on throat.

F. c. africana Levaillant, 1850 (NW Africa east to NW & W Tunisia; sedentary). A somewhat variable race, not least in size, but typically similar size to *coelebs*. ♂ often has pinkish-whitish throat but rather dark pink rest of underparts and quite dark lead-grey hood with much black on forehead and lores, latter contrasting with broken white eye-ring. However, it should be noted that some ♂♂ are paler pink below, approaching *spodiogenys*, and for such birds all characters need to be considered in combination. Mantle usually all green, but a few have faint traces of chestnut-brown admixed, in a very few a little more brown (more frequent at western end of range). Blue-grey nape usually uniformly dark, only rarely (in 19%) with faint white spot at centre. Wing markings usually white (has some pale yellow only in fresh plumage, and then mainly on some inner greater coverts and tertials). Vocalisations as described for *spodiogenys*. **L** 14.5–17.5 cm; **W** ♂ 81–97 mm (*n* 128, *m* 89.8), ♀ 80–88 mm (*n* 16, *m* 84.2); **T** ♂ 60–78.5 mm (*n* 128, *m* 70.4), ♀ 61.5–68.5 mm (*n* 16, *m* 65.3); **T/W** *m* 78.2; **B** 13.5–16.6 mm (*n* 150, *m* 15.1); **B**(f) 11.0–13.6 mm (*n* 150, *m* 12.2); **BD** 7.3–9.7 mm (*n* 146, *m* 8.3); **Ts** 17.5–20.5 mm (*n* 146, *m* 19.3). **Wing formula: p1** minute; **p2** 1–5.5 mm < wt, =5 or 5/6; **pp3–5** about equal and longest; **p6** < wt 2.5–7 mm; **p7** < wt 9–16 mm; **p10** < wt 18–24 mm; **s1** < wt 20–25 mm. Emarg. pp3–6. (Syn. *koenigi*.)

F. c. spodiogenys Bonaparte, 1841 (Tunisia except north-west, NW Libya; sedentary). Very similar to *africana*, but fractionally larger and slightly paler above and below, underparts in ♂ plumage being more diluted pink with a cream-white undertone. The paler bluish crown creates greater contrast against black forehead. Mantle usually pure green. White pattern in wing as a rule more extensive. Still, a few are probably inseparable. About 25% of ♂♂ have a prominent white nuchal spot and another 15% have a faint hint of one. Song short and 'primitive' compared to Eurasian populations, terminal flourish often omitted or shortened. A metallic *chip-chip-chip* often inserted between songs. **W** ♂ 86–97 mm (*n* 50, *m* 91.0), ♀ 81–93 mm (*n* 18, *m* 85.3); **T** ♂ 66–79 mm (*n* 49, *m* 71.3), ♀ 63–72.5 mm (*n* 18, *m* 66.6); **T/W** *m* 78.3; **B** 13.6–16.4 mm (*n* 66, *m* 14.9); **B**(f) 10.6–13.4 mm (*n* 67, *m* 12.1); **BD** 7.5–9.5 mm (*n* 67, *m* 8.4); **Ts** 17.2–19.9 mm (*n* 65, *m* 18.9).

F. c. harterti Svensson, 2015 (N Cyrenaica, NE Libya; presumably sedentary). Differs from similarly large *spodiogenys* of Tunisia in that ♂ plumage is darker above and below, equally dark as *africana* but warmer above, less cold blue and green, green back more tinged golden-brown (more so when fresh than when worn); further by having a heavier bill than both, invariably having a small white nuchal patch

F. c. africana, 1stW ♂, Spain, Feb: as ad, but in winter or early spring broad pale tips somewhat obscure patterns. Primary-coverts and apparently one outermost greater covert retained juv, being slightly paler and greyer than moulted inner greater coverts. (avesdeceuta.com)

F. c. spodiogenys, ad ♂, Tunisia, Apr: slightly paler above and below than *africana*, with pink underparts pale and slightly tinged cream-white. White wing markings on average more extensive. Only ad wing-feathers visible. (D. Occhiato)

(only in 19% of *africana* and 40% of *spodiogenys*, and then generally fainter and smaller). Median coverts and tertial edges tend to be purer pale lemon-yellow (whiter in both other races), yellow colour remaining even in more worn spring plumage. Compared to male *africana*, lores and cheeks tend to be less neatly blue-grey, more sullied brown-grey. Long-tailed and long-legged. **W** ♂ 89–95 mm (*n* 18, *m* 91.6), ♀ 84–87 mm (*n* 8, *m* 85.3); **T** ♂ 67–79 mm (*n* 18, *m* 73.6), ♀ 67–68 mm (*n* 18, *m* 67.8); **T/W** *m* 80.1; **B** 14.3–18.0 mm (*n* 26, *m* 16.2); **B**(f) 12.2–15.2 mm (*n* 26, *m* 13.5); **BD** 8.2–9.7 mm (*n* 25, *m* 9.1); **Ts** 17.8–21.6 mm (*n* 26, *m* 19.3).

ATLANTIC ISLANDS GROUP (Azores, Canaries, Madeira) comprises five races. In general dark bluish-grey above with green usually limited to rump. Long-legged. Note that *moreletti* of the Azores and *maderensis* of Madeira differ a little from the others in plumage, but less in structure, and show some affinities with the African taxa, rather than with populations breeding in the Canaries. These two races form a link between the Eurasian populations and those on the Canaries and in North Africa.

F. c. moreletti Pucheran, 1859 (Azores; resident). Upperparts rather similar to *africana*, with forehead black, crown/nape bluish-grey, mantle green (sometimes sullied or blotched grey), back and rump less bright green. White portions on rr5–6 more restricted than in rest of complex, mainly confined to edges on inner webs. Bill heavy. Proportionately rather short-tailed. Song often lacks terminal flourish, trill being slightly longer and more tentative in delivery. Distinctive alarm call, *chee-chee-chee*, a bit tit-like. **L** 15–16 cm; **W** ♂ 81–88 mm (*n* 15, *m* 84.9), ♀ 76–82 mm (*n* 12, *m* 78.5); **T** ♂ 64–70.5 mm (*n* 15, *m* 67.1), ♀ 58–66 mm (*n* 12, *m* 61.5); **T/W** *m* 78.7; **B** 16.3–19.7 mm (*n* 27, *m* 17.6); **B**(f) 13.5–16.0 mm (*n* 27, *m* 14.5); **BD** 8.4–10.2 mm (*n* 27, *m* 9.4); **Ts** 19.7–22.5 mm (*n* 27, *m* 21.4). **Wing formula: p1** minute; **p2** 1–4 mm < wt, =6 or 6/7 (85%) or = 5/6 (15%); **pp3–5** about equal and longest; **p6** < wt 2–4 mm; **p7** < wt 7–11.5 mm; **p10** < wt 14–20 mm; **s1** < wt 16–21 mm. Emarg. pp3–6.

○ **F. c. ombriosa** Hartert, 1913 (El Hierro, W Canaries; resident). Very similar to *palmae*, differing only subtly and on average in that ♂♂ usually have slightly darker cinnamon-buff throat/chest, and presence of faint green tinge on rump in some (apparently absent in *palmae*). Glossy metallic blue-grey crown distinctly set off from pinkish-white lores. **W** ♂ 84–89 mm (*n* 8, *m* 86.8), ♀ 80.5–85 mm (*n* 6, *m* 81.9); **T** ♂ 69–75 mm (*n* 8, *m* 72.6), ♀ 65–69.5 mm (*n* 6, *m* 67.4); **T/W** *m* 83.1; **B** 15.0–17.5 mm (*n* 14, *m* 16.8); **B**(f) 12.5–14.3 mm (*n* 14, *m* 13.8); **BD** 8.3–9.5 mm (*n* 14, *m* 8.8); **Ts** 20.7–22.7 mm (*n* 14, *m* 21.4).

F. c. palmae Tristram, 1889 (La Palma, W Canaries; resident). Very similar to *ombriosa* but on average chest in ♂ paler cinnamon-buff, with no green on rump. On average less greenish-yellow edges to primaries compared to *ombriosa*. Song and calls like *canariensis*. **W** ♂ 84–90 mm (*n* 13, *m* 87.5), ♀ 81–85 mm (*n* 11, *m* 83.0); **T** ♂ 68–75 mm (*n* 13, *m* 72.5), ♀ 64–70 mm (*n* 11, *m* 67.1); **T/W** *m* 82.0; **B** 15.6–17.9 mm (*n* 24, *m* 17.0); **B**(f) 13.0–15.1 mm (*n* 24, *m* 14.1); **BD** 8.2–9.4 mm (*n* 24, *m* 8.8); **Ts** 20.5–22.5 mm (*n* 23, *m* 21.7).

F. c. canariensis Vieillot, 1817 (Gran Canaria, La Gomera and Tenerife, C Canaries; resident). Throat, breast, upper belly and flanks light cinnamon-buff (restricted to throat and breast in *palmae* and *ombriosa*). Crown dark slate to bluish-black, nape and mantle dark grey with bluish-green suffusion, rump green. ♀♀ appear inseparable from other races in Canary Is. Song as *moreletti*, more tentative in delivery, trill longer and flourish shorter or omitted. One common call is more chirping, like House Sparrow. Also has double call, *che-chelee*, a little like Crimson-winged Finch or Woodlark in tone. **W** ♂ 81.5–92 mm (*n* 16, *m* 85.7), ♀ 77–84 mm (*n* 12, *m* 80.2); **T** ♂ 68–76 mm (*n* 16, *m* 71.1), ♀ 63–67 mm (*n* 12, *m* 65.0); **T/W** *m* 82.2; **B** 15.5–18.6 mm (*n* 28, *m* 17.1); **B**(f) 13.0–15.0 mm (*n* 28, *m* 13.9); **BD** 8.0–9.3 mm (*n* 28, *m* 8.6); **Ts** 20.5–22.7 mm (*n* 28, *m* 21.7). (Syn. *tintillon*.)

F. c. maderensis Sharpe, 1888 (Madeira; resident). Differs from *canariensis*, which it otherwise resembles, in darker upperparts (darker even than *moreletti* of Azores), largely green mantle, predominantly blue-grey back (rarely with brown tinge) and green rump, and tends to have breast-sides and flanks more suffused lead-grey. Bill heavy. ♀ like *canariensis*, but cap and nape sometimes have more distinct sepia stripes, and mantle and scapulars are greener when

F. c. moreletti, ♂, presumed ad, Azores, Jul: dark blue-grey cap, often well-developed pale supercilium (white when worn) and dusky olive-green mantle. Note clear dark lores and bicoloured ear-coverts. White in wing rather restricted. Wing seems evenly feathered. (J. Normaja)

F. c. moreletti, ♀, Azores, Jul: bill markedly long and heavy-based, almost like in Blue Chaffinch. Olive tones of back and rump rather dull, giving fairly uniform impression of upperparts. Possibly ad by seemingly fresh and evenly-feathered wing, although important parts invisible, thus best left un-aged. (J. Normaja)

F. c. moreletti, ad ♂, Azores, Sep: large-billed with characteristic pink-buff throat, fore ear-coverts and breast, thus like other Atlantic Islands' races. However, unlike Canarian taxa it has greenish-tinged mantle (obvious in fresh plumage), more like N African taxa. Note on-going post-nuptial remiges moult, inferring age. (H. Shirihai)

F. c. moreletti, 1stW ♂, Azores, Nov: young autumn ♂ often duller and less boldly patterned due to extensive mainly rufous-buff tips, though main racial features still visible. (D. Occhiato)

F. c. moreletti, ♀, presumed ad, Azores, Sep: Azorean ♀♀ often appear dusky, especially when fresh. Note proportionately large-headed, heavy-billed, strong-legged and short-tailed jizz. Primary-coverts seemingly ad, and all greater coverts of ad type, but certain ageing impossible without a better view. (H. Shirihai)

fresh. Has only one (not two) white outer tail-feathers. **W** ♂ 81–87 mm (*n* 12, *m* 83.3), ♀ 74–81 mm (*n* 12, *m* 77.3); **T** ♂ 67–74 mm (*n* 12, *m* 70.0), ♀ 58–68 mm (*n* 12, *m* 63.5); **T/W** *m* 83.1; **B** 14.5–18.8 mm (*n* 24, *m* 16.3); **B**(f) 12.2–14.5 mm (*n* 24, *m* 13.2); **BD** 7.4–9.1 mm (*n* 24, *m* 8.1); **Ts** 20.0–22.5 mm (*n* 24, *m* 21.5).

TAXONOMIC NOTE Although the rather large morphological differences between the three subspecies groups may invite a split into two or three species, genetic analyses (Marshall & Baker 1999; cf. Collinson 2001) has shown *africana* to be more closely related to European *coelebs* than to *spodiogenys*, a rather unexpected result. Ssp. *spodiogenys* proved to be sister to all other sampled Chaffinch taxa. Against this background it is hardly defensible to split the complex; it is better kept together as one species showing intricate variation.

REFERENCES VAN DEN BERG, A. (2005) *DB*, 295–301; 395. – COLLINSON, M. (2001) *BB*, 94: 121–124. – GRISWOLD, C. K. & BAKER, A. J. (2002) *Evolution*, 56: 143–153. – MARSHALL, H. D. & BAKER, A. J. (1999) *Mol. Phyl. & Evol.*, 11: 201–212. – STANFORD, J. K. (1954) *Ibis*, 96: 449–622. – SVENSSON, L. (2015) *BBOC*, 135: 69–76.

F. c. ombriosa, ad ♂, El Hierro, Canary Is, Jul: generally like *palmae*, but upperparts including rump concolorous bluish. Chest dark cinnamon-buff, rest of underparts white. Metallic blue-grey crown and rear ear-coverts, contrasting with pinkish-buff lores and fore ear-coverts as in most ♂♂ of Macaronesian populations. In full post-nuptial moult. (B. Rodríguez)

F. c. canariensis, ad ♂, La Gomera, Canary Is, Jan: ♂ has bluish-black crown and extensive pinkish cinnamon-buff underparts. Wing-feathers evenly fresh without hint of moult limits. (M. Strazds)

F. c. palmae, ♂, presumed ad, La Palma, Canary Is, Nov: small and pale cinnamon-buff area on face and chest, smaller on chest than in *palmae*. Otherwise, like other Canarian races, has deep slate-blue upperparts. Wing-feathers seem evenly fresh indicating age. (H. Nussbaumer)

F. c. palmae, ♀, La Palma, Canary Is, Nov: heavy-billed as compared to European taxa. ♀ to some degree mirrors paler lores and fore ear-coverts of ♂. Ageing uncertain in this case. (H. Nussbaumer)

F. c. maderensis, ad ♂, Madeira, Apr: more like Azorean population in pinkish ochre-yellow face and chest, greenish tinge on mantle and (slightly) on rump. However, pinkish face typically encompasses larger area of ear-coverts and around eyes. Also typical are contrasting grey breast-sides and flanks. Note heavy bill. Evenly fresh wing with very dark primary-coverts infers age. (H. Shirihai)

F. c. maderensis, ♀, Madeira, Apr: generally ♀♀ are difficult to assign to race, but ♀ *maderensis* is often quite characteristic in having pale face like ♂, with contrasting light eye-surround, lores, fore ear-coverts and throat. Ageing uncertain. (H. Shirihai)

BLUE CHAFFINCH
Fringilla teydea Webb, Berthelot & Moquin-Tandon, 1836

Fr. – Pinson bleu; Ger. – Teydefink
Sp. – Pinzón azul; Swe. – Blåfink

A resident of pinewoods on the C Canary Islands, the Blue Chaffinch is restricted to Tenerife (where it is rather numerous) and SW Gran Canaria (where it is extremely rare), usually at moderate elevations (300–1800 m). It may descend to lower elevations during snow or periods of unusually low temperatures, and there is one extralimital record, from Lanzarote, in late autumn.

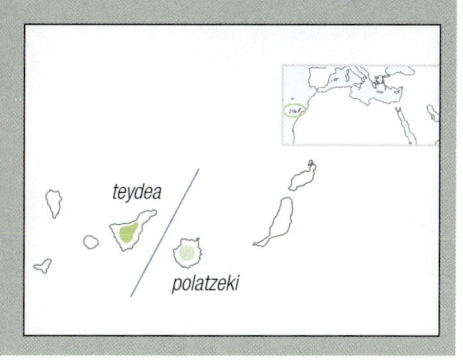

F. t. teydea, ad ♀, Tenerife, Canary Is, Sep: predominantly uniform drab brown-grey (mantle and scapulars washed slightly warmer). Belly and undertail-coverts are palest areas, but wing-bars are pale cream-white and rather distinct. Bill greyish-horn. At end of post-nuptial moult with only ad wing-feathers visible. (H. Shirihai)

F. t. teydea, ad ♂, Tenerife, Canary Is, Mar: often appears very blue, especially in early spring when still very fresh but with fewer pale tips, affording more uniform overall impression. Obvious yet broken white eye-ring, blacker lores and whitish undertail-coverts. Bill is robust, mostly metallic steel-blue, making the species altogether unmistakable. Wing evenly fresh. (V. Legrand)

IDENTIFICATION A large, heavy, dark and long-billed chaffinch, notably more robust than Chaffinch, with comparatively uniform plumage. ♂ is largely *metallic slate-blue to blue-grey*, becoming white on centre of belly, vent and undertail (on Gran Canaria white on much of belly and lower flanks also). Head and mantle often darker, with a *narrow black loral band* (extending to forehead on Gran Canaria), but most distinctive are the *small white eye-crescents*. Chin and throat slightly paler grey than rest of head. Wings black with *bluish-white median-coverts wing-bar and pale bluish-grey fringes and tips to greater coverts*. Tail dull black with broad grey-blue fringes to central feathers, narrower on rest (thus lacks white on outer feathers). Bill pale metallic steel-blue, with black tip in breeding season. Legs grey-brown. ♀ resembles ♂ but is *predominantly drab bluish-grey with variable brownish tone above* (sometimes has slightly deeper earth-brown wash to mantle and scapulars), paler, ash-grey underparts, and *rather more obvious whitish wing-bars*. Bill duller with pale pink base to lower mandible. Habits much as Chaffinch, but most often found in pine forests, and the heavier, stouter bill is mainly used to prise seeds from cones, but it also persistently hawks and searches bark for insects, and is considerably less gregarious. Generally at higher altitudes than Chaffinch, with little overlap.

VOCALISATIONS (*teydea*) Song resembles Chaffinch but is shorter and simpler, the tone slightly harsher and more low-pitched, and the terminal flourish is slightly decelerating, more drawn-out and often repeated, e.g. *sitt-sitt-sitt rüah-rruaah, rruuaah* or a simpler *chet-chet-chat-chut-churreeah*. – Commonest call a disyllabic loud, nasal and squeaky *che-cheeve* or *whee-chay*, at times recalling parrot or bulbul in tone.

SIMILAR SPECIES Unmistakable, as even ♂♂ of blue-backed Canarian races of *Chaffinch* have prominent white in wings and tail, a vinaceous or pink-red breast and extensive white belly. In addition, Blue Chaffinch is noticeably larger, with a substantially larger bill, and often appears to have a more bulging breast and larger head. Even in flight, the larger size is obvious, and the flight pattern is slower and more undulating. ♀ Blue Chaffinch, while more similar to Chaffinch, still differs structurally, and has plainer upperparts, almost entirely greyish-white underparts and less complex wing markings.

AGEING & SEXING (*teydea*) Ageing possible when seen close or handled based on presence of moult contrast or not in wing, and on shape and wear of tail-feathers. Sexes moderately differentiated. – Moults. Complete post-nuptial moult and partial post-juv mostly Jul–Nov (probably more protracted and at times suspended towards winter in 1stW). Post-juv renewal involves head, body, all lesser and variable number of median and greater coverts. Pre-nuptial moult apparently absent or limited. – **SPRING Ad** ♂ See Identification. Plumage wears even brighter, more lead-blue, especially on head. **Ad** ♀ See Identification. Upperparts deeper grey than in autumn. Both sexes reliably aged by lack of moult limits in wing. Primaries, primary-coverts and tail-feathers only moderately worn and bleached. **1stS** Separated by relatively more worn juv primaries, primary-coverts and tail-feathers, usually some outer greater coverts and all or at least most tertials, creating quite noticeable moult limits. Both sexes often somewhat less pure grey and blue, tinged brownish. Some duller 1stS ♂♂ may resemble bluer and older ♀♀. – **AUTUMN Ad** ♂ Fresh dark olive-brown tips sometimes present on upperparts. Bill slightly duller, otherwise as breeding plumage. **Ad** ♀ Head and upperparts browner, underparts more buffish-grey. Both sexes, prior to completing moult, may retain some heavily-worn unmoulted remiges, while once moult is finished they are evenly fresh and lack moult contrast in wing. **1stW** Both sexes aged by juv remiges, primary-coverts and some (few outer to all) greater coverts, which are weaker textured, duller, at least initially fringed slightly buff-brown and form moult limits; retained tail-feathers also more pointed, especially central pair. 1stW ♂ can closely recall fresh ad ♀ (or is intermediate between sexes), with slate-grey upperparts, head- and neck-sides washed dark olive-brown, paler lores, and chin to breast tinged buff-brown. 1stW ♀ like same-sex ad but much less pure grey exposed, including fringes to wing-feathers, thus close to juv. **Juv** Resembles 1stW ♀ but has soft, fluffy

F. t. teydea, ad ♂, Tenerife, Canary Is, Sep: at end of post-nuptial moult and very fresh. Same individual under different lighting, showing how this affects coloration—in shade (left) often appears much bluer, whereas in sunlight looks greyer. Head, upperparts and wing uniform slate-blue to blue-grey, underparts paler. Median and greater coverts have broad ash-grey tips, producing slight wing-bars. Note white inner webs to outer tail-feathers, visible from below (left). (H. Shirihai)

body-feathers, drab greyish-brown upperparts, buffish-white underparts and pale fringes to wing-feathers. Bill initially blackish rather than grey (Garcia-del-Rey & Gosler 2005).

BIOMETRICS (*teydea*) **L** 16.5–18 cm; **W** ♂ 94–106 mm (*n* 19, *m* 101.8), ♀ 91–103 mm (*n* 14, *m* 96.3); **T** ♂ 71–86 mm (*n* 19, *m* 80.0), ♀ 71–80 mm (*n* 14, *m* 74.8); **T/W** *m* 78.2; **B** 17.5–20.9 mm (*n* 33, *m* 19.6); **B**(f) 15.0–18.2 mm (*n* 33, *m* 16.7); **BD** 9.5–11.9 mm (*n* 33, *m* 10.7); **Ts** 20.5–24.1 mm (*n* 32, *m* 22.9). **Wing formula: p1** minute; **p2** 3.5–9 mm < wt, =6 or 6/7; **p3** < wt 1–3 mm; **pp4–5** about equal and longest; **p6** < wt 3–5 mm; **p7** < wt 10–14 mm; **p10** < wt 19–28 mm; **s1** < wt 21–29.5 mm. Emarg. pp3–6 (sometimes faintly on p7 too). – In a study of live birds on Tenerife, Garcia-del-Rey & Gosler (2005) reported similar biometrics except that ♀♀ were clearly smaller, and bill (both sexes) slightly longer: **W** ad ♂ 98–106 mm (*n* 76, *m* 102.8), 1stY ♂ 93–102 mm (*n* 64, *m* 97.5), ad ♀ 89–97 mm (*n* 60, *m* 93.5), 1stY ♀ 86–95 mm (*n* 42, *m* 90.7); **T** ad ♂ 80–87 mm (*n* 52, *m* 83.5), ad ♀ 72–82 mm (*n* 37, *m* 75.8); **B** 18.8–23.0 mm (*n* 242, *m* 20.6); **BD** 9.4–11.3 mm (*n* 242, *m* 10.4).

GEOGRAPHICAL VARIATION & RANGE Two races recognised, differing mainly in size and prominence of pale wing-bars in ♂ plumage, subtly also in vocalisations and genetically. All populations sedentary.

F. t. teydea Webb, Berthelot & Moquin-Tandon, 1841 (Tenerife; common). Treated above. Ad ♂ lacks obvious black frontal band (can have narrow hint at most), while wing-bars are pale blue-grey rather than white, and are often narrow and insignificant. However, note that imm ♂ and ♀♀ usually have whitish wing-bars.

F. t. polatzeki Hartert, 1905 (Gran Canaria; rare and endangered). Subtly smaller than *teydea*. ♂ differs in being whiter on belly and lower flanks, and has diffuse but obvious black frontal band and broad double white or grey-white wing-bars (in *teydea*, wing-bars are blue-grey and usually narrower). ♀ paler with less extensive brown-grey on chin to breast, and more extensive off-white belly. Song on average more reminiscent of Common Chaffinch, but variable and some overlap. Call more whistling, a little Chiffchaff-like, *hueet*, or disyllabic *hueet-hue*. **W** ♂ 95–98.5 mm (*n* 11, *m* 96.8), ♀ 88–95 mm (*n* 5, *m* 90.5); **T** ♂ 74–78 mm (*n* 11, *m* 76.0), ♀ 68–75 mm (*n* 5, *m* 70.8); **T/W** *m* 78.4; **B** 18.2–20.4 mm (*n* 16, *m* 19.0); **B**(f) 15.3–18.3 mm (*n* 10, *m* 16.1); **BD** 10.1–10.8 mm (*n* 16, *m* 10.5); **Ts** 21.6–23.0 mm (*n* 16, *m* 22.2).

TAXONOMIC NOTE Considering the combination of noted differences in genetics, sperm morphology, plumage and size morphology, and vocalisations, separate species status has been suggested (Lifjeld *et al.* 2016, Sangster *et al.* 2016). Not denying these (comparatively fine) differences between the two taxa, we feel that on the whole it is an arbitrary decision whether to regard them as species or subspecies, and we favour the last-mentioned option. In particular when judging island populations it is easy to observe better separation of data than for continental parapatric populations, and one should allow somewhat for this difference when making a judgement. Any conservation issues should be equally important regardless of taxonomic status.

REFERENCES Garcia-del-Rey, E. & Gosler, A. G. (2005) *Ring. & Migr.*, 22: 177–184. – Lifjeld J. T. *et al.* (2016) *BMC Zool.*, Open Access. – Sangster, G. *et al.* (2016) *J. Avian Biol.* 47: 159–166.

F. t. teydea, 1stW ♂, Tenerife, Canary Is, Sep: post-juv moult pattern, intermediate between ♀ and ad ♂, with brownish-grey cast to grey-blue face, upperparts, breast and flanks. Note contrasting ad-type bluish-grey median coverts and rump and uppertail-coverts. (H. Shirihai)

F. t. teydea, juv/1stW ♀, Tenerife, Canary Is, Sep: browner above, less pure grey than ad ♀; fringes to wing-feathers, being juv, are whitish-tipped thus wing-bars more contrasting. Note robust bill. At least until mid-autumn young retain many juv feathers, especially ♀♀. (H. Shirihai)

F. t. polatzeki, ad ♂, Gran Canaria, Canary Is, Oct: smaller and more delicate than *teydea*, including bill. ♂ slightly duller and greyer above, and paler grey below with whiter belly (thus more contrast between upper- and underparts than in *teydea*, which appears more evenly bluer). Diffuse but obvious black frontal band and whiter, grey-white wing-bars. At end of post-nuptial moult. (H. Shirihai)

F. t. polatzeki, ad ♀, Gran Canaria, Canary Is, Oct: very similar to *teydea* but has on average paler underparts, including more extensive off-white belly. At end of post-nuptial moult. (H. Shirihai)

F. t. polatzeki, 1stW ♂, Gran Canaria, Canary Is, Dec: plumage rather intermediate between the two sexes, but combination of some bluish upperparts feathers and young age (note retained juv outer greater coverts) infers age and sex. The new ad-like greater coverts show the diagnostic better-defined, narrower and whiter wing-bars of this race. (M. Angel Peña)

BRAMBLING
Fringilla montifringilla L., 1758

Fr. – Pinson du Nord; Ger. – Bergfink
Sp. – Pinzón real; Swe. – Bergfink

One of the characteristic breeding birds of the taiga, from N Scandinavia across Siberia. Throughout its coniferous forests or alpine birch stands you will hear the Brambling's monotonous reeling or buzzing song. In winter it moves south, although it usually does not reach further than the N Mediterranean. At this season it is perhaps most familiar to the majority of European birders, although its winter occurrences are somewhat irregular, in some years there are large influxes, and small numbers may even reach the northern Sahara and Middle East, but in mild winters many remain further north. At this season, enormous flocks may gather around beech trees, feeding on the seeds from the fallen beechnuts.

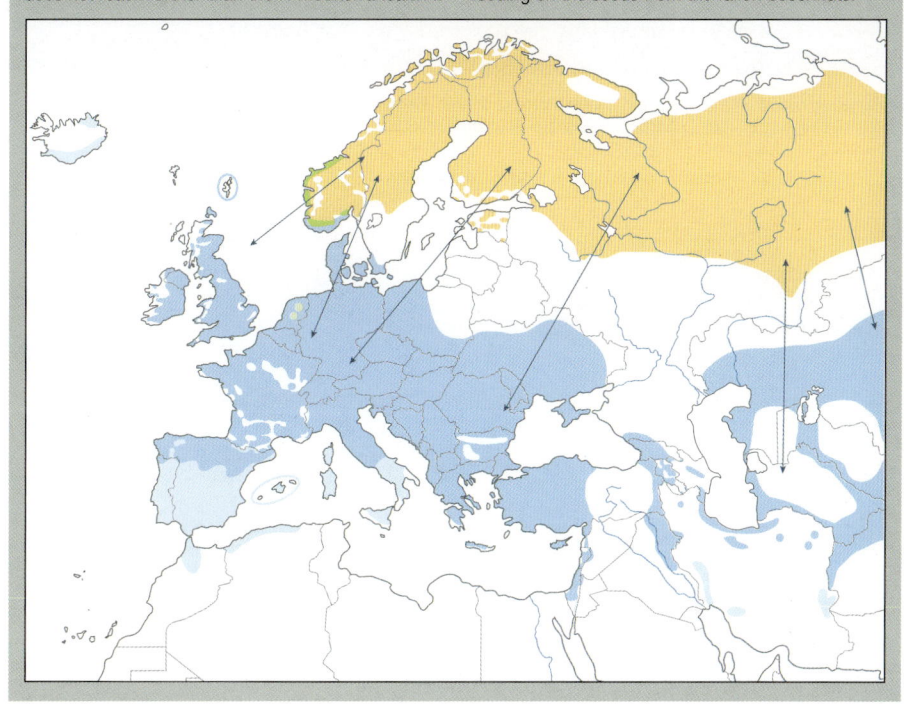

IDENTIFICATION Square-headed finch, with ♂ in breeding plumage being particularly attractive. All plumages have *long clean white rump to uppertail-coverts* (and all-dark tail), especially conspicuous in escape-flight when diagnostic call is usually given. Most also have some *orange-rufous on breast*, making the species even more unmistakable. Recalls Chaffinch in size and structure, but tail is proportionately slightly shorter. *Spring ♂ has glossy blue-black head, mantle, scapulars and wings*, contrasting with extensive bright *orange-rufous lesser coverts* ('wing-bend'), *throat to breast and flanks*. There is a white median-coverts bar (if not concealed by the rufous feathers), a *narrow rufous-white greater coverts bar* and a white spot at base of innermost primaries. Across the rufous flanks there are variable bold, round black spots, while *rest of underparts are almost clean white*. Bill blue-black when breeding. ♀ has dull silvery grey head-sides with bold dark crown- and nape-sides and upper-rear ear-covert stripes. Wing pattern similar to ♂ but duller, and has *much paler and more restricted orange-buff on breast and flanks*. Bill yellow with black tip. Non-breeding ♂ characterised by its blotchy black, buff and grey mantle and head (black bases partly concealed by greyish and rusty-buff feather fringes), and thus approaches ♀ in head pattern. Forms compact flocks (up to several 100,000!) in winter, which mingle freely with Chaffinches, feeding mainly on the ground, especially on beech mast.

VOCALISATIONS Song a grating, monotonous, slowly repeated *rrrrrhüh*, like a distant wood-cutting saw. The song can be mistaken for Greenfinch's alternative song type, but each note is much straighter and more machine-like. – Most characteristic call is a nasal and rather croaking *teh-ehp* (uttered in flight or when perched). In flight, often from migrating flocks, a subdued *yeck*, often repeated, harder than similar call of Chaffinch, more akin to flight call of Twite. In anxiety a repeated, clear, penetrating *slitt, slitt, slitt,...*

SIMILAR SPECIES Owing to similar general characters, behaviour and wing pattern, and duller ♀ plumage, confusion with *Chaffinch* may arise in poor views. However, all plumages possess diagnostic combination of long, oval whitish rump patch and almost all-black tail. Furthermore, breeding ♂ has glossy black head and mantle, with broad orange blaze across forebody, and, in winter, a black-speckled face, crown and mantle. ♀ and juvenile Brambling closer to Chaffinch, but still readily distinguished by dark brown crown- and nape-sides, broad buff supercilium, and grey neck-sides and nape centre.

AGEING & SEXING Ageing possible when seen close or handled based on presence of moult contrast or not in wing, and on shape and wear of tail-feathers. Sexes differ clearly after post-juv moult. Seasonal variation due to feather wear, especially affecting ♂ plumage. – Moults. Complete post-nuptial moult in ad and partial post-juv moult in late Jun–mid Sep. Post-juv moult involves head, body, all lesser and median coverts, most or all greater coverts, rarely some alula-feathers and tertials, central tail-feathers, but no primary-coverts or primaries. Pre-nuptial moult absent. – **SPRING Ad** Wear of feather-tips heightens sexual differences (see Identification), ♂ becoming striking with black hood and back. Aged by lack of moult limits in wing, though some ♀♀ difficult to age due to at times subtle contrast in young birds. Primaries, primary-coverts and tail-feathers still relatively fresh. **1stS** Relatively more worn, and duller

Ad ♂, Finland, Jun: with wear, head and back become solidly black, contrasting with white belly and rump patch, orange breast and shoulders, and black wings (with distinct wing-bars), making this plumage unmistakable. Wing-feathers ad. (D. Occhiato)

Ad ♀, Finland, Apr: diffuse dark stripe either side of crown and darkish ear-coverts gives characteristic 'square' look to head-sides, while orange is duller than in ♂. Dark areas grey- or olive-brown, giving mottled pattern above. Note black dots on flanks and thighs. Bill clearly bicoloured in ♀. Only ad wing-feathers visible. (M. Varesvuo)

1stS ♂, Netherlands, Mar: in transition to 1stS plumage when broadly scaled above and bill still mostly yellowish. Contrastingly worn brownish juv alula and primary-coverts clinch ageing. (R. Pop)

1stS ♂, Finland, Jun: as ad, but plumage less solidly patterned, while browner and worn juv feathers in wing form moult limits, most easily seen in ♂. (D. Occhiato)

juv feathers in wing and tail form moult limits or contrast (readily detected in ♂, less conspicuous in ♀). Both sexes sometimes retain traces of immaturity until late spring, and are overall less immaculate (in ♂♂ black and orange less intense). – AUTUMN **Ad** Both sexes are evenly fresh following complete moult and lack moult limits in wing. **Ad ♂** Ageing of ♂♂ best based on shape of tail-feathers and blackness of primary-coverts. (Numerous 1stY have unspotted rufous-white lesser coverts, not only ad!) Extensive fresh greyish-buff fringes and tips to dark areas (thus somewhat ♀-like), but retains obvious black centres to crown, nape and mantle-feathers, and is deeper orange-buff on breast and lesser coverts. Bill orange-yellow with black tip, legs brownish-flesh. Almost uniform black bases to primaries, alula, primary-coverts and greater coverts (cf. 1stW ♂). **Ad ♀** Compared to fresh ad ♂ or duller 1stW ♂, essentially greyish-buff above with more grey on head. Crown-, nape- and eye-stripes narrower and even less clearly defined than in worn plumage; generally no black feather bases visible on face and mantle, and paler orange breast. ♀♀ can have rather many blackish centres to nape and mantle, approaching 1stY ♂, but note that cheeks are dull grey-brown, lacking black feather-bases, and chest is paler and duller, not as rusty-coloured as in ♂. Very rarely, ad ♀ develops almost ♂-like plumage with well-developed black hood and much black on centre of mantle-feathers. Rump unstreaked pale yellow. Upper wing-bar with tiny black spots on lesser coverts, and colour mix of yellow, rufous and white. **1stW**

Ad ♂ (left), Finland, Apr, and ♀, Israel, Jan: in all plumages long, clean white rump to uppertail-coverts, which together with wing pattern and orange shoulder and breast make even distant flying birds distinctive. ♂ is ad by evenly-fresh wing, while on ♀ apparent moult limits in tail and greater coverts infer young. (Left: A. Below; right: L. Kislev)

Best aged (both sexes) by retained juv remiges, primary-coverts, usually some greater coverts and usually all tertials, alula and tail-feathers, which are weaker textured and duller, creating moult contrast with new, blacker or darker-centred lesser, median and inner greater coverts (but moult limits in ♀ more difficult to detect). Tail-feathers narrower and more pointed. **1stW ♂** Intermediate between fresh ad ♂ and ad ♀, but unlike latter has more black on head and mantle, and more extensive rufous-orange on lesser coverts and breast. Besides ageing criteria like moult limits and shape of tail-feathers (see 1stW ♀), bases to retained juv primaries, alula, primary-coverts and unmoulted greater coverts paler, more greyish-black when fresh or greyish-brown with wear, and black centres to mantle-feathers (visible when raised) are

1stS ♀, Finland, Apr: dark frame to grey neck and nape-sides can become more contrasting with wear, and together with marbled mantle and orange breast make identification and sexing straightforward. Remiges, alula, outer greater coverts and primary-coverts all retained juv. (A. Below)

Finland, Oct: flock of mixed age and sex, though all show diagnostic white rump. (M. Varesvuo)

Ad ♂, England, Feb: extensive fresh greyish-buff fringes and tips to dark areas obscure pattern, but centres to crown, nape and mantle-feathers, as well as wing-coverts, are black (ad), as are remiges and primary-coverts. Unlike summer, bill is orange-yellow with black tip. (G. Thoburn)

Ad ♀, Netherlands, Dec: compared even to least-advanced fresh 1stW ♂, winter ad ♀ usually readily identified. Note rufous-tinged greyish-buff mantle and more grey on head. No black bases visible on face, and orange on breast diluted. If aged correctly first, here an ad (by evenly-fresh wing), then this pattern equals ♀. (N. D. van Swelm)

more pointed, less rounded than in ad (see Svensson 1992). Bases of median and longer lesser coverts sometimes more extensively black, partly visible among cinnamon-brown of lesser coverts, which are less rusty-orange and yellow than ad, and lesser and median coverts often partly speckled or streaked black, being less uniform than most ad. Buff fringes to crown, mantle and scapulars average broader than ad in same state of wear. Black areas also duller, and some may show whitish chin. **1stW ♀** Very similar to fresh ad ♀. (Many have decidedly paler breast, more buff-cinnamon, less rufous. Also greyer, less strongly marked head.) See general ageing characters for 1stW. **Juv** Resembles fresh ♀ but has darker-mottled crown and mantle, narrow buffish-white tips to greater and median coverts, and yellowish cast to belly.

BIOMETRICS L 14–15.5 cm; **W** ♂ 87–96 mm (n 24, m 92.9), ♀ 85–91 mm (n 18, m 87.1); **T** ♂ 57–67 mm (n 24, m 63.5), ♀ 56–63 mm (n 18, m 58.9); **T/W** m 68.0; **B** 13.0–15.6 mm (n 28, m 14.4); **B(f)** 10.7–13.2 mm (n 28, m 11.6); **BD** 7.2–9.2 mm (n 28, m 8.2); **Ts** 17.5–20.0 mm (n 29, m 18.8). **Wing formula: p1** minute; **p2** 0–2 mm < wt; **pp** (2)3–4 about equal and longest; **p5** < wt 2–5 mm; **p6** < wt 10–15 mm; **p7** < wt 16.5–20 mm; **p10** < wt 26–31 mm; **s1** < wt 27–34 mm. Emarg. pp3–5.

GEOGRAPHICAL VARIATION & RANGE Monotypic. – Fenno-Scandia, Estonia, N Russia, N Siberia east to Kamchatka and Sakhalin; winters C & S Europe, rarely N Africa, Turkey, Levant, and S Asia east to Japan. (Syn. *subcuneolata*.)

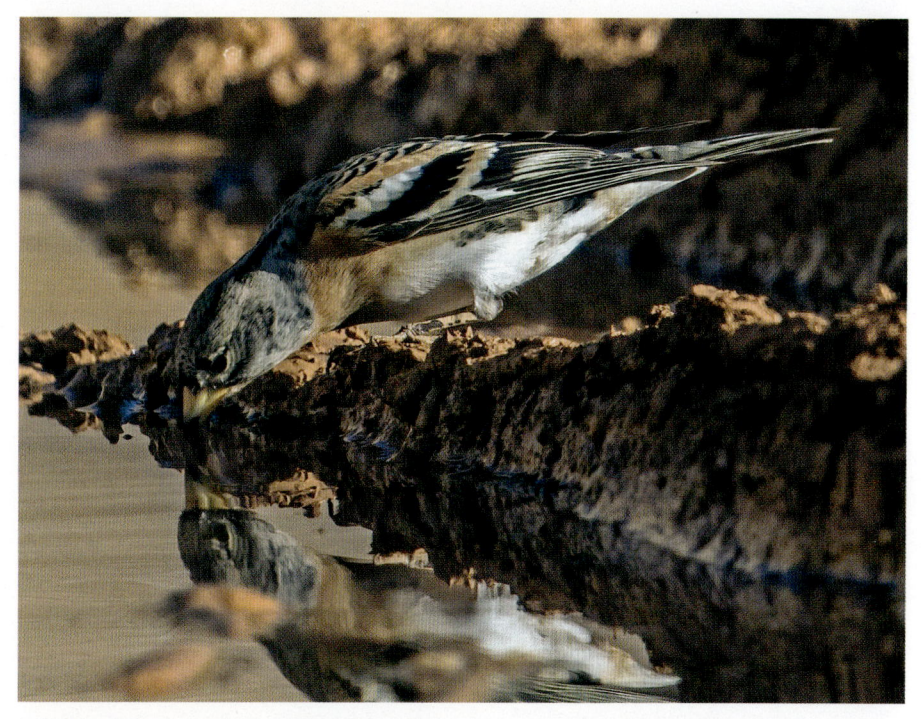

1stW ♂, Israel, Nov: much variation in extent of the exposed of black centres and intensity of orange among winter ♂♂, thus ageing depends on other clues, e.g., brownish juv alula and primary-coverts, with juv secondaries having buffish (not yellowish) edges. (H. Shirihai)

1stW ♀, Israel, Nov: unlike fresh ad ♀, has orange-cinnamon breast, head-sides less pure grey, more buff and brown, and moult limits in wing, including juv alula and primary-coverts and outer greater coverts. Note typical flanks spotting. (H. Shirihai)

Juv, Finland, Jul: fluffy plumage with dark lateral crown and dark-mottled mantle, narrow buffish-white tips to greater and median coverts, and yellowish-buff wash below. (M. Sciborski)

RED-FRONTED SERIN
Serinus pusillus (Pallas, 1811)

Fr. – Serin à front rouge; Ger. – Rotstirngirlitz
Sp. – Serín frentirrojo
Swe. – Rödpannad gulhämpling

A small, attractive Asian finch that is usually seen in mountains near the tree-line or on alpine meadows, feeding on the ground or in low vegetation. The westernmost breeding sites are in W Turkey (Uludağ). A social bird that often is seen in small flocks even in the breeding season, and which in autumn can form very large flocks. Makes chiefly altitudinal or short-range movements in winter. Often noticed by characteristic calls emanating from over-flying parties. Flock formation dense, flight undulating. Nests in dense conifers, sometimes at considerable height above ground.

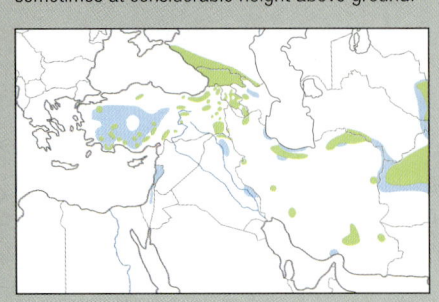

♂ (left) and ♀, Turkey, May: a pretty finch, of which ♂ in abraded spring/summer plumage has head and bib on average blacker, breast-streaking often bolder, and red and yellow on head and body brighter and more extensive than in ♀, but some ♀♀ similarly attractive, and sexing best practised when pair seen together. ♂ is ad (evenly-feathered wing), ♀ a duller individual and probably 1stS (alula and primary-coverts apparently juv). (M. Schäf)

IDENTIFICATION Closely related to Serin of Europe, about as small or only slightly larger, but with a proportionately somewhat longer tail. It has a similarly *short and stubby, rounded bill* as Serin, but the *bill is generally darker*. Red-fronted Serin is rather a *dark and boldly streaked* bird that in adult plumage has a *sooty-black head and breast*, a *bright red or orange-red forecrown patch* and mustard-*yellow wing-bend and edges to wing- and tail-feathers*. Mantle and back buffish-brown strongly streaked dark, and *flanks and lower breast also boldly streaked*, in worn plumage becoming quite dark overall. Buff-white tips to median and greater coverts, apparent both on perched and flying birds, form *two variably distinct pale wing-bars* depending on plumage wear. There is a white patch on secondaries visible when wing is folded. *Rump*, well visible in flight, is characteristically *orange-yellow* (like pale copper or mustard), often unstreaked on centre but can be diffusely streaked throughout. Sexes very similar and only a few extremes possible to separate, or when pair seen together. ♂ has more extensive and deeper black on head and throat, ♀ less. Juvenile and many first-winters readily separated from adults (and early moulters among 1stW) by rufous-cinnamon head (at least on face, cheeks and upper throat) with very little if any black, and no red forecrown patch). – Often encountered in flocks, taking off in dense formation with bounding, somewhat erratic flight, calling frequently. Feeds a lot on the ground in open alpine habitats.

VOCALISATIONS Song recalls both Goldfinch and Citril Finch (more than Serin), an elaborate, prolonged twittering and energetic series of trills and high-pitched notes run together without pause. Contains short wheezing sounds just like the song of the two above-mentioned species. Often delivered in flight, but also from high song-post or from atop a rock. – Call, often heard from birds flying past, a ringing, high-pitched trill that often falls slightly in pitch at the end, *ti-ti-ti-ti-te* or *tvir'r'r'r*, a cheery, soft and spirited sound that is easy to pick out once learnt. Also an upwards-inflected *djuee*, a bit like Redpoll or Greenfinch, perhaps in anxiety.

SIMILAR SPECIES Difficult to confuse with any other species if seen well. In fleeting, distant views, juveniles can perhaps be mistaken for Asian races of *Twite*, but these have dark-streaked forecrown (not near-uniformly rufous-cinnamon as in Red-fronted Serin), a hint of a pale supercilium not seen in juvenile Red-fronted Serin, bold streaking of underparts concentrated on breast-sides/upper flanks (rather than all of breast and flanks boldly streaked), and pale edges to wing- and tail-feathers white, not yellowish.

AGEING & SEXING Ages differ clearly in 1stW, but differences very slight once head and throat moulted to ad appearance (protracted process, Aug–Apr, some still in full juv plumage in Feb), requiring close views or handling. Sexes very similar, but a few extremes can be separated after post-juv moult (from about Aug–Sep, sometimes not until early spring). Size difference gives some guidance. – Moults. Complete post-nuptial moult of ad in (late Jul) Aug–Oct. Partial post-juv moult in late summer or early autumn does not involve flight-feathers, tail-feathers, primary-coverts or outer greater coverts. No pre-nuptial moult other than odd birds, presumably 1stW, replacing scattered feathers on head, central tail-feathers and odd tertials. – **SPRING** ♂♀ Sexing usually difficult (except least-neat birds being ♀♀). Lower rump chrome-yellow, partially unstreaked. Lesser coverts near wing-bend rather uniformly orange-yellow, with only small dark spots visible. (Most attractive birds not only ♂♂.) Size gives some guidance for handled birds. ♀ A very few are separable: dark greyish rear crown streaked black; red patch on forecrown small; lower rump boldly streaked dark; lesser coverts dull yellowish with prominent dark spotting. **F.g.** No contrast in wear or colour visible among greater coverts, tertials or tail-feathers. Tips to tail-feathers slightly rounded, neatly edged. **1stS** Not always separable, but some have remnants of rufous-brown hood of juv plumage, and can show slight contrast in wear and colour of inner moulted and outer unmoulted greater coverts, and rarely a mixture of retained juv (more worn) and odd new (freshly-moulted) tertials and tail-feathers. Tips of tail-feathers on average narrower and more pointed than in ad, but intermediates occur, and wear makes difference progressively less obvious. – **AUTUMN Ad** Throat and sides of head black or blackish, forecrown with red patch of variable size. Sexes

Ad, presumed ♂, Turkey, Jul: solid black hood, boldly black-streaked breast and flanks, and large, bright red forecrown patch identify to species, and make ♂ sex most plausible. Note mustard-yellow shoulder patch and panel on flight-feathers, and hint of pale wing-bars. Aged by evenly-feathered wing and dark-looking primary-coverts. (A. Halley)

FINCHES

Ad ♀, **Turkey**, Jun: sexed by combination of being ad with rather small red forecrown patch, browner and less well-defined hood, browner and less distinct underparts-streaking on duller ground, less bold wing markings, and having dark-streaked shoulder patch with limited mustard-yellow. (S. Olofson)

1stS, presumed ♂, **Armenia**, May: once aged as 1stS (note obvious moult limit in greater coverts), then blackness of throat, degree of unstreaked mustard-yellow on shoulder patch, and chrome-yellow fringes of primaries and tail-base indicate ♂, although the forehead patch is rather small, and indeed most single birds cannot be safely sexed. (S. J. M. Gantlett)

Ad ♀, **Iran**, Apr: many keep fresh winter plumage well into spring, especially the broader fringes above and very spotty lower throat and breast. All visible wing-feathers are evenly fresh, thus an ad, and knowing this, the sooty-brown head, the small forecrown patch and the underparts pattern fit only with ♀. A 1stS ♀ is visible in the background. (M. Nemati)

Ad, presumed ♂, **Kazakhstan**, Jan: when fresh, buffish-grey fringes partly conceal dark centres above, obscuring also bold streaking on chest, while black of head shows fine greyish tips. Ageing straightforward by ad greater coverts and primary-coverts, but sex can only be guessed based on dark-looking and extensive hood. (A. Isabekov)

Presumed 1stW ♂, **India**, Jan: some 1stY can be very similar to ad, as here. The rather large, reddish forecrown patch and extensive dark hood, the bold breast-streaking as well as the quite obvious mustard-yellow on wing and undertail-coverts seem to best fit 1stW ♂. However, closer inspection of the wing would be required for confirmation. (D. Ash)

365

1stW, Israel, Nov: least-advanced young birds in winter have brown hood with variable amount of black admixed on throat and nape, and still lack the red patch on forecrown. Aged—apart from the typical brown head—by juv primary-coverts and pointed tail-feathers, as well as by obvious moult limit in greater coverts. (A. Ben Dov)

Presumed 1stW ♀, Turkey, Sep: another poorly advanced 1stW, also with rufous-tinged head of juv, lacking the red forecrown patch. Although sexing of young birds is normally impossible, this one is at the nondescript end of variation, hence presumably a ♀. (G. S. Bhardwaj)

alike, although a few ♀♀ separable having poorly-developed black and red head/throat pattern, being greyish and dark-streaked rather than uniformly black on rear crown, and having small red patch on forecrown. Wings and tail worn in late summer, in moult by early autumn, then mint (fresh) during much of winter. **Juv** Easily aged by lack of black and red on head/throat, these colours largely replaced by a rufous-ochre hood. Wings and tail fresh in summer, then progressively more abraded in winter.

BIOMETRICS L 12–12.5 cm; **W** ♂ 74–83 mm (n 18, m 77.6), ♀ 70–78.5 mm (n 14, m 74.2); **T** ♂ 52–61 mm (n 18, m 56.4), ♀ 51–57 mm (n 14, m 54.0); **T/W** m 72.4; **B** 7.2–8.8 mm (n 35, m 7.8); **BD** 5.3–6.7 mm (n 34, m 5.9); **Ts** 13.5–15.4 mm (n 31, m 14.2). **Wing formula: p1** minute; **pp2–4**(5) about equal and longest; **p5** < wt 0–1.5 mm; **p6** < wt 5–6.5 mm; **p7** < wt 10–12 mm; **p10** < wt 18–21.5 mm; **s1** < wt 20–22.5 mm. Emarg. pp3–5.

GEOGRAPHICAL VARIATION & RANGE Monotypic. – Turkey, Caucasus, Transcaucasia, Iran, mountains of Central Asia; mainly sedentary but some altitudinal movements in winter to lower levels or other short-range movements.

Juv, Kyrgyzstan, Jul: rufous-brown head and buffish-yellow wing panel on primaries, typical broad cinnamon-buff wing-bars and rather dusky, profusely and diffusely dark-streaked underparts. Fluffy body-feathers obvious in particular on shoulders and underparts. (T. Lindroos)

(EUROPEAN) SERIN
Serinus serinus (L., 1766)

Fr. – Serin cini; Ger. – Girlitz
Sp. – Serín verdecillo; Swe. – Gulhämpling

A small finch that is easiest to see in the Mediterranean area, but which is widely distributed over much of Europe except Fenno-Scandia. It may look like a newly-fledged young bird all its life because of its cutely rounded head and small, stubby bill. Typical bird of churchyards and gardens with ornamental cypresses, where it sings its characteristic song from a high post or embarks on a song-flight with slow-motion wingbeats, revealing its yellow rump patch.

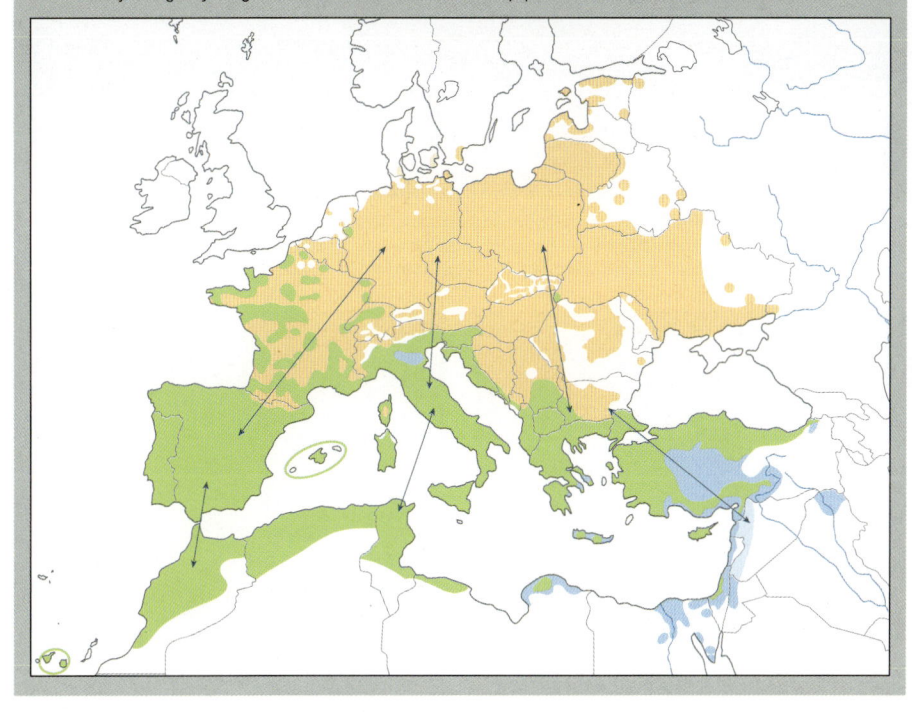

Ad ♂, Spain, May: often perched with slightly drooped wings, e.g. when singing, to show bright yellow rump. With wear, dark areas of head fade to olive, making it look even yellower. Evenly-feathered wings, and rather broad and still fresh rectrices infer age. (A. Torés Sanchez)

IDENTIFICATION A *small, heavily streaked* finch with at least some yellow in plumage. *Bill is horn-coloured, short but deep*, appearing stubby and slightly rounded in outline. Legs pinkish. In all plumages has a *uniformly coloured cheek patch*, olive-green or greyish, *with an isolated pale spot in its lower part. Tail deeply forked*, like in most finches, and also *all dark*, lacking white or yellow sides or bases (as in Linnet, Siskin and Greenfinch). Adult ♂ has *clear yellow forecrown, supercilium, neck-sides, throat, breast and rump*, whereas *hindcrown is rather uniform olive-green* with only faint dark streaks. The ♀ is less clear yellow, more pale greenish-yellow, and both crown and breast tend to be more prominently streaked than in ♂. The juvenile is more buff-brown and has very little yellow in plumage, but recognised by general size and shape, bill shape, streaking and head pattern. Rump is buffish-white and streaked, but streaking is usually a bit lighter than on back. – A lively bird, always dashing off in bounding flight. Often seen in twos or small family parties (but rarely also in larger flocks) like other closely-related small finches. Sings from elevated song post in tall tree, on a wire or in butterfly-like song-flight with slow-motion wingbeats.

VOCALISATIONS Vocal. Song is a peculiar, very fast and high-pitched jingle of sharp, squeaky and 'electric' (sizzling) notes, like the sound from walking on crushed glass over a marble floor. It is characteristic that the pitch is rather even throughout the song (unlike in Siskin and Citril Finch), without any inserted drawn-out squeaky, nasal notes. – Call is a 'bouncing', high-pitched trill with 'crystal-clear' voice reminiscent of the song, *zir'r'r'rl*, often uttered in flight or just after alighting. Other calls are an upwards-inflected, nasal *dvueeh*, sometimes with a short initial note added, *de-dvueeh*. Many other calls described but they are more rarely heard, are variable and add little useful to the identification process.

SIMILAR SPECIES General size and colours resemble Siskin, but that species differs in its longer and more pointed bill, yellowish sides to tail, in ♂ plumage all-dark crown

♂, presumed ad, Italy, Apr: stubby bill and round head, latter with rather complex plumage pattern; note well-defined yellow forehead, supercilium and ear-coverts surround, and dusky moustachial-stripe. Also vividly yellow throat, crown and breast, and double wing-bars. Diagnostic yellow rump just visible. Seemingly evenly-feathered wing suggests ad. (L. Sebastiani)

♀, Spain, Apr: unlike ♂, yellow much reduced and less vivid on face and central breast, and underparts streaking is browner, but streaking covers greater area of breast and flanks; crown also clearly streaked. (R. Debruyne)

(lacking the yellow forecrown) and in most plumages a rather striking yellow wing-bar. – In Levant must be separated from *Syrian Serin*. The latter is much less streaked (can appear nearly unstreaked) and has yellowish sides to tail. It also lacks pale surround to the darker cheek patch so typical of Serin. – Overlaps with *Citril Finch* in many parts of S Europe, but this is a more slim-shaped and less heavily-streaked bird, with usually much broader and more prominent yellowish-green double wing-bars.

AGEING & SEXING Ages differ in summer–autumn, and often in spring too if seen well or handled. Sexes differ following post-juv moult. Size difference very slight. – Moults. Complete post-nuptial moult of ad mainly in late Jul–Oct (Nov). Partial post-juv moult in late summer or autumn (protracted due to long breeding season) usually does not involve flight-feathers, tail-feathers, primary-coverts or outer greater coverts, but some early-hatched birds in S Europe may moult some primaries, most or all greater coverts, and some tail-feathers (commonly only r1) and tertials. (Claims of completed full moult in 1stY require confirmation.) Extent of pre-nuptial moult not well known, apparently usually no moult at all. – **SPRING** ♂ Forecrown and supercilium unstreaked bright yellow (or yellow with slight greenish

Ad ♀, Spain, May: some ♀♀ wear much duller, barely showing any yellow except on face and breast, but instead whole plumage heavily and densely streaked; yellowish dusky-white wing-bars more contrasting. Compared to Siskin, Serin has much shorter and stubbier bill, and lacks bright yellow-and-black wing-bars. Evenly-feathered wing infers age. (C. N. G. Bocos)

♂, presumed 1stS, Spain, Jun: like ad ♂ but this bird approaches ad ♀ in having slightly smaller yellow areas, and some bold breast streaking. Nevertheless, the amount of yellow on face and rump excludes any ♀, but certain ageing as 1stS would require handling (still, apparent moult limit in wing with juv alula and primary-coverts). (C. N. G. Bocos)

1stS ♀, Spain, May: best ageing characters are retained juv primary-coverts and tail (abraded and pointed). However, differences between ad and 1stY ♀♀ less obvious. Note ill-defined yellowish frame to ear-coverts, isolated pale spot on lower cheeks, spotty marks on lower throat and upper breast, longitudinal flank-streaks, and greenish-tinged rump/uppertail-coverts. Wing-bars quite pronounced. (C. N. G. Bocos)

Ad ♂, Spain, Dec: when very fresh, yellow face and throat markings reduced due to obscure olive-grey tips, with dark streaks both above and below. Still unstreaked yellow forecrown and throat of ♂ spring plumage readily visible inferring sex. All wing ad. (M. Varesvuo)

tinge), yellow supercilium continuing around mainly greyish-green ear-coverts. Throat and breast same, unstreaked. Rest of crown grey-green, unstreaked or only faintly streaked. ♀ Crown, throat and breast greenish-white with yellowish tinge (but some have stronger yellow colour, approaching least-yellow ♂♂). Whole crown prominently streaked dark (very rarely streaks thinner, less prominent). **F.g.** No contrast in wear or colour visible among greater coverts, tertials or tail-feathers. Many central greater coverts have broad yellowish-green tips. Tips to tail-feathers rather broad and slightly rounded, neatly edged until early spring, later moderately abraded. **1stS** Not always separable, but some can show slight contrast in wear and colour of inner moulted greater coverts (tipped yellowish-green) and outer unmoulted (more narrowly tipped off-white). Tips of tail-feathers on average narrower and more pointed than in ad, often worn and pointed from late winter and usually heavily abraded from May, but wear makes difference between age categories progressively less obvious. – **AUTUMN** Sexing as in spring, but yellow on forecrown of both sexes completely or partly concealed by olive-grey feather-tips. Amount of streaking on crown best clue. **Ad** No contrast in wear or colour visible among greater coverts, these being similarly dark-centred and tipped/edged yellowish-green. Tips to tail-feathers rather broad and slightly rounded, neatly edged. **1stW** Slight contrast in wear and colour of some (often many) inner moulted greater coverts with slightly darker brown-grey centres and yellowish-green tips/edges, and outer unmoulted subtly paler-centred and edged brownish-white or off-white (rarely yellowish-buff). Tips to tail-feathers on average narrower and more pointed than in ad. **Juv** Easily separated by lack of nearly all yellow and green colours on head and body, these being largely replaced by buff-white, grey-brown and rufous, with only faint yellow cast to breast and belly. Also, tail-feathers edged greenish-yellow. Whole plumage prominently streaked, including rump and breast. All greater coverts tipped and edged rufous-buff.

BIOMETRICS L 11–12 cm; **W** ♂ 68–77 mm (n 23, m 73.2), ♀ 67–74 mm (n 15, m 68.8); **T** ♂ 45–53 mm (n 22, m 48.8), ♀ 43.5–50 mm (n 15, m 46.2); **T/W** m 66.9; **B**(f) 6.5–8.3 mm (n 34, m 7.4); **BD** 5.4–6.9 mm (n 31, m 6.0); **Ts** 13.0–14.5 mm (n 31, m 13.7). (Bill to skull often difficult to measure and therefore not included.) **Wing formula:** p1 minute; p2 < wt 0.5–1.5 mm; pp3–4(5) about equal and longest; p5 < wt 0–2.5 mm; p6 < wt 6–10 mm; p7 < wt 10–14.5 mm; p10 < wt 16–21 mm; s1 < wt 18–23 mm. Emarg. pp3–5.

GEOGRAPHICAL VARIATION & RANGE Monotypic. — Europe (except Britain and much of Fenno-Scandia), including Baltic States, Belarus, Ukraine, W Romania, Bulgaria; Turkey, Cyprus, N Africa; sedentary. — (Syn. *germanicus* [or '*woltersi*' depending on generic arrangement], which was separated on account of very minor size differences, but is too subtle to be accepted. Another synonym is *meridionalis*.)

REFERENCES ROHNER, C. (1981) *Orn. Beob.*, 78: 1–11.

Ad ♂, Italy, Feb: by late winter, upperparts of ♂ already abraded but yellow face and rump not very bright, so many look predominantly greyish. Aged by evenly-feathered wing. (D. Occhiato)

Ad ♀, Spain, Dec: autumn ♀♀ have dullest but most profusely-streaked plumage. Identification rests on stubby bill, typical head profile, plus hint of typical head pattern with isolated pale spot on lower cheeks. Whitish-cream wing-bars, greenish-yellow rump (barely visible) and bold flanks-streaks also typical. Ad by evenly-feathered wing. (M. Varesvuo)

1stW ♂, Spain, Dec: very similar to ad ♂, and young that have moulted all their greater coverts, as seems the case here, are difficult to age, but primary-coverts are juv. Broad dusky streaking on buffish flanks characterise this age. (M. Varesvuo)

1stW ♀, Israel, Nov: young ♀ could be mistaken for similar-plumaged Siskin, but bill never as pointed, and wing-bars not bright yellow-and-black. With experience, face pattern can separate the two at a glance. Combination of dullest plumage, with fewest yellowish elements, obvious brownish tinge, and moult limits among greater coverts infer sex and age. (A. Ben Dov)

Juv, Spain, Jun: yellow parts replaced by ochre-buff, and lacks unstreaked yellow rump. Already a hint of characteristic head pattern discernible. (A. M. Domínguez)

SYRIAN SERIN
Serinus syriacus Bonaparte, 1850

Fr. – Serin syriaque; Ger. – Zederngirlitz
Sp. – Serin sirio; Swe. – Levantsiska

Possibly one of the least-known species in the covered region. It has a very limited range, confined to vegetated lower mountains and slopes in Syria, Lebanon, N Israel and Jordan. Many are sedentary, but a few move south to S Israel and Sinai. Rare or at least rather scarce, which together with its seemingly erratic movements, makes this a rather difficult species to get to know.

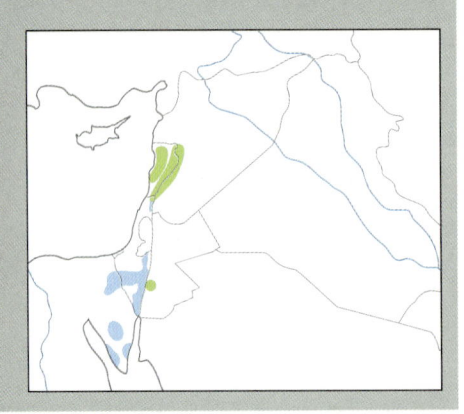

Ad ♂, Israel, Jun: compared to European Serin plumage slightly duller yellow-green and greyer, but above all almost unstreaked. ♂ in summer has rich yellow forehead and deep yellow throat, accentuating grey-tinged hindneck and cheeks. Wing-feathers extensively edged greenish-yellow. Primary-coverts edged bright greenish-yellow infer age. (A. Ben Dov)

Ad ♀, Jordan, Mar: note (individually variable) yellow patch encircling dark eye. Combination of pale brown-grey upperparts with ill-defined dark streaking, some faint streaking on ear-coverts and crown, limited yellow on forehead indicate ♀. Ad tail-feathers with rather rounded tips, and all visible wing-feathers are ad too. (H. Nussbaumer)

IDENTIFICATION Somewhat *larger and slimmer than Serin* with proportionately *slightly longer tail* (thus roughly like Red-fronted Serin in size and structure). *Bill slightly longer than in Serin*, dark grey, and *culmen is convex*. A mainly greenish-yellow bird with elements of purer yellow and subdued brown and grey. Could be compared with Citril Finch, but *lacks the contrasting broad yellow-green double wing-bars* of that species (much of the wings are greenish), and *tail has yellowish sides* lacking in Citril Finch. Most notable feature in both sexes, but more prominent in ♂, is *deep yellow forecrown and eye-surround*. Behind greyish cheeks there is often a *yellowish-ochre patch on neck-side*. Throat pale yellow in ♂, more dull greenish-yellow in ♀. Much of wing-bend and wing-coverts greenish-yellow in ♂. *Rump unstreaked pale yellow-green*, obvious in flight. *Mantle dull greyish-olive and very lightly streaked dark* in ♂, more prominently streaked in ♀. *Flanks tinged greyish*, unstreaked in ♂ and diffusely streaked dark in ♀. Juvenile is tinged rufous-cinnamon on wing-coverts, neck and head, but has the greenish-yellow on wings and tail of adults.

VOCALISATIONS Resembles those of Citril Finch and Red-fronted Serin, but appear to be consistently lower-pitched and perhaps slightly slower than either. Song structure can recall passages of any of Citril Finch, Corsican Finch and Goldfinch, a continuous hurried series of trills, musical whistles and jingling notes. Just as in Goldfinch song, one or two strained and nasal, buzzing notes are inserted into the strophe. The voice is clearly slower, lower-pitched and less sharp than Serin's. – Call is a 'bouncing', merry, short trill, on the whole lower-pitched than Serin or Red-fronted Serin, *pe-re-ret* or *ter'r'r*, sometimes with initial note subtly higher. There is some variation in delivery, some trills being more nasal, others higher-pitched and metallic. A nasal *pe-tchee* is also sometimes heard.

SIMILAR SPECIES Considering its range, the only realistic confusion species is *European Serin*. Syrian Serin is much less boldly streaked, especially on the underparts, and lacks the obvious pale cheek patch-surround of European Serin (most obvious in ♂, but a hint of this pattern visible in most Europeans). Syrian Serin has yellowish tail-sides, while European has all-dark tail-feathers. Syrian Serin is a slightly larger species with a marginally longer tail.

AGEING & SEXING Ages differ in summer–autumn, and sometimes in spring too if seen well or trapped. Sexes differ following post-juv moult. Size difference slight. – Moults. Complete post-nuptial moult of ad mainly in Jul–Sep. Partial post-juv moult in early autumn usually does not involve flight-feathers, tail-feathers or primary-coverts; most or all tertials and some inner greater coverts are normally replaced, at times all greater coverts, and sometimes r1. No pre-nuptial moult recorded. – **SPRING** ♂ Forehead and eye-surround deep (golden-) yellow. Throat, breast and belly rather clear yellow-green. Flanks and belly unstreaked (or very nearly so). Mantle moderately streaked dark. ♀ Like ♂ but rather paler and duller yellow-green on head, throat and breast, with pale brown cast on breast. Lower flanks and belly usually thinly and diffusely streaked. Streaking on mantle heavier than in ♂. **F.g.** No contrast in wear or colour visible among greater coverts, tertials or tail-feathers. Tips to tail-feathers rather broad and slightly rounded, neatly edged. **1stS** Frequently not separable, but some may show slight contrast in wear and colour of inner moulted (broadly tipped and edged yellowish-green) and outer unmoulted greater coverts (more narrowly tipped and edged off-white). Tips to tail-feathers on average narrower and more pointed than in ad, but wear makes differences progressively less obvious, and r1 might be moulted to round-tipped ad type. – **AUTUMN** Sexing as in spring, but yellow or yellow-green on head of both sexes subdued and partially concealed by sandy-grey feather-tips. **Ad** No contrast in wear or colour visible among greater coverts, these being largely yellowish-green with pale ash-grey tips. Dark primaries neat and rather broadly tipped whitish. Primary-coverts rather broad and dark, edged bright yellowish-green. Central alula broadly edged yellow-green (cf. Lehnardt et al. 2012). Tips to tail-feathers rather broad and slightly rounded, neatly edged. **1stW** Slight contrast in wear and colour of some inner moulted mainly yellowish-green greater coverts, and outer unmoulted edged buff-white or off-white. Primary tips worn and bleached brownish with minimal remains of

1stS ♂, Syria, May: unstreaked rich yellow rump. Best to age first, since 1stS ♂ often appears intermediate between the sexes. Here, condition and shape of tips of primaries and tail-feathers suggest young age. Also, age secured by differences in pattern of freshly moulted tertials (right wing), and retained juv ones (left). (A. Audevard)

— 370 —

FINCHES

1stS ♀, Syria, May: least amount of yellowish-green in spring and summer. Unlike ♂, tinged more grey or dull greyish-brown, with more distinctly dark-streaked mantle, scapulars and flanks, and more limited yellow on face. Also limited and paler yellow-green in wings, and no or subdued shoulder patch, greater coverts bar and remiges panel. Aged by juv tertials and primary-coverts. (A. Audevard)

Ad ♂, Israel, Dec: in winter, yellow areas partly concealed, with variable narrow streaks on flanks, and greyish tips to wing-coverts. Sexing can be difficult at this season, but here large and bright yellow face infers ♂; evenly-feathered wing (with broad whitish primary-tips) indicates age. (H. Shirihai)

Ad ♀, Israel, Dec: visibly drabber with more green-tinged yellowish areas than ad ♂ (except bright lemon face), and many also have stronger flanks streaking. Aged by evenly-feathered wing and broad whitish primary-tips. (H. Shirihai)

1stW ♂, Israel, Dec: as fresh ad ♂ but approaches ad ♀ in reduced yellow on face and greener wing and rump, but separable when aged first (tips of retained juv primaries already worn, without broad whitish edges of ad, and only tiny greyish tips left). (H. Shirihai)

1stW ♀, Israel, Dec: overall dull and greyish with strong upperparts streaking and limited yellow face sex this bird, which is reliably aged by retained juv and bleached brownish wing, with hardly any whitish tips. Interestingly, this bird has moulted all of its tail. (H. Shirihai)

Juv, Israel, May: very distinctively tinged rufous-cinnamon, especially on wing-coverts, shoulders and face, but wings and tail edged greenish-yellow. Otherwise mostly brownish; tertials edged drab-brown. Head- and body-feathers rather loose and fluffy. (L. Kislev)

paler tips. Primary-coverts and alula brown-grey, former rather narrow, both with limited dull green edges. Tips to tail-feathers on average narrower and more pointed than in ad. **Juv** Easily separated by being tinged brown on head and body. Tertials edged brown. All greater coverts edged rufous and broadly tipped buff-brown. Body-feathers rather loose and 'fluffy', especially on nape and vent.

BIOMETRICS L 12–13 cm; **W** ♂ 72.5–80.5 mm (n 12, m 77.7), ♀ 71–77 mm (n 8, m 74.6); **T** ♂ 50–59 mm (n 12, m 54.6), ♀ 50–55 mm (n 8, m 53.1); **T/W** m 70.9; **B** 8.6–11.0 mm (n 21, m 10.1); **B**(f) 7.1–8.0 mm (n 12, m 7.7); **BD** 5.9–7.0 mm (n 21, m 6.3); **Ts** 13.5–15.0 mm (n 19, m 14.3). **Wing formula: p1** minute; **pp2–4**(5) about equal and longest; **p5** < wt 0–2.5 mm; **p6** < wt 6–9.5 mm; **p7** < wt 12–14.5 mm; **p10** < wt 19–24 mm; **s1** < wt 21–25 mm. Emarg. pp3–5.

GEOGRAPHICAL VARIATION & RANGE Monotypic. – Syria, Lebanon, N Israel, Jordan; resident, or makes short-range movements only, recorded in Sinai. (Claimed records from Iraq based on misidentified specimens; Porter 2014.)

REFERENCES Khoury, F. (1998) *Sandgrouse*, 20: 87–93. – Lehnardt, Y., Yosef, R. & Perlman, G. (2012) *DB*, 34: 77–84. – Porter, R. F. (2014) *Sandgrouse*, 36: 58–60.

(ATLANTIC) CANARY
Serinus canaria (L., 1758)

Fr. – Serin des Canaries; Ger. – Kanariengirlitz
Sp. – Serin canario; Swe. – Kanariesiska

Everybody knows what a canary is, and that the original wild bird from which the cagebirds are derived lives on the Canary Islands seems logical. It is also found on Madeira and the Azores, but nowhere in Africa or Europe. It is thus a true island bird, being resident all year. Found in all kinds of open habitats with scattered bushes, trees or woods on mountain slopes but also at lower levels right down to the sea. Often first noticed by virtue of its well-known and pleasing song being delivered from high vantage point.

♂, **Madeira, Apr**: quite Serin-like face markings with yellowish forehead and supercilium, and sometimes, like here, greyish-green surround to ear-coverts. Always has rather clear dark moustachial-stripe and isolated yellow lower-cheeks mark like Serin. (H. Shirihai)

Ad ♂, **Canary Is, Aug**: larger and longer-tailed, with stronger bill than Serin, if superficially quite similar. Unlike Serin, note greyer upperparts and yellow extending further onto belly. Pattern of head and wing vary greatly. Only some older ♂♂ have yellow shoulder patch solid and contrasting, but diagnostic if visible. All wing-feathers ad. (B. Rodríguez)

IDENTIFICATION The Canary can be described as a mixture of a Serin and a Citril Finch. It is somewhat *larger and longer-tailed*, and more elongated in shape than a Serin, more like a Citril Finch, but its colours most resemble those of Serin. Cagebird canaries are often colour manipulated and can be clear yellow or white. But the real thing is a rather *nondescript yellowish-green* bird with *moderate streaking on back and flanks*. *Bill is strong and stout*, not quite as short as in Serin though, and often slightly *pinkish-tinged*. Legs, too, are dull pinkish. *The long tail is all dark* like Serin. *The rump can flash yellowish* in flight, but is more streaked and dull compared to Serin (especially dull in ♀ Canary, often showing no or only very little contrast with the back and uppertail). Double wing-bars rather narrow, both tinged yellowish. *Head pattern* mirrors that of Serin, only is *less contrasting*, more greyish-green and subdued, and ♂ lacks the strong contrast of Serin between dark crown and yellow forehead. Juvenile lacks all or nearly all yellow tones, are more buff-brown and evenly streaked, still identified by shape and calls, in addition to range.

VOCALISATIONS Song is loud, pleasing and clearly structured, resembling that of Goldfinch with same pace and a similar mixture of muffled trills, nasal discordant notes, rapid twitter and high-pitched whistles. – Has a variety of calls. Commonly heard from overflying parties, or from an alighting bird, is a slightly descending and tremulous *ti-ti-türr*. Another is an upwards-inflected Greenfinch-like *djueet*. Sometimes utters a more Serin-like 'silvery' trill, *svivivivi*, or a Siskin-like desolate *teeü*, falling in pitch.

SIMILAR SPECIES Considering that the Canary is resident on the North Atlantic islands, with no related finches within hundreds of miles, it can hardly be confused with any other species. The hypothetical vagrant *Serin* would be readily separated by its shorter tail and bill, its more compact appearance, stronger streaking on back and flanks, and by its calls.

AGEING & SEXING Ages differ in summer–autumn, and often in spring too if seen well or trapped. Sexes generally differ following post-juv moult, but a few are difficult. – Moults. Complete post-nuptial moult of ad in late spring or summer, after early breeding. Partial post-juv moult in summer or early autumn (details poorly known) usually does

1stS ♂, **Madeira, Mar**: on some, head may appear largely greenish-yellow, with less clear pattern. Note also quite greyish mantle and back, distinctly streaked dark. Outer three greater coverts retained juv, more whitish-edged than moulted inner, and tail-feathers narrow and pointed inferring age.

not involve flight-feathers, tail-feathers (except perhaps r1) or primary-coverts. Apparently no pre-nuptial moult. – **SPRING** ♂ Ground colour of much of head (forehead, forecrown, supercilium, eye-surround, base of bill, chin) bright yellow-green. Throat/breast same, unstreaked. Rest of crown and nape grey-green or brownish-grey, unstreaked or only faintly streaked. Rump rather unstreaked and bright yellowish-green. Undertail-coverts white with some clear yellow tones, unstreaked (or at most with diffuse smudge on longest). ♀ As ♂ but yellow-green duller and usually with faint brown-grey tinge, nearly always lightly streaked. Crown and nape rather prominently streaked dark. Rump duller and more streaked than in ♂. Undertail-coverts dirty-white with limited yellow tinge and invariably has prominent dark streak on longest. **F.g.** No contrast in wear or colour visible among greater coverts, tertials or tail-feathers. **1stS** Not always separable, but some may show slight contrast in wear and colour of inner moulted (more broadly tipped yellowish-green) and outer unmoulted greater coverts (more narrowly tipped off-white), though wear in spring often makes field assessment difficult. Tips to unmoulted outer tail-feathers on average narrower and more pointed than in ad, but wear makes difference progressively less obvious. – **AUTUMN** Sexing as in spring, but yellow on forecrown of

♂, presumed 1stS, Madeira, Apr: separated by range from Serin, but note that some duller ♂ Canaries can be more similar to that species. However, unlike Serin, facial pattern less bold, tail is proportionately longer, and bill is stronger and appears to be paler, almost pinkish-grey. Ageing impossible in this view, but whiter belly usually indicates 1stY. (H. Shirihai)

♀♀, Canary Is, Jul (left) and Jan: ♀ usually clearly greyer, especially on crown, neck and upper breast, with less extensive or bright yellow on face and underparts, heavier grey-brown streaking on upperparts, and duller, greener and more streaked rump. Much variation in face pattern and underparts streaking, irrespective of age. Ageing impossible without closer views of wings. (Left: J.-M. Fenerole; right: M. Strazds)

Ad, presumed ♂, Tenerife, Canary Is, Sep: autumn plumage seems perfectly fresh without moult limits, and primary-coverts are dark and neatly edged, hence an ad. Sexing slightly more difficult, but amount of yellow on forehead, throat and centre of underparts infer a ♂. (H. Shirihai)

Ad ♂, Madeira, Nov: when very fresh, yellow on face limited to supercilium, forehead, fore ear-coverts and chin, with instead much grey on unstreaked crown, rear ear-coverts, nape and even upper breast. Rump is most obvious yellow area, while yellow on underparts is mixed with grey. Evenly-fresh wing and tertial pattern with neat dark centres. (D. Occhiato)

1stW ♂, Azores, Oct: as ad in autumn, but has browner upperparts, more grey on face and some broad blotchy dark marks on lower flanks. Also aged by clear moult limits in greater coverts. Amount of yellow, including on lower breast and flanks, should eliminate ♀. (D. Occhiato)

Ad ♀, Tenerife, Canary Is, Sep: there is some variation in ♀ plumage, even among ad, and some, like here, can appear rather dull. Still, whole wing is neat and fresh, and primary-coverts are rather broad and rounded at tips with dark centres inferring ad. (H. Shirihai)

1stW ♀, Azores, Oct: young autumn ♀ has dullest plumage, with least amount of yellow on face and below, and has some fine streaking on crown. Unlike equivalent age and sex of Serin, head, face and underparts are plainer and less streaked. Rump on the other hand can show less greenish and yellow, but more streaks. Age confirmed by moult limit in greater coverts. (D. Occhiato)

both sexes slightly duller or partially concealed by olive-grey feather-tips. **F.g.** No contrast in wear or colour visible among greater coverts, these being tipped/edged yellowish-green. Tips to tail-feathers rather broad and slightly rounded, neatly edged. **1stW** Often a slight contrast in wear and colour of several inner moulted greater coverts edged yellowish-green, and a few outer unmoulted edged brownish-white or off-white. Tips to tail-feathers on average narrower and more pointed than in ad. **Juv** Easily separated by lack of nearly any yellow and green on head and body, these being largely replaced by buff-white and grey-brown, with no or only faint yellow cast. Upperparts prominently streaked, including rump, and breast finely streaked. All greater coverts tipped and edged buff.

BIOMETRICS L 12–14 cm; **W** ♂ 71–78 mm (n 31, m 73.7), ♀ 69–74 mm (n 30, m 71.0); **T** ♂ 53–60 mm (n 31, m 56.0), ♀ 52–58 mm (n 30, m 54.8); **T/W** m 76.6; **B** 8.2–10.2 mm (n 60, m 9.0); **BD** 6.4–7.5 mm (n 52, m 6.9); **Ts** 15.9–18.0 mm (n 60, m 17.0). **Wing formula: p1** minute; **p2** < wt 0.5–1.5 mm; **pp3–5** longest, but p5 sometimes < wt 0.5–1 mm; **p6** < wt 4–6.5 mm; **p7** < wt 9–11.5 mm; **p10** < wt 16–19 mm; **s1** < wt 16.5–20.5 mm. Emarg. pp3–5, also usually slightly or clearly on p6.

GEOGRAPHICAL VARIATION & RANGE Monotypic. – Canary Is, Madeira, Azores; resident.

1stW ♀, Azores, Oct: as image above, but body not fully moulted, thus no yellow visible yet. Rather nondescript, but perhaps best recognised on general shape and size, and considering locality. All wing-feathers juv. (D. Occhiato)

Juv, Canary Is, Aug: drabber than later plumages with rufous-tinged grey-brown upperparts, head and chest, lacking any yellowish or greenish tones, with brownish-white belly and extensive dark streaking. (B. Rodríguez)

ARABIAN SERIN
Crithagra rothschildi (Ogilvie-Grant, 1902)

Alternative name: Olive-rumped Serin

Fr. – Serin d'Arabie; Ger. – Olivbürzelgirlitz
Sp. – Serín árabe; Swe. – Arabsiska

Breeds in bushy montane areas from W Saudi Arabia south to Yemen, from about 1000 m (or somewhat lower) to at least 2800 m. Unlike its close relative and most likely confusion species, Yemen Serin, this species is almost never found in towns and villages, and prefers to perch on bushes, not on stony ground. Previously commonly included in the genus *Serinus*, recent research has shown that arrangement to be polyphyletic and in need of revision, thus together with Yemen Serin the species is now placed in *Crithagra*.

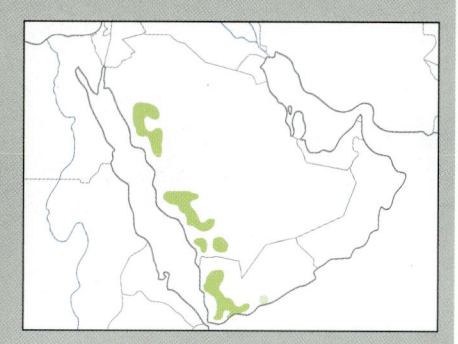

Yemen, Nov: overall drab-coloured, with no sexual differences, but some variation in rump coloration and streaking. Within range could only be confused with Yemen Serin but has stouter, more decurved bill, decidedly darker, smokier grey-brown upperparts, plainer face and yellowish-olive rump. Seemingly all wing ad. (W. Müller)

IDENTIFICATION Very nondescript, overall brownish and rather dingy-looking, relieved only by the *greenish-yellow rump*. It has a very faint *pale supercilium*, only slightly *mottled ear-coverts and throat-sides*, and slightly *buffish chin and throat*, contrasting very slightly with the otherwise greyish-white underparts. *Throat, chest and flanks are finely but densely streaked dark*. A relatively short wing, and the longish tail is (sometimes) accentuated by the species' habit of wagging it. Head almost unstreaked (or very faintly streaked) and somewhat greyer, while rest of upperparts are streaked dull brownish. Subdued pale or whitish wing-bars and tertial edgings are moderately marked but never distinctly patterned. *Deep-based, conical bill usually greyish or horn-coloured* and rather bulbous, giving head a Bullfinch-like appearance. Small eye is dark, and legs generally pinkish. Prefers cultivated areas in highlands, with reasonable cover, as well as rocky areas with grassy ground.

VOCALISATIONS Song described as variable but usually has a plaintive ring, has been compared to that of Linnet, but also with a weak Common Rosefinch due to its melodious quality, short strophes with brief pauses, e.g. *zi juu chi-chi chichi… ti-tiiu-tui-tsu…*, etc. (Clement et al. 1993). Variation applies mostly to number of syllables per strophe, and longer song transcribed *seeoo tee-teacher, seeoo teete-teacher, seeoo tee tee seeoo*, occasionally culminating in a long and varied musical jingle with a warbled quality. – Call is a quiet *tsee-tsee* or *dlit-dlit*, often given in flight (Phillips 1982, Everett 1987, Hollom et al. 1988).

SIMILAR SPECIES Closely recalls Yemen Serin but has larger, stouter and more bulbous bill, which is more uniformly greyish, lacking contrast between dark upper and pale lower mandibles. Overall a slightly darker bird, more smoky grey-brown, with darker cheeks and breast area, and has (at least in ♂♂) yellowish-olive rump (absent in Yemen Serin, but can be absent or difficult to see even in present species). Upperparts (including nape and upper mantle) are more uniformly dull grey-brown (streaking often difficult to see unless close), whereas Yemen Serin has slightly paler nape and upper mantle with visibly more contrasting streaking. Underparts are darker and more sullied mustard-drab, whereas Yemen Serin is slightly whiter with narrower and better-defined streaking. Ear-coverts tend to be plainer (lacking pale mark below and behind eye of Yemen Serin). Absence of moustachial stripe also useful. Usually possesses only a faint pale supercilium (somewhat more marked in Yemen Serin). Primary projection is distinctly shorter, *c.* 1/3 of tertial length (*c.* 2/3 in Yemen Serin). Less gregarious than Yemen Serin. Frequently perches in trees. Often flicks wings and wags tail. Overall much less flighty and active than Yemen Serin.

AGEING & SEXING Ageing probably possible using moult limits in autumn but little studied; at least juv separable. Sexes alike. (Sometimes stated that ♀♀ have buffier, less olive, upperparts, but this could not be confirmed on available specimens.) Limited seasonal plumage variation. – Moults. Ad undertakes complete post-nuptial moult, while post-juv moult seems mostly partial (based on limited and seasonally incomplete material at NHM). No record of partial pre-nuptial moult. Judging from specimens, ad becomes progressively more worn from late Jan, and are very worn by Apr; birds in post-nuptial moult collected Jun–Aug; birds from Sep–Dec rather fresh. Breeding and moult apparently strongly linked to occurrence of seasonal rains twice per year (mostly Mar–Apr and Jul–Sep), hence breeding could occur both in spring and late summer, at least in some years, complicating understanding of moult and plumage variation. – **ALL YEAR** Aging difficult, and possible only with practice and bird in the hand. **Ad** As described under Identification, but note some individual variation mainly in breast streaking, some being rather blotchy and diffuse, not distinctly streaked. Bill greyish but sometimes dull pink, especially lower mandible,

Yemen, Jan: evenly fresh with no sign of moult limits in wing. Greenish rump seems to be lacking. Specific characteristics include lack of pale supercilium, or dark moustachial streak and pale lower cheeks mark of Yemen Serin. Bill almost uniform, predominately greyish with indistinct pinkish-straw hue. (H. & J. Eriksen)

with tiny pale spot at base. When fresh shows reduced (partially concealed) greenish-yellow rump and slightly warmer tinge to pale parts (less pure whitish). More extensively pale-fringed above and on upperwing than in worn plumage. **1stW/S** Hardly separable from fresh ad, though after post-juv moult at least some retain (or suspend moult of) some juv remiges, primary-coverts, possibly outer greater coverts and tail-feathers, which are weaker textured, less fresh and browner, compared to neighbouring new ad-like feathers. Greenish-yellow hue is generally even further reduced, but this is inconsistent as ageing criterion. **Juv** Body plumage soft, fluffy and more heavily streaked than in ad, otherwise similar.

BIOMETRICS L 11.5–13 cm; **W** ♂ 65.5–70 mm (n 15, m 67.2), ♀ 64–68.5 mm (n 14, m 66.0); **T** ♂ 47–55 mm (n 15, m 50.4), ♀ 47.5–52.5 mm (n 14, m 50.4); **B** 10.7–12.9 mm (n 29, m 11.8); **B**(f) 8.4–10.3 mm (n 29, m 9.3); **BD** 6.9–8.1 mm (n 28, m 7.6); **Ts** 13.7–15.7 mm (n 26, m 14.7). **Wing formula: p1** minute; **p2** < wt 1.5–4 mm; **pp3–5**(6) about equal and longest (p5 to 2.0 <); **p6** < wt 0–3 mm; **p7** < wt 1–5.5 mm; **p8** < wt 3–7 mm; **p10** < wt 8–11.5 mm; **s1** < wt 9.5–13.5 mm. Emarg. pp3–6, also often indistinctly on p7.

GEOGRAPHICAL RANGE & VARIATION Monotypic, but see below. – W & S Arabia south of c. 26°N, W Yemen; sedentary. (Syn. *uropygialis*.)

Yemen, May: note strong bill. Besides narrow and subdued underparts streaking, very featureless with bland face, just a hint of paler supercilium and crescent-shaped whitish mark below eye, and no wing-bars. (W. Müller)

TAXONOMIC NOTES Previously, usually referred to *Serinus*, but both Nguembock *et al.* (2009) and Zuccon *et al.* (2012) showed that *Serinus* was polyphyletic and needed to be divided into more than one genus to avoid this. Most of the Arabian and African serins are therefore better moved to genus *Crithagra*. – Has been considered a race of extralimital Yellow-rumped Seedeater *C. atrogularis* of E & S Africa (e.g. in Peters 1968), but not followed here. – Sometimes treated as polytypic; apart from *rothschildi* (S Yemen), '*philbyi*' also recognised (N Yemen and Saudi Arabia), though Peters (1968) placed the latter in synonymy. An independent check of the status of the latter proved difficult due to different condition of specimens in NHM, most *rothschildi* being heavily worn, whereas those of '*philbyi*' were fresh. The question cannot be resolved without comparable material.

REFERENCES Everett, M. J. (1987) *Sandgrouse*, 9: 102–105. – Nguembock, B. *et al.* (2009) *Mol. Phyl. & Evol.*, 51: 169–181. – Phillips, N. R. (1982) *Sandgrouse*, 4: 37–59. – Zuccon, D. *et al.* (2012) *Mol. Phyl. & Evol.*, 62: 581–596.

Yemen, Jan: wing too heavily worn to enable ageing, but buffish edges to greater coverts and softer-textured remiges and tail suggest juv after limited moult. Streaking above almost lacking, and below very narrow and less profuse than on fresh birds in previous images. Greenish rump the only notable character. (H. & J. Eriksen)

YEMEN SERIN
Crithagra menachensis (Ogilvie-Grant, 1913)

Alternative name: Menacha Seed-eater

Fr. – Serin du Yémen; Ger. – Jemen-Girlitz
Sp. – Serín yemení; Swe. – Jemensiska

A drab, endemic small finch of alpine, rocky slopes in the SW Arabian highlands (SW Saudi Arabia and Yemen), often found above 2000 m (and can reach higher than 3600 m). This unobtrusive species often occurs in the same areas as its very similar close relative, the Arabian Serin. Associated with barren montane areas almost devoid of vegetation, but is also found in villages and towns. Recently discovered in extreme SW Oman. Formerly traditionally referred to the genus *Serinus*, but recent research has shown this and many African serin species to be better placed in a separate genus *Crithagra*.

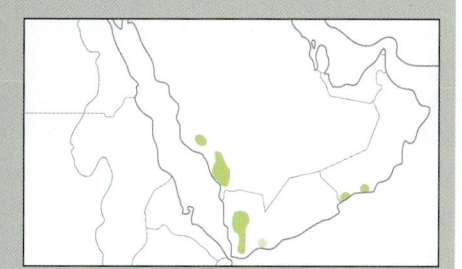

IDENTIFICATION Plumage *dull grey-brown and white*, streaked brown throughout, with any pattern washed-out. Similar shape as Arabian Serin, but is proportionately *longer-winged and shorter-tailed*. Bill is subtly *smaller*, and *upper mandible is usually clearly darker than pale lower*. Indistinct moustachial stripe accentuated by the whitish, *thinly dark-streaked throat*, whereas *breast and flanks are boldly streaked brownish*. A hint of a pale, slightly streaky supercilium. *Upperparts uniform without yellowish-green rump patch*, but crown and nape finely streaked darker, and mantle and rest of upperparts more broadly but diffusely so. *Underparts cleaner whitish*, lacking Arabian Serin's warmer and subtly darker suffusion on chest and flanks. Pale wingbars and tertial edgings rather obscured. Small eye is dark, and leg colour varies from bright to duller flesh. Very sociable and typically feeds on ground, but if disturbed usually flushes into cover or to a small tree. Song is often given from atop a 3–4 m tall tree.

VOCALISATIONS Song recorded as *chew-chee-chee-chwee* (Clement et al. 1993), and a *wit-wit-wit* or *wit-wit-witcha*, etc., given both in dipping flight and perched. – Common call is a rolling, *chi-chi-chi*, often noted as having Redpoll-like quality, with various other musical twitters and sparrow-like chirps also given (mostly in flight), e.g. *crrrreeeet-creeet* and softer *churrt* or *churrut churrut seer*. Also a Yellow Wagtail-like *tseep* and Siskin-like *tsee-oo* with the first note drawn-out, and feeding flocks utter a whispering *tleeit-tleeit*, and in other situations a *twi-twi-twi-twi* (ending in *tuu*), also *cheep-chhp* and *cheir-vrip cheir-virp*, or sonorous *triurp-triurp* (Phillips 1982, Everett 1987, Hollom et al. 1988).

SIMILAR SPECIES For separation from *Arabian Serin*, see that species.

AGEING & SEXING Ageing probably possible using moult limits in autumn but little studied; at least juv separable. Sexes alike. Limited seasonal plumage variation. – Moults. Ad undertakes complete post-nuptial moult, often in late summer or autumn, while post-juv moult seems mostly partial (based on limited and seasonally incomplete material at NHM and AMNH). No record of partial pre-nuptial moult. Also, like Arabian Serin, quite possible that breeding takes place twice per year, both in spring and late summer, with consequences for moult seasons and ageing. – **ALL YEAR** Generally worn while breeding, but fresh after moult and before onset of breeding when more extensively pale-fringed above and in wing. **Ad** As described under Identification. Additionally, upperparts coloration varies from

Yemen, May: like Arabian Serin but not as obviously dull and washed-out as the latter, with especially stronger face pattern, and note boldly brown-streaked breast. Rather smaller-billed with less curved culmen and slightly shorter tail. Apparent moult limits in greater coverts imply 1stY. (W. Müller)

Yemen, Nov: more variegated head pattern than Arabian Serin, including distinct moustachial-stripe contrasting with whitish fore-cheeks, while breast streaking is usually more obvious. Also lacks yellowish-green rump shown by many (but not all) Arabian. Seemingly evenly-fresh wing of left-hand bird suggest ad, but safe ageing not possible. (W. Müller)

Yemen, Oct: small, conical and bicoloured bill. When very fresh, face pattern less bold and upperparts appear more uniform cold grey-brown, but crown and head-sides always clearly dark-streaked. Underparts rather diffuse. All visible wing-feathers seem uniformly fresh, suggesting ad, or a young that moulted completely. (U. Ståhle)

1stW, Oman, Nov: after some body-feathers replaced in post-juv moult this is technically a 1stW. Very blotchy streaking below, with some shaft-streaks and some fluffy fringes. Crown streaking and face pattern provide specific diagnostics, not least prominent dark moustachial-stripe. (T. Lindroos)

earth-brown to greyish-brown, while underparts are off-white to greyish-white. Prominence of dark moustachial stripe varies individually, sometimes being distinct and in others much less well-marked. Lower mandible sometimes brownish-pink or, exceptionally, almost orange-yellow. **1stY** Hardly separable from fresh ad and, if possible, best aged by retained juv remiges and primary-coverts (latter more worn and pointed, creating more contrast versus fresher and newer coverts). **Juv** Body plumage appears soft and fluffy, and more heavily streaked than in ad, otherwise similar. Further work with live birds required to advance on these basic differences.

BIOMETRICS L 12–13 cm; Size of sexes nearly the same. **W** 69–77.5 mm (n 13, m 72.3); **T** 46–54 mm (n 14, m 49.4); **B** 10.3–11.8 mm (n 14, m 10.9); **B** 10.3–11.8 mm (n 14, m 10.9); **B**(f) 7.9–9.2 mm (n 14, m 8.5); **BD** 6.0–7.4 mm (n 14, m 6.4); **Ts** 14.2–16.0 mm (n 14, m 15.1). **Wing formula: p1** minute; **p2** < wt 0.5–1.5 mm; **pp3–5** about equal and longest; **p6** < wt 2–4 mm; **p7** < wt 7–9.5 mm; **p10** < wt 14–18.5 mm; **s1** < wt 15–19.5 mm. Emarg. pp3–6, though less distinctly on p6.

GEOGRAPHICAL RANGE & VARIATION Monotypic. – SW & S Arabia south of *c.* 20°N, W Yemen; sedentary.

TAXONOMIC NOTES Formerly treated as conspecific with extralimital Ankober Serin *C. ankoberensis* of Ethiopia. – See also notes on generic treatment under Arabian Serin.

Juv, Oman, Nov: very similar to image above (top right), but this might still wear its full juv plumage. Coarse brown streaking over much of underparts. (A. Blomdahl)

GOLDEN-WINGED GROSBEAK
Rhynchostruthus socotranus Sclater & Hartlaub, 1881

Fr. – Pinson à ailes d'or; Ger. – Goldflügelgimpel
Sp. – Picogordo árabe; Swe. – Guldvingad fink

Stunningly attractive, robust finch of N Somalia, Socotra, and in Arabia from SW Oman west to SW Saudi Arabia (up to 3100 m). Formerly considered closest to some Himalayan finches based on morphology, but more recently genetic research (e.g. Zuccon *et al.* 2012) has shown the species to be sister to Desert Finch and basal to Greenfinch and allies. Generally a resident breeder that normally frequents arid *Acacia* and *Euphorbia* woodland in the highlands. Rather shy and elusive, not being easy to see or get close to.

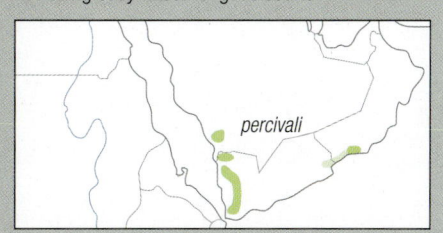

R. s. percivali, ad ♂, Oman, May: stocky finch with large conical bill, black triangular mask, milk chocolate-brown crown, nape, sides of neck and breast, in striking contrast to whitish cheeks. Wing and tail show diagnostic yellow patterns. (H. & J. Eriksen)

IDENTIFICATION Stocky, large sparrow-sized finch with a *massive conical, almost blackish bill, with black surround* that extends and merges with dark eye, and in ♂♂ extends to upper throat, creating *triangular facial mask* in head-on view. *Crown, nape and throat pale chocolate-brown*, contrasting with striking *whitish cheek patches*. Chest and neck-sides *light cinnamon-buff*, and lower nape to back and rest of underparts variously cold greyish or greyish tinged slightly brown. Wings and tail largely black, with *prominent yellow panels on wing*, covering median and greater coverts, also edge of longest tertial, much of secondaries and basal parts of tail-feathers. Also has broad yellowish-green tips to lesser coverts and narrower grey-white fringes to otherwise black tertials. Legs flesh-coloured. Perches on flower heads and low bushes, but is often quite shy, disappearing some distance in bounding flight if disturbed. Usually in pairs or small, loosely formed groups.

VOCALISATIONS Its liquid, rather discordant but quite loud and musical song contains a number of Goldfinch and Linnet-like elements, and has been rendered *whit-whee-oo, sit-eeee-did-ee did-oo-ee,* or *tviit-te-vyt-te-viit* or *whee wee-ooo wheee* and *swip ooo wee wip oo*; it is often given in a fluttering display flight, not unlike that of Greenfinch. – Calls include a Goldfinch-like *tlyit-tlyit* and a soft *wink*. Also a rich rippling *tut-tut-tut-tut* or *dy-dy-dy* followed by a dry trill *drrrt*, and a wheezing *tzee* (Martins 1987, Hollom *et al.* 1988).

SIMILAR SPECIES Unmistakable within range.

AGEING & SEXING (*percivali*) Ageing probably possible using moult limits in autumn but little studied; at least juv separable. Sexes slightly differentiated year-round, both in size and pattern. Seasonal plumage variation limited. – Moults. Limited information on period and extent of moult. Breeding cycle, and hence moult, apparently complicated by the extent of two annual rainy seasons (see Arabian and Yemen Serins). In NHM rather fresh juv collected in Dec (Saudi Arabia) and heavily-worn birds in Jul; ad apparently undertakes complete post-nuptial moult, principally in late spring–summer, while juv moults partly (leaving some inner primaries and several secondaries unmoulted) or completely (both of which may occasionally be suspended, hence a variety of different moult patterns are encountered). Partial pre-nuptial moult unconfirmed. Moult of live birds noted in Apr and Jul. (See also Kirwan & Grieve 2007.) – ALL YEAR **Ad ♂** Black chin and upper throat (on some even slightly extending to upper breast) 8–25 mm long; usually narrow black forehead. Brown cap and cinnamon-buff on neck and chest more saturated and warmer than in ♀, forming greater contrast with dull grey-brown rest of upperparts and greyer (more ash-grey) rest of underparts, respectively. Whitish cheek patches as a rule cleaner white and better marked than in ♀. Furthermore, ♂ generally has deeper (heavier) bill and longer tarsi than ♀, and see general size. **Ad ♀** Unlike ♂, has restricted black feathering below bill only on chin, 2–7 mm long, and no or virtually no black on forehead. Cap is duller and upperparts browner. Grey underparts less pure (sullied pale buffish-brown). White cheek patch tends to be smaller and less well marked, while (diluted) cinnamon-buff of chest extends to throat, which is never black. Smaller on average than ♂, with limited overlap. **1stY** Largely as ad unless retained juv wing-feathers are detected, including all or at least some remiges and especially primary-coverts, which are contrastingly worn and browner (not black like ad); also primary tips relatively more worn than ad at same season. Beware, however, that some moult all flight-feathers and primary-coverts to appear confusingly black and fresh as ad, but most birds retain at least a few juv primary-coverts. Birds that moult completely apparently inseparable from ad. No consistent plumage differences found to separate from ad (both sexes). **Juv** Overall soft fluffy appearance to body plumage, which is brownish olive-buff, with entire head and underparts heavily streaked darker, and upperparts diffusely streaked. Yellow in wing restricted, especially on

R. s. percivali, ad ♂, Yemen, May: two images of the same bird. Relatively large black bib indicates ♂, which below cinnamon-brown breast usually has quite cold greyish rest of underparts. All wing-feathers ad, including fresh and jet-black primary-coverts. (W. Müller)

wing-coverts where it is reduced to tips. Greater and median coverts edged rusty-buff.

BIOMETRICS (*percivali*) L 14–15 cm; **W** ♂ 90–96 mm (*n* 15, *m* 93.4), ♀ 85.5–90 mm (*n* 13, *m* 87.5); **T** ♂ 53–59 mm (*n* 14, *m* 56.2), ♀ 50–54 mm (*n* 13 *m* 52.2); **B** ♂ 17.3–19.3 mm (*n* 13, *m* 18.2), ♀ 15.2–17.1 mm (*n* 12, *m* 16.0); **B(f)** ♂ 14.2–16.3 mm (*n* 13, *m* 15.0), ♀ 11.7–14.0 mm (*n* 13, *m* 13.1); **BD** ♂ 11.3–13.4 mm (*n* 15, *m* 12.8), ♀ 10.7–12.5 mm (*n* 13, *m* 11.5); **Ts** 16.3–19.0 mm (*n* 28, *m* 17.5). **Wing formula: p1** minute; **p2** < wt 0–3 mm; **pp3–4** about equal and longest; **p5** < wt 1–4.5 mm; **p6** < wt 6.5–12 mm; **p7** < wt 13.5–17 mm; **p10** < wt 22–27 mm; **s1** < wt 23–28 mm. Emarg. pp3–5.

GEOGRAPHICAL RANGE & VARIATION Rather well-marked variation between three recognised races: *louisae* in Somalia, *socotranus* in Socotra and *percivali* in S Arabia. The race on Socotra has a black crown, nape and bib. Only *percivali* occurs within the covered region, the other two being extralimital and therefore not treated.

R. s. percivali Ogilvie-Grant, 1900 (SW Arabia, W Yemen, Dhofar in S Oman; resident). Has brown crown and upper nape, thus black on head restricted to lores, narrow band on forehead, and chin and upper throat. (Syn. *yemenensis*.)

R. s. percivali, ad ♀, Yemen, Mar: Unlike ♂, no black bib and triangular dark mask smaller, almost restricted to lores, and not as jet-black. Crown also duller and a little lighter brown, and white cheek patch often smaller and less well defined at rear. Wing seems evenly fresh. (P. Ryan)

R. s. percivali, ♂, Oman, Jan: black chin and extensive mask seemingly including lores, and whitish cheek patch prominent, hence deemed a ♂. Although there could be new central tail-feathers, the tail is mostly covered by a branch, and primary-coverts are in shadow, therefore age best left undecided. (F. Trabalon)

R. s. percivali, 1stY ♀, Oman, May: showing moult contrast (e.g. primaries, primary-coverts and most of tail juv, browner than newer tertials, alula and central rectrices). Very dull grey-brown crown, and little indication of black bib add up to young ♀. (H. & J. Eriksen)

TAXONOMIC NOTE Due to rather well-marked subspecific differences a possible split into three species was discussed by Kirwan & Grieve (2007). Our view is that such differences are not uncommon between widely separated subspecies, and we prefer to await further studies before any change is adopted.

REFERENCES Kirwan G. M. & Grieve, A. (2007) *Bull. ABC*, 14: 159–169. – Martins, R. P. (1987) *Sandgrouse*, 9: 106–110. – Zuccon, D. et al. (2012) *Mol. Phyl. & Evol.*, 62: 581–596.

R. s. percivali, juv, Saudi Arabia, Jul: blackish wings with prominent yellow panel in combination with Hawfinch-sized bill makes identification simple. Juv plumage dull and streaked on off-white ground. (J. Babbington)

R. s. percivali, juv, Yemen, Jan: very distinctive with entire underparts heavily streaked and blotched dark. Species revealed also by bill size, shape and general pattern. (J. Hornbuckle)

(EUROPEAN) GREENFINCH
Chloris chloris (L., 1758)

Fr. – Verdier d'Europe; Ger. – Grünfink
Sp. – Verderón común; Swe. – Grönfink

A widespread and well-known finch with attractive green and yellow colours. Being both common in parks and gardens and having a loud and pleasant song, it is one of those birds that people notice, even without being a birdwatcher. Its strong conical bill reveals that it is mainly a seed-eater, and therefore a resident over much of its range, and only northerly populations migrate south in winter. A frequent guest at bird feeders in gardens, constantly fighting with fluttering wings and open bill over the best positions. Builds a cup-shaped nest in bushes or low dense trees.

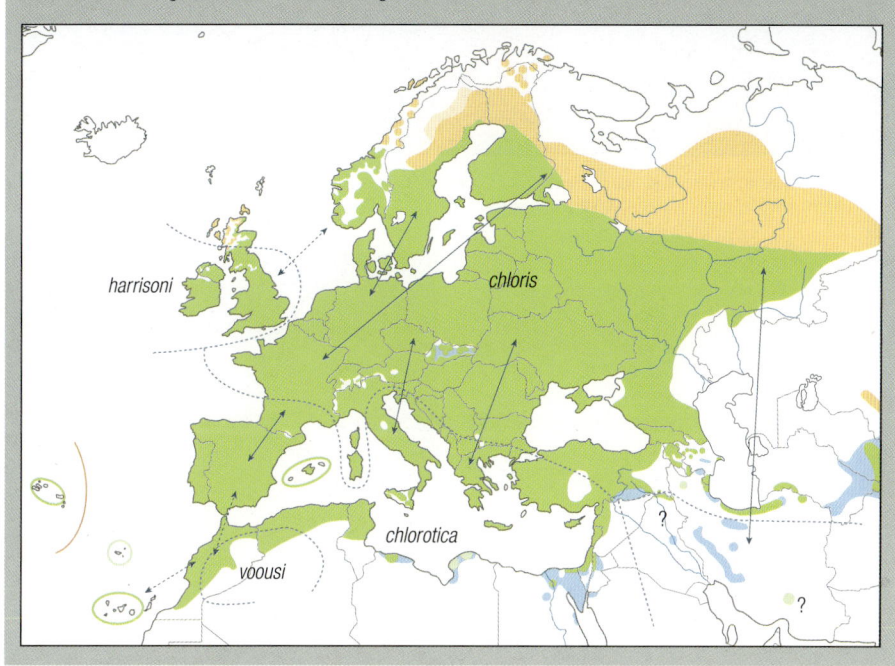

IDENTIFICATION About the same size as House Sparrow but differently shaped, with a slimmer outline, longer and more pointed wings, a slightly shorter tail and shorter legs. Compared to other common finches like Siskin and Redpoll, it has a *larger head* and *heavier bill*, looking somewhat front-heavy. The rather pale *pinkish-grey* (can look simply pale grey) *bill is conical* with straight culmen, and the *legs are pinkish*. The plumage is greyish, green and yellow, duller in the ♀, brighter in the ♂. Adult ♂ is *greenish* on much of head and body with a variable amount of *yellow on belly* and a yellowish-*green throat and rump*. Most characteristic are the bright *yellow edges to the wings* and *yellow bases to outer tail-feathers*. The ♀ is duller, greyish with some diffuse dark streaking on back, and a more limited amount of greenish and yellow. Juveniles resemble the ♀ but are slightly paler greyish and are more prominently streaked. The Greenfinch is a strong and fast flier, the flight path being rather straight and only slightly undulating. Quite wary, taking off nervously at first sign of danger. Lands somewhat clumsily with fluttering wingbeats. Often assumes high perch in tall trees, but also in bushes and hedgerows. Sings both from perch in tree, on wire or TV aerial, and in 'wandering' song-flight with high, slow, butterfly-like wingbeats.

VOCALISATIONS The normal song is a fast, loud and pleasing, almost Canary-like series of trills and repeated whistles, each brief motif on the same pitch but motifs alternate between a few different ones. A common song could be rendered *jup-jup-jup-jup türrrrrrr tuy-tuy-tuy-tuy dyuuwee jip-jip-jip-jip-jip dürdürdürdür turrrrrr twee-twee-twee-twee*. There is also an alternative song type, a 'primitive' nasal wheeze with rather hoarse voice, *dschyeeesch*, or these two are combined to produce a varied and entertaining mixture. Songs can be fairly brief and delivered with short pauses between them, but agitated males give prolonged song that can last several minutes, and in song-flight is always long, varied and pleasing. – Call is a strong brief twitter, *jup-up-up* or *chüp-üp-üp-üp*, in the mixed chorus from migrants discernible by its strength and rather musical voice, familiar from the song. Sometimes gives single *jup* or *chüp*, stronger and more resounding than similar but softer flight-call of Chaffinch. Alarm is an upwards-inflected whistling *dyuuwee* (or *tooee*), also used now and then in the song, rather Canary-like and even possible to confuse with Common Rosefinch and Redpoll if heard only briefly and at some distance. Young give a resounding *chew-chew-chew*, frequently repeated and often heard from feeding parties feeding on dog-rose hips in Aug.

SIMILAR SPECIES From *Siskin* by larger size and more front-heavy look caused by bigger head and heavier bill with much deeper base. Both have yellow on sides of tail-base in ♂ plumage, but Greenfinch has a yellow line along the edge of the wing and greyish tertials, whereas Siskin has a prominent yellow wing-bar across the wing, and pale-edged dark tertials. – *Serin* is also greenish and yellow, but is much smaller with a tiny, rounded bill and very bold dark streaking above and below. – *Citril Finch* is more delicately proportioned and has bold double yellowish-green wing-bars. – *Canary* differs in its slimmer shape, shorter and stubbier bill, and its lack of yellow edges to primaries.

AGEING & SEXING (*chloris*) Ages differ subtly in both autumn and spring, but reliable ageing requires close views. Sexes separable at least after post-juv moult; separable also in juv plumage if seen close or handled. – Moults. Complete post-nuptial moult of ad in late summer, mainly mid Jul–Sep (Oct). Partial post-juv moult at about same time does not usually involve flight-feathers, tertials or primary-coverts, but birds from early broods, especially in C & S Europe, can moult some or even most of these feathers (see also below); in a few cases, moult of all tail and of most primaries and several secondaries has been recorded. In C & S Europe, usually all greater coverts moulted, but in N Europe fairly common for a slight moult contrast to be discernible between 1–3 outer unmoulted and remaining inner moulted. No pre-nuptial moult in winter. – **SPRING Ad ♂** Much of head, throat and breast rather bright yellowish-green. Mantle unstreaked greyish-green with slight brown tinge. Extensive bright yellow on outer webs of outer long primaries reaching shaft or nearly so. Base of outer three tail-feathers yellow across whole feathers, leaving only shaft dark. Alula edged yellow and white, tipped ash-grey. Primary-coverts edged green and tipped ash-grey, tips rounded (although there is some variation, some 1stS being nearly as bright). **1stS ♂** As ad ♂, and some inseparable if not seen well, but usually possible to recognise by slightly less bright green colours, alula with less or no yellow-white on outer web, and primary-coverts more pointed with less bright green edges and less pure ash-grey tips. Also has slightly more pointed and worn tail-feathers (unless not moulted to ad type), and sometimes two generations of tail-feathers. ♀ Duller and less green than ♂, with mottled or faintly-streaked mantle and often crown/nape. Yellow on long outer primaries paler yellow (even whitish) and covers only about half of outer webs. Some yellowish-white

C. c. chloris, ad ♂, Germany, Apr: a rather stocky, large-headed finch with heavy conical pale bill. Plumage generally unstreaked yellow, green and grey, with diagnostic bright yellow primary panel and basal edges to outer tail, all obvious in ad ♂. Generally, plumage becomes brighter yellowish-green by late spring. Ad by bright colours and pure grey-tipped primary-coverts. (A. Noeske)

C. c. harrisoni, ad ♂♂, England, Apr: ♂ is relatively brighter olive-green than ♀ with more extensive and vivid yellow wing and tail markings, characteristic ill-defined greyish cheeks and breast-sides, and large grey wing panel. Two ♂♂ showing individual variation. Neat green-edged and grey-tipped primary-coverts infer age. (G. Thoburn)

basally on three outer tail-feathers, but generally this does not reach shafts, being separated by a dark zone. (Very rarely ♂-like ♀♀ appear with more yellow on outer tail-feathers, and especially r5 can have pale yellow all the way to the shaft, but then not on r4 and r6. The border between yellow and dark is also more irregular than in ♂.) Ageing sometimes possible using wear of tail-feathers, 1stS usually being more abraded than ad. – AUTUMN **Ad** ♂ Much as ad ♂ in spring, although a variable number of fine grey or dull feather-tips often make the green colours less vivid. All greater coverts of same generation, tipped grey. Alula has broad yellow-white outer edge. Primary-coverts mint fresh, tail-feathers fresh, rather broad with rounded tips. **1stW** ♂ Much as 1stS ♂. Tail-feathers narrower and more pointed than ad, tips slightly abraded. Sometimes a few outer greater coverts unmoulted, being juv type with brown-tinged off-white tips. ♀ Much as spring. Ageing as ♂. **Juv** Resembles ♀ but is somewhat paler greyish and less green-tinged with much more prominent dark streaking above and below. Tail pattern sexes any bird seen well or handled. In a study in Galicia, NW Spain, ♂♂ had dark tip of r5 along shaft shorter than 31.5 mm, ♀♀ longer than this (Arenas & Senar 2004).

BIOMETRICS (*chloris*) **L** 14.5–16 cm; **W** ♂ 85–95 mm (n 64, m 89.6), ♀ 82–90 mm (n 42, m 86.9); **T** ♂ 52–61.5 mm (n 65, m 56.2), ♀ 51–57 mm (n 42, m 54.4); **T/W** m 62.7; **B** 14.0–16.8 mm (n 60, m 15.5); **B**(f) 11.0–14.0 mm (n 93, m 12.8); **BD** 9.2–10.9 mm (n 91, m 10.0); **BW** 8.5–10.0 mm (n 60, m 9.2); **Ts** 16.7–18.7 mm (n 83, m 17.7). **Wing formula: p1** minute; **p2** 0–2 mm < wt; **pp**(2)**3–4** longest; **p5** < wt 3–6.5 mm; **p6** < wt 12–16 mm; **p7** < wt 18–20 mm; **p10** < wt 25–30.5 mm; **s1** < wt 27–33 mm. Emarg. pp3–5.

GEOGRAPHICAL VARIATION & RANGE Mostly very slight variation, affecting size, structure and colour saturation. Extremes are fairly different, but most races are connected by wide zones of intergradation and are therefore far from distinct. Four races accepted here, but several more have been described. However, when comparing long series of all taxa, it proved impossible to discern claimed fine differences for some of them, or to ascertain a sufficient level of consistency, meaning that they have been lumped with other more distinct races. – Introduced to Cape Verdes, but the subspecies concerned is unknown.

○ ***C. c. harrisoni*** Clancey, 1940 (S Scotland, Ireland, much of England; mainly resident). Resembles *chloris* with same brown tinge on mantle, but is very slightly smaller, somewhat darker and duller, breast being dull dark greyish olive-green, and throat and belly dirty yellowish-green, the yellow colours being much more subdued than in other races. Still, some birds approach *chloris*. Intergrades with *chloris* both in N Scotland and possibly S England. Stated (e.g. in *BWP*) to have yellow-green forehead contrasting with olive-green crown in worn plumage, but such birds rare, most having a largely uniformly dark crown even when worn. **W** ♂ 85–92 mm (n 18, m 88.4), ♀ 83–89 mm (n 12, m 85.1); **T** ♂ 53–59 mm (n 17, m 55.2), ♀ 51–57 mm (n 12, m 53.3); **T/W** m 62.5; **B** 14.3–17.0 mm (n 30, m 15.6); **B**(f) 11.5–14.4 mm (n 30, m 12.5); **BD** 9.3–11.3 mm (n 30, m 10.0); **BW** 8.3–9.5 mm (n 39, m 9.1); **Ts** 16.8–18.8 mm (n 29, m 17.9). (Syn. *restricta*.)

C. c. chloris (L., 1758) (N Scotland, N Europe east to Urals and W Siberia, south to C & SE France, Corsica, Sardinia, Alps and N Italy, Hungary, N Balkans, Bulgaria, Black Sea region, E Turkey, Caucasus, N Iran and Central Asia; northern and eastern populations move west and south in winter, southern being resident). Described above. Rather large, in ♂ plumage with moderate amount of green and yellow, but some are brighter. Mantle practically invariably tinged brown, olive-brown or greyish-brown, but a few have so few brown elements that they look green-backed. – Birds of Corsica and Sardinia ('*madaraszi*') tend to be slightly brighter and smaller than typical *chloris*, but are still very close to it, nearly always darker than typical *chlorotica* and have brown tinge on back, thus included here in *chloris*. Birds

C. c. chloris, ad ♀, Germany, Apr: the ♀ is usually duller and drabber than the ♂, more grey-green and/or brownish in general tone, with less yellow on wings and tail. Evenly fresh ad wing and tail, and primary-coverts neatly grey-tipped. (A. Noeske)

C. c. chloris, ♂, presumed ad, Finland, Nov: fresh following complete post-nuptial moult, washed dull olive-brown, thus greenish and yellow less vivid than in spring/summer. Racial designation by range and slightly dusky appearance. All visible wing-feathers seemingly ad and very fresh. (M. Varesvuo)

C. c. chloris, ♀, presumed ad, Finland, Jan: just as the autumn ♂, the ♀ is browner-tinged and a little duller when fresh. Rump can appear rather contrastingly greenish, though, and note typical head and bill profile and wing markings. Racial designation by range and brownish plumage. Evenly-fresh wing-feathers. (M. Varesvuo)

C. c. chloris, juv, Switzerland, Jun: streaked darker above and below, with fluffy body plumage fringed pale buffish to paler olive grey-brown, and has duller bill. Presumed ♂ as entire outer web of primaries seemingly bright yellow. (H. Shirihai)

in Bulgaria and Romania are similarly subtly brighter than typical *chloris*, but again this is more of a clinal transition into *chlorotica* of Balkans than meriting formal separation. Birds in N & E Turkey, Caucasus, Transcaucasia and N Iran ('*bilkevitchi*'), and in Central Asia ('*turkestanica*'), are large like *chloris* but on average a little paler and greyer, while individual variation and subtle differences make it difficult to uphold these as separate races. (Syn. *bilkevitchi*; *kaukasica*; *madaraszi*; *menzbieri*; *smithae*; *turkestanica*.)

C. c. chlorotica (Bonaparte, 1850) (S France, Iberia including Balearics, Italy except in the north, S Balkans, W & S Turkey, Levant, NW Africa except Atlas Mts, east to Egypt; mainly resident). Brighter green above in ♂ plumage, lacking brown tinge on mantle of *chloris* and *harrisoni*, and is brighter and more extensively yellow on belly, with strong contrast versus olive-green chest. Often, but variably, forehead and fore supercilium show yellow (especially on sides, making dark lores prominent), and throat and rump are bright greenish-yellow. Somewhat smaller than *chloris*. **W** ♂ 80–90 mm (*n* 93, *m* 85.1), ♀ 78–87 mm (*n* 36, *m* 82.7); **T** ♂ 49–56 mm (*n* 93, *m* 52.5), ♀ 48–55 mm (*n* 36, *m* 51.4); **T/W** *m* 61.8; **B** ♂ 13.9–16.9 mm (*n* 81, *m* 15.6), ♀ 13.5–16.6 mm (*n* 26, *m* 15.3); **B(f)** ♂ 10.3–14.1 mm (*n* 93, *m* 12.5), ♀ 11.2–13.1 mm (*n* 36, *m* 12.2); **BD** 9.1–11.5 mm (*n* 129, *m* 10.0); **BW** 8.5–10.1 mm (*n* 111, *m* 9.2); **Ts** 15.9–19.0 mm (*n* 126, *m* 17.2). Post-juv moult on average more extensive than in *chloris*, nearly invariably all greater coverts, a few tertials and some remiges (generally some outer, but rather irregularly) and part of or whole tail moulted. – Birds of S France and E & S Spain, mainland Italy and W Balkans have been described as '*aurantiiventris*', of Corsica and Sardinia as '*madaraszi*', of NW Spain, Portugal and NW Morocco north of Atlas as '*vanmarli*', and of SE Europe and W Turkey as '*muehlei*', but much variation within these populations and claimed differences very slight and not sufficiently consistent to meet a 75% requirement, thus included here in *chlorotica*, which has priority. Birds in W Iberia ('*vanmarli*') and Levant ('*chlorotica sensu stricto*') are smallest, but size difference very slight and not sufficient grounds for separation. (Syn. *aurantiiventris*; *chlorantica*; *mallorcae*; *muehlei*; *vanmarli*.)

○ **C. c. voousi** Roselaar, 1993 (Atlas Mts, from Morocco to C Algeria; resident). Same bright plumage as *chlorotica*, or is on average subtly paler and less deep green and yellow, while ♀♀ allegedly often more ash-grey (Roselaar 1993), but bill on average larger. A subtle race and some overlap with *chlorotica*, but sufficiently distinct according to 75% rule. **W** ♂ 82–91 mm (*n* 26, *m* 86.0); **T** ♂ 50–57 mm (*n* 26, *m* 53.5); **B** ♂ 15.7–18.2 mm (*n* 26, *m* 16.7); **B(f)** ♂ 12.5–14.6 mm (*n* 26, *m* 13.5); **BD** ♂ 10.0–12.1 mm

C. c. chlorotica, ad ♂, Italy, Mar: subtly smaller than *chloris* and more vividly tinged greenish-yellow, with grey areas lighter and purer ash-grey when still fresh. Fringes of flight-feathers edged bright yellow. Mantle and back green, lacking brown tinge in ad ♂. Whole wing fresh, and neat green-edged and grey-tipped primary-coverts infer age. (D. Occhiato)

C. c. chlorotica, 1stS ♀, Spain, May: ♀ of this race on average paler and yellower, but difference slight and require direct comparison to be appreciated. Primary-coverts juv, dull and brownish-tinged. (C. N. G. Bocos)

Presumed *C. c. chlorotica*, ♀, Italy, Jan: in winter, especially ♀♀ of this race often appear distinctly paler than *chloris*, tinged greyish-brown and less dusky. Note better-marked paler areas on face. Without handling ageing difficult, could be ad (most likely) or young following extensive post-juv moult. (L. Sebastiani)

C. c. chlorotica, 1stW ♂, **Israel, Nov**: in all subspecies when very fresh, 1stW ♂ can appear intermediate between the two sexes. Combination of overall pale plumage, some still ongoing moult in remiges (southern races have more extensive post-juv moult) and extensive yellow underparts indicate *chlorotica*. (A. Ben Dov)

1stW ♂, Italy, Dec: in winter some *chloris* may move south into Italy, and brownish mantle seems to infer this race, still best to avoid subspecific identification in this case. Age inferred by moult contrast between juv tertials and moulted purer grey-edged greater coverts, and by somewhat pointed juv primary-coverts. (D. Occhiato)

Presumed *C. c. chlorotica*, 1stW ♀, **Italy, Nov**: young ♀♀ can be very dull with no yellow or greenish (except on primaries and rump, respectively), and often are finely streaked below (ageing confirmed by worn tips to juv primaries). Overall pale plumage and larger bill in relation to body best fit *chlorotica*, but safe racial designation uncertain. (D. Occhiato)

C. c. voousi, ♂, presumed ad, **S Morocco, Mar**: very similar to *chlorotica* but with slightly heavier bill (almost Hawfinch-like in shape here), and often subtly paler ♂ plumage, with upperparts saturated olive-green, contrasting with paler yellowish-green underparts. Visible wing-feathers seem ad, but angle of view prevents safe ageing. (T. Grüner)

(*n* 26, *m* 10.9); **BW** ♂ 9.3–10.4 mm (*n* 25, *m* 10.0); **Ts** ♂ 16.5–18.3 mm (*n* 26, *m* 17.6).

TAXONOMIC NOTE Recent research (Nguembock *et al.* 2009, Zuccon *et al.* 2012; and references therein) has shown, among other things, that both *Serinus* and *Carduelis* are polyphyletic and require division into multiple genera, respectively. This affects the Greenfinch, which should therefore retain its old genus *Chloris*.

REFERENCES Arenas, M. & Senar, J. C. (2004) *Ring. & Migr.*, 22: 1–3. – Nguembock, B. *et al.* (2009) *Mol. Phyl. & Evol.*, 51: 169–181. – Roselaar, C. S. (1993) *DB*, 15: 258–262. – Zuccon, D. *et al.* (2012) *Mol. Phyl. & Evol.*, 62: 581–596.

C. c. voousi, 1stW ♂ (left) and 1stW ♀, **Oued Massa, S Morocco, Oct**: heavy bill, fairly pale overall plumage and locality support subspecific identification. Both are 1stW, the ♂ by unmoulted juv primary-coverts (but note ad-shaped rounded tail-feathers, obviously moulted), the ♀ by several brownish and worn juv central tail-feathers. (D. Bigas)

(EUROPEAN) GOLDFINCH
Carduelis carduelis (L., 1758)

Fr. – Chardonneret élégant; Ger. – Stieglitz
Sp. – Jilguero europeo; Swe. – Steglits

The colourful plumage of the Goldfinch, together with its entertaining song, made it a favourite cagebird in times now long past. Although its plumage would hardly make headlines in South America, where so many species possess strikingly colourful plumage, it is rather unusual in a European context. Not surprisingly its colours have been charmingly explained in an anecdote, wherein God paints all the birds but forgets one little grey one until the end. God then makes good by wiping off all of his brushes on this last bird—the Goldfinch! Widely distributed in open landscapes, where small parties are often seen feeding on weeds and thistles.

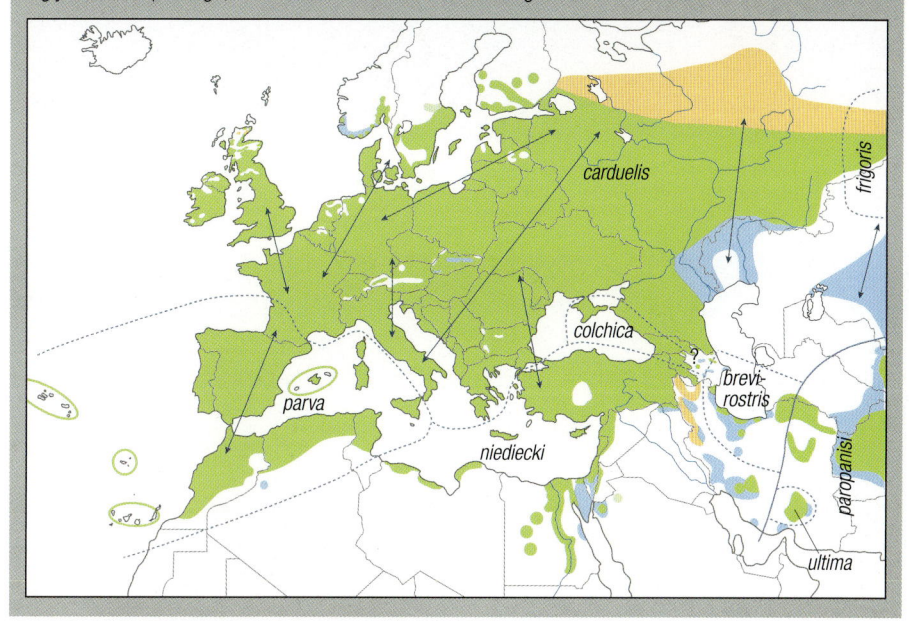

IDENTIFICATION There can be few birds more characteristic and easy to identify than the Goldfinch, at least when confronted by an adult at reasonable range. The boldly-patterned *head in red, black and white*, the *acutely-pointed pale pinkish-white bill*, the striking *wings in yellow, black and white*, and *pale cinnamon-brown back and flanks* combine to make the Goldfinch look exotic, like an imported cagebird from the tropics. In flight, note first on upperwing the *broad yellow wing-bar on a black background* and the *white rump patch*, more visible than the tiny white patches on the otherwise black tail-feathers. Sexes very similar and generally not separable in the field (although ♂ has slightly more extensive red mask than ♀). Juvenile differs more clearly by lacking the bold head pattern of adult, the head and nape being buff-white or pale brown, diffusely streaked darker. Breast, too, is finely spotted, rather than uniformly pale cinnamon. Flight light and bounding. Often seen in twos or small family parties, in off-season also in larger flocks where pasture land holds good stands of thistles or other weeds with plenty of seeds. Clings to these plants agilely, even upside-down, when feeding.

VOCALISATIONS Song a typical hurried, high-pitched cardueline strophe, continuous series of buzzing trills, liquid twittering and fine whistling notes. Can recall Siskin but is generally recognised by typical inserted calls (see below) and absence of the Barn Swallow-like choking or wheezing notes so typical of Siskin song. – Frequently first noted by its calls, either a trisyllabic fine, skipping *tickelitt* or mono- or disyllabic related calls, *telitt* and *litt*. From feeding parties, among the skipping calls, sometimes also a rasping *tschrre*, often repeated in short series, almost like a quarrelling Sand Martin. On one recording also very fine, sharp notes, *ziit*.

SIMILAR SPECIES Can hardly be confused with any other species in the W Palearctic!

AGEING & SEXING (*carduelis*) Ages differ in summer, sometimes in autumn and even in spring if seen well or trapped. Sexes similar but differ subtly in plumage following post-juv moult. Size difference very slight. – Moults. Complete post-nuptial moult of ad mainly in Jul–Oct, but southern populations may start earlier. Partial post-juv moult in late summer or early autumn does not involve flight-feathers or primary-coverts, and in N Europe often 1–4, rarely more, outer juv greater coverts retained, whereas southern population frequently moult all juv greater coverts. Tail-feathers generally not moulted in N Europe (except sometimes r1), but tendency in south to moult tail-feathers partly or, rarely, completely. Very rarely complete post-juv moult has been reported from S Europe (cf. Jenni & Winkler 1994 for some references), but many or all of these might involve arrested moult before all outer primaries and primary-coverts had been moulted. No pre-nuptial moult. – **SPRING** ♂ Red on face averages slightly more extensive than in ♀, above eye

C. c. carduelis, ♂, presumed ad, Switzerland, Dec: red face and black wings with large yellow panels. Sexes almost alike but most ad ♂♂ have red on face extending slightly further behind eye and further onto upper throat; also, nasal hairs are usually jet-black. Ad by seemingly evenly-feathered wing, but impossible to judge for sure. (H. Shirihai)

C. c. carduelis, ♀, presumed ad, Switzerland, Dec: smaller red face, below eye extending shorter than rear edge, and on upper throat averages less extensive; black on loral area and around base of bill on average slightly less extensive than in ♂, while nasal-hairs often grey rather than black. Tentatively aged as previous bird. (H. Shirihai)

often extending nearly 1 mm behind rear edge of eye, below eye reaching rear edge of eye (but some are intermediate). Red on throat extensive, reaching 7–10 mm below base of lower mandible. Nostril-feathers blackish. Lesser coverts very dark (black or brownish-black), sometimes finely tipped brown when fresh. ♀ Red on face on averages less extensive than in ♂, above eye hardly extending to its rear edge, below eye reaching just inside (shorter than) rear edge of eye. Red on upper throat on average less extensive, reaching 5–8 mm from base of lower mandible (thus some overlap with ♂). Nostril-feathers often pale grey, sometimes slightly or much darker grey. Lesser coverts variable, often broadly tipped grey-brown, less commonly with thinner grey-brown tips to dark grey bases (thus approaching ♂ pattern). **F.g.** No contrast in wear or colour visible among greater coverts, tertials or tail-feathers. Tips to tail-feathers slightly rounded, neatly edged. **1stS** Not always separable, but those that have retained few outer juv greater coverts often show slight contrast in wear and colour between these, being shorter and tipped buffish- or off-white, and inner moulted that are longer and more extensively tipped yellow or yellow-white. Tips of tail-feathers on average narrower and more pointed than ad, but intermediates occur, and wear makes difference

C. c. carduelis, 1stW, presumed ♂, Switzerland, Dec: after post-juv moult becomes one of the most attractive finches with head in red, white and black, wings black, white and yellow, and with cinnamon-brown back. Here a 1stW (note moulted central tail-feathers in contrast to browner rest) tentatively sexed by the restricted but deeply red face and greyish-black nasal hairs. (H. Shirihai)

C. c. carduelis, 1stW ♂ (left) and ♀, England, Oct: here two 1stW aged by moult limits between new glossy black wing-feathers and duller juv ones (on left, best seen between new tertials and juv primaries and primary-coverts; on right, limits within tertials). Face pattern including colour of nasal hairs helpful for sexing, but only after ageing is done to prevent confusion with ad ♀. (P. Lathbury)

progressively less obvious. – **AUTUMN** Ageing and sexing as in spring. Those 1stW moulting all greater coverts often still identified by slight contrast between blackish base of greater coverts and subtly greyer juv primary-coverts (requires practice). **Juv** Easily separated by lack of red on face and black on crown and nape, these parts being buff-white and pale brown diffusely streaked or spotted darker. Pale brown breast and flanks also diffusely mottled.

BIOMETRICS (*carduelis*) **L** 13–14 cm. Sexes nearly the same size: **W** ♂ 77–88 mm (n 35, m 82.1), ♀ 77.5–85 mm (n 18, m 81.3); **T** ♂ 47.5–55 mm (n 35, m 51.0), ♀ 48–55 mm (n 18, m 50.8); **T/W** m 62.1; **B**(f) 11.3–14.5 mm (n 53, m 13.0); **BD** 6.6–8.4 mm (n 53, m 7.4); **Ts** 12.5–16.1 mm (n 51, m 14.7). **Wing formula: p1** minute; **pp2–4** longest, p2 and p4 sometimes 0.5–1 mm < wt; **p5** < wt 2–4.5 mm; **p6** < wt 8–14 mm; **p7** < wt 15–19 mm; **p10** < wt 24.5–29 mm; **s1** < wt 27–30 mm. Emarg. pp3–5, but on p5 rarely slightly less well marked (and very rarely a faint emarg. also on p6).

GEOGRAPHICAL VARIATION & RANGE Over much of its range variation is slight and clinal. Only in extreme south-east of the treated region is a markedly different race group found with different head and wing pattern, possibly meriting species status (but see Taxonomic note). In Europe, southern

C. c. carduelis, Denmark, Oct: note broad yellow wing-bar and white rump, making the species unmistakable in flight. Flock formation often dense like this. (S. E. Jensen)

FINCHES

C. c. carduelis, juv, Netherlands, Jul: generally drab buff-brown above and creamy-white below, with diffuse fine dark streaks and spots, and has an almost unmarked head. However, wings as ad, but has buff-brown tips to coverts. (N. D. van Swelm)

C. c. niediecki, ♂, presumed ad, Israel, Nov: similar to ssp. *carduelis*, but upperparts and breast-side patches average more drab-brown, less vivid cinnamon (at least when series are compared). Possibly ad by evenly and very freshly-feathered wing, but difficult to say without handling or a better view. (A. Ben Dov)

populations of the so-called '*carduelis* group' are smaller than northern, but no consistent plumage differences can be found. Here we provisionally acknowledge a total of six races within this group, versus nine by Vaurie (1959) and ten in BWP. This reflects a slightly different approach to subspecies taxonomy, where only more distinct races are accepted. Although we prefer in general to compare adequate series of described races, this has not always been possible. For the Goldfinch for instance, we have been unable to examine sufficiently extensive material of all relevant taxa from the Black Sea and Caspian Sea regions, and rather few *colchica*. – Northern and eastern populations move west or south in winter, at least to southern parts of breeding range, whereas southern populations are mainly resident.

'CARDUELIS GROUP'
C. c. carduelis (L., 1758) (much of N, W & C Europe, including France except south-west, mainland Italy, Balkans and mainland Greece, Romania, N & SE Ukraine except Crimea, east to middle Volga, Kalmuk steppe, N Caucasus). Treated above. Rather large. Mantle, back and flanks saturated cinnamon. Cheeks have limited brown, becoming whitish in worn plumage. Grades into *parva* in S France. – Breeders in Britain & Ireland ('*britannica*') are subtly smaller on average, **W** 73.5–84 mm (n 44, m 78.2), **T** 44–52 mm (n 44, m 48.2), **B**(f) 10.9–13.2 mm (n 44, m 11.9), and cheeks sometimes have more brown, becoming less whitish with wear; still, much overlap and majority cannot be safely separated, thus included in *carduelis* here. Birds in Balkans and mainland Greece ('*balcanica*') often separated on account of paler and more grey-tinged upperparts, but intermediate between *carduelis* and *niediecki* of Turkey, and usually very close to or inseparable from former, therefore included in that. When series of breeders from S Ukraine ('*volgensis*') or SE Russia ('*elegans*') were compared with series of *carduelis*, no differences could be detected, hence included here. (Syn. *balcanica*; *britannica*; *elegans*; *hortensis*; *volgensis*.)

C. c. parva Tschusi, 1901 (SW France, Balearics, Iberia, Atlantic Islands, N Africa east to Tunisia, large Mediterranean islands; presumably this subspecies introduced to Cape Verdes). Very similar to *carduelis* but slightly smaller and has proportionately subtly larger bill. **W** ♂ 71–80 mm (n 49, m 76.5), ♀ 72–79 mm (n 26, m 73.8); **T** ♂ 42–51 mm (n 48, m 47.4), ♀ 43–48 mm (n 27, m 45.6); **T/W** m 62.0; **B**(f) 10.4–14.5 mm (n 82, m 12.1); **BD** 6.0–7.6 mm (n 82, m 7.6); **Ts** 13.0–15.0 mm (n 80, m 14.0). **Wing formula: p5** < wt 0–4.5 mm; **p6** < wt 8–12 mm; **p7** < wt 12.5–16.5 mm; **p10** < wt 21–27 mm; **s1** < wt 21–28 mm. – Note that birds in Libya are intermediate between this and *niediecki*, but closer to *niediecki* (being similarly large), hence included therein. Birds of large islands in W Mediterranean often separated ('*tschusii*') on account of smaller bill than *parva* and slight olive tinge in brown of underparts, but biometric differences are minute (e.g., in *tschusii* BD on average only 0.2 mm less than in *parva*), and fit main clinal pattern; when series of specimens of *parva* and *tschusii* with covered labels were mixed they proved impossible to sort according

C. c. paropanisi, ad ♂, S Kazakhstan, May: very distinctive, with smaller red face bordered by ill-defined white ear-coverts, while rest of head and upperparts drab grey lacking any black. Tips of flight-feathers black, without white spots, while tertials have long white stripe on outer web; note also much longer bill. Ad by evenly-feathered wing. (A. Isabekov)

C. c. paropanisi, ♀ (right) and ♂, presumed ad (right), Kazakhstan, Mar: note difference in amount of red on face, reduced in ♀. At least right-hand bird seems ad by wing-feathers having uniformly deep black portions. (A. Isabekov)

to the above-mentioned plumage criterion. (Syn. *africanus*; *bruniventris*; *propeparva*; *tschusii*; *weigoldi*.)

C. c. niediecki Reichenow, 1907 (Crete, Turkey except perhaps north-east, Cyprus, Libya, Egypt, Levant, Transcaucasia and NW Iran except Talysh and Caspian coast; possibly also W Zagros in W Iran). Smaller than *carduelis* and a trifle duller brown, less pure cinnamon, above and on flanks, and is subtly paler on average than preceding two races. **W** ♂ 77–83 mm (*n* 31, *m* 79.7), ♀ 74–82 mm (*n* 23, *m* 76.8); **T** ♂ 47–52 mm (*n* 31, *m* 50.0), ♀ 44–51 mm (*n* 23, *m* 48.5); **T/W** *m* 62.9; **B**(f) 10.9–14.5 mm (*n* 54, *m* 12.4); **BD** 6.4–8.0 mm (*n* 54, *m* 7.4); **Ts** 13.5–15.3 mm (*n* 50, *m* 14.4). (Syn. *iranensis*; *schiebeli*.)

○ **C. c. colchica** Koudashev, 1915 (Crimea, SE Ukraine, W Caucasus and Transcaucasia; possibly NE Turkey). Very similar to *niediecki*, with same pale cinnamon colour above and on flanks, only is subtly larger with a little heavier bill, and same size as *carduelis*. **W** ♂ 83–87.5 mm (*n* 9, *m* 84.8), ♀ 78–82 mm (*n* 7, *m* 80.2); **T** ♂ 49–56 mm (*n* 9, *m* 52.7), ♀ 47–53 mm (*n* 7, *m* 50.8); **T/W** *m* 62.6; **B**(f) 10.5–14.7 mm (*n* 16, *m* 12.7); **BD** 7.4–9.0 mm (*n* 16, *m* 8.0); **Ts** 14.0–15.5 mm (*n* 15, *m* 14.9). Wing formula: **s1** < wt 25.5–29.5 mm. – Note that birds in much of NC Caucasus are darker brown and inseparable from *carduelis*, in which they are best included. (Syn. *nikolskii*.)

C. c. frigoris Wolters, 1953 (W & C Siberia). Similar to *carduelis* but slightly paler overall, with cleaner white underparts, and all measurements are slightly larger, making it the largest race. Grades into *carduelis* in S Ural region. **L** 13.5–14.5 cm. **W** ♂ 84–95 mm (*n* 30, *m* 87.8), ♀ 82.5–89 mm (*n* 9, *m* 85.5); **T** ♂ 50–60 mm (*n* 30, *m* 54.3), ♀ 51 mm (*n* 9); **T/W** *m* 62.2; **B**(f) 12.2–15.6 mm (*n* 40, *m* 13.8); **BD** 7.6–9.2 mm (*n* 40, *m* 8.3). Wing formula: **p5** < wt 3–6 mm; **p6** < wt 11–15 mm; **p7** < wt 16.5–21 mm; **p10** < wt 27–32 mm; **s1** < wt 28–34.5 mm. (Syn. *major* Taczanowski, 1879—preoccupied.)

C. c. brevirostris Zarudny, 1889 (Talysh and SE Azerbaijan, coastal Iran in north between Lenkoran and Gorgan, south to Zagros). Compared to *niediecki* somewhat darker brown above, on breast-sides and flanks, and is slightly larger but with proportionately a little shorter tail and stronger bill. **W** ♂ 79–86 mm (*n* 16, *m* 82.1), ♀ 77–83 mm (*n* 6, *m* 80.2); **T** ♂ 47–53 mm (*n* 16, *m* 50.5), ♀ 47–52 mm (*n* 6, *m* 49.9); **T/W** *m* 61.7; **B**(f) 11.3–14.9 mm (*n* 21, *m* 12.8); **BD** 7.1–8.2 mm (*n* 22, *m* 7.6); **Ts** 13.8–15.7 mm (*n* 21, *m* 14.7). – Known to intergrade with brown-headed *paropanisi* where they meet in NE Iran. (Syn. *loudoni*.)

'Caniceps Group'

C. c. paropanisi Kollibay, 1910 (NE Iran, S Turkmenistan, SE Uzbekistan, Tien Shan Mts, SE Kazakhstan). Differs markedly from previous group in the lack of a black patch on crown and nape. Only black on head is loral mark. Differs also in having uniformly drab grey crown to back (lacking rufous or tawny-brown mantle and back), a greyish (rather than rufous or ochre-tinged) breast-band and upper flanks, pure white outer webs to tertials, and lacks white tips to primaries. Furthermore, has more extensive white patch on rump/uppertail-coverts, and more extensive white on inner webs of outer tail-feathers. Bill usually markedly longer with a more attenuated point. Sexes alike, or are so similar that sexing is usually impossible. ♂♂ have on average more extensive red on head than ♀♀, but much more overlap than in *carduelis* group, making the character of limited use. Hybridises with *brevirostris* where they meet in Iran, but extent not known. **L** 12.5–14 cm. **W** ♂ 79–92 mm (*n* 26, *m* 84.0), ♀ 77–87 mm (*n* 22, *m* 81.8); **T** ♂ 49–57 mm (*n* 26, *m* 52.9), ♀ 48–57 mm (*n* 21, *m* 51.8); **T/W** *m* 63.1; **B**(f) 12.3–17.0 mm (*n* 51, *m* 14.5); **BD** 7.0–8.4 mm (*n* 50, *m* 7.7); **Ts** 13.0–15.5 mm (*n* 50, *m* 14.4). Wing formula very similar to that of *carduelis*. (Syn. *orientalis*; *subcaniceps*.)

? **C. c. ultima** Koelz, 1949 (Fars district in SC Iran). Very similar to *paropanisi*, only said to have a longer bill (by 'about 1.5 mm'; Vaurie 1959). No material seen.

Extralimital races: **C. c. caniceps** Vigors, 1831 (W Himalayas from Gilgit east; resident or short-range movements). Similar to *paropanisi* but is darker overall, not as pale sandy brown-grey on upperparts and flanks, more tawny.

C. c. subulata (Gloger, 1833) (Altai, W & N Mongolia, Baikal region; winters SW & C Asia). Resembles *caniceps* but is decidedly paler throughout.

TAXONOMIC NOTE Although the two racial groups look sufficiently different morphologically to merit a split into two species, hybridisation where they meet (SW Siberia, NE Iran) appears to be common and is well known; more research into the relations between the two groups is required before the case can be settled.

REFERENCES Robertson, D. (1995) *Tay Ringing Group Report 1996–97*, p. 4–6.

C. c. paropanisi, ♀, presumed ad, Iran, Jan: on some birds possible to see hint of pinkish hue admixed in grey flanks. Wing is seemingly evenly-feathered ad. Note small red face and whitish nasal hairs. (E. Winkel)

C. c. paropanisi, ♂, presumed 1stS, Kazakhstan, Jul: strikingly long and pointed bill in this race. Also, often appears smudgy dusky-grey below, as here on breast, and paler grey on flanks. A ♂ considering amount of red on face, with worn tips to tail-feathers which might indicate 1stS, but best to refrain from certain ageing. (A. Belyaev)

C. c. paropanisi, 1stS ♀, Kazakhstan, Mar: like previous, with contrastingly paler primary-coverts, but being a young ♀ has much smaller red mask that is broken up and paler red or even tinged orange. Note distinctly black-tipped bill. (N. Kim)

CITRIL FINCH
Carduelis citrinella (Pallas, 1764)

Fr. – Venturon montagnard; Ger. – Zitronengirlitz
Sp. – Verderón serrano; Swe. – Citronsiska

A small finch that is found only in mountains of C and SW Europe. The best chance to locate the species is to search alpine meadows or clearings in the upper forest zone, just below the tree-line, and to look for stands of thistles or dandelions that have ceased flowering. There, with a little luck, a small party of Citril Finches will be found feeding. Once the call is learnt, they can also be picked out flying overhead, often seemingly restless on their way somewhere else. Shorter migrations are noted in Germany, NE France, Austria and Switzerland, whereas birds in the Pyrenees and in Iberia only make altitudinal movements.

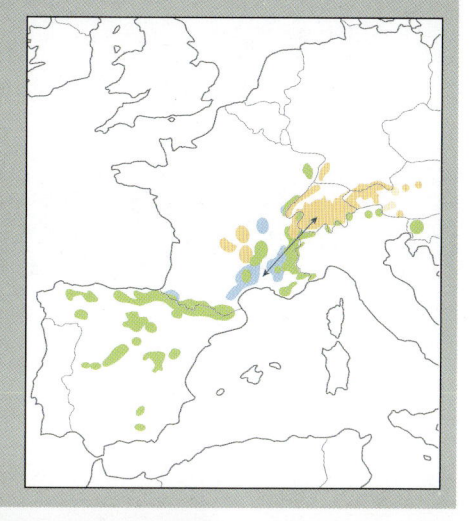

VOCALISATIONS Song is rather similar to that of Goldfinch, but differs slightly in being feebler and often containing Redpoll-like buzzing trills. Some phrases can recall parts of Chaffinch song (the flourish). A song can be built up using several phrases with different motifs separated by brief pauses, but at other times the song is more continuous. – Call is a brief, monosyllabic, nasal and metallic *teh*, either uttered singly or quickly repeated in a trembling trill, *tehehe*, when resembles the call of Red-fronted Serin. Other calls include a sharp *ziit*, an upwards-inflected *tooee* and a nasal piping *tüüt*. Some of these calls, especially when uttered alternatively, can recall flight-calls of Siskin.

SIMILAR SPECIES After the recent split of birds on Corsica and Sardinia as a separate species, the *Corsican Finch*, this has become the major pitfall when identifying a Citril Finch. Note that Citril Finch does not have hazelnut-brown back like Corsican but is mainly greyish-green above. However, this can be difficult to see in poor light, as can the subtly less distinct streaking on the back. Supplementary characters are therefore often useful. Citril Finch has clearer yellow rump patch, but subtly less strong yellow on face and underparts, whereas the opposite is true for Corsican Finch, which has a more subdued greenish rump patch and quite strong yellow on face, eye-surround and

Ad ♂, Spain, Feb: note dark wing with broad yellow-green wing-bar and remiges panel; whole wing-bend also concolorous yellow-green. Ad ♂ is unstreaked greenish-yellow (brightest on face and rump) with ash-grey nape and breast-sides. Note that centre of breast is unbroken greenish-yellow. Evenly-feathered wing further confirms age. (A. Torés Sanchez)

1stS ♂, Switzerland, Apr: very similar to ad ♂, but differs by having outer three greater coverts retained juv, whitish-tipped and more brown. Primaries and primary-coverts, too, are juv, brown-tinged and slightly worn at tips. Breast is greenish-yellow, lacking ash-grey band (perhaps just visible here), thus plumage typical for ♂♂. (P. Hürlimann)

IDENTIFICATION Clearly smaller and slimmer than Greenfinch, more the size of Siskin but subtly *longer-tailed and more elongated*. Plumage predominantly greenish-yellow. *Two broad yellowish-green wing-bars* characteristic of all plumages except juveniles, which have narrower buff-white wing-bars. Note *greenish-yellow unstreaked rump patch*, easily seen in flight, but *all-dark tail-feathers*. Adult ♂ has *unstreaked ash-grey crown and nape*, continuing onto neck-sides and often to breast-sides, *greenish-yellow face and eye-surround*. Underparts unstreaked greenish-yellow. Mantle and back are duller olive-tinged greyish-brown faintly streaked dark. ♀ similar but slightly duller with a faint brown-grey cast on nape, which is diffusely streaked or mottled. Mantle and back slightly more distinctly streaked than in ♂. Furthermore, ♀ has *grey across breast* forming a diffuse or obvious breast-band, while ♂ has no grey, or grey only on sides, leaving *central breast greenish-yellow*. Juvenile is more streaked, especially on breast, and has less green and yellow in plumage, being more tinged buff-brown. Often encountered in small family parties. Like most small finches, rather restless and constantly on the move. Generally noticed by the nasal, discreet but characteristic calls.

♀, Spain, Jun: ♀♀ are duller than ad ♂, with reduced yellowish-green and has grey breast-band. Mantle and back suffused brown-grey and diffusely streaked. Note grey body-sides, often faintly streaked. Yellow face duller and less extensive, and head partly streaked, especially on crown. Wing markings as in ♂, but duller and more subdued. Ageing uncertain. (C. Bradshaw)

entire underparts. Long uppertail-coverts in Citril Finch are greenish-tinged but greyish in Corsican Finch. – *Siskin* is shorter-tailed and compact-looking with a slightly longer bill. It differs clearly by having yellowish tail-sides and a better-marked wing pattern, with a broad yellow or yellowish wing-bar surrounded by blackish. ♂ Siskin has dark crown and chin. – *Serin* can theoretically be confused with Citril Finch but is slightly smaller with a short and stubby bill, is more boldly streaked dark and has pale surround to the dark cheek patch. – *Greenfinch* is altogether much larger and heavier with a heavier bill.

AGEING & SEXING Ages differ readily in summer, and also nearly always in autumn–spring. Sexes differ following post-juv moult. Size difference between sexes insignificant. – *Moults*. Complete post-nuptial moult of ad in late Jul–Oct (early Nov). Partial post-juv moult in late summer or autumn does not involve flight-feathers, tail-feathers, primary-coverts or usually outer greater coverts (only very few moult all greater coverts). No pre-nuptial moult other than by odd birds, presumably 1stW, replacing scattered feathers on head, central tail-feathers and odd tertials. – **SPRING** ♂ Throat

1stS ♀, Switzerland, May: irrespective of age, some ♀♀ are neater and yellower, approaching less colourful young ♂, still recognised by unbroken grey across upper breast. Aged by retained juv outer greater coverts. (C. Plummer)

Ad ♀, Spain, Sep: p2 apparently still growing, indicating ad in post-nuptial moult, and less yellow and greenish in plumage inferring ♀. Combination of yellow face but otherwise grey head, neck and breast, and greenish-yellow tinge to rump and uppertail-coverts, and also wing pattern, identify such dull birds to species. (G. Reszeter)

Ad ♀, Spain, Nov: when fresh, upperparts can be tinged brown and streaked dark, thus more like Corsican Finch. Nevertheless, still has clearer yellow rump (with greenish, not greyish, uppertail-coverts) but less obvious yellow on face and underparts (the opposite in Corsican). Citril also tends to lack Corsican's clearer yellow eye-surround. Evenly fresh wing, rounded primary-tips and heart-shaped secondary-tips confirm ageing. (F. Trabalon)

Ad ♂, Germany, Nov: following post-nuptial moult evenly very fresh, and ad ♂ typically bright and rather extensively yellowish, including on wing-bars, face, breast and rump, and pale and pure ash-grey on nape and neck-sides. Yellow colour more dull and greenish-tinged than in spring and summer. (T. Grüner)

1stW ♂, Spain, Nov: yellow areas on average somewhat duller and more restricted than in ad ♂, and grey maybe a bit more extensive on head and flanks, and even forms hint of narrow semicollar (though never a broad breast-band as in ♀). Primary-coverts juv, lacking any yellowish-green edges, and tail-feathers seem pointed. (F. Trabalon)

and centre of breast unbroken yellowish-green, lacking grey band on lower throat/upper breast. (Sometimes ash-grey breast-sides extensive, creating impression of grey breast-band at many angles.) Lower breast/belly rather bright yellowish, invariably unstreaked. Undertail-coverts usually rather prominently tinged yellow. Mantle/back greenish-grey with only faint dark streaks. Rump usually unstreaked bright yellowish-green. Forecrown and face yellowish-green, rear-crown/nape and neck-sides ash-grey, unstreaked. ♀ Overall darker and duller than ♂. Chin/upper throat and lower breast dull yellowish-green with some grey tones, separated by greyish band on lower throat/upper breast. (Very rarely grey breast-band narrower in centre, but never acquires ♂-like appearance.) Lower breast/belly dull yellowish-green with grey tinge, in some with faint streaks on flanks and sides of belly. Undertail-coverts usually largely whitish, but can have faint yellow tinge basally. Mantle/back greyish-green with slight brown tinge, diffusely streaked dark. Rump rather dull yellowish-green, less bright and contrasting than ♂. Forecrown (dull) yellowish-green, less extensively so than ♂, rear crown/nape greyish (sometimes with slight brown hue), finely streaked. **F.g.** No contrast in wear or colour visible among greater coverts, tertials or tail-feathers. Tips to tail-feathers slightly rounded, neatly edged. **1stS** Not always separable, but some may show slight contrast in wear and colour of inner moulted (tipped and edged green) and outer unmoulted greater coverts (tipped and edged greyish-white). Tips of tail-feathers on average narrower and more pointed than in ad, but intermediates occur, and wear makes difference progressively less obvious. – AUTUMN Sexing as in spring. **Ad** Double wing-bars broad and yellowish-green (duller in ♀) with no moult contrast among greater coverts or tertials. Edges to flight-feathers yellowish-green or dull greenish. Tail-feathers first in moult, then fresh with neat greenish-white edges, tips slightly rounded. **1stW** On average somewhat duller and more heavily streaked above than ad, though some are quite similar. A variable number (usually 2–5) of unmoulted juv outer greater coverts tipped buffish- or off-white, contrasting markedly with moulted inner (broadly tipped yellowish-green). Some show contrast between juv and new tertials. Flight-feathers thinly edged whitish (lacking green tinge). Tail-feathers on average slightly narrower with more pointed tips, becoming progressively more abraded in late autumn. **Juv** Easily separated by lack of nearly all yellow and green on head and by having underparts buff-white with only faint yellow tinge, streaked grey-brown. Upperparts rather prominently streaked, including rump. Wings and tail fresh in summer, progressively more abraded in winter, edges pale yellowish-white, soon bleaching to off-white, lacking clear yellowish-green element. Double wing-bars rusty-buff with only slight yellow tinge.

Presumed 1stW ♂, Switzerland, Dec: combination of indistinct dark mantle streaking and only narrow grey semicollar, and yellow belly reaching lower breast, suggest 1stW ♂. The young age, with all remiges and primary-coverts retained juv, and moult limit in tail, further support this sex. (C. Plummer)

1stW ♀, Spain, Dec: on average paler, streakier and shows less yellow than autumn ad ♀. Aged by moult limit between renewed tertials and retained abraded brownish primaries without whitish tips. (J. Blasco-Zumeta & G.-M. Heinze)

Juv, Switzerland, Jul: lacks yellow and green, with streaked buff-white underparts. Double wing-bars tinged rusty-buff. (H. Shirihai)

BIOMETRICS L 11.5–12.5 cm. Sexes of same size: **W** 75–80 mm (n 43, m 77.7); **T** 51–57.5 mm (n 43, m 54.2); **T/W** m 69.7; **B** 7.3–10.1 mm (n 43, m 8.8); **BD** 5.0–6.6 mm (n 43, m 6.1); **Ts** 13.4–15.3 mm (n 41, m 14.3). **Wing formula: p1** minute; **pp2–4** longest, p2 and p4 sometimes 1 mm < wt; **p5** < wt 1–2.5 mm; **p6** < wt 6.5–10 mm; **p7** < wt 12–15 mm; **p10** < wt 17.5–23.5 mm; **s1** < wt 21–25 mm. Emarg. pp3–5.

GEOGRAPHICAL VARIATION & RANGE Monotypic. – Iberia except south-west, S France, Alps, S Germany, NW Balkans; mainly resident.

REFERENCES HYNDMAN, T. (2008) *BW*, 21: 243–249.

CORSICAN FINCH
Carduelis corsicana (Koenig, 1899)

Fr. – Venturon corse; Ger. – Korsengirlitz
Sp. – Verderón corso; Swe. – Korsikansk siska

A close relative of the Citril Finch, and was until fairly recently regarded as a local race of that, but is now commonly treated as a separate species. It is a resident endemic on Corsica and Sardinia, occurring only on these islands, where it replaces Citril Finch. Found in similar habitat as Citril Finch, including open country with scattered trees and bushes near the tree-line, but also frequently occurs at lower levels, will enter gardens and can be seen in a variety of habitats down to sea level.

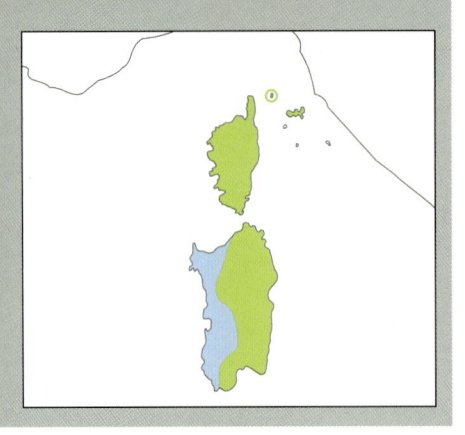

♂, presumed ad, Corsica, Jun: mantle and scapulars brown with bold dark streaking (not washed olive-grey with only thin streaks, as in Citril Finch). Also yellow on face and underparts more extensive and brighter, but rump greenish-yellow and long uppertail-coverts grey. Grey areas around head and neck limited in comparison. All wing seemingly ad. (I. Merrill)

IDENTIFICATION Very similar to its close relative, Citril Finch, with nearly the same size and proportions (only slightly smaller and shorter-winged), and much the same habits. Shares most plumage characters, like *broad double greenish wing-bars*, a *yellowish-green rump patch*, *all-dark tail* and *greyish nape and neck-sides*. Differs subtly in vocalisations and in the following plumage characters: in ♂ plumage, *back is hazelnut-brown* with *more bold dark streaks* than in Citril Finch; *rump more subdued or dull greenish* (Citril: clearer yellowish-green); *underparts* on average *clearer yellow* (Citril: slightly less bright greenish-yellow); and *face and forecrown more extensively yellow*. A supplementary character is provided by the *long uppertail-coverts* being largely *greyish* (Citril: greenish). The ♀ is less pure brown on back than ♂, more greyish-brown, and yellow in plumage is more restricted and subdued. It differs from ♀ Citril Finch in having *slightly more and clearer yellow* on underparts and face. *Does not develop such a clear grey breast-band* as ♀ Citril Finch, is more diffusely greyish-green with some yellowish tones and just a hint of more grey on breast than rest of underparts. Grey nape is diffusely streaked as in ♀ Citril Finch.

VOCALISATIONS Very similar to those of Citril Finch. Song as Citril but averages longer and is more structured, thereby being somewhat reminiscent of Goldfinch and even Wren. – Call like Citril, but has been perceived as being a little feebler with a 'trembling' quality. Both simple monosyllabic calls and multisyllabic brief, nasal trills are heard from flocks in flight.

SIMILAR SPECIES On Corsica, both Siskin and Serin also breed and must be eliminated. On Sardinia, only Serin can cause confusion. Corsican Finch separated from *Siskin* by longer, all-dark tail and slimmer shape, less striking wing-bars and unstreaked underparts. ♂ Siskin has blackish crown and chin. – *Serin* eliminated by its shorter bill, pale surround to darker cheek patch, and boldly streaked underparts.

AGEING & SEXING Ages differ in summer, sometimes in autumn and even in spring if seen well or trapped. Sexes differ in plumage following post-juv moult. Size difference very slight. – Moults. Complete post-nuptial moult of ad mainly in Jul–Sep. Partial post-juv moult in late summer or autumn (protracted due to long breeding season) does not involve flight-feathers, tail-feathers, primary-coverts or outer greater coverts. No pre-nuptial moult. – **SPRING** ♂ Forecrown, face and all of underparts (except undertail-coverts) rather bright yellow (only limited olive-grey tinge to breast-sides and flanks in many). Rear-crown/nape and neck-sides unstreaked ash-grey (or dull olive-grey; with very subdued and fine streaks only on nape). Mantle/back rather bright hazelnut-brown, prominently streaked dark. Rump usually unstreaked dull yellowish-green. ♀ Overall slightly darker and much duller than ♂. Chin/upper throat and lower breast dull yellowish-green with some grey tinge and often a few diffuse streaks on sides and belly. Lower throat/upper breast a mixture of dull yellowish-green and brownish-grey, often having a rather untidy, 'dirty' appearance. Mantle/back greyish-brown (less bright brown than ♂), diffusely streaked dark. Rump dull and dark yellowish-green, less bright and contrasting than in ♂, and often diffusely mottled or streaked. **F.g.** No contrast in wear or colour visible among greater coverts, tertials or tail-feathers. Tips to tail-feathers slightly rounded, neatly edged. **1stS** Not always separable, but some may show slight contrast in wear and colour of inner moulted (tipped and edged green) and outer unmoulted greater coverts (tipped and edged greyish-white). Tips of tail-feathers on average narrower and more pointed than in ad, but intermediates occur, and wear makes difference progressively less obvious. – **AUTUMN** Sexing as in spring. **Ad** Double wing-bars broad and yellowish-green (duller in ♀). Edges to flight-feathers yellowish-green or dull greenish. Tail-feathers first in moult, then fresh with neat greenish-white edges, tips slightly rounded. **1stW** On average somewhat duller colours and more heavily streaked above than ad, though some are quite similar. A variable number of unmoulted juv outer greater coverts tipped buffish- or greyish-white contrast with moulted inner tipped yellowish-green. Flight-feathers thinly edged whitish (lacking green tinge). Tail-feathers on average slightly narrower with more pointed tips, becoming progressively more abraded in late autumn. **Juv** Easily separated by lack of nearly all yellow and green on head and body, these being largely replaced by buff-white and grey-brown, with only faint yellow cast. Both upperparts and underparts prominently streaked, including rump and breast. Wings and tail fresh in summer, then progressively more abraded in winter, edges pale yellowish-white, soon bleaching to off-white, lacking clear yellowish-green element. Double wing-bars rusty-buff with only slight yellow tinge.

BIOMETRICS L 11.5–12.5 cm. Sexes nearly the same size: **W** ♂ 70–76 mm (n 24, m 73.9), ♀ 69–75 mm (n 15, m 72.1); **T** ♂ 49–54.5 mm (n 24, m 52.0), ♀ 47–53 mm (n 15, m 50.5); **T/W** m 70.2; **B** 6.9–9.6 mm (n 39, m 8.4); **BD** 5.2–6.4 mm (n 38, m 5.8); **Ts** 13.8–15.7 mm (n 36, m 14.7). Wing formula: **p1** minute; **pp2–4** longest (p2 and p4 sometimes 0.5–1 mm < wt); **p5** < wt 0.5–3 mm; **p6** < wt 7–10 mm; **p7** < wt 11–14 mm; **p10** < wt 18–21 mm; **s1** < wt 19–23.5 mm. Emarg. pp3–5.

GEOGRAPHICAL VARIATION & RANGE Monotypic. – Corsica, Sardinia; resident.

TAXONOMIC NOTE Previously regarded as a subspecies of Citril Finch, but nowadays generally treated as a separate species, mainly due to consistent differences in plumage

Ad ♀♀, Corsica, May (left) and Apr: compared to ♂, drab greyish-brown above, and yellow on face more restricted, thus more like ♀ or 1stY ♂ Citril Finch. Does not develop clear grey breast-band like ♀ Citril, and long uppertail-coverts diagnostically grey. More and clearer yellow on underparts and face, rather untidy and 'dirty' (and faintly streaked) body-sides are also features of this species. Aged by evenly-feathered wing. (Left: M. Zekhuis; right: G. Howard)

FINCHES

and vocalisations, partially supported also by differences in habitat preferences. There is also a genetic difference, which varies between 2.7% and 3.2% depending on the methods applied (Pasquet & Thibault 1997, Förschler et al. 2009).

REFERENCES FÖRSCHLER, M. I. et al. (2009) Mol. Phyl. & Evol., 52: 234–240. – PASQUET, E. & THIBAULT, J.-C. (1997) Ibis, 139: 679–684.

♂, presumed 1stS, Corsica, May: note that in ♂♂ underparts are more purely yellow and neatly unstreaked, though flanks often have ash-grey tinge. Yellow on face is more extensive than in Citril Finch, and mantle and back is dark-streaked nut-brown. Wing pattern is identical to Citril, though. Tips to primaries fairly abraded, and central tail-feathers newer than rest of tail suggesting age. (F. Pelsey)

Presumed 1stS ♀, Corsica, Jul: while dull yellow underparts with diffuse streaking or flecking is safest sexing criterion for ♀, dull and dusky yellow face, dusky yellowish wing-bars, brownish primary-coverts and very dull brown-grey dark-streaked mantle favour 1stS. In fact, such olive-grey mantle-colour invites confusion with Citril Finch, but note grey long uppertail-coverts and extensive yellowish face. (J.-M. Fenerole)

Ad ♂, Corsica, Aug: in post-nuptial moult, central primaries and outer tail-feathers growing. When fresh, similar to worn summer ad ♂, but plumage less distinct due to some broad pale feather-tips. Note especially that border to yellow face is rather diffuse, and yellow below is pale. (L. Khil)

♀, presumed ad, Corsica, Aug: like fresh ad ♂ but overall much duller, in particular on face. Breast pattern of ♀♀ variable, some being patchy and streaked, others have more grey on sides of breast. Primaries and tertials appear quite fresh and dark-centred, hence presumed ad. (L. Khil)

Presumed 1stW ♂, Corsica, Oct: primaries, primary-coverts and outermost greater covert are probably juv, but image does not permit safe ageing. Degree of yellow on face and underparts, very faint streaking on crown and mantle, but none on flanks seem to infer young ♂. This plumage can be very similar to Citril, but at this season Corsican lacks Citril's greenish-yellow rump and uppertail-coverts. (G. Jenkins)

♀, presumed 1stW, Corsica, Oct: sex inferred by dull and subdued plumage pattern, yellowish flanks tinged brown, and grey of nape and neck-sides blotched and dusky. New central tail-feathers growing while rest appear to be juv. (A. Easton)

Juv, Corsica, Aug: mostly buff-white and grey-brown, and prominently streaked throughout. Separated from other juv Carduelis by cream-buff wing-bars accentuated by black feather bases. (U. Paal)

(EURASIAN) SISKIN
Spinus spinus (L., 1758)

Fr. – Tarin des aulnes; Ger. – Erlenzeisig
Sp. – Jilguero lúgano; Swe. – Grönsiska

A small elegant finch of tall and extensive coniferous forests, but which also feeds in stands of alders or birches, and in winter can visit feeders in gardens. Often seen in pairs or quite small parties in summer, but in autumn and winter the Siskin forms large flocks in favoured areas. A migrant in the north and east, but short-range movements are more common in the south and west, and some birds are resident. Being particularly fond of spruce seeds, it follows that the species will be forced to make mass movements in some years, at least locally, when spruce crops are poor.

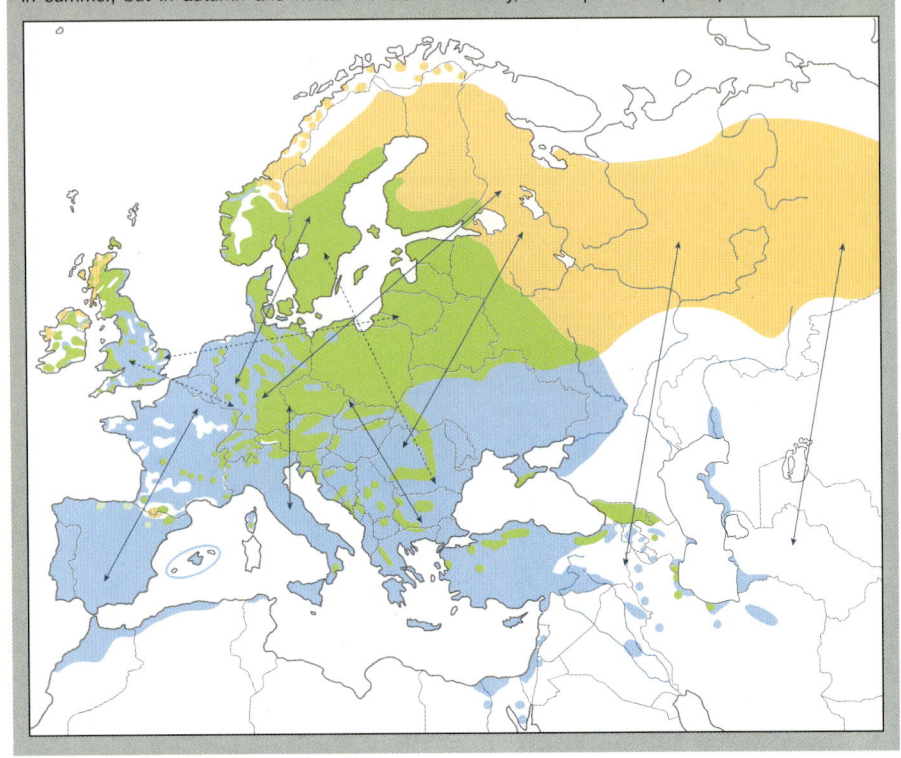

IDENTIFICATION This *long-winged* but rather *short-tailed* finch has the size of a redpoll but is rather more *compact*. It has a seemingly *small head* with a *pointed, conical and rather long bill*, and *short legs*. The tail is forked, a common characteristic of most finches. In flight readily identified by combination of *bold yellow wing-bar on rather pointed, dark wings, yellowish rump* and *some yellow at base of outertail*. When seen well, a perched bird is characteristic enough: *the dark folded wing has two broad yellow wing-bars*, and often visible is *a smaller yellow patch angled to the wing-bar* formed by yellow bases to the inner primaries (or at least their edges are basally yellowish-white). Tertials blackish with contrasting whitish outer edges. Finally, there is the pointed, conical bill, which is rather pale pink-brown (tip and culmen can be darker), and the rather short tail. Summer ♂ is *yellowish-green* on head and breast, brightest on supercilium and upper breast, and has *black crown* and usually *a small black bib*, both of which lacking in the somewhat duller ♀, which has a pale chin and grey-green streaked crown. In all plumages *rather prominently streaked dark on flanks and belly* (though adult ♂ less so), but only *finely streaked on grey-green mantle*. Juvenile even duller than ♀, with yellow washed out, but identified by shape and wing pattern. Usually seen in small parties or larger flocks when feeding, sometimes with redpolls in winter. Very agile and will climb the thinnest branches in birch or alder, sometimes hanging upside-down like a small tit. Flight is fast and undulating, rather lighter and more 'erratic' than Goldfinch, and clearly lighter and more skipping than Greenfinch. Often shy or wary, and seems restless, the whole flock suddenly taking off without obvious reason, either settling again after circling for a while, or flying off far. Nevertheless, can enter gardens and be rather tame at feeders.

VOCALISATIONS The song is a fast twitter, rather similar to that of Goldfinch albeit slightly faster and longer, now and then relieved by a drawn-out choking or wheezing sound (peculiarly, the same song structure occurs in Barn Swallow). Many Siskins include fragments of mimicry of other bird species in the song, and a few are highly accomplished at this. – Flight call, often the best means of detecting a Siskin when it passes overhead, is a disyllabic downward-inflected *teeluh* alternating with upwards-inflected *tluee*. The calls, uttered singly, usually have a rather desolate ring. Feeding flocks utter low, dry *te-te-te* contact calls.

SIMILAR SPECIES The *Serin* is similarly sized, also yellow-rumped and streaked, has yellowish wing-bars, and the ♂ is greenish-yellow on head, breast with some yellow on tail-base, so could be confused with ♂ Siskin. However, Serin has a proportionately longer tail, the bill is very short, and is much more boldly dark-streaked, especially on back. Also, ♂ Serin has yellow forehead and dark cheeks, whereas ♂ Siskin has black crown and rather pale yellowish-green cheeks. The yellowish wing-bars in Serin are also much narrower, and there is no yellow patch on the folded primaries like in Siskin. – *Citril Finch* and *Corsican Finch* differ in their longer all-dark tail, slightly shorter and darker bill, less prominent yellow wing-bars that do not extend to base of primaries, and the more diffuse streaking even in juvenile plumage. – Only *Goldfinch* has anything like such prominent yellow wing-bars as Siskin, but differs in a number of obvious

Ad ♂, Italy, Mar: in all plumages, readily identified by compact shape and short tail, long wings and pointed conical bill. Boldly banded yellow-and-black wings. Yellow base to outertail and yellow rump also act as clues. In ♂, note black crown, with throat and breast largely unstreaked, and often small black bib. Aged by evenly-feathered wing. (D. Occhiato)

Ad ♀, England, Mar: rather finely dark-streaked greyish olive-green upperparts, lacking strong head pattern of ad ♂. Note ill-defined streaky supercilium, and whitish underparts often show dull dark streaking. Wing pattern as ♂, but bands narrower and yellow duller, with basal primary patch much smaller. Evenly-feathered wing indicates age. (G. Thoburn)

ways. Even in a fleeting view of a bird flying away, note Siskin's yellow rump, some yellow at tail-base and rather dark, greenish-tinged streaked back, while Goldfinch has a white rump, black-and-white tail and rather pale uniform tawny-brown back. – *Common Redpoll* shares Siskin's autumn and winter habit of feeding in dense, large flocks in alder and birch, and keeps the same dense flock formation, with a similarly undulating, light and 'nervous' flight, but seen close lacks any green or yellow tones, and have a red forecrown patch.

AGEING & SEXING Ages differ subtly in autumn and spring, but reliable ageing requires close views. Moderate seasonal variation. Sexes separable at least after post-juv moult; if seen close or handled also in juv plumage. – Moults. Complete post-nuptial moult of ad in late summer and early autumn, mainly mid Jul–early Oct. Partial post-juv moult at about same time does not usually involve flight-feathers, tail-feathers, tertials or primary-coverts, but birds from early broods, especially in C & S Europe, can moult some, most or even all tertials and tail-feathers, and even some odd central primaries. In C & S Europe, often all greater coverts moulted, but in N Europe fairly common for a slight moult contrast between (1) 2–5 outer unmoulted (on average fewer in ♂♂) and remaining inner coverts, which are moulted. No pre-nuptial moult in late winter. – **SPRING Ad** ♂ Crown black. Throat and breast largely unstreaked greenish-yellow, chin often (but not invariably) has small black bib. No moult contrast among greater coverts. Outer three tail-feathers have extensive deep yellow bases, tipped dark grey, with sharp and rather straight division between dark tip and yellow base. Inner web of r6 yellow to shaft; only part of shaft and outer web dark; rr4–5 yellow across both webs (Svensson 1995). Shape of tail-feathers slightly more rounded at tips than in imm, but difference less obvious than in many finches, and wear in spring makes difference in shape progressively more difficult to use. **1stS** ♂ As ad ♂, and many inseparable if not seen well or handled, but often possible to recognise by moult contrast between some retained outer greater coverts (narrowly tipped paler yellowish-white) and inner (more broadly tipped greenish). Also, yellow on bases of rr4–6 is paler, and their dark tips are less dark grey and more extensive, with a usually uneven, irregular division to yellow bases. Tail-feathers slightly narrower and more pointed at tips than in ad. **Ad** ♀ Crown green-grey streaked dark. Throat and breast off-white with yellowish-green tinge, breast and belly usually prominently streaked; never has black bib. Outer three tail-feathers have pale yellow bases, tipped rather extensively dark. Inner web

1stS ♂, Finland, Jun: duller than ad ♂, and readily recognised by retained outer greater coverts, which are narrowly tipped paler yellowish-white. Plumage usually much less vivid green and yellow than ad, with yellowish primary patch and tail-edges smaller. However, some individuals are more like ad ♂ and less easily aged without handling. (M. Varesvuo)

1stS ♀, Scotland, Feb: as ad ♀, but note that new ad-like inner greater coverts are broadly tipped yellowish, whereas juv outer ones are both shorter and more tipped buffish-white. Note also pointed tips to juv tail-feathers. Some ♀♀ have stronger yellowish supercilium and ear-coverts surround than others, but apparently not age-related. (S. Round)

of r6 not yellow to shaft, sometimes not even yellow to shaft on r5 (Svensson 1995). Shape of tail-feathers as ad ♂. **1stS** ♀ As ad ♀, but sometimes possible to recognise by moult contrast among greater coverts (see ♂). Also, pale on bases of rr4–6 is very narrow and restricted to broad inner edges of inner web on all three, with very little yellow tinge at all. Shape of tail-feathers as 1stS ♂. – **AUTUMN Ad** ♂ Much as ad ♂ in spring, although black crown-feathers finely tipped grey. Easier to use shape of tail-feathers than in spring. **1stW** ♂ Much as 1stS ♂. Tail-feathers narrower and more pointed than in ad, tips slightly abraded from late autumn. ♀ Much as in spring. Ageing as ♂. **Juv** Resembles ♀ but is somewhat paler greyish, less green-tinged, with more prominent dark streaking above and below. All greater coverts tipped buffish, soon bleaching whitish. Tail pattern helps sex any bird seen well or handled.

BIOMETRICS L 11–11.5 cm. Sexes nearly similar size: **W** ♂ 70–76 mm (*n* 35, *m* 73.1), ♀ 69–74 mm (*n* 35, *m* 71.2); **T** ♂ 42–47.5 mm (*n* 35, *m* 44.6), ♀ 42–46.5 mm (*n* 35, *m* 43.7); **T/W** *m* 61.2; **B** ♂ 11.0–13.0 mm (*n* 18, *m* 12.0), ♀ 10.2–12.1 mm (*n* 25, *m* 11.1); **BD** 6.1–7.4 mm (*n* 39, *m* 6.6); **Ts** 13.0–14.4 mm (*n* 41, *m* 13.6). **Wing formula: p1** minute; **p2** 0–1.5 mm < wt; **pp3–4** longest; **p5** < wt 4–6.5 mm; **p6** < wt 10.5–13 mm; **p7** < wt 15–17.5 mm;

Sweden, Sep: in flight often told by bold yellow wing-bars, yellowish rump and yellow base of outertail, but also readily by characteristic calls once learnt. (M. Varesvuo)

p10 < wt 22.5–28 mm; **s1** < wt 25–29 mm. Emarg. pp3–4, also sometimes slightly on p5.

GEOGRAPHICAL VARIATION & RANGE Monotypic. – Mainly N & C Europe, local in Iberia, France, Italy, Alps, N Balkans, N Turkey, Caucasus, Transcaucasia, east through Russia and Siberia to Sakhalin and Ussuriland; northern breeders move south to S Sweden and Baltic States, C & S Russia; some reach coast of N Africa, Levant; partly nomadic and numbers fluctuate in response to food supply.

REFERENCES Svensson, L. (1995) *Ringers' Bull.*, 9(3): 18.

Ad ♂, Spain, Jan: evenly very fresh plumage, and hence black crown-feathers finely tipped grey. Yellow parts also washed duller or somewhat more greenish-tinged when fresh. (C. N. G. Bocos)

Ad ♀, Spain, Jan: evenly-feathered wing, overall very fresh including neat dark tail-feathers with pale fringes, and body plumage more diffusely streaked than in spring or summer. (C. N. G. Bocos)

1stW ♂, France, Oct: young ♂ in fresh plumage can appear intermediate between the sexes, as black crown is more concealed and black bib hardly indicated. Also often has weaker wing markings. Nevertheless, unlike ♀ usually has dark crown, unstreaked face, breast and rump, with more yellow tones, especially on supercilium. (A. Audevard)

1stW ♀, Germany, Oct: might be confused with young ♀ Serin, but that species lacks bright yellow-and-black wing-bars, and bill is never as pointed, and Siskin lacks a dusky moustachial-stripe. With experience, jizz and markings separate them almost instantly. Combination of dullest plumage, with moult limit in outer greater coverts determines sex and age. (T. Krüger)

Juv ♂, England, Aug: heavily streaked above and below, and has least yellow-tinged plumage. Also rufous-buff wash to head and wing-bars (soon bleaches to off-white). Relatively obvious yellow on rather large primary patch and edges of flight-feathers and tail-base indicate juv ♂. (M. Goodey)

(COMMON) LINNET
Linaria cannabina (L., 1758)

Fr. – Linotte mélodieuse; Ger. – Bluthänfling
Sp. – Pardillo común; Swe. – Hämpling

On a seashore or heath with junipers or brambles, or from a hedge in an open agricultural landscape, you suddenly flush two small birds that take off together in fast, skipping flight; chances are that you have come across the Linnet. In fact, there are few passerines that seem to possess such a close pair-bond as Linnets, since the two adults are always seen together and the one cannot bear to be left behind, so always follows the other. The Linnet is mainly a bird of open landscapes, but will breed in gardens and open woodland with glades and bushes. It is a migrant in N and E Europe.

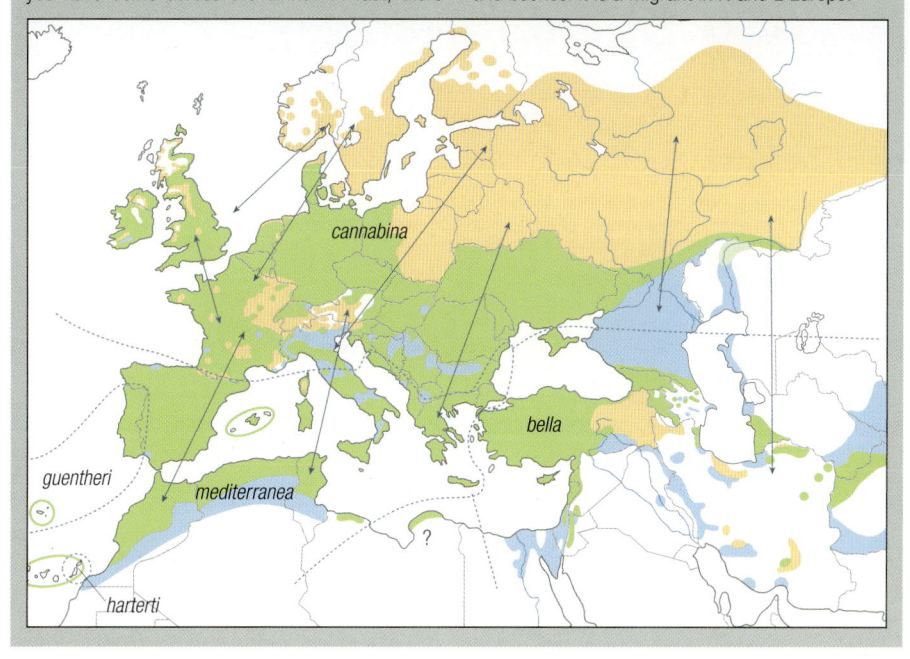

L. c. cannabina, 1stS ♂, England, Jun: greyish head with typical isolated pale central cheek patch, chestnut-brown back and in ♂♂ crimson-red forecrown and breast patches. Note also pure white primary fringes and outer tail-feathers. Occasionally, 1stY moult some inner and/or (as here) central primaries in post-juv moult. (J. Richardson)

IDENTIFICATION A *slim, rather long-tailed* finch with a *stout, greyish bill*. Like most finches the tail is notched at the tip. Both sexes have *warm brown back*, the ♂ brighter and unmarked, the ♀ a little duller and lightly dark-streaked. Both have white edges to bases of primaries and tail-feathers, which form narrow *white panels on the folded wing and tail*. ♂ has *red forecrown* and a *red breast patch*, the latter semi-divided in the centre. Note that while the red is a bright crimson or poppy-red in summer, it is much more subdued and brownish in fresh autumn plumage, and first-summer ♂♂ can remain in a less attractive plumage for longer. Both sexes have *greyish-tinged heads with paler eye-surround and an isolated cheek patch*, a basic pattern that they share with Twite and Serin. Head of the ♂ is slightly purer grey than that of the ♀. In all plumages the pale throat has tiny dark blotches or fine streaking on the centre. Flight is fast and undulating, the birds frequently taking off in towering flight. Before landing in a bush, they usually zigzag or make short jerky movements in the air before selecting a perch. Juveniles can be slightly problematic to identify, especially ♀♀ since these have least white visible on edges of flight-feathers, but bill shape and lack of any yellow or green help, and the typical pattern of paler centre to cheeks surrounded by dark is already discernible, albeit more diffuse.

VOCALISATIONS The song, delivered from visible post atop a bush, is one of the loveliest among the finches due to its sweet voice and varied structure. Rapid trills and unstructured twitters are mixed with drawn-out (almost 'contemplating') pleasant whistles. Variably long strophes are interrupted by brief pauses. There is a fair amount of variation from prolonged and frenzied songs, when the bird is agitated, to brief and 'dreamy' fragments of song with longer pauses. A characteristic phrase with a nice ringing quality is *tuh-ke-yüüh*, another typical one opens with a quickly repeated single note followed by a mellow whistle, *si-si-si-si-sooeeehya*. – Call, commonly heard from birds just before taking off and in flight, is a dry and almost 'bouncing' disyllabic or trisyllabic *tig-itt* or *tig-itt-itt*, with last syllable stressed, and these calls can become an excited, quick and longer series on taking off. The call of Linnet is drier and less hard than the similar call of Twite, and less nasal than that of Tree Sparrow, with the syllables uttered faster (Tree Sparrow gives 'single calls in a series').

SIMILAR SPECIES Whereas the ♂ can hardly be confused with other species, the ♀ and juvenile share many characteristics with *Twite*, including size, shape, the white on edges of primaries and tail-feathers (though narrower) and basic head pattern. Still, Twite differs, even from juvenile Linnet, in its bolder streaking on back, breast and flanks, lack of streaking on centre of throat, stronger ochre-buff tinge in the entire plumage, and by having in fresh plumage a broader and more prominent ochre-buff wing-bar on tips of greater coverts. – Juvenile *Serin* usually lacks any visible pure yellow and can therefore resemble juvenile Linnet but is smaller and more compact, the bill is shorter and slightly more bulbous, and flight-feathers lack any trace of white edges. Also, juvenile Serin tends to be overall more heavily streaked than Linnet (although there is perhaps less of a difference when extremes are compared).

AGEING & SEXING Ages differ subtly in autumn and spring, but reliable ageing requires close views. Moderate seasonal variation. Sexes separable after post-juv moult. –

L. c. cannabina, ♂ (left) and ♀, Denmark, Jul: small, slim and rather long-tailed, with small head and short bill, brown back, and ♀♀ and young generally less heavily streaked than redpolls and Twite. In summer, ♂ has greyer head and crimson-red forecrown and breast patch, while ♀ is colourless and streaked in these areas. (M. Hansen)

L. c. cannabina, presumed 1stS ♂, Netherlands, Jun: some young ♂♂ breed in a more ♀-like plumage. Such retarded ♂ plumage not that rare, and separated from ♀ by amount of brownish-pink blotches over large area on breast and forecrown, and bright chestnut and largely unstreaked back. Ageing in spring problematic, but hint of paler ochre wing-bar, and retarded red development, suggests 1stS. (N. D. van Swelm)

L. c. cannabina, ♀, presumed 1stS, Bulgaria, May: young ♀ has the dullest plumage in spring and summer, with rather weak head pattern. Difficult to judge state of wing-feathers properly, but general dullness of plumage infers age.

Moults. Complete post-nuptial moult of ad in late summer and early autumn, mainly early Jul–mid Oct. Partial post-juv moult at about same time (Jul–Nov) does not usually involve flight-feathers, tail-feathers, tertials or primary-coverts, but birds in S Europe can moult some, most or even all tertials and tail-feathers, and even some odd central (rarely also innermost) primaries. Many inner greater coverts commonly moulted, leaving a variable number of outer unmoulted, but some moult all. No pre-nuptial moult in late winter. – SPRING ♂ Forecrown and breast have red patch in nearly all birds. Those lacking red instead have brown spots (not dark grey streaks). A supporting but not infallible character: white on inner visible 5 mm of outer webs of pp7–9 reaches shafts, or is usually < 0.5 mm from shafts (very few exceptions). ♀ No red on forecrown and practically never any on breast (if any on breast limited to few stains, not a full patch). Breast and crown invariably streaked dark. A supporting but not infallible character: white on inner visible 5 mm of outer webs of pp7–9 separated from shafts by a dark zone, 0.5 mm wide or more, usually 1 mm wide (half the width of the webs). **Ad** No contrast in wear or colour visible among greater coverts, tertials or tail-feathers. Tips to tail-feathers slightly rounded, neatly edged or moderately worn. Primary-coverts blackish (♂) or dark grey (♀), longest neatly edged buff-brown. Note that from Apr many are intermediates, difficult to age safely. **1stS** Among those that retain two or preferably three or more outer juv greater coverts, innermost of these show slight contrast in wear and colour being subtly shorter and tipped/edged slightly paler or duller brown (inner moulted ones longer, a little darker and more rufous-edged). Primary-coverts compared to ad on average subtly paler and browner with narrower and more indistinct paler edges (but many are intermediate in appearance, difficult to age). Sometimes a slight moult contrast visible within alula, but evaluation might require handling. Tips of tail-feathers on average narrower and more pointed and worn than ad, but intermediates occur, and wear makes difference progressively less obvious. A very few (apparently more frequently in C & S Europe) have moult contrast also among primaries, some central and/or inner ones being newer and darker. – AUTUMN **Ad** ♂ Much as ♂ in spring, although red on forecrown and breast often subdued and more pinkish-brown. Easier to use moult contrast among greater coverts and shape of tail-feathers than in spring. **1stW** ♂ As ad ♂ but tail-feathers narrower and more pointed than in ad, tips slightly abraded from late autumn. Often a slight moult contrast visible among greater coverts (see Spring). ♀ Much as in spring. Ageing as ♂. **Juv** Resembles ♀ but somewhat duller and browner overall, and streaking on breast is denser and more diffusely blotched, not as well defined. Look for remnant yellow gape flanges and more woolly plumage on vent and nape.

BIOMETRICS (cannabina) **L** 13–14.5 cm. **W** ♂ 80–86 mm (*n* 37, *m* 82.5), ♀ 78–82 mm (*n* 27, *m* 80.2); **T** ♂ 52–58 mm (*n* 36, *m* 55.1), ♀ 51–57 mm (*n* 27, *m* 53.6); **T/W** *m* 66.8; **B** 10.0–11.8 mm (*n* 35, *m* 10.9); **B**(f) 8.6–10.4 mm (*n* 35, *m* 9.4); **BD** 6.7–7.9 mm (*n* 35, *m* 7.3); **Ts** 15.3–17.1 mm (*n* 35, *m* 16.1). **Wing formula: p1** minute; **pp2–3** longest; **p4** < wt 0.5–1.5 mm; **p5** < wt 4–7 mm; **p6** < wt 10–14 mm; **p7** < wt 16–19 mm; **p10** <

L. c. cannabina, ad ♂, Netherlands, Sep: in fresh winter plumage sexes resemble each other (red of ♂ partly concealed, only wearing to crimson-red by spring), though ♂ still has hint of summer plumage shining through, and more blotchy brown-red flecking of breast. Outermost primaries and tail-feathers growing in last stage of complete moult, inferring ad. (C. van Rijswijk)

L. c. cannabina, ad ♀, Netherlands, Aug: at final stage of complete moult, with several tail-feathers and inner secondaries growing, thus definitely ad. Given age, the boldly streaked back together with dark-streaked crown and much of underparts infer sex. (N. D. van Swelm)

L. c. cannabina, 1stW ♂, England, Nov: 1stW ♂ resembles ad ♀, but note budding red on breast in the form of diffuse blotches, still partly concealed by brown and buff tips (♀ is streaked dark on breast). Rather prominent white edges to outer primaries and tail, and diffuse and limited flanks streaking, also suggest ♂. Typical for younger ♂♂ to be diffusely streaked on back rather than uniform chestnut. (S. Litten)

L. c. cannabina, ♀, presumed 1stW, Netherlands, Oct: dull and heavily-streaked plumage, including head, which is also less pure grey, indicating young ♀. Although outermost few greater coverts seem paler-tipped juv, moult pattern cannot be safely evaluated. (N. D. van Swelm)

wt 25–29 mm; **s1** < wt 27–31 mm. Emarg. pp3–5, on p5 sometimes slightly less prominently.

GEOGRAPHICAL VARIATION & RANGE Throughout its range variation is slight and clinal. Five races here recognised, two of which are very subtle, versus six by Vaurie (1959) and seven (plus one unnamed) in *BWP*. The number depends on how individual plumage variation within any one population is interpreted, a variation which is rather obvious in Linnet. A few ad ♂♂ of one race may differ somewhat from average ♂♂ of another, e.g. in having a deeper chestnut or tawny-brown back, while the majority in the same two populations appear alike. Strictly, subspecies taxonomy should not be based on the appearance of a few extremes, hence our reluctance to accept all described taxa. There is a more uncontroversial cline in size, with largest birds in north and east, and smaller in the south. Bill and tarsus are of the same size throughout the species' entire range, thus become proportionately slightly larger in the south (though much less difference than claimed by some authors).

L. c. cannabina (L., 1758) (N & C Europe including British Isles, south to Pyrenees, N Italy, N Balkans except coasts, W & C Romania east to W Siberia but not Black Sea coasts and Crimea; winters in south of range, departing Fenno-Scandia and N Russia). Rather long-winged and large, back in ad ♂ dark tawny, usually with some dark streaking, breast dark crimson even in summer, not bleaching to pale red. A fair degree of variation as to amount of streaking and darkness. – Scottish birds sometimes separated as '*autochthona*'; inadequate material examined, but judging from 12 Scottish specimens in NHM no consistent difference from *cannabina* detectable, and probably best included therein. (Syn. *autochthona*; *sejuncta*.)

○ **L. c. mediterranea** (Tschusi, 1903) (outer Canaries, N Africa from Morocco to Libya, Spain, all large Mediterranean islands except Cyprus, C & S Italy, Adriatic coast of Balkans, Greece, E Bulgaria, W Black Sea coasts; resident). Very similar to *cannabina* (and included therein by Vaurie 1959) but is nearly always somewhat paler (subtle, and requires long series for comparison!), and has slightly shorter wing and marginally smaller general size. A very subtle race. **W** ♂ 75–83.5 mm (*n* 55, *m* 78.8), ♀ 73–80.5 mm (*n* 34, *m* 76.6); **T** ♂ 48–56 mm (*n* 55, *m* 52.8), ♀ 48–53.5 mm (*n* 34, *m* 51.3); **B** same size as *cannabina*; **Ts** 14.8–16.9 mm (*n* 85, *m* 15.7). – Note that all populations of N Africa and those of the outer Canary Is are included. N African birds are unproblematic since they have the same biometrics and are inseparable by plumage. As to birds of the outer Canaries, Hartert, normally so reliable, claimed that these had a '2 mm longer bill than nominate', hence described them as a new subspecies, '*meadewaldoi*'; however, we found the bill to be

L. c. cannabina, juv, Sweden, Aug: duller overall than later plumages, streaking on breast dense but diffuse, upperparts blotched or diffusely streaked, and body-feathers rather fluffy, especially on breast and flanks. (S. Hage)

L. c. cannabina, England, Dec: in flight note that white edges to primaries and tail-feathers form distinctive patches on spread wings and tail, and also how the chestnut back and innerwing is well visible in the right light. (R. Tidman)

L. c. mediterranea, ♂, presumed ad, Spain, May: compared to *cannabina*, mantle and scapulars of ♂ in spring often slightly paler and more orange-brown, dark head pattern slightly more contrasting, and flanks and belly often tinged darker rufous-cinnamon. Possibly ad by seemingly evenly-feathered and still very fresh wing. (C. N. G. Bocos)

L. c. mediterranea, ♀, presumed ad, Spain, May: ♀ plumage of this race is also slightly paler than *cannabina*. No evidence of moult limits and rather less worn wing indicate ad, but ageing often impossible in this species in spring. (C. N. G. Bocos)

the very same length as in *cannabina* and N African birds. Plumage inseparable between '*meadewaldoi*' and *mediterranea*. (Syn. *meadewaldoi*.)

L. c. guentheri (Wolters, 1953) (Portugal, Madeira; resident). The smallest and darkest race, ad ♂ often (but not invariably) has deeper rufous-tawny back, and ♀ has on average slightly bolder dark streaking. Small, but size difference not marked; mean length of wing and tail 2–3 mm shorter than in rest of Iberia. Birds in Galicia and Andalucia apparently *mediterranea*. **W** ♂ 73.5–79 mm (*n* 16, *m* 75.6), ♀ 73–77 mm (*n* 11, *m* 75.2); **T** ♂ 47–54 mm (*n* 16, *m* 51.1), ♀ 48–52 mm (*n* 11, *m* 50.7); **T/W** *m* 67.5; **B** similar size to *cannabina*; **Ts** 14.2–15.8 mm (*n* 28, *m* 15.0). (Syn. *nana*.)

○ **L. c. harterti** (Bannerman, 1913) (Fuerteventura, Lanzarote; resident). Differs only very subtly from *mediterranea* in being a trifle smaller and in ♂♂ on average very slightly paler cinnamon-brown on back. A very subtle race. It has been claimed that ♂♂ less frequently develop a full red breast patch, but many specimens in collections examined by us are the same as in other races. **W** ♂ 74–81 mm (*n* 22, *m* 76.7), ♀ 72.5–78 mm (*n* 9, *m* 74.6); **T** ♂ 48–54 mm (*n* 22, *m* 50.9), ♀ 45–52.5 mm (*n* 9, *m* 48.9); **T/W** *m* 66.1; **B** similar size as *cannabina*; **Ts** 14.7–16.7 mm (*n* 31, *m* 15.4).

L. c. bella (C. L. Brehm, 1845) (Turkey, Crimea, Levant, Caucasus east through N Iran and Central Asia; mainly resident). About as large as *cannabina* and has a subtly even longer wing and especially tail, but differs most obviously in

L. c. bella, ♂♂, presumed 1stS (left) and ad, Armenia, Jul: rump and uppertail-coverts mainly white, narrowly streaked dark, sometimes with rosy-red admixed. 1stS ♂ has duller/smaller red forehead, slightly paler back, less pure/smaller area of white in tail, and more abraded wingtips, forming moult limits with replaced inner primaries and secondaries. (V. Ananian)

L. c. bella, ♂, presumed ad, Iran, Jun: the brightest race, with most extensive white in wing, head paler ash-grey (distinctly paler than *cannabina*, slightly paler even than *mediterranea*), mantle and scapulars dull cinnamon-brown (not chestnut-brown), and red forehead and breast paler. Bill rather thick. No sign of moult limits in rather fresh wing suggests ad. (E. Winkel)

L. c. bella, ♀, Israel, Jul: subspecies inferred from date and locality, but note also proportionately heavy bill and fairly light general plumage colours. Already in post-nuptial moult, inner primaries replaced. (A. Ben Dov)

being paler overall, particularly in worn summer plumage, when ad ♂ has paler crown/nape and head-sides, and a rather light and bright tawny back; breast is clear rose-red rather than deep crimson. Lower rump/uppertail-coverts of ♂ on average have more white and thinner black streaks; often also some pink on rump region. ♀ only subtly paler than *cannabina*, some being difficult to separate. Fractionally deeper bill on average than *cannabina*, but much overlap. Grades into *mediterranea* in E Balkans. **W** ♂ 79–86 mm (*n* 21, *m* 83.6), ♀ 76–82 mm (*n* 10, *m* 80.7); **T** ♂ 52–60 mm (*n* 21, *m* 56.9), ♀ 51–58.5 mm (*n* 10, *m* 55.1); **T/W** *m* 68.1; **B** 10.7–12.5 mm (*n* 31, *m* 11.7); **B(f)** 8.9–10.4 mm (*n* 31, *m* 9.6); **BD** 6.8–8.5 mm (*n* 31, *m* 7.5); **Ts** 15.1–16.6 mm (*n* 31, *m* 15.8). (Syn. *fringillirostris*; *merzbacheri*; *persica*; *taurica*.)

L. c. bella, presumed 1stW ♂, Israel, Dec: in winter can appear quite pale in strong sunlight. Presumed young bird after extensive moult. Rather pure grey head and rusty-tinged back in combination with subdued and restricted streaking of breast and flanks suggest ♂. (L. Kislev)

L. c. guentheri, ad ♂, Madeira, May: darkest race, with dark rufous-brown back and shoulders. Flanks flecked deep rufous-cinnamon. Red of forecrown and chest bright ruby-red, not dark crimson as in *cannabina*. Ad by evenly-feathered wing, all greater coverts being of ad type, and primaries looking blackish and little worn. (F. Coimbra)

L. c. guentheri, ad ♀, Madeira, Oct: ♀ of this race also dark, with notably heavy streaking both of upperparts and underparts. Wing evenly feathered, and note rather broad and fresh tail-feathers with neat pale fringes. (R. Pop)

YEMEN LINNET
Linaria yemenensis (Ogilvie-Grant, 1913)

Fr. – Linotte du Yemen; Ger. – Jemenhänfling
Sp. – Pardillo yemení; Swe. – Jemenhämpling

An Arabian endemic, confined to Yemen and SW Saudi Arabia, in arid terrain on rocky plains or mountain slopes above 1800 m altitude, with minimal ground vegetation plus scattered bushes and trees. It follows that it is one of the least-encountered or studied species in the entire region. The Yemen Linnet is a resident bird that is fairly common locally and which breeds semicolonially.

Behaviour and appearance are much like Common Linnet, and it is often seen in pairs or small parties. May descend to lower levels in winter.

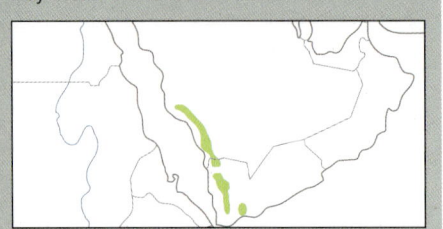

♂, presumed ad, Yemen, May: no range overlap with Common Linnet, and ♂ readily identified by uniform pale ash-grey hood (no dark-and-pale head pattern, and no red on forecrown) and chestnut blotches on body-sides. Wing fairly fresh, lacking any indication of moult limits, which suggests ad. (W. Müller)

IDENTIFICATION About same size as Linnet (only slightly smaller) with same shape and flight mode. Perched birds are very similar to Linnet but differ immediately if seen in flight due to the *striking white wing patches* like in some shrikes. These are confined to the primaries on upperwing, but extend also to the secondaries on underwing. Other differences from Linnet are: (i) *white sides to base of tail* visible in flight, formed by white patches on inner webs of outer two tail-feathers; (ii) *no red on forecrown or breast* in ♂ plumage, *instead sides of breast chestnut*; (iii) *grey on head more uniform*, lacking pale patch on lower cheeks of Linnet; (iv) *more extensively chestnut flanks*; and (v) more subdued or *faint streaking* above and below in ♀ and juvenile compared to Linnet.

VOCALISATIONS Song has been described as 'musical, tinkling, lively and varied; a more drawn-out note, often with an upward inflection, may be introduced at intervals' (Bowden & Brooks 1987). It is most similar to Linnet both generally speaking and in details. – Call in flight has been rendered as a twittering, musical *tirrrit* or *wid-ee*, or as 'a soft liquid *vliet* recalling Goldfinch' (Porter et al. 1996). Some flight-calls might recall Snow Bunting, a more trilling *pr'ruitt*, only clearer and less ringing (M. Ullman pers. comm.).

SIMILAR SPECIES Resembles only *Linnet*, but this species is unknown in SW Arabia, even in winter. In any case, easily separated from Linnet by pure white wing patch and by lack of red on breast and forecrown in ♂ plumage. Has more rufous-brown on flanks than Linnet in all plumages, and head is more uniformly grey.

AGEING & SEXING Ages differ subtly at least in autumn, but reliable ageing requires close views. Ageing not studied in detail, but presumed to be same as for Linnet (which is supported by those specimens examined). Sexes similar but differ subtly after post-juv moult. – Moults. Complete post-nuptial moult of ad apparently in late summer (one ad heavily worn but moult not started 1 Jul). Partial post-juv moult in late summer presumed not to involve flight-feathers, tail-feathers, tertials or primary-coverts, but requires confirmation. No pre-nuptial moult. – **SPRING** ♂ Head and nape rather pure medium grey, crown very finely streaked, grey colour neatly demarcated from partly chestnut mantle. Scapulars and most secondary-coverts chestnut. Mantle and back grey-brown with some chestnut streaks or blotches, appearing unstreaked or very moderately streaked or mottled. Sides of breast and upper flanks paler chestnut. Greater coverts broadly edged chestnut, merging with colour of mantle/back when wing folded. Blackish primaries broadly white-based (except outer). ♀ Head and nape greyish tinged brown, crown thinly streaked, grey colour less sharply demarcated from brown mantle than in ♂. Scapulars bright chestnut, but this colour does not extend to wing-coverts and mantle as in ♂. Mantle and back dull brown (not as bright chestnut or prominent as ♂), invariably lightly streaked or mottled. Sides of breast and upper flanks have more restricted chestnut than ♂. Greater coverts have dark centres, edged/tipped dull chestnut, not appearing as extension of mantle/back like ♂. White bases to primaries less extensive than ♂. – **AUTUMN** Sexing as in Spring. **Ad** Initially heavily worn or in active moult, then freshly moulted. Tips to tail-feathers following moult slightly rounded, neatly edged. **1stW** Perhaps not always separable, but tips of tail-feathers on average narrower and more pointed than ad, though wear makes difference progressively less obvious. **Juv** Resembles ♀ but is overall somewhat duller and browner, including entire head and nape, and has streaked upperparts and breast. Any remnant yellow gape flanges and more woolly plumage on vent and nape will help ageing.

BIOMETRICS L 12–13 cm. **W** ♂ 76–80.5 mm (n 15, m 78.6), ♀ 70–76.5 mm (n 12, m 74.2); **T** ♂ 52–59 mm (n 15, m 55.3), ♀ 48–54 mm (n 12, m 51.4); **T/W** m 69.9; **B** 9.3–12.0 mm (n 23, m 10.5); **B**(f) 7.1–9.4 mm (n 23,

♀, presumed 1stS, Yemen, May: duller than ♂, and readily identified to sex and species by upperparts coloration and pattern of folded wing, including buffish cinnamon-brown shoulder patch and tips to inner greater coverts, leaving broad dark wedge-shaped panel across bases of greater coverts. Juv primaries and outer greater coverts. (W. Müller)

m 8.0); **BD** 5.8–7.0 mm (*n* 25, *m* 6.5); **Ts** 13.5–15.2 mm (*n* 23, *m* 14.3). **Wing formula: p1** minute; **pp2–5** longest (p5 sometimes 0.5–1 mm < wt); **p6** < wt 4–6 mm; **p7** < wt 9–13 mm; **p10** < wt 17–21 mm; **s1** < wt 18–22 mm. Emarg. pp3–6, on p6 sometimes slightly less prominently.

GEOGRAPHICAL VARIATION & RANGE Monotypic. – Yemen, SW Saudi Arabia; mainly resident, making local altitudinal movements at most.

REFERENCES Bowden, C. G. R. & Brooks, D. J. (1987) *Sandgrouse*, 9: 111–114.

♂, presumed ad, Yemen, Jan: almost uniform ash-grey head and neck reaching down to breast, contrasting with chestnut-brown back and wings, greyish-white rump and cinnamon-buff underparts. Rather reduced white in wings and tail compared to Common Linnet. Fresh wing and bright plumage support ageing. (H. & J. Eriksen)

♂, presumed 1stW, Yemen, Jan: combination of solid chestnut greater coverts panel typical of ♂ and duller overall plumage, including greyer and slightly streakier upperparts infer sex and age, though confirmation through handling required for safe ageing. (H. & J. Eriksen)

♀, Yemen, Jan: in fresh plumage, sexing possible, with ♀ having black wedge-shaped panel between chestnut shoulder and tips of inner greater coverts, brown-tinged head and breast, and streaky underparts. (H. & J. Eriksen)

Juv, Yemen, Oct: lacks any grey on head, being also much duller and buff-brown above, and quite heavily streaked on throat, chest and flanks. (U. Ståhle)

TWITE
Linaria flavirostris (L., 1758)

Fr. – Linotte à bec jaune; Ger. – Berghänfling
Sp. – Pardillo piquigualdo; Swe. – Vinterhämpling

The Twite takes its name from its characteristic nasal, twanging call, and this also helps to alert you to its presence. Twites breed on alpine heaths in the north and, unless you visit the species' homeland in summer, it is not until autumn that the species is normally encountered. Then, along a seashore or over an inland meadow, a flock in rather dense formation flies past in skipping flight, giving the characteristic nasal call mixed with clicking notes in chorus. From feeding flocks a low murmur of song fragments is occasionally heard. Often rather shy and restless, soon taking off again.

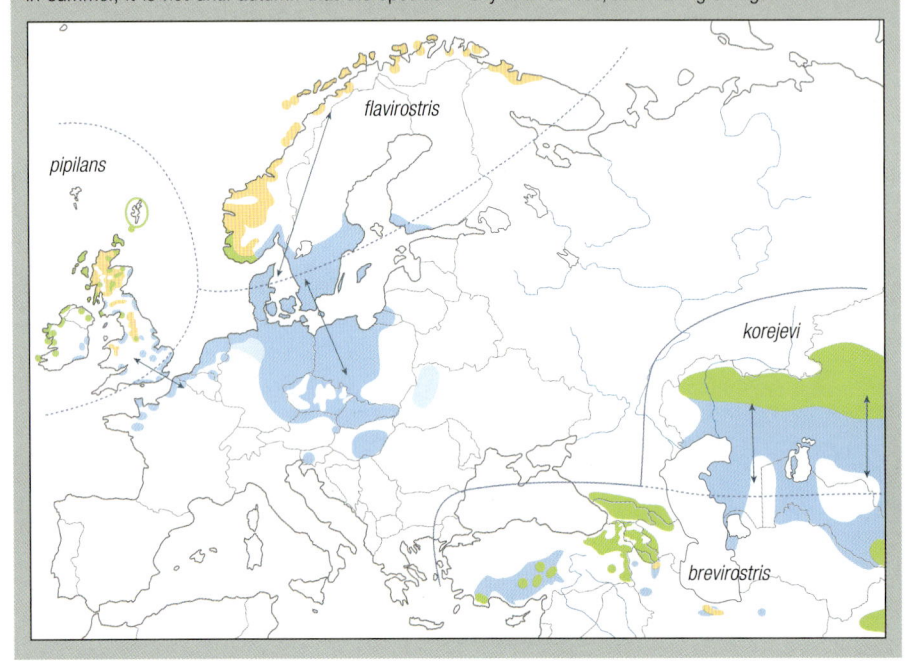

L. f. flavirostris, ad ♂, Finland, Jan: pinkish-red of rump in ♂ partly obscured in fresh plumage by brownish-white feather-tips, but plumage suffused warmer buff, especially around face. Also, bill in winter is yellowish with dark tip. Entire wing is evenly fresh ad, and note broad and rather rounded, fresh tips of tail-feathers. (M. Varesvuo)

IDENTIFICATION Sized like Linnet or Common Redpoll but has proportionately *longer tail*, notched at the tip as in nearly all finches. Plumage *ochre or buff-tinged and white, heavily streaked dark*, thus more nondescript than Linnet or Common Redpoll, lacking any red on crown and breast, grey on head and uniform tawny-brown on back. In fresh plumage fairly pale and buff-tinged, in worn plumage darker and more *heavily streaked*, especially above. Flight-feathers and tail-feathers have white edges like Linnet, only thinner and less obvious. ♂ has *pink tinge on rump*, most obvious in summer when worn, more brownish-pink and subdued in fresh autumn plumage. In fresh plumage a rather *prominent buffish wing-bar* is formed by pale tips to greater coverts, enhanced by *solid dark bases to the secondaries*; in worn summer plumage the wing-bar is narrower, whiter and less obvious. Bill is dark and greyish when breeding, but yellowish in autumn (hence the scientific name *flavirostris* meaning yellow-billed). Head diffusely streaked with a paler area around the eye, although there is also a hint of a darker line through it. Juvenile very similar to ♀, only on average duller and more streaked. – Eastern populations (NE Turkey, Caucasus, Central Asia) differ rather markedly in that breast-sides are densely streaked or blotched, forming a dark gorget contrasting with the unmarked whitish throat, and in their overall paler plumage and more contrasting wing pattern.

VOCALISATIONS Song a mixture of fast trills, passages of unstructured twitters and buzzing notes, often recognised by the interwoven bleating *tveiiht* call and the generally hard and rather scratchy voice, which can even recall song of Dipper. In comparison, Linnet sounds pleasant and softer. Although excited birds sometimes extend the song's length, it is more

L. f. flavirostris, ♂, presumed ad, Norway, Jun: brownish-and-white, heavily dark-streaked above and below, with ochre-buff face and prominent buff-white wing-bars. Distinguished from redpolls by lack of red forecrown or black bib. Characteristic of ♂ in summer is pinkish-tinged rump. Possibly ad by still little worn tail with neat white edges. (A. Below)

L. f. flavirostris, ♀, presumed ad, Norway, Jun: ♀ similar to ♂ but lacks any pink on rump, and is overall darker with less rich rufous-buff face. Also has relatively little white in wing. Typical for the Twite is pale mid-section on folded secondaries. Although tail-feathers are rather pointed they are fairly fresh still in June indicating ad. (M. Varesvuo)

L. f. flavirostris, Netherlands, Dec: thinner and less obvious white edges to wing- and tail-feathers than Linnet. Streaky plumage, buffish face and pinkish-red rump (on at least one bird, top centre, a ♂), typical flight-calls and dense formation usually readily identify the species when seen in flocks. (R. Debruyne)

L. f. flavirostris, ♂, Finland, Dec: dark-streaked buff-brown plumage, with warmer face, supercilium and throat, pale ochre-buff greater-coverts bar (just visible), plus yellowish bill identify this species. Pink-tinged and almost unstreaked rump shows it to be a ♂. Since wing is partly covered by flank-feathers ageing is impossible. (M. Varesvuo)

typically composed, just as in Linnet, of rather short strophes, each slightly different, separated by 1–3 sec pauses. — Main call is a dry, hard clicking Brambling-like *djet* or *jeck*, often given in quick series of about three, forming a loud buzz or noisy chorus from larger flocks. The alternative call is the more characteristic and that which gives the bird its English name, the high-pitched bleating or almost whistling, drawn-out and slightly rising *tveiiht* (or *twe-ee*). This is the call that usually alerts you to an approaching flock of Twites, whereas the *djet* call is easily drowned-out among other calls at a migration spot. A few other minor calls exist, but they have less importance for identification.

SIMILAR SPECIES Could be mistaken for a juvenile Linnet but differs in its longer tail, more ochre or buff plumage tones, the more prominent buffish wing-bar with a dark area on the secondaries. White edges to primaries and tail-feathers much narrower than in Linnet. Note also unstreaked buff throat. — There is a superficial resemblance to juvenile Common Redpoll of the race *cabaret*, but this is a smaller and not as long-tailed bird, which usually has a small dark chin patch before the characteristic red on forecrown starts to develop. Its colours are also deeper rufous-tinged, less ochre-buff than in Twite.

AGEING & SEXING (*flavirostris*) Ages differ subtly at least in autumn, sometimes in spring, but reliable ageing invariably requires close views. Sexes nearly always separable after post-juv moult. — Moults. Complete post-nuptial moult of ad in late summer and early autumn, mainly mid Jul–Oct. Partial post-juv moult at about same time does not involve remiges, tail-feathers or primary-coverts. Sometimes, odd tertials and r1 also moulted. Several inner greater coverts commonly moulted, leaving a variable number of outer ones unmoulted, creating subtle moult contrast. No pre-nuptial moult in late winter. — **SPRING** ♂ Most feathers on rump pink, tipped buff or whitish in autumn, appearing stronger pinkish-scarlet in summer. Rump occasionally pinkish-brown, without visible pink but importantly never streaked dark all over as in ♀. ♀ Rump streaked dark brown-grey on off-white or buffish-white ground. Sometimes one or a few small pinkish spots mixed in, or faint pinkish-brown edges to a few rump-feathers (very rarely a little more pink-red on rump, making sexing problematic). **F.g.** No contrast in wear or colour visible among greater coverts, tertials or tail-feathers. Tips to tail-feathers slightly rounded, neatly edged at least until mid spring. **1stS** Often not separable, but those with moult contrast among greater coverts and heavily worn and pointed tail-feathers can be identified if seen well or handled. — **AUTUMN** Sexing as in spring, only pink on rump of ♂ now more subdued and tinged buff-brown. Absence of prominent streaking over rump best clue. Ageing as in spring, only moult contrast among greater coverts in 1stW now usually easier to detect. **Juv** Resembles ♀ but is somewhat duller and browner overall, and streaking on breast is denser and more diffusely blotched, less well defined. Look for remnant yellow gape flanges and more woolly plumage on vent and nape.

BIOMETRICS (*flavirostris*) **L** 13–14.5 cm; **W** ♂ 76–84 mm (n 20, m 79.2), ♀ 75–78.5 mm (n 16, m 76.6); **T** ♂ 56–65 mm (n 20, m 60.2), ♀ 55–61 mm (n 16, m 58.6); **T/W** m 76.3; **B**(f) 7.3–9.1 mm (n 27, m 8.0); **BD** 5.4–6.7 mm (n 27, m 6.1); **Ts** 15.2–16.7 mm (n 27, m 16.0). **Wing formula: p1** minute; **pp2–4** often longest, but both p2 and p4 sometimes to 1 mm <; **p5** < wt 2–5 mm; **p6** < wt 9–14 mm; **p7** < wt 14–18.5 mm; **p10** < wt 23–27 mm; **s1** < wt 24–29.5 mm. Emarg. pp3–5.

GEOGRAPHICAL VARIATION & RANGE Two racial groups: in NW Europe moderate variation and rather evenly streaked underparts, whereas populations in the Caucasus and Central Asia, separated as they are by wide gap from

L. f. flavirostris, ♀, Finland, Jan: in winter all have yellowish bill with dark tip. General jizz and ochre-buffish face plus streaking above and below identify. All visible wing-feathers seem intermediate in wear, while primary-coverts and greater coverts are hidden, making ageing impossible. Rump lacks any pink, sexing the bird as ♀. (M. Varesvuo)

L. f. flavirostris, 1stW, Finland, Dec: obvious wing-bars (here only greater coverts bar visible), general colouring and yellow bill give the species. Age inferred by seemingly pointed tail-feathers and primary-coverts, and by moult limit in greater coverts. Sex very difficult to decide from this image alone, where rump is invisible. (M. Varesvuo)

L. f. flavirostris, juv, Norway, Aug: like ♀, but more heavily and diffusely streaked body in juv plumage, which is soft and fluffy, with stronger buff fringes and tips, including wing-bars. Note narrow and pointed tail-feathers. (M. Lefevere)

L. f. pipilans, ♂, presumed ad, Scotland, May: darker dorsally than *flavirostris*, streaks on mantle averaging slightly broader. Rufous-buff face and reddish rump often darker, and streaks on breast slightly more marked and extend further down; wing pattern generally darker. Whole wing seems evenly fresh, thus is most likely ad. (G. Hofmann)

those in Britain and Fenno-Scandia, differ more markedly in plumage, having dark breast-side streaking, and size. Several more subspecies described from Asia, but extra-limital and not treated here.

'FLAVIROSTRIS GROUP'

L. f. flavirostris (L., 1758) (NW Fenno-Scandia, mainly in Norway; winters S Fenno-Scandia, NW & C Europe south to N France, Balkans, Bulgaria). Treated above. Rather large with small bill. Compared to *pipilans* in fresh plumage paler and buffier brown, and has more extensive white belly in better contrast to buff-brown breast. Nape often clearly paler brown and lighter streaked, contrasting with mantle.

○ **L. f. pipilans** (Latham, 1787) (British Isles; mainly resident but moving to coasts and lower levels). Very similar to *flavirostris* but on average slightly smaller, darker and browner, with a little bolder and more extensive streaking, apparent when series of same age and season are compared. Wings on average 2 mm shorter, while bill is *c.* 0.4 mm deeper. **L** 12.5–14 cm. **W** ♂ 73.5–81 mm (*n* 21, *m* 76.8), ♀ 71–81 mm (*n* 23, *m* 75.3); **T** ♂ 54–61 mm (*n* 21, *m* 58.5), ♀ 52–62 mm (*n* 23, *m* 56.4); **B**(f) 7.2–9.0 mm (*n* 45, *m* 8.1); **BD** 5.8–7.3 mm (*n* 44, *m* 6.4); **Ts** 15.0–16.5 mm (*n* 44, *m* 15.9). – That birds from Outer Hebrides ('*bensonorum*') should be 'readily separable' as duller brown above and having darker streaking could not be confirmed on the limited material available (NHM). (Syn. *bensonorum*.)

L. f. pipilans, England, Mar: compared to *flavirostris*, upperparts and breast heavier streaked. Wing pattern averages darker. Ochre-tinged face often deeper and more extensive. All visible wing-feathers rather fresh and without handling best left unaged. Similarly, sex difficult to decide without a better view of the rump. (G. Thoburn)

L. f. pipilans, presumed ad ♂, England, Nov: pink just visible on side of rump inferring sex, and neat head pattern with strong ochre-buff tinge, and prominent white edges to inner primaries seem to further confirm this. No obvious moult limit visible in greater coverts suggesting ad, but confirmed ageing would require a better view or tail-feathers. On average very slightly darker and browner than *flavirostris*, but safe separation often requires comparison of series. (P. Sawer)

L. f. brevirostris, ♂, presumed ad, Azerbaijan, May: dense and bold dark streaking on breast-sides, often coalescing to a solid patch. Rather dark brown upperparts, and pale pink rump with some white giving a much paler impression than in *flavirostris*. Only ad wing-feathers apparently visible, but difficult to say for sure. (M. Heiß)

L. f. brevirostris, presumed ad ♀, Georgia, Apr: extensive ochre-buff face to upper breast, contrastingly dark crown and dark-streaked breast-sides are typical features of this race. Sexing difficult unless rump is clearly pink, but ♀ often told by boldly-streaked rump. Possibly ad by apparently uniform wing-feathers, and tail seems still neatly white-edged in spring. (K. Drissner)

'BREVIROSTRIS GROUP'

L. f. brevirostris (Bonaparte, 1855) (S & E Turkey, Caucasus, Transcaucasia, NW Iran; winters in south of range, often on coasts and at lower levels). Differs readily from preceding two races in having throat unstreaked, and dense, bold dark streaking on upper breast (usually concentrated on breast-sides) forming a gorget. Upperparts have paler ground colour, but dark streaking still results in greater contrast. Rump in ♂ pale pink with some white visible, much paler than in *flavirostris*. ♀♀, too, frequently have limited pink on rump. Bases of secondaries edged white forming a white wing patch, and entire wing pattern more contrasting than in European birds. Subtly smaller than *flavirostris*, with long tail and marginally longer bill. **L** 12.5–13.5 cm. **W** ♂ 76–80 mm (*n* 15, *m* 77.6), ♀ 71–77 mm (*n* 14, *m* 74.1); **T** ♂ 58–65 mm (*n* 15, *m* 61.6), ♀ 55–62.5 mm (*n* 14, *m* 58.8); **T/W** *m* 79.4; **B**(f) 7.8–9.5 mm (*n* 29, *m* 8.8); **BD** 5.9–6.8 mm (*n* 18, *m* 6.3); **Ts** 14.0–16.3 mm (*n* 28, *m* 15.4). **Wing formula: p5** < wt 0.5–2 mm; **p6** < wt 5.5–8.5 mm; **p7** < wt 10–13.5 mm; **p10** < wt 17–21.5 mm; **s1** < wt 20–24 mm. Emarg. pp3–5.

○ **L. f. korejevi** (Zarudny & Härms, 1914) (Kirghiz Steppe from west of Ural R. and Transcaspia east to Tarbagatay and Dzungaria; resident or nomadic within same range in winter). Similar to *brevirostris* in spring and summer plumage, but averages slightly smaller and paler, streaking appears less dark and distinct, and streaks both on mantle and breast narrower, not tending to partly coalesce in blotches. In fresh plumage feathers have slightly broader and whiter tips giving it a clearly paler plumage. Birds of Transcaspia and W Kirghiz steppe ('*kirghizorum*') said to differ in being slightly paler, but no difference detected by us when series were compared. **L** 12–13 cm. Sexes appear to be of similar size. **W** ♂ 72–78.5 mm (*n* 14, *m* 75.6), ♀ 71–78 mm (*n* 13, *m* 74.5); **T** ♂ 56–65 mm (*n* 14, *m* 61.2), ♀ 57–65 mm (*n* 13, *m* 60.2); **T/W** *m* 80.6; **B**(f) 7.4–9.4 mm (*n* 30, *m* 8.3); **BD** 5.7–7.0 mm (*n* 27, *m* 6.3); **Ts** 14.4–16.4 mm (*n* 30, *m* 15.4). Wing formula very similar to *brevirostris*. (Syn. *kirghizorum*.)

TAXONOMIC NOTE Considering the well-marked morphological differences between European taxa ('Flavirostris Group') and widely allopatric populations in the Caucasus eastwards ('Brevirostris Group'), a split into two species is not unlikely once the situation has been better studied.

REFERENCES BUB, H. (1978) *Sterna*, 17: 21–23. – MCLOUGHLIN, D. T. *et al*. (2012) *Ringers' Bull*., 27: 76–82.

L. f. brevirostris, ♀, presumed 1stS, Turkey, May: in *brevirostris*, broader white bases to secondaries and white edges to primaries produce larger wing patches, and entire wing is more contrasting than in European birds. Also, tail-feathers are more broadly fringed buffish-white. Tail has very pointed tips suggesting 1stS. (D. Occhiato)

COMMON REDPOLL
Acanthis flammea (L., 1758)

Fr. – Sizerin flammé; Ger. – Birkenzeisig
Sp. – Pardillo norteño; Swe. – Gråsiska

A widespread (circumpolar) small finch common in the boreal forests of N Europe. It breeds mainly in the birch zone just below the tree-line, but may also enter taller willows at the border between forest and tundra. Nests also in spruce or pine forests. Further south in Europe it breeds in many habitats, but especially in mountains and along seashores, wherever there are suitable copses of trees or tall, dense bushes to nest in. As so often among the smaller finches, redpolls are restless and seemingly nervous, constantly taking off and moving far in light, undulating flight, and are often seen in pairs or small parties. Even feeding and nest-building seem to be best performed in the company of the mate.

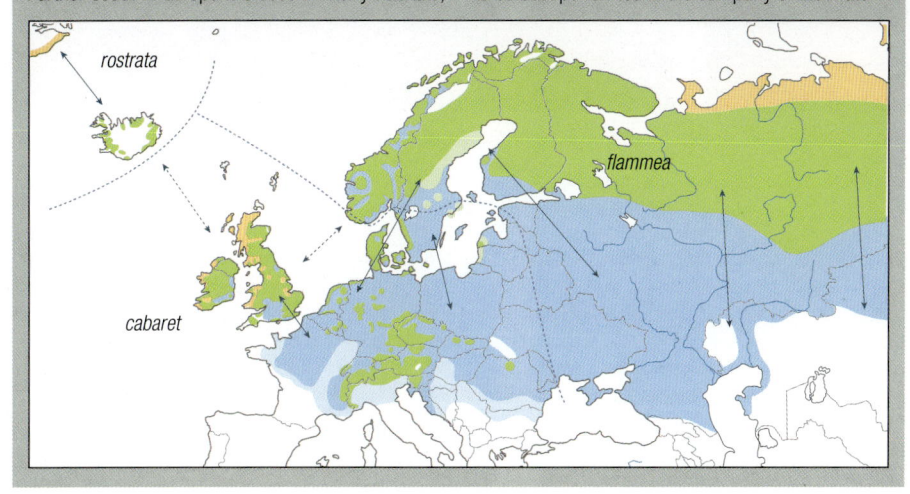

A. f. rostrata, Iceland, Apr: heavily streaked dark on body-sides, with small red forecrown patch and small dark bib perhaps indicating a ♀, but since ageing impossible when unable to assess wear and shape of tail or moult pattern in wing both age and sex best left undecided; only ad ♂ can be excluded. Note the deep bill base. (P. Mugridge)

IDENTIFICATION A small bull-necked finch with a clearly notched tail, mainly *grey and white with dark streaking* and a *pink- or crimson-red rounded forecrown patch*. In fresh autumn plumage, though somewhat varying geographically, *tinged ochre or buff-brown on head, neck, upperparts, breast and flanks*, this colour progressively bleaching in winter and spring to greyish-white. A *small blackish bib* and *pale wing-bars* are typical of all plumages (though bib is poorly marked in some recently-fledged juveniles). There is usually a blackish frontal band across forehead in front of the red patch, the lores and eye-surround are also dark, but there are usually contrasting white 'eyelids' above and below the eye. *White edges to tertials* prominent before plumage becomes heavily worn in late summer. Adult ♂ has *pink-red breast and often rump*, and there may be some pink spots on cheeks and flanks, the reddish colour darkening with wear in summer. At the same time the entire plumage often darkens due to abrasion of pale edges, especially above but also below. *Bill is largely yellowish* with darker tip and culmen *in winter* but *darkens in summer to grey-black*, or is dark with only a trace of yellowish on cutting edges or base of lower mandible. Very similar to Arctic Redpoll and at times inseparable (see that species for details). Often seen feeding in flocks in forests, taking seeds in trees. Climbs agilely on thin branches, frequently in company of Siskins.

VOCALISATIONS Song, usually given in undulating, wide-ranging song-flight, only rarely from perch, consists of a mixture of call notes, the hard and almost echoing, metallic *chet-chet-chet-chet*, and drawn-out, dry reeling trills, *serrrrrrrr*. More 'accomplished' singers alternate the *chet-chet-...* series with a slower and deeper-pitched *chutt chutt chutt* variation, thereby creating a faint similarity with Greenfinch song. – Common flight-call the echoing, metallic *chet-chet-chet* from the song. When alarmed, or excited near the ♀, gives a Greenfinch-like upward-inflected *jueet* (or *dvoee*), often accompanied by *chet-chet-chet* calls. In close contact between two breeding adults, a fine squeaky *tsii-e* can be heard.

SIMILAR SPECIES Very similar to darker plumages of Arctic Redpoll, and a few are difficult or even impossible to separate in the field. Both species vary in darkness and amount of streaking, with the palest and least-streaked Common Redpolls approaching the darkest and most-streaked Arctic Redpolls very closely. See Identification and Similar species under Arctic Redpoll for detailed advice.

AGEING & SEXING (*flammea*) Ages usually differ in 1stY. In ♂♂, breast and rump colour is usually a very obvious ageing character, at least in late winter–summer. Moderate seasonal variation. Sexes usually separable after post-juv moult. – Moults. Complete post-nuptial moult of ad in late summer and early autumn, mainly mid Jul–Oct. Partial post-juv moult at about same time does not involve flight-feathers, tail-feathers or primary-coverts. However, rarely odd inner tertials and r1 also moulted. Usually, only some inner greater coverts moulted. No pre-nuptial moult

A. f. rostrata, presumed 1stS ♂, Iceland, Jun: subspecies by locality, date and bill size. Combination of juv primary-coverts, and faint red dots on fore cheeks suggest age and sex. Larger than *flammea* with heavier bill, usually more heavy streaking. Pinkish-red on lower face and breast variable in summer ♂♂, with some 1stS lacking this altogether. (A. Torés Sanchez)

A. f. rostrata, ad ♂, Canada, Jan: all ad ♂♂ of this taxon have pinkish-red breast, partially concealed by pale tips in winter. There is also often some reddish flecks on cheeks and flanks. Note heavy dark streaking. (J. Iron)

A. f. rostrata, presumed 1stW, Faeroe Is, Oct: note characteristic dusky appearance of this race when fresh, with brownish tinge to much of plumage. Underparts heavily streaked blackish. Relatively large black bib and swarthy face. Bill is mostly yellowish at this season. Apparent moult limit in greater coverts suggests age, whereas sexing is not possible. (S. Olofson)

A. f. flammea, ad ♂, Finland, Jun: Mealy Redpoll (ssp. *flammea*) is generally the greyest and palest race. Breeding ♂ still can be quite dark-streaked, with red forecrown, lower face, breast and rump. Note pointed bill with straight culmen, less deep-based than *rostrata*. Extensive red areas and not heavily worn wing and tail indicate ad ♂. (J. Peltomäki)

in late winter. – **SPRING Ad** Tail-feathers rather broad and tips slightly rounded, neatly edged at least until mid spring, all of same generation. Somewhat variable though, and a few have slightly more pointed tips, especially to outer tail-feathers. **1stS** Can be inseparable if not seen sufficiently well, but those with narrow, heavily worn and pointed tail-feathers can be identified, or those few with moult contrast between r1, ad-type with more rounded tips, and rest of tail-feathers, which are retained juv feathers, narrower with pointed and abraded tips. Note that 1stS ♂ does not develop red breast; this emerges after moult to 2ndW **Ad ♂** Most feathers on breast pink-red, in late winter–early spring somewhat subdued and can have buff-white tips or 'buffish overlay', but strongly pink-red or crimson from late spring. Cheeks and flanks frequently show some pink-red spots. Rump usually also mainly pink-red like breast (and can be completely unstreaked), but a few have reduced pink, being dark-streaked greyish-white with just a few pink spots. **Ad ♀** Breast lacks any, or has very little, pink-red. Flanks without any pink. Rump and cheeks with no or at most only traces of pink spots. All tail-feathers of same generation, rather broad with neatly rounded tips, and moderately worn. **1stS ♂♀** As ad ♀ but tail-feathers pointed and worn or, rarely, show moult contrast (see below). 1stS ♀ never has any pink on cheeks, breast or rump, only on forecrown. Size of red forecrown patch variable and a poor indicator of sex. – **AUTUMN** Sexing as in spring, only pink-red colour on breast and rump now much more subdued and tinged, or 'concealed' by, buff-brown, and can be difficult to discern in the field. Ageing as in spring, only shape and wear of tail-feathers now easier to assess. Sometimes a slight contrast between inner moulted greater coverts (darker-centred, slightly more buff-tipped) and retained outer juv (subtly paler centres, more whitish tips). **Juv** Resembles ♀ but readily recognised by lack of red patch on forecrown.

BIOMETRICS (*flammea*, var. *holboellii* excluded) **L** 12.5–13.5 cm; **W** ♂ 72–80 mm (*n* 97, *m* 76.2), ♀ 71–77.5 mm (*n* 57, *m* 74.0); **T** ♂ 52–61 mm (*n* 96, *m* 56.6), ♀ 51–60 mm (*n* 57, *m* 55.6); **T/W** *m* 74.7; **B** 9.3–13.3 mm (*n* 124, *m* 11.1); **B**(f) 7.6–10.4 mm (*n* 168, *m* 8.8); **BD** 5.6–7.3 mm (*n* 163, *m* 6.5); **BD/B**(f) *m* 74.8; **B**(f)/**T** 13.2–18.3 (*n* 167, *m* 15.6); **Ts** 12.8–15.9 mm (*n* 113, *m* 14.4). **Wing formula: p1** minute; **pp2–4** often longest, but p2 and p4 sometimes 0.5–1.5 mm <; **p5** < wt 2–5 mm; **p6** < wt 9–14 mm; **p7** < wt 13–18 mm; **p10** < wt 22–27 mm; **s1** < wt 23–28.5 mm. Emarg. pp3–5. – Often easier or more reliable to measure bill to feathers than to skull. Disregard nostril-feathers or nasal hairs.

GEOGRAPHICAL VARIATION & RANGE Slight or almost non-existent variation over circumpolar boreal forest and taiga areas, but more marked in isolates of Iceland and Greenland, where differences affect general size, darkness, prominence of streaking and bill-size. The mainly montane and coastal form in W & C Europe (*cabaret*) differs in smaller size and darker and warmer brown colours. Over whole range rather extensive individual variation, and variation related to sex, age and wear of plumage, often making identification problematic, including the safe elimination of Arctic Redpoll. – Most populations resident or nomadic, making only irregular, food-related, winter movements.

A. f. rostrata (Coues, 1861) 'GREATER REDPOLL' (Baffin

A. f. flammea, 1stS, N Norway, Jul: one difficulty in separating Mealy from Arctic Redpoll is that individual variation produces birds with whitish lower back and rump contrasting with darker rest of upperparts. But here entire rump clearly dark-streaked, and head and mantle too dark for Arctic. Two feather generations in tail in Jul infer age, whereas sex is indeterminable. (H. Eskonen)

A. f. flammea, ad ♂, N Finland, Apr: small and elegant finch, heavily streaked with red forecrown patch and small black bib. Note also whitish double wing-bars. Short, sharply pointed bill is usually straw-coloured with dark culmen and tip. Ad ♂ by extensive red throat, breast, fore cheeks and rump. (A. Seppä)

A. f. flammea, ad ♀, **Finland, Apr**: breast is boldly dark-streaked and lacks any pink-red, while cheeks have just a hint of red spotting showing, and these traits in combination with it being ad by evenly-feathered wing and neat dark primary-coverts fit only ad ♀. Note that outer greater coverts are normally more buff-tinged than inner white-tipped, hence this is not a true moult limit. Bold black centres to undertail-coverts firmly eliminate Arctic Redpoll, which otherwise can be rather similar in 1stW plumage. (A. Seppä)

A. f. flammea, ad ♂, **Netherlands, Nov**: non-breeders are considerably paler than in summer, with whitish-grey rump variably dark-streaked, mantle and flanks more greyish, and red parts pale crimson-pink. Birds with much pink on rump, like here, frequently have reduced streaking there. Wing evenly feathered and fresh. (N. D. van Swelm)

A. f. flammea, presumed ad ♀, **Finland, Jan**: some are overall paler making separation from Arctic Redpoll more challenging. Possible ad, since wing is evenly feathered, with ad-like dark, fresh and round-tipped primary-coverts. If ad, the absence of any pinkish-red on breast, and smallness of red forecrown patch, might indicate ♀. (M. Varesvuo)

A. f. flammea, 1stS, presumed ♂, **Finland, Apr**: 1stS (moult limits in wing) and possibly ♂ (underparts clean white with relatively limited streaking) can be similar to darker Arctic Redpolls. This bird seems to lack dark streaks on undertail-coverts, also like Arctic. However, rump seems to be too dark, head pattern a bit too darkish and bill size more typical of Common. (M. Varesvuo)

Island, S Greenland, N Iceland; has bred Scotland; often winters Iceland and straggles to Scotland). Differs from *flammea* by somewhat larger size, heavier and at times basally more bulbous bill, and heavier streaking usually affords a slightly darker overall appearance in comparable plumages. Note that culmen usually is straight (at least in outer part) giving pointed bill-tip, but depth large and base of bill usually slightly more rounded (convex) than in the other taxa. Upperparts including rump usually rather uniformly dark and rufous-tinged brown (autumn) or grey (summer) with dense, bold streaking. Dark bib tends to be large (but variable and overlaps with *flammea*). Note long tarsus, large bill-depth and ratio bill-depth/bill-length to feathering (expressed as a percentage), often helpful with borderline *rostrata*/*flammea*. It has been claimed that not all ad ♂ *rostrata* attain pink-red breast and rump, but frequency of 'retarded' plumage unknown, and should perhaps be reassessed on a larger sample of summer birds. (All ad ♂♂ examined by us had pink-red breast. Odd 1stY ♂♂ in collections with replaced tails, appearing like ad, might provide some explanation, if not all.) **L** 13.5–14.5 cm. **W** ♂ 76–87 mm (n 37, m 81.5), ♀ 75.5–85 mm (n 22, m 79.3); **T** ♂ 57–66 mm (n 36, m 60.3), ♀ 55–63 mm (n 21, m 59.3); **T/W** m 74.4; **B** 10.5–13.9 mm (n 52, m 11.9); **B(f)** 8.1–11.0 mm (n 62, m 9.3); **BD** 6.7–9.2 mm (n 59, m 7.5); **BD/B(f)** m 80.4; **B(f)/T** 12.9–17.9 (n 60, m 15.6); **Ts** 15.2–17.5 mm (n 53, m 16.2). **Wing formula: p5** < wt 3–6 mm; **p6** < wt 10.5–14 mm; **p7** < wt 15–18 mm; **p10** < wt 24–29 mm; **s1** < wt 25–31 mm. – Breeders in N Iceland sometimes separated as '*islandica*' on account of smaller general size and bill, and subtly paler and greyer plumage versus classic *rostrata*. However, plumage differences between dark Icelandic breeders and *rostrata* slight and often non-existent (mainly limited to on average smaller dark patch on longest undertail-coverts), and near-complete overlap exists in measurements, hence dark Icelandic breeders best included here. Strongest difference is tarsus length, but this still only separates 18.5%. **W** ♂ 75–85 mm (n 33, m 80.0), ♀ 75–82 mm (n 25, m 78.2); **T** ♂ 55.5–63 mm (n 33, m 60.0), ♀ 55–64 mm (n 26, m 59.3); **T/W** m 75.4; **B** 9.5–12.8 mm (n 53, m 11.4); **B(f)** 7.7–10.1 mm (n 58, m 8.9); **BD** 6.3–8.2 mm (n 59, m 7.2); **BD/B(f)** m 81.9; **B(f)/T** 12.6–17.7 (n 58, m 14.9); **Ts** 14.5–17.0 mm (n 54, m 15.6). – There is substantial variation among Icelandic breeders, some being considerably paler grey with white or whitish rump and unmarked white or near-white undertail-coverts with only narrow dark streak, but these pale birds are not included in *rostrate* here (instead see Arctic Redpoll). It is possible that some intermediates difficult to assign to taxon are hybrids with Arctic Redpoll, but no proof exists, and it is just as likely that they represent extreme variation of either taxon. (Syn. *islandica*.)

A. f. flammea (L., 1758) 'MEALY REDPOLL' (N & C Fenno-Scandia, Russia, Siberia, North America except Baffin I; possibly also NW Scotland). Treated above under Identification, etc. Medium-large, usually rather prominently streaked including on rump (although some ad ♂♂ attain unstreaked pink-red rump), culmen straight, tip rather slender and pointed. In fresh plumage head, throat and sides of upper breast tinged ochre (most marked in imm), mantle and upper-wing brownish, rest of plumage more grey and white. Back of neck usually rather pale with dark streaking, but some are darker, more like *cabaret* (see below). Rump variable,

A. f. flammea, presumed 1stS, Finland, Apr: judging from wear and pointed shape of tail-feathers this is most likely a 1stS. Sex perhaps ♀ based on small red forecrown patch, but underparts quite white, and streaking limited, more typical of ♂, hence sexing unsure and best left undecided. (M. Varesvuo)

A. f. flammea, juv, Norway, Aug: compared with ♀ tinged slightly more buff-brown, including wing-bars, and more blotchy below, lacking any pink or red in plumage, and black bib small and hardly visible. Fluffy, loose plumage. (M. Lefevere)

too, usually pale grey boldly streaked dark, but some are warmer buff-brown and others are quite whitish and lightly streaked inviting confusion with Arctic Redpoll. With wear in spring, ochre bleaches and becomes greyish-white; breeders in summer darken due to loss of whitish feather-edges and are very different from autumn. (Syn. *fuscescens*; *linaria*.) – Includes a long-billed and larger variety, var. *holboellii*, which presumably habitually breeds in the range of normal *flammea*, possibly more commonly in far north and east (but details poorly known), and even mixed pairs have been recorded (Svensson 1992; specimens in NRM). Plumage very similar or identical to *flammea*, although there is slight tendency for *holboellii* to be rather dark and boldly patterned. Note that not only is bill longer and therefore more pointed, but also somewhat deeper-based than *flammea*. We follow Herremans (1990) in defining as var. *holboellii* any *flammea* with B(f) in ♂♂ ≥ 10.5 mm, in ♀♀ ≥ 10.0 mm. Biometrics for var. *holboellii*: **L** 13–14 cm; **W** ♂ 76–83 mm (*n* 26, *m* 78.5), ♀ 73–79 mm (*n* 9, *m* 75.4); **T** ♂ 55–63 mm (*n* 26, *m* 58.1), ♀ 53–58 mm (*n* 9, *m* 55.8); **B** ♂ 11.5–14.8 mm (*n* 24, *m* 13.3), ♀ 12.1–14.9 mm (*n* 5, *m* 12.9); **B(f)** ♂ 10.5–12.5 mm (*n* 26, *m* 11.1), ♀ 10.0–12.6 mm (*n* 9, *m* 10.5); **BD** ♂ 6.8–8.0 mm (*n* 26, *m* 7.4), ♀ 6.3–7.3 mm (*n* 9, *m* 7.0); **BD/B(f)** *m* 66.8; **B(f)/T** 17.4–22.9 (*n* 35, *m* 19.0); **Ts** 13.8–15.5 mm (*n* 26, *m* 14.7). Wing formula as *flammea sensu stricto*.

A. f. cabaret (Statius Müller, 1776) 'LESSER REDPOLL' (British Isles, continental Europe from N France through Low Countries, Germany and Baltic States, S Scandinavia; also in Alps and Czech mountains). Resembles *flammea* but usually rather distinctly smaller, darker and warmer rufous or tawny-brown above, with slightly bolder and more extensive streaking, usually apparent with practice and can be established when series of same age and season compared. Can be described as a 'dwarf *rostrata*'. Back of neck and sides of head nearly as dark as mantle and hindcrown, thus darker than normal *flammea*. In fresh plumage throat and sides of breast ochre-buff (richest in imm) with marked contrast to whitish belly (usually more contrast than in *flammea*). Rump usually buff-tinged and well streaked (rarely more whitish rump like some *flammea*). Bill small but actually of similar size as *flammea*, hence proportionately subtly larger than latter. Calls and song appears to be on average slightly higher-pitched and faster than *flammea* and other Common Redpoll taxa, more similar to those of Arctic Redpoll *exilipes*, but probably both overlap and variation make this an unreliable distinction. **L** 12–12.5 cm; **W** ♂ 69–74.5 mm (*n* 32, *m* 71.7), ♀ 67.5–73 mm (*n* 21, *m* 69.9); **T** ♂ 49–56 mm (*n* 32, *m* 52.6), ♀ 49–55.5 mm (*n* 21, *m* 51.5); **T/W** *m* 73.6;

A. f. cabaret, ad ♂, England, Feb: smallest and brownest race, and ad ♂ usually has, as here, extensive pinkish-red underparts. Amount of pinkish-red in plumage makes examination of moult and wear of wing-feathers superfluous, only ad ♂ looks like this. (S. Round)

A. f. cabaret, ad ♂, England, Mar: evenly rather fresh wing and tail (note broad and rounded tips), and obvious red throat (breast cannot be judged) infer age and sex. Subspecific identification based on small size, tawny-brown ground colour and bold streaking. Rump unstreaked pinkish-red and white, a pattern sometimes found in ad ♂♂. (P. Lathbury)

A. f. cabaret, ad ♀, Austria, May: subspecies by locality and date, but also note dark brown upperparts and heavily streaked body-sides. Ad due to wing evenly and not heavily worn, tail-feathers fairly well kept and neatly fringed pale. ♀ due to being ad with no reddish-pink on cheeks (save the odd spot), breast or flanks. (L. Khil)

A. f. cabaret, 1stS, presumed ♀, **England**, Apr: aged by apparent pointed, worn and bleached juv tail-feathers. Outer greater coverts, primary-coverts and remiges also juv. No scarlet-pink on cheeks, throat or breast, with upperparts and flanks broadly and densely streaked and tinged tawny, are by and large features of young ♀ *cabaret*. (I. Clarke)

A. f. cabaret, ad ♂, **Netherlands**, Jan: scarlet-pink on cheeks, throat and chest extensive but partially concealed in fresh plumage. Ad by evenly-feathered wing- and tail-feathers with rounded tips. (C. van Rijswijk)

A. f. cabaret, 1stW, presumed ♂, **Ireland**, Sep: in early autumn, a young bird (pointed tips to tail-feathers and narrow brownish primary-coverts) with some reddish on lower cheeks is probably ♂. (S. Cronin)

B 9.0–11.7 mm (*n* 53, *m* 10.6); **B**(f) 7.1–9.8 mm (*n* 57, *m* 8.3); **BD** 5.6–7.1 mm (*n* 58, *m* 6.5); **BD/B**(f) *m* 78.6; **B**(f)/**T** 13.3–17.8 (*n* 57, *m* 15.8); **Ts** 13.2–15.4 mm (*n* 52, *m* 14.1). **Wing formula: p5** < wt 2–4 mm; **p6** < wt 8–11 mm; **p7** < wt 12–14.5 mm; **p10** < wt 19.5–23.5 mm; **s1** < wt 21.5–25 mm. (Syn. *britannica*; *disruptis*; *rufescens*.) – Sometimes treated as separate species (notably by BOU), but not followed here due to frequent reports of apparent overlap between *cabaret* and *flammea* (birds with warm brown colours and bold streaking, but with measurements exceeding normal range of *cabaret*, or greyish-white more lightly streaked birds that are smaller than normal *flammea*), intermediates that seem to be the effect of hybridisation, as the two races are now in contact in Norway and Sweden, and possibly elsewhere. See Taxonomic notes.

TAXONOMIC NOTES The redpolls as a group have long been the subject of taxonomic controversy. Depending on how overlap in morphological characters between taxa are interpreted, the group has been variously treated as comprising one, two or several species. All taxa are quite similar genetically, indicating a recent radiation (Marthinsen *et al*. 2008). We follow Vaurie (1959) and Knox (1988) in accepting two species, and refer readers to the latter for convincing arguments to keep polytypic Arctic and Common Redpolls separate. Within Common Redpoll, the Icelandic situation is particularly complex and still poorly understood.

A few comments on *flammea* taxonomy are warranted. – Zuccon *et al*. (2012) showed that *Carduelis* as traditionally defined is polyphyletic. To avoid this, they recommended that *Acanthis* is resurrected for Common and Arctic Redpolls (and *Linaria* for Linnet and Twite), which is followed here. – Knox *et al*. (2001) recommended splitting *cabaret* as a separate species. Their main argument, sympatric breeding in S Norway without mixed pairs, was based on a preliminary report (Lifjeld & Bjerke 1996) of a study of 11 breeding pairs over one season, but sympatric breeding there ceased subsequently. Mating may well have occurred before arriving at the site. Ottvall *et al*. (2001) showed that there was no significant genetic difference between *cabaret* and *flammea*, and reported increasing numbers of intermediates (*c*. 10%) which could best be interpreted as the result of mixed pairs; consequently, they recommended conspecific treatment. The findings of Marthinsen *et al*. (2008) were similar. With substantial numbers of indistinguishable birds, the requirement of diagnosability is apparently not fulfilled, and *cabaret* is therefore best treated as a race of *flammea*. – The occurrence of large and particularly long-billed birds (at times also heavy-based) (var. *holboellii*) within the range of *flammea*, skewing the normal distribution of measurements at the upper end, is difficult to explain. Perhaps ecological factors affecting natural selection in the northernmost taiga eliminates small birds, but not to the same extent larger birds? We find it practical to present biometrics for this variety separately, but acknowledge that the chosen limit (for practical reasons following Herremans 1990) is arbitrary. – The breeding population on Iceland is puzzling. Both dark and pale birds occur, but judging from specimen material (and supported by taxonomic reviews of the complex, notably Knox 1988) dark and pale birds are much commoner than are intermediates; apparently assortative mating keeps the two types intact. We have not found sufficient differences between the dark Icelandic birds and *rostrata*, hence have included them in the latter (*contra* Herremans 1990, who upheld *islandica*). The pale birds we are unable to separate from Arctic Redpoll, and we fail to see any reason to call them anything else (see that species.).

REFERENCES Boddy, M. (1981) *Ring. & Migr.*, 3: 193–202. – Herremans, M. (1990) *Ardea*, 78: 441–458. – Knox, A. (1988) *Ardea*, 76: 1–26. – Knox, A. *et al*. (2001) *BB*, 94: 260–267. – Landsdown, P., Riddiford, N. & Knox, A. (1991) *BB*, 84: 41–56. – Lifjeld, J. T. & Bjerke, B. A. (1996) *Fauna Norv.*, Ser. C (*Cinclus*), 19: 1–8. – Lindström, Å., Ottosson, U. & Pettersson, J. (1984) *Vår Fågelv.*, 43: 525–530. – Marthinsen, G., Wennerberg, L. & Lifjeld, J. T. (2008) *Mol. Phyl. & Evol.*, 47: 1005–1017. – Molau, U. (1985) *Vår Fågelv.*, 44: 5–20. – Nyström, B. & Nyström, H. (1987) *Vår Fågelv.*, 46: 119–128. – Ottvall, R. *et al*. (2002) *Avian Science*, 2: 237–244. – Reid, J. M. & Riddington, R. (1998) *DB*, 20: 261–271. – Williamson, K. (1961) *BB*, 54: 238–241.

A. f. cabaret, 1stW ♀, **Netherlands**, Nov: red and black marks on head rather weak, and heavily streaked overall on rather brownish ground should eliminate a wintering *flammea*. Age confirmed by retained juv outer greater covert. (J. A. van den Bosch)

ARCTIC REDPOLL
Acanthis hornemanni (Holboell, 1843)

Alternative name: Hoary Redpoll (Am.)

Fr. – Sizerin blanchâtre; Ger. – Polarbirkenzeisig
Sp. – Pardillo de Hornemann; Swe. – Snösiska

Closely related to the Common Redpoll, the Arctic Redpoll is in every way a copy of it, only paler and breeding further north. It prefers lower willow scrub on tundra or mountain heaths rather than taller bushes or birch forests, but from time to time breeds in the closed birch zone, too. The two redpoll species have very nearly the same DNA so obviously they have only very recently diverged from a common ancestor. Arctic Redpoll is well feathered and can generally sustain the cold winters in the north, where it is mainly resident, but in some winters small parties move south and please birdwatchers always on the lookout for something unusual.

IDENTIFICATION A paler version of Common Redpoll, thus a small, seemingly neck-less finch with a notched tail, basically grey and white with some streaking, a pale wing-bar and red forecrown. *Bill yellowish with dark tip* but darkens somewhat during breeding season. The palest Arctic Redpolls, generally adult ♂♂ irrespective of race, are *strikingly white with reduced streaking* and are unproblematic, especially if the *large unstreaked white, square rump patch* and *unmarked white undertail-coverts* are seen. *Head pattern* is usually *clearly paler* than in Common Redpoll with a less prominent or narrower dark frontal band and paler loral area, while *back has paler, often whitish ground colour*, tertials have *broader white edges* and the *white wing-bar is broader*. They also differ clearly from adult ♂ Common Redpoll in that *breast is invariably rosy-pink* even in summer, never becoming darker crimson-red with wear. – Problems start with slightly darker and more streaked non-adult ♂ Arctic, and for these there is a full cline of variation running into Common Redpoll without any clear step. Identification must then rely on a combination of as many criteria as possible, none of them conclusive alone. Since a few adults (mainly ♀♀), and a substantial proportion of first-year Arctics of either sex, show a variable amount of dark streaking on the white rump, usually just a little and partly hidden by overlaying white edges to other feathers, but in some tricky cases quite as much as a lightly streaked Common Redpoll, making separation of Arctic and Common Redpolls using the rump feature alone unsafe. *Ground colour of the rump is usually whiter* in Arctic, more tinged grey or buff in Common, but a few of the latter are almost as white as Arctic. There is often a hint of a distinct and straight division between the more heavily streaked back and the lightly streaked rump of a difficult Arctic, whereas a lightly streaked rump of a controversial Common usually blends diffusely into the darker and more boldly streaked back. However, this difference is both slight and variable, and can only serve as supporting evidence. – The presence or not of a dark central streak or patch on the longest undertail-coverts is useful when extremes are encountered: a bird with *pure white and unmarked undertail-coverts* can only be an Arctic, and a bird with a very broad and extensive dark central patch can only be a Common. Regrettably quite a few are intermediates, and could be either species. Still, in Common the narrowest streaks are at least *c.* 2 mm broad, while the majority have streaks about 3–4 mm wide. In Arctic that shows a streak on longest undertail-coverts, it is often quite thin, 0.5–1.5 mm wide, and birds with broader streaks are very rare indeed. Also, the streak is usually confined mainly to the outer part of the feather but fades away towards the base (admittedly an in-hand character). – The third most important character to consider is bill-size. A typical *exilipes* Arctic has a *short and acutely pointed bill*, whereas most Common Redpolls have a proportionately slightly longer but still pointed bill. All other characters are more subtle and variable. However, in ssp. *hornemanni* the bill is about as long and heavy as in Common Redpolls of corresponding body size, so the usefulness of bill-size is generally limited to comparison between *exilipes* and *flammea*.

VOCALISATIONS Both song and calls of ssp. *exilipes* are very similar to those of 'Mealy Redpoll' (*flammea*), only on average higher-pitched, softer and faster, thereby closely resembling those of Lesser Redpoll (*cabaret*, and doubtfully separable from that). In particular, the reeling trills in the song seem finer and more 'lisping', and have been noted to be slightly shorter but more frequently repeated. The chattering series, too, are often a little feebler, less hard and 'more clattering than clicking'. Individual and acoustic variations unfortunately limit the use of these tentative differences in the field. A detailed comparison of calls of ssp. *hornemanni* with, e.g., *rostrata* or *flammea* is desirable but apparently lacking.

SIMILAR SPECIES Very similar to paler plumages of *Common Redpoll*, and a few are difficult or even impossible to separate in the field. Differences have already been covered under Identification, but the main points are repeated here. (i) Common has streaked rump, usually on a pale greyish or buffish ground, but some fresh autumn birds can appear to have a small area of the rump unstreaked white due to the broad white feather-edges. As some ♀ and 1stY ♂ Arctic Redpolls can have a lightly streaked whitish rump there is an obvious risk that relying on this feature alone can lead

A. h. hornemanni, ad ♂, Greenland, Mar: note relatively pale head pattern, overall whiter plumage and weaker streaking than Common Redpoll, and also broad white wing-bars. Note unstreaked white, faintly pink-tinged rump, plus pale pink on breast, which sex this bird as ♂. Ad by pink breast and evenly-feathered wing. (C. van Rijswijk)

A. h. hornemanni, ad ♂, Netherlands, Oct: broad white wing-bar and very pale, nearly whitish ground colour above. Being fresh, tinged pinkish-ochre on face and breast. Quite clean white, thinly-streaked plumage, and only fresh ad wing-feathers infer age and sex. Any pink on breast probably not visible in this angle. (A. B. van den Berg)

you astray. Arctic with a lightly streaked rump generally has a white or almost white ground colour to the entire rump and a hint of a distinct border to the greyish and more heavily streaked back, whereas Common Redpolls with a hint of a paler rump patch, thereby recalling some Arctic, have more diffuse borders to the pale centre of the rump. Also note that when the rump of adult ♂ Common Redpoll is reddish it is often totally unstreaked. If the red pigment is worn off, bleached (as can happen in summer) or poorly developed such birds can appear to have an unstreaked white rump strongly suggesting Arctic Redpoll. (ii) Longest undertail-coverts in Common invariably have dark central streaks or patches, only rarely slightly narrower streaks, and never so thin as to only represent a shaft-streak. Classic Arctic Redpolls (mostly adult ♂♂) have unmarked white undertail-coverts, a conclusive character of that species. However, many Arctics have a thin dark shaft-streak or even a narrow patch on the centre of the coverts, and there is some overlap in pattern between the two species, although this probably only affects *c.* 10% of birds, or even less. (iii) Common Redpoll has a medium-large bill, conical and pointed in most. In comparison, Arctic Redpoll of the race *exilipes* has an on average slightly shorter bill, invariably finely pointed, sometimes almost 'invisible' among the nostril-feathers and rictal bristles. Other differences to consider are paleness of the head pattern, with most Arctic having an incomplete or narrow dark frontal band and less dark lores, giving the faces a paler expression than the invariably rather swarthy-looking Common Redpoll. (iv) Further supportive but far from conclusive average differences include the slightly shorter tail in Common, the darker overall look with narrower wing-bars and darker-looking back, and the less 'fluffy' plumage (some Arctic can look 'richly and loosely feathered'). As to adult ♂♂, remember that whereas a summer Common Redpoll becomes deeply pink-red or crimson on breast and rump, a summer Arctic remains pale pink on these areas in summer. You must consider all of the above in combination when encountering a tricky bird.

AGEING & SEXING (*exilipes*) Ages usually differ in 1stY. In ♂♂, breast and rump colour is usually a useful ageing character, at least from late winter to summer. Sexes separable after post-juv moult. – Moults. Complete post-nuptial moult of ad in late summer and early autumn, mainly Jul–Sep (Oct). Partial post-juv moult at about same time does not involve flight-feathers, tail-feathers or primary-coverts. Rarely odd inner tertials and r1 also moulted. Usually only a few inner greater coverts moulted. No pre-nuptial moult

A. h. hornemanni, ad ♀, Scotland, Oct: breast and unstreaked white rump lack pink inferring sex. Broad flight- and tail-feathers and primary-coverts with rounded tips indicate ad. Differs from *exilipes* in larger size, deeper bill and is on average whiter with overall less extensive streaking. Colour depends on age and sex and is affected by bleaching and wear, and worn *exilipes* can appear whiter than fresh *hornemanni*. (H. Harrop)

in late winter. – **SPRING Ad ♂** Most feathers on breast pale pink, in late winter–early spring pink somewhat subdued and may have buff-white or white tips, or 'buffish concealment', but becomes purer pink from late spring. Flanks frequently show some pink. Rump usually pure white, but can have variable number of pink spots, at times many. Tail-feathers rather broad and tips slightly rounded, neatly edged at least until mid spring, all of same generation. Somewhat variable and a few have slightly more pointed tips, especially to outer tail-feathers. **Ad ♀** Breast lacks any pink, rump usually too. All tail-feathers of same generation, rather broad with neatly rounded tips, moderately worn. **1stS** Sometimes inseparable if not seen sufficiently well, but those with narrow, heavily-worn and pointed tail-feathers can be identified, or those few with moult contrast between r1, being ad-type with more rounded tips, and rest of tail-feathers, being retained juv feathers, narrower with pointed and abraded tips. Note that 1stS ♂ does not develop a full pink breast; this emerges after moult to 2ndW. – **AUTUMN** Sexing as in spring, but pink on breast and rump in ad ♂♂ now much more subdued and tinged, or 'concealed' by, ochre-buff and white. Ageing as in spring, but now shape and wear of tail-feathers easier to assess. Sometimes a slight contrast between inner moulted greater coverts (darker-centred, slightly more buff-tipped) and retained outer juv (subtly paler centres, more whitish tips). **Juv** Resembles ♀ but readily recognised by lack of red patch on forecrown.

BIOMETRICS (*exilipes*) **L** 12–13 cm; **W** ♂ 73–80 mm (n 72, m 76.5), ♀ 70.5–79 mm (n 49, m 74.3); **T** ♂ 53–62 mm (n 71, m 58.0), ♀ 52–61.5 mm (n 49, m 56.6); **T/W** m 75.9; **B** 8.8–11.0 mm (n 57, m 10.0); **B**(f) 6.9–9.2 mm (n 134, m 7.9); **BD** 5.8–7.0 mm (n 123, m 6.5); **BD/B**(f) m 82.4; **B**(f)**/T** 11.4–15.9 mm (n 133, m 13.8); **Ts** 13.0–15.2 mm (n 116, m 14.2). **Wing formula**: **p1** minute; **p2** < wt 0.5–2 mm; **pp3–4** often longest, but p4 rarely < wt 0.5–1 mm; **p5** < wt 1.5–4 mm; **p6** < wt 7.5–11.5 mm; **p7** < wt 12–17 mm; **p10** < wt 19.5–25 mm; **s1** < wt 21–27 mm. Emarg. pp3–5. – Often easier or more reliable to measure bill to feathers than to skull. Disregard nostril feathering or nasal hairs.

GEOGRAPHICAL VARIATION & RANGE Moderate variation, mainly affecting size, slightly also paleness of plumage. Only two subspecies (but apparently one more undescribed;

A. h. hornemanni, 1stW, presumed ♂, Scotland, Oct: moult limit in greater coverts infers age, while very white plumage, narrow flanks streaking, and well-developed red and black markings on head might indicate ♂. (H. Harrop)

A. h. hornemanni, 1stW, presumed ♀, Scotland, Oct: like previous bird, but here pointed and somewhat abraded tail-feathers can be judged, and thicker streaking below might indicate young ♀. (H. Harrop)

A. h. exilipes, ♀ (left) and ad ♂, Finland, Mar: smaller than *hornemanni*, with slighter bill. ♂ shows some pale pink on breast and flanks from late spring, and well-developed red forecrown and black bib; never strongly streaked on flanks. Amount of pink on right bird automatically ages it as ad. Sex of ♀ inferred by association with ♂ but also by very strong flanks streaking and reduced dark face and bib. (H. Sørensen)

see below). All populations sedentary or irregularly nomadic linked to food supply.

C. h. hornemanni (Holboell, 1843) 'HORNEMANN'S REDPOLL' (N Greenland, Ellesmere & Baffin Is; apparently Iceland, or a closely-related unnamed taxon). Differs from *exilipes* in considerably larger size, heavier and basally deeper bill. Plumage very similar to *exilipes* with much overlap, still on average slightly paler greyish-white ground colour, and palest ad ♂♂ attain very large white rump patch and pale head and upperparts. Longest undertail-coverts as *exilipes*, thus never show broad dark streak as *flammea* (but a narrow one can occur). Note long tarsus, large bill-depth and ratio bill-depth/bill-length as a percentage, often helpful with borderline *hornemanni/exilipes*. **L** 14–15 cm. **W** ♂ 81–91 mm (*n* 25, *m* 86.6), ♀ 79–88.5 mm (*n* 22, *m* 82.8); **T** ♂ 62–69 mm (*n* 25, *m* 65.3), ♀ 59–67 mm (*n* 22, *m* 63.5); **T/W** *m* 75.5; **B** 10.1–12.5 mm (*n* 47, *m* 11.2); **B**(f) 7.8–10.8 mm (*n* 85, *m* 9.1); **BD** 6.9–8.4 mm (*n* 48, *m* 7.7); **BD/B**(f) *m* 87.7; **B**(f)/**T** 11.8–16.4 (*n* 48, *m* 13.9); **Ts** 15.0–16.6 mm (*n* 40, *m* 16.0). **Wing formula: p6** < wt 10.5–13 mm; **p7** < wt 16–18.5 mm; **p10** < wt 24.5–27.5 mm; **s1** < wt 25–32 mm. – Situation in Iceland unresolved. Breeding population includes pale birds that appear identical to Arctic Redpoll and have biometrics intermediate between the two other races. Pale Icelandic birds provisionally treated here as *hornemanni* based on rather similar (only marginally smaller) bill-size and very pale ad ♂ plumage (including unstreaked rump and often all-white undertail-coverts), but as wing and tail are markedly shorter, and bill-depth a little smaller and tarsus somewhat shorter, they may require naming (as first noted by Herremans 1990). **W** ♂ 75.5–84.5 mm (*n* 29, *m* 79.6), ♀ 77.5–80 mm (*n* 4, *m* 78.5); **T** ♂ 55–65.5 mm (*n* 29, *m* 60.8), ♀ 58–62 mm (*n* 4, *m* 60.3); **T/W** *m* 76.4; **B**(f) 8.0–9.5 mm (*n* 33, *m* 8.6); **BD** 6.4–8.0 mm (*n* 33, *m* 7.2); **BD/B**(f) *m* 83.5; **B**(f)/**T** 12.9–15.3 (*n* 33, *m* 14.2); **Ts** 15.0–16.6 mm (*n* 26, *m* 15.6). **Wing formula: p7** < wt 14–17 mm; **p10** < wt 23–28 mm; **s1** < wt 24–30 mm.

C. h. exilipes (Coues, 1861) (N Fenno-Scandia, N Russia, N Siberia, much of N North America except northeast). So-called 'COUES' REDPOLL' is treated above under Identification, etc. Same size as Mealy Redpoll (*A. f. flammea*), only has on average subtly shorter bill and longer tail. Usually rather lightly streaked and therefore paler than Mealy Redpoll of corresponding age and sex, rump either unstreaked white (ad ♂♂ at times with pink spots) or with variable streaking, generally faint but rarely as prominently as Mealy Redpoll (generally 1stY ♀♀). Longest undertail-coverts unmarked white or have only rather narrow and sometimes short dark streak. Bill short, culmen straight, tip pointed. In fresh plumage head, throat and sides of upper breast tinged ochre as Mealy Redpoll (most marked in imm), rest of plumage more pale grey and white. With wear in spring, ochre colour bleaches and becomes greyish-white. (Syn. *pallescens*; although originally a name given to intermediate birds between this and *A. f. flammea*, in reality nearly always referring to imm *exilipes*.)

TAXONOMIC NOTES For general notes on the redpoll complex, see Common Redpoll. – A few redpolls can prove to be very difficult to refer to either Common or Arctic Redpolls, supporting the idea that they might be better treated as conspecific. This idea acquires further support from the knowledge that there is very little genetic variation among the entire redpoll complex (e.g., Marthinsen et al. 2008). Speciation among redpolls is obviously a relatively recent event, and morphological characters appear to develop faster than genetic change. Difficult or not, the two redpolls can nearly always be identified from a careful morphological analysis, although field views are not always sufficient. Despite wide-

A. h. exilipes, ad ♀, Sweden, Nov: compared to ad ♂ washed duskier and stronger streaked on head and face, with no pink below, and being ad (evenly-feathered wing) confirms sex, too. Large unstreaked white rump, but more extensive streaks on flanks than ♂. Red crown rather small. (T. Holmgren)

A. h. exilipes, presumed 1stS ♂, Denmark, Mar: greyish, brownish and buffish plumage of young ♂ often intermediate between the two sexes, though still normally paler and more narrowly streaked than ♀ and usually easily separated from *flammea* Common Redpoll. Tertials seem juv; safe sexing and ageing difficult. (H. H. Larsen)

FINCHES

Arctic Redpoll, ssp. *exilipes* | Common Redpoll, ssp. *flammea*

spring ad ♂♂

spring ad ♀♀

autumn/winter presumed ♂♂

autumn/winter presumed ♀♀

Arctic Redpoll, *A. h. exilipes* (left birds: first two Finland, Apr; third Sweden, Jan; fourth Finland, Mar) versus **Common Redpolls,** *A. f. flammea* (right: first Finland, Apr; second Denmark, Mar; third Denmark, Nov; fourth Finland, Mar): first row spring ad ♂♂ of the two species, second row ♀♀; third and fourth rows autumn/winter presumed ♂♂ and ♀♀, respectively. In all plumages, Arctic has more uniform head with weaker streaking and when fresh ochre-yellowish or cinnamon-buff ground colour to face and breast. Mantle and scapular fringes average paler than in *flammea*, which are browner, though darkest Arctic can be extremely similar to pale *flammea*. Streaking on mantle and back narrower, but against paler ground colour more conspicuous than on most *flammea*. Arctic also has larger and cleaner unstreaked white rump (although 1stY, especially ♀♀, can show slight blotching). Pale tips/fringes to wing-coverts and tertials average broader and whiter in Arctic. 'Standard' Arctic has thinner and sparser underparts streaking than Common when corresponding age and sex are compared. Longest undertail-coverts in Arctic white and unmarked, or have very narrow dark shaft-streaks only (in Common usually broad dark central streaks, though a few are intermediates). Bill of Arctic averages shorter, and in fresh plumage can look even shorter due to bushier feathering at bill base. Birds of both species with red or pink on breast, flanks and rump are ♂♂ (Arctic invariably pink, Common in worn summer plumage red), while ad spring birds that lack this are ♀♀; in autumn/winter sexing difficult, but ♂♂ and ad average paler with less streaking. The most similar plumages of the two species are brownest *exilipes* (probably young ♀: bottom left) and palest Common (probably fresh ♂: third right). Only birds with clear moult limits in wing and narrow and pointed juv tail-feathers can safely be aged as young. (Top left: J. Normaja; second: M. Varesvuo; third: T. Lundquist; bottom: H. Taavetti; top right: A. Seppä; second: M. S. Jensen; third: P. S. Christensen; bottom: H. Harrop)

A. h. exilipes, **1stS, Finland, Apr**: pointed tail-tips and moult limit in greater coverts age this bird, but many 1stS cannot be sexed since young ♂ in spring as a rule lacks pink below. Degree of streaking can give hint of sex but here intermediate. Such young *exilipes* can be difficult to separate from palest *flammea* Common Redpoll, but note prominent clean white rump. (J. Normaja)

A. h. exilipes, **1stW, presumed ♂, Denmark, Dec**: fresh young Arctic Redpolls often sullied buff and yellowish-brown, contrasting with purer white areas, including broad wing-bars. This bird seems to have unusually heavy bill for the race *exilipes*. Aged by juv outer greater coverts, while limited amount of streaking on flanks in a young bird best fits ♂. (R. S. Neergaard)

spread sympatry there is no proof of actual interbreeding; birds with intermediate appearance are more likely extreme variations of either taxon rather than hybrids. When breeding Arctic Redpolls are encountered, both parent birds are invariably white-rumped. On direct comparison, song and calls are subtly finer and higher-pitched than Common Redpoll ssp. *flammea*. Based on available data the two are therefore best retained as different species. – The pale Icelandic breeders are provisionally referred to Arctic Redpoll rather than Common, for three reasons: (i) they are usually inseparable by plumage from Arctic; (ii) they live sympatrically with dark redpolls apparently without mixing; and (iii) it is consistent with recognising Common and Arctic Redpolls as two species elsewhere. More research is desirable. As noted above, Arctic Redpolls on Iceland are intermediate between the two existing races and appear to require naming.

REFERENCES Jännes, H. (1995) *Limicola*, 9: 49–71. – Herremans, M. (1990) *Ardea*, 78: 441–458. – Knox, A. (1988) *Ardea*, 76: 1–26. – Landsdown, P., Riddiford, N. & Knox, A. (1991) *BB*, 84: 41–56. – Marthinsen, G., Wennerberg, L. & Lifjeld, J. T. (2008) *Mol. Phyl. & Evol.*, 47: 1005–1017. – Molau, U. (1985) *Vår Fågelv.*, 44: 5–20. – Nyström, B. & Nyström, H. (1987) *Vår Fågelv.*, 46: 119–128. – Pennington, M. K. & Maher, M. (2005) *BW*, 18: 66–78. – Stoddart, A. (1991) *BW*, 4: 18–23. – Votier, S. C. et al. (2000) *BB*, 93: 68–84. – Williamson, K. (1961) *BB*, 54: 238–241.

A. h. exilipes (top) with two Common Redpolls, **Denmark, Dec**: note pale plumage of Arctic, with much-reduced flanks and cheeks streaking, and less black on face. Head and upperparts paler with broader whitish wing-bars. Cinnamon-buff tinge on head and breast often purer and more striking than in fresh Common Redpoll. Also appears larger due to fluffy plumage. (E. Ødegaard)

TWO-BARRED CROSSBILL
Loxia leucoptera J. F. Gmelin, 1789

Alternative name: White-winged Crossbill (Am.)

Fr. – Bec-croisé bifascié; Ger. – Bindenkreuzschnabel
Sp. – Piquituerto aliblanco; Swe. – Bändelkorsnäbb

A widespread crossbill of the boreal taiga, also occurring in North America. Resident breeder in N Russia and Finland, rarely also in Sweden. Apparently nowhere common west of the Ural Mountains, but breeding occurrences are rather erratic, albeit locally more common in some years. Makes irruptive movements to the south and west, often associating with Common Crossbills, at irregular intervals, generally on a small scale, but more massive irruptions have occurred (e.g. in 1990, 2008 and 2013). Breeds in open coniferous taiga with spruce, pine and larch. A specialist on larch seeds. Also eats seeds of spruce and various berries (e.g. rowan).

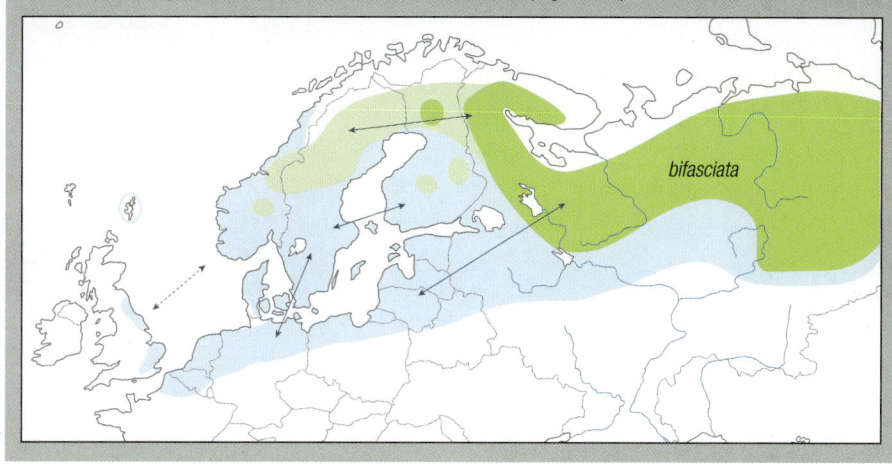

IDENTIFICATION Generally best detected by its *calls*. The song differs slightly, too, compared to its relatives. Unless in heavily worn plumage, the *broad and well-defined white double wing-bars* and extensive pure *white tips to the tertials* of adults should confirm a preliminary identification based on calls. Juveniles have much narrower wing-bars and come close to some aberrant wing-barred Common Crossbills (see Similar species), but luckily young birds frequently appear in the company of one or more adults. Two-barred Crossbill is on average slightly smaller, and has a finer bill, than Common Crossbill, but this is often difficult to discern in the field, and some are the same size, with very nearly as large bills. The red colour of the ♂ is generally somewhat *brighter and more rosy-red*, more the colour of strawberries or geranium (scarlet) than brick-red or crimson; however, again some of the two have very nearly the same red colour. Adult ♀ is yellowish-green and greyish, and resembles ♀ Common Crossbill apart from the white marks, and has on average *bolder dark streaking* and *paler yellow rump*. Very rarely aberrant ♀♀ occur, which have the yellow and green replaced by dull orange or pale red on rump and underparts. Juvenile greyish and heavily streaked dark. Two-barred Crossbill is a somewhat lighter and slimmer bird than the Common (but they are still close in size), and this is reflected in a somewhat lighter flight with perhaps less deep undulations, and also in more agile movements when feeding. Proportionately slightly longer-tailed than Common, augmenting the slimmer-shaped impression. When feeding, often hangs in tit-fashion under a cone, rather than removing it and taking it to a perch. Will also feed on seeds falling down on the snow or ground beneath feeding flocks of Common or Parrot Crossbills.

VOCALISATIONS The song resembles that of other crossbills, comprising call notes mixed with fairly rapid and drawn-out twittering trills and chattering sounds with some variation, but is feebler and at least sometimes a little quicker and more drawn-out, bearing a certain resemblance to the song of, e.g., Goldfinch or Siskin. Listen for the dry but 'full' or slightly clattering common call, *chip-chip-chip*, in the song! – The call from a bird taking off, or from a flock in flight, is a repeated *chip-chip-chip...*, related to that of Common Crossbill but somewhat weaker, drier and less metallic or 'echoing' (often likened to the normal call of Common Redpoll), and often given in quicker succession. As with other crossbills, this call varies in strength, some sounding a little drier and weaker, others being closer to those of Common Crossbill. Another call, immediately identifying the species, heard both from feeding birds and in flight, is a peculiar, nasal piping, *eehp*, like a muted toy trumpet. When heard close from feeding birds it is often slightly upslurred, *eeh-ih*. (Beware of similar-sounding easterly Bullfinches.)

SIMILAR SPECIES If seen briefly and only in flight, could be confused with *Chaffinch*, but has heavier head, all-dark tail, more undulating and purposeful flight, and different calls. – Main risk of confusion is between juveniles/immatures and some aberrant wing-barred *Common Crossbills* (var. '*rubrifasciata*'). Such Common Crossbills can often be eliminated by having: (i) rosy-tinged, buff-white or off-white wing-bars rather than pure white; (ii) diffusely rather than sharply-defined white tips to wing-barred coverts; (iii) white tips of coverts and tertials concentrated on outer webs, rather than almost equally prominent on both webs; (iv) brown uppertail-coverts with insignificantly paler narrow tips, rather than blackish uppertail-coverts with broad (1–2 mm) white tips; and (v) proportionately somewhat shorter tail. A few birds with intermediate characters in all or most of these respects could be hybrids.

AGEING & SEXING Ages differ during 1stCY. Sexes differ at least after post-juv moult. – Moults. Just one, complete post-nuptial moult, mainly in late autumn (Sep–Nov), but post-juv moult more protracted (Jun–Dec) due to variation in fledging time. – **SPRING Ad ♂** Bright red over much of head, breast, flanks, back and rump. Wings and tail blackish, the former with broad white double wing-bar, the lower usually broadest and 'club-shaped', with thickest end pointing at tertials. Tertials have extensive wedge- or V-shaped white tips, typically with much white on both webs, and on some a black point protruding halfway through the white along the shaft. Little seasonal variation, but summer or early-autumn birds can have the white wing-bars somewhat reduced and the white on tertials entirely worn off. Rarely red colour is replaced by greenish-yellow (presumably food-related, but possibly combined with aberration), these told from ♀ by largely unstreaked crown and nape. **Ad ♀** Resembles ♀♀ of

L. l. bifasciata, ad ♂, Finland, Jul: told from other crossbills by bold white wing-bars, emphasised by black covert bases, while tertials are typically white-tipped. Breeding ♂ often strawberry-red (not as crimson as most other crossbills). Immaculate coloration, broad wing-bars and no visible juv feathers in wing indicate age and sex. (H. J. Lehto)

L. l. bifasciata, ad ♀, Shetland, Jul: aside of double white wing-bars and rather small size, note slenderer bill, and listen for typical calls. Mainly dull olive-yellow or grey-green with yellower rump, finely dark-streaked forehead to crown, but more diffuse streaking on mantle and flanks (usually absent in other crossbills). Only ad-like feathers in wing. (H. Harrop)

the larger crossbill species, with an unstreaked lemon-yellow rump and a greenish tinge on the rest of the body, being diffusely streaked dark. Seasonal variation as ♂. Very rarely yellowish and green replaced by orange and pale pinkish-red, but only rump is prominently pink-orange, rest of plumage appears mainly greyish. (Such birds often mistaken for young ♂♂.) **1stS** As ad, and often inseparable unless the more worn and pointed tips to retained juv tail-feathers are seen. Many, however, have narrower wing-bars and less white on tertials, the latter usually being wholly or partly-retained juv-feathers on which all white tips can wear off. — AUTUMN **Ad** As in spring. **1stW ♂** Attains red colour and broader white wing-bars at a variable pace, usually from (Jun) Jul to early Nov (or even later), and to a variable extent, some being red with some orange, buff and yellow admixed through the winter, others becoming virtually all red like ad. **1stW ♀** As in ♂, replaces juv plumage at a variable pace and to different extent. Similar to ad ♀, but those with retained juv, narrow, wing-bars can usually be separated. **Juv** Largely greyish and off-white without yellow or green, boldly streaked brown-grey. Narrow wing-bars.

BIOMETRICS (*bifasciata*) **L** 15–16 cm; **W** ♂ 87–96 mm (n 19, m 93.4), ♀ 84–93.5 mm (n 18, m 90.2); **T** ♂ 56–65 mm (n 18, m 60.3), ♀ 53–62 mm (n 18, m 58.9); **T/W** m 65.8; **B**(f) 16.3–18.0 mm (n 37, m 17.0); **BD** 9.0–10.3 mm (n 36, m 9.8); **Skull+bill** 34.0–38.5 mm (n 21, m 35.6); **Ts** 14.5–16.3 mm (n 29, m 15.1). **Wing formula: p1** minute; **p2** < wt 0.5–4 mm; **p3** longest; **p4** < wt 0.5–2 mm; **p5** < wt 3–5 mm; **p6** < wt 11.5–16 mm; **p10** < wt 29–35 mm; **s1** < wt 31–37 mm. Emarg. pp3–5.

GEOGRAPHICAL VARIATION & RANGE Two subspecies widely separated, possibly better treated as separate species, but kept together until better studied (see below). Mainly resident, but some winter movements most years to west and south, and in years with poor food supplies larger movements reach further.

L. l. bifasciata (C. L. Brehm, 1827) (N Eurasia). Described above. Rather heavy bill. Plumage of ♂ somewhat variable, generally rather warm pinkish-scarlet (not that different from Common Crossbill) but can be more raspberry-red ('poppy red').

Extralimital: **L. l. leucoptera** (J. F. Gmelin, 1789) (North America). Small, with rather fine bill (**BD** c. 8.0–9.0 mm) and streaking is both heavier and darker than in *bifasciata*. Also, plumage of ♂ rather darker red and not as warm scarlet, invariably more carmine (colder bluish-tinged red). It is said to have a different song, but this requires further study; available recordings convey fairly similar songs.

TAXONOMIC NOTES Various authors (e.g. Elmberg 1993, Robb & van den Berg 2002) have suggested that the North American taxon might form a separate species, based on differences in bill structure, plumage colours, and song. If this view is adopted the Eurasian form becomes monotypic *L. bifasciata*. Although such a change would probably be reasonable, morphological and vocal differences are relatively small, and we prefer to await the results of a genetic analysis and a more comprehensive study of vocalisations and behaviour. — Extralimital, widely allopatric but still closely related *L. megaplaga* (Hispaniola) is resident in the Caribbean. Formerly treated as a race of Two-barred Crossbill but now generally afforded species status based on its much heavier bill, shorter wings and tail, narrower wing-bars and different vocalisation (Smith 1997, Boon et al. 2006).

REFERENCES Boon, L. J. R., Ebels, E. B. & Robb, M. S. (2006) *DB*, 28: 99–105. – Ebels, E. B., van Beusekom, R. F. J. & Robb, M. S. (1999) *DB*, 21: 82–96. – Elmberg, J. (1993) *Auk*, 110: 385. – Harrop, A. H. J., Knox, A. G. & McGowan, R. Y. (2007) *BB*, 100: 650–657. – Harrop, H. & Fray, R. (2008) *BW*, 21: 329–339. – Harrop, S & Millington, R. (1991) *BW*, 4: 55–59. – Robb, M. (2000) *DB*, 22: 61–107. – Robb, M. & van den Berg, A. B. (2002) *DB*, 24: 215–218. – Smith, P. W. (1997) *BBOC*, 117: 264–271.

L. l. bifasciata, Denmark, Jan: in flight visibly slimmer than Common Crossbill, but diagnostic very broad white wing-bars and orange-tinged or greenish-yellow rump ensure identification. Three ♂♂ and two ♀♀ dash off from the top of a spruce. (H. Pedersen)

L. l. bifasciata, ad ♂, Finland, Apr: bright red with bold white double wing-bars and blackish wing-feathers. Although there is some overlap, many Two-barred have a somewhat smaller bill with finer tips than Common. (M. Varesvuo)

L. l. bifasciata, ♂, presumed ad, England, Mar: much variation in the red plumage among ♂♂, irrespective of age, showing mixed reddish, orange, yellow and greenish-olive plumage; this presumed ad has more crimson tone. Broad white wing-bars and tertial tips, and no juv (browner) remiges visible infer age. (M. West)

FINCHES

L. l. bifasciata, ♀, England, Dec: fresh winter ♀ lacks strong streaking below of juv. Wing-bars broad and ad-like. Moult limit in greater coverts (outermost ones seem buff-tipped) does not necessarily indicate age. (P. Walkden)

L. l. bifasciata, ♂, presumed ad, Scotland, Aug: a typical ♂ with strawberry-red colour and bold white wing-bars and tertial tips. Age inferred by rather broad tail-feathers, central ones with reddish edge, though these could have been moulted. Some ad ♂♂ can have scattered greenish feathers, as shown here on mantle, thus this is not necessarily a sign of immaturity. Definite ageing unsafe. (H. Harrop)

L. l. bifasciata, ad ♂, England, Jan: some ♂♂ more scarlet or brick-red like Common Crossbill. Moult of tertials can be asymmetric, with old worn ad-type feathers retained on left wing (most of white abraded), thus bird is ad. Compared to image to the left, this bird has maximum size of upper white wing-bar and more extensive black scapulars. (J. Almond)

L. l. bifasciata, 1stW ♂, Finland, Oct: some 1stW moult into reddish ad-like plumage, only with broader pale tips. Still, this bird can be aged by clear moult limit in greater coverts and sharply pointed tail-feathers, and note thin whitish or green-tinged edges to all flight-feathers showing these to be juv in spite of general darkness. (M. Varesvuo)

L. l. bifasciata, 1stY ♂, Sweden, Feb: least-advanced young ♂♂ very distinctive by ragged plumage pattern, with broadly pale-fringed orange and greenish-yellow plumage, and generally obvious moult limits in wing (although latter difficult to evaluate on this image). (T. Lundquist)

L. l. bifasciata, ♀, Finland, Jan: some young ♀♀ have a slight orange tinge in otherwise greyish and yellowish-green colours, and most extreme ones can be difficult to sex. Note how new scapulars, median coverts and many greater coverts are ad-like, latter with blackish bases and broad white tips in contrast to rest of wing being juv. (M. Varesvuo)

L. l. bifasciata, juv, Scotland, Jul: duller than ♀, lacking any green or yellow, and heavily streaked dark, but already has white wing-bars and tertial tips (narrower than ad). To separate from aberrant wing-barred Common Crossbill note that white tips are pure white (not tinged green or red) and concentrated at tips (not continuing narrowly along outer edges). (H. Harrop)

COMMON CROSSBILL
Loxia curvirostra L., 1758

Alternative name: Red Crossbill (Am.)

Fr. – Bec-croisé des sapins; Ger. – Fichtenkreuzschnabel
Sp. – Piquituerto común; Swe. – Mindre korsnäbb

The Common Crossbill is of Holarctic distribution, and the commonest *Loxia* in the treated region. Breeds in coniferous forests over much of Europe and locally in NW Africa, the breeding range being continuous in Fenno-Scandia and Russia, but in rest of range more disjunct and confined to forested mountains. Numbers of breeding pairs in any one area vary substantially depending on food supplies and population growth after good years. Makes large-scale irruptions outside the breeding season to the south and west at irregular intervals. A specialist on seeds of *Picea* and *Abies*, but will exploit *Pinus* or *Larix* where the others are missing. The crossed mandible tips serve as cone-opening 'pliers' when feeding.

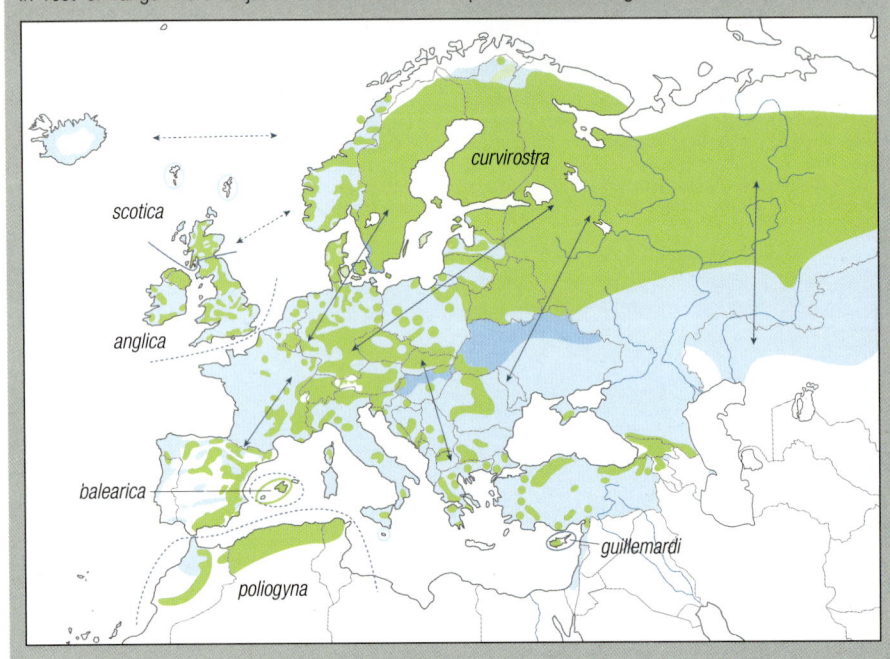

IDENTIFICATION The relative heaviness of a Common Crossbill (between Chaffinch and Bullfinch) is helpful when sorting out a flock of crossbills at long range, as is their habit of alighting in the top of a spruce or pine, the members of a flock rather close together, and quietly but energetically starting to feed on the cones. These are first bitten off by the birds, which move like small parrots among the branches, bending forward and even hanging upside-down, and the cones are then taken to a nearby perch and systematically opened. At closer range recognised as a crossbill by plump build with *large head* and often slightly bull-necked outline, plus heavy *bill with crossed mandible tips*. ♂ is brick-red (scarlet), ♀ greyish with *greenish-yellow rump* (and sometimes green-tinged also on head, breast and centre of belly). Mostly encountered in noisy family parties or larger flocks. In flight, notice dense flock formation, often at considerable altitude, pronounced undulating flight path and (due to altitude) seemingly slow progression. The repeated, metallic, far-reaching calls are frequently uttered before take off and in flight, and serve as the easiest means of detection and identification. Separating Common Crossbill from Parrot requires close views and careful assessment of bill, head and neck size, plus knowledge of calls, and may still not always be possible. (See Similar species.)

VOCALISATIONS Song, delivered from atop a spruce or pine, is rather 'musing': fairly short, clearly-spaced phrases are delivered at a leisurely pace, not unlike the song of Bullfinch. The phrases contain both strained trills and soft twitters, most motifs being repeated, such as a slightly Great Tit-like *tee tee tee tee* or a typical rolling *cherrr*, and best recognised by the interwoven series of *glip* calls. – The common call from a bird taking off, or a flock in flight, is a repeated, strong and metallic *glip glip glip...*, varying somewhat in strength, pitch and details depending on age, individual, mood or population (including subpopulations or even families). Thus, it can sound as *glip*, *chip* or *jip*, at times even from the same flock, but quite commonly it has a 'clipping' quality, indicated by the inserted 'l' in *glip*. The anxiety/alarm call is a deeper, nasal, short, repeated *tohp*. Young beg with a rather high-pitched, shrill, almost trilling *zriit* (Robb 2000).

SIMILAR SPECIES Common and Parrot Crossbills share the same plumage, only differing in general size, head size, bill size/shape, and calls. Common differs from *Parrot Crossbill* by its on average somewhat smaller head and bill, and slightly higher-pitched calls. On a perched crossbill, it is essential to examine the bill in profile: Common as a rule has a slightly longer bill than it is deep, often somewhat longer and more attenuated tips to both mandibles, less strong curvature to both mandibles, the lower mandible somewhat less deep than the upper, and often the tip of the lower mandible is visible well outside the culmen of the upper mandible in side view. To the attuned ear, calls usually provide a good clue. For separation of aberrant wing-barred immatures (var. '*rubrifasciata*') from same-age Two-barred Crossbill, see under latter. – For in-hand identification, particularly important to measure bill depth and lower mandible width at feathering, lower mandible depth across gonys angle, and wing (see Svensson 1992). There is apparently no overlap in bill depth, and very little in lower mandible width, between post-juvenile Common and Parrot (though largest *scotica* of Common come close and require care), and wing length and some other measurements will also help.

AGEING & SEXING Ages differ from fledging and for a period thereafter (between some months to one year depending on hatching and moult). Sexes differ from fledging, but more clearly after post-juv moult. – Moults.

L. c. curvirostra, ♂, presumed 1stW, Finland, Nov: a fairly large stocky finch with large head, stout bill with crossed mandible-tips, and rather long wings. The colourful ♂ is red of varying tone, brightest on the rump, and with characteristic dark frame to ear-coverts. All tertials on left wing retained juv, primaries edged off-white rather than red, and tail-feathers pointed inferring age. (M. Varesvuo)

L. c. curvirostra, ♂, presumed 1stS, Spain, Apr: in some populations, ad ♂♂ occasionally do not become fully red, making it important to age birds before sexing them. Here remiges white-edged (rather than red or yellow) inferring juv, thus most likely 1stS. Degree of orange-red fits only ♂ (some ♀♀ can be quite orange-yellow, but not as red). (C. N. G. Bocos)

L. c. curvirostra, ♀, Spain, May: largely greyish-green with some diffuse grey streaking, especially above, unstreaked greenish-yellow rump and faintly streaked belly. Wing rather uniform, but outermost greater covert seems shorter, older and pale-tipped and could be retained juv, possibly indicating 1stS, but safe ageing impossible from just this image. (C. N. G. Bocos)

L. c. curvirostra, ♀, presumed ad, Netherlands, Nov: rather bright ♀, with quite obvious yellow on crown, belly and rump. When very fresh, ad ♀ can have more marbled olive, grey and greenish-yellow underparts. Whole wing fresh, feathers green-edged inferring age. (C. van Rijswijk)

Generally just one, complete moult, in late summer and autumn (mainly Jul–Oct), but due to protracted breeding season, and irruptive habits, some start to moult in spring (May–early Jun), arrest in summer and resume in autumn. Most 1stY moult partly, a few nearly completely, exceptionally even completely. Advanced post-juv moult can involve all primaries but usually not secondaries or primary-coverts (or at least not inner primary-coverts). Feathers of ♂♂ moulted in summer and autumn (from late Jun) become red, those in winter and spring yellow. Yellow feathers thus no sign of immaturity in ♂♂, but can be food-related (e.g. Newton 1972) or endochrinally governed. – **SPRING Ad ♂** Scarlet or orange-red over much of head, breast, flanks, back and rump. Rarely, some yellow feathers admixed. Wings and tail dark brown-grey, when fresh with very narrow red, rufous or greenish fringes. A variable amount of grey admixed in the red parts. Ear-coverts and scapulars show some brown. Uppertail-coverts dark brown, fringed rufous when fresh. Centre of belly and undertail-coverts white, the latter boldly streaked or spotted dark. Little seasonal variation except wear. **Ad ♀** Except dark brown-grey wings and tail, largely greyish-green with some diffuse grey streaking, especially above, and unstreaked greenish-yellow or bright green rump. **1stS** As ad, and often inseparable unless the more worn and pointed tips to retained juv tail-feathers are seen, or if some narrowly white-tipped juv wing-coverts are retained. – **AUTUMN Ad** As in spring. **1stW ♂** Attains red colour at a variable pace, usually (Jun) Jul–early Nov (or even later), and to a variable extent, some (having a partial moult in late spring and resuming in summer/autumn) being red with some orange, buff and yellow admixed through the winter, others (moulting only in late summer and autumn) becoming virtually all red. **1stW ♀** As in ♂, replaces juv plumage at a variable pace and extent. Resembles ad ♀ as soon as all heavily-streaked juv body-feathers have been replaced. **Juv** Largely greyish and off-white without yellow or green, boldly streaked dark brown-grey. Wing-coverts sometimes finely tipped whitish. Sexes differ in that ♂ has small unstreaked patch where throat meets breast, and on average thinner dark streaks on breast (streaks about as broad as pale background), while ♀ has no such patch on throat/breast, and has broader breast streaking than pale background between streaks (Castell 1983, Edelaar et al. 2005).

BIOMETRICS (*curvirostra*) **L** 15.5–17.5 cm; **W** ♂ 91–105 mm (n 150, m 97.6), ♀ 90–101 mm (n 94, m 95.0); **T** ♂ 55–64 mm (n 66, m 58.3), ♀ 52–60 mm (n 33, m 56.2); **T/W** m 59.1; **B**(f) ♂ 17.9–21.2 mm (n 141, m 19.2), ♀ 17.0–21.4 mm (n 86, m 18.7); **BD** 9.8–12.0 mm (n 227, m 11.0); **BD/B**(f) 51.4–68.7 (n 224, m 58.1); **LM** 13.4–16.8 mm (n 161, m 15.1); **LMD** 4.1–5.7 mm (n 167, m 4.9); **LMD/LM** 27.4–39.6 (n 160, m 32.6); **LMW** 9.6–11.6 mm (once 12.4 and once 9.3; n 107, m 10.7); **Skull+bill** 35.5–41.9 mm (n 139, m 39.6); **Ts** 16.3–18.3 mm (n 77, m 17.3). **Wing formula: p1** minute; **pp2–3** often equal and longest, but sometimes p2 < wt 0.5–2.5 mm; **p4** < wt 1–4 mm; **p5** < wt 6–10 mm; **p6** < wt 14–21 mm; **p10** < wt 31–39 mm; **s1** < wt 35–43.5 mm. Emarg. pp3–4, but often has slight emarg. also on p5.

(LM = lower mandible length, measured on side to where mandible tip is pointed; LMD = lower mandible depth; LMW = lower mandible width; all measurements taken as explained in Svensson 1992; recently-fledged juv have smaller bills.)

GEOGRAPHICAL VARIATION & RANGE Slight variation, mostly involving general size, size of bill and saturation of plumage colours. Five subspecies generally recognised within the treated region. A few others have been described from E Asia, and several from North America, not treated here. All populations mainly sedentary, but irruptions west and south occur depending on food supply in breeding range, more frequently in north of range.

L. c. curvirostra L., 1758 (much of N Eurasia, south to Iberian mainland, Italy and Greece, Crimea, Turkey, Caucasus, Transcaucasia, Russia, W Siberia). Described above. Within its large range slight clinal variations in general size and bill size have been documented (e.g. Alonso et al. 2006). – Birds in Iberia ('*hispana*') said to differ by their longer and thinner bill, but, on the contrary, bill is a fraction shorter

L. c. curvirostra, ♀ (left) and two ♂♂, Finland, Sep: large-headed but short-tailed in flight, which is obviously undulating. Often flies at considerable height but still often noticed by characteristic calls. (M. Varesvuo)

L. c. curvirostra, 1stW ♂, Finland, Feb: amount of yellow, orange or red cannot be used for ageing ♂♂, since this is dependent on food and moult season, thus ♂♂ best left un-aged unless feather wear and moult limits help. Here whole wing seems retained juv, and some dark streaking remains on back and lower belly. (M. Varesvuo)

(in practice same length) and bill depth and width are only 0.6 mm lesser with much overlap, hence here included in *curvirostra*. Note that both bill length and curvature of mandibles vary slightly locally; birds in Italy have slimmer bills with on average longer upper mandibles, like in Iberia, whereas birds in Greece have slightly shorter and heavier bills with more curved mandibles. It is our belief that all these subtle local tendencies (with much overlap) cannot be recognised and named. – Birds in Corsica sometimes separated ('*corsicana*') by being 'less colourful than *curvirostra*', but although this is true for some birds in collections, the majority do not differ much or at all from *curvirostra*. Also incorrectly said to have 'broader tail-feathers'. Due to their feeding habits, many Corsican birds (just as breeders in North Africa and SE Europe) sometimes develop slightly stronger tip of upper mandible, a little different from typically finely pointed tip of average *curvirostra* feeding mainly on the softer spruce cones; still, shape and size of bills of crossbills are known to adapt in response to food choice, hence this slight difference appears insufficient ground for subspecific separation. – Birds in Crimea (*mariae*) said to differ by ♀♀ being paler and ♂♂ paler and brighter red, but no such differences discernible in admittedly scant material examined. Bill averages slightly heavier, but difference small and shows much overlap, hence best included. – Birds in Caucasus ('*caucasica*') have been described as being brighter red in ♂ plumage, but this difference seems variable and insignificant if at all true. – Birds in much of Turkey sometimes separated ('*vasvarii*') by being dark and having a heavy bill, but according to admittedly rather small material any plumage difference seems too variable and faint to warrant formal naming, while the bill has much the same size as in *curvirostra*. (Syn. *caucasica*; *corsicana*; *hispana*; *mariae*; *nidificans*; *taurica*; *vasvarii*.)

L. c. scotica Hartert, 1904 (Scotland). Differs from similarly plumaged *curvirostra* by a little larger size and slightly heavier bill, latter being intermediate between *anglica* and Parrot Crossbill. There is overlap regarding several measurements. Often helpful to note that the *bill is square-looking with rather strong curvature on both mandibles* like in Parrot, only being somewhat smaller. Any bird with flatter-curved mandibles and far-protruding crossed tips, making the bird look slightly longer-billed, is probably a *curvirostra* or *anglica*. **W** ♂ 96.5–105 mm (*n* 35, *m* 100.7), ♀ (once 91) 93–101 mm (*n* 32, *m* 97.5); **T** ♂ 56.5–64 mm (*n* 35, *m* 59.8), ♀ 54–61 mm (*n* 32, *m* 57.5); **T/W** *m* 59.2; **B**(f) ♂ 17.5–20.4 mm (*n* 35, *m* 19.0), ♀ 17.2–20.1 mm (*n* 32, *m* 18.5); **BD** 11.3–13.1 mm (*n* 60, *m* 12.3); **LM** 13.9–16.6 mm (*n* 67, *m* 15.1); **LMD** 4.9–6.4 mm (93% > 5.1; *n* 67, *m* 5.6); **LMD/LM** 31.5–43.4 (*n* 66, *m* 37.2); **LMW** (once 10.4) 11.0–12.7 mm; **Skull+bill** 36.5–44.0 mm (*n* 62, *m* 39.5); **Ts** 16.5–18.8 mm (*n* 52, *m* 17.8). – Calls sound intermediate between *curvirostra* and Parrot Crossbill.

L. c. curvirostra, 1stW ♂, Finland, Feb: some duller 1stY ♂♂ are easily identified in their rather retarded first post-juv plumage, especially if some juv dark streaking on flanks retained, and if coverts have pale tips forming subtle wing-bars. Beware that such pale wing-bars wear whiter and could create resemblance with Two-barred Crossbill, though wing-bars never as broad or as pure white as in that species. (M. Varesvuo)

L. c. curvirostra, 1stS ♀, Spain, Apr: as ad but note moult limit in greater coverts with outer four juv with whitish tips. Rather dull and greyish, with hardly any greenish-yellow on crown and belly. (C. N. G. Bocos)

L. c. curvirostra, 1stW ♀, Finland, Nov: mainly grey and olive, with bright greenish-yellow breast and rump indicating ♀, while streaky plumage and mostly juv wing and tail age the bird. (M. Varesvuo)

L. c. curvirostra, juv, Spain, Mar: heavily streaked black on mainly greyish-white ground, with just slight brownish tinge on back and flanks. Wing-coverts tipped buff-white forming double wing-bars. Although not visible here, rump has often first sign of greenish or yellowish colour. (C. N. G. Bocos)

A careful study of sonograms of some of the calls can apparently separate *scotica*, at least according to current thinking, but this technique and required knowledge is only available to a few. The common call from a bird taking off, and a flock in flight, is a repeated strong and deep, 'echoing' *chup chup chup…*, intermediate between *curvirostra/anglica* and Parrot. On one recording (Robb 2000) a softly downcurved *küup* can be heard, sounding fairly different. The anxiety/alarm call is a subdued, muffled variation of the call, *chu*, repeated in long series.

○ **L. c. anglica** Hartert, 1904 (England, Ireland, probably also parts of S Scotland). When series are compared with *curvirostra* or *scotica*, both sexes are slightly duller, red of ♂♂ is duskier and often admixed with greyish or greenish feathers. Bill rather heavy, short and compact, extremes in upper end overlapping with *scotica*. **W** ♂ 97–102 mm (*n* 12, *m* 99.3), ♀ 94–99 mm (*n* 13, *m* 96.9); **T** ♂ 53–63 mm (*n* 12, *m* 58.3), ♀ 52–60 mm (*n* 13, *m* 56.1); **T/W** *m* 58.3; **B**(f) 17.3–19.8 mm (*n* 25, *m* 18.5); **BD** 10.6–12.0 mm (*n* 25, *m* 11.3); **LM** 13.9–16.5 mm (*n* 25, *m* 15.1); **LMD** 4.7–5.4 mm; **LMD/LM** 29.1–36.0 (*n* 25, *m* 33.0); **LMW**

10.3–11.8 mm (*n* 25, *m* 11.0). A subtle race which will often be difficult or impossible to separate from both *curvirostra* and *scotica* when single birds are encountered in the field.

L. c. balearica (Homeyer, 1862) (Balearic Is). On average smaller than *curvirostra*, the smallest race in the treated range. It is also often paler, ♂♂ frequently being pinkish-red, while ♀♀ are rather dull and pale grey, less green. Bill fairly small with more curved mandibles, but only a slight difference and much overlap (our data; Alonso *et al.* 2006). **W** ♂ 91–97 mm (*n* 19, *m* 94.5), ♀ 89–96 mm (*n* 9, *m* 92.7); **T** 51–59 mm (*n* 28, *m* 55.3); **T/W** *m* 58.8; **B**(f) 16.1–18.7 mm (*n* 28, *m* 17.6); **BD** 10.2–11.8 mm (*n* 27, *m* 10.9); **LMW** 9.4–10.9 mm (*n* 28, *m* 10.3). Alonso *et al.* (2006) noted slightly shorter wing and longer tail than the above values. The flight call can sound rather weak, often lacking the strong 'clipped' quality of N European *curvirostra*.

L. c. poliogyna Whitaker, 1898 (Morocco, N Algeria, N Tunisia). Similar general size as *curvirostra*. ♂♂ resemble *curvirostra* but are generally duller and duskier, often with much grey admixed. ♀♀ lack green on crown and back, have grey wash to greenish rump, and breast is dull greyish with green tinge much subdued; upperparts quite plain, any darker streaking or blotching absent or much reduced. **W** ♂ 95–102 mm (*n* 33, *m* 98.6), ♀ 91–101 mm (*n* 21, *m* 96.0); **T** ♂ 54–60 mm (*n* 31, *m* 57.4), ♀ 53–58 mm (*n* 21, *m* 55.7); **T/W** *m* 58.2; **B**(f) 17.7–21.0 mm (*n* 51, *m* 19.0); **BD** 10.4–12.0 mm (*n* 52, *m* 11.2); **LMW** 9.6–11.3 mm (*n* 43, *m* 10.4). It is noteworthy that this race closely resembles *anglica* both as to biometrics and plumage characters, although ♀♀ are paler and more greyish.

L. c. guillemardi Madarász, 1903 (Cyprus). Differs rather clearly from other races of Common Crossbill in that ♂ does not develop much red, any reddish colour being orange (not scarlet) and is confined to rump and centre of breast and belly, with only slight tinge also on crown and throat; some ad ♂♂ have virtually no orange and are close to inseparable from ♀♀ on plumage (only avering brighter green than typical ♀). ♀ is rather dull and lacks green tinge on crown, or has only very little. Bill-shape similar to *scotica* with rather broad and deep lower mandible. Same general size as *curvirostra*. **W** ♂ 96–103 mm (*n* 24, *m* 99.4), ♀ 93–98.5 mm (*n* 14, *m* 95.9); **T** ♂ 53–64 mm (*n* 24, *m* 58.1), ♀ 53–60 mm (*n* 14, *m* 55.9); **T/W** *m* 58.4; **B**(f) 18.2–21.7 mm (*n* 35, *m* 19.9); **BD** 11.0–12.7 mm (*n* 32, *m* 11.9); **LMW** 10.3–12.9 mm (*n* 36, *m* 11.3).

TAXONOMIC NOTES Since the 1970s, the 'Scottish Crossbill', ssp. *scotica*, has generally been treated as a separate species, but is for consistency here included as a race of Common Crossbill, since at least the Cyprus race *guillemardi* is equally 'different', if not even more distinct. Although sympatric breeding has been reported for *curvirostra* and *scotica* in Scotland, at least in some years and in some areas—usually a good argument for separate species status—this may be due to irruptive populations of the former staying to breed within the range of the more resident latter, and may not be of lasting nature. Genetic differences between *scotica*, *curvirostra* and even Parrot Crossbill are virtually non-existent (Summers & Piertney 2003), indicating recent divergence or gene-flow (e.g. mixed pair of Scottish and Parrot confirmed in Scotland; Summers *et al.* 2007), and it is not known to what extent vocalisation can be learned rather than is inherited, or how quickly bill-size can change in response to available food. Future research will hopefully resolve whether *scotica* is a poorly-defined emerging species, a hybrid swarm as a result of periodic irruptions by Parrot Crossbills to the resident Commons, or—as here treated—a fairly distinct subspecies of Common. – Summers & Jardine (2005) reported strong similarities between the vocalisations of several of the subspecies listed above, but also some minor differences. – Taxonomy of the North American populations has been extensively studied (Groth 1993); although the various taxa seem to be genetically very close, differences in bill size and calls seemed to coincide in different populations, and a division into several cryptic ('sibling') species was proposed. A similar situation for Europe was tentatively suggested by Robb (2000), Parchman *et al.* (2006), Edelaar (2008) and Edelaar *et al.* (2008). However, speciation seems anyway be in its infancy, and like Förschler

Presumed *L. c. anglica*, ad ♂, England, Nov: English breeders average duller and a little darker than *curvirostra*, differences which are best established in direct comparison with museum series at hand. Both primaries and primary-coverts look ad, and immaculate red plumage supports ageing. This bird looks rather heavy-billed, but a drop of water on lower mandible adds to this. (N. Appleton)

L. c. scotica, ♂, Scotland, Feb: differs from *curvirostra* and *anglica* only in on average larger bill, where in particular bill width is broader and lower mandible is heavier, halfway to bill of Parrot Crossbill. Safe separation requires close views or handling to secure detailed measurements. (F. Desmette)

L. c. scotica, ♀, Scotland, May: race *scotica* has an intermediate-sized bill, normally visibly stronger than English or Continental breeders, but some ♀♀, as shown here, can have a slightly smaller bill which only differs marginally from the others. Safe ageing would require handling. (D. Whitaker)

Variation in bill size of crossbills: Common Crossbill ssp. *anglica* (left: England, Mar), ssp. *scotica* (centre: Scotland, Feb) and Parrot Crossbill (right: Finland, Mar), ♂♂: from left to right, general size, thickness of lower mandible and curvature of culmen increase with the effect that tips of mandibles become blunter and less protruding. (Left: J. Richardson; middle: D. Eades; M. Varesvuo)

L. c. scotica, ♀, presumed ad, Scotland, Apr: drop of water on lower mandible makes bill appear very chunky, but lower mandible still seems rather strong, and culmen is quite sharply curved. Although amount of yellowish-green in plumage would indicate ad, as do seemingly greenish edges to most wing-feathers, safe ageing without handling unwise. (F. Desmette)

L. c. poliogyna, 1stY ♂, Morocco, Mar: ♂♂ average somewhat duller red than *curvirostra* as can be seen from this extreme example photographed in the Atlas Mts. However, it is an imm ♂, and some older ♂♂ attain a somewhat redder plumage than this. Biometrics by and large the same as for W & N European populations. (A. B. van den Berg)

& Kalko (2009) we prefer to await further research before evaluating these findings. Keenan & Benkman (2008) found that adults can alter their calls to match those of their mates; although apparently rare, this complicates the suggestion of several cryptic species with different call types. It seems a fair assumption that radiation among Common Crossbill into further species is still at a very early stage and that treating it in the traditional way as one species is not unreasonable.

REFERENCES Alonso, D. *et al.* (2006) *Ardea*, 94: 99–107. – van den Berg, A. B. & Blankert, J. J. (1980) *DB*, 2: 33–35. – Berthold, P. & Schlenker, R. (1982) *DB*, 4: 100–102. – Castell, P. (1983) *Cage & Aviary Birds*, (May): 9–10. – Edelaar, P. (2008) *J. Avian Biol.*, 39: 9–12. – Edelaar, P., van Eerde, K. & Terpstra, K. (2008) *J. Avian Biol.*, 39: 108–115. – Edelaar, P., Phillips, R. E. & Knops, P. (2005) *Wilson Bull.*, 117: 390–393. – Förschler, M. I. & Kalko, E. K. V. (2009) *J. of Orn.*, 150: 17–27. – Groth, J. G. (1993) *Univ. of Calif. Publ. Zool.*, 127: 1–144. – Herremans, M. (1982) *Gerfaut*, 72: 243–254. – Herremans, M. (1988) *Gerfaut*, 78: 243–260. – Jardine, D. C. (1994) *Ring. & Migr.*, 15: 98–100. – Keenan, P. C. & Benkman, C. W. (2008) *Condor*, 110: 93–101. – Knox, A. G. (1990) *BB*, 83: 89–94. – Newton, I. (1972) *Finches*. Collins, London. – Parchman, T. L., Benkman, C. W. & Britch, S. C. (2006) *Molecular Ecology*, 15: 1873–1887. – Proctor, B. & Fairhurst, D. (1993) *BW*, 6: 145–146. – Robb, M. (2000) *DB*, 22: 61–107. – Sewall, K. B. & Hahn, T. P. (2009) *Animal Behaviour*, 77: 123–128. – Snowberg, L. K. & Benkman, C. W. (2008) *J. Evol. Biol.*, 20: 1924–1932. – Summers, R. W. & Piertney, S. B. (2003) *BB*, 96: 100–111. – Summers, R. W. & Jardine, D. C. (2005) *Ardeola*, 52: 269–278. – Summers, R. W., Dawson, R. J. G. & Phillips, R. E. (2007) *J. Avian Biol.*, 38: 153–162. – Summers, R. W., Dawson, R. J. G. & Proctor, R. (2010) *J. Avian Biol.*, 41: 219–228.

L. c. guillemardi, ad ♂, Cyprus, Oct: bill in this race slightly stronger and deeper than in *curvirostra*, with more strongly curved culmen. Never becomes as red as other European races, with ad ♂ usually being bright greenish-yellow, at most having orange-tinged crown, underparts and rump. Wing ad. (H. Shirihai)

L. c. poliogyna, ♀, presumed ad, Morocco, Oct: ♀♀ of the North African race are typically duller and greyer, less bright green than ssp. *curvirostra*, in particular on upperparts lacking green on crown and having subdued green wash only on rump. (A. B. van den Berg)

L. c. guillemardi, ad ♀, Cyprus, Oct: duller race, which in ♀ plumage has characteristically greyish head and upperparts. Wing ad. (H. Shirihai)

L. c. guillemardi, 1stW ♂, Cyprus, Oct: as ad ♂, but much of wing juv, contrasting with moulted lesser, median and inner greater coverts (latter white-tipped). (H. Shirihai)

PARROT CROSSBILL
Loxia pytyopsittacus Borkhausen, 1793

Fr. – Bec-croisé des perroquet
Ger. – Kiefernkreuzschnabel
Sp. – Piquituerto lorito; Swe. – Större korsnäbb

A larger cousin of the Common Crossbill, with a proportionately heavier bill adapted to opening the harder cones of *Pinus* species. However, it also feeds on seeds of other conifers too, including spruce. Range usually restricted to N Europe and NW Siberia, but after irruptions may stay and breed elsewhere, e.g. in Scotland. Often associates with Common Crossbill, making similar irruptions though less numerous.

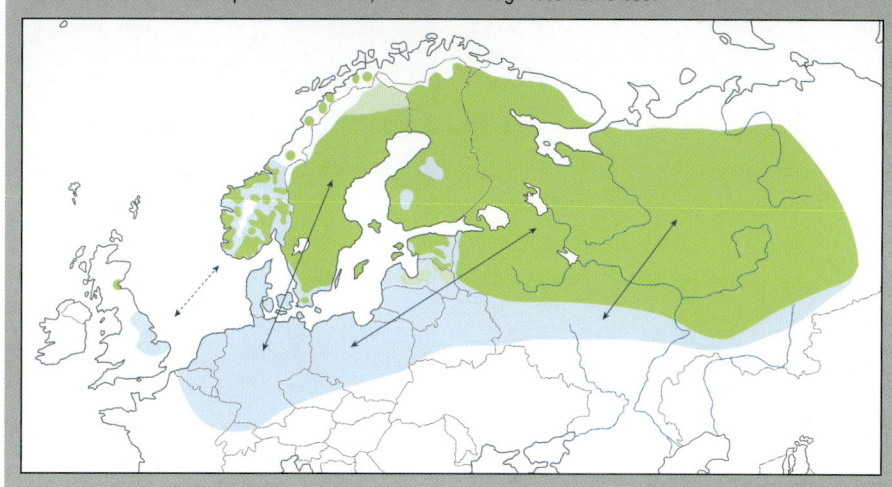

1stS ♂, Finland, Mar: strongly curved mandibles, and depth of lower mandible *c.* 45% of upper, while tip of lower in profile projects at most marginally. Post-juv moult tends to be rather limited, with 1stY often showing many old feathers. Note moult limit in greater coverts. (M. Varesvuo)

IDENTIFICATION See Common Crossbill for general description. Contrary to some claims, plumage is identical to Common Crossbill (thus, ♂♂ not on average greyer or paler orange-red). For reliable identification a reasonably close observation is necessary, and the bill must be studied in profile. Compared to Common Crossbill has somewhat *larger head* and often slightly more *bull-necked* outline, heavier bill with a 'squarer' look, the *depth of the bill being roughly equal to its length*. Adding to this impression is the *stronger curvature of both mandibles*, with almost as deep lower mandible as upper, the *lower mandible having a 'bulging' outline*. A supporting character is the much *broader bill-base* than in Common when seen head-on. To the trained ear on the Continent (outside the British Isles) the *deeper, harder calls* are the best means of both detection and identification of Parrot Crossbill, but even with practice it may be difficult to identify all birds.

VOCALISATIONS Song quite similar to that of Common Crossbill, a repetition of simple notes, whistles, squeaky notes and interwoven series of calls, delivered unhurriedly, and with well-spaced segments. Possibly only recognised by the interwoven call notes. – The common call from a bird taking off, and a flock in flight, is a repeated strong and deep, 'echoing' *tüpp tüpp tüpp…* (or *küp küp küp…*), compared to Common Crossbill with harder initial consonant and sounding slightly deeper (more echoing) and straighter, less 'clipped'. Rather limited individual variation in the call, and any identification problem is due to greater variation among Common Crossbills, of which some populations can approach Parrot's call rather closely, but probably never match it entirely. The anxiety/alarm call is a subdued, rather deep variation of the call, *chu*, repeated in long series. Young beg with rather high-pitched *chit-too* (Robb 2000).

SIMILAR SPECIES Plumage as for *Common Crossbill*, but that species has a somewhat smaller head and bill, and slightly higher-pitched calls (see Common for details, and above under Identification). – For in-hand identification particularly important to measure bill depth and lower mandible width at feathering, lower mandible depth across gonys angle, and wing (cf. Svensson 1992). There is apparently no overlap in bill depth, and very little in lower mandible width, between Parrot and Common. A few controversial and reportedly heavy-billed Common of ssp. *scotica* could actually well have been Parrot.

AGEING & SEXING Ages and sexes differ, much as described for Common Crossbill. – Moults. As in Common Crossbill, generally only one, complete moult, in late summer and autumn (mainly Jul–Oct), but due to protracted breeding season, and irruptive habits, some start to moult in spring, arrest in summer and resume in autumn. Most 1stY moult partly, a few nearly completely, exceptionally even virtually completely. Advanced post-juv moult can involve all primaries but usually not secondaries or primary-coverts. Feathers of ♂♂ moulted in summer and autumn (from late Jun) become red, those in winter and spring yellow. – **SPRING Ad ♂** Scarlet or orange-red over much of head, breast, flanks, back and rump. Rarely, some yellow feathers admixed. Wings and tail dark brown-grey, when fresh with very narrow red, rufous or

Ad ♂, Sweden, Jan: averages larger and bulkier than Common Crossbill, with larger head, thicker neck and massive parallel-sided bill, with bulging, strongly curved gonys and blunter tip. Can show extensive greyish cutting edges. Predominantly bright scarlet, but some grey feathers admixed. Wing evenly fresh. (M. Nord)

♀, presumed ad, Finland, Mar: unlike ♂, mainly greyish-green with grey streaking, especially above, partly greenish-yellow on crown, rump and sometimes belly. Wing apparently ad. Note size of bill, and that mandible curvature is strong resulting in that tip of lower mandible is hardly visible in profile. (M. Varesvuo)

greenish fringes. **Ad ♀** Except dark brown-grey wings and tail, largely greyish-green with some diffuse grey streaking, especially above, and unstreaked greenish-yellow or bright green rump. **1stS** As ad, and often inseparable unless the more worn and pointed tips to retained juv tail-feathers can be seen, or if some narrowly white-tipped juv wing-coverts are retained. – AUTUMN **Ad** As in spring. **1stW ♂** Attains red colour at a variable pace in summer–autumn, and to a variable extent, some being red with orange, buff and yellow admixed through the winter, others becoming virtually all red. **1stW ♀** As in ♂, replaces juv plumage at a variable pace and extent. Resembles ad ♀ as soon as all heavily-streaked juv body-feathers have been replaced. **Juv** Largely greyish and off-white without yellow or green, boldly streaked dark brown-grey. Wing-coverts sometimes finely tipped whitish. Sexing of juv not studied but could probably be the same as described for Common Crossbill.

BIOMETRICS L 17–18.5 cm; **W ♂** 99–111 mm (*n* 72, *m* 105.0), ♀ 97–108 mm (*n* 57, *m* 102.1); **T ♂** 57–68 mm (*n* 72, *m* 63.0), ♀ 58–66 mm (*n* 57, *m* 61.6); **T/W** *m* 60.2; **B** 20.0–26.0 mm (*n* 128, *m* 22.9); **B**(f) 17.6–22.2 mm (*n* 125, *m* 20.2); **BD** 13.1–14.9 mm (*n* 124, *m* 13.9); **BD/B**(f) 61.8–81.3 (*n* 123, *m* 69.4); **LM** 14.0–17.4 mm (*n* 124, *m* 15.6); **LMD** 5.8–7.2 mm (*n* 57, *m* 6.5); **LMD/LM ♂** 35.9–47.9 (*n* 124, *m* 41.9); **LMW** 11.3–13.7 mm (*n* 124, *m* 12.7); **Skull+bill** 39.0–45.6 mm (*n* 124, *m* 41.9); **Ts** 17.6–19.8 (*n* 126, *m* 18.6). **Wing formula: p1** minute; **pp2–4** often equal and longest, but sometimes p2 and p4 < wt 0.5–2.5 mm; **p5** < wt 5–11 mm; **p6** < wt 15–21 mm; **p7** < wt 20–28 mm; **p10** < wt 34–43 mm; **s1** < wt (37) 39–45 mm. Emarg. pp3–4(5).

(LM =lower mandible length, measured on side to where mandible tip is pointed; LMD =lower mandible depth; LMW =lower mandible width; all measurements taken as explained in Svensson 1992; recently-fledged juv have smaller bills.)

GEOGRAPHICAL VARIATION & RANGE Monotypic. – N Europe from Scotland and Fenno-Scandia, east through Estonia, Russia to N Ural Mts; sedentary or nomadic, in some years making longer movements far outside breeding range.

REFERENCES Bowey, K. & Westerberg, S. S. (1994) *BB*, 87: 398–401. – Millington, R. & Harrap, S. (1991) *BW*, 4: 52–54. – Robb, M. (2000) *DB*, 22: 61–107.

♂, Denmark, Oct: fresh and colourful plumage suggests ad male, but too few wing-feathers visible to confirm age using moult. Size and proportions of bill exclude all races of Common Crossbill. (E. Ødegaard)

1stW ♂, Denmark, Oct: somewhat variable, with scarlet-red plumage admixed with orange and yellow, but still readily identified as ♂. Just as in Common Crossbill, some younger birds have whitish wing-bars of variable prominence, these being at first buff-tinged, then bleach to white before they wear off. (E. Ødegaard)

♂ (left) and ♀, Denmark, Feb: even in flight, it can be possible to see heavy and almost 'square-shaped' parrot-like bill with especially chunky lower mandible. (M. S. Jensen)

1stW ♀, Denmark, Nov: showing the powerful bill required to handle the hard pine cones. Once all of the juv streaked body-feathers are moulted appearance becomes very much like ad ♀, but most of wing retained. (S. Olofson)

1stW ♀, Sweden, Dec: some young ♀♀ become very dull and greyish, and somewhat streaky. Note moult limits among greater coverts with only inner two moulted to ad type. Massive and parallel-sided bill. (S. Hage)

Juv, Sweden, Jun: all juv crossbills are greyish and off-white with boldly dark-streaked plumage, at first without any yellow or greenish tinges. Only size and shape of bill discriminates a young Parrot from Common Crossbill (apart from some calls). Note depth of lower mandible, not quite but almost as thick as upper mandible. (L. Petersson)

ASIAN CRIMSON-WINGED FINCH
Rhodopechys sanguineus (Gould, 1838)

Fr. – Roselin à ailes roses; Ger. – Rotflügelgimpel
Sp. – Camachuelo alirrojo asiático
Swe. – Bergsökenfink

A rather scarce finch of rocky mountains and foothills over a wide area from Turkey to Central Asia, wintering at lower elevations or making short-range movements to the south, in the east occasionally reaching N Pakistan and NW India. Usually a frustratingly shy and mobile species, elusive and difficult to get close to. A typical encounter is a sudden appearance, arriving in fast, undulating flight with unmistakable calls, a brief stop to feed, and then all too soon a nervous departure for a distant hill.

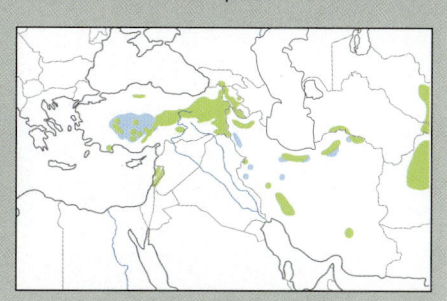

IDENTIFICATION Relatively large, *heavy-billed, robust finch*. Usually appears nondescript dark brown at distance, but at close range displays *large pink wing-panels* (striking in ♂), a *dark crown* which can appear as a broad blackish band along top of head, *brown mantle blotched or diffusely streaked blackish*, and variably *boldly dark-blotched rufous-brown breast and upper flanks* (especially prominent in ♂), some *dark pink-red on face and lower rump*, and *pink or whitish basal patches and white tips to tail*. ♀ and winter birds generally duller. In most plumages, *chin to breast and flanks are mainly warm buff-brown*, with *a pale crescent on the lower breast*, and lower nape and belly are also pale. The combination of all these characters results in a rather characteristic and boldly-patterned plumage when seen close. Stance noticeably horizontal, with heavy *pale bill* and head held up. When encountered, usually solitary, in pairs or small parties, but occasionally gregarious, especially post-breeding when sometimes forms mixed flocks with other finches. *Bounding strong flight*. In breeding season generally restricted to rocky slopes above 1500 m, but may forage or drink at lower levels.

Ad ♂, **Turkey, May**: superficially like other pink-winged finches, but bulky with proportionately short tail, long wings, diagnostic pink-and-black wing pattern, stout bill, crimson mask, blackish crown and pale crescent-like breast-band. Note distinctive pink tinge to lower rump. (D. Occhiato)

Ad ♂, **Turkey, Jun**: ♂ has black cap, brighter reddish-pink face, warm pinkish cinnamon-buff supercilium above brownish-streaked ear-coverts, brighter chestnut mantle and pink in wing-coverts and flight-feathers. Crescent-like breast-band and chestnut-brown-specked flanks that contrast with rest of underparts. Wing seems ad and fresh, with blacker tips and centres. (V. Legrand)

VOCALISATIONS Song is a short, rhythmic and almost melodious strophe, often given in deeply undulating flight, a rapid, jerky *diddle-diddle-de-dii-u* or *chuke-chuke-chu-che-tleeeh-chuke*. Typically one note near end is higher-pitched, drawn-out and stressed. – Flight-call a soft fluting *chiv-lee* or a resounding, melodious, trisyllabic *chu-che-litt*. A rhythmic yodelling *dy-lit-dy-lit* has been likened to Woodlark. A 'querying', timid *tveeh* also noted.

SIMILAR SPECIES Unmistakable if seen well, given size, heavy pale bill, dark cap, extensive pink (and sometimes partly white) coloration in wings, tail, on rump and face. Note also white underwing-coverts. – Very similar to *African Crimson-winged Finch* but latter is extremely unlikely to be encountered in the range of Asian Crimson-winged. Difference between them is covered in some detail under African Crimson-winged, but the main clues are repeated here. Asian Crimson-winged lacks paler bib (or, rarely, has only hint of one), has bolder streaking or dark blotching on mantle, breast and upper flanks, more dark pink on fore-face, tends to have dark-spotted rufous cheeks, subtly heavier (deeper) bill, in adult ♂ more extensive pink on folded wing and slightly shorter tail. Furthermore, has on average blacker and laterally better-defined dark crown patch. – Considerably larger than *Trumpeter* and *Mongolian Finches*, with proportionately larger head and bill, and shorter tail. Both overlap with Asian Crimson-winged Finch in E Turkey, and Asian Crimson-winged has been widely confused with Mongolian Finch there. – Flight fast and powerful, and deeply undulating, being more akin to *Rock Sparrow* (which overlaps in range) than to other finches, but Crimson-winged Finch and Rock Sparrow differ considerably in head and wing patterns, as well as other characters.

AGEING & SEXING Ageing possible, at least in 1stW if seen close or handled using moult limits in upperwing. Sexes similar but moderately differentiated, mainly in spring and summer. Correct sexing, especially in autumn and winter, may first require ageing using moult pattern in wing. Seasonal plumage variation limited and mostly due to feather wear. – Moults. Complete post-nuptial moult of ad and partial post-juv moult mainly in Jul–Sep (can start late Jun). Post-juv moult includes head, body, lesser and median coverts, none or some inner greater coverts and tertials, and a few may moult more extensively (rarely even few inner primaries, all primary-coverts, some outer secondaries and r1, then appearing much as ad in field views). Pre-nuptial moult apparently absent. – **SPRING Ad** ♂ Wear increases contrast of black cap, face pattern (especially pale supercilium and collar), rosy tone to lower back and rump, and red of face (near bill-base, on lores and around eye). Also,

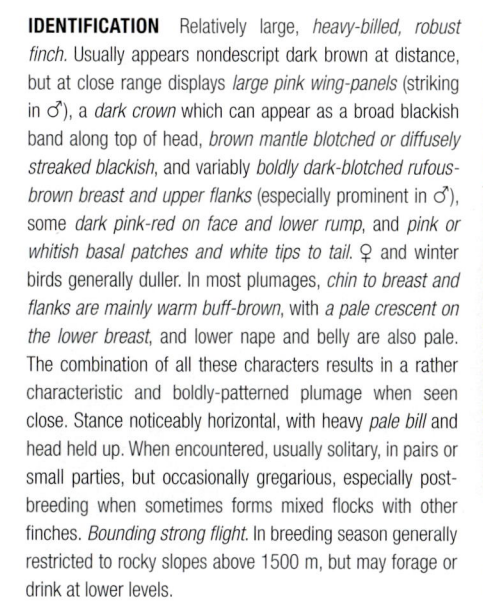

♂, presumed 1stS, **Turkey, Jun**: 1stY ♂ variable, some moult almost into ad-like plumage (though variation poorly known, and an ad which did not finish complete moult is an alternative interpretation; note unmoulted outer three primaries). Compared to African Crimson-winged has slightly deeper bill, usually lacks paler bib, has blacker and better-defined cap, brighter reddish-pink around eye, bolder streaking on mantle, breast and upper flanks, dark-spotted rufous cheeks, and more extensive pink in wing. (D. Barnes)

Presumed 1stS ♂, NW Pakistan, Jun: most likely a 1stS ♂ with bleached brownish primaries and abraded tertials. Nevertheless, dark cap and some stronger reddish elements near the eye already developed, and wing panel more pink than white, so should be a ♂. Such a bird if not correctly aged first could be mistaken as ♀. (I. Shah)

♀, presumed ad, Turkey, May: duller than ♂, with much less pink or white in wings, and no pink visible in basal tail. Also less strongly patterned below with paler chin and throat. Upperparts much less heavily streaked, pale area on rump to uppertail-coverts almost lacks pink, and lores and eye-surround light cinnamon-buff. Despite unusually strong feather wear, all visible wing-feathers seem ad, with still darker remiges and round ad-like primary-tips, pointing to ad, as does some pinkish in greater coverts. (D. Occhiato)

breast-sides become more rufous-cinnamon, black spots on central breast and flanks are more contrasting, and red basal areas on wings and tail become more conspicuous, but white tips to remiges are virtually lost. A few develop pale pinkish cream-brown throat, but not to the extent of some 1stS ♂♂. Bill dull yellow when breeding. **Ad ♀** Duller with reduced dark cap (can be reduced to brown-grey centres to crown-feathers), more white and duller pink in wings, virtually no pink in tail, and whiter, less rufous and less spotted underparts (chin and throat variable, usually with some pale cream-brown or white). Mantle and scapulars browner and much less heavily streaked, lower back to uppertail-coverts paler grey-brown, tinged isabelline, and lores and eye-surround greyish cinnamon-buff (at most faintly reddish), with paler ear-coverts and less contrasting supercilium. Bill greyish-horn. In hand, pink-red edges to remiges and primary-coverts paler and narrower (also has browner and less well-defined centres). **1stS** Much individual variation, especially in 1stS ♂♂, some approaching ad ♂ in overall brightness (but most still safely sexed using same criteria as for ad). Some have variable pale throat reaching upper breast and almost no streaks on underparts, thus approaching African Crimson-winged. Best aged by contrastingly worn retained juv wing- and tail-feathers; extent of subterminal black areas on rr5–6 as 1stW. — **AUTUMN Ad** Both sexes differ from 1stW in being evenly fresh with broader and whiter primary tips. **Ad ♂** Fresh cinnamon-buff fringes to black crown patch. Also, duller carmine-red facial areas, and upperparts more buff, less heavily streaked but broadly fringed, while rump and uppertail-coverts are pale rosy-pink. Whitish band between breast and flanks on the one hand, and upper belly on the other washed pale pink, with yellowish-buff breast and flanks tipped whitish (giving slightly marbled appearance), being unstreaked or nearly so. **Ad ♀** Much as in spring, but broad greyish-buff fringes to crown and upperparts make this plumage even duller. In both sexes pale fringes to wings and tail broader, with carmine-pink and red basal area partially concealed. **1stW** Both sexes differ from ad by moult limits among greater coverts, and by retained juv primaries and tertials, being browner basally and less fresh, with narrower and less sharply-defined white tips. Tail-feathers usually rather abraded at tips. **1stW ♂** Plumage may approach ♀. Striking head and underparts patterns as seen in ad much reduced, with pale flecking more extensive (chin and throat pattern variable, as 1stS ♂).

1stS ♀, Turkey, May: as ad ♀ but duller and least marked. Best aged by dull brown and rather worn primaries, and juv primary-coverts lacking bold black tips and pink outer webs of ad. Beware that such imm plumage can show vague pale bib, and this combined with reduced streaking of underparts can recall African Crimson-winged Finch. (D. Occhiato)

Ad ♂♂, Turkey, Sep: ♂ autumn plumage is more ♀-like due to many pale feather-tips attained through post-nuptial moult. Still, already in Sep sexing possible since black crown is partly visible, as is pink loral area (left bird) and blackish blotches on mantle and flanks (right bird). (Left: P. Thomsen; right: G. Ekström)

1stW ♂ (left) and 1stW ♀, Turkey, Sep: first-winter plumage usually clearly duller than respective ad, but as to 1stW ♂, sex inferred from blackish cap being extensive (although broken up by some pale tips), loral area having much pink and at least some juv wing-feathers retained, which together separate from similar ad ♀. 1stW ♀ resembles the young ♂ but has duller and paler pink in wing and less extensive and even more broken-up black cap. (M. Uyar)

Upperparts essentially warm brown, with paler rump than ad. Pink and white areas in remiges and tail-feathers duller and reduced. Retained juv greater coverts have blackish-brown inner webs and are brown-buff over most of outer web except a thin pink fringe (ad greater coverts have blacker inner webs and almost entire outer webs pink). Retained primary-coverts mostly dull brown with narrow fringes (lacking contrasting blackish inner web and pinkish outer web of ad). Tail has more extensive dark areas than ad ♂, approaching ad ♀. **1stW ♀** Plumage palest and least patterned, with much-reduced pink in wing, and fringes to greater coverts mainly buff. Terminal half of r5 black except diffuse white tip on inner web, and r6 also has slightly more extensive dark areas than ad ♀. **Juv** Soft, fluffy body-feathers generally rather sandy or sandy-brown with few dark centres, very little pink on closed wing, and bill dark horn-yellow becoming brownish at tip.

BIOMETRICS L 15–16 cm; **W** ♂ 101.5–111 mm (n 30, m 106.2), ♀ 97–109 mm (n 16, m 101.7); **T** ♂ 54.5–63.5 mm (n 30, m 58.0), ♀ 53–59 mm (n 16, m 56.1); **T/W** m 54.8; **B** 14.5–17.8 mm (n 46, m 15.8); **B(f)** 11.5–14.0 mm (n 46, m 12.7); **BD** 9.2–11.7 mm (n 45, m 10.9); **Ts** 17.7–20.7 mm (n 43, m 19.5). **Wing formula: p1** minute; **pp2–3**(4) about equal and longest; **p4** < wt 0–6 mm; **p5** < wt 8–12 mm; **p6** < wt 15–20 mm; **p7** < wt 20.5–27 mm; **p10** < wt 31–40.5 mm; **s1** < wt 34–43 mm. Emarg. pp3–5.

GEOGRAPHICAL VARIATION & RANGE Monotypic. – C, S & E Turkey east to mountains of Central Asia and Xinjiang, NW China; sedentary or nomadic, with local and irregular movements in non-breeding season.

Juv/1stW ♀, Turkey Aug: in advanced post-juv moult (juv primaries and coverts). 1stW ♀ is most nondescript plumage, with bland head and wing patterns. Superficially similar to Trumpeter and Mongolian Finches, but has larger head and bill, slightly shorter tail and different wing pattern. (M. Sözen)

TAXONOMIC NOTE The relationship between *sanguineus* and *alienus* was discussed based on morphology (Kirwan et al. 2006), wherein several fine but clear differences were listed indicating separate species status for each. However, no comprehensive genetic examination has been performed, and since vocalisations apparently are the same, further research seems desirable. Still, the split is tentatively adopted here.

REFERENCES Kirwan, G. M. et al. (2006) *Bull. African Bird Club*, 13: 136–146.

AFRICAN CRIMSON-WINGED FINCH
Rhodopechys alienus Whitaker, 1897

Fr. – Roselin de l'Atlas; Ger. – Atlasgimpel
Sp. – Camachuelo alirrojo bereber; Swe. – Atlasökenfink

This close relative of the Asian Crimson-winged Finch is often regarded as a subspecies of that, but is treated here as a separate species considering its widely allopatric range and some clear and consistent morphological differences. Its rather fragmented range is restricted to higher areas of the Atlas Mountains of Morocco and apparently also to a very small area in NE Algeria. It frequents open rocky areas with sparse vegetation, wintering at somewhat lower elevations.

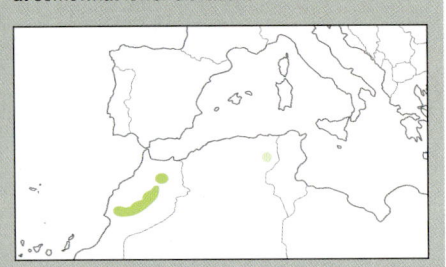

Ad ♂, Morocco, Dec: differs from Asian Crimson-winged by pale brownish-white throat bordered by tawny-brown collar, while brownish body-sides lack blotchy streaks. Black cap of ♂ smaller, bordered by greyish nape and upper neck-sides. Upperparts and head-sides dark grey-brown, less rufous-brown than Asian. Only ad wing-feathers visible, while relatively colourful plumage in Dec also indicative of age. (F. Trabalon)

Ad ♂, Morocco, Feb: compared to Asian Crimson-winged has much-reduced red mask, and hardly any pink on rump, wing-coverts or tertials, thus wing shows smaller panel. Wing evenly feathered and tips of tail-feathers fairly broad, inferring age. (D. Occhiato)

Ad ♀, Morocco, Feb: unlike ♂, cap duller and only blotched black, and rest of plumage less boldly marked, with drabber upperparts, rump/uppertail-coverts less pink, body-sides pale buff-brown and dark streaking very thin. Only ad wing-feathers visible. (D. Occhiato)

IDENTIFICATION Very similar to Asian Crimson-winged Finch: a relatively large, *robust finch with heavy pale bill, strikingly patterned wings* and tail with usually *extensive pink portions, warm rufous-brown breast and flanks, variable pale crescent across lower chest* (usually ill-defined, edges broken) and pale belly. However, is subtly larger and longer-tailed, slightly paler and duller in all plumages, and note the following additional distinctions: (i) invariably a *pale buff-white throat bib* surrounded by darker brown (Asian: paler bib lacking or only rarely present); (ii) breast-sides and flanks rather uniform tawny-brown or reddish-ochre, *without blackish blotching* (Asian: breast-sides and upper flanks more rufous and boldly dark-blotched); (iii) less extensive and less saturated pink on face (Asian: ♂♂ often have deep pink-red lores, eye-surround and fore supercilium); (iv) rump brown, lacking pink (Asian: ♂ often pink on rump); (v) mantle rather uniform brown, darker streaking or mottling subdued (Asian: mantle dark brown diffusely striped blackish); (vi) cheeks rather plain brown (Asian: tends to have dark brown cheeks blotched blackish); (vii) on average less extensive pink on folded wing, both primary-coverts and secondaries rather prominently tipped dark (Asian: secondaries usually all pink, primary-coverts either all pink or have smaller dark tips); (viii) tends to have purer grey nape (Asian: usually greyish-white with buff hue); (ix) dark crown (most prominent in ♂) less black and less well defined (Asian: on average blacker and laterally better-defined crown patch); and (x) tends to have subtly less deep, more rounded, less pointed bill, though overlap (Asian: usually slightly deeper bill with straighter culmen). ♀ more similar to Asian Crimson-winged Finch but duller, and upperparts distinctly more uniform and greyer (less streaked and buff-brown). Shares many behavioural characteristics, and has similar flight pattern and habitat choice to Asian Crimson-winged Finch.

VOCALISATIONS See Asian Crimson-winged Finch; no consistent differences between these two closely-related species are known, and there appears to be much individual variation in the breeding season as to song, further precluding the detection of any slight difference.

SIMILAR SPECIES Separation of Asian and African Crimson-winged Finches is discussed above in some detail, but they are also extremely unlikely to come into contact. When comparing specimens in museum collections it is essential to compare birds of the same age and sex (and season). Some young ♂ Asian Crimson-winged Finches lack the diagnostic blackish breast markings and have a white throat patch, thus superficially recalling African Crimson-winged, but remain separable using other differences mentioned under Identification. African also has a relatively smaller and less pointed, on average more 'chubby' bill. Other confusion risks are the same as mentioned under Asian Crimson-winged, but African is much less likely to be confused with other finches in its home range.

AGEING & SEXING Ageing possible, at least in 1stW if seen close or handled, using moult limits in upperwing. Sexes very similar but slightly differentiated, mainly in spring and summer. Correct sexing, especially in autumn and winter, may first require ageing using moult pattern in wing. Seasonal plumage variation limited and mostly due to feather wear. – Moults. Complete post-nuptial moult of ad and partial post-juv moult mainly in Aug–Sep. Post-juv moult includes head, body, lesser and median coverts, none or some inner greater coverts and tertials, and a few may moult more extensively, rarely even including a few inner primaries, primary-coverts, some secondaries and r1. Pre-nuptial moult apparently absent. – SPRING Ad ♂ Wear somewhat increases contrast of black cap, face pattern (especially pale supercilium and collar), and rosy tone to cheeks (if any present), but seasonal change rather less marked than in Asian Crimson-winged. Also, breast-sides become more rufous-cinnamon, and red basal areas on wings and tail more conspicuous, but white tips to remiges are virtually lost. A few develop pale pinkish cream-brown throat but not to extent of some 1stS ♂♂. Bill dull yellow when breeding. In hand, r5 mostly dark (near black except whitish tip), thus lacking dark subterminal patch on inner web. R6 also almost all dark, including outer web, except usually sharp white wedge on inner web. **Ad ♀** Very similar to ♂ (many probably cannot be sexed in the field) but overall duller with paler crown patch and much-reduced greyish collar; also white throat slightly less well defined, and cheeks usually lack pinkish. Pale fringes to remiges (especially middle) and primary-coverts narrower and paler pink. Tail has rather conspicuously-reduced pale tips (both tips and wedge on inner web of r6 diffuse and sullied pale buff-brown). Bill generally less yellow, more bone-grey. **1stS** Best differentiated by retained juv wing- and tail-feathers with moult limits as 1stW. Extent of subterminal black areas in rr5–6 also as 1stW. Both sexes less strongly patterned than ad and, generally, less easily sexed; especially some 1stS ♂♂ approach ♀ (mainly ad). – AUTUMN **Ad** Both sexes differ from 1stW in being evenly fresh with broader and whiter primary tips. Compared to ad Asian Crimson-winged, pale primary tips narrower and whitish-cream with pale buff-brown cast. **Ad ♂** Feathers on black crown patch narrowly fringed cinnamon-buff, pinkish facial areas duller or lacking, and upperparts slightly more buffish, with brown-buff breast and flanks more obviously tipped whitish. **Ad ♀** Much as in spring. **1stW** Both sexes differ from ad in having juv primaries and tertials browner basally and less fresh, with considerably more diffuse and buffier (less white) tips; birds with moult limits among greater coverts (renewed ad-like inner coverts) more distinctive. Tail-feathers rather pointed. **1stW ♂** Sexual dimorphism obscured, some strongly approach ♀ (especially ad ♀), though unlike most 1stW

1stY ♀, Morocco, Mar: 1stY ♀ most nondescript, with bland face and smaller pink-red wing-panel. Very similar to Asian Crimson-winged, but has less obvious cap and weaker bill. Only likely to be confused with Trumpeter Finch, though readily differentiated by size, jizz, bill shape, wing and head patterns, and also lacks Trumpeter's white eye-ring. Aged by juv primary-coverts. (M. Holck)

♀♀, crown darker, some have hint of grey on nape (none in ♀) and pink in wing much more obvious. Retained juv greater coverts have brown inner web and mostly brown-buff outer web (ad-type greater coverts have darker inner webs and almost entire outer web cinnamon-brown with some pink hue); retained primary-coverts mostly blackish-brown except thin pinkish edge on outer web (ad-type primary-coverts have blackish-brown inner web, extending across 50% of outer web, leaving relatively broad and brighter pink edge). Tail-feathers (retained juv) more extensively dark than ad ♂, pattern approaching ad ♀ with often even more diffuse and buff tips, and reduced (or virtually lacking) pale wedge on inner web of r6. **1stW ♀** Palest and least patterned plumage, with much-reduced dark cap, no grey on nape, reduced pink in wing, and fringes to greater coverts mainly buff. Tail pattern as 1stW ♂ or has even more obscure pale areas. **Juv** Probably similar to Asian Crimson-winged Finch.

BIOMETRICS L 15–16.5 cm; **W** ♂ 104–112.5 mm (n 18, m 108.6), ♀ 101–105 mm (n 8, m 103.6); **T** ♂ 60–65 mm (n 18, m 61.9), ♀ 57–62 mm (n 8, m 60.1); **T/W** m 57.5; **B** 14.6–16.9 mm (n 26, m 15.7); **B**(f) 11.5–13.2 mm (n 28, m 12.3); **BD** 8.4–11.2 mm (n 27, m 10.0); **Ts** 17.5–20.0 mm (n 28, m 18.9). **Wing formula: p1** minute; **p2** < wt 0–2 mm; **p3** (pp2–3) longest; **p4** < wt 0.5–3 mm; **p5** < wt 8–11 mm; **p6** < wt 17–21 mm; **p7** < wt 22.5–28 mm; **p10** < wt 33–42.5 mm; **s1** < wt 37–45 mm. Emarg. pp3–5.

GEOGRAPHICAL VARIATION & RANGE Monotypic. – Mainly at higher elevations in Moroccan Atlas Mts, but apparently also an isolated tiny population in NE Algeria; mainly nomadic or moves to lower altitudes in winter.

TAXONOMIC NOTE See Asian Crimson-winged Finch.

REFERENCES KIRWAN, G. M. et. al. (2006) Bull. African Bird Club, 13: 136–146.

1stW ♂, Morocco, Oct: due to late and prolonged post-juv moult, young ♂♂ can appear ♀-like and undeveloped even in late autumn. The new black feathers emerging on crown infer ♂. Extensive portion of wing still unmoulted, while pointed tips of tail-feathers further confirm young age. (M. Pitt)

Ad ♂, Morocco, Dec: still fresh winter ♂, with pale fringes above and broad primary-tips. Specific characters include greyish nape, small dark cap, narrower pink wing-panel and dark blotching on breast-sides. Note that in straight profile the pale bib can be invisible. Wing evenly ad. (T. Kolaas)

♀-like, Morocco, Dec: combination of hint of moult limits in wing, with seemingly still unmoulted outer primaries suggest 1stW, but safe ageing difficult, and sexing young birds not easy, especially in fresh plumage. (A. Faustino)

DESERT FINCH
Rhodospiza obsoleta (M. H. C. Lichtenstein, 1823)

Fr. – Roselin de Lichtenstein; Ger. – Weißflügelgimpel
Sp. – Camachuelo desertícola; Swe. – Ökenfink

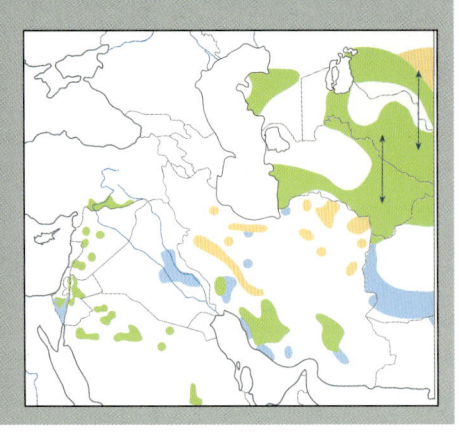

One of a quartet of so-called 'desert finches', comprising the genera *Rhodopechys*, *Rhodospiza* and *Bucanetes*, Desert Finch typically not being a desert dweller—despite its name. It ranges from SE Turkey and the Levant through Iran to Pakistan, and in Central Asia to Gansu (NC China) and Mongolia, occurring in lowland plains, semidesert areas and open country with scattered trees, such as orchards, gardens and oases, and is rather arboreal, often perching in trees though feeds on ground.

IDENTIFICATION A rather pale, predominantly *light sandy-brown* and long-winged finch (although can appear dumpy and rotund). The *stubby bill is quite thick at the base*, almost like in Bullfinch but not as rounded, more straight triangular. Wing pattern characteristic, with *contrasting pink and white wing panels*, most obvious on the folded wing, but usually looks almost translucent white in flight. Also conspicuous is the *whitish outertail* with blackish centre and tips. Breeding adult ♂ has *contrasting black bill and lores*; non-breeding ♂, ♀ and first-winter lack black lores (or black is much reduced) and have pale straw-coloured, dark-tipped bill (though breeding ♀ may acquire dark horn-coloured bill that appears near black). ♀ and young are also rather drabber with less striking plumage, including duller and less extensive pink fringes to coverts and secondaries. Underparts have sandy-buff wash to breast and flanks, deepest in ♂. In all plumages, iris dark brown and tarsus dull brown. Rather sociable, especially post-breeding and in winter, and strongly attracted to cultivation. Its purring flight-calls are diagnostic.

VOCALISATIONS The rather quiet song is a repetitive twittering, somewhat like several *Carduelis* finches, comprising calls interspersed with more nasal notes (not unlike Trumpeter Finch) and others that are more grating. – Most characteristic call a soft *prrut* or *drrr'r* (not unlike a distant European Bee-eater *Merops apiaster*) given in flight or as contact call from feeding parties, a *tvoi* (described as a cross between a Tree Sparrow and Common Rosefinch) and lilting *dveüüt* recalling Twite and Rock Sparrow.

SIMILAR SPECIES Could be confused with any of the other 'desert finches'. However, Desert Finch is distinctly smaller and less bulky than either of the two *Crimson-winged Finches*. In flight, Desert Finch has a largely white (not pink) tail, and at rest confusion should not occur due to different wing patterns and the much plainer, sandy-grey plumage of Desert Finch, and in summer blackish bill and lores of ♂ Desert. In practice, they are relatively unlikely to come into contact due to different habitat preferences. – The same characters should prevent confusion with the two slightly smaller *Mongolian* and *Trumpeter Finches*, which also have similar (yet differing) wing patterns and underparts coloration (pink in ♂♂ of both other species). – The profile may resemble that of *Greenfinch*, including the clearly-forked tail, and some wintering Greenfinches in the Levant are greyer and frequently join flocks of Desert Finch. However, Desert Finch coloration is sandy-buff and grey-white, and the diagnostic wing and tail patterns should prevent confusion even of distant birds.

Ad-like ♂, **Turkey, May**: elegant finch with contrasting black mask and bill, bright pink wing-panel, sharply-patterned black alula and centres to inner greater coverts (invisible here), tertials and primaries. Flank-feathers of ♂ tinged rufous. Evenly fresh ad-type wing and generally immaculate plumage best fit 2ndS (or older) ♂, but theoretically could be 1stS after complete moult. (D. Occhiato)

Ad-like ♀, **Jordan, Jan**: like ♂ but duller with no black on lores, and with alula and tertials centred brownish and bill less intense black. Pink in wing less pure. Evenly fresh ad-type wing is no safe indication of age, without visible moult limits. (J. Tenovuo)

♂, **Turkey, May**: diagnostic wing pattern, with broad diffuse pink-white wing bands along whole length of wings. After extensive post-juvenile moult many 1stS ♂♂ keep 1–3 outer primaries (p2 and p3 here). (K. Malling Olsen)

AGEING & SEXING Ageing requires close inspection of moult pattern and shape and wear of tail-feathers, but due to sizeable proportion of 1stY that undergo complete post-juv moult only young birds which moulted partially can be recognised. Bill coloration often useful for ageing and sexing, at least in early autumn. Sexes readily differentiated in breeding plumage (late winter or early spring to summer). – Moults. Complete post-nuptial moult in late summer and early autumn (Jul–Oct). Post-juv moult at the same time, but highly variable in extent ranging from birds with limited moult (head, body and all lesser and median coverts, and some or all greater coverts, plus some or all tertials renewed) to those with very extensive or even complete moult. Pre-nuptial moult apparently mostly absent, but some mainly 1stY renew some secondary-coverts and remiges (especially secondaries) in winter or early spring. – **SPRING F.g.** (ad and such 1stS with complete post-juv moult) No moult limits in wing. On average less worn compared to 1stS with retained juv feathers, and dark portions of wing darker in each sex, respectively; also, pink and white patterns on average neater and more extensive. **F.g. ♂** Black lores solid, sometimes extending to upper chin. Bill black from early spring. Tertials black-centred, fringed whitish. With wear in summer, plumage becomes slightly greyer above and paler below (but never as dull as ♀). White tertial fringes almost wear off in summer. **F.g. ♀** Duller than ♂ with no black on lores (at most slightly mottled brownish on central area), and tertials pale brown diffusely fringed paler. Bill paler, but often misleadingly dusky grey-black, can appear black at distance or in poor light. Overall more uniform greyish sand-coloured (less buffish-cinnamon and pink), and unlike ♂ has diluted chestnut-buff hue to rump and uppertail-coverts and flanks. White and pink areas in wing less pure or bright (pink especially reduced in greater coverts, being more greyish-pink and restricted to narrower, much less contrasting fringes on outer webs). As with tertials, other dark areas of wings, including alula and tips of primary-coverts, duller brown-grey and less contrasting. **1stS** (mainly birds with partial post-juv moult) As respective full grown sex, but note relatively more worn and abraded retained juv wing- and tail-feathers (see 1stW for moult pattern). 1stY with complete post-juv moult usually inseparable from ad using moult but with practice often visibly less bright and neat, 1stS ♂ often having less black or even pale-mottled black loral mask. Some young ♀♀ very pale and sometimes retain a wholly or partly pale bill. – **AUTUMN Ad** After moult, both sexes evenly fresh and lack moult limits sometimes visible in 1stY (also, prior to completion of moult often still has heavily-worn remiges and wing-coverts). Alula often rather rounded in shape (juv alula more pointed, but some are quite similar). **Ad ♂** Black loral mask as in summer but may have fine pale tips to black feathers when fresh. Most keep all-black or extensively blackish bill into late autumn or early winter. Tertials black neatly fringed whitish-buff and primaries tipped boldly white. Plumage generally as spring but overall slightly warmer toned. **Ad ♀** As in spring, including tertial pattern with brown-grey centres and diffuse sand-coloured fringes, which is best sexing character. Lores even plainer and cleaner pale when fresh. At least until late autumn mostly dark grey-black bill being a key feature to separate from the much paler-billed young birds (see 1stW). **1stW** Both sexes reliably aged in case any juv remiges, primary-coverts, some outer greater coverts, and any tertials, alula or tail-feathers retained (weaker textured and duller, creating moult limits in contrast with recently renewed feathers, best seen in greater coverts). Unmoulted tertials (sometimes kept through Oct) centred browner and fringed sandier. Juv tail-feathers also on average narrower and more pointed (but some overlap). Whitish areas in remiges and tail less pure, washed cream-buff, and pink on secondaries, primary-coverts and greater coverts paler and more restricted (retained greater coverts have pale sandy outer fringes and tips), respectively. When post-juv moult complete no moult contrast visible, still

1stS ♂, Israel, Mar: much variation in post-juv moult, here a young ♂ with juv primary-coverts and alula, thus already duller wing-panel appears even more limited. Nevertheless, black loral mask and bill almost like summer ad ♂, and these and tertial pattern separate it from ♀. (R. Pop)

♂, presumed 1stS, Israel, Mar: no discernible moult limits in wing suggest a bird after complete moult, thus ageing impossible considering the occurrence of complete post-juv in some young birds. Combination of duller pink in wing with partial loral mask indicate 1stS. This ♂ is to some extent nearer ad ♀, but unlike latter tertials black-centred. (H. Shirihai)

1stS ♀, Israel, Jun: the dullest and scruffiest plumage, with patchily-moulted (post-juv) inner wing, with outer juv primaries the most worn. Indication of pink wing panel is limited to new secondaries and primary-coverts. Bill often rather pale, as here, and loral area almost plain sand-coloured. (G. Shon)

Ad ♂, Kazakhstan, Oct: evenly fresh following complete post-nuptial moult, though some tail-feathers are still growing. Differs from any other plumage at this season, as mask and bill are black as in spring–summer. (G. Dyakin)

1stW ♂, Kazakhstan, Oct: following post-juv moult, with ad-type blackish-centred tertials, making sexing easy. Unlike ad ♂ note moulted tertials and primaries are broader-fringed and tinged more extensively buff. Yellowish bill with dark base useful for ageing in early autumn. Pinkish wing-panel and tertial pattern distinguish it from any potential confusion species. (V. Fedorenko)

some can be aged using paler bill until early or even late autumn. **1stW ♂** Once juv tertials have been moulted to ad type (blackish-centred, whitish-fringed) sexing is easy. Rest of plumage intermediate between fresh ad ♂ and ad ♀, but unlike latter has more pink in wings and darker lores. At least until late autumn, bill is yellowish-horn distinctly blackish-tipped and with dusky base. (In winter bill colour darkens and becomes progressively less useful as reliable ageing criterion.) **1stW ♀** Tertials brownish with diffusely paler edges. Duller plumage than 1stW ♂, rather similar to ad ♀. Unlike ad ♀ has pale bill, rather plain pale greyish-horn (can even appear whitish) at least until early autumn, after which colour slowly darkens. **Juv** Like ♀ but has soft fluffy body-feathers, and lacks any pink on median and greater coverts, which are instead broadly and diffusely tipped pale buff. ♂ often told from ♀ by having dark grey lores, lacking in latter.

BIOMETRICS L 14–15.5 cm. **W** ♂ 84–92 (once 95) mm (n 32, m 88.4), ♀ 80–88.5 mm (n 24, m 85.4); **T** ♂ 57.5–65 mm (n 32, m 61.3), ♀ 54–62 mm (n 24, m 59.2); **T/W** m 69.3; **B**(f) 9.8–12.2 mm (n 52, m 11.1); **BD** 9.3–10.8 mm (n 52, m 10.0); **Ts** 15.8–17.8 mm (n 48, m 16.9). **Wing formula**: p1 minute; pp2–3(4) about equal and longest; p4 < wt 0–1.5 mm; p5 < wt 2–6 mm; p6 < wt 9–15 mm; p7 < wt 16–21.5 mm; p10 < wt 26–33 mm; s1 < wt 27–34 mm. Emarg. pp3–5, p6 also near tip.

GEOGRAPHICAL VARIATION & RANGE Monotypic. – SE Turkey, Levant, NW Arabia, much of Iran, Afghanistan, Pakistan, E Central Asia to Gansu (NC China), Mongolia; mainly resident, but local or shorter movements noted, with some birds moving to Iraq and Sinai in winter.

1stW ♀, Israel, Sep: at end of complete post-juv moult, with ivory-coloured bill the only ageing criterion from similar-plumaged ad ♀. Unlike ♂, tertials and alula brownish and lores pale and almost plain. (E. Bartov)

1stW ♀, Israel, Sep: in post-juv moult (some feathers still fluffy, of juv type) and bill still almost whitish-grey. Limited and pale pink in wing, with buffish-brown new tertials, sex this bird. (A. Ben Dov)

MONGOLIAN FINCH
Bucanetes mongolicus (Swinhoe, 1870)

Fr. – Roselin de Mongolie; Ger. – Mongolengimpel
Sp. – Camachuelo mongol; Swe. – Mongolfink

Closely related to the Trumpeter Finch, a well-known species of the Canaries, N Africa and the Middle East, the Mongolian Finch is a more easterly distributed bird of deserts and mountains. It is mainly found in Central Asia, but in fairly recent times has been shown to also breed in easternmost Turkey. Like some other barren mountain and desert birds, not least sandgrouse, it is easiest to see when it comes to drink, so a strategic wait in the desert near fresh water in the morning or evening can be productive. Compared to Trumpeter Finch it favours on average higher altitudes.

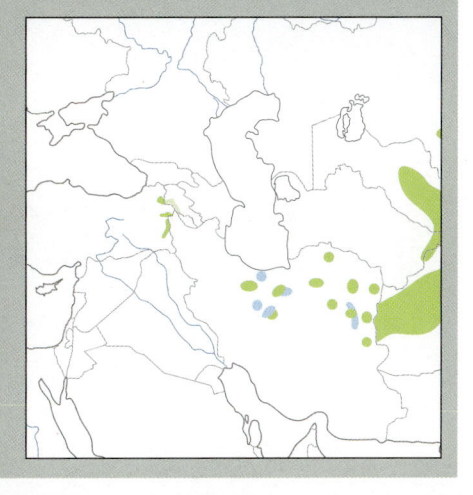

IDENTIFICATION Of the same size and of similar shape to the Trumpeter Finch, but differs in having a proportionately somewhat *smaller and less bulbous bill* (which is never reddish or pink but *pale yellowish-brown* or duller grey-brown), and a more striking wing pattern with *two clearly-separated large white patches, one on bases of greater coverts and the other on secondaries*. Since wear alters the plumage rather dramatically, it is convenient to describe worn and fresh plumages separately. Worn plumage (summer): ♂ has a contrasting *square pink rump patch* (smaller and more concentrated than in ♂ Trumpeter Finch), back being dull brown without any pink. *Conspicuous wing pattern of white–red–white*, secondaries having white bases to outer webs, greater coverts red edges and white bases. Outer part of secondaries dark grey, tipped white. Bases of outer tail-feathers edged white with faint pink tinge in adult ♂. Hint of red supercilium formed by red lateral crown-stripes. *Chin, throat, upper breast and flanks pink*, centre of lower breast and belly whitish. ♀ resembles ♂ but pink on rump lacking, or at least more subdued. Wing pattern similar but less contrasting and bright. No pink on supercilium and sides of crown, or only traces of it. Outer tail-feathers thinly edged white, usually without pink. Fresh plumage (autumn–winter): much more similar to Trumpeter Finch, especially to

Ad ♂, Mongolia, Jun: a rather stocky, thick-billed and pink-winged finch, identified by exposed black centres to greater coverts and secondaries, contrasting white to pinkish-white double wing-bars, highly distinctive both when perched and in flight. ♂ by bright and extensive pink face, underparts, rump and edges to wing-feathers. Wing evenly rather fresh, including pink-fringed primary-coverts, while tail-feathers have fairly rounded tips. (H. Shirihai)

♀ of that species, due to buffish feather-edges and a buffish sheen covering much of white, pink and red. Note *smaller bill* and hint of characteristic summer wing pattern with *broader pale area on central part of outer webs of secondaries*, although not yet white (instead pink and buffish). There is a tendency for feathers on chin and upper throat to have dark grey bases (pale-based in Trumpeter Finch). Juvenile is similar to ♀ but even duller sandy-brown and diffusely mottled darker on upperparts. From very similar juvenile Trumpeter Finch by bill size and tendency to have slightly more contrast on secondaries.

VOCALISATIONS Song given both from ground and in flight is a brief series of soft whistles, upwards-inflected or downslurred in a mixture, and usually including one or two short chirping trills, mainly towards the end. – Call notes short, rather soft but clear *chik* or *chip*, which have been likened to Linnet (Beddard et al. 2002). There is also a two-note flight-call with second note drawn-out and upwards-inflected, *tu… tweet*, sometimes described as being similar to calls of Rock Sparrow. Other variations include a *piuh* with piping, desolate ring and falling pitch, and a more nasal, strained *tweeih* (although nothing like the 'toy trumpet' sound of Trumpeter Finch). The latter notes resemble those of the song.

Presumed ad ♀, Kazakhstan, May: sexes very similar, but ♀ usually has less extensive pink on rump, supercilium, breast and flanks, less contrasting wing pattern (white more restricted, pink outer fringes to greater coverts and primaries narrower) and no or minimal reddish-pink on face. All wing-feathers appear ad, including broadly pink-fringed primary-coverts. (V. & S. Ashby)

1stS ♂, Turkey, Jun: largely as ad ♂, but juv wing pattern less bold and attractive, and primary tips rather brown and abraded. Note obvious pink on side or forecrown. Bill slightly less swollen at base, and slightly more pointed at tip, than Trumpeter Finch, though straw-coloured and lacking pink. (K. Dabak)

♂♂, Kazakhstan, May: showing species-specific bold, double whitish-and-dark wing panels, but without closer views of wing it is impossible to age the birds. The amount of pink and boldness of wing pattern best fit ♂. (R. Porter)

SIMILAR SPECIES As stated under Identification, very similar to *Trumpeter Finch* in fresh autumn plumage. The white and pink-red pattern on wing is then much more subdued and buffish-tinged, and care is needed when evaluating the finer differences between the two species. Bill size is always useful, and the slightly less bulbous-shaped and near-straight culmen in Mongolian Finch is useful. In all plumages there is a tendency for the centre of lower breast and belly to be contrastingly paler in Mongolian, whereas in Trumpeter Finch underparts are more uniformly coloured. White tail-sides strongly indicate Mongolian Finch (Trumpeter having pink or brown tail-sides), as does a pink supercilium/lateral crown-stripe. – Differs from *Asian Crimson-winged Finch* by smaller size and proportionately smaller head and bill, less deeply notched tail-tip, lack of dark crown, and white and pink wing patches divided, rather than united as one large patch.

AGEING & SEXING Ageing sometimes possible after post-juv moult, at least if seen well. Sexes usually differ following post-juv moult, more obviously in 1stS after some bleaching and wear. – Moults. Complete post-nuptial moult of ad mainly in Jul–Oct (Nov). Partial post-juv moult in summer or early autumn. Apparently invariably replaces all secondary-coverts, but no moult of outer primaries or tail-feathers as often occurs in 1stY *B. githagineus*. No pre-nuptial moult. – **SPRING ♂** Contrasting bright pink rump patch, back dull sandy-brown without pink. Striking wing pattern on folded wing with white–red–white, secondaries having white bases (or large inner central part) to outer webs, greater coverts red edges and white bases; median coverts broadly tipped white. Outer part of secondaries dark grey (almost blackish), tipped white. Red lateral crown-stripes (invading sides of forehead and fore supercilium). Chin, throat, upper breast and flanks pink, centre of lower breast/belly whitish. ♀ Resembles ♂, but pink on rump lacking or more subdued. Wing pattern

1stS ♀, China, Apr: least-advanced 1stY can look dull and indistinctly patterned. Diagnostic pattern on greater coverts barely visible, but the broad whitish panels and dark centres on secondaries confirm the identification. Bill shape diagnostic, but shares with Trumpeter Finch white eye-ring, whereas streaky upperparts are not. Dull plumage suggests ♀, and mostly juv wing the age. (J. & J. Holmes)

Juv, China, Aug: this poorly known plumage differs from Trumpeter Finch in less uniform wing, with pale buff tips to median and greater coverts, and dark-centred greater coverts. When secondaries visible, pattern of folded wing also diagnostic. (M. Lagerqvist)

Mongolian Finch (Mongolia, Jun), versus Trumpeter Finch (Jordan, Mar), Desert Finch (Israel, Jul) and Asian Crimson-winged Finch (Turkey, Jun): wing patterns always diagnostic, with Mongolian having bold double whitish-and-dark wing panels, Trumpeter any pale restricted to secondaries, Desert having Linnet-like white-edged primaries, and Asian Crimson-winged (even in dullest autumn bird) has pink wing-panel covering all remiges or at least bases of primaries. Size, jizz and bill shape can be useful too. Note also that some ♂ Mongolian have face to chest tinged pink, Trumpeter can have orange bill, Desert often has black bill and sometimes (in ♂♂) black loral mask, and Asian Crimson-winged a dark cap. Based on moult and general plumage, the Mongolian and Trumpeter Finches are probably 1stS ♂♂, Desert 1stS ♀, and Asian Crimson-winged presumed 1stW ♂. (Top left and right: H. Shirihai; bottom left: E. Bartov; bottom right: H. Shirihai)

similar but less contrasting and bright, white more restricted. No red on crown-sides, or only faint pink traces at most. A few borderline cases may be difficult to sex. **Ad** Tips to tail-feathers rather rounded and neat. Primary-coverts and alula usually have neat white fringes with some pink. **1stS** Tips to tail-feathers rather pointed and abraded. Primary-coverts and alula usually have more diffuse off-white fringes (no pink). — **AUTUMN** Sexing after post-juv moult same as in spring, only in fresh plumage much of pink and red concealed by buffish tips, and broad pale area on bases of secondaries pink or buffish instead of white. Ageing sometimes possible, too, using shape of tail-feathers (see spring). **Juv** Resembles post-juv ♀, but underparts lack any pink, and greater coverts are sandy-brown with paler buff tips.

BIOMETRICS L 13–14.5 cm. **W** ♂ 89–96 mm (n 31, m 92.5), ♀ 85–94 mm (n 16, m 88.5). **T** ♂ 52–60 mm (n 28, m 56.0), ♀ 50–59 mm (n 15, m 54.5); **T/W** m 60.7; **B** 10.8–12.4 mm (n 13, m 11.6); **B(f)** 8.0–10.0 mm (n 48, m 9.2); **BD** 6.7–8.6 mm (n 47, m 7.8); **BW** 6.1–8.4 mm (n 47, m 6.9); **Ts** 16.0–17.8 mm (n 47, m 17.0). **Wing formula: p1** minute; **p2** 0–1 mm < wt; **p3** (pp2–3) longest; **p4** < wt 1–3 mm; **p5** < wt 8–10.5 mm; **p6** < wt 13.5–16 mm; **p7** < wt 18–22 mm; **p10** < wt 27–34 mm; **s1** < wt 29.5–36.5 mm. Emarg. pp3–4 (though on p4 sometimes a little less clear).

GEOGRAPHICAL VARIATION & RANGE Monotypic. – E Turkey, W Armenia, Iran, Central Asia; resident or nomadic, and may descend from higher altitudes.

REFERENCES Barthel, P. H. et al. (1992) Limicola, 6: 265–286. – Beddard, R., Ananian, V. & Finn, M. (2002) Sandgrouse, 24: 144–147. – Kirwan, G. & Konrad, V. (1995) BW, 8: 139–144.

TRUMPETER FINCH
Bucanetes githagineus (M. H. C. Lichtenstein, 1823)

Fr. – Roselin githagine; Ger. – Wüstengimpel
Sp. – Camachuelo trompetero; Swe. – Ökentrumpetare

As its name might imply, this little finch of deserts, arid plains and mountain slopes emits a sound like a plastic toy trumpet when it sings, a nasal, straight, buzzing tone about two seconds long. It breeds mainly in N Africa and the Middle East, but is probably easiest to see on one of the eastern Canary Islands. It has recently spread to Spain and has accidentally occurred as far north as Denmark and Sweden. Often seen singly or in small parties, but larger flocks also occur, especially in winter. A bit unpredictable, encounters can happen suddenly or with long intervals.

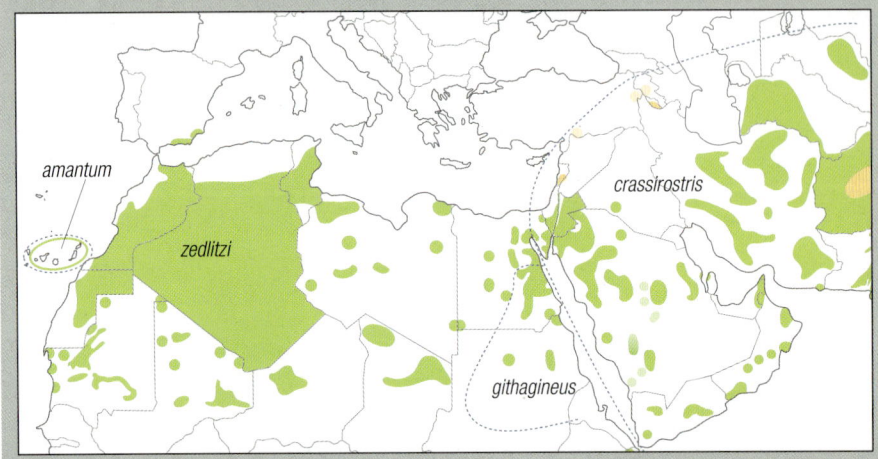

B. g. zedlitzi, ad-like ♂, Morocco, Jan: a pretty, distinctive finch with a short thick bill, short tail, and rather long wings. Plain sandy grey-brown with a rosy-pink wash and conspicuous pale eye-ring in close views. ♂ in spring/summer has ash-grey crown and ear-coverts, vivid rosy-pink face and underparts, and diagnostic orange-red or coral-red bill. Whole wing evenly feathered, but as some 1stY moult completely ageing impossible. (M. Schäf)

IDENTIFICATION A *small* finch with *rather large head* and a hefty bill. The plumage is comparatively featureless, mainly *sandy grey-brown* with some pink in adult ♂. Best character is the *large, bulbous bill* that is either *pale yellowish-brown or pinkish-red to orange-red* (breeding adult ♂), often being *contrastingly pale* against the rather dark face. ♂ has a *pink flush on forehead, throat, belly, flanks, rump, wings and tail, and is rather pure ash-grey on crown, nape, cheeks and sometimes breast*, whereas the ♀ is duller and browner overall, lacking any of the grey and most of the pink tinges of the ♂. A few ♀♀ have more pink and approach younger ♂♂ in appearance. Feet reddish-brown, *tarsi often appearing pinkish*. Normally encountered feeding on the ground, usually in pairs, small parties or less often in larger flocks. Adopts quite upright stance when alert, otherwise more hunched. Undulating or skipping flight is light but fast.

VOCALISATIONS Song is a peculiar, toy trumpet-like, drawn-out and slightly nasal, buzzing sound, *aaaaaahp*, on one pitch and lasting c. 2 sec. The buzzing note is often combined with a few initial twittering or short whistling sounds (but in essence still the same song, hardly a different type). – Most characteristic call is a short, jolting, nasal *aahp*, sometimes doubled or repeated a few times, clearly reminiscent of the song in tone but lower-pitched. Other calls include a buzzing *dzzuh* and a variety of rather subdued and anonymous notes, e.g. *chu*, *zit* and *chu-wit*.

SIMILAR SPECIES Most obvious risk of confusion is with *Mongolian Finch*, especially young or autumn ♀♀, which can have a very subdued wing pattern and can be very similar to Trumpeter Finch. Therefore always best to base identification on size, shape and colour of bill (see Identification) combined with plumage criteria. Even when wing pattern of Mongolian is less characteristic, there is usually a rather obvious diffuse pale oblong patch on the centre of the folded secondaries lacking in Trumpeter Finch. – Both *Crimson-winged Finches* differ in being larger and having a proportionately bigger head with a dark central crown-band prominent in most plumages. Also, show strong contrast between dark flanks and white belly. – Occurs in same range as *Sinai Rosefinch*, and ♂♂ of both have some pink in plumage, but this species is somewhat larger and slimmer, and has a less bulky bill that is never reddish.

AGEING & SEXING (*zedlitzi*) Ageing often possible after post-juv moult, at least if seen well, but due to variation in extent of this moult some 1stW apparently become inseparable from ad. Sexes differ slightly after post-juv moult, more obviously in 1stS after some bleaching and wear. – Moults. Complete post-nuptial moult of ad mainly in Jun–Sep. Most commonly, 1stY moult primaries partly in summer or early autumn, renewing a number of central or all outer, leaving only 3–4 inner primaries unmoulted. Others leave all juv primaries unmoulted, and yet others seem to moult completely. No pre-nuptial moult. – **SPRING** ♂ Underparts pale pinkish including belly and undertail-coverts (though breast frequently greyish). Rump and uppertail-coverts predominantly pink. Forehead and face show pinkish-red spots or short streaks. Bill when breeding pink-red or coral-red. Flight-feathers and tail-feathers edged pink. ♀ Underparts pale buff-brown, sometimes with a faint pink tinge on chin and throat/upper breast (but not continuing prominently onto belly or undertail-coverts; at the most a hint on belly in some). Undertail-coverts buff-brown finely streaked dark (only very rarely show faint pink-buff hue). Forehead and face generally buff-brown, rarely with very slight traces of pink admixed. Flight-feathers and tail-feathers edged pale brown or pinkish-buff (but not bright pink). **F.g.** No moult contrast among primaries, tail-feathers, tertials or wing-coverts. Tips to tail-feathers rather rounded and neat. **1stS** If seen close and well, a moult contrast is often visible among primaries, sometimes also among tail-feathers, tertials or wing-coverts. Tips to any unmoulted tail-feathers rather pointed and

B. g. amantum, ad-like ♂, Canary Is, Mar: richer-coloured than *zedlitzi*, being more pink-red, especially on rump, uppertail-coverts and fringes of wing-feathers. Grey of head and throat in ♂ slightly darker. Age and sex as in previous image. (B. Baston)

B. g. githagineus, ad-like ♂, Egypt, Feb: a rather small race, like *amantum*, but closer to *zedlitzi* in its slightly drabber plumage, though not as extreme as *crassirostris*. Age and sex inferred from bright colours and neat plumage in combination with only ad wing-feathers. (S. Berg Pedersen)

B. g. crassirostris, ♂♂, 1stS (left) and ad-like, Jordan, Mar: during spring/summer ♂♂ of all ages have diagnostic bright orange-red bill, but at least some 1stY possess combination of variable number of juv wing-feathers (e.g. outer primary-coverts in this case) and to some extent drabber plumage, with less reddish-pink, making this race look even more washed-out. (H. Shirihai)

abraded. – AUTUMN Sexing after post-juv moult same as in spring. Ageing often possible with ♂♂, sometimes with ♀♀ too. **Ad** Primaries initially heavily worn, then in active moult and from late Aug or Sep fresh without any moult contrast. Primaries and primary-coverts partly edged pink. Tips to tail-feathers rather rounded. **1stW** Sometimes shows moult contrast among primaries and wing-coverts, moulted feathers often being a little darker, quite fresh and partly edged pink or pinkish-buff, unmoulted slightly paler and becoming abraded as autumn progresses, while edges have no pink tinge. Any unmoulted outer greater coverts are buff-edged and broadly buff-tipped, contrasting with pink-edged inner. **Juv** All ochre-buff lacking any pink (except faintly on outer edges of primaries). Flight-feathers broadly tipped pale buff, tertials very broadly tipped and edged pale, appearing rather uniformly pale as body (unlike in ad).

BIOMETRICS (*zedlitzi*) L 13–14 cm. Size difference between sexes rather slight: **W** ♂ 83.5–93 mm (n 20, m 88.7), ♀ 82–88.5 mm (n 12, m 85.2); **T** ♂ 46–57 mm (n 20, m 52.2), ♀ 46–56 mm (n 12, m 51.3); **T/W** m 59.4; **B** 11.1–13.0 mm (n 24, m 12.2); **B(f)** 8.6–11.3 mm (n 33, m 10.0); **BD** 8.0–10.2 mm (n 33, m 9.3); **BW** 7.0–8.6 mm (n 31, m 7.8); **Ts** 16.2–18.7 mm (n 28, m 17.6). **Wing formula: p1** minute; **p2** 0–1 mm < wt; **p3** longest; **p4** < wt 0.5–3 mm; **p5** < wt 5.5–8.5 mm; **p6** < wt 11.5–16 mm; **p7** < wt 16–20.5 mm; **p10** < wt 27.5–33 mm; **s1** < wt 28–36 mm. Emarg. pp3–4.

GEOGRAPHICAL VARIATION & RANGE Fairly prominent variation, mainly affecting saturation of colours, but also size.

B. g. zedlitzi (Neumann, 1907) (S Spain, Morocco, Western Sahara, Algeria, Tunisia, Libya, N Egypt). Described above. Somewhat larger and paler than *amantum*. Compared to similarly large *crassirostris* a trifle more pinkish-red, especially ♂♂ in worn plumage, which show rather obvious reddish tinge also on upperparts (whereas *crassirostris* is largely grey-brown above). ♂♂ usually develop dark red edges to outer tail-feathers, tertials and greater coverts, whereas these edges are pale pink in *crassirostris*. (Syn. *theresae*.)

B. g. amantum (Hartert, 1903) (C & E Canary Is). Smaller than *zedlitzi* (but same size as *githagineus*) with a proportionately slightly stronger bill, and stronger red pigment in both sexes than any of the other races, worn ♂♂ often developing quite reddish plumage, including upperparts. Several juv seem to perform complete moult, estimated to take place

B. g. crassirostris, ad-like ♀, Jordan, Mar: ♀ of all races more uniform drab brown above, without grey on cap, and with bill dull yellowish-brown, rump and underparts paler, and the pink on face largely restricted to supercilium, cheeks and throat. Also pink fringes in wing and tail narrower than in ♂. Racial differences in ♀♀ less pronounced. Only ad-like wing-feathers visible. (H. Shirihai)

B. g. crassirostris, ad ♂, Israel, Sep: in fresh plumage, rosy-pink mostly overlaid by drab brown feather-fringes, but note still clear pinkish hue on face, scapulars, belly, vent and fringes of remiges and tail-feathers. Bill is dull orange, has stronger 'warm' colour than usually ♀; beware that some ad ♂♂ have duller more ♀-like bill in early autumn. Bulging bill shape exaggerated by hanging drop of water. (H. Shirihai)

Jul–Sep. Fledged young seen from mid Apr. **L** 12–13.5 cm. **W** ♂ 82–88 mm (n 33, m 84.8), ♀ 79–88 mm (n 20, m 82.0); **T** ♂ 46–53 mm (n 33, m 49.4), ♀ 45–52 mm (n 20, m 48.1); **T/W** m 58.4; **B** 9.6–13.5 mm (n 53, m 11.5); **B**(f) 7.0–11.3 mm (n 53, m 9.1); **BD** same as zedlitzi; **BW** 7.6–9.0 mm (n 26, m 8.2); **Ts** 15.5–18.0 mm (n 46, m 17.1). **Wing formula: p10** < wt 24.5–31 mm; **s1** < wt 26–34.5 mm.

B. g. githagineus (M. H. C. Lichtenstein, 1823) (C & S Egypt, Sudan). Considerably smaller than zedlitzi (size as amantum), with proportionately slightly shorter tail, and same pale colours. **W** ♂ 81–87 mm (n 21, m 85.0), ♀ 80–85 mm (n 18, m 82.7); **T** ♂ 45–53 mm (n 21, m 48.5), ♀ 44–50 mm (n 18, m 47.0); **T/W** m 57.0; **B** 10.5–12.5 mm (n 29, m 11.6); **B**(f) 8.7–10.5 mm (n 39, m 9.6); **BD** 8.1–9.4 mm (n 38, m 8.7); **BW** 6.6–8.1 mm (n 31, m 7.4); **Ts** 15.8–17.5 mm (n 28, m 16.7). Wing formula as zedlitzi.

B. g. crassirostris (Blyth, 1847) (Sinai, W Arabia, Levant, SC & E Turkey, Transcaucasia, Iran, Central Asia and east to Sind and Punjab in NW India). Resembles zedlitzi (has same general size, bill size and bill shape) but is on average slightly paler and sandier, less pink. In worn plumage, ♂♂ are grey-brown above with very little pink (whereas zedlitzi consistently shows much pink or red). Larger than githagineus. **W** ♂ 85–92 mm (n 26, m 88.3), ♀ 82–92 mm (n 12, m 86.1); **T** ♂ 47–55 mm (n 25, m 50.9), ♀ 46–53 mm (n 11, m 49.5); **T/W** m 57.5; **Ts** 16.0–19.0 mm (n 32, m 17.2). Wing formula very similar to that of zedlitzi. (Syn. bilkewitschi.)

B. g. crassirostris, ad-like ♀, Israel, Sep: as ad ♀ in spring and summer, but any pinkish tinge reduced, and bill also dull straw-coloured, thus hardly differs from dullest examples of non-breeding ♂. Because some 1stW moult completely, age is impossible to ascertain. (H. Shirihai)

B. g. crassirostris, 1stW, presumed ♂, Israel, Sep: at least in Jul–Oct, young plumage very distinctive with variable cream-white face, as these feathers are last to be replaced, and becomes highly abraded and bleached. Note retained juv primary-coverts with very broad buff-white fringes. Rather pinkish edges to new remiges indicate ♂. (H. Shirihai)

B. g. crassirostris, presumed 1stW ♀, Israel, Oct: dullest example, presumably ♀, but sexing during winter, especially of young birds, usually impossible, and ageing is also uncertain if moult is complete. However, the extreme greyish-tinged plumage and brownish bill indicate young. (H. Shirihai)

B. g. amantum, juv, Canary Is, Jul: juv of all races differ from ♀ in being largely sandy-cinnamon or buff-brown with dark centres to flight- and tail-feathers. Lack of any pink and pinkish-horn bill also indicate age. (J. Érard)

COMMON ROSEFINCH
Carpodacus erythrinus (Pallas, 1770)

Fr. – Roselin cramoisi; Ger. – Karmingimpel
Sp. – Camachuelo carminoso; Swe. – Rosenfink

This medium-sized, mainly Asiatic, finch has spread west in modern times, expanding its range across much of N and C Europe. Although it means a longer migration to its S Asian winter quarters, the species has remained faithful to the latter despite now breeding further west, and is one of rather few European breeders that departs in autumn to the east and south-east. From May, its simple but characteristic whistling song sounds again across open habitats like lakeshores, agricultural areas interspersed with humid meadows and rich shrubbery, and in re-growth in old forest clearings

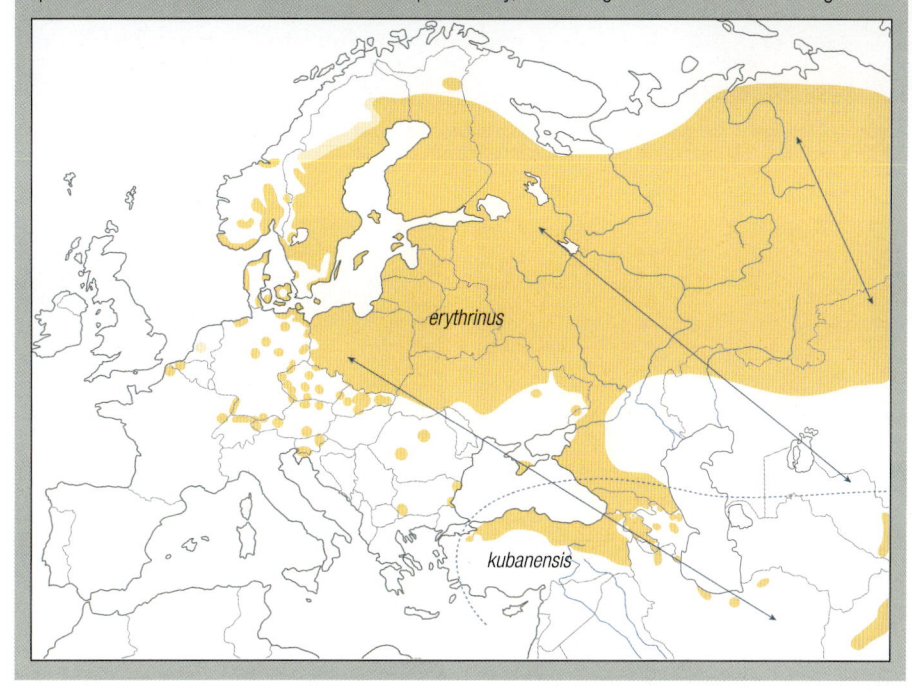

C. e. erythrinus, 1stS ♂, Belarus, May: ♂ usually does not attain fully ad plumage until autumn of 2CY, thus majority of 1stS ♂ (which often sing on breeding grounds) are ♀-like, brownish-grey with no or very limited red on face. (D. Yakubovitch)

VOCALISATIONS Song by ♂ (allegedly very exceptionally also by ♀) a simple, mechanically repeated but sweet and pleasing whistling *weedye wu-weedyah* (often rendered as 'pleased to meet you!'). There is some limited geographical variation in the song, but these small differences do not seem to be strictly linked to existing races, more with local subpopulations. Irrespective of small differences from the song described above, there should be no problem to recognise the species by its voice and general song structure. Each ♂ sticks to one variation (exceptions insignificant). – Calls are basically variations of one rather Greenfinch-like up-curled, clear, whistling *dyooee*, used both in flight, including from migrants, and perched. Alarmed birds use a very similar and related call, only more emphatic and sometimes slightly more strained in tone, not so sweetly whistling. Other minor calls exist but have very limited significance for identification.

IDENTIFICATION In size and shape, between House Sparrow and Corn Bunting, and in many plumages superficially resembles these two. *Tip of tail notched* like most finches. Bill is dark, stout and bulbous-shaped and the legs are dark. The adult ♂ is characteristic, being *deep red over much of head, breast and rump*, whereas belly, flanks and undertail are whitish (upper belly and flanks sometimes with pink tinge). Cheeks, mantle and wings brownish, mantle diffusely streaked darker. This plumage is usually attained in second summer, but already first-summer ♂ can rarely show limited red on forehead, rump, throat and upper breast (so-called 'advanced plumage'). (Very rarely, and presumably food-related, red colour wholly or partly replaced by yellow.) ♀ and most first-summer ♂♂ are inseparable, being rather nondescript grey-brown (sometimes with faint olive cast) and no particular feature to note except two narrow and indistinct paler wing-bars and the *very plain and uniformly grey-brown face in which the dark beady eye is obvious*. Juvenile resembles ♀ but is a touch more brownish, more boldly streaked and has more prominent buffish wing-bars.

SIMILAR SPECIES If size is not immediately evident, might be confused with *Great Rosefinch*, but latter is longer-tailed with a heavier and usually paler bill. – Whereas an adult ♂ is not easily confused with other European birds, the ♀, immature ♂ and juvenile potentially can; without the aid of voice, and in more fleeting views, these plumages can be confused with *Corn Bunting*. However, this species is larger and bulkier, has a proportionately slightly shorter and narrower tail, a somewhat smaller bill with a pale pinkish-yellow base to lower

C. e. erythrinus, ad ♂, Finland, Jun: stocky-bodied, round-headed and stout-billed. Breeding ad ♂ suffused bright red on head to breast/mantle, and paler red on rump and wing-bars. Extensive scarlet plumage infers ad; while a few 1stS develop similar plumage, this bird is aged as ad by reddish-edged remiges and primary-coverts. (D. Occhiato)

C. e. kubanensis, ad ♂, Turkey, May: compared to *erythrinus*, deeper red on head and throat, and pink-red colour reaches further down on belly. Sex obvious from red colour, and age inferred as for previous bird. (D. Occhiato)

Variation in ♂♂, presumed 1stS (left: *C. e. erythrinus*, Finland, Jun; right: *C. e. kubanensis*, Armenia, Aug): some ♂♂ acquire partial red on crown, breast and rump (left bird), while some are more advanced (right) with more extensive red, though still a bit mottled. Both birds show signs of immaturity, e.g. greyish-white upper wing-bars and generally juv feathers in wing. (Left: J. Peltomäki; right: O. Z. Göller)

C. e. erythrinus, presumed ad ♀, Finland, Jun: some ♀-like are very nondescript, with nearly no streaking, and when these also have an olive tinge to edges of wing-feathers it is a good sign of being ad ♀. Note also round-tipped alula with neat pale fringe. Still, safe ageing would require an assessment also of shape and wear of tail-feathers. (D. Occhiato)

Presumed *C. e. erythrinus*, ♀-like, Kazakhstan, May: ♀-like birds (both sexes) which cannot be aged with certainty are usually impossible to separate from ad ♀. ♀-like birds are rather nondescript, with whitish wing-bars, and dark eye in unpatterned head the most conspicuous features. Note streaked underparts with pale greyish-white ground colour, and hefty bill. (J. Normaja)

mandible, and slightly stronger, pinkish-tinged legs. There is often a dark patch on upper breast of Corn Bunting, and the species' entire plumage is more distinctly streaked. – Could theoretically be confused with a dull juvenile *Yellowhammer*, but this is slimmer and a little longer-tailed, the tail has white sides (that of Common Rosefinch is all dark), and there is more contrast between dark cheek and pale submoustachial stripe and throat. Also, the bill is smaller and not as stout and bulbous as Common Rosefinch.

AGEING & SEXING (*erythrinus*) Ages differ in autumn and spring, but reliable ageing usually requires close views. Sexes readily separable after first post-nuptial moult, sometimes already following post-juv moult. – Moults. Complete post-nuptial moult of ad after autumn migration, soon after reaching winter grounds, mainly Oct–Nov (sometimes starting in Sep on presumed stopover in Central Asia and finishing in Dec; Stach *et al.* 2016). Post-juv moult variable and still not fully understood; often starts with body moult in Sep (even before migration) and is then suspended to be completed in winter; some 1stW judging from brown and abraded primaries apparently do not moult any flight-feathers and only a few tail-feathers and tertials, while primary-coverts are retained. An unknown but sizeable percentage replace a few outer primaries along with all tail-feathers and tertials

C. e. erythrinus, ad ♀, Scotland, Sep: during autumn migration and prior to moult, ad ♀ are heavily worn and bleached greyish, differing from completely fresh and slightly olive-tinged 1stW of either sex. (J. Nichols)

1stW, India, Sep: two images of 1stW already in winter quarters, possibly of the same bird. Subspecific determination unsure. Young birds differ in early autumn from ad ♀ by their fresher and slightly olive-tinged plumage, and also often by slightly stronger streaking of underparts and head. Later in autumn, when ad has moulted completely, some of these differences disappear. (C. N. G. Bocos)

and thus show a slight moult limit among primaries. Judging from very neat plumage of some apparent 1stS, an unknown proportion of those might moult more extensively in their first winter, and complete moult in some cannot be ruled out. – **SPRING Ad ♂** Crown, nape, rump, throat and upper breast rather deep carmine-red (exceptionally yellow; see above). Lower breast and sides of belly sometimes possess a pinkish flush, but often dull off-white without any pink. Wings and tail fresh without any moult contrast, most wing-coverts edged/tipped pink. Tips to tail-feathers rather rounded and fresh. **Ad ♀** Upperparts olive-brown, underparts off-white, both rather diffusely streaked dark. Tail-feathers rather broad with rounded tips, edges usually still rather fresh. **1stS ♂ advanced variety** A minority attain quite ad-like plumage with varying amount of red on head, rump and upper breast (Ström 1991). Usually aged as 1stS by incomplete winter moult of wing, showing some moult contrast or slightly more abraded edges/tips to tertials and wing-coverts, latter with no pink. Unmoulted outer tail-feathers more worn and pointed than ad. **1stS ♂ normal variety/1stS ♀** As ad ♀, lacking any red or pink, but tail-feathers slightly narrower and tips more pointed, often with quite abraded edges. Sometimes a moult contrast is visible among wing-coverts, tertials or tail-feathers. Those which moulted only few inner greater coverts can have remnants of off-white wing-bar of juv type. If ageing difficult, as a rule impossible to tell 1stS ♂♀ from ad ♀. (Note that there is no proof or even indication for 2ndS ♂ to be ♀-coloured, *contra* Clement et al. 1993.) – **AUTUMN Ad ♂** As in spring, but wings and tail now heavily worn, and colours a little darker. **Ad ♀ / 2ndW ♂** Same as under autumn ad ♂ apply. Among 2ndW ♂♂ a minority are advanced (variably red), whereas majority resemble ♀♀. Wing-bars narrow or absent due to wear. Tips of some or all tail-feathers and primaries in 2ndW can be especially abraded. Upperparts dull olive-brown or brownish-grey, ad ♀ mostly greyish with no moult, whereas 1stW can start to replace some body-feathers in Europe. **1stW ♂♀** As ad ♀ but whole plumage fresh and tinged olive, tail-feathers not yet heavily abraded. Some have rather prominent whitish wing-bars. **Juv** Very similar to 1stW♂♀ but somewhat duller and paler overall, streaks more diffuse and not as dark. Wing-bars usually slightly tinged buff or cream. Plumage of slightly looser structure than 1stW, especially on vent and nape.

BIOMETRICS (*erythrinus*) **L** 14–15 cm; **W** ♂ 81–88.5 mm (n 29, m 84.4), ♀ 77.5–83 mm (n 11, m 81.1); **T** ♂ 54–63 mm (n 29, m 57.0), ♀ 51–58 mm (n 11, m 54.2); **T/W** m 67.3; **B** 11.5–13.7 mm (n 35, m 13.0); **B**(f) 9.2–11.0 mm (n 36, m 10.2); **BD** 7.5–9.8 mm (n 34, m 8.5); **Ts** 17.1–18.7 mm (n 35, m 18.0). **Wing formula: p1** minute; **p2** < wt 0–1 mm; **pp**(2)3–4 about equal and longest; **p5** < wt 2.5–6.5 mm; **p6** < wt 8–13 mm; **p7** < wt 11.5–18 mm; **p10** < wt 21–26 mm; **s1** < wt 22.5–29 mm. Emarg. pp3–5.

GEOGRAPHICAL VARIATION & RANGE Fairly slight variation, mainly involving darkness or paleness of plumage and amount of red in ad ♂. Two sufficiently distinct races within treated region and two more in E Asia (not treated). Several more races have been described but in our opinion do not meet a 75% rule.

C. e. erythrinus (Pallas, 1770) (N & E Europe, W & C Siberia, N Mongolia; winters mainly India). Treated above. The 'dullest' race, ad ♂ having slightly less extensive carmine-red than the following race, red on underparts often mainly restricted to throat and upper breast, belly being off-white with mere tinge of pink in some. (Syn. *kistjakovskii; pallidorosa*.)

C. e. kubanensis Laubmann, 1915 (N Turkey, Caucasus, Transcaucasia east through Elburz Mts in N Iran, Kopet Dag, W Turkestan, Afghanistan east to Tien Shan, north-east to Dzungaria; winters India and adjacent areas). Differs from *erythrinus* in that ad ♂ usually has more extensive dark pink below the deep scarlet-red throat/upper breast, often covering all or much of belly and flanks. Furthermore, ad ♂♂ in fresh plumage (winter–early spring) often have more saturated rufous-brown on mantle and scapulars, with more red admixed. However, some variation and overlap with neighbouring races. Wing and bill a trifle longer than in *erythrinus*, but rest of biometrics and wing formula nearly identical to *erythrinus*. **W** ♂ 83.5–90 mm (n 14, m 86.4), ♀ 78–86 mm (n 11, m 82.9); **T** ♂ 54–60 mm (n 14, m 57.1), ♀ 52–58 mm (n 11, m 54.7); **T/W** m 66.1; **B** 12.5–14.4 mm (n 25, m 13.3); **B**(f) 9.4–11.9 mm (n 25, m 10.5); **BD** 7.9–9.5 mm (n 24, m 8.7); **Ts** 17.3–19.3 mm (n 24, m 18.1). – When examining long series from the Caucasus with birds from Central Asia (mainly in ZMMU), we concluded that claimed differences between *kubanensis* and *ferghanensis* are extremely subtle if not virtually non-existent, with individual variation being more noteworthy than geographical, hence the latter is lumped with the former, which has priority. (Syn. *ferghanensis*.)

TAXONOMIC NOTE We follow Tietze et al. (2013) in keeping this well-known and widespread species in *Carpodacus*, achieved by including a few more species in this genus than did, e.g., Zuccon et al. (2012).

REFERENCES Stach, R. *et al.* (2016) *J. of Orn.*, 157: 671–679. – Ström, K. (1991) *Ornis Svecica*, 1: 119–120. – Tietze, D. T. *et al.* (2013) *Linn. Soc. Zool. J.*, 169: 215–234. – Zuccon, D. *et al.* (2012) *Mol. Phyl. & Evol.*, 62: 581–596.

C. e. erythrinus, Scotland, Sep: typical 1stW in early autumn, with olive-green tinge and prominent whitish double wing-bars and tertial edges. All plumage juv and fresh. Compared to autumn ad, tail-feathers rather narrow and pointed, and not yet heavily abraded. One outer tail-feather growing, presumably replacing an accidentally lost feather. (H. Harrop)

SINAI ROSEFINCH
Carpodacus synoicus (Temminck, 1825)

Fr. – Roselin du Sinaï; Ger. – Sinaigimpel
Sp. – Camachuelo del Sinaí; Swe. – Sinairosenfink

A specialised rosefinch found only in Sinai, Jordan, NW Arabia and Israel, where it mainly inhabits red granite and sandstone mountains with cliffs, including the famous Petra and Mt Moses or St Katherine's monastery sites. Often seen for the first time by W European birdwatchers in the mountains around Eilat in S Israel in winter, for example at Amram's Pillars, where the birds can be quite tame. Closely related to extralimital Pale Rosefinch *C. stoliczkae* of Central Asia and China.

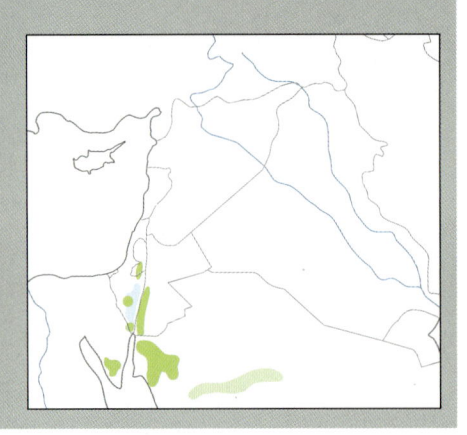

1stY ♂, Israel, Mar: maturation of young ♂ prolonged, with most 1stY like ♀, but is sometimes rather darker around head and may show crimson tinge to parts of head, mantle and other areas (only attains full ad plumage in second autumn, after first post-nuptial moult). Apparent juv primaries and primary-coverts seem to support age. (D. Bar Zakay)

Ad ♂, Israel, Feb: proportionately a somewhat long-tailed, slim and unstreaked rosefinch. Beautiful ad ♂ is reddish crimson-pink (with glistening silvery-white crown) on head, although face usually looks darker and more poppy-red. Upperparts lack obvious pattern, e.g. no pale wing-bars. (A. Ben Dov)

IDENTIFICATION Medium-sized, unobtrusive, ground- and cliff-haunting rosefinch. Adult ♂ largely pink, with *forecrown paler silvery-pink*, and *face, rump and entire underparts pinkish-red* becoming *deeper carmine-red on face and upper breast*, whereas hindneck and rest of upperbody are dull pink with brown suffusion (diffusely dark-streaked), while upperwing is dull brownish-grey like ♀. Tail all dark in both sexes. ♀-like plumages generally nondescript *pale fawn*, with very subdued streaking on upper- and underparts (one of the palest and most poorly-marked finches in the covered region). Slightly darker flight-feathers and tertial centres, dusky grey-brown bill, dark eye and pink-brown legs are the only other striking features of these plumages. Flight light and bounding, ascending quickly. Feeds on ground and is reasonably gregarious, forming small parties post-breeding and in winter. Often nervous, escaping to cliff faces and flying ahead of observer along wadis. Frequently, its high-pitched call is heard long before the birds are seen, but in other situations can be tame.

VOCALISATIONS Song rarely heard, reportedly a varied and melodious prolonged medley. ♂ in display also has buzzing note (not wholly dissimilar to Trumpeter Finch). – Contact and alarm call an emphatic, straight, high-pitched and squeaky *tweet* or *cheek* with minor variations in pitch and length. Flight-call characteristic: a fairly metallic, hard *chip* that is surprisingly sparrow-like.

SIMILAR SPECIES Occurs only in deserts of S Levant and NW Arabia, where impossible to confuse with other *Carpodacus* finches, except perhaps with same-sized Common Rosefinch. However, this has different shape, a marginally heavier bill, and adult ♂♂ differ by having more extensive and deep carmine-red rather than pink parts of the plumage, while ♀/immatures are readily separated by their overall paler tone and subdued streaking. – Superficially similar to Trumpeter Finch, which occurs in same breeding areas, but latter is smaller and more compact, with a thicker bill, shorter, deeper-forked tail, more obvious pale eye-ring and different wing pattern. Also, breeding ♂ has pink-red bill and ♀/non-breeding ♂ a paler one. – ♀/immature could be confused, if seen poorly, with similar-sized Pale Rock Sparrow, which can appear in the same habitat on migration, but this species has unstreaked plumage, different wing pattern and bill shape/colour. – Differs from extralimital Pale Rosefinch (*C. stoliczkae*; see Taxonomic note) ♂ plumage in lacking in obvious border between pale silvery-pink forecrown and much darker drab-brown hindcrown. Instead whole crown and nape is largely pinkish, and this suffusion continues onto mantle and back (Pale Rosefinch: dark brown nape, mantle and back).

AGEING & SEXING Ageing of all plumages possible using moult pattern if seen close or handled. Sexes differ strongly in ad, but 1stY ♂ is ♀-like (and usually breeds in such plumage), but some imm ♂♂ distinctive by intermediate appearance. Seasonal variation of ad rather limited. – Moults. Complete post-nuptial moult of ad in Jul–Sep, and at same time partial post-juv moult including head, body, lesser and median coverts, and most or all greater coverts. Pre-nuptial moult apparently absent. – **SPRING Ad ♂** Largely

♀-like (left: Sinai, Egypt, Apr; right: Israel, Mar): bill rather small (still of course a strong finch bill). ♀-type plumage nondescript, being predominantly drab sandy-brown (slightly gingery when fresh, especially on face) and only vaguely streaked. Black eye obvious. Left bird apparently ad and thus a ♀, but without closer inspection of wing ageing the other bird is impossible; it could be ad ♀ or 1stS of either sex. (Left: D. Monticelli; right: M. Varesvuo)

pink below, as well as on head and rump, mantle and back being streaked or mottled rufous-brown. Forehead deep red, forecrown silvery-pink grading smoothly into pink-red rear crown and upper nape. Sides of head and underparts deep pink-red or carmine from throat to upper breast, with a paler pink colour on lower breast and flanks. Vent and undertail-coverts mainly whitish with at most only a faint pink hue. With wear, red of face darkens, in some almost to blackish wine-red, contrasting more sharply with rest of underparts. Hindcrown, nape, mantle and wings wear to duller, more grey-brown pink, and streaks are reduced. **Ad ♀** At this season, often more dun-coloured with clearly gingery face, isabelline-brown to drab-grey upperparts, browner streaks on crown and mantle, slightly paler uniform rump. Wings have reduced and paler greyish-buff fringes to coverts and tertials, and underparts are rather pale ochre with neat brown streaks on breast and flanks. Slight pink wash to face/breast barely visible in the field. **1stS** Both sexes resemble ♀, but some ♂♂ develop partial reddish-pink feathering on face and chest. Browner retained juv wing-feathers discernible (moult as 1stW). – AUTUMN **Ad ♂** Little noticeable change in overall plumage when fresh, except pinkish-red of face and underparts more extensively tipped silvery-carmine, and dark face less marked. Slightly more reddish-pink wash to mantle. **Ad ♀** Broadly fringed paler above and on wings. Overall more grey, and streaking more diffuse. **1stW** Both sexes resemble fresh ad ♀; unlike latter has retained juv primary-coverts and remiges, and perhaps some outer greater coverts and alula, which are narrower, more pointed, less firmly textured and not as fresh, with slightly more creamy-buff fringes than in ad ♀. **Juv** Like ♀ but has soft fluffy body-feathers and is overall more buff-brown with distinctly paler, cream-tinged rump and even less distinct streaking.

BIOMETRICS L 14–15 cm; **W** ♂ 83–90 mm (*n* 22, *m* 86.7), ♀ 80–86.5 mm (*n* 8, *m* 84.3); **T** ♂ 58–66 mm (*n* 21 *m* 62.0), ♀ 58–64 mm (*n* 8, *m* 60.9); **T/W** *m* 71.7; **B** 12.1–15.1 mm (*n* 30, *m* 13.3); **B**(f) 9.0–11.1 mm (*n* 30, *m* 10.0); **BD** 7.1–8.2 mm (*n* 29, *m* 7.8); **Ts** 18.3–21.8 mm (*n* 30, *m* 19.7). **Wing formula: p1** minute; **p2** < wt 0.5–4 mm; **pp3–4**(5) about equal and longest; **p5** < wt 0–2.5 mm; **p6** < wt 2.5–5.5 mm; **p7** < wt 8.5–13 mm; **p10** < wt 18–23.5 mm; **s1** < wt 20–26 mm. Emarg. pp3–6.

GEOGRAPHICAL VARIATION & RANGE Monotypic. – Sinai, S Israel, S Jordan, NW Arabia; sedentary. (Syn. *kistjakovskii*; *pallidorosa*.)

TAXONOMIC NOTE Very marked variation between widely allopatric Sinai Rosefinch in Levant and N Arabia (*synoicus*) on the one hand, and extralimital Pale Rosefinch *C. stoliczkae* in C & E Asia (with races *salimalii*, *stoliczkae* and *beicki*), strongly suggests two species involved, with strong support from mtDNA, differing by 4.1% (Tietze *et al.* 2013). Birds from C Afghanistan (*salimalii*), E Pamirs and SW Tarim Basin in W China (*stoliczkae*) and NE Qinghai and adjacent Gansu (*beicki*) share distinguishing characters in ♂ plumage of a sharp border between pale pink ('silvery') forecrown and darker drab-brown hindcrown, with nape to back basically brown, whereas *synoicus* has whole crown and nape smoothly pink, and mantle and back are also predominantly pink with some dull brown streaking.

REFERENCES Shirihai, H. (1989) *BB*,82: 52–55. – Tietze, D. T. *et al.* (2013) *Linn. Soc. Zool. J.*, 169: 215–234.

Ad ♂, Israel, Sep: a lovely fresh-plumaged ♂ just after complete post-nuptial moult, when reddish areas appear pinker due to pale tips, but face still more dark pink-red. No other rosefinch frequents such arid areas, where often first detected by its high-pitched calls. (H. Shirihai)

Ad ♀, Israel, Sep: at end of complete post-nuptial moult but still showing sharp contrast between old/worn and new remiges, separating it from a 1stW. (H. Shirihai)

1stW, Israel, Sep: in late summer/early autumn young still in more juv-like plumage, initially evenly fresh and greyer, with cream-buff fringes to wing-coverts. (H. Shirihai)

1stW, Sinai, Egypt, Oct: 1stW (by moult limit in greater coverts), but ♀-like birds cannot be sexed. Unlikely to be confused with any other rosefinch, or with young or winter Trumpeter Finch, which is stockier, with stubby bill and shorter tail, has pale eye-ring and pale fringes to tail- and flight-feathers. (D. Occhiato)

♀-like, Sinai, Oct: some ♀-like birds are virtually unstreaked below, here an extreme example. (D. Occhiato)

GREAT ROSEFINCH
Carpodacus rubicilla (Güldenstädt, 1775)

Fr. – Roselin tacheté; Ger. – Berggimpel
Sp. – Camachuelo grande; Swe. – Större rosenfink

Any serious 'twitcher' who wants to build up a long list of bird species seen in the W Palearctic will sooner or later consider visiting the Caucasus to see one of its most sought-after birds: the Great Rosefinch. A 'lazy' person elects to visit the village of Kazbegi in Georgia at 900 m altitude in early April, when it is still winter and Great Rosefinch is often something of a garden bird visiting bird feeders. Perhaps more 'sportsmanlike' is it to arrive in late May or June to find the bird in its alpine breeding habitat. But that requires a climb to at least 3000 m, frequently even more! The reward is not only to see the beautifully red male but also to hear it sing.

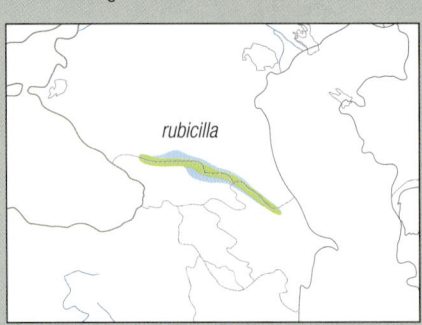

C. r. rubicilla, ♂, presumed ad, Georgia, Apr: individual variation includes paler ♂♂. Especially when whitish tips still fresh and large, these can coalesce to form almost whitish ear-coverts patch and strongly marbled underparts. This bird also has greyer upperparts. Looks best for ad, but safe ageing from this image alone impossible. (W. Müller)

C. r. rubicilla, ad ♂, Georgia, May: a large rosefinch with heavy ivory-coloured bill, longish wings and tail, and almost unstreaked mantle and scapulars. ♂ is red with white spots on head and underparts, making it unmistakable in region. ♂ only attains full red plumage after first post-nuptial moult, i.e. such a ♂ in spring/summer is 2ndS or older. (S. Olofson)

IDENTIFICATION A *large and robust, long-tailed* finch, almost the size of a small thrush, with a characteristic *straw-coloured, heavy, conical bill*. Adult ♂ largely *raspberry-red with whitish spots on crown and entire underparts*. Only the back, wings and tail are mainly dark brownish, the back streaked dark. The paleness of the bill is enhanced by the contrasting swarthy face. ♀ and first-summer ♂ are rather dull grey-brown streaked dark, strongest below, lacking any particular field marks apart from the *hefty, usually pale bill*, the *large size* and the *long and notched tail*. There is a hint of narrow, slightly paler wing-bars. In winter or spring the bill is sometimes darker grey in some ♀-coloured birds, but this could also be individual variation at any time of year. Legs in all sexes and ages dark grey. Flight, at least over longer distances, strong and undulating. Moves on the ground with hops.

C. r. rubicilla, ad ♂, Georgia, May: upperparts dull red, contrasting with pinkish-red rump and uppertail-coverts, red of face and cap deeper, becoming almost blackish-red around eyes. Note ill-defined grey-pink tertial fringes and edges to greater coverts. Plumage evaluation as with image above. (J. S. Hansen)

C. r. rubicilla, presumed 1stS ♂, Georgia, May: ♀-like being drab grey-brown and off-white with ill-defined dusky streaks above and below, but fine spotting on crown and cheeks pinkish-brown, and mantle and scapulars also with faint pink hue inferring young ♂ in its first summer. (J. S. Hansen)

C. r. rubicilla, ♀-like, Georgia, Apr: rather chunky and large-headed jizz, with some heavily-spotted birds appearing overall dappled. Some seem to have greyer breast-sides, flanks and neck-sides. Clear whitish fringes to wing-coverts and pale secondary panel also typical. Safe ageing difficult. (W. Müller)

VOCALISATIONS Song a series of clear notes, usually one (or a few) introductory note(s) and a hint of a pause followed by a quick series of simple, slightly accelerating notes falling subtly in pitch, *chuu, fyu fyu fyu-fyu-fyu-fyu*. A certain variation in length and details of delivery between ♂♂ noticeable, reputedly even from same bird. — Call a high-pitched, cheerful up-curled *tvoee*, often given in flight or when alert on the ground. Also a discordant, 'thick' and somewhat sparrow-like *chick* with some variation in loudness and details.

SIMILAR SPECIES While the adult ♂ can hardly be mistaken for other species within the treated region, the ♀ and young are more nondescript and could easily be mistaken for Common Rosefinch if size is not evident (as often possible in alpine habitat—distant birds and nothing familiar nearby to compare). Normally though, a non-red Great Rosefinch differs in being more clearly streaked on the crown, head-sides and below. The tail is proportionately longer and the bill larger, usually paler and more straw-coloured. — There are at least two other very similar rosefinches in E Asia: extralimital *Streaked* (or *Eastern Great*) *Rosefinch* (*C. rubicilloides*; not treated) is equally large but differs mainly in its much more prominently-streaked upperparts, and *Red-mantled Rosefinch* (*C. rhodochlamys*; not treated) has very similar ♀ plumage and heavy bill but differs in its much shorter wing (c. 80–90 mm) and shorter primary projection (c. 14–16 mm).

AGEING & SEXING (*rubicilla*) Ages differ subtly in autumn and spring, but reliable ageing requires close views. Sexes readily separable after first post-nuptial moult. — **Moults.** Complete post-nuptial moult of ad in late summer and early autumn, mainly early Jul–Oct. Partial post-juv moult at about same time does not normally involve flight-feathers, tail-feathers, tertials, greater coverts or primary-coverts. No pre-nuptial moult in late winter. — **SPRING Ad ♂** Much of plumage except mantle, back, scapulars, flight-feathers and tail raspberry-red, crown and underparts white-spotted. Mantle, back and scapulars tawny-brown with reddish tinge and fine streaking. Tail-feathers rather broad with rounded tips, edges usually still rather fresh. **Ad ♀** Much of plumage, including mantle, back and scapulars greyish-brown prominently streaked dark. No red in plumage. Greater coverts have greyish-white outer edges diffusely demarcated from dark centres. Tail-feathers rather broad with rounded tips, edges usually still rather fresh. **1stS ♂♀** As ad ♀ but tail-feathers slightly narrower and tips more pointed with often quite abraded edges. Greater coverts have slightly narrower and more distinct whitish edges, and these usually continue on tip of inner web. If tail-feathers or greater coverts criteria are difficult to assess, it is as a rule impossible to separate 1stS ♂♀ from ad ♀. — **AUTUMN Ad ♂** Raspberry-red as ad ♂ in spring. **Ad ♀** As in spring. Important to establish ad-type greater coverts and tail-feathers. **1stW ♂♀** As 1stS ♂♀ in spring but tail-feathers not yet heavily abraded. **Juv** Very similar to ad ♀ but somewhat duller and paler overall, streaks more diffuse and not as dark, and belly on average less streaked. Tail-feathers and greater coverts as 1stS (see spring).

BIOMETRICS (*rubicilla*) **L** 19.5–21 cm. **W** ♂ 117–125 mm (*n* 16, *m* 119.2), ♀ 111–118 mm (*n* 8, *m* 114.3); **T** ♂ 87–96 mm (*n* 15, *m* 90.5), ♀ 81–87 mm (*n* 8, *m* 85.4); **T/W** *m* 75.6; **B**(f) 13.6–15.3 mm (*n* 24, *m* 14.3); **BD** 11.3–13.6 mm (*n* 24, *m* 12.3); **Ts** 22.0–24.5 mm (*n* 24, *m* 23.4). **Wing formula: p1** minute; **pp2–4** about equal and longest (p2 sometimes to 2 mm <); **p5** < wt 1–4 mm; **p6** < wt 8–14 mm; **p7** < wt 15–29 mm; **p10** < wt 24–39 mm; **s1** < wt 33–41 mm. Emarg. pp3–6. (Bill to skull difficult to measure accurately in this species, at least on skins.)

GEOGRAPHICAL VARIATION & RANGE Fairly slight. Four races usually accepted, all allopatric in different Asian

C. r. rubicilla, ♀-like, Georgia, Apr: note extremely heavily-streaked underparts, possibly indicating younger age but partly also depending on angle of view. (W. Müller)

mountain ranges. Only one occurs within the treated region.

C. r. rubicilla (Güldenstädt, 1775) (Caucasus; winters at lower elevations but remains close to breeding range). Treated above. The darkest race, ad ♂ being deep scarlet-red with rather small whitish spots on head and underparts. Dark tawny-brown on scapulars and mantle/back.

REFERENCES Loskot, V. M. (1991) *Trudy Zool. Inst. Akad. Nauk SSSR*, 231: 43–116.

C. r. rubicilla, ad ♂, Georgia, Feb: rather fresh winter plumage with slightly broader pale tips and appears at most marginally duller than in spring/summer. (G. Darchiashvili)

C. r. rubicilla, ad ♀, Georgia, Feb: when fresh often overall greyish with streaking below softer looking. Wing evenly fresh, suggesting ad ♀ rather than similar-looking 1stY of either sex. (G. Darchiashvili)

LONG-TAILED ROSEFINCH
Carpodacus sibiricus (Pallas, 1773)

Fr. – Roselin à longue queue; Ger. – Meisengimpel
Sp. – Camachuelo colilargo
Swe. – Långstjärtad rosenfink

A taiga rosefinch often found in dense riverine vegetation (e.g. Salix, Betula). It is mainly a Siberian species, but it was recently discovered breeding locally in European Russia, and is thus a member of the W Palearctic fauna. Long-tailed—as the name says—with a rounded head and, for a rosefinch, small, rounded bill, features that combine to give it a slightly tit-like appearance (picked up in its German name). These features have been used to place it in a separate genus, Uragus, but recent research based on genetic markers has shown the species to be closely related to the majority of rosefinches, in Carpodacus. Mostly sedentary, or moves short distances to the south in severe winters.

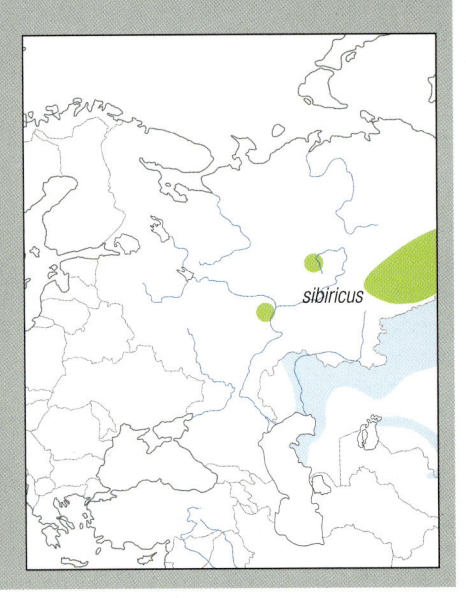

IDENTIFICATION Medium-small, *long-tailed* and *round-headed* with a *short, stubby bill*. Rather pale general colours with *prominent white double wing-bars* in all plumages, in ♂♂ often so extensive as to form *one large white patch*. Outer tail-feathers white in contrast to dark centre. Tertials broadly edged white, as are secondaries forming hint of a panel on the closed wing. Plumage soft and fluffy. Adult ♂ is *glossy pale greyish-white on crown, nape* and *cheeks* (can look silvery or pure white!) faintly tinged pink, and is *bronzy crimson-pink on forehead and face*, with *purer pink on sides of neck, breast, belly and rump*. Throat tinged crimson but densely blotched glossy silvery-white. Mantle, shoulders and back pinkish-buff streaked brown-black. In late summer, pink becomes redder, and white portions decrease with wear. First-year ♂ variable, depending mainly on amount of pink attained in post-juvenile moult, some being almost as dull as ♀♀, others attaining more pink in post-juv moult; with wear during winter, and especially in late spring, concealed pink becomes more visible, and most immature ♂♂ recognised as such by reduced but obvious pink on breast and belly, and traces of pink on upperparts. A few are nearly as neat as adult ♂ (thus safe separation may require handling to examine pattern of primary-coverts and large alula). ♀ lacks crimson on head and upperparts, has only *rump warm orange-buff or pink-orange*, and is otherwise greyish with subdued ochre tones, best told by general shape and the broad double white wing-bar. Juvenile, rarely encountered outside breeding range, resembles ♀ but is a touch duller, brownish and more boldly streaked, with more buffish-tinged wing-bars. Fluttering flight on rounded wings produces sound not unlike that of House Sparrow. Climbs agilely on thin branches like a tit.

VOCALISATIONS Song a brief, fast warble of similarly high-pitched silvery and mellow notes, often without clear pattern, still characteristic, frequently a few notes repeated almost in Coal Tit fashion, *seevitcha-seevitcha-seevitcha-seewee-seewee-seewee*. Contrary to some oft-repeated statements bears little resemblance to song of Common Crossbill. – Commonest call is a Chaffinch-like *fink, fink, fink*, often uttered by anxious birds around nest, hence might be primarily an alarm. There is also a more melodious multisyllabic whistle, variable in details and repeated a few times, perhaps with a contact function more than alarm.

SIMILAR SPECIES Can hardly be confused with other species due to its stubby Bullfinch-like bill, long tail with white sides and prominent double white wing-bars. – It is somewhat larger than *Long-tailed Tit* with thicker and pale bill, and is further readily separated by the double white wing-bars. – In China, a remote confusion risk exists with extralimital Three-banded Rosefinch (*C. trifasciatus*; not treated), but this is much darker, with narrower pale wing-bars and stronger bill.

AGEING & SEXING (*sibiricus*) Ages differ subtly, but reliable ageing usually requires close views or handling. Sexes readily separable after first post-nuptial moult, often already following post-juv moult. – Moults. Complete post-nuptial moult of ad after breeding, mainly Jul–Sep (Oct). Partial post-juv moult often starts with body moult in Jul; 1stW do not moult any flight-feathers and only a few tail-feathers and tertials; primary-coverts and large alula never moulted. Number of greater coverts moulted varies from many inner to all. Pre-nuptial moult in late winter lacking or at least very restricted. – **SPRING Ad** No obvious moult contrast in plumage, all wing-coverts of same generation. Tail-feathers comparatively broad with rounded tips, edges usually still reasonably fresh, at least in early spring. **Ad ♂** Forehead, face and eye-stripe deep crimson-pink with faint brownish tone, sides of neck and throat purer pink with glossy pink-white streaks or blotches. Breast, flanks and rump crimson-pink. Mantle and back pinkish-grey, streaked brown-black. Primary-coverts and large alula rather dark (often almost blackish) with distinct white or whitish outer edges. **Ad ♀** Head, breast, flanks greyish-brown with some faint ochre tinge (lacking any crimson), diffusely or rather finely streaked dark. Crimson or orange confined to rump. **1stS** At times shows moult contrast in wings or tail (but can be difficult to establish due to amount of wear). Tail-feathers on average slightly narrower than ad and tips more pointed with often quite abraded edges. (Beware of similarity with more heavily-worn ad.) **1stS ♂** Either rather similar to ad ♀, having only scant red or pink in plumage away from rump, or has much pink on head, sides of neck, breast and flanks, and some glossy whitish blotches on crown, sides of neck and throat, approaching ad ♂ (pink feathers all attained in post-juv moult, but most initially obscured by brownish or greyish tips and surfaces later abraded). Primary-coverts and large

C. s. sibiricus, ad ♂, Mongolia, May: long-tailed as name says, with short, thick bill, prominent double white wing-bars, white-fringed tertials, whitish secondary panel and broad white edges to tail. Ad ♂ has rose-pink to vinaceous-red body, whitish cap, ear-coverts, throat and flanks, almost blood-red forehead and eye-stripe, and uniform pinkish-red rump. (H. Shirihai)

C. s. sibiricus, ♀, presumed ad, Kazakhstan, Mar: a grey-and-white version of the ♂, with any pink or orange-buff restricted to rump. Most obvious feature is double wing-bars (but upper one can be hidden by scapulars, as here), streaked dusky brown-grey above, with some vague streaking on throat and breast. Wing seems evenly feathered indicating ad, although large alula is rather brownish. (V. Vorobyov)

C. s. sibiricus, ad ♂, Kazakhstan, Dec: when fresh, silvery-white tips broader and conceal parts of the reddish plumage, which therefore appears more pinkish. Wings evenly fresh without visible moult limits, indicating ad (and 1stW ♂♂ do not develop such extensive reddish plumage anyway). (V. Federenko)

C. s. sibiricus, 1stW ♂, Siberia, Jan: some fresh, least-advanced young ♂♂ are very like ♀, but pinkish rather than orange-tinged rump and also faint pink on face, mantle, vent and belly are visible. Subdued amount of pink reveals sex and age. (Y. Malkov)

alula dull brown lacking contrasting whitish outer edges. – **AUTUMN Ad ♂** As in spring, but wings and tail now fresh following complete post-nuptial moult, and pink and silvery-white colours a little duller, partly obscured by brownish and greyish tips, but still much pink visible on face, neck and breast/flanks. Whole plumage of same generation lacking any moult contrast. Note contrasting pattern of primary-coverts and large alula. **Ad ♀** As in spring, but fresh wings and tail without moult contrast, and tail-feathers neat with rounded tips. **1stW ♂** As ad ♀ but shows at least some pink-buff traces on breast, flanks or back of neck. Those with much pink separated from ad ♂ by dull brown primary-coverts and large alula (less dark and lacking contrasting whitish outer edges of ad ♂). At times a slight moult contrast detectable between variable numbers of replaced inner greater coverts and some or all tertials (ad type) and any juv outer unmoulted greater coverts (tipped and edged more buffish). **1stW ♀** Difficult to separate from ad ♀, but sometimes possible on slightly heavier streaking on breast and flanks, narrower tail-feathers with more abraded tips and on average browner and more pointed primary-coverts (though some overlap). **Juv** Very similar to 1stW ♀ but is somewhat duller and brown-tinged overall, streaks being more diffuse and not as dark. Wing-bars usually slightly tinged buff or cream. Plumage of slightly looser structure than 1stW, especially on vent and nape.

BIOMETRICS (*sibiricus*) L 15–16.5 cm; **W** ♂ 72–80 mm (*n* 25, *m* 76.0), ♀ 71–75.5 mm (*n* 11, *m* 73.2); **T** ♂ 77–92 mm (*n* 25, *m* 83.4), ♀ 78–84 mm (*n* 11, *m* 81.2); **T/W** *m* 110.2; **B** 10.3–12.0 mm (*n* 34, *m* 11.0); **B(f)** 6.8–9.0 mm (*n* 33, *m* 8.1); **BD** 7.1–8.3 mm (*n* 34, *m* 7.5); **Ts** 14.9–17.0 mm (*n* 30, *m* 16.1). **Wing formula:** p1 minute; **p2** < wt 5.5–10 mm; **p3** < wt 1–3.5 mm; **pp4–6** about equal and longest (p6 sometimes to 2.5 mm <); **p7** < wt 3–6 mm; **p8** < wt 6–10 mm; **p10** < wt 10–17 mm; **s1** < wt 12–17 mm. Emarg. pp3–7.

GEOGRAPHICAL VARIATION & RANGE Moderate variation within taiga belt, mainly involving darkness or paleness of plumage; usually three subspecies recognised. Isolated from these three races are two darker and shorter-tailed subspecies in C China. Only one subspecies occurs in the treated region.

C. s. sibiricus (Pallas, 1773) (locally C European Russia, Ural region, N Mongolia, Siberia east to Transbaikalia, N Manchuria and lower Amur; sedentary or makes short-range movements in severe winters). Treated above. Rather pale and long-tailed, ♂ having much white in wing. (Syn. *fumigatus*; *stegmanni*.)

TAXONOMIC NOTE Arnaiz-Villena *et al.* (2007) showed that Long-tailed Rosefinch, previously habitually placed in a monotypic genus *Uragus*, was nested in the *Carpodacus* clade, and this was further demonstrated by Liang *et al.* (2008). Then, in their more complete phylogenies of the Fringillidae, this finding was replicated by Zuccon *et al.* (2012) and Tietze *et al.* (2013). Obviously, a short bill, long tail and prominent wing-bars are easily-adapted traits and not proof of long-different life history.

REFERENCES Arnaiz-Villena, A. *et al.* (2007) *Acta Zool. Sinica*, 53: 826–834. – Liang, G. *et al.* (2008) *Zoological Research*, 29: 465–475. – Tietze, D. T. *et al.* (2013) *Linn. Soc. Zool. J.*, 169: 215–234. – Zuccon, D. *et al.* (2012) *Mol. Phyl. & Evol.*, 62: 581–596.

C. s. sibiricus, presumed 1stW ♀, Kazakhstan, Dec–Jan: the lack of any reddish or pinkish on both birds, and also more orange-cinnamon (not pinkish) rump on the right-hand bird strongly suggest 1stW ♀. The obvious contrast between wingtip and tertials also supports this being a young bird. Note the typical fine breast streaking on the left-hand bird. (V. Fedorenko)

PINE GROSBEAK
Pinicola enucleator (L., 1758)

Fr. – Durbec des sapins; Ger. – Hakengimpel
Sp. – Camachuelo picogrueso; Swe. – Tallbit

A very large, rather long-tailed finch that is generally confined to the northernmost taiga and is only rarely seen further south in Europe. Still, in some years (at irregular intervals) quite large numbers evacuate their home forests and move south-west, some exceptionally even reaching the British Isles. Since the birds are not used to humans, they are often surprisingly tame and cling to branches laden with rowan berries overhanging pavements in towns, or in gardens near houses. In earlier times, when birds were commonly caught for food or entertainment, they even acquired the nickname 'nitwits' due to their inability to avoid being easily caught.

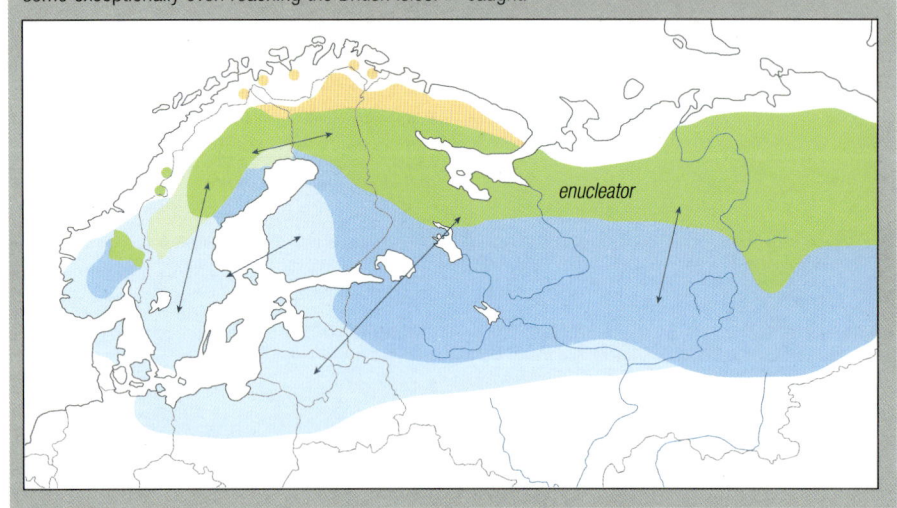

P. e. enucleator, presumed ad ♀, Finland, Apr: clearly duller than ad ♂, but difficult to separate from similar 1stY of both sexes, though has evenly-feathered wing and not heavily worn tail and primary-coverts suggesting an ad. Head, breast and rump orange-yellow and grey, thus ♀-like plumage. (A. Ouwerkerk)

P. e. enucleator, ad ♂, Finland, Apr: a large finch with stubby bill and decurved culmen ending in a hint of a hook tip, and double white wing-bars. ♂ develops rosy-red plumage only after first complete post-nuptial moult. Grey lower flanks, vent and undertail-coverts, with darker lores and feathers around eye. (A. Ouwerkerk)

IDENTIFICATION A *large, bulky finch with long tail* and *dark bulbous bill*. Often seen feeding in small parties or larger flocks in berry-laden trees, climbing nimbly like crossbills or oversized redpolls. Also, often quietly feeds on the ground on berries in taiga shrubbery. In all plumages has *thin double white wing-bars* and *whitish edges to the greater coverts and tertials*. Plumage greyish with variable amount of *raspberry-red* (adult ♂) or *yellow-green or orange-yellow* (♀, young ♂) on head, underparts, lower back and rump. Tail-coverts and lower scapulars diffusely dark-centred, giving a darkish and blotched appearance. Although it may be tempting to believe that non-red birds with an orange tinge to the yellow on head and breast are only immature ♂♂, whereas all ♀♀ are yellowish-green, this is not true. You will find both sex categories represented among both colour categories; it is a matter of individual variation alone, possibly also food-related.

VOCALISATIONS Song is a clear, fast, short yodelling, *c.* 2–3 sec long, has a desolate ring to it and shuttles irregularly over a few clear notes with only small tonal steps, *düdeli-didelü-deedelu-düdelüh*. Peculiarly, it sounds almost like Wood Sandpiper's *Tringa glareola* display, only not quite as monotonous and mechanical, but with a less resolute pattern and a clearer, more 'silvery' voice. – Call is a disyllabic clear *dlüit* or *dee-düh* or other variations, like elements from the song. When alarmed or before taking off, calls are more intense and often contain more syllables, *düll-de-djüh* or *düll-de-dü-djüh*. From feeding flocks come discreet conversational clicking notes, *bitt bitt bitt...*

SIMILAR SPECIES Can hardly be confused with any other species in the treated region. Similar-sized *Great Rosefinch* occurs only in the Caucasus, has larger conical and usually straw-coloured bill, and less prominent pale wing-bars. Other rosefinches are either smaller, lack obvious wing-bars or breed only in E Siberia or China.

AGEING & SEXING Ages differ subtly in both autumn and spring, but reliable ageing requires close views. Sexes often—but not always—separable after first post-nuptial moult. – Moults. Complete post-nuptial moult of ad in late summer and early autumn, mainly early Jul–mid Oct. Partial post-juv moult at about same time does not usually involve flight-feathers, tail-feathers, tertials, primary-coverts or greater coverts. A few, however, moult some inner greater coverts (commonly 2–4), and odd birds appear to replace all, but timing not clear. Apparently no pre-nuptial moult in late winter. – **SPRING** Sex: apart from fully raspberry-red ad ♂♂, or all yellowish-green adult ♀♀ that have first been safely aged, often very difficult to identify. Both ♀♀ (of any age) and 1stY ♂ can either lack any orange tones or have ochre-yellow and greenish plumage with rather extensive orange, even a touch of orange-red on crown, head-sides and, less often, on throat, breast and upper flanks. Subtle difference in number of dark feather-centres on crown,

P. e. enucleator, ad ♂, Finland, Feb: contrastingly white-edged coverts and tertials and fine greyish outer fringes to tail-feathers. Rather distinct scaly pattern and quite deep red upperparts, and somewhat more grey on sides of this bird. Red plumage (2ndS or older), evenly-feathered wing and broad and round-tipped tail-feathers age the bird. (M. Varesvuo)

mantle and back (♂♂ with many and strong contrast, ♀♀ with no or little and less contrast) is only apparent as a slight tendency with substantial overlap and of very limited use. **Ad ♂** Head, back, rump, breast and flanks raspberry-red mixed with varying amount of grey. Tail-feathers rather broad with rounded tips, edges still rather fresh. **Ad ♀** Head and breast vary from yellowish green-grey to greenish-yellow with grey and subdued orange admixed (those with orange-red should only be determined if first safely aged as ad). Rump and flanks yellowish-green or greyish-green. Tail-feathers as ad ♂. **1stS ♂♀** As ad ♀ but tail-feathers slightly narrower and tips more pointed with quite abraded edges. If tail-feather criterion difficult to assess, it is generally impossible to tell 1stS ♂♀ from ad ♀. – AUTUMN **Ad ♂** Raspberry-red as ad ♂ in spring. **Ad ♀** As in spring. Important to establish ad type of tail-feathers. Greater coverts have full, rather broad and pure white, sharply delineated tips and outer edges. **1stW ♂♀** As 1stS ♂♀ in spring but tail-feathers not yet heavily abraded. Greater coverts (if not moulted) rather short, tipped and edged rather thinly and diffusely buffish-white or dusky-white. Any moulted inner greater covert is contrastingly longer and broader-tipped purer white. **Juv** Resembles ♀ but is somewhat duller and greyer overall, breast slightly tinged buff-yellow, and greater coverts are tipped buffish grey-white or yellowish-buff.

BIOMETRICS (*enucleator*) **L** 20–21 cm. Sexes nearly the same size: **W** ♂ 106–117 mm (*n* 39, *m* 110.8), ♀ 104–112.5 mm (*n* 27, *m* 109.1); **T** ♂ 80–92 mm (*n* 39, *m* 86.4), ♀ 81.5–90 mm (*n* 27, *m* 85.5); **T/W** *m* 78.1; **B** 17.2–21.5 mm (*n* 56, *m* 19.1); **B**(f) 14.0–16.7 mm (*n* 61, *m* 15.4); **BD** 10.4–12.5 mm (*n* 59, *m* 11.2); **Ts** 20.0–22.8 mm (*n* 56, *m* 21.5). **Wing formula: p1** minute; **p2** < wt 2–4 mm; **pp3–4**(5) about equal and longest; **p5** < wt 0–3 mm; **p6** < wt 6.5–10 mm; **p7** < wt 17–20 mm; **p10** < wt 28.5–33 mm; **s1** < wt 31.5–37 mm. Emarg. pp3–6, on p6 often slightly less prominently.

GEOGRAPHICAL VARIATION & RANGE Slight and presumably clinal variation, mainly involving size and shape of bill. Only one race occurs within treated region, but three more have been described from C Siberia and E Asia, and several more from North America, none of which is treated here.

P. e. enucleator (L., 1758) (N Fenno-Scandia, N Russia, NW Siberia east to Altai and Yenisei). Described above. Differs from extralimital Asian races only subtly in size and shape of bill. (Syn. *stschur*.)

P. e. enucleator, 1stS ♂, Finland, Mar: much less colourful than ad ♂, or intermediate between plumage of fully adult sexes, with clear moult limits between juv primaries and primary-coverts on the one hand and renewed tertials and greater coverts on the other. (M. Varesvuo)

P. e. enucleator, 1stS ♂, Lapland, Jun: least-advanced young ♂, rather ♀-like, except some reddish feathers developing on lower cheeks, and some red stains on breast. Note brownish, rather pointed juv-type primary-coverts (as far as they can be judged at this angle). (F. Trabalon)

P. e. enucleator, ad ♂, Finland, Nov: winter plumage as summer but now very fresh, and plumage generally appears more marbled. Note rather strong face pattern with dark eye-stripe in this individual. (M. Varesvuo)

P. e. enucleator, ad ♀, Finland, Nov: evenly-fresh wing infers age and in combination with orange-yellow plumage confirms sex as ♀. Large, chunky jizz, yellowish-orange plumage, double wing-bars, clear white tertial fringes and bill structure make even ♀ unmistakable. (M. Varesvuo)

P. e. enucleator, 1stW, Finland, Oct: generally ♀-like, and aged by moult contrast between retained primaries and tertials, and within greater coverts; note also pointed tips to tail-feathers. However, sexing impossible of young birds with this appearance. (M. Varesvuo)

(COMMON) BULLFINCH
Pyrrhula pyrrhula (L., 1758)

Fr. – Bouvreuil pivoine; Ger. – Gimpel
Sp. – Camachuelo común; Swe. – Domherre

Two birds are perhaps better known from Christmas cards and pretty paintings than in real life, both are red-breasted: the Robin and the Bullfinch. In N Europe, Bullfinch is regarded as a winter visitor, coming to visit bird feeders. It is thereby associated with snow, but in fact it is present all year, only leading a quiet, discreet life in the nearest wood during summer. Pairs remain together year-round and apparently mate for life. Consequently, Bullfinches are mainly seen singly or in twos, but small flocks of up to a dozen birds may form outside the breeding season.

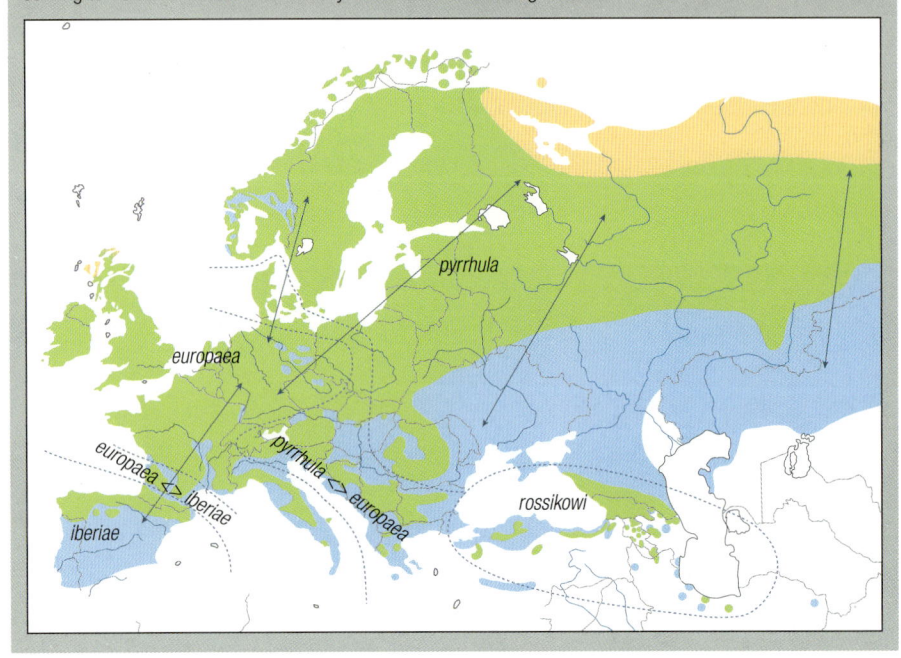

IDENTIFICATION The size of a Hawfinch or slightly larger (i.e. roughly between a sparrow and Blackbird), but size varies geographically, northern and eastern birds being considerably larger than breeders in the west and south. The Bullfinch has a rather *plump body*, of which the head seems to form the upper part (with no visible neck, thus really 'bull-headed'), and a *short but heavy, bulbous bill*. Both sexes have *bill, cap and face black*, with black wings and tail, and they share the *large white rump patch* and *greyish-white wingbar*, both often clearly visible in flight, not least when a bird is flushed and flies away. Despite these similarities the sexes are readily separable. The ♂ has attractive *rosy-red cheeks, throat, breast and upper belly*, and *ash-grey back*, whereas the ♀ has dull pinkish-grey breast (tinged buffish or sandy in some populations) and a greyish-ochre back. Lower belly and vent are white in both. The juvenile differs clearly (but is rarely seen) in its boldly olive-spotted plumage and in lacking the black cap and face. Bullfinches are not shy but wary, flying off a fair distance if approached too closely. Flight is undulating and powerful. When feeding on buds they climb nimbly in trees despite their size. In cold weather they may fluff up their rich feathering into the *rounded form* so often depicted by artists.

VOCALISATIONS Song is rather quiet and low-key, seemingly a recited, slightly tentative and stuttering mixture of scratchy, strained notes, brief fluty whistles (like the call!—see below), brief trills and hard, warbling notes somewhat reminiscent of Dipper song. Typically, the call-like fluty sound is given in pairs, a calmly repeated *phew phew*, which sound carries furthest in the forest. Sometimes, apparently more regularly in C than N Europe, the song is more prolonged and varied. – The contact call is a short low-pitched, ocarina-like, often straight whistle, *phew* (or *teu*), but it can be downwards-inflected, *phew-uh*. This call is also used in alarm when an owl is spotted or when potential nest predators are around. As already stated, it is also frequently included in the song. There are several call variations, partly of a local nature, the best known being the peculiar so-called 'trumpet call', somewhat recalling that of Two-barred Crossbill, a cracked note with a nasal quality, *pjheeh*. The significance of this is still unknown, whether it is used by a little-studied race, or is a call variation that all populations are capable of giving. (It should be noted that proof exists that birds uttering the 'trumpet call' are inseparable from ssp. *pyrrhula* both in morphology and mtDNA.) Another secondary contact call, often heard from feeding flocks or from migrants, is a rather subdued clicking sound, *bt bt bt...* or *tet tet tet...*

SIMILAR SPECIES Should be impossible to confuse with any other species in the covered area if seen reasonably well. The ♀ could perhaps be mistaken for a Hawfinch in a poor and fleeting view, but note black bill and crown, and the white rump patch of Bullfinch. – Azores Bullfinch is very similar to ♀ Bullfinch but lacks white rump and vent, the rump being diffusely paler buff-brown. Also, it has a proportionately heavier bill. – For museum workers a real challenge is to separate ♀ of eastern race *cassinii* from Grey Bullfinch (*P. cineracea*, extralimital, not treated). Grey Bullfinch is on average subtly paler and cleaner grey, lacks the pink-buff, warm cast of *cassinii*, especially apparent on underparts. Grey Bullfinch also has narrower whitish wingbar (only narrow tips of some greater coverts whitish, bases greyish, whereas *cassinii* has broad white or whitish tips to many greater coverts), usually lacks any white on inner web of r6 (at most a narrow stripe in some) and has on average more extensive black on chin and at base of bill. Size of bill said to be smaller, but this difference, if true, is very minor and of limited use.

AGEING & SEXING (*pyrrhula*) Ages differ subtly in autumn and spring, but reliable ageing in the field requires close views. Sexes separable after post-juv moult. – Moults. Complete post-nuptial moult of ad in late summer into early autumn, mainly late Jul–early Oct, but late breeders will moult well into Nov. Partial post-juv moult at same time does not usually involve flight-feathers, longest tertial(s), tail-feathers, outer greater coverts, carpal covert, alula coverts or any primary-coverts. No pre-nuptial moult in winter. – **SPRING** ♂ Cheeks and underparts from throat to belly deep pinkish- or rosy-red. ♀ Cheeks and underparts from throat to belly buffish brown-grey or vinaceous drab-grey. Ageing as in autumn, only any unmoulted juv wing-coverts now tipped off-white, any buff or brown tinges bleached away. – **AUTUMN** Sexing straightforward as in spring. **Ad** All greater coverts of same generation, no moult limits. No buff tips to outer greater coverts. Carpal covert often dark grey with well-marked whitish edge/tip. Primary-coverts on average slightly broader with rounded tips and glossy surface. **1stW** Usually a slight moult limit between inner (commonly 3–5) moulted greater coverts and outer retained juv-feathers, latter subtly less blackish and glossy on basal part, and tips might show faint buff tinge to pale grey tips. Carpal covert nearly always unmoulted, often has faint buff tip, and is less contrasting dark grey with whitish edge/tip like ad, more diffusely grey-edged. Primary-coverts

P. p. pyrrhula, ad ♂, Finland, Mar: bull-necked, plump-bodied, and easily identified (and sexed) by black crown and bib, contrasting with grey upperparts and rosy-red cheeks and underparts. Also note white rump and dark bluish-black wings with single but broad whitish bar. Evenly-feathered wing infers age. (M. Varesvuo)

AZORES BULLFINCH
Pyrrhula murina Godman, 1866

Fr. – Bouvreuil des Açores; Ger. – Azorengimpel
Sp. – Camachuelo de las Azores
Swe. – Azorisk domherre

One of the rarest birds in the entire region is this relative of the Bullfinch that lives only on São Miguel in the Azores, where the estimated total population is fewer than 200 pairs. Although numbers have recently been reported to be slightly higher, it is still regarded as severely threatened. Formerly treated as a local subspecies of the common European Bullfinch, in the light of knowledge of its genetic distance compared to other Bullfinch populations, and in line with several similar cases of isolated and distinct taxa, the Azores Bullfinch is now generally treated as a separate species. Its favourite habitat is low laurel scrub on steep, inaccessible mountain slopes above 500 m, and the remaining birds are restricted to a tiny area.

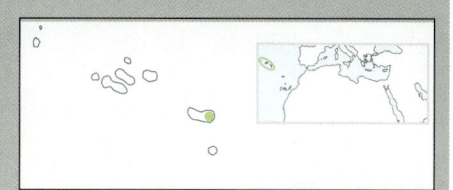

1stW, presumed ♂, Azores, Sep: aged by juv outer greater coverts and (already abraded) primaries, and less intense black head. Tentatively sexed by warmer underparts and in being relatively fresh at end of breeding season. (H. Shirihai)

Ad, presumed ♂, Azores, Oct: told from Common Bullfinch by lack of sexual dimorphism, as both sexes resemble ♀ of latter, and also by lack of white rump. Both sexes mostly drab grey, with cinnamon-buff cheeks and underparts, and steel-blue to purplish gloss to tail and wings. Evenly-fresh wing and rounded tail-tips indicate ad; presumably ♂ by darker and warmer underparts. (D. Occhiato)

IDENTIFICATION Differs from Bullfinch in *lack of white rump*, its *much deeper and heavier bill* and *darker, more uniform and warmer cinnamon-tinged grey-brown underparts* in both ♂ and ♀ plumage (sexes alike in specimens examined). Upperparts are dull earth-brown with somewhat paler buff-grey rump and pinkish-grey broad wing-bar. Wings and tail are black with strong bluish gloss. Wings are slightly shorter and more rounded than in Bullfinch, but general size is the same or even a little larger.

VOCALISATIONS There is one recording of claimed song (T. Linjama, Xeno-Canto), according to which it consists of sparsely emitted single whistling notes of varying type, a few seconds pause between each, *tee-uh…..whey….. tee-uh…..tee-uh…..weeo…*, etc., pensive and with desolate ring. – Calls are compared to Common Bullfinch more drawn-out and clearly bent (down-slurred or up-slurred), e.g. downwards-inflected whistling *pe-hew* or up-bent *tu-wee*. They usually sound plaintive and can vaguely recall such different species as Ringed Plover *Charadrius hiaticula*, Tawny Owl *Strix aluco* or Blue Jay *Cyanocitta cristata*.

SIMILAR SPECIES Differs from *Bullfinch* in heavier bill and by lacking the white rump patch and white vent. Resident on a remote island in the Azores, thus hardly a candidate for confusion with Bullfinch in Europe.

AGEING & SEXING After post-juv moult ages appear to be alike, but not sufficiently studied. Sexes alike or at least very similar. – Moults. Complete post-nuptial moult of ad apparently in late summer, mainly Aug–Oct (Ramos 1998), although two specimens in NHM offer conflicting evidence having just started primary moult in Apr and late May, respectively; the specimens are either both incorrectly dated or indicate a much more protracted or different moult period. Partial post-juv moult apparently does not involve flight-feathers, tail-feathers or primary-coverts. No pre-nuptial moult in late winter. – **SUMMER Ad** Cap and bill-surround black. Plumage neat and smooth. ♂ on average slightly darker and warmer brown below, not quite as pale and pinkish-grey as ♀, but many exceptions and extensive overlap makes sexing of single birds difficult. (♂ said to be slightly more warm brown, or even pink, on belly and flanks than ♀, but this appears to be a very vague tendency and, if confirmed, probably has little value in the field unless a breeding pair is seen together. Another claim is that ♂ sometimes has a buffish-orange tinge to upper breast and ear-coverts. This could be true, but there is much variation and overlap.) **Juv** Differs from later plumages in same ways as described for Bullfinch.

Presumed ♀, Azores, Sep: although sexes are very similar and usually inseparable, there is a tendency for ♀♀ to be slightly paler and less tinged pinkish-brown below than ♂♂, hence this bird might be a ♀. Wing seems neat and very dark, but ageing at this angle still best avoided.

BIOMETRICS **L** 15.5–17 cm; **W** ♂ 88.5–91.5 mm (*n* 16, *m* 89.5), ♀ 85–89 mm (*n* 10, *m* 87.6); **T** ♂ 63–72 mm (*n* 16, *m* 69.5), ♀ 64–71 mm (*n* 9, *m* 67.8); **T/W** *m* 77.4; **B**(f) 10.3–13.8 mm (*n* 28, *m* 12.1); **BD** 9.7–11.7 mm (*n* 28, *m* 10.6); **BW** 9.7–11.2 mm (*n* 28, *m* 10.5); **Ts** 20.0–22.8 mm (*n* 28, *m* 21.0). **Wing formula:** p1 minute; p2 3–6 mm < wt; **pp3–4** longest; p5 < wt 0–1.5 mm; p6 < wt 2.5–4 mm; p7 < wt 8–11 mm; p10 < wt 16.5–19.5 mm; s1 < wt 19–22 mm. Emarg. pp3–6. **White of r6** (on inner web) invariably absent. – Wing-length from a larger sample of live birds (Ramos 1998): **W** ♂ 86–93 mm (*n* 22, *m* 90.0), ♀ 85–92 mm (*n* 28, *m* 88.7).

GEOGRAPHICAL VARIATION & RANGE Monotypic. – São Miguel, Azores; resident.

REFERENCES Bibby, C. J., Charlton, T. D. & Ramos, J. A. (1992) *BB*, 85: 677–680. – Ramos, J. A. (1998) *Ring. & Migr.*, 19: 17–22. – Ramos, J. A. (2000) *Bull. Afr. Bird Club*, 7: 31–33.

Presumed ad ♂, Azores, Aug: bill relatively longer and heavier than in Common Bullfinch. Combination of darker and warmer tinged underparts, rather evenly-fresh wing and tail, and rounded tips to tail-feathers indicate ad ♂ (or young bird that has moulted completely, a pattern as yet unknown in the species). (R. Lowe)

1stS ♀ (right) and juv, Azores, Oct: at end of breeding season ♀ has more worn plumage than ♂, with face and cap heavily abraded, and whiter underparts. Based on moult pattern in wings (noted in the field) this ♀ was aged as 1stS. Juv generally looks fluffy, with buff feathers, except dusky crown-sides and paler bill. (H. Shirihai)

Juv, Azores, Aug: unlike ad, juv lacks black cap and face, and has more fluffy plumage. (R. Lowe)

P. p. pyrrhula, ♀, presumed ad, Finland, Mar: same size, shape and general pattern as ♂ but ash-grey upperparts replaced by drab-grey, and pink-red underparts by dull pinkish-grey. Also, black parts are slightly less glossy, though not a big difference. Wings evenly feathered inferring age, though ageing without handling always difficult. (M. Varesvuo)

P. p. pyrrhula, 1stW ♂, Sweden, Jan: a young bird now in ad-like plumage, differing from ad only in subtly paler greyish primary-coverts and the odd unmoulted juv outer greater coverts being basally greyer and having more diffusely marked whitish tips. (L. Petersson)

on average slightly narrower with more pointed tips in some, and often have slightly less gloss. Still, some so similar that ageing becomes impossible. **Juv** Differs clearly from later plumages by lack of black cap and bill surround, and by having all wing-coverts buff-tipped. Sexes alike.

BIOMETRICS (*pyrrhula*) **L** 15–17.5 cm; **W** ♂ 84.5–99 mm (*n* 151, *m* 93.8), ♀ 86–96.5 mm (*n* 63, *m* 91.8); **T** ♂ 64–75 mm (*n* 151, *m* 69.1), ♀ 62–73 mm (*n* 63, *m* 67.7); **T/W** *m* 73.7; **B**(f) 9.0–11.9 mm (*n* 159, *m* 10.5); **BD** 8.7–11.4 mm (*n* 148, *m* 9.9); **BW** 9.2–11.5 mm (*n* 159, *m* 10.3); **Ts** 16.7–19.0 mm (*n* 140, *m* 17.9). **Wing formula: p1** minute; **p2** 2.5–5 mm < wt; **pp3–5** about equal and longest (though either of p3 and p5 to 2 mm <); **p6** < wt 4–7 mm; **p7** < wt 11–15.5 mm; **p10** < wt 20–25.5 mm; **s1** < wt 21–28 mm. Emarg. pp3–6. **Length of white rump** usually 15–25 mm (extremes 13–30 mm); **white of r6** (along shaft on inner web) generally absent, but 17% had 5–23 mm, including apparent breeders of Fenno-Scandia, Hungary, Germany.

GEOGRAPHICAL VARIATION & RANGE Moderate variation within treated region, mainly affecting size and saturation of colours, whereas extralimital variation in E Asia is quite complex (not covered here). Several described subspecies proved difficult to confirm using examined material and are thus only listed as synonyms. All populations mainly sedentary, but depending on food supplies make irregular irruptions or nomadic movements in winter, mainly to west and south. Intermediate populations link more distinct subspecies, thus large areas inhabited by birds showing mixed or less distinct traits.

P. p. pyrrhula (L., 1758) (Fenno-Scandia, Baltic States, Poland except south-west, Belarus, Slovakia, Austria, Balkans south to N Bulgaria, east through Ukraine and Russia to C Siberia, NW Mongolia). Described above. Rather large with pure colours, in ♂ rosy-red underparts and pure ash-grey back, in ♀ buffish-grey underparts. Wing-bar broad (5–12 mm) and grey-white to nearly white. Outer web of short tertial has some rust-red on outer web. Populations in W Poland, much of E Europe and Balkans tend towards *europaea* in being slightly smaller and duller, but are generally closest to *pyrrhula*. Birds in C Siberia and NW Mongolia tend to be slightly smaller. (Syn. *germanica*; *macedonica*; *major*; *rubicilla*; *vulgaris*.)

P. p. europaea Vieillot, 1816 (British Isles, Netherlands, Germany, SW Poland, west of *pyrrhula* and south to France and Italy; perhaps N & W Switzerland). Resembles *pyrrhula* but markedly smaller (by *c.* 11%) and darker, ♂ having slightly duller red below, ♀ subtly more olive-brown above (many are inseparable). Birds in Britain and Ireland are very slightly duller than continental birds and more tinged olive-brown, but difference very subtle and far from consistent, and more than half of compared birds of the two appeared identical. Note that darker and duller colours of specimens can to some extent be explained by staining (prevailing coal-heating in W Europe during times of collection?) as cleaning a dull British specimen (in NHM) produced lighter and brighter colours. Birds in Denmark, N Germany and SW Poland tend towards *pyrrhula* being subtly larger and brighter, but are generally closest to *europaea*, and hence included therein. Birds in S France (Gers and northern slopes of Pyrenees) tend towards *iberiae*, having slightly brighter colours. **L** 14–15.5 cm; **W** ♂ 77–88 mm (*n* 52, *m* 83.0), ♀ 78–91.5 mm (*n* 43, *m* 81.8); **T** ♂ 57–67 mm (*n* 52, *m* 61.7), ♀ 55–67 mm (*n* 43 *m* 60.6); **B**(f) 8.4–10.5 mm (*n* 71, *m* 9.3); **BD** 8.3–10.5 mm (*n* 67, *m* 9.2); **BW** 8.5–10.3 mm (*n* 71, *m* 9.2); **Ts** 15.5–18.5 mm (*n* 68, *m* 16.9). **Length of white rump** usually 11–18 mm (extremes 6–22 mm); **white of r6** (along shaft on inner web) generally absent, but 12% had 8–15 mm. (Syn. *coccinea*; *hauseri*; *minor*; *nesa*; *orientalis*; *pileata*; *pusilla*; *rufa*; *vieilloti*; *wardlawi*.)

P. p. iberiae Voous, 1952 (N Iberia including Pyrenees). Has same small size as *europaea*, but differs in having clearer, less dull colours, more saturated red in ♂♂, slightly deeper red even than most *pyrrhula* (but some are similar). The red is uniformly strong down to lower belly (in both *pyrrhula* and *europaea* often slightly paler and less evenly strong on belly). Grey of upperparts in ♂ subtly paler and purer grey than *europaea* (but requires direct comparison of series). Birds on N slopes of Pyrenees and in adjacent SW France are

P. p. europaea, ad ♂, England, Apr: the W European race tends to be subtly darker grey above than *pyrrhula*, and underparts slightly darker and more pink-tinged, while the wing-bar is more tinged light grey, less whitish. Note glossy black crown and surround to bill. Age based on very black primary-coverts and neat greater coverts with broad well-marked grey-white tips. (B. Nield)

P. p. europaea, ad ♀, England, Mar: much duller and browner than ♂, being brown-grey above, dull pinkish-drab below, and has black parts less glossy. Wings evenly feathered inferring age. (B. Nield)

P. p. europaea, 1stW ♀, England, Dec: at late stage of post-juv moult with renewed tertials and median and most inner greater coverts, thus except primaries, primary-coverts and most secondaries, and is now closer to ad ♀ in appearance. (B. Garrett)

P. p. europaea, juv, England, Jul: unlike ad lacks black cap and bill surround, rump is buff-tinged and upperparts brownish-buff, with buff (not white) wing-bars. (S. Round)

intermediate between this race and *europaea*. **L** 13.5–15 cm; **W** ♂ 80–84 mm (*n* 15, *m* 82.3), ♀ 77.5–83 mm (*n* 8, *m* 79.9); **T** ♂ 58–63 mm (*n* 15, *m* 61.3), ♀ 55–62 mm (*n* 8, *m* 59.4); **B**(f) 8.6–10.8 mm (*n* 22, *m* 9.6); **BD** 8.3–9.6 mm (*n* 23, *m* 8.8); **BW** 8.7–9.6 mm (*n* 23, *m* 9.2); **Ts** 15.4–17.6 mm (*n* 23, *m* 16.9). **Length of white rump** usually 10–16 mm (extremes 8–21 mm); **white of r6** (along shaft on inner web) invariably absent.

○ ***P. p. rossikowi*** Derjugin & Bianchi, 1900 (N Turkey, Caucasus and adjacent areas on both sides, in north-west to SE Ukraine, in south-east to NW Iran). Medium-sized, slightly smaller than *pyrrhula*. ♂ is bright red below, similar to *pyrrhula*, but on average perhaps a trifle darker red, but many are inseparable. Some ♂♂ have a little red admixed in grey of back, and reddish tinge to base of outer webs of inner greater coverts. Said to have broader bill-base, and in ♂ paler grey back than *pyrrhula*, but neither of these claims could be confirmed. A very subtle race that could be lumped with *pyrrhula*. **W** 87–95 mm (*n* 15, *m* 90.8); **T** 64–71 mm (*n* 15, *m* 67.3); **B**(f) 10.0–11.8 mm (*n* 15, *m* 10.8); **BD** 9.5–11.5 mm (*n* 15, *m* 10.2); **BW** 9.4–11.1 mm (*n* 15, *m* 10.3). **Length of white rump** 14–25 mm; **white of r6** (along shaft on inner web) generally absent, but one from Jan had 21 mm. – Breeders in Mazandaran and on adjacent S Caspian coast ('*caspica*') claimed to be brighter red below in ♂♂ and a little darker grey on back, but the scant material available for examination did not differ from *rossikowi*. – We follow Kirwan (2006) in not recognising a separate race ('*paphlagoniae*') in NW Turkey. Birds from this area may have on average subtly smaller size and paler colours, but differences seem inconsistent and very small. (Syn. *caspica*; *paphlagoniae*.)

Extralimital: ***P. p. cassinii*** S. F. Baird, 1869 (Kamchatka, N Kurils, coast of Sea Okhotsk, Sakhalin, Ussuriland; possibly also E Siberia, from Irkutsk and Yeniseisk and east, but westerly birds might be migrants or intergrades with *pyrrhula*). Same size as *pyrrhula*, but differs in nearly always having a prominent white oblong patch on inner web of r6. Also has on average a larger white rump patch, on average broader greyish-white wingbar (8–14 mm), though much overlap. Shortest tertial in ♂ has grey outer web, or only faint rufous tinge. ♂ has on average subtly paler and more pinkish-red cheeks and throat, and paler and purer grey back than *pyrrhula* (but some are very close or similar). ♀ somewhat paler, especially on upperparts. Back is also purer grey, having less brown cast. **L** 16–17.5 cm; **W** ♂ 90.5–96.5 mm (*n* 11, *m* 95.1), ♀ 87–96 mm (*n* 12, *m* 90.7); **T** ♂ 67–76.5 mm (*n* 11, *m* 71.8), ♀ 65.5–71 mm (*n* 12, *m* 69.1); **T/W** *m* 75.5; **B**(f) 8.7–11.5 mm (*n* 23, *m* 10.2); **BD** 8.7–11.5 mm (*n* 22, *m* 10.1); **BW** 8.8–11.2 mm (*n* 23,

P. p. iberiae, ad ♂, Spain, Aug: a rather subtle race, with ♂ differing only in being slightly paler grey above and having a stronger and more even red hue below, from throat to belly, thus not becoming subtly paler on belly. All wing-feathers ad. (C. N. G. Bocos)

m 10.1); **Ts** 16.5–19.0 mm (*n* 23, *m* 18.2). **Length of white rump** usually 26–33 mm (extremes 20–39 mm); **white of r6** (along shaft on inner web) 14–33 mm in 92%. (Syn. *exorientis*; *kamtchatica*; *rosacea*.)

REFERENCES Fox, T. (2006) *BB*, 99: 370–371. – Kirwan,

Presumed *P. p. iberiae*, 1stW ♀, Spain, Jan: based on locality this should be the Iberian race, but since it is very similar to W European *europaea*, and some local winter movements cannot be ruled out, it is best to avoid a certain race label. Note faintly buff-tipped retained juv outer greater coverts revealing the age. (C. N. G. Bocos)

G. M. (2006) *Sandgrouse*, 28: 10–21. – Newton, I. (1966) *BB*, 59: 89–100. – Newton, I. (1966) *Ibis*, 108: 41–67. – Pennington, M. G. & Meek, E. R. (2006) *BB*, 99: 2–24. – Roselaar, C. S. (1993) *DB*, 15: 258–262. – Vaurie, C. (1949) *Amer. Mus. Novit.*, 1424.

P. p. rossikowi, ad ♂, Turkey, Mar: very similar to *pyrrhula* and doubtfully separable in the field except perhaps that wing-bar is grey-tinged like in *europaea*. On average is slightly smaller and has slightly deeper red underparts in ♂ plumage. All wing-feathers ad. (H. Kahraman)

WHITE-WINGED GROSBEAK
Mycerobas carnipes (Hodgson, 1836)

Fr. – Grosbec à ailes blanches
Ger. – Wacholderkernbeisser
Sp. – Picogordo aliblanco; Swe. – Tujastenknäck

A large Asian finch which, within the treated region breeds only in N Iran, but is widely distributed across Central Asia and reaches well into China. The White-winged Grosbeak breeds in mountains and on high plateaux with open ground and bushes, often junipers or related species. Despite its large size, the species often remains hidden in low scrub and undergrowth while feeding. It is mainly resident but will descend in winter from its breeding quarters, which are near the tree-line or a little above, often at 3000 m altitude.

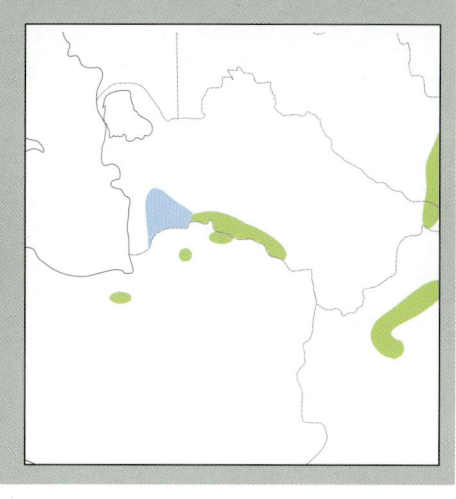

Ad ♂, Kazakhstan, Feb: a large-headed finch with a Hawfinch-like bill. ♂ has dark slate-grey head, upperparts and breast, as well as much of wing and tail, but rump, flanks and belly to undertail-coverts dull mustard-yellow. Wing pattern diagnostic, especially white primary patch, mustard-yellow tips to tertials and inner greater coverts bar. (A. Isabekov)

IDENTIFICATION A *large, long-tailed finch* with a *heavy bill* that is gregarious and often encountered in small parties in open alpine habitats. Often found in low junipers and other scrub, feeding low down and out of sight, and only rarely perches in tall trees or in dense forest. The ♂ is *blackish or dark slate on head, breast, upper belly and back*, and *mustard-yellow on lower belly and rump*. Tertials are partly yellow, too, with a *yellowish wing-bar*. A *white patch on primary bases* is part of the typical wing pattern. The ♀ is similar, but much duller with less clean colours, lacking clear demarcation between greyish breast and yellowish-grey belly, and has *diffuse whitish spotting on throat and breast*, whereas *cheeks are diffusely striped* off-white and dark grey. The *massive bill is steel-grey* with blackish tip in both sexes; the short *legs are pinkish*. Juvenile resembles ♀ but is tinged brownish, and thin pale feather-tips create a scalloped effect on dark parts. Both greater and median coverts tipped pale. Flight strong and slightly undulating, sometimes with long glides when descending to lower levels. Due to long tail and large white primary patches can briefly recall a shrike.

VOCALISATIONS Song poorly studied but supposed to comprise well-spaced call notes, often uttered singly, *kwuee*, but at times quickly doubled, *kwueeh-kwe*, and sometimes a subdued chatter without any clear pattern (subsong?). – Calls include a song-like disyllabic, Canary-like *kwuee*, Magpie-like chatters and squeaky notes, e.g. a nasal note running into a short rattle, *kwee-cha-cha*. In flight, but also from perched birds, often gives a squeaky, drawn-out note with slightly nasal and strained voice, sometimes rendered *schweeup*.

SIMILAR SPECIES Can hardly be confused with any other species in the treated region, but extralimitally in the Himalayas and SE Asia a closely related twin species occurs, the *Spot-winged Grosbeak* (*M. melanozanthos*; not treated), which differs in smaller white primary patch (just a spot), black instead of yellow rump and less extensive black on breast, covering mainly throat and upper breast rather than reaching onto belly.

AGEING & SEXING Ages similar after post-juv moult but often separable if seen well, at least in autumn. Sexes differ after post-juv moult (about late Aug–Oct). – Moults. Complete post-nuptial moult of ad in (Jul) Aug–Oct (Nov). Partial post-juv moult in late summer or early autumn does not involve flight-feathers or primary-coverts. Greater coverts are apparently completely replaced as a rule, but pattern of 1stW greater coverts intermediate between juv and ad types (full width of yellowish-green tips not acquired until after next moult) making it difficult to age 1stW using wing-bar alone. No pre-nuptial moult. – **SPRING** ♂ Head, breast, upper belly, mantle and scapulars sooty-black. Lower back/rump greenish-yellow. Wing-feathers blackish. ♀ Head, breast, upper belly, mantle and scapulars dark grey, not black-looking. Cheeks irregularly striped dirty white and dark grey. Lower back/rump dull greenish-yellow, less bright than ♂. Wing-feathers dark olive-grey. Less obvious contrast between sooty-black breast/upper belly and yellowish lower belly than ♂, often diffuse border between the two, and yellow tinge sullied grey and subdued. – **AUTUMN Ad** Freshly moulted. Tail-feathers fresh and rather broad, and tips rather rounded. Many greater coverts broadly tipped yellowish-green. Median coverts and outer greater coverts lack pale tips. No scalloping created by pale tips to dark feathers. Sexes differ as in spring. **1stW** As ad but tail-feathers narrower and tips more pointed, worn from mid autumn. Often juv greater coverts kept in autumn, recognised by narrower and less green tips than ad. **Juv** Resembles ad ♀ but slightly duller grey above, even tinged brownish, with faint, thin whitish feather-tips creating scalloped pattern. Double wing-bars formed by olive-yellow to off-white tips to median and greater coverts.

BIOMETRICS L 20.5–22 cm; **W** ♂ 117.5–123 mm (*n* 14, *m* 121.3), ♀ 111–124.5 mm (*n* 13, *m* 117.9); **T** ♂

Ad ♂, Kazakhstan, Dec: when very fresh, slate-coloured areas often washed paler greyish, and yellow sometimes brighter. Occasionally the scapulars are also tipped yellowish, as here, and bill can be partly pinkish-tinged. (A. Isabekov)

90–100 mm (*n* 14, *m* 95.1), ♀ 87–98 mm (*n* 13, *m* 92.5); **T/W** *m* 78.4; **B** 19.0–23.1 mm (*n* 27, *m* 21.6); **B**(f) 15.8–18.7 mm (*n* 26, *m* 17.3); **BD** 14.5–17.3 mm (*n* 27, *m* 16.1); **Ts** 23.6–27.0 mm (*n* 27, *m* 25.8). **Wing formula: p1** minute; **p2** < wt 2–6; **pp3–4** longest; **p5** < wt 0–3 mm; **p6** < wt 5–9 mm; **p7** < wt 14–19 mm; **p10** < wt 27–33.5 mm; **s1** < wt 29–37 mm. Emarg. pp3–6.

GEOGRAPHICAL VARIATION & RANGE Monotypic. – N & SE Iran, mountains in Central Asia from W Pakistan and N Afghanistan via Tien Shan to SE Kazakhstan, Kopet Dag in S Turkmenistan; sedentary but can be nomadic or move to lower altitudes. – According to Vaurie (1959), birds in N Iran, S Turkmenistan east to Baluchistan and Afghanistan, '*speculigerus*', differ subtly from *carnipes* of further north and east in Central Asia in being less dark black above in ♂♂, more tinged slaty, and a trifle smaller on average. However, we fail to detect any differences between birds from Baluchistan and those further east and north-east, being of same darkness in ♂ plumage, and of same size. (Syn. *speculigerus*.)

Ad ♀, Kazakhstan, Feb: unlike ♂, dark slate-grey replaced by paler olive-grey, variably mottled whitish, with duller and more restricted yellow in wing and on lower belly. ♀ is easily confused with young ♂, but here evenly-fresh ad wing helps confirm the sex. (A. Vilyayev)

1stW ♂, Kyrgyzstan, Jul: young ♂ gradually attains dark slate-coloured feathers, and, especially if it retains contrastingly worn and browner wing-feathers (here alula, primary-coverts and tertials), they can be aged and sexed with confidence. (T. Lindroos)

1stY ♀, Kazakhstan, Feb: as ad ♀, but differs in having largely juv wing, with contrastingly new inner greater coverts having ad ♀-like pattern, inferring age and sex. Note relatively small white primary patch. (A. Isabekov)

HAWFINCH
Coccothraustes coccothraustes (L., 1758)

Fr. – Grosbec casse-noyaux; Ger. – Kernbeisser
Sp. – Picogordo común; Swe. – Stenknäck

The Hawfinch is the most distinctive of European passerines in having such a disproportionately heavy, conical bill in relation to its body size. The bill is an effective, powerful tool to crush seeds of cherries and other fruit-bearing trees, in order to get at the kernel, but the Hawfinch also feeds on larger insects in summer. Most of the time, it leads a secretive and rather quiet life in small family parties high in the foliage of tall trees (often beech, elm and oak). At times it descends to the ground to feed on seeds or insects. Often noticed only by its sharp clicking call that sounds like an amplified Robin. winter and in young birds. Usually keeps to upper foliage of tall deciduous trees, suddenly dashing off in fast, undulating flight to nearby trees. Often keeps still for longer periods, thus difficult to spot (unless calls are heard). Juvenile clearly differs in being barred or blotched dark on more dusky yellowish-white ground.

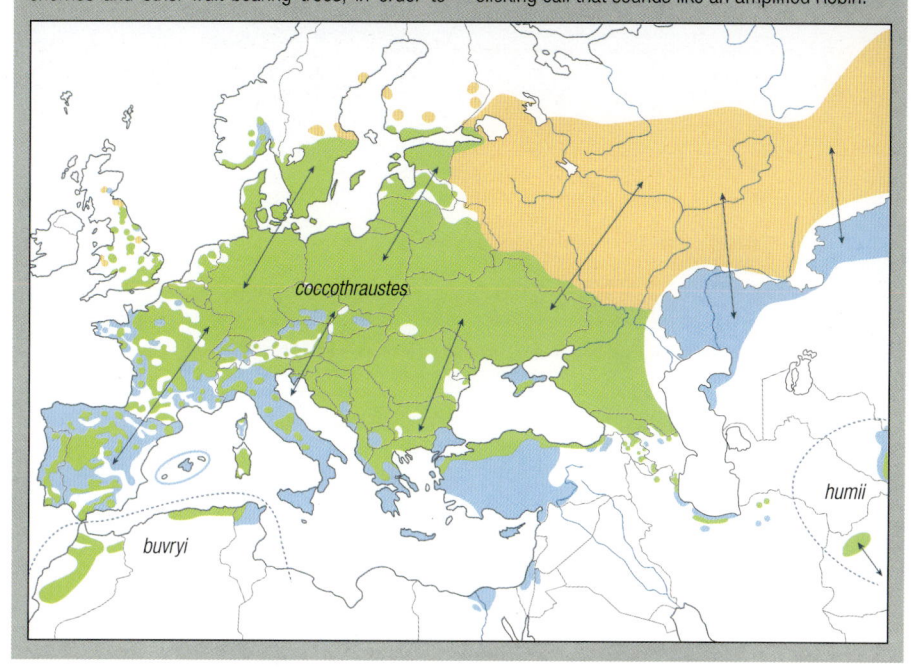

VOCALISATIONS Song discreet and easily overlooked as calls of various other birds or song of Spotted Flycatcher, a series of thin, sharp notes at slow, uneven pace, containing both straight *ziih* and buzzing *zrri*. – Most characteristic call is a very sharp and explosive, clicking *pix!*, most similar to single clicking call of Robin, only much fiercer and louder. Often uttered in flight, sometimes one call per undulation. Other calls include a fine, thin, sharp *ziih* and a more buzzing or rolling *zrri*, both calls like notes from the song.

SIMILAR SPECIES Impossible to confuse with other species if seen reasonably close and well, but if seen from below among leaves and branches could be mistaken for a *Waxwing* as both have rufous-buff or pinkish-brown underparts, a black bib and pale band on tail-tip. However, once the bill is seen or the white wing-bar, the lack of a crest, etc., the Hawfinch is easily identified.

AGEING & SEXING (*coccothraustes*) Ages very similar after post-juv moult, but at least some ♂♂ differ in 1stY, or at least in 1stW. Sexes differ on fledging (or even before). – Moults. Complete post-nuptial moult of ad mainly in Jul–Oct. Partial post-juv moult in summer or early autumn does not involve flight-feathers or primary-coverts. Many moult r1 and a few other tail-feathers, rarely all. All greater coverts moulted, often all tertials as well. No pre-nuptial moult. – **SPRING** ♂ All flight-feathers black with purple or blue gloss to outer webs (sometimes less glossed on outer 1–2 long primaries). Hindcrown darker and more rufous-tinged tawny (forecrown being paler, more ochre). Rump, uppertail-coverts and cheeks warmer rufous. Back darker earth-brown. Underparts greyish-buff with some vinous-pink tinge. Ageing of ♂♂ sometimes possible; see autumn, although wear progressively makes moult differences less visible. ♀ Outer webs of secondaries, and partly on some inner primaries, ash-grey without any gloss. Whole crown rather uniformly coloured, greyish-brown or brown with faint olive tinge (rufous element lacking or much subdued). Rump, uppertail-coverts and cheeks rather uniform with crown, greyish-ochre or dull grey-brown without rufous. Back somewhat paler earth-brown than ♂. Underparts dusky-buff, lacking pink

IDENTIFICATION A stocky, *large-headed, heavy-billed* and *short-tailed* finch, slightly smaller than a Bullfinch. The size of the head and massive *conical bill* in combination with the mainly *buff-brown and rufous colours* instantly identify the species, while the *broad white wing-bar* and *short, broadly white-tipped tail* further confirm it. Adults have a *black bib* on chin/upper throat and *black lores*. Inner primaries are blue-glossed black with *peculiarly extended, wedge-shaped* or 'club-shaped' *tips*. ♂ has all wing-feathers edged glossy black, whereas the ♀ has pale grey outer webs to the secondaries forming *a pale panel on folded wing*. Nape is greyish in contrast to the dark rufous-brown mantle. Bill in breeding adults is bluish-black with paler blue-grey base and cutting edges, whereas it is more yellowish-brown in

C. c. coccothraustes, ad ♂, Sweden, May: thick-headed, short-tailed finch with massive, conical bill, with neat black margin around bill and eye. Note variable grey half-collar, greyish-vinaceous underparts with white undertail-coverts, and tail tipped white. Wing evenly feathered, while black secondary panel and large 'club-heads' show it to be ♂. (D. Pettersson)

C. c. coccothraustes, ad ♂, Spain: hindcrown and cheeks deep tawny-rufous with paler forehead. Remiges black glossed purple or blue, back earth-brown and rump warmer rufous, along with pale brown underparts tinged vinaceous-pink. Note blue-glossed 'club-shaped' tips to inner primaries, otherwise black secondary panel and evenly-feathered wing infer age and sex. (C. N. G. Bocos)

hue. Ageing of ♀♀ appears very difficult if even possible.
– AUTUMN Sexing same as in spring. **Ad** ♂ Whitish greater and median coverts tinged rufous, densely textured, edges neat. Secondaries broadly edged with purplish gloss, and outer webs have wedge-shaped extensions distally. **1stW** ♂ Whitish greater and median coverts tipped yellowish-ochre, rather loosely textured. Outer webs of secondaries variable as to degree of gloss, some with narrow rim and others without. Outer webs of secondaries have rounded tip or only slight hint of wedge-shaped extension. Ageing of ♀♀ appears very difficult if possible, limited to few birds with moult contrast among greater coverts, alula or tail-feathers. **Juv** Heavily barred or blotched dark below on dusky-white ground. Throat and much of head tinged yellowish.

BIOMETRICS (*coccothraustes*) **L** 16.5–18.5 cm. Size difference between sexes slight: **W** ♂ 98–109 mm (n 32, m 104.6), ♀ 97–108 mm (n 26, m 102.2); **T** ♂ 51–58 mm (n 31, m 55.1), ♀ 50.5–57 mm (n 26, m 53.3); **T/W** m 52.5; **B**(f) 17.3–21.7 mm (n 53, m 19.8); **BD** 13.4–17.4 mm (n 53, m 16.0); **Ts** 20.0–23.0 mm (n 58, m 21.4). **Wing formula: p1** minute; **pp2–4** about equal and longest (p2 sometimes to 2 mm <); **p5** < wt 3–5 mm; **p6** < wt 13–17 mm; **p7** < wt 19–22.5 mm; **p10** < wt 30–36 mm; **s1** < wt 32–38 mm. Emarg. pp3–5. **Width of extended tip outer web p7** 4.5–9.0 mm. **White on r6** (inner web) ♂ 18–28.5 mm, ♀ 14–21 mm.

GEOGRAPHICAL VARIATION & RANGE Mostly very slight and clinal variation, mainly affecting saturation of colours, but also size in south-west of range. The described races are both very similar and in a few cases doubtfully separable, and are connected by wide zones of intergradation. Some of these taxa could be lumped, as proposed here. All populations mainly sedentary, but some northern populations make short-range winter movements.

C. c. coccothraustes (L., 1758) (Europe, much of Siberia east to Amur, NW Turkey, Caucasus, Transcaucasia, N Iran, wooded parts of N Central Asia). Described above. Rather saturated colours. – Birds in Crimea, Caucasus east to N Iran ('*nigricans*') have perhaps on average a slightly shorter primary projection and are said to be slightly darker on mantle and more tinged pink below, but if true certainly only a very minor tendency and difficult to confirm in studied samples. Birds in Siberia ('*verticalis*') tend to have slightly heavier bill, but difference clinal and very minor. (Syn. *boehmei*; *insularis*; *loennbergi*; *nigricans*; *schulpini*; *tatjanae*; *verticalis*.)

C. c. buvryi Cabanis, 1862 (NW Africa). Differs in being smaller and having somewhat paler and duller colours, and greyer rump (subtle). **L** 15–17 cm. **W** ♂ 96–103 mm (n 17, m 98.8), ♀ 94.5–100 mm (n 21, m 97.1); **T** ♂ 51–56 mm (n 17, m 53.4), ♀ 49–55 mm (n 21, m 54.0); **T/W** m 54.0; **B**(f) 16.0–20.7 mm (n 38, m 18.7); **BD** 13.2–15.9 mm (n 38, m 14.4); **Ts** 19.5–23.4 mm (n 38, m 20.9). **Wing**

C. c. coccothraustes, ad ♀, Germany, Feb: duller than ♂, with less glossy black tracts and paler rufous and brown ones but, most importantly, tips to secondaries and terminal half of outer web of primaries are contrastingly ash-grey, forming a broad pale panel. Only ad wing-feathers visible. (M. Schäf)

C. c. coccothraustes, in flight, Jul, Germany: note wing pattern and general jizz that secure identification in flight, even if diagnostic calls are not heard. All three images possibly of the same bird, but at least the left hand one shows a ♀, whereas the other two cannot be sexed. (M. Putze)

C. c. coccothraustes, 1stW ♂, Hungary, Jul: combination of being ♂ (dark secondary panel, black edges to primaries) with duller, washed-out plumage, juv primary-coverts, alula and tertials, greyish iris and pinker bill, confirm age. (M. Varesvuo)

formula: **p10** < wt 29–31 mm; **s1** < wt 30.5–35 mm. White on **r6** (inner web) ♂ 20–25 mm, ♀ 13.5–20 mm. (Syn. *theresae*.)

Extralimital races: **C. c. humii** Sharpe, 1886 (Central Asia in N Afghanistan, SE Uzbekistan north to Dzungaria). Large like *coccothraustes* but decidedly paler overall and slightly more yellow-tinged ochre, less pinkish-grey. Doubtful if it ever straggles to our region.

C. c. japonicus Temminck & Schlegel, 1848 (Sakhalin, Japan). Similar to *coccothraustes* but averages slightly paler on back and head, and white tail-tip a little narrower, usually 15–20 mm.

REFERENCES Fornasari, L., Pianezza, F. & Carabella, M. (1994) *Ring. & Migr.*, 15: 50–55. – Herremans, M. L. J. (1990) *Ring. & Migr.*, 11: 86–89.

C. c. coccothraustes, 1stW ♀, Jan: dull plumage and greyish secondary panel indicate ♀, while seemingly juv primary-coverts and certainly juv alula, and greyish-tinged iris confirm age. (C. N. G. Bocos)

C. c. coccothraustes, 1stS ♀, Mongolia, May: a particularly dull and greyish-tinged bird. Pale grey secondary panel and edges to primaries confirm sex, while bleached brownish juv alula suggests age. (H. Shirihai)

C. c. coccothraustes, juv, Hungary, Jul: easily differentiated from ad by yellowish-buff head, with contrasting black bars, spots or scaly patches on chest and flanks. (M. Varesvuo)

C. c. buvryi, ♂, presumed ad, Morocco, April: breeders in NW Africa differ only subtly in being somewhat smaller and slightly duller and greyer. Sexed by glossy blue-black flight-feathers, primary tips with 'club-head extensions', while seemingly uniform ad-type wing infers age. (M. Buckland)

C. c. buvryi, ♀, Morocco, April: slightly duller and greyer than ♀ *coccothraustes* but sharing with that race the grey secondary panel, readily indicating sex. Ageing of ♀♀ without handling extremely difficult. (M. Buckland)

OVENBIRD
Seiurus aurocapilla (L., 1766)

Fr. – Paruline couronnée; Ger. – Pieperwaldsänger
Sp. – Reinita hornera
Swe. – Rödkronad piplärksångare

This North American ground-dwelling thrush-like warbler is an extremely rare vagrant to Britain and Ireland, where all records are in late autumn or early winter. Its name pertains to the shape of its nest, which to some recalls that of a Dutch oven. But Ovenbirds in North America should not be confused with a group of sedentary birds living in Middle and South America, which as a group are also called ovenbirds. The Ovenbird breeds in tall deciduous woodland mainly in Canada and winters in Mexico, Central America, the West Indies and N South America. Look out for a migrant even in urban areas—it is e.g. commonly seen in Central Park, New York City. Or heard; listen for its rhythmic *tea-cher tea-cher tea-cher*.

IDENTIFICATION A readily identified Nearctic passerine due to its combination of *plump body, orange-rufous median crown-stripe* bordered by black, *unbroken buff-white eye-ring* making it look somewhat startled, olive upperparts and *very clean white underparts heavily streaked black on chest and flanks*, with a sometimes well-marked lateral throat-stripe. Feeds unobtrusively on ground, usually within cover, *moving with very deliberate, but continuous steps* recalling a pipit.

VOCALISATIONS Song (sometimes heard from migrants) is a loud, characteristic *tea-cher tea-cher tea-cher tea-cher*…, increasing in loudness towards end. – Common call is a dry *chip* given from ground. Flight-call described as 'a short, high piercing, rising *seek*' (Sibley 2000).

SIMILAR SPECIES If seen well, can only be confused with a waterthrush, but rather easily separated from any species by presence of orange crown patch (though not always easy to see in the field), presence of white eye-ring but lack of supercilium and eye-stripe, and overall very different structure and jizz, being much squatter and plumper, less attenuated, with striking habit of cocking and flicking tail deliberately upwards (quite unlike waterthrushes) while walking, which is a continuous and slightly strutting 'stroll'.

AGEING & SEXING Aged reliably only by close examination of moult pattern and feather wear in wing. No sexual dimorphism, and seasonal plumage variation very limited. – Moults. Complete post-nuptial moult of ad and partial post-juv moult mostly in Jun–Aug (Sep). Post-juv moult includes head, body, all median and usually all greater coverts (but no remiges or alula). Pre-nuptial moult absent. – SPRING **Ad** May show pale yellow tone to underparts, and crown patch more obvious than in autumn/winter. **1stS** Indistinguishable from ad in the field, and separation in the hand not easy (moult limits as in autumn but retained juv remiges, primary-coverts and rectrices have even more abraded tips). – AUTUMN **Ad** Much as in spring (now evenly fresh), but stronger olive wash to upperparts, with buff or tawny wash to neck-sides, ear-coverts and flanks in some. Underparts streaking may appear broader but more diffuse than in spring, and grey feather-tips obscure crown stripes. **1stW** Much as ad, but when still fresh has rather noticeable rusty tips to tertials; also check for retained juv primary-coverts, primaries and tail-feathers (narrower, more pointed, less fresh and duller, with more distinct rusty-buff fringes, i.e. with

Presumed *S. a. aurocapilla*, 1stS, Cuba, Mar: highly terrestrial, plump-bodied parulid. White eye-ring, orange-rufous median crown-stripe bordered black, olive upperparts and heavily streaked black-on-white underparts identify the species. Two outermost greater coverts are buff-tipped (retained juv), confirming age. (H. Shirihai)

S. a. aurocapilla, USA, Apr: faint yellow tinge to breast and vivid rufous median crown-stripe suggestive of ad, but primary-coverts seem slightly browner and more worn than greater coverts indicative of 1stS, and knowing there is much overlap between ages in spring this bird is best left unaged. The difference in length among greater coverts, with inner ones seemingly longer, is a 'false moult limit', as such irregularities occur at all ages. (D. Monticelli)

S. a. aurocapilla, 1stW, Azores: often walks with very deliberate but continuous steps, even over bare ground, displaying characteristic squat body, square-shaped head, and short but frequently cocked tail. Note pale fringes to juv outer median and greater coverts. Primary-coverts are brown and rather pointed. (V. Legrand)

S. a. aurocapilla, 1stW, Azores: same bird as image to the left and besides pale-tipped tertials (highly indicative of juv feathers), the primary-coverts and alula are much browner than more olive greater coverts, and primary tips appear more pointed. (V. Legrand)

S. a. furvior, Newfoundland, E Canada, Jun: this race is also a potential vagrant to the treated region since it breeds far east in North America. Similar to *aurocapilla* but differs by on average slightly bolder underparts streaking and more brown-tinged, less greenish upperparts. (J. Clarke)

less diffuse olive fringes of ad). Moult limits best seen against renewed ad-like greater coverts; r6 further tapered and lacks or has indistinct pale edges at tip (clearer and whiter tip in ad). Crown patch usually duller. **Juv** Totally different from later plumages, much of head and body, except belly, tinged pinkish-rufous, sparsely marked with small distinct black streaks. Not expected to occur in Europe.

BIOMETRICS (*aurocapilla*) **L** 13–14.5 cm; **W** ♂ 73–83 mm (*n* 15, *m* 77.9), ♀ 71–79 mm (*n* 12, *m* 74.7); **T** ♂ 50.5–58.5 mm (*n* 15, *m* 53.9), ♀ 48.5–55 mm (*n* 12, *m* 51.3); **T/W** *m* 69.0; **B** 13.3–16.0 mm (*n* 26, *m* 14.8); **B**(f) 10.7–13.0 mm (*n* 26, *m* 11.7); **BD** 4.0–5.2 mm (*n* 26, *m* 4.5). **Wing formula: p1** vestigial; **pp2–4** about equal and longest (p2 sometimes < wt 0.5–2 mm); **p5** < wt 0.5–2.5 mm; **p6** < wt 6–9 mm; **p7** < wt 9–12.5 mm; **p10** < wt 14–20 mm; **s1** < wt 16–21 mm. Emarg. pp3–5.

GEOGRAPHICAL VARIATION & RANGE Polytypic with usually three subspecies accepted. Race *aurocapilla* (over most of range including E USA and much of Canada except Newfoundland) is the race that is presumed in the first place to have occurred in Europe as a vagrant, the alternative being *furvior* breeding in E Canada.

S. a. aurocapilla (L., 1758) (C & SE Canada, E USA; winters West Indies, Mexico, adjacent Central America, N South America). Treated above.

S. a. furvior Batchelder, 1918 (C & S Newfoundland, E Canada; winters in Bahamas, Cuba and in adjacent areas). Similar to *aurocapilla* but differs by on average slightly bolder underparts streaking and a little darker and more brown-tinged, less greenish upperparts. Orange median crown-stripe averages narrower, whereas black lateral crown-stripes are often bolder.

NORTHERN WATERTHRUSH
Parkesia noveboracensis (J. F. Gmelin, 1789)

Fr. – Paruline des ruisseaux; Ger. – Drosselwaldsänger
Sp. – Reinita charquera norteña
Swe. – Nordlig piplärksångare

In spite of its English name, this species is a member of the American wood warblers, or parulids. Habitually, this terrestrial warbler lives in thickets near water and swamps across Alaska and much of Canada, locally into the NE USA and Pacific Northwest. It is an extremely rare vagrant to NW Europe, where it has occurred both at reservoirs and other inland habitats, and on seashores.

Presumed 1stS, Mexico, Apr: ground-dwelling pipit-like wood warbler. Rather dusky with intense dark streaking below. Long off-white supercilium (variably tinged buffish-yellow) emphasised by dark eye-stripe. Wing appears brown and worn, including primary-coverts, contrasting slightly with new greater coverts, suggesting 1stS, but best left unaged without handling. (H. Shirihai)

IDENTIFICATION A small, robust and short-tailed pipit-like bird with dense dark streaking below on a pale background. Easily separated from all other North American wood warblers by uniform dark olive-brown upperparts, *heavily dark-streaked whitish to pale yellowish underparts*, its *long off-white supercilium* (which can have a slight buffish-yellow tinge) emphasized by the *dark eye-stripe* (but see also Similar species regarding Louisiana Waterthrush). Has habit of *feeding on ground*, strutting about on comparatively *long, brownish legs* and *constantly wagging, waving and pumping tail* and rear body (in a manner recalling pipits or Common Sandpiper *Actitis hypoleucos*).

VOCALISATIONS Song is rather like a short Chaffinch song, but is unlikely to be heard in the W Palearctic. – Call is a loud metallic *chink*. There is also a high, buzzy note given in flight when flushed.

SIMILAR SPECIES For differences from Ovenbird see that species. – Separated from closely related extralimital *Louisiana Waterthrush* (not treated) by smaller size, narrower supercilium (Louisiana: broad), darker brownish legs (Louisiana: paler pink) and more yellowish- or olive-sullied underparts (Louisiana: purer white).

AGEING & SEXING Ageing usually possible in autumn using thin rusty tips to tertials and some coverts present in 1stW, supported by degree of feather wear in hand. Sexes alike and seasonal plumage variation very limited. – Moults. Complete post-nuptial moult in ad and partial post-juv moult mostly Jun–Sep. Post-juv moult includes head, body and all median and greater coverts, but no remiges or alula. Pre-nuptial moult in Oct–May is absent or variable in extent, mostly limited to body-feathers. – **SPRING Ad** Supercilium may taper slightly behind eye, and ground colour of supercilium and underparts varies from whitish to pale yellow-buff or buffish-olive. Tips of 1–4 outer tail-feathers often have white fringe or shallow white triangular spot on inner web, but much individual variation and white fringes rapidly reduced by wear. **1stS** Indistinguishable from ad in the field, often also in the hand; handled birds possible to age by same moult criteria as in autumn, but feather abrasion must be kept in mind (in general, from c. Mar–Apr tips of outer primaries and tail-feathers frayed and browner, whereas still neat and dark in ad). – **AUTUMN Ad** Almost identical to spring, but stronger buff or yellow wash to supercilium and underparts, overall darker upperparts and stronger throat spotting. Often inseparable from 1stW without solid experience of variation in wear at different seasons. **1stW** As ad, but at least until Nov–Dec tertials show narrow rusty tips (lacking in ad). Juv primary-coverts, central alula and r1 can have narrow rufous fringe on tip (indistinct in some, and can wear off quickly). Juv outer tail-feathers also more pointed and usually lack obvious pale edges at tip (often whiter and bolder in ad, but much overlap). **Juv** Upperparts feathering tipped rufous or ochre, tips to median and greater coverts forming buff wingbars. Underparts tinged buff with less distinct streaking. Not expected to occur in the covered region.

BIOMETRICS L 13–14 cm; **W** ♂ 72.5–80 mm (*n* 12, *m* 76.5), ♀ 72–77 mm (*n* 12, *m* 74.5); **T** ♂ 48–56 mm (*n* 12, *m* 51.0), ♀ 46–53 mm (*n* 12, *m* 48.8); **T/W** *m* 66.1; **B** 14.4–16.7 mm (*n* 24, *m* 15.5); **B**(f) 10.0–13.9 mm (*n* 24, *m* 11.6); **BD** 3.3–4.3 mm (*n* 24, *m* 4.0). **Wing formula:** p1 vestigial; pp2–4 about equal and longest (p2 sometimes < wt 0.5–1.5 mm); p5 < wt 1.5–4 mm; p6 < wt 7–10 mm; p7 < wt 9–15 mm; p10 < wt 16.5–20 mm; s1 < wt 18–23 mm. Emarg. pp3–5.

GEOGRAPHICAL VARIATION & RANGE Monotypic. – Alaska, Canada, E & extreme NW USA; winters south to Florida, Bermuda, Middle America, West Indies and N South America.

1stS, Puerto Rico, Mar: when feeding on ground, constantly wags tail and rear body, producing Common Sandpiper-like impression. Only liable to be confused with Louisiana Waterthrush (see Vagrants), but smaller bill, narrower supercilium, yellowish-sullied underparts and lack of pink-buff wash on flanks eliminate that species. Body-feathers and greater coverts fresh, and latter more olive than flight-feathers, while pale-tipped tertials and primary-coverts confirm age. (H. Shirihai)

NEW WORLD WARBLERS

TAXONOMIC NOTE Dunn & Garrett (1997) already predicted that Northern Waterthrush was not closely related to Ovenbird and 'perhaps deserves generic rank'. Sangster (2008) described a new genus, *Parkesia*, for Northern and Louisiana Waterthrushes. In a comprehensive genetic-based phylogenetic study of nearly the entire Parulidae (Lovette *et al.* 2010), this change was adopted to avoid polyphyly of *Seiurus*. Whereas Ovenbird is basal and sister to all other parulids, and remains in *Seiurus* being its type species, Northern Waterthrush is better transferred to the new genus *Parkesia*.

REFERENCES Lovette, I. J. *et al.* (2010) *Mol. Phyl. & Ecol.*, 57: 753–770. – Sangster, G. (2008) *BBOC*, 128: 212–215.

1stW, Azores, Oct: often draws attention while feeding with its loud metallic call and by its odd rhythmic bobbing movements. Pale tips to tertials, worn and pointed tips to tail-feathers and brown-tinged primary-coverts contrasting with greater coverts (though not as much as in spring) indicate age. (D. Monticelli)

1stW, Azores, Oct: moult and ageing as the other two birds on this page, but also note that bill can be mostly pale in 1stW. Smaller bill, narrower supercilium, yellowish-sullied underparts and lack of buff wash on flanks separate Northern from Louisiana Waterthrush. (V. Legrand)

1stW, Azores, Oct: some show bolder underparts streaking. Pale-tipped tertials, browner-looking primaries, primary-coverts and alula are all unmoulted juv feathers. (D. Monticelli)

BLACK-AND-WHITE WARBLER
Mniotilta varia (L., 1766)

Fr. – Paruline noir et blanc; Ger. – Kletterwaldsänger
Sp. – Reinita trepadora; Swe. – Svartvit skogssångare

Infrequently recorded in W Europe in late autumn, although several records in the covered region are from winter, this charismatic and acrobatic bird is a guaranteed crowd-pleaser due its smart plumage and nuthatch-like movements on tree-trunks and branches. It breeds in deciduous woodland across North America, and winters in the extreme SE USA to NW South America.

Ad ♂, USA, Apr: head striped black-and-white, upperparts heavily streaked, together with white wing-bars and heavily dark-streaked white underparts make the species unmistakable. Note dark wings and tail, blacker cheeks and black primary-coverts of ad ♂—the most attractive plumage with the boldest pattern. (S. Elsom)

IDENTIFICATION More or less automatic, given a clear view of its *black-and-white striped head*, heavily-streaked upperparts, *bold double white wing-bars* and tertial fringes, with *mostly white underparts* which are *also heavily streaked*. Rather slender and slightly curved bill, eye and legs mostly dark/blackish-grey. Moves on and around main trunks and larger branches, at times with head pointing downwards, its behaviour even *recalling a Nuthatch*.

VOCALISATIONS Most frequent calls are buzzy clicking *chizz* notes, or a high-pitched, thin *ziit-ziit* (sometimes uttered singly), and a harder clicking *tuc*.

SIMILAR SPECIES Only if seen very poorly should this species pose an identification problem. Could perhaps vaguely recall *Blackpoll Warbler*, but given a reasonable view is unmistakable.

AGEING & SEXING Ageing requires consideration of head and flank plumage and a close check of moult and feather wear in wing and tail. Sexes clearly differentiated in ad, chiefly in spring and summer. – Moults. Complete post-nuptial moult of ad and partial post-juv moult mostly Jun–Aug. Post-juv moult generally involves head, body, lesser, median and greater coverts. Pre-nuptial moult (in both ad and 1stY) partial and limited, occurring in late autumn to Feb–Apr (mostly involving some tertials and sometimes central tail-feathers). – **SPRING Ad** Both sexes best aged by being evenly fresh. **Ad ♂** Apparently unrecorded within the treated region. The strongest and boldest-patterned plumage, with all-black ear-coverts and extensive black chin and throat, while some have mottled white or mostly white chin. Outer tail-feathers purer grey to blackish with more extensive white patches. Whitish areas cleaner (lacking obvious buff or cream suffusion), including below, with bolder flanks streaking. **Ad ♀** Like ad ♂, apparently unknown in W Palearctic. Compared to ad ♂, ear-coverts and throat paler and lack black, black eye-stripe less pronounced and underparts less heavily streaked with slight buffish wash to flanks, which are streaked duskier. Outer tail-feathers also have less pure and smaller white area. **1stS** Best aged by presence of juv primary-coverts and primaries, which are heavily abraded and more tinged brownish-grey, not so blackish and white. In some ♂♂, dark on throat and ear-coverts reduced or almost absent due to white mottling. – **AUTUMN Ad ♂** Rather variable; some have black throat (black of chin concealed by white tips), others mostly white, but black ear-coverts (variable, can be only mottled dark) separate ad ♂ from all other plumages at this season. **Ad ♀** As in spring. **1stW ♂** Closely approaches ad ♀ due to lack of black cheeks and mostly whitish throat. Best aged by juv primary-coverts, which are narrow, tapered and already slightly abraded and browner, with much-reduced pale fringes (in ad, primary-coverts are broad, truncate, relatively fresh and dusky with narrow but distinct pale fringes). Remiges, especially primaries, weaker textured and somewhat less fresh than ad at this season. 1stW ♂ also tends to lack any wash to whitish flanks (streaked boldly blackish, but often more blurred than ad ♂). Moult limits (with retained juv remiges and primary-coverts) more difficult to discern in the field than in spring. **1stW ♀** Shows less bold, indistinctly greyish-streaked underparts and always has buff-brown wash to ear-coverts, flanks and undertail-coverts; from 1stW ♂ primarily by duller flanks with indistinct blurry streaking. Ageing by moult limits as in 1stW ♂. **Juv** Soft, fluffy body plumage, slightly sullied buff. Head and body largely diffusely

Ad ♀, USA, Apr: like ♂, but no black bib, paler ear-coverts (thus black eye-stripe stands out better) and ground colour of underparts less pure white. For a ♀ still quite heavily streaked underparts, while dark and fresh wings and tail, including blackish primary-coverts, are indicative of ad. (S. Arlow)

1stS ♂, USA, May: like ad ♂ with obvious black bib and ear-coverts, but retained brownish-tinged primary-coverts, large alula and tips to primaries age it as 1stS. (B. van den Boogaard)

NEW WORLD WARBLERS

streaked rufous and blackish. Not expected to occur in the covered region.

BIOMETRICS L 11.5–13 cm; **W** ♂ 69–74 mm (*n* 16, *m* 71.3), ♀ 64–69 mm (*n* 12, *m* 67.4); **T** ♂ 48–51 mm (*n* 16, *m* 49.3), ♀ 44–50 mm (*n* 12, *m* 46.9); **T/W** *m* 69.4; **B** 12.9–15.3 mm (*n* 27, *m* 14.2); **B(f)** 9.0–12.0 mm (*n* 26, *m* 10.9); **BD** 2.9–3.9 mm (*n* 25, *m* 3.4); **Ts** 16.0–18.5 mm (*n* 25, *m* 17.1); **white patch r6** ♂ 13.5–21 mm, ♀ 8–16 mm. **Wing formula: p1** vestigial; **pp2–4** about equal and longest (p2 sometimes 0.5–3 mm <); **p5** < wt 0.5–2 mm; **p6** < wt 5–9 mm; **p7** < wt 9–12.5 mm; **p10** < wt 13–19 mm; **s1** < wt 14.5–21 mm. Emarg. pp3–5.

GEOGRAPHICAL VARIATION & RANGE Monotypic. – Canada except NE, C & E USA; winters West Indies, Central America, N South America.

1stS ♀, Dominican Republic, Apr: instantly identified as young ♀, being dullest and least streaked below in spring, although dullest ad ♀ only fully eliminated by presence of juv primary-coverts and primaries being slightly brownish. Shape and plumage pattern remain highly characteristic, especially head and wing markings. (H. Shirihai)

Ad ♂♂, Costa Rica (left), Venezuela (centre) and Mexico (right), Dec: typical postures, especially bird in centre with head pointing down while moving along branch. All three show immaculate plumage striped black-and-white, even when fresh, although black bib largely concealed (a few black bases just visible on left bird). All three show mainly black cheeks of ♂. All are further aged by uniform wing-feathers and seemingly rather broad tail-feathers. (Left & centre: H. Shirihai; right: S. N. G. Howell)

Ad ♀, Mexico, Oct: note pale cheeks (with thin dark eye-stripe) and less heavily streaked underparts, plus dark flight-feathers with greyish fringes and whitish tips, these characters together age and sex this bird. (S. N. G. Howell)

1stW ♂, Azores, Oct: diagnostic black lateral crown- and eye-stripes bordering white median crown-stripe and supercilium. The general degree of streaking is intermediate between ad of either sex, but closer to ad ♂, with bolder and blacker flanks streaking. Nevertheless, if aged first (by brown juv primary-coverts and primary tips), sexing safer. (D. Monticelli)

1stW ♀, USA, Sep: overall dullest autumn plumage, with indistinct streaking on breast-sides, and pale grey cheeks, indicating young ♀. Still, 1stW ♀ at times quite similar to 1stW ♂, and rest of plumage not that far from latter, although weak streaking on underparts should infer a ♀. (T. Shimba)

— 467 —

TENNESSEE WARBLER
Oreothlypis peregrina (A. Wilson, 1811)

Fr. – Paruline obscure; Ger. – Tennessee-Waldsänger
Sp. – Reinita de Tennessee; Swe. – Tennesseesångare

This comparatively poorly-marked wood warbler is a very rare vagrant to the W Palearctic, with five September records from Britain and the Faeroes, one from Iceland in October and four from the Azores in October and November. It breeds in coniferous and deciduous woodlands in N North America from SE Alaska, across much of Canada south to extreme NE USA. Active and restless, moving around in the canopy when foraging, often high in the trees and on outer branches (though can be seen lower down). Like many other wood warblers can hang upside-down in the thinnest branches.

IDENTIFICATION In plumage, shape and size quite similar to Common Chiffchaff, with (variably marked) *short dark eye-stripe and vague pale supercilium*, bright *olive-green upperparts*, *rump paler and brighter green*, and pale underparts. Latter have variable (depending on season, age and sex) pale *yellow wash*, but are more whitish on centre of belly and undertail-coverts. Often diffusely and irregularly sullied greyish on chest and flanks, and longest undertail-coverts have diffuse greyish centres. Entire plumage invariably *unstreaked*. Especially in first-winter and fresh adult plumage, faint pale wing-bars may be visible. Breeders have greyer head, nape and neck-sides (noticeably purer and more bluish-grey in ♂), and whiter underparts, but this plumage has not been recorded in Europe. Fine, sharp bill. Iris and legs mostly dark (blackish-grey). Long, fairly pointed wings with *good primary projection*. When handled or seen close, note lack of reduced outer primary of *Phylloscopus* and tendency to have thin whitish edges distally on inner webs of outer tail-feathers, while some have a large patch on outermost 1–2 pairs; this surprisingly large variation seems unrelated to sex or age.

VOCALISATIONS Call is a thin and soft *tsip* or *tsit* or *tsee*. These, and a sweet *chip* note, are most frequently heard.

SIMILAR SPECIES Tennessee Warbler superficially resembles some of the wing-barred Eurasian *Phylloscopus* warblers, though its long and sharply-pointed bill with a paler grey base to lower mandible is diagnostic. Also, the relatively short tail with white-edged inner webs of outer tail-feathers and distinctly white-tipped exposed primaries generally eliminates all *Phylloscopus* species in a closer look. Note dark legs. – In the Americas, Tennessee Warbler is sometimes subject to confusion with other parulids (see, e.g., Dunn & Garrett 1997), but most of these have not yet been recorded in Europe. It recalls Orange-crowned Warbler (*O. celata*, a potential future vagrant to the W Palearctic; not treated); the shorter tail, greater contrast between the dorsal and ventral parts (underparts being more whitish), generally better-marked face and, in fresh plumage, pale wing-bars and primary tips, and white undertail-coverts are important distinguishing features from Orange-crowned. In fresh autumn birds, chest usually sullied yellowish, without subdued streaking on breast-sides of most Orange-crowned. Tennessee Warbler also lacks any streaking below, and their calls are quite different.

Ad ♂, USA, Apr: unstreaked with pointed bill, bright olive-green upperparts and whitish underparts including undertail-coverts (slight yellow wash on flanks, and diffusely sullied greyish on chest). Bluish-grey head, nape and neck-sides a feature of ♂, while evenly-feathered wing with indistinct pale-fringed greater and median coverts (lacking whitish wing-bar of some 1stY birds) confirm age. (A. Murphy)

1stS ♂, USA, Apr: long, rather pointed wings with obvious primary projection. Bluish-grey head, nape and neck-sides of ♂, while brownish alula, worn and bleached primary-coverts, primaries and tail indicate young age. All greater coverts replaced, but these vary in colour (here many outer form grey panel), wear and number of pale tips, but in general whitish wing-bar is more obvious in first-spring birds than in ad (A. Murphy)

NEW WORLD WARBLERS

AGEING & SEXING Age and sex criteria must be combined, and individual variation must be taken into account. Sexes virtually alike in autumn (differing to some degree in spring and summer), and seasonal plumage variation rather limited. – Moults. Usually complete post-nuptial moult in ad in late summer, but may suspend and leave some flight-feathers unmoulted during migration. Partial post-juv moult (involving head, body, and wing-coverts) mostly Jun–Sep. Pre-nuptial moult partial or limited, mostly confined to some head- and body-feathers, Dec–Apr. – **SPRING Ad** Unrecorded in the treated region. Extent and shade of grey on crown and nape, and yellowish wash on throat, vary individually, but ♀ usually has crown and nape washed olive, mantle duller grey-green and some pale yellow on throat and upper breast, whereas at least most typical ♂ has contrasting darker and greyer bluish crown/nape, whiter throat, purer green mantle and back, and more prominent whitish patches to outer two rectrices. **1stS** Very like respective ad (only slightly duller), but possible to age by worn and browner juv remiges, and subtly more pointed and rather bleached primary-coverts and tail-feathers. Moult contrast best seen against slightly fresher post-juv median and greater coverts. – **AUTUMN Ad** Sexes similar, but ♂ tends to have larger white tail spots, a greyer (or grey-green) crown, and averages less yellow below (belly whitish with greyish-sullied flanks). Fresh ad ♀ (as 1stW ♂) extensively but variably washed yellowish below. Some are intermediates, impossible to sex reliably. Usually evenly fresh without moult limits in wing and tail (cf. 1stW) but those with few suspended (heavily-worn) remiges distinctive. Many are intermediates, difficult or impossible to age reliably. **1stW** Very similar to fresh ad, but has on average more yellowish tinge to underparts and face, sometimes extending to undertail-coverts (young ♂ has relatively less yellow below than young ♀, the same sexual difference as in ad). Narrow wing-bars tend to be slightly more apparent than in fresh ad. Can sometimes be aged by retained juv primary-coverts (slightly narrower, less fresh and duller), primaries and tail-feathers, which are also fringed narrower olive-buff (forming slight moult contrast with renewed median and greater coverts). In both sexes r6 lacks or has only slight whitish smudges at tip (i.e. almost as ad ♀ but much less than fresh ad ♂). Narrow pale tips to primaries in all ages/plumages at this season. **Juv** Rather like 1stW ♀ but pattern duller, less neat, and wing-bars very prominent. Extremely unlikely to occur in Europe.

BIOMETRICS L 10.5–11.5 cm; **W** ♂ 62–70 mm (*n* 12, *m* 65.4), ♀ 60–64 mm (*n* 13, *m* 62.1); **T** ♂ 42–46 mm (*n* 12, *m* 43.5), ♀ 37–43 mm (*n* 13, *m* 39.6); **T/W** *m* 65.1; **B** 11.7–13.7 mm (*n* 24, *m* 12.8); **B(f)** 9.3–10.9 mm (*n* 23, *m* 10.1); **BD** 3.5–4.1 mm (*n* 23, *m* 3.7); **Ts** 16.0–18.7 mm

♀, presumed 1stS, USA, Apr: ♀ usually has crown and nape washed olive, mantle duller grey-green and some pale yellow on throat and upper breast. In Apr, due to wear, it is not easy to decide age, but worn and bleached primaries and tail-feathers might indicate 1stS. Still, some ad in spring can be similar, and age remains unsafe. (A. Murphy)

Ad ♂, USA, Sep: more difficult to sex in autumn, as bluish-grey head, nape and neck-sides of ♂ can be only just visible. Evenly feathered wing with blackish and white-tipped primaries and indistinct pale-fringed greater and median coverts confirm age. (R. Brewka)

Ad ♀, Canada, Sep: fresh ♀ usually lacks any trace of grey on head and is mostly greenish above and yellow below. When age confirmed as ad (by evenly fresh wing-feathers with no moult limits and blackish primaries with whitish tips) separation from 1stW ♂ is easier. (P. Eriksson)

1stW ♂, USA, Sep: a bright autumn plumage indicates either ad ♀ or 1stW ♂, but when tentatively aged as 1stW (by seemingly pointed juv primary-coverts and hint of wing-bar on greater coverts), in combination with some grey on crown and cheeks, and white rather than yellow belly, it can be identified as 1stW ♂. (L. Spitalnik)

1stW ♀, Azores, Oct: combination of mostly juv wing and duller plumage infer age and sex. Beware possible confusion with Orange-crowned Warbler (as yet unrecorded in Europe), mainly due to rather unusual subdued grey streaking on breast of this individual, but note whitish undertail-coverts, better-marked face and pale wing-bar of Tennessee Warbler. (V. Legrand)

Presumed 1stW ♀♀, USA, Sep: plumage variation makes some birds difficult to age and therefore also sex. Left bird apparently has juv primary-coverts (brownish and slightly pointed, contrasting with fresh yellowish-green greater coverts) indicating 1stW, while yellowish underparts best fit ♀. Right bird, despite primary-coverts, primaries and tail appearing ad-like, has greyish-tinged greater coverts panel with broad, well-defined pale tips of 1stW, while richer yellow underparts and no grey on head suggest ♀. (Left: N. McKown; right: S. N. G. Howell)

(n 25, m 16.8). **Wing formula: p1** minute; **p2** < wt 0–1 mm, = 4/5 or =wt; **pp(2)3–4** about equal and longest; **p5** < wt 0.5–2.5 mm; **p6** < wt 5–8 mm; **p7** < wt 8–11.5 mm; **p10** < wt 12.5–18.5 mm; **s1** < wt 15–19.5 mm. Emarg. pp3–5.

GEOGRAPHICAL VARIATION & RANGE Monotypic. – S Alaska, Canada, E USA; winters SE Mexico to NW South America.

TAXONOMIC NOTE A recent comprehensive genetic-based phylogenetic study of nearly the entire Parulidae (Lovette *et al.* 2010) confirmed the results of previous studies (e.g. Avise *et al.* 1980), which suggested that several genera as traditionally treated were polyphyletic and therefore requiring revision. In the case of Tennessee Warbler, formerly usually named *Vermivora peregrina*, this was transferred to the genus *Oreothlypis*, which is followed here. For an alternative arrangement, see Sangster (2008).

REFERENCES Avise. J. C., Patton, J. C. & Aquadro, C. F. (1980) *Journal of Heredity*, 71: 303–310. – Lovette, I. J. *et al.* (2010) *Mol. Phyl. & Ecol.*, 57: 753–770. – Sangster, G. (2008) *BBOC*, 128: 207–211.

(COMMON) YELLOWTHROAT
Geothlypis trichas (L., 1766)

Fr. – Paruline masquée; Ger. – Weidengelbkelchen
Sp. – Mascarita común; Swe. – Nordlig gulhake

Very rare vagrant to Britain, with about a dozen records and single, somewhat unusual, overwintering and spring records. There is also the odd record from Iceland. Typical habitat includes marshes, reedbeds, brushy pastures and abandoned fields in North America, where widespread from SE Alaska across much of Canada and south over most of the USA (except the south-west), as well as in Mexico south to Puebla. Vagrants can occur in a variety of wooded habitats, often low down in underbrush.

IDENTIFICATION A *short-winged*, *longish-tailed* North American warbler, with adult ♂ automatically identified by *bold black mask* that seems to *droop at rear* (reaching from forehead through eye to ear-coverts) outlined by pale grey upper border, olive-green upperparts and yellow underparts, intense on throat to undertail-coverts. ♀ duller, lacks black mask, has *slight supercilium* in front of eye and *reduced yellow below*, but *first-winter* ♂ being similar to ♀ *has mask partly developed*, reduced in extent, brownish and mottled buff-white. Unmistakable within the treated region, but beware of somewhat featureless young ♀ in autumn. Despite longish tail has rather *Wren-like shape*, enhanced by *frequently-cocked tail* being distinctive.

VOCALISATIONS Most frequently heard vocalisation in non-breeding season is a dry, rather sharp *chip*. There is also a harsher *tchek* slowly repeated.

SIMILAR SPECIES ♂ unmistakable, and even ♀ and non-adults should be readily distinguished from other Nearctic wood warblers in the W Palearctic.

AGEING & SEXING Ages similar, but ♂♂ generally separable in autumn. Clear sexual dimorphism, while seasonal plumage variation is often age-related, with 1stW reminiscent of ♀. 1stW ♂♂ almost invariably have partial mask, therefore easy to tell from ♀♀. – Moults. Complete post-nuptial moult of ad and partial post-juv moult mostly Jun–Sep. Post-juv moult includes head, body, all median and greater coverts, sometimes greater alula and some tertials and central tail-feathers. This moult is quite variable both individually and geographically, sometimes including several outermost primaries and innermost secondaries, and even most or all tail-feathers, but such extensive moult appears to be unusual in northern and more migratory populations (thus, 1stW with extensive moult unlikely to reach W Palearctic). Partial pre-nuptial moult variable, absent or mainly limited to body-feathers, Oct–May. – **SPRING Ad** ♂ See above for a general description. Bold black mask with pale grey upper border. May have crown tinged rufous. **Ad** ♀ Much duller than ad ♂ and lacks mask, though occasionally, presumably older birds, may show 'ghost' of ad ♂ pattern with suggestion of dark mask, especially enhanced by wear late

G. t. trichas, ad ♂, USA, May: unmistakable, ad ♂ has full, glossy black mask, broad square rectrices and black bill (inferring age). Ssp. *trichas* has rather limited and subdued brown wash on flanks, while yellow of rest of underparts is not so intense, as shown here and in image to right. (D. Monticelli)

G. t. trichas, 1stS ♂, USA, May: bold black mask and yellow underparts automatically identify this short-winged, longish-tailed parulid. Mask includes forehead, eye-surround and ear-coverts and appears to droop at rear, with pale grey-white upper border. Olive-green upperparts with only moderate brownish hue supports racial assignation. Aged by brown-tinged primary-coverts and one alula feather, mask being less glossy black and flecked pale, brown rectrices with worn tips and paler bill. (D. Monticelli)

G. t. trichas, ad ♀, USA, Apr: dull grey and greenish in general, lacking black mask of ♂, but still has rich yellowish throat and breast. Latter, plus fact that all wing-feathers apparently of same generation, with neat dark primary-coverts and ad-looking tail, confirm age. (A. Murphy)

Presumed *G. t. trichas*, 1stS ♀, USA, Apr: compared to the bird at left, even duller, with juv primary-coverts (worn and brown), pale yellow throat and seemingly dull brownish tips to primaries and rectrices. Such pale and bland birds more typical of races in interior W USA, though locality (Galveston, Texas) should fit *trichas*, and ♀♀ often difficult to assign to subspecies anyway. (A. Murphy)

in spring. **1stS** In the field, both sexes difficult to distinguish from ad, but ♂ often shows some brown in mask and trace of buff eye-ring, while ♀ can be very faded and worn, and has less yellow on throat and breast than same-sex ad. Best aged by more worn, retained juv wing-feathers and tail-feathers (moult pattern as 1stW but wear and bleaching must be taken into account). – AUTUMN **Ad ♂** Darker and browner upperparts, with buff wash to underparts (intense on flanks) and develops some pale tips to mask feathers. **Ad ♀** More brownish-tinged upperparts, and flanks strongly washed brown. Unlike 1stW, both sexes evenly fresh. **1stW** Sometimes possible to age by paying attention to slight moult contrast created by retained juv primary-coverts, primaries and tail-feathers, which are generally narrower, weaker textured, browner, somewhat less fresh and duller – moult limits best seen against renewed greater coverts and tertials. Note that many are difficult to age without practice and comparative material at hand. **1stW ♂** Duller than respective ad, with at least a few scattered black feathers in mask or more extensively blackish mask (and therefore easily separated from ad ♀) and may retain some buff feathers in eye-ring from juv plumage. **1stW ♀** Extremely dull and uniformly plumaged. Upperparts entirely olive-brown and underparts largely buff. **Juv** Similar to 1stW ♀ but generally has duller, less yellow throat, and ochre greater coverts wing-bar more prominent.

BIOMETRICS (*trichas*) **L** 11–13 cm; **W** ♂ 56–60 mm (*n* 12, *m* 57.5), ♀ 50–56 mm (*n* 12, *m* 52.8); **T** ♂ 44.5–54 mm (*n* 12, *m* 49.3), ♀ 45–53 mm (*n* 12, *m* 48.1); **T/W** *m* 88.4; **B** 11.7–14.3 mm (*n* 24, *m* 13.3); **B(f)** 9.0–11.7 mm (*n* 23, *m* 10.7); **BD** 3.5–4.1 mm (*n* 23, *m* 3.7); **Ts** 17.7–21.0 mm (*n* 23, *m* 19.7). **Wing formula: p1** minute; **p2** < wt 2–4 mm, =p6 or 7; **pp3–5** about equal and longest; **p6** < wt 0.5–3 mm; **p7** <wt 2–5 mm; **p8** < wt 4–6.5 mm; **p10** < wt 7–11 mm; **s1** < wt 7.5–12.5 mm. Emarg. pp3–5.

GEOGRAPHICAL VARIATION & RANGE Polytypic, over a dozen races described, but differences generally slight and clinal. Only ssp. *trichas* thought to have occurred in Europe. This and some similar species of *Geothlypis* are sometimes considered conspecific, most having allopatric ranges.

G. t. trichas (L., 1766) (mainly E USA and E Canada; winters West Indies, Mexico, N South America). Treated above. Comparatively small. Less tinged brown above and on flanks than some other races, not as brightly yellow on underparts as others. Reliable racial identification usually requires comparison with museum material.

G. t. trichas, ad ♂, USA, Oct: following complete post-nuptial moult, plumage resembles breeding ♂ (although black mask has white tips on forehead and ear-coverts, and intense yellow underparts are restricted to throat and upper breast). Wing appears rather evenly fresh. Bill has pale cutting edges at this season. (E. Myles)

G. t. trichas, 1stW ♂, Azores, Oct: besides strongly pale-flecked blackish mask and reduced yellow below, juv primary-coverts are already bleached and slightly brownish, contrasting with greener greater coverts, and tips to primaries—as far as can be judged—look brownish, too. (V. Legrand)

G. t. trichas, 1stW ♂, Azores, Oct: near total absence of dark mask, with only a few vaguely exposed dark feather bases on cheeks, yet this in combination with rich yellow throat indicates a 1stW ♂. Additionally aged by moult pattern as in previous bird. (D. Monticelli)

G. t. trichas, 1stW ♀, USA, Sep: drabbest plumage in autumn, but still unlikely to be confused with any Palearctic warbler or chat. Indistinct (sometimes broken) whitish eye-ring, olive-brown upperparts (brighter on rump/uppertail-coverts and tail fringes) and hint of yellow on throat reveal identity. Aged by moult pattern as in previous 1stW ♂♂. (F. Jacobsen)

AMERICAN REDSTART
Setophaga ruticilla (L., 1758)

Fr. – Paruline flamboyante
Ger. – Schnäpperwaldsänger
Sp. – Candelita norteña
Swe. – Rödstjärtsskogssångare

A habitual catcher of flies taken during aerial sallies (unusual among New World wood warblers), this distinctive bird is a very scarce vagrant to the treated region (mostly recorded in Britain and Ireland), mainly in October–November; also accidental in, e.g., Azores, Iceland, France and Madeira. Inhabits open woodland in North America. Typically droops wings slightly and, especially, semi-spreads tail even when feeding, usually only quickly but at times for a few seconds, so the striking tail pattern is likely to be seen on a bird under prolonged observation.

1stS ♂, Cuba, Mar: compared to similar ♀ generally darker and browner, sometimes with variable blackish lores and frequently with scattered black patches on face to breast, or even as here most of head brownish-black, also blackish-tinged uppertail-coverts and often richer yellow (at times even orange) patches on breast-sides. Note retained and bleached brownish juv primaries, primary-coverts, greater coverts and tertials. (H. Shirihai)

Ad ♂, USA, Apr: moves restlessly, fanning its tail and 'drooping' wings, revealing diagnostic brilliant orange breast-sides, basal tail-sides and bases to remiges. Belly to undertail-coverts white, otherwise black, indicative in spring of ad ♂ (cf. 1stS ♂ above right). Entirely ad wing. (A. Murphy)

IDENTIFICATION Unmistakable adult ♂ plumage (strikingly black, white and bright orange) unrecorded in Europe, but other plumages also readily identified by *orange-yellow* (1stW ♂) or *yellow* (either sex) *bases to outer tail-feathers* (recalling similar pattern in white of Red-breasted Flycatcher) and *orange* (♂) or *yellow* (either sex) *patch on breast-side and upper flanks*. Otherwise, ♀/1stY ♂ has grey head, olive upperparts and dull off-white underparts. In spring, immature ♂♂ appear first in ♀-like plumage but attain some black feathering of adult plumage. Stubby but pointed dark bill and thin black legs. General shape also characteristic, not least due to *proportionately long tail*. It is habitually very active, moving around restlessly in the canopy, often fanning the tail and 'drooping' the wings.

VOCALISATIONS Common contact calls are a 'wet', stone-clicking *chip* and a slightly higher-pitched *tseet*, vaguely recalling call of Song Thrush, but longer; once learnt these calls are, in an American context, identifiable as belonging to this species.

SIMILAR SPECIES Can only be confused with Red-breasted Flycatcher, given a brief view, and shares some behavioural characteristics, flycatching from at least mid height in trees, and at other times holding wings drooped and tail cocked. But easily distinguished by larger size and longer tail, which is habitually kept fanned (cocked but folded in Red-breasted), while colour of basal corners to tail and presence of wing flashes and breast-side patches, among other plumage features, provide ready distinctions.

AGEING & SEXING Ageing requires close check of moult pattern and feather wear in wing and tail-feathers, and size of yellow bases to remiges (in non-ad ♂ plumages). Sexes differ dramatically from 2ndCY autumn plumage, while in 1stW and 1stS, ♂ has ♀-like plumage (thus until first post-nuptial moult); seasonal plumage variation mainly involves moderate feather wear. – Moults. Complete post-nuptial moult of ad and partial post-juv moult mostly Jun–Aug (Sep). Post-juv moult includes head, body, all median and greater coverts, and usually greater alula feather. Partial pre-nuptial moult variable, may involve some innermost greater coverts. – SPRING Ad ♂ Head, upperparts and breast all black. Brilliant orange breast-sides, basal tail-sides and bases

Ad ♀, USA, Apr: wing, tail and breast-side patches yellow, and eye-ring and loral area finely flecked narrowly white on otherwise brownish-grey head and olive-tinged upperparts. Whole wing ad, and note dark primary-coverts and extensive yellow base to fourth outermost tail-feather (r3). (A. Murphy)

1stS ♀, USA, Apr: dullest plumage in spring, with yellow wing patches much reduced, often largely concealed (but these vary individually, not with age and sex). Most obvious are retained alula and primary-coverts, but this bird has replaced nearly all rectrices (accidentally?) to ad pattern with extensive yellow base on r3 (only outermost left tail-feather unmoulted, duller with paler brown tip). (A. Murphy)

1stW ♂♂, Venezuela, Dec (left) and Mexico, Oct: individual variation in wing markings occur, with left bird having broad yellow bases to flight-feathers, while right bird has very limited and dull yellow bases. Both share worn and slightly browner remiges and tail-feathers, orange breast-side patches and rather greyish head. Sometimes, orange patches in tail and on body-sides can appear rather small, especially when tail tightly folded and wings drooped. (Left: H. Shirihai; right: S. N. G. Howell)

to remiges, and belly to undertail-coverts white. Extent of black and orange variable. **Ad ♀** Head greyish-brown and upperparts tinged olive; uppertail-coverts and rump tinged greyish (only indistinctly contrasting with back). Wing, tail and breast-side patches yellow, and eye-ring and lores flecked narrowly white. **1stS ♂♀** Like ad ♀. Sometimes possible to age by retained (subtly more worn and bleached browner) juv primaries, primary-coverts and tail-feathers, and rarely some outer greater coverts, but in general differences are subtle, and many cannot be aged on wear. 1stS ♂ compared to 1stS ♀ generally darker and browner, usually with variable blackish lores and frequently with scattered black patches on head, neck and chest, blackish or dark bronze-grey upper-tail-coverts and rump (contrasting with paler olive-grey rump and back), and most have broader richer yellow (at times even orange) patches on breast-sides; ♀ duller, with yellow patches much reduced, often almost lacking or concealed. Size of yellow bases to remiges differ between sexes and from ad ♀ in the same way as in autumn (see 1stW). — **AUTUMN Ad ♂** As spring, though may possess buff fringes to black feathers. **Ad ♀** Much as spring, but upperparts browner, throat and breast tinged dull buff, with less contrast between head and mantle. Differs from 1stW in being evenly fresh (no moult contrast), but handling or exceptionally close views needed to discern this. Tail-feathers average subtly broader and more rounded at tips, again a difficult character to use. Yellowish bases to secondaries usually visible outside tips of greater coverts, sometimes also narrowly on primaries outside primary-coverts. Olive fringes to tertials rather distinctly set off compared to more sullied pattern of juv feathers. Fine white tips to r6 generally a good sign of ad ♀. **1stW** As ad ♀, and frequently difficult to separate. As a rule, should average slightly more worn on tips of remiges and tail-feathers, and these also average slightly browner, but differences often marginal and many are intermediates. Size and coloration of breast-side patches, and size of yellow wing patches, variable, but orange (rather than lemon-yellow) breast-sides very strongly indicative of ♂ (many such birds labelled '♀' in collections are probably wrongly sexed). Very limited yellow visible in wing and reduced amount of yellow to fourth outermost rectrix (r3) are fairly strong indications of 1stW ♀. Any bird with dark grey or blackish patches on head, throat or breast, preferably combined with contrastingly dark uppertail-coverts, should be 1stY ♂ (although rarely, older ♀♀ can show such dark patches on head and underparts and richer yellow or orange-tinged breast patches). **Juv** Rather brownish upperparts and greyish underparts. Double yellowish-white wing-bars. Not expected to occur in the covered region.

BIOMETRICS L 11.5–13 cm; **W** ♂ 63–68 mm (n 22, m 64.6), ♀ 59–65 mm (n 18, m 61.7); **T** ♂ 54–59 mm

Ad ♂, USA, Sep: the old ♂ in autumn has same attractive black plumage with bold orange pattern and white belly as in spring making it unmistakable. (G. Koziara)

1stW ♀, Scotland, Sep: even in more subdued 1stW ♀ plumage, immediately identified to species by yellow-based tail-sides. Sex indicated by overall dull plumage and lack of orange on body-sides, and age inferred by subtle contrast between brown-tinged juv primary-coverts and renewed greater coverts, and also by rather worn and pointed tail-feathers. (M. Rayment)

(n 22, m 56.3), ♀ 52–59 mm (n 18, m 54.6; **T/W** m 87.7; **B** 11.5–13.0 mm (n 21, m 12.2); **B**(f) 8.2–9.3 mm (n 19, m 8.8); **BD** 3.0–3.4 mm (n 17, m 3.2); **Ts** 14.8–16.9 mm (n 19, m 16.3); **yellow on pp > pc** 1stY ♂ 0.5–8 mm (n 10, m 4.0), ♀ 0–5 mm (n 18, m 1.2); **yellow on ss > gc** 1stY ♂ 4–10 mm (n 7, m 7.6), ♀ 1–11 mm (n 9, m 6.3). **Wing formula: p1** minute; **p2** < wt 0.5–4 mm, =5 or 5/6 (=6); **pp3–4**(5) about equal and longest; **p5** < wt 0–3 mm; **p6** < wt 4–8 mm; **p7** < wt 7–10.5 mm; **p10** < wt 11–16 mm; **s1** < wt 13–17 mm. Emarg. pp3–5.

GEOGRAPHICAL VARIATION & RANGE Monotypic. — SE Alaska across much of Canada and south over E USA; winters in parts of California and Florida, in Mexico, Central America, West Indies and W & N South America. (Syn. *tricolora*.)

NEW WORLD WARBLERS

NORTHERN PARULA
Setophaga americana (L., 1758)

Fr. – Paruline à collier; Ger. – Meisenwaldsänger
Sp. – Parula norteña; Swe. – Messångare

One of the most attractive transatlantic vagrants, which irregularly makes landfall in the treated region, most frequently in Britain in October. When feeding in canopy active and makes twitching movements of wings, constantly altering perch and acrobatically dangling upside-down like a tit. Feeds at moderate to high levels in foliage, but—mainly in autumn—is also attracted to weedy patches and scrubby thickets. Breeds in deciduous woodland but also in coniferous forests, frequently near water, most commonly in E North America. It winters in Central America, West Indies and S Florida.

Ad ♂, USA, Apr: in all plumages, bold white wing-bars and yellow throat and breast make this vagrant unmistakable. In spring, blue-grey upperparts of ♂ purer, emphasising reduced but brighter yellow-green mantle patch, while tawny-rufous admixed with black within bright yellow bib is variable but usually visible. White wing-bars to ad wing-coverts often appear as broad white patches. Near-black lores and eye-surround distinctive, as are white eye-crescents. Age secured by evenly-feathered ad wing and blue-edged primary-coverts. (D. Monticelli)

IDENTIFICATION A *small*, *short-tailed* and therefore compact-looking parulid that generally presents few identification problems. ♂ is *blue-grey above with two bold white wing-bars*, prominent *white patches in outertail* (largest in ad ♂), *triangular-shaped bronze-green (or yellowish-green) mantle patch*, *yellow throat and breast*, and *bold but separate white eye-crescents*. ♂ further has broad *dark, mainly rufous band across breast* at all seasons, while ♀ has no breast-band or only a hint of rufous and grey spots on predominantly yellow breast. Grey of head in ♂♂ darkens to *near-black on lores, eye-surround* and narrowly on forehead. ♀ further has upperparts more greenish-grey than bluish-grey (though some overlap), and has reduced bronze-coloured saddle effect on mantle. Pink-brown ('flesh-coloured') legs. *Two-coloured bill* with upper mandible dark, lower pale pink-yellow.

VOCALISATIONS Common call is a liquid but sharp *chip* or *tsip*, which is often repeated. In flight gives a descending softer *tseef*, also often repeated.

SIMILAR SPECIES Even poorly marked 1stW are unmistakable within the treated region due to the bold double wing-bar and vivid lemon throat and breast.

AGEING & SEXING Ageing as a rule possible by plumage and moult pattern in both autumn and spring, although some ♀♀ in particular can be tricky. Sexes differ more obviously in breeding plumage. Rather little seasonal variation. – Moults. Complete post-nuptial moult generally in Jun–Aug, and partial post-juv moult (including head, body and most wing-coverts, sometimes greater alula but almost never any tertials—see below) in the same period as ad, or starts slightly earlier. Pre-nuptial moult absent (or limited, in Jan–May). – **SPRING Ad ♂** Not yet recorded within the treated region; variable tawny-rufous breast-band bordered by black band (also varies in prominence). Lores and eye-surround near-black. **Ad ♀** Duller and lacks breast-band, though some may have a trace of rufous spots, mainly on breast-sides. Crown green-tinged and lores pale. **1stS** Very like respective ad. Can be aged by retained, more strongly bleached juv wing-feathers (moult pattern as 1stW) and green fringes to primary-coverts. Both sexes have duller (less pure grey) upperparts, and young ♂ usually has less extensive breast-band. – **AUTUMN Ad** No moult contrast in wing or tail. Primary-coverts blackish with blue (♂) or bluish-green (♀). Sexes more alike, but despite greenish wash, upperparts of ♂ still predominantly blue-grey, and hint of breast-band visible (though strongly obscured by yellow tips). Lores in ♂ less dark than in spring, still clearly darker than in ♀. **1stW** Resembles ad of respective sex: ♂ has duller breast-band, more extensive greenish tinge to head (lost with wear) and brown primary-coverts edged greenish. Lores darkish. ♀ even duller, lacks any trace of breast-band, has even more greenish-tinged upperparts, primary-coverts fringes and has greenish wash to wing-bars. Unlike ad, retains juv remiges (less fresh, browner and fringed greenish, less grey) and primary-coverts, all or at least most tertials and alula, and occasionally outer greater coverts, as well as tail-feathers (usually more pointed). Note, in particular, greenish fringes to primary-coverts, secondaries and tertials. **Juv** Lacks any yellow, is more dull greyish. Not expected to ever occur within the covered region.

BIOMETRICS L 10–11.5 cm; W ♂ 57.5–65.5 mm (*n* 12, *m* 60.9), ♀ 56–63 mm (*n* 12, *m* 58.3); T ♂ 40–45 mm (*n* 12, *m* 42.2), ♀ 39–43 mm (*n* 12, *m* 40.9); T/W *m* 69.8; B 11.1–13.3 mm (*n* 24, *m* 12.3); B(f) 9.0–11.0 mm (*n* 24, *m* 9.8); BD 3.1–3.7 mm (*n* 23, *m* 3.4); Ts 15.5–17.3 mm (*n* 23, *m* 16.5). **Wing formula: p1** minute; **p2** < wt 0–1.5 mm, =4/5; **pp**(2)3–4(5) about equal and longest; **p5** <

Ad ♀, USA, Apr: duller version of ♂, with paler lores and no breast-band (or just slight rusty flecks at sides) while limited bluish in crown is green-tinged; yellowish-green mantle duller and more extensive. Aged as ad ♂ above (but especially note dusky primary-coverts with bluish edges). (D Monticelli)

1stS ♂, Cuba, Mar: some 1stS ♂♂ exceed ad ♂ in their colourful plumage, especially in having a near-complete rufous breast-band with black upper border, so ageing must rely on moult pattern—note here juv brown large alula and primary-coverts (latter green-edged), and also brownish primary tips. (H. Shirihai)

HANDBOOK OF WESTERN PALEARCTIC BIRDS

wt 0–1.5 mm; **p6** < wt 4–6 mm; **p7** < wt 7–10 mm; **p10** < wt 11–15.5 mm; **s1** < wt 13–17 mm. Emarg. pp3–5.

GEOGRAPHICAL VARIATION & RANGE Monotypic. (Sometimes a western subspecies 'ludoviciana' is recognised as being subtly smaller, with ♂ having more prominent breast pattern, but not followed here.) – Much of E Canada and E USA, very sporadically also in the west; winters Mexico, Central America, S Florida, West Indies.

TAXONOMIC NOTE Previously known as *Parula americana*, this species was moved to *Setophaga* by Lovette *et al.* (2010) to avoid polyphyly. In this comprehensive study of the Parulidae, it was shown that the genera *Setophaga* and *Dendroica*, plus *Wilsonia citrina* and two species of *Parula*, were closely related. Therefore, all these species were proposed to constitute a single genus, which takes the oldest available name *Setophaga*, a proposal followed here.

REFERENCES Lovette, I. J. et al. (2010) *Mol. Phyl. & Ecol.*, 57: 753–770.

1stS ♀, **Cuba, Mar**: dullest plumage in spring/summer with less pure grey (more olive) upperparts and usually duller yellow throat and breast, but always best aged by more strongly bleached juv wing-feathers. Note moult limit in greater coverts (found only in *c.* 10% of birds), and brown primary-coverts with green or virtually colourless fringes. (S. Bury)

Ad ♂, **Canada, Sep–Oct**: mainly blue-grey upperparts and well-marked breast coloration, while lores solid black near eye (but closer to bill concealed by pale tips), indicating age and sex. (T. Shimba)

Ad ♀, **USA, Sep**: combination of evenly-feathered wing, large area of white in tail-base, fresh blue-edged primary-coverts, and duller overall plumage, with pale lores and no breast-band, infer age and sex. (I. Davies)

Variation in presumed 1stW ♂♂, **USA, Sep (left) and France, Oct**: sexing in autumn not always straightforward, even when aged first, as some presumed young ♂♂ can be relatively dull. Nevertheless, the rather bright yellow bib with some rusty tones indicates both are probably young ♂♂. Retained primary-coverts and remiges fringed olive-brown and dull bluish infer age. (Left: S. N. G. Howell; right: R. Armada)

1stW ♀, **USA, Oct**: duller still, lacking any breast-band, even more greenish-tinged upperparts and crown, and has least obvious wing-bars. Moult pattern as 1stW ♂, but especially note subtly duller juv primary-coverts and apparently unmoulted juv outer greater coverts. (F. Jacobsen)

(AMERICAN) YELLOW WARBLER
Setophaga petechia (L., 1766)

Fr. – Paruline jaune; Ger. – Goldwaldsänger
Sp. – Reinita de manglar; Swe. – Gul skogssångare

Very rare vagrant on this side of the Atlantic with about a dozen records in Britain, where it is unusual in having appeared several times in August, well before most transatlantic passerine vagrants are usually spotted. A few records are also known from Iceland and other Atlantic islands. Breeds in scrub, second growth and thickets of North America from Alaska and Canada south virtually throughout USA and parts of Mexico; winters from extreme S USA and West Indies south to C South America. A polytypic species, but all vagrants believed to be of the subspecies *aestiva*.

Presumed *S. p. aestiva*, ad ♂, USA, May: striking golden-yellow face and underparts, sometimes with chestnut stains on cap, and well-developed (but age-related variable) rufous streaking below. Contrasting yellow edges to tertials. Inner webs of tail-feathers diagnostically all yellow. Note hint of yellowish wing-bars. Evenly-feathered wing supports age. (R. Royse)

IDENTIFICATION An easily identified North American wood warbler. Adult ♂ is very striking with *bright yellow head, underparts and uppertail-coverts*, the yellow body *finely streaked rufous*. Rest of upperparts yellowish olive-green, nearly plain or only lightly streaked darker. Crown sometimes has faint rufous hue. The *black eye in an otherwise plain yellow face* is another typical feature, as is the plump, short-tailed appearance and *yellow undertail*. Note *contrasting yellow edges to tertials*. Inner webs of tail-feathers (save all-dark central pair) *diagnostically mostly to completely yellow*. At close range, pale eye-ring visible, as are hint of yellowish double wing-bars. ♀ and young in autumn are duller and plainer, with much less noticeable streaking, or none at all, and yellow is often paler. Especially in these plumages, the pale yellow eye-ring is quite conspicuous. Very active when hunting for insects, and habitually bobs tail.

VOCALISATIONS Often rather vocal, giving a frequently repeated 'tongue-clicking', loud *tschip*.

SIMILAR SPECIES Only likely to be confused with other North American wood warblers in non-breeding plumage, e.g. young *Tennessee Warbler* (which see), but whitish vent and undertail-coverts, better-marked face pattern and plainer wings of latter should readily prevent confusion. — Yellow Warbler might also be confused with *Hooded Warbler*, which has white (not yellow) outertail-feather patches, paler legs and a less yellow head. — From *Wilson's Warbler* by yellow in tail and lack of prominent yellow lores (or entire supercilium in ♂ of latter). Both Wilson's and Hooded Warblers are also proportionately much longer-tailed, have less obvious pale wing-feather fringes and wing-bars than Yellow Warbler, and generally appear pale-faced and dark-capped, respectively. — Differs from *Common Yellowthroat* by quite different jizz and structure, and combination of same plumage features mentioned for previous species. — Yellow Warbler's distinctive shape (including relatively short tail), in many birds rufous underparts streaking, pale eye-ring, unique yellow pattern to outer rectrices, and overall uniform plumage should eliminate confusion with other superficially similar and potential vagrant parulids (see Vagrants section, p. 591), even in dullest autumn plumages.

AGEING & SEXING Ageing often requires a close check of moult and feather wear in wings and tail. Sexes differ mainly in breeding plumage. Seasonal plumage variation limited. — Moults. Complete post-nuptial moult of ad (occasionally suspended during migration) in Jun–Sep and partial post-juv moult (including head, body, some to all median and greater coverts, and usually greater alula and some tertials) generally May–Sep. Pre-nuptial moult (mostly partial, may include some to all greater coverts, tertials and even some secondaries) in Dec–Apr. — **SPRING Ad ♂** Unrecorded in Europe. Bright gold-yellow face and underparts, sometimes with chestnut wash to cap, and well-developed (but variable) rufous streaking. **Ad ♀** Like ♂, but underparts less brilliant yellow, rufous streaking much reduced or occasionally entirely lacking, and olive-green upperparts includes nape and crown. **1stS** Very like respective ad, but typical birds sometimes possible to age by retained juv wing-feathers (especially contrastingly worn and bleached primaries, primary-coverts and perhaps some unmoulted outer greater coverts, alula and tertials), and more worn, tapered and less yellow tail-feathers; ♂ has slightly greener crown than ad and better-developed flanks-streaking than ad ♀. Beware that ad may show confusing moult contrast in wings following pre-nuptial moult. — **AUTUMN Ad** Sexes more alike, both being less strongly patterned, with (especially) fewer chestnut streaks below and more olive hue above, ♂ otherwise much as spring. **1stW ♂** Very like ad ♀, being overall greener (crown only slightly yellower than upperparts, and less vivid yellow underparts which lack or have reduced rufous streaking), with some whitish (not yellowish) fringes to inner wing-feathers. **1stW ♀** Rather uniform yellowish-green with utterly 'plain' olive appearance (slightly brighter rump and undertail-coverts), mostly lacking streaking below. Tips of greater and median coverts dull yellow, retained tertial fringes whitish. For reliable separation from ad, in both sexes look for retained juv primaries, primary-coverts, sometimes some outer greater and median coverts, and some alula and tertials (narrower and weaker textured, less fresh with narrower dull yellowish or greenish-buff fringes). Also, note more pointed juv tail-feathers with smaller yellow patches in outermost feathers (restricted to inner fringes). **Juv** Dull and brownish-washed overall. Not expected to ever reach the covered region.

BIOMETRICS (*aestiva*) L 10.5–12 cm; **W** ♂ 64–70 mm (*n* 12 *m* 66.2), ♀ 58–65 mm (*n* 12, *m* 61.9); **T** ♂ 44–49 mm (*n* 12, *m* 46.4), ♀ 42–46.5 mm (*n* 12, *m* 43.8); **T/W** *m* 70.4; **B** 12.2–14.0 mm (*n* 24, *m* 13.2); **B**(f) 9.2–11.0 mm (*n* 23, *m* 10.0); **BD** 3.2–4.0 mm (*n* 23, *m* 3.6); **Ts** 17.3–19.6 mm (*n* 22, *m* 18.5). **Wing formula**: p1 minute; **p2** < wt 0–1.5 mm, =4/5 or =wt; **pp**(2)**3–4** about equal and longest; **p5** < wt 0.5–3 mm; **p6** < wt 4–7 mm; **p7** < wt 7–10 mm **p10** < wt 13–17.5 mm; **s1** < wt 15–19 mm. Emarg. pp3–5.

GEOGRAPHICAL VARIATION & RANGE Complex taxonomy. Three groups of subspecies recognised: the migratory (northern) *aestiva* group (described above), Golden Yellow Warbler (the *petechia* group, confined to mangroves in the Caribbean region) and 'Mangrove Yellow Warbler' (*erithachorides* group, which is also confined to mangroves, from N Central America to W South America). While the *aestiva*

Presumed *S. p. aestiva*, ♀, USA, Apr: underparts less deep yellow than ♂, with rufous streaking much reduced or lacking, and olive-green upperparts extending to nape and crown. Not easy to age, especially in spring since both age-groups renew coverts and tertials in winter, and this bird is best left unassigned. The bright yellow plumage suggests ssp. *aestiva*, but ♀♀ difficult to assign to race. (A. Murphy)

Presumed *S. p. aestiva*, ad ♂, Madeira, Aug: when fresh in autumn less strongly patterned, more ♀-coloured with more subdued chestnut streaks below (note how their appearance on this same bird changes with angle) and more olive above. All wing seems evenly fresh, and amount of chestnut streaks already in Aug supports ageing. (T. Olsen)

and *petechia* groups are rather similar, ♂ Mangrove Yellow Warbler usually has completely brick-red hood. As might be expected, birds recorded in the W Palearctic all represent the *aestiva* group, either (perhaps most likely) the brighter and more vividly-plumaged *S. p. aestiva* (S Canada, NC & NE USA, east from the Great Plains south to *c.* 33°N) or the duller and weakly differentiated *amnicola* (breeds north of *aestiva*, from Newfoundland west to NE British Columbia, the Yukon and C Alaska), but autumn birds are difficult to label reliably except to subspecies group.

S. p. aestiva (J. F. Gmelin, 1789) (S Canada, NC & NE USA, east from the Great Plains south to *c.* 33°N; winters Central and N South America). Described above.

TAXONOMIC NOTE See Northern Parula for reasons to move Yellow Warbler from *Dendroica* to *Setophaga*.

REFERENCES LOVETTE, I. J. *et al.* (2010) *Mol. Phyl. & Ecol.*, 57: 753–770.

Presumed *S. p. aestiva*, presumed ad ♀, USA, Sep: combination of rather rich yellow plumage (but duller upperparts and no rufous streaking below), and being possibly ad (apparently evenly-feathered wing with dark and yellow-edged primary-coverts) suggests ad ♀. Racial identification of ♀♀ in autumn often difficult, but strong yellow colour might favour *aestiva*. (A. Murphy)

Scotland, Aug: especially in autumn, pale yellow eye-ring can be quite conspicuous. Rather greener-tinged upperparts, quite vivid yellow underparts and seemingly juv primary-coverts suggest 1stW ♂, while very dark and glossy and neatly pale-tipped primaries would fit ad ♀, thus best left undecided. With such uncertainty with a vagrant, subspecies assignment best not attempted either. (H. Harrop)

1stW, presumed ♀, Azores, Oct: rather uniform dull olive upperparts (slightly brighter on rump/uppertail-coverts, though not visible here), creamy yellowish underparts with no clear streaking, and retained tertials with whitish fringes make it a young bird. Primaries, primary-coverts, and large alula also juv. Dull plumage suggests ♀, but small percentage of 1stW ♂♂ can also appear this dull. (D. Monticelli)

BLACKPOLL WARBLER
Setophaga striata (J. R. Forster, 1772)

Fr. – Paruline rayée; Ger. – Streifenwaldsänger
Sp. – Reinita estriada; Swe. – Vitkindad skogssångare

Of the North American wood warblers, this is the most frequent vagrant to W Europe, presumably due to its typically very long migrations: it breeds as far north as Alaska and is occasionally observed south to Argentina and Chile. The Blackpoll Warbler breeds in boreal forests of North America from Alaska and across much of Canada to the NE USA; and mainly winters in N South America. Most records in Europe have been made in late autumn, with one fairly recent (2000) spring occurrence in Britain.

1stS ♂, USA, May: unmistakable as ♂, with black cap, lateral throat-stripe bordering white cheeks. Note black-streaked white underparts, grey upperparts, and prominent white wing-bars and tertial fringes. Rather robust (and relatively large) with proportionately long wings but short tail and bill. Worn and browner wing (except moulted and fresh median and inner greater coverts, latter forming contrast with outer ones) defines age. (A. Murphy)

IDENTIFICATION Unmistakable in adult plumages, especially breeding ones, but it is first-winters that are most frequently recorded in Europe, and these require careful separation from very similar Bay-breasted Warbler (*S. castanea*, recently recorded in Britain; see Vagrants section p. 594). In breeding plumage sexes are readily separated in that ♂ has a *black cap and boldly black-streaked white underparts*, whereas ♀ is duller with finely streaked olive-brown crown and nape, and is washed yellowish-olive or dusky grey below with subdued streaking. Non-breeding Blackpoll Warbler resembles adult ♀, has *olive-green, faintly streaked upperparts, two variably broad yellow-white wing-bars*, olive-tinged face with *faint dull yellow supercilium* (slightly emphasized by short dark eye-stripe), *throat and breast*, whitening on rear underparts and being *narrowly streaked*, especially on sides of breast and flanks. Relatively large with proportionately *long wings* and undertail-coverts. *Thin, short bill* and *short tail* with *large white patches distally* on outer three pairs of tail-feathers characteristic.

VOCALISATIONS Vagrants in Britain have been recorded giving a thin Goldcrest-like *zree* call. Otherwise, usual call in America is a sharp clicking *chip*, apparently identical to that of Bay-breasted Warbler.

SIMILAR SPECIES Adult may briefly recall much more boldly-patterned *Black-and-white Warbler*, but given good view they are not confusable. – Non-breeders may require careful separation from very similar plumages of *Bay-breasted Warbler* (these two generally exhibit similar structure, behaviour and calls). In comparison to latter, Blackpoll exhibits greater contrast between yellow of breast and white rear underparts, especially undertail-coverts, and also lacks the richer buff tinge to flanks, vent and undertail-coverts of most, but not all, Bay-breasted. Blackpoll also has more noticeably (but still relatively weakly) streaked breast-sides and flanks, duller, more olive-green fringes to remiges, darker olive (less lime-green) upperparts and more strongly marked eye-stripe and supercilium. Colour of tarsus and toes is generally most diagnostic character, Blackpoll having paler, yellower legs (with yellow soles and toes, often extending to tarsus), whereas Bay-breasted has greyish legs without yellow. Also, bill is generally thinner and shorter, and whitish primary tips more distinct in Blackpoll.

AGEING & SEXING Handling birds to check moult pattern and feather wear in wing often essential for reliable ageing, especially in autumn when all age and sex classes are quite similar. Note also that in spring, both ad and 1stS can show moult contrasts in wing. Certainly, sexing of most 1stW impossible unless wing-length assists. Sexes differ almost exclusively in breeding plumage. Seasonal plumage variation in ♂ rather marked. – Moults. Complete post-nuptial moult of ad and partial post-juv moult (including head, body, all median and greater coverts, and usually greater alula and some tertials) mostly in Jun–Sep. Partial pre-nuptial moult, chiefly Feb–Apr (May), extensive in ♂ to permit rather dramatic change of appearance between winter and spring. In both sexes and age categories may include substantial number of greater coverts and tertials (then creating moult limits), sometimes all. – **SPRING Ad** ♂ Distinctive: black cap and lateral throat-stripe bordering white cheek, otherwise grey upperparts streaked black, mainly white below with black-streaked sides, and quite pure white wing-bars and tertial fringes. Some have duller black cap and more olive upperparts. **Ad** ♀ Lacks dark cap, overall duller and more olive-green (less grey) above with paler and narrower streaks below. **1stS** Generally like respective ad, best aged by more worn, browner juv wing-feathers and tapered, subtly narrower tail-feathers. ♂ duller on average and has narrower lateral throat-stripe. – **AUTUMN Ad** ♂ Basically like ad ♀ being olive grey-green above mottled or streaked dark. Usually the most heavily streaked below (including at base of lateral throat-stripe) and above (including black centres to back feathers and sometimes with some black in ear-coverts) at this season. Ground colour of underparts can be quite yellow. **Ad ♀/1stW** Like ad ♂ in autumn, thus greenish above and tinged yellowish below. Indistinct underparts streaking, and little streaking or none on crown. Ad ♀ sometimes possible to identify by subtly more rounded tail-feathers with larger white patches subterminally on outer rectrices, and by on average more diffuse dark streaking or mottling on mantle, but inseparable even in the hand. Ageing of 1stW sometimes possible as tail-feathers are relatively narrower and more pointed at the tips, and outer rectrices usually have smaller or narrower white patches. Also, streaking of mantle within each sex averages more distinct than in ad (see Pyle 1997). Sexing of 1stW mostly impossible except extreme birds, and by using biometrics. Some 1stW ♂♂ better marked, approaching some ad ♂♂ at this

♀, presumed ad, USA, May: ♀ by duller and more olive-green (less grey) head and upperparts and paler (yellowish-olive or dusky) underparts with fewer black streaks, and no black cap. Compared to bird in image above the wing is neater and more evenly feathered perhaps indicating ad, still primary-coverts seem a bit brown, and a firm ageing best avoided. (A. Murphy)

Ad ♂, USA, Sep: during autumn in fresh winter plumage, ♂ becomes similar to ♀, being olive grey-green above streaked dark, but note black patches at base of lateral throat, which is diagnostic of age and sex at this season, while ground colour to underparts can be quite yellowish-tinged as in a ♀. Wings and tail are ad. (L. Spitalnik)

England, Oct: especially in autumn, ageing often difficult, and sexing harder still (except some extreme birds). This individual illustrates well the challenge: entire wing appears quite fresh and uniform with dark large alula suggesting ad, while probability for a European vagrant strongly favours 1stW. Since ageing unsure, sexing cannot be attempted. (G. Thoburn)

Presumed 1stW ♂, USA, Sep: being presumably 1stW (seemingly duller and rather pointed juv primary-coverts) in combination with prominently streaked back, greenish head and clear yellow tinge on throat, suggest young ♂. Pay attention to the characteristic broad whitish wing-bars but short and indistinct pale supercilium. (R. Brewka)

1stW, presumed ♀♀, Iceland (left) and Azores, Oct: autumn birds can require careful separation from several other North American warblers (see Vagrants). Look for greater contrast between yellow of throat/breast and white rear underparts, lack of rich buff on flanks, dusky-olive and faintly streaked upperparts, broad whitish wing-bars, bold whitish primary tips, short and indistinct pale supercilium (slightly emphasised by hint of dark eye-stripe) and, especially, bright yellow feet. Juv primary-coverts and lack of heavier streaks below indicate young, while rather dull plumage without strong streaking suggests ♀, but sex uncertain as only extreme examples can be separated. (Left: Y. Kolbeinsson; right: D. Monticelli)

season, but still show less markedly-streaked flanks and lack lateral throat-spotting, though some exhibit black marks on breast. **Juv** Duller than 1stW and diffusely mottled darker, both above and below. Not expected to be recorded this side the Atlantic.

BIOMETRICS L 12–13 cm; **W** ♂ 74–81 mm (n 14, m 77.1), ♀ 71–75 mm (n 12, m 73.1); **T** ♂ 47.5–55 mm (n 14, m 51.0), ♀ 47–51 mm (n 12, m 49.0); **T/W** m 66.6; **B** 12.8–14.0 mm (n 25, m 13.4); **B**(f) 9.3–11.0 mm (n 25, m 10.1); **BD** 3.5–4.2 mm (n 23, m 3.8); **Ts** 17.2–19.9 mm (n 23, m 18.7); **white patch r6** ♂ 14–22 mm, ♀ 12–19 mm. Wing formula: **p1** vestigial; **pp2–4** about equal and longest (p4 sometimes to 1.5 mm <); **p5** < wt 4–7.5 mm; **p6** < wt 8–12 mm; **p7** < wt 11–16 mm; **p10** < wt 19–24.5 mm; **s1** < wt 21–27 mm. Emarg. pp3–4.

GEOGRAPHICAL VARIATION & RANGE Monotypic. – Alaska and across much of Canada to NW and NE USA; winters in N & C South America. (Syn. *breviunguis*; *lurida*.)

TAXONOMIC NOTE See Northern Parula for reasons to move Blackpoll Warbler from *Dendroica* to *Setophaga*.

REFERENCES Lovette, I. J. et al. (2010) *Mol. Phyl. & Ecol.*, 57: 753–770.

YELLOW-RUMPED WARBLER
Setophaga coronata (L., 1766)

Alternative name: Myrtle Warbler, Audubon's Warbler

Fr. – Paruline à croupion jaune; Ger. – Kronwaldsänger
Sp. – Reinita coronada
Swe. – Gulgumpad skogssångare

Also commonly known as the Myrtle Warbler (then referring only to the eastern races, notably *coronata*), this distinctive and widespread American wood warbler is a rather rare vagrant to W Europe, almost always found in October. It breeds in coniferous and mixed woodlands of North America, from Alaska and over much of Canada, the NE and W USA south to NW Mexico. It is a hardy bird wintering rather far north for a parulid. Geographical variation quite well marked, sometimes leading to the view that it is best split into two species, the eastern one being the Myrtle Warbler, which is that recorded in Europe.

S. c. coronata, 1stS ♂, Canada, Jun: conspicuously patterned with yellow rump, throat and narrow median crown-stripe and striking yellow breast-side patches. Typical also are bold black markings below which sometimes merge into a solid patch. Upperparts largely bluish-grey streaked dark, with two whitish wing-bars. Mainly heavily abraded and brownish juv primary-coverts and remiges (only outer one or two have been renewed). (A. Murphy)

IDENTIFICATION Conspicuous *yellow rump, crown and breast-side patches* make adults, especially breeding ♂, easily identified, whereas first-winters, in particular the young ♀, are more nondescript, but all post-juv plumages have the conspicuous yellow rump. Upperparts of breeding adults largely grey *streaked dark*, more brownish-olive and less prominently streaked in immature and non-breeding plumages. Head and facial features variably well-marked. Two narrow but *clear white or buff-white wing-bars* and usually rather *boldly dark-streaked white underparts* (both wing-bars and streaks weakest in first-winter ♀) are all useful field marks. Eastern 'Myrtle Warbler', which has turned up in Britain, has invariably white (or buff-white) throat, and adults have dark cheeks and a white supercilium. Note also white patches subterminally in outer tail-feathers.

VOCALISATIONS (*coronata*) An emphatic *chek* or dull *chup* are the most common contact calls. Flight-call has been described as a high-pitched, short *tsip*.

SIMILAR SPECIES Easily distinguished from all congeners by virtue of rather dark and well-streaked, largely grey-and-yellow plumage combined with yellow rump. In the Americas, especially young ♀♀ may occasionally be subject to confusion with several congeners in same plumages (see, e.g., Dunn & Garrett 1997 for good guidance among these), but most of these are as yet unrecorded in Europe. The extremely rare vagrants Cape May Warbler (*S. tigrina*) and Palm Warbler (*S. palmarum*; see Vagrants section) may represent potential confusion risks versus young autumn ♀♀. Cape May Warbler, however, is smaller, with a more slender bill and shorter tail, a dull greenish (not yellow) rump, lacks the yellow breast-side patches, is more finely and uniformly streaked below, and often has a yellowish wash to throat (lacking in 'Myrtle Warbler'). Palm Warbler has a yellowish rump, but this is duller and more ill-defined. Furthermore, Palm Warbler diagnostically pumps the tail constantly and has bright yellowish undertail-coverts, among other separating characters.

AGEING & SEXING (*coronata*) Correct ageing (which should precede sexing, especially in autumn when examining ♀ and young) requires assessment of moult and feather wear in wing. Sexes differ mainly in breeding plumage. Seasonal plumage variation moderate, though quite marked maturation, especially in ♂♂. (In the Americas ageing and sexing complicated due to intergradation with *S. c. auduboni*; see Taxonomic notes.) Partial pre-nuptial moult in both age categories involves replacement of feathers on head and body plus smaller secondary-coverts and a varying number of inner greater coverts. – Moults. Complete post-nuptial moult in ad mostly in Jun–Sep, and partial post-juv moult in same period. Post-juv moult includes head, body, all median and most or all greater coverts, and often greater alula but no tertials. Partial pre-nuptial moult in Dec–May, often includes a few or even many greater coverts. – SPRING **Ad** ♂ Unrecorded in Europe. Readily sexed by black ear-coverts and bright white throat. Both sexes have broken white supercilium, which is brighter and both broader and bolder in ♂. Bold black markings on breast and flanks variable, at times forming an almost solid black patch. **Ad** ♀ Much duller than ad ♂, with brown ear-coverts, browner upperparts, yellow areas paler and black underparts more spotted than boldly streaked. **1stS** Generally like respective ad, but aged by strongly abraded juv wing-feathers (especially tips to primaries and primary-coverts) and worn, more tapered tail-feathers. Both ad and 1stS can show moult contrast in spring as substantial number of greater coverts replaced by both age categories in winter. – AUTUMN In the hand, ad of both sexes separated from 1stW by having on average broader and fresher tail-feathers with more rounded tips (especially valid for the outer pairs; beware individual variation and that not all birds can be aged using tail-feather shape). Some support offered by neat plumage without moult contrast in wing, but many

S. c. coronata, ♀, presumed 1stS, USA, Apr: rather nondescript compared to ♂, with brown ear-coverts, browner upperparts, more spotted breast and ill-defined browner flanks streaking, but still shows species' diagnostic (albeit duller) yellow rump and (smaller) breast-side patches. Probably 1stS by apparently juv (browner) primaries and primary-coverts and alula, with more tapered tail-feathers. Pre-nuptial moult of body-feathers not completed yet, hence plumage transitional. (A. Murphy)

S. c. coronata, 1stY ♀, Grand Cayman, Mar: as late as Mar still in 'winter plumage' (very pointed tips to tail-feathers, brown primary-coverts and dull brown plumage lacking blue fringes confirm age). Yellowish-buff breast-side patches can be very difficult to see in dullest ♀♀, and rump can be completely concealed by wings. Such birds may be confused with same plumages of Cape May and Palm Warblers (see Vagrants), but rump—when seen!—always brighter yellow. (H. Shirihai)

sides varies extensively in both sexes. **Juv** Lacks any yellow in plumage, and is streaked all over, including rump. Similar to some congeners at same age but not expected to occur in Europe.

BIOMETRICS (*coronata*) **L** 12.5–13.5 cm; **W** ♂ 71–79 mm (n 26, m 74.7), ♀ 70–76 mm (n 22, m 72.4); **T** ♂ 51–58 mm (n 25, m 55.2), ♀ 52–57 mm (n 21, m 54.2); **T/W** m 78.5; **B** 11.0–13.8 mm (n 24, m 12.5); **B**(f) 7.0–9.6 mm (n 24, m 8.9); **BD** 2.9–3.7 mm (n 23, m 3.3); **Ts** 16.2–20.0 mm (n 24, m 18.4); **white patch r6** ♂ 17–23 mm, ♀ 17–21 mm. **Wing formula: p1** minute; **p2** < wt 0–3.5 mm, =3–5/6; **pp**(2)**3–5** about equal and longest (p5 sometimes to 3 mm <); **p6** < wt 4.5–8 mm; **p7** < wt 8.5–12 mm; **p10** < wt 15.5–19 mm; **s1** < wt 15.5–22 mm. Emarg. pp3–5 (rarely also faintly on p6).

GEOGRAPHICAL VARIATION & RANGE Polytypic with rather clear differences between the two main groups, eastern 'Myrtle Warbler' comprising *coronata* and *hooveri*, having in common a white throat and in adults slate-grey or blackish cheeks and white supercilium, and western and southern 'Audubon's Warbler' comprising three further races, which in ad have in common yellow throat and plainer grey head-sides. Only the first-mentioned subspecies is believed to have straggled to Europe, and is the only one treated here.

    S. c. coronata (L., 1766) 'MYRTLE WARBLER' (Canada, C & E USA; winters S North America, West Indies, Central America). Treated above. Very similar to more westerly ssp. *hooveri* but subtly smaller and has on average bolder black markings below and often a little more brownish suffusion in plumage.

TAXONOMIC NOTES *S. c. coronata* (Myrtle Warbler) and *S. c. auduboni* ('Audubon's Warbler') were formerly considered separate species, but they intergrade on the east slope of the Canadian Rockies. They are therefore perhaps better treated as one species with two distinct race groups. – It has

1stW are almost as fresh looking with no easily detectable moult contrast. In the field, note that ad (both sexes) has on average subtly broader outer edges and tips to tertials which are evenly rufous-tinged (1stW: on average narrower and more whitish edges or at least tips), darker black centres to back feathers, and bolder dark streaking on underparts (at least compared to 1stW ♀). **Ad** ♂ Upperparts browner than in spring (though usually still some grey admixed, especially on scapulars, back and uppertail-coverts), and large visible black centres to blue-grey uppertail-coverts. Head pattern more like ♀, but occasionally some black on lores and in ear-coverts patch, and yellow rump patch intense, whereas patch on crown partly or even wholly concealed by brown tips. Amount of yellow on breast-sides variable. Usually quite bold black streaks on breast and flanks. **Ad** ♀ Overall browner than ad ♂, including uppertail-coverts, with duller and smaller dark centres to back feathers and smaller and less bright yellow marks, but otherwise often inseparable from ad and 1stW ♂♂ in the field. **1stW** Aged in the hand (with difficulty) by on average narrower, more pointed and generally more abraded or 'frayed' juv tail-feathers with subtly smaller white patches on outermost feathers. Also on average narrower and more pointed primary-coverts (with on average narrower pale fringes) and sometimes visible moult contrast between the odd unmoulted outer greater coverts and moulted inner. Browner uppertail-coverts with relatively narrower and more indistinct dark central streaks. (See also ad in autumn.) **1stW** ♂♀ Usually inseparable, but any that lack any yellow on crown (even hidden under brown tips) should invariably be a ♀, and any with blackish feathers on lores and ear-coverts are probably ♂. Uppertail-coverts differ on average in that most ♂♂ have these bluish with larger black central streaks, whereas most ♀♀ have them brownish with diffuse or no dark central streak, but many are intermediates. Note that amount of yellow on breast-

S. c. coronata, 1stW ♂, USA, Oct: much individual variation in post-juv moult can produce well-marked imm ♂♂, with yellow and blue-grey emerging on crown, lower back and on uppertail-coverts, but note that this bird still has largely juv wing with alula, primary-coverts and remiges brownish and contrasting with moulted greater coverts (blue-grey but, interestingly, with rusty tips). Note blackish lores and hint of eye-stripe. (L. Spitalnik)

NEW WORLD WARBLERS

S. c. coronata, ad ♂, USA, Oct: upperparts and head browner than in spring, still usually some bluish-grey visible, especially on scapulars and uppertail-coverts. Large black centres to blue-grey uppertail-coverts. Head pattern more like ♀, but occasionally as here has a little black on lores and hint of dark eye-stripe. Yellow rump bright. Rather black streaks on breast and flanks, while yellow on breast-sides often rather pale, but brighter than in other autumn plumages. (L. Spitalnik)

S. c. coronata, ad ♀, USA, Dec: being ad (evenly fresh wing) in combination with more brownish-olive upperparts than ad ♂, thinner and less distinct dark streaks above, smaller and duller yellow breast-side patches, and having edges to remiges and rectrices less tinged bluish. Otherwise very similar to some dull 1stW ♂♂ (these two often inseparable if not safely aged first). (A. Murphy)

S. c. coronata, 1stW ♂, Scotland, Sep: black-centred blue uppertail-coverts together with 1stW wing pattern (only visibly moulted feathers are three innermost greater coverts, rest are juv) and pointed tail-feathers confirm age and sex. Some young ♂♂ can have ♀-like brownish head and upperparts, reduced dark streaking on flanks and buffish breast-sides. Such individual variation in autumn plumage must always be borne in mind. (H. Harrop)

S. c. coronata, 1stW, presumed ♂, Scotland, Sep: juv primary-coverts (more pointed, brownish with narrow pale fringes) and more pointed tail-feathers confirm age. Moderately narrow dark streaks on underparts, with hardly any yellow (buffish) on breast-sides, more uniformly brown and less dark-streaked above and on head (no blackish lores) could suggest least advanced young ♂. A better view of uppertail-coverts required to confirm sex. (H. Harrop)

also been suggested that Audubon's Warbler is the product of hybridisation between Myrtle Warbler and the closely-related Black-fronted Warbler *S. (c.) nigrifrons* (Brelsford *et al.* 2011). – See Northern Parula for reasons to move Yellow-rumped Warbler from *Dendroica* to *Setophaga*.

REFERENCES Brelsford, A., Milá, B. & Irwin, D. E. (2011) *Molecular Ecology*, 20: 2380–2389. – Lovette, I. J. *et al.* (2010) *Mol. Phyl. & Ecol.*, 57: 753–770.

S. c. coronata, 1stW ♀, Azures, Oct: combination of being young (moult limit among outer greater coverts and fairly dull-looking primary-coverts), absence of any yellow on crown and having greyish lores suggest ♀. Note also very limited yellow wash on breast-sides, and duller overall plumage. (D. Mitchell)

SCARLET TANAGER
Piranga olivacea (J. F. Gmelin, 1789)

Fr. – Piranga écarlate; Ger. – Scharlachtangare
Sp. – Piranga escarlata; Swe. – Scharlakanstangara

An American passerine, until recently considered a member of the family Thraupidae, but now classified among the cardinals, family Cardinalidae. However, tanagers are largely insectivores, thereby lacking the heavy, conical bill of true cardinals adapted for seed-eating. Scarlet Tanager breeds in E North America. It is a highly migratory species, wintering primarily in South America, in the eastern foothills of the Andes and in W Amazonia, as far south as Bolivia. Most records in the covered region have been made in autumn, mainly in the Azores, Britain and Ireland.

IDENTIFICATION Larger than Common Rosefinch or Bullfinch, but compared to the latter *outline is rather slender*, with *relatively long wings and tail*. Bill fairly short and stout with clearly curved upper mandible profile. Unmistakable summer ♂ *predominantly bright red* (flame scarlet) *with contrasting jet-black wings and tail*. Vagrants to this side of the Atlantic, however, can be expected only in ♀-like plumages, *dull greenish above, dusky yellowish below*, with *dark brownish or blackish wings*. In moult from late summer to autumn, ♂♂ show blotchy green and red to mostly yellow-green plumage in winter, when some only separable from ♀ by still-blackish scapulars, wings and tail. First-year ♂ midway between ♀ and winter adult ♂ (cf. Ageing & Sexing). Wings generally uniform, but when fresh some pale fringes to wing-coverts and tertials can form a hint of wing-bars. Adults of both sexes have medium grey to pale horn-coloured bill and legs. Often forages high in trees or sallies to catch insects in flight, though vagrants reported to exploit a variety of habitats.

VOCALISATIONS Call (given by either sex, including alarm or in territorial dispute) is quite characteristic, a hard chip followed by a louder, drawn-out buzzy note, *chik-drrreey*; the chip can be given alone or as double *chik* note without the *drrreey*. Flight-call a clear whistle, *puwi*.

SIMILAR SPECIES Given reasonable views unlikely to be confused with any native species in our region, even in ♀-like plumages. The ♂'s red coloration is even more intense and deep red than ♂ *Summer Tanager* (see Vagrants section), which also lacks black wings of Scarlet. Duller ♀♀, immature and non-breeding ♂♂ must be distinguished with care from corresponding plumages of Summer Tanager, which are generally more brownish overall with more yellowish (rather than whitish) wing fringes when fresh. Summer Tanager is usually also notably larger and heavier with a longer bill, although these differences require familiarity with both species.

AGEING & SEXING Rather complex seasonal plumage variation, pronounced in ♂♂, but reliable ageing (and often sexing) requires close inspection of moult patterns. In spring–summer, sexes easily separable, but following post-nuptial moult all birds generally become ♀-like, though some ♂♂ can be distinguished (see below). – Moults. Complete post-nuptial moult of ad in late summer and early autumn, mainly early Jul–Sep. Partial post-juv moult at about same time does not normally involve flight-feathers, tail-feathers, tertials and primary-coverts, but a variable number of inner (none to all) greater coverts may be replaced. Partial pre-nuptial moult (Jan–May) when summer plumage attained,

Ad ♂, Mexico, Apr: striking scarlet-red plumage with jet-black (variably metallic blue) wings and tail make ad ♂ unmistakable. Rather elongated jizz with comparatively narrow but quite powerful bill. Red even more intense and deeper than ♂ Summer Tanager, which also lacks black wings. (H. Shirihai)

♀, USA, Apr: ♀ is olive-green above, with brownish-grey wings and tail, tertials having thin and subdued yellow-olive fringes, and yellowish underparts. Whole wing seems of similar age and comparatively fresh, still ageing without handling in this case difficult. (A. Murphy)

1stS ♂, USA, May: often mixed red, orange and greenish-yellow, appearing intermediate between two sexes in ad plumage. Ageing usually easy, but much variation with some young ♂♂ more olive and ♀-like, thus less obviously different. Nevertheless, in close views all young ♂♂ should show strong moult contrast in wing, with at least some black wing-coverts versus heavily-abraded and browner juv feathers. (L. Spitalnik)

involves in ad head, body, 8 or all greater coverts and occasionally 1–3 tertials and central tail-feathers. Pre-nuptial moult in 1stY is similar but often more extensive, seemingly also involves some secondaries, and more (at times all) tail-feathers. – SPRING Sex sometimes difficult to decide, leaving aside very obvious scarlet ad ♂♂ and dull yellowish-green ad ♀♀ which have first been aged using moult contrast and feather wear. Both ♀♀ (of any age, especially older) and some least advanced 1stY ♂ can be mostly yellowish-green with variable number of scarlet or reddish-orange feathers on crown, head-sides and, less often, on rest of throat, breast and upper flanks. **Ad ♂** Scarlet-red with contrasting jet-black wings and tail, and edges of tail-feathers still rather fresh. **Ad ♀** Yellowish underparts and olive above, with brownish wings and tail, and yellow-olive feather fringes. For reliable ageing of both sexes attempt to confirm that only ad wing-feathers are present, with only moderately worn primaries and primary-coverts. **1stS ♂** Variable, generally all red, or red, green and yellow forming complex variegated plumage intermediate between the two sexes in ad plumage. Ageing of such birds usually easy due to strong moult contrast in wing, with more heavily-abraded and browner juv primaries, primary-coverts and outer greater-coverts. **1stS ♀** As ad ♀ but note moult limit in greater coverts, outer unmoulted juv being slightly duller with narrower olive edges and with yellow tips in contrast to moulted inner with broader and more diffuse olive edges and no yellow tips. Further, most remiges and primary-coverts, sometimes even tail-feathers (slightly narrower with more pointed tips) are juv, being already quite abraded compared to ad, but moult contrast can be less clear than in young ♂♂. – AUTUMN **Ad ♂** Body plumage generally ♀-like yellowish olive-green, occasionally with some orange on breast and/or rump, or with some isolated red feathers. Ad remiges and tail-feathers blacker than in other plumages. Scapulars usually blackish. Several have a hint of a dusky mask. **Ad ♀** As spring but overall fresher. For reliable ageing, establish presence of only ad-type remiges, tail-feathers and wing-coverts, these being firmer textured, fresher and darker, and tail-feathers have on average rounder tips. **1stW ♂** Overall ♀-like. Like winter ad ♂ has blackish scapulars (though can also have less conspicuous dark grey). Lacks dusky mask. The safest clue for ageing is clear moult contrasts in wing, with new inner wing-coverts being fresher and blackish in contrast to retained duller and less fresh juv primaries, primary-coverts and outer greater coverts, and with juv tail-feathers being narrower and more pointed at tips. Body plumage moderately bright olive-green. **1stW ♀** As 1stW ♂ but body plumage on average a little duller and duskier (though many are alike!), and moulted inner wing-coverts dull olive-grey, not blackish. **Juv** Resembles ♀ but plumage duller and greyer, with dusky streaking below, otherwise body and coverts more fluffy in structure.

Ad ♂, USA, Sep: following post-nuptial and post-juv moults, all birds generally ♀-like, though ad ♂ distinguished by blacker remiges and tail-feathers (no contrast with blackish wing-coverts and scapulars). (L. Spitalnik)

1stW ♀, USA, Sep: vagrants only likely in ♀-like plumages, i.e. dull greenish above and dusky-yellowish below; here a 1stW ♀ with moderately dark wing, and moult limit in greater coverts, outer unmoulted juv coverts yellow-tipped and forming slight wing-bar. Duller ♀-like plumages must be carefully separated from same plumages of larger and heavier Summer Tanager. (L. Spitalnik)

1stW ♂, Azores, Oct: like winter ad ♂, blackish scapulars and secondary-coverts form clear moult limits with duller and less fresh juv primaries, primary-coverts, outer greater coverts and alula, while juv tail-feathers narrower with more pointed tips. (D. Monticelli)

1stW ♀, Azores, Oct: unlike rather similar ad ♀ note juv (already quite abraded brownish) alula, primary-coverts and remiges; also lacks blackish scapulars and secondary-coverts of 1stW ♂. (Left: V. Legrand; right: D. Monticelli)

BIOMETRICS L 16–17 cm. **W** ♂ 92–102 mm (n 29, m 97.2), ♀ 91–98 mm (n 18, m 94.3); **T** ♂ 61–68 mm (n 24, m 65.3), ♀ 61–67 mm (n 16, m 64.0); **T/W** m 71.3; **B** 16.8–19.9 mm (n 21, m 18.3); **B**(f) 13.0–16.5 mm (n 23, m 14.4); **BD** 7.4–9.1 mm (n 24, m 8.0); **Ts** 17.9–20.6 mm (n 23, m 19.3). **Wing formula: p1** minute; **p2** < wt 0–2.5 mm; **p3** longest (sometimes pp2–3); **p4** < wt 1–2 mm; **p5** < wt 2.5–7 mm; **p6** < wt 8.5–15 mm; **p7** < wt 13–20 mm; **p10** < wt 23–30 mm; **s1** < wt 25–33 mm. Emarg. pp3–4 (on p5 often too, albeit slightly less prominently).

GEOGRAPHICAL VARIATION & RANGE Monotypic. – SE Canada and E USA; winters mainly in W South America.

SONG SPARROW
Melospiza melodia (A. Wilson, 1810)

Fr. – Bruant chanteur; Ger. – Singammer
Sp. – Chingolo cantor; Swe. – Sångsparv

A long-tailed mostly ground-dwelling abundant Nearctic sparrow which has been recorded in a few NW European countries, mainly in spring but also twice in autumn. It breeds across Canada and from E USA to W Alaska and south to California, resident in the south but short-range migrant in the north. Often seen feeding on the ground in open forest floor thickets, in weeds along hedges or pond edges, moving with short hops or running longer stretches.

M. m. melodia, presumed ad, USA, Mar: streaky, large-headed and long-tailed American sparrow with rounded tail lacking white sides. Strong conical bill. Rufous crown with narrow greyish median stripe, and dark rufous stripe from eye to nape. Whitish underparts coarsely streaked rufous-black. Upperparts boldly dark-streaked, while rufous-toned wings have indistinct slightly paler wing-bars. Difficult to age but overall good plumage in Mar suggests ad. (R. Royse)

IDENTIFICATION A medium-sized streaky American sparrow with ash-grey, off-white and rufous-brown colours. Characteristic proportions with a head that can appear rather large (due to slightly bull-necked outline) and *long and well rounded tail, lacking any white on sides*. Conical bill grey (but can be partly pink-tinged), *legs brown*. Rufous and black-streaked lateral crown-stripes on each side of suggested narrow ash-grey median crown-stripe. A dark rufous-brown stripe from eye to side of nape and another along lower edge of cheeks (submoustachial stripe) on otherwise ash-grey sides of head including lores. Throat white with bold rufous and black lateral throat-stripes forming triangular patches. Rufous-tinged wings have hint of paler double wing-bars. Breast and belly white distinctly and coarsely streaked rufous-black (except at centre of belly), *streaks sometimes coalescing to a dark patch on centre of breast*. Flanks olive-brown boldly streaked rufous-black. Sexes alike. *In flight pumps up and down with tail* in peculiar manner (a habit shared with Lincoln's Sparrow (see vagrants section, p. 601).

VOCALISATIONS Main call often described as a 'nasal, hollow *chimp*', to a European perhaps best described as a muffled call of a Great Spotted Woodpecker *Dendrocopos major*, *chipp*. Often (irregularly) repeated.

SIMILAR SPECIES Must first of all be separated from *Savannah Sparrow* (see vagrants section, p. 600), but this has clearly shorter and slightly notched tail, each tail-feather being pointed at tip, white rather than grey median crown-stripe, a little yellow on fore supercilium, pink legs and largely pinkish bill. Its plumage is also more brown and buff-white, not as rufous and grey as Song Sparrow. – *Lincoln's Sparrow* is also fairly similar and shares the habit of often pumping tail in flight but differs by finer bill, less bold underparts streaking and slightly shorter tail. It has also less rufous, more greyish-brown upperwing and tail. – *Swamp Sparrow* (*Melospiza georgiana*; unrecorded within treated region) is smaller, has finer bill with a little yellow at base of lower mandible, darker lateral crown-stripes and much less streaking on underparts. – Could theoretically and in a poor view be mistaken for an immature *White-throated Sparrow*, a similarly abundant species with similar habits, but this has finer bill, unstreaked white or whitish throat, a grey-brown wash over dark-mottled rather than black-streaked breast, a buff or yellowish tinge to forepart of supercilium and a square or even notched tail-tip.

AGEING & SEXING Ageing mostly possible in autumn and in the hand. No sexual dimorphism. Age-related seasonal variation (1stW and ad differ). – Moults. Complete post-nuptial moult in ad, and partial post-juv moult, generally Jul–Oct. Post-juv moult is variable in extent, always includes head, body, usually all secondary-coverts and some tertials, and often central tail-feathers, sometimes more tail-feathers (rarely all). Sometimes some innermost secondaries and up to seven outer primaries moulted, too, but no primary-coverts. Pre-nuptial moult absent. (Partly from Pyle 1997.) – **SPRING Ad** Pale grey sides of head with dark rufous stripes, crown has narrow pale greyish median stripe. Bill varies from dark grey to grey with a faint pink tinge to lower mandible. Centre of underparts white with neat distinct streaking. **1stS** As ad, and sometimes inseparable but averages slightly duller with less neat head pattern and buff-tinged off-white underparts with slightly more blurred streaking. With practice, some birds in the hand can be reliably aged by stronger wear to retained juv primaries, primary-coverts and tail-feathers. – **AUTUMN Ad** Differs from spring plumage only by generally fresher appearance. Tail-feathers broader and with firmer structure, outer ones more obtuse (less pointed) at tips. Greater coverts rather evenly fringed buff. **1stW** Differs from ad as in spring. Any retained juv tail-feathers somewhat narrower and of less firm structure, tips often slightly abraded early on. (Note that outer tail-feathers have rather rounded tips.) Outer unmoulted greater coverts (and often tertials) have more whitish tips (less buff) divided at centre (along shaft) by blackish streak along shaft. Juv primary-

M. m. melodia, 1stW, USA, Nov: a dark autumn bird, differentiated from superficially similar Savannah Sparrow (see Vagrants) by longer-tailed appearance, more rufous and buff-white (not brown-and-grey) upperparts, white (versus grey) median crown-stripe, slightly buffish fore supercilium (no yellow supraloral spot), and largely greyish (not pinkish) bill. Aged by juv, narrower, more pointed and slightly abraded tail and apparently unmoulted tertials. (R. Royse)

coverts subtly narrower with more indistinct pale fringes, but experience required before all (or even most) can be recognised. **Juv** Not expected to occur in the treated region.

BIOMETRICS (*melodia*) **L** 14–15.5 cm; **W** ♂ 65–71 mm (*n* 17, *m* 68.1), ♀ 63–69 mm (*n* 13, *m* 64.3); **T** ♂ 63–70 mm (*n* 17, *m* 66.6), ♀ 58–67 mm (*n* 13, *m* 64.3); **T/W** *m* 98.2; **TG** 7–14 mm; **B** 11.5–14.6 mm (*n* 31, *m* 13.7); **B**(f) 10.0–12.4 mm (*n* 30, *m* 10.9); **BD** 6.2–7.9 mm (*n* 29, *m* 7.3); **Ts** 20.4–23.7 mm (*n* 31, *m* 21.7). **Wing formula: p1** minute; **p2** < wt 5.5–9 mm, =8, =8/10 or 10/ss; **p3** < wt 0–3 mm; **pp**(3)**4–5**(6) about equal and longest; **p6** < wt 0–3 mm; **p7** < wt 1.5–5 mm; **p8** < wt 4–7.5 mm; **p10** < wt 7–12 mm; **s1** < wt 9–14 mm. Emarg. pp3–6.

GEOGRAPHICAL VARIATION & RANGE Polytypic with about 25 subspecies recognised in most modern handbooks (and about twice as many described). Extremes rather different but neighbouring subspecies quite similar, connected clinally making exact racial identification difficult. Vagrants to W Palearctic not firmly identified as to subspecies, but reasonably only *melodia* is involved.

M. m. melodia (A. Wilson, 1810) (WC Canada east to Newfoundland and much of E USA; winters S USA in Maryland to Florida and west to S Texas). A rather variable race as to darkness and prominence of streaking, birds in coastal areas from Long I. south to C North Carolina ('*atlantica*') tending to be a touch darker and stronger streaked but much overlap and variation (D. Sibley, pers. comm.), hence here provisionally included. (Syn. *atlantica*.)

REFERENCES PATTEN, M. A. & PRUETT, C. L. (2009) *Systematics and Biodiversity*, 7: 33–62. – VAUGHAN, T. (1994) *BW*, 7: 407–409.

M. m. melodia, 1stW, USA, Nov: paler autumn example. Relatively dull with buffy submoustachial and slightly greyish supercilium, superficially like Lincoln's Sparrow (see Vagrants), but differs by slightly longer tail, chunkier bill, much whiter breast with much bolder underparts streaking, and more rufous upperwing and tail. Ageing as previous bird, although here the tertials have been moulted. (R. Royse)

M. m. melodia, USA, Jan: another pale but strongly greyish-tinged bird, which could also be mistaken for Lincoln's Sparrow, especially given greyish supercilium. Without close view of entire wing, ageing impossible. (R. Bonser)

Germany, Apr: apparently one of interior W North American subspecies, e.g. *montana*, *fallax*, or an intergrade (P. Pyle, pers. comm.). Given this, it could be an escaped cagebird, as these taxa are at most short-distance migrants. Difficult to age, but the bird seems rather fresh. (M. Gottschling)

WHITE-CROWNED SPARROW
Zonotrichia leucophrys (J. R. Forster, 1772)

Fr. – Bruant à couronne blanche; Ger. – Dachsammer
Sp. – Chingolo coroniblanco; Swe. – Vitkronad sparv

This attractive Nearctic sparrow is an extremely rare vagrant to several NW European countries, mainly in autumn but also in spring. It frequents shrubby areas across N North America from Alaska to Newfoundland, but in the west extending south to California, being less widespread in the east. It winters widely across the USA and in N Mexico. Often seen feeding on the ground, moving with short hops.

Z. l. leucophrys, presumed ad, USA, Apr: bold black-and-white head, mainly reddish-pink bill, plain greyish cheeks to chest, buffish-brown belly, heavily streaked upperparts and rusty-brown wings with white-spotted wing-bars identify this vagrant. Both age classes can show moult limits in greater coverts in spring, thus despite outer greater coverts looking worn and bleached, these could be strongly 'sun-bleached' ad feathers. This is ssp. *leucophrys* by black lores and pinkish (not tawny) bill. (J. Normaja)

IDENTIFICATION Usually straightforward, the size and shape roughly like a Reed Bunting *Emberiza schoeniclus* and with general plumage pattern broadly recalling Rock Bunting *E. cia*. But identification involves primarily the safe separation from other American species, not from European. Sexes similar. Adults have *bold black-and-white crown and head-sides pattern*, *mainly pinkish bill*, largely *plain greyish face and underparts*, boldly dark-streaked upperparts and brownish wings with *white-spotted double wing-bars*. The long dark tail lacks any white. Note that *throat is grey or pale buffish-grey with no or only slight contrast to cheeks*, never contrastingly white as in White-throated Sparrow *Z. albicollis*. Similarly, it lacks the yellow on fore supercilium so typical of White-throated Sparrow; instead replaced by black in adults, narrow and dull white in immatures. Even first-winters should not pose a problem, being similar to adults, but are in general duller, and lateral crown-stripes and eye-stripe are reddish-brown, not black, while the white supercilium of adult is more subdued, and median crown-stripe is buff-brown, not white. Also, the bill is duller.

VOCALISATIONS Song a rather slow strophe of high-pitched, fine notes, being rather variable in structure but usually commencing with some clear whistles, continuing with a series of buzzy trills, e.g. *eeh uuh-uuhtree-tree-tuu*. – Main calls rather similar to White-throated Sparrow, a metallic *pzit* and a thin high-pitched, drawn-out *tsiiip*. Conversational calls from feeding birds a Yellowhammer-like *bt bt bt...*

SIMILAR SPECIES Only significant confusion risk in Europe is *White-throated Sparrow*, but lacks clearly-defined whitish throat with dark border, never shows yellow on fore supercilium, has broader and striking white median crown-stripe, greyer upperparts and plain ash-grey ear-coverts and neck-sides. First-winter White-crowned separated from White-throated by neater brown and buffish head markings (including indistinct lateral crown-stripe, eye-stripe and lateral throat-stripes), greyer ear-coverts, unstreaked grey neck-sides, breast and flanks, whiter tips to median and greater coverts, and paler bill. White-crowned is also slightly larger and more slender looking (with more erect posture) than White-throated. – Immatures could be confused with several other American sparrows. Differs from *Fox Sparrow* with similar head and neck pattern by its unstreaked underparts (Fox: boldly streaked rufous) and more prominent white wing-bars. Fox Sparrow also averages larger. – *Song Sparrow* likewise has similar head and neck pattern as immature White-crowned but is readily separated by its boldly streaked underparts and prominent dark lateral throat-stripe.

AGEING & SEXING Ageing mostly possible in autumn and in the hand. No sexual dimorphism. Age-related seasonal variation quite pronounced (1stW and ad differ). – Moults. Complete post-nuptial moult in ad, and partial post-juv, generally Jul–Oct. Post-juv moult includes head, body, usually all wing-coverts and some tertials, and often central tail-feathers. Pre-nuptial moult in Jan–May variable, mostly involving just head, part of body, none to most of greater coverts, some tertials and often central tail-feathers. – **SPRING Ad** Pale grey head with crown striped black and white, and white supercilium. Bill varies from yellowish to reddish-pink. **1stS** As ad, and both age categories can show moult contrast in this season. With practice, some birds in the hand can be reliably aged by stronger wear to retained juv primaries, primary-coverts and tail-feathers. – **AUTUMN Ad** Differs from spring plumage only by generally

Z. l. leucophrys, ad, England, Jan: immaculate black-and-white head and evenly fresh wing infer age. Little black in lores perhaps suggests an intergrade *leucophrys* × *gambelii* but pinkish bill and white breast indicate *leucophrys*. Note plain buffish grey-brown lower back to uppertail-coverts (latter only slightly brighter) and long dark tail without any white. (I. H. Leach)

Z. l. gambelii, ad, Azores, Oct: race *gambelii* by pale lores, tawny bill and crisp grey head and breast. This race breeds in Alaska and W Canada and winters S to Mexico, but is only a rare migrant in E North America, thus probably less likely to reach Europe, despite larger population than nominate (P. Pyle, pers. comm.). Still, this one reached the Azores! Full ad plumage in autumn and evenly fresh wing infer age. (D. Monticelli)

fresher appearance. **1stW** Differs from ad by rufous-brown lateral crown-stripes (rather than jet-black), pale buff-brown median crown-stripe (ad: pure white) and sullied eye-stripe. Juv primary-coverts subtly narrower with more indistinct pale fringes. **Juv** Not expected to occur in the treated region.

BIOMETRICS (*leucophrys*) **L** 16–17 cm; **W** ♂ 78–86 mm (*n* 12, *m* 82.5), ♀ 75–81 mm (*n* 12, *m* 78.1); **T** ♂ 71–78 mm (*n* 12, *m* 74.5), ♀ 69–75 mm (*n* 12, *m* 71.2); **T/W** *m* 90.8; **B** 12.2–14.2 mm (*n* 26, *m* 13.3); **B**(f) 9.3–11.7 mm (*n* 26, *m* 10.8); **BD** 6.6–7.9 mm (*n* 24, *m* 7.1); **Ts** 21.5–24.5 mm (*n* 26, *m* 23.3). **Wing formula: p1** minute; **p2** < wt 3–6 mm, =5/6, =6 or 6/7; **pp3–5** about equal and longest; **p6** < wt 1.5–4 mm; **p7** < wt 4.5–10 mm; **p10** < wt 12–18 mm; **s1** < wt 14–21 mm. Emarg. pp3–6.

GEOGRAPHICAL VARIATION & RANGE Polytypic, but apparently only ssp. *leucophrys* has been positively identified in Europe. Some western populations, notably *gambelii* (Alaska, W Canada) differ markedly in plumage, moult and vocalisations but are extremely unlikely to occur in W Europe.

Z. l. leucophrys (J. R. Forster, 1772) (E Canada west to NC Ontario; winters SE USA, Cuba, N Mexico). Treated above. Has black or dark brown supra-loral streak and on average rather dark orange-brown bill with blackish tip. (Syn. *nigrilora*.)

Z. l. gambelii (Nuttall, 1840) (Alaska, W Canada; winters SW Canada to N Mexico). Differs primarily by having paler, often whitish) lores without much contrast to supercilium (lores darker in *leucophrys*), and breast usually slightly paler. Bill on average paler pinkish.

REFERENCES Bending, R. & Bending, S. (2008) *BW*, 21: 14–18.

Z. l. leucophrys, 1stW, Azores, Oct: the young is more nondescript, still easily identified by striking head pattern (rufous-brown lateral crown-stripe and eye-stripe, and grey-tinged pale buff-brown median crown-stripe), white wing-bars and long dark tail without white edges. Note grey or pale buffish-grey throat only slightly contrasting with cheeks. Dark through upper lores, contrasting head pattern and whitish breast indicate *leucophrys*. All greater coverts and tertials renewed, with broad dark rufous edges. (K. Haataja)

Z. l. gambelii, 1stW, USA, Feb: some 1stW retain rusty-brown head markings and dull orange bill into late winter at least, but subspecific characteristics still obvious—note pale lores of *gambelii*. 'False moult limits' in greater coverts with three outermost fringed more rufous, but these are just differently patterned, not different feather generations. (C. Bradshaw)

Z. l. gambelii, ad, USA, Jan: ad by head pattern (young retain full brown-and grey cap until at least Mar) and reddish eye, but some rufous-brown feathers on rear crown-stripes suggest 2ndW (Pyle 1997). Primary-coverts, remiges and rectrices ad with 'false moult limit' among inner greater coverts. Pale lores, tawny bill and grey breast indicate *gambelii*. (J. Normaja)

WHITE-THROATED SPARROW
Zonotrichia albicollis (J. F. Gmelin, 1789)

Fr. – Bruant à gorge blanche; Ger. – Weißkehlammer
Sp. – Chingolo gorjiblanco; Swe. – Vitstrupig sparv

The most frequently recorded North American sparrow in Europe, mostly found in spring in the treated region, mainly in Britain, although it is not quite annual. Breeds in mixed forest and thickets across much of Canada and in the extreme NE USA, and winters over much of the E and S USA and in the coastal west. It is a short- to medium-range migrant in America, but birds thrown off course are obviously capable of crossing the Atlantic, even though it is suspected that most or all are to some extent ship-assisted.

IDENTIFICATION Chunky, well-marked sparrow, readily identified by contrasting *white throat patch* with pencil-thin *black lower border and moustachial stripe*, *white median crown-stripe and broad supercilium* that is characteristically *yellow in front of eye*. *Black crown-sides and eye-stripe*, and *grey ear-coverts and breast*. Upperparts largely rufous-brown and buff, boldly streaked darker, rufous wings with white tips to median and greater coverts forming *spotty white wing-bars*. Tail rather long and lacks any white. Sexes very similar with much overlap in characters. Some adults are duller and close to first-winter birds, which have streaked breast, browner general appearance with brown head-stripes and white of head tinged buffish. Conical *bill mostly dull greyish-horn*, almost black on tip.

VOCALISATIONS Song (sometimes heard in Europe) a few high-pitched clear straight whistles, given in slow, rhythmic pattern and often followed by some shorter notes, e.g. *sooooo seeeeee süüüh dididi dididi*. – Calls include a sharp metallic *chink*, and in flight a high *seeet*, generally similar to White-crowned Sparrow.

SIMILAR SPECIES Can be confused with *White-crowned Sparrow*, but note contrasting white throat bib outlined narrowly black. Also note yellow tint to fore supercilium, lacking in White-crowned. Supporting characters are slightly darker pink-brown or grey-brown bill, darker cheeks, on average narrower or partly broken white median crown-stripe, and in first-winter birds diffusely streaked breast (White-crowned: plain greyish-buff). – See also *Song Sparrow* for differences from that.

AGEING & SEXING Quite pronounced age-related seasonal plumage variation (i.e. 1stW and ad often differ in plumage); combination of several plumage criteria might prove useful for ageing of some known-sex birds and vice versa, but experience required. Sexes almost alike. – Moults. Complete post-nuptial moult and partial post-juv moult including head, body, usually all secondary-coverts but no tertials or tail-feathers, in Jul–Sep. Very limited pre-nuptial moult chiefly involves forebody including head in Mar–May. – SPRING **Ad** Sexes very similar or inseparable. Polymorphic with 'white-striped' and 'tan-striped' morphs, and some

Ad, Scotland, May: unmistakable chunky build and well-marked white throat lined with black, usually broader below, as well as striking ad head pattern with white median crown-stripe and broad supercilium characteristically bright yellow in front of eye, bordered by black crown-sides and eye-stripe. Grey ear-coverts and breast. Upperparts rufous-brown boldly streaked black, with rufous wings and spotty white wing-bars. Tail rather long and lacks white. Ad by dark primary-coverts and very neat wingtips and rectrices. Perhaps bright enough to be a ♂ but caution required in spring, even with well-marked birds. (H. Harrop)

intermediates: white-striped has paler, less contrasting head markings, with browner dark and buffish pale stripes, less pronounced yellow fore supercilium, often has variable mottling on breast and occasionally even a very narrow lateral throat-stripe. ♀ averages duller, quite streaky on chest with browner head-stripes, less yellow on fore supercilium and lores, and has duller legs. Much overlap and intermediates (both sexes and morphs). **1stS** Much as ad but in the hand aged with experience by relatively more abraded retained juv primaries, primary-coverts and outer tail-feathers, while least-advanced birds occasionally still show some or even more breast-streaking and duller greyish-brown iris (reddish-brown in ad). – AUTUMN **Ad** Much as in spring, but overall fresher; wings evenly fresh with broad and relatively well-fringed primary-coverts and tail-feathers. Iris reddish-brown. Breast uniform greyish or only weakly streaked. **1stW** Unlike ad has rather extensively streaked neck-sides, breast and flanks. Also note retained juv remiges, tail-feathers and primary-coverts (less fresh and weaker textured), and greyish-olive iris. Some, particularly ♀♀, very dull and lack any yellow on supercilium. Usually white throat patch less sharply delimited, with quite pronounced dark moustachial and lateral throat-stripes. **Juv** Not expected to occur in Europe.

1stS, England, Jun: unlike similar ad, note vestiges of immaturity. Still partly brown-feathered black crown-stripes, dusky streaking on breast-sides and flanks, and duller greyish-brown iris. However, moult is main ageing criterion; note worn brownish juv primary tips and outer rectrices, which are also pointed. Inner three greater coverts have been moulted. Sexing not possible. (C. Griffin)

1stS, USA, Mar: very streaky fore parts and browner crown-stripes in spring could suggest 1stS ♀, but sexing often best not attempted. Note greyish-sullied supercilium and juv (worn brownish and pointed) primary-coverts and primary-tips, and outer greater coverts contrasting with renewed innermost three, all indicating age. Sexing often best not attempted. (S. Sharma)

Ad, USA, Mar: 'tan-striped morph' with buffish-sullied supercilium. Ad-type primary-coverts and head pattern in autumn clearly indicate age, and also supported by bright head pattern in autumn, but given much overlap, as well as morph variation, reliable sexing extremely difficult. (I. Davies)

BIOMETRICS L 15–16 cm; W ♂ 74–80 mm (n 13, m 76.4), ♀ 70.5–78.5 mm (n 13, m 74.4); T ♂ 68.5–76.5 mm (n 13, m 72.5), ♀ 68–75 mm (n 13, m 71.4); T/W m 95.5; B 12.5–14.5 mm (n 24, m 13.5); B(f) 10.0–11.8 mm (n 24, m 10.8); BD 7.0–8.4 mm (n 23, m 7.6); Ts 21.0–24.3 mm (n 24, m 22.8). **Wing formula**: **p1** minute; **p2** < wt 6–10 mm, =p8/10; **p3** < wt 0.5–3 mm; **pp4–6** about equal and longest (though p6 often 1 mm <); **p7** < wt 3–5 mm; **p8** < wt 6–10.5 mm; **p10** < wt 11–14.5 mm; **s1** < wt 11.5–16 mm. Emarg. pp3–6 (rarely faintly also on p7).

GEOGRAPHICAL VARIATION & RANGE Monotypic. – Canada, NE USA; winters S USA, E Mexico.

Ad, Azores, Oct: all wing and tail ad, and ageing further suggested by overall colourful plumage in autumn. Probably ad ♀ (given strongly streaked underparts) of 'tan-striped morph' (or 'intermediate colour' as buff wash confined to rear of crown-stripes), but sexing of these problematic due to variation. (D. Monticelli)

1stW, Ireland (left: Oct) and USA (right: Dec): note heavily streaked neck- and body-sides, without any yellow on supercilium, while wings (left bird) mainly juv, including remiges and primary-coverts. Quite pronounced dark moustachial and lateral throat-stripes. Sexing probably impossible, but overall dull and very streaky underparts suggest ♀♀ (some extremely dull tan-morph 1stW ♂♂ can look very similar). From White-crowned Sparrow, even in young plumages the contrasting white throat narrowly outlined black readily separates them. Also useful are diffusely streaked breast (plain greyish-buff in White-crowned), darker cheeks, slightly darker pink-brown bill, and on average narrower or partly broken white median crown-stripe. (Left: S. Cronin; right: C. Grande Flores)

DARK-EYED JUNCO
Junco hyemalis (L., 1758)

Fr. – Junco ardoisé; Ger. – Junko
Sp. – Junco pizarroso; Swe. – Mörkögd junco

Despite their slightly odd name, juncos of North and Central America are closely related to buntings of the Palearctic, generally referred to the family Passerellidae, New World sparrows. Winter and spring occurrences in Europe of this vagrant from across the Atlantic are the norm. Like other American sparrows, most vagrants to Europe have occurred in Britain. Breeds in forest and brushy areas of North America, wintering widely across virtually the entire USA and N Mexico.

J. h. hyemalis, ad ♂, England, Apr: medium-small, long-tailed, bunting-like bird, easily identified by dark slate-grey plumage except for white belly and outer tail-feathers. Note blackish face, making pinkish-tinged bill even more striking. Wing evenly-feathered ad. (S. Gray)

IDENTIFICATION A medium-small, rather long-tailed passerine with conical bill. Readily identified on combination of *plain dark grey plumage contrasting with clear-cut white belly*. Undertail also white, as are *prominent white outer tail-feathers*. Note *pale pinkish bill with a dark tip* and *reddish iris*. It is thus impossible to confuse the species with any other species (including escapees). Sexes very similar, and only when series are compared in collections, or when a typical bird is seen close, can they be confidently separated. ♂♂ average purer white on belly and darker slate-grey on hood and chest, whereas ♀♀ tend to be a little more sullied buff-grey on the white belly, and have less dark grey hood and chest, both often subtly brown-tinged. Upperparts of a ♂ usually cleaner grey, not as brown-tinged as in ♀, but age differences must be kept in mind: first-winter birds of both sexes in autumn are more brownish-grey with duller iris. Spends much time on the ground when feeding.

VOCALISATIONS Song is a fast rattling trill, occasionally interspersed with warbles, twitters and *chip* notes, generally somewhat like Yellowhammer, without final wheeze. – Calls include a distinctive high-pitched twittering, a liquid *chek* and sharp *dit* or *chip*.

SIMILAR SPECIES Wholly distinctive.

AGEING & SEXING (*hyemalis*) Assessment of moult pattern often essential for reliable ageing. Sexes almost alike, but differ in measurements and in colour and pattern of tail-feathers. 1stY and ad differ chiefly in autumn, otherwise rather limited seasonal plumage variation. – Moults. Complete post-nuptial moult and partial post-juv moult generally Jul–Oct. Post-juv moult mostly involves head, body, most or all median coverts and tertials, a variable number of greater coverts (from a few inner to all) and often central tail-feathers. Pre-nuptial moult is mostly limited, involving head and forebody, in Feb–Apr. (See Pyle 1997 for some geographical and individual variation in moult.) – **SPRING Ad** ♂ Dark slate-grey except contrasting white belly. Feathering on lores, chin and throat may appear blackish. Pale pink bill with tiny black tip. **Ad** ♀ Differs from ♂ in having grey parts slightly paler and subtly washed brownish (especially upperparts). Also, flanks less clearly ash-grey and centre of belly sullied off-white. **1stS** In the hand aged by relatively more abraded retained juv primaries and primary-coverts, and subtly more pointed and weakly pale-fringed primary-coverts, while least-advanced birds may have duller greyish-brown iris (reddish-brown to red in ad). Birds showing moult limits among median and greater coverts, with contrasting new tertials, more easily aged. Some duller ♂♂ somewhat approach ad ♀ plumage, thus ageing should precede sexing. – **AUTUMN Ad** Overall fresher, otherwise as spring. **1stW** Much like ad, but washed browner and buffier, with some retained juv tertials (browner fringes and whitish tips), and outer median and greater coverts have buffish or whitish tips. Iris greyish-brown to brown (deep red in ad). ♀ usually noticeably browner than ♂, with broad brown tertial fringes. **Juv** Duller colours, plumage largely streaked. Not expected to occur within the treated region.

BIOMETRICS (*hyemalis*) **L** 14–15 cm; **W** a 75.5–83 mm (n 12, m 79.7), o 72–81 mm (n 12, m 75.3); **T** a 64–70 mm (n 12, m 67.0), o 57–68 mm (n 12, m 62.5); **T/W** m 83.5; **B** 11.5–13.3 mm (n 24, m 12.5); **B**(f) 9.0–11.1 mm (n 24, m 10.1); **BD** 6.0–6.7 mm (n 24 m 6.4); **Ts** 19.0–21.7 mm (n 23, m 20.7). **Wing formula: p1** minute; **p2** < wt 5–7 mm, =p6/7 or 7; **pp3–5** about equal and longest; **p6** < wt

J. h. hyemalis, 1stS ♂, Scotland, May: abraded juv primaries and especially contrasting brownish alula and primary-coverts, being duller and somewhat brownish-tinged. Moult limits among greater coverts, and contrasting new central tertials. (H. Harrop)

J. h. hyemalis, 1stS ♀, USA, Jun: about dullest plumage at this season, with less dark grey hood and chest, brown-tinged upperparts and sullied buff-grey around white belly, still general pattern of ad discernible. Ageing as in image to the left. (B. Hubick)

0.5–3 mm; **p7** < wt 5.5–8 mm; **p10** < wt 11.5–18 mm; **s1** < wt 13–18.5 mm. Emarg. pp3–6.

GEOGRAPHICAL VARIATION & RANGE Polytypic, but generally only ssp. *hyemalis* thought to have reached Europe, although it is reasonably probable that *carolinensis* (Appalachian Mts, W Virginia to N Georgia) could also have reached this side of Atlantic occasionally.

J. h. hyemalis (L., 1758) (Alaska, Canada, NE USA to Massachusetts; winters S USA, N Mexico). Treated above.

TAXONOMIC NOTES Formerly considered to represent four (strikingly marked) species, Slate-coloured Junco (*hyemalis*, the only form recorded in the W Palearctic), Oregon Junco (*oreganus*; including Pink-sided Junco *mearnsi*), White-winged Junco (*aikeni*) and Grey-headed Junco (*caniceps*), but they intergrade freely, sometimes over broad areas of overlap, except where suitable nesting habitat is restricted. It is accordingly better to treat them as four racial groups of a single species.

REFERENCES Eritzsøe, J. Svenningsen, H. (1996) *DB*, 18: 1–5.

J. h. hyemalis, 1stY ♀, Netherlands, Feb: apparent moult limits in greater coverts and possibly juv primary-coverts and primaries suggest age. Species identification straightforward even with dull ♀, given bright pinkish bill contrasting with darker face. Some ♀♀ have rather strongly olive-tinged brownish upperparts. (P. Cools)

J. h. hyemalis, ad ♂, USA, Oct: evenly fresh with very slight pale feather tips to dark parts, and pale-fringed uppertail-coverts, but otherwise as spring. No brownish tinges, upperparts pure bluish-grey, belly clean white. Wing apparently ad, including bluish-grey tinge to pale fringes to remiges. (R. Brewka)

J. h. hyemalis, ♂, presumed ad, USA, Oct: some difficult to age. Plumage already advanced to almost full ad ♂, and fresh body plumage including pale-tipped breast indicate ad, but apparent brownish primary-tips suggest 1stW (primary-coverts insufficiently visible to be used for ageing). (P. Post)

J. h. hyemalis, ♀, USA, Oct: in autumn, some very fresh ♀♀ are tinged rufous-brown. (I. Davies)

SNOW BUNTING
Plectrophenax nivalis (L., 1758)

Fr. – Bruant des neiges; Ger. – Schneeammer
Sp. – Escribano nival; Swe. – Snösparv

To most European birdwatchers, the Snow Bunting is a rather exotic breeding bird, requiring a long trip to the north to see, but in late autumn many are delighted by the sight of a restless flock in whirling flight over the sea like the winter's first snowflakes. Flocks often amount to 15–50 birds, but 100–200 can occur. The Snow Bunting breeds above the tree-line on rocky slopes, on tundra with boulders and rocky Arctic coasts. A tiny population also breeds in the Scottish mountains. It is a hardy species, some wintering as far north as on SW Norwegian coasts and in S Iceland, but the majority migrate to Britain, Netherlands and Germany. Fair numbers can also be found along the shores of S Baltic.

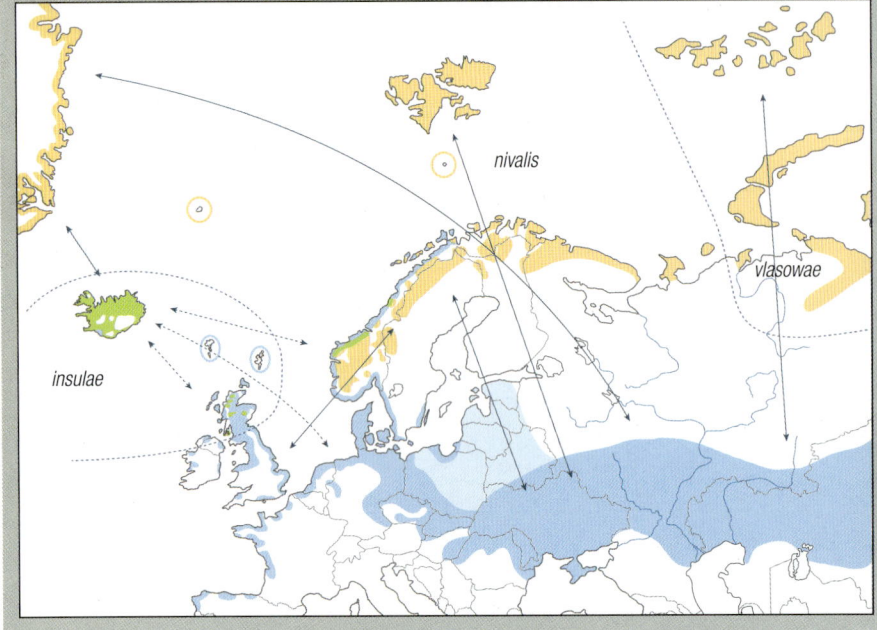

P. n. nivalis, 1stS ♂, Greenland, Jul: as ad but more than half length of primary-coverts dark brown with uneven border to white bases. (C. Siems)

IDENTIFICATION A fairly large, *long-winged* bunting with characteristic white patches or *extensive white portions on wings and tail*. The ♂ is *all white* in summer except for black back, wingtips and centre to the tail. The bill is black in summer but turns *yellowish with dark tip* in autumn, when the white plumage is also partly obscured by rufous and buff feather-tips. Legs black. The ♀ is less extensively white and has some *grey patches or streaking on crown and cheeks* even in summer, and back is not solidly black but streaked greyish. A few very neat ♀♀ approach ♂ plumage closely and require careful observation before they can be sexed correctly. First-winter birds are on average browner than adults and show less white in the wing, the young ♂ is therefore quite similar to the adult ♀, while the young ♀ has white on the wing reduced to a narrow wing-bar. In flight, adult ♂ has *entire innerwing white with contrasting black tip and alula*. ♂♂ irrespective of age show sharp contrast on underwing between blackish tip and white inner parts, whereas ♀♀ have diffuse border between rather pale grey tip and white innerwing. Juvenile plumage, which is not seen away from the breeding grounds, differs in being sullied brownish-grey and mottled darker over much of head, breast and upperparts, while underparts are diffusely streaked. In autumn migrants, the white is most exposed in flight, and a flying flock of Snow Buntings is therefore very characteristic. Migratory flight often at high altitude (for a small mainly diurnal passerine), but can also fly low over sea or fields. Flight is fast with long undulations. Runs on ground much like a lark. Behaves seemingly nervously, rushing ahead and making brief stops when feeding, taking off for short flights without apparent reason, then settling again.

VOCALISATIONS Song, usually delivered from perch on a rock but sometimes also in spectacular gliding song-flight, is a clear, rhythmic warble lasting *c.* 2–2.5 sec, often comprising some 10–12 syllables The song has a desolate ring to it, recalling both Rustic and Lapland Buntings, but a little harder than the former and without the trilling sounds of the latter. The phrase is repeated without much variation with pauses of 5–10 sec, sometimes longer. – Calls in flight

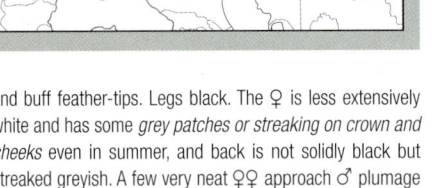

P. n. nivalis, ad ♂, Norway, May: plumage acquired entirely via abrasion (not moult). Combination of all-white primary-coverts (or, at most, tiny black tips), black primary tips and alula, and white head. Nearly all-black mantle indicative of age and sex in summer. (M. Varesvuo)

P. n. nivalis, ad ♂, Norway, May: note white innerwing with contrasting black tip and alula. Age and sex as for image to the left. (M. Varesvuo)

BUNTINGS

P. n. nivalis, ad ♀ (left) and 1stS ♀, Bear I, Jun: less contrastingly marked than ♂, with diffuse greyish streaking on crown, cheeks, nape and mantle, and less extensive white in wing, being usually restricted to panel on secondaries and tips to greater coverts. Note brownish juv alula, pointed primary-coverts, less white head and more yellowish bill-base in 1stS. (H. Shirihai)

are mainly of two types, a single whistling *pew* somewhat recalling Little Ringed Plover *Charadrius dubius*, repeated a few times within hearing distance, and a trilling *per'r'r'ret*, peculiarly similar to the common call of Crested Tit. These two can also be given in combination. Another call is sometimes heard from flying or feeding flocks, a harsh *bersch*, a little like that given from some wheatear species in S Europe.

SIMILAR SPECIES The only species within the treated range sharing the wing and tail patterns of Snow Bunting is the *Snow Finch*, but this is restricted to alpine habitats of S & C Europe, and the probability of encountering both species together anywhere is minimal (though see p.335). Snow Finch has a proportionately longer tail and in most plumages a black bib (or at least some grey mottling on chin) lacking in Snow Bunting.

AGEING & SEXING (*nivalis*) Ages differ in autumn, and can usually also be separated in spring if seen well. Sexes usually differ on fledging. – Moults. Rapid complete post-nuptial moult of ad on breeding grounds after breeding finished, mainly Jul–Sep (Oct). Partial post-juv moult at same time, usually includes head, body, median coverts, sometimes some inner greater coverts, odd tertials and rarely r1, but not primaries, rest of tail-feathers or any primary-coverts. Apparently no pre-nuptial moult. – **SPRING** Nearly always possible to sex immediately using plumage. A few less typical birds can often be sexed if aged first, ad being ♀, 1stS being ♂. ♂ Entire head and neck, and whole or most of rump white (though rear crown can have dark tips into May or even

P. n. nivalis, 1stS ♂, Finland, Apr: during spring and early summer whitish fringes above not yet lost, producing much variability in transitional ♂ plumages. Age based on broad dark tips to primary-coverts with uneven border between dark tips and white bases, plus quite pointed tips to tail-feathers, while sex inferred from black wingtip and alula. (A. Juvonen)

P. n. nivalis, ad ♂♂, Finland, Mar (left) and Dec: as summer but fresh tips to upperparts still conceal much of the black bases. Winter plumage characteristically tinged rusty yellow-brown on crown/nape, cheeks, breast-sides and shoulders, while rump and uppertail-coverts are white with variable orange-brown blotching. Bill yellowish with tiny black tip. (M. Varesvuo)

Jun, 1stS on average more so than ad). Mantle, back and scapulars black, either with whitish and ochre-buff tips or, in summer when heavily worn, nearly uniformly black. Generally much white basally on long primary-coverts (75% or more of visible length), and some ad have all-white primary-coverts. (A claim by Salomonsen 1947 that a percentage do not develop ad ♂ pattern of primary-coverts, but instead have dark ♀-type pattern, could not be confirmed in material at ZMC, NHM and NRM.) On handled birds, diagnostic sharp contrast on underwing between blackish (or at least very dark grey) outer part of primaries and inner pure white part possible to establish. (Ambiguous birds as to this character extremely rare. For such birds check all other criteria.) ♀ Crown and ear-coverts usually partly white but feathers basally dark and often broadly tipped rufous. Nape whitish striated grey-buff. Mantle, scapulars and most of rump brown-black edged whitish and rufous-buff, rump usually has more white admixed (but rarely dominating). Bases of long primary-coverts dark or mixture of dark and off-white, never all white. On handled birds, diagnostic diffuse border on underwing between rather pale grey outer part of primaries and inner pure white part possible to establish. **Ad** Tips of tail-feathers

P. n. nivalis, 1stW ♂♂, Finland, Dec (left) and Scotland, Oct: unlike ad ♂ has rather broad dark tips to primary-coverts with uneven border between the dark tips and extensive white bases. Typically suffused extensively buff-brown on upperparts (still usually with broader visible black bases than most ♀♀), especially on top of head. Underparts whitish with buff breast-sides. Bill pale yellowish with dark tip. (Left: M. Varesvuo; right: H. Harrop)

P. n. nivalis, ad winter ♀ (left: England, Nov) and 1stW ♀ (right: Germany, Oct): unlike any ♂, no or very little whitish visible at base of primaries, and primary-coverts mostly dark with narrow pale edges when fresh. Upperparts more narrowly streaked. No or very little white on rump/uppertail-coverts. Young ♀ has least white in wings (confined to wing-bars, with very narrow secondaries panel). Tips of tail-feathers and primary-coverts in young birds pointed. (Left: R. Brooks; right: T. Krüger)

P. n. nivalis, Finland, Mar: flock showing range of age and sex variation in flight. Ad ♂ has whitest wing, with all-white greater coverts and primary-coverts, isolating black alula, and all-white secondaries, while at other extreme 1stW ♀ (none visible here) has mostly dark greater coverts and primary-coverts plus broadly dark-tipped secondaries, creating a much darker wing with limited white. (M. Varesvuo)

rounded in early spring (but later in summer often too worn to be reliably judged), black centre at tip of r1 rather wide and obtuse (often visible even on worn feathers with practice). In ♂: long primary-coverts all white, or white with small dark tip; alula black; outer part of primaries jet-black, in fresh plumage (sometimes even in spring) finely but neatly edged white; tips of outer tail-feathers have rather restricted dark markings, usually not reaching far along shafts towards base (on r5 8–18 mm, on r6 5–17 mm); often entire secondaries white, or with very little dark on tips of ss5–6. In ♀: fairly rounded tips to tail-feathers, with much white on inner webs of rr4–6; inner webs of all secondaries whitish (except sometimes a small dark patch near tip of s1); whitish base to inner web of long primary-coverts (in-hand character); rather much white visible on sides and outer part of rump/ inner uppertail-coverts. **1stS** Tips of tail-feathers somewhat pointed, dark centre at tip of r1 rather pointed, too. In ♂: long primary-coverts either white with small dark tip (like some ad) or with half or whole of visible part of primary-coverts dark; alula dark grey or blackish-grey (but invariably subtly paler than typical ad); outer part of primaries dark grey or blackish, in fresh plumage (sometimes even in early spring) with pale fringes present, such fringes variable but often less distinct than in ad; tips of outer tail-feathers with rather extensive dark markings, often covering tip of both webs and reaching far along shafts towards base (on r5 13–23

P. n. nivalis, juv, presumed ♂, Norway, Jul: juv has very different plumage and look compared to later plumages. Fluffy body-feathers sullied brown-grey, with blackish-centred tertials broadly edged rufous. Quite blackish wing-feathers indicate a ♂. (M. Varesvuo)

P. n. insulae, Iceland, ad summer ♂ (left: Jul) and 1stS ♂ (right: Jun): ad usually has small black tips to primary-coverts (sometimes large, but coverts apparently never all white as often in *nivalis*), while in 1stY juv coverts can be all black, augmenting the more extensive black on wings and tail in this race. (Left: R. Martin; right: A. Tores Sanchez)

mm, on r6 10–27 mm, but thin black line on shafts often much longer, not included in these measurements); usually extensive dark patches or markings on outer webs of many outer secondaries. In ♀: rather pointed tips to tail-feathers, quite large dark areas on outer parts of rr4–6; inner webs of outer three secondaries usually have large dark portions (but rarely same pattern as ad ♀); long primary-coverts all dark.
— AUTUMN Although summer pattern obscured by extensive whitish, ochre-buff and rufous tips or fringes, sexing often possible if seen well. Essential to attempt ageing in combination with sexing. **Ad ♂** All tips of tail-feathers rounded and fresh, black centre of r1 wide and obtuse at tip. Much white on primary-coverts and secondaries (see spring). Wing-feathers including alula jet-black, primaries neatly edged pure white. Rump often have some visible white at sides of rump and on short uppertail-coverts. Much black on mantle/back/scapulars, overall impression not streaked but blackish with rufous-ochre and white fringes. **Ad ♀** All tips of tail-feathers rounded and fresh, black centre of r1 wide and obtuse at tip. No or only very little visible whitish at base of primary-coverts. Upperparts more streaked than solidly dark-looking. Generally no or only very little white on rump/uppertail-coverts. **1stW ♂** Tail-feathers pointed. Compared to ad ♂, often reduced white on primary-coverts and secondaries (see spring). Wing-feathers including alula

P. n. insulae, ♀, possibly ad, Iceland, Jun: dark and rounded primary-covert tips suggest ad. Crown, scapulars, rump and uppertail-coverts more frequently tinged rufous than in ♀ *nivalis*. (M. Varesvuo)

dark grey or blackish, primaries edged diffusely whitish. No or very little visible white on sides of rump/uppertail-coverts. Blackish or dark grey centres to feathers of mantle/back/scapulars somewhat less prominent than ad ♂, dark centres pointed. **1stW** ♀ Tail-feathers pointed. Primary-coverts all dark, secondaries with strongly reduced white. No white patch visible on rump, upperparts being streaked dark admixed rufous, ochre and white. **Juv** Distinct plumage, only seen on breeding grounds. Head, nape, mantle and breast sullied brown-grey, finely mottled or streaked dark. Centres of tertials blackish with broad rufous edges/tips. Greater coverts edged dusky-white with some rufous tinge. Most can be sexed if seen well by using amount of white visible on primary-coverts and secondaries, and (when handled) darkness and contrast of underside of primaries to white innerwing (as outlined above).

BIOMETRICS (*nivalis*) **L** 16–18 cm; **W** ♂ 106.5–117 mm (*n* 58, *m* 110.7), ♀ 100–111 mm (*n* 42, *m* 105.2); **T** ♂ 60–73 mm (*n* 58, *m* 67.3), ♀ 58–70 mm (*n* 42 *m* 63.0); **T/W** *m* 60.4; **B** 11.5–15.1 mm (*n* 94, *m* 13.1); **B**(f) 8.5–11.5 mm (*n* 98, *m* 10.1); **BD** 5.0–7.0 mm (*n* 84, *m* 6.3); **Ts** 19.5–22.3 mm (*n* 92, *m* 21.0). **Wing formula: p1** minute; **pp2–3** longest, or p3 < wt 0.5–1 mm; **p4** < wt 2.5–4 mm; **p5** < wt 11–17 mm; **p6** < wt 17–25 mm; **p7** < wt 25–31 mm; **p10** < wt 37–46 mm; **s1** < wt 40–48 mm. Emarg. pp3–4.

GEOGRAPHICAL VARIATION & RANGE Rather slight variation despite large breeding range, although Icelandic population is rather clearly darker. At least one more extralimital race has been described (not treated).

P. n. nivalis (L., 1758) (N Fenno-Scandia, NW Russia, Svalbard, Greenland, N North America; in Palearctic winters in S Scandinavia, W, C & SE Europe). Described above. Grades into *vlasowae* at least from lower Pechora in NE Russia, but already around Archangel starts to become paler and whiter on average. Salomonsen (1931) discerned the Greenland population ('*subnivalis*') as different from Fenno-Scandian breeders, but if any difference at all it is too small to warrant separation. (Syn. *borealis*; *hiemalis*; *montanus*; *mustelinus*; *subnivalis*.)

P. n. insulae Salomonsen, 1931 (Iceland, N Scotland; partly resident, partly migrant to Ireland, British Isles, North Sea coasts). Rather clearly darker than *nivalis* and *vlasowae*, feathers of upperparts and flanks in fresh plumage fringed/tipped rather dark rufous-ochre, and rufous-brown band across chest darker and more prominent on average. Both sexes in worn plumage tend to retain more dark areas than *nivalis*, most notable difference in both sexes being the dark rump (in *nivalis*, ♂ has all-white or largely white rump, ♀ partly white rump). Primary-coverts of ♂ have on average slightly more extensive dark tips than *nivalis*, and all-white primary-coverts are quite rare, while ♀ is all dark even in ad. Bill is proportionately stronger than in other races. **W** ♂ 105–116 mm (*n* 29, *m* 110.4), ♀ 99–109 mm (*n* 26, *m* 103.4); **T** ♂ 61–73 mm (*n* 29, *m* 65.5), ♀ 57–67 mm (*n* 26 *m* 60.8); **T/W** *m* 59.1; **B** 12.0–15.9 mm (*n* 52, *m* 14.0); **B**(f) 9.4–13.0 mm (*n* 55, *m* 11.0); **BD** 5.7–7.0 mm (*n* 49, *m* 6.5); **Ts** 20.0–22.5 mm (*n* 53, *m* 21.2).

○ *P. n. vlasowae* Portenko, 1937 (NE Russia, N Siberia; winters S Russia, Kirghiz Steppe in Kazakhstan). Differs by being a trifle paler and more extensively white, having a slightly larger white rump patch than *nivalis*. In fresh plumage feathers tipped more buff-white, less rufous- or ochre-tinged. Many are difficult to separate though. Not a distinct race, but perhaps just enough different to be warranted. Already around Archangel, birds tend to be paler and whiter than typical *nivalis*, and this tendency runs as a cline eastward. **W** ♂ 106–118 mm (*n* 28, *m* 112.5), ♀ 102–108 mm (*n* 12, *m* 105.9); **T** ♂ 62–72 mm (*n* 28, *m* 67.9), ♀ 61–67 mm (*n* 12 *m* 63.4); **B** 11.6–14.0 mm (*n* 35, *m* 12.9); **B**(f) 9.3–10.7 mm (*n* 34, *m* 9.9); **BD** 6.0–7.0 mm (*n* 32, *m* 6.6). (Syn. *pallidior*.)

TAXONOMIC NOTE Traditionally the Snow Bunting has been afforded a separate genus, *Plectrophenax*, together with its close relative McKay's Bunting *P. hyperboreus*, largely based on external characters. Recent analyses of the relationships of most taxa within Emberizidae based on mtDNA (e.g. Carson & Spicer 2003, Klicka *et al*. 2003) showed that *Plectrophenax* can be placed in the same clade as most *Calcarius* species, and therefore is perhaps better merged with *Calcarius* to avoid paraphyly in the latter. However, other solutions to this problem exist, and we prefer to await a more general agreement before making a change.

REFERENCES Carson, R. J. & Spicer, G. S. (2003) *Mol. Phyl. & Evol*., 29: 45–57. – Klicka, J., Zink, R. M. & Winker, K. (2003) *Mol. Phyl. & Evol*., 26: 165–175. – Salomonsen, F. (1931) *Ibis*, 73: 57–70. – Salomonsen, F. (1947) *Dansk Orn. For. Tidsskr*., 41: 138. – Smith, R. D. (1992) *Ring. & Migr*., 13: 43–51.

P. n. insulae, ♀, presumed 1stS, Iceland, Jul: typically duskier, with broad dark bases to upperwing-coverts, and lower mantle and scapulars more extensively dark than *nivalis*. Pale areas of supercilium, neck-sides and rump are more mottled and streaked grey-brown than in worn ♀ *nivalis*. Worn and pointed primary-coverts suggest age. (R. Martin)

P. n. insulae, ad ♂, Jan Mayen, Jun: birds at this locality tend to be intermediate between *nivalis* and *insulae* in amount of white on rump and black on primary-coverts, although they are much closer to, and included in, *insulae*. (H. Shirihai)

P. n. insulae, ♀, presumed ad, Jan Mayen, Jun: most ♀♀ on this island are much like Icelandic *insulae*, with rusty-tinged individuals being common. Relatively fresh plumage in Jun suggests ad. (H. Shirihai)

P. n. insulae, presumed ad ♀, Iceland, Mar: the overall dark plumage, including rump, of this race is sometimes appreciable in winter too. Quite dark primaries with distinct white tips indicate ad, while narrow black centres to scapulars, extensive dark tips to secondaries, no visible white in primary-coverts and mainly yellow bill indicate ♀. (Ó. Runólfsson)

LAPLAND BUNTING
Calcarius lapponicus (L., 1758)

Alternative name: Lapland Longspur (Am.)

Fr. – Bruant lapon; Ger. – Spornammer
Sp. – Escribano lapón; Swe. – Lappsparv

Together with species like the Golden Plover *Pluvialis apricaria* and Meadow Pipit, the Lapland Bunting gives character and spirit to the open heaths and damp osiers of northern fells and tundra. As soon as you leave the birch zone on a mountain trek for the open slopes and plains, the peculiar ringing song of the Lapland Bunting can be heard everywhere. The attractive ♂ is easily spotted, perched atop a willow or embarking on a short song-flight. In autumn, the species moves south to winter in W and C Europe. Spring migration occurs mainly in April and the first half of May.

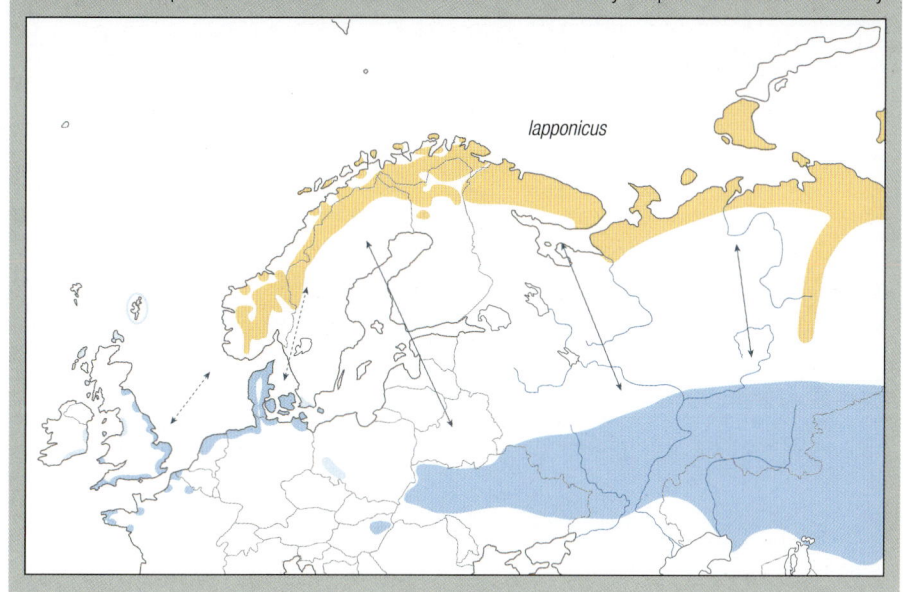

C. l. lapponicus, 1stS ♂, Norway, May: only safely differentiated from ad ♂ by moult: note pointed and less dark-centred juv primary-coverts, as well as possible moult limit in tertials (at least longest is juv). (D. Occhiato)

IDENTIFICATION A very neat-looking, fairly large and long-winged bunting of open alpine plains and valleys with willows and scrub. The ♂ has *jet-black crown, face and upper breast* and *vivid chestnut nape*. The black continues onto sides of breast, with a prominent *pale yellow supercilium* which becomes white and *bends down across side of neck*, dividing black face and chestnut nape. The *bill is yellowish with a dark tip*. Lower breast and belly are white with a few black streaks on flanks. Back is rather dark grey-brown streaked black. Dark tail has narrow white sides. The ♀ differs in lacking the black *head, throat and upper breast*, these parts being *brown or grey mottled dark, with throat and submoustachial stripes whitish. The pale throat is often encircled by a dark grey, patchy gorget on upper breast*. Furthermore, there is no distinct pale supercilium or continuation of this on the neck-side like the ♂. Bill is pinkish-buff, less yellow than ♂. A very few ♀♀ are more advanced and approach ♂ appearance, but they can invariably be separated (see Ageing & sexing). In autumn, Lapland Buntings are more discreet-looking and could be mistaken for Reed Buntings. Still, they are usually recognised using a combination of features: (i) *pale ochre-buff head-sides with a dark line encircling the ear-coverts*; (ii) *dark lateral crown-stripes with a paler brown median crown-stripe* (these two first points together create an appearance vaguely recalling Little Bunting, but size, bill colour and many other differences separate); (iii) *double buff-white or white wing-bars* and *chestnut-edged greater coverts*; (iv) *chestnut edges to tertials*; and (v) often *rufous-tinged nape*, most obvious in adults. Juvenile, only seen on breeding grounds, differs in that entire head, neck, breast and upperparts are buffish-white or yellowish-ochre mottled or streaked darker. Flight and behaviour much as in Snow Bunting.

VOCALISATIONS Song generally given from exposed perch in top of a willow or from lower shrub on a bog, but also in high song-flight, from which the bird glides down with spread wings and fanned tail, delivering an extended strophe. Most characteristic is the inclusion of trilling or jingling notes and the mix of shorter sounds and one or two drawn-out notes, *kretle-krleee-tr kliitre-kretle-tree*. It can slightly recall song of Horned Lark due to the same high-pitched notes and jingling elements (and often the same habitat!) but differs by its more even pace, lacking the faltering opening of each phrase as in Horned Lark. – Calls are many and frequent. Flight-calls on migration rather similar to those of Snow Bunting, a short, soft, whistling *chu* (or even more Snow Bunting-like, *pyu*) and a dry rattling *pr'r'r'rt*, harder and drier than corresponding call of Snow Bunting. Another call used

C. l. lapponicus, ad ♂, Norway, Jun: mid-sized bunting, characteristic of Arctic tundra. In most plumages unmistakable, especially the immaculate ad summer ♂. Aged by uniformly ad wing, including primary-coverts. (H. Harrop)

C. l. lapponicus, ♀, presumed ad, Norway, Jul: note variegated black-and-white face markings, pale lores and central ear-coverts bordered by dark lateral crown-, eye-, and moustachial-stripes. Whitish underparts have dense blackish streaks on breast and flanks, often forming gorget or irregular breast patch. Without closer inspection of tail-feathers and primary-coverts ageing is difficult, but overall rather colourful plumage suggests ad. (M. Varesvuo)

by migrants (also by night) is a *chüp*, clearly related to the first-mentioned call but slightly hoarse and more emphatic. When anxious near the nest often gives a disyllabic feeble, almost trembling *tihü*.

SIMILAR SPECIES Differs from slightly smaller *Rustic Bunting* by brown-grey rather than rufous rump- and flanks-streaking, more vividly chestnut-edged greater coverts and ochre-tinged head-sides. – From about same-sized *Reed Bunting* by more distinct whitish double wing-bars, more rufous-tinged greater coverts, and yellowish or pinkish bill with dark tip, rather than all-dark bill. – If size is not fully appreciated, as can easily happen when trying to identify birds in unfamiliar terrain, can be confused with clearly smaller *Little Bunting*, sharing dark lateral crown-stripes leaving centre of crown paler, ochre-tinged head-sides with rather distinct black lines around ear-coverts, and even a hint of a pale buff or whitish eye-ring. However, note pinkish or yellowish dark-tipped rather strong and bulbous bill of non-breeding Lapland Bunting, and much longer primary projection.

C. l. lapponicus, ♀, presumed 1stW, England, Feb: dullest plumage (any rufous on nape very subdued, and breast-sides only sparsely streaked), with remiges deemed to be juv, as are also just visible primary-coverts having worn and rather pointed tips. (R. Chittenden)

C. l. lapponicus, 1stS ♀, Norway, Jul: duller than ad ♀, including less extensive and more washed-out rufous nape, but close inspection of tail-feathers and primary-coverts essential for reliable ageing (here, primary-coverts clearly juv, and pointed rectrices confirmed in another image). (H. Shirihai)

C. l. lapponicus, spring ♂♂ (left ad; right presumed 1stS), England, Mar: irrespective of age, much variation in development of summer plumage during late winter to mid spring. At same time and location, ad ♂ has on average more solid black facial areas, while age is most safely established when wear and shape of tail-feathers can be assessed. (Left: J. Theobald; right: N. Appleton)

AGEING & SEXING (*lapponicus*) Ages differ in autumn, but generally not safely separated in spring due to wear except for very typical birds recognised after some practice. Sexes usually differ after post-juv moult. – Moults. Rapid complete post-nuptial moult of ad on breeding grounds after breeding finished, mainly late Jun–Aug (early Sep), sometimes final stages concluded during or even after autumn migration. Partial post-juv moult at same time, usually includes head, body, lesser and median coverts, but not greater coverts, primary-coverts, primaries or tail-feathers. Both age categories have a partial pre-nuptial moult confined to head and throat. – **SPRING** Nearly always possible to sex straight away on plumage. ♂ Lores, ear-coverts and throat black, rarely with a few white tips admixed. Prominent supercilium pale lemon, connected to white line across side of neck. Nape unstreaked deep chestnut. (Very rarely, more ♂-like ♀♀ occur approaching this pattern, though never attain fully jet-black lores and ear-coverts, and buffish-white supercilium of ♀ usually reaches bill, whereas in ♂ is separated by 2 mm of black.) ♀ Lores and throat mixed off-white and black (some with black dominating), ear-coverts pale brown admixed with some black patches or spotting (ear-coverts never nearly all black). **Ad** Those with reasonably little worn tips of tail-feathers, tips being fairly broadly rounded, are safely ad. Intermediates will occur best left unaged. **1stS** Those with rather narrow central tail-feathers which are heavily worn are safely 1stS. – **AUTUMN** Although summer pattern obscured by extensive ochre-buff and buff-white tips or fringes, sexing often still possible if seen well. **Ad** Tips of tail-feathers often broad and well rounded, but rarely rather pointed, though extreme tips invariably neatly rounded, and especially r1 has broad and rounded tip. Much unstreaked chestnut visible on nape despite being partly concealed by some buff-ochre tips. **1stW** Tips of tail-feathers usually discernibly narrow and pointed compared to ad type, tips sometimes 'frayed' at extreme tips; r1 tapers off towards tip, which is not as broad and rounded as ad. Colour of nape variable, either like ad (some ♂♂) or has reduced chestnut tinge, streaked or mottled dark. ♂ Crown-feathers have extensive black centres, ochre-buff tips limited to small spot on each web (ad), often possible to see at close range or on photographs, but in 1stW more intermediate with broad black bases less extensive, approaching some ♀♀. Feathers of upper breast have extensive black rounded centres partially concealed by buff-white tips. Much chestnut visible on nape. ♀ Crown-feathers have blackish central streaks, ochre-buff edges to each web extensive. Feathers of upper breast streaked, dark centres mainly pointed, extensively edged buff-white. Nape either like ♂ (ad) or has much-reduced chestnut (even without any) dark streaks or mottling (1stW). **Juv** Resembles 1stW ♀, but entire head and neck including super-cilium, nape and lower throat yellowish-buff, evenly mottled or

C. l. lapponicus, autumn ♂♂, presumed ad, Oct (left: Faeroe Is; right: England): both have seemingly uniformly fresh ad wing and fairly round-tipped tail-feathers, supported by ad-like larger and rounder marks on flanks and breast-sides (left bird) and strong face pattern and large bright rufous nape (right). (Left: S. Olofson; right: G. Thoburn)

C. l. lapponicus, ad ♀, Finland, Sep: evenly fresh ad wing (very broad white fringes to primaries and primary-coverts), narrow dark crown-streaking, plus only slight rufous on nape infer age and sex. Winter ♀ can superficially resemble smaller Little Bunting, but has stronger, dark-tipped pinkish-yellow bill, and the brown-grey (rather than rufous) rump and brighter chestnut greater coverts panel prevent confusion with Rustic Bunting. Wing panel (and whitish double wing-bars) also eliminate same-sized Reed Bunting. (A. Seppä)

C. l. lapponicus, ♂, presumed 1stW, England, Oct: certainly a ♂ based on broadly black centres on feathers of lower throat and upper breast combined with deep chestnut nape. Age more difficult to decide, and not that wrong for an ad if it was not for one seemingly moulted inner primary-covert in contrast to browner and juv-looking outer strongly inferring 1stW. (G. Thoburn)

spotted dark. Upper breast and flanks boldly streaked black. Feathers of neck, nape and vent often more 'loose' and fluffy (admittedly requiring close observation).

BIOMETRICS (*lapponicus*) **L** 14.5–16.5 cm; **W** ♂ 91–100 mm (*n* 56, *m* 95.1), ♀ 86–92.5 mm (*n* 31, *m* 89.7); **T** ♂ 60–68 mm (*n* 57, *m* 63.6), ♀ 56–64 mm (*n* 31 *m* 60.0); **T/W** *m* 66.8; **B** 11.2–13.6 mm (*n* 51, *m* 12.4); **B**(f) 8.2–11.0 mm (*n* 50, *m* 9.8); **BD** 5.7–7.0 mm (*n* 49, *m* 6.4); **Ts** 18.0–21.3 mm (*n* 49, *m* 20.1); **HC** 7.9–13.2 mm (*n* 31, *m* 9.8). **Wing formula: p1** minute; **pp2–4** about equal and longest, or p4 < wt 0.5–3 mm; **p5** < wt 6–9 mm; **p6** < wt 14–18 mm; **p7** < wt 18–24 mm; **p10** < wt 25–35 mm; **s1** < wt 28–37 mm. Emarg. pp3–4, often slightly on p5.

GEOGRAPHICAL VARIATION & RANGE Rather slight variation, mainly involving colour saturation, but also size. One race breeds within the treated region, another is extra-limital in summer but may occur in winter on a rare but regular basis. Three additional extralimital races are not likely to ever occur and are therefore not treated (*alascensis, coloratus, kamtschaticus*).

C. l. lapponicus (L., 1758) (N Fenno-Scandia, N Russia, N Siberia; winters W, S & E Europe, Central Asia). Described above.

○ *C. l. subcalcaratus* (C. L. Brehm, 1826) (Greenland, N North America; some breeders in Greenland winter NW Europe). Differs from *lapponicus* by being very subtly paler, less dark chestnut on nape and perhaps slightly paler on mantle in summer plumage. Very slightly larger, although much overlap with *lapponicus* in wing and tail measurements. However, bill and tarsus are proportionately rather clearly longer, and tail a little shorter. A combination of all these slight average differences might enable a few typical birds to be identified, even away from breeding areas. **L** 15–17 cm. **W** ♂ 92–102 mm (*n* 29, *m* 96.8), ♀ 87–97 mm (*n* 19, *m* 92.7); **T** ♂ 58.5–68 mm (*n* 29, *m* 62.8), ♀ 55–68 mm (*n* 19 *m* 59.7); **T/W** *m* 64.7; **B** 12.0–14.8 mm (*n* 47, *m* 13.3); **B**(f) 9.1–12.0 mm (*n* 46, *m* 10.6); **BD** 5.7–7.0 mm (*n* 46, *m* 6.5); **Ts** 19.8–22.3 mm (*n* 46, *m* 21.0); **HC** 8.3–12.5 mm (*n* 18, *m* 10.3).

TAXONOMIC NOTE See note on relationship with *Plectrophenax* under Snow Bunting.

REFERENCES Alström, P. *et al*. (2008) *Mol. Phyl. & Evol.*, 47: 960–973. – Carson, R. J. & Spicer, G. S. (2003) *Mol. Phyl. & Evol.*, 29: 45–57. – Garner, M. (2007) *BW*, 20: 203–204. – Klicka, J., Zink, R. M. & Winker, K. (2003) *Mol. Phyl. & Evol.*, 26: 165–175. – Mjøs, A. T. (2007) *BW*, 20: 348.

C. l. lapponicus, presumed 1stW, England, winter: much of wing apparently juv, including narrow pointed primary-coverts, suggesting a young bird. Many young cannot be sexed, especially if age not established, being intermediate in pattern and brightness. (M. Lane)

C. l. lapponicus, 1stW ♀, Finland, Jan: most wing-feathers and tail-feathers juv, the latter being discernibly narrow and pointed. Note weak face pattern and no rufous on nape, as well as narrower, ill-defined and browner streaks below. (M. Varesvuo)

C. l. lapponicus, juv, Finland, Jun: almost evenly mottled or spotted dark, with rather obvious fluffy body-feathers. Still, the characteristic chestnut greater coverts panel bordered by two white wing-bars already well developed. (M. Varesvuo)

BLACK-FACED BUNTING
Emberiza spodocephala Pallas, 1776

Fr. – Bruant masqué; Ger. – Maskenammer
Sp. – Escribano carinegro; Swe. – Gråhuvad sparv

One of the many taiga-dwelling buntings, the Black-faced Bunting prefers rather tall and dense vegetation. In Siberia it often occurs on the dampest patches of closed taiga (especially overgrown forest bogs) but also frequents moist edges of rivers and pools with a mix of tall bushes and trees. Often spotted by its peculiarly low-pitched, simple song, performed from a side branch, rather than a treetop. It is a migrant, wintering mainly in S Korea and S China, returning to its most north-westerly breeding sites in the Altai and SW Siberia (upper Yenisei) only in early June. It has straggled several times to W Europe.

E. s. spodocephala, ♂, Mongolia, May: unmistakable due to blackish mask on lores and greenish-grey throat, grading into yellowish underparts. Thin white wing-bars and white wedges on outer tail-feathers can also be obvious. Relatively thick conical bill. Without close inspection of primary-coverts and tail-feathers, ageing in spring usually impossible. (H. Shirihai)

E. s. spodocephala, ♀, presumed ad, China, May: pale bill, grey to green-tinged head and neck-sides, unstreaked olive-brown rump (here invisible), cream-white to yellowish-tinged submoustachial stripe and underparts, with narrow black lateral throat-stripe and boldly black-streaked flanks and upper breast. Ageing often uncertain (unless bird handled) but rounder tail-tips and purer grey, olive and yellow indicate ad. (M. Parker)

IDENTIFICATION Of Reed Bunting size with similar proportions. Usually gives a rather dark and slightly dull first impression in the field, without striking features. In summer plumage the ♂ in Siberia (ssp. *spodocephala*) has *lead-grey head, throat and chest* with swarthy, *blackish face and chin*. Belly is pale yellowish (can look washed out and nearly white at western end of range) with some *dark streaks on flanks*. Upperparts tawny-brown streaked black, but *rump unstreaked grey-brown*. There is some *white on tail-sides*, and two rather indistinct pale ochre wing-bars. Bill strong with *straight culmen, grey with pinkish lower part*. Legs pink. In fresh autumn plumage the grey head and chest is partly obscured by brown tips, but the main pattern of the summer plumage is still usually evident. The ♀ is much more featureless lacking the swarthy face and the uniformly grey head and chest of the ♂, having instead an *off-white throat* and *whitish submoustachial stripe* with indistinct darker lateral throat-stripe. Crown is streaked brown-grey, and there is a dull and indistinct paler supercilium (the *submoustachial stripe is the palest part of the head-sides*, paler than the supercilium). While the back is rather warm brown streaked dark, the *head, neck and breast-sides usually look rather greyish-tinged*. Underparts dusky-white with a subtle creamy-yellow cast, flanks and breast-sides boldly streaked dark. Rump grey-brown with some subdued darker mottling (but lacks rufous or chestnut tones). Limited seasonal changes in ♀. Juvenile is like a dull and slightly more brownish ♀, lacking most of the greyish cast on head and neck.

VOCALISATIONS (*spodocephala*) Song a short (1.5–2 sec) low-pitched, metallic, jingling phrase containing many trilling notes. Rather extensive variation in details, but each ♂ largely sticks to one theme. Compared to Reed Bunting lower-pitched and delivered at more even, fluent pace, and tonal steps more marked. Song often terminated with a couple of similar trilling notes. Repeated every 4–6 sec when fully worked up. – Call a sharp, fine *zit* or *zrit*, rather like call of Song Thrush. When agitated several subdued calls can be uttered in rapid series.

SIMILAR SPECIES The adult ♂ should be distinctive enough, but ♀ and young are rather nondescript and could be mistaken for several other bunting species of similar age and sex. From *Pine Bunting* by slightly smaller size, lack of chestnut rump and slight yellowish tinge in plumage visible in at least some. Also, tail is proportionately shorter, and lower part of bill is pink. – *Reed Bunting* is similar but has better-marked dark lateral throat-stripe with heavier blackish patch at lower end, lacks greyish cast on neck and head, rufous lesser coverts and wing-bend, and more strongly-marked cheeks. Furthermore, bill of Black-faced has pink lower part (Reed: all grey) and straight culmen (Reed: usually slightly convex). – *Yellowhammer* has at least some slight yellowish cast to head and breast, chestnut rump, apart from larger size, longer tail and many different calls. – *Cirl Bunting* is fairly similar but has even duller, more grey-tinged rump, generally a chestnut tinge on scapulars and more contrasting cheek pattern with darker lines on upper and lower borders. Also, underparts in Cirl are more extensively but also more finely streaked dark.

AGEING & SEXING (*spodocephala*) Ages differ in autumn, sometimes also in spring; reliable ageing in the field usually requires close views. Sexes usually differ after post-juv moult, but some extreme variations are very similar. – Moults. Complete post-nuptial moult of ad after breeding, mainly Jul–Sep. Partial post-juv moult late Jul–early Oct

E. s. spodocephala, 1stS ♂, France, May: as ad but note pointed, worn and brownish tail-feathers being retained juv. (M. Thibault)

— 502 —

E. s. spodocephala, presumed 1stS ♀, Mongolia, May: paler wing-bars of ♀ often weak or as here (three images of the same bird) near absent. Irrespective of age, some ♀♀ are duller, with almost no greenish-olive or yellow on head or underparts. The apparently pointed, worn and brownish tail-feathers and juv-like primary-coverts suggest 1stS. (H. Shirihai)

affects head and many (often all) wing-coverts and tertials, but does not involve any primaries, secondaries or primary-coverts. Tail-feathers are either not moulted at all, or, more commonly, r1 is moulted, rarely more tail-feathers (though apparently never all). Both age categories have a partial pre-nuptial moult in winter (Nov–Mar) confined to parts of head and throat. – **SPRING** ♂ (The following criteria separate most birds, but some must be left un-sexed in the field due to intermediate characters.) Head, neck and throat and upper breast uniform grey tinged olive-green. Face (lores, feathers at base of bill and below eye) blackish (dark grey in some). ♀ Chin and partly lores and throat and upper breast pale yellow. A few have extensive olive-grey on throat and upper breast but not quite as much as ♂. **Ad** Those with reasonably little worn tips of tail-feathers, tips being fairly broadly rounded, are safely ad. Intermediates will occur best left unaged. Ad ♂ has throat and upper breast uniform grey tinged olive-green (faintly white-tipped in a few). Primary-coverts still fresh and rounded, fringed brown or olive-brown. Tips of longest primaries and tail-feathers reasonably fresh. A few ad ♀♀ possible to separate if aged first (rather fresh longest primaries, tail-feathers and primary-coverts), by having much olive-grey on sides of neck and throat and upper breast. **1stS** Those with rather narrow central tail-feathers which are heavily worn are safely 1stS. 1stS ♂ has throat and upper breast with some yellowish patches or spots mixed in the olive-grey (although some are nearly as neat as ad ♂). Primary-coverts worn, rather pointed with slightly frayed tips. Longest primaries and tail-feathers often heavily abraded. 1stS ♀ tends to be more boldly streaked, lack olive-grey and have heavily-abraded longest primaries. – **AUTUMN** Sexing often difficult due to head pattern being partially concealed by fresh brown-grey feather-tips, but also because some young ♂♂ do not develop ♂ characters until after pre-nuptial moult. If ageing possible most categories can usually be separated. **Ad** After completion of moult, longest primaries and all tail-feathers quite fresh, tips of tail-feathers slightly rounded, edges neat. Primary-coverts have rather glossy surface, rounded tips and neat brown fringes in contrast to dark centres. Iris rufous-brown. **1stW** Tips of longest primaries and of tail-feathers slightly worn, tips of tail-feathers pointed. Primary-coverts rather dull grey-brown with less neat paler brown edges. Iris grey-brown. **1stW** ♂ Either very similar to ad ♂ or more like ♀♀, lacking grey-green on throat/upper breast and having many brown tips to crown-feathers. Latter category often possible to separate from 1stW ♀ by fewer dark streaks on breast (difficult and requires practice). **1stW** ♀ A bird positively aged as 1stW that lacks virtually any concealed olive-grey on head or throat, and with much dark streaking on throat/upper breast, should be a ♀. **Juv** Resembles a dull ♀ but often has warmer brown (rufous-tinged) crown and nape, lacking any olive hue.

E. s. spodocephala, ♂, China, Oct: some ♂♂ in autumn attain summer-like plumage, including some advanced 1stW birds. Thus, without close inspection of primary-coverts and tail-feathers age is best left uncertain. Note blackish mask on lores and largely greyish head and throat. (I. Fisher)

E. s. spodocephala, 1stW ♂, China, Oct: greyish collar (sides of neck) and much brown on crown eliminate ♀. Note narrow dark lateral throat-stripe and boldly black-streaked flanks. Primary-coverts and remiges apparently juv and less fresh than moulted rest of wings, further supporting ageing and sexing. (I. Fisher)

BIOMETRICS (*spodocephala*) **L** 14–16 cm; **W** ♂ 68–77 mm (*n* 24, *m* 72.2), ♀ 65–71.5 mm (*n* 18, *m* 67.9); **T** ♂ 56–69 mm (*n* 24, *m* 63.6), ♀ 57–66 mm (*n* 18, *m* 60.6); **T/W** *m* 88.7; **B** 11.1–13.5 mm (*n* 38, *m* 12.5); **B**(f) 9.1–11.0 mm (*n* 37, *m* 10.0); **BD** 5.6–6.9 mm (*n* 36, *m* 6.3); **Ts** 17.5–19.8 mm (*n* 36, *m* 18.9). **Wing formula: p1** minute; **p2** < wt 2–4 mm; **pp3–4** longest; **p5** < wt 1.5–3.5 mm; **p6** < wt 5–8.5 mm; **p7** < wt 7.5–11 mm; **p10** < wt 12–16 mm; **s1** < wt 13.5–16.5 mm. Emarg. pp3–6 (though somewhat less prominently on p6).

GEOGRAPHICAL VARIATION & RANGE Fairly well-marked variation in east of range. Only three races, all extra-limital, seem sufficiently distinct to be warranted.

E. s. spodocephala Pallas, 1776 (Siberia, Altai, east through N Mongolia to Amur, Ussuriland, Manchuria, N Korea;

E. s. spodocephala, 1stW ♀, **Germany, Dec**: dull-plumaged 1stW ♀ can recall a ♀ House Sparrow, though obviously differs by dark-streaked breast and flanks. Apparently juv-like primary-coverts support ageing. Such 1stW ♀♀ lack obvious distinguishing characters, and could be confused with several other bunting species. (V. Legrand)

E. s. spodocephala, 1stW ♀, **China, Oct**: if aged as 1stW (note apparently faded and worn juv remiges and pointed tail-feathers), the complete lack of olive-grey on head or throat will safely sex this bird as ♀. (I. Fisher)

E. s. personata, ♂ (left: Feb) and ♀ (Apr), **Japan**: this eastern race, with clearly different morphology and song, is a potential vagrant to the treated region. Note especially the extensive bright yellow underparts, heavy and long bill, reduced white in tail (invisible here), and (in breeding ♂) a yellow chin and throat, sharply contrasting with olive-grey cheeks. ♀ has diagnostic fine arrowhead streaking on upper flanks (cf. Yellow-breasted Bunting). Ageing in spring is difficult even when the bird is handled. (Left: I. Fisher right: J. Ponces)

winters S & E China). Described above. Yellow of underparts generally pale and varies clinally in that least-yellow birds occur in the west, some even being nearly whitish on belly, while those with strongest yellow breed in Amur and Ussuriland. Grey of head and throat in ♂ more lead-grey, less greenish-tinged than *personata*. (Syn. *extremiorientis*; *flaviventris*; *oligoxantha*.)

E. s. personata Temminck, 1836 (Sakhalin south to N Japan; winters S Japan). Differs in being greener on crown and nape, less lead-grey, and has centre of throat yellow instead of (greyish-)green. Bill somewhat stronger and wing longer than *spodocephala*. Tendency for black underparts streaking to be arrow-shaped (more straight streaks at least in *spodocephala*). Song appears to be consistently more uneven in pace, containing more fleeting halts like Reed Bunting.

E. s. sordida Blyth, 1845 (C and SW China; winters SE Asia west to Nepal). Resembles *spodocephala* in having a dark chin and throat, but all colours darker and more saturated. Black on face (chin, around bill-base, lores) darker and more extensive. (Syn. *melanops*.)

REFERENCES Alker, P. J. (1997) *BB*, 90: 549–561. – Bradshaw, C. (1991) *BW*, 85: 653–665. – Hough, J. (1994) *BW*, 7: 98–101.

E. s. personata, presumed 1stW ♂, **Japan, Dec**: the yellower underparts and bolder dark face markings of ♀ and young autumn birds could invite confusion with same plumages of Yellow-breasted Bunting, but note the arrowhead-shaped dark breast streaking. Primary-coverts and remiges seemingly juv hint at age and sex. (R. Bonser)

PINE BUNTING
Emberiza leucocephalos S. G. Gmelin, 1771

Fr. – Bruant à calotte blanche; Ger. – Fichtenammer
Sp. – Escribano cabeciblanco; Swe. – Tallsparv

The Pine Bunting is an eastern counterpart of the common European Yellowhammer. They are genetically closely related, and even hybridise to some extent where they overlap in W Siberia. They occupy much the same habitats, and their songs are very similar if not even identical, but they differ dramatically in plumage. The Pine Bunting breeds from easternmost European Russia across much of Siberia, south to the large steppes in Kazakhstan and Mongolia, with an isolated population in NE Tibet and Gansu. Northern populations are migrants to S Asia while southern birds are resident or move only shorter distances. Oddly, a few winter regularly in S Europe and in the Middle East.

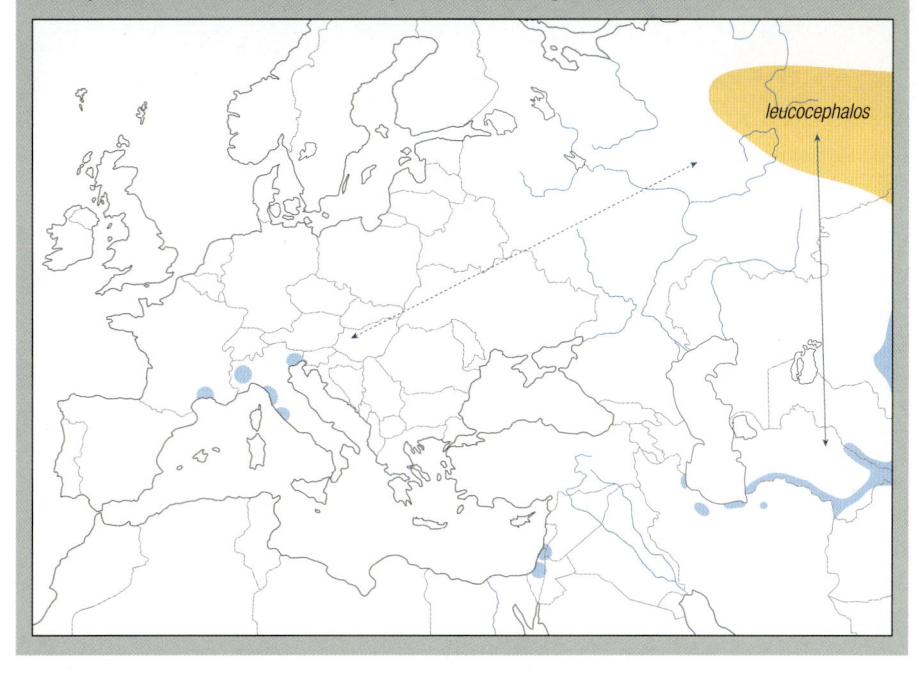

IDENTIFICATION One of the larger buntings. Similar in shape to Yellowhammer, thus rather long-tailed with *conical and slightly bulbous bill* adapted for mostly seed-eating habits. Adult ♂ in summer attractive, with mainly *deep chestnut-red head* including throat, except *white crown and cheek patches outlined in black. Chestnut rump* and some chestnut streaking or blotching on lower breast and flanks contrasting with white upper breast patch and belly, and some *rufous on scapulars* and greater coverts. On particularly neat ♂♂, the chestnut on breast forms a solid area. *Tail-sides white.* Bill grey, upper mandible darker than lower. Winter plumage (attained from late summer) is similar, but chestnut colours partly concealed by pale tips, and white crown patch partly obscured by grey tips. ♀ and immature ♂ are less neat than adult ♂, with more subdued colours, and crown is mainly brown and streaked dark. Throat is off-white lined by dark lateral throat-stripes, and there is some rufous mottling or spotting on lower throat (rarely rather extensive rufous spotting and can closely resemble immature ♂ plumage). Note *lack of any yellowish or olive*, not even on edges of primaries, colours which would indicate Yellowhammer rather than Pine Bunting, or at least a hybrid between them. Juvenile and first-winter ♀ have practically no reddish-brown in their plumage except on rump and a little on scapulars and tertials, and they are more streaked on head and breast than other plumages. Like all buntings, feeds mainly on ground but when alarmed, or when resting or singing, takes elevated perch in bush or tree. Flight powerful with deep undulations.

VOCALISATIONS (*leucocephalos*) Song very similar or identical to Yellowhammer, and safe separation usually impossible given individual variation in both. Song comprises a usually quick repetition of 5–7 short, high-pitched notes ending with a different and more drawn-out note, *ze-ze-ze-ze-ze sreeee* or *zri-zri-zri-zri-zri-zri seeh*. – Calls also by and large similar to those of Yellowhammer, but the common hoarse *chüff* call of the latter is either shorter and finer, more like *chick*, or softer and slightly downwards-inflected, *psheu*. Other calls include a fine clicking *petelit*, and strident, thin, sharp *zee*.

SIMILAR SPECIES There is really only one major pitfall to consider, pale or aberrant first-winter ♀ Yellowhammer lacking yellow pigmentation, or hybrids between that species and Pine Bunting. Note that Pine Bunting is subtly larger and more long-tailed (differences very slight and rarely of real help) and lacks any trace of yellow or olive, including on edges to primaries, tail-feathers and on belly/vent (or when handled, on axillaries/underwing-coverts). Most birds have rather prominently grey and rufous tones in their plumage and look both 'cleaner' and slightly paler than Yellowhammer. Usually, the supercilium appears broader giving a more 'open-faced' expression, and streaking above and below is slightly thinner and sparser giving a somewhat paler look. Hybrids, which are fairly frequent in zone of overlap between the two species in W Siberia, pose a special problem. There is no safe way of eliminating a hybrid that closely resembles a Pine Bunting (back-crosses with pure Pine Buntings look even more similar to the real thing), but a careful check of the entire plumage for any yellow tones helps. (Cf. Hellquist 2015.)

AGEING & SEXING (*leucocephalos*) Ages differ in autumn, rarely also in spring if handled or seen close. Sexes differ after post-juv moult. – Moults. Complete post-nuptial moult of ad on breeding grounds post-breeding, mainly late Jul–Sep. Partial post-juv moult at same time, usually includes most or all median and greater coverts, sometimes also odd tertials and r1, but not primaries, rest of tail-feathers or any primary-coverts. Both age categories appear to undergo partial pre-nuptial moult in late winter limited to parts of head and body. – **SPRING** ♂ Crown white, chin and throat uniform rufous, and unstreaked white patch below rufous throat. ♀ Crown brown (rarely a little white visible) streaked dark. Chin, throat and upper breast off-white or buff-white (sometimes with a little rufous admixed). **Ad** Tail-feathers rather broad with rounded tips, usually still fairly moderately abraded (but some are ambiguous due to more wear than usual); beware that a very few might renew some central tail-feathers and show moult contrast in tail (otherwise an indication of 1stS!). ♂ has much pure white and chestnut in plumage. **1stS** Tail-feathers narrower than typical ad, and tips more pointed and abraded; beware that a few renew r1 (or further central tail-feathers) either in autumn or late winter and thus have rounded tips on the most visible tail-feathers. ♀ generally heavily streaked and lacks any extensive pure white on head

E. l. leucocephalos, ♂, presumed ad, Mongolia, Jun: Yellowhammer-like in behaviour, size and jizz. Breeding ♂ unmistakable: rufous head, throat and breast relieved by white crown, cheeks and narrow upper breast-band. Centre of breast and belly also white. Safe ageing requires close look at tail and primary-coverts, but general handsome plumage seems to indicate ad. (H. Shirihai)

E. l. leucocephalos, ad ♀, Mongolia, Jun: an old bird based on rather broad tips to outer tail-feathers. Age supported by extensive rufous in plumage and very little black streaking below. Note rufous in supercilium and lateral throat-stripe, off-white throat, upper breast streaked dark brown, and dark brown lateral crown-stripes. (H. Shirihai)

E. l. leucocephalos, 1stS ♂, Mongolia, Jul: ♂-like but head and breast pattern less neatly developed. Ad ♀ is excluded by large, pure white crown patch, but an advanced ad ♀ rarely might otherwise look much like this (P. Dubois)

E. l. leucocephalos, 1stS ♀, Mongolia, Jun: unlike ad ♀, note very worn juv wing and tail, the latter with pointed tips. Unlike older ♀ lacks visible rufous in supercilium but has bright patches in submoustachial stripe and breast-sides. Limited dark streaking below. (H. Shirihai)

E. l. leucocephalos, ad ♂, Italy, Nov: in winter head pattern becomes duller, but still is obvious, with supercilium and throat clearly chestnut, but whitish patches on crown and cheeks less contrasting (pale fringes to head-feathers wear to reveal stronger white, black and chestnut breeding plumage). Uniformly fresh wing and tail further confirm age. (D. Occhiato)

or underparts, or extensive chestnut on breast. – **AUTUMN ♂** Crown white, only partly concealed by brown-grey tips. Chin and throat rufous, finely tipped white. ♀ Crown brownish-grey streaked dark brown, sometimes with a little white on feather-bases. Chin and throat off-white or buff-white, sometimes with some rufous on feather-bases, partly concealed by pale tips. **Ad ♂** Crown-feathers white tipped greyish-buff. Tips of tail-feathers rounded and neat. **1stW ♂** Crown-feathers white tipped brownish-grey with blackish central streaks. Tips of unmoulted tail-feathers somewhat pointed, slightly abraded by late autumn. **Juv** Very similar to ♀ but overall more buff, dull brown and cinnamon, less off-white and rufous in plumage. Crown duller brown streaked dark, rather more uniform than ♀.

BIOMETRICS (*leucocephalos*) **L** 16–18 cm; **W** ♂ 90.5–100 mm (n 26, m 95.2), ♀ 87–94 mm (n 15, m 89.9); **T** ♂ 73–84 mm (n 26, m 78.9), ♀ 72–78 mm (n 15, m 75.8); **T/W** m 83.4; **B** 11.9–13.8 mm (n 41, m 12.9); **B**(f) 9.0–11.5 mm (n 41, m 10.4); **BD** 5.5–6.9 mm (n 38, m 6.4); **Ts** 18.0–20.3 mm (n 40, m 19.1). **Wing formula: p1** minute; **pp2–4** longest (or p2 and/or p4 < wt 0.5–2 mm); **p5** < wt 0.5–2 mm; **p6** < wt 6–9 mm;

BUNTINGS

E. l. leucocephalos, ad ♀, Israel, Dec: while sex is obvious from general plumage pattern, age is not so easy to decide when bird is seen head-on (though rather distinct head pattern with hint of rufous at rear supercilium, and a trace of white visible on rear crown indicate ad), ageing becomes easier when same bird is seen from rear; note ad tail-feathers, rather broad with rounded and neat tips, outer feathers white-tipped. (A. Ben Dov)

E. l. leucocephalos, ♀, presumed 1stW, England, Oct: sex inferred by dullest end of plumage variation, where rufous breast-side patches are merely faintly hinted, streaking of crown, breast and flanks prominent and head pattern subdued in grey-brown and off-white colours. Age more difficult to decide at this angle, but general dullness might indicate 1stW. (R. Bonser)

E. l. leucocephalos, 1stW ♂♂, Italy, Nov: as fresh-plumaged ad, but broader pale fringes obscuring rufous head pattern to a greater degree. Pointed tail-feathers further enable ageing. Young ♀♀ have generally more diffuse head pattern than these with less whitish cheek patch and not as blackish mark around. (D. Occhiato)

p7 < wt 13–16.5 mm; **p10** < wt 21.5–25 mm; **s1** < wt 23–26.5 mm. Emarg. pp3–5.

GEOGRAPHICAL VARIATION & RANGE Slight variation only. Two sufficiently distinct races described, but still differ only moderately.

E. l. leucocephalos S. G. Gmelin, 1771 (E Russia, Siberia, Transbaikalia, N Mongolia, Amur, Sakhalin, N China; winters rarely S & SE Europe, more commonly Afghanistan, Pakistan, NW India, Himalayas, China). Treated above. Black bands on forecrown and crown-sides moderately broad, rufous in plumage medium dark. (Syn. *karpovi; stachanowi*.)

Extralimital: *E. l. fronto* Stresemann, 1930 (NE Qinghai, N Gansu; winters south of range and at lower elevations). Differs in that ad ♂ has broader black bands on forehead and crown-sides, and slightly deeper rufous on head and rump. Also, wing said to be slightly longer on average. Few specimens examined by us. Song very similar but perhaps slightly slows towards end, and calls include at least the soft, down-curled, *psheu*. (Syn. *kamtschatica*.)

TAXONOMIC NOTE Hybridisation between Pine Bunting and Yellowhammer is well known and apparently fairly frequent within the large area in mainly W Siberia where the

1stS ♀, either Yellowhammer or hybrid Pine Bunting × Yellowhammer, England, Mar: dull non-yellow plumage, little rufous below, no visible rufous above eye and bluish cast to lower mandible suggest a hybrid. Note worn juv wing and tail, the latter with pointed tips. (R. Chittenden)

— 507 —

Autumn/winter ♀ Pine Bunting versus Yellowhammer (top left: 1stW ♀ 'pure' Pine, Netherlands, winter; top right: 1stW ♀, presumed hybrid Pine × Yellowhammer, Kazakhstan, Oct; bottom left: 1stW ♀ Yellowhammer, Sweden, Oct; bottom right: 1stW ♀ Yellowhammer, Finland, Jan): top left a reasonably typical Pine Bunting, with some rufous above eye (visible in other photos of same bird), extensive rufous and weak dark streaking below. Primaries and tail juv. Top right a presumed hybrid based on combination of clean white belly and yellow primary edges; possible 1stW by apparent moult limit in greater coverts. Bottom left is a tricky little-yellow Yellowhammer, with only faint yellow on primaries and supercilium (overall less contrasting than typical Pine). Apparent moult limit in wing supports ageing. Bottom right another tricky Yellowhammer, with olive-yellow hue to neck-sides among other evidence of yellow below in otherwise rather Pine Bunting-like plumage. (Top left: F. Jiguet; top right: A. Hellquist; bottom left: F. Ström; bottom right: A. Below)

two species overlap (e.g. Panov *et al.* 2003, and references therein). The genetic distance is very small, just 0.4% in one study of cyt b of mtDNA (Alström *et al.* 2008), and practically non-existent in another (Irwin *et al.* 2009). Still, it is generally thought that hybrids are fewer than if interbreeding was totally unhindered (although Panov *et al.* 2003 postulated complete fusion with time in areas of overlap), there are some slight structural differences, and vast allopatric areas exist where each species is both genetically pure and morphologically distinct. The two also differ more in their nuclear DNA (Irwin *et al.* 2009), which could indicate recent introgression of mtDNA. Therefore, it seems best to maintain the common treatment as two different species.

REFERENCES Alström, P. *et al.* (2008) *Mol. Phyl. & Evol.*, 47: 960–973. – Ayé, R. & Schweizer, M. (2003) *DB*, 25: 40–43. – Bradshaw, C. & Gray, M. (1993) *BB*, 86: 378–386. – Hellquist, A. (2015) *Vår Fågelv.*, 2015/6: 34–43. – Irwin, D. E., Rubtsov, A. S. & Panov, E. N. (2009) *Biol. J. Linn. Soc.*, 98: 422–438. – Lewington, I. (1990) *BW*, 3: 89–90. – Occhiato, D. (2003) *DB*, 25: 1–16. – Occhiato, D. (2003) *DB*, 25: 32–39. – Panov, E. N., Roubtsov, A. S. & Monzikov, D. G. (2003) *DB*, 25: 17–31.

Hybrid Pine Bunting × Yellowhammer, ♂ (left: Kazakhstan, Jun) and ♀ (right: Russia, Jun): in summer, some hybrid ♂♂ possess extraordinary and rather confusing colour patterns, while some ♀♀ can have yellow-tinged primary edges and throat, and an olive cast on upper breast, sufficient to be confidently labelled as hybrids. (Left: V. & S. Ashby; right: P. Parkhaev)

YELLOWHAMMER
Emberiza citrinella L., 1758

Fr. – Bruant jaune; Ger. – Goldammer
Sp. – Escribano cerillo; Swe. – Gulsparv

Any bird that popularly has its song transcribed as a 'little-bit-of-bread-and-no-cheeeese' is bound to be fairly common and well known, or both. Still, probably few outside the ranks of birdwatchers are really familiar with the Yellowhammer and can identify it instantly by sight. It is all around us, yet not quite part of common knowledge. The Yellowhammer breeds in open habitats with scattered bushes and trees, is common on arable fields and pastures with hedgerows and scattered dense copses for cover and nesting. It is resident in Britain and France, but a migrant further north and east. In winter quite large flocks can be found where food is abundant.

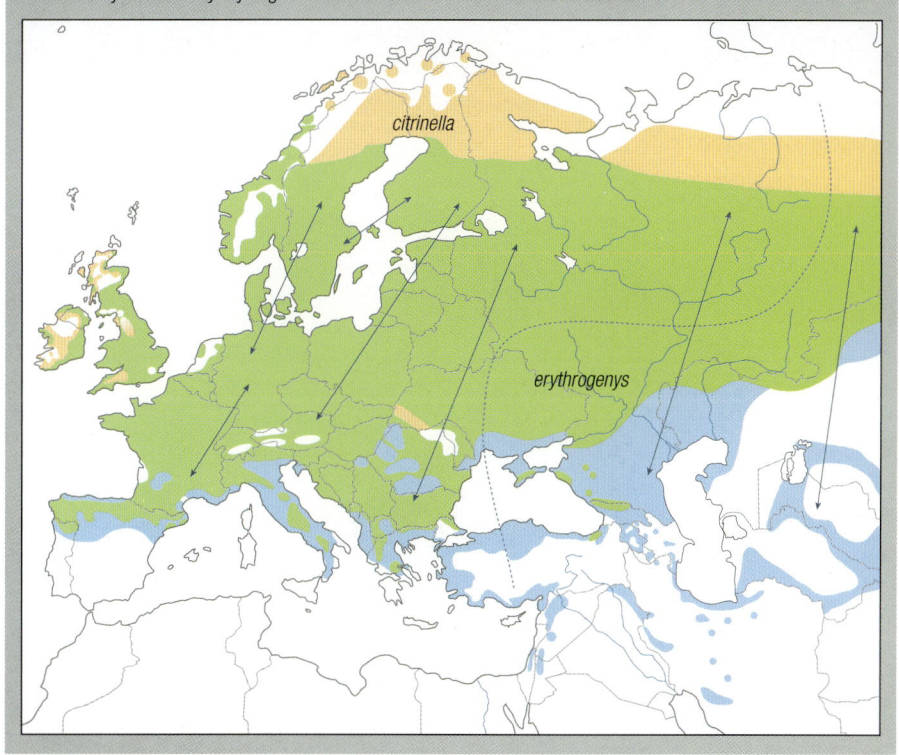

IDENTIFICATION A rather *large, long-tailed bunting* that often appears as a *brown, dark-streaked* bird with *rufous rump* and *white tail-sides*. In a closer look *some yellow* in the plumage should also be evident, even in duller birds. Bill is grey with darker culmen. The ♂ in summer plumage is an attractive bird with *bright yellow head* (with some olive-grey marks like stripes above and below cheeks and along crown-sides, olive mottling on nape, etc.), *yellow on much of underparts* and has some *rufous and olive-green on breast and flanks*. Mantle and shoulders are tawny-brown streaked black, and tertials have rufous edges unless heavily worn. The brightest birds acquire a *completely yellow head* in summer, which can contrast very strongly against the darker brown body. The ♀ is much less bright and more streaked, even in worn summer plumage, the head being a mix of olive, grey and brown with at most a hint of a pale yellow crown patch (finely streaked), supercilium, submoustachial stripe and throat. Autumn plumages, especially young birds, are less neat and more olive grey-brown with more prominent streaking, so ♂♂ appear rather similar to adult ♀, whereas first-winter ♀ is very dull and olive-brown with just a faint tinge of olive-yellow on belly and edges of flight-feathers, and is best identified by the red-brown rump, long tail with white sides, calls and behaviour. Often encountered feeding in flocks on the ground (e.g. on spilt grain) and flushes before any of the sparrows (Yellowhammers are rather wary), escaping in strong but skipping, slightly undulating flight, 'collapsing' with turgid wings and tail at last second to perch in a high bush or tall tree at some distance, waiting until the 'coast is clear'. Often flicks tail upwards or, less frequently, to sides when uneasy, and raises crown-feathers slightly.

VOCALISATIONS Sings from exposed high perch on bush, tree or wire, a mechanically-repeated simple song, the well-known 'cheese-pleading' or 'counting-to-seven song', six (5–7) similar short notes in quick succession followed by either one drawn-out, thin note, *si-si-si-si-si-si seeee*, or two different notes or other variation at the end. The initial series of notes invariably involves a repetition of the same note, but the note itself can vary, from a thin, sharp *si* or *tse* to a more strident or buzzing *dzre* (River Warbler voice!) or even a disyllabic *zri-e*. – Call from perched or flying birds a discordant *chüff*. Also a fine clicking conversational *pt... pt, pt-pt... peti-litt* etc., and fine, thin *zee*. A sharp *tsit* from alarmed birds.

SIMILAR SPECIES First-winter ♀ can be confused in particular with corresponding plumage of *Pine Bunting*, a species that shares the rufous rump, size and general shape with Yellowhammer. On Yellowhammer, note at least a faint yellowish hue on centre of lower belly and edges to primaries, a more olive-brown tinge, not as cold greyish and rufous-tinged as in Pine Bunting. Yellowhammer is also usually a slightly darker and more densely and heavily-streaked bird than Pine Bunting, although extremes are close.

AGEING & SEXING (*citrinella*) Ages differ in autumn, rarely also in spring; reliable ageing in the field usually requires close views. Sexes differ after post-juv moult, but sexing can be difficult in autumn, especially of 1stW birds, then easier in spring. – Moults. Complete post-nuptial moult of ad on breeding grounds after breeding, mainly Aug–Oct.

E. c. citrinella, ad ♂, Switzerland, Jul: yellow, especially on head and underparts. With wear in summer, dusky ear-coverts surround, mottled olive-green and red-brown breast, and streaky flanks create bold pattern. Darker olive-brown upperparts with yellowish wing-bars also obvious. Very fresh feathers in Jul indicate an ad, an assessment supported by extensively yellow head and tail-feathers looking broad and rounded. (H. Shirihai)

E. c. citrinella, ♀, presumed ad, England, Apr: unlike ♂, heavily streaked, duller and less yellow overall. Yellowish tinge to moustachial stripe, sometimes also to supercilium and central crown and always to belly and primary edges separate from similar Pine Bunting. Fresh exposed primary-tips and richer coloration indicate ad. (P. Blanchard)

E. c. citrinella, 1stS ♂, Finland, May: aged as 1stS by moult limits in tertials (longest is juv) and lesser coverts (some dull juv feathers retained). Note that apparent moult limit between inner three more rufous-edged greater coverts and rest is an artefact; normally all greater coverts are renewed in post-juv moult, none in pre-nuptial. (D. Occhiato)

E. c. citrinella, ♀, presumed 1stS, Hungary, Jun: dull with limited yellow, streaking narrow but bold below. Apparently retained juv primary-coverts (narrow and rather pointed) and alula suggest 1stS. (M. Varesvuo)

Partial post-juv moult at same time, usually includes most or all median and greater coverts, sometimes also odd tertials and r1 (rarely more tail-feathers), but not primaries, rest of outer tail-feathers or any primary-coverts. For both age categories, in at least some birds there appears to be a partial pre-nuptial moult in late winter limited to parts of head and body, and renewal of some central tail-feathers. — **SPRING** Sexing generally possible using plumage characters. For the few ambiguous birds ageing will help, as the two most similar are ad ♀ and 1stS ♂. ♂ Has a variably large clear yellow central crown patch, sometimes with thin dark streaking at its edges or all over. Throat usually unstreaked clear yellow, or has some rufous spots or streaks laterally, but many 1stS have some dark brown or black streaks or spots on sides, or even over nearly all throat. Tends to have an unstreaked or diffusely mottled olive-green breast-band and many rufous patches or streaking on lower breast and flanks. ♀ Crown variable, either grey-brown and rather prominently streaked dark or has some pale yellow visible in centre among bold dark streaks and some grey-brown feather-edges. Pale yellow throat nearly invariably streaked, at least partly. Birds with most of throat unstreaked are rare and have pale yellow colour, paler than in ♂. Breast is generally rather boldly streaked (in at least some ad, streaking more reduced) and has reduced amount of olive-green. Lower breast and flanks either have some subdued and limited rufous streaking (though consistently less than ♂) or has no rufous, instead being dull greenish-yellow streaked grey-brown. **Ad** Tail-feathers rather broad with rounded tips, usually still fairly moderately abraded (but some are ambiguous due to more wear than usual); beware that a very few might renew some central tail-feathers and show moult contrast in tail (otherwise an indication of 1stS!). ♂ has much clear yellow and olive-green in plumage. **1stS** Tail-feathers narrower than typical ad, and tips more pointed and abraded; beware that a few renew r1 (or further central tail-feathers) either in autumn or late winter and thus have rounded tips on the most visible tail-feathers. ♀ generally heavily streaked and lacks any clear yellow on head or underparts, or green on breast. — **AUTUMN Ad** ♂ Tips of primaries and tail-feathers fresh (after moult finished), latter usually rather broad and tips rounded (but some variation, and odd 1stW can replace some rectrices in autumn, new feathers being ad-like in shape). Same plumage criteria for sexing as spring, but yellow of crown-feathers now partly concealed by grey-brown tips; note that these tips are thin and that any shaft-streaks on crown-feathers are usually thin, short and not prominent. However, Dunn & Wright (2009) showed that a few ad ♂♂ do not develop typical crown pattern with much yellow partly hidden and only thin dark shaft-streaks, thus all characters should be used in combination for any ambiguous

E. c. citrinella, ad ♂, Netherlands, Jan: in fresh plumage visible yellow reduced by grey-green feather-tips, but becomes progressively brighter as these abrade. Combination of relatively rich yellow plumage, with yellow crown patch already in Jan, and uniformly fresh wing and tail (latter with seemingly rather broad and rounded feathers) indicative of ad. (R. Schols)

E. c. citrinella, ad ♀, Israel, Dec: being a ♀ based on general plumage pattern, the combination of relatively extensive yellow on head and underparts, and evenly fresh wing infer age. (A. Ben Dov)

bird, including wing length. Nearly always some clear yellow above and behind eye. Much olive-green usually visible on breast. Outer two tail-feathers have nearly invariably about same-sized, large white wedges, and rump-feather shafts of same colour as rest of feather or paler; longest uppertail-coverts have partly or wholly black shafts (Dunn & Wright 2009). **Ad ♀** Tips of primaries and tail-feathers fresh and rounded. Same plumage criteria apply as spring, but due to dark feather-tips some are more difficult to sex reliably. However, if first correctly aged, ad ♀ should be possible to separate from sometimes similar-looking 1stW ♂. Only rarely some yellow above and behind eye. No or limited and subdued olive on breast. As a rule, white wedge on r5 considerably smaller than that on r6, and rump-feather shaft black or blackish (Dunn & Wright 2009). **1stW ♂** Tips of primaries and tail-feathers slightly abraded from Sep, unmoulted tail-feathers often narrower and tips more pointed (rarely slightly broader and more rounded in a few difficult 1stW). Plumage characters as ad ♂ or a little duller with less prominent yellow and green, then approaching ad ♀. No black on shaft of longest uppertail-coverts (Dunn & Wright 2009). **1stW ♀** Tips of primaries and tail-feathers as 1stW ♂. Plumage characters as ad ♀ or a little duller with heavier streaking and no or very little apparent yellow away from throat and centre of belly (still pale yellow), and no green. **Juv** Very similar to 1stW ♀ but overall more buff and dull brown, less pale yellow in plumage. Entire head on average duller and more uniformly brown streaked dark. Lateral throat-stripes less developed, whole throat diffusely mottled or streaked.

BIOMETRICS (*citrinella*) **L** 16.5–17.5 cm; **W** ♂ 87–97 mm (*n* 64, *m* 91.6), ♀ 83–90 mm (*n* 48, *m* 86.8); **T** ♂ 71–81 mm (*n* 64, *m* 75.3), ♀ 67–77 mm (*n* 48, *m* 71.5); **T/W** *m* 82.3; **B** 12.0–14.5 mm (*n* 105, *m* 13.2); **B**(f) 9.6–12.0 mm (*n* 104, *m* 10.6); **BD** 5.7–7.0 mm (*n* 83, *m* 6.3); **Ts** 18.5–21.2 mm (*n* 102, *m* 19.7). **Wing formula: p1** minute; **pp2–4** about equal and longest (or p2 < wt 0.5–3 mm); **p5** < wt 0.5–2.5 mm; **p6** < wt 6.5–10.5 mm; **p7** < wt 14–18 mm; **p10** < wt 21–25 mm; **s1** < wt 22–26.5 mm. Emarg. pp3–5.

GEOGRAPHICAL VARIATION & RANGE Very slight and clinal variation, mainly affecting colour saturation. Only two subspecies accepted here.

E. c. citrinella L., 1758 (Europe except south-east; northernmost populations winter further south and south-west, southern sedentary). Described above. Grades without sharp borders into *erythrogenys* of Baltic States, Poland, Belarus, S Russia, and perhaps NE Balkans, thus birds over large area difficult to assign to race even with series for comparison. – A separate race sometimes recognised from Scotland, N England, Wales and Ireland ('*caliginosa*'), being subtly smaller and duller with marginally bolder streaking, but differences very slight indeed. Probably more than 50% are alike, and '*caliginosa*' thus does not seem to come close to passing a 75% rule. (Syn. *caliginosa*; *nebulosa*; *sylvestris*.)

○ **E. c. erythrogenys** C. L. Brehm, 1855 (C & E Ukraine, NE Turkey, SE Russia, W Caucasus, W & C Siberia; largely sedentary, but moves to lower elevations in winter, or is short-range migrant). Differs only slightly from *citrinella* in ♂ plumage, which is slightly brighter and paler yellow, and has subtly paler and more greyish-brown or cinnamon-tinged mantle/back (less olive and rufous) with slightly less heavy dark streaking. Wing-bars often appear a little clearer whitish, not as buffish-yellow as *citrinella*. In fresh plumage the chestnut rump-feathers have more whitish tips than *citrinella*. Grades into *citrinella* over wide zone (see *citrinella*). **W** ♂ 85–97 mm (*n* 16, *m* 91.1), ♀ 83–89 mm (*n* 15, *m* 87.0); **T** ♂ 67–80.5 mm (*n* 16, *m* 74.8), ♀ 66–77 mm (*n* 15, *m* 71.9). Rest of biometrics and wing formula very nearly identical to *citrinella*.

TAXONOMIC NOTE Panov *et al.* (2003) and Irwin *et al.* (2009) suggested that ssp. *erythrogenys* could represent a hybrid swarm, the result of hybridisation by *citrinella* with Pine Bunting *E. leucocephalos*, and back-crossing by

E. c. citrinella, 1stW ♂♂, Netherlands, Nov (left) and Germany, Oct: sexed by relatively extensive yellow and less streaky head, but not as bright as most ad, with juv primary-coverts, and worn and pointed tail-feathers to support ageing. Nevertheless, there is much individual variation, with least-advanced 1stW ♂♂ (right bird) not that different from richest yellow ad ♀, and if not aged correctly could easily be mistaken for one. (Left: R. Schols; right: M. Schmitz)

hybrids, a hypothesis we find unlikely considering first of all the close similarity between *erythrogenys* and *citrinella*, and the consistent morphological appearance of *erythrogenys* over a broad area.

REFERENCES Alström, P. *et al.* (2008) *Mol. Phyl. & Evol.*, 47: 960–973. Dunn, J. C. & Wright, C. (2009) –*Ring. & Migr.*, 24: 240–252. – Hellquist, A. (2015) *Vår Fågelv.*, 2015/6: 34–43. – Irwin, D. E., Rubtsov, A. S. & Panov, E. N. (2009) *Biol. J. Linn. Soc.*, 98: 422–438. – Panov, E. N., Roubtsov, A. S. & Monzikov, D. G. (2003) *DB*, 25: 17–31.

E. c. citrinella, 1stW ♀, Israel, Dec: dull with heavy streaking and only very pale yellow tinge (limited to throat, supercilium, belly and primary edges) are often first clues for 1stW ♀, but juv primaries and primary-coverts represent only sure-fire ageing criterion (this bird has replaced many central tail-feathers to ad-like, perhaps accidentally?). (A. Ben Dov)

E. c. citrinella, juv, England, Aug: overall fluffy feathering and rather dull brown plumage, tinged buff and with hardly any hint of yellow in plumage, and entire head streaked and throat diffusely and narrowly streaked, too. Typical for the species note rufous rump and uppertail-coverts. (C. Fleming)

CIRL BUNTING
Emberiza cirlus L., 1766

Fr. – Bruant zizi; Ger. – Zaunammer
Sp. – Escribano soteño; Swe. – Häcksparv

Compared to Yellowhammer—one of its closest relatives—the Cirl Bunting differs in having a more south-westerly range and being slightly more arboreal. It breeds from SW England to NW Africa and east to E Turkey and is typically found in open agricultural areas with copses, tall hedges, vineyards and orchards, at woodland edges or in woods with clearings. When singing its simple, monotonous song, it usually takes a post in top of tallest trees. It is mainly resident except in the northern parts of its range from where it moves south a moderate distance in winter.

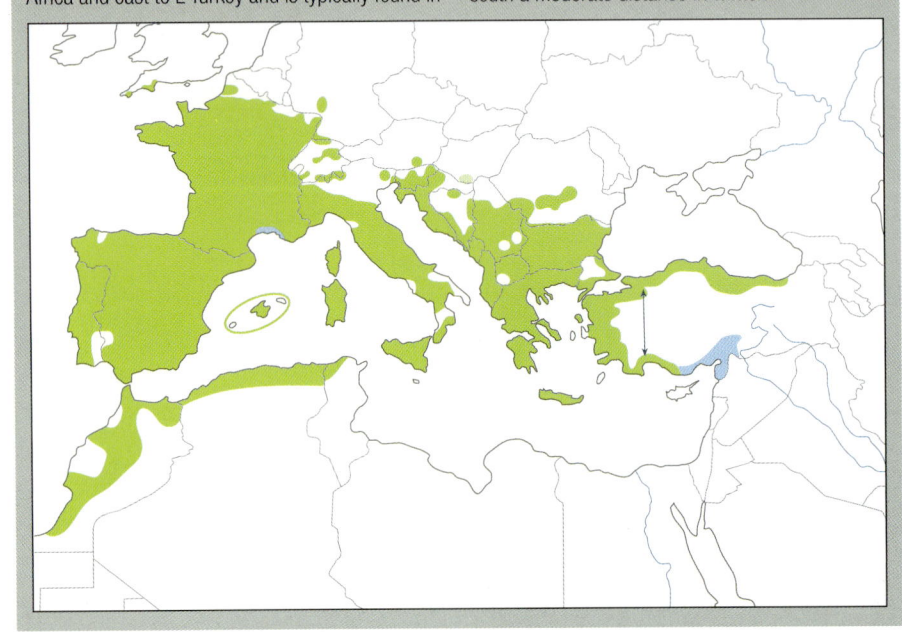

IDENTIFICATION On average a trifle smaller than Yellowhammer with proportionately slightly shorter and more rounded wings, otherwise resembles that species, especially in ♀ and immature plumages. (That the tail is also slightly shorter, as often stated, is hard to see on photographs or skins and, if correct, is probably of little significance for identification.) Adult summer ♂ is characteristic given its *black throat*, narrow *pale yellow upper breast-band* and *pale yellow face with a broad blackish eye-stripe*, its *olive-green crown* (thinly streaked black), largely *unstreaked olive-green neck and chest, rufous patches either side of breast* and *rufous scapulars*. Note that the *rump is olive-grey* in all plumages, rather than rufous as in Yellowhammer and Pine Bunting. To separate ♀ and first-winters from similar-looking Yellowhammer, note in addition to the olive-tinged grey-brown rump, (i) the two on average *more prominent and contrasting broad dark bands on the pale greyish-yellow face* bordering the ear-coverts above and below, and (ii) the *rufous scapulars* which in Cirl Bunting contrast more against grey-brown rest of upperparts than in Yellowhammer. Juvenile is even more nondescript and buff-brown than later plumages. Note tendency even in juvenile to have rufous tinge to scapulars and breast-sides (unlike juvenile Yellowhammer), and fewer and thinner streaks on flanks and belly.

VOCALISATIONS Song differs clearly from that of Yellowhammer by lacking a different end note, a monotonous 1.5–2 sec-long flat, dry or slightly metallic trill of similar structure and pitch as the song of Arctic Warbler (albeit faster, and differs in other fine details, too), *sre'sre'sre'sre'sre'sre'sre'sre*. Sometimes shifts between two similar variations on different pitch. There are also many variants in tone and voice, some buzzing or scratchy, others clear. – Calls include a fine short *tsit* that can recall both Song Thrush and some bunting species in Siberia (but is less hard and clicking than latter). There is also a piercing, slightly downslurred *seeeu* like alarm from a Blackbird. A lower-pitched slightly 'bent' *cheu* can recall the autumn call of Reed Bunting. When agitated in territorial disputes, or when ♂ courts ♀, can utter explosive trills almost like Long-tailed Tit, *zir'r'r*.

SIMILAR SPECIES The risk of confusion with *Yellowhammer* is treated under that species and Identification above. – Young ♀ can theoretically resemble young ♀ *Reed Bunting*, but Cirl Bunting is slightly larger and usually has some faint yellow tones on face and underparts (mainly on breast), lacking in Reed Bunting, and scapulars and breast-sides tend to show more vivid rufous tones than back and wing-coverts, whereas in Reed Bunting the lesser coverts are warmest brown but the scapulars do not differ from surrounding coverts, and there is never any rufous on the breast-sides. Also, Reed Bunting usually has dark legs, while those of Cirl are pinkish.

AGEING & SEXING Ages differ in autumn, but reliable ageing usually requires close views. Sexes differ after post-juv moult. – Moults. Complete post-nuptial moult of ad on breeding grounds after breeding finished, mainly mid or late Jul–Oct. Partial post-juv moult at same time, usually includes

Ad ♂, England, Apr: jizz recalls Yellowhammer, but slightly smaller with proportionately stronger bill. Unmistakable in ♂ plumage, especially in breeding plumage. Note olive-green head and upper breast, with contrasting black eye-stripe and bib, sharply delimited from bright yellow supercilium, stripe below eye, and lower throat, as well as partial rusty breast-band. Unworn wing and tail, and with seemingly broad and rounded tips to tail-feathers, confirm age. (D. Rickards)

♂, presumed ad, Italy, May: with wear in late spring and summer, ♂ becomes even more boldly patterned, with brighter yellow parts, and breast-band often increasingly green-tinged. Entire wing evenly feathered and looking quite fresh indicating ad, but safe ageing would require a closer look at primary-coverts and shape and wear of tail-feathers. (L. Sebastiani)

most or all median and greater coverts, sometimes also odd tertials and r1, but not primaries, rest of tail-feathers or any primary-coverts. Both age categories appear to undergo limited partial pre-nuptial moult in late winter, limited to parts of head and throat. – SPRING Ageing sometimes possible with typical birds based on shape and wear of unmoulted tail-feathers, those with broader and at tips slightly rounder feathers with only moderate abrasion being ad, those with narrower and pointed and heavily abraded tips being 1stS. Many birds will appear as intermediates and are best left unaged until practice and experience help to age more birds. ♂ Chin and throat greyish-black. Supercilium, band over ear-coverts and patch between throat and breast pale lemon-yellow. Breast-band uniform olive-green. Mantle-feathers partly rufous. ♀ Chin and throat dusky yellowish-white, throat spotted or streaked dark. Supercilium and band over ear-coverts pale yellowish-white, partly spotted or streaked dark. Breast dusky yellowish-white tinged dull olive, streaked dark. Mantle more dull brown, less rufous-tinged than ♂. – AUTUMN Sexing as in spring, only head pattern more blurred, blackish bib of ♂ partly concealed by pale tips

♀, Greece, May: unlike ♂, heavily streaked, duller and less yellow overall. Distinguished from ♀ Yellowhammer by dull brownish olive-grey tinge to upperparts, including rump (without rufous) and by rufous on scapulars. Underparts washed buffish and cream, and flanks narrowly streaked dark. Wing appears evenly fresh suggesting ad, but safe ageing impossible without a closer look. (P. Cools)

1stS ♂, Greece, May: similar to ad ♂, but worn central tail-feathers and primary-tips confirm age, while plumage is also duller and less well patterned in many first-year ♂♂. (K. Mauer)

1stS ♀, Morocco, Feb: very similar to ad ♀ but worn juv tail-feathers and wing confirm age, and whole plumage averages slightly duller. Classic dark framing of ear-coverts surrounding bold whitish spot. Olive-tinged grey-brown rump just visible, and rufous scapulars important species criterion. (D. Occhiato)

1stS ♀, Spain, Mar: dull-plumaged bird, with no yellow, but still shows diagnostic rufous scapulars, as well as grey-brown rump (with broad, dark streaking, a vestige of immaturity, as is heavy streaking on crown). Ageing clinched by moult limit among tertials (central one contrastingly post-juv), with worn juv primaries and tail. (M. Varesvuo)

Ad ♂, England, Dec: as breeding ♂, but black of head and rufous of mantle, scapulars and upper belly partly obscured by pale olive-grey fringes. Intense chestnut edges to secondaries and greater coverts form a red-brown panel. Pure black-centred primary-coverts and broad-tipped outer tail-feathers support ageing. (S. Ray)

Ad ♀, England, Feb: entire plumage evenly fresh and in good condition (no contrast between tertials and secondaries or between greater coverts and tertials, as sometimes shown by 1stY), with dark-centred primary-coverts and fresh chestnut edges to inner secondaries. Some ad ♀♀ can have rather intense yellow supercilium. (T. Hovell)

1stW ♂, England, Jan: like ad ♂, but often duller with dark area on head duskier instead of black, and yellow diluted. Note juv (brownish) alula and primary-coverts, and paler edges to flight-feathers subtly contrasting with tertials. Vaguely olive-tinged grey breast more boldly streaked dark than in normal ad winter ♂. (C. Fleming)

and yellow parts partially by grey tips. Some 1stW ♂♂ have poorly developed breast-band and some streaking on breast, thus resemble ad ♀; start with ageing to separate these. **Ad** After completion of moult, longest primaries and all tail-feathers quite fresh, tips of tail-feathers slightly rounded, edges neat. **1stW** Tips of longest primaries and tail-feathers slightly worn from mid autumn, tips of tail-feathers pointed, except birds that have moulted r1 (rarely more inner pairs of tail-feathers). Birds with moult contrast between retained juv and new tail-feathers easy to recognise. **Juv** Resembles a dull ♀ but usually has rather densely but diffusely streaked or blotched breast and rump. Also, crown is duller brownish-grey with rather heavy streaking. Look for loose plumage on vent and hindneck.

BIOMETRICS L 15–17 cm; **W** ♂ 75.5–84.5 mm (n 16, m 81.8), ♀ 75–82.5 mm (n 14, m 78.0); **T** ♂ 68–74 mm (n 16, m 71.1), ♀ 67–73 mm (n 14, m 69.8); **T/W** m 88.1; **B** 12.0–14.0 mm (n 30, m 13.2); **B(f)** 9.9–11.3 mm (n 30, m 10.4); **BD** 5.5–7.3 mm (n 27, m 6.5); **Ts** 17.9–19.0 mm (n 26, m 18.3). **Wing formula: p1** minute; **p2** < wt 1–3 mm; **pp3–5** about equal and longest; **p6** < wt 1.5–3.5 mm; **p7** < wt 7–10 mm; **p10** < wt 14.5–18 mm; **s1** < wt 15–19 mm. Emarg. pp3–6, sometimes p7 very slightly.

1stW ♀, England, Jul: irrespective of age, some ♀♀ show hint of ♂-like head pattern, but always lack black bib and eye-stripe, while the grey-green breast and partially rufous upper belly are very subdued, with much stronger streaking than any ♂. In post-juv moult with missing tertials and fresh but juv flight-feathers. (B. Nield)

♀, presumed 1stW, England, Oct: dull overall plumage indicate sex, and could infer young age, too. However, reliable ageing depends on seeing well any juv (pointed) tail-feathers (not really possible here), primaries and primary-coverts (again, poor angle of view to assess these). Especially in such dull plumage, and if rump or scapulars not seen, confusion with dull Yellowhammer becomes more likely. (J. Davenport)

Juv, Turkey, Aug: overall buff-brown and, unlike juv Yellowhammer, has already a hint of rufous-tinged scapulars and breast-sides, and fewer dark streaks on flanks and belly. (S. Bekir)

♀ Cirl Bunting (top and bottom left: Spain, Mar) versus Yellowhammer (top right: Israel, Dec; bottom right: Scotland, Apr): this pitfall is frequent in S Europe, especially as diagnostic lack of rufous on rump and uppertail-coverts, and presence of rufous scapulars, in Cirl Bunting are often difficult to see. Luckily, other subsidiary clues can be employed: Cirl has (i) whitish rear ear-coverts spot larger and better defined (in general, the face pattern in Yellowhammer is usually less distinct), (ii) stronger streaking on crown to neck-sides and nape (Yellowhammer lacks streaks on neck-sides and nape, or they are very subtle, while central crown is often unstreaked), and (iii) Cirl almost always lacks Yellowhammer's yellow-fringed primaries. With experience, subtly different jizz can also be used, with (iv) the bill often visibly heavier in Cirl. – Only the top left Cirl can be aged as 1stS by moult limit between tertials and dull-edged secondaries, worn (brown) primary-coverts and by an apparently retained juv largest tertial. (Top and bottom left: M. Varesvuo; top right: A. Ben Dov; bottom right: D. Jacobsen)

GEOGRAPHICAL VARIATION & RANGE Monotypic. – S & C Europe north to France, S Britain, S Germany; Turkey, NW Africa; sedentary. – Birds from Corsica and Sardinia ('*nigrostriata*') claimed to be more heavily streaked on flanks and duller on mantle, but impossible to confirm any difference on upperparts in NHM with adequate series at hand. A few ♂♂ from Corsica have slightly more black streaking on flanks than average *cirlus* of Iberia, France or Greece, but odd, similarly-streaked *cirlus* match this. If there really is an average difference it must be minimal, and is deemed insufficient grounds for separation. (Syn. *nigrostriata*.)

REFERENCES GUTIÉRREZ, R. (1997) *Alula*, 3: 174–180.

ROCK BUNTING
Emberiza cia L., 1766

Fr. – Bruant fou; Ger. – Zippammer
Sp. – Escribano montesino; Swe. – Klippsparv

This long-tailed bunting of rocky slopes near or above tree-line in the Mediterranean region, Turkey, Levant and Central Asia is peculiarly discreet in its habits and often evades detection unless you make an effort to find it. The song, too, is low and modest and does not help greatly, but when the bird is seen it is quite attractive with rufous and light grey colours marked in black. And the quiet song, if you pay attention to it, is quite pleasing, clear and jingling. The Rock Bunting is mainly resident but some move to lower elevations in winter.

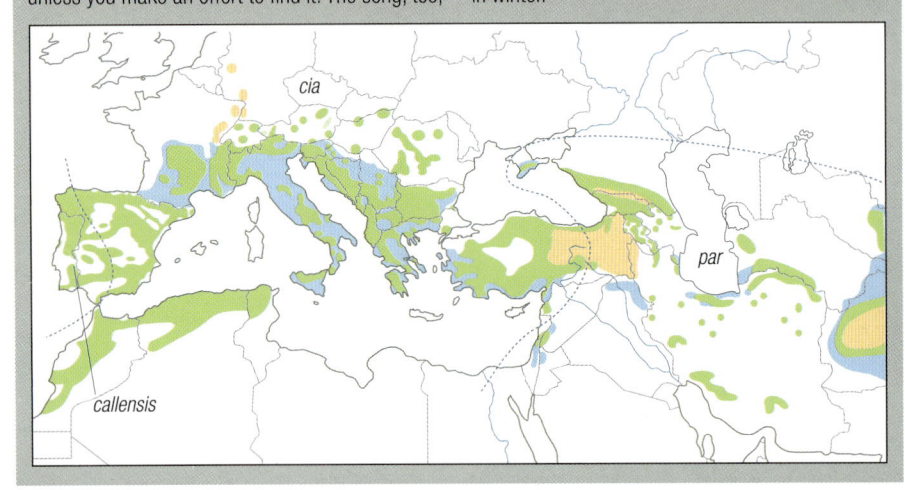

IDENTIFICATION A slim, *long-tailed bunting*, medium-large to large, in post-juvenile plumages with characteristic *head pattern, pale grey with distinct black stripes*. *Rump, belly and vent are warm rufous-brown*, rump darkest. The *ash-grey colour* of the head *extends onto the breast and neck*, both being unstreaked. There are *black lateral crown-stripes* (creating a pale grey median crown-stripe), a *black eye-stripe* and an *angled black stripe below the ear-coverts*, separated from the rear end of the eye-stripe by only a narrow gap. *Mantle tawny-brown streaked dark*, tertials and greater coverts edged tawny-brown with dark centres, whereas median coverts are largely black with broad buffish tips (soon bleaching whitish). *Lesser coverts and wing-bend grey* (often concealed when the bird is perched). *Tail-sides white*, clearly visible in flight at least. *Bill* usually *all grey* with darker upper mandible (exceptions with pinkish-yellow tinge on lower mandible or whole bill very rare), *legs pinkish*. Sexes are similar, ♀ being only somewhat duller in colours and head pattern, crown often being slightly darker and more streaked than ♂, and breast-sides and flanks can have some streaking, invariably lacking in ♂. Juvenile very different and potentially difficult to identify: much duller and more brown-grey, lacking all or most of the head pattern of older birds, and all or most of the rufous on belly and rump. Usually head, neck, mantle, breast and flanks are dull grey-brown or buff-brown with diffuse dense dark mottling or streaking. Rump and centre of belly can show hint of rufous, and there may be traces of the dark lines on the face of later plumages, but generally the juvenile must be identified using size, shape, tertial pattern (evenly-broad narrow buff-brown edges to otherwise blackish tertials), tail pattern, greyish lesser coverts, grey bill and calls. Habits, gait and flight much as Yellowhammer.

VOCALISATIONS Song is a high-pitched clear warble of 2–2.5 sec, often recognised by its faltering opening, uneven rhythm and inclusion of a few softer notes in the middle. The clear, high-pitched voice and somewhat irresolute warbling structure are reminiscent of Dunnock, while some phrases can recall Wren due to the quick shift between trills and series of twittering notes. However, both these other species sing at an even pace without brief pauses, unlike Rock Bunting. A combination of range, habitat and the clear, jingling notes delivered at uneven pace and the hesitant start generally identify the species if the singing bird has not yet been seen. An average song, usually introduced by one or two short *tsit* notes, could be rendered *tsit, zit-itt, svee cha-cha-sivi-seea-seea zitt-seeviserrr zri*. – Calls resemble those of Cirl Bunting, the most commonly heard being a short, sharp *tsi*, a slightly longer and fuller *tsip*, and a drawn-out thin and very high-pitched *tseeee*, sometimes faintly downslurred at the end and recalling, apart from Cirl Bunting, both Robin and Blackbird alarm calls, and when downwards-inflected even Penduline Tit a little. When agitated brief, sharp trills are uttered, a *zerr* or more drawn-out *zir'r'r*.

SIMILAR SPECIES Must be distinguished from similar *Striolated Bunting* (these two only narrowly come into contact in Levant, being otherwise well separated by range). Striolated shares general plumage pattern of Rock Bunting but has dark-streaked grey throat/breast (Rock: unstreaked), whiter-looking supercilium and submoustachial stripe but darker grey cheeks than Rock, lacks a white median coverts wing-bar, has pinkish-yellow lower mandible (Rock: grey), rufous-tinged lesser coverts (Rock: greyish) and duller grey-brown rump (Rock: vivid rufous). – Overlaps widely with *House Bunting* in NW Africa but differs in paler grey head with more contrasting black stripes, prominently dark-streaked mantle/back (House: unstreaked rufous) and grey bill (House: pinkish-yellow lower mandible). – *Godlewski's Bunting* (*E. godlewskii*; extralimital, not treated) is superficially similar to Rock Bunting but has chestnut lateral crown-stripes, chestnut line on cheeks around ear-coverts, not black, and rufous scapulars. – Although widely separated in range, it should be noted that extralimital sub-Saharan ssp. *goslingi* of *Cinnamon-breasted Bunting* resembles Rock Bunting to a surprising degree, but differs by lack of unstreaked rufous rump, grey of throat extending much less far onto upper breast, and face pattern is more contrasting and boldly striped.

AGEING & SEXING (*cia*) Ages differ in autumn, rarely also in spring; reliable ageing usually requires close views. Sexes differ after post-juv moult. – Moults. Complete post-nuptial moult of ad on breeding grounds after breeding, mainly late Jul–Sep (Oct). Partial post-juv moult at same time, usually includes most or all median and greater coverts, sometimes also odd tertials and r1, but not primaries, rest of tail-feathers or any primary-coverts. Apparently no pre-nuptial moult. – **SPRING** Ageing sometimes possible with typical birds based on shape and wear of unmoulted tail-feathers, and pattern of r4 (see under Autumn), those with broader and at tips slightly rounder feathers with only moderate abrasion, and with r4 white-tipped, being ad, those with narrower and pointed and heavily abraded tips, and no trace of white on tip of r4, being 1stS. Due to wear many birds will appear as intermediates

E. c. cia, ♂, presumed ad (left) and 1stS, Turkey, Jun (left) & May: a largish, long-tailed bunting, with pale grey and black stripes on head, and grey breast contrasting with warm chestnut-brown rest of underparts. Rufous-brown upperparts, back black-streaked, rump not. Large white tail corners often highly distinctive from below or when tail spread, and rather thin white wing-bars also typical. Here the bold black head-stripes (very wide over lores) with pure grey throat and breast indicate a ♂. Ageing on the other hand is difficult: if anything, width of tail-feather tips of left bird seem broad and rounded (thus ad-like), and primary-coverts seem dark (though viewing angle not perfect); the right bird appears to have heavily worn primary-tips and heavily worn central tail-feathers suggesting 1stS. (Left: H. Shirihai; E. Yoğurtcuoğlu)

and are best left unaged until practice and experience help to age more birds. A few are also difficult to sex due to intermediate characters. ♂ Lores and head stripes jet-black and well defined. Flanks invariably unstreaked rufous. A small area on centre of crown is pure grey with no more than very fine black streaking. (Rear crown/nape often slightly more prominently streaked black.) Throat/upper breast pure ash-grey. Division between ash-grey throat and chestnut-ochre breast/belly usually distinct and even. ♀ Lores blackish or greyish-brown, though a few have darker lores and approach ♂ pattern. Dark bands on head often diffuse (or even lacking) and a mixture of black and brown, on average less dark and distinct than ♂. (Birds with least-neat head pattern have whole crown grey-brown streaked dark.) Flanks sometimes (but not always) diffusely streaked. Centre of crown grey or brownish-grey, usually with slightly heavier streaking than ♂. Throat/upper breast ash-grey with faint buff-brown tinge, sometimes also with diffuse mottling. Division between greyish throat and chestnut breast/belly often rather diffuse. — AUTUMN Sexing much as in spring. **Ad** Tips of tail-feathers rather rounded, r4 tipped pure white. Iris rufous-brown. Note that grey head with black lines of ♂ now largely concealed by buffish-rufous tips. **1stW** Tips of tail-feathers rather pointed, r4 all dark (no white tip). Iris sepia (brown-grey). **Juv** Readily separated from ♀ by lack of distinctive dark-and-pale head pattern of post-juv, whole head, neck, mantle, throat, breast and flanks being dull greyish-white with faint brown hue or buff-brown streaked dark. No or only very little rufous on rump or underparts.

BIOMETRICS (*cia*) **L** 16–17.5 cm; **W** ♂ 80–89 mm (*n* 23, *m* 84.4), ♀ 75–88 mm (*n* 21, *m* 81.1); **T** ♂ 70.5–81 mm (*n* 23, *m* 75.8), ♀ 67–80 mm (*n* 21, *m* 73.6); **T/W** *m* 90.3; **B** 12.0–15.0 mm (*n* 39, *m* 13.3); **B**(f) 9.3–11.6 mm (*n* 42, *m* 10.5); **BD** 5.5–7.2 mm (*n* 36, *m* 6.5); **Ts** 18.3–20.5 mm (*n* 42, *m* 19.6). **Wing formula: p1** minute; **p2** < wt 1.5–6 mm; **pp3–4**(5) longest; **p5** < wt 0–2 mm; **p6** < wt 1.5–3 mm; **p7** < wt 6–8.5 mm; **p10** < wt 15–19 mm; **s1** < wt 16–21 mm. Emarg. pp3–7, but on p7 sometimes less clear and confined to tip.

GEOGRAPHICAL VARIATION & RANGE Slight and clinal variation, largely blurred by extensive individual variation, evident when sufficiently large series are directly compared. Three races in treated region and one extralimital could be regarded as reasonably distinct, but a more drastic

E. c. cia, ad ♀, Turkey, Jun: unlike ♂, head and breast patterns less striking with crown streaked and some brownish admixed. Also much less grey over throat and breast. Uniformly rather fresh wing, including blackish primary-coverts and primaries (latter with neat whitish-grey fringes), and seemingly broad and rounded tail-feathers confirm age. (H. Shirihai)

E. c. cia, 1stS ♂, Germany, Apr: some 1stY ♂♂ show clear vestiges of immaturity, with remnant brownish feathers on black crown-stripes, thus very similar to ad ♀♀ except tendency to have broader and blacker head-stripes (usually with broad black lores), and more extensive and clearer grey breast. Sexing is easier if young ♂♂ are aged first (here, primary-coverts and alula are juv, being contrastingly brown and worn). (M. Schäf)

E. c. cia/callensis, ad ♂, Spain, Sep: at end of post-nuptial moult, plumage already mostly fresh. Sexing can be complicated by some ♂♂ having less intense black crown-stripes, but brownish tones in crown of this bird do not match a ♀ (given an ad) while broad black lores and sharp border to extensive grey breast also indicate a ♂. (H. Shirihai)

E. c. cia, presumed 1stW ♂, Italy, Oct: ageing always a priority, especially when in autumn attempting to separate 1stW ♂ from ad ♀, in particular as crown in both can be variably streaked and extensively washed brownish. Although rather advanced grey-and-black head pattern and very limited breast streaking seem to suggest sex as ♂, whole plumage seems fresh lacking obvious moult limits, and safe ageing and sexing would require handling. (D. Occhiato)

treatment lumping all, or nearly all, is defensible considering the minute differences and large overlap. All populations mainly sedentary.

E. c. cia L., 1766 (C & S Europe except W Iberia; also in NW Africa, W & C Turkey, N Levant). Described above. The few specimens examined from N Africa ('*africana*'), indicate that this population is on average slightly paler below and perhaps smaller than *cia*, but retained in *cia* due to extensive variation within latter. (Syn. *africana*; *barbata*; *hordei*; *meridionalis*.)

E. c. callensis Ticehurst & Whistler, 1938 (Portugal, W Spain). Smaller than *cia* with a finer bill, shorter legs, etc. Slightly darker above when series are compared, and rump/uppertail-coverts are deeper rufous. Sexes generally more similar than in other races, ♀ often very close to ♂ in plumage. **L** 15–15.5 cm. **W** ♂ 74–80 mm (*n* 16, *m* 77.7), ♀ 72.5–77.5 mm (*n* 9, *m* 75.1); **T** ♂ 65–75 mm (*n* 16, *m* 71.3), ♀ 65.5–71 mm (*n* 9, *m* 68.0); **T/W** *m* 91.1; **B** 11.4–13.0 mm (*n* 25, *m* 12.4); **B(f)** 9.3–11.0 mm (*n* 26, *m* 10.0); **BD** 5.0–6.2 mm (*n* 26, *m* 5.6); **Ts** 17.5–20.0 mm (*n* 25, *m* 18.6).

○ *E. c. par* Hartert, 1904 (Crimea, NE Turkey, Caucasus,

E. c. cia/callensis, 1stW ♂, Spain, Oct: much variation in 1stW ♂♂, some retaining an overall dull plumage and bill colour, and even a streaky lower breast, thus appearing ♀-like, but unlike ♀ have blacker broad lores and at least hint of extensive grey throat, and upper breast almost clean of streaking. Ageing difficult at such angles, but overall plumage pattern is age diagnostic. (A. M. Domínguez)

Rock Bunting (left: Spain, ssp. *cia/callensis*, Aug) versus Striolated Bunting (right: Israel, ssp. *striolata*, Sep): a potential identification pitfall that is not widely appreciated, but still likely at a few localities in Middle East, mostly in autumn and winter. Usually readily separated by calls and the more contrasting white tail edges of Rock Bunting, but with silent birds look for (i) Rock's thin white wing-bars, most prominent on median coverts, lacking the solidly rufous wing including lesser coverts of Striolated (grey in Rock); also, (ii) lower mandible in Rock is grey, not pinkish-yellow as in Striolated; (iii) Rock has prominent dark lateral throat-stripe, but lacks dark moustachial stripe of Striolated; and (iv) Rock has predominantly rufous rump and uppertail-coverts, not greyish-brown as in Striolated. Both are ad at end of post-nuptial moult; the Rock is ♀ and Striolated a ♂. (Left: A. M. Domínguez; right: H. Shirihai)

E. c. cia, 1stW ♀, Israel, Dec: there is much variation among young ♀♀ in autumn and into winter, with dullest birds almost lacking any rufous on underparts, rump and uppertail-coverts, with extensive but narrow streaking on lower throat to chest and flanks. Such birds can be challenging without previous experience and if call not heard, but note diagnostic combination of emerging dark eye-stripe (including lores) and lower border to ear-coverts, broad dark scapular centres with rufous edges, and clear whitish median coverts wing-bar. Juv wing and tail secure age. (L. Kislev)

E. c. callensis, 1stS ♂, Portugal, Apr: a rather small-billed, darkish race with especially deep rufous rump and uppertail-coverts. Pure grey head with bold, broad black head-stripes, but heavily worn juv central tail-feathers and primary tips infer age and sex. (M. Lefevre)

BUNTINGS

E. c. par, ♂, presumed ad, Iran, Jun: compared to *cia* more narrowly black-streaked and slightly paler sandy-cinnamon above, including rump, but wing-bars buff and less obvious. On average paler and more uniform pinkish rufous-cinnamon below, but much individual variation with age, sex, bleaching and wear. Pure grey, and bold, wide black head-stripes infer ad but difficult to be sure due to abrasion to wing and tail. (E. Winkel)

Transcaucasia, C & S Levant, Iraq, Iran, S Central Asia). A variable race, some being very similar in plumage to *cia*, others being deeper rufous on belly and flanks. Fractionally larger than *cia*, with slightly longer wing and tail, but tarsus same and bill-depth smaller. Contrary to statements elsewhere, it is not paler above than *cia*, and streaking on mantle is equal. Could well be lumped under *cia*. **L** 15.5–17 cm. **W** ♂ 82.5–90 mm (*n* 25, *m* 86.7), ♀ 77–88 mm (*n* 21, *m* 82.0); **T** ♂ 73–82 mm (*n* 25, *m* 77.9), ♀ 69–79.5 mm (*n* 21, *m* 74.5); **T/W** *m* 90.4; **B** 12.1–14.7 mm (*n* 43, *m* 13.5); **B**(f) 9.3–12.2 mm (*n* 46, *m* 10.7); **BD** 5.3–7.0 mm (*n* 31, *m* 6.4); **Ts** 18.0–21.0 mm (*n* 43, *m* 19.4). – Birds from Crimea through Caucasus to NW Turkey and Transcaucasia ('*prageri*') sometimes differ in having a faint brown or buff hue on the grey throat, and have been described as being subtly darker and larger than *par*, but these differences are insignificant and rather variable, and best included in *par*. (Syn. *lasdini*; *mokrzeckyi*; *prageri*; *serebrowskii*.)

Extralimital: **E. c. stracheyi** Moore, 1856 (W Himalayas) is similar to *cia* but has rufous-buff wing-bar (greater coverts) rather than white or buff-white as in the others.

E. c. par, ♀, presumed ad, Azerbaijan, Jun: overall pinkish rufous-cinnamon underparts in this race. Predominantly dusky head-stripes confirm sex and, despite being rather worn in Jun, it seems to have only ad-like wing and tail. (M. Heiß)

E. c. par, 1stW, presumed ♂, Iran, Jan: comparatively extensive streaking of underparts infer age, and this is supported by rather brownish primary-coverts. Sex more difficult to decide, though quite distinct and bold head pattern in a young bird point at ♂. (E. Winkel)

E. c. par, juv, Iran, Jun: heavily streaked with very weak facial stripes, but look for diagnostic dark lower border to ear-coverts. The proportionately long tail in this bunting almost comically obvious on this juv. (E. Winkel)

— 519 —

STRIOLATED BUNTING
Emberiza striolata (M. H. C. Lichtenstein, 1823)

Alternative names: Mountain Bunting, Striated Bunting

Fr. – Bruant striolé; Ger. – Wüstenammer
Sp. – Escribano estriado; Swe. – Bergsparv

Formerly often considered conspecific with the House Bunting, but treated here as a separate species. Unlike House Bunting, the Striolated Bunting is usually much less confiding in man, preferring unspoilt, desolate areas, stony and scrubby hillsides, although it is attracted to drinking water around habitation. It breeds in NE Africa and the Levant north to S Israel, east through Arabia to N and C India. Largely resident but may seek lower elevations in winter.

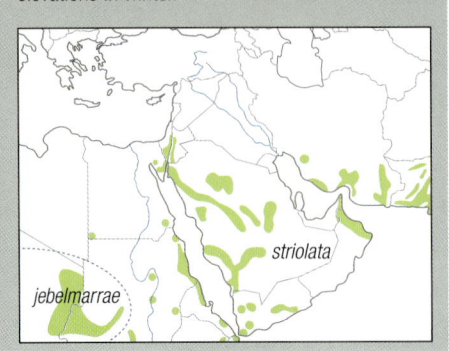

IDENTIFICATION Small, *slim* and rather nondescript *rusty-brown* bunting with *two-toned bill* (dull horn-yellow to bright yellow-orange lower mandible) and *dark tail* indistinctly edged rufous. ♂ (especially in worn plumage) has *rather dark greyish head to upper breast* and *striped face* with quite pronounced dark eye-stripe and moustachial stripe, usually a distinct pale supercilium and often a quite noticeable pale submoustachial stripe and lower cheek-stripe. Black crown-streaking rather uniform, but usually denser on sides, with tendency to show pale median stripe, while dark lateral throat-stripes are narrow or poorly developed. Throat and upper breast somewhat paler grey, irregularly blotched black. Rest of *underparts pale pinkish-cinnamon*. Upperparts sandy-grey, *mantle* prominently *dark-streaked* contrasting with *bright rufous shoulder and upperwing* (dark centres to tertials well exposed). Greater and median coverts only subtly paler-tipped, never forming obvious wing-bars. Fresh (especially immature) ♂ less grey around head with less striking facial stripes. ♀ generally like duller version of ♂, with a browner, indistinctly patterned head and less strongly russet plumage. Stance low and rather horizontal, appearing short-necked with relatively short wings and a long slim tail. Often well camouflaged in desert habitats and only its calls can betray its presence. Normally in pairs or small parties.

VOCALISATIONS Song is simpler than that of House Bunting, a rather even, short phrase with a jerky rhythm and further differs from House Bunting in that all notes are rather clear whistles (rather few trilling or buzzing sounds), hence lacks any similarity to Chaffinch. A common phrase could be rendered *chewy-chowee-chewy-cheyawee*. – Common call is a nasal rather sharp and loud *chuet* or *twett*, when more subdued a bit like Tree Sparrow. Other calls include a clearly disyllabic *sweee-doo*, lower-pitched and much less nasal, with the second part never as drawn-out as in House Bunting.

SIMILAR SPECIES Striolated Bunting is one of several similar-sized buntings with a dark-striped head, rather dark colours and largely streaked plumage. Its dark rufous-edged tail is shared only by *Cinnamon-breasted Bunting*, which has a considerably bolder head pattern with obvious crown-stripes and has less rufous wings, among other characteristics. – Very similar to *House Bunting*, but they are unlikely to meet, at least in the covered region. House Bunting has same size but a proportionately slightly longer tail. Striolated Bunting shows less contrast between grey dark-streaked throat and cinnamon-pink unstreaked breast and belly compared to House Bunting, and upperparts are more streaked, not as plainly rufous-tinged. Also, dark horizontal stripe across pale cheeks is more contrasting in Striolated, not as subdued as in House. – Somewhat smaller and slighter than *Rock Bunting* (which is otherwise separable by its all-greyish bill, no or limited breast-streaking, contrastingly rufous rump, white outer tail-feathers, narrower and blacker eye- and moustachial stripes that tend to meet on ear-coverts, and combination of clearer dark lateral crown-stripes and pale median crown-stripe). Rock Bunting also has narrow white wing-bars, lacking in Striolated. – *Ortolan* and *Cretzschmar's Buntings* should rarely pose a problem, especially given their characteristic reddish-pink bills, bright pale eye-ring, more heavily-streaked upperparts and wings, and white outer tail-feathers.

AGEING & SEXING (*striolata*) Ages differ in autumn; reliable ageing usually requires close views and experience. Sexes differ after post-juv moult, but beware individual and some seasonal variation in strength of head pattern (especially ♂). In general, rather limited seasonal variation. – Moults. Complete post-nuptial moult of ad on breeding grounds after breeding, mainly late Jun–Sep. Partial post-juv moult at same time, usually includes head, body, all lesser and median coverts and most or all greater coverts, sometimes also odd tertials and all or most tail-feathers. Apparently only rarely pre-nuptial moult, then usually in 1stY. – **SPRING Ad** ♂ Wears greyer on head to breast with clearer-cut head-stripes, purer white supercilium, whitish-grey submoustachial stripe, and sometimes partly white ear-coverts and median crown-stripe. Compared to fresh autumn plumage, upperparts bleach paler, sandy-rufous to sandy-grey, with more prominent dark streaks. **Ad** ♀ Usually duller and paler than ♂, with noticeably more obscure head markings, the dark stripes clearly browner and more diffuse, and pale areas, including supercilium, buffier. More diffuse streaking above and on breast. **1stS** As respective ad but has generally more worn retained juv wing-feathers (but beware that ad can wear quite strongly, too), and some show moult limit among greater coverts and tertials (apparently only a few undertake limited winter moult, and these are usually 1stY). Some 1stY ♂♂ have head markings duller and less bold, approaching ♀ pattern. – **AUTUMN Ad** ♂ After Aug–Oct has broader pale fringes and may appear more evenly grey- and black-streaked on head (pale mark on ear-coverts and median crown-stripe especially obscured), with slightly broader mantle-streaking and more uniform rufous wings. **Ad** ♀ As in spring, but broader rufous fringes to wing-feathers and some buffish on head. Head and breast less clearly patterned. Seasonal variation less obvious than in

E. s. striolata, ♂, presumed ad, Israel, Mar: bright yellow lower mandible and conspicuous head pattern, with off-white and dark stripes, and dark-streaked lower throat and breast, all patterns augmented by wear. Tips of primaries and tail-feathers fresh suggesting ad, latter broad and rounded (though beware that odd 1stY replace whole tail). (H. Shirihai)

♂, especially when very fresh, sufficient to make separation of brightest older ♀♀ and dullest (especially 1stW) ♂♂ less straightforward. Both sexes best aged by being evenly fresh without moult limits in wings. **1stW** After some feather wear (usually Oct onwards) general plumage similar to fresh ad and, as many moult all greater coverts, tertials and tail-feathers, ageing not easy, even in the hand. Retained juv remiges and primary-coverts are only slightly weaker textured and indistinctly more worn, thus not always easily detected. Those that retain some outer greater coverts and perhaps tertials (juv coverts have paler and better-defined paler fringes) can be aged without difficulty. Some also retain several outer tail-feathers, which are slightly more worn and more pointed with more diffuse rusty tips and fringes. Lower mandible usually duller than in ad, and iris more dark olive (but by late autumn can be almost pure sepia-brown like ad). When still very fresh, just after post-juv moult, many 1stW ♂♂ have head markings and coloration less advanced, midway between ad ♂ and ♀. **Juv** Overall browner with soft, fluffy body-feathers (generally paler and more nondescript than ♀), and only hint of head pattern. Bill duller, although similar to 1stW ♀.

BIOMETRICS (*striolata*) **L** 12–13 cm; **W** ♂ 73–80 mm (*n* 12, *m* 76.3), ♀ 70–76 mm (*n* 12, *m* 73.0); **T** ♂ 54–64 mm (*n* 12, *m* 58.8), ♀ 53–59 mm (*n* 12, *m* 55.8); **T/W** *m* 76.8; **B**(f) 8.2–9.7 mm (*n* 24, *m* 9.0); **BD** 5.0–6.1 mm (*n* 22, *m* 5.4); **Ts** 14.5–16.0 mm (*n* 24, *m* 15.4). **Wing formula: p1** minute; **pp2–5** about equal and longest (but either or both of p2 and p5 sometimes to 2.5 mm <); **p6** < wt 2–4 mm; **p7** < wt 6–10 mm; **p10** < wt 13.5–17 mm; **s1** < wt 14.5–18 mm; **tert. tip** < wt 10–15 mm. Emarg. pp3–6.

GEOGRAPHICAL VARIATION & RANGE Only one subspecies breeds within the covered region, *striolata*. A couple more subspecies have been described from sub-Saharan E Africa, but these are extralimital and not treated here.

E. s. striolata M. H. C. Lichtenstein, 1823 (Levant north to Dead Sea, Sinai, Egypt, N Sudan, Red Sea coast south to N Somalia, W & C Saudi Arabia, Yemen, Oman, S Iran, Pakistan, NW India; mainly resident). Described above. Relatively pale and moderately dark-streaked compared to other sub-Saharan populations. Possibly a tendecy for Central Asian breeders to average a touch paler, but not distinct enough to warrant separation. (Syn. *dankali*, *kovacsi*, *tescicola*.)

Extralimital: ***E. s. jebelmarrae*** (Lynes, 1920) (SW Sudan, apparently also E Chad). A darker race with bolder streaking both above and on breast. Chestnut of belly deeper than in *striolata*. (See also Taxonomic notes below.)

E. s. striolata, ♀, presumed ad, Israel, Mar: differences between sexes more obvious with wear in spring. Especially note weaker head-stripes than ♂, with pale stripes more buff-tinged, and dark ones browner and less contrasting. Rest of plumage duller, with fainter streaking above and on breast, and the bright rufous wings often the only contrasting feature. Wing condition and feather generations point to ad. (J. Normaja)

E. s. striolata, ♂, United Arab Emirates, Feb: a neat and attractive ♂ in winter, although head pattern slightly less contrasting before all pale feather-tips worn off from dark stripes. The very fresh wing and tail could indicate an ad, but best left unaged without handling. (V. & S. Ashby)

E. s. striolata, ad ♂, Israel, Sep: in fresh autumn plumage, with main differences compared to worn plumage that dark streaking on breast is to some extent concealed, and the striped head pattern subtly less contrasting. Blotchier (less streaky) grey breast in combination with bold head pattern indicative of ad ♂. (H. Shirihai)

E. s. striolata, ad ♀, Israel, Sep: at end of complete post-nuptial moult, otherwise rather similar to fresh plumage. Note subdued head pattern and duller wing compared to ♂. (H. Shirihai)

E. s. striolata, two ♂♂ (2nd and 4th birds from left) and two ♀♀, **India, Jan**: regardless of sex or age, species-specific characters obvious, like contrast between rufous-tinged wings and duller brown-grey rest of plumage, longitudinally striped heads and two-coloured bill with yellow lower mandible. Lack of paler wing-bars and somewhat streaked or blotched breast supplement the description. (P. Ganpule)

TAXONOMIC NOTES Striolated Bunting and House Bunting are closely related and often treated as conspecific. However, we follow Kirwan & Shirihai (2007) who argued that morphological and vocal differences are more typical of two different species. Byers *et al.* (1995) stated that populations in SE Sahara were intermediate and form an apparent link between the two, but according to Kirwan & Shirihai (2007), material from this area available for examination is scant, and existing specimens can rather comfortably be referred to one or the other of the two species. More recently, a phylogeny of the entire 'Brown-rumped Bunting' complex of Africa (Olsson *et al.* 2013) confirmed that a split between *striolata* and *sahari* is well supported genetically. – According to preliminary genetic data (M. Schweizer, G. Kirwan & H. Shirihai, pers. comm.), *jebelmarrae* is more closely related to House Bunting *E. sahari* than to Striolated Bunting. However, we prefer to await a more comprehensive study before making a change.

REFERENCES Kirwan, G. M. & Shirihai, H. (2007) *DB*, 29: 1–19. – Olsson, U., Reuven, Y. & Alström, P. (2013) *Ibis*, 155: 534–543.

E. s. striolata, 1stW ♂, **Iran, Jan**: wing and tail juv and already somewhat bleached, including dull and rather narrow primary-coverts, and tail shows clear moult limit. Otherwise rather similar to ad ♂, only has weaker head-stripes and less bright rufous wing. Birds in the east tend to average slightly paler. (A. Ouwerkerk)

E. s. striolata, ♀, presumed 1stW, **United Arab Emirates, Sep**: young ♀♀ have drabbest plumage, still show hint of striped head pattern and rufous parts on wing apart from streaked back and breast. Age mainly inferred from general drabness, since whole wing seems evenly fresh and no obvious moult limits can be detected. (M. Barth)

E. s. striolata, juv, **Israel, Sep**: rather nondescript with soft, fluffy body-feathers that are predominantly brownish-grey and almost uniformly diffusely streaked, striped head pattern as yet only hinted, leaving rufous wing as most striking feature. Bill, too, has duller colours. (H. Shirihai)

HOUSE BUNTING
Emberiza sahari Levaillant, 1850

Fr. – Bruant du Sahara; Ger. – Hausammer
Sp. – Escribano sahariano; Swe. – Hussparv

The House Bunting constitutes the NW African component of what has hitherto usually been regarded as a single, widespread Saharo-Sindian species. Recently, distinctive morphological, vocal and ecological differences have been taken to justify the recognition of two species (as here). It occupies similar habitats to Striolated Bunting, but differs in its habit of much more frequently living close to man (in some places freely entering houses to feed or nest). It breeds in the Atlas of NW Africa, Western Sahara and extends to S Mauritania and Mali. It has steadily spread north during the last century, even colonising cities.

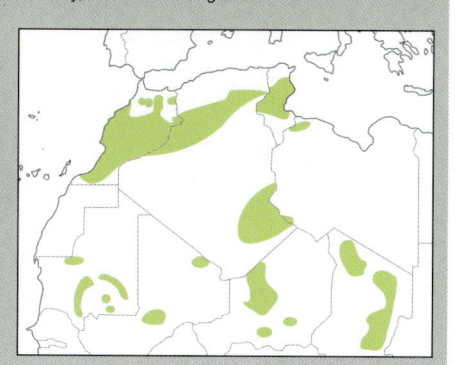

IDENTIFICATION Similar to Striolated Bunting, of same size (including the bill) but with proportionately a *somewhat longer tail* and *longer legs*. It also differs in having plumage overall less streaked. Especially ♂ differs by *more uniform head and breast*, which generally appear more evenly and narrowly streaked (although some come close). Has rather more reduced dark eye-stripe (when worn) or almost lacks it (when fresh); also, moustachial stripe and supercilium narrower, less bold and less pure white, and paler median crown-stripe and stripe on ear-coverts diffuse or almost absent. Dark streaking on throat and breast narrower and ill-defined, and thus inconspicuous. Typically has *darker, richer rusty-brown mantle, scapulars* and *rump*, with very *poorly-defined shaft-streaks confined to mantle* (sometimes appears almost unstreaked, whereas Striolated Bunting is heavily streaked). Also unlike Striolated Bunting, *wings more uniformly and extensively rufous* (almost concolorous with mantle). Dark tertial centres fringed rufous. No trace of any wing-bars. Underparts generally more uniform than Striolated Bunting, rather distinctly set-off from grey breast. ♀ generally duller than ♂ and, despite strong individual variation, usually differs from ♀ Striolated by *plainer head/breast* and *more evenly rufous upperparts and upperwing*, as in ♂.

VOCALISATIONS Song is a brief, consistently and oft-repeated, jerky phrase that can be likened to the efforts of an immature ♂ Chaffinch, which has not yet learnt to sing properly. It lacks the acceleration and drop in pitch of Chaffinch, but can have rather similar voice and include a hint of that species' flourish, e.g. *chippy cherr-pitchy cherr-pitchy* (Byers *et al.* 1995) or *tve-tvi tve **vitt**ya-**trrüü**oo-tvetvittyü*. – Calls variable but usually higher-pitched and squeakier than those of Striolated, e.g. a nasal and noisy double-noted *sai-wai* (also rendered *dzwe-wee* or *che-**veeh**-e*), the second part often slightly prolonged, a call type that can recall Rock Sparrow at a distance. A variant is a downslurred glissando ***chee**yaa*. More clicking, subdued and insignificant notes occur too, like *tvoit* and *che*.

SIMILAR SPECIES Very similar to *Striolated Bunting*: the main differences are outlined above, although in any case the two are unlikely to come into contact in the W Palearctic. It is important to remember that there is much individual variation in both groups, mainly due to seasonal feather wear and age differences, and there is some overlap between them in head pattern of ♂♂. The dullest Striolated (mostly when fresh and apparently 1stY) and brightest House (mostly when worn and apparently ad) can appear quite similar. When dealing with trapped birds or specimens in museums, biometrics are very useful, but it is important to always compare birds of equivalent sex and age, and separation is best based on the more constant differences in the upperparts. See Striolated Bunting for other identification issues, especially separation from *Rock Bunting* (they winter in the same areas).

AGEING & SEXING Ages differ in autumn; reliable ageing usually requires close views. Sexes differ after post-juv moult, but beware individual and some seasonal variation in strength of head pattern (especially ♂). In general, rather limited seasonal variation. – Moults. Complete post-nuptial moult of ad on breeding grounds after breeding, mainly late Jun–Sep. Partial post-juv moult at same time, usually includes head, body, all lesser and median coverts and most or all greater coverts, sometimes also odd tertials and all or most tail-feathers. Apparently no pre-nuptial moult in ad, whereas some 1stY can moult some or much body-feathers (e.g. one from 30 Mar had mostly retained juv wing-feathers but freshly moulted tertials, all lesser and some median coverts and one inner greater covert, as well as parts of the forebody and head). – **Spring Ad ♂** Compared to ♀ especially note greyer (sometimes almost ash-grey) cast to head and breast, broader and blacker crown-streaking, and more pronounced lateral crown-stripes. **Ad ♀** Duller and paler than ♂ with head generally pale buffish-brown, including supercilium, narrower and denser crown-streaking, and throat and upper breast hardly contrast with rest of underparts (which are paler rufous). Upperparts and upperwing also duller. **1stS** As respective ad but has more worn

♂, presumed ad, Morocco, Jan: characteristically rather weakly streaked, with ash-grey head and upper breast and rest of plumage chestnut-brown, affording bicoloured effect, especially in attractive ♂. Presumed to be ad by seemingly uniformly fresh wing, including as far as can be judged primary-coverts. (M. Schäf)

Ad ♂ House Bunting (left: Morocco, Mar) versus ad ♂ Striolated Bunting (right: ssp. *striolata*, United Arab Emirates, Jan): although differences between these two species most striking in ad ♂♂, the following differences hold to some extent for all plumages: compared to Striolated, House's (i) paler grey head and upper breast appear rather marbled with (ii) more ill-defined face pattern (including less pronounced dark eye-stripe behind eye, often less pure white supercilium and submoustachial stripe, the latter bordered by only vague dark moustachial and lateral throat-stripe); (iii) brighter and almost uniform chestnut-tinged upperparts and wing lacking Striolated's stronger dark streaking (breast) and larger dark centres (wing). Both ♂♂ still in fresh plumage; above-mentioned differences will become more accentuated by wear. Evenly feathered ad wing (House) and tail (Striolated) evident. (Left: E. F. Henriksen; right: M. Barth)

♂ (left) and ♀, latter presumed ad, Morocco, Feb: some ♀♀, especially when still fresh, very nondescript and predominantly buff-brown with very obscure dark streaks, almost lacking any obvious head pattern, and head and breast coloration hardly contrasts with warmer underparts. Only the ♂ can here be properly judged as to wear, and this appears to have evenly fresh ad wing, including primary-coverts. (M. Varesvuo)

♀, presumed ad, Morocco, Mar: in worn plumage, ♀'s whitish supercilium and dark-streaked crown become more apparent, but unlike any ♂, ground colour of head and breast drabber and more brownish-tinged. With wear ♀ head pattern becomes slightly stronger, and hence more similar to poorly streaked ♀ *striolata*. Tips to tail-feathers seem round and ad-like. (D. Monticelli)

retained juv primaries and primary-coverts, less frequently some outer greater coverts and tertials. Some 1stS ♂♂ have less advanced head colour, approaching ad ♀. Ageing not easy as most or all juv median and greater coverts, usually all tertials and even tail replaced, while ad can have almost equally-worn primaries and primary-coverts.. – **Autumn Ad ♂** Much as in spring but broader pale fringes and thus more diffuse and uniform head pattern, with paler grey crown and breast. **Ad ♀** As in spring, but due to broader brownish fringes, especially to head- and breast-feathers, appears plainer. Both sexes differ from 1stW in being evenly fresh. **1stW** As respective fresh ad but retains primaries and primary-coverts, which are more weakly textured and somewhat less fresh and browner. Less frequently retains some outer greater coverts and tertials, then more easily aged (juv coverts have paler and better-defined fringes). Also sometimes retains outer tail-feathers (slightly more worn and more pointed with more diffuse rusty tips and fringes). Sexing not always easy, some ♂♂ being almost midway between ad ♂ and ♀ in head markings and general coloration. **Juv** Soft, fluffy body-feathers. More nondescript than ♀, with rather rufous wings and browner head. Bill duller.

BIOMETRICS L 13–14.5 cm; **W** ♂ 73–80 mm (*n* 12, *m* 76.5), ♀ 71–76 mm (*n* 12, *m* 73.8); **T** ♂ 58–68 mm (*n* 12, *m* 63.4), ♀ 55–65 mm (*n* 12, *m* 61.2); **T/W** *m* 82.9; **B**(f) 8.5–10.0 mm (*n* 24, *m* 9.1); **BD** 4.9–5.9 mm (*n* 22, *m* 5.5); **Ts** 15.7–18.0 mm (*n* 22, *m* 17.0). **Wing formula: p1** minute; **p2** < wt 0.5–5 mm; **pp3–6** about equal and longest (but p5 and p6 sometimes to 3 mm <); **p7** < wt 4–7 mm; **p10** < wt 12–16 mm; **s1** < wt 11.5–16 mm; **tert. tip** < wt 9–13 mm. Emarg. pp3–6.

GEOGRAPHICAL VARIATION & RANGE Monotypic (but see below, Taxonomic notes). – Morocco, Algeria, Tunisia, W Tripolitania, Libya, Mauritania, N Senegal, S Mali; mainly sedentary. – Birds in S Morocco ('*theresae*') and Mauritania, N Senegal and S Mali ('*sanghae*') have all been described as being darker and more saturated in colour, but we have compared available types and long series in museums, and agree with Kirwan & Shirihai (2007) in treating both races as insufficiently distinct or documented. (Syn. *sanghae*; *theresae*.)

TAXONOMIC NOTES Often treated as conspecific with Striolated Bunting, but treated as separate species here (see Taxonomic note under Striolated). – According to preliminary genetic data (M. Schweizer, G. Kirwan & H. Shirihai, pers. comm.), the extralimital race *jebelmarrae* (SW Sudan, E Chad) is more closely related to House Bunting than to Striolated Bunting *E. striolata*, to which it is currently referred. Future taxonomic change within this complex seems therefore inevitable.

REFERENCES Kirwan, G. M. & Shirihai, H. (2007) *DB*, 29: 1–19. – Olsson, U., Reuven, Y. & Alström, P. (2013) *Ibis*, 155: 534–543.

♀, presumed 1stS, Morocco, Apr: a poorly marked ♀ with almost no obvious face pattern, rufous-brown upperparts less neat, with some thin dark streaks on mantle and exposed dark centres to scapulars and tertials. Nevertheless, overall rusty tones above and below are conspicuous. Tail appears to have moult limits, while primary-coverts and primaries seemingly are juv in quality and degree of wear. (R. Armada)

♂, Tunisia, Dec: irrespective of age, some ♂♂ have a stronger face pattern closer to *striolata*, but dark stripes (except perhaps lores) are dark grey, not black, pale stripes less bold, and dark streaking on breast limited. Furthermore, the diagnostically bright rusty plumage with unstreaked upperparts and no dark centres on wing show well. Impossible to age from this image alone. (P. Casali)

♀, Morocco, Dec: ♀♀ of this species can appear very bland. A particularly uniform and unstreaked bird. (B. van Elegem)

CINNAMON-BREASTED BUNTING
Emberiza tahapisi A. Smith, 1836

Fr. – Bruant cannelle; Ger. – Bergammer
Sp. – Escribano camelo; Swe. – Afrikansk klippsparv

A widespread African bunting that recalls the more familiar Rock Bunting, only with stronger colours and patterns. The Cinnamon-breasted Bunting reaches north across sub-Saharan Africa to Sudan, and occurs on both Socotra and in extreme S Arabia, as well as south all the way to South Africa. Many birders have probably made its acquaintance in S Oman. A claim from Sinai is now rejected. It usually frequents dry, rocky areas with scattered bushes, but will of course regularly come to drink at wells or creeks. It is largely resident, with some seasonal movements reported.

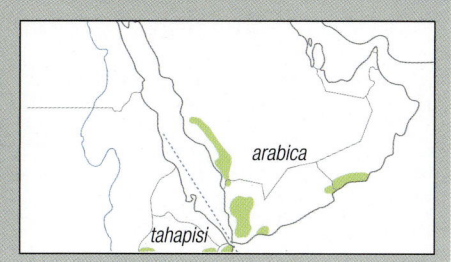

E. t. arabica, ♀ (left) and ♂, Oman, Sep: classic image illustrating both sexes. Despite that both seem to show moult limits in wing or tail, we are reluctant to even tentatively identify them as 1stY because of our limited experience with the moult strategy in this species, which breeds twice per year (with a potentially complex moult cycle) linked to the local rainy seasons. (H. & J. Eriksen)

IDENTIFICATION The ♂ has immaculately *striped black-and-white hood*, with *blackish throat and upper breast*, white median crown-stripe, cheek-stripe and submoustachial stripe, warm brown, heavily-streaked upperparts, *mainly cinnamon-brown underparts* and blackish-brown *tail, lacking any white*. ♀ generally paler, with less contrasting head pattern, and paler throat and breast, both densely mottled blackish. Shares bicoloured bill with House and Striolated Buntings, and has similarly-coloured legs. Occurs singly, in pairs or small groups, and, like most buntings, typically feeds on ground, with a shuffling, unobtrusive motion.

VOCALISATIONS Song short, often delivered from prominent perch, a short, accelerating, twittering trill, *dzüt dzit-dzirerit* or *try-tri, tve-rerir*, the second and last notes higher pitched, sometimes culminating in an excited twitter. Some commence with a longer trill but others emit more of a staccato phrase and can recall Yellowhammer. – Calls include a nasal *per-wee-e* or *daav* in contact, an unobtrusive *dwee* in flight and a very thin *tsiii* in alarm.

SIMILAR SPECIES *House Bunting* of NW Africa is readily separated from the present species, not least because within the covered region they are well separated geographically. – Only in S Arabia, where Cinnamon-breasted overlaps with *Striolated Bunting*, is there a real confusion risk. Striolated tends to have a more evenly dark-streaked crown, with usually reduced pale median and darker lateral crown-stripes, and head is generally paler and greyer than Cinnamon-breasted with better-streaked throat and breast. Also, unlike Cinnamon-breasted, the pale head-side stripes are usually broader than the dark ones. Furthermore, Striolated is distinctly less boldly and more narrowly streaked above, with more rufous or sandy mantle/scapulars, and the rump/uppertail-coverts are more weakly streaked or even unstreaked, thus more concolorous with rest. Striolated Bunting has an overall more rufous wing with centres to all wing-coverts less dark or contrasting. Also, there are no dark centres exposed (or only faint-shaft streaks) on lesser and median coverts, whereas in Cinnamon-breasted the opposite is true, thus offering a consistent diagnostic character separating the two species, even for the otherwise most difficult individuals. – Following the split of 'Gosling's Bunting' *E. goslingi*, its easternmost subspecies (of possible hybrid origin) *septemstriata* occurring in NE & E Sudan south to N Ethiopia is a real confusion risk with *tahapisi* of S Ethiopia and further south. The former usually has more extensive rufous on wing-feathers and upper wing-coverts, and throat is slightly paler and mottled grey-black and white, thus lacking solid black bib of ♂ *tahapisi*.

AGEING & SEXING (*arabica*) Reliable aging requires close check of moult pattern and feather wear in wing and tail. Sexes usually well differentiated. Rather limited seasonal variation. – Moults. Complete post-nuptial moult of ad in summer after breeding, perhaps mainly May–July. However, moult period seems to vary with breeding season (which varies in response to rainy season or annual precipitation). Post-juv moult at the same time is mainly partial, mostly of head, body, all lesser and usually most or all median and greater coverts, some tertials, all or most tail-feathers and sometimes inner primaries, primary-coverts and outer secondaries. Pre-nuptial moult absent or limited. – **BREEDING Ad** ♂ See Identification. **Ad** ♀ Despite some variation (e.g. better-marked ad ♀ superficially resembles least-advanced, duller imm ♂ with less strongly-marked head), most have dark pattern of head much paler and browner (with stronger dark streaking) than ♂, and dark throat less extensive. Pale head-stripes often tinged buffish and less clearly defined, especially median one is obscured by heavy streaking. Dark upperparts streaking duller and less bold than ♂, and flanks/breast-sides more cinnamon. Unlike ♂, tends to have yellowish (rather than pinkish-yellow) lower mandible. **1stS** As respective ad, but usually retains some heavily-bleached juv wing- and

E. t. arabica, ♂, presumed 1stY, Oman, Nov: large, unstreaked dark bib and four grey-white stripes over dusky head identify. Appears to show three generations of wing-coverts, and even if we disregard the single fresh median covert (perhaps replaced due to an accident), most lesser and median coverts seem to be juv and post-juv. Primary-coverts and primaries might also be juv. (D. Occhiato)

tail-feathers. Some ♂♂ further distinguished by less bold pattern, including on head. The few recorded birds with limited pre-nuptial moult (e.g. with renewed tertials) are of this age class. – **NON-BREEDING Ad** Largely as breeding, but plumage overall paler-fringed. Best aged by being evenly fresh. **1stW ♂** Because many moult all median and greater coverts, tertials and tail-feathers, separation by moult limits not always straightforward from fresh ad ♂. However, those few with moult limits among these feathers should be reliably separated with experience (juv remiges and primary-coverts only slightly weaker textured and more worn in early autumn). Some appear less advanced in head pattern, but still readily separated from most ♀♀. **1stW ♀** Dullest of all, with head even duller/streakier than ad ♀, more uniform grey throat and breast, and has even buffier head-stripes. Moult limits among wing- and tail-feathers best ageing clues. Both sexes, at least in early stages, tend to have dark grey-brown iris (dark chestnut-brown in ad) and slightly darker legs (fleshy straw-coloured). Lower mandible flesh (not yellow or pink-yellow of ad). Also, tail-feathers narrower and more pointed than in ad, with rusty tips to outermost pair. **Juv** Most like

E. t. arabica, presumed 1stY ♂, Oman, Nov: on present knowledge presumed to be young ♂ due to slight brownish tinge to black head, but not clear why it lacks the large dark bib of most ♂♂ (including young). Attempting ageing based on visible moult pattern would be arbitrary. (D. Occhiato)

E. t. arabica, left to right, ♀, ♂ (presumed 1stY), and ♂, Oman, Nov: ♀ on left has relatively mottled and brownish plumage with ill-defined face pattern (but ageing impossible as wing invisible). Central bird particularly interesting as it shows ♂ pattern but dark of head dark grey rather than black suggesting young age (and obvious moult limit in tertials seems to confirm this). ♂ on right appears to be in very attractive plumage, perhaps hinting at ad, but it is impossible to be sure about its age, and seemingly very freshly moulted tertials suggest 1stY. Note in all three striking bicoloured bill (black upper mandible and yellow lower). (D. Occhiato)

♀ but has soft, fluffy body-feathers, head and throat grey-brown flecked darker, with even more indistinct and buffier pale head-stripes.

BIOMETRICS (*arabica*) **L** 14–15 cm; **W** ♂ 71–77.5 mm (n 13, m 74.9), ♀ 70–76 mm (n 7, m 73.0); **T** ♂ 60–66 mm (n 13, m 62.5), ♀ 54.5–63 mm (n 7, m 60.0); **T/W** m 83.3; **B** 10.3–11.7 mm (n 22, m 11.1); **B**(f) 8.2–9.9 mm (n 22, m 9.1); **BD** 5.3–6.8 mm (n 12, m 6.0); **Ts** 14.5–17.7 mm (n 21, m 16.1). **Wing formula:** p1 minute; pp2–4 about equal and longest (but p2 sometimes to 2 mm <); p5 < wt 0.5–2 mm; p6 < wt 2–6.5 mm; p7 < wt 7–11.5 mm; p10 < wt 11–15.5 mm; s1 < wt 10.5–15.5 mm; tert. tip < wt 2–13.5 mm. Emarg. pp3–6.

GEOGRAPHICAL VARIATION & RANGE A recent study of all sub-Saharan and Arabian brown buntings (Olsson *et al.* 2013) showed that *tahapisi* and *arabica* on the one hand, and *goslingi* and *septemstriata* on the other, differed genetically (mtDNA cyt b) by as much as >5%. This, together with rather clear morphological differences, are reasons enough to treat the two groups as separate species, Cinnamon-breasted Bunting and 'Gosling's Bunting'. Taxon *septemstriata* (E Sudan east of the Nile south to N & C Ethiopia) is morphologically intermediate and somewhat variable, but

E. t. arabica, presumed 1stW ♂, Oman, Nov: like ♀, being duller than ad ♂ with buffish-cream and dark grey head pattern. Nevertheless, dusky areas of crown, ear-coverts, cheeks and bib are still rather solid with little of streaking and mottling suggesting a young ♂. Moult limits in wing support ageing. Note the characteristic rufous-edged outer tail-feathers of the species. (D. Occhiato)

E. t. arabica, presumed juv and 1stY ♀♀, Oman, Nov: perhaps throughout year, there is much variation among 'transitional' young plumages, from birds still with mostly juv feathers (bird on left) to those that already completed post-juv moult much earlier. Strategy and seasonality of moult never studied, and is especially poorly known in race *arabica*. Mainly differs from ad ♀ or young ♂ in being even duller with streakier crown and breast. (T. Langenberg)

♀ Cinnamon-breasted Bunting (left: Oman, Sep) versus Striolated Bunting (right: ad, Israel, Sep): generally, corresponding ages/sexes of the two species are readily separated by stronger face pattern but less rufous wing and less blotchy breast of Cinnamon-breasted. However, least marked young ♀ can almost match head-stripes pattern of ♂ Striolated, while fresh ♀♀ of both species can show rather similar unmarked breast. The clincher in all plumages is pattern of median coverts, especially outermost two, which invariably differ in pattern of dark centres: very large in Cinnamon-breasted, but narrow shaft-streaks in Striolated. Age of former unsure despite appearing to show very fresh wing and rest of plumage, although a few outer greater coverts are unmoulted, but the Striolated is an ad in post-nuptial moult. (Left: A. Al-Sirhan; right: H. Shirihai)

genetically very close to *goslingi*, thus on present knowledge best included in Gosling's Bunting. To conclude, just one subspecies occurs within the treated region, *arabica* of S Arabia.

E. t. arabica (Lorenz & Hellmayr, 1902) (SW Saudi Arabia, Yemen, S Oman; sedentary). Described above. Subtly paler underparts than ssp. *tahapisi* of S Ethiopia, E & S Africa.

REFERENCES Olsson, U., Yosef, R. & Alström, P. (2013) *Ibis*, 155: 534–543.

E. t. arabica, juv, Oman, Nov: most like ♀, but head grey-brown with buffier pale stripes; warm buff underparts flecked darker; all wing and body plumage juv. Note absence of rufous on wing, which separates it from any Striolated Bunting. (A. Audevard)

CINEREOUS BUNTING
Emberiza cineracea C. L. Brehm, 1855

Fr. – Bruant cendré; Ger. – Türkenammer
Sp. – Escribano cinéreo; Swe. – Gulgrå sparv

A rare and local summer visitor over much of its breeding range, the Cinereous Bunting is one of the most-coveted Middle Eastern specialities. There are two rather different forms, and separate species status for these could be considered, but we cautiously retain them as one in want of a comprehensive study. Breeds on bushy, stony arid slopes on the Greek islands of Lesbos and Chios in the west and through Turkey and Middle East to SW Iran. In winter and on migration frequents dry open country with short grass, semi-desert, rocky hills, scrubby or cultivated areas. Often associates with congeners, feeding on ground.

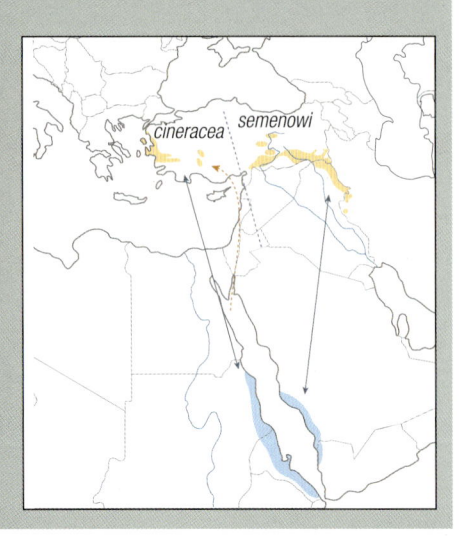

E. c. cineracea, ♂, presumed ad, Israel, Mar: yellow-green head contrasts with brown-grey, finely dark-streaked mantle and scapulars and off-white underparts. Eye-ring tinged lemon-yellow. Wing and scapulars of same generation, in perfect condition, inferring an ad, but viewing angle prevents a firm conclusion. (L. Kislev)

IDENTIFICATION Rather large and long-bodied bunting with fairly plain plumage, lacking strong patterns. Bill typically *pale bluish-grey and appears long* compared to some congeners. Predominant plumage colours grey, white and pale yellow (with some olive and brownish tones). Note *white outer tail-feathers*, hint of double wing-bars and a *discreet pale eye-ring*. *Dark tertial centres* generally rather *ill-defined* when fresh (with wear, centres become more dark-looking); 'L-shaped' notch on outer webs of tertials varies from absent to moderately developed. ♂ of ssp. *cineracea* (E Greece, W Turkey) has *pure yellow throat, unstreaked yellowish greyish-olive head, yellowish-white eye-ring and variable yellow-green tone to face*, a vague yellow submoustachial stripe above olive-grey lateral throat-stripe, and sometimes a short pale yellow supercilium. *Neck and breast olive-grey*, grading into *white or grey-white belly* and undertail-coverts. Upperparts rather plain and inconspicuously streaked, including *ash-grey rump and uppertail-coverts*. ♂ of ssp. *semenowi* (E Turkey, Middle East, Central Asia) is similar but has *yellow throat tinged olive*, and *olive-yellow colour of throat continues below olive-grey breast-band onto lower breast and belly*, leaving only undertail-coverts whitish. Further, rump and uppertail-coverts are slightly olive-tinged. Double wing-bars tend to be slightly less prominent compared to *cineracea*. ♀ is duller grey-brown lacking any strong yellow, and are more obviously dark-streaked above. ♀ *cineracea* averages slightly more brown-tinged grey above, while ♀ *semenowi* is a trifle more olive-tinged grey, less brownish. Greyish-brown breast variably finely streaked or spotted dark. Unstreaked greyish rump and uppertail-coverts and cream-white belly as ♂. Less advanced immature ♂ in first spring similar to ♀, and not always separable, but lateral throat-stripe and breast-streaking usually less well developed or—more often—lacking, and streaks on head and back reduced compared to ♀; note also that head of immature ♂ is often tinged yellowish or greenish, and breast-band is purer grey. Immature ♀ is least colourfull; note presence of *fine triangular-shaped dark spots on pale cream-buff throat and greyer breast*. Mainly terrestrial, but will perch in low trees. Wary but generally not shy, and migrants can be rather tame.

VOCALISATIONS No study has been undertaken of song differences between the two races, although a seemingly fairly consistent difference appears to be the following: (1) *cineracea* song often recalls 'southern Ortolan song', thus after a few short accelerating notes there is a buzzing drawn-out note (often) at lower pitch followed by a very brief note; (2) *semenowi* song opens with a few brief notes followed by a half-halt then a little flourish at the end, the latter often consisting of two similar complex notes (e.g. *seewich-seewich*). Whether this holds true on a larger sample remains to be established. The *semenowi* song type can closely recall Grey-headed Bunting song and often requires a check of the bird to safely identify the songster. – Calls are probably more similar, or even the same, for both subspecies, and since they are often varied a comparison is not straightforward. Commonest call (not that different from Cretzschmar's or Grey-necked Bunting) is either a short, rasping *tschrip* or a full and more melodious *chülp* (or shorter *chü* or *dju*), the rasping and the melodious often uttered alternatively for some period (in Ortolan fashion).

SIMILAR SPECIES In a fleeting view, head colour and yellow throat of ♂ may suggest *Ortolan Bunting*, but grey bill and predominantly greyer plumage, lacking any rufous on underparts, permit easy distinction. (The same features and at least faintly yellow throat are also useful in separating *Cretzschmar's* and *Grey-necked Buntings*.) ♀ / first-autumn browner and streakier, hence closer to same plumages of several similar buntings, especially first-autumn Ortolan Bunting (and related species), but are paler and more featureless, with diagnostic pale horn-grey bill (rather than pinkish or flesh-coloured), less patterned head-sides, and greyer and relatively less heavily-streaked upperparts and

E. c. cineracea, 1stS ♂, Greece, Apr: irrespective of age, a few ♂♂ have browner (less pure grey) and stronger dark streaking on mantle, but like all others show clean grey rump and uppertail-coverts. Pointed tips to worn and bleached juv primary-coverts, brown and abraded primary tips and tail-feathers infer age. (R. Pop)

crown, any markings being ill-defined and narrower (also lacks any strong rufous coloration above, often, but not always, present in the other species). First-autumn plumages of the Ortolan group rather more distinctly and uniformly streaked dark above, although there is some variation (Grey-necked least-prominently streaked). The fresh autumn ♀ and first-winter Cinereous are also typically paler or even whiter below (lacking any obvious rufous or yellow). Tertials have more ill-defined centres than Ortolan and Cretzschmar's Buntings, but better developed than Grey-necked. In addition, rump and uppertail-coverts are usually plainer, being less heavily streaked than Ortolan and lack rufous-brown tinge often shown by Cretzschmar's, but again more variation and overlap in these features exist than usually acknowledged in the literature. – In all plumages, extensive white on outer tail-feathers instantly excludes Black-headed and Red-headed Buntings; unlike these two, undertail-coverts never brighter yellow than belly, rump more concolorous with upperparts, and tertials have broad, yet diffuse, fringes.

AGEING & SEXING (*cineracea*) Ageing requires a close check of moult pattern and shape of outer tail-feathers. Age related to seasonal variation generally quite evident, especially the transformation from young to ad ♂ and fresh autumn ad ♀ to breeding ♀. Sexes separable (but mostly of ad, especially in spring; ♂ also considerably larger—see Biometrics). – Moults. Complete post-nuptial moult and partial post-juv moult mostly in Jul–Aug (Sep). Post-juv moult includes head, body, all lesser coverts, all or most median coverts and at least some inner greater coverts; possibly no alula-feathers or tertials moulted, nor central tail-feathers, but limited material examined. Apparently a partial pre-nuptial moult, but winter birds rarely studied, hence extent little known. – **SPRING Ad ♂** Always purer yellow and greyer than ♀, with head largely unstreaked and strongly suffused yellow, ashy or olive-grey (amount of yellow varies individually, but generally birds in early spring are less bright). Throat bright yellow (unstreaked) and grades into variable brownish/greenish-grey breast-band, while lower breast tinged yellow, flanks washed grey and rest of underparts cream-white, often with clean ash-grey nape. Mantle and scapulars plainer grey-brown, with relatively narrow or ill-defined dark streaking. **Ad ♀** Drabber than ♂ with streakier plumage: head browner with only limited green and yellow wash, and has narrow but distinct dark shaft-streaks on crown. Cheeks mottled drab grey, and yellow of face and throat fainter (usually tinged pale buffish-white or buffish-yellow), and faint brownish-grey lateral throat-stripe meets grey-brown breast and nape, which are both slightly dark-streaked; eye-ring also indistinctly washed yellow. Rest of upperparts slightly darker than in ♂ with better-defined streaking, especially on lower mantle and scapulars; rump and uppertail-coverts slightly browner and variably streaked. **1stS** Aged by more heavily worn and browner retained juv wing-feathers. Usually shows moult contrast between unmoulted primaries and primary-coverts on the one hand, and fresh median and inner greater coverts on the other. Plumage resembles respective ad, but ♂ tends to have less yellow than ad ♂ and often some diffuse streaking below (see Identification); 1stS ♀ especially dull, having little or no yellow on head. – **AUTUMN Ad** Evenly fresh lacking any moult contrast in wings or tail. **Ad ♂** Very much as spring, but head greyer or more olive, yellow of throat diluted, upperparts tinged browner, and especially scapulars are warmer fringed. Remiges thinly fringed cream or pale cinnamon. Upper breast sometimes streaked slightly darker. **Ad ♀** Much as spring but limited material examined (or birds observed). Those seen suggest that after moult is completed there is no yellow on throat, not even concealed at base of feathers (*contra* common statements in the literature). **1stW** Difficult to sex, overall as fresh ad ♀ but usually browner and even streakier, including on throat and breast, as well as above, while throat has practically no yellow or entirely lacks it. Best aged by retained juv remiges (less fresh and weaker textured), primary-coverts and at least some outer greater coverts, forming moult contrasts with recently moulted ones. Also, tail-feathers slightly more worn and slightly narrower than ad, with white wedges on inner webs of rr5–6 less square-cut, more

E. c. cineracea, ♀♀, Israel, Apr (left) and Mar: head grey-brown finely streaked darker, often (right-hand bird) with hint of yellowish-green to face, enhancing greyish nape and neck-sides. Throat and moustachial stripe pale yellow (often admixed with small but fine dark streaks, as on right bird). Underparts as ♂ but often less pure grey-white, suffused pale buff-brown, with diffuse dusky streaks on breast and flanks. Left-hand bird has limited yellow on face and apparently juv primary-coverts and alula, suggesting 1stS. Right bird, although extensively yellow on head and state of primary-coverts appear ad-like, apparent moult limit in tertials contradict this, so best left unaged. (Left: A. Ouwerkerk; right: T. Krumenacker)

E. c. cineracea, 1stS ♀, Kuwait, Apr: young ♀ in spring has drabbest plumage, having least yellow on face, but unrelated to age some are very weakly streaked overall, as shown here. All plumages show pale wing-bars. Worn juv wing, including narrow, pointed primary-coverts. (G. Brown)

E. c. semenowi, ad ♂, E Turkey, May: when encountering a maximally yellow ♂ like this, one may start to consider separate species status for the strikingly different eastern race *semenowi*. Evenly fresh wing, including neat and greenish-tipped primary-coverts, infer age. (L. Svensson)

E. c. semenowi, ad ♂, Turkey, May: most ♂♂ differ from corresponding plumages of *cineracea*, mainly by their greenish-olive head, chest and flanks surrounding bright sulphur-yellow face and throat to upper breast. Upperparts brown-grey, often with richer chestnut cast. Vent and undertail-coverts whitish-grey admixed with yellow. Uniformly fresh wing in May infer age. (A. Below)

E. c. semenowi, ♂, Turkey, Apr: extensive white outer corners on tail visible from below. Note olive-tinged breast and warm yellow belly, very different from ssp. *cineracea*. Ageing unsure. (H. Nussbaumer)

E. c. semenowi, ♀, Turkey, May: unlike ♀ *cineracea*, dusky underparts extensively tinged yellow. Insufficient wing-feathers visible to age with certainty. (T. Luiten)

narrowly V-shaped; tips of tail-feathers often more pointed, less broadly truncate. **Juv** Compared to ♀ has warmer and browner upperparts, heavier-streaked mantle, yellowish-grey crown streaked dark, forehead mottled darker, but head rather indistinctly patterned. Underparts tinged buff and quite heavily streaked on throat to flanks. Not unlike juv Ortolan, Cretzschmar's and Grey-necked Buntings, and perhaps most easily separated by greyish, rather than pale pinkish, bill.

BIOMETRICS (*cineracea*) **L** 16–17 cm; **W** ♂ 89–98 mm (*n* 20, *m* 93.3), ♀ 85–89.5 mm (*n* 6, *m* 87.6); **T** ♂ 68–78 mm (*n* 20, *m* 73.1), ♀ 68–70 mm (*n* 6, *m* 68.5); **T/W** *m* 78.4; **B** 13.0–15.1 mm (*n* 27, *m* 14.0); **B(f)** 10.3–11.9 mm (*n* 27, *m* 11.2); **BD** 6.3–7.3 mm (*n* 25, *m* 6.9); **Ts** 19.3–21.2 mm (*n* 27, *m* 19.9); **white on r6** ♂ 25–35 mm, ♀ 25.5–29 mm. Wing formula: **p1** minute; **pp2–3** about equal and longest (p2 sometimes to 1.5 mm <); **p4** < wt 0.5–2 mm; **p5** < wt 2.5–4 mm; **p6** < wt 8.5–12 mm; **p7** < wt 13–17 mm; **p10** < wt 21–26 mm; **s1** < wt 22–26 mm. Emarg. pp3–5.

GEOGRAPHICAL VARIATION & RANGE Two well-differentiated races; intermediates apparently rare or non-existent due to small population or even a gap in range between E Taurus Mts (eastern limit of *cineracea*) and the Gaziantep region (western limit of *semenowi*). Reports of intermediate ♂♂ regularly observed on migration in Israel (Shirihai 1996) are not supported by museum specimens, and may be more due to individual or age-related variation, probably involving 1stS *semenowi*.

E. c. cineracea C. L. Brehm, 1855, 'Smyrna Bunting' (Lesbos and Chios in E Greece, W & SC Turkey east to E Taurus Mts; winters mainly Eritrea, possibly also Sudan). Described above.

E. c. semenowi Zarudny, 1904, 'Kurdish Bunting' (SE Turkey from Gaziantep east through Iraq to SW Iran; winters S Arabia including Yemen, and NE Africa at least in Eritrea). Differs from *cineracea* by more extensive yellow including lower breast and belly in ♂ plumage, and faint yellow tinge below also in ♀♀. Grey parts (chest-band, neck-sides, nape, flanks) have olive tinge, more so on average than *cineracea*. Also, yellow throat and belly have slight olive hue (in *cineracea* throat is pure lemon-yellow). Grey of lower back and rump tinged brown (purer grey in *cineracea*). Same size as *cineracea* but has a trifle longer tail and thinner bill. **W** ♂ 88–98.5 mm (*n* 18, *m* 93.6), ♀ 85–91 mm (*n* 10, *m* 87.8); **T** ♂ 70–78 mm (*n* 18, *m* 73.4), ♀ 67–73.5 mm (*n* 10, *m* 70.5); **T/W** *m* 79.1; **BD** 5.8–7.3 mm (*n* 26, *m* 6.5); **white on r6** ♂ 21–30 mm, ♀ 24–27 mm.

TAXONOMIC NOTE Considering the rather well-marked differences and apparent gap in range of the two subspecies a case could be made for treating them as separate monotypic species. However, a comprehensive genetic study has not yet been conducted, and vocal differences need to be better studied and confirmed, thus for now it is best to continue to treat them as one species.

E. c. semenowi, 1stS ♂, Turkey, May: seemingly worn and brownish juv primary-coverts, primaries and tail (with narrow, pointed and worn feathers) infer age, but this image also demonstrates that young ♂ in spring can be as colourful as ad. (D. Occhiato)

REFERENCES Chappuis, C., Heim de Balsac, H. & Vielliard, J. (1973) *Bonn. zool. Beitr.*, 24: 302–316. – de Knijff, P. (1991) *BW*, 4: 384–391. – Walther, B. A. (2006) *Sandgrouse*, 28: 52–57.

E. c. semenowi, ad ♂♂, Israel, Sep: separation of races in autumn and winter, when yellow areas of *semenowi* can be partly concealed, is sometimes less straightforward than in spring and summer. The amount of yellow tinge indicates sex; both birds seem already to to have moulted extensively during completed post-nuptial moult in late summer, and tail-feathers appear broad and rounded at tips. (R. Segali)

E. c. semenowi, ad ♀, Oman, Sep: classic image of a poorly-known plumage, in early autumn, when still in uniformly fresh plumage after recently completed post-nuptial moult. Otherwise similar to ♀ in spring, including yellow-tinged belly that confirms race. (H. & J. Eriksen)

ORTOLAN BUNTING
Emberiza hortulana L., 1758

Fr. – Bruant ortolan; Ger. – Ortolan
Sp. – Escribano hortelano; Swe. – Ortolansparv

A familiar bunting with a pleasant song, regrettably heavily hunted in some W and S European countries for its few grams of meat claimed to be a delicacy, despite being generally protected. A summer visitor to Europe with a range that stretches east to Central Asia and W Mongolia, the Ortolan breeds in open country with scattered copses and scrub, often in agricultural land or grazed meadows with scattered small woods. It winters across N sub-Saharan Africa and in the S Middle East and Arabia.

IDENTIFICATION Relatively round-headed, plump but long-tailed bunting, very similar to both Grey-necked and Cretzschmar's Buntings, and separation requires care, especially with non-adult ♂ plumages. Always displays a rather *conspicuous pale eye-ring* (pale yellow, can look white), a *pale yellow submoustachial stripe* contrasting with dark (olive-grey) lateral throat-stripe, *rufous or warm buffish underparts*, and is typically *heavily streaked above*. Scapulars fringed rufous, and pale double wing-bars rather prominent. Bill rather longish with straight culmen, and *bill and legs pink to dull orange-red*. ♂ easily recognised by *greyish-green head and breast*, sharply set-off from *yellow throat* and *rusty peach-brown belly*. Both *submoustachial stripe and eye-ring* are *lemon-yellow*. Duller and streakier ♀ and immature plumages, however, are much less distinctive and require more careful separation from some congeners. Stance half-upright to horizontal and typically hops on ground, but may appear almost to creep when feeding. More insectivorous than most buntings, often foraging in trees and even catches insects in flight. Typically feeds on fallen caterpillars on the ground under large oaks. Flight superficially recalls Tree Pipit in speed and undulations. Often elusive when breeding, but can be highly gregarious on migration and in winter.

VOCALISATIONS Song (repeated endlessly) consists of a 'ringing' first part (3–4 units) and lower-pitched, more melancholy second part (usually 2–3 units), the second part almost as if echoing the first, *srü-srü-srü-srü-dru-dru-dru* or *sia sia sia drü drü* (the echoed notes often lower-pitched), or the song is a little more strained in tone, less ringing, *dzii dzii dzii dzii hüü hüü*; in south of range the last part often cut short, last note drawn-out in Cretzschmar's-like fashion, *sür-srü-srü-srü-chuuüü* or *zree-zree-zree züüü*. Thus, in SE Europe and Turkey, where both species occur, until voice and variation of both have been learnt always best to see the songster for reliable identification. – Commonest calls, often heard from migrants in flight, are a metallic, almost disyllabic *sli-e* and short *chu*, frequently given alternately at 2–3 sec-intervals. Variants of these two calls, like *tseeu* and *plett*, are not so important for identification.

SIMILAR SPECIES Only marginally smaller than *Yellowhammer* which rarely causes a significant confusion risk. – In W Europe pinkish bill and bright lemon-yellow eye-ring are instantly useful characteristics in all plumages, but are shared by the much more local Middle Eastern *Grey-necked Bunting*. However, this has proportionately a little longer tail, more subdued streaking on mantle and lacks a dark breast-band (though in fresh adult plumage colours of Ortolan breast-band duller, especially in ♀, which often has

Ad ♂, Finland, May: uniform greyish-green head and breast-band with conspicuous yellow eye-ring, submoustachial stripe and throat. Belly rufous-orange, while undertail-coverts yellowish. Distinctly dark-streaked olive-brown and rusty-fringed mantle and scapulars. Lack of dark streaks on breast or lateral throat-stripe as often shown by 1stS ♂. Age supported by seemingly rounded tail-feather tips and ad wing. (M. Varesvuo)

Ad ♂♂, Israel, Mar: irrespective of age, two types of ♂ occur (though not in same frequency across range), greyish-and-cream (left) and bright green-and-yellow (right), the former caused by a deficiency (and is diet-related), rather than due to a mutation. Degree of yellow on throat (and head) is caused by carotenoids. (M. Varesvuo)

♀, United Arab Emirates, Apr: duller than ♂, with less pure green-and-yellow head, has fine dark streaking on crown and lateral throat-stripes. Note also dark-streaked off-white breast, yellowish throat (can be flecked in centre, as here) and unstreaked (or slightly mottled) rufous belly. Ageing never easy, sometimes not even in the hand, and best avoided here. (H. Roberts)

1stS ♂, Turkey, Jun: note vestiges of immaturity in tiny dark markings on crown and overall rather dull colours, as often shown by 1stS ♂. Aged also by juv tail (pointed tips, visible in other images of same individual) and wing, including rather pointed primary-coverts. (H. Shirihai)

breast-streaking rather than a solid band and can lack clear-cut demarcation between greyish breast and rufous lower underparts). Also, ♂ has more lead-grey head and nape, not olive-green as Ortolan, and pink bill is thinner (and can therefore appear more pointed). Note also slightly different tertial pattern in Grey-necked, with marginally narrower and more uniformly wide warm brown outer edges (Ortolan: more brightly rufous outer edges to tertials widen halfway, forming wider rufous panel). – In SE Europe and Turkey confusion can occur with *Cretzschmar's Bunting*, which may require careful separation, especially in non-adult plumages. Only Ortolan has extensive yellow on the throat and underwing or a green tone to the head. It also has a dull brown rump, not rufous like Cretzschmar's. Ortolan averages slightly larger (but difference subtle), and in adult plumages unmistakable if seen well, but still easily confused if head colours not seen. Tertial pattern in Ortolan same as Cretzschmar's. Juvenile/first-winter Ortolan far less distinctive (see the other two species for details, especially Cretzschmar's Bunting for a detailed account). Both have broad white tail corners, but in the hand the detailed pattern of the outer pair of rectrices is a useful distinguishing feature. – See *Cinereous Bunting* for separation from that species.

AGEING & SEXING Ageing requires close scrutiny of moult and feather wear in wing and tail. Sexes often differ subtly after post-juv moult, more readily distinguished from 1stS. Neither exhibits significant seasonal variation. – Moults.

Complete or suspended post-nuptial moult of ad (when suspended, all or some secondaries left until after autumn migration), and partial post-juv moult in early Jul–early Sep. Post-juv moult includes head, body, all lesser coverts and most or all median coverts, sometimes no greater coverts but usually a few innermost to all, and very occasionally some tertials and alula. Partial pre-nuptial moult in ad involves any retained secondaries, often head, body, usually part of or all greater and median coverts (more coverts moulted in 1stY), all tertials, some tail-feathers, but also rarely part of alula, mostly late Dec–late Mar. Unlike many buntings, acquires a less advanced, to some degree juv-like 1stW plumage, compensated by an extensive pre-nuptial moult in winter quarters, where 1stW plumage is replaced by a more ad-like plumage. – **SPRING Ad ♂** See Identification. Eye-ring, submoustachial stripe and throat lemon-yellow. Crown, nape and breast-band vary, some being olive-green, others much greyer and less greenish; others yet again even slightly mottled on breast. Belly usually bright orange-rufous. **Ad ♀** Like ♂ but head greyish-olive, usually tinged brownish and more streaked on crown. Lores and throat essentially buffish-yellow, with more noticeable lateral throat-stripe and duller ill-defined greyish breast with variable fine streaking, particularly on centre. Upperparts and wings duller than ♂, and underparts more buffish yellow-brown, less uniform and less rufous, appearing mottled when seen close. However, sexing not always straightforward (especially involving duller young ♂♂). **1stS** Differs from similar ad by retained and contrastingly more worn juv primaries, primary-coverts, perhaps some median and greater coverts, and some tail-feathers (but differences often difficult to detect even in the hand). Some ♂♂ have head, neck and breast-band on average greyer than ad, throat and submoustachial stripe paler yellow and, often, some diffuse streaking below. ♀ especially dull, with even less yellow and is more streaked (very similar to Cretzschmar's, but usually shows traces of greenish on crown and nape). Both 1stS and ad can show moult limit among greater and median coverts in spring, and it is not known whether only ad can have new secondaries (some ad apparently do not moult any secondaries in winter). – **AUTUMN Ad** Both sexes reliably aged by being evenly fresh, and most birds are even more distinctive during autumn migration by having several unmoulted secondaries (often all, and occasionally other feathers) due to suspension. All greater coverts and tertials new, fringed pale rufous. Tail-feathers rounded, rr5–6 with white tips. **Ad ♂** As spring, though head has variable greyish cast, slightly more streaky crown, darker moustachial and lateral throat-stripes, and lores, eye-ring and throat vary from pale lemon to buffish-yellow, and breast from olive-grey to olive-green with perhaps some dark mottling and spots. **Ad ♀** Much as spring but overall duller with more breast-streaking, and unlike fresh ♂ tends to lack clear-cut division between yellowish throat, only slightly greyish upper breast and rufous-buff of rear under-

1stS ♀, Israel, Mar: dullest plumage in spring, with extreme examples keeping streaky 1stW-like plumage (especially bold blackish lateral throat-stripe and breast streaking), olive-grey breast and rufous-tinged belly subdued, hardly any yellow on eye-ring or submoustachial stripe. Ageing based on strong moult limit in greater coverts (unmoulted outer feathers bleached juv) and on generally undeveloped plumage. (R. Pop)

Ad ♂, Spain, Aug: at end of complete post-nuptial moult with many remiges and rectrices still growing. Greyish cast to greenish head, at the most narrowly streaked crown, dark lateral throat-stripe, rather pale yellow throat, and breast and flanks can show scattered dark streaking, otherwise as spring. (C. Requena Aznar)

parts, while underparts below chest diluted cinnamon-buff, less deep rufous. Crown and head-sides, lateral throat-stripe and chest more heavily spotted or streaked black. **1stW** Generally recalls juv and, unlike fresh ad ♀, plumage duller with even browner head (lacking green tinge), less yellow on throat and submoustachial stripe (but some), and has heavily-streaked lateral throat-stripe and boldly streaked breast and flanks, with ground colour pale rufous-buff and gingery-yellow; also more extensively streaked upperparts with fewer warm brown tones. Retained juv greater coverts (most or all) and tertials differ from ad pattern and are fringed whitish, much of the white abraded during autumn, but occasional 1stW shows completely renewed greater coverts, and some ad can show moult contrast due to suspended moult in these tracts. Primaries, primary-coverts, tertials and tail-feathers also slightly worn, tips of tail-feathers pointed, and rr5–6 lack whitish tips or have only small area on r6 (Svensson 1992). Sexing of 1stW usually impossible, but ♀ often more heavily streaked below, with less yellowish tinge, while advanced ♂ can be slightly brighter and approach ad ♀ (see also Biometrics). See Cretzschmar's Bunting for comparison of 1stW plumages. **Juv** As 1stW but has soft fluffy feathering, heavily streaked and sometimes scalloped, and pale parts virtually lack yellow pigments.

BIOMETRICS L 15–16 cm; **W** ♂ 83–95.5 mm (n 28, m 89.0), ♀ 80–88.5 mm (n 20, m 84.0); **T** ♂ 63–75 mm (n 28, m 67.8), ♀ 60–67 mm (n 20, m 63.3); **T/W** m 75.9; **white on r6** (to tip) 27.5–38 mm (n 47, m 33.0); **B** 12.0–14.6 mm (n 47, m 13.3); **B**(f) 9.1–12.2 mm (n 50, m 10.5); **BD** 5.2–6.7 mm (n 48, m 6.0); **Ts** 17.0–20.0 mm (n 47, m 18.7). **Wing formula: p1** minute; **pp2–4** about equal and longest (though p2 sometimes to 2.5 mm <); **p5** < wt 2–6 mm; **p6** < wt 6–13 mm; **p7** < wt 11–19 mm; **p10** < wt 21–30 mm; **s1** < wt 23–31.5 mm; **tert. tip** < wt 9.5–19.5 mm. Emarg. pp3–5 (sometimes faintly on p6).

GEOGRAPHICAL VARIATION & RANGE Monotypic. – Europe (though absent over much of W & C, including Britain), Turkey, Middle East, Central Asia east to Altai and NW Mongolia; winters sub-Saharan Africa, scattered occurrences from Sierra Leone to Sudan and Eritrea. (Syn. *elisabethae*; *shah*.)

REFERENCES SMALL, B (1992) *BW*, 5: 223–228.

Ad ♀, Bulgaria, Jul: very fresh flight-feathers with still growing primaries (at least one unmoulted very worn outer primary). All greater coverts moulted. Overall duller than autumn ad ♂ with streaky breast and solid lateral throat-stripe, but no clear-cut yellowish throat, only slight greyish wash on breast and rufous-buff underparts. Still, compared to 1stW, streaks on crown and breast and dark lateral throat-stripe far less developed. (I. Hristova)

1stW ♂, Scotland, Aug: ageing and sexing in autumn can be challenging, with 1stW ♂ often being very similar to ad ♀. Since ageing here straightforward due to obvious moult limits with several central greater coverts whitish-fringed retained juv, and primary-coverts pointed, young ♂ seems best option, supported by strong yellow tinge on face. (H. Harrop)

1stW, presumed ♀, Azerbaijan, Sep: dullest example in autumn, with retained overall juv-like pattern, especially much reduced (sometimes almost lacking) greenish-olive and yellow on head, throat and eye-ring, no grey or rufous below, but dense blackish streaks on chest strongly suggest 1stW ♀ (M. Heiß)

Juv, France, Jul: most like 1stW ♀ but soft fluffy body-feathers, heavily streaked, and note that pale plumage parts virtually lack any yellow. (D. Occhiato)

GREY-NECKED BUNTING
Emberiza buchanani Blyth, 1845

Fr. – Bruant à cou gris; Ger. – Steinortolan
Sp. – Escribano cabecigrís; Swe. – Bergortolan

This mainly Central Asian, elegant and long-tailed bunting typically occurs on open, rocky slopes or hillsides with boulders, low scrub and patches of grass, a habitat it shares with, among others, Cinereous and Rock Buntings. It breeds in easternmost Turkey but is perhaps commoner further east in Central Asia, and can be found at over 3000 m altitude. There is a small isolated population in Mongolia, but the centre of the range is from Iran to Kazakhstan. It winters in India and returns to its breeding sites from late April into May.

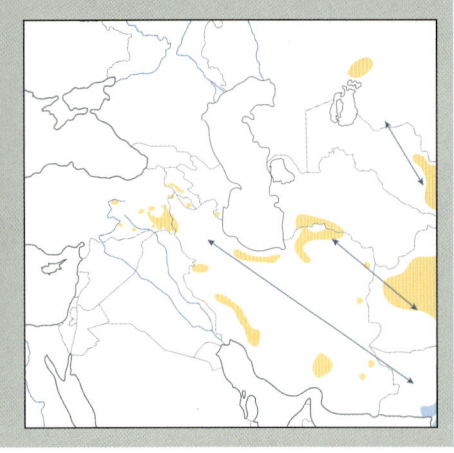

IDENTIFICATION A slim, medium-sized bunting with *long, white-sided tail* and, for a bunting, rather a *long but delicate, all-pink bill* (without darker culmen or tip). The Grey-necked Bunting is one of a trio of rather similar-looking buntings with a basically greyish head, pale (yellow or orange-buff) throat and submoustachial stripe, pale eye-ring, brown dark-streaked back and uniform reddish-brown lower breast and belly, the other two being Ortolan and Cretzschmar's Buntings. Grey-necked differs by *lacking a greyish chest-band* and having rather *subdued streaking on the mantle*. Furthermore, the *scapulars are rufous* in contrast to the *paler earth-brown or brown-grey mantle* and *greyish lesser coverts*. Undertail-coverts are *whitish*, and black-centred *tertials have usually evenly-broad cinnamon-brown edges* (less rufous-edged, lacking the step between narrower pale edge basally but abruptly broader pale edge distally on longest two tertials of Ortolan and most other buntings; some exceptions noted). *Head, nape, neck-sides and lateral throat-stripe are bluish-grey*, while *throat and submoustachial stripe are pale yellow*. There is a *complete whitish eye-ring*. The *cold red-brown breast and belly* are finely vermiculated white, at least before becoming worn in summer. Sexes similar but ♂ neater with clearer colours, ♀ slightly duller. Juvenile is more uniform brown-grey and more streaked on head and below. Look for rather long but thin, *all-pinkish bill* with *straight culmen*, a hint of rufous on scapulars (less than in adult, but still usually some traces) and greyish lesser coverts/wing-bend. The relatively lightly streaked mantle, also in juveniles, is a supporting character.

VOCALISATIONS Song a mechanically-repeated fast, short and rather plain phrase with slightly shrill or hoarse tone. In structure it broadly resembles Ortolan Bunting, with a few high-pitched or ascending notes, then a sudden change to lower-pitched concluding ones. It often opens with 3–4 very fast or accelerating high-pitched notes, after which pitch drops and pace slows a little, and the second half varies in details, e.g. *srisrisrisri sru-sru-**sreh** sru*. Other song variations have been rendered *sresresre-su**sreeh**-sreea* or initially rising, then falling *tru-trü-**tri** tre-tra*. Often the second or third last note is a little more stressed. Song varies somewhat between ♂♂, and some variations recall the partly sympatric Cinereous Bunting, found in similar habitats, so care and practice are needed to eliminate this species. (Cinereous Bunting song opens at more leisurely pace and ends quickly with a few downslurred notes.) – Calls quite similar to those of Cinereous and Cretzschmar's Buntings, including an emphatic *chüpp* and a high-pitched, sharp *zrip*. These two calls can be combined and alternated, and, as with so many birds, the finer details vary a little depending on the individual and circumstances.

SIMILAR SPECIES Separated from adult *Ortolan Bunting* by vocalisations, bluish-grey head (not greenish-tinged), whitish eye-ring (not yellowish), gradually paler reddish-brown lower belly and whitish vent and undertail-coverts (not uniformly orange-brown), lack of complete olive-grey chest-band (bluish-grey neck-sides at most), thinner and slightly longer all-pink bill with straight culmen, finer-streaked mantle, rufous scapulars but grey lesser coverts, and usually evenly broad (rather narrow) pale brown edges to longest two tertials (not broader red-brown edges distally, forming a step halfway along feathers). Juveniles are more similar, both being brownish and streaked, but differences relating to bill, prominence of mantle-streaking, colour of lesser coverts, scapulars and undertail still apply. – Differs from adult *Cretzschmar's Bunting* by song, mainly whitish throat and submoustachial stripe (red-brown spotting at most, not all rusty-buff), cold brown-grey mantle thinly streaked in contrast to rufous scapulars (mantle and scapulars uniformly reddish- or tawny-brown and prominently streaked), paler vent and undertail-coverts, thinner and slightly longer all-pink bill with straight culmen, and tertial pattern (see above). Juveniles, again, are more similar but most differences detailed under comparison with Ortolan apply. – Separation from first-winter Cretzschmar's and Ortolan Buntings, especially young ♀♀, can be less straightforward. Note: (i) breast-streaking noticeably narrower and shorter, and generally reduced in extent, hardly reaching lower breast and invariably not extending to flanks (differences more obvious compared to Ortolan); (ii) upperparts tinged greyer and rump/uppertail-coverts the same tone but unstreaked (stronger and broader dark upperparts streaking in Cretzschmar's and Ortolan, and streaking often extends to rump/uppertail-coverts in Ortolan, while they are rustier in Cretzschmar's); (iii) lateral throat-stripe of Grey-necked narrower and less solid, with whiter and more extensive throat patch; and finally, (iv) check for diagnostic evenly dark-centred (or almost so) longest two tertials, lacking clear angled 'L' notch of the other two species. – Juvenile separated from juvenile Cinereous Bunting by the pink, slenderer

Ad ♂♂, Kazakhstan, May (left) and Turkey, Jun: slender jizz with long tail and narrow, finely pointed pinkish-orange bill, but whitish eye-ring also characteristic. Dark tertial centres lack sharp step on outer web of Ortolan and Cretzschmar's Buntings. Combination of blue-grey head and lateral throat-stripe, contrasting with cream-white submoustachial stripe and throat, plus vinaceous rufous-brown underparts, while weakly streaked mantle emphasises conspicuous rufous scapular panel, provide both specific and sexing clues. White wedge on outer tail-feathers, and dull greyish-brown rump. Both have evenly worn wings (on right bird innermost median covert looks different, being more greyish without specific pattern, as often the case with this feather). (Left: C. Bradshaw; right: V. Legrand)

bill with straight culmen (Cinereous: grey, slightly stockier bill) and the faint rusty tinge on underparts (Cinereous: off-white on belly, without brown or rufous tinges).

AGEING & SEXING Ages differ in autumn, sometimes in spring; reliable ageing usually requires close views or handling. Sexes usually differ slightly after post-juv moult, but many are very similar and can be inseparable. – Moults. Complete post-nuptial moult of ad after breeding, mainly Jul–Oct. Partial post-juv moult at same time but usually somewhat later affects head, body and many wing-coverts but does not involve any primaries, secondaries, tail-feathers (except r1 in some, rarely more) or primary-coverts. Variable number of greater coverts moulted, from none to all. 1stY usually has limited partial pre-nuptial moult in late winter confined to parts of head, some or all tertials and r1 (if these not replaced in autumn). – **SPRING Ad ♂** (The following criteria separate many birds, but quite a few must be left un-sexed due to intermediate characters. Wing-length often helpful for sexing.) Breast and belly distinctly deep rufous without dark spots or streaks. Crown/nape unstreaked grey, or show only very faint streaks in some. Scapulars often rather vivid rufous. **Ad ♀** Breast shows varying amount of rufous, streaked or spotted dark. Crown greyish-brown, finely streaked dark (never pure ash-grey, or unstreaked). Check wing-length. **1stS** Sometimes differs from similar ad by retained and contrastingly more worn juv primaries, primary-coverts, perhaps some median and greater coverts, and some tail-feathers being narrower and more worn and pointed at tips (but differences often difficult to detect even in the hand). Some birds are intermediates, difficult to determine, hence best left unaged. – **AUTUMN** Sexing as in spring but rufous of underparts partly concealed by whitish tips. **Ad** After completion of moult, longest primaries and all tail-feathers quite fresh, tips of tail-feathers slightly rounded, edges neat. All greater coverts fresh and edged rufous-brown. **1stW** Tips of longest primaries and tail-feathers slightly worn, tips of latter pointed, except birds that moulted r1, rarely more inner pairs of tail-feathers. In birds with some outer or all juv greater coverts retained, these are slightly worn and edged pale buff (rather than being fresh and edged rufous). Birds with moult contrast between retained juv and new tail-feathers also easily recognised. **Juv** Resembles a dull ♀ but usually easily separated by rather densely streaked or blotched breast and rump (largely unstreaked in post-juv). Also, crown duller brownish with rather heavy streaking.

BIOMETRICS L 15–16 cm; **W** ♂ 82–94 mm (n 37, m 87.4), ♀ 77–87 mm (n 23, m 82.7); **T** ♂ 67–79 mm (n 37, m 71.1), ♀ 60–72 mm (n 23, m 67.2); **T/W** m 81.4;

♀, Mongolia, Jun: closely recalls ♂, but shows fine dark shaft-streaks on crown, duskier lateral throat-stripe and very fine dusky streaking on breast-sides to upper flanks, and rufous on underparts and scapulars diluted and less extensive. ♀ can develop rufous on breast, then almost indistinguishable from 1stY ♂. Tail-feathers rather narrow and worn, but best left unaged. (H. Shirihai)

1stS ♂, Kazakhstan, Jun: like ad ♂ but duller overall and in many respects midway between the sexes. Nevertheless, grey head usually purer with hardly any dark crown streaking, rufous on breast deeper and on scapulars more solid and brighter than any ♀. Differences even clearer if bird aged first: here, much of wing, including primary-coverts, is juv. (H. Shirihai)

Ad ♀, India, Oct: age and sex due to fine dark shaft-streaks on crown, less pure grey lateral throat-stripe (some dark mottling admixed), diluted and smaller area of rufous on underparts, clear dark centres to rufous scapulars and overall less saturated wing, despite all feathers being ad. (A. Deomurari)

Ad ♂, Oman, Jan: overall plumage freshness and neat pattern without any dark streaking indicates ad ♂ (no obvious juv feathers). It is interesting to note that the innermost long greater covert is very recently moulted, as to be expected by Jan. That the innermost median covert looks different, being more greyish without specific pattern, is normal and not moult related. (H. & J. Eriksen)

B 12.4–15.1 mm (*n* 34, *m* 13.6); **B**(f) 10.0–12.7 mm (*n* 59, *m* 11.1); **BD** 4.9–6.2 mm (*n* 58, *m* 5.4); **Ts** 18.0–20.2 mm (*n* 31, *m* 18.9). **Wing formula: p1** minute; **p2** < wt 0.5–2 mm; **pp3–4** longest; **p5** < wt 0.5–4 mm; **p6** < wt 6–9.5 mm; **p7** < wt 12–15 mm; **p10** < wt 19–23 mm; **s1** < wt 20–24 mm. Emarg. pp3–5, p6 sometimes very slightly emarg. too.

GEOGRAPHICAL VARIATION & RANGE Monotypic. – E Turkey, Transcaucasia, Iran, Central Asia, Mongolia. – Claimed differences in darkness and prominence of streaking above, which have given rise to three described subspecies apart from *buchanani*, could not be corroborated on material examined by us (NHM, AMNH, ZMC, NRM, MNHN), individual variation being more noticeable than geographical. (Syn. *cerrutii*; *huttoni*; *neobscura*; *obscura*.)

REFERENCES Roselaar, C. S. & Castricum, V. (2006) *DB*, 28: 284–291.

1stW ♂, Germany, Oct: like ad ♂, but grey above and rufous below less pure, but this (relatively advanced) individual is still brighter than any ♀. Greater coverts are juv. See also below for plumage variation of 1stW ♂. (G. Schuler)

1stW, ♂ (left) and 1stW ♀, India, Oct: least advanced young autumn plumages (still with many juv feathers, while moulting others). However, both can be sexed, with ♂ having unstreaked greyer head, warmer pinkish-brown underparts and weaker lateral throat-stripe. (A. Deomurari)

1stW ♀, India, Nov: duller plumage combined with clear dark streaks on crown, mantle and breast, narrow dark brown lateral throat-stripe and broad dark centres to scapulars diagnostic of 1stW ♀. Remiges, tertials and (the just visible) primary-coverts are juv, while all wing-coverts are post-juv and very fresh. Juv tertials have dark centres showing only a shallow step different from Ortolan and Cretzschmar's Buntings. (S. Singhal)

Juv, Kazakhstan, Jul: soft fluffy feathering, and heavily streaked overall. Long tail well developed, less so long pointed bill, and latter is slightly dusky at tip and culmen. (M. Westerbjerg Andersen)

CRETZSCHMAR'S BUNTING
Emberiza caesia Cretzschmar, 1827

Fr. – Bruant cendrillard; Ger. – Grauortolan
Sp. – Escribano ceniciento; Swe. – Rostsparv

A local E Mediterranean speciality that summers on barren, stony slopes with sparse, scrubby vegetation. Cretzschmar's and the more widespread Ortolan Buntings can be a challenge to identify, especially on migration where they can occur together. However, in the breeding season where they meet, Cretzschmar's is generally found at lower altitudes than Ortolan. Cretzschmar's winters mainly in NE Africa, in Sudan and S Egypt south to Eritrea, but possibly also in S Arabia.

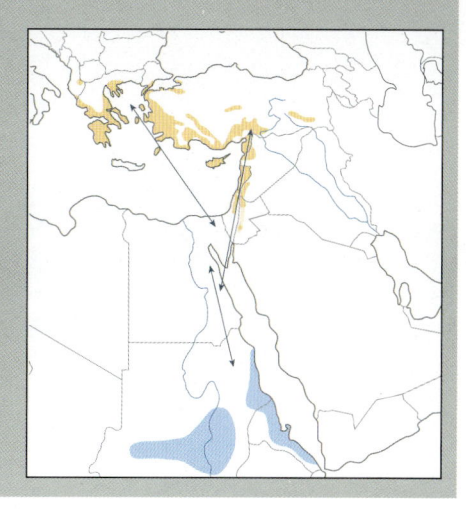

IDENTIFICATION Slightly smaller than closely related Ortolan Bunting, with similar structure and almost identical plumage pattern (though not colours). Legs and bill as Ortolan, bright reddish pink-brown, but *bill slightly shorter and more slender*. Prominent *whitish or pale buff eye-ring* in all plumages. Adult ♂ has *rusty-orange throat, submoustachial stripe and lores*, contrasting with pale *bluish-grey head and breast-band* sharply demarcated from *rufous-brown rest of underparts*, while boldly black-streaked upperparts are overall rather drab rufous-brown, and *rump unstreaked rufous*. ♀ duller and streakier than ♂, usually with *much-reduced grey on head and a poorly-defined and slightly streaked breast-band*. Note clear-cut notched rufous outer fringes to longest two tertials, and rusty tips to median and greater coverts, all richer on average in ♂. Young browner and even streakier, and more difficult to separate from Ortolan (see Similar species). Flight, behaviour and actions very similar to Ortolan, but perhaps less shy. Markedly terrestrial, though often flushes to trees if disturbed.

VOCALISATIONS Song similar to Ortolan in same regions with 2–3 repeated notes, but the drawn-out penultimate (or last) one is thinner, and lacks the pleasant ringing quality of the first part, *zwiie-zwiie-zwiie ziüüü* or *che-che-che-cheee* and faster *si si-siüüü* or *ziii-ziii-ziii-ziiii*; some birds alternate this with a hoarse (at times almost asthmatic!) variant. – Contact call also like Ortolan (or Yellowhammer) but sharper, sounding slightly rasping, *chitt* (also rendered *spit* or *tsrip* or *tchipp*), and a melancholy, low-pitched, full whistle with a faint downslur, *chüu*. Also a drier *plett* and a disyllabic *chittlet*, a bit like Ortolan.

SIMILAR SPECIES Adult ♂ unmistakable if seen well, especially combination of blue-grey head and breast-band, rusty-orange submoustachial stripe and throat, plus lack of any pure lemon-yellow or green; mantle also averages slightly more reddish, and belly deeper rufous-brown, while rump is diagnostically unstreaked rufous-brown. – Adult ♀ usually readily differentiated from similar ♀ Ortolan Bunting (and Grey-necked Bunting, which see) by usually having at least some hint of ♂ characters. – First-autumn birds very similar to Ortolan Bunting in same plumage, but the following characters in Cretzschmar's, used in combination, are useful: (i) all (or nearly all; may leave 1–3 central) greater coverts moulted early, before Sep, ad-type greater coverts tipped and edged rusty-cinnamon creating a rusty wing-bar that blends in with rest of wing, whereas first-year Ortolan retains all or most juv greater coverts through autumn, these being slightly worn and rather whitish, creating a whitish more narrow (still more contrasting) wing-bar; any moulted to ad type in Ortolan are not especially rusty-tipped, thus differ only moderately from juv greater coverts; (ii) median coverts in Cretzschmar's tipped similarly rusty-cinnamon as greater coverts, whereas Ortolan has median coverts tipped warm sandy or dull ochre-tawny, thus on average a little duller and less rusty than Cretzschmar's; (iii) rump in Cretzschmar's almost unstreaked and tawny-cinnamon, with more chestnut feather-centres appearing at most as diffuse blotches, while in Ortolan rump is more clearly streaked and ground colour is more greyish-brown, less rufous or chestnut-tinged; (iv) on average (though some variation and near overlap of extremes) Cretzschmar's is more chestnut on belly in first autumn (function of earlier and more extensive moult), but some Ortolan can show a little chestnut, and all have a warm ochre-buff colour; (v) on average, dark streaking of upperparts (notably of scapulars), chest and flanks of Cretzschmar's a little less bold, more narrow than in Ortolan (some overlap); (vi) blackish centres to two longest tertials on average (again, some overlap) more restricted, narrower than Ortolan; (vii) white on outer two tail-feathers on average less extensive, and proximal border more obtuse (running a little more across the feather than along it), while Ortolan often has slightly more extensive white and more oblique (diagonal) proximal border; some are similar though; (viii) due to slightly more extensive pale loral area, Cretzschmar's appears to have broader supercilium in front of eye, sometimes almost meeting above bill, while Ortolan usually has narrower supercilium in front of eye and darker forehead; and furthermore, (ix) bill averages shorter and more conical, and (x) tail subtly longer in Cretzschmar's.

Ad ♂, Turkey, May: slightly smaller than Ortolan, which it otherwise resembles in shape and general plumage, but grey areas purer and tinged blue rather than green, and yellow on face replaced by rufous. Also, prominent eye-ring whitish, lacking yellow tinge. In spring, ♂ looks smart with its 'clean' bluish-grey head and breast-band, framing rusty-orange face and throat. All wing ad. (D. Occhiato)

♀, Israel, Mar: irrespective of age, some breeding ♀♀ can resemble ♂, being faintly washed bluish-grey on head and breast, even more than shown here. Nevertheless, grey colours never as bright or extensive, and dark markings on crown, lateral throat-stripe and breast invariably better developed than in even least advanced 1stS ♂. Seemingly abraded brownish remiges and primary-coverts could suggest 1stS, whereas amount of grey on head and breast indicates ad, hence best left unaged. (H. Shirihai)

In the hand, usually has more white and cream-buff (not yellowish) underwing-coverts and axillaries. – Juveniles of Cretzschmar's and *Ortolan Bunting* are often confused on the breeding grounds (some are perhaps only separable if seen with adults). Both show extensive, distinctly triangular streaks below, but in Cretzschmar's these are browner and concentrated on the central upper breast (less obvious on breast-sides, with very few on upper flanks and usually none on rear flanks); and juvenile Cretzschmar's also tends to have a warmer cinnamon or pink-buff lower breast to undertail-coverts, contrasting with much paler upper breast, throat and submoustachial stripe, while the dark lateral throat-stripe is relatively weak and browner (Ortolan: more extensive and broad blackish-brown streaks, on some covering the entire breast and reaching rear flanks, and ground colour almost invariably pale gingery-yellow; lateral throat-stripe broader, forming a prominent patch and gorget, somewhat recalling Tree Pipit). Juvenile Cretzschmar's has a warmer brownish-grey mantle and rustier rump and uppertail-coverts (never olive brown-grey like Ortolan), with mantle and scapulars more finely streaked. – For separation from *Grey-necked Bunting* see that species.

AGEING & SEXING Ages differ in autumn, sometimes in spring; reliable ageing usually requires close views or handling, when wear and moult contrast in wing, and shape

1stS ♂, Israel, Apr: as ad ♂ but less pure bluish-grey head (ear-coverts washed brownish) and breast, with some streaks on crown and breast, both being faintly streaked, and thin dark lateral throat-stripe. Age confirmed by moult limits in wing, especially the renewed inner greater covert versus new innermost scapular. Also note uniform rufous submoustachial stripe. (G. Shon)

Spring comparison of ♀♀ Cretzschmar's Bunting (left Israel, Mar) versus Ortolan Bunting (right: Israel, May): difficulties occur with dull ♀ Ortolans (usually young) that almost lack any yellow in plumage, but such birds in optimal light conditions usually show at least some greenish-olive tinge to head. Any hint of bluish-grey on breast-sides can sometimes clinch Cretzschmar's in spring, but beware that grey is often obscured (usually in young birds), and conversely some dull grey feathering maybe seen on breast-sides of Ortolan. Most reliable difference, but requiring experience, is the subtly smaller (shorter, more pointed) bill of Cretzschmar's. Both birds most likely 1stS with subtly contrasting juv remiges and primary-coverts. (Left: H. Shirihai; right: L. Kislev)

Ad ♂, Israel, Jul: a bird in post-nuptial moult, secondaries and some outer primaries still growing. Compared to breeding plumage, lead-grey of head and breast less pure, partly concealed by fresh rufous-buff tips. (L. Kislev)

Ad ♀, Greece, Aug: wing evenly fresh and ad-like (especially having blackish-centred primary-coverts). Compared to breeding plumage overall more mottled, with notably buffish tips below. Note almost pure white conspicuous eye-ring. (P. Petrou)

BUNTINGS

and wear of outer tail-feathers can help. Sexes usually distinguishable. Seasonal variation rather limited. – Moults. Complete post-nuptial moult of ad after breeding, mainly Jul–Sep. Partial post-juv moult at same time involves head, body, all lesser and median coverts, and usually all or most greater coverts, tertials and central tail-feathers (in ♂♂ occasionally all tail-feathers). Pre-nuptial moult partial, usually in winter quarters but highly variable in extent, typically including head and body, all or most median coverts, inner greater coverts, all tertials, and perhaps central tail-feathers, most extensive in ♂♂ (especially among 1stY). – SPRING **Ad** Entire wing uniformly moderately worn, lacking strong moult contrast created by any retained heavily-worn and brown juv feathers. Any detectable moult contrast only slight. **Ad ♂** See Identification for general description. **Ad ♀** Unlike ♂, has brown-tinged, streaked crown and nape, albeit usually with some ash-grey, at least on neck- and breast-sides, and paler, buff-white throat with dark-flecked lateral throat-stripe. Greyish breast buffish, always but variably streaked dark brown. Demarcation between nape and mantle less obvious, and underparts paler than ♂. Still, sexing not always straightforward. **1stS** Resembles respective ad, but often has some retained, brown-bleached juv wing-feathers. (However, both ad and 1stS show moult contrast in spring and, as usually no juv greater coverts are retained, it is recommended to attempt to establish that primaries and primary-coverts are definitely juv, being distinctly worn and bleached—with practice most can be aged in the hand, but some must be left undetermined.) Many ♂♂ less advanced, approaching ♀, especially by having any grey less pure, and on average more distinct streaking below. – AUTUMN **Ad** Both sexes aged by being evenly fresh. **Ad ♂** Generally duller and less clearly patterned than spring (greyish areas less pure), crown/nape may have stronger streaking, and underparts tipped paler. **Ad ♀** Much as spring but overall duller, with pale brownish and faintly-streaked head-sides, underparts rather mottled pinkish-buff and overall paler, and breast streaking stronger and more extensive. **1stW** Both sexes resemble ad ♀ though duller, browner and more extensively streaked below. Those that have retained some outer greater coverts easily aged, but others moult all greater coverts, making it advisable to confirm that remiges, primaries and primary-coverts are juv (less glossy and tips slightly more worn). Many ♂♂ have at least some grey on head and breast, and rusty-cinnamon below (some being very close to ad ♀, or even approach fresh ad ♂), but ♀ usually has only hint of these colours, or nearly always lacks grey on crown, with much heavier-streaked throat, breast and flanks. Combination of plumage and size (see Biometrics) reliable for sexing many in autumn, except a few intermediates. **Juv** As 1stW but has soft fluffy feathering, and pale parts cream-toned. Generally heavily streaked, especially on breast/upper flanks.

Spring comparison of ♀♀ Cretzschmar's Bunting (top left ad, centre & bottom left 1stW: Israel, Sep) and Ortolan Bunting (top right: Kuwait, Sep; centre and bottom right: Israel, Sep): separation can be more tricky than in spring, and especially of dull 1stW ♀ Ortolan lacking yellow tinge in face and greenish-olive hue on head and upperparts. Also, note that some ♀ Cretzschmar's in autumn hardly show any bluish-grey on neck- and breast-sides. Nevertheless, and regardless of age, any autumn ♀ Cretzschmar's (after late summer moults) is more smoothly patterned, as this species moults more completely, unlike 1stW Ortolan which usually retains more juv-like heavily streaked underparts, often keeps juv median and greater coverts being on average paler tipped, and has obvious moult limits in wing. Note that central streaks of scapulars in Cretzschmar's tend to be more pointed than in Ortolan, although there is much overlap. The slightly smaller bill of Cretzschmar's, being shorter and more pointed (with less bulging lower mandible outline) is often a useful clue, too. Only the top left Cretzschmar's is ad (evenly fresh after complete post-nuptial moult), while the rest are 1stW by retained juv remiges and primary-coverts. (Top & central left: I. Drob; bottom left: I. S. Tov; top right: A. Audevard; central & bottom right: E. Hadad)

Juv/1stW, Turkey, Jun: a young bird in early stages of post-juv moult. Body plumage still loose and fluffy, and underparts streaking extensive. (H. Shirihai)

BIOMETRICS L 14.5–15.5 cm; **W** ♂ 81–90 mm (*n* 22, *m* 84.6), ♀ 78–84 mm (*n* 16, *m* 81.0); **T** ♂ 61–69 mm (*n* 22, *m* 65.3), ♀ 60.5–66 mm (*n* 16, *m* 62.8); **white on r6** (to tip) 23–31 mm (*n* 32, *m* 27.0); **T/W** *m* 77.3; **B** 11.7–14.1 mm (*n* 35, *m* 12.7); **B**(f) 9.0–11.3 mm (*n* 38, *m* 10.2); **BD** 5.2–6.4 mm (*n* 33, *m* 6.0); **Ts** 17.5–19.6 mm (*n* 35, *m* 18.5). Wing formula: **p1** minute; **pp2–4** about equal and longest (but p2 sometimes to 2 mm <); **p5** < wt 2.5–7 mm; **p6** < wt 7–12 mm; **p7** < wt 12–16 mm; **p10** < wt 19.5–24 mm; **s1** < wt 21–26 mm; **tert. tip** < wt 7–19 mm. Emarg. pp3–5.

GEOGRAPHICAL VARIATION & RANGE Monotypic. – Greece, W & S Turkey, Cyprus, Levant; winters Sudan, S Egypt south to Eritrea; possibly also in S Arabia.

REFERENCES MILD, K. (1990) *Bird Songs of Israel and the Middle East*. Stockholm. – SMALL, B (1992) *BW*, 5: 223–228.

YELLOW-BROWED BUNTING
Emberiza chrysophrys Pallas, 1776

Fr. – Bruant à sourcils jaunes
Ger. – Gelbbrauenammer
Sp. – Escribano cejigualdo; Swe. – Gulbrynad sparv

Yet another of the many bunting species breeding in the Siberian taiga that occurs in the covered region only as a vagrant, the Yellow-browed Bunting has been recorded in Britain, Netherlands and Sweden. It has a more easterly range than most treated in this book, breeding from Lake Baikal eastwards. It is primarily a forest bird that can be encountered even in rather dense woodland. Like all buntings it spends much time on the ground when feeding. Winters in SE Asia, mainly in China. It returns late in spring, in May or early June.

♂, South Korea, Apr: although overall plumage, especially head pattern, secondary-coverts and tertials, are like that of ad ♂, quality of remiges is slightly uneven or ambiguous to alternatively suggest an 'advanced' 1stS ♂. Tertials strikingly patterned black and rufous, but wing-bars and olive-grey lesser coverts in this bird rather weakly expressed. (A. Audevard)

IDENTIFICATION The size of Reed Bunting but perhaps somewhat more compact and short-tailed. The conical *bill is rather strong and long* with a *straight culmen*. Sexes are quite similar but ♂ is often slightly neater and has a more contrasting head pattern, the ♀ being slightly duller with less pure head colours. Adult ♂ (only minor seasonal variation) has a largely *black head with pure lemon-yellow eyebrows* becoming white at rear, a *narrow white median crown-stripe* and an isolated *white spot on rear part of blackish cheeks*. Some ♂♂ have slightly less black lateral crown-stripes, while cheek has some dark brown at the centre, and these are very similar to neatest ♀♀. Throat whitish with prominent black lateral stripes, and *pure white submoustachial stripes*. Breast and flanks pale brownish-white (or light brownish-grey), finely but *distinctly streaked black*, centre of belly white. Mantle medium grey-brown streaked black, with a tendency to have *colder grey-brown sides to mantle* but more reddish tinge on its centre, scapulars and tertial-edges. *Rump is unstreaked chestnut* with slightly paler edges when fresh, much as Rustic Bunting. *Two narrow whitish wing-bars*, the upper on median coverts purest white and most prominent. Upper mandible largely dark grey, *lower mandible pink*. Legs pinkish. Juvenile, a rather unlikely straggler to Europe, lacks the bold head pattern of older birds, has brownish lateral crown-stripes and entire crown and much of cheeks streaked dark.

VOCALISATIONS Song a short, pleasant phrase, repeated with little variation. Often recognised by the initial two or three softly 'bent', drawn-out, slow, whistling notes, the first one low-keyed and the following one or two a single octave higher (if two, these are identical), followed by a few shorter and quicker notes, often ending in a Redstart-like softly whistled 'diphthong'. Typical song can be rendered *twaooh tweeh tweeh-tse-tse-tse tsuih*. Could possibly be confused with song of Little Bunting (usually longer and more varied in pipit fashion) or Chestnut Bunting (has more scratchy or trilling notes, and often opens with three similar notes). – Call, like so many Siberian buntings, a short, sharp clicking *zick*.

SIMILAR SPECIES From *Little Bunting* by larger size, much heavier, longer bill with pink lower mandible and strong yellow tinge on eyebrows. Note also whitish submoustachial stripes and quite dark brown cheeks with black outline, if not even uniformly black cheeks. – Differs from *Rustic Bunting* even in rather similar immature plumages by having at least some yellow on fore supercilium, and black streaking to slightly brown-tinged breast and flanks, not chestnut streaks and blotches on pure white ground. Also, note hint of chestnut necklace in Rustic, whereas Yellow-browed has greyish nape and upper mantle. – Among several extralimital relatives in the Emberizidae, ♀ *Tristram's Bunting* (*E. tristrami*; not treated) resembles ♀ and immature Yellow-browed, but differs by lack of yellow on fore supercilium and buffier submoustachial stripe. Also has a longer tail. – There is a superficial resemblance between Yellow-browed Bunting and extralimital adult North American *White-throated Sparrow*, but note the latter's larger size, lack of underparts streaking and brown-grey rather than chestnut rump.

AGEING & SEXING Ages differ only in autumn; reliable ageing in the field usually requires close views. Sexes rather similar, at times inseparable, but most differ following post-juv moult. Size often helpful when plumage is intermediate. – Moults. Poorly understood and mainly inferred from specimens not actively moulting, but a few moulting birds examined. Complete post-nuptial moult of ad after autumn migration, mainly (Aug?) Sep–Oct (Nov?). Some birds in Sep had still not started and were heavily worn. First birds with completed moult in Oct. Partial post-juv moult at same time (or earlier?) affects head, body, tertials, a varying number of central tail-feathers and some wing-coverts, but apparently does not involve any primaries, secondaries (except perhaps sometimes 1–2 innermost) or primary-coverts. Whether all or only some inner greater coverts are replaced requires further investigation; both strategies seem to occur. There does not seem to be any extensive pre-nuptial moult in late winter, but specimens in Jan and Mar are still very fresh, and one (in NHM) had growing s6 in Mar (moult aberration?).

1stY ♂, Sweden, Jan: note on left image 'large-headed' jizz of this smallish bunting, plus strong conical bill with dusky-grey culmen and pink lower mandible, but general shape dependent on posture, as right image demonstrates. What makes ♂ unmistakable is broad bright yellow supercilium, progressively becoming whiter at rear; there is also some yellow on chin and upper submoustachial stripe. Black crown has narrow but distinct whitish median stripe, and dark ear-coverts brownish-tinged. Retained juv wing-feathers and pointed tail-feathers confirm the age of this bird. (M. Bergman)

According to *BWP*, there is a partial moult affecting head and forebody. – **SPRING** Ageing difficult except for very neat birds without much wear to wings and tail being ad, those that are heavily worn being 1stS. Requires practice and a careful approach. ♂ Sides of crown jet-black. (If any brown present in the black, only scattered small tips.) Ear-coverts mainly black too, but some variation (perhaps linked to age) in that some have black admixed with much brown at centre. Eye-stripe nearly always black. Upperparts on average more rufous-tinged (but some overlap with ♀). Entire underparts usually white, or very nearly so (some having a faint brown-buff cast on breast and flanks), streaked black. Yellow of supercilium clear lemon. ♀ Sides of crown rufous streaked black (about equal proportions of black and brown). Ear-coverts brown with some paler brown and dark brown-black admixed. Eye-stripe dark brown (not black). Upperparts on average less rufous, more earth-brown. Most of underparts white, but at least flanks and breast-sides brown-tinged, sometimes whole breast. Supercilium either pale yellow-white or equally clear yellow as ♂. – **AUTUMN** Sexing as in spring, but 1stW ♂ often difficult to separate from ♀. Some, probably mainly ♂♂, have faint yellow cast on forepart of median crown-stripe, but this bleaches and becomes white in spring. **Ad** Iris deep rufous-brown. Longest primaries and all tail-feathers quite fresh once complete moult finished, tips of tail-feathers slightly rounded. Rump and uppertail-coverts invariably unstreaked rufous. **Ad ♂** Much black visible on crown-sides and ear-coverts. Median crown-stripe white (often admixed with a few thin dark streaks). **1stW** Iris dark grey-brown. Tips of longest primaries and tail-feathers often slightly worn, tips of latter pointed. Rump and uppertail-coverts either unstreaked rufous as in ad, or streaked and less intense rufous (like juv plumage). **1stW ♂/♀♀** Limited black visible on head. Central crown-stripe not pure white, streaked dark throughout. Crown-sides a mix of black and brown. Ear-coverts more brown than black. Throat and breast off-white, latter usually tinged buff-brown, densely streaked dark. Ad ♀ can be distinguished if ageing possible. **Juv** Resembles dull ♀ but often separated by less contrasting head pattern with brown lateral crown-stripes and entire crown streaked dark brown. Rump always streaked dark. All tail-feathers pointed.

BIOMETRICS L 14–15.5 cm; **W** ♂ 75–84 mm (*n* 26, *m* 79.7), ♀ 71–78 mm (*n* 20, *m* 75.1); **T** ♂ 57–66 mm (*n* 24, *m* 62.6), ♀ 55–64 mm (*n* 18, *m* 60.1); **T/W** *m* 79.4; **B** 12.3–15.0 mm (*n* 33, *m* 13.4); **B(f)** 9.3–11.5 mm (*n* 38, *m* 10.4); **BD** 6.9–8.0 mm (*n* 33, *m* 7.4); **Ts** 18.0–21.0 mm (*n* 27, *m* 19.2). **Wing formula: p1** minute; **p2** < wt 0.5–2 mm; **pp3–5** longest, or p5 < wt 0.5–2 mm; **p6** < wt 2.5–5.5 mm; **p7** < wt 7.5–11.5 mm; **p10** < wt 16–20 mm; **s1** < wt 17–21.5 mm. Emarg. pp3–6.

GEOGRAPHICAL VARIATION & RANGE Monotypic. – C & E Siberia, from Baikal north to *c.* 65° and east to upper and middle Lena; winters SE China.

♀, China, May: similar to ♂, only duller and less bright, supercilium being paler yellowish, and ear-coverts and lateral crown-stripes much paler and browner. Note typical white rear spot on ear-coverts. Mottled buffish rust-brown and black-streaked crown may show pale central stripe. Wear to wing difficult to safely assess, hence best left unaged. (M. Parker)

Variation in 1stS ♀♀ (left: South Korea, May; right: China, Apr): 'advanced' young bird on left very similar to ad ♀ in spring and summer, including head pattern, but moult limits (older feathers juv) in alula, greater coverts (inner three being post-juv), odd median coverts and tail; also, primary-coverts appear to be juv being brown, abraded and rather pointed. 'Less advanced' bird (right) lacks obvious yellow pigments to supercilium. Note dull rufous-brown rump and uppertail-coverts found in this species. (Left: A. Audevard; right: J. Martinez)

Ad ♀, China, Nov: tail, tertials, visible median and greater coverts, and primary-tips all ad-like in shape and quality. Long thick bicoloured supercilium and bold white rear ear-coverts spot in otherwise brownish head indicate both species and age. Note diffusely dull brownish rump and uppertail-coverts. (M. & P. Wong)

Presumed 1stW ♂, China, Oct: although ageing as 1stW here tentative and mainly based on seemingly juv type of primary-coverts, general plumage fits best with young autumn ♂. Note fairly contrasting head pattern with quite yellow supercilium enhanced by black markings. Often gives a compact impression with peaked crown and heavy bill. (I. Fisher)

RUSTIC BUNTING
Emberiza rustica Pallas, 1766

Fr. – Bruant rustique; Ger. – Waldammer
Sp. – Escribano rústico; Swe. – Videsparv

One of the many 'taiga buntings', with a vast range from Fenno-Scandia to the Pacific, the Rustic Bunting favours wet patches of closed forest, often with sparse spruces mixed with birch and willow and rich shrubs of marsh tea and lingonberry sprigs. It is a discreet bird, often first noticed by its softly-whistled song, which has a ring of its own and is easily picked out. Once breeding is finished in NW Europe it embarks on its long migration, first due east through Siberia, keeping to wooded habitats, then south through E China to reach its winter quarters in SE Asia.

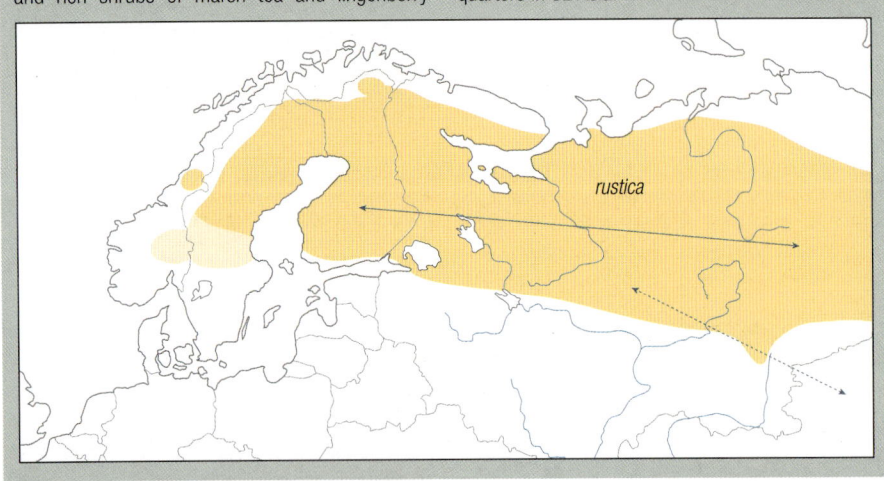

E. r. rustica, ♂, presumed ad, Finland, May: quite robust bunting with hint of crest raised when singing and characteristic black-and-white head and bold chestnut breast and flanks streaking, and by and large unstreaked chestnut rump and scapulars. Ageing of this bird uncertain as primary-coverts and tail cannot be properly assessed. (D. Occhiato)

IDENTIFICATION A small to medium-sized bunting, compact (like Reed and Little Buntings) with a rather long conical *bill with straight culmen*. The bill is usually *bicoloured with dark grey upper mandible and pink lower*. *Legs pinkish*. Feathers of crown are slightly elongated and frequently raised to form a *hint of a crest*, rendering the bird a *flat-crowned* impression when relaxed. In all plumages, *rump, uppertail-coverts and flanks-streaking chestnut*, and adults and first-winter ♂ also have dense chestnut breast-streaking, the streaks on breast often merging into a *chestnut band or patch*. The chestnut breast-band and flanks-streaking (latter broad and lacking any black elements) typically contrast strongly against *pure white rest of underparts*. Adult ♂ has a striking head pattern, *black with prominent white supercilium and narrow median crown-stripe*. A small white spot is usually isolated at the bottom rear part of the black cheeks. Most have *narrow chestnut lateral throat-stripes*, separating the white throat from *white submoustachial-stripes*, but a few lack them and look particularly white-throated. Some adult ♂♂ are *very reddish-brown on upper mantle*, this becoming nearly uniform chestnut, with only a hint of black and pale brown streaking. Adult ♀ has less black head; only lateral crown-stripes are blackish or very dark grey, *cheeks being brown-grey* with darker margins, supercilium and crown-stripe are less pure white and less distinct, lores not all dark, and mantle is more tawny-brown (boldly streaked black), less rufous. Still, beware of odd, more ♂-like, ♀♀ (and conversely ♂♂ that are slightly less developed and neat, approaching ♀ plumage); sexing requires care. In autumn all plumages rather similar since the smart head pattern and chestnut breast-band and nape of adults are partly concealed by ochre-buff or brown margins. Typically, has *yellowish-buff supercilium and submoustachial stripe*, a *white wing-bar on median coverts* and a less conspicuous buffish second wing-bar on greater coverts. As in spring, chestnut streaking on breast and flanks contrasts sharply with pure white rest of underparts. Adult ♂ shows most traces of the brighter summer plumage, while first-winter ♀ has very little chestnut at all. Flight rather strong, slightly twitching or skipping. Restless and active, often flicking wings and tail.

VOCALISATIONS Song a very pleasant jaunty 'yodelling' or irresolute warble, a rapid series of mellow whistled notes on a rather low pitch that can actually recall Garden Warbler a little. The phrase is repeated with no or only very minor changes and only brief pauses. An average song can be rendered *duu-delee-diidu-deluu-delee-dee*do. – Call is a fine, sharp, short, high-pitched but not very loud *tsee* or *zit*, a bit like the call of Song Thrush.

SIMILAR SPECIES Roughly same size and shape as *Reed Bunting* but differs—also in autumn when their plumage patterns are more similar—in neater and more contrasting general pattern; rufous, partly coalescing, streaks on breast and flanks in contrast to white belly (Reed can have partly rufous streaking on flanks but they are narrower and admixed with some more blackish streaks); bill has straight culmen and pink lower mandible (Reed: nearly uniform greyish bill with usually slightly convex culmen); two pure white wing-bars (Reed: invariably rufous-brown or ochre tips to median and greater coverts, not forming very prominent wing-bars); hint of chestnut nape and/or necklace (never on Reed); chestnut-tinged rump and uppertail-coverts (Reed: grey-brown or tawny-brown, heavily streaked); and often visible small crest. Sides of crown in Rustic brown-streaked or blotched blackish, never tinged reddish-brown like some Reed Buntings. A couple of particularly strongly marked paler buff-brown central stripes along mantle/back is a sign of Reed; Rustic is also striped on upperparts, but contrast not as strong. Strong rufous-ochre edges to folded secondaries is another sign of Reed Bunting, whereas Rustic has less reddish and thinner edges. – *Little Bunting* is a smaller bird, but this is not always obvious, making it useful to note pink lower mandible (Little: all-grey bill), duller brown-grey cheeks (Little: ochre or rufous) and chestnut flanks-streaking (Little:

E. r. rustica, ♂, presumed 1stS, Finland, Jul: ♂ in summer has striking black crown and ear-coverts, white supercilium and throat, rufous nape, breast-band and flanks streaking, and silky-white underparts. Crown may show slight pale median stripe and variable whitish spot on rear ear-coverts. Alula and primary-coverts appear quite brown and worn suggesting 1stS, but ageing in spring often difficult. (M. Varesvuo)

distinct black streaks). – From similar-sized *Yellow-browed Bunting* by lack of any lemon-yellow on supercilium (but can be warm buffish), presence of at least a hint of rufous on neck/nape and rufous streaking on breast and flanks.

AGEING & SEXING (*rustica*) Ages differ in autumn, but normally not in spring; reliable ageing in the field usually requires close views. Sexes usually differ after post-juv moult, but extremes are very similar and can even be almost inseparable. Seasonal variation through fresh feather-tips being brown or buff partly concealing brighter breeding plumage. – Moults. Complete post-nuptial moult of ad after breeding but before autumn migration, mainly Jul–Sep (rarely end Jun). Partial post-juv moult at same time (or starting slightly earlier) affects head, body and many wing-coverts but does not involve any primaries, secondaries, tail-feathers (except r1 in some) or primary-coverts. Both age categories usually have a partial pre-nuptial moult in late winter confined to parts of head and throat. – **SPRING** ♂ (The following criteria separate most birds, but a few must be left un-sexed in the field due to intermediate characters.) Crown and ear-coverts usually unmarked jet-black. Lores and forehead black. Chestnut lateral throat-stripe narrow, ending well short of base of lower mandible (rarely missing altogether). Chestnut band across breast broad and dark, widening in centre. (Rarely, black on head less uniform, approaching ♀ pattern.) ♀ (A few attain more ♂-like plumage, difficult to sex reliably.) Crown and ear-coverts brown-grey, usually with black streaks or patches. Lores and forehead brown with rufous and/or buff tinge and often some black spots. Lateral throat-stripe mix of rufous, brown and dark grey, usually rather prominent and reaching close or virtually to base of lower mandible. Breast-band like ♂ but less dark chestnut and often not as broad and well defined, partly more streaked. – **AUTUMN** Sexing and ageing often difficult due to head pattern partly concealed by fresh ochre-buff feather-tips. If ageing possible, most typical ad ♂ and 1stW ♀ can often be identified. **Ad** After completion of moult, longest primaries and all tail-feathers quite fresh, tips of tail-feathers slightly rounded, edges neat. Much rufous on lower nape (though partly concealed by buff tips). Those with much black (partly concealed by thin ochre-buff fringes) on sides and rear of crown are ad ♂♂. **1stW** Tips of longest primaries and tail-feathers slightly worn, tips of latter pointed, extreme tips sometimes 'frayed'. Some pale rufous on nape in some, none in others, concealed by broad buff tips and admixed with some dark streaking. Those with very little black on sides and rear of crown limited to narrow central streak on

E. r. rustica, ♀, Finland, May: a dull bird with dark areas of head browner and fringed buff. ♀ usually has thin dark lateral throat-stripe, and reddish on nape, rump and breast less bright, and less extensive than most ♂♂. Ageing difficult in this view. (M. Varesvuo)

E. r. rustica, 1stS ♀, Finland, Jun: the least striking plumage, which combined with heavily bleached juv primaries, primary-coverts, alula and tail infer 1stS ♀. Still, underparts white with same bold chestnut streaking as in other plumages. Note unstreaked chestnut rump just visible. (J. Peltomäki)

E. r. rustica, 1stW ♂, England, Nov: combination of clearly juv primaries, primary-coverts, alula and most of greater coverts with rather black frame to crown and ear-coverts, and rather deep and extensive rufous above, indicative of age and sex. (S. Ray)

E. r. rustica, 1stW, presumed ♀, Netherlands, Oct: aged as 1stW on narrow and pointed tail-feathers. Reduced black on head, and duller and less extensive chestnut above (including uppertail-coverts) and below, suggest 1stW ♀. (R. Pop)

each feather, and with no (or practically no) rufous visible on nape, are 1stW ♀♀. **Juv** Resembles dull ♀ but often has diffuse blotches or streaks on breast (rather than more uniform bib), and streaks are mix of blackish and rufous, not pure chestnut. Head pattern slightly more blurred than in post-juv plumages.

BIOMETRICS (*rustica*) **L** 13.5–15 cm; **W** ♂ 75–81 mm (*n* 19, *m* 79.2), ♀ 74–78 mm (*n* 12, *m* 76.5); **T** ♂ 55–61.5 mm (*n* 19, *m* 58.8), ♀ 55–61 mm (*n* 12, *m* 57.8); **T/W** *m* 74.8; **B** 11.9–13.2 mm (*n* 32, *m* 12.6); **B**(f) 9.0–10.8 mm (*n* 32, *m* 9.8); **BD** 5.2–6.7 mm (*n* 32, *m* 6.0); **Ts** 17.0–19.2 mm (*n* 29, *m* 18.3). **Wing formula: p1** minute; **p2** < wt 0.5–2 mm; **pp3–4** about equal and longest; **p5** < wt 1–2.5 mm; **p6** < wt 5–8 mm; **p7** < wt 11.5–14.5 mm; **p10** < wt 19–22 mm; **s1** < wt 20–23 mm. Emarg. pp3–5.

GEOGRAPHICAL VARIATION & RANGE Variation slight. Two races described, differing only subtly.

E. r. rustica Pallas, 1776 (N Fenno-Scandia, N Russia, Siberia except extreme east; winters SE Asia, mainly E China). Described above. Separation from following race generally possible only with neat ♂♂ and direct comparison. (Syn. *borealis*.)

Extralimital: *E. r. latifascia* Portenko, 1930 (Anadyr, Kamchatka, possibly south to N Amur; winters E Asia, perhaps mainly Japan and South Korea). Differs in ♂ summer plumage by being slightly darker and more contrasting, whole crown from base of bill to nape solidly jet-black, ear-coverts the same, chestnut breast-band on average more complete and darker, sometimes with some black blotches at upper edge, and has a subtly stronger bill. Streaking of back averages bolder. Size appears to average subtly larger, but much overlap. **W** ♂ 77–83 mm (*n* 12, *m* 79.5), ♀ 74–82 mm (*n* 5, *m* 78.0); **T** ♂ 55–63 mm (*n* 12, *m* 59.0), ♀ 55–62 mm (*n* 5, *m* 57.9); **B** 11.7–14.8 mm (*n* 17, *m* 13.1); **B**(f) 9.5–12.0 mm (*n* 17, *m* 10.4); **BD** 6.0–6.8 mm (*n* 17, *m* 6.3).

REFERENCES Bradshaw, C. (1991) *BW*, 4: 309–313.

E. r. rustica/latifascia, ad ♂, China, Oct: evenly fresh following post-nuptial moult, with remiges and rectrices darker, more firmly textured, tail-feathers having neat, rounded tips. Generally has ♀-like head pattern, but black on sides and rear of crown less concealed by ochre-buff fringes. All plumages show rather complex tertial pattern and prominent pale wing-bars (that on median coverts usually whiter and more obvious). (I. Fisher)

E. r. rustica, ♀, presumed ad, Finland, Sep: in mint plumage after apparent complete post-nuptial moult inferring age, and since crown is only narrowly black-streaked, and nape has very little rufous visible, it is fair to conclude that the bird is ♀. Note prominent rufous flank-streaking. (H. Taavetti)

E. r. latifascia, presumed ♂, Japan, Nov: darker and more contrasting than ssp. *rustica*, with on average bolder overall pattern and deeper and more extensive chestnut. Too few wing-feathers visible to safely age this bird, making sex also less straightforward; still, relatively bright plumage and boldly patterned head strongly indicate ♂. (T. Shimba)

E. r. rustica, 1stW, Scotland, Oct: typical autumn view of the species, crown feathers slightly raised to form hint of crest, bold chestnut breast and flank streaking contrasts with unmarked white centre of belly. Buff-white supercilium and submoustachial stripe prominent, and double whitish wing-bars also striking. (H. Harrop)

LITTLE BUNTING
Emberiza pusilla Pallas, 1776

Fr. – Bruant nain; Ger. – Zwergammer
Sp. – Escribano pigmeo; Swe. – Dvärgsparv

One of several taiga-living buntings that can brighten up your walk in these remote forests, and one of the smallest. It has a delightful song and a discreet but attractive plumage. Together with the Arctic Warbler and Rustic Bunting, this small bird performs one of the longest migrations of all, from N Fenno-Scandia to SE Asia. Winter quarters in S China, Myanmar, Vietnam and India are reached after first flying mainly east, turning south when reaching C Siberia and Mongolia. It arrives at its most distant breeding sites only in June.

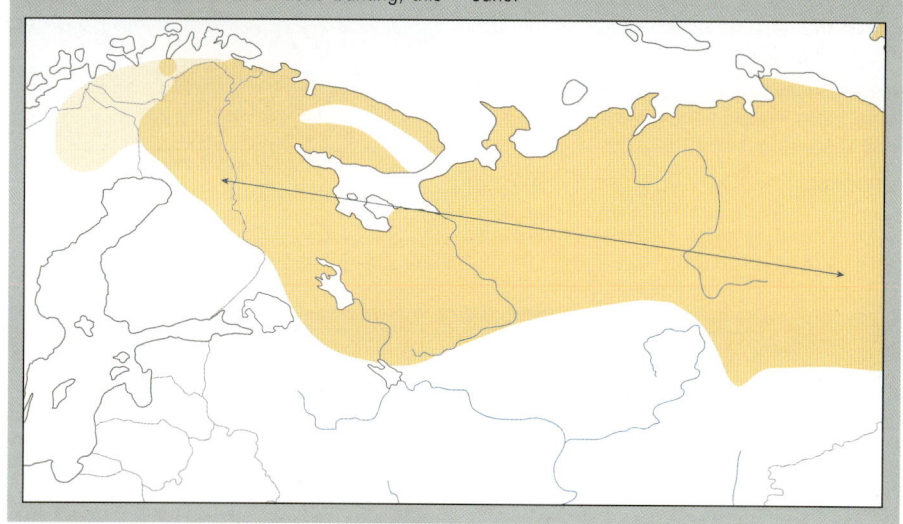

Presumed ad ♂, Finland, May: tiny bunting, in all plumages characterised by chestnut cheeks and pale eye-ring, with black lateral crown-stripes and black upper and rear borders to ear-coverts. Has contrasting whitish underparts prominently streaked blackish on breast and flanks. Sexes very similar or indistinguishable except when seen together. Rounded tips to tail-feathers infer age. (H. Harrop)

IDENTIFICATION A *small* bunting the size of a Goldfinch, appearing rather *compact* (not especially long-tailed) with a *small grey, pointed bill with straight culmen*. Upperparts including rump grey-brown, boldly streaked black on mantle, lightly on rump. Median coverts tipped buff-white (bleaching whitish) forming *a wing-bar*. Underparts buff-white with *narrow, distinct black streaking*. In all summer plumages *chestnut on face, crown and sometimes chin/upper throat*. Note that both sexes can have chestnut on chin/throat, but those with most extensive and deepest colour appear to be always ♂♂. Broad jet-black (♂) or dark grey (♂/♀) *lateral crown-stripes border the chestnut median crown-stripe*. Chestnut ear-coverts encircled by narrow black stripe, but latter does not reach bill-base or to eye. At lower rear of chestnut cheeks there is often a small paler spot. Submoustachial stripes whitish (or pale buff), well-marked lateral throat-stripes black. An *unbroken whitish eye-ring* gives the facial expression a neat appearance. *Legs pink.* Autumn adults appear very similar to summer birds, but chestnut is more ochre-tinged (due to ochre tips to many feathers). First-winters similar and difficult to age in the field; on average slightly duller and less smart, the median crown-stripe is less solid, more streaked, and chestnut is more subdued and perhaps on average a little more concealed by ochre-buff tips than adults. Flight fast and light, at times skipping on take-off or before alighting. Often feeds on the ground but freely perches in bushes and trees, and invariably sings from high (but usually concealed) perch. Acts 'nervously' and quickly, but does not appear shy.

VOCALISATIONS Song is a pleasant short strophe, high-pitched and melodious with a flow of rather different motifs, rather like a slow Wren, e.g. *setru-setru-setru zrizrizri sveeu-sveeu sisi-sürrr*. Some strophes are longer, and it is actually one of the most accomplished singers among the *Emberiza* buntings, and apart from Wren the song recalls Linnet, Goldfinch and a pipit, but most of all Chestnut Bunting, including softer whistles among more buzzing or trilling notes. – Call is a high-pitched hard clicking *zick* like miniature Hawfinch. The call is very similar, and difficult to separate from, several other Asian buntings. When agitated utters several such calls in series and of differing strength.

SIMILAR SPECIES Differs from ♀ and young *Reed Bunting* by smaller size, pink legs, straight culmen (giving more pointed look to the bill), better-marked paler median crown-stripe (but Reed Bunting can have a hint), duller grey-brown lesser coverts (Reed Bunting: reddish-brown, but can be hidden in wing pockets) and calls. Dark lateral throat-stripe never reaches base of lower mandible (in Reed Bunting invariably so). Whitish eye-ring neater and more obvious than Reed Bunting. – From *Rustic Bunting* by smaller size, fine black streaks on breast and flanks (Rustic: large diffuse chestnut streaks), finer all-grey bill, paler and more chestnut or ochre-tinged cheeks, and lateral throat-stripe never reaches base of bill (Rustic: lateral throat-stripe meets bill). Rustic also tends to have at least part of the rump dark, dull chestnut with no or only indistinct streaks. – *Yellow-browed Bunting* can be eliminated by its white (or at least whitish) median crown-stripe and pink lower mandible, if yellow fore supercilium is not evident (can be nearly missing in first-winter ♀). – Like young autumn *Lapland Bunting* has pale median crown-stripe and ochre-tinged cheeks, there is a remote risk of confusion with Little Bunting, but the considerable size difference, dark-tipped pinkish-yellow bill, longer primary projection and more vividly rufous-edged greater coverts of Lapland Bunting should quickly resolve any problem.

AGEING & SEXING Ages differ in autumn, but usually not in spring; in the field, reliable ageing usually requires close views. Sexes very similar and often inseparable, but at least extremes differ after post-juv moult. – Moults. Complete

♂, Finland, May: head-on, chestnut median crown-stripe visible. Chestnut ear-coverts show pale spot at rear. Especially ♂ can show buffish-rufous submoustachial stripe and throat, but dark lateral throat-stripe missing or subdued. Colourful plumage could suggest ad, but feather generations or possible moult limits difficult to assess, so best left unaged. (J. Tenovuo)

post-nuptial moult of ad after breeding, mainly Jul–Sep. Partial post-juv moult at same time affects head, body and some wing-coverts but does not involve any primaries, secondaries, tail-feathers, primary-coverts or outer greater coverts. A variable number of tertials can be replaced, too. Both age categories have a partial pre-nuptial moult in late winter confined to parts of head and body. – **SPRING** ♂ (The following criteria separate only a minority, most birds being indeterminate as to sex.) Centre of crown and ear-coverts unmarked deep chestnut. Broad lateral crown-stripes rather well marked and solid jet-black. ♀ Only the dullest birds can safely be sexed as ♀♀. Centre of crown dull chestnut, streaked dark. Broad lateral crown-stripes diffuse and brown-grey, often with hint of darker streaking. Ear-coverts dull chestnut, sometimes with some paler brown admixed. Note that both sexes can have some chestnut on chin/upper throat, not only ♂. – **AUTUMN** Sexing as in spring, but even more difficult as head pattern partly concealed by fresh buff feather-tips. **Ad** Longest primaries and all tail-feathers quite fresh, tips of latter slightly rounded. Much rufous on head (but partly concealed by buff tips). **1stW** Tips of longest primaries and tail-feathers slightly worn, tips of tail-feathers pointed. Some pale rufous on head in some, none in others, concealed by broad ochre or buff tips and admixed with some dark streaking. **Juv** Resembles a dull ♀ but often separated by more white-tipped median coverts with pointed black central marks (rather than rufous-buff or ochre-buff tips, and less prominent and less pointed dark central spots). Streaking on underparts usually rather broad, diffuse and brown-grey (post-juv has blackish, distinct and narrow streaking).

BIOMETRICS **L** 12–13.5 cm; **W** ♂ 68–77 mm (n 30, m 72.9), ♀ 67.5–71.5 mm (n 10, m 68.8); **T** ♂ 52–61 mm (n 30, m 56.5), ♀ 52–57 mm (n 10, m 54.4); **T/W** m 77.9; **B** 10.0–12.2 mm (n 40, m 11.3); **B**(f) 8.0–9.9 mm (n 40, m 8.7); **BD** 4.8–5.5 mm (n 32, m 5.2); **Ts** 16.4–18.3 mm (n 36, m 17.3). **Wing formula:** p1 minute; p2 < wt 0.5–2 mm; **pp3–5** longest, or p5 < wt 0.5–2 mm; **p6** < wt 4–8 mm; **p7** < wt 10–12 mm; **p10** < wt 16–19 mm; **s1** < wt 17.5–20.5 mm. Emarg. pp3–5, sometimes slightly on p6.

GEOGRAPHICAL VARIATION & RANGE Monotypic. – N Fenno-Scandia, N Russia, N Siberia east to Anadyr; winters E Nepal, Myanmar east through S & SE China.

REFERENCES BRADSHAW, C. (1991) *BW*, 4: 309–313. – STODDART, A. (2008) *Birdwatch*, 197 (Nov): 29–31. – SWANBERG, P. O. (1954) *Vår Fågelv.*, 13: 213–240. –SVENSSON, L. (1975) *Vår Fågelv.*, 34: 311–318.

Ad ♀, China, Apr: combination of relatively dull plumage, including head pattern, especially for an ad (based on all tail-feathers being comparatively broad with little visible wear) safely sex this bird as ♀. (S. Laukkanen)

1stS ♂, Finland, Jun: brighter chestnut face and scapulars, and bold black face markings suggest ♂. Other images of same bird confirm that outer three greater coverts, alula, primary-coverts and flight- and tail-feathers are juv. Note typical whitish underparts with narrow but black streaks on breast and flanks. (T. Muukkonen)

1stY ♀, Netherlands, Feb (left) and England, Apr: rather dull plumage with limited chestnut colour on head-sides and slightly streaked or mottled median crown-stripe, with dark brown-black and broken-up lateral crown-stripes rather than these being solid black, indicate ♀. Primary-coverts and wing- and tail-feathers juv, latter clearly pointed and somewhat abraded. (Left: A. B. van den Berg; right: D. Robson)

BUNTINGS

Ad ♂, Netherlands, Dec: in autumn, black feathers on head-sides fringed rufous, other feathers of upperparts with buff or grey-olive, partly obscuring sexual differences, but brighter ♂ still identifiable. Buffish-rufous submoustachial stripe, and heavily reduced dark lateral throat-stripe. Wing evenly fresh in Dec infers age. In fresh plumage, underparts, especially flanks, sullied warmer. (A. Ouwerkerk)

Ad ♀, China, Oct: ageing as image to the left (note broad fringes to primary-coverts and very fresh primaries with distinct white tips), but being a ♀ has streaky crown without obvious black element, and greyer rest of upperparts. (I. Fisher)

1stW ♂♂, Germany, Oct (left) and Belgium, Jan: fresh bird in Oct (left) and another bleached duller in midwinter (right). Blacker head markings and relatively bright chestnut cheeks (at least left bird) indicate ♂♂, while age confirmed by pointed tail-feathers (left) and moult limit in tertials (right). (Left: A. Halley; right: V. Legrand)

1stW ♀, England, Oct: dullest plumage in autumn, with browner lateral crown-stripes and buffier central crown. Also, chestnut of cheeks dull and subdued. Most if not all of greater coverts juv. (M. Goodey)

— 547 —

CHESTNUT BUNTING
Emberiza rutila Pallas, 1776

Fr. – Bruant roux; Ger. – Rötelammer
Sp. – Escribano herrumbroso; Swe. – Rödbrun sparv

The size of Little Bunting, and also a taiga species, preferring mixed woodland. It has a more easterly and southerly distribution than Little Bunting, being confined to SC and SE Siberia and Transbaikalia, not breeding further west than the Krasnoyarsk region. Its range is probably the reason why it is only an extremely rare vagrant to Europe, but there are now several records from NW and S Europe, which are mostly presumed to involve genuine stragglers from Asia. In ♀-like plumages superficially similar to the more widespread Yellow-breasted Bunting.

Ad ♂, China, May: very distinctive in breeding plumage, being basically bicoloured, with uniform chestnut head to upper breast, upperparts and wing-coverts, and pale yellow lower breast to undertail-coverts, with diffusely dusky-streaked flanks. Black-centred tertials just visible. All wing evenly feathered and rather fresh. (R. Schols)

IDENTIFICATION A small, fairly short-tailed bunting in which the adult ♂ is most distinctive, having a *bright rufous hood, breast, back, rump, wing-coverts, tertials and inner secondaries*, in contrast to *bright yellow underparts* with flanks streaked olive-grey. Flight-feathers and tail blackish. Bill is rather *pale brown-grey* with pinkish base to lower mandible, *pointed* and has straight slightly darker culmen. ♀-like plumages on the other hand can be difficult to separate from corresponding plumages of Yellow-breasted Bunting (see Similar species). Head and upperparts generally greyish-buff, the head being rather streaky, but note contrasting *rufous rump* and more ill-defined head pattern. ♀ has *yellowish* (finely streaked) *underparts*, though *throat is buffish or whitish*, never yellow. Dark frame to ear-coverts, lateral crown-stripe, pale supercilium and wing-bars are all rather subdued, while *lateral throat-stripe is relatively stronger*. Unobtrusive when feeding on ground, but escapes to tree canopy if disturbed, and sings from high perch.

VOCALISATIONS Song perhaps unlikely to be heard in the covered region, but is a short strophe recalling a section from Olive-backed Pipit, nearly invariably starting with three slow, pure whistling notes (or double notes) followed by a quicker series of scratchy notes and ending in a soft, mellow flourish in Redstart fashion, e.g. *tvee*a *tvee*a *tvee*a *sre-sre-sra-seesichawey*. Must be separated in the first place from songs of Little and Yellow-browed Buntings. – Commonest call very similar to Little Bunting, a short monosyllabic, clicking *zic* or *zit*. Also gives a thin high *tseep*.

SIMILAR SPECIES Adult ♂ unmistakable, but ♀-like plumages recall *Yellow-breasted Bunting* and may be hard to separate if diagnostic unstreaked, bright rufous rump or all-dark tail-feathers not seen (at most a small white patch on inner web of r6). Yellow-breasted has duller (less rufous) well-streaked rump, and quite obvious white tail-sides (although white may be concealed on a folded tail). Furthermore, Chestnut Bunting has a duller head, with less bold and contrasting lateral crown-stripes, less clean, dusky yellow-buff supercilium, dull, rather uniform dusky-olive ear-coverts (with indistinct dark border). Thus, pale facial surround less marked, though pale submoustachial stripe is emphasized by relatively prominent, short blackish lateral throat-stripe. In Yellow-breasted crown-stripes and supercilium, and dark frame to pale ear-coverts, are prominent, but lateral throat-stripe is often weaker. In Chestnut, throat is usually creamy-white or buff (only rarely tinged yellow) and often slightly paler than rest of underparts including pale yellow undertail-coverts (which are usually whitish or buffish, only rarely yellow in Yellow-breasted). Breast usually mottled rufous-brown, and generally warmer sides have soft dusky-olive streaks on upper flanks (streaks on Yellow-breasted usually more extensive, darker and bolder). Upperparts of Chestnut less boldly striped dark, and lack distinctive yellowish-buff lining, while wings are duller without bright wing-bars (especially median covert bar highly distinct in Yellow-breasted). – Chestnut Bunting's small size matches *Black-faced*, *Little* and *Pallas's Reed* Buntings, but rufous rump, yellowish belly and almost uniformly dark tail are unlike any of these. – Inexperienced observers should bear in mind that ♀ *Yellowhammer* can appear superficially similar, but ♀ Chestnut is smaller, with a shorter tail, different bill shape, less streaked underparts, and almost completely lacks white in the tail. Also, the throat is never yellow.

Spring ♀♀, presumed ad, Korea, May (left) and India, Mar: relatively featureless, but still has diagnostic and contrasting rufous rump and uppertail-coverts, with vivid yellow to pale buffish-yellow underparts diffusely streaked dark. Rest of upperparts and head brown- or greyish-olive streaked blackish-brown. Rather weak head pattern with darker head-sides, ill-defined pale median crown-stripe and usually rear cheek spot (right). Variable amount of rufous on head and scapulars, but in ♀♀ none or very little rufous on chin and throat. Relatively fresh wing of left bird plus apparently strongly yellow underparts indicate ad, but moult pattern and feather generations impossible to evaluate for reliable ageing. Note that in this species even ad can have quite pointed tail-feathers (left). (Left: A. Audeward; right: J. Kuriakose)

BUNTINGS

AGEING & SEXING Ages differ in autumn and for most ♂♂ in spring; reliable ageing of ♀♀ usually requires close views and experience. Sexes sometimes differ in autumn after post-juv moult, at least when handled. Sexes often differ in spring, although many ♀♀ and 1stS ♂♂ cannot be separated. Rather obvious seasonal variation. – Moults. Complete post-nuptial moult in ad after breeding, usually late Jul–Sep; some reputedly commence migration before moult is finished (Olsson in Byers *et al.* 1995). Partial post-juv moult at the same time includes head, body, all lesser coverts and most or all median coverts, usually a few innermost greater coverts and some tertials and alula. Ad ♂ apparently attains breeding plumage through wear of feather-tips, without pre-breeding moult; occasionally though, perhaps a limited partial pre-nuptial moult in winter quarters in imm ♂♂ (including at least head and nape). – **SPRING Ad ♂** See Identification for general description. May have restricted brown mantle markings and, occasionally, a vestigial white patch on outermost pair of tail-feathers. **Ad ♀** Occasionally some chestnut in wings (mainly on lesser coverts, tertials and secondary fringes), crown and mantle. Ad ♀ with chestnut above separated from similar-looking 1stS ♂ if first correctly aged. Note that usually chin and throat are buff-white in ♀, while 1stS ♂ is more rufous. Large alula moulted to more rounded tip, often being partly or entirely chestnut on outer web. **2ndS ♂** Variable, some almost halfway between ad ♂ and ♀, with streaked mantle, partially chestnut head and breast, paler throat and supercilium, and less uniform rufous scapulars and upperwings, with dark-centred median and greater coverts (median and at least outer greater coverts have pale tips forming faint wing-bars). **1stS ♂** Like ad ♀ with streaked head and upperparts, and similar upperwing pattern (some probably inseparable), but throat and breast of 1stS ♂ often have some chestnut, sometimes quite extensively so on crown. 1stS ♂ usually differs from ad ♀ in retaining at least some juv wing- and tail-feathers, forming clear moult limits, but some birds moult extensively and only leave a few outer greater coverts, these being white-tipped (ad can have paler, buffish, tips to outer webs of otherwise chestnut greater coverts). Large alula unmoulted, pointed and brown-grey. **1stS ♀** Like ad ♀ but usually less rufous visible. Differs from ad ♀ in retained juv wing- and tail-feathers, forming clear moult limits. – **AUTUMN Ad ♂** Fresh off-white fringes to chestnut tracts create mottled appearance, most noticeable on supercilium and throat/breast, but ♂ plumage still obvious. **Ad ♀** Upperparts and breast browner than spring, and overall plumage much paler due to broad pale fringes. **2ndW ♂** Intermediate between ad ♂ and 1stW ♂, with broader fringes more extensively obscuring chestnut. Central throat rufous-buff or pale chestnut. Median coverts have dark bases, and greater coverts more extensive dark centres (faint wing-bars, absent in ad). Lesser coverts chestnut but sometimes tinged olive, as in ♀. In some, mantle and scapulars faintly streaked black. **1stW ♂** Unlike similar ad ♀, wing has predominantly juv feathers. Unlike 1stW ♀ often has some chestnut (mostly concealed) on crown, perhaps also on ear-coverts and breast, and occasionally some chestnut among renewed upperwing-feathers. Differs from 2ndW ♂ in lacking chestnut bases to central throat-feathers, but beware a few more advanced 1stW ♂ (and check age first). Juv tertials often retained (blackish-centred with rufous fringes to outer webs), and any early-renewed feathers are not chestnut over entire outer web as in 2ndW (or older) ♂. Differs from ad ♂ by lack of extensive chestnut in plumage. **1stW ♀** Dullest of all and resembles ad ♀, but lacks any rufous, except on rump, and always has grey-brown lesser coverts. Juv tertials mostly retained. **1stW** In both sexes retained primaries and primary-coverts more weakly textured and slightly worn. Some outer greater coverts also retained, forming moult contrast. Ageing by shape of tail-feathers difficult, as relatively pointed in ad too. **Juv** Not expected to occur in the treated region. More heavily streaked than ♀, especially above and on breast. Rump less bright and head pattern more striking, with more prominent supercilium.

1stS ♂, China, May: advanced 1stS ♂ resembles ad ♂ but rufous of head and breast grizzled with buffish-grey. Mantle and scapulars more like ♀, but show variable number of rufous feathers on mantle, and especially on scapulars. Least advanced young ♂ can even more strongly resemble ♀. Note many old wing- and tail-feathers, the latter vey pointed. (B. van den Boogaard)

Ad ♀, Hong Kong, May: Plumage variation among ♀♀ include those with more olive-tinged upperparts and quite restricted chestnut away from rump, here only a hint on lateral crown-stripes and scapulars. The acutely pointed tail-feathers might lead you to age this as 1stS ♀, but oddly in this species both ad and younger birds have such tail-feather shape. (M. & P. Wong)

Ad ♂, China, Oct: when freshly moulted, chestnut areas are fringed buff, and yellow below partly washed pale olive-grey, thus autumn plumage partly conceals ♂'s very distinctive pattern (I. Fisher)

1stW ♂♂, Korea, Oct (left) and Taiwan, winter: extent of post-juv moult variable, some moulting less extensively (left) before migration, others more so (right). Young ♂ on left shows extensive rufous mainly on rump (here extending down towards vent) and first signs on rear crown, nape and shoulders, and also rather pale yellow underparts. Bird on right, however, is easily sexed on much rufous visible on head, throat, upperparts, rump and uppertail-coverts, and bright yellow underparts. Left bird has retained all juv wing-coverts. (A. Lewis)

1stW ♀, China, Nov: in this plumage most likely to be confused with same sex and age Yellow-breasted Bunting, but smaller, has finer bill and less patterned head with indistinct supercilium, less contrasting dark lateral crown-stripes and less prominent dark border to ear-coverts, whereas lateral throat-stripe relatively obvious. Throat usually cream-white (not yellow as in Yellow-breasted) and contrasts with rest of underparts (softly streaked on flanks). With difficult birds, confirm that tail either shows no white or at the most just a tiny white wedge on outermost feather. (M. & P. Wong)

BIOMETRICS L 13–14 cm; **W** ♂ 70–79 mm (*n* 36, *m* 75.3), ♀ 67–76 mm (*n* 22, *m* 70.7); **T** ♂ 51–62 mm (*n* 36, *m* 56.4), ♀ 49–57 mm (*n* 22, *m* 52.9); **T/W** *m* 74.9; **B** 11.1–14.5 mm (*n* 41, *m* 12.5); **B**(f) 9.5–12.1 mm (*n* 31, *m* 10.2); **BD** 5.8–6.7 mm (*n* 28, *m* 6.1); **Ts** 16.8–19.2 mm (*n* 31, *m* 18.0). **Wing formula: p1** minute; **pp2–5** about equal and longest (but p2 and/or p5 sometimes to 3 mm <); **p6** < wt 6–10 mm; **p7** < wt 9.5–14 mm; **p10** < wt 16.5–21 mm; **s1** < wt 18–22 mm; Emarg. pp3–5.

GEOGRAPHICAL VARIATION & RANGE Monotypic. – SC & SE Siberia, Transbaikalia; winters SE Asia. (Syn. *pamirensis*.)

REFERENCES Peltomäki, J. & Jantunen, J. (2000) *DB*, 22: 187–203. – Votier, S. & Bradshaw, C. (1996) *BB*, 89: 437–449.

YELLOW-BREASTED BUNTING
Emberiza aureola Pallas, 1773

Fr. – Bruant auréole; Ger. – Weidenammer
Sp. – Escribano aureolado; Swe. – Gyllensparv

This was formerly one of the most abundant species in taiga bogs and damp willows, birch copses and thickets along lakes and rivers from Russia eastwards. It also had a westerly outpost in N Finland, but this is now gone, and the species has also decreased dramatically across many areas of European Russia and Siberia, although it is still possible to hear its characteristic and persistently repeated song in many places. The Yellow-breasted Bunting is a migrant, spending the winter mainly in India and S China, and massive hunting in winter may be part of the explanation for the decline. It returns in May or, in the case of birds breeding in NW Russia, not until late May or June.

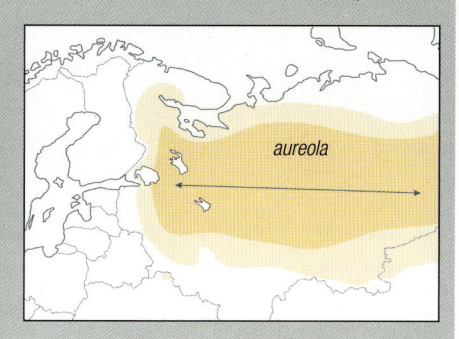

E. a. aureola, 1stS ♂, Mongolia, Jun: 1stS has only partly developed, pale-mottled black face, but usually quite uniform chestnut crown, scapulars and rump, underparts paler yellowish, chestnut breast-band broken and often less extensive or less streaked white shoulder panel. Remiges, primary-coverts and alula are juv, brownish and worn. (H. Shirihai)

IDENTIFICATION Reed Bunting-sized but has a proportionately longer and stronger bill. The adult ♂ is an attractive bird with *bright yellow breast and belly* (undertail-coverts white), the yellow breast with a *chestnut band across it*. The *throat is black*, and the black *continues on the cheeks, lores and forehead*, while *crown and nape are deep chestnut* like the breast-band. *Rump and scapulars unstreaked chestnut*. Mantle and back are generally chestnut, too, but streaked black. Amount of chestnut and streaking on mantle/back varies individually to some degree even in adult ♂, some being slightly duller brown with bolder streaking, others almost unstreaked deep chestnut. There is a *large pure white wing patch* formed by largely white median coverts, and below this a *narrow white wing-bar on tips of greater coverts*. These features suffice to separate the adult ♂ from any other bunting species. First-summer ♂ is similar, only somewhat less neat, lacking (or having a much reduced) white wing patch, black and chestnut on head usually streaked or broken up by a varying number of brown, grey and white feathers, and the chestnut breast-band is narrower and less distinct, even broken in the centre. As a rule, one-year-old ♂ also has hint of a pale supercilium behind eye and less chestnut on mantle/back, where often duller brown. Identification problems are usually limited to ♀♀ and immatures in autumn. These can vaguely recall Yellowhammer being *pale yellow and streaked below*, streaked also on mantle, and rump is tinged reddish-brown at least in some adults. However, note tendency to have a *paler median crown-stripe* due to darker lateral crown-stripes being dark brown streaked black, but central crown paler grey-brown and only lightly streaked. Also *much finer and more limited streaking below*, a fairly prominent *white wing-bar on tips of median coverts* and a narrower and less prominent second bar on greater coverts, a largely *pinkish lower mandible* and pinkish cutting edges to upper mandible (only culmen and tip dark grey) and a *broad buffish or off-white supercilium*. Throat and breast have a buff hue, whereas belly is clearer pale yellowish. *Ear-coverts are rather pale grey-brown encircled by a dark line*. There can be a *hint of a paler spot on rear part of cheeks*. In autumn both adults and first-winters attain a more buffish-tinged, less clear yellow, plumage. Juvenile similar to ♀, only invariably lacks chestnut, being duller brown and more streaked from mantle to rump.

VOCALISATIONS Song quite characteristic due to its structure and 'ringing' quality, most notes being uttered in pairs and the song's 'building steps' having a wide tonal range, commonly working up the ladder until a final drop, *tru-tru trüa-trüa **tri-tri** tra*, or seesawing up and down with a final rise in pitch, *trü-trü tra tro-tro **triih***, with numerous other variations on same theme. Delivered at a steady, rather leisurely pace and all notes slightly trilling or 'scratchy'. – Call a short, sharp clicking *zick*.

SIMILAR SPECIES The remote risk of confusion with *Yellowhammer* is mentioned under Identification. Yellowhammer is a larger and more long-tailed bird that usually shows some olive tones in its plumage and lacks a prominent white median coverts wing-bar. – The smaller ♀ *Chestnut Bunting* is another potential pitfall, but apart from size differs in its lack of whitish wing-bars, much more restricted white on outertail (sometimes none), more prominent chestnut rump including in immature ♀ (duller and more streaked in immature Yellow-breasted), and much duller head pattern without prominent pale supercilium or dark lines around brownish cheeks.

AGEING & SEXING (*aureola*) Ages differ in autumn, and at least in ♂♂ often in spring, but reliable ageing in the field usually requires close views. Individual variation makes some ♂♂ difficult to age reliably. Sexes differ after post-juv moult. – Moults. Complete post-nuptial moult of ad after onset of autumn migration, on stopover (C China) before reaching final winter quarters, mainly Aug–early Oct. Partial post-juv moult at same time and together with ad, usually includes most or all median and greater coverts, and tertials, but not primaries, tail-feathers or primary-coverts. Both age categories undergo limited partial pre-nuptial moult in late winter, at least in ♂♂, mainly confined to head and breast. – SPRING Ad ♂ Chin, ear-coverts, lores and forehead jet-black. Crown to rump deep chestnut, mantle rather finely streaked dark, crown and rump unstreaked. Throat to belly lemon-yellow apart from narrow chestnut breast-band. Median and lower lesser coverts white (or white with very limited dark bases or speckles in some). Wear of tail-feathers helpful if seen well: although ad shape can be quite pointed, wear is usually only moderate during first part of spring. **1stS** ♂ Chin, ear-coverts, lores and forehead dark brownish-grey, or mix of black, grey and white. Sometimes an ill-defined white supercilium behind eye. Crown and rump chestnut, more or less streaked dark. Mantle duller brown than ad, with only some rufous tones and heavily streaked dark. Underparts

E. a. aureola, summer ad ♂♂, Russia, Jun: unmistakable by deep chestnut crown and upperparts, contrasting with jet-black rest of head. Underparts deep yellow, with narrow chestnut bar on upper breast, broader at sides. Diagnostic wing pattern includes prominent white shoulder, and white-tipped greater coverts forming a bar. Entire wing ad in both birds, but bird on right has tail abraded (misleadingly like retained juv). (Left: P. Parkhaev; right: D. Occhiato).

as in ad or somewhat paler, yellowish-white. Breast-band rarely as prominent as ad, more commonly thinner or even broken in centre. Median coverts usually dark brown tipped white or, rarely, almost all white. Lower lesser coverts grey-brown, sometimes speckled white. For difficult borderline cases between ad and 1stS as to upperparts and wing pattern, check abrasion of tail-feathers, these usually being quite heavily abraded, and r1 strongly pointed and narrow, in 1stS. Some ♂♂ best left un-aged. Furthermore, least colourful 1stS ♂ can be extremely difficult to separate from advanced ad ♀, and without positive ageing such birds must also be left un-sexed. ♀ Chin buffish-white (rarely with a little yellow). Throat to belly pale yellow-white, tinged buff on throat and breast. Breast usually has a few ill-defined, scattered streaks (or rarely some chestnut spots). Lores and supercilium dusky-white. Upperparts vary from greyish-brown and heavily streaked, crown and rump sometimes with rufous tinge, to more ♂-like pattern with clear chestnut on rump and some on crown (probably ad ♀♀). Typically, ♀ has crown rufous-tinged and heavily streaked dark laterally, but pale brown-grey on centre with finer streaks forming a diffuse paler median crown-stripe. Ageing should only be attempted with typical plumage variations and if supported by wear of tail-feathers; many intermediate birds are impossible to age. — AUTUMN Sexing largely as in spring. Ad ♂ While still in breeding areas, as in spring, only now heavily worn. After moult on stopover started, much as spring but lemon-yellow underparts now tinged ochre-buff, and black and chestnut on head more blurred and slightly resembles ♀ pattern. Also, largely white lesser and median coverts tipped buff. Feathers of mantle/back typically chestnut with thin buff tips and thin dark central streaks stopping short of buff tips. 1stW ♂ Underparts pale or deep yellow with buff-brown hue on lower throat/breast, no or practically no streaking on breast (any streaks diffuse). Median coverts

E. a. aureola, ♀, Russia, Jul: broad pale supercilium and pale median crown-stripe (just visible) separated by dark brown lateral crown-stripes. Pale grey-brown ear-coverts with darker frame. Upperparts grey-brown dark-streaked and with pale 'braces'. Underparts pale yellowish, streaked dark on flanks but more faintly on breast. Ageing unsure without close views of wing. (P. Parkhaev)

E. a. aureola, ad ♂, India, Dec: when freshly moulted in autumn or early winter, all feathers show grey-buff fringes, partly obscuring bright colours, and plumage, especially head, therefore gives more ♀-like appearance. However, yellow underparts brightest at this season, and breast usually still shows partial chestnut bar, and white wing patches just visible. (A. Das)

E. a. aureola/ornata, ♀, presumed ad, China, Oct: note buffish-tinged broad supercilium, dark frame to ear-coverts. Combination of well-streaked crown with barely evident rufous on rump (rufous not so deep, buff fringes broad) indicates ♀, while all wing- and tail-feathers appear ad-like. Bright yellow throat and duller buffish-yellow underparts with finer streaking at sides support age/sex designation. (M. Hale)

E. a. aureola/ornata, presumed 1stW ♂, India, Dec: ♀-like plumage but deeper rufous on rump/uppertail-coverts indicates young ♂. Age very difficult to assess (especially as post-juv moult is very variable in timing and extent), but perhaps 1stW given already very pointed tail-feathers, and the rather broad white tip to one exposed median covert confirms the sex. (A. Das)

E. a. aureola, 1stW, Scotland, Sep: young ♂ initially often inseparable from ♀, but during course of first winter some develop rufous on breast-sides and brighter yellowish underparts. Nevertheless, this bird still has mostly unmoulted rump/uppertail-coverts, with as yet no indication of rufous of ♂, thus to be safe best left unsexed. (H. Harrop)

dark, tipped buffish-white. Note typical pattern of feathers on mantle/back, which are dark brown, or brown with moderate rufous tinge, and bold black central streak penetrating buff tips and edges. Tail-feathers rather narrow and pointed, in early autumn fresh. ♀♀ Underparts pale yellow, breast and flanks finely streaked dark. Ad ♀ differs from 1stW ♀ in having heavily worn flight-feathers initially, then moulting these on stopover to completely fresh ones, tail-feathers with rounded, neat tips. **Juv** Very similar to ♀ but overall more buff and less clear yellowish. Supercilium yellowish-buff (not off-white) and entire head pattern less contrasting.

BIOMETRICS (*aureola*) **L** 14–15 cm; **W** ♂ 74.5–82 mm (n 32, m 78.4), ♀ 72–77 mm (n 14, m 74.2); **T** ♂ 56–64 mm (n 32, m 59.5), ♀ 54–59 mm (n 14, m 56.4); **T/W** m 75.9; **B** 12.0–15.0 mm (n 43, m 13.4); **B(f)** 9.7–11.5 mm (n 42, m 10.6); **BD** 6.0–7.4 mm (n 40, m 6.7); **Ts** 18.7–21.2 mm (n 43, m 20.1). **Wing formula: p1** minute; **pp2–4** longest, or p2 and p4 < wt 0.5–2 mm; **p5** < wt 0.5–2 mm; **p6** < wt 7–9 mm; **p7** < wt 11.5–14.5 mm; **p10** < wt 19–22 mm; **s1** < wt 19–23.5 mm. Emarg. pp3–5.

GEOGRAPHICAL VARIATION & RANGE Slight and clinal variation. ♂♂ tend to be slightly darker and more saturated in the east. Only two races deemed warranted.

E. a. aureola Pallas, 1773 (NE Europe, W & C Siberia, N Mongolia except extreme north-east, east to Anadyr; winters SE Asia, from Nepal to SW China). Described above. Upperparts of ad ♂ chestnut-brown, throat to breast bright yellow with narrow chestnut breast-band. (Syn. *suschkini*.)

Extralimital: *E. a. ornata* Shulpin, 1928 (E Siberia from Chita east, N Manchuria, Ussuriland, Amur, Kamchatka, Sakhalin, Japan; winters SE Asia, mainly S China). Differs in that ad ♂ has slightly darker chestnut upperparts and deeper yellow underparts. Often, the breast-band is more prominent and blacker. Size appears very similar. Unlike *aureola*, ad moults completely on breeding grounds prior to autumn migration, with a partial pre-nuptial moult in late winter. (Syn. *insulanus*; *kamtschatica*.)

REFERENCES Harrop, H. (1993) *BW*, 6: 317–319. – Peltomäki, J. & Jantunen, J. (2000) *DB*, 22: 187–203.

E. a. aureola, 1stW ♀, Hong Kong, Nov: note combination of pointed tail-feathers and ♀-like plumage indicating age and sex. Note very typical bold head pattern, with hint of paler median crown-stripe just visible, much chestnut on rump and strong bill with typical pink lower mandible and cutting edges. (M. & P. Wong)

E. a. aureola, juv, Russia, Jul: generally ♀-like, but ground colour buffier and streaking on upperparts, breast and flanks more prominent. Post-juv moult has apparently just started. (P. Parkhaev)

♀ Yellow-breasted Bunting (top left: China, May), Yellow-browed Bunting (top right: China, Apr), Chestnut Bunting (bottom left: India, Mar) and Rustic Bunting (bottom right: Japan, Mar): the closest species to Yellow-breasted is Chestnut Bunting, both having yellowish underparts, but latter lacks clear whitish wing-bars and outer tail-corners, and has more prominent chestnut rump (all of the other species can to some extent develop a chestnut rump, but should already show other distinctive ♂-like characters; cf. Rustic image). If Yellow-breasted is seen well, confusion with Yellow-browed or Rustic Buntings unlikely. Beware 'washed-out' Yellow-browed with hardly any yellow on fore supercilium, or conversely, a Yellow-breasted with little yellow on underparts. Rustic less likely to be confused with Yellow-breasted due to Rustic's strong rufous streaks on flanks (but not all ♀-like Rustic have obvious such streaking). At least first three birds apparently 1stS, while Rustic is perhaps ad. (Top left: V. & S. Ashby; top right: J. Martinez; bottom left: J. Kuriakose; bottom-right: A. Halley)

(COMMON) REED BUNTING
Emberiza schoeniclus (L., 1758)

Fr. – Bruant des roseaux; Ger. – Rohrammer
Sp. – Escribano palustre; Swe. – Sävsparv

The Reed Bunting is a characteristic sparrow-sized bird of damp habitats with low vegetation. It is found in reed marshes, lakeside sedges with scattered bushes and Salix bogs in alpine habitats. It is widespread and has developed a rich variety of subspecies that differ markedly in size, bill shape and general colours. Most populations in the north and east are migrants, moving to warmer climes in winter, mainly in W and S Europe, whereas southern birds are more or less resident or only short-range migrants. It is quite an abundant breeder in N Europe and is accordingly a commonly seen migrant in autumn further south. Being hardy it returns early, usually in March and early April, to its favoured breeding sites.

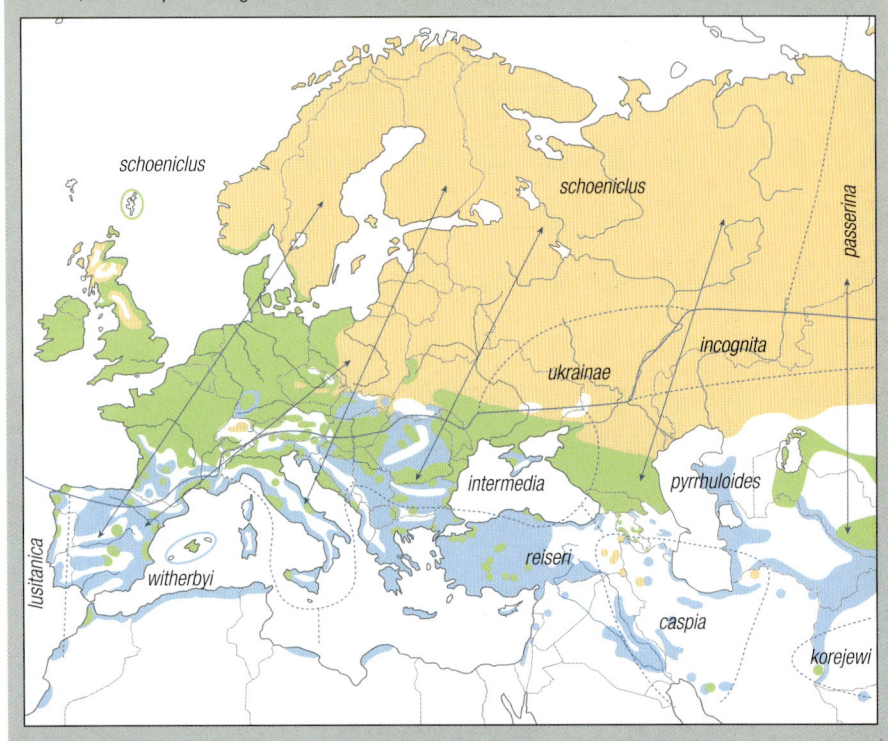

IDENTIFICATION Medium-sized to large bunting with *greyish or dull brown rump* and some *white on tail-sides*. European adult ♂ in summer unmistakable with *black head and bib* and *white collar and submoustachial stripe*. Upperparts brown-grey and rufous streaked black, although rump is an unstreaked mouse-grey (very pale buff-white or grey-white in some eastern populations). Wings are slightly warmer tawny or rufous-tinged brown, and *wing-bend and lesser coverts* (if visible) *are vivid rufous*. Tips to median and greater coverts are ochre, bleaching pale buff in summer, but usually inconspicuous, i.e. Reed Bunting has *no prominent pale wing-bars*. Underparts are dusky off-white with a faint brownish cast, nearly unstreaked with *only a few streaks on flanks* (virtually unstreaked in eastern populations). *Legs* variable in colour, anything from pinkish-brown to red-brown or even brown-black, but *invariably look rather or very dark* in the field (unlike e.g. Rustic or Little Buntings). *Bill* of variable size and shape, but in NW Europe *rather small and mainly dark greyish*, black in breeding season. In S and E Europe has heavier, more bulbous-shaped bill, with those in SE Europe and Turkey, and breeders in Central Asia, very different because of large rounded bill, large size and pale plumage. In autumn, when freshly moulted, much of the neat head and throat pattern is concealed by brown or buff-white feather-fringes, but such birds often have visible, broad and extensive (rounded) black feather-centres on crown and lower throat. ♀♀ (year-round) and many immature ♂♂ in autumn lack the striking head pattern of summer ♂, having *crown brown streaked dark*. Brown-grey or *rufous-tinged brown cheeks have darker margins*, and there is a *whitish submoustachial stripe* and a *conspicuous blackish lateral throat-stripe* that broadens on side of upper breast into a dark patch. Supercilium off-white in summer but ochre-buff in autumn. Flanks more heavily streaked dark than in adult ♂, and rump is tinged brownish, less grey, and lightly streaked or mottled dark. More advanced immature ♂ already has emerging ♂ head pattern. Juvenile similar to autumn ♀. When flushed from ground or low vegetation, often towers to some height in skipping, energetic flight and drops into cover after some distance, usually with sudden steep dive.

VOCALISATIONS Song a brief, slow verse of rather shrill and metallic notes. It opens with a few simple notes at slow pace, then makes a half-stop and concludes with a slight acceleration and a little flourish, or a quick repetition of a few

E. s. schoeniclus, ♂♂, Switzerland, May (left) and Finland, Jun: typical breeding plumage with blackish head and bib, prominent long white submoustachial stripe, and conspicuous white hind-collar. Upperparts generally streaked blackish-brown and fringed buff or rufous, except greyish rump. Wings fringed deep rufous (no pale wing-bars in any plumage). Underparts mainly whitish, with ill-defined streaks on flanks varying in width and extent. Right-hand bird appears ad by rather broad tail-feathers and only moderate wear. (Left: H. Shirihai; right: D. Occhiato.)

shrill notes, e.g. *sripp, sripp, sreeaa, srisrisri* or a more even-paced *srri, sru, chuveet, srisri, serr*. The song is repeated without much variation and short pauses, delivered from top of *Carex* or *Salix* bush. – Call is a soft down-curled whistling note, *tseeoo*, less fine and drawn-out than Penduline Tit and slightly lower-pitched. Another call frequently heard from autumn migrants, often in flight, is a slightly hoarse and 'raw' *chreh* or *brze*. Also a short fine, sharp *zi* when anxious at nest or young, often repeated to become a fine twitter, *zit-it-it-it-it-it*.

SIMILAR SPECIES Of same size and general plumage pattern as Rustic Bunting in autumn, but differs by more blackish than rufous flanks-streaking, less pure white belly, has neither a rufous necklace nor white double wing-bars as Rustic, darker (not pure pink) legs and an all-grey bill with a usually convex culmen (Rustic straight culmen and partly pink lower mandible). – Pallas's Reed Bunting is similar in general plumage but is smaller (not always apparent) and has a proportionately slimmer body and usually longer tail. The safest difference is the rufous-brown lesser coverts (mainly grey or blackish in Pallas's Reed), supplemented by the greyish or brown and prominently streaked rump (unstreaked buff or white in Pallas's Reed, but note that the large-sized and large-billed Reed Bunting ssp. *pyrrhuloides* in Central Asia also has pale, unstreaked rump). – Little Bunting is smaller and has rufous-ochre cheeks against which the neat white eye-ring stands out. Also, dark lateral throat-stripe does not reach bill-base. Bill invariably has straight culmen. Both submoustachial stripe and supercilium are at least partly tinged rufous-ochre in Little, but pale buff or whitish in Reed. – Cirl Bunting is about same size and also has grey-brown or greyish rather unstreaked rump like Reed Bunting, but differs in contrasting, striped cheeks, less prominent lateral throat-stripe, finer flanks-streaking and often has hint of rufous on breast-sides and yellowish cast to belly. Also, Cirl has a whitish forewing-bar and duller brown-grey mantle in contrast to warmer rufous-tinged scapulars. – Young Lapland Bunting has similarly streaked and striped tawny-brown upperparts and streaked flanks on pale buff-brown ground, but readily differs in yellowish dark-tipped bill, more uniformly rufous-tinged cheeks and two white and well-marked wing-bars. – Yellowhammer is hardly an option given its larger size, longer tail, chestnut-tinged rump and at least some yellow in the plumage. – Young Pine

E. s. schoeniclus, ad ♂, Netherlands, Mar: in early spring, ♂ can still show remnants of pale tips on sides of head and throat. Rounded tips and rufous edges to primary-coverts and dark tips to primaries with distinct white fringes indicate age. (N. D. van Swelm)

E. s. schoeniclus, 1stS ♂, Finland, May: note heavily abraded tertials, worn tips to juv primaries, and pointed tips to primary-coverts and tail-feathers, while extensive pale tips to black areas of head and bib also indicate age. (M. Varesvuo)

E. s. schoeniclus, ♀, Germany, Apr: typical appearance in spring and summer, when differences from ♂ very clear, showing distinct buff supercilium and submoustachial stripe, brown to grey-buff cheeks framed by dark border that reaches bill base, and blackish lateral throat-stripe. Scapulars have rich rusty tone, lesser coverts uniform rufous. Without close views of wing and tail, ageing impossible. (T. Grüner)

E. s. schoeniclus, ♀, presumed 1stS, Netherlands, Mar: irrespective of age, some ♀♀ in spring and summer have much black mottling on head-sides, and as here a broader black lateral throat-stripe, somewhat recalling head pattern of winter ♂, but rest of plumage is typical ♀. Tail-feathers seem all pointed and abraded, hence presumably retained juv. (N. D. van Swelm)

E. s. schoeniclus, presumed ad ♂, Netherlands, Sep: when very fresh, pale brown tips conceal black of head and throat, thus recalls young and ♀, though often separable by exposed white feathers on nape and more extensive black mottling on head. Fresh tail- and wing-feathers suggest ad, but bear in mind possibility of young bird with moulted tail, so definite ageing is not possible. In all plumages outermost tail-feathers show extensive white. (A. Ouwerkerk)

Bunting could perhaps be more confusing, but again is a larger and longer-tailed bird with the same chestnut rump as Yellowhammer.

AGEING & SEXING (*schoeniclus*) Ages can differ in autumn if all, or at least some, juv tail-feathers are retained, but only rarely in spring; reliable ageing in the field probably as a rule impossible and invariably requires close views if attempted. Sexes differ in spring, usually also in autumn after post-juv moult, but some autumn birds are very similar and difficult to sex in the field. – Moults. Complete post-nuptial moult of ad after breeding, mainly mid Jul–mid Sep. Partial post-juv moult in same period affecting head, body, lesser, median and greater coverts, but does not involve any primaries, secondaries or primary-coverts. A varying percentage of birds (higher in S & W Europe) replace some or all tertials and tail-feathers. Depending on study or population, 5–20% renew whole tail making ageing impossible or at least very difficult. A partial pre-nuptial moult in late winter affects mainly head and throat of ♂. – **SPRING** ♂ Head and throat black (which before they are abraded in late spring can have some whitish tips). Usually well-defined white

E. s. schoeniclus, ad ♀, England, Jan: ad-like and still rather fresh remiges, primary-coverts and tail confirm age (note false moult limit among greater coverts, innermost few naturally broader-tipped rufous, most obvious when fresh). Being ad with no trace of black areas on head confirms sex as ♀. (G. Reszeter)

E. s. schoeniclus, presumed 1stW ♂, Italy, Jan: combination of juv remiges and primary-coverts being rather brownish and, as to coverts, pointed, already extensive white on nape and rather much black visible on crown and ear-coverts indicate age and sex, but still some uncertainty remains for both. (D. Occhiato)

E. s. schoeniclus, 1stW ♀, Italy, Dec: aged as young bird by same features as previous bird, plus visibly narrow tail-feathers with very pointed tips, but here the lack of any black areas on head confirm sex as ♀. (D. Occhiato)

E. s. schoeniclus, juv, Netherlands, Jun: densely streaked dull black on warm buff breast and flanks, with dusky ear-coverts, bold black lateral throat-stripe, and loose, fluffy body plumage. Yellowish-white gape-flanges typical of juv. Very dark head pattern could indicate ♂, but best left unsexed. (S. Hendriks)

E. s. ukrainae, summer ♂, Russia, May: culmen clearly curved, but bill still fine and pointed compared to thick-billed group. Note long-tailed impression. Rump, uppertail-coverts and flanks in this race have on average finer and less extensive streaking. Note rufous lesser coverts, an important species criterion. Without handling ageing difficult. (P. Parkhaev)

submoustachial stripe joining white half-necklace around nape. Birds with retained juv tail-feathers in spring that are very heavily abraded and pointed, identifiable as 1stS. (Those with all tail-feathers rather fresh and with rounded tips are either ad or 1stS.) ♀ Head (rufous-)brown, crown streaked dark, cheeks mottled. Broad, diffuse supercilium off-white or buffish-white. Chin, throat and submoustachial stripe off-white, separated by a brown-black lateral throat-stripe. Ageing as ♂. – **AUTUMN** Sexing often possible but more difficult than in spring due to characteristic head pattern being partly concealed by fresh pale feather-tips. If ageing possible sexes can usually be identified. **Ad** After completion of moult, longest primaries and all tail-feathers quite fresh, tips of latter slightly rounded, edges neat. Note that a minority of 1stW also replace whole tail making ageing in the field impossible. **1stW** Tips of longest primaries and tail-feathers slightly worn, tips of any unmoulted juv tail-feathers pointed. Many moult r1 and/or r6, and a minority replace whole tail, thus using tail to separate 1stW from ad not always possible in the field, sometimes not even in the hand. **Juv** Resembles a dull ♀ but usually has more yellowish-tinged head and neck, rather evenly streaked or spotted dark. Lateral throat-stripes blackish-grey (rather than black) and both less distinct and more extensive than post-juv ♀. Greater coverts edged/tipped somewhat narrower and more yellowish-buff, with black penetrating pale edge at shaft (post-juv edged more rufous and lacks black penetrating rufous edge at shaft).

BIOMETRICS (*schoeniclus*) L 14–15 cm; **W** ♂ 76–85 mm (n 45, m 81.8), ♀ 72–80 mm (n 50, m 76.1); **T** ♂ 62–72 mm (n 45, m 67.1), ♀ 59–67 mm (n 50, m 63.4); **T/W** m 82.7; **B** 10.0–12.6 mm (n 80, m 11.1); **B**(f) 7.4–9.8 mm (n 92, m 8.7); **BD** 4.7–6.1 mm (n 88, m 5.2); **BW** 4.5–5.9 mm (n 78, m 5.2); **Ts** 17.5–20.5 mm (n 85, m 19.0). **Wing formula: p1** minute; **p2** < wt 0.5–4.5 mm; **pp3–5** about equal and longest (p5 sometimes to 1 mm <); **p6** < wt 0.5–3.5 mm; **p7** < wt 5–10 mm; **p8** < wt 9–13.5 mm; **p10** < wt 12–19 mm; **s1** < wt 12.5–20 mm. Emarg. pp3–6, sometimes slightly on p7 near tip.

GEOGRAPHICAL VARIATION & RANGE Rather well-marked and diverse variation with *c.* 30 races described (though probably only about little over half of them are warranted, many being apparently based on small series or inadequate material, or the product of a different approach to subspecies taxonomy). Races form two groups, northern and southern, but with intermediate populations between them, which are sometimes (but not here) treated as a third group. It should be clearly stated that borders between groups (whether two or three) are not sharp, that groupings are somewhat arbitrary and mainly serve as a practical but rough first sorting. Apart from the subspecies listed below, there are several extralimital ones in E Asia, not treated here. – Races in the northern group tend to be smaller with a finer bill and thin tarsi, while the southern group includes taxa that are larger, slightly longer-tailed and have heavier, more bulbous bills and stronger feet. Among the large-billed group, those taxa with heaviest bills occur in the east, whereas those at the western end have comparatively smaller bills (although this is only a broad pattern with many irregularities, rather than a simple, smooth cline). Within each group a cline runs from west to east in that general colours become paler. There is a considerable difference in size between the smallest races in N Europe and NW Siberia, and the largest in Central Asia. The (generally diffuse) border between the northern small-billed and southern large-billed groups runs from W & N Iberia south of the Alps, through C Balkans, just north of the Black Sea and lower Volga, and east at the transition between taiga and steppe. Birds in the intermediate or contact zone between these two main groups are rather more variable in the size and shape of bill, which makes defining borders for the involved subspecies more difficult than elsewhere. The 11 races treated here regularly breed within the treated region (or are probable migrants) and are recognised as being sufficiently distinct. Apart from these, there are a few extralimital similarly distinct races in Asia (not treated). – We have received much valuable data on variation in Russia from E. Koblik and Y. Redkin (pers. comm.), but since the tradition in Russia is to recognise very fine shades of variation and subdivide smooth clines into several separate taxa—invariably a difficult task and not attempted by us—we have drawn our own conclusions from their data.

'NORTHERN SMALL-BILLED GROUP'

E. s. schoeniclus (L., 1758) (N & W Europe including Britain, Fenno-Scandia, Russia and the Ural Mts, south to N Iberia and S France, the Alps including Italian foothills, Carpathians, Belarus except in south; northern populations make short- or medium-range winter movements towards south-west or south, reaching W & S Europe, Turkey). Described above. Culmen of bill straight or only very slightly convex, forming a subtle cline, with northernmost birds having on average smallest bill with straightest culmen ('steinbacheri') and southern birds (e.g. 'goplanae') with convex culmen and slightly longer bill. However, we find the difference both subtle and far from consistent, or insufficiently clear-cut to warrant subspecific division. Plumage rather dark, although on average a little darker near the Atlantic and paler in the east, and paler birds in E Europe ('wotiakorum') can be considerably paler above than those in Scandinavia, Germany or France; nevertheless, it is a stepless cline with much variation, and included here in schoeniclus. Grades into very similar passerina between the Ural region and lower Ob. (Syn. goplanae; mackenziei; septentrionalis; steinbacheri; turonensis; wotiakorum.)

○ **E. s. passerina** Pallas, 1771 (much of W Siberian taiga between Ob and Yenisei; winters mainly Central Asia, partly in Iran, less commonly Iraq). Differs only very subtly from schoeniclus in being on average slightly larger, a tad paler above and having slightly finer or sparser streaking on flanks and rump. Many cannot be separated even in direct comparison. Bill subtly heavier and tarsus longer than schoeniclus. Birds from the southern taiga sometimes separated as 'pallidior' on account of paler and more buff tones and heavier bill, but impossible to confirm as a consistent difference when fair series compared. Similarly, C Siberian breeders often separated as 'parvirostris' due to claimed paler plumage and finer streaking, but appear identical to us in long series. **W** ♂ 78–88 mm (n 24, m 82.4), ♀ 75–85 mm (n 16, m 78.8); **T** ♂ 62.5–72 mm (n 23, m 67.6), ♀ 63–72 mm (n 16, m 67.0); **B** 10.5–12.2 mm (n 35, m 11.6); **B**(f) 7.7–10.0 mm (n 39, m 8.9); **BD** 4.4–6.2 mm (n 40, m 5.6); **BW** 4.7–6.2 mm (n 31, m 5.5); **Ts** 18.5–21.5 mm (n 38, m 19.6). (Syn. pallidior; pallidissima; parvirostris; tazensis.)

○ **E. s. ukrainae** (Zarudny, 1917) (S Belarus, N & C

E. s. witherbyi, ♂♂, presumed ad, Spain, (left: Jul, right: May): this thick-billed race (albeit less deep-billed than some races from further east) is also typically rather dark-looking. Upperparts fringed rather dark rufous, even when worn, and rump and uppertail-coverts dark grey, latter often strongly streaked. Flanks rather extensively streaked rufous-black. Some ♂♂ have shorter white submoustachial stripe, not reaching bill (right-hand bird), adding to dark appearance of this race. Immaculate plumage and lack of obvious juv feathers in wing and tail suggest ad. (E. Ayala)

E. s. witherbyi, juv, Spain, Jun: a juv with yellow gape flanges, just starting to moult into 1stW plumage, all median coverts dropped and body plumage a bit disorderly. (E. Ayala)

Presumed *E. s. intermedia*, ♂♂, presumed 1stY, Italy, Feb: both show approximate bill dimensions of intermedia, or are intermediate between that race and schoeniclus. Nicely illustrating late winter transition into breeding plumage. Upperparts typically slightly paler than in schoeniclus. Both show worn, bleached and pointed longest primary-tips in Feb, thus juv (?), and very scruffy head pattern with many brownish feathers admixed in the black supports this. (D. Occhiato)

Ukraine, adjacent S Russia east to Volga and Orenburg in S Urals; winters SE Europe, Black Sea and S Caspian Sea regions). Very similar to *schoeniclus*, only subtly larger and longer-tailed, slightly paler and on average less streaked below and over more greyish rump. The bill is slightly stronger than in *schoeniclus* with a convex culmen, but is not as strong as *intermedia*. Grades into *schoeniclus* in the north and west, and *intermedia* in the south. Subtle, and perhaps a borderline case whether best separated or included within *intermedia*. Scant material examined. Biometrics provided by Y. Redkin (pers. comm.): **W** ♂ 80–87 mm (*n* 32, *m* 82.5), ♀ 74–82 mm (*n* 13, *m* 78.1); **T** ♂ 66–72 mm (*n* 32, *m* 68.9), ♀ 62–69 mm (*n* 13, *m* 65.8); **T/W** *m* 83.7; **B** 11.8–13.3 mm (*n* 44, *m* 12.5); **B**(f) 7.8–8.9 mm (*n* 44, *m* 8.3); **BD** 5.4–6.9 mm (*n* 38, *m* 6.0); **BW** 5.4–6.8 mm (*n* 46, *m* 6.0); **Ts** 19.0–21.4 mm (*n* 45, *m* 19.9).

E. s. lusitanica Steinbacher, 1930 (Portugal, possibly Galicia; resident). Very similar to *schoeniclus* but on average even darker, wearing in summer to nearly all brown-black upperparts, and has a somewhat heavier bill with more clearly convex culmen. The white collar in ♂ plumage averages narrower than *schoeniclus* and *witherbyi*. Close to *schoeniclus* of N Iberia and SW France but just separable. Scant material examined. **W** ♂ 73–79 mm (*n* 7, *m* 77.5), one ♀ 70 mm; **T** ♂ 61–68 mm (*n* 7, *m* 65.0), ♀ 63 mm; **B** 11.9–12.4 mm (*n* 8, *m* 12.1); **B**(f) 9.0–10.2 mm (*n* 8, *m* 9.6); **BD** 5.4–6.3 mm (*n* 8, *m* 5.9); **BW** 5.4–6.2 mm (*n* 8, *m* 5.9); **Ts** 18.4–19.7 mm (*n* 8, *m* 19.2).

'SOUTHERN LARGE-BILLED GROUP'

E. s. witherbyi von Jordans, 1923 (Iberia except west and north, E Pyrenees, Mediterranean coast of France, Balearics, probably also Morocco; resident). Similar to *schoeniclus* but has a markedly heavier bill with convex culmen. In the west, grades into subtly darker and smaller-billed birds in Portugal (*lusitanica*). On average perhaps a little darker than *schoeniclus*, and some ♂♂ have shorter white submoustachial stripe not reaching bill. **W** ♂ 73–86 mm (*n* 18, *m* 78.9), ♀ 70–83.5 mm (*n* 7, *m* 76.4); **T** ♂ 63–69 mm (*n* 18, *m* 66.0), ♀ 58–66 mm (*n* 7, *m* 63.5); **B** 11.1–12.7 mm (*n* 26, *m* 12.1); **B**(f) 8.8–10.7 mm (*n* 26, *m* 9.8); **BD** 6.2–7.4 mm (*n* 26, *m* 6.7); **BW** 5.9–7.9 mm (*n* 26, *m* 6.6); **Ts** 19.0–21.4 mm (*n* 25, *m* 20.4). **Wing formula: p7** < wt 4–7 mm; **p8** < wt 7–11 mm; **p10** < wt 12–16.5 mm; **s1** < wt 13.5–17.5 mm. – In a study of Iberian birds from Catalonia, Castilla-La Mancha, Valencia and the Balearics (Belda *et al.* 2009) with subspecific identity confirmed by mtDNA, a bill depth of > 5.9 mm classified 95% correctly, whereas 15 Finnish birds all had < 5.9 mm.

E. s. intermedia Degland, 1849 (Italy south of Piedmont and foothills to Alps, N Balkans, E Austria, Hungary, Romania, S Ukraine, W Caucasus, apparently also NE coast of Turkey, possibly also interior NE Turkey; mainly resident, but some populations make short-range movements in winter). Resembles *witherbyi* in bill-size but differs in being usually slightly paler above. Apart from having a rather strong bill with clearly convex culmen, differs from *schoeniclus* in having paler upperparts with ochre and grey-white portions more dominant, and in often (but not invariably) having rufous rather than blackish streaking on flanks. Rump averages paler and more greyish-tinged. Still, a rather variable race as to bill-size and paleness of plumage, possibly partly due to secondary intergradation between small-billed and large-billed populations. **W** ♂ 76.5–85 mm (*n* 15, *m* 82.0), ♀ 72–82.5 mm (*n* 9, *m* 77.6); **T** ♂ 66–72 mm (*n* 15, *m* 68.7), ♀ 63.5–71 mm (*n* 9, *m* 67.3); **T/W** *m* 84.9; **B** 10.7–13.9 mm (*n* 23, *m* 12.2); **B**(f) 8.5–11.4 mm (*n* 24, *m* 9.7); **BD** 5.6–8.2 mm (usually at least 5.8; *n* 23, *m* 6.6); **BW** 5.2–8.0 mm (*n* 21, *m* 6.4); **Ts** 18.5–22.6 mm (*n* 19, *m* 20.4). (Syn. *compilator*; *harmsi*; *palustris*; *stresemanni*; *tschusii*.)

E. s. reiseri Hartert, 1904 (S Balkans including Greece and Albania, possibly also Bulgaria, W & C Turkey; mainly resident, but short-range movements in winter noted). Resembles *intermedia* in plumage but somewhat larger

E. s. intermedia, ♀, Italy, May: showing typical bill dimensions. Unclear in this image if tips to tail-feathers are abraded or just wet, while wing is reasonably (though not entirely) fresh. Thus, ageing unsure. Beware inclination to consider ♀-like birds with blotchy black feathers on head in spring/summer as being 1stS ♂♂. Our field experience indicates some ♀♀ (perhaps especially older birds) develop substantial black on crown, ear-coverts and lores, but such plumage never shown by ♂♂ of any age at this season. (F. Ballanti)

on average with a strikingly heavy and more bulbous bill, roughly as large as *caspia* and *pyrrhuloides* (see below), and possibly the most bulbous-shaped of all taxa (except perhaps *korejewi*). Plumage rather dark and boldly streaked above with tendency to have much ash-grey on nape/upper mantle, scapulars and rump. Underparts in ♂ unstreaked, or nearly unstreaked, in ♀ thinly but prominently streaked on breast and flanks. **W** ♂ 84–91 mm (*n* 11, *m* 88.3), ♀ 78–83 mm (*n* 8, *m* 81.1); **T** ♂ 72–81 mm (*n* 11, *m* 76.8), ♀ 63.5–74 mm (*n* 8, *m* 70.2); **T/W** *m* 86.8; **B** 12.8–14.4 mm (*n* 19, *m* 13.7); **B**(f) 10.3–12.0 mm (*n* 19, *m* 11.2); **BD** 7.2–9.1 mm (*n* 19, *m* 8.5); **BW** 7.2–8.7 mm (*n* 19, *m* 8.2); **Ts** 20.9–22.8 mm (*n* 17, *m* 21.8). **Wing formula: p7** < wt 6–10 mm; **p8** < wt 10–13 mm; **s1** < wt 15–19 mm.

? **E. s. incognita** (Zarudny, 1917) (NW Kazakhstan east of lower Volga, east through Kirghiz Steppe to Astana region; partially migratory, wintering in S Central Asia). Described as having a slightly smaller bill than *pyrrhuloides* (still quite strong and bulbous) and very subtly darker plumage (still almost as pale as *pyrrhuloides*). Scant material examined. Two ♂♂ had W 80–81 mm and T 67–68.5 mm. BD 6.2 mm.

E. s. pyrrhuloides Pallas, 1811 (N & NW Caspian region including Kalmuk Steppe and lower Volga, Aral Sea, Central Asia south of Kirghiz Steppe east to Balkhash region; mainly resident but short-range winter movements possible). The largest and palest race with a heavy, bulbous-shaped bill (same average size as *caspia* and *reiseri*). Quite long-tailed

E. s. reiseri, ♂♂, ad (left: Turkey, Mar) and 1stS (Greece, May): generally like *intermedia* but tends to be somewhat darker, and bill usually much heavier. In particular the right bird has a typically very deep and bulbous-shaped bill. Left-hand bird is ad by evenly feathered wing, and right-hand is 1stS with pointed and abraded juv tail-feathers and wingtips. (Left: A. Atik; right: P. Dougalis)

E. s. reiseri, ad ♂ (left) and ad ♀, Turkey, Nov: bill thicker than in *intermedia*, similar in size to *caspia* but on average more bulbous than in either. Both birds have broad tail-feathers, and left bird fresh wing, indicating ad. ♂ at left has fresh pale-tipped feathering on black head and bib, still centre of throat and top of head clearly mainly black, whereas ♀ is characteristic as such. (Left: Z. Kurnuç; right: G. Hatipoğlu)

with strong feet. Entire underparts of ♂ unstreaked white (except black bib). ♀ has some very thin and faint rufous streaks on flanks in fresh plumage. Rump unstreaked pale grey-white with buff-brown tinge, often wearing to near white in summer. – Birds in lower Volga region, Sarepta ('*volgae*'), were described as having slightly smaller bill and marginally broader dark streaks on subtly darker upperparts, but those we have examined had just as heavy bill as birds from the core range of *pyrrhuloides*. The streaking varies a little in all races and claimed differences for '*volgae*' do not seem distinct enough in direct comparison. Birds described as '*centralasiae*' of Chinese Turkestan tend to have slightly smaller bill and perhaps a little bolder streaking on still very pale upperparts, but a slight and inconsistent difference, thus included here. **L** 15.5–17.5 cm; **W** ♂ (once 79) 81–94 mm (*n* 32, *m* 87.6); ♀ 77–85.5 mm (*n* 17, *m* 81.4); **T** ♂ 69–86 mm (*n* 32, *m* 77.6); ♀ 68.5–79 mm (*n* 17, *m* 74.1); **T/W** *m* 89.7; **B** 11.3–14.8 mm (*n* 47, *m* 12.9); **B**(f) 9.0–11.7 mm (*n* 46, *m* 10.3); **BD** 6.3–9.8 mm (*n* 48, *m* 8.1); **BW** 6.2–8.3 mm (*n* 48, *m* 7.5); **Ts** 19.5–22.3 mm (*n* 49, *m* 20.9). **Wing formula: p2** < wt 1.5–6 mm; **p10** 13–20 mm; **s1** 13–21 mm. (Syn. *centralasiae*; *harterti*; *volgae*; *zaidamensis*; *zaissanensis*.)

E. s. caspia Ménétries, 1832 (SE Turkey, Transcaucasia, E Iraq, Kuwait, S Caspian region, W Iran; mainly resident). Differs from *intermedia* in being slightly paler and more greyish-tinged with thinner dark streaking, and has much heavier bill. Bill about as heavy as in *reiseri* and *pyrrhuloides*. Plumage resembles *pyrrhuloides* but clearly darker above, and is streaked (mainly rufous and thinly) below. Limited material examined. **W** ♂ 85–87 mm (*n* 4, *m* 86.1), ♀ 77–83 mm (*n* 5, *m* 79.9); **T** ♂ 71–78 mm (*n* 4, *m* 74.16), ♀ 66–74 mm (*n* 5, *m* 70.3); **T/W** *m* 86.4; **B** 11.7–14.6 mm (*n* 18, *m* 13.1); **B**(f) 9.7–12.0 mm (*n* 18, *m* 10.6); **BD** 6.7–8.5 mm (*n* 17, *m* 7.7); **BW** 6.3–7.5 mm (*n* 18, *m* 7.0); **Ts** 20.3–22.3 mm (*n* 18, *m* 21.4). **Wing formula: p10** < wt 14–19 mm; **s1** < wt 15–21 mm.

○ **E. s. korejewi** (Zarudny, 1907) (SE Iran, Seistan, SW Afghanistan; resident or short-range movements in winter). Very similar to *caspia*, equally pale, large, long-tailed and heavy-billed, but from limited material available differs subtly in its more bulbous bill with more rounded culmen and less pointed bill-tip. Claimed to be darker and less greyish on mantle than *caspia*, but examined material identical in these respects. **W** 76–85 mm (*n* 4, *m* 81.5); **T** 70–76 mm (*n* 4, *m* 72.3); **T/W** *m* 88.7; **B** 12.7–14.2 mm (*n* 4, *m* 13.4); **B**(f) 10.7–11.6 mm (*n* 4, *m* 11.0); **BD** 8.3–9.0 mm (*n* 4, *m* 8.7); **BW** 7.3–7.8 mm (*n* 4, *m* 7.6); **Ts** 20.3–21.4 mm (*n* 4, *m* 20.8).

TAXONOMIC NOTES Regarding the correct position of ssp. *minor*, sometimes claimed to belong to Common Reed Bunting, see Pallas's Reed Bunting. – Defining geographic variation based solely on material available in visited collections (albeit often in fair numbers) is fraught by the paucity in these of some taxa, and this is particularly true for Reed Bunting. There is also the problem of separating local breeders from wintering or migrant birds. The interpretation offered here should be reasonably accurate, but is inevitably limited by the above-mentioned problems and reflects the slightly broader subspecies concept applied. We believe that finer shades of variation are better described under fewer formally named taxa than subdivided into a mosaic of subspecies, which in some literature approaches the level of populations or even subpopulations. We note the occurrence of both larger-billed and comparatively smaller-billed birds collected in late spring and summer in Italy, parts of N Balkans, in Turkey and the lower Volga region, to name a few areas. Some of these discrepancies might be explained by smaller-billed late migrants lingering among larger-billed local breeders, but in some cases these might just as well indicate heterogeneous populations due to the mixing of small-billed and large-billed populations in the zone of intergradation, or mean that some subspecies are simply more morphologically variable, adding to the above-mentioned difficulties. The existence of two separate Reed Bunting species, one small-billed and one large-billed, has been suggested. – There is certainly still room for improvement in subspecies definitions and general taxonomy of this species through targeted field research during the breeding season.

REFERENCES Belda, E. J. et al. (2009) *Ardeola*, 56: 85–94. – Bell, B. D. (1970) *Bird Study*, 17: 269–281. – Kasparek, M. (1979) *J. f. Orn.*, 120: 247–264. – Kvist, L. et al. (2011) *J. of Orn.*, 152: 681–693. – Neto, J. et al. (2013) *PLoS One*, 8(5): e63248. – Svensson, L. (1975) *Vår Fågelv.*, 34: 311–318.

E. s. pyrrhuloides, 1stS ♂♂, Mongolia, Jun: bill very deep and bulbous in shape but averages blunter (shorter) than in other races. Large and proportionally long-tailed race with upperparts typically overall pale. Mantle and scapulars worn with relatively narrow black streaks augmenting pale impression. Rump unstreaked pale ash-grey, flanks unstreaked too and purer white when worn. Remiges fringed buff-white when worn, contrasting with rustier tips and fringes to wing-coverts. Both are 1stS by juv primaries and tail. (H. Shirihai)

E. s. pyrrhuloides, 1stS ♀, Mongolia, Jun: observed paired with right-hand ♂ above. Heavy bill not always obvious. Also apparently 1stS by many worn (juv) wing- and tail-feathers. (H. Shirihai)

E. s. caspia, ♂ (left: Turkey, May) and ♀ (Armenia, Oct): thick-billed race, but bill averages less heavy and plumage darker than *pyrrhuloides*. Note extremely broad white necklace of ♂ when plumage fluffed up, perhaps in display. Without close views of wing and tail, ageing impossible. (Left: D. Occhiato; right: V. Ananian)

PALLAS'S REED BUNTING
Emberiza pallasi (Cabanis, 1851)

Fr. – Bruant de Pallas; Ger. – Pallasammer
Sp. – Escribano de Pallas; Swe. – Dvärgsävsparv

Closely related to Reed Bunting, the Pallas's Reed Bunting is a smaller version that lives in two separate areas, one large in the N Siberian tundra from the northern end of the Urals eastwards, the other in steppe, lakeside meadows and other open habitats, including alpine heaths on mountains, in N Mongolia and adjacent parts of S Siberia. It may actually extend its breeding range to just west of the Urals, and is thus afforded full coverage here, but the details are poorly known. It has straggled to Sweden and Britain. The winter is spent in SE Asia, from where it returns in May and June.

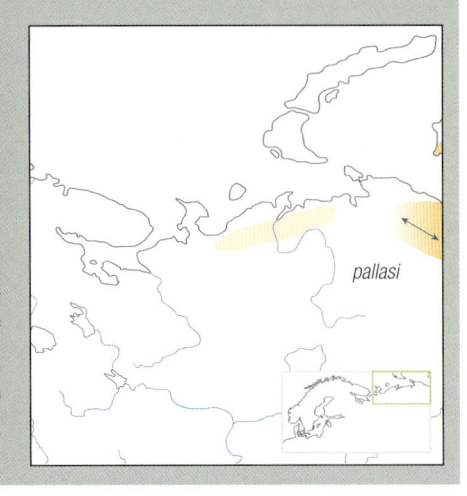

E. p. pallasi, ♂ (left) and ♀, Kazakhstan, May: all plumages like Common Reed Bunting, but note finer bill (all black when breeding), and relatively longer tail compared to *schoeniclus*. Also grey to greyish-brown lesser wing-coverts (rufous-brown to cinnamon in Common Reed) and buffish (less rufous) tips to median coverts. Feathers of mantle and scapulars typically have broad dark centres, fringed buffish (wearing whitish). ♀, in this the darkest race, often has obvious dark lateral throat-stripes ending in necklace on breast, but unlike Common Reed flanks almost unstreaked. Also tends to lack (or almost lack) dark lateral crown-stripes. Pointed and heavily bleached primary-coverts, primary-tips and tail-feathers of ♂ indicate possible 1stS, while ♀ appears rather fresh suggesting ad. (W. Müller)

IDENTIFICATION Like a small Reed Bunting but has a *large unstreaked pale rump patch* and very restricted streaking on the underparts in post-juvenile plumages. As *small* as Little Bunting or Chestnut Bunting, but compared to these two the tail is proportionately slightly longer and fuller. *Bill* medium-sized, *with straight culmen* (or very slightly curved) and outside breeding season (unlike Little) shows some *pink on lower mandible*. However, at least breeding ♂ acquires an all-black bill. Legs have same colour as Reed Bunting, i.e. variably dull pink with darker toes, all pink-brown or darker brown-grey. Adult ♂ in summer is like a small long-tailed Reed Bunting, with a black head and throat, a *broad white half-necklace* and *white submoustachial stripe*, latter usually more broadly triangular in shape (not as narrow as in Reed Bunting; H.-D. Altmann *in litt.*). In spring, nape can be tinged yellow-buff, but in summer the whole collar is white. Underparts slightly paler than in Reed Bunting due to *fewer streaks on flanks*, and has a *large unstreaked whitish rump patch*. There is also a more pronounced *contrast between dark back and pale rump* than in any eastern population of Reed Bunting. Furthermore, compared to Reed, *upperparts less rufous* or warm tawny-brown, more boldly black-streaked on grey-brown ground, feathers on back fringed yellowish-grey or buff. Like Reed Bunting usually two prominent pale stripes are formed on centre of mantle/back, but these are on average less conspicuous compared to Reed. The folded wing is less rufous than Reed, and very characteristically ♂ has *greyish* (when worn *almost slate-coloured*) *lesser coverts* (however, see ssp. *minor*). Median coverts are black with narrow off-white edges and slightly broader buff-white tips. In fresh autumn plumage the neat head and neck pattern are partly concealed by brown and buff-white tips, the rump and entire underparts being rufous-buff or ochre-tinged, and the upperparts are less dark due to broad yellowish-buff edges. Adult ♀ is similar to Reed Bunting but as underparts are nearly unstreaked, the *prominent blackish lateral throat-stripe* stands out much more as a *dark patch or angled mark on sides of lower throat*. On some ♀♀ the two angled marks almost meet on centre of breast forming outline of a bib. Subdued streaks or spots on breast and flanks are rufous, not blackish or brown-grey as in Reed Bunting. *Wing-bars slightly more prominent* since tips to median and greater coverts are yellowish-white rather than rufous-ochre as in Reed Bunting (some autumn birds are rather similar). Lesser coverts are brown-grey or grey-brown lacking rufous tinge, and median coverts, too, lack the rufous edges of Reed Bunting. Crown uniformly warm brown finely streaked dark. Cheeks rufous-tinged brown with a darker border at lower edge, but usually *lacks dark eye-stripe* present in Reed, and this makes the buff-white supercilium stand out less clearly. In fresh autumn plumage the main difference is, as in ♂, the *rufous-ochre tinge on rump* (which can have faint rufous streaks or spots) *and underparts* (stronger rusty tinge than Reed), and the in general *somewhat paler and warmer upperparts*. First-winters resemble adults and can be difficult to separate. Juvenile resembles first-winter ♀ but more heavily streaked blackish (including on breast and grey-tinged rump) with narrower and more off-white wing-bars (just like juvenile Reed). Still, separated as Pallas's Reed by combination of size, shape, bill with straight culmen and pink on lower mandible, and grey-tinged dull brown lesser coverts. Flight hopping or skipping, often low over ground or vegetation. Spends much time on ground, cleverly hiding. Twitches tail to sides nervously on exposed perch.

VOCALISATIONS Song is a very simple, rather monotonous series of a shrill or rasping note, repeated 4–6 times, *srih-srih-srih-srih-srih*. Can sound somewhat like song of Grey Wagtail or White-capped Bunting (*E. stewarti*; extralimital, see vagrants). – Call has been likened to Tree Sparrow, a soft or fine *chleep* or *tsilip*. There is also a more Reed Bunting-like hoarse note, which has been described as a weaker variant of the flight-call of Richard's Pipit, *brzyh*.

SIMILAR SPECIES Differences from Reed Bunting are largely covered under Identification. Note on post-juvenile birds absence of rufous on upperparts and wings (there can be a narrow rufous area on bases of wing-feathers immediately outside tips of greater coverts), pale and largely unstreaked rump, and greyish lesser coverts (most obvious in ♂, slightly brown-tinged in ♀). Juvenile can have quite rufous edges to tertials, flight-feathers and scapulars, recalling a Reed Bunting, but note grey tinge to streaked rump, and paler brown, less contrasting head pattern. Lesser coverts are brownish, but not rufous as in Reed. – Japanese Reed Bunting (*E. yessoensis*; extralimital, not treated, but see p. 563) can closely resemble Pallas's Reed in ♀ and immature plumages. However, Japanese Reed has: (i) contrasting dark crown (bold black streaking) but unstreaked rufous-tinged nape/upper mantle (Pallas's: rufous or tawny crown and nape, both rather finely streaked dark); (ii) rather uniform rufous-tinged central tail-feathers (Pallas's: strong contrast between blackish-brown central streak and paler brown sides); (iii) a much narrower and pointed white wedge on inner web of penultimate tail-feather (Pallas's: white wedge of variable shape, but generally broad and obtuse); (iv) unstreaked rufous rump (Pallas's: rump cinnamon or

Presumed *E. s. pallasi*, 1stS ♀, E China, Apr: showing well the proportionately long tail. ♀ in spring usually shows cinnamon pigments on crown and ear-coverts, and pale (almost whitish) rump, and is thereby distinctive. Note weak head pattern, with especially weak dark eye-stripe and lateral crown-stripe. Aged by juv tail-feathers. (M. Parker)

BUNTINGS

E. p. pallasi, ♀, presumed 1stS, Mongolia, May: spring plumage rather contrasting, upperparts with sandy-white, cinnamon-brown and black pattern when pale tips of winter plumage have worn off. Unstreaked cream-buff rump, two-coloured bill and long tail good field marks when restricted streaking of underparts cannot be seen. (A. Audevard)

Presumed *E. p. pallasi*, 1stW ♀, E China, Nov: in fresh winter plumage streaking of upperparts and head pattern much subdued due to pale buff-brown tips and fringes. Told from Common Reed Bunting by almost unstreaked rump and underparts and pale pinkish-tinged lower mandible. Age inferred by narrow and pointed retained juv tail-feathers. Racial determination tentative. (M. Parker)

whitish, variably streaked or blotched rufous); and (v) median coverts broadly tipped chestnut (Pallas's: median coverts tipped ochre-white with rufous central streak). – *Little Bunting* can resemble juvenile Pallas's Reed Bunting but has much darker crown-sides leaving a fairly obvious paler median crown-stripe lacking in Pallas's Reed. Also, rufous-ochre cheeks in Little have a neat, almost complete dark border (lacking only on lores and near lower mandible), but Pallas's Reed has a (broader and more ill-defined) dark line at the lower edge, and sometimes a dark patch at the lower rear corner, but no dark line at the upper edge. Furthermore, Little has no pink lower mandible (juvenile can have a paler, diffusely yellowish-grey base) and rump is uniform with back, both being streaked dark.

AGEING & SEXING (*pallasi*) Ages differ in autumn, sometimes in spring; reliable ageing in the field usually requires close views. Seasonal change quite marked. Sexes differ in spring, usually also in autumn after post-juv moult, but some autumn birds are very similar and difficult to separate in the field. – Moults. Rapid complete post-nuptial moult of ad after breeding, mainly Aug–Sep but can start in late Jul.

Pallas's Reed Bunting (top left: China, Jan; bottom left: Hong Kong, Nov) versus Common Reed Bunting (top right: Netherlands, Sep; bottom right: Italy, Nov) in ♀-like winter plumage: both left images show relatively 'easy' examples of Pallas's Reed, both having nearly unstreaked underparts and bicoloured bill with pinkish-tinged lower mandible (most obvious on top one), typically very little streaking on underparts, upperparts boldly black-streaked and fringed rufous and yellowish-grey. Also blackish lateral throat-stripe obvious. As often the case, diagnostic brownish-grey lesser coverts are concealed by breast-sides, the large unstreaked pale rump is only partly visible on lower bird, and proportionately longer tail not always apparent. The lower right image shows a more challenging Common Reed, being rather poorly marked on the head. Nevertheless, it has obviously dark-streaked flanks and rufous lesser coverts. By range and general coloration all seem to be *pallasi* and *schoeniclus*, respectively. Those with visibly pointed juv tail-feathers can be aged as 1stY. (Top left: M. Hale; top right: A. Ouwerkerk; bottom left: J. Holmes; bottom right: D. Occhiato)

E. p. lydiae, ♂♂, Mongolia, Jun: unlike race *pallasi*, a distinctly pale form, largely sand-coloured grey-buff above, also on wing-coverts, and largely white below. Dark-streaked pale brown mantle and browner scapulars. Ageing difficult without handling. (H. Shirihai)

Partial post-juv moult in about same period affecting head, body, lesser and median coverts, but does not involve any primaries, secondaries, tail-feathers or primary-coverts. Apparently no pre-nuptial moult in winter, but perhaps poorly known. – **SPRING** ♂ Head and throat black (before late spring may have some whitish tips admixed). Usually well-defined white submoustachial stripe joining white (or sometimes more yellowish-buff) half-necklace around nape. Underparts unstreaked white or buffish-white. Ageing sometimes possible if bird seen near and well: ad has rather broad and rounded tips to tail-feathers, only moderately worn, whereas 1stS often has narrower tail-feathers with r1 quite pointed and heavily abraded. Note, only extremes should be aged; many are intermediate and best left un-aged. ♀ Head basically (rufous-)brown, crown streaked dark, cheeks warm brown faintly mottled darker, with dark lower border and may have blackish mark at lower rear corner. Broad, diffuse supercilium off-white or buffish-white. Chin/throat and submoustachial area off-white, separated by a brown-black lateral throat-stripe. Upper breast and flanks lightly and sparsely streaked rufous (but streaks may be nearly absent or covered by pale tips in early spring). Ageing as in ♂. – **AUTUMN** Sexing often possible but more difficult than in spring due to characteristic head pattern being partly concealed by fresh pale feather-tips. If ageing possible most categories can usually be separated. **Ad** After completing moult, longest primaries and all tail-feathers quite fresh, tips of latter slightly rounded, edges neat. **1stW** Tips of longest primaries and tail-feathers slightly worn, tips of latter pointed. **Ad ♂** As spring but black-and-white parts of head and throat partially concealed by ochre or buff-white tips. **1stW ♂** As ad ♂ but tail-feathers pointed and slightly worn. **Ad ♀** As spring but warmer buffish-ochre. Tips to tail-feathers fresh and rounded. **1stW ♀** As ad ♀ but tail-feathers pointed and slightly worn. **Juv** Differs clearly from post-juv ♀ in heavily black-streaked throat, breast and rump; typically has rather uniform rufous-tinged cheeks with a blackish patch at rear and prominent white submoustachial stripe.

BIOMETRICS (*pallasi*) **L** 12.5–14 cm; **W** ♂ 69–77.5 mm (*n* 43, *m* 73.8), ♀ 66–74 mm (*n* 28, *m* 68.9); **T** ♂ 58–69 mm (*n* 43, *m* 62.5), ♀ 55–64 mm (*n* 28, *m* 59.4); **T/W** *m* 85.3; **B** 9.8–12.4 mm (*n* 67, *m* 11.0); **B**(f) 7.5–9.9 mm (*n* 67, *m* 8.7); **BD** 4.6–5.9 mm (*n* 64, *m* 5.1); **Ts** 16.9–19.5 mm (*n* 63, *m* 18.0). **Wing formula: p1** minute; **p2** < wt 0.5–4 mm; **pp3–5** about equal and longest; **p6** < wt 0.5–5 mm; **p7** < wt 4–9 mm; **p8** < wt 7.5–12 mm; **p10** < wt 12–17 mm; **s1** < wt 13–18 mm. Emarg. pp3–6 (very rarely slightly also on p7).

GEOGRAPHICAL VARIATION & RANGE Variation mainly slight, but also perhaps insufficiently well known. Museum material includes many likely migrants of unknown populations, which makes reliable examination difficult. Birds in fresh plumage differ substantially from those in worn, more so than in many other species, adding to the complications: only specimens from comparable seasons and plumage states can be used. One extralimital race might represent a separate species, but more research is desirable before this issue can be decided. We have benefited from valuable input by Y. Redkin and E. Koblik, ZMMU (Moscow), but the

E. p. lydiae, 1stS ♀, Mongolia, Jun: ♀ of this race often appears distinctly washed-out, with weaker head and beast markings. Underparts almost unstreaked, but blackish lateral throat-stripe relatively obvious. Compared to Common Reed Bunting, more prominent and whiter wing-bars, while lesser coverts (not always visible though) are grey-brown lacking rufous tones. Pointed and heavily bleached primary-tips and tail-feathers indicate age. (S. Pfützke)

subspecific taxonomy presented below represents our own view.

E. p. pallasi (Cabanis, 1851) (tundra habitats and taiga bogs in N Siberia, apparently also extreme NE Russia and pockets of alpine habitats in SC Siberia south to Krasnoyarsk, Irkutsk, Altai, Sayan and N Mongolia, extending east to Anadyr and south-east to SE Transbaikalia, Amur and Ussuriland; possibly also locally in Tien Shan; winters S & E Mongolia, S Ussuriland, Manchuria, E China, Korea). Described above. Includes '*polaris*' of the northern tundra, described as being slightly smaller on average (though the limited material examined contradicts this) and a trace darker with slightly heavier streaking above. However, material from N Siberia seems identical to *pallasi*, and the two are therefore lumped. **W** 67.5–76 mm (*n* 11, *m* 71.9); **T** 58–65 mm (*n* 11, *m* 61.4); **T/W** *m* 85.5; **B** 9.8–12.2 mm (*n* 10, *m* 11.1); **B**(f) 7.8–9.3 mm (*n* 10, *m* 8.6); **BD** 4.6–5.3 mm (*n* 10, *m* 4.9); **Ts** 17.0–19.1 mm (*n* 9, *m* 17.7). (Syn. *latolineata*; *montana*; *passerina*; *polaris*.)

Extralimital subspecies: ***E. p. lydiae*** Portenko, 1929 (N Mongolia, Transbaikalia; thought to be resident or short-range migrant reaching NE China, but poorly known). Rather markedly paler above than *pallasi* with broader yellow-buff feather-edges in fresh plumage and narrower and browner,

Presumed *E. p. lydiae*, ad ♂, **China, Nov**: overall pale plumage suggests race *lydiae*. Following complete post-nuptial moult, when evenly fresh and in non-breeding plumage, generally ♀-like but black feather bases on crown and throat and white on neck-sides sometimes visible. In fresh plumage markedly paler than most Common Reed. Underparts tinged cream-buff with very little or no streaking. (T. Beeke)

1stW, presumed ♂ (left: Oct) and presumed ♀ (right: Nov), **China**: racial identification perhaps impossible outside breeding range, especially in winter when individual variation and overlap is greater, even between palest and darkest races. But, overall pale plumage of both suggests race *lydiae*. Both aged as 1stW by juv primaries, primary-coverts and tail (also confirmed via other images), but sexing is far more challenging. The better-exposed black facial markings on left bird suggest a ♂, unlike that on the right. (Left: I. Fisher; right: M. Hale)

less black, streaks. It also appears to be proportionately subtly longer-tailed than *pallasi*. Scant material examined, but studied in the field. When flushed, it habitually flies very low over grass and low tussock vegetation. Call noted as rather Reed Bunting-like, a somewhat slurred *dziu*, recalling also a common type of *flava* Yellow Wagtail call. Size of 3 ♂♂: **W** 65.5–75 mm (*m* 69.8); **T** 59–65 mm (*m* 61.3); **T/W** *m* 87.9; **B** 9.9–10.7 mm (*m* 10.3); **B**(f) 8.2–8.7 mm (*m* 8.4); **BD** 4.8–5.1 mm (*m* 5.0); **Ts** 17.2–18.1 mm (*m* 17.6).

E. p. minor Middendorff, 1851 (SE Siberia, Manchuria, perhaps W Ussuriland; short-range winter movements probably limited). Differs in having lesser coverts/wing-bend more brown than grey, and originally described by Middendorff as a race of Reed Bunting, subsequently supported by Vaurie (1956). However, its structure and size fit better with Pallas's Reed, and E. Koblik (pers. comm.) refers to it as a definite race of *pallasi* based on song, behaviour and ecology, in agreement with Stepanyan (1990). Only scant material seen. **W** ♂ 71–74 mm (*n* 4, *m* 72.8), one ♀ 68 mm; **T** 60–66 mm (*m* 62.9); **T/W** *m* 87.6; **B** 10.7–11.2 mm (*n* 5, *m* 10.9); **B**(f) 8.2–8.6 mm (*n* 5, *m* 8.4); **BD** 5.2–5.3 mm (*n* 4, *m* 5.3); **Ts** 17.0–18.3 mm (*n* 5, *m* 17.6). Hardly a candidate for vagrancy to the covered area.

TAXONOMIC NOTE The apparently ecologically separated southern *lydiae* breeds in grassy tussocks at lakesides,

Japanese Reed Bunting *E. yessoensis*, 1stW ♀, **Korea, Jan**: a potential vagrant, in some plumages only distinguished from Pallas's Reed Bunting with care. In all plumages note its contrasting dark-streaked crown and cheeks (especially rear cheeks), unstreaked pinkish-buff hind-collar and paler buff-brown central tail-feathers, while not visible in this image is the unstreaked rufous-buff rump and broadly chestnut-tipped median coverts. Worn and narrow tail-feathers indicate 1stW, general plumage ♀. (M. Poll)

on tall-grass steppe and other open lowland habitats. It has been claimed to occur sympatrically with *pallasi*, although this contradicts observed habitat differences and requires confirmation. As the calls apparently differ, and *lydiae* is a morphologically rather distinct form, it might merit separate species status. Further study required.

REFERENCES Haldén, P. (2010) *Roadrunner*, 18(3): 28–31. – Loskot, V. M. (1986) *Proc. Acad. Sci. Zool. Inst. Leningrad*, 150: 147–170. – Vaurie, C. (1956) *Amer. Mus. Novit.*, 1795: 85–94.

RED-HEADED BUNTING
Emberiza bruniceps Brandt, 1841

Fr. – Bruant à tête rousse; Ger. – Braunkopfammer
Sp. – Escribano carirrojo; Swe. – Stäppsparv

The Red-headed Bunting is abundant and widespread in steppe, semi-deserts and cultivated areas of Central Asia wherever there are bushes, low trees or thickets. Its simple song, persistently repeated, is an integral part of the Asian plains and lower mountains. Despite clearly different ♂ plumages, Red-headed and Black-headed Buntings are closely related and often mix where they meet, e.g. north and south of Caspian Sea. Red-headed is a migrant, moving south-east in autumn to winter in India. It has occurred many times in Europe, but all of these records are traditionally thought to involve escaped cagebirds (though genuine vagrancy in a few cases is very likely).

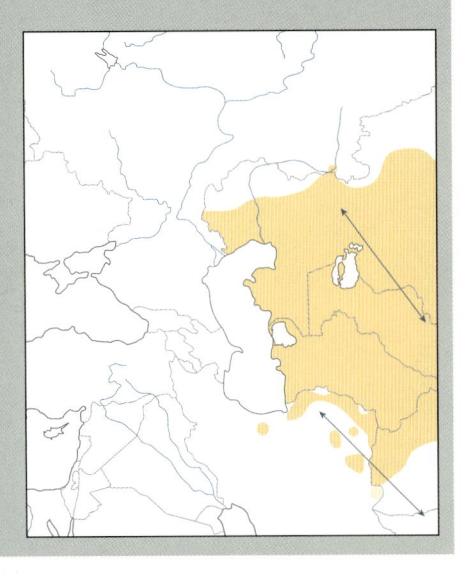

♂, Kazakhstan, May: tail without any white, and in breeding plumage golden-brown to chestnut head and bib, contrasting with bright yellow nuchal collar, rump and belly, and dark-streaked moss-green mantle. Complete winter moult in both age classes means spring birds cannot be aged. (C. Bradshaw)

♂ (top) and ♀, Iran, May: one of few bunting species where sexes differ distinctly. The drab ♀ is very similar to ♀ Black-headed Bunting, separation requiring care. Intense yellow on undertail-coverts, as here, plus lack of white in tail eliminate ♀ Cinereous Bunting, whereas any clear streaking on breast and flanks would have excluded post-juv Red-headed. (A. Sadr)

IDENTIFICATION A medium-sized to large bunting with a *stout, conical all-grey bill* and *no white in tail*. Adult ♂ in summer is a strikingly beautiful bird with *red-brown head and throat, bright yellow rest of underparts and rump*, and a yellowish-tinged *grey-green back streaked blackish*. The yellow usually forms a necklace, broken only on nape. The extension of red-brown on the head varies individually, and a few have crown partly or all yellow instead. In late autumn much of the red-brown is concealed by buffish tips, but still partly visible on throat and cheeks. Legs brown. The ♀ is much duller, *pale grey-brown* without any characteristic features except *pale yellow undertail-coverts*, although some have more yellow on underparts, the yellow invading the belly, breast and throat. Note that the *head is very plain*, lacking any dark or pale marks. A few have a little rufous on forehead and a *yellow-green hue to the rump*, while others have a stronger yellow tinge on rump and rufous on forecrown and head-sides, although not to the extent to make sexing difficult. Wing-coverts and tertials are dark with pale brown distinct edges. Two indistinct and hardly contrasting wing-bars are formed by buff-brown tips to coverts. Mantle/back greyish-brown, rarely with a slight olive hue (never rufous), streaked dark, and crown very finely spotted or streaked. Juvenile and first-winter ♀ are very similar, sandy grey-brown with just the pale yellow tinge on undertail-coverts of later plumages. They are often inseparable from corresponding plumages of Black-headed Bunting (Similar species). Red-headed Bunting is a rather stocky bird with a sparrow-like flight, a little clumsy and fluttering when alighting or taking off.

VOCALISATIONS Song a short, monotonously repeated, loud phrase with rather shrill tone, opening with a few staccato-like sharp notes, then accelerating and falling in pitch, e.g. *zrit... zrit... srüt, srüt-srüt setteri-sett sütterreh*. Contains mainly trilling and scratchy sounds. Very similar if not identical to song of closely related Black-headed Bunting. Sometimes gets stuck and for a period repeats the same shrill, sharp introductory note, *zrit... zrit... zrit...*, etc., almost sounding like Zitting Cisticola. An alternative song is heard at dawn, a 'pensive' or clearly-paused series (at walking pace) of individual notes of varied pitch and structure, all sharp and shrill, *srit... tsih... trett... zrieh... tsrit... zri-zri... zih...* etc. – Calls include Ortolan-like *chüpp* and Cretzschmar's-like *zrit*. Others resemble common call of Yellowhammer, *chüf* or *chüh*, and there is also, like in Yellowhammer, a fine, sharp *zit* that can be repeated in a fast clicking trill, *zir'r'r*.

BUNTINGS

SIMILAR SPECIES ♀ and immature very similar to corresponding plumages of *Black-headed Bunting*, but at least some are separated by (i) on average smaller size; (ii) slightly smaller and shorter bill; (iii) duller, less conspicuous pale wing-bars (depends on wear and season, and of little help in fresh autumn plumage); (iv) on average less streaked crown but more distinctly streaked mantle; and (v) never has rufous tinge to mantle (can be slightly yellowish-green). – Can recall similar-sized non-adult ♂ Cinereous Bunting, but latter has some white on tail-sides. – Differs readily from *Chestnut Bunting* in much larger size, greenish-tinged upperparts, hint of yellow necklace, yellow instead of chestnut rump, etc.

AGEING & SEXING Ages differ in autumn; reliable ageing in the field usually requires close views. Sexes differ after post-juv moult. – Moults. Partial post-nuptial moult after breeding but before autumn migration (Jul–Aug), more extensive in ♂ (most or all feathers of head, nape, upperparts, tertials, parts of underparts, a few tail-feathers and some wing-coverts) than in ♀ (no moult, or replaces only scattered feathers on head and body). Near-complete post-nuptial moult of ad including flight-feathers and tail-feathers immediately after reaching winter quarters, mainly Aug–early Nov. Partial post-juv moult Jul–Aug usually limited to head, body, tertials and many or all wing-coverts. Near-complete moult of 1stW, including flight-feathers and tail-feathers, immediately after reaching winter quarters. No pre-nuptial moult in ad or 1stS later in winter. –

SPRING Ageing impossible. ♂ Head and throat (and often upper breast) rufous or (rarely) golden-brown, rest of underparts bright yellow. Mantle/back yellowish-green, usually with some dark streaking. Rump bright yellow. ♀ Variable, perhaps most having crown and cheeks greyish-brown with some faint spots of yellow, crown streaked dark, while probably fewer show some rufous on forecrown and yellow on rump. Mantle/back 'medium' brown or olive-brown, lacking rufous tinge, generally clearly streaked dark. Underparts dusky yellowish-white, often with some buff on breast, but sometimes rather rich yellow from chin to undertail-coverts. – **AUTUMN** Sexing of ad largely as in spring. Sexing of 1stW while still on breeding grounds usually impossible. **Ad** ♂ Throat a mix of rufous and yellow, feathers tipped white. Crown brown, streaked. Longest primaries and tail-feathers heavily abraded while still on breeding grounds. **Ad** ♀ As spring, but primaries and tail-feathers heavily worn while still on breeding grounds. Tertials and wing-coverts rather worn, pale (dull whitish) fringes narrow and uneven. **1stW** Resembles ad ♀ but tips of long primaries and tail-feathers still quite fresh. Upperparts buffish-brown streaked dark. Underparts buffish-white with some yellow, invariably including undertail-coverts. Tertials and median coverts have prominent pale buff fringes. **Juv** Very similar to plainer ad ♀ and 1stW but overall more buff, less greyish and clear yellowish, and crown/nape and rump more heavily streaked dark. Tertials evenly bordered

♀, Kazakhstan, May: like a poorly marked *Passer* sparrow, with pale grey-brown or sandy-buff upperparts, and buff-white underparts, dark-streaked crown and mantle, and clear-cut pale wing-covert and tertial fringing the only obvious features. With wear, many ♀♀ show little to no yellow on vent and undertail-coverts, as shown here. (M. Westerbjerg Andersen)

Ad ♂, India, Oct: seems to be near end of complete post-nuptial moult having reached its winter grounds. Chestnut of head and bib now partly concealed by yellow or grey-buff tips on freshly grown feathers. (S. Garg)

1stW ♂, India, Sep: early in autumn moult (no wing feathers yet replaced) with new yellow feathers on underparts, and some reddish on face just visible separating it from ♀, but also from often similar-looking Black-headed Bunting. This bird has accidentally lost its entire tail. (C. Abhinav)

Juv/1stW, Kazakhstan, Jul: after some body moult, but much of wing still juv. Note dark streaking and well-defined buffish fringes on upperparts and wing, and whitish and buff underparts with some yellow, most intense on undertail-coverts. Very similar or inseparable from Black-headed Bunting in same plumage. Here, identification is based on location/date and apparently shorter bill. (K. Haataja)

buffish-white with sharp border to dark centres (in post-juv more diffuse border between dark centre and pale fringes). Upperparts buffish-brown streaked or spotted dark, including faintly on nape. Primaries and tail-feathers fresh. Breast buffish-white, ♂ usually has clear yellow on rump, belly and undertail-coverts, ♀ only on undertail-coverts. Sides of breast lightly streaked dark.

BIOMETRICS L 15–16.5 cm; W ♂ 84–94.5 mm (n 29, m 88.7), ♀ 79–88 mm (n 11, m 82.7); T ♂ 66–75 mm (n 29, m 70.1), ♀ 62–70 mm (n 11, m 65.5); T/W m 79.1; B 13.5–16.3 mm (n 37, m 15.1); B(f) 11.1–14.0 mm (n 40, m 12.5); BD 6.7–8.2 mm (n 36, m 7.2); Ts 19.2–22.2 mm (n 38, m 20.5). Wing formula: **p1** minute; **pp2–4** about equal and longest (rarely p2 to 1.5 mm <); **p5** < wt 1–5 mm; **p6** < wt 6–11 mm; **p7** < wt 10–16 mm; **p10** < wt 18–24 mm; **s1** < wt 21–27 mm. Emarg. pp3–5, rarely slight emarg. on p6 as well.

GEOGRAPHICAL VARIATION & RANGE Monotypic. – Extreme SE Russia from lower Volga east, Kazakhstan east to Zaysan and extreme W Mongolia and NW China, Transcaspia, N Iran east to N Afghanistan, foothills of Tien Shan; winters India.

♂, India, Nov: just after complete moult, when bright rufous colours partly concealed by yellow tips and to some extent appears mottled or washed-out, but otherwise unmistakable. Note pale long bill and bright yellow underparts. (S. Singhal)

♀ Red-headed Bunting (left: China, Jan) and Black-headed Bunting (right: India, Mar): further emphasising their extreme similarity and overlap in characters; no single diagnostic character exists! The ♀ Red-headed Bunting is often obviously smaller, with shorter bill and overall drabber plumage, almost cream-brown or tinged olive-brown above (lacking any sign of rufous). Only birds showing all these characters can sometimes be separated with certainty from ♀ Black-headed. (Left: M. Hale; right: N. Sant)

REFERENCES Dernjatin, P. & Vattulainen, M. (2007) *Alula*, 13: 50–54. – Vinicombe, K. (2003) *Birdwatch*, 137: 32. – Wilson, K. (2011) *BW*, 24: 342–352.

Red-headed Bunting 1stY ♂ (right) and presumed Black-headed Bunting, ♀, India, Nov: ♂ is 1stY (still shows only first signs of red bib and head in Nov), while bird on left appears to be a ♀ Black-headed (probably ad by seemingly early-completed moult with only ad type wing-feathers). Note on left bird longer and stronger bill suggesting Black-headed, but might also be a hybrid between the two. (S. Singhal)

BLACK-HEADED BUNTING
Emberiza melanocephala Scopoli, 1769

Fr. – Bruant mélanocéphale; Ger. – Kappenammer
Sp. – Escribano cabecinegro
Swe. – Svarthuvad sparv

This close relative of the Red-headed Bunting breeds in SE Europe including the Balkans, SE Russia, Turkey, the Levant, Transcaucasia and NW Iran. Like the Red-headed Bunting, it winters in India, leaving its European breeding grounds early, at the end of summer (sometimes in July), returning fairly late, in late April or May. Its favoured breeding habitats are open, cultivated land with hedgerows, copses, shrubbery and orchards, in on average more lush and vegetated areas and at slightly lower altitudes than Red-headed Bunting, although in many respects their preferences are similar.

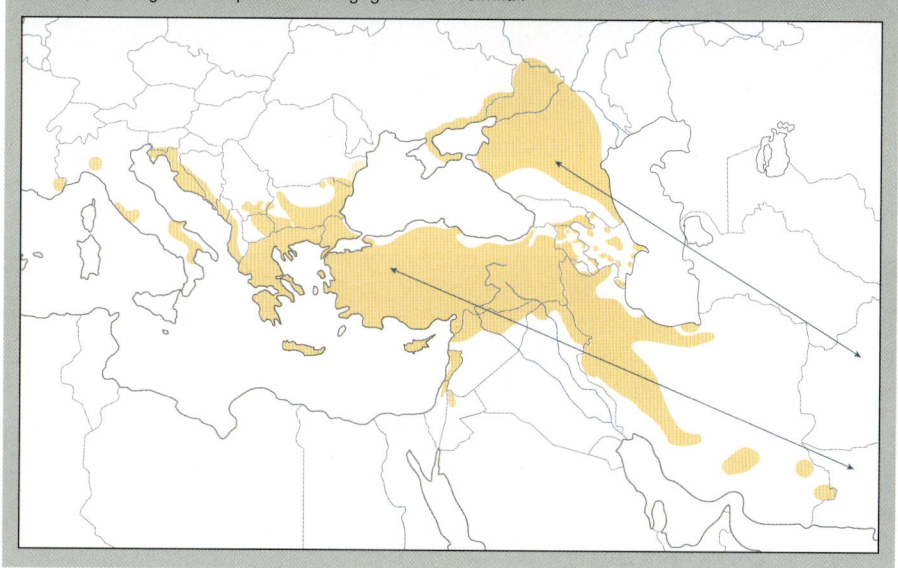

IDENTIFICATION A *large, long-tailed* bunting with a *heavy, rather long, conical all-grey bill*. Lacks white in tail, a feature that reveals its close relationship with Red-headed Bunting. Adult ♂ in summer easily identified by *black head, bright yellow underparts* and largely *unstreaked red-brown back and rump*. In late autumn much of the black is concealed by buffish tips, but is still partly visible. Legs brown. ♀ much duller and very closely resembles ♀ Red-headed Bunting, being usually rather pale *grey-brown* without any characteristic features save the *pale yellow undertail-coverts*, although some have pale yellow over most or entire underparts. A few worn summer birds can attain a plumage that shows traces of ♂ plumage, with a somewhat darker grey head and better contrast to pale yellow throat, a hint of rufous on upperparts, especially on rump but often also on breast-sides. Wing-coverts and tertials are dark with distinct pale brown edges. Two indistinct and only slightly contrasting wing-bars are formed by buff-white tips to coverts, on average a little marginally more contrasting than in Red-headed Bunting. Mantle/back is greyish-brown, rarely with a slight rufous tinge (never any yellow-olive), very lightly streaked dark (on average less distinctly streaked than Red-headed). Juvenile and first-winter ♀ very similar, sandy grey-brown with just the pale yellow tinge on undertail-coverts of later plumages. They are often inseparable from corresponding plumages of Red-headed Bunting (cf. Similar species under Red-headed). Habits and behaviour as Red-headed Bunting.

VOCALISATIONS Song and calls virtually identical to Red-headed Bunting (which see).

SIMILAR SPECIES Main confusion risk is *Red-headed Bunting* in any plumage, apart from adult ♂. See Identification and under Red-headed Bunting for separating characters.

AGEING & SEXING Sexes differ after post-juv moult. Ages differ in autumn; reliable ageing in the field usually requires close views. – Moults. Partial post-nuptial moult after breeding but before autumn migration (Jul–Aug), more extensive in ♂ (most or all feathers of head, nape, upperparts, tertials, parts of underparts, a few tail-feathers and some wing-coverts) than in ♀ (no moult or replaces only scattered feathers on head and body). Near-complete post-nuptial moult of ad including flight-feathers and tail-feathers after reaching winter quarters, mainly Oct–early Dec. Partial post-juv moult Jul–Aug usually limited to head, body, tertials and many or all wing-coverts. Near-complete moult of 1stW, including flight-feathers and tail-feathers, after reaching winter quarters. No pre-nuptial moult in ad or 1stS later in winter. – **SPRING** Ageing impossible. ♂ Crown and cheeks black. Entire underparts yellow. Mantle/back unstreaked rufous-brown, rump rufous with yellow cast. A few have black on head mixed with some grey-brown and yellow, but cheeks nearly always virtually black. ♀ Crown and cheeks greyish-brown with some faint spots of yellow, crown streaked dark. Underparts dusky yellowish-white, often with some buff on breast, but sometimes rather rich yellow from chin to undertail-coverts. Mantle/back greyish-brown with some rufous admixed, generally streaked dark. A few have rather extensive black on crown, especially in summer when plumage worn, but cheeks are grey-brown, not blackish. – **AUTUMN** Sexing of ad largely as spring. Sexing of juv while still on breeding grounds usually impossible. **Ad ♂** Black, yellow and rufous partially concealed by broad greyish-buff and white tips, yet the underlying colours are readily visible. Tips of longest primaries and tail-feathers heavily worn. **Ad ♀** As spring, but primaries and tail-feathers heavily worn while still on breeding grounds. **1stW** Resembles ad ♀ but

♂, **Turkey, Jun**: ♂ unmistakable in breeding plumage, by black hood, yellow underparts and largely unstreaked chestnut upperparts. Otherwise a heavy-bodied bunting with stout bill and long tail that shows no white. A trifle larger and slightly longer-billed than Red-headed Bunting. (H. Shirihai)

♂, **Israel, May**: often sings from exposed perch atop a bush. Chestnut above extends to rump, but most uppertail-coverts greyish-olive. Due to complete winter moult by both age classes cannot be aged in spring. (A. Ben Dov)

♀♀, **Iran**, Jun (left), and **Israel**, Jul: pale wing-covert and tertial fringes often obvious. Note extreme variation in summer ♀♀, especially depth of yellow below. The duller bird on right could be confused with ♀ Red-headed Bunting. Nevertheless, bill longer and stronger. Still, ♀♀ of the two species always difficult to separate needing care. (Left: E. Winkel; right: A. Ben Dov)

Ad ♂♂, **Israel**, Jul (left), and **Netherlands**, Oct: during partial post-nuptial moult (left) or before complete pre-nuptial moult (right), these two birds show some of the variation in ♂♂ at these seasons. (Left: A. Ben Dov; right: C. van Rijswijk)

tips of long primaries and tail-feathers still quite fresh. Upperparts buffish-brown streaked dark. Underparts buffish-white with some yellow, invariably including undertail-coverts. **Juv** Very similar to ad ♀ and 1stW but overall more buff and less clear yellowish. Upperparts buffish-brown streaked dark, including faintly on nape. Underparts buffish-white lacking any clear yellow except on undertail-coverts. Breast lightly streaked dark.

BIOMETRICS L 15.5–17 cm; **W** ♂ 91.5–98.5 mm (n 22, m 95.1), ♀ 82.5–91 mm (n 11, m 86.2); **T** ♂ 68.5–77 mm (n 22, m 72.2), ♀ 63–71.5 mm (n 11, m 66.4); **T/W** m 76.2; **B** 13.6–17.7 mm (n 32, m 16.2); **B(f)** 11.8–15.9 mm (n 33, m 13.5); **BD** 7.0–8.2 mm (n 33, m 7.5); **Ts** 19.7–23.4 mm (n 33, m 21.5). **Wing formula:** p1 minute; pp2–4 about equal and longest (sometimes p2 and/or p4 to 2 mm <); **p5** < wt 3.5–6.5 mm; **p6** < wt 8–13 mm; **p7** < wt 12–17.5 mm; **p10** < wt 23–28.5 mm; **s1** < wt 22–31 mm. Emarg. pp3–5, but at times only slightly on p5.

GEOGRAPHICAL VARIATION & RANGE Monotypic. – SE Europe from Italy (local SE France) through Balkans, W

Ad ♀, **China**, Nov: very worn plumage prior to pre-nuptial moult, which strangely has not started yet in this bird (normally moult is concluded early Dec, but this may be disrupted in vagrants). Only the yellow vent sticks out in otherwise nondescript plumage. (M. & P. Wong)

BUNTINGS

Black Sea coasts, Turkey, Crimea, SE Ukraine, S Russia, Transcaucasia, NW & SW Iran, Baluchistan; winters Pakistan, N & C India.

REFERENCES Baxter, P. & Shaw, D. (2007) *BW*, 20: 299–302. – Olsen, K. M. (1999) *Roadrunner*, 2: 10–14. – Shirihai, H. & Gantlett, S. (1993) *BW*, 6: 194–197. – Wilson, K. (2011) *BW*, 24: 342–352.

Ad ♀ (centre) with two juv/1stW, Azerbaijan, Aug: note variation in breast streaking between two juv in same clutch! Also that juv has faint pinkish hue to pale area of bill, not bluish-grey like ad. (M. Heiß)

Juv/1stW, China, Oct (left) and Scotland, Sep: birds with mostly juv plumage, including richly streaked underparts, can be seen on autumn migration (e.g. left-hand bird). Nevertheless, some young moult earlier and more of body plumage, but usually leave wings until complete pre-nuptial moult. Renewed body plumage of right-hand bird duller, with strongly streaked upperparts and almost lacks yellow below, i.e. very similar to fresh ad ♀ Red-headed Bunting. Therefore, ageing first is very important. Mantle-feathers of bird on right show dark centres and hint of warmer brownish fringes of Black-headed Bunting, while long bill supports species identification. (Left: M. & P. Wong; right: H. Harrop)

♂♂, India, Dec: two different birds in last stage of complete winter moult, with apparently some inner secondaries still growing. Impossible to age either once last juv feather shed, but fair to guess that at least right bird is ad ♂ since it has developed slightly more of the breeding plumage (though this can also just be individual variation). (Left: R. Mallya; right: A. Deomurari)

♀, India, Dec: very fresh after complete winter moult, when ageing impossible using moult and feather pattern, but perhaps amount of yellow tinge below (at least on belly) suggests a young bird. In this plumage a rather featureless olive-buffish bird with a plain head relieved only by pale eye-ring, and note pale double wing-bars. (N. Sant)

CORN BUNTING
Emberiza calandra L., 1758

Fr. – Bruant proyer; Ger. – Grauammer
Sp. – Escribano triguero; Swe. – Kornsparv

The odd one out among the buntings, being both rather thickset and 'clumsy' in flight and movements, and having rather dull greyish-brown plumage with coarse streaking, the same for both sexes. Along with a few larks, it is one of the most widespread and abundant species in southern agricultural districts, typically encountered singing from a high and visible song post, such as a telephone wire, its characteristic jangling, creaking song. It is everywhere a resident. In its southern range it tends to also adapt to open and barren slopes.

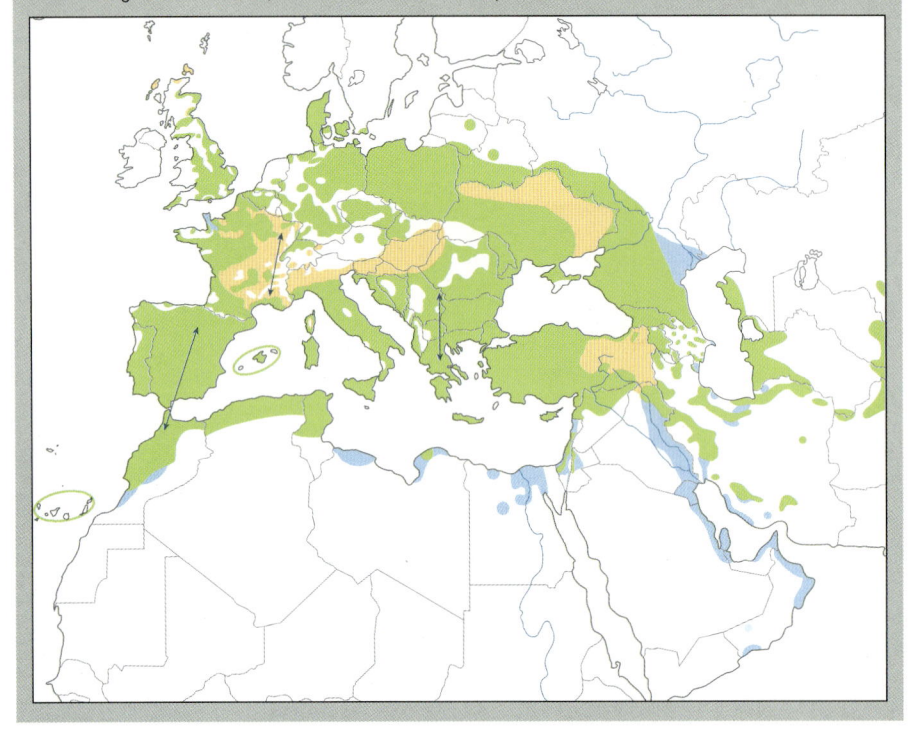

Israel, Apr: unique among Old World buntings by lack of sexual dimorphism and complete rather than partial post-juv moult. Large head with broad dark line on lower ear-coverts and another at upper rear corner, separated by white spot, as well as pale supercilium. Dark streaks on breast often coalesce into ill-defined patch in centre. Powerful bill with decurved culmen. (H. Shirihai)

VOCALISATIONS Song characteristic, monotonously and eagerly repeated without variation, a shrill, brief phrase, the opening halting, then accelerating into a quick, sharp jingling, *teck teck-zick-zik-zkzkzkzrississs*. No geographical variation noted. – Call a metallic, sharp *zrit*, a drier discreet clicking *bitt* (or very short *bt*), sometimes repeated in fast series, *bt't't't*.

SIMILAR SPECIES Due to its large size, lack of white in the tail and streaked grey-brown plumage, rather characteristic. Could perhaps be confused with young *Cirl Bunting*, but that is smaller, with a slightly smaller all-grey bill and often has a slight rufous tinge to scapulars and breast-sides, and a little white on tail-sides. – ♀♀ of *Black-headed* and *Red-headed Buntings* are slimmer, nearly unstreaked below and invariably have some pale yellow on vent/undertail-coverts. – A non-adult ♂ *Common Rosefinch* is similarly dull grey-brown and streaked all over like Corn Bunting, but is smaller and slimmer, has a smaller head with a seemingly heavier bill, and the pale wing-bars are more prominent as a rule. – Various larks, e.g. *Skylark*, can be a confusion risk mainly due to colours, behaviour and shared habitat, but Skylark is eliminated by pale rear edge to wings, white tail-sides and thinner bill.

IDENTIFICATION A *large, stoutly-shaped* bunting with *comparatively large head* and *short tail, the tail lacking white sides*. The plumage is colourless, dull *grey-brown above and off-white below* streaked dark throughout. The dark streaks often coalesce to form an *ill-defined larger dark patch on centre of breast*. Blackish lateral throat-stripes of varying prominence, sometimes quite thin or short. On finely-streaked cheeks often a tendency for *brown darker margin on lower edge of ear-coverts*, and a dark brown patch in upper rear corner. An unbroken pale eye-ring affords a rather 'mild' look. The breast and flanks usually have a faint brownish tinge. Tertials dark with evenly broad paler edges. The *bill, straw-yellow* with darker grey culmen, has a *rather small upper mandible* with a peculiarly 'curvy' cutting edge, visible when the ♂ bends its head back and sings with open bill. Legs rather strong and pinkish. Sexes and ages alike. Often seen flying off in *low fluttering flight with dangling legs*. Often drops clumsily into crop fields in lark fashion. As wings are rather broad, similarity with Skylark is enhanced, but note *absence of pale rear edge to wings*.

AGEING & SEXING Ages inseparable after post-juv moult. Sexes alike in plumage, but size helpful for majority; see Biometrics. – Moults. Complete post-nuptial moult of

Scotland, Jun (left), and England, May: typical singing posture (left) and diagnostic display flight with drooping legs. (Left: I. H. Leach; right: R. Hayes)

England, Apr: a large bunting with stocky body and almost lark-like plumage. Pale greyish-brown upperparts rather evenly dark-streaked, off-white underparts with bold dark streaks on upper breast and finer black spots or streaks on throat, lower breast and flanks, and no white in tail. Pale wing-bars prominent, especially that on median coverts. (S. J. M. Gantlett)

Turkey, May: from late spring on wears browner, dark streaks often bleach to paler brown. (D. Occhiato)

ad mainly in Aug–Oct. Complete post-juv moult at the same time, which possibly makes it the only species of bunting with a complete moult in both ad and juv. No pre-nuptial moult. — AUTUMN **Ad/1stW** Ground colour of plumage off-white or greyish-white with no or only limited brown or buff tinge. Tertials and greater coverts edged and tipped pale buff-brown, edges grading into dark centres without sharp borders. **Juv** Ground colour of plumage slightly yellowish-buff, less grey-white. Uniformly broad buff-white fringes to tertials and greater coverts more sharply defined than ad, sharply contrasting against uniformly dark brown feather-centres.

BIOMETRICS L 16–19.5 cm. Marked size difference between sexes (mix of birds from S Scandinavia, Germany, Baltic States, Poland, Hungary, Spain, Cyprus, N Africa): **W** ♂ 99–107.5 mm (n 36, m 103.1), ♀ (once 87.5) 90–101 mm (n 22, m 93.6); **T** ♂ 69–80 mm (n 36, m 74.0), ♀ 63–73 mm (n 22, m 67.8); **T/W** m 72.0; **B** 13.5–17.3 mm (n 57, m 15.6); **B**(f) 11.1–13.8 mm (n 57, m 12.6); **BD** 7.8–10.0 mm (n 44, m 8.9); **Ts** ♂ 23.3–26.7 mm (n 35, m 25.2); ♀ 22.0–25.3 mm (n 20, m 24.1). **Wing formula: p1** minute; **p2** < wt 0–2.5 mm; **pp**(2)**3–4** longest; **p5** < wt 2–6 mm; **p6** < wt 8–15 mm; **p7** < wt 14–22 mm; **p10** < wt 24–30 mm; **s1** < wt 24–32 mm. Emarg. pp3–5, also slightly near tip of p6.

GEOGRAPHICAL VARIATION & RANGE Monotypic. — Europe except most of Fenno-Scandia and Ireland; also in N Africa, Turkey, Caucasus, Transcaucasia, N & W Iran, Central Asian mountains to SE Kazakhstan. — Several subspecies apart from *calandra* have been described, for example '*clanceyi*' (W Ireland, W Scotland), '*germanica*' (Germany), '*algeriensis*' (NW Africa), '*thanneri*' (Canary Is) and '*buturlini*' (SE Turkey, much of Middle East, S Central Asia). There is a very subtle cline running from west to east in that plumage becomes slightly paler and greyer, and less prominently streaked. Birds on Canary Is may also be a trifle smaller on average. However, these differences are quite variable and occur also within subpopulations as individual variation, so any definition of racial boundaries creates insoluble borderline problems, hence the species is treated here as monotypic. (Syn. *algeriensis*; *buturlini*; *clanceyi*; *germanica*; *thanneri*.)

REFERENCES CAMPOS, F. *et al.* (2005) *Ring. & Migr.*, 22: 159–162. – MORGAN, J. H. (2006) *Ring. & Migr.*, 23: 125.

Israel, Nov (left) and Saudi Arabia, Jan: variation in fresh plumage, but ageing impossible after complete post-nuptial or post-juv moult. Note individual variation in streaking below. When neck stretched white ear-coverts spot looks larger (left-hand bird), and sometimes vague lateral crown-stripes may be apparent. (Left: A. Ben Dov; right: P. Roberts)

Denmark, Sep: complete post-nuptial or post-juv moult just finished, or nearly so. Possibly there are still some unmoulted feathers at sides of neck and throat, but since these cannot be assigned to generation, ageing is impossible. Note here surprisingly long-legged appearance. (K. Dichmann)

Juv, Spain, Jun: slightly white-scalloped upperparts, more spotted (not so streaked) breast and better-marked dark lower border to ear-coverts. Note also ochre-buff tinge to head and breast not seen in later plumages. (M. Schäf)

ROSE-BREASTED GROSBEAK
Pheucticus ludovicianus (L., 1766)

Fr. – Casse-graine à poitrine rose
Ger. – Rosenbrust-Kernknacker
Sp. – Picogrueso pechirrosado; Swe. – Brokig kardinal

Almost annual autumn vagrant to W Europe, mainly Britain, where it is most frequent in SW England in October. Some records outside Britain are regarded as escapes. Rose-breasted Grosbeak, a member of the family Cardinalidae, is a migrant breeder in North America, where it frequents deciduous forest and second growth. Vagrants to Europe often feed on berries. It is not particularly shy, often permitting rather close approach.

IDENTIFICATION A large, *chunky passerine with a hefty pale bill of almost Hawfinch-like proportions*. The bill of adult ♂ is usually pale pinkish bone-coloured (with only slightly darker tip), whereas ♀ and young have somewhat darker bills. Apparently all genuine records in Europe have involved individuals in first-winter plumage, which has a striking head pattern involving *long broad whitish supercilium and median crown-stripe*, and a *whitish submoustachial stripe* partly surrounding the dark brown cheeks. It also has a complex wing pattern (distinctive in flight) with *two white wing-bars and white basal remiges patch* (very small in young ♀♀, though) *and white tertial tips*. Rest of plumage, especially upperparts, largely brown, streaked darker, with off-white and less heavily streaked underparts, often with a yellow-buff tone. There is a characteristic *chrome-yellow patch near wing-bend*, but this can be concealed when wings are tucked in properly. Underwing-coverts are lemon-yellow. Legs dark brown.

VOCALISATIONS Only call heard in Europe is a hard *chick* note.

SIMILAR SPECIES Quite distinctive and should not cause identification problems in Europe.

Ad ♀, USA, May: note strong conical bill and bold whitish supercilium that, together with whitish submoustachial stripe, borders dark brown cheeks. Dark-streaked upperparts brownish, while pale underparts, breast and flanks being buffish, are boldly streaked. Spotty white wing-bars, and compared to ♂ much smaller basal white primary patch and almost uniform tail. Aged by still fresh dark ad primary-coverts. In-hand character is yellow underwing-coverts. (B. Coster)

AGEING & SEXING Reliable ageing and sexing require check of moult and feather patterns, especially in autumn. Sexual dimorphism striking in ad. Strong seasonal and age-related plumage variation. – Moults. Generally complete post-nuptial moult of ad mostly in Jun–Sep, though some suspend and leave a few remiges until after autumn migration. Partial post-juv moult, including head, body, usually some to all median and greater coverts, also occasionally isolated secondaries and r1, but no alula and tertials, mostly in Jun–Sep. Partial pre-nuptial moult in Oct–May of variable extent, may involve most of body, some median and some to all greater coverts, tertials and tail-feathers, but seems more extensive in 1stY and is to some extent complementary (i.e. a continuation of post-breeding and post-juv moults). – **SPRING Ad ♂** (plumage develops both through abrasion of feather tips and moult). Striking, with largely black head to upper breast and upperparts, clean white belly, rump and wing markings (largest/boldest), with large rose-red breast and underwing-coverts patches; rr4–6 have large white tips. A bird in this plumage in W Palearctic is probably an escape. **Ad ♀** Largely brownish above, mantle heavily streaked black, as are scapulars, breast and flanks. Rather bold whitish supercilium, with reduced white primary patch and lacks white distal patch on outer tail-feathers. Also note buffish underwing-coverts. **1stS ♂** As ad ♂ but many

Ad ♂♂, Mexico, Apr (left & right): unmistakable ♂ especially attractive in breeding plumage due to black head, bright red breast, and white rest of underparts, rump and large spots on uppertail-coverts. Has black wings with spotty white wing-bars (upper bar more like shoulder patch) and tips to tertials, and note substantial basal white primary patch. Rather long, slightly notched black tail has white patches visible from below, or when spread. Whole wing evenly feathered. In-hand characters include bright pink underwing-coverts. (H. Shirihai)

1stS ♂, Mexico, Apr: like ad but black areas browner, with bleached and very worn brown primaries, primary-coverts and secondaries, and much smaller basal white primary patch. Partial white supercilium behind eye, like in winter plumage (just visible here). Tail already moulted. (H. Shirihai)

1stS ♀, Cuba, Apr: unlike ad, primary-coverts (tips just visible), alula and a few outer greater coverts are juv (worn and bleached browner), while any basal white primary patch is usually covered by wing-coverts. Age confirmed by moult limit in tail. Note often visibly pale rear ear-coverts patch, while fore supercilium often appears wider and buffier. (H. Shirihai)

CARDINALS AND ICTERIDS

Ad ♂♂, Costa Rica (left) and Panama, Dec: following complete post-nuptial moult, in partly barred or spotted winter plumage due to pale brownish edges to black head and upperparts, but whitish areas also appear smudged, with dark-streaked breast-sides and flanks. Pale supercilium and submoustachial stripe surround mottled cheeks. But wings and tail still black and white like in summer, and large white primary patch is still best age criterion. (Left: H. Shirihai; right: V. & S. Ashby)

least-advanced birds intermediate between non-breeding and breeding plumages. Most show mix of worn and bleached brown (juv) and contrastingly black or black-and-white (ad-like) wing- and tail-feathers. **1stS ♀** As ad ♀ but shows mix of feather generations as 1stS ♂. – AUTUMN **Ad ♂** Breeding plumage concealed by fresh buffish-olive fringes and thus recalls ♀♀, with brownish and streaky appearance above, and whitish supercilium. Still, unlike ♀, typically white markings in wings and tail most extensive, and has blacker centres to mantle-feathers and scapulars, dark-barred white back and rump, more pink-red on throat and breast, lesser coverts and underwing-coverts. Also unlike breeding plumage, throat mainly pinkish, breast streaky and some white areas washed cream-buff. **Ad ♀** As in spring but broadly fringed pale. Both sexes evenly fresh or have distinctive unmoulted and very worn remiges (due to suspension). As a rule no white distal patches on outer tail-feathers as in ♂♂. **1stW ♂** Much like ad ♀ (or intermediate between ad ♀ and non-breeding ad ♂), but remiges, primary-coverts, alula and tail juv, slightly weaker textured, less fresh and browner, while tail-feathers are subtly narrower and more pointed (with reduced off-white distal patches compared to ad ♂). Size of white basal primary patch intermediate, larger than in ad ♀ and distinctly smaller than in ad ♂; tips of retained greater coverts narrower and buffier than ad ♂-type. Also unlike ♀, less heavily streaked on breast, but usually has some pink, and always has pink underwing-coverts (not buffish). **1stW ♀** Dullest plumage, but generally as fresh ad ♀, from which distinguished by moult limits in wings similar to those in 1stW ♂. White basal primary patch small and ill-defined (even absent). Wing-bars narrower, and pink suffusion to plumage lacking, or virtually so. **Juv** Not recorded in the covered region.

BIOMETRICS L 17.5–19 cm; W ♂ 99–110 mm (n 18, m 103.9), ♀ 97–107 mm (n 12, m 101.1); T ♂ 69–77 mm (n 18, m 73.6), ♀ 69–79 mm (n 12, m 73.3); T/W m 71.5; B 17.5–21.6 mm (n 30, m 19.6); B(f) 14.0–18.6 mm (n 30, m 16.2); BD 11.0–15.0 mm (n 29, m 13.3); Ts 20.6–23.5 mm (n 24, m 22.4); **white on r6** ad ♂ 30–36 mm, 1stY ♂ 0–27 mm, ♀♀ 0 mm; **white primary patch** (> pc) ad ♂ 15–29 mm, ad ♀ 0–3 mm, 1stW ♂ 2.5–9 mm, 1stW ♀ 0 mm. **Wing formula: p1** minute; **p2** < wt 0.5–4.5 mm, =5 or 5/6; **pp3–4** about equal and longest; **p5** < wt 0.5–4 mm; **p6** < wt 6.5–11 mm; **p7** < wt 12–19 mm; **p10** < wt 22–30 mm; **s1** < wt 24–32 mm. Emarg. pp3–5.

GEOGRAPHICAL VARIATION & RANGE Monotypic. – S Canada, E USA; winters from S Mexico to NW South America.

1stW ♂♂, Canada, Sep (left) and Azores, Oct: overall, 1stW birds are ♀-like. Remiges, primary-coverts and most tail-feathers are juv, forming strong moult limits in wings of young ♂♂ as soon as new ad-like coverts grow. On left bird, broadly white median coverts already moulted, and in right bird new black greater coverts are very contrasting. Also unlike ad ♀, white primary patch slightly larger and bolder, with primaries basally blacker enhancing the white patch. Not visible here are yellow rather than pink underwing-coverts. (Left: M. Gauthier; right: R. Ek)

1stW ♀, Azores, Oct: unlike ad ♀ and 1stW ♂, white primary patch usually wholly lacking. Aged by juv primary-coverts, but juv feathers sometimes more easily detected in alula. In all ♀-like plumages, visibility of pale median crown-stripe varies with angle and feather position, while dark lateral crown-stripes appear almost like a cap in profile. (V. Legrand)

INDIGO BUNTING
Passerina cyanea (L., 1766)

Fr. – Passerin indigo; Ger. – Indigofink
Sp. – Azulillo índigo; Swe. – Indigofink

In spite of its name this attractive bird is more of a finch than a bunting, being related to the Red Cardinal and some of the New World grosbeaks. The completely blue adult breeding ♂ is very handsome, whereas the ♀ is nondescript like a (small) ♀ House Sparrow. It breeds in SE Canada and in much of E USA in brushy areas with scattered trees and taller bushes, often feeding in weeds and tall grasses, while winter is mainly spent in Central America. Recorded in a number of European countries in spring, summer and autumn, but some of these records might relate to escapes.

Ad ♂, USA, Jun: a small, stocky finch with conical bill and rather short tail that is unlikely to be confused, especially in near all-blue ad ♂ plumage. In close views, the deeply blue head, black loral area and some exposed dark-centred upperwing-coverts should be visible. (D. Monticelli)

IDENTIFICATION A well-proportioned Linnet-sized finch with conical dark bill with slightly curved culmen, dark legs and dark eye. Adult breeding ♂ is entirely blue with darker lores and centres to wing-feathers, whereas non-breeding and immature ♂♂ are patchily blue mixed with brown. ♀ and first-winter are rather plain grey-brown with faint paler buff-brown double wing-bars and a short and very indistinct paler supercilium. The breeding ♀ develops a slightly paler throat and more obvious dark streaking on breast, and uppertail wears to more prominent blue (being more obscure in fresh autumn plumage). First-summer ♂♂ are rather variable as to amount of blue, some being quite similar to adults, only less brilliant blue, while others have much brown among the emerging blue. Those with less developed blue have the pale double wing-bars like in autumn.

VOCALISATIONS Call is a sharp and liquid clicking *tsick*, rather like a 'muted Hawfinch call', repeated when alarmed. Also a drier variation, *tett*.

SIMILAR SPECIES ♀ and immature *Lazuli Bunting* resemble corresponding plumages of Indigo Bunting but differ by unstreaked more cinnamon-tinged breast and more prominent pale buff-white double wing-bars. – *Blue Grosbeak* is larger with proportionately heavier bill, and ♀/immature have prominent rufous-buff double wing-bars.

AGEING & SEXING Ageing using moult status, and development of blue in plumage for ♂♂, possible, but some uncertainty as to extent of moult in 1stY limits technique. Sexual dimorphism striking in breeding plumage and also obvious in non-breeding adults. Strong (also age-related) seasonal plumage variation. – Moults. Complete post-nuptial moult in ad, generally Aug–Oct. Partial post-juv moult in Jun–Dec includes head, body, usually all median and most inner to all greater coverts, occasionally some tertials but no remiges or primary-coverts. Pre-nuptial moult in winter (Jan–May) variable and perhaps partly a continuation of post-juv moult for imm, often includes many, sometimes all greater coverts, a few or several outer primaries, rarely 1–2 outer primary-coverts, a few or several inner secondaries (Pyle 1997) and odd tertials. More research required to better map moult variation in the two age groups. – SPRING **Ad ♂** All blue, head, nape, throat, breast and upper belly glossy ultramarine, rest more cobalt or cerulean-blue. Lores and centres to flight-feathers black, partly visible. Bill strikingly two-coloured, upper mandible black, lower ivory-white. All remiges, primary-coverts, greater coverts and alula uniform, edged blue. **Ad ♀** Brown and off-white, slightly rufous-tinged on head, back, upperwing and flanks, olive-tinged on chest and rump. Throat whitest, chest diffusely streaked or mottled. Hint of dull and faint ochre double wing-bar. Tail-feathers often edged blue in abraded summer plumage, and rarely some purplish-blue tinge on rump. Bill similarly two-coloured as ♂, only duller. **1stS ♂** Variable amount of blue developed, some close to adult ♂ appearance, others have much brown still. Remiges, primary-coverts and alula edged mainly brown (only little blue tinge) and more worn, often also some outer greater coverts retained juv feathers in contrast to blue-edged inner. Sometimes mixture of all-brown retained inner primaries and outer secondaries contrast to darker and fresher rest being finely edged bluish. **1stS ♀** As ad ♀ but sometimes, before too worn, identified by mixture of old and new remiges and primary-coverts. – AUTUMN **Ad ♂** Variable, some remaining almost as blue and neat as in spring but more often much blue replaced by brown or covered by brown tips, resulting in chequered plumage. **Ad ♀** As in spring. All primary-coverts fairly broad and rounded, fresh and rather dark, usually faintly blue-edged. Remiges uniformly fresh and densely textured, often thinly edged pale blue. **1stW ♂** As ad ♀, but primary-coverts slightly narrower and more pointed, slightly paler brown and not so glossy, from Oct first blue feathers or patches begin to appear, often first on breast, later all over. **1stW ♀** As ad ♀, but primary-coverts slightly narrower and more pointed, slightly paler brown and not so glossy. No blue in plumage, but may show faint green tinge on rump and lesser coverts. **Juv** Similar to ♀/imm but plumage less neat, more soft, and feathers of upperparts tipped grey and underparts more heavily streaked dark. Not expected to occur in Europe.

1stS ♂, Mexico, Mar: unlike ad ♂, 1stS ♂ has more brown feathers above and less pure blue feathering. Note also juv brownish greater coverts and primary-coverts, but recently renewed blue-edged outer primaries, which is a typical moult pattern of this age. (S. N. G. Howell)

Ad ♂, USA, Apr: despite distinctive blue coloration, moult and plumage variation in relation to age and sex are highly complex. Ad ♂, here apparently after incomplete post-nuptial moult, and which theoretically therefore could be in its 2ndS (or is just normal variation of ad), still with some retained winter coverts and tertials with brownish fringes. Note bluish-edged ad-type primary-coverts, primaries and tail. (S. Elsom)

CARDINALS AND ICTERIDS

BIOMETRICS **L** 12–13 cm; **W** ♂ 66–72 mm (n 13, m 69.0), ♀ 61–68 mm (n 16, m 64.5); **T** ♂ 49–54 mm (n 13, m 51.6), ♀ 42–50 mm (n 16, m 47.5); **T/W** m 74.2; **B** 12.3–14.3 mm (n 29, m 13.0); **B**(f) 9.0–11.2 mm (n 29, m 10.1); **BD** 6.1–7.2 mm (n 28, m 6.6); **Ts** 15.2–18.0 mm (n 28, m 17.1). **Wing formula: p1** minute; **p2** < wt 0–7 mm; **pp**(2)**3–4**(5) about equal and longest; **p5** < wt 0–2.5 mm; **p6** < wt 3–6.5 mm; **p7** < wt 6.5–10.5 mm; **p10** < wt 10–17 mm; **s1** < wt 10–18 mm. Emarg. pp3–5 (rarely faintly on p6 too).

GEOGRAPHICAL VARIATION & RANGE Monotypic. – SE Canada, E & SC USA; winters Mexico, Panama, West Indies, less commonly Florida.

REFERENCES Wolfe, J. D. & Pyle, P. (2011) *Western Birds*, 42: 257–262.

♀, Azores, Oct: ♀♀ are often sullied cinnamon-buff below. This bird from Jul has obvious moult limits among greater coverts and tertials, with outer greater coverts being shorter and more whitish-tipped than inner suggesting these are juv and the bird is *c.* 1 year old, before starting complete post-nuptial moult. However, both moult and development of any blue is variable, and safe ageing of 1stS best accomplished by noting browner and more frayed juv primary-coverts and replaced outer but retained inner primaries. This often requires handling, and this bird best left unaged. (V. Legrand)

1stW ♂♂, Azores, Oct: visible greater coverts (at least most), remiges, primary-coverts, alula and rectrices are juv, being already slightly worn. Further, some blue feathers have already grown on rump, lesser coverts and side of rump indicating ♂. (Left: V. Legrand; right: D. Monticelli)

♀♀, presumed 1stW, Azores, Oct: there is some variation in underparts streaking of ♀♀, some being more streaked (right bird), and both types could occur at same season, differences which may relate to age or timing of moult. Note characteristic jizz, bill shape, diffuse pale eye-ring and rather plain brown head. (Left: V. Legrand; right: D. Spittle)

BOBOLINK
Dolichonyx oryzivorus (L., 1758)

Fr. – Goglu des prés; Ger. – Bobolink
Sp. – Tordo charlatán; Swe. – Bobolink

This member of the New World blackbirds is a vagrant to NW Europe, most frequently recorded in Britain, usually in late September or October on the Shetlands or Isles of Scilly, where it is almost annual.

A migrant breeder that frequents tall grass, flooded meadows and farmland across much of S Canada and the N USA, Bobolinks winter in marshes, rice fields and pastures of N Argentina, and are a widespread transient east of the Andes and in Amazonia. It is thus a strong flier with long pointed wings.

♂, USA, May: this New World icterid is remarkably similar to Afrotropical bishops (weavers) in both appearance and use of grassy habitats. Full ♂ breeding plumage unmistakable with its buff nape and largely black-and-white rest of plumage, but not recorded in the treated region. Bobolinks undergo a complete pre-nuptial moult in both ad and 1stW, hence ageing not possible in spring and summer. (E. Orf)

IDENTIFICATION Beautifully plumaged ad ♂ unmistakable but not yet recorded in Europe, where apparently only first-winters have been observed. Latter (and similar ♀ and non-breeding ♂) typically have strikingly *yellowish-buff plumage* and well-marked head pattern of *dark eye-stripe and crown-sides, yellow-buff supercilia and median crown-stripe*, and pale lores and cheeks. *Upperparts, breast and flanks heavily streaked dark, wing-coverts and tertials boldly fringed paler*. Very characteristic are two *striking buff tramlines on mantle. Strong pinkish-straw conical bill.* Fairly *short tail* with narrow, *pointed rectrices*. Frequently walks on ground, with *lark-like gait* and constant forward-stretching of neck and head; often pauses briefly, raising head to look around. *Flight* pattern useful for identification, as seemingly cardueline appearance is belied by *direct* (not undulating) path typical of all icterids.

VOCALISATIONS Calls are mainly monosyllabic and clicking or grating. Commonly gives a repeated, hard *chek*. Another call is softer and more buzzing, *cheez*, also often repeated.

SIMILAR SPECIES Given good view, unmistakable. Only significant confusion risk is *Yellow-breasted Bunting* (rare breeder in extreme NE Europe and vagrant to W Europe), but Bobolink should be easily separated by being larger, with a paler, slightly larger bill, having no dark lower ear-coverts border, its more striking and yellower head pattern, and pointed tail-feathers. – Greatest confusion risk in Europe probably lies with *escaped weavers* in ♀-like plumages, as many species share Bobolink's streaky appearance, as well as bill and tail shape, and behaviour on ground. For these, consult suitable handbooks.

AGEING & SEXING Understanding of moult pattern and seasonal change essential for reliable ageing. Sexual dimorphism striking in breeding plumage. Complex and strong (also age-related) seasonal plumage variation. – Moults. Two complete moults per year (exceptional strategy, known only from few other species): post-nuptial rapid (e.g., all tail-feathers sometimes replaced concurrently) but can be suspended during migration, while pre-nuptial also swift, commenced on wintering grounds but again can be suspended during migration if not finished in time. Partial post-juv moult (to unknown extent sometimes suspended), includes head, body, usually all median and most to all greater coverts, occasionally some tertials but apparently no alula or tail-feathers. In general, moult on breeding grounds occurs mostly Jul–Oct and in winter quarters Jan–mid Apr. – **SPRING Ad ♂** All-black face and underparts (when still very fresh, tipped and fringed pale yellow), with pale buff nape and shawl, buffish-white fringes to otherwise black wings, and striking white rump and scapulars. Birds with extensive buff tipping still in late spring thought to represent individual variation rather than 1stS, but details still poorly known. **Ad ♀** Buff median crown-stripe outlined by broad blackish lateral crown-stripes. Pale buff broad supercilium, narrow dark eye-stripe. Upperparts and flanks heavily streaked dark on pale buffish ground. Plumage fresh by Apr–May. **1stS** Indistinguishable from ad once plumage completely replaced. Some (both ages, but apparently chiefly 1stS) return to breeding areas having suspended before some wing-feathers are replaced; ageing still difficult due to similarity between the two age groups. – **AUTUMN Ad ♂** After post-nuptial moult completed inseparable from ♀, though some have blackish throat-feathers (or distinctive black centres); black streaks on flanks usually bolder and more extensive. See also Biometrics. **Ad ♀** Rather bright and distinctly fringed buff-orange following moult. Often inseparable from ad ♂. During autumn migration both sexes differ from 1stW by being evenly fresh. Birds that have suspended moult and possess some retained heavily-worn remiges (frequent!) are highly distinctive too. **1stW** (Sexes indistinguishable but see Biometrics.) Like fresh ad but remiges, primary-coverts, alula, tail and occasionally 1–2 outer greater coverts and tertials juv, these being slightly weaker textured and browner. The sharply-pointed tail-feathers gradually taper off (in ad, inner web has marked bulge subterminally, especially on outer feathers; cf. Pyle 1997). Typically has bright, more yellowish plumage, and tends to have sharper off-white fringes to wing-feathers (more buffish and less sharp or less even-width in ad). Pale areas of upperparts tinged yellowish-buff (more cinnamon-buff and olive in ad). Underparts strongly tinged pale yellow, less warm buff and white, and chest usually lacks small dark spots. Black streaks on flanks generally less heavy and lower down end in a point (ad: rounded blob). Some ♂♂ develop a few partly or all-black feathers on cheeks, throat or chest during autumn/winter. Midwinter ♂♂ can look variegated, with isolated black feathers on underparts and wing emerging. **Juv** Not expected to occur in Europe, as soon achieves 1stW plumage.

♀, USA, Jun: also colourful though lacking strong pattern of ♂, and readily distinguished from any European passerine, with pinkish-straw conical bill (dusky tip and culmen), long wings and mostly ochre-buff head with dark stripe behind eye and dark lateral crown with narrow median crown-stripe. Note boldly streaked creamy-buff flanks. (M. Costina)

♂, Canada, May: some spring ♂♂ differ in that black feathers on face and below are still extensively tipped buff, and more brownish wing-coverts being fringed and tipped buff, pale edges remaining well into summer. This is believed to represent individual variation rather than such birds being 1stS (though proof still lacking). Upperparts and wings have been moulted once but new feathers more intermediate (centres browner with broad pale fringes). (D. Beadle)

CARDINALS AND ICTERIDS

Winter to spring ♂♂, Argentina, Mar: with European records as late as Dec, and once in Apr, a bird in this plumage on this side the Atlantic is a possibility. The bird on the left has completed moult and has less brown fringing remaining, while the bird on the right is just finishing moult (outer primaries and some rectrices growing) and has broader brown fringes on back and wing-feathers. Neither have retained juv or ad winter feathers and are therefore not possible to age. (R. Güller)

England, Oct: either ad ♀ or 1stW, most of which are impossible to sex (though see images below). Like spring/summer ♀, but brighter, as feathers above fringed richer buff, and face and underparts also warmer, while bill brighter pink. Remiges, primary-coverts and alula may be ad but could be fresh juv, so best to refrain from ageing or sexing. Pale median crown-stripe and stripes on back shown nicely here. (G. Reszeter)

BIOMETRICS L 16–18 cm; **W** ♂ 92–105 mm (n 24, m 97.8), ♀ 83–92 mm (n 12, m 87.6); **T** ♂ 58–70 mm (n 24, m 65.2), ♀ 57–66 mm (n 12, m 61.0); **T/W** m 67.6; **B** 15.1–18.0 mm (n 35, m 16.3); **B(f)** 12.0–15.5 mm (n 34, m 13.6); **BD** 8.0–10.4 mm (n 33, m 9.3); **Ts** ♂ 25.5–28.3 mm (n 23, m 27.1), ♀ 24.5–28.0 mm (n 12, m 26.0). **Wing formula: p1** minute; **pp2–3** about equal and longest; **p4** < wt 3–7 mm; **p5** < wt 7.5–13 mm; **p6** < wt 12–18.5 mm; **p10** < wt 24.5–35 mm; **s1** < wt 27–37 mm. Emarg. pp3–4.

GEOGRAPHICAL VARIATION & RANGE Monotypic. – S Canada, N USA; winters N Argentina, E Bolivia, S Brazil and adjacent areas.

1stW ♂, Azores, Oct: black bases on upper breast-sides suggest a ♂. Already worn juv remiges, primary-coverts, alula, outermost greater coverts and tertials age as 1stW. Head pattern characterised by long, broad buff supercilium bordered by dark eye-stripe and lateral crown-stripe, and well-demarcated pale median crown-stripe (F. Lopez)

1stW, presumed ♀, Azores, Oct: ageing as bird at left, but much plainer face and underparts suggest ♀. (D. Monticelli)

BALTIMORE ORIOLE
Icterus galbula (L., 1758)

Fr. – Oriole de Baltimore; Ger. – Baltimore-Trupial
Sp. – Turpial de Baltimore; Swe. – Baltimoretrupial

Another rare North American autumn vagrant to W Europe, most frequent in Britain in autumn (September–October), with some winter and even spring records. A migrant breeder in riparian woodland, orchards and savanna from SW and S Canada south over much of E USA. It rarely winters in SE USA, but usually withdraws in winter to Middle and N South America. Vagrants can occur in a variety of bushy habitats with copses or scattered trees, including gardens, where the species even has visited feeders.

Ad ♂, USA, Apr: mid-sized icterid with a pointed long bill. Plumage of breeding ♂ would be unmistakable if ever recorded in the treated region (so far no such records). Evenly very fresh ad wing and solid black and bright orange plumage infer age and sex. (A. Murphy)

IDENTIFICATION The *size of a small, slim-shaped starling* with a *rather long, sharply-pointed bill*. Adult ♂ unmistakable due to its jet-black and bright orange plumage, so far unrecorded among vagrants to Europe, which have all been first-winter birds in ♀-like plumage. These are also easily identified by their *pale grey-brown, long conical, pointed bill*, *prominent double white wing-bars* and a certain amount of *orange-yellow in the plumage*, mostly on the head, breast, uppertail-coverts, uppertail, undertail and vent. Note also the variably dark olive-brown mantle and *dark eye*, the latter prominent in an otherwise rather featureless face.

VOCALISATIONS Peculiar call variation, one type almost resembling the call of a Greenshank *Tringa nebularia*, a strong, resounding, deep whistle *cheu-cheu*, another a harsh chattering *che-che-che-che-...*, recalling a quarrelling Magpie.

SIMILAR SPECIES No other species of *Icterus* has been recorded in Europe, and it should be very difficult to confuse this species with any native European species. However, there is a remote chance that in future, the very similar Bullock's Oriole (*Icterus bullockii*; not treated) is confirmed as a vagrant, requiring careful check of immature plumages (most likely to occur) before these two can be separated. Bullock's has on average paler and purer yellowish cheeks and less prominent lower wing-bar (greater coverts more narrowly tipped white). Also, Bullock's tends to have slightly more prominent darkish eye-stripe.

AGEING & SEXING For correct ageing check moult contrast, feather wear and pattern in wing and tail. Strong sexual dimorphism, but except development of 1stY plumages, there is limited seasonal plumage variation. Use of biometrics recommended for sexing, especially with intermediate plumages (especially ♀♀ and 1stY plumages). – Moults. Complete post-nuptial moult of ad and partial post-juv moult mostly Jun–Sep. Post-juv moult includes head, body, and a variable number of median coverts (none to some) and sometimes a few inner greater coverts. Partial pre-nuptial moult in Nov–May (most significant in 1stY, but variable) includes some or all greater coverts and tail-feathers, and some tertials. – **SPRING Ad** ♂ Head, throat, mantle, back, much of wings and central tail-feathers black, with a black basal bar on outer tail-feathers; rest of underparts, rump, uppertail-coverts, wing-bend, scapulars, median coverts and broad tips to outer tail-feathers deep yellow-orange. Contrasting white greater coverts wing-bar and white panel to secondaries. **Ad** ♀ Much duller than ad ♂, crown and nape golden-brown, mantle and back orange-brown blotched blackish; throat yellow-orange like breast (no black bib), and tail rather plain olive-yellow, darker above, more pale yellow below (thus no black in tail). Wing-bars and panel on inner secondaries reduced, and wing-bend and scapulars blackish with golden-brown tips; median coverts blackish with broad yellow-white tips forming upper wing-bar. **1stS** From ad by retained juv primaries (heavily worn and bleached browner), primary-coverts and perhaps some other coverts, tertials and tail-feathers (but differences in the field not as obvious as in ♂). Retained outer greater coverts have thin (abraded) pale edges contrasting with fresh inner new ones, which have

Ad ♀, USA, May: in profile, note long, still thick-based, pointed bill; in ad summer, pale area of bill is bluish-grey. Some ♀♀ mirror ad ♂ plumage, being dark-headed but upperparts are browner and to some extent mottled (rather than solid black), still can be difficult to separate from dull 1stS ♂. If aged first (by evenly very fresh wing, and round-tipped primary-coverts and tail-feathers), such plumage must be ad ♀. Tends to lack yellow-orange shoulder patch (wing-bend) and has brownish rather than black central rectrices. (A. Murphy)

Ad ♀, USA, May: a pale bird, readily separated from any dull young ♂ by its lack of blackish head and its wing and tail pattern. Moderate wear and evenly-feathered ad wing, and rounder-tipped primary-coverts and tail-feathers are best means of ageing. (A. Murphy)

CARDINALS AND ICTERIDS

broad edges to dark centres. **1stS ♂** Generally intermediate between ad ♂ and ad ♀, though usually clearly brighter than latter (e.g. has all-black rather than dark-blotched mantle, though beware that some pale fringes to dark mantle-feathers can persist until Jun in some young ♂♂), and often has more yellowish-orange (rather than the bright and deep orange of ad ♂) on rump and underparts. A few least-advanced birds not readily sexed in the field. Ageing of spring/summer ♂ usually unproblematic since new inner black ad greater coverts with wide white tips and edges are highly contrasting against the outer brown retained juv ones. **1stS ♀** Head to mantle dark olive-brown (dusky centres smaller and less deep than ad ♀), bib only faintly suggested and shorter, or sometimes lacking, thus in some chin and throat mostly yellow (with only indistinct and irregular dark smudging). Pale wing markings reduced. – AUTUMN **Ad ♂** Much as spring, but white wing fringes broader, in some birds black of mantle and scapulars narrowly tipped orange, and orange rump and underparts can be faintly tinged olive. **Ad ♀** As breeding but head, upperparts and wings broadly fringed pale. Unlike any 1stW, at least in early to mid autumn there are at least some (often numerous) dusky centres to crown and mantle, giving marbled pattern, and some (not all) have dark bib. Only later in autumn or winter, some 1stY ♂♂ develop more dark feathers above and on bib, and can be difficult to separate from strongly-marked ad ♀, if not aged first. **1stW** Rather uniform and dull with black or dark feather-centres or blotches above largely reduced or concealed. Beware that wear, pattern and colour of primary-coverts rarely differ from ad ♀, thus look for more pointed and narrower tail-feathers, as well as relatively less fresh condition of primaries, which are more reliable for ageing in early autumn. With experience, early autumn 1stW ♂ may be sexed by relatively broader pale tips and fringes to retained juv greater coverts (sometimes twice as wide), as well as deeper orange-yellow wash to chest, duskier face and more exposed dark centres on mantle, but much variation and overlap between the sexes in young plumages, rendering differences of limited use. Nevertheless, dullest individuals, with entire plumage more olive-grey, less yellowish-orange below, are most probably always 1stW ♀♀. Later in winter, 1stW ♂ acquires more black on head, upperparts, wings and throat, and is easier to sex. **Juv** Not expected to occur in Europe.

BIOMETRICS L 17–22 cm; **W** ♂ 91–101 mm (*n* 15, *m* 97.2), ♀ 87.5–96 mm (*n* 12, *m* 91.9); **T** ♂ 68–77 mm (*n* 15, *m* 72.5), ♀ 67–75 mm (*n* 12, *m* 70.8); **T/W** *m* 73.0; **B** 18.6–22.3 mm (*n* 27, *m* 20.4); **B(f)** 16.0–19.2 mm (*n* 26, *m* 17.4); **BD** 6.0–8.7 mm (*n* 25, *m* 7.7); **Ts** 21.1–25.5 mm (*n* 25, *m* 23.8). **Wing formula: p1** minute; **p2** < wt 0–3 mm, =4/5 or =5; **pp(2)3–5** about equal and longest; **p6** < wt 4.5–8 mm; **p7** < wt 9.5–15 mm; **p8** < wt 13–20 mm; **p10** < wt 20–26 mm; **s1** < wt 23–29.5 mm. Emarg. pp3–6.

GEOGRAPHICAL VARIATION & RANGE Monotypic. – S Canada, much of USA except south-east; winters Mexico south through Central America to N South America, including Trinidad & Tobago and NW Ecuador, and the Greater Antilles.

TAXONOMIC NOTE The Baltimore Oriole has enjoyed a slightly chequered taxonomic history, long considered a monotypic species it was thereafter 'lumped' with Bullock's Oriole (*I. bullockii*, which mainly breeds in the W USA) under the name Northern Oriole, but they are now again regarded as separate based on robust mtDNA evidence and some obvious morphological differences in ♂ plumage. Degree of hybridisation is apparently limited and stable.

1stS ♂♂, USA, Apr: note extreme variation in first ♂ breeding plumage, but both still have heavily worn and abraded, brownish juv remiges and primary-coverts. Least-advanced young ♂ on right has deeper orange rump and darker (recently renewed) central rectrices, which are important to check versus dark-headed ad ♀. (A. Murphy)

Ad ♂, Costa Rica, Dec: as summer ad ♂, but evenly very fresh after complete post-nuptial moult. Ad ♂♂ in fresh plumage may have deeper and more reddish-orange underparts. (H. Shirihai)

1stS ♀, USA, May: unlike similar pale ad ♀ (left), note heavily worn and abraded brownish juv primaries and moult limits between tertials and secondaries, and in tail. (A. Murphy)

1stW ♂, Netherlands, Jan: aged as 1stW by largely juv wing, and by strong orange pigments rather than yellow signalling ♂ (without the black head and upperparts of an ad ♂). Compare with young ♀♀ at left and right, and with ad ♂ above right. (T. Luiten)

1stW ♀, USA, Sep: even in dullest autumn plumage unlikely to be confused with any European bird, given long, thick-based but sharply pointed bill, prominent double white wing-bars, and yellowish-tinged head, breast, uppertail-coverts and uppertail, while black eye can be contrasting. Upperparts often pale greyish olive-brown and belly whiter. Age and moult as in previous two images. (A. Murphy)

VAGRANTS TO THE REGION

This section covers those very rare passerine vagrants which have seldom been recorded in the treated region (see Introduction for geographical coverage of the handbook). In general, until the end of 2016, fewer than ten individuals of each species included here are believed to have reached the covered region unassisted and in a wild state. Records of rare species are made continuously, and for some the limit of ten birds has already been exceeded. Although lists of rare birds often focus on the number of occasions that a certain species has been found, i.e. 'records', we feel that it is more interesting to know the number of birds involved. One record can involve more than one bird.

Descriptions in this section are kept short, and the number of photographs low, to save space and because most of these species are neither part of the normal avifauna of the region, nor even of the entire Palearctic. They have often arrived from distant regions due to extreme weather conditions or imperfect navigation abilities in young and inexperienced birds. Some may even have been partly or wholly ship-assisted, but details of this are understandably largely unknown, and all records listed are thought to involve natural movements of wild birds (i.e. excluding known or strongly suspected introductions and escapes, often referred to as 'category D' species). It is hoped that these descriptions will suffice as a first introduction to identification, as well as summarising known records in the covered region.

The texts are not always of the same length, some East Palearctic species being afforded somewhat more detailed accounts reflecting the anticipated greater interest in identification and taxonomy of these birds. Also, species more challenging to identify due to their similarity to related species have often received more attention than, say, a fairly straightforward Red-breasted Nuthatch or Common Grackle.

We gratefully acknowledge the valuable help in compiling the list of records for each species and country, largely done by José Luis Copete and Marcel Haas. They have in turn corresponded with the various national Rarity & Record Committees or similar bodies to as far as possible secure that all records are officially accepted. Others who in various ways helped assemble and check the data, in particular regarding the Middle East, namely Guy Kirwan, Nigel Redman and Magnus Ullman, are also gratefully thanked.

Vagrancy, and its underlying reasons, has always attracted interest both among keen amateur ornithologists (notably by 'twitchers' focussing on rare birds) and by professional ornithologists or scientist trying to understand the mechanisms behind these movements. By listing all of these rare species and their records and numbers within the covered range, an up-to-date tool is provided for further study and analysis. For those wanting to delve deeper into the subject we recommend reading introductory sections in, e.g., Alström *et al.* (1991–), Haas (2012) and Howell *et al.* (2014), and a useful paper by Thorup (2004).

BLUE-AND-WHITE FLYCATCHER *Cyanoptila cyanomelana* (Temminck, 1829)

United Arab Emirates two (Nov 1980, Feb 1999), Oman two (Jan 1982, Nov 2006). A summer visitor to E Asia, wintering in mainland SE Asia and the Philippines south to the Greater Sunda Is. Favours forested mountains, generally in damp mature mixed broadleaf forest with dense undergrowth, often near open places with water and rocky walls, preferring wooded canyons. **IDENTIFICATION** A relatively large (**L** 14.5–16 cm) flycatcher with a stunning ♂ plumage that combines *glossy ultramarine or cobalt-blue crown* becoming more *violet-blue on nape and upper mantle*, and also on *wing-bend, rump and tail-sides* (sometimes entire mantle and back, too); lower mantle, back and upperwing sometimes duller blue, even tinged greenish (individual variation rather than age-related). *Forehead may appear silvery-blue* depending on light and angle. The *black of face, breast and flanks contrasts strongly with the white belly* and *white lateral tail-base*. ♀ is much duller and *characteristically plain*, being generally light olive- or grey-brown, though shows a distinct white eye-ring, and often has a characteristic *pale buff-grey central throat patch on otherwise dark olive-brown throat chest*. Fringes of wings and tail slightly tinged rufous and contrast moderately with the brown of upperparts. ♀ has no white at the tail-base. Bill and legs black. Song, typically delivered from high perch, varied with short fluty warbles and some cracked and melancholic notes. Call a strong clicking *tchuck*. **AGEING & SEXING** **Ad** Sexes differ markedly as described above. **1stW** ♂ resembles ♀, but has blue wings, rump and tail. Young ♀ can be aged on moult limits in wing with retained juv feathers, including some outer greater coverts tipped buffish-rufous. In spring, 1stS ♂ is like ad ♂ but still has some juv outer greater coverts contrastingly pale-tipped and dull greenish, not glossy ultramarine. Rarely all greater coverts moulted to ad type, but primary-coverts still juv, dull bluish-green with little or no gloss. **GEOGRAPHICAL VARIATION & RANGE** Polytypic, with two subspecies often recognised, but the race to have reached the treated region has not been established. *C. c. cyanomelana* (S Kurile Is, Japan and S Korea) is described above, while the other race *cumatilis* ('Zappey's Flycatcher', by some treated as a separate species; NE China, SE Russia and N Korea) differs in ♂ plumage on generally paler bluish coloration, with crown and upperparts turquoise or azure-blue (instead of ultramarine), and face, breast and flanks deep verditer-blue (instead of black), and throat often shows greenish gloss, while nape and upper mantle lack violet tinge of ssp. *cyanomelana*, and tail-base has more white at the centre than sides. ♀ *cumatilis* is generally darker or more rufous-brown with less white on throat. Further, irrespective of age, ♂ appears to have two colour morphs, one with darker glossy blue, another with paler blues; intermediates between them are not infrequent.

C. c. cyanomelana, ad ♂, South Korea, Apr: unmistakable elegant plumage with variable gloss and shades of blue on upperparts, forming contrast with black face and breast and pure white belly. (R. Newlin)

C. c. cyanomelana, ad ♀, China, Sep: rather nondescript, though note the distinct unbroken white eye-ring and narrow dusky-white throat patch. Breast olive-brown, belly whiter sullied light cream-brown, wing and tail tinged more rufous. Seemingly evenly fresh ad wing, and only limited bluish bases visible on crown, infer age and sex. (D. Thirunavukkarasu)

C. c. cumatilis, ad ♂, Hong Kong, Oct: distinctive, with blue parts being paler than in *cyanomelana*, including face to breast, lacking the black facial patch of the other race. (M. & P. Wong)

C. c. cumatilis, 1stW ♂, Malaysia, Dec: has variable degree of blue on back and rump and in wings and tail. Note retained juv outer greater coverts, tipped rufous-buff. (Amar-Singh HSS)

(ASIAN) VERDITER FLYCATCHER *Eumyias thalassinus* (Swainson, 1838)

Iran one (Mar 2001). Summer visitor through Himalayas west to N Pakistan and NE India, and through C & S China in SE Asia to Sumatra and Borneo, usually found in open deciduous, evergreen or mixed forest adjacent to open country; winters on lower elevations and in south of range, where it also frequents plantations and parkland habitats. **IDENTIFICATION** A conspicuous *rather large* (L 14.5–16 cm) and *long-tailed* but small-headed flycatcher, easily identified by its *uniform turquoise-blue colour*. Head, rump and uppertail-coverts often notably brighter, and *remiges edged cobalt-blue*. Has *conspicuous black lores* but is otherwise wholly glossy greenish-blue (forehead often shinier). *Undertail-coverts broadly tipped whitish*, creating a chevron pattern. ♀ resembles ♂ but is subtly *duller greenish-blue and has dark grey* (instead of black) *lores*. Both sexes become glossier blue-green in worn plumage. Legs and short bill black. Confiding habits. Song a series of rapid undulating notes, strident but musical, gradually descending. Call a plaintive *pseeut*. Also utters multisyllabic, drier *tze-ju-jui* or *tzeju-jui*. **AGEING & SEXING Ad** Sexes quite similar, ♀ only slightly duller and greener. **1stY** Very similar to ad but at least in first autumn often has remnants of barred juv feathers among buff-white and grey on chin, upper throat and below gape, and if still some unmoulted juv wing-coverts kept (lighter and brownish-tinged with hint of pale tips), with moult contrast most notable in young ♂, aging is possible. **GEOGRAPHICAL VARIATION & RANGE** Polytypic, with two races recognised, ssp. *thalassinus* (N Pakistan, N India, continental SE Asia, C & S China) being the paler race, with ssp. *thalassoides* (S Myanmar, Thai-Malay Peninsula, Borneo, Sumatra) slightly darker.

Far left: *E. t. thalassinus*, ad ♂, India, Apr: the normally rather long-tailed and small-headed jizz less evident here, but uniform verditer-blue colour (pale greenish-blue) obvious, and wings notably brighter still. There is also a conspicuous black loral patch or mask; all characters taken together make it unmistakable. (M. Kulkarni)

Left: *E. t. thalassinus*, ♀, India, Apr: a duller and paler version of ♂ with less black on lores. Here the broadly white-tipped undertail-coverts can be seen well. (B. Sarkar)

DARK-SIDED FLYCATCHER (SOOTY FLYCATCHER) *Muscicapa sibirica* J. F. Gmelin, 1789

Iceland one (Oct 2012). Breeds in C & SE Siberia east to Kamchatka, Japan and N Korea, further south from E Afghanistan through Himalayas through China to NW Vietnam; winters in SE Asia. Breeds in mixed broadleaf forests in lowlands and mountain foothills, but in non-breeding season found in all types of wooded habitats. **IDENTIFICATION** A quite small (**L** 11–12.5 cm), dark brown-grey and white flycatcher with long dark wings, *very small black bill*, differing from related species by its *extensively sooty-brown breast and flanks*, leaving only *centre of throat and belly white*. Primary projection long, somewhat exceeding visible tertial length. Compared to Brown Flycatcher (also from Siberia) has *reduced pale area on lores*. When seen well, *chin, throat-sides and breast are boldly streaked dark grey-brown*, but streaks often coalesce to form *near-solid grey-brown breast-sides*. *White lower throat* widens to rounded patch, often extending narrowly to neck-sides. Longest *undertail-coverts often show diffusely darker centres*. Off-white eye-ring rather obvious, slightly broader behind eye. Base of lower mandible yellowish-horn, but pale portion rarely visible in field views. Hunts flying insect in dashing sallies, often well visible outside or above tall trees, and characteristically often returns to same favourite perch. The high-pitched, piercing song is probably not heard away from breeding grounds, a brief stanza of mouse-high piping or whistling notes, easy to overlook and difficult to hear for elderly people. Also some interwoven lower-pitched, harder notes. Call outside breeding season also rarely heard, a very high-pitched squeaky *zi-zi-zi*, rather like the sound of a shrew or Goldcrest call. **AGEING & SEXING Ad** Sexes similar. Limited age-related and seasonal plumage variation, apart from the effect of wear and bleaching. In spring and early summer, birds with both very fresh and slightly worn flight-feathers occur; not known whether such differences are due to different moult strategies applied in different breeding ranges or represent individual variation as to moult and wear. In autumn, ad is evenly fresh and lacks pale spotting above. Tertials either plain dark or with thin paler fringes. Breeders in Siberia and Russian Far East are thought to perform partial post-nuptial moult and have a complete winter moult, but more study of this required. **1stW** Upperparts finely spotted cream-white, including tips to median coverts and scapulars, such tips (from juv plumage) often still present in Oct and sometimes in Nov. Greater coverts tipped buff or have cream-white tips shaped as shallow wedges, innermost few usually moulted to ad type in autumn (late post-juv moult). Tertials variably broadly fringed whitish. Underparts more spotted or vermiculated than streaked as ad. **GEOGRAPHICAL VARIATION** Polytypic; four subspecies described. Although the Icelandic bird was not subspecifically identified, it most likely belonged to the ssp. *sibirica*, which breeds in Siberia.

Above: Mongolia, Jun: both birds show the characteristic extensively blotched or diffusely streaked breast and flanks with sooty-brown (see Asian Brown Flycatcher). Beware of some variation (individual or wear-related, or depending on posture). Note also almost entirely dark pointed bill, while undertail-coverts are blotched dark (unmarked white in Brown). (H. Shirihai)

Left: 1stW, Hong Kong, Oct: when evenly fresh in autumn, 1stW often has more extensive dusky breast and sides below, appearing plainer and not so streaky. Note reduced pale area on lores, and off-white eye-ring rather obvious, slightly broader at rear of eye. Bill small and pointed with only limited paler base to lower mandible. (M. & P. Wong)

ULTRAMARINE FLYCATCHER *Ficedula superciliaris* (Jerdon, 1840)

Iran one (Apr 2009). Breeds in the Himalayas, mountains of SC China and neighbouring parts of SE Asia; western populations reach through the Himalayas as far as E Afghanistan and N Pakistan (and move south in non-breeding season to C & S India). Frequents the canopy of open deciduous, coniferous, or mixed mountain forests with well-developed understorey of scrub, but on dispersal and in winter habitat choice is more varied. **IDENTIFICATION** Rather small arboreal *Ficedula* flycatcher (L 10.5–11.5 cm). Ad ♂ is strikingly bicoloured with *bright blue upperparts, head and breast-side patches* which contrast with *silky-white throat and rest of underparts*. *Lores black* and flight-feathers, wing-coverts, and tail black with bright dark blue fringes. ♂ of the westernmost population has a *short white supercilium above the ear-coverts* and a variable amount of *silvery-white edges to rump-feathers*, also white corners to the tail-base (often hard to see in the field), but these white areas increasingly replaced by blue towards the east of the range (see below). Wear has some influence on ♂ plumage, too, as the blue of the upperparts and side of breast is non-glossy azure-blue when fresh, but glossier cerulean-blue when worn, especially on the more exposed parts like forehead, lateral crown, and uppertail-coverts. The ♀, in comparison, is rather nondescript, being mainly pale brown-grey above and off-white below, including on throat. Note ill-defined ash-grey to dusky lateral breast-patches (mirroring ♂ pattern), with some black specks on crown and a cerulean-blue wash from lower mantle or rump to tail. Flight-feathers and tertials have grey-white fringes, and there may also be a vague buffy supercilium. Bill relatively narrow at base for a flycatcher, and both bill and legs blackish or very dark. Feeds among foliage and branches, and frequently sallying for flying insects. Song feeble, high-pitched and rather disjointed, with short trills and chirps. Calls include a short, staccato, rising squeaky note, often followed by a rapid hard rattling, *chee-tr-r-r-r*, or just a low, rattling *t-r-r-r-t*. **AGEING & SEXING Ad** Sexes differ markedly as described above. **1stY** ♂ like ♀ but has a much stronger bluish cast on lower mantle and deep blue rump and uppertail-coverts, blue edges on wings and tail, and distinct pale edges to blackish tertials and buff greater coverts wing-bar. Young ♀ similar to ad ♀ but is even duller and lacks blue on rump and tail, wing-coverts tipped rufous or buff-brown, while underparts are sullied light buff. **1stS** ♂ Resembles ad apart from less pure and deep blue (more cerulean) on upperparts and pale tips to the retained outer juv greater coverts. Remiges and primary-coverts visibly more worn. **GEOGRAPHICAL VARIATION** Polytypic, with two (in ♂ plumage distinct) subspecies, the western one (*superciliaris*, described above) showing white on the rear supercilium and at the bases of rump- and tail-feathers, the eastern one not. Apparently the source of vagrancy to the treated region is the western race *superciliaris*.

Above left: *F. s. superciliaris*, ad ♂, India, May: attractive blue upperparts and head, upperwing fringes and pectoral patch on breast-side. In contrast to this a short white supercilium behind eye, pure white throat and rest of underparts. (H. Punjabi)

Above right: *F. s. superciliaris*, ad ♀, India, May: rather pale brown-grey above and off-white below, but often with a faint cerulean-blue wash on mantle and scapulars. Throat and centre of upper breast usually form a slightly whiter bib, bordered by brown-grey. Note also relatively narrow and blackish bill. (D. Laishram)

Left: *F. s. superciliaris*, 1stW ♂, India, Dec: shares general appearance with ♀, including the vague whitish bib, but compared to any ♀ shows stronger bluish cast on shoulders, wings and tail-edges. Retained juv greater coverts tipped buff forming a narrow wing-bar. (S. Damle)

MUGIMAKI FLYCATCHER *Ficedula mugimaki* (Temminck, 1836)

Russia one (Aug 2007), Italy one (Oct 2011; a record from the 1950s is under review); a British record (Nov 1991) is currently under Cat. D (genuine vagrancy of a wild bird doubted). A summer visitor in moist taiga, usually in spruce, fir and pine forests in valleys, from upper Ob, middle Yenisei, and N Mongolia east to Sakhalin, Russian Far East, N Korea and NE China; winters in S China, SE Asia, the Greater Sunda Is and the Philippines. **IDENTIFICATION** A small *Ficedula* flycatcher (**L** 11.5–12.5 cm) that in general size and structure resembles Red-breasted Flycatcher. ♂ breeding plumage is unmistakable by combination of blackish or *slate-coloured upperparts and head*, a *small white patch behind the eye*, a *large white panel on the inner wing-coverts* extending to the fringes of the tertials, and a *deep cadmium-orange throat, breast* and belly that contrasts with the white vent and undertail-coverts. There is usually also some white laterally at the tail-base (but beware that white at the tail-base of ♂ is somewhat variable, and some have less even if adult). The ♀ is far less conspicuous, having uniform dull greyish olive-brown upperparts including upperwing and head, a narrow whitish-buff eye-ring, sometimes a hint of a *light olive-buff supercilium behind the eye* (absent in some 1stY), *narrow whitish double wing-bars* (that on the median coverts weaker or lacking in some, or is the first to wear off), and narrow pale fringes to tertials. *Throat and breast are dull mustard-orange* (much less saturated than in ad ♂), grading to pale buff on belly and white on vent. Tail grey-brown, virtually without white at base (if any, just extreme basal edge of one or two feathers, mainly hidden). 1stY ♂ is ♀-like but separated in close views by *stronger orange of underparts, grey-black or black uppertail-coverts* and at least *some obvious white at base of outer tail-feathers* (usually only on outer webs). Bill fine, narrow compared to most other flycatchers, black. Tarsus black, rather short. Song simple and mostly consists of segments of loud trills and whistling phrases, which fall stepwise in pitch. Calls include a rattling *tur'r'r'rt*, a low *chuck* and a soft disyllabic *tyyuh*. **AGEING & SEXING Ad** Sexes differ markedly from 2ndW, as described above. **1stW** Both sexes closely resemble ad ♀ but note many retained juv wing-coverts distinctly tipped white or buff-white, and 1stW usually has more acutely pointed tail-feathers; note also that 1stW ♂ usually has deeper orange (less mustard-tinged) breast, blackish uppertail-coverts and some (largely hidden) white at base of blacker tail. **1stS** ♂ As 1stW ♂, thus still mainly ♀-like, told on same criteria as 1stW ♂. After complete post-nuptial moult inseparable from ad ♂. **GEOGRAPHICAL VARIATION** Monotypic.

Ad ♂, Thailand, Mar: unambiguously patterned with dark slate-grey upperparts and well-marked white supercilium behind the eye, and a large white panel on the inner wing-coverts, as well as deep orange throat and breast, leaving lower belly vent and undertail-coverts white. (P. Ruangjan)

1stS ♂, Mongolia, May: most 1stS ♂♂ show only partly developed white supercilium, and have ♀-like white double wing-bars. The blackest area on head is the moustachial stripe. Not visible here are the white bases to outer tail-feathers recalling pattern of Red-breasted Flycatcher, a potential confusion risk. (H. Shirihai)

♀, Hong Kong, Nov: unlike Red-breasted or Taiga Flycatchers usually has narrow whitish double wing-bars, and throat and breast a warmer and more extensive orange-buff. Whitish portions on bases of outer tail-feathers not visible here but when seen usually clearly smaller than in Red-breasted. (M. & P. Wong)

INDIAN PARADISE FLYCATCHER (ASIAN PARADISE FLYCATCHER) *Terpsiphone paradisi* (L., 1758)

United Arab Emirates one (Oct 2011). Widespread in the Oriental region and extends well north into the mountains of W Central Asia, roughly to Afghanistan and eastward into E Asia. Favours open deciduous and evergreen forest with shady canopy, often near water, locally also in groves, plantations and gardens; mostly in lowlands and foothills below 1600 m, in the Himalayas occasionally up to 2400 m. Tropical subspecies are mostly sedentary, but the W Central Asian *leucogaster* winters in W & S India. **IDENTIFICATION** A large (**L** ♂ with tail-streamers 42–48 cm; ♀ 18–21cm) and distinctly plumaged monarch flycatcher, with ♂ *having elongated central tail-streamers* (projecting up to *c.* 27 cm). Adult ♂ is either of white morph, then *mostly white with a glossy bluish-black head and neck*, and black on some of the exposed centres to flight-feathers, or of rufous morph with *chestnut-rufous upperparts, wings and tail*, still with a *black head*. The proportion of white and rufous morph ♂♂ varies between populations, as does the presence of ♂♂ *with mixed morph plumage* (mixing mainly affecting wing pattern). All ♂♂ have a narrow but conspicuous *blue eye-ring* and a *blue bill*. Many have elongated crown-feathers forming a *shaggy crest*, but the development of a crest varies both individually and geographically. ♀♀ and imm *resemble the rufous ad ♂ but have no or less developed tail-streamers*, and have the *black cap grading into grey sides of head, neck, throat and sides of breast*. The tail is short, but older ♀♀ have the central tail-feathers often elongated for a few centimetres. Song a mixture of rasping, fluty and whistled notes, or a simpler descending warble, *chu-wu-wu-wu-wu-wu*. Rather noisy calls, uttering sharp *skreek* or coarse, nasal, trisyllabic *gii-geh-guh*, falling in pitch. **AGEING & SEXING** Ad Sexes differ markedly as described above. **1stY** ♂ resembles ad ♀ but the rear and side of neck is darker grey, blackish or glossy black. **2ndY** Resembles the rufous-morph ad ♂, but have tail-streamers of only half length compared to ad. **GEOGRAPHICAL VARIATION & RANGE** Polytypic with at least four subspecies recognised, mostly distributed in the Oriental region, with the westernmost ssp. *leucogaster* (valleys of Tien Shan and Pamir-Alai ranges, E Afghanistan, W Pakistan, NW Himalayas from N Pakistan to C Nepal) characteristic in showing a distinct pointed crest and a bright pale orange or rufous tinge to upperparts, wings and tail at all ages and in both sexes (except white-morph ♂♂). Most ad ♂♂ of the rufous morph have breast and belly white, sharply contrasting with the glossy black throat, without the grey breast and upper flanks of some eastern races. White ad ♂♂ are absent north of the Hindu Kush (Afghanistan), but rufous ♂♂ with white feathers intermixed in the rufous of tail and wings are frequent there. Further south and east, white ♂♂ are common or even predominant.

T. p. leucogaster, ad ♂ white morph, India, Jul: mostly white, including the elongated central tail-streamers, contrasting with glossy bluish-black head, shaggy crest and neck. Note also blue eye-ring and bill. A white morph does not occur in the Arabian race *harterti* of African Paradise Flycatcher. (D. Laishram)

T. p. leucogaster, ad ♂ rufous morph, India, Apr: rufous morph has the white of upperparts replaced by rufous. Crest in respective age and sex much longer and shaggier than in the Arabian race *harterti* of African Paradise Flycatcher, and underparts are whiter. (H. Punjabi)

Above: *T. p. leucogaster*, ad ♀, India, Jun: resembles rufous ad ♂♂ but has no elongated tail-streamers, and in comparison with young ♀ head and crest is bluer, has more gloss and crest is longer, but above all throat and neck-sides are grey, not blackish. (P. J. Saikia)

Left: *T. p. leucogaster*, 1stY ♂, India, May: unlike similar ♀, has a dark throat, creating a 'hooded' effect, and the tail-streamers are only partly grown. (D. Laishram)

T. p. leucogaster, ad ♂ mixed rufous/white-morph, India, Apr: in intermediate birds white and rufous colours mix in a variable way, mostly affecting wing pattern. The crest can point slightly forward when fully erected. (A. Arya)

BLACK-NAPED MONARCH *Hypothymis azurea* (Boddaert, 1783)

Iran one (Feb 2011). Widespread in the Oriental region, from India to Indonesia, north to S China. Favours thick forests and other well-wooded habitats, but found also in scrub and overgrown plantations, normally in the lowlands, but sometimes up to 1300 m; also breeds in secondary forest and urban parks. **IDENTIFICATION** A medium-sized (**L** 14.5–16 cm), slender and attractively plumaged monarch flycatcher, the ♂ being almost entirely *deep cobalt-blue or ultramarine* except three unique *black marks: a rim at the base of forehead, a patch on the rear crown, and a narrow black half collar* ('necklace') on upper breast. Underparts grade from blue upper breast to bluish-grey lower breast and upper belly, and to white abdomen and undertail-coverts. Dark flight-feathers fringed blue. The ♀ is duller having blue restricted to head, neck and upper breast, and lacks the black markings, rest of upperparts being brown. Bill strong with broad base, black. Iris and tarsi black. Song a monotonous, clear ringing *wii-wii-wii-wii-wii-wii*. Contact calls a sharp and abrupt *skrip*, a harsh *chee chee chee*, or a series of three whistled notes, *treet-treet-treet*. **AGEING & SEXING Ad** Sexes differ markedly as described above. **1stY** Apparently no plumage difference from ad, all autumn and winter birds similarly neatly plumaged with in ♂♂ blue wing-coverts without visible contrast or differences. **GEOGRAPHICAL VARIATION & RANGE** The monarch flycatcher with the largest number of races recognised (e.g., no fewer than 23 in Dickinson & Christidis 2014!). Judging by range, the western race *styani* (much of India, Nepal and into S China and east to Vietnam) is the most likely source of vagrancy to the treated region. It has distinct black markings on head and breast (see above).

H. a. styani, ad ♂, China, Jul: unmistakable and highly attractive Asian speciality. Note in particular the small black tuft on rear crown, as well as a narrow black half collar on upper breast in overall deep cobalt blue. The blue becomes duller on lower breast and grades into light grey-white on underparts. (J. Martinez)

♀, Thailand, winter: duller than the ♂ and lacks the black markings, but still unmistakable due to bluish-tinged grey head and breast, forming contrast with grey-brown upperparts and grey-sullied white underparts. Subspecies undetermined.

RED-BREASTED NUTHATCH *Sitta canadensis* L., 1766

Iceland one (May 1970), Britain one (Oct 1989–May 1990). Resident in northern and subalpine coniferous forests, generally from Alaska and W Canada, through USA to Mexico; irruptive migrant, with numbers and winter range varying annually, but in the south-east of the range mostly resident or only partial migrant. Closely related to geographically distant Chinese Nuthatch (*S. villosa*; extralimital, not treated), Yunnan Nuthatch (*S. yunnanensis*; extralimital, not treated) and Corsican Nuthatch (p. 117). **IDENTIFICATION** A small (**L** 11–12 cm), strikingly patterned nuthatch, where ♂ has a *black cap* and a *long and broad black eye-stripe*, enhancing a *conspicuous white supercilium*. Furthermore has white chin, throat and lower cheeks, while *rest of underparts are a rich buff to ochre-cinnamon*. Black eye-stripe broadens on sides of neck. Upperparts grey-blue, and has white tail-spots rather like Eurasian Nuthatch. Straightforwardly distinguished from other North American nuthatches by head pattern and warm underparts coloration. Note that *white supercilia meet on forehead* (in Corsican not). ♂ readily told from Corsican by its more richly coloured underparts (Corsican ♂ at most is greyish pink-buff below). Red-breasted ♀ and young ♂ have paler underparts than ♂, but still have warmer underparts coloration (in particular on flanks and vent) than any Corsican. Feeds on small branches and outer twigs. Song a series of repeated far-carrying, high-pitched, nasal notes, e.g. *ehk, ehk, ehk, ehk,…* or *yeh, yeh, yeh, yeh,…*, varying somewhat in pitch and speed. Call often a single note, and in alarm a monotonous, fast series of nasal, hard notes, *neh neh neh neh…*, or a trilled *triiiiir'r'r'r'r*. **AGEING & SEXING Ad** Undergoes complete post-nuptial moult. Sexes rather similar, but ♀'s underparts average paler, and black facial parts less intense, thus the blackish-grey of cap and eye-stripe does not contrast so strongly with grey mantle as in ♂. **1stY** Performs partial post-juv moult, and in fresh autumn plumage the cap varies from mid grey-blue to blackish-grey, though usually the forecrown is still dull black, and the eye-stripe to some degree narrower and mottled paler than in respective ad; unmoulted wing-feathers tinged brownish-grey with limited bluish edging, and if not completely moulted may show traces of buff tips to retained juv greater coverts. **GEOGRAPHICAL VARIATION** Monotypic.

♂, Canada, Jan: Superficially similar to Corsican Nuthatch due to bold head pattern, but has longer, broader and better-defined black eye-stripe. Being a ♂ also has more richly coloured underparts. Without handling ageing difficult. (D. Beadle)

Ad ♀, Alaska, USA, Dec: underparts less deeply saturated than ♂, and black eye-stripe not quite as black; cap dark grey with almost no contrast to rest of upperparts. Whole wing evenly fresh, primary-coverts neatly edged blue-grey inferring age. (F. Cezus)

BLACK-NAPED ORIOLE *Oriolus chinensis* L., 1766

Oman one (Dec 2011), United Arab Emirates one (Feb 2012). A migratory breeder in E Siberia, Ussuriland, E China and Korea, and a mostly resident breeder south through SE Asia and parts of India to the Philippines, where it also winters. Habitat much as Golden Oriole, favouring tall deciduous or evergreen trees with dense canopy in mature or open forest, parkland, plantations, and parks; mostly in lowlands but also in hills up to at least 1600 m. **IDENTIFICATION** A thrush-sized (**L** 24–26 cm) bright black-and-yellow oriole, which in all plumages except juv is characterised by a *very heavy pink-red bill* and *black bar across the nape connected to the black eye-mask on each side*. Ad ♂ of the E Palearctic ssp. *diffusus* has *bright yellow head, body, and wing coverts*, with a rather narrow black stripe through the eye and a black nape bar up to 2 cm wide. Much of

O. c. diffusus, ad ♂, Thailand, winter: unlike the equivalent plumaged *O. oriolus* has a rather broad black band through the eye that continues onto nape where it widens (though black does not extend to crown, as it does in some more striking races in SE part of the range). Black primary-coverts with moderately wide yellow patch at the tips, and tertials diagnostically (but variably) have black inner webs and yellow outer webs. (B. Promjiam)

wing contrastingly black, but primary-coverts tipped yellow (creating a moderately large patch) and, when fresh, primaries narrowly fringed and tipped off-white to yellow. *Tertials diagnostically (but variably!) have black inner webs and yellow outer webs*, and black secondaries are broadly edged and tipped yellow, all together giving a rather *more variegated wing pattern* than Golden Oriole. Ad ♀ variable, some closely resemble ♂, but lores are greyish, the black nape bar is less sharply defined, and yellow of mantle, scapulars, upper wing-coverts and central uppertail tinged greener. Underparts either unstreaked deep yellow as in ad ♂ or, sometimes, paler yellow with traces of dark shaft-streaks on breast. The tail in adult ♂ is *black, with broad bright yellow corners*, whereas these corners are smaller and tail-feathers have largely dark outer webs in ♀. Young resemble ♀, but differ on greater coverts being narrowly tipped pale yellow. Iris red in ad, bill dark reddish-grey with paler base (imm) to bright red (ad), and legs dark flesh-brown to greyish-black. Differs from Golden Oriole and Indian Golden Oriole (p. 163) by thicker bill and (in most plumages) by black or dark nape bar, and in wing pattern (at least with well-marked birds); also differs by wing formula. Song a clear fluty whistling, *lywee wee wee-leeow*. Call described as a harsh, nasal and somewhat jay- or cat-like *niiie*, *myaa* or *gyaa*. **AGEING & SEXING Ad** Rather weak sexual differences, as described above (complicated by extensive individual variation and by more or less ♀-like 1stY plumage in ♂). **1stY** Resembles ad ♀, but dark eye-stripe and nape bar are more dusky olive than blackish, and note pale-tipped retained juv greater coverts. Sexing often very difficult of

O. c. diffusus, ad ♀, Sri Lanka, Feb: duller than ♂, and lores are greyish, and the black nape bar is less sharply defined but still present on most, separating them from *O. oriolus* and *O. kundoo*. (M. Mandal)

1stW birds, but ♂ averages darker on uppertail and inner webs of tertials, less greenish-grey. Underparts variably yellowish-white or more yellow on breast and flanks, boldly black-streaked. Undertail-coverts bright yellow. In 1stS, eye and bill still only partly red. **GEOGRAPHICAL VARIATION & RANGE** Polytypic, with as many as 20 subspecies described, of which the single Palearctic one, *diffusus*, occurs in summer in SE Siberia from Lake Baikal east to the Russian Far East and south through NE & E China to N Laos, N Vietnam, and possibly SW & C Myanmar).

Above: *O. c. diffusus*, 1stW, presumed ♀, Hong Kong, Sep: differs from *O. oriolus* and *O. kundoo* by thicker bill and by at least an indication of dark nape-bar, even in such dull young autumn plumage. Sexes rather similar in 1stW, but ♂♂ usually have somewhat more yellow on primary-coverts. (M. & P. Wong)

BLACK DRONGO *Dicrurus macrocercus* (Vieillot, 1817)

Oman seven (Nov 1991, Dec 1998, Jan 1999, Nov 2006, Nov 2007, Dec 2008, Apr 2013), United Arab Emirates nine (Nov 2005–Jan 2006, Jan–Feb 2012, Mar–Apr 2012, Mar 2014, Nov 2014, Apr 2015, Jun 2015, Oct-Nov 2015, two Dec 2015-Feb 2016), Kuwait one (Nov–Dec 2015). Found in much of tropical S Asia from S Pakistan and through India and Sri Lanka east to S China and Indonesia. Formerly apparently bred SE Iran. A bird of open habitats with sparse tree cover, often in cultivated countryside and parkland areas; western and northern populations are migratory, others short-range migrants or sedentary. (Slightly more records than stipulated for being treated as a vagrant, still basically an East Asian species that only recently have been repeatedly found in the covered region.) **IDENTIFICATION** A medium-sized (L 29.5–32 cm), uniformly glossy black drongo, characterised by its *very long and deeply forked tail*, narrow at base but widening at tip (tips of outer tail-feathers curved outwards and often slightly twisted, too). Black bill moderately strong. Depending on light, black plumage can show dull steel-blue or greenish sheen, most obvious on head, upperparts and breast grading into duller black rest of underparts. At least some birds show *diagnostic small white mark behind rictal bristles at the gape*. Underwing-coverts and inner webs of flight-feathers paler, and rump sometimes paler greyish, too. Easily confused with Ashy Drongo (which see), though Black Drongo has a glossy black throat and breast, while Ashy is more bluish-slate above

and paler slate-grey below lacking gloss. Shape of tail and white mark at gape are important key distinctions (but beware moulting birds with tail appearing less forked), but also in ad the much darker red-brown iris (instead of bright red as in Ashy, but beware again age-related variation as to iris colour, as juv/1stY Ashy also can have darker eyes). Calls include both hoarse Jackdaw-like notes and clear short whistles. **AGEING & SEXING Ad** Sexes similar. **1stY** As ad but the black feathers of breast to undertail-coverts, as well as the longest uppertail-coverts, show pale grey to white tips, giving mottled appearance, while moult limits in wing and shallower tail-fork usually indicate young age. **Juv** Initially almost as glossy as ad, but the plumage is fluffier, and due to wear the feathering gradually becomes sooty at first and browner later. **GEOGRAPHICAL VARIATION & RANGE** Polytypic, with several races described, possibly forming two main groups, a western *macrocercus* group and an eastern *cathoecus* group. The relationship between these groups is poorly understood. The race found as a vagrant in the treated region is probably the westernmost *D. m. albirictus* (E Afghanistan east through NW & N Pakistan and in the Himalayan foothills, also S Pakistan and apparently at least formerly also in SE Iran), a large subspecies with a long and deeply forked tail. Black of plumage fairly strongly glossed green-blue. A small white rictal spot frequently present. Underwing with well-developed medium silvery-grey feather-edges.

D. m. albirictus, Oman, Apr: a bluish-black bird with very long and deeply forked tail, the tail being rather narrow at base but widens somewhat towards tip with outermost feathers curved outward and twisted. Black plumage has green-blue sheen, most obvious above. Note also diagnostic, if present, a small white mark behind the rictal bristles at the gape. Also unlike Ashy Drongo, has darker red-brown (instead of bright red) iris. (H. & J. Eriksen)

ASHY DRONGO *Dicrurus leucophaeus* Vieillot, 1817

United Arab Emirates ten (Dec 2006, Feb–Mar 2008, Dec 2009, Jan–Mar 2010, Oct 2011, Feb–Mar 2012, Jan–Mar 2014, Dec 2014, Jan–Feb 2015, Oct–Nov 2015), Kuwait four (Apr 2010, two Dec 2010–Jan 2011, Feb–Apr 2013), Iran two (Feb 2014, Jan 2016), Oman one (Nov 2014), Israel one (Dec 2014). Breeds across S & SE Asia, with populations from N & C China migratory, those of Afghanistan, the Himalayas, N Myanmar and S China partly migratory or moving altitudinally, wintering south to Sri Lanka and mainland SE Asia. A bird of open broadleaved or mixed forest, forest edges, glades and clearings, as well as in cultivation with little tree cover. Can reach altitudes up to 3300 m. (Again more records than the stipulated nine for treatment as a vagrant, but several of these have been only in the last few years.) **IDENTIFICATION** Medium-sized (L 26–30 cm) drongo with slender jizz, characterised by varying *ashy or greyish hues to predominantly black plumage*, and by its *deeply forked long tail* (outer feather flaring outward but without much curvature at tip; cf. Black

Far left: *D. l. longicaudatus*, presumed ad, Gujarat, India, Mar: slender jizz, characterised by varying degree of ash-grey hue mostly below to otherwise largely black plumage, and by its deeply forked long tail (outer feathers flaring outward but without curvature at tip; see Black Drongo). The bright red iris suggests ad, and provides further differentiation from Black Drongo. (P. Ganpule)

Left: *D. l. longicaudatus*, 1stY, Kuwait, Jan: long tail with pointed outer feathers and ash-grey hue below but with attractive white crescent-shaped tips to undertail-coverts characteristic. Beware, however, that iris is blackish in young birds, and this could be a cause of misidentification as Black Drongo. (V. Legrand)

Drongo). Beware that western race *longicaudatus* (see below) is blacker and thus far more difficult to identify, being very similar in colour and tail-shape to Black Drongo from the same region. Still, in close views, the plumage is slate-black to dark slate-grey with a sky-blue lustre rather than being deep black with a dark metallic blue gloss, and especially the underparts appear duller, greyer and less glossy than Black Drongo. Also has a more striking *red iris*, and it *lacks a white rictal spot*. The *entire underwing is white* at all ages. Calls resemble those of Black Drongo. **AGEING & SEXING Ad** Sexes similar. **1stY** Closely resembles ad, but may show clear moult limits in wing, and/or reduced white area on underwing, limited to white tips on otherwise blackish underwing-coverts, and slightly also on belly and undertail-coverts. **Juv** Shallower tail fork and rather unglossed grey-black fluffy feathering. **GEOGRAPHICAL VARIATION & RANGE** Polytypic with 15+ subspecies often recognised, conveniently arranged in three groups according to colour and size. Western and northern *longicaudatus* ('Dark Ashy Drongo'; E Afghanistan, Pakistan, in foothills of the Himalayas east to E Nepal) characterised by being large and slate-black with a long deeply forked tail, and a rather short bill.

LARGE-BILLED CROW (JUNGLE CROW) *Corvus macrorhynchos* Wagler, 1827

Iran several older records (1896–1916, spring and late autumn). From W Pakistan and Afghanistan east through the Himalayas, Tibet and the Indian Subcontinent to the Philippines, E Asia and Japan, and north to SE Siberia. Highly adaptable, its habitat ranging from ice-covered coasts of E Siberia to tropical lowland forests of SE Asia, but largely shuns dry country; often lives close to man and habitation; in Central Asia inhabits high mountains; largely sedentary. **IDENTIFICATION** Rather large (**L** 44–56 cm) and *strongly built crow*, but size quite variable within wide range, with larger-sized taxa in, e.g., Tibet, and in north-east of range almost as large as Raven, with the smaller southern taxa as Carrion Crow in size. Most*ly dull black, with distinct purple sheen to wings and tail*, slightly also on cap, mantle and scapulars, and sometimes has some green or purple gloss on throat and breast (less glossy than Raven and Carrion Crow, and far less so than Rook). Variable grey tinge (or just softer feathering) around the neck, rear ear-coverts and behind and above the eye, contrasting with blacker forehead and throat. Feathering of crown forms domed cap, with the *vertical forehead* often emphasized when these feathers erected (top of head not as flat and gradually sloping to bill-base as in Raven and Carrion Crow). House Crow (p. 252) is remarkably similar in proportions, feather structure and bill shape (especially the distinctly arched upper mandible and steep forehead), but is considerably smaller than even the smaller southern taxa of Large-billed Crow and, unlike in House Crow, the contrast between blacker head and throat and seemingly paler grey nape and breast is much less obvious. Feathers at bill-base short and velvety, and the lanceolate throat-feathering not long, but sometimes raised, giving *shaggy-throated impression*. Bill heavy, with strongly decurved and highly ridged culmen, with nasal bristles dense and long, extending along about half of culmen but not covering its top (in Raven and Carrion Crow, bristles cover about 1/3 of bill, including basal part of culmen). Tail-tip protrudes behind wingtips at rest. When calling, tail usually raised slightly and body often crutched horizontally. Iris dark brown, bill and legs black. Call a raucous, almost laughing *awa-awa-awa* or a harsh, clear *kaaw*, *gwarr* or *kaa kaa*, hoarser than Carrion Crow, sometimes intermixed with throaty rattling sounds. **AGEING & SEXING Ad** Sexes similar as to plumage. **1stY** Like ad but tail-feathers are narrower and more rounded at tip (less broadly rounded or truncate in shape). As a rule very difficult to see a difference in gloss or wear between moulted and retained juv greater coverts. **GEOGRAPHICAL VARIATION & RANGE** Highly complex variation with many races described, but within the treated region ssp. *intermedius* (often called 'Tibetan Crow'; Afghanistan, W & N Pakistan, the Himalayas, SW Xizang east to NC Nepal, but probably also this race in S Tajikistan, S Turkmenistan and W Xinjiang) is described above and is likely to be the race involved in vagrancy to Iran. It has a relatively short and slender bill.

C. m. intermedius, presumed ad, India, Oct: glossy plumage with green and purple sheen often giving silky impression, and vertical forehead often emphasised, otherwise mostly a dull black large crow. Markedly larger than House Crow though similar in proportions, especially bill shape. However, contrast between blacker head and throat and paler grey nape and breast is much less developed. (K. Dasgupta)

DAURIAN STARLING *Agropsar sturninus* (Pallas, 1776)

Netherlands one (Oct 2005); a Norwegian record (Sep 1985) has been moved recently to Cat. D (natural origin questioned). A summer breeder from Transbaikalia to NE China, winters to SE Asia and Greater Sunda Is. In its native range highly gregarious, breeding in lightly wooded country, often in second growth, at forest edge, and in cultivated lowlands. **IDENTIFICATION** Comparatively small stocky starling (**L** 17–19 cm) with short tail and small, stout and slender, dark bill. Rather remarkably variegated, almost pied, plumage. *Dark upperwing has pronounced broad white wing-bar on median coverts*, and rear scapulars form prominent brown-white 'V'. Whitish tips to inner greater coverts and wedge-shaped tips to tertials, as well as a *pale brown-grey or cinnamon-buff panel on secondaries*. Sexually dimorphic, with ad ♂ *glossy blackish* (iridescent purple and green) *above* including the dark areas of wing, merging with pale, whitish to pale cream-grey or pinkish-violet hue of head, upper mantle and breast. Further, there is a purplish-black patch on hindcrown and nape (often difficult to see), and *rump is dirty ochre grey-white* contrasting with glossy green-black tail. ♀ patterned as ♂, but duller, with glossy black of wing and tail replaced by brown, crown duller, mantle dull grey-brown and has a less contrasting pale rump. ♀ also has narrower pale wing markings, and a diffuser brown nape spot. Iris dark brown, bill almost wholly black in breeding ♂, but at other times and in other plumages greyer with extensive horn-coloured base. Tarsus greenish-grey. Song a rapid and complex series of mixed whistles, trills and chatters, including some mimicry. Said to utter a harsh, explosive *chelee* and a throaty *gyuru gyuru*. When flushed also a harsh, drawn-out *chirrup*, resembling similar call of Common Starling. **AGEING & SEXING Ad** Sexes differ as described above. **1stY** Like ad of respective sex, but as long as moult is not completed (during prolonged post-juv moult), some inner secondaries and outermost tertials are often retained till midwinter. Plumage predominantly brownish, and when still fresh, coverts distinctly pale-fringed, though by spring all birds are similarly plumaged and can no longer be aged. **Juv** Generally closer to ♀ with its duller plumage, being pale brown above, but also diffusely mottled brown-grey below, and has browner wings, hardly any white on scapulars and only narrow grey tips to median coverts. **GEOGRAPHICAL VARIATION** Monotypic.

♂, China, Apr: white wing-bars form prominent white V when seen from behind. Being a ♂ has iridescent purple and green and blacker upperparts, including dark areas of wing, and head and underparts purer whitish- to pale cream-grey with pinkish hue. Purplish-black patch on hindcrown and conspicuous whitish rump are just visible in this image. (J. Martinez)

♀, India, Jan: duller, with pale areas greyer and tinged brownish, and with narrower and less solid pale wing markings. (N. Sant)

CHESTNUT-TAILED STARLING *Sturnia malabarica* (J. F. Gmelin, 1789)

Oman one (Nov 2010). A resident or partly migratory species of wooded habitats in India and SE Asia; known to perform some poorly understood movements, e.g. ssp. *malabarica* (apparently the source of vagrancy to the treated region—see below) has been recorded in Pakistan and C & S India. **IDENTIFICATION** Rather small to mid-sized starling (**L** 17–20 cm), appearing thick-necked with a stout-based but pointed bill, and rounded short wings, *grey-brown upperparts and blackish remiges*. The colours of the rest of the plumage vary between subspecies. In ssp. *malabarica*, *underparts are mostly rufous or dirty buff-brown*, while the *head is paler, greyish-white with whitish hackles of loose, pointed plumes* giving a streaked effect (especially on rear crown, neck and breast). Tail has chestnut outer feathers. *Iris white*, and *bill yellow with a pale greenish-blue base*. Legs pinkish-yellow. Typically omnivorous, found in open woodland and cultivation, like most starlings, eating fruit, plant material and insects, and flies in tight flocks which often rapidly change direction with great synchrony. Song comprises a series of short hard notes and low squeaky churrs, also rapid subdued outbursts of rambling warbles, chortles and squeaky churrs from the song. Calls include a metallic or harsh rasping note, *trreh*, and a tremulous downslurred whistle, *trreeahl*, a little like White-breasted Kingfisher. **AGEING & SEXING Ad** Sexes rather similar, but in breeding ♂, head becomes purer white and underparts deeper chestnut. **1stY** Once post-juv moult completed inseparable from ad. **Juv** Overall *plainer and duller, with grey-brown upperparts, and rufous-brown sides and tips to tail*. **GEOGRAPHICAL VARIATION & RANGE** Polytypic, but variation not fully understood. Three races usually recognised, *malabarica* (NE India, Nepal, Bhutan, Bangladesh, NW Myanmar, and in non-breeding period from Pakistan to C & S India), which could be the source of vagrancy to Oman, but considering the proximity of the breeding range of *nemoricola* (S China, SE Asia), this race might equally likely reach the treated region.

S. m. malabarica, ad, Nepal, India, summer: grey upperparts and mostly rufous underparts, and the head is mainly a light silvery-grey with whitish hackles giving streaky effect. Tail has chestnut outer feathers. Iris is bluish-white, bill greenish-blue with yellowish tip, and legs are pinkish-yellow, much the same colour as belly. Here an ad at nest.

S. m. malabarica, NW India, summer: striking plumage with near-white head and breast in contrast to rufous rest of underparts (including feet). Multi-coloured bill with orange-yellow tip adds to personality. (G. Bagda)

RED-BILLED OXPECKER *Buphagus erythrorhynchus* (Stanley, 1814)

Yemen one (Apr 1998). Breeds in E & S Africa, from Eritrea, Ethiopia, Djibouti, extreme SE Sudan and parts of Somalia south to NE Namibia and N & E South Africa. Resident, occurring in association with game mammals and livestock in savanna, thornbush and montane grassland. **IDENTIFICATION** Rather large (**L** 19–22 cm), *slim, long and slender-tailed* starling with chunky short bill, unmistakable by its unique shape and behaviour of constantly perching and nimbly clambering about on mammals to feed on insects. It is mostly *dark brown, except buffy-yellow underparts*, all-red bill and reddish iris, eye surrounded by *broad yellow orbital ring* and wattle. The only two oxpeckers are distinctive birds and are only likely to be confused with one another. When perched the two should be readily separable using bare-parts coloration: Yellow-billed Oxpecker (*B. africanus*; unrecorded in the treated region) lacks a yellow orbital ring but has an extensive bright yellow base to the bill, and even in flight they should prove identifiable, given the large pale rump and uppertail-coverts in Yellow-billed (concolorous with the back and tail in Red-billed). Young are rather less easily separated, but Red-billed often shows a *dull reddish base to the mandible* and *lacks any rump contrast*, although note that the latter feature is much reduced in young Yellow-billed. Red-billed calls consist of spat clicking notes and a monotonous hissing *ssshhhhhh*. **AGEING & SEXING Ad** Sexes alike with very little seasonal variation. **1stY** Probably much like ad following post-juv moult. **Juv** Differs from ad in having duller iris, no coloured orbital ring and a dark bill, often with a very pale reddish base only, which along with the lack of contrast between the rump and rest of upperparts is the best feature separating it from Yellow-billed Oxpecker at same age. **GEOGRAPHICAL VARIATION** Monotypic.

Ethiopia, Sep: habitually feeds on mammal parasites, and often first spotted due to bright coloration of bare parts. (H. Shirihai)

VILLAGE WEAVER (BLACK-HEADED WEAVER) *Ploceus cucullatus* (Statius Müller, 1776)

Egypt one (May 2006); an Israeli record (summer 1987) could also be of a genuine vagrant. Widespread and abundant resident in much of sub-Saharan and tropical Arica, thriving in lush savanna, along watercourses, in moist thornbush and cultivation, rural settlements, also in hillsides and lower mountains to *c.* 2000 m, locally even higher. **IDENTIFICATION** (*abyssinicus*) Rather large (**L** 14–17 cm), strongly built weaver. ♂ readily separated in breeding plumage from congeners by combination of large size, mainly *yellow plumage with black head*, and *black wings with broad yellow fringes, yellow reaching onto mantle, nape and rump*, further a *very strong black bill* and *red iris*. Size and iris colour also useful for separating ♀♀ (and non-breeding ♂♂), and further note *narrow dull yellow supercilium* and *broad dusky yellowish-white fringes to upperwing feathers*. Among other larger weavers in the treated region or NE Africa, no other species shows all these features, but note that other subspecies

P. c. abyssinicus, ad ♂, Ethiopia, Sep: black face extending to mid crown and to upper breast at centre, narrowly edged rich chestnut, with rest of head, mantle and underparts yellow. Largely black upperparts with broad yellow fringes, and yellow rump. Yellow neck reaching onto the mantle. Black wings with broad yellow fringes forming wing-bars, and has red iris. (H. Shirihai)

differ. Calls many, often heard is a constant chattering, with short *churrs* and higher-pitched squeaks being uttered more occasionally. In flight from flocks shorter *chik* notes. ♂ song consists of buzzing or wheezing notes, terminating in a short twitter. **AGEING & SEXING Ad** Sexes differ. ♂ has distinct breeding and non-breeding plumages. Breeding ♂ has black face extending to mid crown and well onto upper breast, sometimes slightly edged chestnut (but chestnut more extensive on underparts and nape in other races), with rest of head, mantle, rump and underparts yellow. Largely black upperparts, wings and tail with broad yellow fringes to most feathers. Non-breeding ♂/♀ have dark greenish-olive head and upperparts, yellowish supercilium, throat and breast, pale yellow fringes to wing-feathers, and largely whitish rear underparts. Breeding ♀ lacks the black head of ♂ and has much paler yellow underparts (sometimes being almost entirely greyish-white) and slightly less dark bill. **1stY/Juv** Very closely recalls ♀ but iris is brown in young ♀ and reddish-brown in young ♂ (ad of both sexes have brighter, more reddish eye). **GEOGRAPHICAL VARIATION** Polytypic, with about eight subspecies described, of which the subspecies *abyssinicus* (Ethiopia, Eritrea and SE Sudan) is the most likely to appear as a vagrant in the treated region.

P. c. abyssinicus, ad ♀, Ethiopia, Sep: dark greenish-olive head and upperparts, yellow supercilium, throat and breast, pale yellow fringes to wing-feathers, and more whitish rear underparts. Note heavy bill and orange-brown iris. (H. Shirihai)

WHITE-EYED VIREO *Vireo griseus* (Boddaert, 1783)

Azores four (Oct–Nov 2005, Oct 2008, Oct 2009, Oct 2012). A North American migratory vireo breeding in dense young second growth, often attracted to areas rich with berries, generally from extreme SE Canada to C Mexico and Florida; winters from SE USA to Greater Antilles and Central America. **IDENTIFICATION** Rather small (**L** 12.5–14 cm) and warbler-like, greyish-olive or olive-green above including on nape and crown (head often appears greyer than more greenish back), paler below, with *yellowish lower chest and flanks* and often belly too. Chin, throat, and upper breast to some degree paler. The most distinctive features are *bold double whitish wing-bars* and facial pattern with *yellow supercilium in front of eye* over *blackish lores* interrupting *broad yellowish eye-ring*. Also distinctive in ad at close range is white iris. Bill quite long, strong and hooked at tip, with upper mandible black, lower grey. Legs dark grey. Calls include a short, sharp ticking, and a harsh miaowing note. Alarm a repeated series of nasal clicking notes, *chet-chet-chet-chet-*… **AGEING & SEXING** Sexes similar, and during autumn ages rather similar, too. **Ad** Following complete post-nuptial moult has evenly fresh wing (no moult limits). **1stW** Aged by brownish iris, and sometimes has visible moult limits among greater coverts and remiges. Post-juv moult eccentric, with outer primaries and some inner primaries replaced. **GEOGRAPHICAL VARIATION & RANGE** Geographic variation moderate, with birds photographed in Azores matching ssp. *griseus* (which breeds throughout most of the range in N America). This race has rather distinctive facial pattern, and also bright yellowish flanks and rich olive tinge above.

V. g. griseus, Azores, 1stW, Oct: note yellow supercilium most pronounced in front of the eye, and blackish lores interrupting yellowish eye-ring. Also, note bold double whitish wing-bars. Dark iris and seemingly dull juv primary-coverts infer age. (V. Legrand)

YELLOW-THROATED VIREO *Vireo flavifrons* Vieillot, 1808

Britain one (Sep 1990), Germany one (Sep 1998), Azores three (two Oct 2008, Oct 2009). A North American migratory vireo that prefers canopy, forest edge, tall second growth, but which on migration often is found in more open areas and low scrub; like other vireos, eats lots of berried on migration and in winter. Breeds from SE Canada, and through C & E USA; most winter from S USA and E Mexico to West Indies and south into Colombia and Venezuela. **IDENTIFICATION** A slightly larger vireo than most (**L** 13.5–15 cm), generally warbler-like but with bigger head, heavier and slightly hooked bill, characteristically *bright yellow spectacles*, and *vividly yellow throat and breast*. Yellow breast contrasts with white belly, and note *striking bold double white wing-bars*. Secondaries broadly edged white. Upperparts olive-green to grey, with *contrasting purer slate-grey rump* and forewing (latter partly visible on folded wing near wing-bend). Flanks tinged grey. Upper mandible dark grey, lower blue-grey with darker tip. Legs dark grey. In North America associates freely with mixed flocks of other birds, and often seen slowly and methodically gleaning foliage, usually high in trees, in summer taking relatively large insects. Often cocks tail. Call a straight or slightly descending series of harsh, grating notes, *cheh cheh cheh*…, very distinctive and to a European recalling both Great Tit and Garden Warbler, often the first clue to the bird's presence. May sing in winter, a slow repetition of buzzy, low-pitched disyllabic or three-note phrases separated by long pauses, often containing a down-slurred *free-way* (or rendered '*three-eight*'). Gives a chattering or scolding series when alarmed. **AGEING & SEXING** Sexes similar. Only limited age-related or seasonal variations. **Ad** During autumn, after complete post-nuptial moult, has evenly fresh wing (no visible moult limits). **1stW** Can often be aged by presence of moult limits among greater coverts, and as all primaries and primary-coverts are retained juv (subtly browner and looser textured) soon becomes a little more abraded than ad. Both age groups can show moult limits in tertials in spring. **GEOGRAPHICAL VARIATION** Monotypic.

Azores, Oct: note rather bright greenish head and upperparts and bright yellow spectacles and supercilium in front of the eye, also yellow throat and breast, and a contrastingly white belly, as well as bold double white wing-bars below greyish shoulder patch. (V. Legrand)

Above: Azores, Oct: when seen head on, characteristic face pattern with distinct lemon-yellow spectacles and throat contrasting with dark olive elsewhere is striking. Broad white double wing-bars add to useful field marks. (V. Legrand)

PHILADELPHIA VIREO *Vireo philadelphicus* (Cassin, 1851)

Ireland two (Oct 1985, Oct 2008), Britain one (Oct 1987), Azores thirteen (2005–2016, Sep–Oct). Breeds in Canada and extreme northern USA; winters Central America to N Colombia. This North American migratory vireo prefers open woodland and forest edge, and is often found in woodland glades with tall second growth, but may visit denser forest canopy, too. (More records than stipulated for being treated as a vagrant, but number of birds found in the Azores almost exploded the last few years.) **IDENTIFICATION** Rather small (**L** 12–13 cm) and somewhat variably coloured, with *yellowish underparts*, palest on belly, more *greenish above*, with contrasting *greyish crown* (only vague or no dark lateral crown-stripes). Lacks *wing-bars* or just shows a hint of one (still, can be pale and quite noticeable) as ill-defined dull greyish-olive edges and tips to greater coverts. Facial pattern usually involves dull white supercilium, dusky eye-stripe and ill-defined whitish patch below the lores and eye. Fresh autumn birds, and especially 1stW, are often brighter yellow below, strongest at centre of throat and breast. Note also jizz with *relatively rounded head* and *thick bill* (usually quite clearly bicoloured). Iris dark, legs grey. On migration solitary but at times joins mixed-species flocks or groups of birds. From superficially similar Red-eyed Vireo (which see) told by combination of small size and shorter whitish supercilia (which are usually more clearly joined on base of forehead), and in having only indistinct dark lateral crown-stripe. Generally more yellow on underparts, and has dark (rather than red) eye. Usually quiet and unobtrusive, gives only very short and subdued *chip*, but occasionally when agitated utters a series of soft descending and nasal notes, *weah weah weah weah...* **AGEING & SEXING** Sexes similar. Seasonal variations slight. **Ad** During autumn (following complete post-nuptial moult) shows evenly fresh wing (no moult limits). **1stW** Can be aged by moult limits among, e.g., the greater coverts, with primary-coverts usually clearly abraded browner and more pointed (retained juv feathers), and is usually richer yellow below. **GEOGRAPHICAL VARIATION** Monotypic.

Azores, 1stW, Oct: generally more yellow on underparts and dark (rather than red) eye, as well as short whitish supercilium running over base of bill on forehead. Has indistinct dark lateral crown-stripe, or at least diffusely darker grey crown-sides, separating the species from similar but larger Red-eyed Vireo. (V. Legrand)

PALLAS'S ROSEFINCH *Carpodacus roseus* (Pallas, 1776)

Hungary one (Dec 1850), Russia at least six (one Nov 1900, one Dec 1995, one Nov 1998, two Nov 2005, one Jan 2011) and Ukraine four (one Nov 1902, two Dec 1902, one Nov 1927), thus just over nine birds recorded but since some records are quite old the species is still placed among the vagrants. Breeds in C & E Siberia, where it favours dense scrub of birch, juniper, larch or dwarf pine, mixed open coniferous forest on low hills in the taiga, from close to sea-level in the N Siberian plain to much higher elevations, e.g. at *c.* 2000 m in the Altai. A short-distance migrant, wintering mostly south of breeding range, but at this season may perform opportunistic movements, depending on local seed crops, occasionally irrupting beyond normal winter range, with some reaching, e.g., W Siberia, and even wandering to the treated region. **IDENTIFICATION** A medium-sized rosefinch (**L** 15–17 cm), rather similar to Scarlet Rosefinch with *bright rosy-red head, breast, rump and uppertail-coverts* but a slenderer bill, slightly longer tail and in ad ♂ *forecrown and throat typically streaked or spangled glossy pink-white*. In most plumages has pale-fringed wing-coverts and tertials, these fringes forming *two prominent pale wing-bars* (broadest on greater coverts, with that on median coverts usually slightly narrower and sometimes even largely concealed). Mantle and scapulars boldly dark-streaked. Ad ♀ brown above streaked blackish, pale buff to off-white below with dark brown streaks on breast, upper belly and flanks, rather like ♀♀ of many other rosefinches, but differs by being distinctly washed pinkish orange-red, with a trace of silvery-white spangles on throat, and with a paler (less streaked) face and narrow buff-grey eye-ring. Tail of ♂ has dull pink and of ♀ buff-white fringes. Iris dark. Bill steel-grey to horn-grey, with paler base chiefly in winter. Legs pinkish-grey. Main call a weak, whistled *fee*, or a sharper and louder *tsuiii*; other calls include a discordant, bunting-like *dzih*. **AGEING & SEXING Ad** In worn plumage,

C. r. roseus, ad ♂, Siberia, Jan: note pink-white 'spangles' on crown and throat, to otherwise extensive bright raspberry-red plumage. Also bold dark streaking above, bright rosy-red rump and uppertail-coverts and largely white vent (difficult to see here). Pale double wing-bars rather prominent. (I. Ukolov)

C. r. roseus, ad ♀, Siberia, Jan: note characteristic heavily streaked grey-brown plumage with pinkish-orange tinge on head, breast and rump. Prominent tertial edges and double wing-bars buff-white. Bill conical, pale horn-coloured with darker culmen. (I. Ukolov)

sexes differ clearly (see above), when fresh in autumn somewhat less. ♂'s feathers in autumn narrowly fringed pale pink-grey, concealing silvery spangles on forecrown and throat, and partly covering rosy-red in remaining plumage. Ad-plumaged ♂♂ with brown-buff (rather than pink-tinged) fringes to freshly moulted upperparts and with pink-brown fringes to wing-feathers are perhaps 2ndY. **1stY** Following partial post-juv moult both sexes generally ♀-like, but differ by moult limits among greater coverts, browner and narrower-fringed flight-feathers and tertials, and narrower and more pointed tips to tail-feathers, while 1stY ♂ to a variable extent may have the reddish plumage of ad, including scarcely streaked pink rump-patch (1stY ♀ lacks any pink tinge on body, except rump). **Juv** Dark streaking blacker and heavier and ground-colour greyer. Lacks largely pink rump and uppertail-coverts, these streaked buff and brown like remaining upperparts, though a slight pink-red wash is often present. Fringes of wing-coverts, flight-feathers, and tail tawny or pinkish-brown. **GEOGRAPHICAL VARIATION & RANGE** Only two subspecies generally recognised. In the treated region the western race *roseus* is found, which is characteristically rather pale, with pinker red plumage, and greyer upperparts.

EVENING GROSBEAK *Hesperiphona vespertina* (W. Cooper, 1825)

Britain two (Mar 1969, Mar 1980), Norway two (May 1973, May 1975). A mainly North American finch, breeding in Canada and W USA south to SW Mexico in mixed forests; in winter frequents both deciduous and coniferous trees, often also visiting gardens and feeders. In some winters performs erratic movements south further into the continental USA. **IDENTIFICATION** Medium-sized to rather large (**L** 16.5–18 cm), stocky but long-winged finch with a thick neck, full chest and relatively short tail. *Bill powerful, conical and pale*, greenish-yellow (spring/summer) to whitish (autumn/winter). In all plumages has *striking white wing pattern* with extensive pure white patch on whole inner wing (ad ♂) or a smaller silvery-white inner wing patch on much of tertials and inner wing-coverts plus a white primary-patch (♀♀). Ad ♂ has *prominent yellow forehead and broad supercilium* being especially contrasting in spring/summer from rest of *dusky-brown head and neck* (which appears blackish at a distance), gradually merging into *golden-yellow rest of plumage*. ♀ is generally *greyish above and paler grey below* with *greenish nape* and *faint yellow tinge on belly and/or flanks*. Wing pattern differs from that of ♂ as explained above in being much less extensive but including a *white patch basally on primaries*. 1stW ♂ resembles ad ♀ (see Ageing & Sexing). Often social and found in flocks, particularly in winter, when feeding on seeds and berries. Noisy, often uttering a loud, strident slightly downslurred call, e.g. *chii-ep* or *chee-er*, rather like a high-pitched and sonorous call by House Sparrow. **AGEING & SEXING Ad** Sexes differ clearly in ad plumage as described above. In all seasons evenly feathered ad wing provide safe aging, and colourful ♂♂ (with no vestiges of immaturity in plumage) can be safely aged as ad, too. When very fresh, ad ♂'s head tipped or sullied greyer and underparts more dusky greenish-brown, thus yellow bases mostly, or at least partly, concealed. **1stY** Both sexes reminiscent of ad of respective sex, but retain variable amount of outer juv greater coverts and tertials (latter more pointed and uniform brownish, lacking contrasting blackish centres and inner webs of ad ♀), as well as most flight-feathers. Basal white primary patch smaller in juv ♂, and absent or almost so in young ♀. **GEOGRAPHICAL VARIATION & RANGE** Geographical variation limited. Ssp. *vespertina* (C & E Canada and NE USA; wintering throughout C & E USA south to Texas and N Florida) is apparently the race that has reached Europe as vagrant.

H. v. vespertina, ♂ and ♀ (right), New York, USA, January: the 'Hawfinch of North America', a large finch equipped with very heavy conical bill and shortish tail. Both sexes have pale yellowish bills with hint of green tinge, and blackish wings with a large white tertial patch in ♀, a smaller wing patch in ♂. Note striking yellow forehead and supercilium in ♂. (M. Read)

GOLDEN-WINGED WARBLER *Vermivora chrysoptera* (L., 1766)

Britain one (Jan–Apr 1989), Azores one (Oct 2012). A summer visitor to a narrow zone in SE Canada and NE USA; winters in SE Mexico and Guatemala south to Panama, with a few reaching N South America. Generally, lives in shrubby, young forest habitats. The species is in sharp decline (with some estimates suggesting that the population has fallen by 98%), the main reasons being habitat loss on both the breeding and wintering grounds, and hybridisation with the closely related Blue-winged Warbler (see below). **IDENTIFICATION** A small (**L** 11.5–12.5 cm) *and delicately built wood warbler*, with highly attractive ad ♂ plumage having *black bib* and *black mask*, the mask *emphasized by whitish surrounds*, but also an eye-catching *yellow cap* and a *yellow wing-coverts panel* on otherwise delicate *plumbeous-grey rest of plumage*. Tail grey, but outer tail-feathers somewhat whiter. ♀ similarly patterned but much duller, with dark facial marks subdued, being mostly greyish-black, and yellow marks less bright being suffused with green. ♂♂ unmistakable, and although ♀♀ and 1stY are more subtly patterned they are still recognised due to at least a hint of dusky cheeks and throat and yellow-olive wing panels and forecrown, while most hybrids with Blue-winged Warbler are distinctive by their mixed characteristics (see below). Often utters a sharp *chip* or *tsik* (but is usually quiet in winter). **AGEING & SEXING** **Ad** Both ad and subsequent imm plumages show clear sexual differences as described above. **1stY** Similar to ad, only a little duller, some probably only separable by presence of retained juv remiges and tail-feathers, tertials and primary-coverts, and some alula-feathers (relatively abraded and more tapered), while grey of plumage to some extent washed brownish, and chin usually whitish, contrasting with the greyish (♀) or black (♂) throat, and with white in the outer tail-feathers reduced or less pure. **GEOGRAPHICAL VARIATION** Monotypic. – Hybridises with Blue-winged Warbler, resulting in two basic hybrid phenotypes, with commonest being birds with white underparts and a thin black eye-stripe, the so-called 'Brewster's Warbler', and the less frequent 'Lawrence's Warbler' showing yellow underparts and a dark throat and dark auricular patch.

1stW ♂, Corvo, Azores, Oct: unmistakable due to attractive dark facial mask and bib, yellow cap and yellow wing panel on otherwise plumbeous-grey rest of plumage. Outer tail-feathers have much white. Dusky rather than black mask and bib, still complete yellow wing patch, and contrastingly brown-tinged primary-coverts and alula infer age and sex. (V. Legrand)

1stY ♀, Costa Rica, Feb: the dull plumage, especially the dark facial marks and poorly developed bib, invite confusion with several species. Mostly greyish-black, with yellow marks duller and suffused with green. Primary-coverts and primaries are retained juv; these with the dull plumage infer age and sex. (M. Danzenbaker)

Close to 'Brewster's Warbler', 1stW, presumed ♀, Azores, Oct: unlike 'classic Brewster's', which should be whiter below, this bird still falls within the wide variation of what some would consider the 'Brewster's type', thus a hybrid Golden-winged × Blue-winged Warbler. The brown-tinged primary-coverts and primaries indicate young age, and probably a ♀ by the overall dull plumage (V. Legrand)

BLUE-WINGED WARBLER *Vermivora cyanoptera* Olson & Reveal, 2009

Ireland one (Oct 2000), Azores two (Oct 2011, Oct 2015). Summer breeder in EC North America inhabiting brushy meadows and second-growth woodlands; winters from SE Mexico south to Costa Rica, occasionally Panama. Breeding range is expanding northwards, where it is gradually replacing Golden-winged, and the two also hybridise. **IDENTIFICATION** Rather small (**L** 11.5–12.5 cm) North American wood warbler, with quite *bright yellow* ad ♂, brightest parts being *forecrown, face and underparts*, while *upperparts are mostly greenish-olive*, but for blue-grey tinge on wings. Other characteristics include *two white wing-bars* and prominent *black eye-stripe*. Undertail-coverts white. ♀ duller overall, with forehead and narrow supercilium yellow, emphasized by dusky eye-stripe, but crown greenish-yellow, and underparts dull yellow. Bill proportionately *long, slender and spike-like*. *Extensive white in tail*, visible when spread or if seen from below. Other North American vagrant wood warblers that are mostly yellow below, such as Yellow Warbler (which see), lack the dark eye-stripe and broad whitish wing-bars of Blue-winged. However, the main pitfall is hybrids and intergrades with Golden-winged Warbler (which see). Often utters squeaky *swik* or *sirk* notes; also a sharp, metallic *tchip* and thin, high-pitched *zzee*. **AGEING & SEXING** **Ad** Sexes differ as described above. All ad undergo complete post-nuptial moult, hence lack any moult contrasts in plumage. **1stY** After partial post-juv moult, 1stY ♂ similar to ad ♀, but crown more yellow in spring and eye-stripe blacker; rectrices, remiges, alula and primary-coverts on average more worn, especially in spring; young ♀ averages duller still than ad ♀, with much of crown more or less concolorous with dull greenish-olive upperparts, and eye-stripe more dusky and indistinct; supercilium often better marked, underparts often more diluted yellow, and wing-bars less clearly defined. **GEOGRAPHICAL VARIATION** Monotypic.

1sW ♂, Azores, Oct: black eye-stripe quite distinctive in otherwise predominantly yellow plumage, brightest on forecrown, face and underparts. Hind-crown to rump greenish, while wings are tinged blue-grey with two white wing-bars. White undertail-coverts useful if visible (though not here). Combination of brown primary-coverts, moderate amount of yellow on forehead and blackish eye-stripe infer age and sex. (V. Legrand)

HOODED WARBLER *Setophaga citrina* (Boddaert, 1783)

Britain two (Sep 1970, Sep 1992), Azores two (Oct 2005, Oct 2008). Summer visitor in SE Canada and E USA to coastal Gulf of Mexico and N Florida favouring swamps and moist woodland; migrates to winter in Central America and the Caribbean. **IDENTIFICATION** A medium-sized (**L** 12–13.5 cm) North American wood warbler, ♂ with strikingly *black hood surrounding yellow face and forecrown*. ♀ usually a clearly duller version of ♂, but some advanced ♀♀ are more ♂-like (showing blackish-admixed olive crown and sides of neck, sometimes also black throat or black spots on breast). All ages to some degree *tinged olive-green above*, with *bright yellow underparts*, *dark lores* and *white outer tail-feathers*. Legs pale pinkish-brown. Dullest 1stY ♀♀ lack clear hooded

1stW ♂, Corvo, Azores, Oct: note strikingly black-hooded ♂ head pattern, with black surrounding yellow face and forecrown. 1stW by duller retained juv primary-coverts (in this species, 1stW and ad autumn ♂ are quite similar in head and body plumage). (M. Bruun)

pattern, are uniform olive-green above and yellow below and can resemble 1stY Yellow Warbler, but that species lacks white in tail and has darker legs, with at least some hint of yellowish wing-bars (no pale wing-bars in Hooded); Yellow never shows clear contrast between olive crown and yellowish face like Hooded. Typically appears *strong-billed* and *large-eyed*. Constantly flicks and spreads tail when feeding in canopy. Call a short, metallic *chink*. **AGEING & SEXING Ad** Sexes differ as described above. Ad undergoes complete post-nuptial moult, hence shows no moult contrasts. **1stY** After partial post-juv moult the sexes resemble respective ad, and especially ♂♂ are difficult to age in autumn. Still, ageing sometimes possible at close range, since rectrices and remiges, large alula and primary-coverts average more worn. **GEOGRAPHICAL VARIATION** Monotypic.

1stW ♀, Corvo, Azores, Oct: there is usually only a hint of the ♂ head pattern, with faintly darker edge to yellow face. 1stW by duller retained juv primary-coverts. (D. Monticelli)

CAPE MAY WARBLER *Setophaga tigrina* (J. F. Gmelin, 1789)

Britain two (Jun 1977, Oct–Nov 2013). Summer breeder in Canada and NE USA; migrates primarily to Caribbean (mainly Greater Antilles), but also regularly to Bermuda. Interestingly, the tongue of this species is exceptionally structured among wood warblers, being curled and semitubular, seemingly an adaption for collecting nectar on wintering grounds. **IDENTIFICATION** Medium-sized (**L** 11.5–13 cm) wood warbler which has *compact* build due to proportionately *short tail*. Also note *thin, slightly downcurved bill*. Attractive breeding ♂ has *yellow face, underparts and rump* (yellow usually extending to sides of neck), *blackish crown* and a variably developed *chestnut patch on ear-coverts and rear supercilium*. *Underparts boldly streaked*, and an *extensive white wing-coverts panel* further makes ad ♂ distinctive. Note greenish edges to flight-feathers, forming a hint of a secondary panel. ♀ much drabber, greyer and greenish-tinged (as a rule lacking the chestnut cheeks and blackish cap of ad ♂), though still distinctive due to brown-streaked yellowish underparts, two narrow white wing-bars, greenish secondary panel, and *greenish-yellow rump*. Note also blackish tail edged olive, and white spots on inner web of outer tail-feathers (largest in ad ♂, smallest in 1stY ♀)—unless the tail is spread, best seen from below. 1stY ♀ can resemble a dull Yellow-rumped Warbler but is smaller, shorter-tailed, and rump is greenish rather than bright yellow. Main call note a high-pitched, thin *tseep*. **AGEING & SEXING Ad** After complete post-nuptial moult plumage lacks moult contrasts. Sexes differ as described above, but in addition, the neatest ♂♂ can have the yellow on sides of neck merge at nape to form almost complete collar, while chestnut may extend onto throat. Ad ♂ in fresh autumn plumage has less solidly black crown due to olive tips, chestnut of face paler and more limited due to yellow-buff tips, and some grey tips on hindneck. The yellow on neck-sides is also slightly obscured due to olive-grey feather tips, while the yellow of rest of underparts is less deep than in spring (but usually still clearly brighter and more extensive than in any ♀ or 1stW ♂). **1stY** Following partial post-juv moult ageing possible using slight moult contrasts. Plumages recall respective ad, but are duller, thus 1stW ♂ resembles ad ♀ and in spring, unlike ad ♂, has chestnut ear patch much less distinct (hence face pattern can appear rather obscured) and has smaller white wing patch. Young ♀ can be extremely drab, with greyer face, only a fraction of yellow below, streaking on breast-sides and flanks washed out, and rump greenish-olive. Young age, however, often must be confirmed by presence of retained juv rectrices and remiges, large alula and primary-coverts (being on average more worn). **GEOGRAPHICAL VARIATION** Monotypic.

♂♂, ad (left: Dominican Republic, Apr) and 1stS (right: Cuba, Mar): in breeding plumage note bright yellow head and underparts, usually extending to sides of neck, boldly streaked underparts, and white patch on median coverts. Note diagnostic chestnut cap and cheeks, but these are duller and more rust-brown on cap and with only a hint of it on the ear-coverts in the young ♂ (right). The two birds are also aged by primary-coverts. (H. Shirihai)

1stW ♂, USA, Sep: young autumn ♂♂ need first of all be separated from ad ♀. Brownish-tinged retained juv primary-coverts here contrast with replaced greater coverts signalling 1stW, and rather a lot of yellow on face and underparts (breast distinctly but finely black-streaked) strongly indicate ♂. Note also yellowish rump. (S. N. G. Howell)

1stS ♀, Cuba, Mar: less bright colours compared to young ♂, with much subdued yellow, and boldly brown-streaked or blotched underparts. Has two narrow white wing-bars, greenish secondary panel, and rump is greenish-yellow (not visible here though). Aged by retained juv brownish primary-coverts, worn tips to primaries and tail-feathers, and overall dull plumage. (H. Shirihai)

1stW ♀, Mexico, Dec: drabbest plumage, with greyish face and just a hint of yellow below, streaks on breast-sides and flanks washed out, and rump dull, greenish-olive rather than yellow. Note retained juv brownish-tinged primary-coverts contrasting with replaced greater coverts. (S. N. G. Howell)

CERULEAN WARBLER *Setophaga cerulea* (A. Wilson, 1810)

Iceland one (Oct 1997). Breeds locally in SE Canada and NE USA (but with noted sharp decline in the core of its range), mainly feeding in the upper canopy of extensive mature deciduous woods, and often found in mixed woodlands near water; migrates to W South America, wintering from N Colombia and Venezuela south to S Peru and W Bolivia. **IDENTIFICATION** Among the smallest of all North American wood warblers (**L** 11–12 cm) and *proportionately short-tailed* (giving rather plump or compact impression), but *primary projection is long*. Bill rather strong and thick. In all plumages shows *broad double white wing-bars*, and to some degree

Ad ♀, USA, Apr: mantle tinged greyish-olive, while crown and rump are blue-green. Supercilium pale cream-white, typically broadening behind eye, and wing-bars narrower than in ad ♂; whitish underparts washed yellowish-cream, with indistinct dark streaking on the sides (here hidden by the wings). Wing evenly feathered ad. (A. Murphy)

characteristic *bluish hue on head and upperparts* (though ♀ and imm are more olive-tinged). Ad ♂ is typically *bright cerulean-blue above* (purest on crown), with some dark streaks on mantle; *white below* with *narrow grey-blue breast-band* and bold but diffuse *blackish blue-grey streaking on flanks*. Ad ♀ has *greenish-tinged mantle*, but still blue-green or *bluish crown and rump* (fairly similar 1stS ♂ shows more bluish colour and dark streaks above than any ♀). ♀ lacks the ♂'s breast-band, or just shows a hint of it. Further, ♀ is *almost unstreaked above*, but has *broad pale supercilium* (typically broader behind the eye). Throat, flanks and breast are pale yellowish-cream, indistinctly streaked dark on sides. In all plumages *white tail-corners* (in form of large spots) prominent but quite small (visible from below, or when tail is spread). Young ♀ in autumn should be distinguishable with care from drab young ♀ Blackburnian and Black-throated Grey Warblers (which see); Cerulean is usually visibly smaller and shorter-tailed than both, and has at least some bluish tinge above and some yellowish-cream tinge below, and is less clearly streaked above. In addition, Cerulean has less strongly marked cheek patch, and its supercilium does not meet the pale neck-side band. Finally, Cerulean has less extensive white on the outer tail-feathers. Main call note a slurred *chip*. **AGEING & SEXING Ad** Sexes differ as described above. Limited seasonal variation. Complete post-nuptial moult in late summer. Bluish becomes purer and any streaking enhanced by wear in spring–summer, when plumage becomes immaculate in ♂. **1stY ♂** intermediate between the two sexes. 1stY ♀ has little if any hint of bluish in plumage. Pale supercilium striking, underparts dull whitish but extensively washed with pale yellowish, especially on sides of throat, across breast, and on flanks, which are only diffusely streaked with greyish-olive. Any safely identified retained juv primary-coverts, remiges or rectrices in autumn should clinch ageing and make sexing of difficult birds more reliable. **GEOGRAPHICAL VARIATION** Monotypic.

1stS ♂, USA, spring: attractive plumage with largely dark-streaked pale blue upperparts with two bold white wing-bars, and pure white underparts with a narrow black breast-band and some black blotches along flanks. Aged by to brownish primaries in stark contrast to black tertials. Slightly green-tinged secondary fringes confirm. (E. Schneider)

1stW ♂, USA, Jul: judged to be 1stW by overall rather dull plumage pattern, and this, combined with having greyish-blue tones in plumage with only slightest green hue above, fits best with young ♂. Note also almost unmarked mantle and diffuse flank-streaking, as well as ♀-like pale supercilium. (G. Koziara).

1stW ♀, USA, Aug: the young ♀ differs rather clearly from ♂ in being greener, less bluish-grey above, and more yellowish, less off-white below. Age further indicated by rather dull primary-coverts. (G. Koziara).

MAGNOLIA WARBLER *Setophaga magnolia* (A. Wilson, 1811)

Britain two (Sep 1981, Sep 2012), Iceland two (Sep–Dec 1995, Oct 1995), Azores two (Oct 2009, Oct 2012). Breeds across S Canada, NE USA, the Great Lakes Region, and the Appalachians; migratory, wintering from Mexico to Central America, as well as in the Caribbean. Breeds mostly in coniferous forests, though on migration favours a variety of woodlands. **IDENTIFICATION** Rather smallish (**L** 12–13.5 cm) wood warbler which in all plumages show diagnostic and conspicuous *large white innertail* (most extensive in ad ♂), *bright yellow throat and breast*, in breeding plumage *a black bar across breast* and *boldly black-streaked sides*. Note *double white wing-bars* (can form large white wing panel in ad ♂). Ad ♂ has *blackish mask, mantle and back*, and a *white supercilium starting from eye and bordering bluish-grey crown*. Note also distinctive *yellow rump patch* and *white undertail-coverts*. Ad ♀ variably duller than any ♂, with dull grey head, whitish lores, faint and narrow pale supercilium, indistinct but complete eye-ring, grey ear-coverts and mottled mantle, latter being more olive-tinged than in ♂. White wing-bars narrower, and streaking on breast and flanks narrower and less bold. The dullest plumages are those of 1stW (see below), which recall ad ♀, only with duller or no dark streaking below, face pattern less distinct, crown and upperparts olive-grey, and white wing-bars narrow. Main contact call a nasal *zic*. **AGEING & SEXING Ad** Sexes differ clearly (as described above), also in fresh autumn plumage (following complete post-nuptial moult). Moult in late summer affects mostly ad ♂ in which black areas become reduced due to broad yellow-olive tips (resulting in a black-spotted appearance above, or nearly all black is concealed), crown and face grey with a thin white eye-ring, and often a hint of white supercilium behind eye; underparts become paler yellow, often with reduced black streaking, and white wing panel less solid. Amount of white in tail greater than in any 1stW. **1stY ♂** intermediate between ad of the two sexes. Young ♀ duller than ad ♀, with grey crown and hindneck slightly tinged olive, olive-yellow upperparts have sparse, inconspicuous dark streaking, rump tipped yellowish olive-grey, while throat, breast and belly are duller yellow, and underparts have partly obscured black streaking, mostly on flanks. Ageing of young birds further confirmed by any moult limits in wing (as a result of partial post-juv moult in late summer). **GEOGRAPHICAL VARIATION** Monotypic.

Ad ♂, Mexico, Apr: here in breeding plumage. Note in particular the head pattern with pale lead-grey crown and extensive white in greater coverts. Aged by neat plumage and ad-type blackish primary-coverts. (H. Shirihai)

1stW ♂, Corvo, Azores, Oct: in all plumages shows conspicuous white band across tail-base, yellow underparts streaked black down sides, as well as double white wing-bars. Note moult limit in greater coverts, and apparent contrast with retained browner primary-coverts and primaries, and quite pointed tail-feathers. (V. Legrand)

♀, presumed ad, Mexico, Sep: note diagnostic large white tail-base. Quite similar to ♂ but subtly duller grey head (but for whitish lores and eye-ring), and olive-tinged and slightly mottled mantle. Also, by comparison, white wing-bars narrower, and streaking on body-sides narrower and less extensive. Wing seems evenly fresh ad. (S. N. G. Howell)

BAY-BREASTED WARBLER *Setophaga castanea* (A. Wilson, 1810)

Britain one (Oct 1995). Summer breeder across S Canada and NE USA, migrating to winter in Central America and NW South America (in autumn migrates earlier than Blackpoll Warbler). Breeds in open coniferous and mixed forests, often in swampy areas, and in winter found in forest edges and second growth with scattered trees, where it often joins mixed-species feeding flocks. **IDENTIFICATION** Rather large (**L** 13–14 cm), *short-tailed* wood warbler with distinctive, almost unique breeding plumage, ad ♂ having *black facial mask extending well onto forehead*, pale *creamy-buff patch on neck-sides, chestnut crown, breast and flanks, greyish dark-streaked upperparts, double white wing-bars* and *creamy-white underparts*. Breeding ad ♀ is usually clearly duller, with *head being mostly grey, mottled black, crown faintly washed chestnut and streaked darker*, and pale neck-side patch subdued, while chestnut on body-sides is diffuse and patchy. Note also dark tail with white patches on inner web of outer three rectrices (largest in ad ♂, smallest in 1stY ♀). Non-breeding ad and 1stW plumages (both sexes) are very different, being yellowish olive-green and somewhat dark-streaked on head and upperparts, with chest and flanks pale brown merging into off-white throat and belly, and has dark legs. Non-breeding ♂♂, and especially ad, are still strongly streaked black above and retain extensive chestnut on flanks, whereas 1stW ♀♀ are the dullest, and will need careful separation from same-plumaged Blackpoll Warbler, but differ in having greyish legs, fairly uniform pale whitish-buff and essentially unstreaked underparts, and weaker-marked eye-stripe and supercilium. Main contact call a sharp, somewhat husky *tchip*, and a rough *zeet*. **AGEING & SEXING** Ad Sexes differ clearly in breeding plumage (spring–summer), but are quite alike following complete post-nuptial moult when in fresh autumn plumage (see above). Non-breeding ♂♂ retain strong buff tinge on underparts with chestnut cast on flanks. They also have larger dark feather-centres above than ♀♀ and young birds. **1stY** During first autumn (following partial post-juv moult to 1stW plumage) resembles non-breeding ad, but following late winter transformation (partial pre-nuptial moult and wear) 1stS patterned like breeding ad, albeit plumage less neat and attractive. Age of 1stY can be confirmed by moult limits (mostly due to partial post-juv moult), with retained juv remiges, rectrices and primary-coverts. **GEOGRAPHICAL VARIATION** Monotypic.

Ad ♂, Canada, spring: aged by full breeding plumage, especially head, and evenly ad wing, including dark primary-coverts. Unmistakable due to black facial mask extending to forecrown, pale creamy-buff patch on neck-sides, bay-coloured crown, bib and (somewhat paler) on flanks. Further, black-streaked pale grey upperparts, and double white wing-bars, as well as creamy-white unstreaked underparts.

Ad ♀, USA, Apr: age and sex inferred by dark primary-coverts, nice wing-edgings, and overall dull, ♀-like head and body plumage. Chiefly note mostly grey head mottled black, crown faintly washed rufous, and paler neck-side patch, while rufous on breast diffuse and patchy, lacking on flanks. (B. Small)

Ad ♂, Colombia, Jan: non-breeding ad is very different, being largely yellowish olive-green, but with relatively extensive dark red-brown tinge to flanks and some on chest, merging with off-white throat and belly. Further aged by blackish primary-coverts and relatively bold white double wing-bars. (H. Shirihai)

1stW, presumed ♂, California, Sep: 1stW by and large are the dullest, and sexes require care to be separated. The retained juv slightly duller primary-coverts and some outer unmoulted greater coverts, and just a hint of chestnut wash on flanks, infer age and sex. (S. N. G. Howell)

BLACKBURNIAN WARBLER *Setophaga fusca* (Statius Müller, 1776)

Britain three (Oct 1961, Oct 1988, Sep 2009), Iceland one (autumn 1987). Breeds in E North America (generally in coniferous or mixed forests, such as pine-oak woodlands); migratory and winters in S Central America and South America, from Colombia, Venezuela and N Brazil south in the Andes to C Bolivia. Occupies wide range of habitats, with preference for tall vegetation, and often dwells in the canopy of trees. **IDENTIFICATION** Rather smallish (**L** 11.5–13 cm) North American wood warbler which in all plumages typically has *pale 'tramlines' on mantle* and *streaked flanks, double white wing-bars* and striking facial pattern, with conspicuous *broad pale supercilium, neck-sides, throat* and *median crown-stripe* (not always visible), and a *pale crescent-shaped patch below eye*, surrounded by triangular-shaped *dusky ear-coverts*, while *darker lores and eye-stripe* can be pronounced in some birds. Breeding ad ♂ has *dark head and dorsal areas solid jet-black* with *broad white wing-bars merging to form quite extensive panel*, plus attractive *orange head pattern, throat and breast*, and *black-streaked flanks*. Breeding ad ♀ is usually clearly duller, with diluted yellow areas (though often quite bright on face), and dark areas olive to dusky-grey (rather than black), narrower white wing-bars, and flanks less boldly streaked. 1stY ♂ resembles ad ♀, with 1stW ♀ the dullest, with yellow colour, facial pattern and upperparts streaking reduced or subdued. The white bases and edges on outer tail-feathers are largest in ad ♂, smaller in ad ♀, and even smaller in 1stY. Main call a sharp *tsip*, and variant is a thin, slightly buzzy *seet*. **AGEING & SEXING** Ad Sexes differ clearly in spring–summer as described above. Undergoes complete post-nuptial moult mostly before autumn migration, but there is also a limited partial pre-nuptial moult in winter. In fresh autumn plumage, only ad ♂ differs markedly being to some degree duller, with less intense yellow-orange on face and underparts, feathers of upperparts tipped greyish, and white wing-bars form less or no solid panel. **1stY** Initially in autumn, 1stW ♂ is like ad ♀ but with more white in wing, while 1stW

Ad ♂, USA, spring: dark areas solidly jet-black with large white wing-bars merging to form quite large wing panel, and pale areas of face bright yellow or brilliant orange (mainly throat), and streaks on flanks black. Aged by having black primary-coverts.

1stS ♀, USA, Jun: clearly duller than ♂, with diluted yellow-orange areas (though often quite bright around the face), with olive-tinged dusky-grey dorsal areas, the two white wing-bars narrower (still quite broad!), and flank-streaking less bold. Aged by brownish-tinged retained juv primary-coverts, and relatively dull head for summer. (D. Monticelli)

Ad ♀, USA, Oct: in fresh autumn ♀♀, pale 'tramlines' on mantle and median crown-stripe often concealed or very weakly patterned, while dark streaks on flanks ill-defined. Further, note double white wing-bars and striking facial pattern (conspicuous broad pale yellow supercilium and neck-side line). Aged by dark primary-coverts not contrasting with rest of wing. (D. Delimont)

♂, presumed 1stW, Azores, Oct: similar to ad ♀, but on average head pattern rather bolder, and yellow-and-orange facial areas brighter, in particular on supercilium. Subtly brown-tinged and duller primary-coverts and remiges contrast slightly with dark-based greater coverts, which further infer age. (V. Legrand)

♀ is much duller than rest. By 1stS both sexes attain plumages similar to their respective ad, though often with vestiges of immaturity. 1stY often best aged by moult limits (due to partial post-juv moult), with retained juv remiges, rectrices and primary-coverts being browner, contrasting to darker moulted coverts. **GEOGRAPHICAL VARIATION** Monotypic.

CHESTNUT-SIDED WARBLER *Setophaga pensylvanica* (L., 1766)

Britain two (Sep 1985, Oct 1995), Azores one (Oct 2009), France one (Oct 2010). Breeds in S Canada south through Great Lakes region to NE USA in second-growth deciduous forest, also forest edges and clearings; migrates to winter grounds, mainly from SE Mexico south to Panama. **IDENTIFICATION** A medium-sized (**L** 12.5–14 cm) and compact North American wood warbler. In breeding plumages unmistakable due to *yellow crown*, a *black eye-stripe connected on lores to black 'moustache'*, leaving *extensive white cheeks*. Underparts white with some *chestnut on breast-sides and flanks*. *Upperparts yellowish olive-green, heavily streaked black*, while wings have distinctive *double yellowish wing-bars*. Breeding ad ♂ is brightest, with purer yellow crown, has strongest face pattern (due to solid black mask) and most extensive and pure chestnut on flanks. Breeding ad ♀ recalls ad ♂ but is usually clearly duller and less boldly marked, even less so than 1stS ♂, with crown more greenish-yellow, black markings on side of head narrower, greyer and often mottled, and chestnut on body-sides heavily reduced, while upperparts are less clearly dark-centred, being more olive-green overall. All non-breeding plumages (ad and 1stW, both sexes) lack black mask on head-sides, with much-reduced chestnut on flanks and dark streaking above, with 1stW ♀ the dullest, but still distinctive, especially due to *bright lime-green crown and upperparts*, *greyish-white face and underparts*, and *prominent unbroken white eye-ring*. In autumn/winter, ad ♂ retains extensive chestnut on flanks, and some 1stW ♂♂ can have rather much chestnut, too, but ad ♀ often shows only a trace, and 1stW ♀ usually lacks it. Habitually, tail often held slightly cocked. Main call a sharp, somewhat husky *tchip* (similar to Yellow Warbler), and a rough *zeet*. **AGEING & SEXING Ad** Sexes differ clearly in spring–summer as described above. Undergoes complete post-nuptial moult mostly before autumn migration, but there is also a limited partial pre-nuptial moult in winter. Beware that least advanced 1stS ♂ can look quite like ad ♀. In fresh autumn plumage sexes more similar but (evenly fresh) ad ♂ still has more chestnut on flanks. 1stW ♂ and non-breeding ad ♀ very similar but if aged first, sexing can be done using upperparts and flank coloration. **1stY** By first spring, both sexes attain plumages similar to their respective ad, although often with vestiges of immaturity, and especially imm ♂♂ can

1stS ♂, USA, spring: even 1stS (primary-coverts retained juv) has rather pure yellow crown, strong face pattern (due to solid black mask), with black eye-stripe connected to 'moustache'. Also note extensive white cheeks, and extensive and solid chestnut on flanks.

Ad ♀, Canada, May: duller and less boldly marked ♀♀, with crown more greenish-yellow, black markings on side of head narrower, greyer and often mottled, and chestnut on body-sides heavily reduced, while wings have narrower double yellowish wing-bars. Wing evenly feathered ad. (S. Gettle)

1stW ♂, USA, Dec: note combination of no black markings on head-sides and lime-green crown contrasting with grey head-sides, while chestnut on flanks can still be obvious. Rest of upperparts also green, diffusely streaked dark. Further aged by retained juv slightly browner primary-coverts and alula, and remiges with duller edges. (D. Miller)

1stW, USA, Oct: dullest plumage, still distinctive, especially due to bright lime-green crown and upperparts, pale grey face and underparts, and prominent white eye-ring. 1stW ♀ usually (though not always) lacks chestnut on flanks, but about half of 1stW ♂♂ also lack chestnut on sides, hence this is not a reliable sex difference. The primary-coverts are retained juv. (B. Zwiebel)

be duller and less boldly patterned. 1stY often best aged by moult limits (due to partial post-juv moult). In non-breeding plumages it is important to age a bird before sexing it, chiefly when attempting to separate 1stW ♂ from ad ♀. However, usually non-breeding ad ♂ and 1stW ♀ distinctive on plumage, as described above. **GEOGRAPHICAL VARIATION** Monotypic.

BLACK-THROATED BLUE WARBLER *Setophaga caerulescens* (J. F. Gmelin, 1789)

Iceland two (Sep 1988, Oct 2003), Azores six (Oct–Nov 2005, two Oct 2006, Oct 2013, two Oct 2015). Breeds in the interior of deciduous and mixed coniferous forests in E North America; migrates to winter mostly in Central America and the Caribbean islands. **IDENTIFICATION** Rather small (**L** 12–13.5 cm) wood warbler with near-diagnostic *white basal primary patch* (though this could be very limited or invisible in some 1stY ♀♀), and attractively patterned ad ♂ plumage with the main features being *black face, throat and flanks*, *bluish-grey to deep blue upperparts* and white remaining underparts. ♀♀ are very different, lacking any black or bluish, or the pure white belly, instead has *brownish-olive upperparts* and *pale buff-white underparts*. Shares with ♂ the distinctive *white primary patch*, only smaller. Has a *narrow buff-white supercilium*. Limited seasonal variation in ad, and 1stW of each sex usually already very similar to respective ad (see below). Call a fairly quiet metallic *twik*, much like Dark-eyed Junco. **AGEING & SEXING Ad** Sexes differ as described above, with winter ad ♂ very similar to breeding plumage in spring or with just a fraction of olive wash above. Undergoes complete post-nuptial moult prior to autumn migration. **1stY** As ad following partial post-juv moult in late summer, but 1stW ♂ is washed more extensively olive above, and black bib is less solid, being mottled with whitish tips, especially near chin and on upper throat. Also recognised by browner-tinged and more pointed primary-coverts, narrower and more pointed tail-feathers, and has on average smaller white basal primary patch. 1stY ♀ usually safely aged on much smaller, if visible, white basal primary patch, but some borderline cases will need to be confirmed with moult limits in wings, with retained juv, slightly worn and browner primary-coverts, and browner primaries with less fresh tips. **GEOGRAPHICAL VARIATION** Monotypic.

1stW ♂, Azores, Oct: note characteristic white basal primary patch, black face, throat and flanks, bluish upperparts and white remaining underparts. The slightly browner primary-coverts and primaries, contrasting with bluer-edged greater coverts, infer age. (V. Legrand)

Ad ♀, Cuba, Mar: note combination of bold and quite extensive (for ♀) white basal primary patch, upperparts being brownish-olive and underparts tinged pale buffy-ochre, as well as narrow, but contrasting buff-white supercilium. Note also evenly greenish-tinged wing feathering, and dark primary-coverts. (H. Shirihai)

1stW ♀, Azores, Oct: as ad ♀ but note tiny, just visible white basal primary patch indicating that remiges are juv. Belly has interesting 'marbled' pattern. (V. Legrand)

PALM WARBLER *Setophaga palmarum* (J. F. Gmelin, 1789)

Iceland one (Oct 1997). Breeds in open coniferous bogs across Canada and NE USA, where shows a preference for open, unwooded habitats more than most other warblers. Rather late autumn migrant, wintering from S USA to the West Indies and south to Panama. During migration and winter found in open brushy woodland and marshes, only occasionally in palms. (Both vernacular and scientific names misrepresentative, as no inclination to use palms during breeding, and little in winter either.) **IDENTIFICATION** Rather large (**L** 13–14.5 cm) North American wood warbler, with little sexual dimorphism restricted to breeding season. In all plumages shows *yellow undertail-coverts*, often *contrasting with the remaining underparts which are paler yellow or whitish*. A ground-dweller *constantly tail-wagging* as it forages, which habit will clinch identification most of the time. Ad in spring–summer has *chestnut cap*, and breast-sides and flanks streaked chestnut. In all plumages *prominent dark eye-stripe* enhances *broad yellowish-white supercilium*. Upperparts olive-grey or olive-brown indistinctly streaked darker, but *rump is dull olive-yellow*. Greater and median coverts narrowly tipped buff-white, forming *ill-defined double wing-bars*. *Tail corners have white spots*. A yellowish-white lower eye-crescent is rather well developed, with a very narrow and indistinct dark lateral throat-stripe. During autumn and winter ad and 1stW lack (or just show a trace of) the chestnut cap typical of breeding plumage, and yellow in plumage much obscured. 1stW (most likely to reach the treated region) have the most nondescript plumages, with crown more or less uniform with upperparts, throat whitish, not contrasting with rest of underparts, but usually a faint yellowish-buff wash on flanks (vaguely streaked darker), supercilium and submoustachial stripe. Crown and mantle thinly and indistinctly streaked darker. Main call described as a sharp *chip* or high-pitched *seep*, mostly given in flight. **AGEING & SEXING Ad** Sexing difficult, although breeding ♂♂ tend to show more and brighter chestnut on crown than ♀♀ in all seasons; in fresh autumn plumage sexing is only possible with some ad ♂♂ that retain substantial chestnut on crown, and aging is possible only by moult pattern, ad wings being evenly fresh (following complete post-nuptial moult prior to autumn migration); chestnut crown and yellow on face and below develop by midwinter (through partial pre-nuptial moult). **1stY** As ad but can be aged by moult limits (due to partial post-juv moult) created by retained juv (slightly worn and browner) primary-coverts, and primaries with less fresh tips. The unmoulted tertials may also form moult contrast with renewed secondary-coverts. **GEOGRAPHICAL VARIATION & RANGE**

D. p. hypochrysea, 1stW, presumed ♂, USA, Oct: typical combination of bright yellow fore supercilium and undertail-coverts, yellowish rest of underparts, lack of prominent wing-bars and slight rufous tinge to crown. Although both sexes and ages quite similar, 1stW inferred by retained juv brownish primary-coverts contrasting with replaced greater coverts, and ♂ guessed by extensive yellow on face and underparts and obvious rufous on crown. (S. Byland)

D. p. hypochrysea, ♀, presumed 1stW, E USA, Sep: no reddish on crown indicates ♀, while relatively extensive yellowish wash on face and underparts for a bird that is seemingly 1stW (although primary-coverts are largely concealed they seem to contrast against newer greater coverts) is characteristic for this eastern race. (S. N. G. Howell)

D. p. palmarum, presumed ♀, Cuba, Nov: the western subspecies *palmarum* is less yellow, more off-white below down to belly, making the bright yellow undertail-coverts stand out. It is more widespread and more numerous than *hypochrysea*, hence could be just as likely to straggle to Europe. No rufous on crown and overall dull colours suggest ♀, but it could also be a very dull ♂. (G. Bartley)

The eastern race *hypochrysea* (SE Canada, E USA; quite possible taxon to reach Europe) has in breeding plumage a brighter chestnut cap, browner upperparts with stronger olive wash, and bright yellow underparts, supercilium and submoustachial stripe, with chestnut streaking on sides of breast. In non-breeding plumage the yellower hue on face and underparts is evident, at least in ad and ♂♂. Western *palmarum* (C Canada; abundant in NE USA in autumn and very likely vagrant to Europe) has on average paler yellow on head and underparts with whitish belly and darker (less chestnut) streaks on sides of breast. The two races are known to intergrade in a broad zone.

YELLOW-THROATED WARBLER *Setophaga dominica* (L., 1766)

Azores one (Oct 2013). Breeds from Iowa to Pennsylvania and New Jersey, south to E Texas and Florida; winters from Georgia and Texas, south to Central America and the Caribbean. Often forages high in the canopy of oak and pine forests, also in swamps and riparian woodland. On migration and in winter found in a variety of woodlands, brushes and thickets, but attracted to pines where available. **IDENTIFICATION** A rather large (**L** 13.5–15 cm) and elongated wood warbler with striking plumage pattern with little variation regardless of sex or age. Typically has *black forecrown*, a *white supercilium* (yellowish at the front), *black eye-stripe which extends onto lower cheeks and neck-sides* becoming *two or three 'tramlines' of bold black blotches on breast-sides and flanks*. A *white crescent under eye* encircled by black, and a *white vertical rear ear-patch*. *Throat and upper breast vividly yellow*, contrasting with *pure white rest of underparts*. Rear crown, nape and rest of *upperparts uniform bluish-grey* except *prominent double white wing-bars*. Outer four tail-feathers have white spots (visible when tail spread or from below). Sexes similar, but ad ♀ is on average duller, with less extensive black on face, mostly grey-streaked black forecrown, and can have a brownish wash on mantle; flank-streaking moderately narrow and indistinct. 1stY similar to ad but slightly duller. *Bill long, pointed and slightly decurved*. Call a high *see* or a sharp *chip*. **AGEING & SEXING Ad** Sexes differ slightly as described above. Undergoes complete post-nuptial moult prior to autumn migration. Fresh autumn ad ♂ patterned as breeding but to some degree less boldly. Beware that some strongly marked ad ♀♀ are rather similar

♂, presumed 1stW, Azores, Oct: note characteristic long, pointed and slightly decurved bill, and striking plumage, especially black facial marks with white supercilium (yellowish-tinged at front), sub-ocular and neck-side patches. Primary-coverts are seemingly retained juv (subtly paler grey, contrasting very slightly with greater coverts) suggesting 1stW, while neat head pattern infers ♂. (V. Legrand)

1stW, presumed ♀, USA, Nov: similar to previous bird, but duller and more weakly patterned. Note lack of black streaking on crown, brown wash to back and mask, and (especially) the lack of bolder black streaks along flanks, which could suggest ♀. Primary-coverts contrast against newer greater coverts. (E. Nielsen)

to least advanced 1stY ♂♂, but if aged first by moult limits, sexing is possible with experience of autumn plumages. **1stY** 1stW best aged by moult pattern (following partial post-juv moult with, e.g., retained slightly worn and browner juv primary-coverts, or rectrices averaging more pointed). 1stW ♂ usually less smartly patterned than ad ♂, while many 1stW ♀♀ distinctly duller with paler and more ill-defined face pattern, weaker flank-streaking and less white in tail and wing-bars. By spring, 1stS of both sexes similar to respective ad, but can be aged by moult limits in wing, and they are often visibly somewhat less advanced than ad. **GEOGRAPHICAL VARIATION** Four subspecies have been described, but we follow Dickinson & Christidis (2014) and treat it as monotypic.

PRAIRIE WARBLER *Setophaga discolor* (Vieillot, 1809)

Azores one (Oct 2012). Found in scrubby and second-growth habitats and mixed forests throughout E & SC USA (thus, not a prairie dweller!), generally from S Maine to S Missouri, south to N Florida and E Texas. Resident in many places in coastal Florida, but is more common as a winter visitor throughout Florida, Bermuda, Bahamas, Greater Antilles and Virgin Islands, but also Belize and Honduran coasts. **IDENTIFICATION** A small (**L** 11–12 cm) *tail-wagging* North American wood warbler with mainly *olive-green upperparts*, *yellow underparts* and neat *black facial marks* (eye-stripe, arc-shaped moustachial stripe plus lower neck-side patch), further emphasized by *yellow supercilium* and broad yellow crescent-shaped ear-covert mark. *Yellow flanks often boldly marked black*. These marks vary considerably with age and sex, and to some extent geographically and seasonally, too: ad ♂ has bolder and blacker marks on brighter yellow ground, ad ♀ and 1stW ♂ are duller with flank markings dusky or greyish-black, while 1stW ♀ is duller still with markings highly subdued, thereby inviting confusion with other species. Amount of *rufous blotching on mantle* also varies, most extensive in ad ♂ (even merging into a larger patch), more reduced in ♀ and young, and in some 1stW ♀♀ it can be lacking altogether. All plumages show *indistinct double* (broad but ill-defined) *olive-yellow wing-bars*, and tail has large white spots on outer feathers (size varies with age and sex). Usually forages in lower branches and brush, *constantly bobbing or wagging its tail up and down*, and especially when distressed it appears to bob the tail quickly downward, then raise it more slowly. Most frequent call a short stone-clicking *chip*. **AGEING & SEXING Ad** Sexes differ slightly as described above. Undergoes complete post-nuptial moult prior to autumn migration, with a partial pre-nuptial moult in late winter and spring. Ad ♂ in fresh autumn plumage similar to breeding, but smart facial pattern and streaking on flanks so obvious in spring to some degree subdued due to feathers tipped greenish. Beware that some strongly marked ad ♀♀ can be rather similar to least advanced 1stW ♂♂ in their dark markings, but if first aged correctly (ad evenly fresh, 1stW with some moult contrast due to retained juv remiges and primary-coverts) sexing often possible. **1stY** 1stW most resembles ad ♀, but aged using moult pattern (following partial post-juv moult with retained slightly worn and browner juv primary-coverts, or on average more pointed rectrices). 1stW ♀ is dullest, separable with practice by less grey on cheeks than 1stW ♂. By spring, 1stS of both sexes similar to their respective ad but can be aged by moult pattern, and they are also somewhat less neat than ad. **GEOGRAPHICAL VARIATION & RANGE** Two races described, with *discolor* (SE Canada, E USA) characterised by on average more striking chestnut on mantle, with deeper yellow underparts and black markings on sides relatively bolder. The single record from the Azores most probably refers to this race.

S. d. discolor, ♂ (left) and ♀, Cuba, Mar: two birds with reduced amount of rufous on mantle, presumably mainly age-related (as both seem to be 1stS by having some retained juv wing- and tail-feathers), but could to some extent also be due to individual variation. The ♀ has the yellow areas duller and tinged greener, and facial and flank markings are duskier and less bold. (H. Shirihai)

1stW ♂ (left: Azores, Oct) and ♀ (right: Canada, Sep): note hint of grey cheeks and rufous on back of ♂, which is also rather bright yellow on head and body plumage, and shows rather boldly black-streaked flanks (unlike ♀). Both these relatively duller 1stW seem to have retained juv primary-coverts. (Left: V. Legrand; right S. N. G. Howell)

BLACK-THROATED GREEN WARBLER *Setophaga virens* (J. F. Gmelin, 1789)

Germany one (Oct 1858), Iceland one (Oct 2003), Azores five (Oct 2008, two Oct 2009, Oct 2013, Oct 2014). Breeds in coniferous or mixed forests in E North America and W Canada, and cypress swamps on South Atlantic coast and in the Appalachian Mountains. Migrates to Mexico, Central America, the West Indies and S Florida. **IDENTIFICATION** Rather small (**L** 12–13.5 cm) and attractive North American wood warbler, not least owing to ad ♂'s *olive-green and yellow head pattern*, *uniform olive upperparts*, solidly *black bib and breast-sides*, and *very broad white double wing-bars*. Although vent is vividly yellow, undertail-coverts are contrastingly pure white. Ad ♀ and 1stW ♂ show similar but duller pattern, with no black on upper throat and instead any black largely restricted to lateral throat-stripe and upper breast continuing as streaks on flanks. 1stW ♀ is generally significantly duller, with only a hint of dark facial marks, and black below restricted to thin streaking on breast-sides. All plumages share white edges to tail. Especially young birds can be confused with corresponding plumage of several congeners, all of which as yet unrecorded in Europe. Main call a soft, flat *tip* or *tsip*. **AGEING & SEXING Ad** Sexes differ slightly as described above. Undergoes complete post-nuptial moult prior to autumn migration, with a partial pre-nuptial moult in late winter and spring. Ad ♂ in fresh autumn plumage resembles breeding ♂, but black areas have narrow olive edges making pattern less neat and clear. Wing-feathers evenly fresh. **1stY** If aged first by moult pattern (following partial post-juv moult with retained juv slightly worn and browner primary-coverts, and browner primaries with less fresh tips the

Ad ♂, Mexico, Apr: distinctive due to yellow face contrasting with solidly black bib, breast and flanks, unstreaked dusky olive-green crown, upperparts, eye-stripe and moustachial stripe. Aged by black primary-coverts and overall neat plumage. (H. Shirihai)

♀, Mexico, Apr: unlike ♂, shows duller pattern, and has black largely restricted to the lateral throat-stripe and upper breast that continues as streaks on flanks. Ageing difficult based on just one image. (H. Shirihai)

Presumed 1stW ♂ (left) and ♀, Azores, Oct: since age of the left bird is unsure, the overall plumage pattern could be either ad ♀ or 1stW ♂ (perhaps to some degree better fitting the latter). Similarly, the right bird is affected by photo flash making it impossible to evaluate edgings to primary-coverts and remiges, but overall plumage pattern best fits 1stW ♀. (V. Legrand)

rather similarly patterned ad ♀ and 1stW ♂ can be separated. The dullest plumage, 1stW ♀, is described above. By spring, both sexes are similar to their respective ad but can be aged by moult pattern, and often by their visibly less crisply patterned plumage. **GEOGRAPHICAL VARIATION** Monotypic.

CANADA WARBLER *Cardellina canadensis* (L., 1766)

Iceland one (Sep 1973), Ireland one (Oct 2006), Azores two (Oct 2009, Oct 2016). Breeds across E & C Canada and NE USA, generally in dense second-growth forests, red maple swamps or high-elevation alpine forests; winters mostly in Central America and N South America. **IDENTIFICATION** Medium-sized (**L** 12.5–14 cm) delicate-looking wood warbler with *fairly long tail*. Combination of *lead-grey* (or olive-grey) *upperparts, distinct pale eye-ring* and *concentrated bold black streaking across chest on yellow underparts* quite characteristic. *Wings and tail uniform* and unmarked (lacking wing-bars or tail spots), and yellow rest of underparts contrasts with whiter undertail-coverts. Ad ♂ *bluish-grey above*, has *black mark on forehead, fore lateral crown, lores and moustachial stripe*, and bold black streaks across breast heavier than in other plumages forming a necklace. Note short and narrow yellow supercilium from eye to bill. Ad ♀ and young autumn birds are generally duller, being less pure grey above, mostly lacking black markings on face, and have duskier breast-streaking (but some distinct black streaks present in necklace of at least some spring ad ♀♀). Although short supercilium and eye-ring are somewhat thinner and duller than in ad ♂ they are still visible. Also, yellow underparts average paler than in ♂. 1stW ♀ has dullest plumage, lacking any obvious dark facial markings and has more greyish-olive upperparts, the whitish eye-ring tinged pale yellow, breast very faintly and diffusely marked with short greyish streaks, while dullest individuals are nearly unmarked on the breast, showing only some inconspicuous markings. Usually forages in undergrowth or low branches, but also seen flycatching. Call a quiet, dry *chick* or *check*. **AGEING & SEXING Ad** Sexes differ slightly as described above. Undergoes complete post-nuptial moult prior to autumn migration, and has partial pre-nuptial moult in late winter and spring. Ad ♂ in fresh autumn plumage resembles breeding ad ♂, but grey tips to some degree obscure black facial pattern, and black breast-streaking is diffuser. **1stY** 1stW described above, but best aged by moult pattern (following partial post-juv moult has slightly worn and browner juv primary-coverts, and browner primaries with less fresh tips). Some advanced 1stW ♂♂ can already show some blackish bases on face.

Ad ♂, Colombia, Feb: note the lead-grey upperparts, black mark on lateral crown to forehead, lores and moustachial stripe, bold black streaking across chest on yellow underparts and distinct pale eye-ring, but no wing-bars or tail spots. The evenly-feathered upperwing with grey primary-coverts together with strong head and breast pattern, infer age and sex. (H. Shirihai)

♀, presumed 1stS, Colombia, Feb: duller version of ♂, being to some degree less pure grey above, with much reduced dark facial and necklace markings, and whitish eye-ring narrower. The combination of dull head and breast plumage, seemingly retained juv brown primary-coverts, worn primaries and rectrices all indicate 1stS ♀. (H. Shirihai)

1stW, Ireland (left: Oct) and Ecuador (right: Nov): both birds are 1stW (e.g. have retained juv primary-coverts) but the left bird also lacks any blackish on head or around neck securing sex as ♀. The right bird could be either sex, a well-marked 1stW ♀ or more likely a poorly marked 1stW ♂, especially considering the bright and rich yellow underparts, blacker loral patch, broader white eye-ring and quite substantial dusky necklace markings. (Left: C. Batty; right: H. Shirihai)

By spring, 1stS of both sexes similar to their respective ad, but can be aged by moult pattern, and often by their visibly less smartly patterned plumage compared to ad. **GEOGRAPHICAL VARIATION** Monotypic.

WILSON'S WARBLER *Cardellina pusilla* (A. Wilson, 1811)

Britain two (Oct 1985, Oct 2015), Ireland one (Sep 2013). Breeds across the N North American boreal zone in shrubby thickets and riparian habitats, also near lakes and in montane overgrown meadows and bogs; winters in tropical evergreen and deciduous forest of Central America, as well as in mangroves, second growth, thorn-scrub and plantations. **IDENTIFICATION** Among the smallest (**L** 11.5–12.5 cm) North American wood warblers, habitually *restless* and active when feeding in canopy, with *rather long tail often held cocked* (or twitched but not spread). Has *mainly olive upperparts* and *yellow underparts*. When seen close, note brighter yellow forehead, lores, supercilium and submoustachial area, giving impression of *yellower face* (but varies individually, geographically, seasonally and with age and sex). In all plumages also has *indistinctly greenish-washed lower flanks*, and *lack of obvious wing-bars* (though when fresh, paler edges to secondary-coverts could hint at double wing-bars). Legs pinkish-flesh. Ad ♂ has small but distinctive *black cap*, while in ♀ of eastern subspecies cap is smaller, patchy and ill-defined, and it can be entirely lacking in 1stW, in particular in ♀. Young without black cap are superficially similar to Hooded Warbler, but usually visibly smaller, with no white in tail, and has pale lores. Smaller and much greener above than dullest Canada Warbler, with yellow (not white) undertail-coverts, but no obvious pale eye-ring as in Canada. Further, a Yellow Warbler with least streaked and uniform yellow underparts could initially cause confusion, but they differ strongly in wing and tail patterns (Yellow has prominent pale edges to wing-coverts and tertials, as well as yellow tail spots), and the two differ considerably in structure and behaviour, too (tail in Yellow Warbler shorter and lacks typical tail waving of Wilson's). ♀ Common Yellowthroat (and related extralimital species) may be superficially similar, but usually clearly larger and differs in structure, head pattern and underparts coloration. Apart from flicking tail nervously, often also flicks wings. Frequently hover-gleans and flycatches. Call a loud, flat, fairly low-pitched *chet*; also a hard *tlik* or a downslurred *tsiep*. **AGEING & SEXING Ad** Sexes differ as described above. Undergoes complete post-nuptial moult prior to autumn migration. Ad ♂ in fresh autumn plumage resembles breeding ♂, but some greenish tips affect black cap (chiefly rear part). However, sexing birds in the field can be difficult because of variability in the extent of the black cap in ♀♀ (some strongly marked ♀♀ rather similar to least advanced 1stY ♂♂ in this feature). **1stY** 1stW described above, but preferably should be aged by moult pattern (following partial post-juv moult with retained slightly worn and browner juv primary-coverts and browner primaries with less fresh tips, often rectrices also more pointed). Advanced 1stW ♂ can already show quite substantial black cap but it is usually uneven (less solid) and smaller, while many 1stW ♀♀ lack it. In autumn, 1stW birds resemble ad of respective sex. Ageing often difficult but at times possible by moult pattern if seen close or handled. **GEOGRAPHICAL VARIATION & RANGE** Three subspecies recognised, differing slightly in size, plumage, vocalisation and ecology; *pusilla* (described above; S & E Canada, NE USA) averages slightly duller and greener than the other two, having duller (less bright yellow or orange-tinged lores and less vividly yellow colour elsewhere), and is the most likely race to have occurred in Europe as a vagrant.

Above left: *C. p. pusilla*, ♂, presumed 1stW, England, Oct: note blackish cap to otherwise mostly yellowish-green upperparts and yellow head and underparts colorations. The possibly retained juv primaries, primary-coverts and tail (but for renewed one central feather) infer 1stW. The not so bright yellow facial area and slightly dusky-tinged lores indicate ssp. *pusilla*. (M. Eade)

Above right: *C. p. pusilla*, 1stW ♀, USA, Sep: dull head plumage with a hint of dark spots on crown, seemingly retained primary-coverts contrasting with greater coverts, and duller grey rectrices infer age and sex. Ssp. *pusilla* by dull dusky olive lores and reduced darkness on cap. (F. Jacobsen)

VAGRANTS TO THE REGION

SUMMER TANAGER *Piranga rubra* (L., 1758)

Britain one (Sep 1957), Azores three (Oct 2006, Oct 2010, Oct 2011). Breeds from S USA to N Mexico; winters from C Mexico to W Ecuador, N Bolivia and Amazonian Brazil. Frequents canopy and edges of evergreen and deciduous forest, but especially outside breeding season found in light woodland and second growth, scattered trees in clearings and shady gardens. **IDENTIFICATION** Medium-sized (**L** 17–19 cm) finch-like New World passerine, by and large unpatterned, but striking reddish ad ♂ is *dusky rose-red on head, upperparts and wing fringes*, and paler rosy-red below. ♀ is olive above and yellow-olive below, with an overall ochraceous or mustard tone (or even a slight orange tinge), brightest on head and breast. The *pale yellowish-buff* (in ad ♀) and whitish-cream (in 1stY ♀) eye-ring can appear rather distinctive in the field. Bill is large and thick, and predominantly *pale yellowish-horn* (lower mandible and/or cutting edges pale yellowish). Legs greyish. Compared to Scarlet Tanager (which see, p. 484), usually visibly larger with stockier, paler bill, and in both red ♂ and olive-yellow ♀ wings are barely darker than their bodies, unlike the sharply contrasting blacker/darker wings of ♂ Scarlet (♀ Scarlet has less contrasts); also lacks the distinct tooth at the edge of the upper mandible of Scarlet, and has yellow underwing-coverts (Scarlet: white). Solitary in winter, often accompanying mixed flocks of other birds, even warblers. Eats fruits and berries of various sorts as well as large insects. May also catch insects in flight. Call a sharp, distinctive staccato *ki-tick, ki-ti-tuk* or *ki-ti-te-tuk*. **AGEING & SEXING Ad** Evenly fresh or worn depending on season, after complete post-nuptial moult in late summer. Sexes differ clearly (year round) as described above. **1stY** During first autumn, both sexes like ♀, although age revealed by any moult limits with retained juv primaries, primary-coverts, some outer greater coverts and most tail-feathers. Some notably dull, greenish above and buff-tinged below, with feathers of wings edged brownish or greyish (lacking red or orange). Young ♂ in spring usually has brighter yellow plumage (sometimes from midwinter), often tinged orange admixed with scattered red feathers, especially on head, breast and back. Pre-nuptial moult generally limited but in the event it is more substantial can serve as an indication of 1stS, especially ♂. **GEOGRAPHICAL VARIATION & RANGE** Two or three races recognised, with *P. r. rubra* breeding in E USA (SE Nebraska, S Iowa, Ohio and Delaware south to C Texas, Gulf Coast and Florida) being that most likely to have crossed the Atlantic.

P. r. rubra, ad ♂, Costa Rica, Dec: racial identification by small darkish bill and dark red plumage. Compared to Scarlet Tanager, bill is paler and stockier (and lacks the latter's distinct tooth at the edge of the upper mandible), and wings are barely darker than the red body. (H. Shirihai)

P. r. rubra / cooperi, 1stS ♀, Mexico, Apr: this bird was encountered on migration in S Mexico suggesting ssp. *rubra*, but here the bill appears somewhat chunky, not so small. Unlike the smaller Scarlet Tanager, the wings are barely darker than the rest of the plumage. 1stS ♀ by dull green plumage, and worn tips to primaries and rectrices. (H. Shirihai)

P. r. rubra, 1stW ♂, Azores, Oct: 1stW ♂ by brownish-tinged retained juv primary-coverts, retained outer 3–4 greater coverts, worn tips to primaries and rectrices, green plumage but some red at base of rectrices and large bill (to rule out 1stW ♀). (V. Legrand)

EASTERN TOWHEE (RUFOUS-SIDED TOWHEE) *Pipilo erythrophthalmus* (L., 1758)

Britain one (Jun 1966). Breeds throughout E USA and SE Canada; partly migratory, with northern birds moving to the southern parts of the breeding areas. Inhabits second growth with dense shrubs and extensive brushy areas, also woodland edges. **IDENTIFICATION** Rather large (**L** 18–21 cm), proportionately long-tailed and attractively patterned New World sparrow, especially ad ♂ with its *black upperparts and hood* that contrast with *rufous sides and white underparts*. *Basal white primary patch* and *white tertial edges* diagnostic. *Extensive white in outer tail-feathers* conspicuous when tail spread, like in flight, or if seen from below when bird is perched. ♀ is similarly patterned, but black areas are brownish-tinged. Most have red eyes. When feeding on ground characteristically scratches with its feet together. Told from Spotted Towhee *P. maculatus* (unrecorded in the treated region) by lack of white-spotted scapulars and lesser coverts, and absence of spotty white wing-bars on median and greater coverts. Main call an emphatic, upslurred *chewink* or *chewriss*, slowly repeated; various chipping calls when agitated. **AGEING & SEXING Ad** Evenly fresh or worn depending on season, age revealed by lack of any moult limits due to complete post-nuptial moult. Sexes differ (year round) as described above. **1stY** Resembles ad but has moult limits with retained browner juv primaries, primary-coverts, sometimes some outer greater coverts and most tail-feathers (but beware more extensive moult in southern populations) contrasting with darker replaced feathers. At least until midwinter, iris is grey-brown to dull red. **GEOGRAPHICAL VARIATION & RANGE** Four faintly to moderately differentiated subspecies usually recognised, with overall size and extent of white in wings and tail declining from north to south. Thus, the larger northern ssp. *erythrophthalmus* (S Canada and E USA) is the most likely vagrant to Europe. It has the most extensive white in tail, and a red iris.

P. e. erythrophthalmus, ad ♂, E USA, spring: long-tailed New World sparrow, with attractive rufous sides to white underparts and large white patches on black wings and white-sided tail. Upperparts and hood also black and sharply delimited. Ad ♂ by smart plumage, all-black wing, red eye, and black head.

LARK SPARROW *Chondestes grammacus* (Say, 1822)

Britain two (Jun–Jul 1981, May 1991). Breeds in S Canada, much of the USA and N Mexico, with southern populations being resident while northern ones are migratory, wintering in the southern breeding ranges and south to Guatemala. Inhabits variety of open habitats including grasslands, brush and cultivation. **IDENTIFICATION** Rather large (**L** 15–16 cm) and strikingly patterned New World sparrow, having especially complex and bold facial pattern with *chestnut crown and cheeks*, a *narrow median crown-stripe* and *broad and long supercilium*, stripes being white or faintly tinged buff-brown, stripes partly or wholly outlined in black. There is also a *crescent-shaped white mark below the eye* and a *broad white submoustachial stripe*. Narrow black eye-stripe and *prominent black wedge-shaped lateral throat mark*. A small *white patch on rear ear-coverts* adds to the characteristic appearance. Also note *long tail with white outer corners* (conspicuous in flight), and grey-brown dark-streaked upperparts (but nape and lower back to rump pale brown-grey and almost unstreaked). *Breast whitish with peculiar, isolated bold black central spot*. Two pale buff to whitish wing-bars, bolder and wider on median coverts. Also a variable *basal white primary patch*. Often gregarious and feeds on the ground. Call a sharp *tsip*, often given in a rapid series. Has more subdued churring short calls, too. **AGEING & SEXING Ad** Evenly fresh or worn (without moult contrasts due to complete post-nuptial moult) depending on season. Sexes similar in plumage. **1stY** Following partial (though can be quite extensive) post-juv moult attains appearance quite like ad, but most birds may retain some juv primaries, primary-coverts, outer greater coverts and tail-feathers. Some 1stW migrate with partial juv plumage retained, having duller brown head pattern. **GEOGRAPHICAL VARIATION & RANGE** Rather limited geographical variation with two subspecies described. The more easterly race *grammacus* (C USA; winters S USA) is the more likely source of vagrancy to Europe. It is overall darker with more distinct head pattern and broader upperparts streaking compared to westerly *strigatus* (C & W USA).

C. g. grammacus / strigatus, presumed ad, W Mexico, Mar: distinctive head pattern striped in pale and dark. Shape of rectrices cannot be judged at this angle but appear broad and blackish, hence possibly an ad. Subspecies not safely distinguished in field; probably *strigatus* based on locality, but perhaps *grammacus* considering distinct head pattern and bold upperpart-streaking. (H. Shirihai)

C. g. grammacus, 1stW, USA, Sep: in spite of variable extent of post-juv moult, some young are clearly differentiated by overall duller and browner plumage, and especially note the limited amount of chestnut on lateral crown-stripes and cheeks. Although partly covered, primary-coverts seem brownish and slightly worn confirming age. (S. N. G. Howell)

SAVANNAH SPARROW *Passerculus sandwichensis* (J. F. Gmelin, 1789)

Britain three (Apr 1982, Sep–Oct 1987, Oct 2003), Azores three (one Oct 2002, two Oct 2009). Breeds in Alaska, Canada, N, C and Pacific coastal USA, Mexico and Guatemala, with southern populations being either resident or partly migratory, while northern ones are migratory, wintering in the southern breeding range and south to across Central America and the Caribbean. Found in tundra, mountain meadows, saltmarshes, grasslands, sandy regions with dense but low vegetation and short-grass prairies. **IDENTIFICATION** A medium-sized (**L** 13–14.5 cm) short-tailed North American sparrow, with *heavily streaked* general appearance, rather like a ♀ Reed or Yellow-browed Bunting, but note *lack of white in tail*, and usually has *pink bill with darker culmen*. There is a *narrow whitish median crown-stripe* (unlike Reed) and *quite dark lateral crown-stripes*, while eye-stripe is less prominent. *Pale buff-white supercilium broad and long*, often lemon-tinged in its forepart (but the yellow hue can be lacking in winter birds, in particular in the easternmost race *princeps*). Cream-buff submoustachial stripe enhanced by *bold black lateral throat-stripe and lower edge of ear-coverts*. Streaking on body is very strong both above and below, and streaks may coalesce on centre of breast to form a hint of a larger patch. Legs pink. Forages on the ground or in low bushes. When flushed takes quickly to cover in grass. Song Sparrow (which see) can be superficially similar but invariably lacks yellow wash to fore supercilium, and streaking on head and body-sides more chestnut-brown, less black. Further, on average has larger size, proportionately longer tail and larger bill. Beware that dark central breast patch, frequently cited as a good field mark for Song Sparrow, is often present on Savannah Sparrow, too. Main call a thin *tzip*; sometimes also an emphatic *chip chip chip*. **AGEING & SEXING Ad** Depending on season, evenly fresh or worn (no moult contrasts due to complete post-nuptial moult). Sexes similar in plumage. **1stY** Following partial post-juv moult appears like ad, but retains juv primaries, primary-coverts, some tertials and tail-feathers, and these form slight moult contrasts in winter (in spring both age groups can have contrast). **GEOGRAPHICAL VARIATION & RANGE** Many races described, traditionally arranged in at least five different groups, some of which have recently been treated as separate species (cf. e.g. HBW). However, it is beyond the scope of this book to try to sort out the complex taxonomy, or to try to identify the race involved for the few records in the treated region. Nevertheless, based on both geography and plumage these should concern *sandwichensis* s.l. (Alaska, Canada, much of USA, Mexico; wintering from S Canada and south).

1stW, Azores, Oct: typically heavily streaked appearance, with striped facial pattern. These are the two birds recorded in 2009 on Corvo Island, Azores. Aged (with the help of other images) by the retained juv primary-coverts and rectrices, and left bird assumed to be ♂ due to brighter saturated and bolder streaking than the presumed ♀ on the right, but due to scant sexual differences and much overlap we recommend leaving Western Palearctic records unsexed. (V. Legrand)

(RED) FOX SPARROW *Passerella iliaca* (Merrem, 1786)

Iceland one (Nov 1944), Ireland one (Jun 1961), Estonia one (Dec 2012), Finland one (Dec 2012–Apr 2013); the Estonian and Finnish records were of the same individual. Breeds in much of N and W North America; in winter migrates towards the Pacific coast, from S British Columbia and south to N Baja California, but also across S USA and N Mexico. In summer inhabits high elevations, especially wet meadows with scattered conifers, but during winter favours more open woodlands with bush cover, attracted to areas rich with blackberry thickets. **IDENTIFICATION** Rather large (**L** 15–17.5 cm) rusty-tinged North American sparrow, with typical mixture of *pure grey and rufous on head and upperparts* (ear-coverts, sides of crown, scapulars, wings, rump and tail being rufous, latter two often conspicuous in flight, whereas supercilium, eye-surround and sides of nape are grey). *Breast and flanks extensively streaked rufous*, with some black streaks also on flanks. Median and greater coverts narrowly tipped buff, forming indistinct double wing-bars. Head pattern includes narrow and indistinct pale median crown-stripe and usually a large dark lateral throat-mark. In some, the rufous of ear-coverts and chest almost coalesce to form ragged rufous central breast patch or necklaced impression. Note that there is much variation in the precise pattern and degree of rufous. Told from Song Sparrow (which see) by larger size and typically more rufous plumage, notably above and on face, and by heavier streaking on breast, and larger bill with some yellow on lower mandible. Feeds on the ground, and often scratches with both feet among the leaves, almost in a towhee-like manner. Alarm a loud clicking *tchek*, or *chick*. **AGEING & SEXING Ad** Depending on season, evenly fresh or worn (no moult contrasts due to complete post-nuptial moult). Sexes similar in plumage. **1stY** Following partial post-juv moult becomes ad-like, but most birds retain juv primaries, primary-coverts, sometimes outermost greater coverts, and all

1stW, Estonia, Dec: characteristic mixture of pure grey and rufous in plumage, and note yellow lower mandible. This vagrant proposed to be ssp. *iliaca* based on reddish plumage and range, but very similar and intergrading *zaboria* cannot be ruled out, hence best if regarded as just 'Red Fox Sparrow'. 1stW by retained juv brownish-tinged primary-coverts and alula. (D. Monticelli)

tertials and tail-feathers, which can create visible (though subtle) moult contrasts to help when ageing. **GEOGRAPHICAL VARIATION & RANGE** At least 15 subspecies described within the Fox Sparrow complex, traditionally arranged in four race groups, some of which by some considered to be separate species (cf. e.g. HBW). If the complex is split in more than one species the one involved as to WP vagrancy is reasonably the 'Red Fox Sparrow' with two very similar and intergrading subspecies, *iliaca* (E Canada) and *zaboria* (Alaska, W Canada). Given mostly subtle and clinal variation, racial determination is seldom possible of single birds, and the occurrence of intergrades makes this even more challenging.

AMERICAN TREE SPARROW *Passerella arborea* (A. Wilson, 1810)

Sweden one (Nov–Dec 2016). Breeding summer visitor across Canada; winters in W & C USA. Found in summer on tundra with low-growing thickets or in shrubby bogs, but during migration and in winter selects open habitats with grassy, weedy and brushy areas. **IDENTIFICATION** Medium-sized (**L** 15–16.5 cm) North American sparrow with a *fairly long tail* and rounded head. *Double white wing-bars*, streaky rufous-tinged upperparts, *ash-grey head* with *rufous crown and eye-stripe*, plus a *rufous patch on breast-side* narrow down choices. Confirmation by *small dark central breast spot* and *bicoloured bill*, upper mandible dark, lower yellowish. Lacks a pale median crown-stripe (at most a hint on forehead only) or any dark lateral throat-stripe. There is some plumage variation in that some are darker and have more rufous on flanks, others are paler and greyer. Often forages on the ground, either openly or in cover of vegetation. Field Sparrow (*Spizella pusilla*; unrecorded and not treated) is smaller and has a narrow grey median crown-stripe through the rufous cap, all-pinkish bill, and lacks the dark central breast spot. Call a hurried, melodious, multisyllabic *see-de-loo*, or in flight a very high-pitched and buzzy, straight, short *zeet*, a little like Song Thrush. **AGEING & SEXING Ad** Depending on season, evenly fresh or worn (no moult contrasts due to complete post-nuptial moult). Tips of tail-feathers rather truncate. Sexes similar in plumage, although ♀♀ average duller, especially on crown. **1stY** Following partial post-juv moult in late summer–early autumn basically ad-like, but most birds retain juv primaries, primary-coverts and most tail-feathers, but replace 1–3 tertials, hence retained feathers often subtly more worn, and tail-feathers more narrow and tapering at tips; moult limits in tertials often best clue. Some have traces of streaks on crown (remnant from juv plumage). As ad, sexes similar though ♀♀ average duller, especially on crown. **GEOGRAPHICAL VARIATION & RANGE** Two subspecies recognised, the subtly smaller, darker and more rufous-tinged eastern *arborea* (NC & NE Canada) being the most likely taxon to reach Europe.

P. a. arborea, 1stW, Sweden, Nov: note ash-grey head with rufous crown and thin eye-stripe. Has double white wing-bars, but here the median coverts wing-bar is concealed by breast-side feathers, and the characteristic rufous patch on breast-sides is also only partly visible. 1stW established from other images and by worn rectrices. Sex unknown but possibly ♂ based on bright plumage for a 1stW bird. Subspecies *arborea* by likelihood and dark plumage. (L. Jensen)

LINCOLN'S SPARROW *Melospiza lincolnii* (Audubon, 1834)

Azores four (one Oct 2010, two Oct 2012, Oct 2016), Iceland one (Dec 2013–Apr 2014). Breeding summer visitor across Canada, Alaska and NE & W USA (less common in eastern parts of its range); migrates to S USA, Mexico and N Central America. Generally a bird of elevated areas with wet thickets or shrubby bogs, but during migration and in winter inhabits grassy, weedy and brushy areas, often near wetlands. **IDENTIFICATION** Rather small (**L** 13.5–15 cm) North American sparrow with a fairly short tail and a *fine conical bill*. It has a characteristic *broad grey supercilium* and a *thin pale median crown-stripe*. Lateral crown-stripes, eye-stripe and moustachial stripes brown streaked black, whereas lateral throat-stripe is very thin and often broken. *Cheeks and neck greyish-tinged.* Wings have rufous edges, and median and greater coverts are narrowly tipped cream, forming indistinct double wing-bars. The breast and flanks are tinged buff and finely streaked black (the streaks often converging to a central breast patch). Rather secretive and unobtrusive, often forages on the ground in dense vegetation. The Song Sparrow (which see) is generally larger and heavier built, with a proportionately longer tail, and its lateral throat-stripe and breast spots are bolder and broader against whiter background, and although Lincoln's Sparrow may have a hint of a central breast patch, this is generally more pronounced in Song Sparrow. Beware that due to general head pattern and square-shaped head with often raised rear crown, can superficially resemble Rustic Bunting (at least initially). Call a light and sharp *chip*, or in flight a high and buzzy *zeet*. **AGEING & SEXING Ad** Depending on season, evenly fresh or worn (no moult contrasts due to complete post-nuptial moult). Sexes similar in plumage. **1stY** Following partial post-juv moult ad-like, but retain juv primaries, primary-coverts, and usually most tertials and all (except often the central) tail-feathers, hence often subtle moult contrasts visible. **GEOGRAPHICAL VARIATION & RANGE** Only slight variation, this being clinal with generally three subspecies recognised, the brighter eastern *lincolnii* (Alaska, Canada, extreme N USA) being the most likely taxon to reach the Atlantic islands or W Europe.

Presumed *M. l. lincolnii*, Azores, Oct: well-marked plumage with bright reddish-brown crown and grey cheeks and crisp black streaking all indicate ssp. *lincolnii*. Note characteristic short tail and fine bill, broad grey supercilium and thin pale grey median crown-stripe. Submoustachial stripe, breast and flanks tinged buff and finely streaked black, while wings are rufous-tinged with weak double pale wing-bars. Although vagrants usually are 1stW, ageing not possible from these images alone. (V. Legrand)

WHITE-CAPPED BUNTING *Emberiza stewartii* (Blyth, 1854)

United Arab Emirates one (Nov 1992). Breeding summer visitor in S Central Asia (S Kazakhstan, Kyrgyzstan, Tajikistan, Uzbekistan, Turkmenistan, Afghanistan, Pakistan, NW India, Nepal) in boreal forests, to temperate grassland, and on dry, barren or rocky hillsides with scattered low bushes, juniper scrub, mostly below tree-line; winters in Pakistan and NW India, but populations from C Afghanistan southward are sedentary or make only short-distance movements, mainly altitudinal. **IDENTIFICATION** A medium-sized bunting (**L** 13.5–15 cm). In all plumages *unstreaked rufous-chestnut rump* and proportionately *long tail with distinctive white wedges on outer feathers*, and a relatively *small fine mainly dark grey bill*. Leg brown or pinkish-brown. ♂ superficially recalls Pine Bunting, but is readily separated by combination of *contrasting broad black eye-stripe and a black throat*, light ash-grey to *whitish cap, cheeks, upper breast* and belly to under tail-coverts. Mantle and scapulars are rufous-chestnut

♂, India, Mar: a most striking plumage with head and breast boldly patterned in pale grey, white and black, in strong contrast to rufous-chestnut breast patch (hint of division at centre), bright rufous rump and pure white undertail-coverts. Primary-coverts seem to be brownish and worn, and this together with many white tips remaining on bib and breast patch might indicate a 1stS. Still safe ageing would require greater familiarity with variation in this species. (M. Mathur)

(concolorous with back to upper tail-coverts), either uniform or narrowly streaked black, and a *broad breast-band and flanks are also rufous-chestnut*. As a rule only one wing-bar moderately developed (on median coverts). In fresh (non-breeding) plumage, ♂'s head and body pattern is partly obscured by grey-olive and buff feather-fringes; chestnut band on lower breast limited to sides. ♀ is rather nondescript, not that dissimilar to larger ♀ Pine Bunting, but is slightly duller grey-brown above, *streaked dark grey on cap* and black on mantle, and often has *some chestnut on outer scapulars*. Facial pattern much plainer than in ♂ (with whitish eye-ring, and much obscured supercilium), but has distinctive *white rear ear-coverts spot* and *pale submoustachial stripe*. Underparts pale cream to off-white with greyer-tinged dark-streaked breast, and many show *hint of rufous patch at side of lower breast*, especially in worn plumage. Some worn ♀♀ show traces of ♂'s black eye-stripe and throat-patch, but less prominently than in 1stY ♂. The ♀ with its finely streaked grey-brown head, broad dark spots on lateral and lower throat, and a broken chestnut breast-band recalls ♀ Chestnut-eared Bunting (see below), but that is smaller, with a thicker bill, has much less white on t5, a paler rufous rump with fine black streaks, and usually has a hint of rufous on ear-coverts. Gathers in small flocks in winter, often with finches and other buntings. Calls described as similar to those of Grey-necked Bunting. Also a 'thin, upslurred, metallic *jweing!*' (Rasmussen & Anderton 2005). **AGEING & SEXING Ad** Sexes differ clearly. Some seasonal variation in ♂ as described above. Complete post-nuptial moult Jul–Sep (early Oct). **1stY** Generally similar to respective ad, but the grey-olive and buff fringes on the freshly moulted body-feathers of ♂ are broader than in ad, taking longer to wear off. 1stS ♂ has more extensive dark centres on median coverts and less rufous outer edges compared to ad ♂. Nearly all 1stS ♂♂ differ from ad ♂ by less solid chestnut breast-band, less chestnut on more boldly streaked mantle and more heavily worn, narrower central tail-feathers (unless replaced in pre-nuptial moult). Young ♀ has less or no chestnut on outer scapulars and lower breast. Both sexes have extensive uniform deep rufous rump and uppertail-coverts, deep white wedges on r5–r6, and may retain some narrowly white-tipped juv outer greater coverts. **GEOGRAPHICAL VARIATION** Monotypic.

♀, **Uzbekistan, May**: note especially rufous colour on scapulars, mantle and unstreaked rump patch, contrasting with the greyish cheeks and neck-sides, and also the very narrow blackish streaks on breast, becoming more diffuse on flanks. (R. Jordan)

1stW ♂, **India, Oct**: plumage of ♂ much obscured and variably patterned in 1stW, with cap washed brownish, and black eye-stripe and bib mottled buff-white. Also has less chestnut on more heavily streaked mantle, mainly obvious only on scapulars. The retained juv tail-feathers are already pointed and worn. Double wing-bars more obvious due to fresh plumage. Unlike superficially similar ♀ Pine Bunting, pale cheek patch much larger. (P. J. Saikia)

1stY ♀, **India, Feb**: the young ♀ has a rather featureless plumage with little or no rufous on scapulars and mantle, but at least the uniform rufous on rump and uppertail-coverts is evident (just discernible on this image). Note the proportionally long tail, and also the lack of a distinct pale supercilium and ear-coverts surround found in Pine Bunting. (S. Singhal)

MEADOW BUNTING *Emberiza cioides* Brandt, 1843

Italy one (1910), Finland one (May 1987), France one (Apr 1993), Spain one (Dec 1994), Denmark one (Jun 1996). Breeds in S Siberia, N & E China, E Kazakhstan, Kyrgyzstan, Mongolia, Korea and Japan; resident or short-distance migrant (moving south as far as to S China and Taiwan). Mostly found in open scrubby habitats or forest edges, usually in hilly or mountainous countryside, descending to lowlands in winter. **IDENTIFICATION** Medium-sized to rather large (**L** 14.5–16 cm) and largely rusty-coloured bunting with bold head pattern. Breeding ♂ has striking facial pattern, with *chestnut crown and cheeks, white supercilium and moustachial stripe* and *bold black lores and submoustachial stripe*, as well as *white throat and lower neck-sides*. Typically, *rear neck-sides are greyish* (but individual and geographical variation in head pattern considerable). Upperparts generally greyish-brown thinly dark-streaked, scapulars and edges of lesser coverts and uppertail-coverts chestnut. Breast, flanks and rump buffish-chestnut. No obvious pale wing-bars, and white tail-edges inconspicuous. Ad ♀ usually a paler version of ♂, with browner and mottled cheeks, streaky crown, and less pure white or well-marked black pattern on face, while breast and body-sides are lighter buff. Indistinct dark streaking on flanks in some, while the *chestnut scapulars* stand out against the otherwise paler upperparts. In fresh non-breeding plumage (both sexes), pale fringes make the whole plumage appear paler, with stronger streaking above, while chestnut on breast is largely obscured or reduced to some spots across the centre. Iris dark brown. Two-toned bill, blue-grey below with dark tip and culmen. Tarsus pinkish-brown. Call a high-pitched, sharp *zit-zit-zit* in rapid series of 3–4 notes. **AGEING & SEXING Ad** Sexes differ clearly. Some seasonal variation in ♂ as described above. Complete post-nuptial moult Jul–Sep. Iris warm brown. **1stY** Generally similar to respective ad, but iris dark grey-brown, and 1stW ♂ approaches ad ♀, while 1stW ♀ is the most nondescript plumage, with less contrasting facial marks and paler underparts. 1stS ♂ has more extensive dark centres on median coverts and less rufous outer edges compared to ad ♂. **GEOGRAPHICAL**

Above: *E. c. cioides*, ad ♂, **Mongolia, Jun**: complex but attractive ♂ plumage with chestnut crown and cheeks, white supercilium and moustachial stripe and bold black lores and submoustachial stripe, as well as white throat forming contrast with warmer chestnut breast patch and greyish neck-sides mark. Although difficult to assess moult and wear of wing and tail, such immaculate plumage infers ad. (H. Shirihai)

Right: *E. c. cioides*, ♀ presumed 1stW, **Mongolia, Mar**: race inferred from locality. Chestnut cheek patch often concentrated behind eye (along upper cheeks), being paler rufous-buff below, unlike any other bunting. Also greater coverts and lower scapulars often typically appear stronger rufous, forming contrast with greyish lesser coverts patch on folded wing. Breast often rufous-tinged. Age inferred from rather dull and brown-tinged primary-coverts. (P. Tsolmonjav)

E. c. tarbagataica, 1stW ♂, **Kazakhstan, Oct**: a freshly moulted 1stW ♂ (aged by pointed tail-feathers with new contrastingly neat central ones) with broad pale ochraceous-buff fringes above making the whole plumage appear paler, while chestnut on crown, cheeks and breast is now largely obscured. This race averages paler in general than *cioides*. (V. Fedorenko)

VAGRANTS TO THE REGION

VARIATION & RANGE Polytypic, five subspecies usually recognised with clinal variation in-between. By range either ssp. *cioides* (SC Siberia, NE Kazakhstan east through Mongolia, Baikal region to Transbaikalia and NC China) or *tarbagataica* (SW Altai, S & C Tien Shan, NW China) are the most likely vagrants to the treated region. Ssp. *cioides* is described above, while *tarbagataica* in ♂ plumage differs by having less rufous on rump, and narrower rufous breast-band.

CHESTNUT-EARED BUNTING *Emberiza fucata* Pallas, 1776

Britain two (Oct 2004, Oct 2012), Sweden one (Oct 2011). Distribution rather disrupt, ranging from NC Mongolia and SE Siberia in the north and patchily through E China reappearing in Himalayas and west to N Pakistan. Inhabits damp open grassy marshland mixed with patches of herbs and scattered deciduous shrub in riverine forest, on coastal plains, and in, e.g., Pakistan up to 2700 m. Southern populations partial short-distance migrants, but those of Russia and Mongolia are largely migratory, wintering mainly in E & S China. **IDENTIFICATION** Rather large to medium-sized bunting (**L** 15–16.5 cm). Ad readily separated from congeners by distinctive *chestnut ear-coverts patch, ash-grey crown and nape* marked with fine black streaks, *uniform ash-grey supercilium* and a conspicuously *white moustachial stripe*, an *extensive black lateral throat-stripe continuing as a gorget* of heavy black streaking on lower throat and breast, contrasting with *uniform white throat and rest of breast*, and below this a *chestnut band across lower breast*. Rest of underparts white with a cream-coloured or pale rufous tinge on flanks. In ♀, the grey cap is tinged olive or light brown, supercilium and moustachial stripe tinged buff (less pure grey and white), black streaks on throat less profuse, white of throat and of rest of underparts tinged buff, and chestnut breast-band more reduced in width and extent and somewhat paler rufous. Birds in fresh autumn plumage are less strongly patterned, especially ♀♀ are duller and more washed out; supercilium and underparts extensively tinged buff. *Rump rufous* in all plumages, *eye-ring whitish*. Although there is some white in outertail (as in several related buntings), the white on inner web of the penultimate tail-feather (r5) is much reduced, present either as a narrow line or totally lacking. *Tail rather short*, feathers about equal in length but each sharply pointed, even in ad. *Bill strong*, conical and deep at base, horn-black above with greyish edges and lower mandible, and a pink base. Iris dark brown. Legs pink-brown. The dullest 1stW Chestnut-eared Bunting (see Ageing & Sexing) with its prominent pale eye-ring may be overlooked as a young Ortolan Bunting, but the former's ear-coverts are always more contrasting rufous, and the supercilium, lateral throat-stripe and streaking on breast are better developed, while the throat lacks significant streaking, and wing-bars are rather ill-defined; also note that rump is usually noticeably rufous in Chestnut-eared (not concolorous with mantle as in Ortolan), and their tail patterns differ, too. Beware also that Little Bunting has superficially similar chestnut ear-coverts, but that species is usually visibly much smaller with the head-stripes better marked (but ending short of bill). Reed Bunting (within the same range chiefly extralimital ssp. *pyrrhulinus*) can appear superficially similar, but should still readily be eliminated by more prominent supercilium and darker lateral crown-stripes, almost unstreaked underparts and paler rump, lacking the rufous tinge of Chestnut-eared. Meadow Bunting (see above) may also show chestnut ear-coverts and rufous rump, but its supercilium is broader and longer, altogether much better marked, while its breast is more indistinctly streaked or even unstreaked by comparison. Calls include an explosive *zick*, a higher-pitched *züi*, and lower-pitched *chutt*. **AGEING & SEXING Ad** Sexes rather similar (but see above for slight differences). Seasonal variation also rather limited. Complete post-nuptial moult Aug–Sep. Beware that ad has acutely pointed tail-feathers. Iris chestnut-brown. **1stY** 1stW can have rather pronounced dark border to ear-coverts and dark lateral crown-stripes similar to many other buntings. Young ♀ is rather nondescript, with well-streaked pale olive- or buff-brown upperparts and a strong buff tinge to supercilium, moustachial stripe and underparts, but with a marked contrast between the colour of upperparts and deep rufous of scapulars, lesser wing-coverts, rump and ear-coverts. Note also whitish eye-ring and heavy blotching or streaking on sides of throat and breast. Lower breast has traces of rufous, and flanks have rather deep cinnamon tinge. 1stW has dark grey-brown iris. **GEOGRAPHICAL VARIATION** Polytypic, with relatively minor differences between the three recognised subspecies, and the main characters for recognition highly influenced by season and degree of wear; subspecific identification outside the breeding ranges difficult or impossible. The comparatively pale race *fucata* (Lake Baikal region, Mongolia east to the Russian Far East, S Sakhalin, S Kurile Is, Transbaikalia, NE China, Japan, Korea) is a long-distance migrant and the one likely to have straggled to NW Europe.

E. f. fucata, ♂, presumed 1stS, Mongolia, May: chestnut ear-coverts patch, ash-grey supercilium, nape and neck-side, and an attractive blotchy black lateral throat-stripe joining across lower throat in a gorget on whitish throat and breast, the latter bordered below by a chestnut band, too, making this bunting unique. Primary-coverts and flight-feathers seem retained juv. (H. Shirihai)

Presumed *E. f. fucata*, 1stW, Shetland, Scotland, Oct: even on dullest 1stW the hint of rufous ear-covert patch is still detectable, as is the grey-tinged rear supercilium, nape and neck-side. This plumage usually shows the strongest blackish stripes on mantle, becoming an important character when faced with a tricky bird. Note also rufous hue on scapulars and neat whitish eye-ring. Flight-feathers and tail are thought to be retained juv. (H. Harrop)

DICKCISSEL *Spiza americana* (J. F. Gmelin, 1789)

Norway one (Jul 1981), Azores five (two Nov 2009, one Sep 2011, two Oct 2012). Breeds in S & C Canada, C & E USA; winters from C Mexico to N South America. Favours open, grassy areas, notably weedy meadows and prairie grasslands, but also attracted to grain fields. Numbers and distribution outside core breeding range vary locally from year to year. **IDENTIFICATION** Rather large (**L** 14–16 cm), a sparrow-like and characteristically *heavy-billed*, streak-backed species. In all plumages *some yellow on breast* and distinctive *yellow or cream-coloured supercilium and submoustachial stripe*, and usually *chestnut median coverts and wing-bend* ('shoulders'). Ad ♂ has head mostly greyish, crown tinged olive and streaked black; yellow supercilium, often becoming white behind eye. *Black gorget on lower throat and upper breast* diagnostic, but variable in size, shape and solidity, partly depending on posture and angle of view (and when fresh can to some degree be concealed by pale greyish fringes that wear off by spring). Upperparts pale brown, streaked black, wing-feathers edged buff, contrasting with chestnut 'shoulders'. *Breast bright yellow*, shading to buff-white on lower belly and *purer white undertail-coverts*. ♀ like ♂ but more heavily streaked above, head more brownish tinged, and 'shoulders' paler chestnut (with dusky feather bases often exposed). Yellow supercilium duller and narrower. Throat white, often bordered by a narrow black lateral throat-stripe (lacks black gorget of ♂). Yellow less extensive and more diluted, and often has thin and ill-

Ad ♂, Venezuela, Dec: note yellow tinge on breast, and bright yellow submoustachial stripe and supercilium. Also rufous on shoulders and wing-bend, and note typical black bib on lower throat and upper breast (chin always white). Being ♂, rest of head mostly greyish. Aged as ad by having rather colourful plumage in Dec. (H. Shirihai)

1stW ♂, Oct, Azores: young ♂ in autumn has rather nondescript plumage, lacking black throat patch, instead the breast and flanks are lightly streaked. But much yellow already developed on supercilium, submoustachial stripe and breast. There is some limited chestnut, too, mainly on shoulders and median coverts. (V. Legrand)

defied dark streaks on flanks. Culmen blackish, rest of bill more greyish. Legs brownish. In non-breeding plumage, the black gorget or lateral throat mark partly concealed by pale feather-tips; yellow below strongly reduced. In its main non-breeding range highly gregarious, appearing in small groups to large flocks. Picks seeds off spike or from ground. A ♀ House Sparrow could perhaps at first glance appear superficially similar to ♀ Dickcissel, but the latter's thin but striking black lateral throat-stripe and flank-streaks, its bluish-tinged paler part of the bill, as well as at least some yellow on supercilium and submoustachial stripe, readily identify. The common call described as low-pitched, rather harsh and grunting *drrt* or *djrrt*, but in flight also often gives a distinctive electric-buzzer *bzrrrrt*. **AGEING & SEXING Ad** Sexes differ, also in size, ♂ being larger with almost no overlap. Some seasonal variation as described above. **1stY** Duller overall than respective ad, breast and flanks lightly streaked, with young ♀ the dullest, which may show almost no yellow or chestnut. Breast and indistinct supercilium and submoustachial stripe tinged yellowish-buff. Some young ♂♂ have more rufous developed on 'shoulders' and have more ad-like facial pattern and underparts coloration. Moult limits also occur between renewed greater coverts and retained primary-coverts, and sometimes among tertials. **GEOGRAPHICAL VARIATION** Monotypic.

BROWN-HEADED COWBIRD *Molothrus ater* (Boddaert, 1783)

Norway two (Jun 1987, May 2010), Britain five (Apr 1988, Apr–May 2009, May 2009, Jul 2009, May 2010), France one (May 2010), Germany one (May 2012). A brood-parasitic icterid of wide range across temperate to subtropical North America, with northern birds migratory, wintering in S USA and Mexico, whereas other populations are permanent residents. Found in woodlands, farmlands, suburbs. **IDENTIFICATION** Moderately sized (**L** 18–20 cm) and compact North American passerine, almost starling-like in general size and shape, and terrestrial foraging behaviour, as well as in its dark plumage, but smaller and slimmer. Being an icterid wholly distinctive in jizz, but in this species also chiefly by its finch-like head and bill with the *bill smaller than most icterids*. ♂ has *brown head contrasting with iridescent greenish blue-black body*. ♀ dusky grey-brown above, paler below with an even paler and greyish- to buffish-tinged throat, rear supercilium and ear-surround, giving paler face impression. Sometimes central breast a bit paler, and very ill-defined streaking on the underparts can be visible in close views, too. Juv is paler above with pale edgings giving a scaly look (including on wing-coverts, forming vague pale wing-bars), and is also more heavily streaked below. Forages on ground (with tail cocked), often around grazing mammals. Calls include a harsh rattle and squeaky whistles. **AGEING & SEXING Ad** Sexes differentiated as described above. **1stY** As juv undertakes near-complete (post-juv, until Oct) moult like ad, the ages are close to indistinguishable in field views once moult completed. Young ♂♂ moulting to ad plumage in late summer shows mixture of buff, brown, and black. **GEOGRAPHICAL VARIATION & RANGE** By range the most likely origin of vagrants to Europe is medium-sized and thick-billed *ater* (SE Canada south throughout E USA, and in E Mexico; winters in Florida and in Mexico south to Oaxaca).

M. a. ater, ♂, presumed 1stY, Mexico, Mar: note finch-like head and bill, with that brown head contrasting with iridescent greenish blue-black body. Probable 1stY male by browner primaries, even though they are replaced during post-juv moult. Subspecies *ater* by stout bill and range. (H.Shirihai)

M. a. ater, presumed 1stY ♀, USA, spring: dusky grey-brown above and on head, paler below with subdued and ill-defined streaking, throat being paler still and slightly tinged buffish. 1stY by seemingly extensive retention of juv wing-feathers. Subspecies *ater* by stout bill and range. (T. Reichner)

Juv/1stW, Canada, May: pale edgings give a scaly pattern above, including on the wing-coverts, forming vague pale wing-bars. The bird is more heavily streaked below compared to post-juv ♀. Much of wing is retained juv feathering. Subspecies *ater* by stout bill and range. (P. Reeves)

COMMON GRACKLE *Quiscalus quiscula* (L., 1758)

Netherlands one (Apr 2013). Abundant, often highly social, found in open and semi-open areas (e.g. fields, marshes, parks, suburban areas and wide range of other modified areas) across North America west to Rocky Mountains; resident in much of its range, but northern populations migrate to south of the range, including to SE USA. **IDENTIFICATION** A large (**L** 27–34 cm) and lanky *all-dark* icterid (=New World blackbirds, orioles, grackles, etc.) with *long, keel-shaped tail*, and strong and sharply pointed bill. Black legs rather long, and the *pale yellow eyes* give an intent expression, usually obvious even at a distance. ♂ has *glossy purplish blue head, neck, and breast*, sometimes with a green or bronzy sheen, chiefly on scapulars, wing-coverts, lower breast and upper flanks. Upperparts blackish-brown and moderately glossed, but blackish-brown belly is the least glossed area. ♀ similar, but usually visibly smaller, with *shorter tail* (and unlike ♂, tail does not appear keel-shaped in flight), is *duller with limited gloss* to plumage, and has browner-tinged mantle and belly, but many are difficult to sex. Both sexes often appear all black at a distance. Juv or still moulting young birds in autumn have dark brown plumages and a dark eye. Habitually forages on the ground, lawns and

Q. q. quiscula / stonei, ad ♂, USA, Apr: age and sex by all-black primaries and glossy, iridescent plumage, with purplish-blue head, neck, and breast, while green or purplish-bronze sheen may be evident on scapulars. Due to distinctive overall shape, bill shape, glossiness, bluish plumage and white eye, it may appear unmistakable, but beware that there are many more grackles and New World blackbirds in Central & South America not yet recorded as vagrants. (F. Jacobsen)

Presumed *Q. q. quiscula / stonei*, ad ♀, New York, USA, Sep: rather glossy plumage, quite black primaries bright yellow eye though not quite as iridescent plumage as in typical ♂, in combination all infer ad ♀. Note quite strong all-dark bill. (J. Martin)

in agricultural fields, pecking for food rather than scratching, and when resting atop trees or on telephone lines, maintain a raucous chattering. Call note a loud *chack*, and a repeated churring *terr* in alarm. **AGEING & SEXING Ad** Sexes separable as described above. **1stY** Undertake near-complete post-juv moult once moult is completed, ageing of ♀♀ in field-views not possible, whereas 1stY ♂♂ show browner wings and average duller overall than ad ♂♂. **GEOGRAPHICAL VARIATION & RANGE** Three races recognised with *stonei* (New Jersey and Pennsylvania south to inland Louisiana, Alabama, Georgia and South Carolina) the most likely taxon to wander to Europe.

Q. q. versicolor, ♂, presumed 1stS, USA, May: quite distinctive even in flight at medium range due to large whitish eye, long tail with narrow base but broad and rounded end, and all-dark iridescent plumage. Wing shape and bill proportion rather like Eurasian Jackdaw. Brownish underwing-coverts, and seemingly brownish primaries, might infer 1stS. (F. Jacobsen)

YELLOW-HEADED BLACKBIRD *Xanthocephalus xanthocephalus* (Bonaparte, 1826)

Netherlands one (May–Jul 1982), Iceland one (Jul 1983). (Several other records in N & W Europe currently regarded as not involving wild or unassisted migrants.) Breeds in SC Canada and W & C USA south to extreme NW Mexico; migrates to winter in S USA and Mexico. Favours marshy habitats and open areas nearby. **IDENTIFICATION** A medium to *rather large* (**L** 22–26 cm), stout-bodied and large-headed blackbird, with moderately long, *conical bill*, the only member of the genus *Xanthocephalus* but in shape and size recalling both orioles and other closely related North American blackbirds. ♂ is highly distinctive due to *golden-yellow head and breast* that *contrast sharply with black facial mask and rest of plumage*, including most of wings. *Characteristic white wing patch*. Ad ♀ quite different, being much smaller and predominantly *dusky-brown with some dull yellow on breast and face*. ♀ lacks the white wing patch; often streaked with white in the centre of belly. Gleans food mostly on the ground, and is highly gregarious year round. Call note described as sounding like a rusty farm gate opening, as a hoarse creak. Also short hard notes in long series. **AGEING & SEXING Ad** Evenly fresh following complete post-nuptial moult. Sexes differ as described above. **1stY** Following partial post-juv moult, 1stY ♂ resembles ♀ but is darker, and has white-tipped wing-coverts. Acquires ad plumage by following autumn. In young ♀ yellow on head and breast usually mottled dark. Both sexes also differ from ad by moult pattern with moult limits due to substantial retention of juv wing-feathers. **GEOGRAPHICAL VARIATION** Monotypic.

♂♂, ad (left) and 1stY, Mexico, Mar: note characteristic orange-tinged golden-yellow head and breast that contrast sharply with rest of plumage. There is also a black facial mask extending around the bill. Wings have a distinctive white patch, mainly located on primary-coverts, though obviously much reduced on the 1stY bird on the right. (H. Shirihai)

♀♀, ad (left: Canada, spring) and 1stS (right: USA, Jun): distinct enough in spite of being predominantly dusky-brown and lacking wing patch, but have variable yellow supercilium and ear-surround, throat and chest. Breast in 1stY birds (right) is streaked with cream-white, forming mottled patch. (Left: T. Zurowski; right: S. N. G. Howell)

1stW ♂, Baja California, Mexico, Oct: 1stW ♂ resembles ♀ but has rather unique pattern, is overall darker, has white-tipped primary-coverts and the black eye mask embedded in more extensive yellowish facial area. Told from ad also by moult limits due to substantial retention of juv wing-feathers. (S. N. G. Howell)

CHECKLIST OF THE BIRDS OF THE WESTERN PALEARCTIC – PASSERINES

This checklist is a complete list of all the species covered in Volumes 1 and 2, including vagrants. The sequence adopted is, for practical reasons, a traditional one, being very similar to the one used in both *The Birds of the Western Palearctic* vol. V–IX (Cramp *et al.* 1988–94) and *Collins Bird Guide* (Svensson, Mullarney & Zetterström 2009). Please see a brief overview on page 27 explaining modern trends in systematics and how handbooks in the future probably will be arranged to better mirror evolution and true relationships.

The delimitation of the Western Palearctic as defined here is explained on page 8. The main difference compared to most previously published works is the inclusion of the whole of Arabia and Iran, resulting in many additional species. The recently published checklist, *Birds of Europe, North Africa and the Middle East* (Mitchell 2017) also includes Arabia and Iran but, unlike that work, we prefer to follow international country borders. This has resulted in the omission of a few African species claimed to have been recorded in the north of several countries bordering the region in the south (in Mauritania, Niger and Sudan). We have also omitted some localised or recently established introductions in several central and southern European countries.

Species limits and nomenclature follow our preferences based on our own studies, as explained in 'Species taxonomy' in the Introduction on pages 9–10.

In order to make the list more useful to readers, volume and page numbers are given for each species, together with a coded general status:

B = breeder, resident or summer visitor

B (I) = established introduced breeder

B (E) = endemic breeder (91 species breed only within the covered area)

M = migrant (or winter visitor) within the treated region

S = scarce visitor, in some parts of the region almost annual, with a total of > 10 records up to the end of 2016

V = vagrant, usually < 10 records ever up to the end of 2016; these are presented in a separate section at the end of each volume

TYRANNIDAE
Eastern Kingbird *Tyrannus tyrannus* **V** – Vol 1: 620
Eastern Phoebe *Sayonis phoebe* **V** – Vol 1: 620
Acadian Flycatcher *Empidonax virescens* **V** – Vol 1: 621
Alder Flycatcher *Empidonax alnorum* **V** – Vol 1: 621
Least Flycatcher *Empidonax minimus* **V** – Vol 1: 621
Eastern Wood Pewee *Contopus virens* **V** – Vol 1: 622

PITTIDAE
Indian Pitta *Pitta brachyura* **V** – Vol 1: 622

ALAUDIDAE
Singing Bush Lark *Mirafra cantillans* **B** – Vol 1: 31
Black-crowned Sparrow Lark *Eremopterix nigriceps* **B** – Vol 1: 33
Chestnut-headed Sparrow Lark *Eremopterix signatus* **V** – Vol 1: 622
Dunn's Lark *Eremalauda dunni* **B** – Vol 1: 36
Arabian Lark *Eremalauda eremodites* **B** (E) – Vol 1: 38
Bar-tailed Lark *Ammomanes cinctura* **B** – Vol 1: 40
Desert Lark *Ammomanes deserti* **B** – Vol 1: 42
(Greater) Hoopoe Lark *Alaemon alaudipes* **B** – Vol 1: 46
Dupont's Lark *Chersophilus duponti* **B** (E) – Vol 1: 49
Thick-billed Lark *Ramphocoris clotbey* **B** – Vol 1: 51
Calandra Lark *Melanocorypha calandra* **B** – Vol 1: 54
Bimaculated Lark *Melanocorypha bimaculata* **B** – Vol 1: 57
Black Lark *Melanocorypha yeltoniensis* **S** – Vol 1: 59
(Greater) Short-toed Lark *Calandrella brachydactyla* **B** – Vol 1: 62
Rufous-capped Lark *Calandrella eremica* **B** – Vol 1: 66
Hume's Short-toed Lark *Calandrella acutirostris* **B** – Vol 1: 68
Lesser Short-toed Lark *Calandrella rufescens* **B** – Vol 1: 70
(Indian) Sand Lark *Calandrella raytal* **B** – Vol 1: 75
Crested Lark *Galerida cristata* **B** – Vol 1: 77
Thekla's Lark *Galerida theklae* **B** – Vol 1: 81
Woodlark *Lullula arborea* **B** (E) – Vol 1: 84
Oriental Skylark *Alauda gulgula* **B** – Vol 1: 86
(Common) Skylark *Alauda arvensis* **B** – Vol 1: 88
White-winged Lark *Alauda leucoptera* **S** – Vol 1: 91
Raso Lark *Alauda razae* **B** (E) – Vol 1: 94
Horned Lark *Eremophila alpestris* **B** – Vol 1: 96
Temminck's Lark *Eremophila bilopha* **B** (E) – Vol 1: 101

HIRUNDINIDAE
Banded Martin *Neophedina cincta* **V** – Vol 1: 623
Brown-throated Martin *Riparia paludicola* **B** – Vol 1: 103
Chinese Martin *Riparia chinensis* **S** – Vol 1: 105
Sand Martin *Riparia riparia* **B** – Vol 1: 107
Pale Martin *Riparia diluta* **S** – Vol 1: 110
Tree Swallow *Tachycineta bicolor* **V** – Vol 1: 623
Purple Martin *Progne subis* **V** – Vol 1: 623
Rock Martin *Ptyonoprogne fuligula* **B** – Vol 1: 113
Crag Martin *Ptyonoprogne rupestris* **B** – Vol 1: 116
Barn Swallow *Hirundo rustica* **B** – Vol 1: 118
Wire-tailed Swallow *Hirundo smithii* **S** – Vol 1: 123
Ethiopian Swallow *Hirundo aethiopica* **V** – Vol 1: 624
Red-rumped Swallow *Cecropis daurica* **B** – Vol 1: 125
Lesser Striped Swallow *Cecropis abyssinica* **V** – Vol 1: 624
Streak-throated Swallow *Petrochelidon fluvicola* **S** – Vol 1: 128
(American) Cliff Swallow *Petrochelidon pyrrhonota* **S** – Vol 1: 129
Asian House Martin *Delichon dasypus* **V** – Vol 1: 624
(Common) House Martin *Delichon urbicum* **B** – Vol 1: 130

MOTACILLIDAE
Richard's Pipit *Anthus richardi* **S** – Vol 1: 132
African Pipit *Anthus cinnamomeus* **B** – Vol 1: 135
Blyth's Pipit *Anthus godlewskii* **S** – Vol 1: 136
Paddyfield Pipit *Anthus rufulus* **V** – Vol 1: 625
Tawny Pipit *Anthus campestris* **B** – Vol 1: 139
Long-billed Pipit *Anthus similis* **B** – Vol 1: 142
Berthelot's Pipit *Anthus berthelotii* **B** (E) – Vol 1: 145
Olive-backed Pipit *Anthus hodgsoni* **S** – Vol 1: 147
Tree Pipit *Anthus trivialis* **B** – Vol 1: 149
Pechora Pipit *Anthus gustavi* **S** – Vol 1: 152
Meadow Pipit *Anthus pratensis* **B** – Vol 1: 154
Red-throated Pipit *Anthus cervinus* **B** – Vol 1: 156
Rock Pipit *Anthus petrosus* **B** (E) – Vol 1: 159
Water Pipit *Anthus spinoletta* **B** – Vol 1: 163
Buff-bellied Pipit *Anthus rubescens* **S** – Vol 1: 166
Golden Pipit *Tmetothylacus tenellus* **V** – Vol 1: 625
Forest Wagtail *Dendronanthus indicus* **S** – Vol 1: 169
Yellow Wagtail *Motacilla flava* **B** – Vol 1: 170

Citrine Wagtail *Motacilla citreola* **B** – Vol 1: 180
Grey Wagtail *Motacilla cinerea* **B** – Vol 1: 185
White / Pied Wagtail *Motacilla alba* **B** – Vol 1: 188
African Pied Wagtail *Motacilla aguimp* **B** – Vol 1: 196

PYCNONOTIDAE
White-eared Bulbul *Pycnonotus leucotis* **B** – Vol 1: 198
White-spectacled Bulbul *Pycnonotus xanthopygos* **B** (E) – Vol 1: 199
Common Bulbul *Pycnonotus barbatus* **B** – Vol 1: 200
Red-vented Bulbul *Pycnonotus cafer* **B** (I) – Vol 1: 201

BOMBYCILLIDAE
Cedar Waxwing *Bombycilla cedrorum* **V** – Vol 1: 626
(Bohemian) Waxwing *Bombycilla garrulus* **B** – Vol 1: 202
(Grey) Hypocolius *Hypocolius ampelinus* **B** – Vol 1: 205

CINCLIDAE
(White-throated) Dipper *Cinclus cinclus* **B** – Vol 1: 207

TROGLODYTIDAE
(Eurasian) Wren *Troglodytes troglodytes* **B** – Vol 1: 210

MIMIDAE
Northern Mockingbird *Mimus polyglottos* **V** – Vol 1: 626
Brown Thrasher *Toxostoma rufum* **V** – Vol 1: 627
Grey Catbird *Dumetella carolinensis* **V** – Vol 1: 627

PRUNELLIDAE
Dunnock *Prunella modularis* **B** – Vol 1: 213
Siberian Accentor *Prunella montanella* **B** – Vol 1: 216
Black-throated Accentor *Prunella atrogularis* **B** – Vol 1: 219
Radde's Accentor *Prunella ocularis* **B** (E) – Vol 1: 221
Alpine Accentor *Prunella collaris* **B** – Vol 1: 224

TURDIDAE
Rufous-tailed Scrub Robin *Cercotrichas galactotes* **B** – Vol 1: 227
Black Scrub Robin *Cercotrichas podobe* **B** – Vol 1: 230
(European) Robin *Erithacus rubecula* **B** – Vol 1: 232
Rufous-tailed Robin *Larvivora sibilans* **V** – Vol 1: 627
Siberian Blue Robin *Larvivora cyane* **V** – Vol 1: 628
White-throated Robin *Irania gutturalis* **B** – Vol 1: 235
Thrush Nightingale *Luscinia luscinia* **B** – Vol 1: 237
(Common) Nightingale *Luscinia megarhynchos* **B** – Vol 1: 240
Bluethroat *Luscinia svecica* **B** – Vol 1: 243
Siberian Rubythroat *Calliope calliope* **B** – Vol 1: 248
Red-flanked Bluetail *Tarsiger cyanurus* **B** – Vol 1: 250
Eversmann's Redstart *Phoenicurus erythronotus* **S** – Vol 1: 253
Black Redstart *Phoenicurus ochruros* **B** – Vol 1: 255
(Common) Redstart *Phoenicurus phoenicurus* **B** – Vol 1: 262
Daurian Redstart *Phoenicurus auroreus* **V** – Vol 1: 628
Moussier's Redstart *Phoenicurus moussieri* **B** (E) – Vol 1: 267
Güldenstädt's Redstart *Phoenicurus erythrogastrus* **B** – Vol 1: 269
Little Rock Thrush *Monticola rufocinereus* **B** – Vol 1: 272
(Common) Rock Thrush *Monticola saxatilis* **B** – Vol 1: 273
Blue Rock Thrush *Monticola solitarius* **B** – Vol 1: 276
Blue Whistling Thrush *Myophonus caeruleus* **V** – Vol 1: 629
Whinchat *Saxicola rubetra* **B** – Vol 1: 279
Fuerteventura Stonechat *Saxicola dacotiae* **B** (E) – Vol 1: 282
(European) Stonechat *Saxicola rubicola* **B** (E) – Vol 1: 284
Eastern Stonechat *Saxicola maurus* **B** – Vol 1: 287
African Stonechat *Saxicola torquatus* **B** – Vol 1: 294
Pied Stonechat *Saxicola caprata* **B** – Vol 1: 295
Isabelline Wheatear *Oenanthe isabellina* **B** – Vol 1: 297
Red-breasted Wheatear *Oenanthe bottae* **B** – Vol 1: 300
(Northern) Wheatear *Oenanthe oenanthe* **B** – Vol 1: 302
Seebohm's Wheatear *Oenanthe seebohmi* **B** (E) – Vol 1: 306
Pied Wheatear *Oenanthe pleschanka* **B** – Vol 1: 309
Cyprus Wheatear *Oenanthe cypriaca* **B** (E) – Vol 1: 315
Black-eared Wheatear *Oenanthe hispanica* **B** (E) – Vol 1: 319
Desert Wheatear *Oenanthe deserti* **B** – Vol 1: 325
Finsch's Wheatear *Oenanthe finschii* **B** – Vol 1: 328
Red-rumped Wheatear *Oenanthe moesta* **B** (E) – Vol 1: 332
Kurdish Wheatear *Oenanthe xanthoprymna* **B** (E) – Vol 1: 334
Persian Wheatear *Oenanthe chrysopygia* **B** – Vol 1: 337
Blyth's Wheatear *Oenanthe picata* **B** – Vol 1: 339
Mourning Wheatear *Oenanthe lugens* **B** – Vol 1: 343
Maghreb Wheatear *Oenanthe halophila* **B** (E) – Vol 1: 346
Basalt Wheatear *Oenanthe warriae* **B** (E) – Vol 1: 348
Arabian Wheatear *Oenanthe lugentoides* **B** (E) – Vol 1: 350
Hooded Wheatear *Oenanthe monacha* **B** – Vol 1: 353
Hume's Wheatear *Oenanthe albonigra* **B** – Vol 1: 355
White-crowned Wheatear *Oenanthe leucopyga* **B** – Vol 1: 356
Black Wheatear *Oenanthe lucura* **B** (E) – Vol 1: 358
Blackstart *Oenanthe melanura* **B** – Vol 1: 360
White's Thrush *Zoothera aurea* **B** – Vol 1: 362
Siberian Thrush *Geokichla sibirica* **S** – Vol 1: 364
Varied Thrush *Ixoreus naevius* **V** – Vol 1: 629
Wood Thrush *Hylocichla mustelina* **V** – Vol 1: 630
Hermit Thrush *Catharus guttatus* **S** – Vol 1: 366
Swainson's Thrush *Catharus ustulatus* **S** – Vol 1: 367
Grey-cheeked Thrush *Catharus minimus* **S** – Vol 1: 368
Veery *Catharus fuscescens* **S** – Vol 1: 369
Yemen Thrush *Turdus menachensis* **B** (E) – Vol 1: 370
Tickell's Thrush *Turdus unicolor* **V** – Vol 1: 630
Ring Ouzel *Turdus torquatus* **B** – Vol 1: 371
(Common) Blackbird *Turdus merula* **B** – Vol 1: 375
Eyebrowed Thrush *Turdus obscurus* **S** – Vol 1: 378
American Robin *Turdus migratorius* **S** – Vol 1: 380
Naumann's Thrush *Turdus naumanni* **S** – Vol 1: 382
Dusky Thrush *Turdus eunomus* **S** – Vol 1: 385
Red-throated Thrush *Turdus ruficollis* **S** – Vol 1: 388
Black-throated Thrush *Turdus atrogularis* **B** – Vol 1: 391
Fieldfare *Turdus pilaris* **B** – Vol 1: 394
Song Thrush *Turdus philomelos* **B** – Vol 1: 397
Redwing *Turdus iliacus* **B** – Vol 1: 400
Mistle Thrush *Turdus viscivorus* **B** – Vol 1: 402

SYLVIIDAE
Cetti's Warbler *Cettia cetti* **B** – Vol 1: 405
Zitting Cisticola *Cisticola juncidis* **B** – Vol 1: 408
Cricket Warbler *Spiloptila clamans* **B** – Vol 1: 411
Graceful Prinia *Prinia gracilis* **B** – Vol 1: 414
Levant Scrub Warbler *Scotocerca inquieta* **B** – Vol 1: 418
Saharan Scrub Warbler *Scotocerca saharae* **B** – Vol 1: 421
Pallas's Grasshopper Warbler *Locustella certhiola* **S** – Vol 1: 423
Lanceolated Warbler *Locustella lanceolata* **B** – Vol 1: 425
(Common) Grasshopper Warbler *Locustella naevia* **B** – Vol 1: 427
River Warbler *Locustella fluviatilis* **B** – Vol 1: 430
Savi's Warbler *Locustella luscinioides* **B** – Vol 1: 433
Gray's Warbler *Locustella fasciolata* **V** – Vol 1: 630
Moustached Warbler *Acrocephalus melanopogon* **B** – Vol 1: 436
Aquatic Warbler *Acrocephalus paludicola* **B** (E) – Vol 1: 439
Sedge Warbler *Acrocephalus schoenobaenus* **B** – Vol 1: 441
Paddyfield Warbler *Acrocephalus agricola* **B** – Vol 1: 443
Blyth's Reed Warbler *Acrocephalus dumetorum* **B** – Vol 1: 445

Marsh Warbler *Acrocephalus palustris* **B** – Vol 1: 448
(Common) Reed Warbler *Acrocephalus scirpaceus* **B** – Vol 1: 452
Cape Verde Warbler *Acrocephalus brevipennis* **B** (E) – Vol 1: 457
Clamorous Reed Warbler *Acrocephalus stentoreus* **B** – Vol 1: 459
Oriental Reed Warbler *Acrocephalus orientalis* **V** – Vol 1: 631
Great Reed Warbler *Acrocephalus arundinaceus* **B** – Vol 1: 463
Basra Reed Warbler *Acrocephalus griseldis* **B** (E) – Vol 1: 465
Thick-billed Warbler *Iduna aedon* **S** – Vol 1: 467
Isabelline Warbler *Iduna opaca* **B** (E) – Vol 1: 469
Olivaceous Warbler *Iduna pallida* **B** – Vol 1: 471
Booted Warbler *Iduna caligata* **B** – Vol 1: 475
Sykes's Warbler *Iduna rama* **B** – Vol 1: 477
Upcher's Warbler *Hippolais languida* **B** (E) – Vol 1: 480
Olive-tree Warbler *Hippolais olivetorum* **B** (E) – Vol 1: 483
Icterine Warbler *Hippolais icterina* **B** – Vol 1: 485
Melodious Warbler *Hippolais polyglotta* **B** (E) – Vol 1: 487
Marmora's Warbler *Sylvia sarda* **B** (E) – Vol 1: 490
Balearic Warbler *Sylvia balearica* **B** (E) – Vol 1: 493
Dartford Warbler *Sylvia undata* **B** (E) – Vol 1: 495
Tristram's Warbler *Sylvia deserticola* **B** (E) – Vol 1: 498
Spectacled Warbler *Sylvia conspicillata* **B** (E) – Vol 1: 501
Western Subalpine Warbler *Sylvia inornata* **B** (E) – Vol 1: 505
Eastern Subalpine Warbler *Sylvia cantillans* **B** (E) – Vol 1: 509
Moltoni's Warbler *Sylvia subalpina* **B** (E) – Vol 1: 513
Ménétries's Warbler *Sylvia mystacea* **B** – Vol 1: 516
Sardinian Warbler *Sylvia melanocephala* **B** (E) – Vol 1: 520
Cyprus Warbler *Sylvia melanothorax* **B** (E) – Vol 1: 524
Rüppell's Warbler *Sylvia ruppeli* **B** (E) – Vol 1: 528
Asian Desert Warbler *Sylvia nana* **B** – Vol 1: 531
African Desert Warbler *Sylvia deserti* **B** – Vol 1: 533
Arabian Warbler *Sylvia leucomelaena* **B** – Vol 1: 535
Yemen Warbler *Sylvia buryi* **B** (E) – Vol 1: 537
Western Orphean Warbler *Sylvia hortensis* **B** (E) – Vol 1: 539
Eastern Orphean Warbler *Sylvia crassirostris* **B** – Vol 1: 542
Barred Warbler *Sylvia nisoria* **B** – Vol 1: 546
Lesser Whitethroat *Sylvia curruca* **B** – Vol 1: 549
Common Whitethroat *Sylvia communis* **B** – Vol 1: 554
Garden Warbler *Sylvia borin* **B** – Vol 1: 558
Blackcap *Sylvia atricapilla* **B** – Vol 1: 560
Eastern Crowned Warbler *Phylloscopus coronatus* **V** – Vol 1: 631
Pale-legged Leaf Warbler *Phylloscopus tenellipes* **V** – Vol 1: 632
Brown Woodland Warbler *Phylloscopus umbrovirens* **B** – Vol 1: 564
Arctic Warbler *Phylloscopus borealis* **B** – Vol 1: 565
Large-billed Leaf Warbler *Phylloscopus magnirostris* **V** – Vol 1: 632
Green Warbler *Phylloscopus nitidus* **B** (E) – Vol 1: 567
Greenish Warbler *Phylloscopus trochiloides* **B** – Vol 1: 569
Two-barred Warbler *Phylloscopus plumbeitarsus* **S** – Vol 1: 572
Pallas's Leaf Warbler *Phylloscopus proregulus* **S** – Vol 1: 574
Yellow-browed Warbler *Phylloscopus inornatus* **B** – Vol 1: 576
Hume's Leaf Warbler *Phylloscopus humei* **S** – Vol 1: 578
Radde's Warbler *Phylloscopus schwarzi* **S** – Vol 1: 581
Dusky Warbler *Phylloscopus fuscatus* **S** – Vol 1: 583
Sulphur-bellied Warbler *Phylloscopus griseolus* **V** – Vol 1: 633
Western Bonelli's Warbler *Phylloscopus bonelli* **B** (E) – Vol 1: 585
Eastern Bonelli's Warbler *Phylloscopus orientalis* **B** (E) – Vol 1: 588
Wood Warbler *Phylloscopus sibilatrix* **B** – Vol 1: 590
Plain Leaf Warbler *Phylloscopus neglectus* **B** – Vol 1: 592
Common Chiffchaff *Phylloscopus collybita* **B** – Vol 1: 594
Mountain Chiffchaff *Phylloscopus lorenzii* **B** (E) – Vol 1: 600

Iberian Chiffchaff *Phylloscopus iberiae* **B** (E) – Vol 1: 603
Canary Islands Chiffchaff *Phylloscopus canariensis* **B** (E) – Vol 1: 606
Willow Warbler *Phylloscopus trochilus* **B** – Vol 1: 608
Ruby-crowned Kinglet *Regulus calendula* **V** – Vol 1: 633
Goldcrest *Regulus regulus* **B** – Vol 1: 611
(Common) Firecrest *Regulus ignicapilla* **B** (E) – Vol 1: 616
Madeira Firecrest *Regulus madeirensis* **B** (E) – Vol 1: 619

MUSCICAPIDAE
Blue-and-white Flycatcher *Cyanoptila cyanomelana* **V** – Vol 2: 551
(Asian) Verditer Flycatcher *Eumyias thalassinus* **V** – Vol 2: 552
Dark-sided Flycatcher *Muscicapa sibirica* **V** – Vol 2: 552
(Asian) Brown Flycatcher *Muscicapa dauurica* **S** – Vol 2: 31
Spotted Flycatcher *Muscicapa striata* **B** – Vol 2: 33
Gambaga Flycatcher *Muscicapa gambagae* **B** – Vol 2: 37
Ultramarine Flycatcher *Ficedula superciliaris* **V** – Vol 2: 552
Mugimaki Flycatcher *Ficedula mugimaki* **V** – Vol 2: 553
Red-breasted Flycatcher *Ficedula parva* **B** – Vol 2: 39
Taiga Flycatcher *Ficedula albicilla* **B** – Vol 2: 42
Semicollared Flycatcher *Ficedula semitorquata* **B** (E) – Vol 2: 45
Pied Flycatcher *Ficedula hypoleuca* **B** – Vol 2: 48
Atlas Flycatcher *Ficedula speculigera* **B** (E) – Vol 2: 52
Collared Flycatcher *Ficedula albicollis* **B** (E) – Vol 2: 55

MONARCHIDAE
African Paradise Flycatcher *Terpsiphone viridis* **B** – Vol 2: 58
Indian Paradise Flycatcher *Terpsiphone paradisi* **V** – Vol 2: 554
Black-naped Monarch *Hypothymis azurea* **V** – Vol 2: 555

TIMALIIDAE
Bearded Reedling *Panurus biarmicus* **B** – Vol 2: 61
Iraq Babbler *Turdoides altirostris* **B** (E) – Vol 2: 64
Common Babbler *Turdoides caudata* **B** – Vol 2: 66
Arabian Babbler *Turdoides squamiceps* **B** (E) – Vol 2: 68
Fulvous Babbler *Turdoides fulva* **B** – Vol 2: 70

AEGITHALIDAE
Long-Tailed Tit *Aegithalos caudatus* **B** – Vol 2: 72

PARIDAE
Marsh Tit *Poecile palustris* **B** – Vol 2: 77
Willow Tit *Poecile montanus* **B** – Vol 2: 80
Caspian Tit *Poecile hyrcanus* **B** (E) – Vol 2: 84
Siberian Tit *Poecile cinctus* **B** – Vol 2: 86
Sombre Tit *Poecile lugubris* **B** (E) – Vol 2: 88
Crested Tit *Lophophanes cristatus* **B** (E) – Vol 2: 91
Coal Tit *Periparus ater* **B** – Vol 2: 93
(European) Blue Tit *Cyanistes caeruleus* **B** – Vol 2: 97
African Blue Tit *Cyanistes teneriffae* **B** (E) – Vol 2: 100
Azure Tit *Cyanistes cyanus* **B** – Vol 2: 103
Great Tit *Parus major* **B** – Vol 2: 106
Cinereous Tit *Parus cinereus* **B** – Vol 2: 111
Turkestan Tit *Parus bokharensis* **B** – Vol 2: 113

SITTIDAE
Krüper's Nuthatch *Sitta krueperi* **B** (E) – Vol 2: 115
Corsican Nuthatch *Sitta whiteheadi* **B** (E) – Vol 2: 117
Algerian Nuthatch *Sitta ledanti* **B** (E) – Vol 2: 119
Red-breasted Nuthatch *Sitta canadensis* **V** – Vol 2: 555
(Eurasian) Nuthatch *Sitta europaea* **B** – Vol 2: 121
Eastern Rock Nuthatch *Sitta tephronota* **B** – Vol 2: 125
(Western) Rock Nuthatch *Sitta neumayer* **B** (E) – Vol 2: 128

TICHODROMIDAE
Wallcreeper *Tichodroma muraria* **B** – Vol 2: 131

CERTHIIDAE
(Eurasian) Treecreeper *Certhia familiaris* **B** – Vol 2: 134
Short-toed Treecreeper *Certhia brachydactyla* **B** (E) – Vol 2: 137

REMIZIDAE
(Eurasian) Penduline Tit *Remiz pendulinus* **B** – Vol 2: 140
Black-headed Penduline Tit *Remiz macronyx* **B** – Vol 2: 145
White-crowned Penduline Tit *Remiz coronatus* **S** – Vol 2: 147

NECTARINIIDAE
Nile Valley Sunbird *Hedydipna metallica* **B** – Vol 2: 149
Purple Sunbird *Cinnyris asiaticus* **B** – Vol 2: 152
Shining Sunbird *Cinnyris habessinicus* **B** – Vol 2: 155
Palestine Sunbird *Cinnyris osea* **B** – Vol 2: 157

ZOSTEROPIDAE
Abyssinian White-Eye *Zosterops abyssinicus* **B** – Vol 2: 160
Oriental White-Eye *Zosterops palpebrosus* **B** – Vol 2: 162

ORIOLIDAE
Black-naped Oriole *Oriolus chinensis* **V** – Vol 2: 555
(Eurasian) Golden Oriole *Oriolus oriolus* **B** – Vol 2: 163

LANIIDAE
Rosy-patched Bush-Shrike *Rhodophoneus cruentus* **B** – Vol 2: 167
Black-crowned Tchagra *Tchagra senegalus* **B** – Vol 2: 169
Brown Shrike *Lanius cristatus* **S** – Vol 2: 171
Turkestan Shrike *Lanius phoenicuroides* **B** – Vol 2: 174
Isabelline Shrike *Lanius isabellinus* M – Vol 2: 178
Red-backed Shrike *Lanius collurio* **B** – Vol 2: 185
Bay-backed Shrike *Lanius vittatus* **B** – Vol 2: 188
Long-tailed Shrike *Lanius schach* **S** – Vol 2: 190
Lesser Grey Shrike *Lanius minor* **B** – Vol 2: 192
Great Grey Shrike *Lanius excubitor* **B** – Vol 2: 195
Northern Shrike *Lanius borealis* **B** – Vol 2: 207
Iberian Grey Shrike *Lanius meridionalis* **B** (E) – Vol 2: 210
Woodchat Shrike *Lanius senator* **B** (E) – Vol 2: 213
Masked Shrike *Lanius nubicus* **B** (E) – Vol 2: 218

DICRURIDAE
Black Drongo *Dicrurus macrocercus* **V** – Vol 2: 556
Ashy Drongo *Dicrurus leucophaeus* **V** – Vol 2: 556

CORVIDAE
(Eurasian) Jay *Garrulus glandarius* **B** – Vol 2: 221
Siberian Jay *Perisoreus infaustus* **B** – Vol 2: 227
Azure-winged Magpie *Cyanopica cyanus* **B** – Vol 2: 229
(Common) Magpie *Pica pica* **B** – Vol 2: 231
Pleske's Ground Jay *Podoces pleskei* **B** (E) – Vol 2: 236
(Spotted) Nutcracker *Nucifraga caryocatactes* **B** – Vol 2: 238
Alpine Chough *Pyrrhocorax graculus* **B** – Vol 2: 241
(Red-billed) Chough *Pyrrhocorax pyrrhocorax* **B** (E) – Vol 2: 244
(Western) Jackdaw *Corvus monedula* **B** – Vol 2: 247
Daurian Jackdaw *Corvus dauuricus* **S** – Vol 2: 250
House Crow *Corvus splendens* **B** (I) – Vol 2: 252
Rook *Corvus frugilegus* **B** – Vol 2: 254
Carrion Crow *Corvus corone* **B** – Vol 2: 257
Hooded Crow *Corvus cornix* **B** – Vol 2: 260
Large-billed Crow *Corvus macrorhynchos* **V** – Vol 2: 557
Pied Crow *Corvus albus* **B** – Vol 2: 264
Brown-necked Raven *Corvus ruficollis* **B** – Vol 2: 265
(Common) Raven *Corvus corax* **B** – Vol 2: 268
Fan-tailed Raven *Corvus rhipidurus* **B** – Vol 2: 272

STURNIDAE
Tristram's Starling *Onychognathus tristramii* **B** (E) – Vol 2: 274
(Common) Starling *Sturnus vulgaris* **B** – Vol 2: 276
Spotless Starling *Sturnus unicolor* **B** (E) – Vol 2: 280
Pied Myna *Gracupica contra* **B** (I) – Vol 2: 283
Rose-coloured Starling *Pastor roseus* **B** – Vol 2: 285
Wattled Starling *Creatophora cinerea* **S** – Vol 2: 289
Brahminy Starling *Sturnia pagodarum* **S** – Vol 2: 291
Daurian Starling *Agropsar sturninus* **V** – Vol 2: 557
Chestnut-tailed Starling *Sturnia malabarica* **V** – Vol 2: 558
Common Myna *Acridotheres tristis* **B** (I) – Vol 2: 292
Bank Myna *Acridotheres ginginianus* **B** (I) – Vol 2: 294
Amethyst Starling *Cinnyricinclus leucogaster* **B** – Vol 2: 296
Red-billed Oxpecker *Buphagus erythrorhynchus* **V** – Vol 2: 558

PASSERIDAE
Saxaul Sparrow *Passer ammodendri* **B** – Vol 2: 298
House Sparrow *Passer domesticus* **B** – Vol 2: 300
Italian Sparrow *Passer italiae* **B** (E) – Vol 2: 305
Spanish Sparrow *Passer hispaniolensis* **B** – Vol 2: 307
Sind Sparrow *Passer pyrrhonotus* **B** – Vol 2: 310
Dead Sea Sparrow *moabiticus* **B** (E) – Vol 2: 311
Iago Sparrow *Passer iagoensis* **B** (E) – Vol 2: 314
Desert Sparrow *Passer simplex* **B** – Vol 2: 316
Zarudny's Sparrow *Passer zarudnyi* **B** – Vol 2: 318
(Eurasian) Tree Sparrow *Passer montanus* **B** – Vol 2: 319
Arabian Golden Sparrow *Passer euchlorus* **B** – Vol 2: 321
Sudan Golden Sparrow *Passer luteus* **B** – Vol 2: 323
Pale Rock Sparrow *Carpospiza brachydactyla* **B** – Vol 2: 325
Yellow-throated Sparrow *Gymnoris xanthocollis* **B** – Vol 2: 327
Bush Sparrow *Gymnoris dentata* **B** – Vol 2: 329
(Common) Rock Sparrow *Petronia petronia* **B** – Vol 2: 330
(White-winged) Snowfinch *Montifringilla nivalis* **B** – Vol 2: 333

PLOCEIDAE
Streaked Weaver *Ploceus manyar* **B** (I) – Vol 2: 336
Rüppell's Weaver *Ploceus galbula* **B** – Vol 2: 338
Village Weaver *Ploceus cucullatus* **V** – Vol 2: 558

ESTRILDIDAE
Common Waxbill *Estrilda astrild* **B** – Vol 2: 340
Arabian Waxbill *Estrilda rufibarba* **B** (E) – Vol 2: 342
Red Avadavat *Amandava amandava* **B** – Vol 2: 343
Zebra Waxbill *Amandava subflava* **B** – Vol 2: 345
Indian Silverbill *Euodice malabarica* **B** – Vol 2: 346
African Silverbill *Euodice cantans* **B** – Vol 2: 348
Red-billed Firefinch *Lagonosticta senegala* **B** – Vol 2: 350

VIREONIDAE
White-eyed Vireo *Vireo griseus* **V** – Vol 2: 589
Yellow-throated Vireo *Vireo flavifrons* **V** – Vol 2: 589
Philadelphia Vireo *Vireo philadelphicus* **V** – Vol 2: 589
Red-eyed Vireo *Vireo olivaceus* **S** – Vol 2: 351

FRINGILLIDAE
(Common) Chaffinch *Fringilla coelebs* **B** – Vol 2: 353
Blue Chaffinch *Fringilla teydea* **B** (E) – Vol 2: 359
Brambling *Fringilla montifringilla* **B** – Vol 2: 361
Red-fronted Serin *Serinus pusillus* **B** – Vol 2: 364
(European) Serin *Serinus serinus* **B** (E) – Vol 2: 367
Syrian Serin *Serinus syriacus* **B** (E) – Vol 2: 370
(Atlantic) Canary *Serinus canaria* **B** (E) – Vol 2: 372
Arabian Serin *Crithagra rothschildi* **B** (E) – Vol 2: 375

Yemen Serin *Crithagra menachensis* **B** (E) – Vol 2: 377
Golden-winged Grosbeak *Rhynchostruthus socotranus* **B** – Vol 2: 379
(European) Greenfinch *Chloris chloris* **B** – Vol 2: 381
(European) Goldfinch *Carduelis carduelis* **B** – Vol 2: 385
Citril Finch *Carduelis citrinella* **B** (E) – Vol 2: 389
Corsican Finch *Carduelis corsicana* **B** (E) – Vol 2: 392
(Eurasian) Siskin *Spinus spinus* **B** – Vol 2: 394
(Common) Linnet *Linaria cannabina* **B** – Vol 2: 397
Yemen Linnet *Linaria yemenensis* **B** (E) – Vol 2: 402
Twite *Linaria flavirostris* **B** – Vol 2: 404
Common Redpoll *Acanthis flammea* **B** – Vol 2: 407
Arctic Redpoll *Acanthis hornemanni* **B** – Vol 2: 412
Two-barred Crossbill *Loxia leucoptera* **B** – Vol 2: 417
Common Crossbill *Loxia curvirostra* **B** – Vol 2: 420
Parrot Crossbill *Loxia pytyopsittacus* **B** (E) – Vol 2: 425
Asian Crimson-winged Finch *Rhodopechys sanguineus* **B** – Vol 2: 427
African Crimson-winged Finch *Rhodopechys alienus* **B** (E) – Vol 2: 430
Desert Finch *Rhodospiza obsoleta* **B** – Vol 2: 432
Mongolian Finch *Bucanetes mongolicus* **B** – Vol 2: 435
Trumpeter Finch *Bucanetes githagineus* **B** – Vol 2: 438
Common Rosefinch *Carpodacus erythrinus* **B** – Vol 2: 441
Sinai Rosefinch *Carpodacus synoicus* **B** (E) – Vol 2: 444
Pallas's Rosefinch *Carpodacus roseus* **V** – Vol 2: 560
Great Rosefinch *Carpodacus rubicilla* **B** – Vol 2: 446
Long-tailed Rosefinch *Carpodacus sibiricus* **B** – Vol 2: 448
Pine Grosbeak *Pinicola enucleator* **B** – Vol 2: 450
(Common) Bullfinch *Pyrrhula pyrrhula* **B** – Vol 2: 452
Azores Bullfinch *Pyrrhula murina* **B** (E) – Vol 2: 455
White-winged Grosbeak *Mycerobas carnipes* **B** – Vol 2: 457
Hawfinch *Coccothraustes coccothraustes* **B** – Vol 2: 459
Evening Grosbeak *Hesperiphona vespertina* **V** – Vol 2: 560

PARULIDAE
Ovenbird *Seiurus aurocapilla* **S** – Vol 2: 462
Northern Waterthrush *Parkesia noveboracensis* **S** – Vol 2: 464
Black-and-white Warbler *Mniotilta varia* **S** – Vol 2: 466
Golden-winged Warbler *Vermivora chrysoptera* **V** – Vol 2: 591
Blue-winged Warbler *Vermivora cyanoptera* **V** – Vol 2: 591
Tennessee Warbler *Oreothlypis peregrina* **S** – Vol 2: 468
(Common) Yellowthroat *Geothlypis trichas* **S** – Vol 2: 471
American Redstart *Setophaga ruticilla* **S** – Vol 2: 473
Northern Parula *Setophaga americana* **S** – Vol 2: 475
(American) Yellow Warbler *Setophaga petechia* **S** – Vol 2: 477
Chestnut-sided Warbler *Setophaga pensylvanica* **V** – Vol 2: 595
Cerulean Warbler *Setophaga cerulea* **V** – Vol 2: 592
Black-throated Blue Warbler *Setophaga caerulescens* **V** – Vol 2: 595
Yellow-throated Warbler *Setophaga dominica* **V** – Vol 2: 596
Black-throated Green Warbler *Setophaga virens* **V** – Vol 2: 597
Blackburnian Warbler *Setophaga fusca* **V** – Vol 2: 594
Prairie Warbler *Setophaga discolor* **V** – Vol 2: 597
Cape May Warbler *Setophaga tigrina* **V** – Vol 2: 592
Magnolia Warbler *Setophaga magnolia* **V** – Vol 2: 593
Yellow-rumped Warbler *Setophaga coronata* **V** – Vol 2: 481
Palm Warbler *Setophaga palmarum* **V** – Vol 2: 596
Blackpoll Warbler *Setophaga striata* **S** – Vol 2: 479
Bay-breasted Warbler *Setophaga castanea* **V** – Vol 2: 594

Hooded Warbler *Setophaga citrina* **V** – Vol 2: 591
Wilson's Warbler *Cardellina pusilla* **V** – Vol 2: 598
Canada Warbler *Cardellina canadensis* **V** – Vol 2: 598

THRAUPIDAE
Summer Tanager *Piranga rubra* **V** – Vol 2: 599
Scarlet Tanager *Piranga olivacea* **S** – Vol 2: 484

EMBERIZIDAE
Eastern Towhee *Pipilo erythrophthalmus* **V** – Vol 2: 599
Lark Sparrow *Chondestes grammacus* **V** – Vol 2: 599
Savannah Sparrow *Passerculus sandwichensis* **V** – Vol 2: 600
Fox Sparrow *Passerella iliaca* **V** – Vol 2: 600
American Tree Sparrow *Passerella arborea* **V** – Vol 2: 601
Song Sparrow *Melospiza melodia* **S** – Vol 2: 486
Lincoln's Sparrow *Melospiza lincolnii* **V** – Vol 2: 571
White-crowned Sparrow *Zonotrichia leucophrys* **S** – Vol 2: 488
White-throated Sparrow *Zonotrichia albicollis* **S** – Vol 2: 490
Dark-eyed Junco *Junco hyemalis* **S** – Vol 2: 492
Snow Bunting *Plectrophenax nivalis* **B** – Vol 2: 494
Lapland Bunting *Calcarius lapponicus* **B** – Vol 2: 499
Black-faced Bunting *Emberiza spodocephala* **S** – Vol 2: 502
Pine Bunting *Emberiza leucocephalos* **B** – Vol 2: 505
Yellowhammer *Emberiza citrinella* **B** – Vol 2: 509
Cirl Bunting *Emberiza cirlus* **B** – Vol 2: 512
White-capped Bunting *Emberiza stewartii* **V** – Vol 2: 601
Rock Bunting *Emberiza cia* **B** – Vol 2: 516
Meadow Bunting *Emberiza cioides* **V** – Vol 2: 602
Striolated Bunting *Emberiza striolata* **B** – Vol 2: 520
House Bunting *Emberiza sahari* **B** – Vol 2: 523
Cinnamon-breasted Bunting *Emberiza tahapisi* **B** – Vol 2: 525
Cinereous Bunting *Emberiza cineracea* **B** (E) – Vol 2: 528
Ortolan Bunting *Emberiza hortulana* **B** – Vol 2: 531
Grey-necked Bunting *Emberiza buchanani* **B** – Vol 2: 534
Cretzschmar's Bunting *Emberiza caesia* **B** (E) – Vol 2: 537
Chestnut-eared Bunting *Emberiza fucata* **V** – Vol 2: 603
Yellow-browed Bunting *Emberiza chrysophrys* **S** – Vol 2: 540
Rustic Bunting *Emberiza rustica* **B** – Vol 2: 542
Little Bunting *Emberiza pusilla* **B** – Vol 2: 545
Chestnut Bunting *Emberiza rutila* **S** – Vol 2: 548
Yellow-breasted Bunting *Emberiza aureola* **B** – Vol 2: 551
(Common) Reed Bunting *Emberiza schoeniclus* **B** – Vol 2: 554
Pallas's Reed Bunting *Emberiza pallasi* **S** – Vol 2: 560
Red-headed Bunting *Emberiza bruniceps* **B** – Vol 2: 564
Black-headed Bunting *Emberiza melanocephala* **B** – Vol 2: 567
Corn Bunting *Emberiza calandra* **B** – Vol 2: 570
Dickcissel *Spiza americana* **V** – Vol 2: 603
Rose-breasted Grosbeak *Pheucticus ludovicianus* **S** – Vol 2: 572
Indigo Bunting *Passerina cyanea* **S** – Vol 2: 574

ICTERIDAE
Bobolink *Dolichonyx oryzivorus* **S** – Vol 2: 576
Brown-headed Cowbird *Molothrus ater* **V** – Vol 2: 604
Common Grackle *Quiscalus quiscula* **V** – Vol 2: 604
Yellow-headed Blackbird *Xanthocephalus xanthocephalus* **V** – Vol 2: 605
Baltimore Oriole *Icterus galbula* **S** – Vol 2: 578

PHOTOGRAPHIC CREDITS

Front cover, main image: Rose-coloured Starling by Daniele Occhiato. Small images clockwise from top left: Algerian Nuthatch (Vincent Legrand), Eurasian Penduline Tit (aaltair/Shutterstock), Corsican Finch (Stefano Turri), Cape Verde Sparrow (Daniele Occhiato), Blue Tit (Carlos N. G. Bocos), Taiga Flycatcher (assoonas/Shutterstock), Collared Flycatcher (Leif Ingvarson/Shutterstock), Eurasian Jay (Wildlife World/Shutterstock), Cirl Bunting (Lev Paraskevopoulos/Shutterstock), Red-backed Shrike (Daniele Occhiato), Tristram's Starling (Mor Ben Zion/Shutterstock)

Spine image: Azure-winged Magpie by Carlos N. G. Bocos.

Back cover, central trio (l-r): Marsh Tit (Markus Varesvuo), Willow Tit (Harri Taavetti), Caspian Tit (Carlos N. G. Bocos). Small images clockwise from top left: Iraq Babbler (Hadoram Shirihai), Rüppell's Weaver (Daniele Occhiato), Short-toed Treecreeper (Daniele Occhiato), Tree Sparrow (janveber/Shutterstock), Yellow Warbler (Tor A. Olsen), Lapland Bunting (Dmytro Pylypenko/Shutterstock), Palestine Sunbird (Hadoram Shirihai), Bearded Reedling (AndrewSproule/Shutterstock), Pine Grosbeak (Jarkko Jokelainen/Shutterstock), White-winged Snowfinch (Thomas Krumenacker).

Photographic credits for images that appear in this book. l = left; r = right; t = top; b = bottom; c = centre.

AbdulRahman Al-Sirhan Alenezi 179cr; 201tr; 203bl; 527cl: **Abhishek Das** 552cl; 552bl: **Abi Warner**:abiwarnerphotography.com 223cr: **Adam Riley** / www.rockjumperbirding.com 113cr: **Aditya "Dicky" Singh** 190tr: **Adrian Drummond-Hill** 177bl; 182tr; 346cl: **Ahmed Al-Aumari** 321tl; 321br: **Aidan G. Kelly** 43tr: **Alan Lewis** 550tl: **Alan Murphy / AlanMurphyPhotography.com** 468tl; 468b; 469tr; 471bl; 471br; 473bl; 473br; 473cl; 477bl; 478cr; 479tl; 479br; 481bl; 481tl; 483tr; 578cl; 578br; 578bl; 579tl; 579tr; 579bl: **Alan Murphy, BIA/Minden Pictures/FLPA** 484bl; 592bb: **Alan Williams/NaturePL** 79b: **Alastair Wilson** www.amwphotos.co.uk 308tr: **Alejandro Torés Sánchez** 51br; 100br; 367tl; 389cl; 407tl; 497cr: **Alexander Fadeev/Shutterstock** 262cr: **Alexander Hellquist** 508cl: **Alexandr Belyaev** 388cr: **Alexey Timoshenko** 197br; 198tl: **Ali Sadr** / http://birdsofiran.com 564bl: **Ali Sangchouli** 237cl: **Alison McArthur** 109bl: **Amano Samarpan** / www.amanosamarpan.com 154bl: **Amar-Singh HSS** 581br: **Amir Ben Dov** 23tl; 23br; 36cr; 41bl; 47tr; 55br; 68tl; 68tr; 90tr; 129bl; 129br; 130tr; 159cl; 159cr; 159bl; 164bl; 165cr; 187cl; 187cr; 220br; 255bl; 256br; 265tr; 267cl; 267bl; 270b; 273tr; 273bl; 279tl; 293br; 308br; 309bl; 325cl; 326br; 332cl; 347c; 354bl; 366br; 369bl; 370cl; 384tl; 387tr; 400br; 434br; 444cl; 507tl; 507tc; 510br; 511cl; 515br; 567br; 568tr; 568cl; 571cl: **Anand Arya, India** / www.anandarya.com 166cr; 310cl; 310cr; 310bl; 344tl; 584cr: **Anders Blomdahl** 378br: **Andreas Noeske** 77bl; 79tl; 98br; 122br; 123cl; 381br; 382cr: **Andreas Uppstu** 208bl; 208br: **Andrés M. Dominguez** 74bc; 210bl; 210br; 212tr; 212b; 214bl; 369bl; 518cl; 518tr: **Andrew Easton** 118tr; 393bc: **Andrey Vilyayev** 200br; 458tl: **Andy & Gill Swash** (WorldWildlifeImages.com) 53br: **Andy Hood** 47cl: **Antonio Martínez Pernas** 74br: **Antti Below** 138bl; 362cl; 362bl; 404bl; 508cl; 529br: **Apostoli Rosella/Getty Images** 305br: **Arash Yekdaneh** / Tarlan Birding & Ornithological Group, Iran 85cl; 162tr; 237cr: **Arend Wassink** 194br: **Ari Seppä** 408br; 409tl; 415tr; 501tl: **Arie Ouwerkerk** 22trb; 56br; 153tr; 154br; 288tr; 307tr; 450br; 450cl; 522cl; 529tl; 547tl; 555br; 561cr: **Arijit Banerjee** 283cl; 303bl: **Arnoud B van den Berg/The Sound Approach** 120tr; 412br; 424tr; 424cr; 546bl: **Arpit Deomurari** 294br; 295br; 535cr; 536bl; 536cr; 569bc: **Arto Juvonen** 107cl; 107tl; 495cr: **Askar Isabekov** / birds.kz 113bl; 114cl; 144tr; 144bl; 144cr; 200cl; 365bl; 387bl; 387br; 457cl; 457br; 458bl: **Assoonas/Shutterstock** 43cl: **Augusto Faustino** 238br; 431cr: **Aurélien AUDEVARD** 22tl; 34tl; 45tl; 81cl; 114tr; 148tr; 174br; 182cl; 215bc; 222tr; 245tl; 245br; 259cl; 270tl; 354br; 370br; 371tl; 396tr; 527cl; 539tl; 540tl; 514cl; 548bl: **avesdeceuta.com** 356bl: **Axel Halley** 53cl; 129tr; 178tr; 180tr; 216bl; 364br; 547cl; 553br: **Ayhan Öztürk- Dörtçeker Wildlife Photography** (www.dortceker.com) 45br: **Aziz Atik** 558cl: **Bill Baston** (www.billbaston.com) 438br: **Barry Nield** 453bl; 453br; 514cl: **Bas van den Boogaard** 44cr; 63t; 191bl; 466br; 549tr: **Benéharo Rodríguez** - www.gohnic.org 245cr; 358tl; 374br: **Bernard Castelein/NaturePL** 303br: **Bernard van Elegem** 524br: **Bernhard Putsch** 167br: **Bill Coster/FLPA** 572tr: **Bill Hubick** / www.billhubick.com 492br: **Biraj Sarkar/Shutterstock** 582tr: **Bird** / Alamy Stock Photo 455br: **Bishan Monappa**, Karnataka, India 166cl: **Björn Anderson** / iGoTerra 84tr; 84br; 126br: **Bob Garrett** 454tl: **BOONCHAY PROMJIAM/Shutterstock** 585b: **Boris Nikolov/ Biota Films** 132cl: **Brendan Marnell** 234bl: **Brian E Small/Alamy Stock Photo** 594tr: **Brian Zwiebel/Sabrewing Nature Tours** 595crt: **Buiten-Beeld / Alamy Stock Photo** 356tr: **Butterfly Hunter/Shutterstock** 585trb: **C. Abhinav** 565cr: **Carlos Grande Flores** 491br: **Carlos N. G. Bocos** 283b; 303cl; 303cr; 306br; 307br; 309br; 313cr; 313bl; 317cl; 320cr; 329bl; 330cl; 331tr; 331cr; 345cl; 353bl; 355bl; 355br; 368tr; 368cl; 368cr; 383cr; 396tr; 396cl; 400tl; 400tr; 420br; 421tl; 422cl; 422bl; 443tl; 443tr; 454cl; 454cr; 459br; 461tr: **Carsten Siems** 494tr: **Charles Fleming** - Website "Wildlife in a Suburban Garden" 511br; 514tr: **Chris Batty** 264cl; 598cl: **Chris Griffin** / www.griffinwildlifephotography.com 490bl: **Chris van Rijswijk** / birdshooting.nl 78tc; 398bl; 411cl; 412bl; 421tr; 568cr: **Christophe Gouraud** 204cr; 226tr: **Christophe Sidamo/Biosphoto/FLPA** 2-3: **Christopher J.G. Plummer**, Switzerland / pbase.com/cplummer 390tr; 391tr: **Clement Francis** 337cr: **Colin Bradshaw** 39br; 105cr; 136tr; 175cr; 275cr; 389br; 489bl; 534bl; 564tr: **Conrado Requena Aznar (Conry)** / conry-conry.blogspot.com.es 532br: **D. Yakubovitch/Shutterstock** 441tr: **Dan Miller** / birdsofthebog.com 595clt: **Daniel Pettersson** / danielpettersson.com 140bl; 459bl: **Danita Delimont/Getty** 594bl: **Danny Laredo** 219bl: **Darran Rickards** / www.e-wildlife.co.uk 512bl: **Darren Robson** 546br: **Darryl Spittle** 575br: **Dave Barnes** / www.pbase.com/davebarnes 54tr; 427br: **Dave Clark** 266br: **Dave Hutton** 196cl: **Dave Pressland/FLPA** 48cl: **David Beadle** 576br; 585cl: **David Bigas** 384bl; 384br: **David Monticelli** pbase.com/david_monticelli 66b; 66cr; 71tl; 115br; 119tr; 119br; 126bl; 150tl; 169br; 204tc; 205tr; 209tr; 216tl; 444bl; 462b; 465tl; 465br; 467bc; 471tl; 471tr; 472tr; 472bc; 475bl; 475tl; 478br; 480br; 485cr; 485br; 489tl; 491cl; 524tr; 574cl; 575cr; 577br; 592t; 594crb; 600b: **David Speiser/www.lilibirds.com** 209cr: **David Tipling** 73br; 88bl; 283tr: **David Whitaker** / Alamy Stock Photo 9bl; 423cr: **Dean Eades BirdMad** / BirdMad.com 423bc: **Dennis Jacobsen/Shutterstock** 34cl; 515br: **Devaram Thirunavukkarasu** / www.gamebirds.me 581tr: **Dibyendu Ash** 365br: **Dick Forsman** / www.dickforsman.com 68cl; 180bl: **Dinah Saluz** 47tl: **Dolly Laishram** 166br; 583tr; 583tlb; 584br; 584bl: **Dominic Mitchell** (www.birdingetc.com) 483br: **Dominic Robinson** / Alamy Stock Photo 283tl: **Dorit Bar Zakay** 444tr: **DP Wildlife Vertebrates/Alamy** 79br: **Dr JA Theobald** 500cl: **Dr. Gaurang Bagda**, Gujarat 588tr: **Dr Malay Mandal**, India 586tl: **Duha a alhashimi** 151cr; 322br: **Earl Orf** / www.earlorfphotos.com 576tl: **Ed Schneider** 593tllt: **Eduardo Ayala** 557tl; 557tr; 557cl: **Edward Myles/FLPA** 472tr: **Edwin Winkel** 22trt; 23tr; 56tr; 95tc; 99br; 127bl; 127br; 183cr; 186br; 193cr; 194br; 262cr; 305cr; 314cr; 315cr; 341bl; 388tr; 400br; 519t; 519bl; 519br; 568tr: **Eigil Ødegaard** 195br; 416br; 426tr; 426cl: **Elena Shnayder** 207cl; 207bl; 207br; 209tl: **Emin Yoğurtcuoğlu** 516br: **Eric Didner** 332tl: **Erich Kuchling/Getty Images** 122br: **Erik Nielsen** 596bb: **Erni/Shutterstock** 94tr; 214bl; 372br: **Eva Foss Henriksen** 22brt; 523bl: **Evgeny N. Panov** 178br: **Eyal Bartov** 434cr; 437bl: **Ezra Hadad** 539cr; 539br: **Fabio Ballanti** 558tr: **Ferran López** 47cl; 102cr; 132cl; 577bl; 577bc: **Fikret Yorgancıoğlu** 56tr; 76cl: **FotoRequest/Shutterstock** 594tl: **Fran Trabalon** 74bl; 95tr; 109cr; 139bl; 170cr; 249br; 270tr; 302cr; 302bl; 338bl; 338br; 380cl; 390cr; 390br; 430cl; 451cl: **Francisco Coimbra** 401bl: **Frank Cezus/Getty** 585cr: **Frédéric Desmette** / www.fredericdesmette.com 423cl; 424tl: **Frédéric Jiguet** 508tl: **Frédéric Pelsy** 393tl: **Fredrik Ström** 508cl: **Friedhelm Adam/Getty Images** 81bl: **Frode Falkenberg** / www.falkefoto.no 173br: **Frode Jacobsen** 472br; 476br; 598bl; 604bl; 605tr: **Gabriel Schuler** 53cr; 536tl: **Gal Shon** 55tl; 433bl; 538tr: **Gary Brown**, Sultan Qaboos University, Muscat, Oman 529cr: **Gary Howard** 392br: **Gary Jenkins** 117br; 118tl; 118cr; 201bl; 261cr; 277cr; 393br: **Gary Thoburn** (GaryTsPhotos) 21cr; 41tl; 73cr; 94cr; 99tr; 99bl; 108cr; 123tr; 135b; 136br; 232br; 276br; 351cl; 363tl; 382ll; 382tr; 394br; 406tl; 480cl; 500br; 501tr: **Geetha Venkataraman** / https://aud-in.academia.edu/GeethaVenkataraman 203cl: **Gene Koziara MD** / https://www.flickr.com/photos/genekdr/ 474cr; 593tr; 593tlb: **Gennadiy Dyakin** 200tr; 434tl: **George Reszeter** 89cr; 173bl; 229br; 230cr; 390bl; 556br; 577cl: **Georges Olioso** http://georges.olioso.oiseaux.net/ 109tr; 226br; 317tr: **Gerhard Hofmann** 405br: **Gianni Conca** pbase.com/birdclick 165cr; 279tr; 306cr: **Gianpiero Ferrari/FLPA** 222br: **Giorgi Darchiashvili** 447bl; 447br: **Glenn Bartley/Getty** 596c: **Glyn Sellors** - www.glynsellorsphotography.com 49tr: **Gobind Sagar Bhardwaj** 366br: **Goran Ekström** 428br: **Graham P Catley** 41tc: **Graham R.**

Lobley 59br: Gunhan Hatipoğlu 558br: Gurhan Sinan Ozgunlu 124tr: Hadoram Shirihai 9tr; 9br; 10cl; 15bl; 24tl; 31tl; 31bc; 31br; 34tr; 35tl; 36tr; 44tl; 46tl; 46tc; 46cl; 47clt; 58tr; 59tl; 64cl; 65bl; 68br; 76cr; 83bl; 89tr; 95br; 96tl; 101cr; 102tl; 105tl; 105tr; 110tl; 110tr; 116tr; 116bl; 116br; 122tl; 126tr; 126cr; 133tr; 139br; 148br; 150br; 155cr; 155br; 157cl; 157br; 158tl; 158tr; 158cl; 159tl; 159tr; 168bl; 168tr; 168tl; 171cl; 171tr; 172tl; 172tr; 173tr; 179cl; 198tr; 202tl; 206tr; 206bl; 211br; 217tr; 217cr; 219cl; 220tl; 223br; 224tl; 230br; 234cr; 240cl; 242bl; 242br; 243cr; 243br; 246b; 249bl; 250cl; 251cl; 251br; 251br; 253cr; 253cl; 253br; 257br; 262tl; 264tr; 265b; 267cl; 267bl; 269bl; 271tr; 271br; 272tl; 272bl; 272br; 274bl; 275tl; 275cl; 275br; 278cr; 279br; 281br; 282bl; 285bl; 285br; 286cr; 289cl; 289br; 290cl; 290cc; 292cl; 295cl; 297cl; 297bl; 298br; 299tr; 299tl; 303tr; 308tl; 311tr; 312tr; 313cl; 314br; 315tl; 315tr; 315br; 320br; 327tr; 327cl; 329cl; 329br; 332cr; 333cl; 333br; 335cr; 338cl; 348cl; 350bl; 350br; 355tr; 357cr; 357br; 358br; 359br; 360tr; 360br; 362bl; 371tr; 371cl; 372tr; 373br; 383tr; 385br; 391cr; 424br; 433cr; 435cr; 437tl; 439tr; 439br; 440cl; 445cl; 445cr; 455tr; 461cl; 464tr; 467tl; 467cc; 474tl; 482cl; 495tl; 498cr; 502cl; 503tc; 505bl; 506tr; 506bl; 517tr; 520bl; 521bl; 522br; 532cr; 535tr; 538cl; 551tr; 559tl; 559cl; 562tr; 570tr; 572cr; 572br; 579cr; 582cr; 588cr; 589tr; 592crt; 593ct; 595bl; 597tl; 597tr; 597clb; 597crb; 598tl; 598tr; 598cr; 599tl; 599tr; 600tl; 602clb; 603tl; 603bl; 604tl; 605cr: Hakan Kahraman 454br: Hakan Yıldırım 46tr: Hanne & Jens Eriksen 43br; 58bl; 58br; 58cl; 68bl; 104cr; 151tr; 153cl; 160br; 161tr; 161br; 162cl; 162bc; 162br; 177cr; 188br; 198cr; 201cl; 202br; 202bl; 273c; 274cl; 287br; 326bl; 329cr; 339bl; 342tr; 375br; 376br; 379cl; 380cr; 403cr; 403cl; 403bl; 525tl; 530br; 535b; 586cr: Hanne & Jens Eriksen/NaturePL 174bl; 176tl: Hannes Nussbaumer 61bl; 63cl; 63cr; 63bl; 73bl; 102bl; 102br; 358cl; 358cr; 370tr; 530tl: Hannu Eskonen / www. luontokuvateskonen.com 408bl: Hans Henrik Larsen 230tr; 414br: Harri Taavetti 43cr; 78cl; 80bl; 87tr; 134br; 415bl: Harri Taavetti/FLPA 228cl; 544cr: Harry J. Lehto 417bl: Harvey van Diek 67bl; 100tr; 237tl: Heiko Schmaljohann / http://www.ifv-vogelwarte.de/index.php?id=176 57cr: Helge Sørensen 414t: Henny van Egdom/ Minden Pictures/FLPA 124br: Henrik Pedersen 418tr: Hervé MICHEL / www. oiseaux-nature.com 258tr; 349c: Hira Punjabi/Alamy 583tlt; 584cl: Hugh Harrop 44br; 86br; 87cr; 140br; 181cl; 181cr; 413tr; 413tl; 413bl; 413br; 415br; 417br; 419tc; 419br; 443br; 478br; 483br; 483cr; 490cl; 490cr; 492bl; 496tr; 499bl; 533cr; 544br; 545cl; 552br; 569cr; 603tr: Huw Roberts 336bl; 337tr; 532tl: Iain H. Leach 276br; 488bl; 488br; 570bl: Iain Lowson Wildlife / Alamy Stock Photo 337cl: Ian Boustead 172br; 181tr; 182bl: Ian Butler Bird / Alamy 274br: Ian Clarke 411tl: Ian Davies 476cr; 491tr; 493br: Ian Fisher/Cahow Photography 248tl; 503cr; 503br; 504tr; 541bl; 544tr; 547tr; 549br; 563cl: Ian Merrill 392cl: Ibrahim Tunca 76br: Igal Siman Tov 539bl: Ignacio Yúfera, www.iyufera.com 167cl: Igor Maiorano / www. pbase.com/igormaiorano 205tc: Ilya Ukolov 95tl; 144tl; 590cl; 590cr: Imran Shah 428tl: Ingo Schultz/GettyImages 308cr: Irit Drob 539tl; 539cl: Istvan Moldován 155cl: Iva Hristova/Biota Films 164tl; 533tr: J. A. van den Bosch 411br: Jacques Érard 440br: Jainy Kuriakose 548br; 553bl: James Hager / robertharding /Getty Images 290cr: James Lidster 213br; 250br: János Oláh / Birdquest 322cl: Jared Clarke -- www.birdtherock.com 463bl: Jari Peltomäki 103br; 108tr; 1896cr; 221bl; 408tr; 442tl; 543cr: Javier Blasco-Zumeta 47cr; 51cl; 53cc: Javier Blasco-Zumeta & Gerd-Michael Heinze 391cl: Jean Iron / www.jeaniron.ca 47br: Jean-Michel Fenerole 373cr; 393tr: Jens Hering 302cr: Jens Søgaard Hansen 446bl; 446br: Jens Thalund 204br; 236b; 237br: Jeremy Babbington 13tr; 160cl; 235br; 296cl; 296bl; 296br; 297tr; 322tl; 380bl: Jerry Jourdan / jerryjourdan.blogspot.com 290bl: Jesús Laborda 234br: Jim Almond 419tr: Jim Martin / @jimspim 604br: Joao Ponces 504cr: Joe Wynn 241br: Johannes Mayer 156br: John & Jemi Holmes 32cr; 203tl; 436bl: John Davenport 514bl: John Hawkins/FLPA 82cr: John Holmes/FLPA 184tl; 184tr; 561bl: John Richardson 196tr; 397tr; 423bl: Jon Hornbuckle 332br; 380br: Jonathan Lewis/Getty Images 79cl: Jonathan Martinez 541cr; 553cr; 585trt; 587bt: Jonathan Meyrav 180tl: Jorma Tenovuo/www.jtenovuo.com 53tl; 432bl; 545bl: José Luis Copete 51cr; 54bl; 54cr: Jose Viana 340tr: Joseph Nichols / www.flickr.com/people/josephbirdphotography/ 442br: Juan José Bazán Hiraldo 216tr: Juan Luis Muñoz 214bl: Jukka Jantunen 209br: Jules Fouarge, Aves, Belgium 61br; 62t; 62bl; 141bl: Jussi Vakkala 40cc: Jyrki Normaja 40tl; 40bl; 42bl; 42br; 172tc; 199cl; 227tl; 256cl; 285tr; 357tl; 357br; 415tl; 416tl; 442cr; 488tl; 489br; 521tr: Kaajal Dasgupta 587cr: Kadir Dabak 436tl: Kai Gauger 216cl: Karel Mauer 513cl: Kari Haataja 175cl; 299br; 489cl; 489cr; 565b: Kevin Lane 334tl: Khalifa Al Dhaheri 326cr; 328bl: Kjell Johansson 197cr: Klaus Dichmann 228tr; 228br; 571bl: Klaus Drissner 406br: Klaus Malling Olsen 126tl; 246cl; 432br: Kris de Rouck 65tr; 350tr: Krzysztof Błachowiak 103tl: Larry Corbett/Alamy 82bl: Lars Gabrielsen 243cl: Lars Jensen 601t: Lars Jonsson 22brb; 49cr; 55tr: Lars Petersson, www.larsfoto.se 280br; 426br; 453tr: Lars Svensson 15br; 15cr; 20tl; 326tl; 529bl: Lars Svensson/Natural History Museum 11tl: Lars Svensson/NRM 20cl; 20tr; 20cr: Lars-Olof Landgren 197tr: Leander Khil, www.leanderkhil.com 393cl; 393cr; 410br: Lefteris Stavrakis 179tl: Leo J. R. Boon/Cursorius 117tl ; 150br : Lesley van Loo - www.LTDphoto.com 128bl: Lior Kislev 158cr; 187tl; 198cl; 198cr; 198tl; 216cc; 217tl; 217bl; 220c; 309tr; 310br; 313tl; 346bl; 346br; 347tr; 362cr; 371cr; 401tr; 518bl; 528cl; 538cr; 538bl: Lloyd Spitalnik / lloydspitalnik. com 470tl; 480tl; 480tr; 482br; 483tl; 484br; 485cl; 485tr: Lubos Mráz / Naturfoto. cz 82tr: Luigi Sebastiani/www.birds.it 33bl; 83tr; 83cr; 93br; 133cl; 185bl; 334cl; 367br; 383br; 512br: Lukas Howald 133tl: Magnus Hellström 50tl; 179br; 183tr: Magnus Hellström / Ottenby Bird Observatory 12cr: Magnus Martinsson/N 44tl: Magnus Ullman 311cl: Magnus Westerlind 236tl: Mandy West 418br: Manjula Mathur 191cr; 601b: Mansur Al Fahad 235bl; 322cr: Manuel Estébanez Ruiz 123bl; 187bl: Marc Raes 318bl; 318br: Marc Thibault 502br: Marcel Gauthier 573cl: Marek Szczepanek / www.marekszczepanek.pl 94tl: Marie Read/Naturepl 590b: Marinco Lefevere / marinco.synology.me/photo 211tl; 214bl; 235cl; 405bl; 410tr; 518br: Māris Strazds / www.putnubildes.lv 358tr; 373cr: Mark Breaks 32tc: Markku Saarinen 137bl: Marko Matešić 135tr: Mark Rayment 474br: Markus Lagerqvist (pbase.com/lagerqvist) 436br: Markus Römhild 60tl; 60tr; 170tr; 213br: Markus Varesvuo 24cl; 33br; 39bl; 47br; 48tr; 62bl; 69cr; 72tr; 72bl; 78tl; 78cr; 81bl; 81tr; 91bl; 91br; 97tl; 106bl; 122tr; 135tl; 138bl; 138br; 163br; 164br; 198br; 216cr; 219br; 221br; 227br; 233tl; 233cl; 233cr; 239tr; 239bl; 240tl; 242cl; 242bl; 242tl; 242tr; 243tr; 245br; 248tl; 248bl; 248br; 249cr; 249tl; 252br; 253br; 253tl; 254bl; 254br; 256tr; 256bl; 256br; 260br; 261bl; 261tr; 268bl; 268tr; 269tr; 269cr; 277br; 278cl; 281br; 281tr; 281bl; 293tl; 293br; 320br; 326br; 347bl; 353br; 354cl; 354cr; 355tr; 361br; 362br; 368br; 369cl; 369cr; 382br; 383br; 395tr; 395bl; 404tr; 404br; 405tr; 405cr; 405cl; 409cl; 409cr; 410cr; 415clt; 418br; 419cr; 419bl; 420bl; 421cr; 421br; 422cr; 422br; 423br; 425tr; 425br; 444br; 450br; 451tc; 451cr; 451bl; 451br; 452bl; 453tl; 460br; 461br; 494bl; 494br; 495bl; 495br; 496tl; 496b; 497tr; 497bl; 499br; 501bl; 501br; 510tr; 513bl; 515bl; 515tl; 524tl; 531br; 531br; 542br; 543tr; 555cl: Martin Fowler/Shutterstock 231bl: Martin Gottschling 487b: Martin Hale 40cr; 552cr; 561cr; 563cr; 566cr: Martin P Goodey 215bl; 396br; 547b: Martin Pitt 431cl; 431br: Martin van der Schalk 161bl: Mateusz Matysiak www. fotomatysiak.pl 134bl; 164tr: Mateusz Sciborski/Shutterstock 363br: Mathias Putze 234tl; 234tr; 251tr; 255tr; 258tl; 258br; 263tl; 263cl; 263cc; 263cr; 460c: Mathias Schäf / living-nature.eu 156cr; 205br; 206tl; 211bl; 218br; 256cr; 263tr; 271br; 282br; 292br; 300bl; 301tr; 311br; 310bl; 348cr; 349bl; 364cl; 438cl; 460tr; 517cr; 523c; 571br: Matt Eade, East Sussex, UK 598bl: Matt Poll / snowyowllost. blogspot.ca 563bc: Matthieu VASLIN 147br: Mattias Ullman / pbase.com/mull 201cr: Menno Hornman 237tr: Michael Bergman 540bl; 540br: Michael Heiß 57tl; 130b; 240tr; 279br; 281br; 286b; 331bl; 331br; 337bl; 406bl; 519cr; 533bl; 569tl: Michael Pope 23cr; 143bl; 143br; 175br; 187br; 193cl; 294br; 309br; 313tr: Michael Schmitz 220bl; 511tr: Michael Westerbjerg Andersen 105br; 536br; 565tr: Michele Mendi 88br; 215tr: Michelle & Peter Wong 32bl; 43bl; 541bl; 541bc; 549cr; 550b; 553tl; 568br; 569cl; 581br; 582b; 583b; 586tr: Miguel Angel Peña 360br: Miguel Angel Rojas Ruiz 340cl: Miguel Rouco 150cr; 170bl: Mika Bruun 591br: Mikael Nord 425bl: Mike Barth 152bc; 153cr; 154tl; 171br; 175tr; 176tr; 188bl; 189bl; 291bl; 295bl; 325br; 336br; 522bl; 523br: Mike Buckland 461bl; 461br: Mike Danzenbaker / avesphoto.com 591tr: Mike Lane 501br: Mike Parker 502br; 541tr; 560br; 561br: Mike Potts/NaturePL 246cr: Mikkel Holck 431tr: Mircea Costina/ Alamy 576bl: Mogens Hansen 397bl: Mohit Kulkarni/Shutterstock 582tl: Mohsen Vejdani 112tr: Morten Scheller Jensen 415crt; 426cc: Morten Winness / www. mortenwinness.no 121bl: Morteza Nemati 90cl; 365c: Murat Uyar 429tr; 429tl: Mustafa Sözen 124tl; 429cr: Nancy McKown / nancybirdphotography.com 470bl: Natali Kim (www.birds.kz) 388br: Neil Bowman/FLPA 146tr; 146bl; 298cl: Neil Loverock 341br: Nick Appleton 423tr; 500cr: Nick Kontonicolas/1000birds.com 89tl: Nigel Blake 92tr: Nigel Redman 318bl: Nikhil Devasar 182br; 188cr; 295bl;

310tr; 310br: **Niranjan Sant** 32cl; 189cr; 566cr; 569br; 587bb: **Norman Deans van Swelm** 50cl; 50cr; 57br; 107bl; 107br; 247br; 301tr; 301tl; 301cr; 301bl; 362tr; 387tl; 398tl; 398br; 399tr; 409tr; 555tr: **Norman West** 223tr: **Ole Krogh** 77tr: **Ole Zoltan Göller** 57bc; 442tr: **Ómar Runólfsson/Getty** 498bl: **Ömer Necipoğlu** 57bl: **Otgonbayar Baatargal** 148cr; 148bl: **Oz Horine/ www.facebook.com/BirdsFamiliesOfTheWorld** 41tr: **Pankaj Maheria / www.escapeintothewild.net** 291tr: **Paolo Casali / http://www.pbase.com/lep** 524bl: **Paschalis Dougalis / https://dougaliswildlifeart.blogspot.de/** 558cr: **Patrick Eriksson** 469b: **Paul Cools / pbase.com/paulcoolsphotography** 493tr; 513tr: **Paul Hürlimann** 389cr: **Paul Lathbury / www.wildlifepictorial.co.uk** 386cr; 386cr; 410bl: **Paul Reeves Photography/Shutterstock** 604cl: **Paul Sawer/FLPA** 406tr: **Paul Vangiersbergen** 206br: **Pavel Parkhaev** 508br; 551bl; 552tr; 553tr; 556bl: **Pekka Fågel** 46cr; 154cr: **Per Schans Christensen** 415crb: **Pete Blanchard / www.flickr.com/photos/flyingfast/** 509br: **Pete Morris** 288b: **Pete Walkden / www.petewalkden.co.uk** 196tl; 419tl: **Peter Adriaens** 205cr: **Peter Arras** 188tl: **Peter Ericsson www.pbase.com/peterericsson** 32tr: **Peter Hvass** 196tl: **Peter Leigh of Firecrest Wildlife Photography** 219cr: **Peter Ryan, FitzPatrick Institute** 290tr; 328cl; 380tr: **Peter Thomsen** 428br: **Peter W. Post** 493bl: **Petros Petrou** 45bl; 129tl; 538br: **Petteri Hytönen** 44tr: **Phil McLean/FLPA** 319bl: **Phil Roberts** 571cr: **Phil Slade / anotherbirdblog.blogspot.co.uk/** 35cr: **Philip Mugridge / Alamy** 407tr: **Philippe Dubois** 278br; 287tr; 506cl: **PhotonCatcher/Shutterstock** 550tr: **Piero Alberti** 177tl; 216br; 304cr: **Pieter Vantieghem** 47cl: **Pranjal J. Saikia** 291br; 584bc; 602tr: **Prasad Ganpule** 522t; 586br: **Purevsuren Tsolmonjav** 602br: **Rafael Armada (www.rafaelarmada.net)** 52tl; 52tr; 53tr; 64tr; 112bl; 204tr; 205tl; 287cl; 312tl; 356tl; 476bl; 476bc; 524cr: **Ralph Martin / www.visual-nature.de** 21bl; 55cl; 89bl; 96tr; 115bl; 116tl; 131tr; 132tl; 138tc; 139tl; 140tr; 354tl; 497cl; 498tl: **Ram Mallya** 166tr; 304tr; 344tr; 347br; 569bl: **Ran Schols** 42cl; 277tl; 510cr; 511bl; 548tl: **Raphaël Jordan** 602tl: **Rashed Al-Hajji** 181bl; 182cl; 183tl; 192br; 194bl; 218bl: **Ray Wilson / Alamy** 49cl: **Raymond Wilson** 131tl; 142br: **Rei Segali** 530bl: **René Pop** 47crb; 75br; 128tr; 176tc; 185br; 186tl; 192br; 204tl; 233br; 240cr; 246tr; 247bl; 259br; 277cl; 278br; 287cr; 300br; 330br; 363tl; 401br; 433tr; 528br; 532bl; 543br: **René van Rossum** 170tl; 317tl: **Ricardo Rodriguez** 137br; 372cl: **Richard Bonser** 487cl; 504br; 507tr: **Richard Brooks/FLPA** 496cl: **Richard Ek** 573cr: **Richard Lowe, Derbyshire** 456tr; 456tl: **Richard Porter** 345tr; 436tr: **Richard Steel / wildlifephotographic.blogspot.com** 223tl; 231tr; 231br; 232bl: **Rob Felix** 65cl; 262bl; 262br: **Robert Newlin** 581tl: **Robert Royse** 477tl; 486bl; 486br; 487tr: **Roberto Güller / www.flickr.com/photos/robertoguller_ok** 577tl; 577tr: **Robin Chittenden (www.robinchittenden.co.uk)** 500tl; 507tr: **Rochdi Nehal** 13tl; 120b: **Roger Ridley** 179bl; 261br: **Roger Tidman** 280tl; 319br: **Roger Tidman/FLPA** 399br; **Rolf Kunz** 132cr; 239bl; 339bl; 349br: **Roman T. Brewka** 469cr; 480cl; 493c: **Ronald Messemaker** 316br: **Rudi Debruyne** 220tr; 367br; 405tl: **Ruedi Aeschlimann, Switzerland / www.vogelwarte.ch/de/voegel/voegel-der-schweiz** 131br: **Rufino Fernández González / www.avesdeburgos.com** 50bl; 51tr; 123br: **Rune Sø Neergard** 416tr: **Russell Hayes / http://birdmanbirds.blogspot.com** 570br: **S. J. M. Gantlett** 193br; 365tr; 571tl: **Sam Northwood / https://samalij.wixsite.com/samsphotopoetry** 173bc: **Sampo Laukkanen** 546tr: **Sander Bot** 133br: **Sanket Nilaxi Hrishikesh Mhatre** 295tl: **Satie Sharma / www.satie.in** 284cl; 491tl: **Saxifraga-Mark Zekhuis** 392bl: **Sean Cronin** 411bl; 491tl: **Sean Gray / grayimages.co.uk** 492tl: **Sebastian Kennerknecht/Minden Pictures/FLPA** 59tr: **Sergey Pisarevskiy** 104b: **Sharon Heald / NaturePL.com** 173tl: **Shivam Tiwari** 152cl; 189br: **Siddharth Damle** 189cl: **Sijmen Hendriks Nature Photography / sijmenhendriks.com** 556cr: **Silas K. K. Olofson/ www.birdingfaroes.wordpress.com** 365tl; 408tl; 426cr; 446cl; 500bl: **Simon Berg Pedersen** 439tl: **Simon Litten / Alamy Stock Photo** 399tl: **Simon Papps/Birdwatchingbreaks.com** 145tr; 145br: **Simonas Minkevicius / Shutterstock** 614, 620; **Skapuka/Shutterstock** 1: **Soner Bekir** 514br: **Stanislav Harvančík www.birdphotoworld.sk** 114b; 132tr; 132bl; 133bc; 138tl; 141tl; 141tr; 200bl: **Steen E. Jensen** 386br: **Stefan Hage, www.birds.se** 181tl; 399cr; 426bl: **Stefan Johansson / www.avesphotography.se** 49tl: **Stefan Pfützke /Green-Lens.de** 98tr; 562bl; 562br: **Steve Arlow (www.birdersplayground.co.uk)** 466bl: **Steve Ashton / www.steveashtonwildlifephotography.co.uk** 197cl: **Steve Bury** 476tr: **Steve Byland/Shutterstock** 596tl: **Steve Fletcher** 211bl: **Steve Gettle/Minden Pictures/FLPA** 595tr: **Steve N. G. Howell** 467cr; 467bl; 470br; 474tr; 574tr; 592clb; 592bt; 593b; 594crt; 596tr; 597crt; 600tr; 605br; 605bc: **Steve Ray / www.flickr.com/photos/26135972@N05/** 513tr; 543bl: **Steve Rooke / www.sunbirdtours.co.uk** 85bl: **Steve Round (steven-round-birdphotography.com)** 35br; 79cr; 92tr; 259tr; 395tr; 410cr; 454tr: **Stuart Elsom LRPS** 466tl; 574tr: **Stubblefield Photography/Shutterstock** 594clb; 595tl; 599br: **Sudhir Garg** 565cl: **Sumit K Sen** 153br; 285cr: **Sunil Singhal** 177tr; 183bl; 184br; 190cl; 190br; 191tl; 191tr; 191cl; 203tr; 343tr; 536bl; 566tr; 533cr; 602clt: **Sylvia Remijn** 335tr: **Tadao Shimba** 251tl; 467tr; 476cl; 544bl: **Tahir Abbas** 112cr: **Terje Kolaas / www.terjekolaas.com** 31bl: **Thierry Helsens** 324b: **Thierry Quelennec** 50tr: **Thom Haslam / EvolvingHumanities.org** 324tr: **Thomas Bernhardsson / www.tbernhardsson.fotosidan.se** 180br: **Thomas Grüner** 138bc; 238bl; 384cr; 390bl; 555cr: **Thomas Krumenacker** 39 tr; 40tr; 40cl; 47cl; 101cl; 334tr; 529tr: **Thomas Langenberg** 288tl; 527tl; 527tl: **Thomas Luiten / www.pbase.com/thomasluiten** 530tr; 579bc: **Thomas Varto Nielsen** 46bl; 46br; 345bl: **Thorsten Krüger / thorsten-krueger.com** 258tl; 396tl; 496cr: **Tim Jones** 323cl; 324cr: **Tim Zurowski/Shutterstock** 605bl: **Timur Çağlar** 320br: **Tom Beeke** 563tr: **Tom Jenner** 71br; 151bl; 345br: **Tom Lindroos** 104tr; 366tr; 378tr; 458cl: **Tom Reichner/Shutterstock** 604tr: **Tomas Lundquist/N / www.tomaslundquist.com** 228tl; 415clb; 419cr: **Tomi Muukonen** 87br; 121br; 546c: **Tommy Holmgren** 414cr; **Tommy P. Pedersen** 152tr; 266br; 295br: **Tony Hovell** 514tl: **Tony Tilford** 165br: **Tor A. Olsen** 478tl; 478tr: **Uku Paal** 215br; 393br: **Ulf Ståhle** 378tl; 403br: **Utopia_88/iStock** 588tl: **Vadim Ivushkin / flickr.com/photos/91544025@N05/** 208tl; 208tr; 208cr: **Vasil Ananian (Armenia)** 400cr; 559br: **Vasilly Fedorenko** 197bl; 201tl; 434tr; 449tl; 449bl; 449br; 602crb: **Vaughan & Svetlana Ashby/Birdfinders** 70tl; 113cl; 147cl; 150tr; 169cl; 225tr; 264br; 289tr; 290tl; 299cl; 337tl; 435bl; 508br; 521cr; 553cl; 573tr: **Veena Nair/ Getty Images** 291bl: **Veer Vaibhav Mishra** 337bl: **Vincent Legrand** 67t; 96tl; 96bl; 101bl; 119cl; 120tl; 151bl; 194cl; 226tl; 264cl; 267tl; 308cl; 314bl; 317b; 320tl; 336cl; 336cr; 351tl; 351br; 352bl; 352tr; 359cl; 427tr; 463tl; 463br; 465tl; 470tr; 472cr; 485br; 504tl; 534br; 547cr; 573br; 575tl; 575cl; 575bl; 586br; 589bl; 589br; 590tr; 591tlt; 591tlb; 591cr; 593cbl; 593cbr; 594br; 595clb; 595crb; 595br; 596bt; 597clt; 597tl; 597br; 599c; 600cl; 600cr; 601cl; 601cr; 603br: **Vincent Legrand/AGAMI** 589cr: **Vincent van der Spek** 324cl: **Vitantonio Dell'Orto/exuviaphoto.com** 228bl: **Vladimir Vorobyov / birds.kz** 224br; 448br: **W. Weenink / Alamy stock photo** 195bl: **Werner Müller** 37tl; 37br; 38tl; 38br; 38bl; 41bl; 76tr; 85tr; 111cr; 111bl; 342cl; 342tr; 342bc; 375tr; 376tr; 377tl; 377b; 379bl; 379br; 402cl; 402br; 446tr; 447tl; 447tr; 560tl; 560tr: **Werner Suter** 109cl: **willridge images / Alamy Stock Photo** 398tr: **Yann Kolbeinsson / Birding Iceland** 480bl: **YAY Media AS / Alamy Stock Photo** 583cl: **Yoav Perlman** 125br; 127t; 328br: **Yosef Kiat, Israeli Bird Ringing Center (IBRC)** 24bl; 24br: **Yuriy Malkov** 449tr: **Zafer KURNUÇ-TURKEY** 558bl: **Zlatan Celebic** 323br.

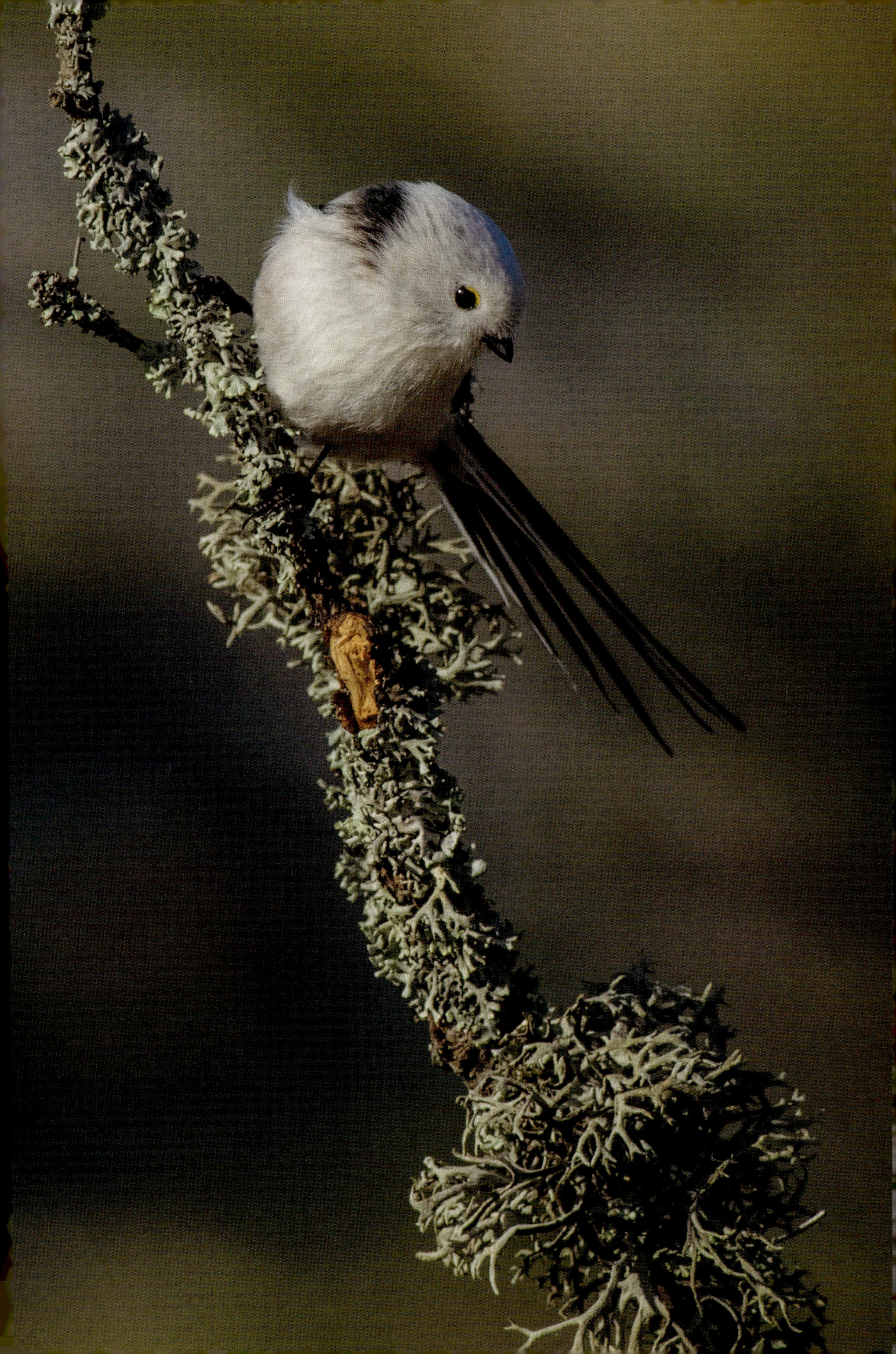

INDEX OF SCIENTIFIC NAMES

A

abyssinicus, Ploceus cucullatus 588
abyssinicus, Zosterops 160
acaciae, Turdoides fulva 71
ACANTHIS 412
ACRIDOTHERES 292
AEGITHALOS 72
aestiva, Setophaga petechia 478
africana, Fringilla coelebs 356
AGROPSAR 587
albicilla, Ficedula 42
albicollis, Ficedula 55
albicollis, Zonotrichia 490
albirictus, Dicrurus macrocercus 586
albus, Corvus 264
algeriensis, Lanius excubitor 204
alienus, Rhodopechys 430
alpicola, Montifringilla nivalis 335
alpinus, Aegithalos caudatus 76
altirostris, Turdoides 64
AMANDAVA 343
amandava, Amandava 343
amandava, Amandava amandava 344
amantum, Bucanetes githagineus 439
americana, Setophaga 475
americana, Spiza 603
ammodendri, Passer 298
ammodendri, Passer ammodendri 299
anatoliae, Poecile lugubris 89
anglica, Loxia curvirostra 423
aphrodite, Parus major 109
arabica, Emberiza tahapisi 527
arabicus, Cinnyricinclus leucogaster 297
arabs, Zosterops abyssinicus 161
arborea, Passerella 601
arborea, Passerella arborea 601
arenarius, Lanius isabellinus 183
asiatica, Sitta europaea 123
asiaticus, Cinnyris 152
asirensis, Pica pica 235
astrild, Estrilda 340
ater, Molothrus 604
ater, Molothrus ater 604
ater, Periparus 93
ater, Periparus ater 94
atlas, Periparus ater 96
atlas, Sitta europaea 124
atricapillus, Garrulus glandarius 225
aucheri, Lanius excubitor 200
aureola, Emberiza 551
aureola, Emberiza aureola 553
aurocapilla, Seiurus 462
aurocapilla, Seiurus aurocapilla 463
azurea, Hypothymis 585

B

bactriana, Pica pica 233
bactrianus, Passer domesticus 304
badius, Lanius senator 215
balearica, Loxia curvirostra 423
balearica, Muscicapa striata 35
balearicus, Cyanistes caeruleus 99
barbara, Petronia petronia 331
barbarus, Pyrrhocorax pyrrhocorax 245
bella, Linaria cannabina 400
bianchii, Lanius borealis 209
biarmicus, Panurus 61
biarmicus, Panurus biarmicus 62
biblicus, Passer domesticus 303
bifasciata, Loxia leucoptera 418
bokharensis, Parus 113
borealis, Lanius 207
borealis, Lanius borealis 209
borealis, Poecile montanus 82
brachydactyla, Carpospiza 325
brachydactyla, Certhia 137
brachydactyla, Certhia brachydactyla 139
brandtii, Garrulus glandarius 224
brevirostris, Carduelis carduelis 388
brevirostris, Cinnyris asiaticus 153
brevirostris, Linaria flavirostris 406
britannica, Certhia familiaris 136
britannicus, Periparus ater 95
bruniceps, Emberiza 564
BUCANETES 435
buchanani, Emberiza 534
buchanani, Turdoides fulva 71
BUPHAGUS 588
buryi, Lanius excubitor 202
buvryi, Coccothraustes coccothraustes 460

C

cabaret, Acanthis flammea 410
cabrerae, Periparus ater 95
caerulescens, Setophaga 595
caeruleus, Cyanistes 97
caeruleus, Cyanistes caeruleus 98
caesia, Emberiza 537
caesia, Sitta europaea 123
calandra, Emberiza 570
CALCARIUS 499
callensis, Emberiza cia 518
canadensis, Cardellina 598
canadensis, Sitta 585
canaria, Serinus 372
canariensis, Fringilla coelebs 357
caniceps, Carduelis carduelis 388
caniceps, Lanius schach 191
cannabina, Linaria 397
cannabina, Linaria cannabina 399
cantans, Euodice 348
cantans, Euodice cantans 349
capellanus, Corvus cornix 263
CARDELLINA 598
CARDUELIS 385
carduelis, Carduelis 385
carduelis, Carduelis carduelis 387
carnipes, Mycerobas 457
CARPODACUS 441, 590
caryocatactes, Nucifraga 238
caryocatactes, Nucifraga caryocatactes 239
caspia, Emberiza schoeniclus 559
caspius, Remiz pendulinus 143
cassinii, Pyrrhula pyrrhula 454
castanea, Setophaga 594
caucasica, Sitta europaea 124
caudata, Turdoides 66
caudatus, Aegithalos 72
caudatus, Aegithalos caudatus 73
CERTHIA 134
cerulea, Setophaga 592
cervicalis, Garrulus glandarius 226
chinensis, Oriolus 585
CHLORIS 381
chloris, Chloris 381
chloris, Chloris chloris 382
chlorotica, Chloris chloris 383
CHONDESTES 599
chrysophrys, Emberiza 540
chrysoptera, Vermivora 591
cia, Emberiza 516
cia, Emberiza cia 518
cinctus, Poecile 86
cinctus, Poecile cinctus 87
cineracea, Emberiza 528
cineracea, Emberiza cineracea 530
cinerea, Creatophora 289
cinereus, Parus 111
CINNYRICINCLUS 296
CINNYRIS 152
cioides, Emberiza 602
cioides, Emberiza cioides 602
cirlus, Emberiza 512
cirtensis, Corvus monedula 249
cisalpina, Sitta europaea 124
citrina, Setophaga 591
citrinella, Carduelis 389
citrinella, Emberiza 509
citrinella, Emberiza citrinella 511
COCCOTHRAUSTES 459
coccothraustes, Coccothraustes 459
coccothraustes, Coccothraustes coccothraustes 460
coelebs, Fringilla 353
coelebs, Fringilla coelebs 355
colchica, Carduelis carduelis 388
collurio, Lanius 185
confusus, Lanius cristatus 173
contra, Gracupica 283
contra, Gracupica contra 284
cooki, Cyanopica cyanus 230
corax, Corvus 268
corax, Corvus corax 269
cornix, Corvus 260
cornix, Corvus cornix 262
coronata, Setophaga 481
coronata, Setophaga coronata 482
coronatus, Remiz 147
coronatus, Remiz coronatus 148

corone, Corvus 257
corone, Corvus corone 258
corsa, Certhia familiaris 136
corsicana, Carduelis 392
corsus, Parus major 109
CORVUS 247, 587
crassirostris, Bucanetes githagineus 440
CREATOPHORA 289
cristatus, Lanius 171
cristatus, Lanius cristatus 173
cristatus, Lophophanes 91
cristatus, Lophophanes cristatus 92
CRITHAGRA 375
cruentus, Rhodophoneus 167
cruentus, Rhodophoneus cruentus 168
cucullatus, Ploceus 588
cucullatus, Tchagra senegalus 170
cumatilis, Cyanoptila cyanomelana 581
curvirostra, Loxia 420
curvirostra, Loxia curvirostra 421
cyanea, Passerina 574
CYANISTES 97
cyanomelana, Cyanoptila 581
cyanomelana, Cyanoptila cyanomelana 581
CYANOPICA 229
cyanoptera, Vermivora 591
CYANOPTILA 581
cyanus, Cyanistes 103
cyanus, Cyanistes cyanus 104
cyanus, Cyanopica 229
cypriotes, Periparus ater 96
cyrenaicae, Cyanistes teneriffae 102

D

dauurica, Muscicapa 31
dauurica, Muscicapa dauurica 32
dauuricus, Corvus 250
decolorans, Parus cinereus 112
degener, Cyanistes teneriffae 102
dentata, Gymnoris 329
DICRURUS 586
diffusus, Oriolus chinensis 585
digitatus, Pyrrhocorax graculus 243
dilutus, Passer montanus 320
discolor, Setophaga 597
discolor, Setophaga discolor 597
docilis, Pyrrhocorax pyrrhocorax 246
DOLICHONYX 576
domesticus, Passer 300
domesticus, Passer domesticus 303
dominica, Setophaga 596
dorotheae, Certhia brachydactyla 139
dresseri, Poecile palustris 78
dresseri, Sitta tephronota 126
dubius, Poecile lugubris 90

E

elegans, Lanius excubitor 203
EMBERIZA 502, 601
enucleator, Pinicola 450
enucleator, Pinicola enucleator 451
erythrinus, Carpodacus 441

erythrinus, Carpodacus erythrinus 443
erythrogenys, Emberiza citrinella 511
erythronotus, Lanius schach 191
erythrophthalmus, Pipilo 599
erythrophthalmus, Pipilo erythrophthalmus 599
erythroramphos, Pyrrhocorax pyrrhocorax 245
erythrorhynchus, Buphagus 588
ESTRILDA 340
euchlorus, Passer 321
EUMYIAS 582
EUODICE 346
europaea, Pyrrhula pyrrhula 453
europaea, Sitta 121
europaea, Sitta europaea 123
europaeus, Aegithalos caudatus 74
excelsus, Parus major 109
excubitor, Lanius 195
excubitor, Lanius excubitor 199
exigua, Petronia petronia 331
exilipes, Acanthis hornemanni 414

F

familiaris, Certhia 134
familiaris, Certhia familiaris 136
faroensis, Sturnus vulgaris 278
fennorum, Pica pica 233
FICEDULA 39, 582
flammea, Acanthis 407
flammea, Acanthis flammea 409
flaviceps, Ploceus manyar 337
flavifrons, Vireo 589
flavipectus, Cyanistes cyanus 105
flavirostris, Linaria 404
flavirostris, Linaria flavirostris 406
frigoris, Carduelis carduelis 388
FRINGILLA 353
fronto, Emberiza leucocephalos 507
frugilegus, Corvus 254
frugilegus, Corvus frugilegus 256
fucata, Emberiza 603
fucata, Emberiza fucata 603
fulva, Turdoides 70
fulva, Turdoides fulva 71
furvior, Seiurus aurocapilla 463
fusca, Setophaga 594

G

galbula, Icterus 578
galbula, Ploceus 338
gambagae, Muscicapa 37
gambelii, Zonotrichia leucophrys 489
GARRULUS 221
GEOTHLYPIS 471
ginginianus, Acridotheres 294
githagineus, Bucanetes 438
githagineus, Bucanetes githagineus 440
glandarius, Garrulus 221
glandarius, Garrulus glandarius 223
glaszneri, Garrulus glandarius 224
graculus, Pyrrhocorax 241
graculus, Pyrrhocorax graculus 243
GRACUPICA 283

grammacus, Chondestes 599
granti, Sturnus vulgaris 278
griseus, Vireo 589
griseus, Vireo griseus 589
guentheri, Linaria cannabina 400
guillemardi, Loxia curvirostra 423
GYMNORIS 327

H

habessinicus, Cinnyris 155
habessinicus, Cinnyris habessinicus 156
harrisoni, Chloris chloris 382
harterti, Fringilla coelebs 356
harterti, Linaria cannabina 400
harterti, Terpsiphone viridis 60
hedwigii, Cyanistes teneriffae 102
HEDYDIPNA 149
hellmayri, Cinnyris habessinicus 156
HESPERIPHONA 590
hibernicus, Garrulus glandarius 224
hibernicus, Periparus ater 95
hilgerti, Rhodophoneus cruentus 168
hispaniensis, Sitta europaea 124
hispaniolensis, Passer 307
hispaniolensis, Passer hispaniolensis 309
holboellii (var.)*, Acanthis flammea* 410
homeyeri, Lanius excubitor 200
hornemanni, Acanthis 412
hornemanni, Acanthis hornemanni 414
hortulana, Emberiza 531
hufufae, Passer domesticus 304
humii, Coccothraustes coccothraustes 461
huttoni, Turdoides caudata 67
hyemalis, Junco 492
hyemalis, Junco hyemalis 493
hyperrhiphaeus, Cyanistes cyanus 104
hypochrysea, Setophaga palmarum 596
hypoleuca, Ficedula 48
hypoleuca, Ficedula hypoleuca 50
HYPOTHYMIS 585
hyrcanus, Garrulus glandarius 224
hyrcanus, Passer domesticus 303
hyrcanus, Poecile 84

I

iagoensis, Passer 314
iberiae, Ficedula hypoleuca 50
iberiae, Pyrrhula pyrrhula 453
ICTERUS 578
iliaca, Passerella 600
incognita, Emberiza schoeniclus 558
indicus, Passer domesticus 304
infaustus, Perisoreus 227
infaustus, Perisoreus infaustus 227
intermedia, Emberiza schoeniclus 558
intermedia, Petronia petronia 331
intermedius, Corvus macrorhynchos 587
intermedius, Parus cinereus 112
irbii, Aegithalos caudatus 75
isabellinus, Lanius 178
isabellinus, Lanius isabellinus 182
italiae, Aegithalos caudatus 75

italiae, Passer 305

J

jagoensis, Estrilda astrild 341
japonicus, Coccothraustes coccothraustes 461
jebelmarrae, Emberiza striolata 521
jebelmarrae, Lanius excubitor 206
JUNCO 492

K

karelini (var.), *Lanius phoenicuroides* 176
kinneari, Cinnyris habessinicus 156
kirmanensis, Poecile lugubris 90
kleinschmidti, Poecile montanus 82
koenigi, Lanius excubitor 205
kordofanicus, Rhodophoneus cruentus 168
korejevi, Linaria flavirostris 406
korejewi, Emberiza schoeniclus 559
kosswigi, Panurus biarmicus 63
krueperi, Sitta 115
krynicki, Garrulus glandarius 225
kubanensis, Carpodacus erythrinus 443
kundoo, Oriolus 166

L

LAGONOSTICTA 350
lahtora, Lanius excubitor 203
LANIUS 171
lapponicus, Calcarius 499
lapponicus, Calcarius lapponicus 501
lapponicus, Poecile cinctus 87
latifascia, Emberiza rustica 544
laurencei, Corvus corax 271
ledanti, Sitta 119
ledouci, Periparus ater 96
leucocephalos, Emberiza 505
leucocephalos, Emberiza leucocephalos 507
leucogaster, Cinnyricinclus 296
leucogaster, Terpsiphone paradisi 584
leucophaeus, Dicrurus 586
leucophrys, Zonotrichia 488
leucophrys, Zonotrichia leucophrys 489
leucoptera, Loxia 417
leucoptera, Loxia leucoptera 418
leucoptera, Pica pica 233
leucopterus (var.), *Lanius excubitor* 200
leucopygos, Lanius excubitor 205
levantina, Sitta europaea 124
LINARIA 397
lincolnii, Melospiza 601
lincolnii, Melospiza lincolnii 601
longicaudatus, Dicrurus leucophaeus 586
lonnbergi, Poecile montanus 82
LOPHOPHANES 91
LOXIA 417
lucionensis, Lanius cristatus 173
ludovicianus, Pheucticus 572
lugubris, Poecile 88
lugubris, Poecile lugubris 89
lusitanica, Emberiza schoeniclus 558
luteus, Passer 323
lydiae, Emberiza pallasi 563

M

macedonicus, Aegithalos caudatus 74
macrocercus, Dicrurus 586
macrodactyla, Certhia familiaris 136
macronyx, Remiz 145
macronyx, Remiz macronyx 146
macrorhynchos, Corvus 587
macrorhynchos, Nucifraga caryocatactes 239
maderensis, Fringilla coelebs 358
magnolia, Setophaga 593
major, Aegithalos caudatus 75
major, Parus 106
major, Parus major 108
malabarica, Euodice 346
malabarica, Sturnia 588
malabarica, Sturnia malabarica 588
manyar, Ploceus 336
maroccana, Turdoides fulva 71
mauritanica, Certhia brachydactyla 139
mauritanica, Pica pica 235
mediterranea, Linaria cannabina 399
melanocephala, Emberiza 567
melanotos, Pica pica 236
melodia, Melospiza melodia 486
MELOSPIZA 486, 601
menachensis, Crithagra 377
menzbieri, Remiz pendulinus 143
meridionalis, Lanius 210
metallica, Hedydipna 149
michalowskii, Periparus ater 96
minor, Emberiza pallasi 563
minor, Garrulus glandarius 226
minor, Lanius 192
mitratus, Lophophanes cristatus 92
MNIOTILTA 466
moabiticus, Passer 311
moabiticus, Passer moabiticus 312
mollis, Lanius borealis 208
moltchanovi, Periparus ater 95
MOLOTHRUS 604
monedula, Corvus 247
monedula, Corvus monedula 248
mongolicus, Bucanetes 435
montanus, Passer 319
montanus, Passer montanus 320
montanus, Poecile 80
montanus, Poecile montanus 83
MONTIFRINGILLA 333
montifringilla, Fringilla 361
moreletti, Fringilla coelebs 357
mugimaki, Ficedula 583
muraria, Tichodroma 131
muraria, Tichodroma muraria 133
murina, Pyrrhula 455
muscatensis, Turdoides squamiceps 69
MUSCICAPA 31, 582
MYCEROBAS 457

N

nargianus, Lanius vittatus 189
nemoricola, Sturnia malabarica 588

nepalensis, Tichodroma muraria 133
neumanni, Muscicapa striata 36
neumayer, Sitta 128
neumayer, Sitta neumayer 129
newtoni, Parus major 108
niediecki, Carduelis carduelis 388
nigricans, Remiz macronyx 146
niloticus, Lanius senator 217
niloticus, Passer domesticus 303
nivalis, Montifringilla 333
nivalis, Montifringilla nivalis 335
nivalis, Plectrophenax 494
nivalis, Plectrophenax nivalis 498
nobilior, Sturnus vulgaris 279
noveboracensis, Parkesia 464
nubicus, Lanius 218
NUCIFRAGA 238

O

obscura, Sitta tephronota 126
obscurus, Cyanistes caeruleus 98
obsoleta, Rhodospiza 432
oligastrae, Cyanistes caeruleus 98
olivacea, Piranga 484
olivaceus, Vireo 351
olivaceus, Vireo olivaceus 352
ombriosa, Fringilla coelebs 357
ombriosus, Cyanistes teneriffae 101
ONYCHOGNATHUS 274
opicus, Perisoreus infaustus 228
OREOTHLYPIS 468
orientalis, Corvus corone 258
orientalis, Cyanistes caeruleus 99
orientalis, Euodice cantans 349
ORIOLUS 163, 585
oriolus, Oriolus 163
ornata, Emberiza aureola 553
oryzivorus, Dolichonyx 576
osea, Cinnyris 157
osea, Cinnyris osea 158

P

pagodarum, Sturnia 291
pallasi, Emberiza 560
pallasi, Emberiza pallasi 563
pallescens, Corvus cornix 263
pallidirostris, Lanius excubitor 201
palmae, Fringilla coelebs 357
palmarum, Setophaga 596
palmarum, Setophaga palmarum 596
palmensis, Cyanistes teneriffae 101
palpebrosus, Zosterops 162
palpebrosus, Zosterops palpebrosus 162
palustris, Poecile 77
palustris, Poecile palustris 78
PANURUS 61
paradisi, Terpsiphone 584
par, Emberiza cia 518
PARKESIA 464
paropanisi, Carduelis carduelis 388
PARUS 106
parva, Carduelis carduelis 387

parva, Ficedula 39
PASSER 298
PASSERELLA 600
PASSERINA 574
passerina, Emberiza schoeniclus 557
PASSERULCUS 600
PASTOR 285
pendulinus, Remiz 140
pendulinus, Remiz pendulinus 143
pensylvanica, Setophaga 595
percivali, Rhynchostruthus socotranus 380
percivali, Tchagra senegalus 170
peregrina, Oreothlypis 468
PERIPARUS 93
PERISOREUS 227
persica, Sitta europaea 124
persicus, Cyanistes caeruleus 99
personata, Emberiza spodocephala 504
petechia, Setophaga 477
PETRONIA 330
petronia, Petronia 330
petronia, Petronia petronia 331
phaeonotus, Periparus ater 96
PHEUCTICUS 572
philadelphicus, Vireo 589
phoenicuroides, Lanius 174
PICA 231
pica, Pica 231
pica, Pica pica 232
PINICOLA 450
pipilans, Linaria flavirostris 406
PIPILO 599
PIRANGA 484, 599
PLECTROPHENAX 494
pleskei, Podoces 236
PLOCEUS 336, 588
plumbea, Sitta neumayer 130
PODOCES 236
POECILE 77
polatzeki, Fringilla teydea 360
poliogyna, Loxia curvirostra 423
poltaratskyi, Sturnus vulgaris 279
purpurascens, Sturnus vulgaris 279
pusilla, Cardellina 598
pusilla, Cardellina pusilla 598
pusilla, Emberiza 545
pusillus, Serinus 364
puteicola, Petronia petronia 331
PYRRHOCORAX 241
pyrrhocorax, Pyrrhocorax 244
pyrrhocorax, Pyrrhocorax pyrrhocorax 245
pyrrhonotus, Passer 310
PYRRHULA 452
pyrrhula, Pyrrhula 452
pyrrhula, Pyrrhula pyrrhula 453
pyrrhuloides, Emberiza schoeniclus 558
pytyopsittacus, Loxia 425

Q

quiscula, Quiscalus 604
QUISCALUS 604

R

raddei, Cyanistes caeruleus 99
reiseri, Emberiza schoeniclus 558
REMIZ 140
rhenanus, Poecile montanus 82
rhipidurus, Corvus 272
rhipidurus, Corvus rhipidurus 273
RHODOPECHYS 427
RHODOPHONEUS 167
rhodopsis, Lagonosticta senegala 350
RHODOSPIZA 432
RHYNCHOSTRUTHUS 379
rogosowi, Perisoreus infaustus 228
rosaceus, Aegithalos caudatus 74
roseus, Carpodacus 590
roseus, Carpodacus roseus 590
roseus, Pastor 285
rossikowi, Pyrrhula pyrrhula 454
rostrata, Acanthis flammea 408
rothschildi, Crithagra 375
rubicilla, Carpodacus 446
rubicilla, Carpodacus rubicilla 447
rubra, Piranga 599
rubra, Piranga rubra 599
rubrifasciata (var.), *Loxia curvirostra* 420
rufibarba, Estrilda 342
ruficollis, Corvus 265
rufitergum, Garrulus glandarius 224
rupicola, Sitta neumayer 129
russicus, Panurus biarmicus 62
rustica, Emberiza 542
rustica, Emberiza rustica 544
ruthenus, Perisoreus infaustus 228
ruticilla, Setophaga 473
rutila, Emberiza 548

S

saharae, Passer simplex 317
sahari, Emberiza 523
salicarius, Poecile montanus 82
salvadorii, Turdoides caudata 66
sandwichensis, Passerculus 600
sandwichensis, Passerculus sandwichensis 600
sanguineus, Rhodopechys 427
sardus, Periparus ater 95
schach, Lanius 190
schach, Lanius schach 191
schiebeli, Fringilla coelebs 356
schoeniclus, Emberiza 554
schoeniclus, Emberiza schoeniclus 557
scotica, Loxia curvirostra 422
scoticus, Lophophanes cristatus 92
SEIURUS 462
semenowi, Emberiza cineracea 530
semitorquata, Ficedula 45
senator, Lanius 213
senator, Lanius senator 215
senegala, Lagonosticta 350
senegala, Lagonosticta senegala 350
senegalus, Tchagra 169
SERINUS 364
serinus, Serinus 367
SETOPHAGA 473, 591
sharpii, Corvus cornix 263
sibirica, Muscicapa 582
sibiricus, Aegithalos caudatus 74
sibiricus, Carpodacus 448
sibiricus, Carpodacus sibiricus 449
sibiricus, Lanius borealis 208
siculus, Aegithalos caudatus 75
simplex, Passer 316
simplex, Passer simplex 317
SITTA 115, 585
socotranus, Rhynchostruthus 379
soemmerringii, Corvus monedula 249
solomkoi, Fringilla coelebs 355
sordida, Emberiza spodocephala 504
speculigera, Ficedula 52
spermologus, Corvus monedula 248
SPINUS 394
spinus, Spinus 394
SPIZA 603
splendens, Corvus 252
splendens, Corvus splendens 253
spodiogenys, Fringilla coelebs 356
spodocephala, Emberiza 502
spodocephala, Emberiza spodocephala 503
squamiceps, Turdoides 68
squamiceps, Turdoides squamiceps 69
stanleyi, Corvus rhipidurus 273
stewartii, Emberiza 601
stoliczkae, Passer ammodendri 299
stoliczkae, Remiz coronatus 148
stonei, Quiscalus quiscula 604
stracheyi, Emberiza cia 519
striata, Muscicapa 33
striata, Muscicapa striata 35
striata, Setophaga 479
striolata, Emberiza 520
striolata, Emberiza striolata 521
STURNIA 291, 588
sturninus, Agropsar 587
STURNUS 276
styani, Hypothymis azurea 585
subcalcaratus, Calcarius lapponicus 501
subflava, Amandava 345
subflava, Amandava subflava 345
subulata, Carduelis carduelis
superciliaris, Ficedula 582
superciliaris, Ficedula superciliaris 583
superciliosus, Lanius cristatus 173
synoicus, Carpodacus 444
syriaca, Sitta neumayer 129
syriacus, Serinus 370

T

tahapisi, Emberiza 525
tarbagataica, Emberiza cioides 602
tauricus, Sturnus vulgaris 279
TCHAGRA 169
teneriffae, Cyanistes 100
teneriffae, Cyanistes teneriffae 101
tephronota, Sitta 125

tephronota, Sitta tephronota 127
TERPSIPHONE 58, 584
terraesanctae, Parus major 110
teydea, Fringilla 359
teydea, Fringilla teydea 360
thalassinus, Eumyias thalassinus 582
thalassoides, Eumyias thalassinus 582
tianschanicus, Cyanistes cyanus 105
TICHODROMA 131
tigrina, Setophaga 592
tingitanus, Corvus corax 271
tingitanus, Passer domesticus 303
tomensis, Ficedula hypoleuca 50
transcaspicus, Passer hispaniolensis 309
transcaucasicus, Passer montanus 320
transfuga, Gymnoris xanthocollis 328
trichas, Geothlypis 471
trichas, Geothlypis trichas 472
tristis, Acridotheres 292
tristis, Acridotheres tristis 293
tristis, Sturnia pagodarum 293
tristramii, Onychognathus 274
tsaidamensis, Lanius isabellinus 184
tschitscherini, Sitta neumayer 130

TURDOIDES 64
tyrrhenica, Muscicapa striata 35

U

ukrainae, Emberiza schoeniclus 557
ultima, Carduelis carduelis 388
ultramarinus, Cyanistes teneriffae 102
uncinatus, Lanius excubitor 203
unicolor, Sturnus 280

V

varia, Mniotilta 466
varius, Corvus corax 269
VERMIVORA 591
verreauxi, Cinnyricinclus leucogaster 297
vespertina, Hesperiphona 590
vespertina, Hesperiphona vespertina 590
vieirae, Periparus ater 95
virens, Setophaga 597
VIREO 351, 589
viridis, Terpsiphone 58
vittatus, Lanius 188
vittatus, Lanius vittatus 189
vlasowae, Plectrophenax nivalis 498
voousi, Chloris chloris 383
vulgaris, Sturnus 276

vulgaris, Sturnus vulgaris 278

W

weigoldi, Lophophanes cristatus 92
whitakeri, Garrulus glandarius 225
whiteheadi, Sitta 117
witherbyi, Emberiza schoeniclus 558

X

XANTHOCEPHALUS 605
xanthocephalus, Xanthocephalus 605
xanthocollis, Gymnoris 327

Y

yatii, Passer moabiticus 312
yemenensis, Linaria 402
yemenensis, Turdoides squamiceps 69

Z

zagrossiensis, Parus major 110
zarudnyi, Passer 318
zedlitzi, Bucanetes githagineus 439
ziaratensis, Parus cinereus 112
ZONOTRICHIA 488
ZOSTEROPS 160
zugmayeri, Corvus splendens 253

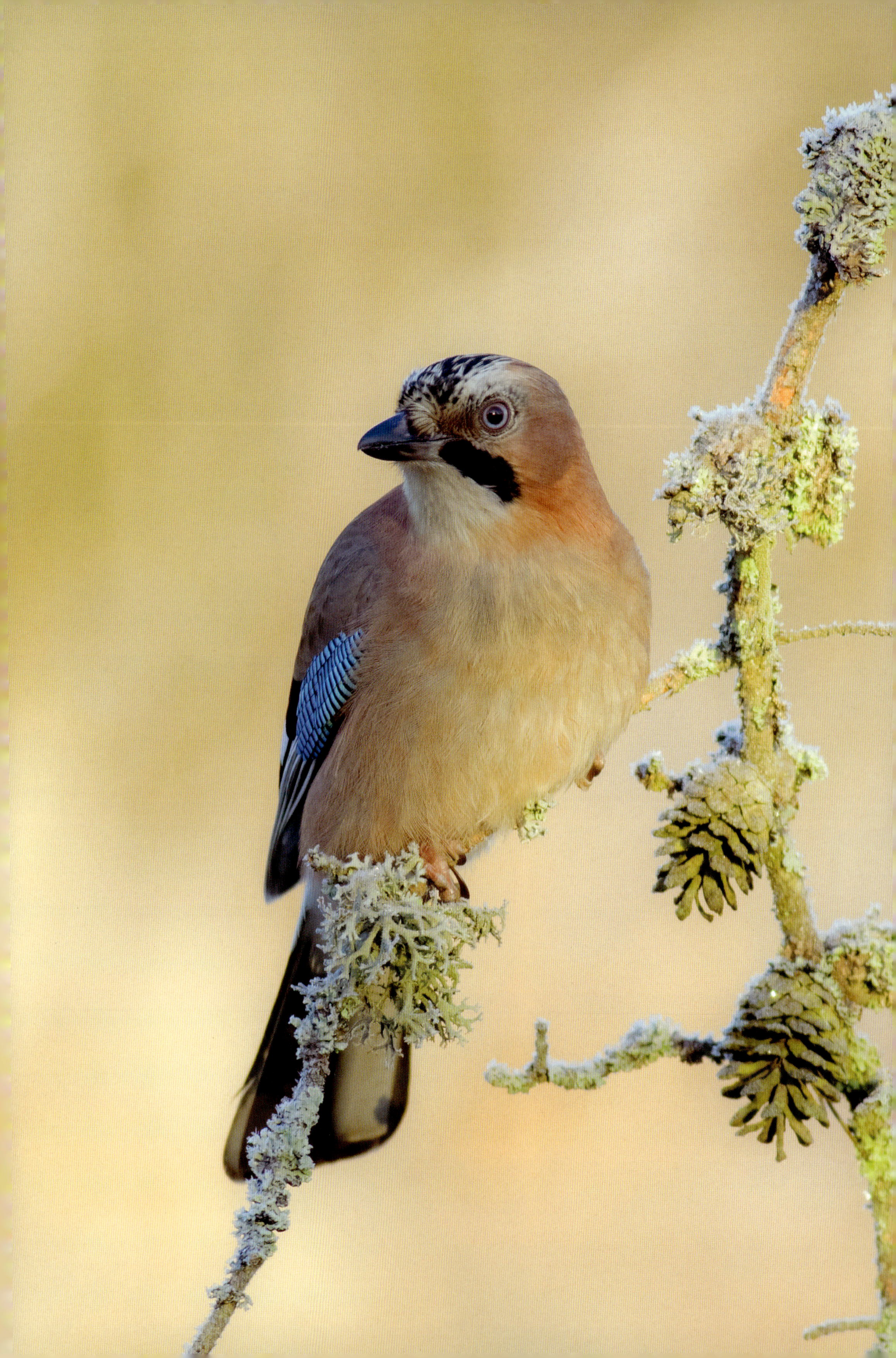

INDEX OF ENGLISH NAMES

A

Avadavat, Red 343

B

Babbler, Afghan 66, 67
　　Arabian 68
　　Common 66
　　Fulvous 70
　　Iraq 64
Blackbird, Yellow-headed 605
Bobolink 576
Brambling 361
Bullfinch, Azores 455
　　Common 452
Bunting, Black-faced 502
　　Black-headed 567
　　Chestnut 548
　　Chestnut-eared 603
　　Cinereous 528
　　Cinnamon-breasted 525
　　Cirl 512
　　Common Reed 554
　　Corn 570
　　Cretzschmar's 537
　　Grey-necked 534
　　House 523
　　Indigo 574
　　'Kurdish' 530
　　Lapland 499
　　Little 545
　　Meadow 602
　　Mountain 520
　　Ortolan 531
　　Pallas's Reed 560
　　Pine 505
　　Red-headed 564
　　Reed 554
　　Rock 516
　　Rustic 542
　　'Smyrna' 530
　　Snow 494
　　Striated 520
　　Striolated 520
　　White-capped 601
　　Yellow-breasted 551
　　Yellow-browed 540
Bush-shrike, Rosy-patched 167
Bush-tit, Long-tailed 72

C

Canary, Atlantic 372
Chaffinch, Blue 359
　　Common 353
Chough, Alpine 241
　　Red-billed 244
Cowbird, Brown-headed 604
Crossbill, Common 420
　　Parrot 425
　　'Scottish' 423
　　Two-barred 417
Crow, Carrion 257
　　Hooded 260
　　House 252
　　Jungle 587
　　Large-billed 587
　　'Mesopotamian' 263
　　Pied 264

D

Dickcissel 603
Drongo, Ashy 586
　　Black 586

F

Finch, African Crimson-winged 430
　　Asian Crimson-winged 427
　　Citril 389
　　Corsican 392
　　Desert 432
　　Mongolian 435
　　Strawberry 343
　　Trumpeter 438
　　White-winged Snow 333
Firefinch, Red-billed 350
Flycatcher, African Paradise 58
　　Asian Brown 31
　　Asian Verditer 582
　　Atlas 52
　　Blue-and-white 581
　　Collared 55
　　Dark-sided 582
　　Gambaga 37
　　Indian Paradise 584
　　Mugimaki 583
　　Pied 48
　　Red-breasted 39
　　Semicollared 45
　　Sooty 582
　　Spotted 33
　　Taiga 42
　　Ultramarine 582
　　Verditer 582

G

Goldfinch, European 385
Grackle, Common 604
　　Tristram's 274
Greenfinch, European 381
Grosbeak, Evening 590
　　Golden-winged 379
　　Pine 450
　　Rose-breasted 572
　　White-winged 457

H

Hawfinch 459

J

Jackdaw, Daurian 250
　　Western 247
Jay, Eurasian 221
　　Pleske's Ground 236
　　Siberian 227
Junco, Dark-eyed 492

L

Linnet, Common 397
　　Yemen 402
Longspur, Lapland 499

M

Magpie, 'African' 235
　　'Arabian' 235
　　Azure-winged 229
　　Common 231
Monarch, Black-naped 585
Munia, Red 343
Myna, Bank 294
　　Common 292
　　Pied 283

N

Nutcracker, Spotted 238
Nuthatch, Algerian 119
　　Corsican 117
　　Eastern Rock 125
　　Eurasian 121
　　Krüper's 115
　　Red-breasted 585
　　'Siberian' 123
　　Western 121
　　Western Rock 128
　　Wood 121

O

Oriole, Baltimore 578
　　Black-naped 585
　　Eurasian Golden 163
　　Indian Golden 166
Ovenbird 462
Oxpecker, Red-billed 588

P

Paradise Flycatcher, African 58
　　Asian 584

Indian 584
Parula, Northern 475
Penduline Tit, Black-headed 145
　　Eurasian 140
　　White-crowned 147
Petronia, Bush 329

R

Raven, Brown-necked 265
　　Common 268
　　Fan-tailed 272
Redpoll, Arctic 412
　　Common 407
　　'Coues' 414
　　'Greater' 408
　　Hoary 412
　　'Hornemann's' 414
　　'Lesser' 410
　　'Mealy' 409
Redstart, American 473
Reedling, Bearded 61
Rockfinch, Pale 325
Rook 254
Rosefinch, Common 441
　　Great 446
　　Long-tailed 448
　　Pallas's 590
　　Sinai 444

S

Seed-eater, Menacha 377
Serin, Arabian 375
　　European 367
　　Olive-rumped 375
　　Red-fronted 364
　　Syrian 370
　　Yemen 377
Shrike, 'Algerian Grey' 204
　　Bay-backed 188
　　Brown 171
　　'Canary Islands Grey' 205
　　'Daurian' 182
　　'Desert Grey' 201
　　'Elegant Grey' 204
　　Great Grey 195
　　'Homeyer's Grey' 200
　　Iberian Grey 210
　　'Indian Grey' 203
　　Isabelline 178
　　Lesser Grey 192
　　'Levant Grey' 200
　　Long-tailed 190
　　Masked 218
　　Northern 207
　　Red-backed 185
　　Red-tailed 174
　　Rosy-patched 167

'Sahel Grey' 206
Siberian Grey 207
'Socotra Grey' 204
'Steppe Grey' 201
'Tarim' 183
'Tsaidam' 184
Turkestan 174
Woodchat 213
'Yemen Grey' 203
Silverbill, African 348
　　Indian 346
Siskin, Eurasian 394
Snow Finch, White-winged 333
Sparrow, 'Afghan' 312
　　American Tree 601
　　Arabian Golden 321
　　Bush 329
　　Cape Verde 314
　　Chestnut-shouldered Bush 327
　　Dead Sea 311
　　Desert 316
　　Fox 600
　　House 300
　　Iago 314
　　Italian 305
　　Jungle 310
　　Lark 599
　　Lincoln's 601
　　Pale Rock 325
　　Red Fox 600
　　Rock 330
　　Rufous-backed 314
　　Savannah 600
　　Saxaul 298
　　Sind 310
　　Sind Jungle 310
　　Song 486
　　Spanish 307
　　Sudan Golden 323
　　Tree 319
　　White-crowned 488
　　White-throated 490
　　Yellow-throated 327
　　Zarudny's 318
Starling, Amethyst 296
　　Asian Pied 283
　　Brahminy 291
　　Chestnut-tailed 588
　　Common 276
　　Daurian 587
　　Rose-coloured 285
　　Rosy 285
　　Spotless 280
　　Tristram's 274
　　Violet-backed 296
　　Wattled 289
Sunbird, Nile Valley 149

Palestine 157
Purple 152
Shining 155

T

Tanager, Scarlet 484
　　Summer 599
Tchagra, Black-crowned 169
Tit, African Blue 100
　　Azure 103
　　Bearded 61
　　Blue 97
　　Caspian 84
　　Cinereous 111
　　Coal 93
　　Crested 91
　　Great 106
　　Long-tailed 72
　　Marsh 77
　　Siberian 86
　　Sombre 88
　　Turkestan 113
　　Willow 80
Towhee, Eastern 599
　　Rufous-sided 599
Treecreeper, Eurasian 134
　　Short-toed 137
Twite 404

V

Vireo, Philadelphia 589
　　Red-eyed 351
　　White-eyed 589
　　Yellow-throated 589

W

Wallcreeper 131
Warbler, Bay-breasted 594
　　Black-and-white 466
　　Blackburnian 594
　　Blackpoll 479
　　Black-throated Blue 595
　　Black-throated Green 597
　　Blue-winged 591
　　Canada 598
　　Cape May 592
　　Cerulean 592
　　Chestnut-sided 595
　　Golden-winged 591
　　Hooded 591
　　Magnolia 593
　　'Myrtle' 482
　　Palm 596
　　Prairie 597
　　Tennessee 468
　　Wilson's 598
　　Yellow 477

    Yellow-rumped 481
    Yellow-throated 596
Waterthrush, Northern 464
Waxbill, Arabian 342
    Common 340
    Orange-breasted 345
    Zebra 345
Weaver, Black-headed 588
    Rüppell's 338
    Streaked 336
    Village 588
White-eye, Abyssinian 160
    Indian 162
    Oriental 162

Y

Yellowhammer 509
Yellowthroat, Common 471